D0164140

ELEMENTARY HYDROLOGY

Vijay P. Singh

Department of Civil Engineering
Louisiana State University

PRENTICE HALL, Englewood Cliffs, NJ 07632

Library of Congress Cataloging-in-Publication Data

Singh, V. P. (Vijay P.)
Elementary hydrology / V.P. Singh.
p. cm.
Includes bibliographical references and index.
ISBN 0-13-249384-5
1. Hydrology. I. Title.
GB661.2.S56 1992
551.48--dc20 91-12242
CIP

Acquisitions editor: ***Doug Humphrey***
Editorial/production supervision
and interior design: ***Richard DeLorenzo***
Copy editor: ***Peter Zurita***
Cover design: ***Joe DiDomenico***
Manufacturing buyer: ***David Dickey***
Prepress buyer: ***Linda Behrens***
Supplements editor: ***Alice Dworkin***
Editorial assistant: ***Jaime Zampino***

Also available from Prentice Hall by Vijay P. Singh:

Hydrologic Systems, Vol. I—Rainfall Modeling
Hydrologic Systems, Vol. II—Watershed Modeling

A Simon & Schuster Company
Englewood Cliffs, New Jersey 07632

Printed in the United States of America

10 9 8 7 6 5 4 3 2 1

ISBN 0-13-249384-5

Prentice-Hall International (UK) Limited, *London*
Prentice-Hall of Australia Pty. Limited, *Sydney*
Prentice-Hall Canada Inc., *Toronto*
Prentice-Hall Hispanoamericana, S.A., *Mexico*
Prentice-Hall of India Private Limited, *New Delhi*
Prentice-Hall of Japan, Inc., *Tokyo*
Simon & Schuster Asia Pte. Ltd., *Singapore*
Editora Prentice-Hall do Brasil, Ltda., *Rio de Janeiro*

To Hydrologists and Water Scientists
Around the Globe

Everywhere water is a thing of beauty, gleaming in the dewdrop;
singing in the summer rain;
shining in the ice-gems till the leaves all seem to turn to living jewels;
spreading a golden veil over the setting sun;
or a white gauze around the midnight moon.

—J. B. Glough, "A Glass of Water"

CONTENTS

PREFACE

There has been an enormous increase in public awareness of the interaction between man and environment. It is, therefore, no surprise that public debates on virtually every facet of the environment have become commonplace before any project, be that construction of a highway or urban development, is proposed. The result of this increased awareness and emphasis on environmental quality has been establishment of higher standards for protection of the environment from humankind's activities, and demand for greater accountability of those responsible for such activities. Environmental consequences of humankind's activities are multifaceted and require an interdisciplinary approach for their assessment. Environmental problems are, therefore, many times more complex today than they were, say, a couple of decades ago.

Air, soil, and water are the three principal components of our environment. Water occupies the central position in the environmental makeup, and so do water resources amongst environmental resources. Environmental protection includes protection of water resources. Hydrology plays a vital role in protection and management of not only water resources, but also other environmental resources. Protection and management of environmental resources involve a broad range of scientific disciplines. As a result, hydrology these days forms a part of undergraduate curricula of such diverse disciplines as agricultural engineering, civil engineering, environmental sciences, climatology and geography, geology and earth sciences, forestry and watershed sciences, and some others. In these curricula, hydrology is usually taught to students who are either at the junior or senior level. Whatever the case may be, students are exposed to hydrology for the first time primarily through this course, and students taking the course have not had an opportunity to be exposed to hydrologic jargon before. And, in most cases, this may be the only course the students will have in hydrology in their undergraduate schooling. Therefore, this hydrology course must be at an elementary level, present basic concepts of hydrology, and develop a flavor for application of hydrology to the solution of a range of environmental problems. It is these considerations that motivated writing this book.

Elementary Hydrology is written at the elementary level, requiring no previous background in hydrology. Indeed, students can understand the book for the most part by themselves. The material for the book came from class notes prepared for an undergraduate course taught to geoscience students at the New Mexico Institute of Mining and Technology, and to civil engineering students at the George Washington University, Mississippi State University, and Louisiana State University. The criticisms and discussions offered by the students influenced, to some extent, the style of presentation in the book. Considerable effort has been made to discuss concepts and illustrate them with examples, even at the expense of increasing size and having a little bit of repetition. This style is deemed desirable for students being exposed to hydrology for the first time. Students learn best by doing.

In an undergraduate hydrology course, students are drawn from a wide range of disciplines and have obviously widely differing backgrounds and interests. The textbook used in teaching such a course must have sufficient material to appeal to a diversified class of students. This consideration also influenced the material covered in the book. Experience indicates that about 70 to 80% of the book may occupy a typical undergraduate course. Inclusion of about 20 to 30% extra material is deemed desirable to provide flexibility to the instructor in the choice of depth and breadth of coverage. Depending upon the particular requirements of a course and the pace of coverage, one can choose material from different chapters of the book. It is hoped that the material covered would also be of some interest to those beginning careers in agricultural sciences, civil engineering, earth sciences, environmental sciences, geography and climatology, forest and range sciences, meteorology, and water and land sciences.

The subject matter of *Elementary Hydrology* is divided into nine parts. Part 1, spanning four chapters, is composed of preliminaries. By introducing the definition of hydrology and its association with other sciences, the concepts of hydrologic cycle and hydrologic budget are presented in Chapter 1. Also introduced are the notion of scales for hydrologic conceptualization and the sources of hydrologic data. Chapter 2 takes stock of water resources available through various sources and develops a perspective for better management of these resources. It brings out the interrelationship amongst atmospheric, land, and oceanic systems. The chapter is concluded with a survey of environmental and water-resources problems for solutions of which hydrology plays a critical role. Hydrologic solutions are obtained within a spatial unit called a watershed, and the techniques of developing these solutions depend upon the type of watershed. A discussion of watershed types is presented in Chapter 3. Hydrologic solutions involve parameters that are characteristic of hydrologic systems being dealt with. Chapter 4 concludes Part 1 with a discussion of methods of determining these parameters.

Part 2, comprised of Chapters 5 to 7, presents hydrologic inputs. Chapter 5 discusses drainage-basin characteristics that have been found useful in developing hydrologic solutions. This chapter also shows close association of hydrology with geomorphology. The other hydrologic input is supplied by the atmosphere in the form of precipitation. A qualitative discussion of weather and precipitation is given in Chapter 6, showing the relation between atmospheric science (meteorology) and hydrology. Chapter 7 presents measurement of precipitation and methods of analyzing the measured precipitation.

Part 3 is on subsurface flow, occupying two chapters. Chapter 8 discusses infiltration and soil moisture. Both the measurement and the methods of prediction of infiltration are presented. The chapter is concluded with a short discussion of soil moisture. It also shows the association of hydrology with agronomy (soil and plant sciences). Chapter 9 deals with groundwater and baseflow. Introducing basic concepts of groundwater hydrology, the discussion focuses on geology of aquifers, well hydraulics, stream–acquifer interaction, recharge, and groundwater runoff. The chapter is concluded with a discussion of baseflow recession and methods for its prediction. It also shows the association of hydrology with geology and other earth sciences.

Part 4, encompassing Chapters 10 to 11, deals with above-surface flow and hydrologic abstractions. The term "abstraction" is used in the book from an engineering design point of view and, hence, has narrow interpretation and limited usefulness. Evaporation and transpiration are treated in Chapter 10. Both the methods of measurement and prediction are included. This chapter also shows the relation of hydrology to aerodynamics and botany. Chapter 11 discusses interception and depression storage, indicating the link between hydrology and agricultural and forest sciences.

Part 5, comprising Chapters 12 and 13, presents streamflow measurement and hydrograph analysis. Techniques of streamflow measurement and sources of streamflow data are presented in Chapter 12. It also points out the association between hydrology and river morphology. The concept of the streamflow hydrograph and its analysis are introduced in Chapter 13. The recession part of the streamflow hydrograph is of particular significance in water supply and pollution abatement. Making use of the discussion in Chapter 9, the methods of hydrograph separation are presented. The chapter is concluded with a short discussion of complex hydrographs and effective rainfall.

Part 6, comprising Chapters 14 to 19, deals with the precipitation–runoff relation. Methods of estimating surface-runoff volume are discussed in Chapter 14. Given the amount of surface runoff, the surface-runoff hydrograph can be estimated in a number of ways. The unit hydrograph method is one of the methods and is discussed in Chapter 15. Empirical methods of synthesizing the unit hydrograph for ungaged watersheds are presented in Chapter 16. Conceptual models of the unit hydrograph are developed in Chapter 17. Frequently, peak discharge is a sufficient statistic for hydrologic design of a multitude of projects. Methods of estimating peak discharge are presented in Chapter 18. The discussion in Chapters 14 to 18 is restricted to rainfall–runoff relation. Chapter 19 introduces snowmelt and the methods of determining runoff from snowmelt.

Part 7, comprising Chapters 20 and 21, deals with flow routing. Chapter 20 discusses methods of flow routing through reservoirs. Methods of flow routing through open channels are discussed in Chapter 21.

Part 8, encompassing Chapters 22 and 23, is on watershed simulation. Chapter 22 presents concepts related to upland erosion and sediment transport, and methods of estimating sediment yield from small watersheds. Chapter 23 presents concepts for building either event-based or continuous streamflow simulation models. This chapter shows how the concepts discussed in the preceding chapters are integrated for development of watershed models.

Part 9, comprising Chapters 24 to 26, is the concluding part and deals with hydrologic design involving statistical methods. Chapter 24 reviews statistical preliminaries needed for performing hydrologic design. Methods of frequency analysis are presented in Chapter 25. Concluding Chapter 26 presents methods for estimating design storm and design flood.

There exists voluminous literature on the topics covered in this elementary book. A comprehensive list of references has advertently been omitted. Only those references that are deemed most pertinent have been cited. This is because the intended audience is the beginner, who, for want of time, background, and experience, would have little appetite to pursue the advanced literature. This, however, in no way reflects a lack of appreciation for either the importance or quality of literature. Indeed this book would not have been completed without reviewing the pertinent literature.

ACKNOWLEDGMENTS

My teachers, colleagues, friends, and students have had a lasting impression on my career. They have broadened my perspective, enhanced my understanding of hydrology, and enriched my life. There are tens of people working in hydrology and related sciences whom I have never met and will probably never meet but whose works have been inspiring. I am grateful they chose to do what they did. Because of their collective efforts, hydrology is a more exciting and challenging science.

Professor W. F. Rogers, formerly of the Department of Civil Engineering, University of Nebraska at Omaha, helped in early versions of Chapters 5 to 8, 12, 13, and 18. Professor A. Kumar, Department of Civil Engineering of Delhi College of Engineering, Delhi, India, reviewed an earlier draft of the book and made many suggestions for its improvement. Professor V. L. Zitta, Department of Civil Engineering of Mississippi State University; Dr. Paul C. Chan, Department of Civil Engineering of the New Jersey Institute of Technology; and Dr. D. K. Borah of the Department of Biological and Agricultural Engineering, Rutgers—The State University of New Jersey, reviewed the entire manuscript and offered a number of suggestions that have been incorporated in the final draft. Amongst those reviewing parts of the manuscript were Mr. R. G. Greene and Dr. J. F. Cruise, Department of Civil Engineering of Louisiana State University; and Messrs. M. R. Jourdan and J. G. Collins, Environmental Laboratory of the U.S. Army Engineer Waterways Experiment Station, Vicksburg, Mississippi. A number of chapter-end exercises were given by Professor K. V. S. Sarma of the Department of Civil Engineering, Federal University of Paraiba, Campina Grande, Brazil; and Professor V. L. Zitta of Mississippi State University. I, of course, accept full responsibility for any omissions, shortcomings, or mistakes that may remain. Ms. S. Sartwell did the typing with skill, care, and untiring devotion. Dr. R. K. Seals, former Chairman, and Dr. R. R. Avent, current Chairman of the Department of Civil Engineering, Louisiana State University, provided encouragement and support whenever needed. My wife, Anita, and children, Vinay and Arti, created an atmosphere for work at home during nights, weekends, and holidays. I owe them my gratitude for their endurance, love, patience, sacrifices, and understanding. My mother, brothers, sisters, who are in India, and friends offered encouragement and were always there when I needed them. Without support from all of these people, this book would not have been completed.

V. P. Singh
Baton Rouge, Louisiana

PART 1
Preliminaries

CHAPTER 1
Introduction

1.1 DEFINITION OF HYDROLOGY

All life on earth is dependent, one way or another, on water. The study of the science of water is, therefore, important. Hydrology deals with some aspects of water as a resource. Specifically, hydrology can be defined as the science that deals with space–time characteristics of the quantity and quality of the waters of the earth, encompassing their occurrence, movement, distribution, circulation, storage, exploration, development, and management. These characteristics are determined by the relation of water to the earth. This definition of hydrology is not unique, but may suffice to indicate its scope. Price and Heindl (1968) have summarized a number of definitions of hydrology that have been reported in the literature.

Customarily, hydrology is partitioned into surface-water hydrology and groundwater hydrology. Surface-water hydrology is confined to the relation between water and the surface of the earth. Groundwater hydrology deals with the relation between water and the lithosphere or the subsurface portion of the earth. This text deals principally with surface-water hydrology, with emphasis on the drainage basin as the origin of surface water.

The definition of hydrology encompasses some aspects of a multitude of disciplines involving agriculture, biology, chemistry, geography, geology, glaciology, meteorology, oceanography, physics, volcanology, and many other disciplines. The involvement of hydrology with these sciences comes about by the reason of the close association of water with the atmosphere and the earth. Many branches of hydrology, therefore, have been distinguished, as shown in Figure 1.1. This association also points out that hydrology is an interdisciplinary science that touches almost all aspects of life. Frequently, hydrology is thought of as an element of agriculture,

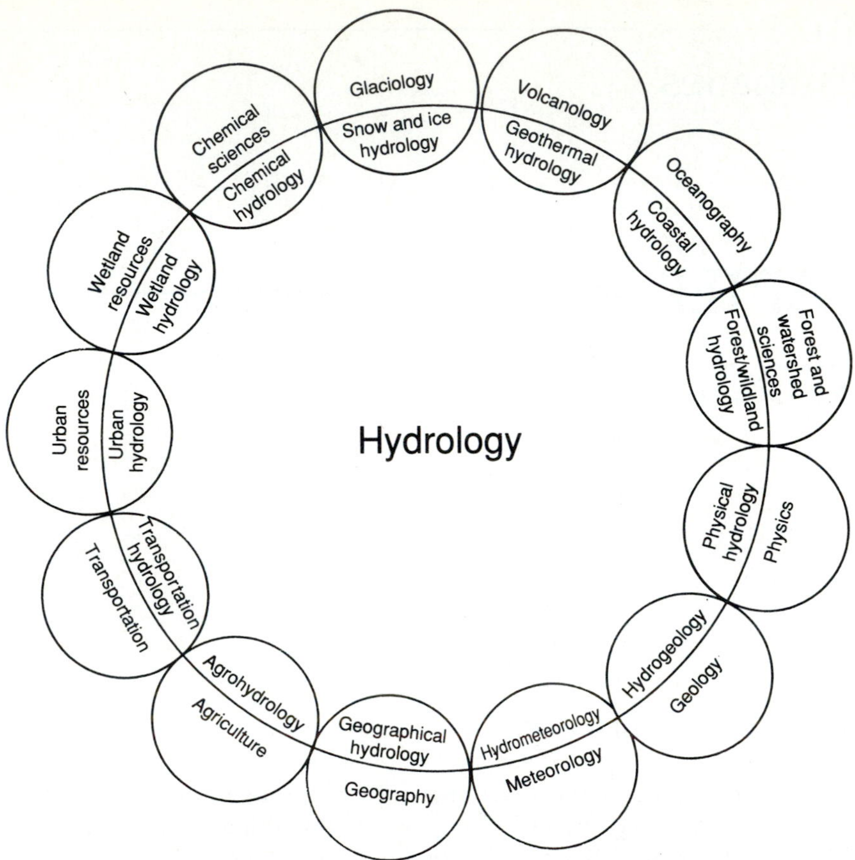

Figure 1.1 Classification of hydrology according to its association with other branches of science.

engineering hydraulics, forestry, geography, or geology. Present sociopolitical culture requires an environmental assessment of all changes in the natural relation of water to the surface of the earth. Therefore, hydrology should be perceived in terms of the entire reaction of water with the environment.

Techniques for solving hydrologic problems are borrowed from several disciplines such as mathematics, statistics, probability theory, operations research, control theory, information theory, and others. Based on the treatment of hydrologic data by these techniques, hydrology can be subdivided into different branches, as shown in Figure 1.2.

1.2 HYDROLOGIC CYCLE

The waters of long geologic history are the waters of the twentieth century; little has been added or lost through the ages since the first clouds formed and the first rains fell. The same water has been transferred time and time again from the oceans into

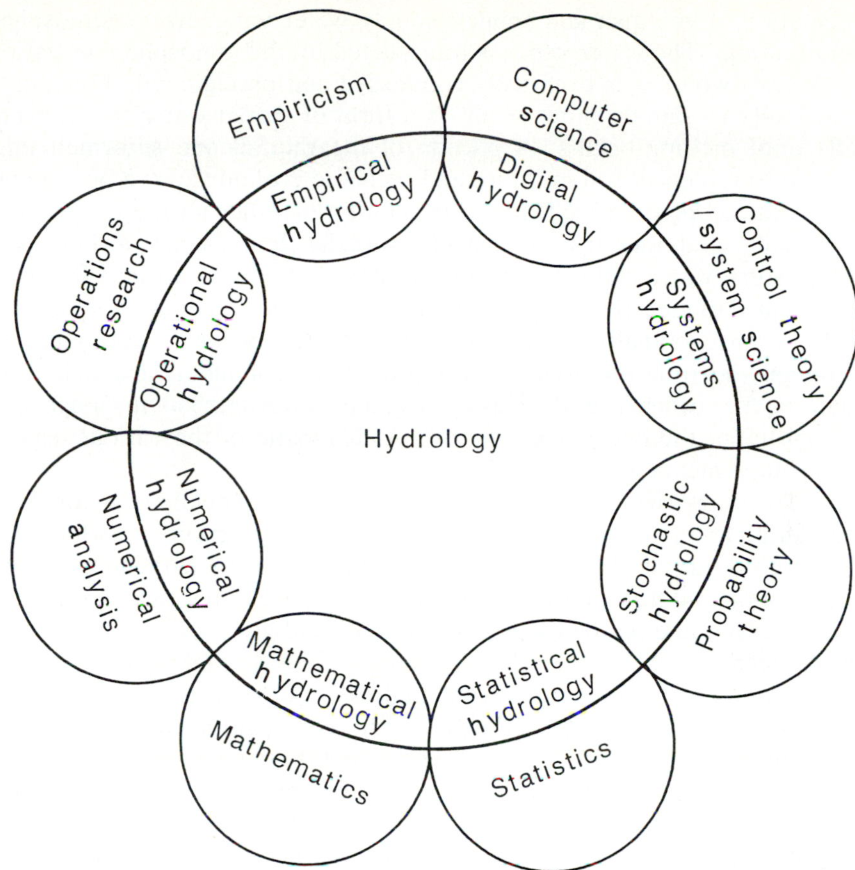

Figure 1.2 Classification of hydrology according to methods of solution.

the atmosphere, dropped upon the land, and transferred back to the sea. This endless circulation is the hydrologic cycle.

Water is evaporated from bodies of water such as lakes, ponds, reservoirs, oceans, and rivers, as well as wet-land surfaces or transpired through plants as vapor to the atmosphere and is transported in the atmosphere to a location where the water vapor is condensed and falls as precipitation on the surface of the earth. At least a portion of this precipitated liquid water runs off over and below the land surface and eventually finds its way back to the ocean where it is reevaporated. This endless circulation has neither beginning nor end. In the geologic past, large alterations in the cyclic roles of the atmosphere and the oceans have produced deserts and ice ages across entire continents. Even now, small alterations of the local patterns of the hydrologic cycle produce floods and droughts.

Evaporation of water from the oceans, lakes, and other free water surfaces throughout the world occurs due to the energy from the sun, thereby providing a supply of vapor to the atmosphere. At the same time, transpiration to the atmosphere of water taken in by plants occurs in forests, prairies, cropland, and other vegetation. Even man and animals supply water vapor to the atmosphere through perspiration. The water vapor is transported by the atmosphere to various parts of the world, where it is eventually condensed and precipitated. Precipitation occurs principally as rain or snow. Snow is a form of stored water that remains where it falls until melting occurs. A portion of the rainfall and snowmelt infiltrates the ground to replenish soil moisture and recharge groundwater; a portion of precipitation is intercepted and adheres to vegetation or other abstract objects, where it is eventually evaporated; a portion fills surficial depressions, forming small ponds, where some water infiltrates and evaporates, and a portion runs off over the surface of the earth to join streams and is eventually transported to the ocean. Some groundwater even eventually finds its way back to the ocean as springflow to streams, although the time required for this process is commonly much longer than that required for surface runoff. This cycle can be interrupted at any stage, but over long enough time, the cycle repeats itself. A schematic of the various aspects of these water movements is shown in Figure 1.3.

To summarize, the hydrologic cycle is a natural machine, a constantly running distillation and pumping system. The sun supplies heat energy, which together with the force of gravity keeps the water moving: from the earth to the atmosphere as evaporation and transpiration, from the atmosphere to the earth as condensation and precipitation, and on the earth as streamflow and groundwater movement, and then to the oceans.

From a global perspective, the hydrologic cycle can be considered to be comprised of three major systems (Singh, 1989): the oceans as the major source of water, the atmosphere as the deliverer of water, and the land as the user of water. In this cycle, there is no water gained or lost, but the amount of water available to the user may fluctuate because of variations at the source, or, more usually, in the delivering system. Clearly, precipitation, runoff, and evaporation are the principal processes that transmit water from one system to the other, as illustrated in Figure 1.4. This illustration encompasses the interactions between the earth (lithosphere), the oceans (hydrosphere), and the atmosphere. The cycle of water movement is a closed system on a global basis, but an open one on a local basis (Dooge, 1973). When the water movement of the earth system is considered, three systems, as shown in Figure 1.5, can be recognized: the land system, the subsurface system, and the aquifer (or geologic) system. Streamflow in a perennial river is derived from these systems, connected through the processes of infiltration, exfiltration, percolation, and upward movement of water. If we focus our attention on the hydrologic cycle of the land system, then precipitation, surface runoff, infiltration, and evapotranspiration are the dominant processes transmitting water, as shown in Figure 1.6. The land system itself can be comprised of three subsystems, as shown in Figure 1.7: vegetation subsystem, structural subsystem, and soil subsystem. These subsystems abstract water through interception, depression, and detention storage, whereby water is either lost to the atmospheric system or subsurface system.

The components of the hydrologic cycle can be summarized as precipitation (rainfall, snowfall, hail, sleet, dew, drizzle, fog, etc.), runoff (surface runoff, subsurface runoff or interflow, and groundwater runoff or baseflow), evaporation, transpiration, infiltration, percolation, seepage, interception, depression storage, and moisture storage over and below the land surface.

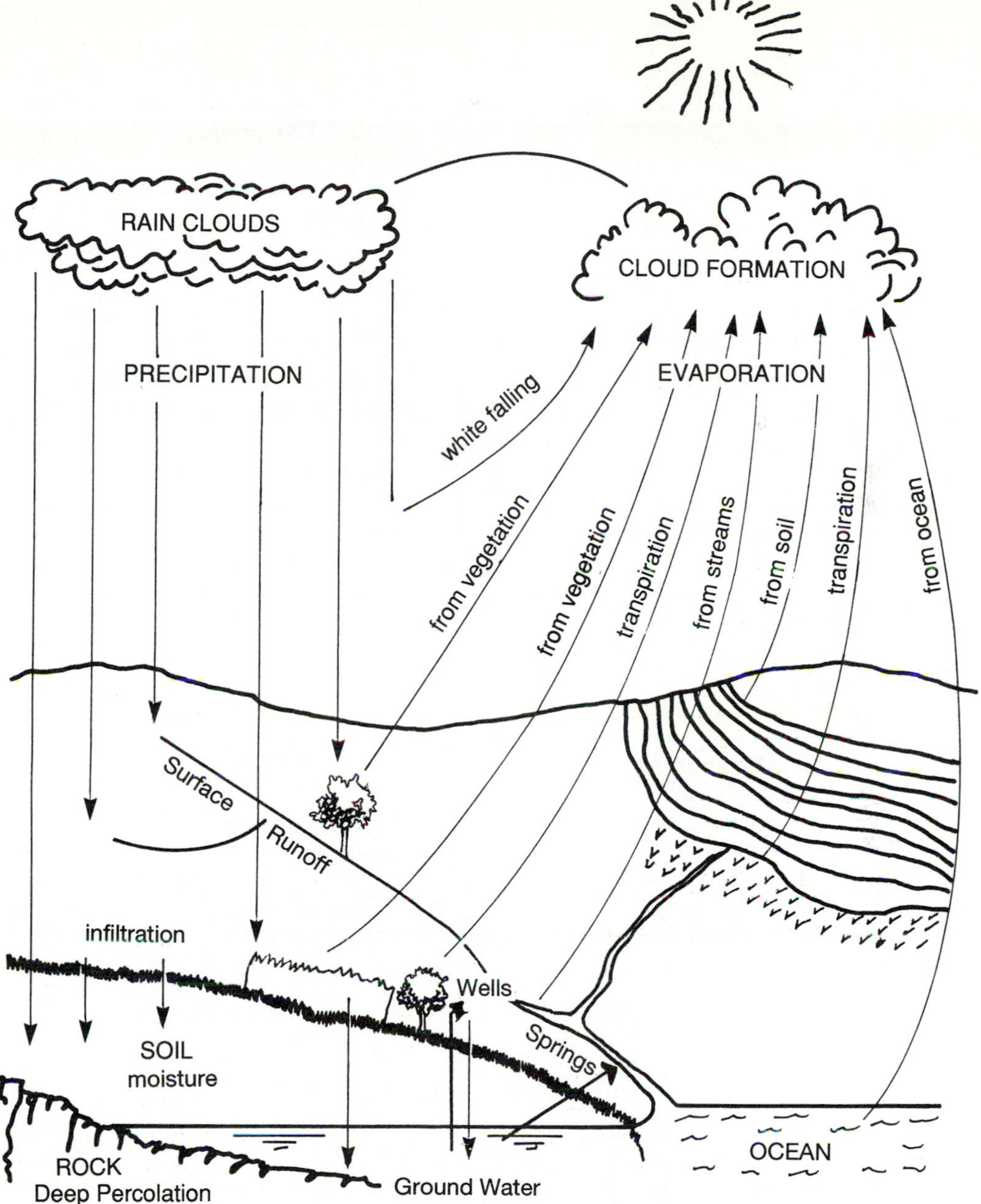

Figure 1.3 A descriptive representation of the hydrologic cycle (after Ackermann et al., 1955). Precip = precipitation, ET = evapotranspiration, SF = streamflow, TF = tidal flow, SR = sea rise, and GW Int. = groundwater intrusion.

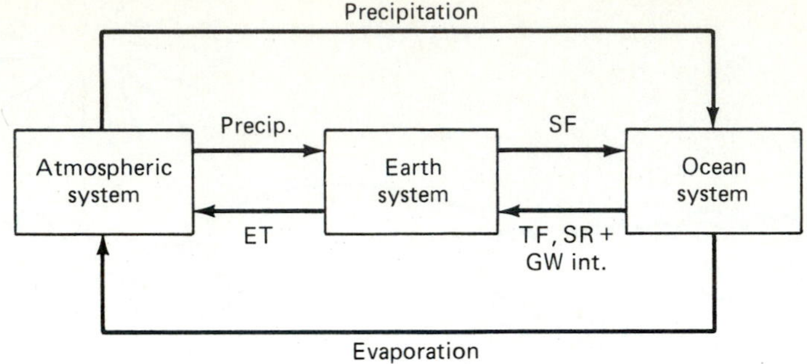

Figure 1.4 A global schematic of the hydrologic cycle.

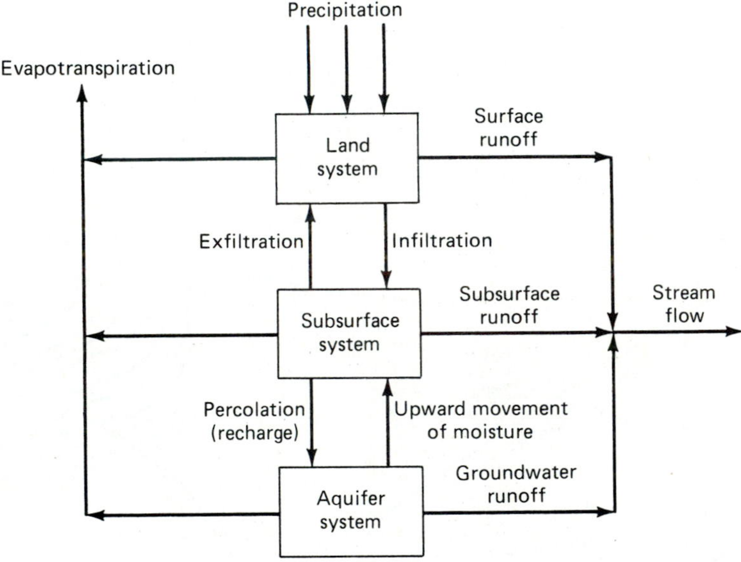

Figure 1.5 A schematic of the hydrologic cycle in the earth system.

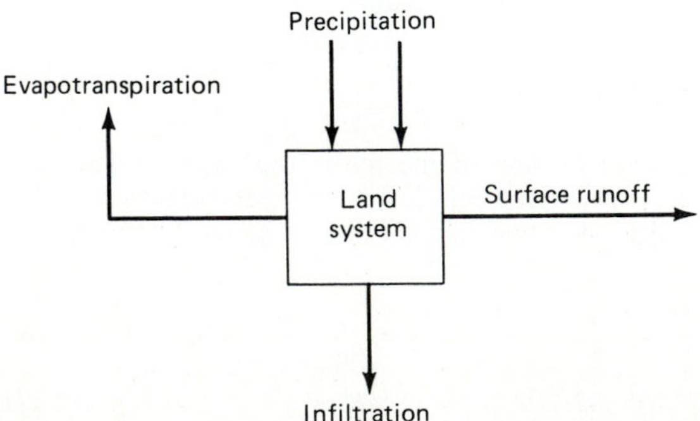

Figure 1.6 A schematic of the hydrologic cycle in the land system.

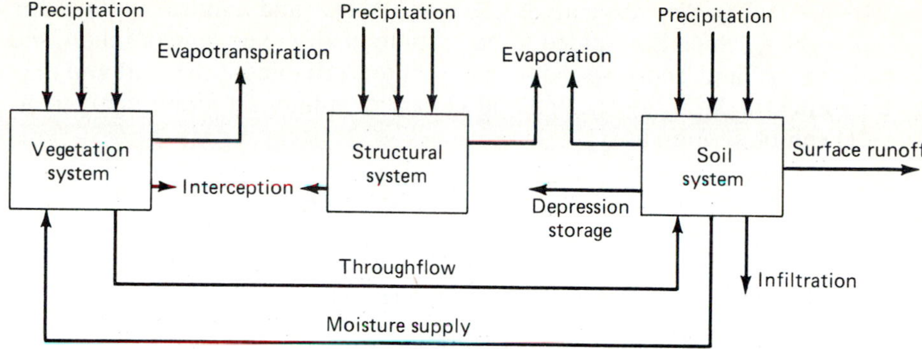

Figure 1.7 A more complete schematic of the hydrologic cycle in the land system.

1.3 HYDROLOGIC BUDGET

The hydrologic budget or water balance of a drainage basin is a mathematical statement of its hydrologic cycle. It is expressed by equating the difference between inflow, I, and outflow, O, of a drainage basin to the rate of change of storage within the basin, ΔS, for a specified period of time, Δt. When the basin is considered as a black-box system, as shown in Figure 1.8(a), or as a reservoir, as shown in Figure 1.8(b), its hydrologic budget can be expressed as

$$\frac{\Delta S}{\Delta t} = \bar{I} - \bar{O} \tag{1.1a}$$

$$\frac{S_2 - S_1}{\Delta t} = \frac{I_1 + I_2}{2} - \frac{O_1 + O_2}{2} \tag{1.1b}$$

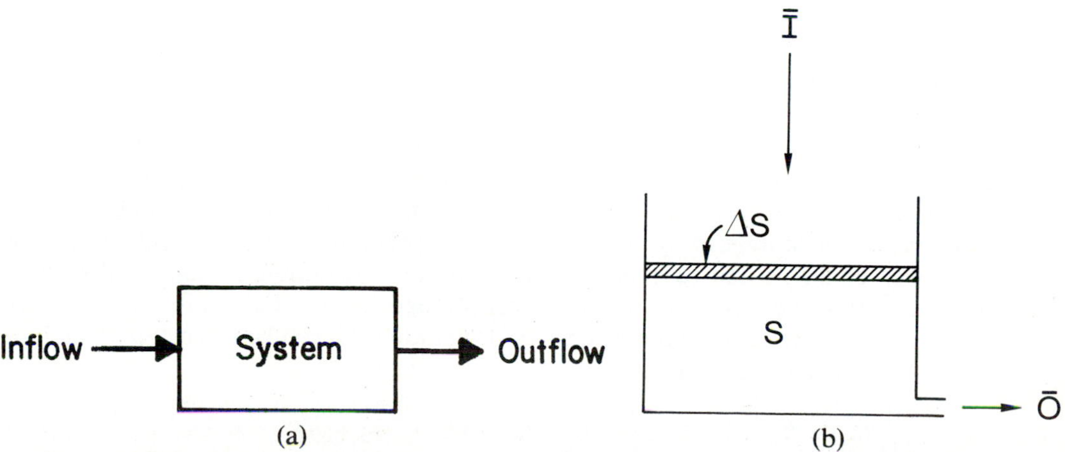

Figure 1.8 (a) The drainage basin as a simple black-box system. (b) The drainage basin as a reservoir receiving average inflow and discharging average outflow.

where $\bar{I}$ and $\bar{O}$ are, respectively, average inflow and average outflow for the time interval Δt, which is assumed to be small to justify averaging of inflow and outflow. Subscripts 1 and 2 correspond to values of the variables at the start and the end of the time interval $\Delta t = t_2 - t_1$. If I and O vary continuously with time t, then Equation (1.1) can be written as

$$\frac{\mathrm{d}S(t)}{\mathrm{d}t} = I(t) - O(t) \tag{1.2}$$

Implied in Equation (1.1) or (1.2) is that I, O, and S do not vary in space or are spatially lumped. Equation (1.2) is also referred to as the spatially lumped continuity equation, or sometimes as the water budget. Equation (1.1) or (1.2) forms the basis of the systems approach in hydrology.

All hydrologic analyses of drainage basins must satisfy Equation (1.2), or else the analysis is incomplete and is, therefore, not reasonable. The appearance of this equation is deceiving in its simplicity. For most hydrologic problems, more than one variable is unknown and, therefore, Equation (1.2) cannot be solved without additional information. For example, both $O(t)$ and $S(t)$ are unknown when a rainfall–runoff relation is desired. Without an extra relation between $S(t)$ and $O(t)$ with or without $I(t)$, $O(t)$ cannot be evaluated. Furthermore, I and O are not known as continuous explicit functions of time. The difficulty of this equation lies in evaluating the component variables. The methods for accomplishing this evaluation are presented in several later chapters.

In Equation (1.2), I and O are expressed as rates having the dimensions of $L^3/T/L^2$ (= L/T) or L^3/T. By integrating, this budget equation can also be written with these quantities expressed in volumetric units as

$$\int \mathrm{d}S(t) = \int [I(t) - O(t)]\, \mathrm{d}t$$

$$S(t) - S(0) = \int_0^t I(t)\, \mathrm{d}t - \int_0^t O(t)\, \mathrm{d}t \tag{1.3}$$

$$= V_I(t) - V_0(t)$$

in which $S(O)$ is the initial storage or storage at $t = 0$, $V_I(t)$ and $V_0(t)$ are volumes of inflow and outflow at time t having the dimensions of L^3/L^2 (= L) or L^3. Equation (1.2) or its variant in Equation (1.1b) or (1.3) is the fundamental governing equation for hydrologic analysis and synthesis.

For a drainage basin, the inflow may be comprised of rainfall, snowfall, hail, and other forms of precipitation. Surface runoff, subsurface runoff, groundwater runoff, evaporation, transpiration, and infiltration may constitute the outflow. The components of storage may include surface storage (over the ground, including storage in channels and reservoirs, depression and detention storage), subsurface storage (within the rootzone), groundwater storage (within the aquifers), and interception (over vegetation, buildings, etc.). Equation (1.2) can be rewritten by including all these components as

$$\frac{\mathrm{d}}{\mathrm{d}t}(S_s + S_m + S_g + S_i) = I_r + I_{sn} - O_{sr} - O_{sb} - O_g - e - e_t - f \tag{1.4}$$

where S with a subscript denotes a component of storage with the subscript s for surface storage, m for soil-moisture storage, g for groundwater storage, and i for interception storage; I_r is rainfall intensity, I_{sn} is rate of snowfall, O_{sr} is surface runoff, O_{sb} is subsurface runoff, O_g is groundwater runoff, e is rate of evaporation, e_t is rate of transpiration, and f is infiltration rate. Equation (1.4) can be written, analogous to Equation (1.3), in volumetric units as

$$\begin{aligned} S_s(t) - S_s(0) + S_m(t) - S_m(0) + S_g(t) - S_g(0) \\ + S_i(t) - S_i(0) = P - V_Q - E - T - F \end{aligned} \tag{1.5}$$

in which $P = \int_0^t (I_r + I_{sn})\,dt$ = volume (or depth) of precipitation, $V_Q = \int_0^t (O_{sr} + O_{sb} + O_g)\,dt$ = volume (or amount) of runoff, $E = \int_0^t e\,dt$ = amount of evaporation, $T = \int_0^t e_t\,dt$ = transpiration (volume of water transpired), and $F = \int_0^t f\,dt$ = cumulative infiltration. The term V_Q provides the estimate of basin yield. The terms E, T, F, and S_i constitute what are frequently known as hydrologic abstractions.

Subsequent chapters will deal with each of these variables appearing in Equation (1.5). Only when the role of each of these variables is completely understood and evaluated can a reasonable solution to a hydrologic problem be determined. Let us consider an example of determining the volume of water that runs off from a drainage basin following a rainfall or snowmelt. This volume is determined by (1) the amount of rainfall or snowmelt, and (2) by the amount of rainfall or snowmelt that is lost to runoff by storage deficiency and abstractions of the drainage basin. Let us assume that there is no change in groundwater storage. The change in soil-moisture storage is due to infiltration, and, hence, the former need not be considered. Then with $S_s(0) = S_i(0) = 0$, Equation (1.5) becomes

$$V_Q = P - F - E - T - S_i - S_s \tag{1.6}$$

Its simplicity belies its importance. For individual rainfall events, the quantities E, T, and S_i are much smaller than P, F, V_Q, and S_s and can be neglected. Equation (1.6), therefore, can be written as

$$V_Q = P - F - S_s \tag{1.7}$$

Equation (1.7) shows that determination of the runoff volume from a drainage basin caused by a precipitation event requires subtraction from total precipitation of the amount of water diverted or used by the drainage basin through abstractions. The units of V_Q and all other variables in this equation are best represented as inches or centimeters of water uniformly distributed over the drainage basin. The reason that these units are desirable is that they conform to familiar inches or centimeters of rainfall terminology. Of course, water units such as cubic feet or acre-feet could be used, but such units would have little meaning to an individual for this application.

EXAMPLE 1.1 The storage in a river reach at a specified time is 3 hectare-meters. At the same instant, the inflow to the reach is 15 cubic meters per second (m^3/s) and the outflow 20 m^3/s. One hour later, the inflow is 20 m^3/s and the outflow is 20.5 m^3/s. Determine the change in storage in the reach that occurred during the hour. Is the storage at the end of the hour greater or less than the initial value? What is the storage at the end of the hour?

Solution $I_1 = 15\ m^3/s$, $I_2 = 20\ m^3/s$, $O_1 = 20\ m^3/s$, $O_2 = 20.5\ m^3/s$, $S_1 = 30{,}000\ m^3$, and $\Delta t = 1\ h = 3600\ s$. Therefore, from Equation (1.1b),

$$\Delta S = \frac{I_1 + I_2}{2}\,\Delta t - \frac{O_1 + O_2}{2}\,\Delta t$$

$$= \frac{15\ \text{m}^3/\text{s} + 20\ \text{m}^3/\text{s}}{2}\,3600\ \text{s} - \frac{20\ \text{m}^3/\text{s} + 20.5\ \text{m}^3/\text{s}}{2}\,3600\ \text{s}$$

$$= -9900\ \text{m}^3$$

Therefore, the storage is less than the initial storage. The storage at the end of the hour = 30,000 m^3 − 9,900 m^3 = 20,100 m^3.

EXAMPLE 1.2 The storage in a river reach is 2 hectare-meters at a given time. Determine the storage 1 hour later if the average rates of inflow ($\bar{I}$) and outflow ($\overline{O}$) during the hour are 21 m^3/s and 18 m^3/s, respectively.

Solution $\bar{I} = 21\ \text{m}^3/\text{s}$, $\overline{O} = 18\ \text{m}^3/\text{s}$, $S_1 = 20{,}000\ \text{m}^3$, and $\Delta t = 3600$ s. Therefore,

$$S_2 = S_1 + (\bar{I} - \overline{O})\,\Delta t$$

$$= 20{,}000\ \text{m}^3 + (21 - 18)\ \text{m}^3/\text{s} \times 3600\ \text{s}$$

$$= 30{,}800\ \text{m}^3 = 3.08\ \text{hectare-meters}$$

Thus, the change in storage during the hour = (30,800 − 20,000) m^3 = 10,800 m^3.

EXAMPLE 1.3 Three centimeters of water evaporated from a 200-hectare vertical walled reservoir during 24 hours. Storm water was added to the reservoir at a constant rate of 3 m^3/s during this period. Determine the volume in ha-cm of water released during the period (through the bottom of the reservoir) if the water level in the reservoir was the same at the beginning and the end of the day.

Solution $I = 30\ \text{m}^3/\text{s} = 0.03$ ha-cm/s, $E = 3\ \text{cm} \times 200\ \text{ha} = 600$ ha-cm, $A = 200$ ha, and $\Delta t = 1$ day = 86,400 s.

$$\frac{dS}{dt} = I - O = I - e - O_{sr}$$

Since, $dS/dt = 0$, $O = I - O_{sr} - e$, or $O_{sr} = I - e$,

$$V_Q = O_{sr}\,\Delta t = (I - e)\,\Delta t$$

$$= I\,\Delta t - E$$

$$V_Q = (0.03\ \text{ha-cm/s})(86{,}400\ \text{s}) - (3\ \text{cm})(200\ \text{ha}) = 2592 - 600$$

$$= 1992\ \text{ha-cm}$$

EXAMPLE 1.4 A 600-acre reservoir receives 100 in. of rainfall during a period of 24 months. During this period, the mean inflow to the stream is 5 cfs, mean outflow from the stream is 4.5 cfs, and an increase in storage is 2000 acre-feet. Compute evapotranspiration in acre-feet as well as in inches.

Solution

$$\frac{dS}{dt} = 2000\ \text{acre-ft},\ I = \frac{100}{12} \times 600\ \text{acre-ft} + \frac{1}{43{,}560} \times 5$$

$$\times 60 \times 60 \times 24 \times 365 \times 2\ \text{acre-ft}$$

$$0 = \frac{1}{43{,}560} \times 4.5 \times 60 \times 60 \times 24 \times 365 \times 2\ \text{acre-ft} + e + e_t$$

where $e + e_t$ represents the sum of evaporation and evapotranspiration.

$$\frac{dS}{dt} = I - 0$$

$$2000 = \frac{100}{12} \times 600 + \frac{1}{43{,}560} \times 5 \times 3600 \times 24 \times 730 - \frac{1}{43{,}560}$$

$$\times 4.5 \times 3600 \times 24 \times 730 - e - e_t$$

Therefore,

$$e_t + e = -2000 + 5000 + \frac{1}{43{,}560} \times 0.5 \times 3600 \times 24 \times 730$$

$$= -2000 + 5000 + 723.97 = 3723.97 \text{ acre-ft}$$

$$= 6.21 \text{ feet} = 74.48 \text{ in.}$$

EXAMPLE 1.5 Rainfall of intensity 0.2 in./h fell on a 1.5-mi^2 drainage area for 6 h. Measured runoff during this period was 2,509,056 ft^3. How much of the total 6-h rainfall was lost?

Solution

$$I = \frac{0.2}{12} \times 6 \times 1.5 \times 5280 \times 5280 \text{ ft}^3$$

$$O = 2{,}509{,}056 \text{ ft}^3$$

The difference between I and O is considered as loss. Therefore,

$$\text{Loss} = 4{,}181{,}760 - 2{,}509{,}056 = 1{,}672{,}704 \text{ ft}^3$$

$$= \frac{1{,}672{,}704}{1.5 \times 5280 \times 5280} = 0.04 \text{ ft.} = 0.48 \text{ in.}$$

$$= \frac{0.48}{6} = 0.08 \text{ in./h}$$

1.4 SPACE–TIME SCALES IN HYDROLOGY

The hydrologic cycle best defines the scope of hydrology. Depending on the hydrologic problem under consideration, the hydrologic cycle or its components can be treated at different scales of time and space. As a consequence, different hydrologic problems may have different space–time scales. The global scale is the largest spatial scale and the watershed, or drainage basin, the smallest spatial scale. A drainage basin, or watershed, is the area that diverts all runoff to the same drainage outlet. In between these two scales lie such scales as continental, regional, and other space scales convenient for hydrologic analysis. Clearly, the watershed, or drainage-basin, scale is the most basic of all; and all other scales can be constructed by building on the drainage basin scale. Most hydrologic problems deal with a drainage basin. This approach distinguishes hydrology from hydraulics, whose spatial scale is a channel or a channel segment. It should be clearly understood that the watershed scale does not usually or necessarily coincide with territorial or jurisdictional boundaries that might be determined by political or economic considerations. A drainage basin can be of almost any size. It might be as small as a small parking lot or as large as or larger than the entire Mississippi River basin, which occupies approximately 41% of the United States. Large watersheds are usually broken down into smaller drainage basins to suit the requirements of a particular problem and to assist in orderly quantitative analysis.

Time scales used in hydrologic studies range from a fraction of an hour to a year or perhaps many years. The time scale used in a hydrologic study depends on the purpose of the study and the problem involved. Hourly, daily, weekly, monthly, or seasonal time scales are common. Sometimes the time interval for the collection of data determines the time scale for hydrologic analysis. Hydrologic time scales often do not coincide with those used in fluid mechanics or in hydraulics and likewise do not coincide with political, environmental, or economic time scales.

This discussion of time scales and space scales becomes more meaningful and significant when considered in association with computations, where interpretation and extrapolation of data are performed. From this point onward, the drainage basin will constitute the spatial scale unit unless otherwise specified. Different time scales will be used for different problems and will be so specified. A comprehensive discussion of scales in hydrology is given by Klemes (1983), and their bearing on development of hydrologic methods by Dooge (1984).

1.5 HYDROLOGIC DATA

An accurate hydrologic evaluation cannot be made without accurate data. Many governmental agencies, private companies, and individuals are engaged in the collection of hydrologic data. This data is usually available to anyone having need of it. The ensuing discussion does not list every source of data or maps useful in hydrology. There are many other agencies—local, state, and national—as well as private companies that maintain and collect useful hydrologic data. A person with reasonable understanding of the basic principles of hydrology will be able to anticipate sources of data by relating the information needed to those agencies or companies that also might need such information.

1.5.1 Streamflow Records

The United States Geological Survey (USGS), in cooperation with state and local agencies, maintains a network of thousands of stream gages. Some of these gages have more than 100 years of records, whereas others have been recently installed. Since 1964, the USGS has maintained a joint effort with appropriate state agencies to collect stream-gage data. As a result, a publication entitled "Water Resources Data, (State Name), Water Year 19—" is published jointly by these agencies each year. This publication contains a summary of the daily water flows (volumetric flow rates) and water-quality analyses. Although this is the largest source of streamflow data, other agencies such as the state Department of Water Resources; Soil Conservation Service, Agricultural Research Service and Forest Service of the U.S. Department of Agriculture; Bureau of Reclamation, Bureau of Land Management, both of the

U.S. Department of Interior; U.S. Army Corps of Engineers; Natural Resource Districts; agencies operating dams; and possibly some state universities operate and maintain stream gages. Private companies are presently furnishing USGS data on compact discs for access by personal computers for a nominal charge.

1.5.2 Precipitation and Climatological Data

Climatological data are collected by the National Weather Service (NWS) of the U.S. Department of Commerce. This agency has been known as the United States Weather Bureau and older records are listed under that name. The climatological data collected by this agency include hourly and daily amounts of rainfall, a station index showing the location of each station, total evaporation and wind movement, soil temperatures, temperatures and temperature extremes, and freeze data. The NWS maintains the largest and most complete network of rain gages in the United States. Many rain gages operated by the NWS have more than a 100 years of record. The complete record of any such gage can be obtained from the NWS Records Center in Asheville, North Carolina. Rain gages are maintained by many other agencies and individuals. Generally, the same agencies that maintain stream gages will also have a rain gage. Local inquiry will usually locate individuals who also maintain a rain gage. Such private rain gages are often well worth the search because they will often supply key data needed to better solve a hydrologic problem.

1.5.3 Topographic Maps

The USGS constructs and distributes $7\frac{1}{2}$-minute quadrangle topographic maps, which are most useful for hydrologic purposes. These maps cover nearly all the United States and are available from the Denver Federal Center, Denver, Colorado, or from state Geological Survey offices. They are also sold in some bookstores and sporting goods stores in larger cities. These maps are useful for determining relevant drainage-basin characteristics such as drainage area, topography, slope, and stream pattern.

1.5.4 Groundwater Data

Information on water tables, water wells, piezometric levels, and aquifer characteristics can be obtained from the USGS, state Geological Survey offices, universities, and local Soil Conservation Service offices. Local well drillers can usually supply valuable groundwater information for the area in which they operate.

1.5.5 Evaporation and Transpiration Data

These data, also known as evapotranspiration (ET) data, are needed for some hydrologic applications. Evaporation is measured by the NWS and evaporation data can be obtained from that agency. This information is published in the Climatological Summary by the NWS. Transpiration data can usually be obtained from the College of Agriculture at most land-grant universities. Other sources might be the Agricultural Research Service, Forest Service, or the local Soil Conservation Service.

1.5.6 Soil Maps

The Soil Conservation Service publishes generalized soil maps for each state and a detailed soil map for each county. These county maps are included in a publication that describes engineering characteristics of the soils found in that county/parish. Each county Soil Conservation Service office can supply the publications they have available for that county. These data are useful for many hydrologic purposes.

1.5.7 Geological Maps

Geological maps show lithology, stratigraphy, and structural properties of geological formations. They indicate the areal distribution of rock outcrops and the nature of those rocks. These maps are available from the USGS and state Geological Survey offices. The maps are useful for determining hydrological parameters of aquifers and groundwater exploration and development.

1.6 UNITS OF MEASUREMENT

The United States is unique in clinging to English units rather than metric units. The metric system of units is used throughout the rest of the world. For this reason, the English as well as metric units will be used freely in this text. Appendix A provides factors for conversion from one system of units to the other. Some of the commonly used hydrologic variables and their units are given in Table 1.1. The most commonly used English units are cubic feet per second (cfs or ft^3/s) for runoff discharge and cubic feet (cf or ft^3), acre-feet (a-f), and inches (in.) of runoff for volume of water.

TABLE 1.1 SOME COMMONLY USED HYDROLOGIC TERMS AND THEIR UNITS

Variable	Characteristics	Units of measurement
Precipitation	Depth	Inches; centimeters
	Intensity	Inches per hour; centimeters per hour
	Duration	Hours
Evaporation	Rate	Inches per day, month, or year; centimeters per day, month, or year
	Amount	Inches; centimeters
Infiltration	Rate	Inches per hour; centimeters per hour
	Depth	Inches; centimeters
Interception	Equivalent depth	Inches per storm duration; centimeters per storm duration
Depression storage	Equivalent depth	Inches per storm duration; centimeters per storm duration
Runoff	Discharge	Cubic feet per second; cubic meters per second
	Volume	Cubic feet; acre-feet, hectare-centimeter
	Equivalent depth	Equivalent inches over drainage basin area; equivalent centimeters over basin area

The corresponding metric units are cubic meters per second (cumec or m^3/s), cubic meters (m^3), hectare-meters (ha-m), and centimeters (cm).

The term acre-foot is defined as an area of 1 acre covered with water 1 foot deep. An acre is a land unit measurement approximately 209 feet by 209 feet square containing 43,560 square feet. This unit is useful and meaningful for irrigation applications. Its adoption throughout hydrology permits the use of manageable numbers for large volumes of water. An acre-foot is equal to 43,560 cubic feet. One inch of runoff is the volume of runoff water that is equivalent to 1 inch of water uniformly distributed over the entire drainage-basin area. This term permits ready comparison with standard terminology for rainfall amounts. Thus, it is convenient and illustrative to relate given inches of rainfall over a drainage basin to the resulting inches of runoff from that same drainage basin produced by that same rainfall event. A term, no longer in use but one that might be encountered in the older literature, the miners inch, is equivalent to a value ranging from 0.020 cfs to 0.28 cfs, depending on the state in which the measurement is made.

EXAMPLE 1.6 The total water supply of the world is about 326 million cubic miles. Express this supply in cubic feet, gallons, cubic meters, acre-feet, cubic kilometers, kilometer-meters, liters, and tons.

Solution

$$\text{Total water supply} = 326 \times 10^6 \text{ mi}^3$$

$$= 326 \times 10^6 \text{ mi}^3 \times \left(\frac{5280}{1 \text{ mile}}\right)^3 = 4.8 \times 10^{19} \text{ ft}^3$$

$$= 4.8 \times 10^{19} \text{ ft}^3 \times \frac{1 \text{ acre-ft}}{43{,}560 \text{ ft}^3} = 1.1 \times 10^{15} \text{ acre-ft}$$

$$= 1.1 \times 10^{15} \text{ acre-ft} \times \frac{3.25 \times 10^5 \text{ U.S. gallons}}{1 \text{ a-f}}$$

$$= 3.59 \times 10^{20} \text{ U.S. gallons}$$

$$= 4.8 \times 10^{19} \text{ ft}^3 \times \frac{1 \text{ km}^3}{(3281 \text{ ft})^3} = 1.36 \times 10^9 \text{ km}^3$$

$$= 1.36 \times 10^9 \text{ km}^3 \times \frac{10^3 \text{ m}}{1 \text{ km}} = 1.36 \times 10^{12} \text{ km}^2\text{-m}$$

$$= 1.36 \times 10^{12} \text{ km}^2\text{-m} \times \frac{10^6 \text{ km}^2}{1 \text{ km}^2} = 1.36 \times 10^{18} \text{ m}^3$$

$$= 1.36 \times 10^{18} \text{ m}^3 \times \frac{1000 \text{ liters}}{1 \text{ m}^3} = 1.36 \times 10^{21} \text{ liters}$$

$$= 3.59 \times 10^{20} \text{ U.S. gallons} \times \frac{9.2 \text{ lb}}{1 \text{ U.S. gallon}} \times \frac{1 \text{ ton}}{2240 \text{ lb}}$$

$$= 1.34 \times 10^{18} \text{ tons}$$

In the United States, 1 ton is 2000 lb. Therefore,

$$\text{Total water supply} = 1.60 \times 10^{18} \text{ tons}$$

EXAMPLE 1.7 If the world's total supply of 326 million cubic miles of water were poured upon the 50 United States, what would be the depth of water by which the land surface would be submerged? The land area of the United States can be taken to be approximately 2274 million acres.

Solution

$$\text{Volume of world's water supply} = 326 \times 10^6 \text{ mi}^3 \times \frac{5280^3 \text{ ft}^3}{1 \text{ mi}^3} \times \frac{1 \text{ acre}}{43{,}560 \text{ ft}^2}$$

$$= 1.102 \times 10^{15} \text{ acre-ft}$$

$$\text{The depth of submergence} = \frac{1.102 \times 10^{15} \text{ acre-ft}}{2274 \times 10^6 \text{ acre}}$$

$$= 485{,}609 \text{ ft}$$

1.7 A SHORT HISTORY OF HYDROLOGY

The concept of the hydrologic cycle is universally accepted. Chow (1964) has classified the historical development of hydrology into eight periods: speculation (ancient to 1400 A.D.), observation (1400 to 1600), measurement (1600 to 1700), experimentation (1700 to 1800), modernization (1800 to 1900), empiricism (1900 to 1930), rationalization (1930 to 1950), and theorization (1950 to 1965). To this classification can be added the period of modeling and computer simulation (1965 to date).

Ancient Chinese engineers were able to drill wells to a depth of 1500 meters as early as 2000 B.C. Persians constructed Quanats as early as 800 B.C. with drainage systems 500 feet below the ground surface. The Egyptians were able to irrigate about 1800 square miles of land around 500 B.C. These are only a few of the many important and complex water resources projects that were accomplished by ancient engineers. Early philosophers such as Thales (circa 650 B.C.), Plato (427–347 B.C.), and Aristotle (384–322 B.C.) taught that water from springs and rivers was driven into the rock by wind, and that fresh water came from a vast cavern called Tartarus or was condensed in a subterranean environment and raised to the surface. With all the speculative theories existing at that time, it is not surprising that experimental research was not undertaken.

Precipitation measurements were made in the fourth century B.C. in India. Practical observations made by Hero of Alexandria (ca. 1 A.D.), the Roman engineer Marcus Vitruvius (ca. 1 A.D.), Leonardo da Vinci (1452–1519), Bernard Palissy (1509–1589), and others provided a basis for the development of a quantitative understanding of the hydrologic cycle. Pierre Perrault (1608–1680) gaged rainfall over a 3-year period and was able to compute the runoff from a portion of the Seine River drainage basin. Crude as his measurements were, he disproved the existing theory that rainfall was inadequate to account for the discharge of springs and rivers. Edmé Mariotte (1620–1684) confirmed Perrault's work independently. He also demonstrated infiltration by experiments using the roof of the cellar of the Paris Observatory. English astronomer Edmund Halley (1656–1742) observed evaporation from the Mediterranean Sea and demonstrated that this evaporation was ample to supply the quantity of water returned to the sea by the streams flowing into it.

These early scientists formed the basis, empirical but quantitative, of hydrology. Subsequent work by later scientists advanced the understanding of hydrology and provided the basis for present knowledge. A more comprehensive historical account has been given by Meinzer (1949), Rouse and Ince (1957), Jones et al. (1963), Chow (1964), and Biswas (1972).

Those who wonder why it took so long for the concept of the hydrologic cycle to be understood would do well to consider the danger of a too familiar association with water. Many people living in modern environments take water for granted. Water arrives at the turn of a tap, dams hold back water, and television broadcasts make many experts on weather. The net result seems to be to ignore the hydrologic cycle. Perhaps more accurately, the effect seems to be to accept only that part of the hydrologic cycle that serves the desired purpose. This attitude seems to be the cause of many errors in hydrologic design. Inadequacy of hydrologic design in many cases can be linked to inadequate consideration of the hydrologic cycle.

EXERCISES

1.1. The annual evaporation from a lake, with a surface area of 1600 hectares, is 3 meters. Determine the average daily evaporation rate in hectare-centimeters per day during the year.

1.2. Rainfall takes place at an average intensity of 1 cm/h over a 250-hectare area for 3 days. Determine the average rate of rainfall in cubic meters per second (m^3/s). Determine the 3-day volume of rainfall in hectare-cm and hectare-meters. Also determine the 3-day volume of rainfall in centimeters of equivalent depth over the 250-hectare area.

1.3. Water is to be supplied from a reservoir fed by a stream with a discharge of 2 m^3/s to meet domestic requirements of an area with a population of 150,000. The average daily consumption is 300 liters per person. The lowest discharge of the stream is 0.25 m^3/s for a period of 15 days. Determine the reservoir size in km^3 and the rate of outflow when the reservoir is full.

1.4. Compute the time required to fill the reservoir in Exercise 1.3 when the demand of the population is being simultaneously fed by the stream and the reservoir is empty after a drought period. The stream discharge is 1.75 m^3/s.

1.5. An area is being irrigated by a stream with a drainage area of 300 km^2. The drainage area contribution is 0.1 $m^3/s/km^2$. Determine the discharge of the channel and the area irrigated if 0.37 m^3/s are required per 1000 hectares.

1.6. The average monthly precipitation in a watershed of 4500 km^2 is 46 cm. If the cumulative losses are 20% of precipitation, determine the area of Exercise 1.5 that can be irrigated with the remaining water. Also calculate the channel discharge.

1.7. Estimate the storage capacity of a reservoir for Exercise 1.6 when the average precipitation is 28 cm for a period of 20 days. The area calculated above is to be continuously supplied with its full demand.

1.8. Water is to be supplied to an area for both domestic and agricultural purposes. The population is 200,000 and the area to be irrigated is 3600 hectares. Water is to be pumped from the river. If the average daily consumption is 320 liters per person and the agricultural demand 0.33 $m^3/s/1000$ hectares, find the number of pumps required when 30% of the pumps are required to be standby. Also calculate the minimum discharge in the river to meet the above demand. The individual pump capacity is 0.1 m^3/s.

1.9. A town has an approximate population of 8000. The daily consumption of water per capita is about 500 liters per day. The annual precipitation is about 30 centimeters, 1% of which is lost by interception. The town plans to construct a storage reservoir for storage and supply of water to meet the demand of water consumption by the town population. It is estimated that reservoir would lose approximately 30% of annual precipitation through the processes of evaporation, infiltration, and deep percolation. Determine the following:

(a) What should be the reservoir volume in cubic meters?

(b) Find the volume in hectare-meters.

(c) What should be the reservoir area if it is to have a depth of 1 meter, 2 meters, 3 meters, 4 meters, and 5 meters?

(d) What should be the extent of the drainage area supplying water to the town?

1.10. There is a reservoir whose surface area is 20,000 hectares. The reservoir is initially dry. The average annual depth of water in the reservoir is 5 meters. There is a drainage basin with an area of 100,000 hectares that supplies water to the reservoir. Evaporation, infiltration, and deep percolation losses approximate 20% of annual precipitation. There is no other source of losing water. Determine the average annual precipitation in centimeters and meters.

1.11. Consider an input, over an area of 100,000 hectares, of 100 cm lasting for a period of 1 day. In the hydrologic budget equation, assume the following relationship between storage S and output Q:

$$S = 2Q$$

The daily demand of water is approximately 4,000,000 liters. Determine the periods of water excess and water deficiency. Take the initial condition as $Q = 0$ at $t = 0$, where t is time in hours.

1.12. Rainfall of intensity 0.5 in./h fell on a 2.0-mi^2 area for 6 h. Measured runoff (sum of hydrograph ordinates) at the end of this period was 3×10^6 ft^3. How much of the total 6-h precipitation was lost?

1.13. It is desired to fill a reservoir by means of a 36-in.-diameter corrugated metal pipe (CMP). The capacity curve of Figure 1.9 gives the volume of storage to be filled between elevations of 50 and 70 ft national geodetic vertical datum (NGVD). The tailwater rating curve of Figure 1.10 gives flow through the pipe for various reservoir elevations. Derive a filling schedule (elevation vs. time) for filling this reservoir by 2-ft increments.

1.14. For a river reach, the average rates of inflow and outflow during 1 hour are 800 cfs and 700 cfs, respectively. At the beginning of the hour, the storage in the reach is 20 acre-feet. Determine the storage in the reach at the end of the hour.

1.15. At a given time, the inflow to a river reach is 300 cfs and the outflow from it is 500 cfs. At the same time, the storage is 12 acre-ft. One hour later, the inflow rate is 500 cfs and the outflow is 550 cfs. Determine the change in storage in the reach that occurred during the hour. Is the storage at the end of the hour greater or less than the original value?

1.16. The storage in a river reach is 2 hectare-meters at a given time. Determine the storage 1 hour later if the average rates of inflow and outflow during the hour are 21 m^3/s and 18 m^3/s, respectively.

1.17. The storage in a river reach at a specified time is 2 hectare-meters. At the same time, the inflow to the reach is 15 m^3/s and the outflow from it is 20 m^3/s. One hour later, the inflow is 20 m^3/s and outflow is 20.5 m^3/s. Determine the change in storage that occurred in the reach. Is the storage at the end of the hour greater or less than the initial value?

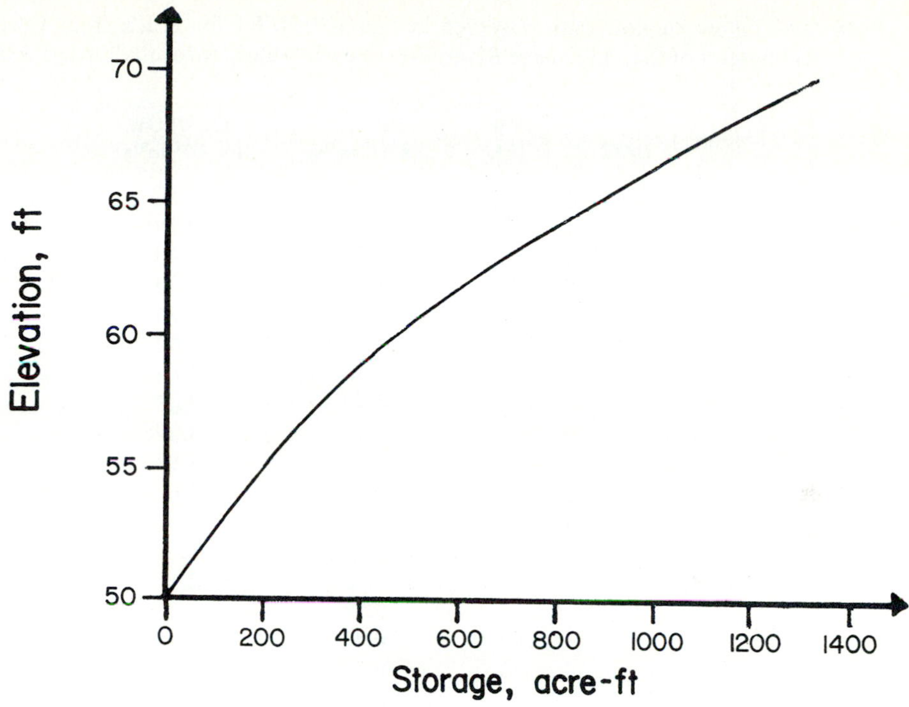

Figure 1.9 Cooper River rediversion: area-capacity curve for area between 50-ft NGVD plug and the powerhouse.

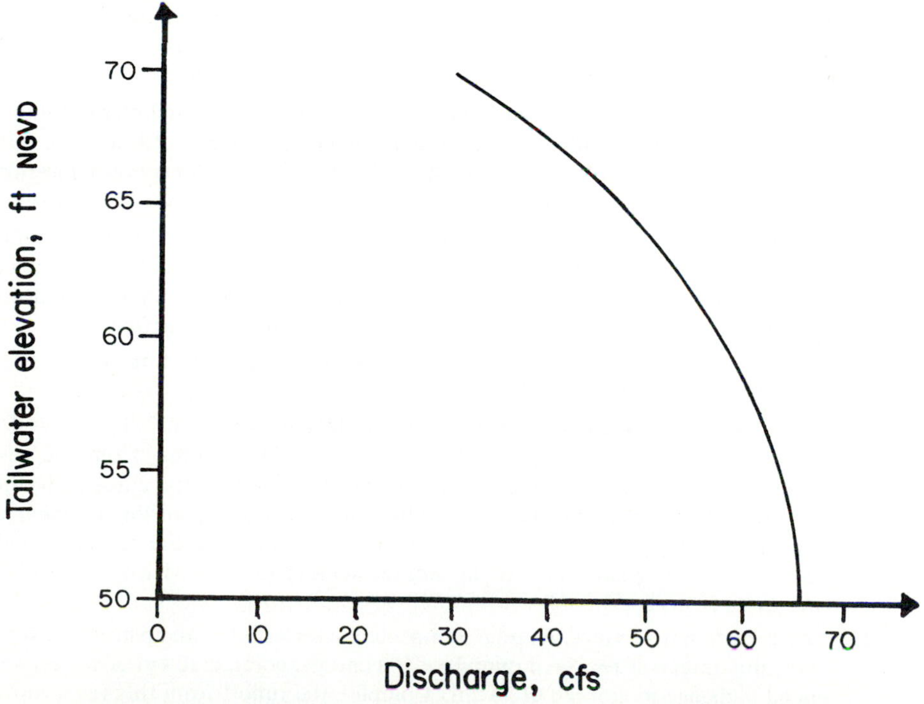

Figure 1.10 Cooper River rediversion: tailwater rating curve for area between 50-ft NGVD plug and the powerhouse.

1.18. The inflow–outflow data observed in March 1926 for the reach from Burrendong to Hellington of the Macquarie River, New South Wales, Australia, are as follows:

DATE	TIME (h)	INFLOW ($\times$ 10 m^3/s)	OUTFLOW ($\times$ 10^3 m^3/s)
24	12	0.18	0.12
	15	0.32	0.20
	18	0.70	0.40
	21	1.70	0.80
	24	3.22	1.36
25	03	4.56	2.00
	06	5.66	2.64
	09	4.40	3.28
	12	3.12	3.80
	15	3.94	3.94
	18	2.16	3.62
	21	1.80	3.06
	24	1.46	2.52
26	03	1.18	2.14
	06	0.98	1.76
	09	0.80	1.44
	12	0.66	1.18
	18	0.48	0.74
	24	0.34	0.50
27	12	0.32	0.30
	24	0.14	0.18
28	24	0.06	0.09

Compute the storage in the reach. Plot storage versus outflow.

1.19. A 1000-acre reservoir receives 100 in. of rainfall during a period of 24 months. During this period, the mean inflow to the stream is 10 cfs, the mean outflow from the stream is 8 cfs, and an increase in storage is 4000 acre-ft. Compute evapotranspiration (ET) in acre-feet as well as in inches.

1.20. A channel reach has an initial storage of 10,000 m^3. This reach receives inflow at the rate of 10 m^3/s for a period of 2 hours. Determine the volume of outflow produced by this channel reach if the final storage of the reach is 10,000 m^3. Assume that there is no loss of water due to seepage, evaporation, or any other abstraction.

1.21. A flow of 5 m^3/s enters a 500-hectare reservoir. Determine the time in hours required to raise its water level by 15 cm.

1.22. Consider a reservoir, initially dry, whose surface area is 1 km^2. It receives, from the beginning, a uniform inflow of 40 cm/h for 5 h. The outflow from the reservoir is observed as 0 at the beginning as well as 10 h later. When time equals 5 h, the outflow equals inflow. For simplicity, assume that the variation in outflow is linear during its rise as well as during its recession. Compute the storage in the reservoir and graph it against time. At what time will the storage achieve its peak? How long will it take for the reservoir to get dried?

1.23. There are two reservoirs, A and B, connected in series. A is an upstream reservoir. In 1 year, this reservoir received rainfall of 100 cm, evaporated 30 cm of water, and experienced a change in storage of 20 cm. Compute the runoff from this reservoir assuming that other abstractions are vanishingly small. Reservoir *A* contributes its runoff to reservoir *B* in which storage remains constant. Compute the runoff from this reservoir if its rainfall and evaporation are the same as for reservoir *A*.

1.24. Consider two reservoirs, *A* and *B*, connected in series. Reservoir *A* is upstream of reservoir *B*. The area of *A* is half of *B*. The area of *B* is 10,000 acres. Reservoirs *A* and *B* are subject to rainfall of 50 cm and evaporation of 30 cm on an annual basis. Reservoir *A* supplies 100 m^3/s of flow to reservoir *B*. A small town of 50,000 is located near reservoir *B*. The per-capita water consumption is 150 gallons per day. The town is contemplating to use reservoir *B* as a water supply reservoir. Determine its annual water storage. For how long can it supply water to the town?

CHAPTER 2 Environmental and Water Resources Problems

Water is essential to life. Without it, human survival could not be possible. The extent to which modern society depends upon water is apparent in the variety of ways in which water is used. Hydrology occupies the central role in development, management, and conservation of this vital resource—water. It may be instructive to survey the world's water resources before discussing the role of hydrology in developing solutions to specific water-related problems. Leeden (1975), L'vovich (1979), and Schendel (1975), among others, have given comprehensive accounts of the world's water resources. The discussion in the following section draws heavily from Morris (1968), and Geological Survey (1969).

2.1 INVENTORY OF THE WORLD'S WATER

The water in the world today has been about the same since the origin of the earth. The earth's estimated 326 million cubic miles (1358.72 million cubic kilometers) of water has remained unchanged in quantity throughout the 4 or 5 billion years of its existence. This water is unevenly distributed over the earth's surface in oceans, rivers, and lakes. An understanding of how uneven this distribution is and where the water occurs provides a clue to water resources development and management.

L'vovich (1979) has given an approximate distribution of the world's water, as given in Table 2.1. The figures in the table are approximate and may differ from those of others. For example, Table 2.2 shows an estimate of the world water balance presented by Nace (1971), which is slightly different from that of Table 2.1. Thus, these figures are to be considered in an approximate sense. Of the estimated 326 million cubic miles ($1{,}358.72 \times 10^6$ cubic kilometers) of water on earth, about 317

TABLE 2.1 WORLD'S ESTIMATED WATER SUPPLY

Source/location	Surface area		Water volume		% of total water
	mi²	km³	[a]mi³	km³	
Surface water					
Freshwater lakes	330,000	854,700	30,000	125,040	0.009
Saline lakes and inland seas	270,000	699,300	25,000	104,200	0.008
Average in stream channels	—		300	1250.4	0.0001
Subsurface water	50,000,000	129,500,000			
Vadose water (includes soil moisture)			16,000	66,688	0.005
Groundwater within depth of half a mile			1,000,000	4,168,000	0.31
Groundwater—deep lying			1,000,000	4,168,000	0.31
Other water locations					
Ice caps and glaciers	6,900,000	17,871,000	7,000,000	29,176,000	2.15
Atmosphere (at sea level)	197,000,000	51,023,000	3,100	12,920.8	0.001
World ocean	139,500,000	361,305,000	317,000,000	1,321,256,000	97.2
Total (rounded)			326,000,000		100

[a] A cubic mile of water equals 1.1 trillion gallons.

TABLE 2.2 ESTIMATE OF THE WATER BALANCE OF THE WORLD (AFTER NACE, 1971)

Source	Surface area ($\times 10^6$ km^2)	Volume ($\times 10^6$ km^3)	Volume (%)	Equivalent depth[a] (m)	Residence time
Oceans and seas	361	1370	94	2500	~4000 years
Lakes and reservoirs	1.55	0.13	<0.01	0.25	~10 years
Swamps	<0.1	<0.01	<0.01	0.007	1–10 years
River channels	<0.1	<0.01	<0.01	0.003	~2 weeks
Soil moisture	130	0.07	<0.01	0.13	2 weeks–1 year
Groundwater	130	60	4	120	2 weeks–10,000 years
Icecaps and glaciers	17.8	30	2	60	10–10,000 years
Atmospheric water	504	0.01	<0.01	0.025	~10 days
Biospheric water	<0.1	<0.01	<0.01	0.001	~1 week

[a] Computed by assuming a uniform distribution of storage over the earth.

million cubic miles (1,321.26 × 10^6 cubic kilometers), or 97%, is contained in the oceans, as shown in Figure 2.1. This is salt water spreading over the approximate oceanic basin area of 139.5 million square miles (361.3 million square kilometers), with an average depth of 12,000 feet (3660 meters). If the oceanic basins were shallow, seas would spread far onto the continents. An additional 1 million cubic

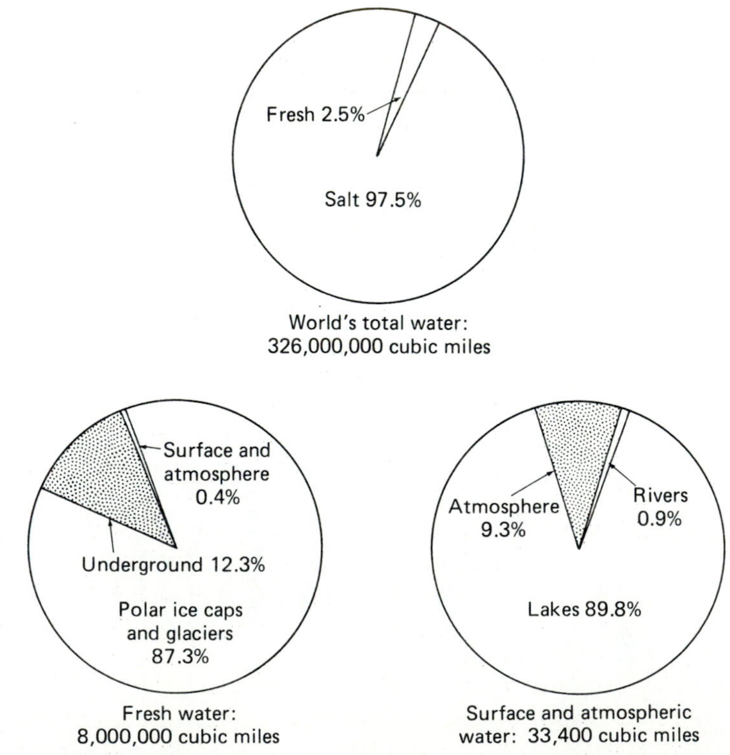

Figure 2.1 The world's water distribution (after Morris, 1968).

TABLE 2.3 APPROXIMATE DISTRIBUTION OF THE FRESH WATER ON THE EARTH

Source	Volume ($\times 10^6$ km^3)	Percentage
Polar ice caps and glaciers	29	77.33
Groundwater in depths up to 800 m	4.15	11.07
Groundwater in depths below 800 m	4.15	11.07
Lakes, rivers, and streams	0.12	0.32
Soil moisture and seepage	0.067	0.18
Atmospheric water vapor	0.013	0.03
Total	37.50	100

miles (4.17 million cubic kilometers) of salt water is buried underground. Thus, the remaining 2.5% (8.15 million cubic miles, or 33.97 million cubic kilometers) makes up the world's total supply of fresh water. Table 2.3 shows an approximate distribution of this fresh water on earth. The largest amount—some 7 million cubic miles (29.18 million cubic kilometers)—is frozen in the polar ice caps and in glaciers in various parts of the world. Ninety percent of this ice is in Antarctica, and the rest is, for the most part, in the Greenland ice cap. Approximately 2 million cubic miles (8.34 million cubic kilometers) of fresh water is stored in the earth, about half within a half mile (0.8 km) depth from the surface. This is more than 35 times the amount held on the surface in lakes, rivers, and inland seas, but relatively small compared to that stored in glaciers and ice caps. There is, however, a relatively small amount—about one hundredth of one percent—of the world's water supply stored in lakes, rivers, soils, and the atmosphere. This small amount is vitally important and is indeed essential to life on earth. This relatively small amount of water is estimated to be 33,400 cubic miles (139,211 cubic kilometers). About 30,000 cubic miles (125,040 cubic kilometers) is stored in lakes in various parts of the world, and the remaining 3400 cubic miles (14,171 cubic kilometers) is present in the world's rivers and in the atmosphere at any time.

2.1.1 The Atmosphere

The atmosphere can be considered as another kind of ocean, but the amount of water contained there is relatively small. The volume of the lower 7 miles of the atmosphere—the realm of the weather phenomena—is roughly four times the volume of the world's oceans, but the atmosphere contains only about 3100 cubic miles (12,921 cubic kilometers) of water, chiefly in the form of invisible water vapor, some of which is transported over land by air currents. If all the water in the atmosphere were to fall at once, the earth would be covered with only 1 inch (2.54 cm) of water. Water is continuously removed from the atmosphere and is precipitated over the surface of the earth. An equivalent amount is returned to the atmosphere through evaporation from land and water bodies and through transpiration by plants. As a result, the atmospheric water content remains practically constant.

Each year the earth receives some 113,000 cubic miles (470,984 cubic kilometers) of precipitation, 89,000 cubic miles (370,952 cubic kilometers) on the oceans

TABLE 2.4 PRECIPITATION AND EVAPORATION ON LAND AND SEA SURFACES

Item	Land	Sea
Area	136×10^6 km^2	374×10^6 km^2
Precipitation (P)	750 mm/year	870 mm/year
Evaporation (E)	545 mm/year	940 mm/year
$P - E$	+205 mm/year	−70 mm/year

and 24,000 cubic miles (100,032 cubic kilometers) on the land, as shown in Table 2.4. The 24,000 cubic miles (100,032 cubic kilometers) of annual land precipitation averages about 27 inches (68.6 cm). The amount of evaporation from the oceans is on the average of about 9% greater than ocean precipitation. This difference causes the water vapor to drift with the winds to fall as precipitation over the land areas. It is this precipitation that sustains the flow of all the world's rivers. The conterminous United States receives an average of 30 inches (76.2 cm) of precipitation every year, or about 1430 cubic miles (5960 cubic kilometers) in total volume. Evapotranspiration returns approximately 21 inches (53.34 cm) of this water to the atmosphere (about 1000 cubic miles, or 4168 cubic kilometers). Of course, some rain is water that was vaporized from the land areas and is being reprecipitated. It must be noted that precipitation is not uniformly distributed on earth. There are vast desert areas where precipitation occurs only rarely; on the other hand, there are areas where the annual precipitation is as much as 400 inches or even more. The importance of atmospheric water lies in its continued replenishment.

2.1.2 Rivers

Limited amount of information on flows of large rivers, as shown in Table 2.5, is compiled by the Military Engineer (1958), whereas the U.S. Geological Survey (1949) computed discharges of the principal rivers of the United States, as given in Table 2.6. The estimated total runoff from all rivers, large and small, measured and unmeasured, is about 9200 cubic miles (38,345.6 cubic kilometers) yearly, or 25 cubic miles (104 cubic kilometers) daily. Each year, some 8400 cubic miles (35,011.2 cubic kilometers) of water is discharged into the sea by the rivers of the world. Langbein and others (1949) have summarized the flow from the United States to the oceans, as given in Table 2.7. This represents an average flow of nearly 40 million cubic feet per second (1.132 million cubic meters per second). Sixty-six principal rivers of the world discharge about 3720 cubic miles (15,505 cubic kilometers) of water yearly. The world's largest river, the Amazon, has an average flow of 7.5 million cubic feet per second (0.212 million m^3/s), nearly one-fifth of the world's total. It is roughly 4 cubic miles per day and some 1300 cubic miles (5418.4 cubic kilometers) per year. The Congo, second to the Amazon, has an average discharge of nearly 1.4 million cubic feet per second (0.043 million m^3/s), or approximately 340 cubic miles (1417 cubic kilometers) per year, less than one-fifth of the flow of the Amazon or nearly 3.5% of the world's total. The estimated annual discharge of all African rivers is about 510 cubic miles (2125.6 cubic km). The North American continent's largest

TABLE 2.5 LARGE RIVERS IN THE WORLD (SUBJECT TO REVISION, AFTER MILITARY ENGINEER, 1958)

Name	Average annual discharge (cfs)	Drainage area (mi^2)	Length (mi)
Amazon (S. Am.)	7,200,000	2,772,000	3,900
LaPlata–Parana (S. Am.)	2,800,000	1,198,000	2,450
Congo (Africa)	2,000,000	1,425,000	2,900
Yangtze (Asia)	770,000	750,000	3,100
Ganges-Brahmaputra (Asia)	707,000	793,000	1,800
Mississippi–Missouri (N. Am.)	620,000	1,243,700	3,892
Yenisei (Asia)	610,000	1,000,000	3,550
Mekong (Asia)	600,000	350,000	2,600
Orinoco (S. Am.)	600,000	570,000	1,600
Mackenzie (N. Am.)	450,000	682,000	2,525
Nile (Africa)	420,000	1,293,000	4,053
St. Lawrence (N. Am.)	400,000	565,000	2,150
Volga (Europe)	350,000	592,000	2,325
Lena (Asia)	325,000	1,169,000	2,860
Ob (Asia)	—	1,000,000	2,800
Danube (Europe)	315,000	347,000	1,725
Zambesi (Africa)	—	513,000	2,200
Indus (Asia)	300,000	372,000	1,700
Amur (Asia)	—	787,000	2,900
Niger (Africa)	—	584,000	2,600
Columbia (N. Am.)	235,000	258,000	1,214
Yukon (N. Am.)	150,000	330,000	2,300
Huang (Asia)	116,000	400,000	2,700
Sao Francisco (S. Am.)	—	252,000	1,811
Euphrates (Asia)	—	430,000	1,700
Murray–Darling (Australia)	13,000	414,000	2,345

river, the Mississippi, has an average flow of roughly 630,000 cubic feet per second (17,829 m^3/s); its drainage area is 1.243 million square miles, or 3.22 million km^2 (about 40% of the total area of the 48 conterminous United States). This amounts to some 133 cubic miles (554.34 cubic km) per year, or approximately 34% of the total discharge from all the rivers of the United States.

The Columbia, the nearest American competitor of the Mississippi, discharges less than 75 cubic miles (312.6 cubic km) per year. The Colorado River discharges only about 5 cubic miles (20.84 cubic km) annually. By comparison, the Amazon is nearly 10 times the size of the Mississippi, and about three times the flow of all U.S. rivers.

Although the amount of water contained in the world's rivers is relatively small, the importance of rivers is related to the flow of water and its residence time. The river water has a turnover time on the order of 2 weeks. Since the river flow is primarily produced by precipitation that is neither continuous nor uniform, the flow of a river is continually fluctuating. Two-orders-of-magnitude variation from minimum flow to peak flow is not uncommon. The flow in a river fluctuates from day to day, month to month, and year to year.

TABLE 2.6 LARGE RIVERS IN THE UNITED STATES IN ORDER OF AVERAGE DISCHARGE AT MOUTH (FIRST-ORDER TRIBUTARIES MARKED T, SECOND-ORDER TRIBUTARIES MARKED TT) (AFTER U.S. GEOLOGICAL SURVEY, 1949)

Rank	River	Length (mi)	Drainage area (mi^2)	Average discharge (1921–1945) (ft^3/s)	Most distant source	Mouth
1	Mississippi	[a]3,892	1,243,700	[b]620,000	Source of Red Rock River, MT	Gulf of Mexico
2	Ohio (T)	1,306	203,900	255,000	Potter Co., PA	Mississippi River
3	Columbia	1,241	258,200	235,000	Columbia Lake, B.C.	Pacific Ocean
4	St. Lawrence	—	[c]302,000	[c]226,000	—	—
5	Mississippi above Missouri River (T)	1,170	171,600	91,300	Lake Itasca, MN	Confluence with Missouri River
6	Missouri (T)	2,714	529,400	70,100	Source of Red Rock River, MT	Mississippi River
7	Tennessee (TT)	900	40,600	63,700	SW Virginia	Ohio River
8	Mobile	758	42,300	59,000	NW Georgia	Mobile Bay
9	Red (T)	1,300	[d]91,400	[d]57,300	Eastern edge of New Mexico	Mississippi River
10	Arkansas (T)	1,450	160,500	45,200	Lake Co., CO	Mississippi River
11	Snake (T)	1,038	109,000	44,500	Ocean Plateau, Teton Co., WY	Columbia River
12	Susquehanna	444	27,570	35,800	Otsego Lake, Otsego Co., NY	Chesapeake Bay
13	Alabama (T)	720	22,600	31,600	Jacks Creek, NW, GA	Mobile River
14	White (T)	690	28,000	—	Madison Co., AR	Mississippi River
15	Willamette (T)	270	11,250	30,700	Tumblebug Creek, Douglas Co., OR	Columbia River
16	Wabash (TT)	475	33,150	30,400	Darke Co., OH	Ohio River
17	Cumberland (TT)	720	18,080	27,800	Port Fork, Letcher Co., KY	Ohio River
18	Illinois (T)	420	27,900	27,400	Source of Kankakee River, Joseph Co., IN	Mississippi River
19	Tombigbee (T)	525	19,500	27,000	NE Mississippi	Mobile River
20	Sacramento	382	[e]27,100	—	Siskiyou Co., CA	Suisan Bay
21	Apalachicola	500	19,500	25,000	Towns Co., GA	Gulf of Mexico
22	Pend Oreille (T)	490	25,820	24,600	Near Butte, MT	Columbia
23	Colorado	[f]1,360	[f]242,900	—	Rocky Mountain National Park, CO	—
24	Hudson	306	13,370	21,500	Essex Co., NY	Upper NY Bay
25	Allegheny (TT)	325	11,700	19,200	Potter Co., PA	Ohio River
26	Delaware	[g]390	[g]12,300	19,000	Source of W. Branch, Schoharie Co., NY	Delaware Bay

[a] The length from mouth to source of the Mississippi River in Minnesota is 2350 miles.

[b] About 25% of the flow occurs in the Atchafalaya River.

[c] At international boundary, lat. 45°.

[d] Flow of Ouachita River has been added.

[e] About

[f] At Arizona–Sonora boundary; natural flow not accurately known because of large depletions for irrigation.

[g] At Deepwater Point on Delaware Bay.

TABLE 2.7 SUMMARY OF ESTIMATED FLOW FROM THE UNITED STATES TO THE OCEANS (AFTER LANGBEIN AND OTHERS, 1949)

Part of drainage system	Description	Area (mi^2)	Mean annual flow (ft^3/s)
1	North Atlantic slope basins	148,000	210,000
2	South Atlantic slope and eastern Gulf of Mexico basins	284,000	325,000
3, 5, 6, 7	Mississippi River basin	1,250,000	620,000
5	Hudson Bay basins	48,000	5,000
4	St. Lawrence River basin	130,000	140,000
8	Western Gulf of Mexico basins	320,000	55,000
9	Colorado River basin	246,000	23,000
10	Great Basin	215,000	0
11	Pacific slope basins in California	117,000	80,000
12, 13, 14	Columbia River basin and coastal streams in Oregon and Washington	262,000	345,000
Total		3,020,000	1,803,000

2.1.3 Lakes

Lakes have been defined as wide places in rivers. This is essentially true of the many small lakes that are impounded by river channels. There is a great variation in lake characteristics and geologic conditions that resulted in lake formation. The earth's land areas are dotted with hundreds of thousands of lakes. Wisconsin and Minnesota contain some tens of thousands each. This same is true of Finland and Canada. The lakes of the world contain practically all the fresh, liquid, surface water in existence, which is about 30,000 cubic miles (125,040 cubic km). The total lake surface area is about 330,000 square miles (854,700 square km). Important as they may be locally, these numerous small lakes hold only a minor amount of the world supply of fresh water, most of which is contained in a relatively few large lakes on three continents, as shown in Figure 2.2. A large lake can be considered to have 5 or more cubic miles of water. Canada has probably more lake area than any other country in the world. The Great Lakes of North America hold nearly one-fifth of the world's total lake storage, which is estimated at 5000 cubic miles (20,840 cubic km). The Great Lakes, and other large lakes in North America (chiefly in 48 United States and Canada), hold about 7800 cubic miles (32,510.4 cubic km) of water, 26% of all liquid fresh water in existence. The large lakes of Africa—Nyasa (Malawai), Tanganyaki, and Victoria—together contain an estimated 8700 cubic miles (36,261.6 cubic km), which is 29% of the world's total. Russia's Lake Baikal, the deepest lake in the world, contains over 6300 cubic miles (26,258.4 cubic km), over 20% of the world's total, and considerably more than the combined water content of the five North American Great Lakes. Asia's large lakes contain about 6400 cubic miles, 21% of the total, nearly all of which is in Lake Baikal. Together, these nine lakes contain about two-thirds of the world's fresh surface water. The remaining third is distributed in hundreds of thousands of smaller lakes in all parts of the world. Lakes of the three continents—North America, Africa, and Asia—account for approximately

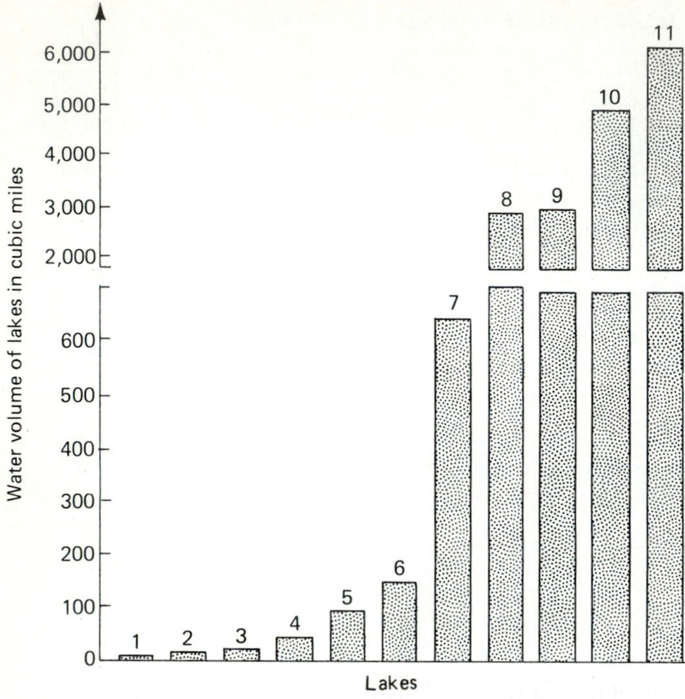

1. Dubawnt, Canada
2. Tungting, China
3. Leman, Switzerland
4. Vanern, Sweden
5. Nipigon, Canada
6. Erie, U.S. & Canada
7. Great Bear, Canada
8. Superior, U.S. & Canada
9. Nyasa, Africa
10. Tahganyika, Africa
11. Baikai, Siberia

Figure 2.2 Distribution of water volume in the largest lakes.

75% of the world's fresh surface water. Lakes on other continents—Europe, South America, and Australia—contain only about 720 cubic miles (3001 cubic km), or roughly about 2% of the total. Thus, the remaining less than one-fourth of the total fresh surface water fills the hundreds of thousands of rivers and smaller lakes found throughout the world.

Whether a lake contains fresh or salt water makes a considerable difference in its usefulness to man. Saline lakes are equivalent in magnitude to fresh water lakes. Their total area is 270,000 square miles (699,300 square km) and their total volume is about 25,000 cubic miles (104,200 cubic km). The distribution, however, is quite different. Nearly 75% of the total saline water (19,240 cubic miles, or 80,192 cubic km) is in the Caspian Sea, and most of the remainder is in Asia. The Great Salt Lake of the United States accounts for only 7 cubic miles, a comparatively insignificant amount.

The importance of lakes lies in their ability to store water, allowing rivers to flow during periods of low precipitation. Lakes also regulate flow pattern of rivers. There are numerous lakes created by man to store water where little or none existed previously. As an example, Lake Mead, in the United States, created by the Hoover Dam on the Colorado River, holds about 9 cubic miles (37.5 cubic km) of water, the lake behind the Kariba Dam in Africa has a volume of 39 cubic miles (162.55 cubic km), and the lake behind the huge Portage Mountain Dam on the Pence River in British Columbia, Canada, has a volume of 25 cubic miles (104.2 cubic km). Storage of water in natural or man-made lakes is extremely valuable, and billions of dollars have been spent to build dams and control works to usefully manage this water.

2.1.4 Subsurface Water

Beneath the surface of the earth, there exists a huge reservoir of fresh water. This subsurface water can be divided into soil moisture, vadose water, groundwater, and deep-lying groundwater. The zones of soil moisture and vadose water are combined into a zone of aeration. The amount of water held as soil moisture at any time is on the order of 6000 cubic miles for the whole world, an insignificant amount by comparison with the earth's total water, but vital to life. This water is crucial to plant life and food production.

Beneath the surface of the earth, there is a zone where the pores of rocks and sediments are completely saturated with water. This water is termed as groundwater. The upper surface of this zone of saturation is called the water table, which is located at varying distances below the ground, and in some cases, right at the land surface. Water in the unsaturated zone above the water table is called vadose water and includes the zone of soil moisture. The volume of the vadose water below the soil moisture zone is estimated to be 10,000 cubic miles (41,680 cubic km) for the world as a whole. Although this water is not extractable, its importance rests with its being potential groundwater recharge. The zone of aeration acts as a filter for water passing through it. Practically all groundwater is derived from precipitation, which filters down through the vadose zone before reaching the groundwater.

Below the water table, to a depth of half a mile of the earth's surface, there is estimated to be about 1 million cubic miles of groundwater. An equal, if not greater, amount is deeply buried, perhaps down to 5 miles (8 km) below the surface. This deep groundwater is not economically recoverable, and much of it is strongly mineralized. The volume of groundwater in the upper half mile (0.8 km) of the continental crust is nearly 3000 times greater than the volume of water in all rivers at any one time, and about 20 times greater than the combined volume of water in all rivers and lakes.

The quantitative importance of groundwater can be further appreciated by comparing it with other components of the hydrologic cycle. Groundwater accounts for nearly two-thirds of the freshwater resources of the world if saline waters of the oceans and seas are excluded from the earth's water. If only the utilizable freshwater resources (excluding the icecaps and glaciers) are considered, groundwater accounts for almost the total volume. If only the most "active" groundwater regimes are considered, which are estimated at 4×10^6 km^3 by L'vovitch (1979), then groundwater accounts for 95% of the volume, and the remaining 5% is distributed with 1.5% as soil moisture and 3.5% in lakes, swamps, reservoirs, and river channels. The volumetric superiority of groundwater is, however, tempered by its slow motion, resulting in long residence time.

Through baseflow, groundwater sustains perennial rivers during rainless periods and droughts. Groundwater flows slowly through moderately to highly permeable strata, called aquifers, at rates of a few inches to perhaps several hundred feet per day; 40 to 50 feet (10 to 15 m) per day would be a rather high rate of flow. Depending upon the distance the groundwater has to travel to reach the surface discharge area, water in shallow to moderately deep zones may remain underground from a few hours to more than 100 years. At great depths, water may take tens or hundreds of thousands of years to pass through an aquifer, and some water is completely stagnant. The volume of groundwater in storage in the United States to a depth of half a mile (0.8 m) is approximately equal to the total of all recharge during about the last 150 years. Crude as this estimate is, it emphasizes the fact that groundwater reserves, although huge, are not wholly self-renewing annually. Whenever these reserves have been depleted by pumpage, it may take many decades to recover even if pumping were stopped completely.

EXAMPLE 2.1 Assume that the world's recoverable groundwater is 4×10^6 km^3. For how long will this water be able to meet the water needs of the world population of, say, 5 billion if per-capita daily consumption (domestic, commercial, etc.) is taken to be 4000 liters?

Solution

$$\text{World population} = 5 \times 10^9$$

$$\text{Per-capita daily consumption} = 4000 \text{ liters}$$

$$\text{Total world daily consumption} = 5 \times 10^9 \times 4000 = 2 \times 10^{13} \text{ liters}$$

$$\text{Total groundwater supply} = 4 \times 10^6 \text{ km}^3 = 4 \times 10^{15} \text{ m}^3$$

$$= 4 \times 10^{18} \text{ liters}$$

$$\text{Time the supply will last for} = \frac{4 \times 10^{18}}{2 \times 10^3} \text{ days} = 2 \times 10^5 \text{ days}$$

$$= \frac{2}{365} \times 10^5 \text{ years} = 548 \text{ years}$$

2.1.5 Ice and Snow

An estimated 7 million cubic miles of fresh water is permanently frozen in the polar ice caps, far greater in quantity than all the liquid fresh water in the world. This fresh water, however, is as unusable as the salt water of the oceans. The Greenland ice cap is about 667,000 square miles (1,727,530 square km) in area, and nearly 5000 feet (1524 m) in thickness; its volume is about 630,000 cubic miles (2,625,840 cubic km). If melted, it would produce enough water to maintain the Mississippi River for more than 4700 years. Nevertheless, this constitutes less than 10% of the total volume of ice caps and glaciers.

By far the largest amount of the world ice is located in the Antarctica ice cap. The area of the ice cap is about 6 million square miles (15.54 million square km) and the total volume is between 6 and 7 cubic miles (between 25 and 29 cubic km), about 90% of the total world ice and about 64% of all water outside the oceans. To visualize the amount of water frozen in Antarctica, if its ice cap were melted at a suitable uniform rate, then it could feed: (1) the Mississippi River for more than 50,000 years, (2) all rivers in the United States for about 17,000 years, (3) all of Canada's rivers for about 13,000 years, (4) the Amazon River for approximately 5000 years, or (5) all the rivers in the world for about 800 years. From a hydrologic standpoint, the volume of ice in the ice caps remains constant. The importance of these polar ice masses lies in their influence on weather patterns. If there were to be extensive melting of ice caps due to climatic changes, the sea level would rise, having disastrous effects in coastal areas.

Mountain glaciers, such as those of the Alps in Europe (alpine glaciers), the Himalayas of Asia, and the Cascades of North America, exert a direct influence on the hydrologic cycle, and are like lakes and groundwater reservoirs. The total volume of all alpine glaciers and small ice caps in the world is only about 50,000 cubic miles (104,200 cubic km), comparable to the combined volume of large saline and fresh water in lakes. These glaciers are important locally, but they contain only an insignificant fraction of the world's water. Water from melting glaciers frequently sustains streamflow through dry seasons where flow would otherwise cease. Analogous to lakes or other reservoirs, the importance of glaciers lies not so much in the actual quantity of water contained in them, but in their ability to help provide more uniform streamflow.

Snow is another form of ice and can store varying amounts of water for varying periods of time. Precipitation stored during winter months in the form of snow exerts a marked effect on the distribution of streamflow throughout the year. Instead of infiltrating the soil or running off in stream channels as rainfall does, this water is stored for periods of as long as several months. The water is released upon melting, which then follows the usual phases of the hydrologic cycle. Much of the total precipitation in Canada may occur in the form of snow.

2.1.6 Oceans

The total amount of precipitation falling on the earth each year is about 0.42 million cubic kilometers (0.32 million cubic kilometers on the ocean and 0.1 million cubic kilometers on the land). As seen from Table 2.4, there is about 9% more evaporation than precipitation over oceans; this deficit is balanced by excess precipitation over the land surface. Thus, the volume of water carried by rivers to the sea each year is about 0.038 million cubic kilometers. Less than 4% of this total river flow is used for irrigation and the remainder, unutilized, flows to the sea.

It is not precisely known as to how much water moves from one ocean to the other, say, from the Pacific to the Atlantic Ocean, by vapor transfer, precipitation, and runoff. Table 2.8 provides an approximate water balance of the oceans. Estimated total flow into the sea from rivers in the 48 United States is about 1,803,000 cubic feet per second (51,025 m^3/s), which amounts to approximately 390 cubic miles (1,625.52 cubic km) per year. Values for runoff (390 cubic miles, or 1,625.52 km^3) plus evaporation (nearly 1000 cubic miles, or 4168 km^3) do not quite equal the precipitation (nearly 1430 cubic miles, or 5960 cubic km), because these values are not precise. The missing 40 cubic miles of water, roughly 10% of the value for streamflow, might represent direct groundwater discharge.

TABLE 2.8 WATER BALANCE OF OCEANS IN mm/year (AFTER SCHENDEL, 1975)

Ocean	Area ($\times 10^6$ km^2)	Precipitation (mm/year)	Inflow from adjacent continents (mm/year)	Evaporation (mm/year)	Water exchange with other oceans (mm/year)
Atlantic	107	780	200	1040	−60
Arctic	12	240	230	120	350
Indian	75	1010	70	1380	−300
Pacific	167	1210	60	1140	130

2.1.7 Continents

An approximate water balance of the various continents is shown in Table 2.9. L'vovich (1979) has computed the world distribution of runoff, as shown in Table 2.10. The Water Resources Council (1968) has presented a comprehensive account of water resources of the United States. Africa, despite its equatorial forest zones, is the driest continent of the world, with only 20% precipitation becoming runoff. Europe and North America have the highest runoff rates. The United States gets about 30 inches (76.2 cm) of precipitation every year, or about 4300 billion gallons per day. Evapotranspiration returns about 21 inches of this water to the atmosphere, or 3100 billion gallons a day—nearly 70% of the total water supply. The total streamflow from surface and groundwater sources is about 8.5 inches (21.6 cm) a year, or about 1200 billion gallons per day. This is the amount available for human use—homes, industry, irrigation, recreation, and so on.

2.2 APPLICATION OF HYDROLOGY

Hydrology touches every human life in some manner. To some, it is simply a need for drinking water, and to others, the need for water might be economic or just for convenience. Modern applications of hydrology are often concerned with floods and flooding along with floodplain management. Changing urban patterns have aggravated flooding, and as a result, flooding is higher and more widespread in some areas

TABLE 2.9 WATER BALANCE OF CONTINENTS IN mm/year (AFTER SCHENDEL, 1975)

Continent	Area ($\times 10^6$ km^2)	Precipitation (mm/year)	Total runoff (mm/year)	Runoff as % of precipitation	Evaporation (mm/year)
Africa	30.3	686	139	20	547
Asia	45.0	726	293	40	433
Australia	8.7	736	226	30	510
Europe	9.8	734	319	43	415
North America	20.7	670	287	43	383
South America	17.8	1648	583	35	1065

TABLE 2.10 WORLD DISTRIBUTION OF RUNOFF (AFTER L'VOVICH, 1979)

Continent (or other area)	Atlantic slope		Pacific slope		Regions of interior drainage		Total land area	
	Area ($\times 10^3$ mi^2)	Runoff (in.)	Area ($\times 10^3$ mi^2)	Runoff (in.)	Area ($\times 10^3$ mi^2)	Runoff (in.)	Area ($\times 10^3$ mi^2)	Runoff (in.)
Europe (including Iceland)	3,073	11.7	—	—	661	4.3	3,734	10.3
Asia (including Japanese and Philippine Islands)	4,626	6.4	6,422	11.8	5,273	0.66	16,321	6.7
Africa (including Madagascar)	5,110	14.0	2,109	8.6	4,291	0.54	11,510	8.0
Australia (including Tasmania and New Zealand)	—	—	1,634	5.5	1,441	0.24	3,075	3.0
South America	6,041	18.7	519	17.5	381	2.6	6,941	17.7
North America (including West Indies and Central America)	5,657	10.8	1,914	19.1	322	0.43	7,893	12.4
Greenland and Canadian Archipelago	1,499	7.1	—	—	—	—	1,499	7.1
Malayan Archipelago	—	—	1,012	63.0	—	—	1,012	63.0
Total or Average	26,006	12.4	13,610	15.5	12,369	0.82	51,985	10.5

than before. Drought is common throughout the world. To those people who depend on water for crops and livestock, this is the most important role of hydrology. Water supply is a common application of hydrology—so common that it is often unappreciated by many people. Increasing populations and the accompanying increase in industry have provided tremendous sources of pollution for our water resources. Hydrologists are deeply involved in attempting to alleviate this serious problem.

Other applications of hydrology apply to fewer people than those mentioned previously. Industry throughout the world has an important concern with hydrology. Navigation of streams, harbors, and seas has always been a basis for commerce. Highways, railroads, and other commercial entities require bridges to span streams and rivers. Agriculture is dependent on irrigation for the production of food and fiber. The irrigated food production is so important that the present world could not be fed without this hydrologic application of water. Water sports are an important part of life of many people. Fortunately, other hydrologic applications such as dams for power and irrigation provide added opportunity for the recreational use of water. The fishing industry and recreational fishermen have a vital interest in providing water compatible with fishing. More and more, modern society demands that the appearance of water development and use be maintained in a manner that is pleasing to view. These and other demands by people, government, and industry provide unlimited opportunities for application of hydrology. Good discussions on the role of hydrology in water resources development are given by Dooge (1988) and Maione (1988).

2.2.1 Flood Control

A flood occurs when a lake, reservoir, or channel is unable to contain the amount of water it receives. It also occurs when an area has inadequate drainage to drain excess precipitation. The result is an inundation of what is usually a dry land. Floods are sometimes caused by the failure of hydraulic structures such as dams, levees, and dykes. Natural floods are, however, more common. The problem of flooding is defined by its areal extent, duration, intensity, and damage. The projects designed to mitigate flooding and flood damage may be structural (e.g., dams, levees, dykes, diversions, floodwalls, and channels), nonstructural (e.g., flood proofing, floodplain management, and relocation), or a combination of both. The hydrologic input needed to design such projects includes: (1) peak discharge and its frequency of occurrence, (2) duration and volume of flood hydrograph and their probabilities of occurrence, and (3) the arrival of the next flooding.

EXAMPLE 2.2 Are floods greater and more frequent today than they were a century or so ago?

Solution The answer to the question is partly yes and partly no. Due to increased urbanization and removal of forests, local flooding may be more frequent and more severe in some areas today than it was in earlier times. However, whether extensive flooding today has increased in magnitude or frequency over large areas is debatable and the empirical evidence does not point to increased flooding. For example, long periods of record on European rivers (800 years on the Danube and 350 years on the Seine) show no increase in either the magnitude or the frequency of floods in these river basins despite deforestation. So the belief that floods are caused by denuded watersheds or the removal of forests has increased the magnitude of large floods does not hold universally.

2.2.2 Drought Mitigation

A drought occurs when there is a shortage of water by comparison with the demand for it. There may not be enough water in lakes, reservoirs, or streams, or precipitation may be deficient. Agricultural, hydrological, and meteorological droughts are usually distinguished. These three types are significantly interrelated, although in the extreme, these may be independent of one another. Analogous to flooding, the problem of drought is defined by its areal extent, duration, severity, and the onset of the next drought. From a hydrological perspective, low discharge (defined over a period) and its frequency of occurrence, duration of this low discharge and volume of low flow, as well as their frequencies, and the probability of occurrence of the next drought are useful to design drought-mitigation projects. A similar type of information is needed for rainfall in case of meteorological drought and for soil moisture in case of agricultural drought. Construction of water impoundments, groundwater pumpage, interbasin transfer, water conservation, and even augmentation of atmospheric precipitation through cloud seeding are some of the ways to mitigate droughts. However, droughts are known to occur over large areas, even as large as continents, and the effectiveness of mitigation measures is only limited.

2.2.3 Water Supply

Water is the key to modern development, supplying the renewable energy required for industrial growth, providing access to raw materials, and playing a vital part in the processing of these materials. In the United States, the per-capita consumption of water is about 1500 gallons a day. Less than 10% of this amount is used by individuals in their own households for drinking, preparing food, washing, lawn and garden sprinkling, etc. The greater part is used by industry and agriculture; the demands for irrigation are higher than those of industry.

A water-supply scheme must provide sufficient water of acceptable quality to serve its intended purpose, be it urban, agricultural, or industrial. The disruption in water supply should be minimum. Hydrology determines the volume of water to be stored to achieve the desired objective and the probability with which this volume of water will not be available. Hydrology also specifies the arrival of the next shortage of water and the frequency of its occurrence. In coastal areas, groundwater aquifers are threatened by saltwater intrusion. This problem is further exacerbated by excessive pumping of groundwater. Hydrological techniques are used to determine a safe yield without encroachment of saltwater.

2.2.4 Pollution Control

Water is an efficient and economical carrier of undesirable materials. It dilutes the waste and to a certain extent, by natural processes, disposes of that waste. However, there is a limit to the amount of waste that can be absorbed by any water-course, including rivers, lakes, reservoirs, and seas. This limitation is too often forgotten in the rush of disposing of waste resulting from growing population and expanding industry. Our polluted water bodies are an ample evidence to attest to this attitude. This, however, is not to suggest prohibition of all water products from watercourses, but to plea for wise water management, economically and socially viable. Hydrology is a key to achieve an acceptable, economic balance that takes into account the many and various services rendered by water bodies. Specifically, it provides information for disposition of water in time and space, both in terms of quantity and quality, in water bodies.

2.2.5 Urban Development

Davis (1976) reported that as of mid-1976, farm land was being swallowed up by urban development at the rate of 2 million acres per year. Land is also being transferred from agriculture to energy and industrial usage, both accentuating urban development. Urban planning and development involve construction of houses or subdivisions, schools, sports and recreation facilities, shopping centers, roads, culverts, bridges, drainage systems, parks, water-supply schemes, waste-disposal facilities, etc. Hydrology gives the design discharge and its probability of occurrence needed for design of hydraulic works. It specifies the extent and severity of flooding needed to ensure building of houses on safe grounds, out of floodplains. It also quantifies, on the other hand, hydrologic consequences upstream and downstream of urban development. For example, hydrology determines if flooding will increase or decrease as a result of urban development.

2.2.6 Industrial Development

Industry has an enormous thirst for water. The largest quantity is used for cooking purposes, but manufacturing processes also use up considerable quantities of water. Plant sanitation is another important area of water use. Just to give a relative indication of industrial water use, 10 gallons of water is needed to refine a gallon of gasoline, 18 gallons of water to refine a barrel of oil, 100 gallons of water to produce a gallon of alcohol, 250 tons of water to produce a ton of sulphate wood pulp, and about 60,000 gallons of water to produce a ton of steel. Therefore, for industrial development to take place, two basic problems have to be resolved: (1) water supply, and (2) disposition of waste. Industrial pollution is a problem in industrialized countries these days. Hydrology assists with addressing these problems, as discussed in part before and to be discussed in part later. However, industrial development also involves roads, land-use change, etc., and hydrology determines consequences of these changes.

2.2.7 Design of Hydraulic Works

Dams, culverts, spillways, bridge crossings, dykes, levees, diversions, channel-improvement works, drainage works, etc., are typical hydraulic works required for water-resources development and management. Design of these works requires an estimate of peak discharge of given frequency. Hydrology produces this estimate. Also estimated using hydrology are the environmental consequences of these works.

2.2.8 Agricultural Production

The largest user of water is agriculture. Efficient water management is essential, especially in dry areas. In humid areas, where irrigation is not generally required, methods of agriculture substantially affect streamflow. Proper agricultural practices conserve precipitation for crop use, prevent the loss of precious soil, and preserve the quality of the streams that drain the land. Today, vast irrigation projects that carry water long distances are common. Crop production involves moisture forecasting, supply of water to farms, management of irrigation water, application of chemicals and fertilizers, drainage of excess waters, soil conservation, etc. Hydrology is used to determine the time history of soil moisture needed for irrigation scheduling, and to dispose of excess waters during flooding. It also is needed to determine soil erosion and sediment transport, migration of chemicals and fertilizers, and their impact on water quality. Hydrology may be used to design a network of wells for a farm, or plan a system of dams, canals, and ditches based on soil properties, land slope, location of the water table, climate, and other factors.

2.2.9 Energy-Resources Development

Thermal, nuclear, and hydropower plants constitute the principal sources of electrical power generation. About 15% of the electricity in the United States today is generated by hydroelectric plants. Hydrology is applied to design these plants safely to avoid flooding and minimize consequent risk of failure. Thermal and nuclear power plants generate waste that needs to be disposed of. Hydrology is applied to determine the water supply needed for cooling purposes and for safe disposition of plant-generated waste. Geothermal energy appears as steam from deep beneath the earth's surface. Hydrology is used to help locate areas where use of the geothermal energy may be feasible and then locate and help design well fields to extract the heated water. Hydrology plays a crucial role in mining and oil exploration. The landscape disturbed, as a result of these activities, should be restored to its original form. Hydrology is applied to design such a landscape.

2.2.10 Land Conservation

Careless farming methods can speed up the runoff of rainfall, resulting in erosion of soil. This increases the danger of flooding downstream and causes streams to become more turbid because of increased concentration of sediments in the stream. Loss of fertile lands due to erosion and of coastal areas has been of growing concern. Not only does hydrology determine the space–time history of erosion, but is also used to develop scenarios for prevention of erosion through, for example, soil conservation, appropriate farm practices, vegetation management, water diversion, afforestation, reduced flooding, and controlled land use.

2.2.11 Environmental-Impact Assessment

Sediment transport, fertilizers, pesticides and feedlot waste, disposal of urban and industrial waste, chemical spills, etc., have major impact on the quality of environment and ecology. With increasing industrialization and urbanization, larger and larger amounts of waste are generated and their disposition, without detrimental effects, is of growing concern. Sediment from eroded fields may choke streams and silt reservoirs. Fertilizers, pesticides and feedlot waste, and disposal of hazardous waste through landfill may leach into groundwater or wash into streams, poisoning plants, fish, and wildlife. Hydrology determines migration of these wastes and their effect on water quality, thereby developing standards for safe and economic disposal of waste through water bodies.

2.2.12 Land-Use Change

The land-use change can be point or nonpoint. Agricultural practices, afforestation and deforestation, urbanization, highway development, channel improvement, and so on are examples of nonpoint change. Dams, culverts, bridges, industrial plants, landfill sites, etc., represent point changes. Hydrologic consequences of these changes are to be determined before a land-use change can be justified. These changes can have significant effect on environment, the quality of life, fish and wildlife, plants and vegetation, etc.

2.2.13 Forest and Wildlife Management

Application of pesticides and chemicals, forest clearing and cutting, forest fires, plantation, logging, road construction, etc., are typical forest-management practices. Preservation of wildlife, animal grazing, animal husbandry, etc., are within the purview of wildlife management. Hydrology determines the consequences of these activities on water quantity and quality. Forest and vegetation cover certainly slow down the rate at which surface water flows to the main channels and spreads runoff over a longer period, and reduces peak flow at the same time. This effect is significant in the case of small streams and small floods, and may not be so for large watersheds and large floods. Great floods overcome the retarding effects of vegetation, and the nature of the land surface becomes of little importance in slowing runoff.

2.2.14 Military Operations

Hydrology plays a crucial role in the planning and conduct of military operations. Military camps are to be located on safe grounds. When the ground is trafficable is of vital importance for movement of military vehicles. A knowledge of river flow ahead of time is required to determine if river crossing would be safe. Dam breaching and the resulting damage are important in planning tactical offenses against enemies as well as adequate defense (Wickham and Jourdan, 1985; Jourdan and Sullivan, 1987). Downstream flooding can be an effective combat multiplier. In addition to damaging structures, the resulting flood wave may create a significant barrier to troop and vehicular movement. Military camping is done at awkward places, and locating water supply quickly is crucial. Hydrology is used to address all of these problems of military tactical environment.

2.2.15 Rural Development

In the so-called third-world countries such as India, China, Pakistan, etc., the bulk of the population lives in rural areas. The key to development of these countries lies, therefore, in development of their rural areas, without massive migration from rural to urban areas. Only then will rural areas be attractive for people to live. Hydrology is needed to properly plan development of activities constituting rural development, such as water-supply schemes, housing, schools and hospitals, roads and drainage systems, sanitary systems, ponds and fisheries development, communication, energy resources, afforestation, and recreation.

2.2.16 Navigation

In the United States, Canada, Europe, and certain other countries, river navigation is widespread, for water provides the most economic means of transportation for bulky raw materials, such as coal, pulp and paper, lumber, minerals, etc. There are nearly 26,000 miles of improved inland waterways for navigation in the United States. In order to maintain navigability, a minimum depth of flow has to be maintained in the river. A system of locks and dams is built on the river, which monitors the river traffic. Hydrology is employed to provide peak discharge and its probability of occurrence for designing these dams. The volume of water that will be available for river flow is estimated using hydrology. Because of siltation, navigable rivers may have to be dredged. The bulk of the sediment received by the river is generated in the upland areas. The upland erosion and the supply of this sediment to the river are determined using hydrology. The study of sedimentation processes is needed to determine the location of jetties and levees so as to minimize future silting problems.

2.2.17 Recreation

Many forms of recreation are common. Some of the popular forms of recreation are based on clean water, such as swimming, fishing, river boating, water skiing, canoeing, and parks. Many rivers and lakes close to urban population centers are highly polluted to the extent of being useless for recreational purposes. The demand for recreation in the western world is such that many reservoirs have been built for the sole purpose of recreation, for the existing facilities are not adequate. Recreational requirements these days are an important consideration in the development of water-use projects. Indeed, the present-day demands for recreation are so strong that they also influence the location of water-resources projects as well as their operation. Recreation is turning into a major commerce industry. Design and operation of these facilities call for adequate availability of water, which is estimated using hydrology. Moreover, measures for protection of the facilities from vagaries of weather and other extremes depend on hydrologic analyses.

2.2.18 Fisheries

This is a significant industry in many parts of the world. Besides the commercial freshwater fisheries, there are thousands of sport fishermen who each year cast their lures into lakes and rivers in all parts of the world. The fish grow if good quality water is available sufficiently. The increasing polluted condition of many lakes and streams has had a serious effect on both the quantity and type of fish available for sports of commerce. To an increasing extent, commercial and sport fishing are receiving important consideration in the preliminary planning and design of water-use projects. This consideration, in some cases, has not only dictated the nature of the projects, but has also influenced the choice of location. Hydrology is used to determine how much, and of what quality, water will be available in streams, ponds, reservoirs, etc., for a specific time period. Thus, hydrology plays a critical role in development of fisheries resources.

A summary of hydrologic inputs needed for addressing the various water-resources problems presented before is given in Table 2.11.

EXERCISES

2.1. The total water supply of the world is about 326 million cubic miles. Express this supply in cubic feet, acre-feet, gallons, cubic kilometers, kilometer-meters, cubic meters, liters, and tons.

2.2. If the world's total supply of water were poured upon the 50 United States, what would be the depth of the water by which the land surface would be submerged? The land area can be taken to be approximately 2274 million acres.

2.3. Find out an approximate distribution of the world's water in oceans, groundwater, ice and glaciers, surface water on land, soil moisture, rivers, and atmospheric vapor. Indicate volumes and percentages.

2.4. Show an approximate distribution of the fresh water on the earth.

2.5. Show an approximate water balance of oceans.

2.6. Show an approximate water balance of continents.

2.7. Show the approximate annual precipitation on and evaporation from land and sea surfaces.

2.8. Show runoff of the 10 largest rivers in the world. Also show runoff of the 10 largest rivers in the United States.

2.9. A multipurpose reservoir has been built on a river and is expected to serve its intended purpose during its expected life. There is concern that the reservoir may not live its expected life due to siltation. What hydrologic parameters are needed to develop measures for saving the reservoir? Present a qualitative but brief discussion.

2.10. A river is being used for navigation throughout the year. What hydrologic parameters are needed to facilitate and maintain navigation in the river?

2.11. It is desired to reclaim a large quantity of land (more than 10,000 km^2). What specific parameters does hydrology provide for undertaking reclamation of this land?

2.12. A waste-water treatment plant is to be constructed for a township. What hydrologic parameters are needed for designing such a plant?

2.13. A lagoon is proposed for disposal of waste from a small town. The lagoon is to be

TABLE 2.11 A SUMMARY OF WATER-RESOURCES PROBLEMS AND NEEDED HYDROLOGIC INFORMATION

Class	Specific problem	Peak discharge	Volume of runoff	Duration of hydrograph	Frequency	Others
Flood protection	(a) Construction of a flood-protection reservoir	X[a]	X		X	
	(b) Flood-damage assessment for flood insurance	X	X	X	X	Damage curves
	(c) Disruption of traffic	X		X	X	
	(d) Floodplain management	X			X	Floodplain mapping
Drought mitigation	(a) Water supply		X		X	
	(b) Water transfer		X		X	
	(c) Damage assessment		X	X	X	
	(d) Saltwater intrusion			X	X	Low-flow saltwater wedge
Water supply	(a) Agricultural irrigation		X		X	
	(b) Municipal water supply		X	X	X	
	(c) Water supply in coastal areas		X	X	X	Groundwater pumping; saltwater wedge
Pollution control	(a) Discharge of waste from lagoons	X		X	X	Period of hydrograph rise
	(b) Dilution of waste		X	X		
	(c) Operation of treatment facilities		X	X		
Urban development	(a) Detention storage	X			X	
	(b) Flood-damage assessment			X	X	Damage curves
	(c) Sewage disposal		X	X	X	
	(d) Drainage system	X			X	

TABLE 2.11 *(Continued)*

Class	Specific problem	Peak discharge	Volume of runoff	Duration of hydrograph	Frequency	Others
Industrial development	(a) Waste disposal		X		X	
	(b) Water supply		X		X	
Design of hydraulic works	(a) Design of dams, culverts, spillways, bridges, etc.	X			X	
	(b) Design of reservoirs		X		X	
	(c) Design of canals		X		X	Sediment transport
Agricultural production	(a) Irrigation scheduling		X			Soil-moisture forecasting
	(b) Drainage of excess waters			X	X	
	(c) Application of fertilizers		X			Transport of chemicals
Energy development	(a) Water supply		X		X	
	(b) Waste disposal		X		X	
	(c) Safety against extremes	X			X	
Land conservation	(a) Construction of terraces		X		X	
	(b) Diversion structures		X			Sediment delivery
	(c) Sediment-control structures		X	X	X	Sediment hydrograph
Environmental impact assessment	(a) Chemical spills		X	X	X	Migration of chemicals
	(b) Waste disposal	X		X	X	
	(c) Vegetation management		X			Migration of waste-water quality determinations

Land-use change	(a) Urbanization	X	X	X	X	
	(b) Deforestation	X	X	X	X	
	(c) Afforestation	X	X	X	X	
	(d) Highway development	X	X	X	X	
Forest and wildlife management	(a) Forest fires		X	X	X	
	(b) Forest clearing		X	X	X	
	(c) Road development	X	X	X	X	
	(d) Forest growing		X			Migration of chemicals
Military operation	(a) Dam breaching			X		Breach evolution
	(b) Reservoir operation		X			
	(c) Water prospecting		X	X		Location of groundwater
	(d) Traffic operation		X	X		Soil-moisture forecasting
Rural development	(a) Water supply		X		X	
	(b) Drainage works	X			X	
	(c) Road construction	X	X	X	X	
Navigation	(a) Design of locks and dams	X			X	
	(b) Design of reservoirs		X	X	X	
	(c) Dredging			X		Sediment transport
Recreation	(a) Fishing		X		X	
	(b) Parks		X		X	
Fisheries	(a) Water supply		X		X	Water-quality determination
	(b) Flow regulation		X	X	X	

[a] X indicates data indicated are required.

located near a stream. It will discharge waste whenever streamflow rises and will store when streamflow is minimal or at steady state. What hydrologic parameters are required for designing the lagoon?

2.14. A detention pond is proposed for stormwater management in an urban area. What hydrologic parameters are needed to design such a pond?

2.15. A rural highway is to be built to connect two cities about 200 km apart. There are two streams between the cities. Discuss the hydrologic parameters to be needed for design of this highway.

2.16. A new small city is to be built. In addition to other facilities, an adequate drainage system is to be constructed. What hydrologic parameters are needed for design of such a system?

2.17. Consider two areas *A* and *B*—*A* having excessive rainfall and *B* having deficient rainfall. A scheme is under consideration for transfer of water from area *A* to area *B*. What hydrologic parameters are needed for designing such a scheme?

2.18. A scheme of canal irrigation is to be designed for an area. What hydrologic parameters are needed for design of such a scheme?

2.19. A nuclear power plant is to be built for energy supply. What hydrologic parameters are needed for design of the plant?

2.20. There is a large natural lake fed by a stream. The lake is used for, among other things, recreation, fishing, and water supply. What role does hydrology play in ensuring the health and life of the lake?

CHAPTER 3
Types of Watersheds

A watershed is like a natural laboratory of hydrology. All hydrologic processes occur in the watershed. Depending upon the type of watershed, these processes vary in space and time within that watershed. Topography, soil, geology, vegetation, land use, and stream network are the principal factors contributing to variability of hydrologic processes. These factors vary with the type of the watershed and within the same watershed. For hydrologic analyses, a watershed constitutes the spatial unit, and hydrologic problems are solved in the context of that spatial unit.

3.1 WATERSHED CHARACTERISTICS

The watershed characteristics pertain to the land and channel elements of the watershed. Size, shape, slope, elevation, density of channels, vegetation, land use, soil type, hydrogeology, lakes, swamps, artificial drainage, and so on are some of the characteristics of basin (or land elements). Channel characteristics are the hydraulic properties of the channel such as size and shape of channel cross-section, slope, roughness, and length of channels in the channel network. The heterogeneity or nonuniformity of a watershed can be described and measured in terms of these characteristics. A watershed may be relatively uniform in terms of one characteristic, but quite heterogeneous in terms of others. This same may be true when any two watersheds are considered. Stated another way, these characteristics may have different combinations of variabilities.

3.2 HYDROLOGICAL INDICES

Simple indices can be defined to illustrate variability of hydrologic behavior with watershed type. Some of these may include runoff volume, peak discharge, timing of runoff, baseflow, infiltration, evaporation, interception, and erosion, to name but a few. The spatially lumped or distributed form of these indices can be used to characterize the watershed type.

Watersheds can be classified based on such characteristics (or criteria) as size, mean slope, length, land use, etc. The watersheds possessing similar characteristics can be grouped together for regionalization. Two hydrologically meaningful criteria are size and land use. In this chapter, we discuss hydrological consequences of watersheds typified by these criteria.

3.3 WATERSHEDS BY SIZE

Three types of watersheds are distinguished according to size: small, medium, and large. This classification is vague, but the implication is in terms of spatial heterogeneity and dampening (averaging) of hydrological processes. For consideration of runoff generation on these watersheds, two phases can be considered: land phase and channel phase. Each phase has its own storage characteristics. Large watersheds have well-developed channel networks and channel phase, and, thus, channel storage is dominant. Such watersheds are less sensitive to high-intensity rainfalls of short duration. On the other hand, small watersheds have dominant land phase and overland flow, have relatively less conspicuous channel phase, and are highly sensitive to high-intensity, short-duration rainfalls. Two watersheds of the same size may behave very differently if they do not have similar land and channel phases. Small watersheds are usually least heterogeneous and large watersheds are most heterogeneous. In other words, spatial variability of watershed characteristics increases with size. The watersheds are considered small if the area is less than 100 mi^2 (or 250 km^2), medium if the area is between 100 and 1000 mi^2 (or between 250 and 2500 km^2), and large if the area is greater than 1000 mi^2 (or >2500 km^2). These numbers should be considered only as rough guidelines and may vary from one geological area to another.

If all three types of watersheds are spatially uniform, then they will behave hydrologically similarly for a specified rainfall. Under this condition, classification of watersheds serves no purpose and becomes moot. As the watershed size increases, storage increases and averaging of hydrologic processes increases as a result. The effect of averaging is to linearize the watershed behavior. On the average, small watersheds are more nonlinear than large watersheds. A small parking lot, for example, is highly nonlinear. If a watershed is complex both in terms of its landscape and land use, then it may be desirable to decompose it into smaller homogeneous subwatersheds. In the computer era, these subwatersheds can be quite small. Or a complex watershed can be represented by a finite grid of elements that may be rectangular, triangular, or both. Each element is considered spatially uniform. Hydrology of large watersheds is more difficult to describe, due to complexity of a multitude of factors operating simultaneously. Nevertheless, their overall hydrologic response is tractable, for interactions of the various phenomena work themselves out.

Under constant conditions, mean discharge must be directly proportional to watershed area (A), and the average flow per unit area must be the same for all watershed sizes. However, constant conditions are never obtained, and watershed behavior is highly variable from one watershed size to another. Small watersheds are, within a given drainage system, represented by upland areas where rainfall and runoff depths are usually greater and an extensive, well-developed channel system is lacking. Lee (1980) has reported that flow rates per unit area, Q (m^3/s-ha, or depth/time), generally follow the relationship

$$\frac{Q}{A} = kA^{x-1} \tag{3.1}$$

where k is an empirical constant, and $x < 1$ for peak flows (typically $x = 0.8$), $x > 1$ for low flows (typically $x = 1.2$), and $x \cong 1$ for average discharge. Figure 3.1 illustrates the typical effect of watershed size. This equation shows that as A increases, Q/A decreases at high flows; but at low flows, Q/A increases with area due to delayed subsurface flow.

The methods of determining hydrologic characteristics vary with the watershed size. Consequently, some workers (Ponce, 1989) prefer to group hydrological analyses into hydrology of small watersheds, hydrology of midsize watersheds, and hydrology of large watersheds. This grouping underscores hydrologic differences of different size watersheds

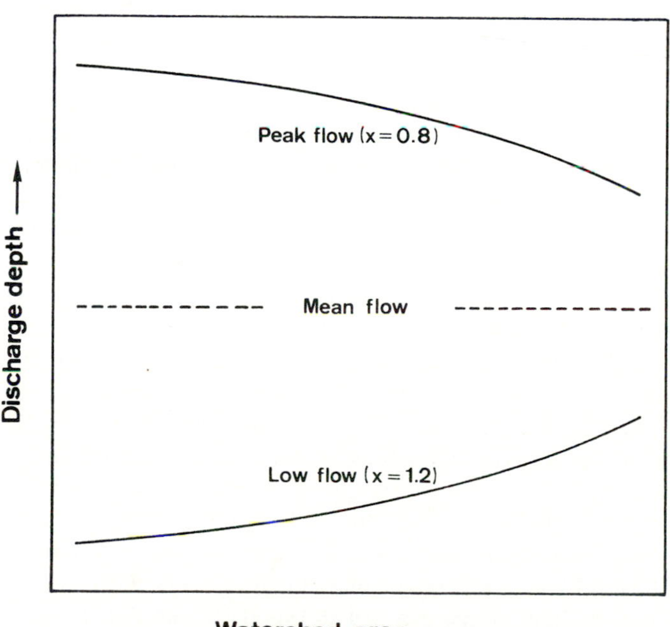

Figure 3.1 Relative discharge and watershed area.

3.4 WATERSHEDS BY LAND USE

Land use defines exploitation of a watershed. Accordingly, watersheds can be classified as agricultural, urban, mountainous, forest, desert, coastal or marsh, or mixed—a combination of two or more of the previous classifications. These watersheds behave hydrologically differently, indeed so differently that they have given rise to different branches of hydrology. For example, treatment of agricultural watersheds constitutes what is called agricultural hydrology. Similarly, there is urban hydrology for urban watersheds, coastal hydrology for coastal watersheds, forest hydrology for forest watersheds, desert hydrology for desert watersheds, wetland hydrology for marshes and wetlands, and mountain hydrology for mountainous watersheds. The ensuing discussion, although brief, brings home the point that hydrology is basic to the study of many scientific areas.

3.4.1 Urban Watersheds

An urban watershed is dominated by buildings, roads, streets, pavements, and parking lots. These features reduce the infiltrating land area and increase imperviousness. Because drainage systems are artificially built, the natural pattern of water flow is substantially altered. For a given rainfall event, interception and depression storage can be significant, but infiltration is considerably reduced and so is the case with evaporation. As a result, there is pronounced increase in runoff and pronounced decrease in soil erosion. Thus, an urban watershed is more vulnerable to flooding if the drainage system is inadequate. It is no surprise that flooding and the resulting flood damage are on the rise in many parts of the United States and are threatening the survivability of the flood insurance program executed by the Federal Emergency Management Administration (FEMA) of the federal government. Depending upon the degree of urbanization, topography, and drainage facility, the production of runoff varies for different parts of the watershed. If lakes, ponds, and parks are numerous in the watershed, evaporation will be significant and may compensate for reduction in evaporation elsewhere due to impervious land surfaces. Once a watershed is urbanized, its land use is almost fixed. Then its hydrologic behavior changes due to changes in precipitation. If a small urbanizing watershed is considered by itself, then runoff peak increases and its time of occurrence decreases with urbanization. This is because as development proceeds, there are more pavements, sidewalks, houses, parking lots, storm sewers, channels, etc., which all decrease infiltration and increase runoff. When an entire complex watershed is considered, then the runoff peak may actually be reduced because its roads, bridges, tunnels, etc., can cause impoundments that dampen the runoff hydrograph.

Urban runoff markedly affects water quality. Cherkauer (1975) compared for a storm (0.87 in., 2.2 cm) water-quality response of an urban watershed (7.5 km^2) with a physiographically and climatically similar rural watershed (9.7 km^2) located a few kilometers from each other. Land-use data of these two watersheds are listed in Table 3.1. Cherkauer observed that the peak flow from the urban watershed was over 250 times that of the rural watershed. Table 3.2 provides total loads of the various water-quality constituents of the two watersheds. Clearly, the urban stream transported vastly greater dissolved material.

TABLE 3.1 LAND USE OF TWO DRAINAGE BASINS (AFTER CHERKAUER, 1975)

	Brown Deer (urban) (Total area = 7.5 km²)		Trinity (rural) (Total area = 9.7 km²)	
Land use	Percent of basin area	Subtotals	Percent of basin area	Subtotals
Residential	56.9		5.0	
Light industry	5.7		0.3	
Parking lots	2.5		0.4	
Total developed		65.1		5.7
Open land	33.4		94.3	
Construction sites	1.5		0.0	
Total undeveloped		34.9		94.3

Urbanization causes climatic change, the most obvious being an increase in temperature. Landsberg (1970) summarized average changes in the elements of climate on a global basis, as shown in Table 3.3. These numbers should, however, be taken only as guidelines. The regional climate and local topography are very important and may override any urban effects.

3.4.2 Agricultural Watersheds

Unlike urban watersheds, an agricultural watershed experiences perhaps the most dynamically significant land-use change. Each year, its land surface is tilled many times and crops are grown as many as three or four times. Cropping pattern and the method of growing crops also change. As a result, land use and the treatment may be distinguished (Soil Conservation Service, 1971) as cultivated land, fallow, row crop, small grain crop, rotation meadow, rotation, straight row, contoured, grassland, meadows, woods and forests, and gardens. All these usually lead to increased infiltration, increased erosion, and /or decreased runoff. Increased infiltration is expected, for farm practices are designed for that purpose, which is in the interest of

TABLE 3.2 TOTAL LOADS OF VARIOUS MATERIALS PER UNIT DRAINAGE AREA CARRIED OUT OF BASIN BETWEEN 05:30, OCTOBER 6, AND 09:00, OCTOBER 8, 1974 (AFTER CHERKAUER, 1975)

Material	Urban load (kg/km²)	Rural load (kg/km²)
Water	86,300	675
Suspended sediment	287	0.8
Total dissolved solids	199	5.6
Sodium	25.7	0.6
Chloride	37.5	0.7
Calcium	23.7	1.0
Magnesium	11.6	0.2
Bicarbonate	61.3	3.4

TABLE 3.3 AVERAGE CHANGES IN GLOBAL CLIMATIC VARIABLES DUE TO URBANIZATION (AFTER LANDSBERG, 1970)

Variable	Change compared with rural area
Temperature	
Annual mean	0.5 to 1°C more
Winter minima (average)	1 to 2°C more
Wind speed	
Annual mean	20 to 30% less
Extreme gusts	10 to 20% less
Calms	5 to 20% more
Contaminants	
Condensation nuclei and particles	10 times more
Gaseous admixtures	5 to 25 times more
Radiation	
Global	15 to 20% less
Shortwave, winter	30% less
Shortwave, summer	5% less
Sunshine duration	5 to 15% less
Cloudiness	
Cloud cover	5 to 10% more
Fog, winter	100% more
Fog, summer	30% more
Precipitation	
Totals	5 to 10% more
Days with less than 5 mm	10% more

crops. Depression storage also is increased by agricultural operations. When the fields are barren, falling raindrops tend to compact the soil and infiltration is, as a result, reduced. There is lesser development of streams in agricultural watersheds, for small channels formed by erosion and runoff are obliterated by tillage operations. The soil texture is altered by regular application of organic and/or inorganic manure. This, in turn, leads to changed infiltration characteristics. Table 3.4 shows the effect of future agricultural land use and treatment on runoff volume due to a specified value of rainfall for three antecedent moisture conditions. There is marked reduction in runoff volume in each case. Evapotranspiration constitutes the principal loss of water from an agricultural point of view. Crops are irrigated on a scheduled basis. The water, in excess of plant rootzone, becomes irrigation return flow or recharges the groundwater. These features are in contrast with those of urban watersheds.

3.4.3 Forest Watersheds

The hydrological behavior of forest watersheds is quite different from that of agricultural or urban watersheds. Interception is significant, and evapotranspiration is a dominant component of the hydrologic cycle. In forest watersheds, the ground is usually littered with leaves, stems, branches, wood, etc. Consequently, when it rains, the water is held by the trees and the ground cover and has greater opportunity

TABLE 3.4 EFFECT OF FUTURE AGRICULTURAL LAND USE AND TREATMENT ON RUNOFF AMOUNT (ADAPTED FROM SOIL CONSERVATION SERVICE, 1971)

Rainfall (cm)	Runoff (cm)					
	AMC I[a]		AMC II[a]		AMC III[a]	
	Present	Future	Present	Future	Present	Future
1.27	0	0	0	0	0.20	0
2.54	0	0	0.05	0	0.89	0.30
5.08	0	0	0.97	0.28	2.92	1.78
7.62	0.58	0.05	2.46	0.35	5.21	3.68
10.61	1.52	0.46	4.27	2.62	7.62	5.84
12.70	2.79	1.09	6.25	4.19	10.03	8.13
Curve number[b]:	57	45	75	65	91	83

[a] AMC: antecedent moisture condition. AMC I corresponds to a dry condition, AMC II corresponds to an average condition, and AMC III corresponds to wet condition.

[b] The curve number is an empirical number between 10 and 100, reflecting the runoff-producing capacity of a soil–vegetation–land–use complex. A higher value of this number corresponds to a higher runoff-producing capacity of the soil.

to infiltrate. The subsurface flow becomes dominant and there are times when there is little to no surface runoff. The subsurface flow may last for quite some time even after the cessation of rainfall. There is greater recharge of groundwater. Because forests resist flow of water, the peak discharge is reduced, although inundation of the ground may be increased. This reduces flooding and flood damage downstream. Due to reduced surface-runoff potential, stream development is much less. Plants and trees provide good protective cover to soil from erosion. It is usually held that the annual water yield is increased by deforestation, as shown in Table 3.5 based on a summary by Anderson et al. (1976). Complete deforestation generally increases the annual water yield between 20 and 40% of normal, but a maximum increase seldom exceeds 40 cm/year.

3.4.4 Mountainous Watersheds

As the name suggests, the landscape of these watersheds is predominantly mountainous. Because of higher altitudes, such watersheds receive considerable snowfall. There is so much snowfall in some watersheds that they are a haven for winter sports. Naturally, hydrology of such watersheds is quite different. By and large, such watersheds have substantial vegetation, such that in some cases, these could be considered as forest watersheds also.

Interception is significant. Due to steep gradient and relatively less porous soil, infiltration is less and surface runoff is dominantly high for a given rainfall event. Flash floods are a common occurrence. The areas downstream of the mountains are vulnerable to flooding whenever there is a heavy rainfall in the mountains. Flooding in valleys downstream may be even more severe when there is rain in mountains on the top of snow. There is little to virtually no change in land use. Erosion is minimal if the mountains are rocky. Sliding and collapsing of slopes are not uncommon

TABLE 3.5 ANNUAL WATER YIELD INCREASE FOLLOWING FOREST REMOVAL (AFTER ANDERSON ET AL., 1976)

Location	Forest type[a]	Removal		Normal yield, Q (mm/yr)	Yield increase by years (mm/yr)			$\Delta Q_1/(Q \cdot f)$ (%)
		Fraction (f)	Method		ΔQ_1	ΔQ_2	ΔQ_3	
North Carolina	MH[b]	1.00	Clear-cutting	792	370	283	279	47
	MH[c]	1.00	Clear-cutting	607	127	95	59	21
	MH	0.50	Strips	1275	198	155	130	31
	MH	0.22	Selection	1222	99	56	71	37
West Virginia	MH	0.85	Clear-cutting	584	130	86	89	26
	MH	0.36	Selection	660	64	36	—	27
	MH	0.22	Selection	762	36	—	—	21
	MH	0.14	Selection	635	8	—	—	9
Colorado	AC	1.00	Clear-cutting	157	34	47	25	22
	PSF	0.40	Clear-cutting	283	86	53	79	76
Oregon	DF	1.00	Clear-cutting	1448	462	457	—	32
	DF	0.30	Clear-cutting	1448	150	163	150	35
Arizona	PP	1.00	Clear-cutting	158[d]	96	23	46	—
	PP	0.32	Strips	170[d]	50	15	9	—
	PP	0.75	Thinning	194[d]	22	37	38	—

[a] MH: mixed hardwoods, AC: aspin–conifer, PSF: pine–spruce–fir, DF: Douglas fir, PP: Ponderosa–pine.

[b] Northeast aspect.

[c] Southeastern aspect.

[d] Winter discharge.

occurrences during periods of heavy precipitation. Snow melting generates surface runoff during spring and summer, which can be used for water supply. Recharge of groundwater is small and evapotranspiration is considerable.

3.4.5 Desert Watersheds

There is little to virtually no vegetation in desert watersheds. The soil is mostly sandy and little annual rainfall occurs. Sand dunes and sand mounds are formed by blowing winds. Stream development is minimal. Whenever there is little rainfall, most of it is absorbed by the porous soil, some of it evaporates, and the remaining runs off—only to be soaked in during its journey. Local flooding may occur due to heavy showers. There is limited opportunity for groundwater recharge due to limited rainfall.

3.4.6 Coastal Watersheds

The watersheds in coastal areas may partly be urban and are in dynamic contact with the sea. Their hydrology is considerably influenced by backwater from wave and tidal action or other controlled streams. Usually, these watersheds receive high rainfall, mostly of cyclonic type, do not have channel control in flow, and are vulnerable to severe, local flooding. Coastal erosion is a continuing problem due to tidal action, and land-use change is common. The water table is high, and saltwater intrusion threatens the health of coastal aquifers, which usually are a source of water supply. The land gradient is small, drainage is slow, and the soil along the coast has a considerable sand component.

3.4.7 Marsh, or Wetland, Watersheds

Wetlands are a significant natural resource in the United States and some other countries. Such lands are almost flat and are comprised of swamps, marshes, water courses, etc. They have rich wildlife and plenty of vegetation. Evaporation is one of the dominant components of the hydrologic cycle, for water is no limiting factor to satisfy evaporative demand. Rainfall is normally high and infiltration is minimal. Most of the rainfall becomes runoff, which discharges slowly for minimal land declivity. Erosion is also minimal, except along the coast. The flood hydrograph peaks gradually and lasts for a long time.

EXAMPLE 3.1 Show the effect of urban development on the peak discharge and runoff hydrograph of a stream.

Solution We consider the study conducted by Stall et al. (1970) on the effect of urbanization. Figure 3.2 shows the effect of urbanization on streamflow in four cases: Case I—rural watershed with no connections to storm drains; Case II—two-thirds of the watershed is urban, one-quarter of which is paved, and half of this is connected to drains; Case III—the watershed is urbanized with 50% paved and half of this connected to the drains; and Case IV—the watershed is intensely urbanized, 75% paved and two-thirds of this is connected to drains. For the same rainfall distribution and the same amount of water as a result, the runoff peak and hydrograph base (duration of runoff)

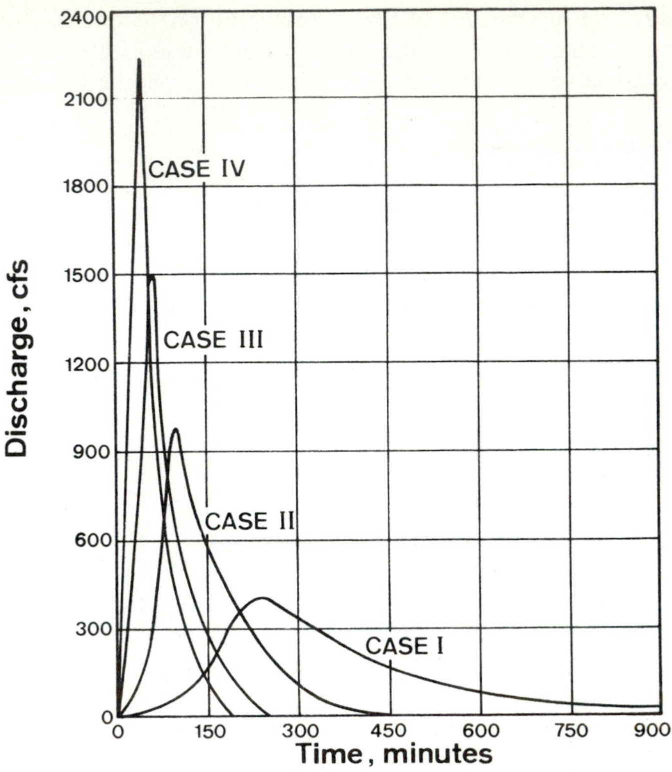

Figure 3.2 Effect of urban development on the peak discharge and runoff hydrograph of a stream (after Stall et al., 1970).

are markedly different for the four cases, depending upon the degree of urbanization. There is a sixfold increase in peak discharge and a fivefold decrease in runoff duration from Case I to Case IV. This case vividly illustrates hydrologic consequences of urbanization.

EXAMPLE 3.2 Briefly show the effect of forest cover on annual sediment yield of a watershed.

Solution Forests exercise a stabilizing influence on soil. Upland forests with mineral soil fully protected by a cover of forest litter and humus contribute little or no sediment to streams. Forest litter protects the soil from raindrop impact and helps maintain high infiltration capacity. As a result, surface erosion is seldom a serious problem in undisturbed forests. Tree roots also bind the soil mass, greatly reducing the hazard of mass soil movements even on steep slopes. Harrold et al. (1974) have reported that annual runoff from small forested watersheds carried almost insignificant amounts of sediment (2 to 4 tons per square mile). For small watersheds, Fleming (1969) developed design curves for estimation of suspended sediment load. General relationships for four cover types are shown in Figure 3.3 for the range of discharges typical of small watersheds. However, stream channels erode during periods of high flood waters. On large watersheds with pronounced channel networks, stream-channel erosion is significant. Harrold et al. (1974) reported that annual sediment yield of large watersheds with miles of

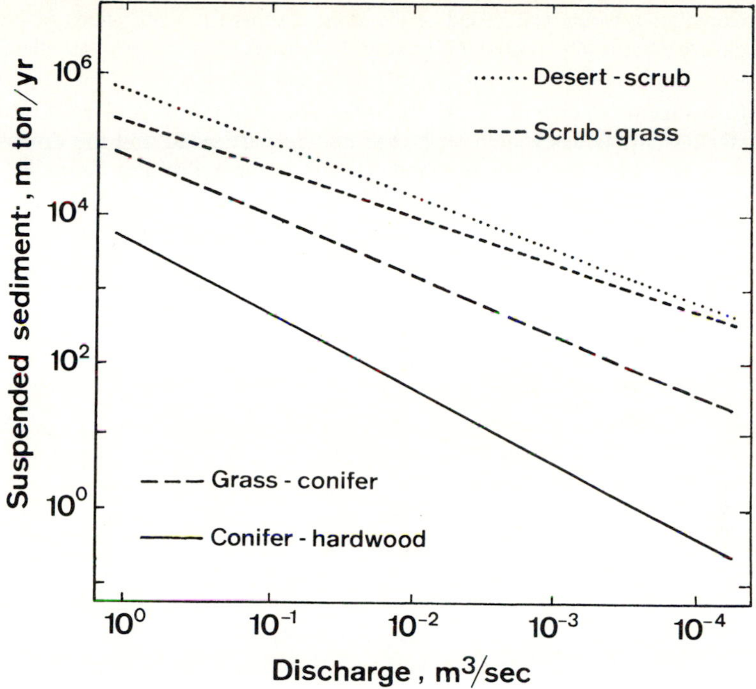

Figure 3.3 Suspended sediment yield as a function of average stream discharge for various cover types (adapted from Fleming, 1969).

stream channels ranged from 22 to 400 tons/mi^2, depending upon the amount of forest cover, as shown in Table 3.6.

TABLE 3.6 FOREST COVER AND ANNUAL SEDIMENT YIELD OF POTOMAC RIVER SUBWATERSHEDS

Forest cover in watershed (%)	Sediment yield from watershed (tons/mi^2)
20	400
40	200
60	95
80	45
100	22

EXERCISES

3.1 Consider two watersheds of the same size—one is urban and the other agricultural. Both watersheds are subjected to the same rainfall. Compare their hydrologic behavior in terms of runoff amount, amount of water infiltrated, and sediment yield.

3.2 Consider a forest watershed of the same size as the watersheds in Exercise 3.1. Compare its hydrologic behavior separately with each of those watersheds. Then compare hydrologic response of the three watersheds.

3.3 Consider a watershed 50 km^2 in area. The watershed is divided into three parts: upstream part is designated as A, the middle part as B, and the downstream part as C. Consider the following cases and compare their hydrologic behavior.

CASE	PART	LAND USE
1	A	Agricultural
	B	Forest
	C	Urban
2	A	Forest
	B	Agricultural
	C	Urban
3	A	Urban
	B	Forest
	C	Agricultural

3.4 Consider a coastal watershed that is urbanizing rapidly. Due to increased urbanization and consequent increase in population, more and more groundwater is being pumped to meet water-supply demand. Should pumping be restricted to save the groundwater from saltwater intrusion? Discuss your answer in qualitative terms.

3.5 Modern agriculture depends heavily on chemical fertilizers. Briefly discuss hydrological consequences of using these fertilizers.

3.6 Take $x = 1$ in Equation (3.1) and then discuss hydrological implications of this equation.

3.7 Which of the following watersheds are most prone to flooding and why? The watersheds are urban, agricultural, mountainous, desert, and forest. Climatic conditions are assumed to be the same.

3.8 What are the two most prominent hydrological processes in desert watersheds?

3.9 Louisiana is losing approximately 50 mi^2 every year of its coast to the Gulf of Mexico. Can you think of any hydrologic reasons for this coastal loss?

3.10 Why is rainfall usually higher along the seacoast than island?

3.11 Discuss the hydrologic consequences of modern agricultural practices.

3.12 Discuss the hydrologic consequences of urban development.

3.13 Land erosion, and consequent loss of productive soil, is a serious problem in some agricultural areas. Soil-conservation measures such as terracing, bunding, etc., are recommended. What are the hydrologic consequences of these measures?

CHAPTER 4
Parameter Estimation

When a relationship between two or more variables is expressed in mathematical form, the form contains certain parameters. For example, annual evaporation (E) can be linearly related to annual rainfall (P) for some watersheds as $E = a + bP$, where a and b are parameters. For given observations of E and P, the parameters a and b must be estimated such that this relationship "best" represents the observations. The meaning of the term "best" will become clear later. There exists a multitude of methods for estimating parameters in hydrology (Singh, 1988). The methods of moments and least squares are two commonly used methods (Dooge, 1973) and are discussed here. Parameter estimation is an important area of mathematics and statistics. The brief discussion in this chapter shows the role these disciplines play in study of hydrology.

4.1 METHOD OF MOMENTS (MOM)

Moments represent certain characteristics of a function such as mean, variance, and skewness, and can be called descriptors of that function. In order to determine parameters of a function, as many moments are determined as the number of parameters in the function. The parameters so determined are such that theoretical moments of the function are equated to the moments computed from data. Thus, the term "best" corresponds to matching of moments and not the matching of the entire function to the observed data.

4.1.1 Definition and Notation

Let x be a continuous variable and $f(x)$ its function, as shown in Figure 4.1, satisfying some necessary conditions. This function is continuous and x varies from $-\infty$ to $+\infty$. Consider an arbitrary point a away from the origin and a strip of thickness dx, which has a height $f(x)$ at location x. The distance between the center of the strip and the point a is $x - a$. Then the rth moment M_r^a of $f(x)$ about point a can be defined as

$$M_r^a(f) = M_r^a(x) = M_r^a = \int_{-\infty}^{\infty} (x - a)^r f(x)\, \mathrm{d}x \tag{4.1a}$$

where the area under the curve $f(x)$ is unity,

$$\int_{-\infty}^{\infty} f(x)\, \mathrm{d}x = 1 \tag{4.1b}$$

This is the definition used normally in engineering. The quantity $f(x)$ dx specifies the weight assigned to the corresponding value of $x - a$. If Equation (4.1b) is not satisfied, then the definition of the rth moment about point a is

$$M_r^a(f) = M_r^a = \frac{\int_{-\infty}^{\infty} (x - a)^r f(x)\, \mathrm{d}x}{\int_{-\infty}^{\infty} f(x)\, \mathrm{d}x} \tag{4.1c}$$

This is the definition used in statistics. Because the denominator in Equation (4.1c) defines the area under the curve, which is usually unity or made to unity by normalization, the two definitions are numerically the same. In this text, we use the definition of Equation (4.1a) with $f(x)$ normalized beforehand.

If the function is discrete, as shown in Figure 4.2, represented as $f(x_j) = f_j$, $j = -\infty, \ldots, -1, 0, 1, 2, \ldots, \infty$, then its rth moment about the origin or any other

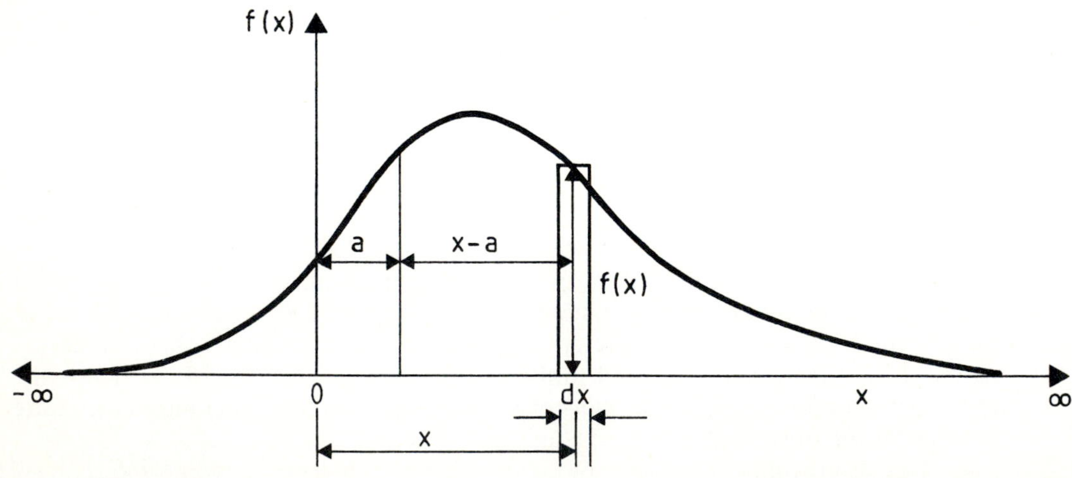

Figure 4.1 Expressing moments of a function $f(x)$ about an arbitrary point a away from the origin.

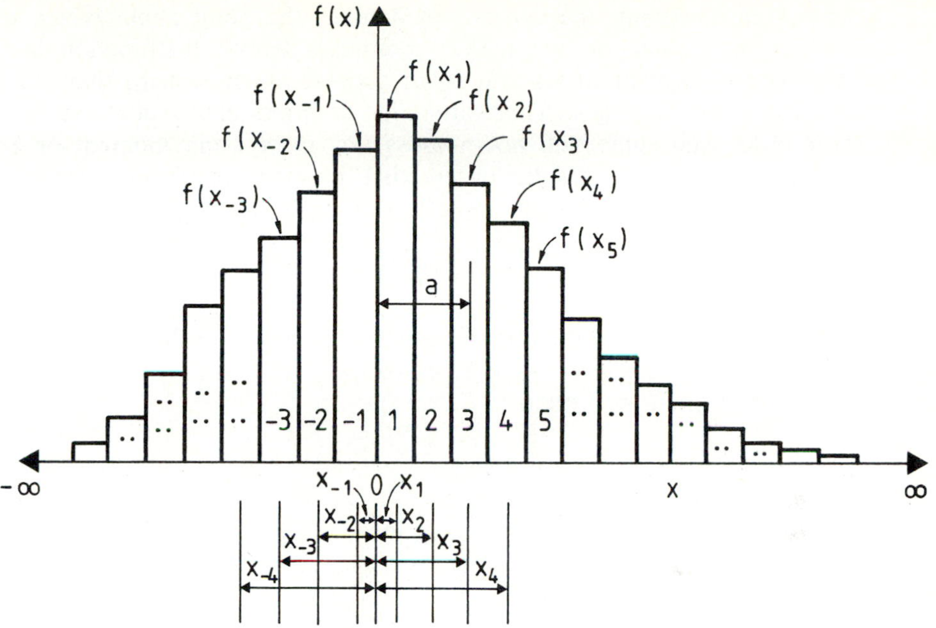

Figure 4.2 Expressing moments of a discrete function $f(x_j)$, $j = -\infty, \ldots, 1, 2, \ldots, +\infty$, about an arbitrary point a away from the origin.

arbitrary point can be defined in an analogous manner. f_j is located at x_j away from the origin. If the arbitrary point is a, then the distance between the location of f_j, x_j, and a is $x_j - a$. Then the rth moment is defined as

$$M_r^a(f) = M^a = \sum_{j=-\infty}^{\infty} (x_j - a)^r f(x_j) \tag{4.2a}$$

where x_j is the point where f is defined. The quantity $f(x_j)$ is the weight assigned to $x_j - a$. It is assumed here that $f(x_j)$ is normalized, that is,

$$\sum_{j=-\infty}^{\infty} f(x_j) = 1 \tag{4.2b}$$

Otherwise

$$M_r^a(f) = M_r^a = \frac{\sum_{j=-\infty}^{\infty} (x_j - a)^r f(x_j)}{\sum_{j=-\infty}^{\infty} f(x_j)} \tag{4.2c}$$

In the previous definitions, the reference axis is the abscissa (x-axis) and the moments are taken about the vertical axis. With equal ease, moments can be taken about the horizontal axis with the ordinate (y-axis) as the reference axis.

Here $r \geq 0$ is the order of the moment. This notation will be employed throughout this chapter. To be explicit, M denotes the moment, the subscript denotes the order of the moment, the superscript denotes the point about which to take the moment, and the quantity within the parentheses denotes the function, in normalized form, whose moment to take. It is, of course, assumed here that the integral in Equation (4.1) converges. There are some functions that will possess moments of lower order and some will not possess any except the moment of zero order. However, if a moment of higher order exists, moments of all lower orders must exist.

If $r = 0$,

$$M_0^a = \int_{-\infty}^{\infty} (x - a)^0 f(x)\, dx = \int_{-\infty}^{\infty} f(x)\, dx = 1$$

Thus, the zero-order moment is the area under the curve defined by $f(x)$ subject to $-\infty < x < \infty$.

If $r = 1$ and $a = 0$, then

$$M_1^0 = \int_{-\infty}^{\infty} x f(x)\, dx = \mu$$

where μ is the centroid of the area or mean. If $a = 0$, the first moment gives the mean.

If $r = 1$ and $a \neq 0$, then

$$M_1^a = \int_{-\infty}^{\infty} (x - a)^1 f(x)\, dx = \mu - a$$

Thus, the first moment is the weighted mean about point a. When $a = \mu$, then the rth moment about the mean is

$$M_r^\mu = \int_{-\infty}^{\infty} (x - \mu)^r f(x)\, dx \tag{4.3}$$

In particular, if $r = 1$, then

$$M_1^\mu = \int_{-\infty}^{\infty} (x - \mu) f(x)\, dx = \mu - \mu = 0$$

Henceforth, we will drop the superscript if the moment is taken about the origin "0." The descriptive properties of the moments with respect to a specific function can be summarized as

M_0 = area under the function

M_1 = mean or lag of the function

M_2^μ = Measure of dispersion of the function about the mean, called variance

M_3^μ = Measure of asymmetry or skewness of the function

M_4^μ = Measure of the peakedness of the function, called kurtosis

For statistical functions, these properties play a special role and, therefore, will be described in Chapter 24.

EXAMPLE 4.1 Consider a function $f(x) = a \exp(-ax)$, where $x > 0$, and $a > 0$. Determine M_r^μ and M_r.

Solution

$$M_r^\mu = a \int_0^\infty (x - \mu)^r \exp(-ax)\, dx$$

Therefore,

$$M_0^\mu = -\exp(-ax)\,|_0^\infty = 1$$

$$M_1^\mu = a \int_0^\infty x \exp(-ax)\, dx - a\mu \int_0^\infty \exp(-ax)\, dx$$

$$= \mu - \mu = 0$$

and so on.

Likewise,

$$M_r = a \int_0^\infty x^r \exp(-ax)\, dx$$

Therefore,

$$M_0 = -\exp(-ax)\,|_0^\infty = 1$$

$$M_1 = a\left[-\frac{x}{a}\exp(-ax) - \frac{1}{a^2}\exp(-ax)\right]\Big|_0^\infty = \frac{1}{a}$$

and so on.

EXAMPLE 4.2 A rainfall event has been observed. The following data represent the histogram of this event:

TIME (min)	RAINFALL INTENSITY (cm/h)	TIME (min)	RAINFALL INTENSITY (cm/h)
0–10	2	10–20	4
20–30	2	30–40	4
40–50	2	50–60	1

Compute the first two moments about the origin.

Solution The histogram of this rainfall event is shown in Figure 4.3. From Equation (4.2c), $f_j = f(x_j)$ is the jth block, as shown in the figure, and not just the jth intensity. For example, $f_1 = 2 \times 10/60$, $f_2 = 4 \times 10/60$, and so on. Furthermore, f_1 takes place at 5 min from the origin, f_2 at 15 min from the origin, and so on. The centers of the various blocks are displayed in the figure.

First, we compute the area under the histogram, which is the same as the amount of rainfall.

$$M_0 = (2 + 4 + 2 + 4 + 2 + 1) \text{ cm/h} \times 10/60 \text{ h}$$

$$= 15 \text{ cm/h} \times 1/6 \text{ h} = 2.5 \text{ cm}$$

Thus, the amount of rainfall = 2.5 cm.

$$M_1 = [5 \text{ min} \times 2 \text{ cm/h} \times 10 \text{ min} + 15 \text{ min} \times 4 \text{ cm/h} \times 10 \text{ min}$$

$$+ 25 \text{ min} \times 2 \text{ cm/h} \times 10 \text{ min} + 35 \text{ min} \times 4 \text{ cm/h} \times 10 \text{ min}$$

$$+ 45 \text{ min} \times 2 \text{ cm/h} \times 10 \text{ min} + 55 \text{ min} \times 1 \text{ cm/h} \times 10 \text{ min}]$$

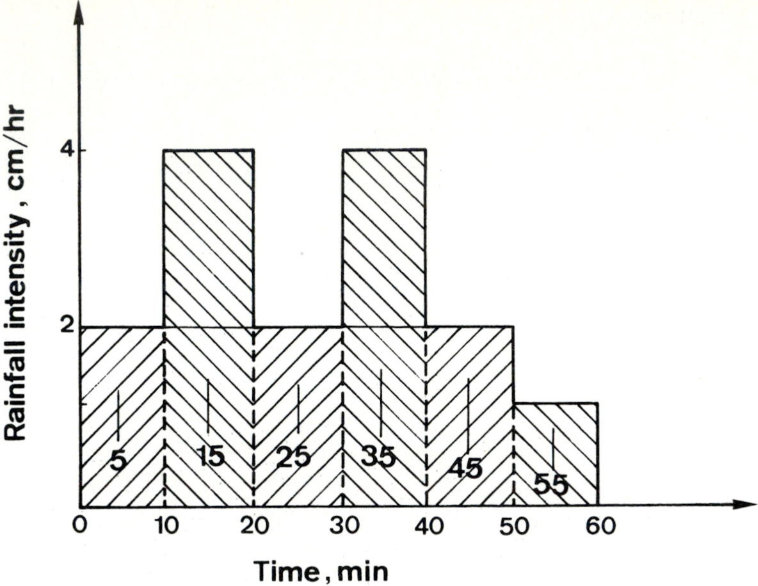

Figure 4.3 Rainfall histogram.

$$\times \frac{1}{2.5 \text{ cm} \times 60 \text{ min/h}}$$

$$= \frac{10 + 60 + 50 + 140 + 90 + 55}{2.5 \times 60} \times 10 \text{ min}$$

$$= \frac{405}{15} = 27 \text{ min}$$

This value represents the centroid μ of the histogram. Stated simply, half of the rainfall amount is to the left of the time = 27 min and the half to its right.

$$M_2 = [(5 \text{ min})^2 \times 2 \text{ cm/h} + (15 \text{ min})^2 \times 4 \text{ cm/h}$$

$$+ (25 \text{ min})^2 \times 2 \text{ cm/h} + (35 \text{ min})^2 \times 4 \text{ cm/h}$$

$$+ (45 \text{ min})^2 \times 2 \text{ cm/h} + (55 \text{ min})^2 \times 1 \text{ cm/h}]$$

$$\times \frac{10 \text{ min}}{2.5 \text{ cm} \times 60 \text{ min/h}}$$

$$= [1 \times 2 + 9 \times 4 + 25 \times 2 + 49 \times 4$$

$$+ 81 \times 2 + 121 \times 1] \times \frac{25 \times 10}{2.5 \times 60}$$

$$= [2 + 36 + 50 + 196 + 162 + 121] \times \frac{5}{3} = 945 \text{ min}^2$$

EXAMPLE 4.3 In Example 4.2, moments are computed about the vertical axis. Using the same data, compute the first two moments about the origin with respect to the horizontal axis.

Solution

$$M_0 = (60 + 50 + 20 + 20)\text{ min} \times \frac{1\text{ h}}{60\text{ min}} \times \frac{1\text{ cm}}{\text{h}} = 2.5\text{ cm}$$

$$\begin{aligned} M_1 &= [(0.5 \times 1 \times 60) + (1.5 \times 1 \times 50) + (2.5 \times 1 \times 20) \\ &\quad + (3.5 \times 1 \times 20)] \times \frac{1}{2.5}\frac{\text{cm}}{\text{h}} \times \frac{1\text{ h}}{60\text{ min/h}} \\ &= [30 + 75 + 50 + 70] \times \frac{1}{2.5} \times \frac{1}{60} = 1.5\text{ cm/h} \end{aligned}$$

$$\begin{aligned} M_2 &= [(0.5^2 \times 60) + (1.5^2 \times 50) + (2.5^2 \times 20) + (3.5^2 \times 20)] \\ &\quad \times \frac{1}{2.5} \times \frac{1}{60} \\ &= [15 + 112.5 + 125 + 245] \times \frac{1}{150.0} = 3.317 \left(\frac{\text{cm}}{\text{h}}\right)^2 \end{aligned}$$

4.1.2 Laplace Transform

For certain functions, it is more convenient to use the Laplace transform. For functions that are zero for $x < 0$, the ordinary Laplace transform can be used, which can be expressed as

$$F(s) = L[f(x)] = \int_0^\infty f(x)\, e^{-sx}\, dx$$

where L denotes the Laplace transform, and s is the Laplacian variable. For functions that have values for $x < 0$, we must use the bilateral Laplace transform given by

$$F(s) = L[f(x)] = \int_{-\infty}^\infty f(x)\, e^{-sx}\, dx$$

Differentiating r times the function $F(s)$ with respect to s and evaluating the derivatives at $s = 0$,

$$\begin{aligned} \frac{d^r}{ds^r} F(s)\big|_{s=0} &= \int_0^\infty (-1)^r x^r f(x)\, dx \\ &= (-1)^r M_r \end{aligned} \tag{4.4}$$

Thus, we see that once the Laplace transform of a function is known, its moments can be obtained by differentiation.

EXAMPLE 4.4 Determine the moments of the function in Example 4.1 by using the Laplace transform.

Solution

$$\begin{aligned} F(s) &= a \int_0^\infty \exp\,[-(a + s)x]\, dx \\ &= \frac{a}{a + s} \end{aligned}$$

$$\frac{d^r}{ds^r} F(s) = (-1)^r \frac{r!a}{(a + s)^{r+1}}$$

$$M_r = (-1)^{-r} \frac{d^r}{ds^r} F(s)|_{s=0} = \frac{r!}{a^r}$$

Therefore,

$$M_0 = 1$$

$$M_1 = \frac{1}{a}$$

$$M_2 = \frac{2}{a^2}$$

and so on.

4.1.3 Change of Reference Points for Moments

Let a and b be two constants and let $c = b - a$. Let us denote the rth moments of a function about a and b, respectively, by M_r^a and M_r^b. Expanding $(x - a)^r$ binomially,

$$\begin{aligned}(x - a)^r &= (x - b + b - a)^r = (x - b + c)^r \\ &= \sum_{j=0}^{r} \binom{r}{j} (x - b)^{r-j} c^j\end{aligned} \tag{4.5}$$

Then we write

$$M_r^a = \int_{-\infty}^{\infty} (x - a)^r f(x)\, dx$$

Substituting Equation (4.5),

$$\begin{aligned}M_r^a &= \int_{-\infty}^{\infty} \sum_{j=0}^{r} \binom{r}{j} (x - b)^{r-j} c^j f(x)\, dx \\ &= \sum_{j=0}^{r} \binom{r}{j} c^j \int_{-\infty}^{\infty} (x - b)^{r-j} f(x)\, dx \\ &= \sum_{j=0}^{r} \binom{r}{j} c^j M_{r-j}^b\end{aligned} \tag{4.6}$$

This gives the rth moment of a function about a in terms of its rth moment and lower moments about b. Writing the suffixes as power indices (without, of course, interpreting them as such except for the purpose of expansion) or the symbolic mnemonic form of the previous relationship,

$$M_r^a = [M^b + c]^r, \text{ for all } r \tag{4.7}$$

in which $[M^b]^r$ is interpreted as M_r^b for all r. If we specialize by taking a as the origin and b as the centroid or the first moment μ, then

$$M_r = \sum_{j=0}^{r} \binom{r}{j} (\mu)^j M^{\mu}_{r-j} \tag{4.8a}$$

The symbolic mnemonic form of the previous relationship is

$$M_r = [M^{\mu} + \mu]^r, \text{ for all } r \tag{4.8b}$$

In particular this leads to

$$\begin{aligned}
M_0 &= M_0^{\mu} = 1 \\
M_1 &= M_1^{\mu} + \mu M_0^{\mu} = \mu \\
M_2 &= M_2^{\mu} + \mu^2 \\
M_3 &= M_3^{\mu} + 3\mu M_2^{\mu} + \mu^3 \\
M_4 &= M_4^{\mu} + 4\mu M_3^{\mu} + 6M_2^{\mu}\mu^2 + \mu^4
\end{aligned} \tag{4.8c}$$

and so on. Equation (4.8b) can be manipulated to express moments about the centroid in terms of moments about the origin. In particular, this yields

$$\begin{aligned}
M_0^{\mu} &= 1 \\
M_1^{\mu} &= 0 \\
M_2^{\mu} &= M_2 - \mu^2 \\
M_3^{\mu} &= M_3 - 3M_2\mu + 2\mu^3 \\
M_4^{\mu} &= M_4 - 4M_3\mu + 6\mu^2 M_2 - 3\mu^4
\end{aligned} \tag{4.9}$$

EXAMPLE 4.5 Compute the first two moments about the centroid of the rainfall histogram of Example 4.2.

Solution From Example 4.2, $M_1 = \mu = 27$ min.

$$\begin{aligned}
M_1^{\mu} = {} & [(5 - 27) \text{ min} \times 2 \text{ cm/h} + (15 - 27) \text{ min} \times 4 \text{ cm/h} \\
& + (25 - 27) \text{ min} \times 2 \text{ cm/h} + (35 - 27) \text{ min} \times 4 \text{ cm/h} \\
& + (45 - 27) \text{ cm/h} \times 2 \text{ cm/h} + (55 - 27) \text{ min} \times 1 \text{ cm/h}] \\
& \times \left(\frac{10}{60}\right) \text{h} \times \frac{1}{2.5} \text{ cm} \\
= {} & [-22 - 48 - 4 + 32 + 36 + 28] \times \frac{1}{6} \times \frac{1}{2.5} = 0
\end{aligned}$$

The first moment about the centroid is zero.

$$\begin{aligned}
M_2^{\mu} = {} & [(5 - 27)^2 \text{ min}^2 \times 2 \text{ cm/h} + (15 - 27)^2 \text{ min}^2 \times 4 \text{ cm/h} \\
& + (25 - 27)^2 \text{ min}^2 \times 2 \text{ cm/h} + (35 - 27)^2 \text{ min}^2 \times 4 \text{ cm/h} \\
& + (45 - 27)^2 \times 2 \text{ cm/h} + (55 - 27)^2 \text{ min}^2 \times 1 \text{ cm/h}] \\
& \times \left(\frac{10}{60}\right) \text{h} \times \frac{1}{2.5} \text{ cm}
\end{aligned}$$

$$= [(-22)^2 \times 2 + (-12)^2 \times 4 + (-2)^2 \times 2 + (8)^2 \times 4$$

$$+ (18)^2 \times 2 + (28)^2 \times 1] \times \frac{1}{15}$$

$$= [3240] \times \frac{1}{15} = 216 \text{ min}^2$$

This gives the variance of 216 min^2.

EXAMPLE 4.6 Verify the relations in Equations (4.8c) and (4.9) for the rainfall histogram in Example 4.2

Solution From Example 4.2, $M_1 = 27$ min, and $M_2 = 945$ min^2; and from Example 4.5, $M_1^\mu = 0.0$, and $M_2^\mu = 216$ min^2.

$$M_1 = M_1^\mu + \mu M_0^\mu = 0.0 \text{ min} + 27 \text{ min} \times 1 = 27 \text{ min}$$

$$M_2 = M_2^\mu + \mu^2 = 216 + (27)^2 = 216 \text{ min}^2 + 729 \text{ min}^2 = 945 \text{ min}^2$$

EXAMPLE 4.7 Using the data of Example 4.2, compute the first two moments about the centroid with respect to the horizontal axis.

Solution The centroid, $\mu = 1.5$ cm/h.

$$M_1^\mu = [(0.5 - 1.5) \times 60 + (1.5 - 1.5) \times 50 + (2.5 - 1.5) \times 20$$

$$+ (3.5 \times 1.5) \times 20] \times \frac{1}{2.5} \times \frac{1}{60}$$

$$= [-60 + 0 + 20 + 40] \times \frac{1}{150} = 0 \text{ cm/h}$$

$$M_2^\mu = [(0.5 - 1.5)^2 \times 60 + (1.5 - 1.5)^2 \times 50 + (2.5 - 1.5)^2 \times 20$$

$$+ (3.5 \times 1.5)^2 \times 20] \times \frac{1}{2.5} \times \frac{1}{60}$$

$$= [60 + 0 + 20 + 80] \times \frac{1}{2.5} \times \frac{1}{60} = 1.07 \text{ cm/h})^2$$

EXAMPLE 4.8 Verify the relations in Equations (4.8c) and (4.9) between the moments about the origin and the moments about the centroid, both taken about the horizontal axis, for the rainfall histogram in Example 4.2.

Solution

$$M_1 = 1.5 \text{ cm/h}, M_2 = 3.32 \text{ (cm/h)}^2, M_1^\mu = 0.0 \text{ cm/h}$$

$$M_2^\mu = 1.07 \text{ (cm/h)}^2$$

$$M_1 = M_1^\mu + \mu = 0 + 1.5 = 1.5 \text{ cm/h}$$

$$M_2 = M_2^\mu + \mu^2 = 1.07 + 1.5 \times 1.5 = 3.32 \text{ (cm/h)}^2$$

4.1.4 Invariance Property

The moments have an invariance property that states that when the variate values are multiplied by a constant, the rth moment M_r is multiplied by a^r. This is evident at once from its definition.

4.2 METHOD OF LEAST SQUARES (MOLS)

Suppose there are n pairs of x and y values plotted in Figure 4.4, and a straight line $y_e = a + bx$ can be fitted to the data points, where a and b are parameters. The value y_e, obtained from the straight line, is the estimate of observed y_0. The parameters must be determined such that the n points lie as close to the line as possible. In other words, the sum of errors between all values of y_0 and y_e, that is, $\Sigma|y_0 - y_e|$ has the smallest possible value. One way to accomplish this objective is to determine a and b (or define the line) such that $\Sigma(y_0 - y_e)^2$ has the smallest value. This procedure is called the method of least squares (MOLS), where the sum is computed for the given n pairs of x and y values. This is the same as done for regression analysis in statistics (Draper and Smith, 1966). Thus, the advantage of using the MOLS for parameter estimation is that statistical inferences can be made pertaining to the goodness of fit to the observed data.

If the relationship between two variables x and y is expressed as $y = f(x; a_1, a_2, \ldots, a_m)$, where a_1, $i = 1, 2, \ldots, m$, are parameters to be estimated, then the MOLS involves estimating the parameters by minimizing the sum of squares of all deviations between observed and computed values of y. Mathematically, this sum S can be expressed as

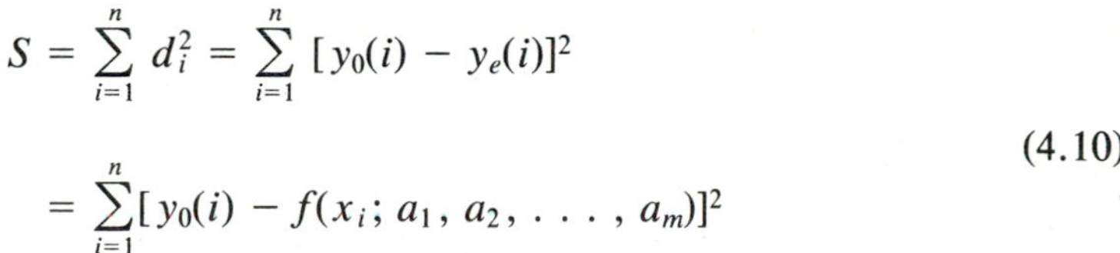

$$\begin{aligned} S &= \sum_{i=1}^{n} d_i^2 = \sum_{i=1}^{n} [y_0(i) - y_e(i)]^2 \\ &= \sum_{i=1}^{n} [y_0(i) - f(x_i; a_1, a_2, \ldots, a_m)]^2 \end{aligned} \tag{4.10}$$

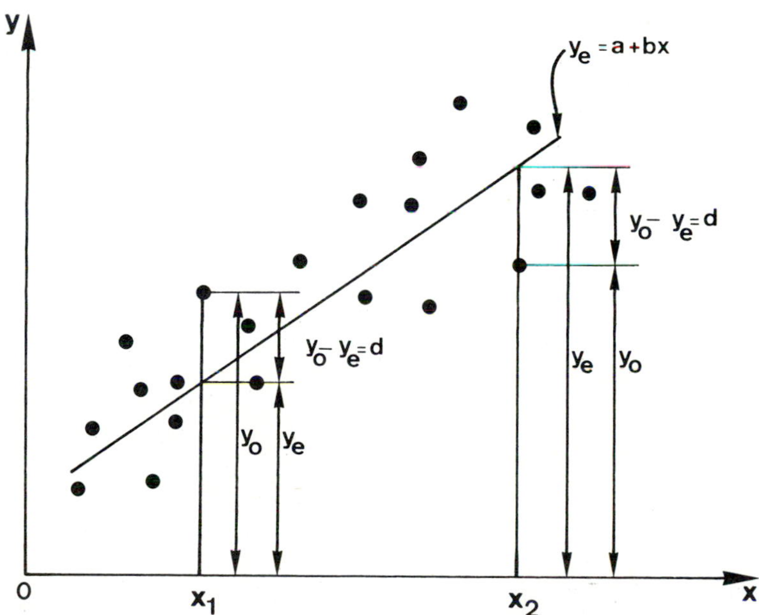

Figure 4.4 Observed values of x and y represented by a straight line, and differences between observed values y and estimated values y_e.

where $y_0(i)$ is the ith observed value of y, $y_e(i)$ is the ith computed value of y, $d_i = y_0(i) - y_e(i)$, and $n > m$ is the number of observations. The minimum of S in Equation (4.10) can be obtained by differentiating S partially with respect to (w.r.t.) each parameter and equating each partial derivative to zero:

$$\frac{\partial \sum_{i=1}^{n} [y_0(i) - f(x_i; a_1, a_2, \ldots, a_m)]^2}{\partial a_1} = 0$$

$$\frac{\partial \sum_{i=1}^{n} [y_0(i) - f(x_i; a_1, a_2, \ldots, a_m)]^2}{\partial a_2} = 0 \tag{4.11}$$

$$\vdots$$

$$\frac{\partial \sum_{i=1}^{n} [y_0(i) - f(x_i; a_1, a_2, \ldots, a_m)]^2}{\partial a_m} = 0$$

The m equations, sometimes called the normal equations, are obtained for estimation of m parameters. The minimum of S corresponds to the ''best'' representation of the data by the function.

The function with the parameters estimated by the MOLS is called the equation of regression of y and /or the equation of prediction for y. It should be pointed out that the MOLS can also be used for nonlinear relationships between x and y, and their parameters can be derived analytically provided these relationships can be linearized using appropriate transformations. For example, if $y = a \exp(ax)$, then it can be linearized by logarithmic transformation as $\ln y = \ln a + ax$, and then parameter a can be estimated.

EXAMPLE 4.9 Let a straight line be expressed as $y = a + bx$. Determine the parameters a and b by the MOLS for a sample of n values of x and y.

Solution

$$S = \sum_{i=1}^{n} (y_0(i) - a - bx_i)^2$$

Differentiating S partially w.r.t. a and b and equating each partial derivative to zero,

$$\frac{\partial S}{\partial a} = -\Sigma\, y_0(i) + na + b\, \Sigma\, x_i = 0$$

$$\frac{\partial S}{\partial b} = -\Sigma\, y_0(i)x_i + a\, \Sigma\, x_i + b\, \Sigma\, x_i^2 = 0$$

The normal equations then are

$$na + b\, \Sigma\, x_i = \Sigma\, y_0(i)$$

$$a\, \Sigma\, x_i + b\, \Sigma\, x_i^2 = \Sigma\, x_i\, y_0(i)$$

In these equations, all the summation quantities are known from the sample. $y_0(i)$ can now be simply written as y_i. Parameters a and b, therefore, can be determined by solving these equations:

$$a = \bar{y} - b\bar{x}$$

$$b = \frac{\Sigma\, x_i y_i - n\bar{x}\bar{y}}{\Sigma\, x_i^2 - n\bar{x}_i^2}$$

$$\bar{y} = \frac{1}{n}\Sigma\, y_i, \qquad \bar{x} = \frac{1}{n}\Sigma\, x_i$$

EXAMPLE 4.10 Let there be a function $y = a + bx + cx^2$. Determine the parameters a, b, and c by the MOLS for a sample of n values of x and y.

Solution Following the same procedure, we construct S as

$$S = \sum_{i=1}^{n} [y_0(i) - a - bx_i - cx_i^2]^2$$

Differentiating S partially w.r.t. a, b, and c, and equating each partial derivative to zero results in three normal equations:

$$\frac{\partial S}{\partial a} = 2\sum_{i=1}^{n} [y_0(i) - a - bx_i - cx_i^2]\,(-1) = 0$$

$$= -\Sigma\, y_0(i) + an + b\,\Sigma\, x_i + c\,\Sigma\, x_i^2 = 0$$

or

$$an + b\,\Sigma\, x_i + c\,\Sigma\, x_i^2 = \Sigma\, y_0(i)$$

$$\frac{\partial S}{\partial b} = 2\sum_{i=1}^{n} [y_0(i) - a - bx_i - cx_i^2]\,(-x_i) = 0$$

$$= -\Sigma\, y_0(i)x_i + a\,\Sigma\, x_i + b\,\Sigma\, x_i^2 + c\,\Sigma\, x_i^3 = 0$$

or

$$a\,\Sigma\, x_i + b\,\Sigma\, x_i^2 + c\,\Sigma\, x_i^3 = \Sigma\, x_i y_0(i)$$

$$\frac{\partial S}{\partial c} = 2\sum_{i=1}^{n} [y_0(i) - a - bx_i - cx_i^2](-x_i^2) = 0$$

$$= -\Sigma\, y_0(i)x_i^2 + a\,\Sigma\, x_i^2 + b\,\Sigma\, x_i^3 + c\,\Sigma\, x_i^4 = 0$$

or

$$a\,\Sigma\, x_i^2 + b\,\Sigma\, x_i^3 + c\,\Sigma\, x_i^4 = \Sigma\, x_i^2 y_0(i)$$

The previous three equations can be solved for a, b, and c using Kramer's rule, since all summation terms are known from the sample.

To that end, one can write the equations in matrix form as

$$\begin{bmatrix} n & \Sigma x_i & \Sigma x_i^2 \\ \Sigma x_i & \Sigma x_i^2 & \Sigma x_i^3 \\ \Sigma x_i^2 & \Sigma x_i^3 & \Sigma x_i^4 \end{bmatrix} \begin{bmatrix} a \\ b \\ c \end{bmatrix} = \begin{bmatrix} \Sigma\, y_0(i) \\ \Sigma x_i y_0(i) \\ \Sigma x_i^2 y_0(i) \end{bmatrix}$$

Therefore, the solution is

$$a = \frac{1}{D_e}\begin{bmatrix} \Sigma\, y_0(i) & \Sigma x_i & \Sigma x_i^2 \\ \Sigma x_i y_0(i) & \Sigma x_i^2 & \Sigma x_i^3 \\ \Sigma x_i^2 y_0(i) & \Sigma x_i^3 & \Sigma x_i^4 \end{bmatrix}$$

$$= \frac{1}{D_e}[\Sigma\, y_0(i)(\Sigma x_i^2\, \Sigma x_i^4 - \Sigma x_i^3\, \Sigma x_i^3) - \Sigma x_i\, (\Sigma x_i y_0(i)\, \Sigma x_i^4$$

$$-\Sigma x_i^2 y_0(i)\, \Sigma x_i^3) + \Sigma x_i^2\, (\Sigma x_i y_0(i)\, \Sigma x_i^3 - \Sigma x_i^2 y_0(i)\, \Sigma x_i^2)]$$

$$b = \frac{1}{D_e}\begin{bmatrix} n & \Sigma\, y_0(i) & \Sigma x_i^2 \\ \Sigma x_i & \Sigma x_i y_0(i) & \Sigma x_i^3 \\ \Sigma x_i^2 & \Sigma x_i^2 y_0(i) & \Sigma x_i^4 \end{bmatrix}$$

$$= \frac{1}{D_e}[n(\Sigma x_i y_0(i)\, \Sigma x_i^4 - \Sigma x_i^2 y_0(i)\, \Sigma x_i^3) - \Sigma\, y_0(i)(\Sigma x_i\, \Sigma x_i^4$$

$$-\Sigma x_i^2\, \Sigma x_i^3) + \Sigma x_i^2\, (\Sigma x_i\, \Sigma x_i^2 y_0(i) - \Sigma x_i^2\, \Sigma x_i y_0(i))]$$

$$c = \frac{1}{D_e}\begin{bmatrix} n & \Sigma x_i & \Sigma\, y_0(i) \\ \Sigma x_i & \Sigma x_i^2 & \Sigma x_i y_0(i) \\ \Sigma x_i^2 & \Sigma x_i^3 & \Sigma x_i^2 y_0(i) \end{bmatrix}$$

$$= \frac{1}{D_e}[n(\Sigma x_i^2\, \Sigma x_i^2 y_0(i) - \Sigma x_i^3\, \Sigma x_i y_0(i)) - \Sigma x_i\, (\Sigma x_i\, \Sigma x_i^2 y_0(i)$$

$$-\Sigma x_i^2\, \Sigma x_i y_0(i)) + \Sigma\, y_0(i)\, (\Sigma x_i\, \Sigma x_i^3 - \Sigma x_i^2\, \Sigma x_i^2)]$$

$$D_e = \begin{bmatrix} n & \Sigma x_i & \Sigma x_i^2 \\ \Sigma x_i & \Sigma x_i^2 & \Sigma x_i^3 \\ \Sigma x_i^2 & \Sigma x_i^3 & \Sigma x_i^4 \end{bmatrix}$$

$$= n(\Sigma x_i^2\, \Sigma x_i^4 - \Sigma x_i^3\, \Sigma x_i^3) - \Sigma x_i\, (\Sigma x_i\, \Sigma x_i^4 - \Sigma x_i^2\, \Sigma x_i^3)$$

$$+ \Sigma x_i^2\, (\Sigma x_i\, \Sigma x_i^3 - \Sigma x_i^2\, \Sigma x_i^2)$$

EXAMPLE 4.11 Let a general function be expressed as $y = a_0 + \Sigma_{i=1}^{m} a_i x_i$. Estimate the parameters a_i, $i = 0, 1, 2, \ldots, m$, from a sample of n values of $y, x_1, x_2, \ldots, x_m$, $n > m$.

Solution: This expression is similar in form to the linear equation in Example 4.9. Therefore, the normal equations of that example can be extended to the general case. By following the same procedure, the normal equations can be written as

$$na_0 + a_1\, \Sigma\, x_i + a_2\, \Sigma\, x_2 + a_3\, \Sigma\, x_3 + \ldots + a_m\, \Sigma\, x_m = \Sigma\, y_0$$

$$a_0\, \Sigma\, x_1 + a_1\, \Sigma\, x_1^2 + a_2\, \Sigma\, x_1 x_2 + a_3\, \Sigma\, x_1 x_3 + \ldots + a_m\, \Sigma\, x_1 x_m = \Sigma\, x_1 y_0(i)$$

$$a_0\, \Sigma\, x_2 + a_1\, \Sigma\, x_1 x_2 + a_2\, \Sigma\, x_2^2 + a_3\, \Sigma\, x_2 x_3 + \ldots + a_m\, \Sigma\, x_2 x_m = \Sigma\, x_2 y_0(i)$$

$$\vdots$$

$$a_0\, \Sigma\, x_m + a_1\, \Sigma\, x_1 x_m + a_2\, \Sigma\, x_2 x_m + a_3\, \Sigma\, x_3 x_m + \ldots + a_m\, \Sigma\, x_m^2 = \Sigma\, x_m y_0(i)$$

Thus, there are as many equations as there are unknowns. These equations, therefore, can be solved to obtain the parameters. It is more convenient to obtain the solution using matrix algebra.

4.2.1 Matrix Representation of the MOLS

The case involving more than one variable can be expressed more efficiently using matrix notation. The normal equations in the MOLS can often be reduced to the form

$$X \cdot A = Y \tag{4.12}$$

Where X is a matrix of $(n + 1, m + 1)$ dimensions, A is a vector of the parameters having $(m + 1, 1)$ dimensions, and Y is a vector having $(n + 1, 1)$ dimensions. In this section, uppercase letters signify matrices or vectors. Superscript T signifies the transpose. The deviations between observed and computed values of y can be written as

$$D = Y - XA$$

where D is a vector of deviations d_i, $i = 1, 2, \ldots, n$.

The sum of squares of deviations can be obtained by using the inner product, which is obtained by multiplying D by its transpose D^T as

$$\begin{aligned} \Sigma\, d_i^2 &= D^T D \\ &= [Y^T - A^T X^T][Y - XA] \\ &= Y^T Y - Y^T XA - A^T X^T Y + A^T X^T XA \end{aligned} \tag{4.13}$$

Since A and Y are column vectors, their transposes will be row vectors. Thus, the second and third terms on the right side of Equation (4.13) will be scalar in form. Since a scalar and its transpose are the same,

$$d_i^2 = Y^T Y - 2A^T X^T Y + A^T X^T XA$$

Differentiating w.r.t. A, and equating to zero,

$$X^T XA = X^T Y$$

The matrix $[X^T X]$ and the vector $[X^T Y]$ are generated from observed values of the variates. The parameters can then be determined as

$$A = [X^T X]^{-1} X^T Y \tag{4.14}$$

Equation (4.14) expresses in compact form a least squares solution.

4.2.2 Error of Fit

In the MOLS method, how well the function represents the data can be analyzed. The sum of squares of errors, S, given by Equation (4.10), is a measure of the dispersion of the observed y_0 values about their corresponding estimated y_e values. To that end, the variance V of y_0 values is defined as

$$V = \sum_{i=1}^{n} (y_0(i) - \bar{y}_0)^2, \qquad \bar{y}_0 = \frac{1}{n} \sum_{i=1}^{n} y_0(i) \tag{4.15}$$

which is a measure of the dispersion of observed y_0 values about their mean $\bar{y}_0$. Similarly, the dispersion of the estimated y_e values about the mean of the observed values, $\bar{y}_0$, is defined as

$$V_e = \sum_{i=1}^{n} (y_e(i) - \bar{y}_0)^2 \tag{4.16}$$

The three sums of squares of deviations given by Equations (4.10), (4.15), and (4.16) are related to each other as

$$V = V_e + S \tag{4.17}$$

Equation (4.17) states that the dispersion of the observed y_o values about their mean is equal to the sum of the dispersion of the estimated y_e values about that mean and the dispersion of the observed y_0 values about their corresponding estimated y_e values from the least squares fit.

For measuring the goodness of fit, two measures are commonly used: the standard error of estimate and the correlation coefficient. The standard error of estimate for a least squares fit to a sample of n pairs of x and y values is defined as

$$S_e = \left\{\frac{1}{n} \sum_{i=1}^{n} [y_0(i) - y_e(i)]^2\right\}^{0.5} = \left[\frac{S}{n}\right]^{0.5} \tag{4.18}$$

where S is given by Equation (4.10). It is the standard deviation of the errors of estimation. Clearly, a value of $S_e = 0$ means that all points lie on the fit. The smaller the value of S_e, the closer the points to the fit.

The correlation coefficient for a sample of n pairs of x and y values is defined as

$$r = \frac{\Sigma xy - \dfrac{\Sigma x \Sigma y}{n}}{\left[\left\{\Sigma x^2 - \dfrac{(\Sigma x)^2}{n}\right\}\left\{\Sigma y^2 - \dfrac{(\Sigma y)^2}{n}\right\}\right]^{0.5}} \tag{4.19}$$

or

$$r = \frac{\Sigma xy - n\bar{x}\bar{y}}{[(\Sigma x^2 - n\bar{x}^2)(\Sigma y^2 - n\bar{y}^2)]^{0.5}} \tag{4.20}$$

It can be shown that

$$r^2 = \frac{V_e}{V} = \frac{\text{explained variance}}{\text{total variance}} \tag{4.21}$$

The square of the correlation coefficient, r^2, is termed as the coefficient of determination. This coefficient is related to the standard error of estimate through Equation (4.17). Equation (4.21) shows thar r^2 is the proportion of the total variation in (or variance of) y that can be explained by the relationship existing between x and y. For example, the correlation coefficient of 0.9 indicates that $(0.9)^2 \times 100 = 81\%$ of the variation is due to the relation between x and y; the remaining 19% of the variation is due to unexplained factors.

If all the data points lie on the fit, then $r = +1$ or -1. This is evident because $S = 0$, and, consequently, $V = V_e$, and $r^2 = 1$ from Equation (4.21). If there exists no relationship between x and y, then $r = 0$. For no relationship, $V_e = 0$, and thus $r^2 = 0$ from Equation (4.21). Since r^2 is always between 0 and 1, r is always between -1 and $+1$.

EXAMPLE 4.12 Annual precipitation at Flagstaff (y) is found to be linearly related to annual precipitation at Fort Valley (x) in Arizona. If the linear relation is assumed to be of the form $y = a + bx$, then determine parameters a and b using the method of least squares. Also compute the correlation coefficient, the amount of explained variance, and the standard error of estimate. Verify if Equation (4.17) is valid.

Solution Pertinent calculations for this problem are shown in Table 4.1. First, the values in columns (2) and (3) are summed to produce their average values respectively as

$$\bar{y} = \frac{\Sigma y}{n} = \frac{1430.96}{30} = 47.70$$

$$\bar{x} = \frac{\Sigma x}{n} = \frac{1770.49}{30} = 59.02$$

Then the parameters of the line are estimated using equations derived in Example 4.9. For this example, calculations are shown in columns (4) and (5).

$$b = \frac{\Sigma\, xy - n\bar{x}\bar{y}}{\Sigma\, n^2 - n\bar{x}^2} = \frac{87{,}261.63 - 30 \times 59.02 \times 47.70}{108{,}188.67 - 30 \times 59.02^2} = 0.76$$

$$a = \bar{y} - b\bar{x} = 47.70 - 0.760 \times 59.02 = 2.85$$

The correlation coefficient is determined from Equation (4.20) for which the calculations are shown in columns (4), (5), and (6).

$$r = \frac{\Sigma\, xy - n\bar{x}\bar{y}}{[(\Sigma x^2 - n\bar{x}^2)(\Sigma y^2 - n\bar{y}^2)]^{0.5}}$$

$$= \frac{87{,}261.63 - 30 \times 59.02 \times 47.70}{[(108{,}188.67 - 30 \times 59.02^2)(71{,}145.79 - 30 \times 47.7^2)]^{0.5}}$$

$$= 0.86$$

This produces

$$r^2 = \text{amount of explained variance} = 0.74$$

The standard error of estimate is given from column (8) as

$$S_e = \left[\frac{1}{n}\Sigma(y - y_e)^2\right]^{0.5} = \left(\frac{753.37}{30}\right)^{0.5} = 5.01$$

For verification of Equation (4.17), we need V, V_e, and S, which are given, respectively, by columns (9), (10), and (8).

$$V = \Sigma\,(y - \bar{y})^2 = 2890.90$$

$$V_e = \Sigma(y_e - \bar{y})^2 = 2137.46$$

$$S = \Sigma\,(y - y_e)^2 = 753.47$$

TABLE 4.1 ANNUAL PRECIPITATION AT FLAGSTAFF AND FORT VALLEY, ARIZONA; AND LEAST SQUARES ANALYSIS

Year (1)	Annual precipitation at Flagstaff, y (cm) (2)	Annual precipitation at Fort Valley, x (cm) (3)	xy (4)	x^2 (5)	y^2 (6)	y_e (7)	$(y - y_e)^2$ (8)	$(y - \bar{y})^2$ (9)	$(y_e - \bar{y})^2$ (10)
1920	49.10	60.73	2,981.84	3,688.13	2,410.81	49.00	0.01	1.96	1.69
1921	56.97	69.64	3,967.39	4,849.73	3,245.58	55.77	1.44	85.93	65.12
1922	63.68	64.06	4,079.34	4,103.68	4,055.14	51.53	147.62	255.36	14.69
1923	53.51	68.81	3,682.02	4,734.81	2,863.32	55.14	2.66	33.76	55.35
1924	42.52	48.26	2,052.02	2,329.03	1,807.95	39.52	9.00	26.83	66.91
1925	48.46	60.33	2,923.59	3,639.71	2,348.37	48.70	0.06	0.58	1.00
1926	42.12	51.87	2,184.76	2,690.50	1,774.09	42.27	0.02	31.14	29.48
1927	60.53	72.64	4,396.90	5,276.57	3,663.88	58.05	6.15	164.61	107.12
1928	37.79	53.72	2,030.08	2,885.84	1,428.08	43.67	34.57	98.21	16.24
1929	39.42	53.29	2,100.69	2,839.82	1,553.94	43.35	15.44	68.56	18.92
1930	54.73	65.16	3,565.66	4,244.52	2,995.37	52.36	5.57	49.42	21.81
1931	51.67	63.24	3,267.61	3,999.30	2,669.79	50.91	0.58	15.76	10.30
1932	55.80	58.42	3,259.84	3,412.90	3,113.64	47.25	73.10	65.61	0.20
1933	39.63	51.06	2,023.51	2,607.12	1,570.54	41.65	4.08	65.13	36.60
1934	37.59	45.97	1,728.01	2,113.24	1,413.07	37.78	0.04	102.21	98.41
1935	42.47	62.05	2,635.26	3,850.20	1,803.70	50.00	56.70	27.35	5.39
1936	49.02	72.37	3,547.58	5,237.41	2,402.96	57.85	77.97	1.74	103.02
1937	49.30	71.73	3,536.29	5,145.19	2,430.49	57.36	64.96	2.56	93.32
1938	52.27	64.59	3,376.12	4,171.87	2,732.15	51.93	0.12	20.89	17.89
1939	32.70	43.38	1,418.53	1,881.82	1,069.29	35.82	9.73	225.00	141.13
1940	53.90	73.25	3,948.17	5,365.56	2,905.21	58.52	21.34	38.44	117.07
1941	63.55	79.40	5,045.87	6,304.36	4,038.60	63.19	0.13	251.22	239.94
1942	25.14	38.79	975.18	1,504.66	632.02	32.33	51.70	508.95	236.24
1943	45.16	59.47	2,685.67	3,536.68	2,039.43	48.04	8.29	6.45	0.12
1944	44.45	55.37	2,461.20	3,065.84	1,975.80	44.93	0.23	10.56	7.67
1945	44.73	51.72	2,313.44	2,674.96	2,000.77	42.15	6.66	8.83	30.80
1946	55.22	59.51	3,286.14	3,541.44	3,049.25	48.07	51.12	56.55	0.14
1947	33.35	34.01	1,134.23	1,156.68	1,112.22	28.69	21.77	205.92	361.38
1948	38.89	44.42	1,727.49	1,973.14	1,512.43	36.61	5.20	77.62	123.99
1949	67.29	73.23	4,927.64	5,362.63	4,527.94	58.50	77.26	383.77	116.64

$\Sigma y = 1{,}430.96$, $\Sigma x = 1{,}770.49$, $\Sigma xy = 87{,}261.63$, $\Sigma x^2 = 108{,}188.67$, $\Sigma y^2 = 71{,}145.79$, $\Sigma(y - y_e)^2 = 753.47$, $\Sigma(y - \bar{y})^2 = 2{,}890.90$, $\Sigma(y_e - \bar{y})^2 = 2{,}137.46$

EXAMPLE 4.13 Assume a linear relation $x = a_1 + b_1 y$ between annual precipitation at Fort Valley denoted by x and annual precipitation at Flagstaff denoted by y. Determine the parameters a_1 and b_1 by the method of least squares using the data of Example 4.12. Also compute the correlation coefficient, the amount of explained variance, and the standard error of estimate. Verify if Equation (4.17) is valid.

Solution By following the same procedure, $\Sigma x = 1770.49$ cm, $\Sigma y = 1430.96$ cm, $\Sigma xy = 87{,}261.63\ \text{cm}^2$, $\Sigma x^2 = 108{,}188.67\ \text{cm}^2$, $\Sigma y^2 = 71{,}145.79\ \text{cm}^2$, $\Sigma \bar{x}^2 = (59.02)^2$, and $\Sigma \bar{y}^2 = (47.4)^2$.

$$b_1 = \frac{\Sigma xy - n\bar{x}\bar{y}}{\Sigma y^2 - n\bar{y}^2} = \frac{87{,}261.63 - 30 \times 47.7 \times 59.02}{71{,}145.79 - 30 \times 47.7 \times 47.7}$$

$$= \frac{2804.07}{2886.99} = 0.97$$

$$a_1 = \bar{x} - b_1\bar{y} = 59.02 - 0.97 \times 47.7 = 12.75$$

$$r = \frac{\Sigma xy - n\bar{x}\bar{y}}{[(\Sigma x^2 - n\bar{x}^2)(\Sigma y^2 - n\bar{y}^2)]^{0.5}}$$

$$= \frac{87{,}261.63 \times 30 \times 47.7 \times 59.02}{\{[71{,}145.79 - 30 \times (47.7)^2][108{,}188.67 - 30 \times (59.02)^2]\}^{0.5}}$$

$$= \frac{2804.07}{[(2886.99)(3686.128)]^{0.5}} = \frac{2804.07}{3262.18} = 0.86$$

$$r^2 = 0.74$$

The standard error of estimate, S_e, is 5.87. Thus,

$$x = 12.75 + 0.97y$$

EXAMPLE 4.14 Compare the values of the parameters obtained in Examples 4.12 and 4.13. Also compare the values of the correlation coefficient, the amount of explained variance, and the standard error of estimate of the two examples.

Solution From the results of Examples 4.12 and 4.13:

PARAMETER	EXAMPLE 4.12 $y = a + bx$	EXAMPLE 4.13 $x = a_1 + b_1 y$
a or a_1	2.85	12.75
b or b_1	0.76	0.973
r	0.86	0.86
r^2	0.74	0.74
S_e	5.01	5.87

EXERCISES

4.1. Consider the normal distribution with mean $\bar{x}$ and variance S^2:

$$f(x) = \frac{1}{S(2\pi)^{0.5}} \exp\left[-\frac{(x - \bar{x})^2}{2S^2}\right], \qquad -\infty < \bar{x} < \infty, \qquad S^2 > 0$$

Determine the first two moments about the origin as well as about the centroid.

4.2. Using the moments derived in Exercise 4.1, show that Equations (4.8c) and (4.9) are valid.

4.3. Consider the two-parameter distribution:

$$f(x) = \frac{1}{a\Gamma\,(b)}\left(\frac{x}{a}\right)^{b-1} \exp\,(-x/a), \qquad a > 0, \qquad b > 0$$

where a and b are parameters. Derive the first two moments about the origin as well as about the centroid.

4.4. Show that Equations (4.8c) and (4.9) are valid for the moments derived in Exercise 4.3.

4.5. Consider the relation between x and y as

$$y = a - bx, \qquad 0 \le x \le 10, \qquad b > 0$$

Determine parameters a and b using the method of moments.

4.6. The amount of runoff, V, and the associated peak discharge, Q, are found to be related for a watershed as $Q = aV^b$, which is linear in log space, $\log Q = \log a + b \log V$. Determine parameters a and b using the method of least squares for the data given in Table 4.2. The logarithmic version is analogous to $y = c + bx$ and should be used for the MOLS.

4.7. Determine the amount of variance explained by the log linear fit in Exercise 4.6, the correlation coefficient, and the standard error of estimate. Verify Equation (4.17).

4.8. Determine the parameters a and b in Exercise 4.6 using the method of moments. Then compute the correlation coefficient, standard error of estimate and the amount of variance explained. Show by using the results from Exercise 4.7 that the MOLS produces smaller sum of squares of errors S than MOM.

4.9. The amount of direct runoff V and the sediment yield S are found to be related as $S_y = aV^b$, where a and b are parameters. Determine the parameters using the MOLS for the data given in Table 4.3. Linearize the relation as $\log S_y = \log a + b \log V$ before doing the analysis.

4.10. Determine the correlation coefficient, standard error of estimate, and the amount of explained variance for the relation in Exercise 4.9.

4.11. A rainfall event has been observed in a histogram form as

TIME (min)	INTENSITY (cm/h)	TIME (min)	INTENSITY (cm/h)
0–10	5	10–20	10
20–30	5	30–40	5

TABLE 4.2 RUNOFF AMOUNTS AND ASSOCIATED PEAK DISCHARGES FOR THE WATERSHED D (4.5 km² AREA), RIESEL (WACO), TEXAS

Runoff amount (cm)	Peak discharge (cm/h)	Runoff amount (cm)	Peak discharge (cm/h)	Runoff amount (cm)	Peak discharge (cm/h)
1.97	1.70	4.12	1.53	1.42	0.18
1.28	0.42	5.74	2.27	3.19	0.66
5.59	1.53	2.29	0.58	1.58	0.57
0.40	0.12	0.43	0.69		

TABLE 4.3 AMOUNT OF RUNOFF AND SEDIMENT YIELD FOR VARIOUS EVENTS ON WATERSHED W-5, NEAR OXFORD, MISSISSIPPI

Runoff amount (cm)	Sediment yield (metric tons)	Runoff amount (cm)	Sediment yield (metric tons)	Runoff amount (cm)	Sediment yield (metric tons)
4.54	482.62	2.50	330.06	5.43	975.10
2.29	843.00	0.96	24.73	2.96	521.00
0.11	7.07	0.41	77.56	1.45	205.81
1.55	374.55	2.57	421.28	1.52	449.89
2.48	728.42				

Use the abscissa as the reference axis.

(a) Compute the first two moments about the origin.

(b) Compute the first two moments about the centroid.

(c) Verify the relation between the moments about the origin and the moments about the centroid using the moment values obtained in (a) and (b).

4.12. Use the data in Exercise 4.11, and use the vertical axis (ordinate) as the reference axis. Then compute the first two moments about the origin as well as the first two moments about the centroid. Verify the relation between these two types of moments using the moment values just obtained.

PART 2
Hydrologic Inputs

CHAPTER 5
Drainage-Basin Characteristics

The drainage basin can be considered as the laboratory of the hydrologic cycle. With the only exception of the atmospheric aspects of precipitation, that is where the hydrologic processes occur and where these processes leave permanent imprints of their effects on the drainage basin itself. The evidence of these imprints is displayed in the surficial topography and exhibited in the stream channels themselves by their size, number, and dimensions. These imprints are related to soil type, geology, and vegetation. It is no exaggeration that understanding the clues displayed in a drainage basin is vital to successful prediction of the hydrologic behavior of that drainage basin.

The drainage basin can be thought of as a system that converts rainfall to runoff. During this process, the drainage basin retains some of the rainfall for its own use. This use includes all the loss elements or abstractions of the runoff equation. The drainage basin also controls the rate at which the runoff will occur and the degree to which the runoff water will be concentrated. In other words, it governs the runoff volume from a given rainfall, and the shape and magnitude of the runoff hydrograph, including the peak discharge resulting form this runoff volume. Thus, the key to predicting the runoff response from any drainage basin is the understanding of the basin itself. This chapter reviews some of the drainage-basin characteristics that are useful in hydrology. It also clarifies the relation between geology and hydrology.

5.1 SURFICIAL ENVIRONMENT

All the factors affecting runoff with which rainfall is in contact constitute the surficial environment. These factors include the land surface, soil, geology, vegetation, and stream network. The surficial environment must also include climatic factors that indirectly affect soil development and vegetation. These factors vary from one drainage basin to another in such a manner that no two drainage basins are alike. A drainage basin is as unique to hydrology as fingerprints are to human beings. The runoff responses from varying combinations of these characteristics are almost always different, but sometimes two basins might respond in nearly the same manner and could be grouped together as similar basins. Grouping basins in a similar category is hazardous unless actual response measurements show that grouping is justified. These responses can be likened to two humans with different fingerprints but like personalities. The surficial environment largely determines personalities of the drainage basins.

5.2 DEFINITION OF A DRAINAGE BASIN

A drainage basin is defined as any portion of the earth's surface within a physical boundary defined by topographic slopes that divert all runoff to the same drainage outlet. This definition permits the selection of any drainage outlet desired. One can move the drainage outlet up the drainage system or down the drainage system to any location of interest. This characteristic is important because many hydrologic problems require an analysis of drainage at a certain point. By definition, any point on the main drainage system can be selected as the basin outlet. Thus, a watershed is defined with respect to the outlet. Figure 5.1 shows watershed Y near Riesel (Waco), Texas, which is composed of six subwatersheds (U.S. Department of Agriculture, 1961). Each subwatershed has its own drainage outlet. As a drainage outlet is moved down along the main drainage channel, more drainage area is included. For example, when the outlet of the subwatershed Y-6 is moved down to the point where the subwatershed Y-4 has its outlet, then the drainage area increases from 16.3 to 79.9 acres. The areas on the map show how the drainage area increases with downward movement of the drainage outlet. The physical boundary of the drainage basin is defined by the direction in which surface runoff water will drain, and follows the ridge line between hydrologic units; on a topographic map, as in Figure 5.1, it appears as an irregular closed traverse that is everywhere normal to the land contour. This boundary is usually called the drainage divide. Any precipitation falling on the drainage divide will either drain into the basin of interest because of the topographic slopes or will drain away from the drainage basin of interest. Thus, the drainage divide is a physical boundary for the drainage basin. The watershed area includes all points that lie above the elevation of the outlet and within the drainage divide that separates adjacent watersheds.

Other terms used by hydrologists and laymen that are synonymous with drainage basin are watershed, catchment, basin, river basin, runoff area, and stream basin. Watershed, basin, and catchment are the terms most commonly used by hydrologists as synonyms for drainage basin.

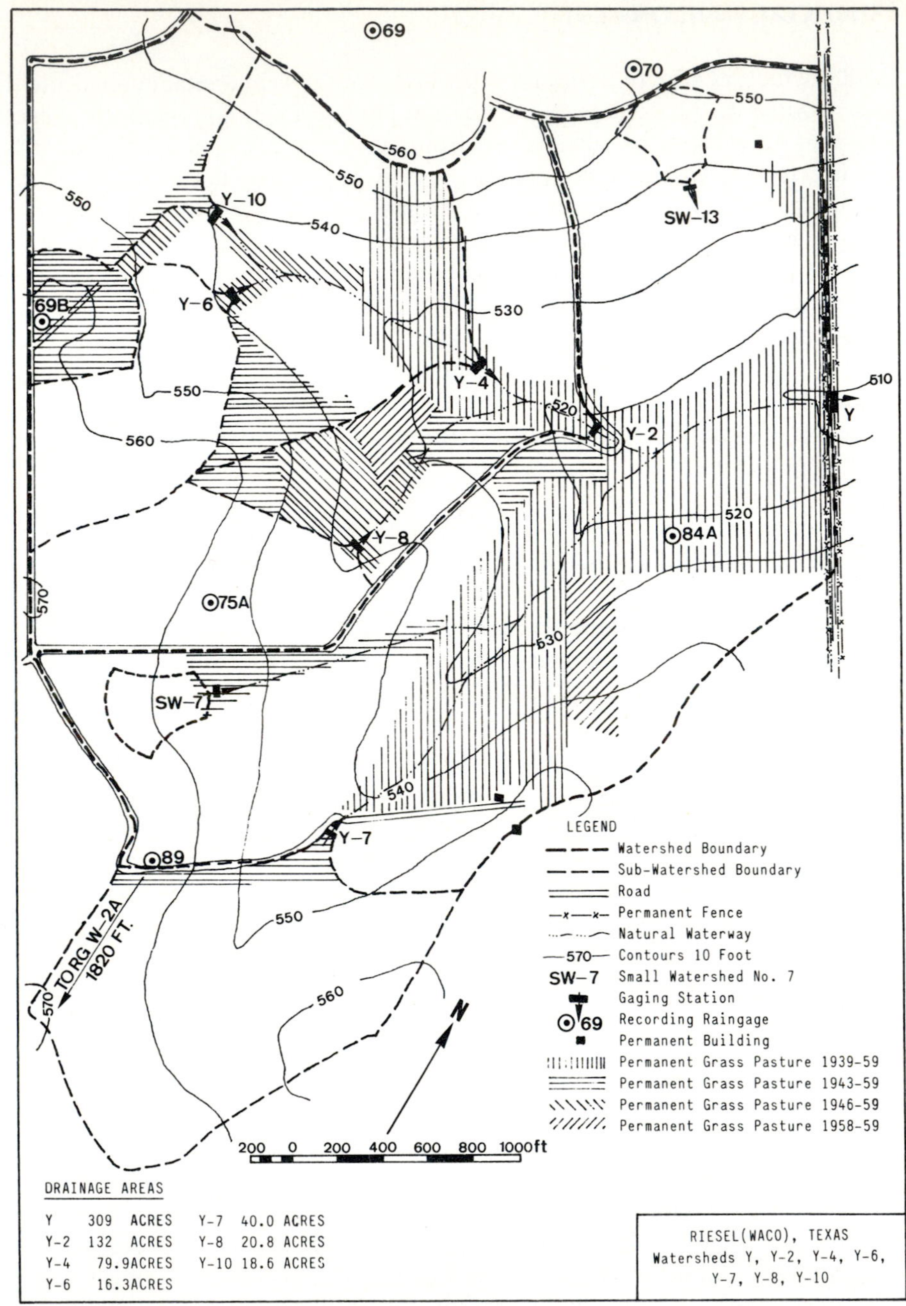

Figure 5.1 Watershed Y and its subwatersheds, Riesel (Waco), Texas (after U.S. Department of Agriculture, 1961).

5.3 QUANTITATIVE CHARACTERISTICS OF DRAINAGE BASINS

Certain characteristics of drainage basins reflect hydrologic behavior and are therefore useful, when quantified, in evaluating the hydrologic response of the basins. These characteristics relate to either the physical drainage basin or to the channels. Physical characteristics of the drainage basin include drainage area, basin shape, ground slope, and centroid, or center of gravity, of the basin. Channel characteristics include channel order, channel length, channel slope, channel profile, and drainage density. Many other characteristics of drainage basins exist and can be quantified but are not commonly identified with diagnostic hydrologic analyses. These and other drainage-basin characteristics are commonly used in geomorphologic studies and to some extent in river engineering. Topographic maps can be digitized these days, and basin characteristics can be conveniently estimated without tedious manual work. Digitizers and simple computer programs are available to determine many of these basin characteristics.

The Irvington drainage basin, located in Little Papillion Creek in the Douglas and Washington counties in Nebraska, will be used as an example basin to illustrate important elements. This basin is shown in Figure 5.2. The stream gage is located near Irvington, Nebraska, just below the junction of Little Papillion and Thomas creeks. The area of this drainage basin above the Irvington gage is approximately 32 mi^2 (83 km^2).

5.3.1 Basin Order and Channel Order

Drainage areas may be characterized in terms of the hierarchy of stream ordering. The order of the basin is the order of its highest-order channel. An inspection of a drainage-basin channel network reveals that as one traces the flow from one of the uppermost channels in the basin toward the outlet, this uppermost channel joins another channel, which in turn joins another channel, and so on. As shown in Figure 5.3, the first-order streams are defined as those channels that have no tributaries. These are the streams whose flow is dependent entirely on surface overland flow to them. The junction of two first-order channels form a second-order channel. A second-order channel receives flow from the two first-order channels that form it, and from overland flow from the ground surface, and it might receive flow from another first-order channel that flows directly into it. Thus, a second-order channel must carry much more flow of water than a first-order channel. A third-order channel is formed by the junction of two second-order channels. A third-order channel receives flow not only from the two second-order channels that form it, but also direct overland flow and possibly from first-order channels that run directly into it and possibly other second-order channels that might join it. Thus, a stream of any order has two or more tributaries of the next lower order. The ordering system continues in the same manner; the junction of two third-order channels form a fourth-order channel and so on. This scheme of stream ordering is referred to as the Horton–Strahler ordering scheme (Horton, 1945; Strahler, 1957). A watershed is described as first-, second-, third-, or higher-order, depending upon the stream order at the outlet (i.e., the highest-order stream within the watershed area). In relatively

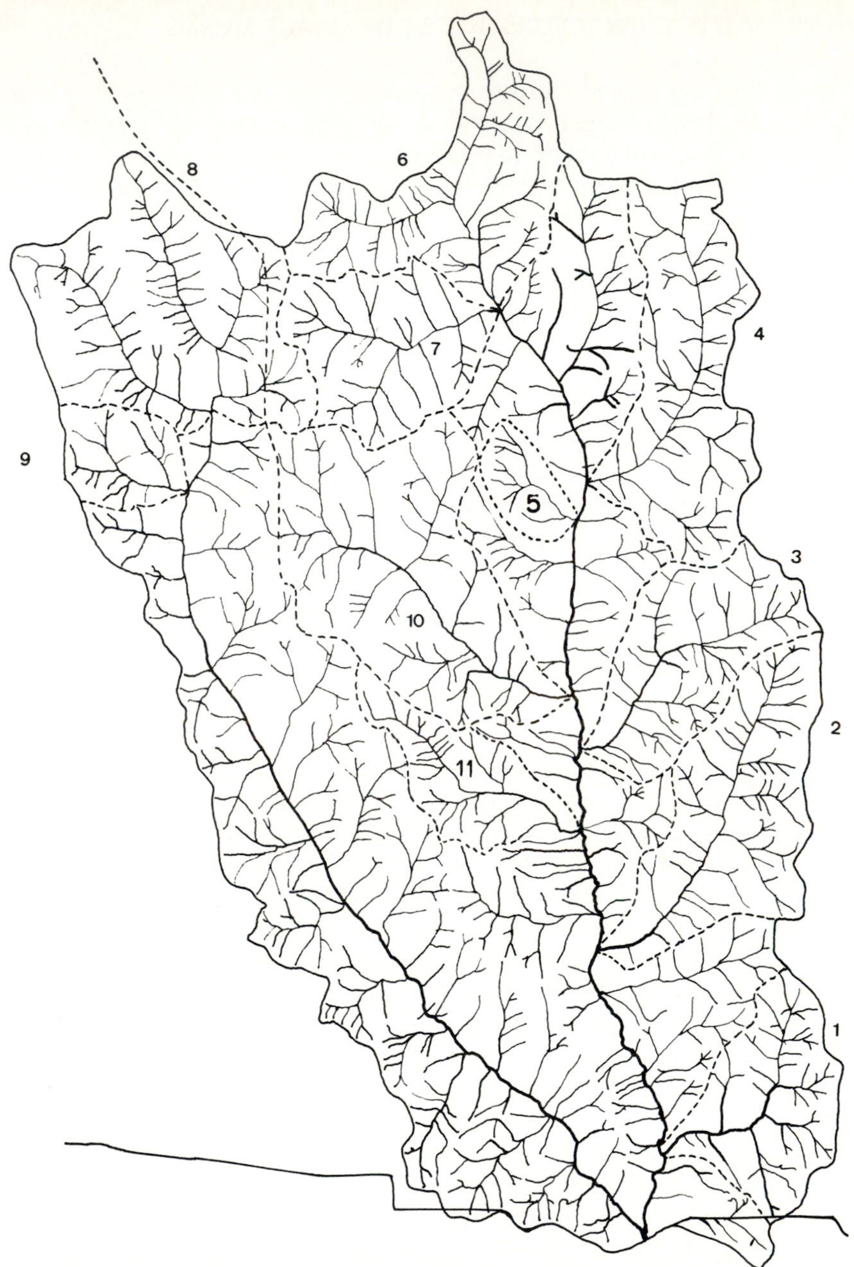

Figure 5.2 The Irvington drainage basin, Nebraska.

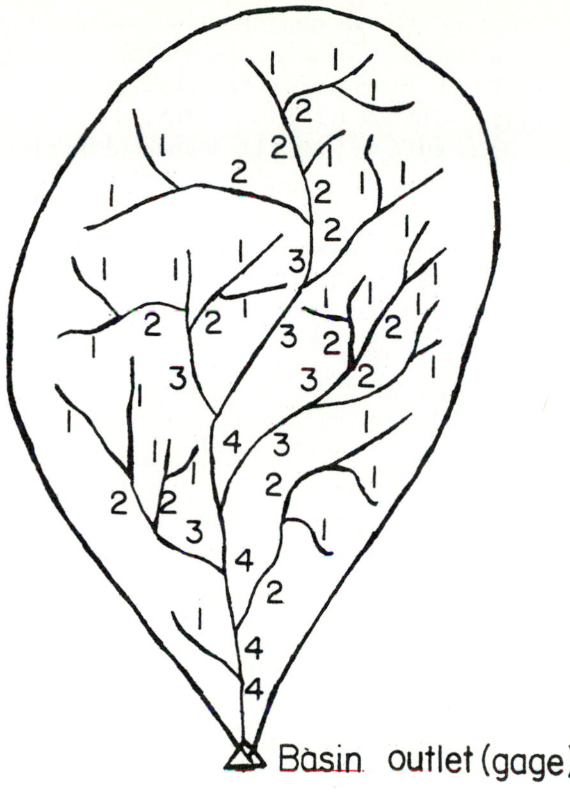

Figure 5.3 Channel ordering by the Strahler method for a fourth-order drainage basin.

homogeneous regions, basin area and other characteristics are found to be highly correlated with basin order.

5.3.2 Basin Area

Basin area is defined as the area contained within the vertical projection of the drainage divide on a horizontal plane. This condition is exhibited on a topographic map, whereon the drainage divide may be traced beginning with the drainage outlet on the stream and following the divide around the drainage basin and back to the outlet. It is apparent that the boundary of the drainage basin can only cross the draining stream at the basin outlet. Drainage area is measured in square miles, square kilometers, acres, or hectares.

The drainage area includes all area within the vertical projection of the drainage boundary. Some drainage basins contain, within their boundaries, areas that do not contribute runoff to the drainage system. These areas are isolated from the drainage system and comprise closed drainages that might form lakes, swamps, or simply areas where drainage infiltrates into the soil or becomes groundwater. Thus, the effective drainage area might be different from the total drainage-basin area. Conditions just described would result in a smaller effective drainage area. It is possible for underground leakage to occur from one drainage basin to another, and this condition would result in a larger effective drainage area for the basin receiving the leakage and a smaller effective drainage area for the basin transmitting the leakage.

The watershed area is comprised of two subcomponents: (1) stream areas and (2) interbasin areas. The interbasin areas are the surface elements contributing flow directly to streams of order higher than 1. Stream areas are those areas that would constitute the area draining to a predetermined point in the stream or outlet. For example, the stream area for first-order streams would be delineated by measuring the drainage area for each first-order channel. This classification of areas is useful in geomorphologic modeling of watershed runoff.

Horton (1945) inferred that mean drainage areas of progressively higher orders might form a geometric sequence. This characteristic was formulated as a law of drainage areas by Schumm (1954), who stated that the mean drainage areas of streams of each order tend to approximate a geometric progression as

$$\bar{A}_w = \bar{A}_1 R_a^{w-1} \tag{5.1a}$$

or

$$\log \bar{A}_w = \log \bar{A}_1 + (w - 1) \log R_a$$

$$\log \bar{A}_w = \log \left(\frac{\bar{A}_1}{R_a}\right) + w \log R_a = a + bw, \qquad a = \log \left(\frac{\bar{A}_1}{R_a}\right), \qquad b = \log R_a \tag{5.1b}$$

where $\bar{A}_w$ is the mean area of basins of order w, $\bar{A}_1$ is the mean area of first-order basins, and R_a is the stream area ratio defined as $R_a = A_w/A_{w-1}$ and normally varies from 3 to 6. Strictly speaking, this relation is valid for basins on uniform soils within a given drainage area.

Drainage area is highly correlated with several hydrologic parameters. Watershed discharge Q has been related to drainage area (Leopold and Miller, 1956; Hack, 1957). Gray and Wigham (1970) have presented a comprehensive list of flow-area relationships. One of the simplest relationships is of the form

$$Q = kA^x \tag{5.2}$$

where k and x are parameters. The magnitude of k depends on rainfall and other watershed parameters, and x varies with Q. The variable Q is some measure of flow, such as mean annual flow, maximum flood, minimum flood, etc. For average flow rates, $x \cong 1$, and $Q/k = A$; but $x < 1$ for high flows and $x > 1$ for low flows. This shows that the average depth of discharge does not depend on the area. However, in larger watersheds, it takes longer for total flow to reach the outlet, which leads to smaller peak flow, and a given flow is sustained longer during drought, which results in greater low flows.

The relation between mean annual discharge and drainage area was derived by Hack (1957), as shown in Figure 5.4. This relation is reasonably good for drainage basins on soils with reasonably similar hydrologic characteristics, but results in considerable scatter for drainage basins located on dissimilar soils and for other dissimilar hydrologic conditions.

Equations (5.1) and (5.2) can be combined to obtain an average flow for a watershed of given order as

$$Q_w = Q_1 R_a^{x(w-1)} \tag{5.3}$$

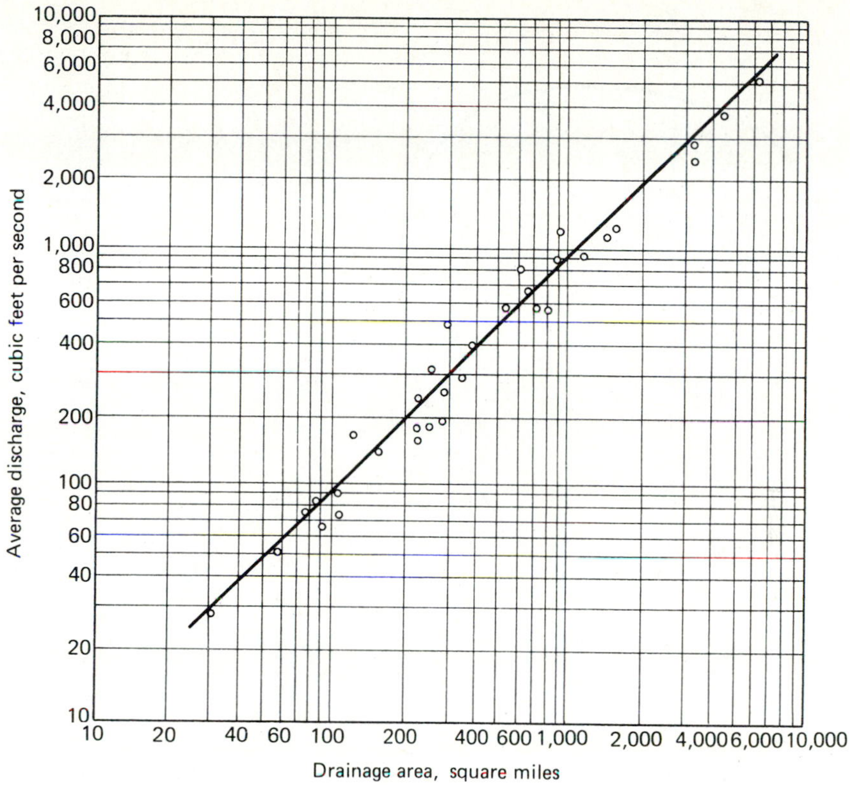

Figure 5.4 Relation between the mean annual discharge and drainage area (after Hack, 1957).

in which Q_w is the average flow for $\bar{A}_w$, and Q_1 is the average flow for $\bar{A}_1$. Equation (5.3) may be useful for planning purposes, but is inadequate for hydrologic design.

EXAMPLE 5.1 Compute the stream-area ratio for the Irvington drainage basin.

Solution The Irvington drainage basin is a sixth-order basin, as shown in Figure 5.2. First, areas of each channel of each order are planimetered. Then areas of all the channels of the same order are summed up, and average area of each order channel is computed. These calculations show that

$$\bar{A}_5 = 41.5 \text{ km}^2, \bar{A}_4 = 4.11 \text{ km}^2, \text{ and } \bar{A}_3 = 0.908 \text{ km}^2$$

Therefore,

$$R_a = \frac{\bar{A}_5}{\bar{A}_4} = \frac{41.5}{4.11} = 10.1; \qquad R_a = \frac{\bar{A}_4}{\bar{A}_3} = \frac{4.11}{0.908} = 4.53$$

Clearly, R_a has much larger variability for this basin.

5.3.3 Basin Shape

Numerous symmetrical and irregular forms of drainage areas are encountered in practice. A frequently occurring shape is a pear shape in plan view, as shown in Figure 5.5(a). The watershed surface, however, is always a tilted concavity that

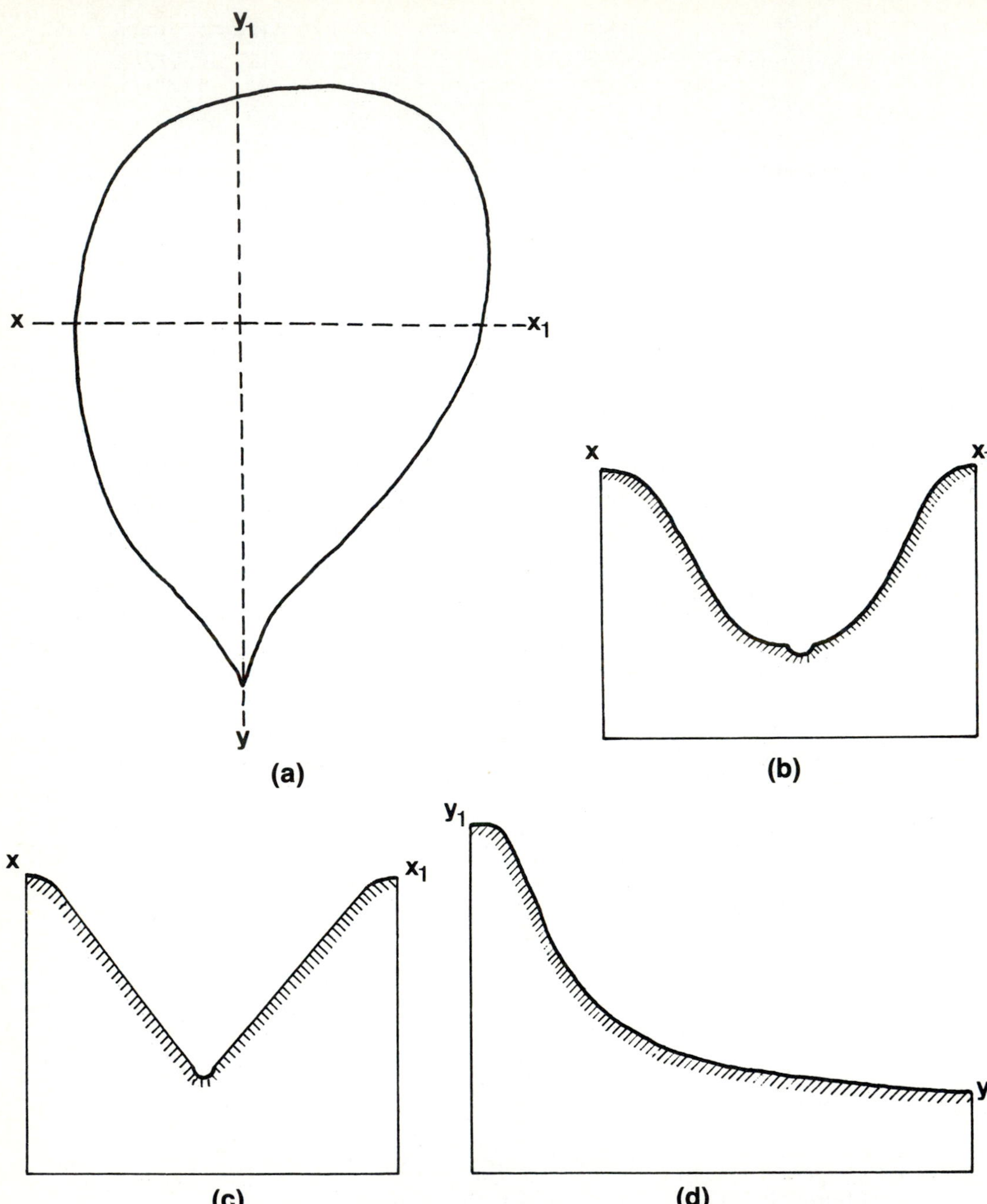

Figure 5.5 (a) Typical watershed shape. (b) Cross-section along x–x_1 approximating a U-shaped valley. (c) Cross-section along x–x_1 approximating a V-shaped valley. (d) Cross-section along y–y_1.

determines the general direction of flow. Depending upon the interaction of climate and geologic processes, the lateral section of a watershed may approximate a U-shaped valley, Figure 5.5(b), or a V-shaped one, Figure 5.5(c); the transverse section, Figure 5.5(d) displays increase in steepness toward the upstream area.

A multitude of dimensionless parameters have been proposed to quantitatively define watershed shape (Morisawa (1958). Some of the commonly used parameters are form factor, shape factor, elongation ratio, circularity ratio, and compactness coefficient, and are given in Table 5.1. These factors involve watershed length L, area, and/or perimeter P_r. Length is defined in more than one way: (1) the greatest straight-line distance between any two points on the perimeter, (2) the greatest distance between the outlet and any point on the perimeter, or (3) the length of the main stream from its source (projected to the perimeter) to the outlet. Clearly, the form factor is less than 1 and its reciprocal, the shape factor, is greater than 1. The elongation ratio, circularity ratio, and compactness coefficient approach 1 as the watershed shape approaches that of a circle.

A square drainage basin would have a shape factor $B_s = 1$, whereas the long narrow drainage basin would have a shape factor $B_s < 1$. The shape parameters can be used to quantify the degree of similarity of drainage-basin shapes.

The watershed shape may influence the hydrograph shape, especially for small watersheds. For example, if a watershed is long and narrow, then it will take longer for water to travel from watershed extremities to the outlet and the resulting runoff hydrograph will be flatter, as shown in Figure 5.6(a). For more compact watersheds, the runoff hydrograph is expected to be sharper with a greater peak and shorter duration, as shown in Figure 5.6(b). A compact watershed is more likely to be covered by the area of maximum rainfall intensity of local storms. For a watershed that is partly long and narrow, and partly compact, the runoff hydrograph is expected to be a complex composite of the aforementioned hydrographs, as shown in Figure 5.6(c). However, other factors have a greater effect on the hydrograph shape than does the basin shape.

TABLE 5.1 WATERSHED-SHAPE PARAMETERS[a]

Parameter (author)	Definition	Formula	Value
Form factor (Horton, 1932)	$\dfrac{\text{Watershed area}}{(\text{Watershed length})^2}$	$\dfrac{A}{L^2}$	<1
Shape factor, B_s (U.S. Army Corps of Engineers, 1954)	$\dfrac{(\text{Watershed length})^2}{\text{Watershed area}}$	$\dfrac{L^2}{A}$	>1
Elongation ratio (Schumm, 1956)	$\dfrac{\text{Diameter or circle of watershed area}}{\text{Watershed length}}$	$\dfrac{1.128A^{0.5}}{L}$	≤1
Circularity ratio (Miller, 1959)	$\dfrac{\text{Watershed area}}{\text{Area of circle of watershed perimeter}}$	$\dfrac{12.57A}{P_r^2}$	≤1
Compactness coefficient (Strahler, 1964)	$\dfrac{\text{Watershed perimeter}}{\text{Perimeter of circle of watershed area}}$	$\dfrac{0.2821P_r}{A^{0.5}}$	≥1

[a] A = watershed area, L = watershed length, and P_r = perimeter.

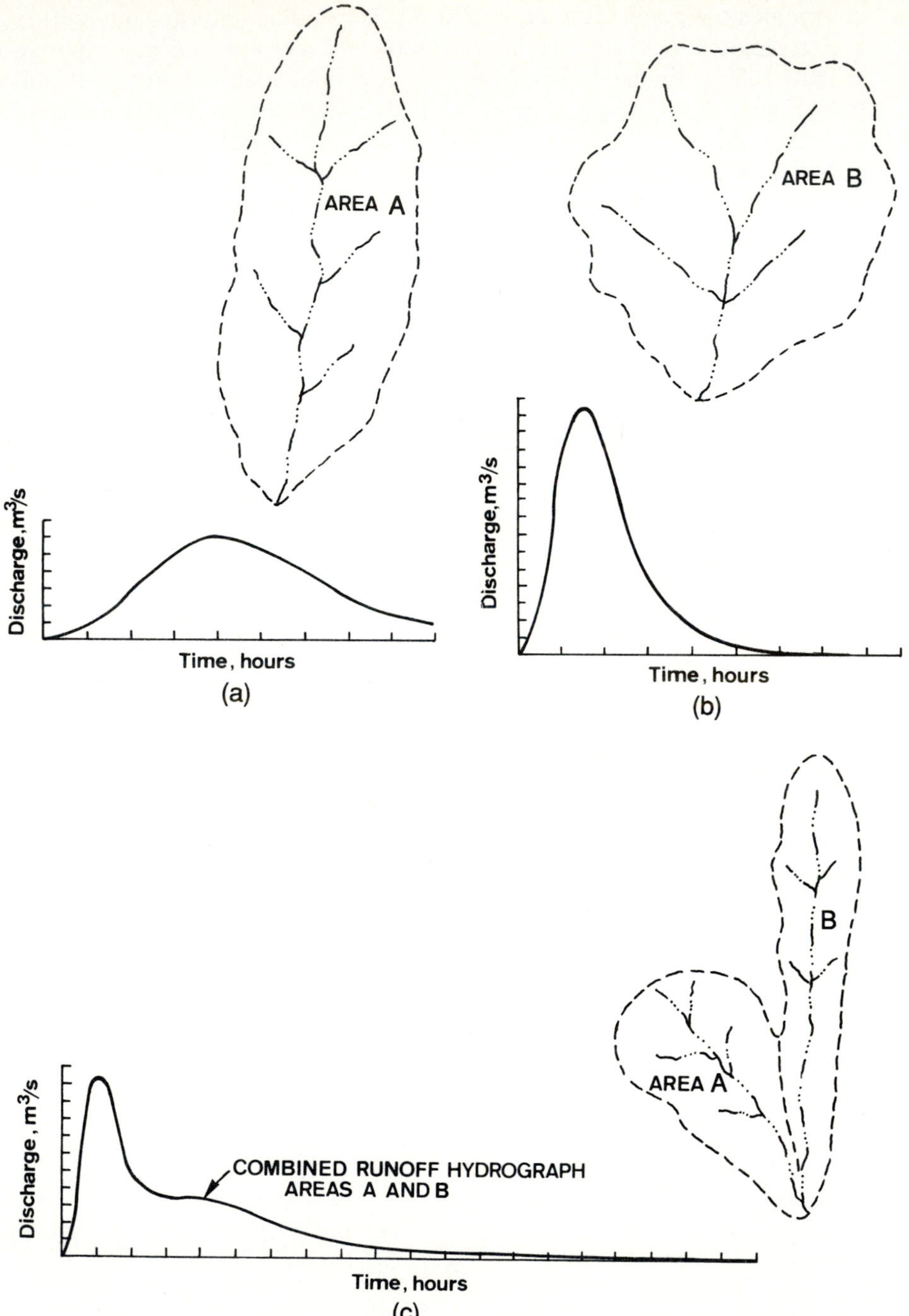

Figure 5.6 Effect of watershed shape on the runoff hydrograph.

EXAMPLE 5.2 Compute the form factor for the Irvington drainage basin.

Solution For this basin, the distance from the outlet to the farthest point on the divide is $L = 14.3$ km, and $A = 83$ km^2. Therefore,

$$B_s = 83/(14.3 \times 14.3)$$
$$= 0.406$$

5.3.4 Basin Slope

Basin slope has a profound effect on the velocity of overland flow, watershed erosion potential, and local wind systems. When coupled with slope orientation, it influences the receipt of solar radiation, and, in turn, microclimate, snowmelt, and distribution of precipitation. Basin slope S is defined as

$$S = h/L \tag{5.4}$$

where h is the fall in feet or meters, and L is the horizontal distance (length) over which the fall occurs. Because ground slope varies greatly from point to point within the drainage basin, Equation (5.4) is not adequate. A better representative value of S can be obtained following a slight modification of the method presented by Horton (1932). The method involves representing the drainage area by a grid system on its topographic map, as shown in Figure 5.7. The elevation contours are assumed to be

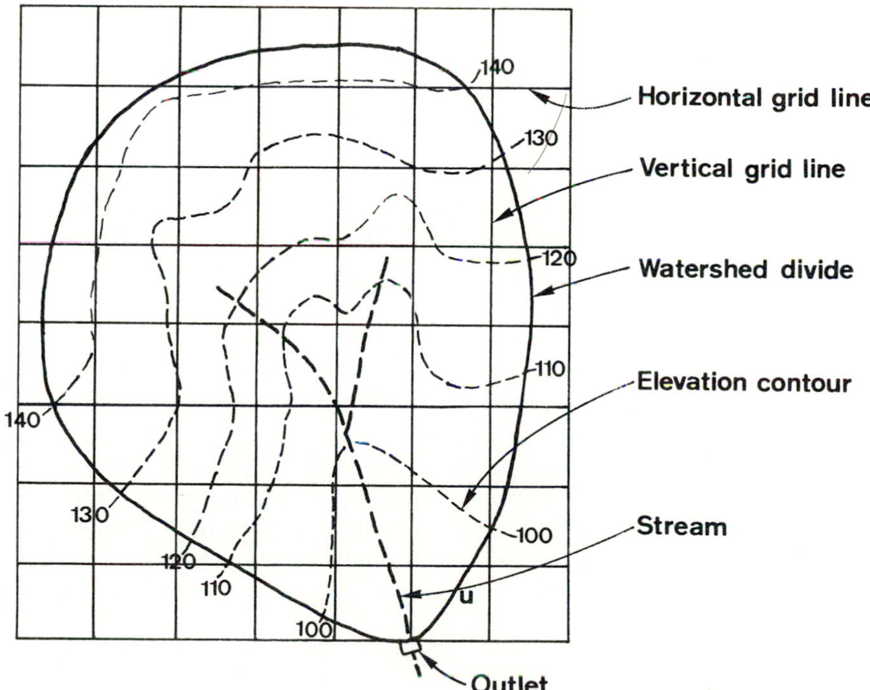

Figure 5.7 Grid system superimposed on the topographic map of a drainage area for the determination of basin slope.

at the same interval h. Each horizontal grid line is measured between its intersections with the watershed boundary, and the total length L of grid line segments is obtained; the same is done for vertical grid lines. Then the number of intersections of each horizontal grid line with contour lines is obtained and the sum of these intersections, N, is obtained; the same is done for vertical grid lines. These data are used to get S as

$$S = \frac{S_V + S_H}{2} \tag{5.4a}$$

where

$$S_x = \frac{Nh}{L}, \qquad x = V \text{ or } H \tag{5.5b}$$

in which S_V is the average slope for the vertical, S_H is the average slope for the horizontal, and N as well as L correspond to S_V or S_H.

EXAMPLE 5.3 Compute the ground slope for the Irvington drainage basin.

Solution For this basin, the farthest point from the outlet to the divide is $L = 14.3$ km, and the difference in elevations of these two points is $h = 0.083$ km. Therefore,

$$S = 0.083/14.3$$
$$= 0.0058$$

This is a gross value of the basin slope.

5.3.5 Centroid (Center of Gravity)

The centroid of a drainage basin is simply the location of the point within the drainage basin that represents the weighted center of the basin. It is the first moment of the area about the origin. If the basin were a perfect circle, then the centroid would coincide with the exact center. Since most basins are not perfect circles, the centroid location will vary greatly. One method of locating the centroid, in the absence of necessary mathematical equations, is to use a cutout tracing of the drainage basin. Trial-and-error balancing of this tracing on the point of a pencil will locate the centroid sufficiently well for hydrologic uses. Another method that is used is to cut the basin map into equal grid sizes and equalize the weights of the cutout grids on a sensitive chemical balance.

The centroid of a drainage basin can be determined more accurately by using the method of moments and representing the basin by a grid system. A grid system can be superimposed on the drainage, as shown in Figure 5.7, such that grids determine equal subareas within the basin, except at the boundary. Because the centroid is the first moment of the area, its coordinates $(\bar{x},\bar{y})$ can be computed as

$$\bar{x} = \frac{\sum_{i=1}^{N} x_i a_i}{\sum_{i=1}^{N} a_i} = \frac{1}{A}\sum_{i=1}^{N} x_i a_i \tag{5.6a}$$

$$\bar{y} = \frac{1}{A} \sum_{i=1}^{M} y_i a_i \tag{5.6b}$$

where x_i is the distance between the y-axis and the center of the ith grid square, a_i is the area of the ith grid square, y_i is the distance between the x-axis and the center of the ith grid square, N is the total number of grid squares in the x direction, and M is the total number of grid squares in the y direction. It is to be noted that the grid is not square at the watershed boundary, and the partial grids within the boundary can be represented by rectangular grids of equivalent areas for purposes of calculations.

For 47 watersheds in midwestern United States, Gray (1961) found that the distance along the channel from the basin outlet to a point opposite the centroid was about one-half the basin length. This distance was used by Snyder (1938) for synthesis of runoff hydrograph. The drainage basins, used by Gray, were relatively well proportioned whatever their shape.

5.3.6 Basin Length

Schumm (1956) defined basin length L_b as the longest dimension of a basin parallel to its principal drainage channel. (This geomorphic definition should be distinguished from the definition used for the time of concentration and other hydrologic purposes. Hydrology is concerned with flow in the drainage channels and, therefore, specifically requires that such measurements are made along the drainage system to a given point.) In a similar vein, basin width can be measured in a direction approximately perpendicular to the length measurement. The relation between mainstream length and drainage-basin area has been investigated by Smart and Surkan (1967). Gray (1961) analyzed data from a number of small watersheds and showed that

$$L_b = 1.4A^{0.568} \tag{5.7}$$

with a 24.8% standard error of estimation, where L_b is in miles and A in square miles. The value 1.4 becomes 1.312 if L_b is in km and A in km^2. Equation (5.7) shows that larger watersheds tend to elongate.

EXAMPLE 5.4 Verify if Equation (5.7) holds for the Irvington drainage basin.

Solution For this basin, 11 drainage areas are identified, as shown in Figure 5.2. For each area, L_b is measured from the map, as well as computed from Equation (5.7). The results are given in Table 5.2. Equation (5.7) is found to be a reasonable approximation for the Irvington drainage basin.

5.3.7 Drainage-Basin Similarity

There can be three types of drainage basin similarity: (1) geometric similarity in terms of basin area, shape, main channel slope, and topography; (2) hydrologic similarity in terms of hydrologic processes such as rainfall, snowfall, infiltration, runoff, and valley storage; and (3) geologic similarity in terms of the properties relating to groundwater flow, soil erosion, porous media, sediment characteristics, and sediment transport, all of which are derived from parent geology. Hydrologic relations can be transferred to similar basins if basin similarity requirements are met.

TABLE 5.2 CALCULATIONS FOR LENGTH–AREA RELATION FOR THE IRVINGTON DRAINAGE BASIN

Area no.	Area (km^2)	L_b (km) Computed	Measured	Error (%)
1	3.10	2.50	2.53	−1.19
2	4.77	3.19	4.42	−27.83
3	4.49	3.08	3.32	−7.23
4	5.15	3.53	3.72	−10.48
5	0.84	1.19	1.46	−18.49
6	4.77	3.19	3.35	−4.78
7	3.49	2.67	2.56	4.30
8	7.59	4.15	3.84	8.07
9	1.33	1.54	1.77	−12.99
10	7.42	4.10	4.57	−10.28
11	2.25	2.08	3.11	−33.12

Strahler (1957) hypothesized that drainage basins for which all corresponding geomorphic quantities having the dimensions of length are in the same ratio can be said to possess exact geometric similarity. No two basins are exactly similar but can be approximately so if their responses, in terms of the similarity desired, are similar. According to Strahler's similarity hypothesis,

$$\frac{A}{L_b^2} = C \tag{5.8}$$

where C is a constant. Clearly, when Equation (5.6) is cast in this form,

$$\frac{A}{L_b^2} = 0.58A^{-0.136} \tag{5.9a}$$

or

$$\frac{A^{1.163}}{L_b^2} = 0.58 \tag{5.9b}$$

which shows that the watersheds analyzed by Gray (1961) did not possess exact geometric similarity.

EXAMPLE 5.5 Show whether or not the Irvington drainage basin possesses exact geometric similarity.

Solution For the 11 drainage areas identified previously (Figure 5.2), the value of C is computed as given in Table 5.3. The value of C ranges from 0.23 to 0.53, and is thus far from constant. Hence, the Irvington drainage basin does not possess exact geometric similarity.

5.3.8 Number of Channels and their Order

Identification of first-order channels is the first step in applying channel order in hydrology. Application of stream ordering requires the use of a topographic map

TABLE 5.3 CALCULATIONS TO DETERMINE GEOMETRIC SIMILARITY OF THE IRVINGTON DRAINAGE BASIN

Area no.	Area (km^2)	L ($\mathrm{k}^{b}_{\mathrm{m}}$)	C
1	3.10	2.53	0.48
2	4.77	4.42	0.24
3	4.49	3.32	0.41
4	5.15	3.72	0.37
5	0.84	1.46	0.40
6	4.77	3.35	0.43
7	3.49	2.56	0.53
8	7.59	3.84	0.51
9	1.33	1.77	0.42
10	7.42	4.57	0.36
11	2.25	3.11	0.23

such as the United States Geological Survey (USGS) quadrangle maps on a 1 : 24,000 scale. An examination of these maps will show that all drainage channels are not shown by the usual stream symbols. For this reason, it is necessary to take the map to the field in order to establish a consistent procedure in identifying first-order channels. Experience has shown that if a drainage basin is examined closely enough, very small channels of only a few inches in width can be identified in certain soils. It is not practical to try to map these channels. Further examination will show that the topographic contours are good indicators of channels. Contours indicate channels by crenulating upward in the channel. The uppermost topographic crenulation shown on the map often coincides with the upper end of a well-defined channel observed in the field. By comparing the topographic map with actual observations in the field, it is possible to consistently select from the map the upper end of a well-defined channel. With this relationship established by actual field comparison with the topographic map, it is then possible to sketch in drainage channels without actually observing each one.

Another method of ordering channels is by use of air photographs. A reasonable scale photograph must be used in order to clearly identify channels. This photograph must be taken to the field and compared with the upper end of the uppermost well-defined channel so that the field and photo relationship may be established. With a little practice and experience, it is possible to interpret channels in a consistent manner from the photographs. When this condition is accomplished, the drainage system can be sketched on the photos and then transferred to a topographic map for further measurements. Actual plotting of the drainage network on a topographic map is necessary in order to use this information for hydrologic purposes.

The number of stream channels of each order is expressed by a mathematical relation, known as Horton's law of channel numbers, in which the number of stream channels of each order forms an inverse geometric sequence with order number,

$$N_w = R_b^{W-w} \tag{5.10a}$$

or

$$\log N_w = W \log R_b - w \log R_b$$
$$= a - bw, \qquad a = W \log R_b, \qquad b = \log R_b \tag{5.10b}$$

where N_w is the number of streams of order w, W is the order of the watershed, and R_b is the bifurcation ratio defined as $R_b = N_w/N_{w+1}$ and varies between 3 and 5. This law is an expression of topological phenomenon, and is a measure of drainage efficiency.

As an example, the Salt Creek drainage basin has a drainage area of 684 mi^2 (1771.6 km^2) above the gage at Lincoln, Nebraska. This watershed is a ninth-order drainage basin; ninth order identifies the highest-order channel in the basin. Ordering the basin shows that there is one ninth-order channel, 2 eighth-order channels, 6 seventh-order channels, 16 sixth-order channels, 76 fifth-order channels, 309 fourth-order channels, and 1418 third-order channels. Because of the large number of channels, the second- and first-order channels were not counted. A regression analysis was made using the counted channels for the higher orders that resulted in

$$\log N_w = 4.60 - 0.533w \tag{5.11}$$

from which the number of second-order channels was determined to be 3410 and first-order channels 11,641. Equation (5.11) is graphed in Figure 5.8. It is apparent that this relation is valid only within this drainage basin and does not represent the relation that might exist in other drainage basins. It is, however, likely that if two drainage basins were located on the same soils in the same area, then the channel number for each order would be reasonably similar.

The number of channels of a given order in a drainage basin is a function of the nature of the surface of that drainage basin. In general, the greater the infiltration of the soil material covering the basin, the fewer will be the number of channels required to carry the remaining runoff water. Moreover, the larger the number of channels of a given order, the smaller is the area drained by each channel order.

EXAMPLE 5.6 Verify the law of stream numbers for the Irvington drainage basin.

Solution Streams of this basin (Figure 5.2) are ordered using the Horton–Strahler ordering scheme, and the basin is found to be a sixth-order basin. Channels of each order are counted as given in Table 5.4. Figure 5.9 shows the relation between number of channels and their order. The best-fit line is obtained by the least squares method with

$$a = 3.598$$

$$b = 0.626$$

$$\text{Correlation coefficient} = 0.996$$

$$\text{Standard error} = 0.121$$

For this basin, Equation (5.10) holds almost perfectly.

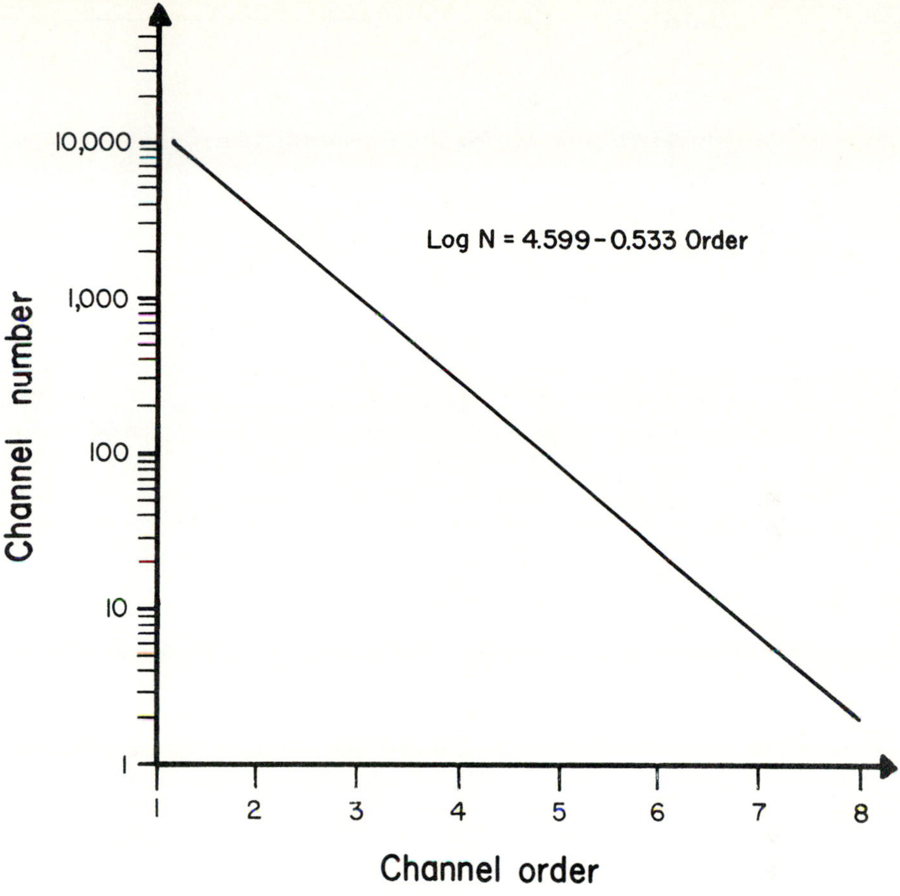

Figure 5.8 Law of channel numbers for Salt Creek at Lincoln, Nebraska.

TABLE 5.4 NUMBER OF STREAMS AND THEIR ORDER FOR THE IRVINGTON DRAINAGE BASIN

Order	N_w	log N_w
1	981	2.992
2	250	2.398
3	51	1.708
4	11	1.041
5	2	0.301
6	1	0

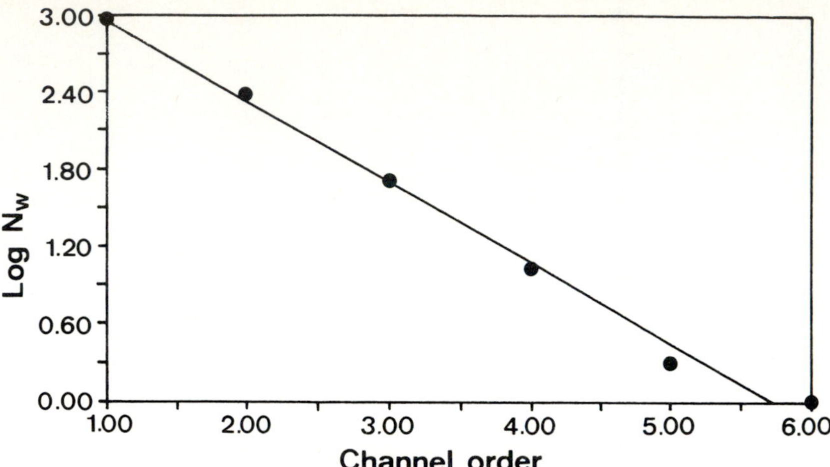

Figure 5.9 Relation between number of channels and their order for the Irvington drainage basin.

5.3.9 Channel Length

This refers to the length of channels of each order. The average length of channels of each higher order increases as a geometric sequence. Thus, the first-order channels are the shortest of all the channels and the length increases geometrically as the order increases, as shown in Figure 5.10. This relation is called Horton's law of channel lengths and can be formulated as

$$\bar{L}_w = \bar{L}_1 R_L^{w-1} \tag{5.12a}$$

or

$$\log \bar{L}_w = \log \bar{L}_1 + (w - 1) \log R_L$$

or

$$\log \bar{L}_w = a - bw, \qquad a = \log (\bar{L}_1/R_L), \qquad b = -\log R_L \tag{5.12b}$$

where

$$\bar{L}_w = \frac{\sum_{i=1}^{N_w} L_{w_i}}{N_w} = \frac{L_w}{N_w} \tag{5.12c}$$

in which L_{w_i} is the length of the ith channel of order w, L_w is the total length of all channels of order w, N_w is the number of channels of order w, $\bar{L}_1$ is the mean channel length of order w, $\bar{L}_1$ is the mean length of the first-order streams, R_L is the stream-length ratio and is generally between 1.5 and 3.5, and is defined as $R_L = \bar{L}_w/\bar{L}_{w-1}$.

Horton (1945) observed that the laws of stream number and length can be combined to produce an expression for the total length of all the streams of a given order. From Equation (5.12c),

$$L_w = N_w \bar{L}_w \tag{5.13}$$

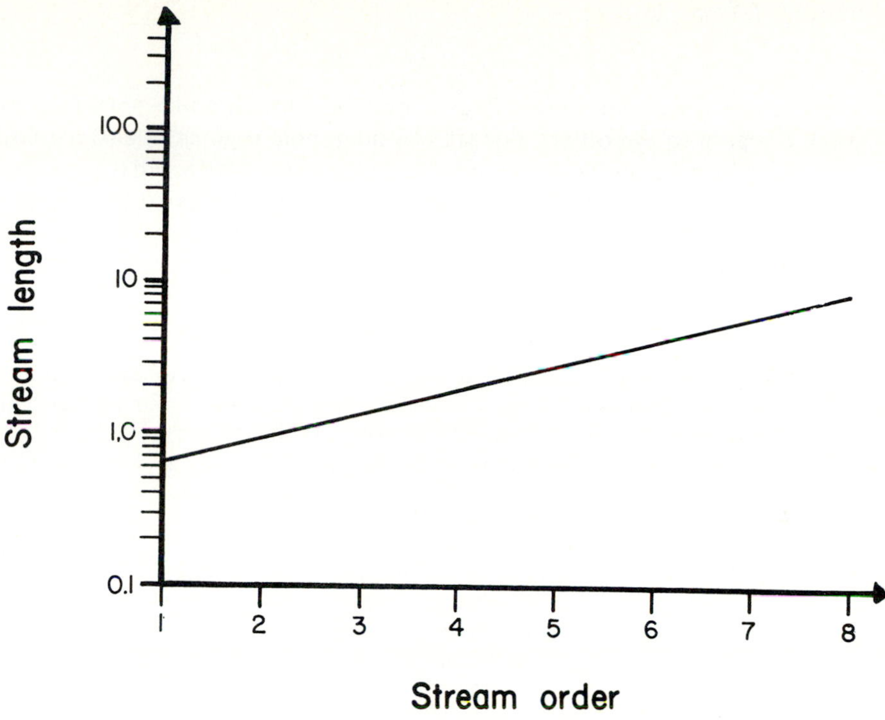

Figure 5.10 Law of stream lengths for Salt Creek at Lincoln, Nebraska.

By replacing N_w and L_w by Equations (5.10), (5.12a), and (5.12b), respectively,

$$L_w = \bar{L}_1 R_b^{W-w} R_L^{w-1} \tag{5.14}$$

Thus, the total length of all the channels of order w can be calculated by knowing the average length of the first-order streams, the bifurcation ratio, and the stream-length ratio. Equation (5.14) can be easily extended to give the total length of all streams of all orders, L_W, within a drainage basin as

$$L_W = \sum_{w=1}^{W} L_w = \sum_{w=1}^{W} \bar{L}_1 R_b^{W-w} R_L^{w-1} \tag{5.15a}$$

$$= \bar{L}_1 \frac{R_b^W}{R_L} \sum_{w=1}^{W} r^w, \qquad r = \frac{R_L}{R_b} \tag{5.15b}$$

$$= \bar{L}_1 \frac{R_b^W}{R_L} \frac{r^W - 1}{r - 1} \tag{5.15c}$$

The lengths of channels of a given order are determined largely by the type of soil covering the drainage basin. Generally, the more pervious the soil, the longer will be the channel length of a given order.

For geometrically similar watersheds, Morisawa (1967) suggested a relation between mean annual discharge, Q(m^3/s), and longest stream length (mouth to source), L (km), as

$$Q = aL^x \tag{5.16}$$

where a is a coefficient, and x is an exponent. Both a and x vary from one physiographic region to the other. For six physiographic regions in eastern United States studied by Morisawa, the value of a varied between 0.0058 and 0.1167, and the value of x between 1.00 and 1.95. Because a is strongly influenced by basin geology and vegetative cover, Equation (5.16) may provide estimates useful for planning purposes only.

EXAMPLE 5.7 Verify the law of channel lengths for the Irvington drainage basin.

Solution First, the length of each channel of each order is measured from the map and lengths of channels of the same order are tabulated. Then, the average length of channels of each order is calculated as shown in Table 5.5. Figure 5.11 shows the relation between channel length and order for the Irvington drainage basin. The best-fit line is

TABLE 5.5 AVERAGE LENGTH OF CHANNELS OF DIFFERENT ORDERS FOR THE IRVINGTON DRAINAGE BASIN

Order	$\bar{L}_w$ (km)	$\log \bar{L}_w$ (km)
1	0.242	−0.616
2	0.363	−0.44
3	0.835	−0.0783
4	1.406	0.148
5	11.89	1.075

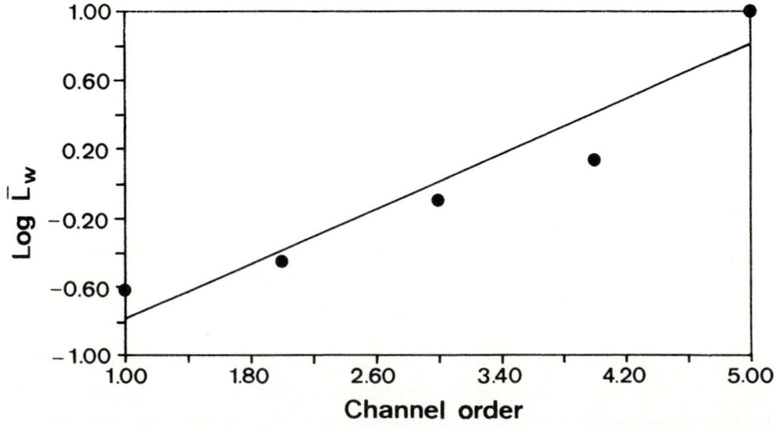

Figure 5.11 Relation between channel length and order for the Irvington drainage basin.

obtained by the least squares method with

$$a = -1.173$$

$$b = +0.397$$

$$\text{Correlation coefficient} = 0.948$$

$$\text{Standard error} = 0.244$$

This relation explains nearly 90% of variance.

5.3.10 Channel Area

The basin area has already been discussed previously. The channel area of order w, A_w, is the area of the watershed that contributes to the channel segment of order w and all lower-order channels. Hack (1957) examined Equation (5.1) and derived an expression for A_w based on $\bar{A}_1$, R_b, and r, which can be written as

$$A_w = \bar{A}_1 R_a^{w-1} \frac{r^w - 1}{r - 1}, \qquad r = \frac{R_L}{R_b} \tag{5.17}$$

The ratios, R_b, R_L, and R_a, are of fundamental significance not only for describing drainage-basin composition, but also for hydrologic synthesis. Rodriguez-Iturbe (1982) has employed these ratios along with stream lengths and stream areas for synthesizing runoff. Many other investigators have since used these and other drainage-basin parameters for hydrologic synthesis (Gupta et al., 1986).

5.3.11 Channel Profile

The longitudinal channel profile represents the relationship between altitude and horizontal distance, and can be determined from a watershed topographic map. The profile changes nonuniformly in a longitudinal direction, that is, it may be made up of a number of segments, each having a uniform slope. Data on channel profile are used to compute the slope of the channel of a given order. Standard curve-fitting techniques can be used to represent the channel profile mathematically (Strahler, 1964). Sribnyi (1961) presented an interesting discussion on analytical generalization of longitudinal profiles of streams. By referring to Figure 5.12, a profile factor b can be defined to describe the channel profile:

$$b = (LH_p/F) - 1 \tag{5.18}$$

where H_p is the height of the divide above the outlet, L is the length of the profile, and F is the area of relief deficiency (shaded area). For a rectangular outline of the area F, $b = 0$; for a parabolic outline, $0 < b < 1$; for a triangular outline, $b = 1$; and for an inverted parabolic outline, $b > 1$.

The channel profile is normally found to be concave upward. The nature of this concavity is a function of the basin geology and precipitation (Hack, 1957). Generally, the upper part of the channel profile is steeper than the lower portion, particularly when the geology of the drainage basin is fairly uniform. Although the general configuration of the stream profile does exhibit a concave-upward appearance, there is a wide range of variations caused by the underlying geology. Where nonuniform

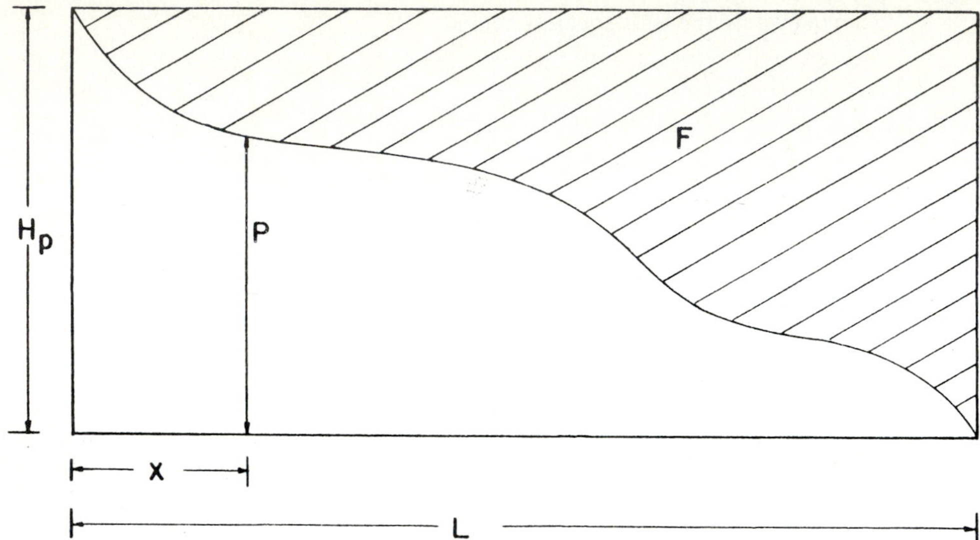

Figure 5.12 Channel profile and relief deficiency.

geology occurs in a drainage basin, rocks more resistant to erosion distort the uniform nature of the concave profile. The result of changes in the stream profile caused by bedrock geology can cause large variations in the flow velocity of a stream channel.

5.3.12 Channel Slope

Channel slope has a profound effect on the velocity of flow in a channel, and, consequently, on the flow characteristics of runoff from a drainage basin. The importance of channel slope lies in two areas: (1) determining discharge and velocity using the Manning or Chezy equation, and (2) as a variable in multivariate analysis to determine the amount of influence accounted for by channel slope. Because the slope varies longitudinally, an average value of slope must be determined.

Method 1. For use in either the Manning or Chezy equation, the channel slope is a local measurement to approximate the energy slope, assuming uniform flow. For this purpose, the channel slope S is computed from Equation (5.4). The fall over the channel reach of interest is measured. However, this measurement is a local measurement and cannot hold for other channel reaches in the drainage basin. Cross-sectional area, channel roughness, and cross-sectional shape also affect flow velocity. Therefore, there is not as much difference in the flow velocity between various reaches of the drainage basin as one might think.

Method 2. When used in multivariate analysis, the arithmetic slope in Equation (5.4) might be determined in many ways. One common method is to compute the fall from the head of the uppermost first-order channel to the basin outlet and divide this fall by its horizontal length.

Method 3. A geometric slope is sometimes used. This slope is determined by locating the median channel-profile elevation on the main channel and computing the fall from this point to the outlet. The length is the horizontal distance between the point of median elevation to the outlet. Slope is then computed using Equation (5.4).

Method 4. Benson (1962) found that the "85–10" slope factor was the most satisfactory in his study of floods in New England. This factor is the slope between 85% (excluding the upper 15%) and 10% (excluding the lower 10%) of the distance along the stream channel from the basin outlet to the divide. It should be noted that the distance is measured to the divide and not to the end of the defined stream channel. The fall and horizontal length between these two points are computed using Equation (5.4).

Method 5. This method is from Johnstone and Cross (1949). The channel can be divided into N number of reaches, each having a uniform slope S_i. Then the equivalent uniform slope S_m is

$$S_m = \left(\frac{\sum_{i=1}^{N} L_i S_i^{1/2}}{\sum_{i=1}^{N} L_i} \right)^2 \tag{5.19}$$

This is designed to estimate the slope that would result in the same total time of travel as the actual stream if length, roughness, channel cross-section, and any other pertinent factors other than slope were unchanged.

Method 6. This is due to Laurenson (1962). Again the stream is divided into N reaches, each of uniform slope. Further, it is assumed that the effects of roughness and hydraulic radius on velocity are the same for all reaches. This assumption is questionable but has been used previously (Taylor and Schwarz, 1952). This assumption is also implied in the first method. The velocity U_i of flow through any reach i can be written as

$$U_i = BS_i^{0.5}$$

where B is a constant. Then the time of flow t_i is

$$t_i = \frac{L_i}{U_i}$$

Therefore, the total time of travel T_c down the main channel is

$$T_c = \frac{1}{B} \sum_{i=1}^{N} \frac{L_i}{S_i^{0.5}}$$

The mean velocity of flow U_m can be written as

$$U_m = \sum_{i=1}^{N} \frac{L_i}{T_c} = \frac{\sum_{i=1}^{N} L_i}{\sum_{i=1}^{N} L_i / S_i^{0.5}}$$

Further,

$$U_m = BS_m^{0.5}$$

Hence,

$$S_m = \left(\frac{\sum_{i=1}^{N} L_i}{\sum_{i=1}^{N} L_i / S_i^{0.5}} \right)^2 \tag{5.20}$$

Method 7. This method is from Gray (1961) and Lane (1975). Gray defined S_c as the slope of a line drawn along the measured profile that has the same area as is under the observed profile. By referring to Figure 5.13, the slope is the slope of the hypotenuse of a right-angle triangle with the same A and length L_c as the observed profile. If A is the area under the observed profile, then

$$A = hL_c/2$$

where h is the ordinate of the right-angle triangle. Therefore,

$$S_c = h/L_c \tag{5.21}$$

or

$$S_c = 2A/L_c^2 \tag{5.22}$$

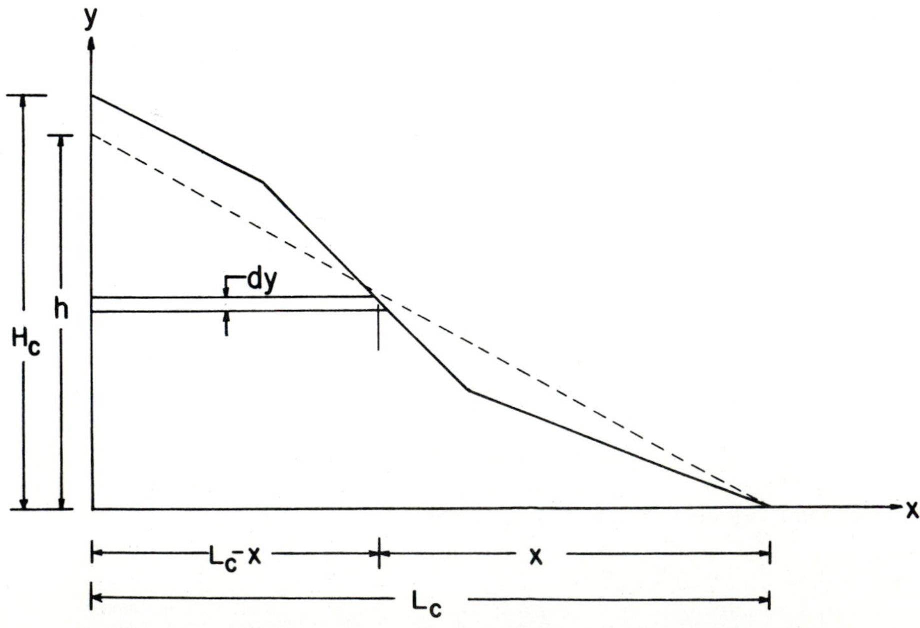

Figure 5.13 Observed channel profile shown by solid line and dashed line.

However, a more accurate measure of S_c can be obtained as follows. We can express

$$A = \int_0^{H_c} (L_c - x)\, dy$$

Since the slope S is continuously changing with x,

$$dy = S(x)\, dx$$

Therefore,

$$A = \int_0^{L_c} (L_c - x) S(x)\, dx$$

Then Equation (5.22) yields

$$S_c = 2 \int_0^{L_c} \frac{L_c - x}{L_c^2} S(x)\, dx \tag{5.23}$$

where $(L_c - x)/L_c^2$ can be considered as a weighting factor. To illustrate, it is 0 at $x = L_c$ and a maximum at $x = 0$ (at the outlet). Thus, Gray's method produces a channel slope that is weighted by distance from the head to the outlet.

The quantity h/H_c can be used as an index of concavity. If it is less than 1, which is normally the case, the stream profile is concave. A value of this quantity greater than 1 would correspond to a convex profile. Further, this quantity can be used as an index of how well the natural channel slope is represented by a straight line.

It can be seen from the foregoing methods of determining channel slope that the main channel is used to determine a slope that is presumed to be representative of all other channels in the drainage basin. One might wonder why so many different slopes are necessary. Although slope is recognized to be important in predicting hydrologic response, results of multivariate analyses using slope have produced less than conclusive results. These and other slope values have been developed in an effort to find a value of slope that is consistently important for multivariate hydrologic analyses. No such slope has yet been discovered.

Law of stream slopes

Horton (1945) introduced the law of stream slopes, which states that the average slope of streams of each order tends to approximate an inverse geometric series.

$$\bar{S}_w = \bar{S}_1 R_s^{W-w} \tag{5.24}$$

where $\bar{S}_w$ is the average slope of streams of order w; $\bar{S}_1$ is the average slope of first-order streams; R_s is the slope ratio, defined as $R_s = \bar{S}_w/\bar{S}_{w-1}$; and W is the order of the basin. The value of R_s is approximately 0.55.

5.3.13 Drainage Density

Drainage density is defined as the length of drainage per unit area. This term was first introduced by Horton (1932) and is expressed as

$$D_d = \frac{L}{A} \tag{5.25}$$

or

$$D_d = \frac{\sum_{w=1}^{W} \sum_{i=1}^{N_w} L_{w_i}}{A} \tag{5.26}$$

which can be simplified using the laws of channel lengths and channel numbers to

$$D_d = \frac{\bar{L}_1}{A} \frac{R_L^W - R_b^W}{R_L - R_b} = \frac{\bar{L}_1 R_b^{W-1}}{A} \frac{r^W - 1}{r - 1}, \qquad r = \frac{R_L}{R_b} \tag{5.27}$$

where D_d is the drainage density of a watershed of order W in miles per square mile or kilometers per square kilometer, L is the total length of all channels of all orders in the drainage basin, and A is the area of the drainage basin in square miles or square kilometers. Clearly, D_d is a measure of the closeness (density) of channel spacing. It is an indication of the drainage efficiency of overland flow and the length of overland flow as well as the index of relative proportions.

The inverse of the drainage density has been characterized by Schumm (1956) as the constant of channel maintenance. Its value increases with the size of the watershed. The constant provides an estimate of the area in square meters of watershed required to maintain a meter of channel. The number of stream channels per unit area is called the stream frequency F (Horton, 1945). Melton (1957) analyzed 156 watersheds covering a broad range in size, relief, cover, and climate, and found that

$$F = 0.694 D_d^2 \qquad \text{or} \qquad \frac{F}{D_d^2} = 0.694 \tag{5.28}$$

where D_d is in miles/square mile.

Drainage density is a fundamental concept in hydrologic analysis. Hydraulics of streams flowing on movable beds, such as silt, sand, or gravel, require that when flow reaches a certain depth, such flow will move sediment. As flow continues or is repeated, enough sediment is moved to form a channel. For this reason, the development of a channel indicates that the relation between rainfall and runoff is sufficient that a channel is required to carry that runoff. The greater the runoff, the more and larger the channels required to carry that given runoff. Thus, drainage basins with high drainage densities indicate that a large proportion of the precipitation runs off; on the other hand, low drainage densities indicate that most rainfall infiltrates the ground and few channels are required to carry the runoff (Rogers, 1971). As shown in Figure 5.14, infiltration capacity is high for soils in sand and granite fragments, and for this material, drainage density is low. Fine-grained soils high in clay and silt content have low infiltration and their runoffs produce high drainage densities. This relation is further demonstrated in Figure 5.15 in which runoff increases as drainage density increases. This result is what would be predicted from Figure 5.14.

Melton (1957) found a strong correlation between drainage density and the ratio of average annual precipitation P to average annual evaporation E. His study showed the effectiveness of vegetal cover (which would vary directly with P/E) in

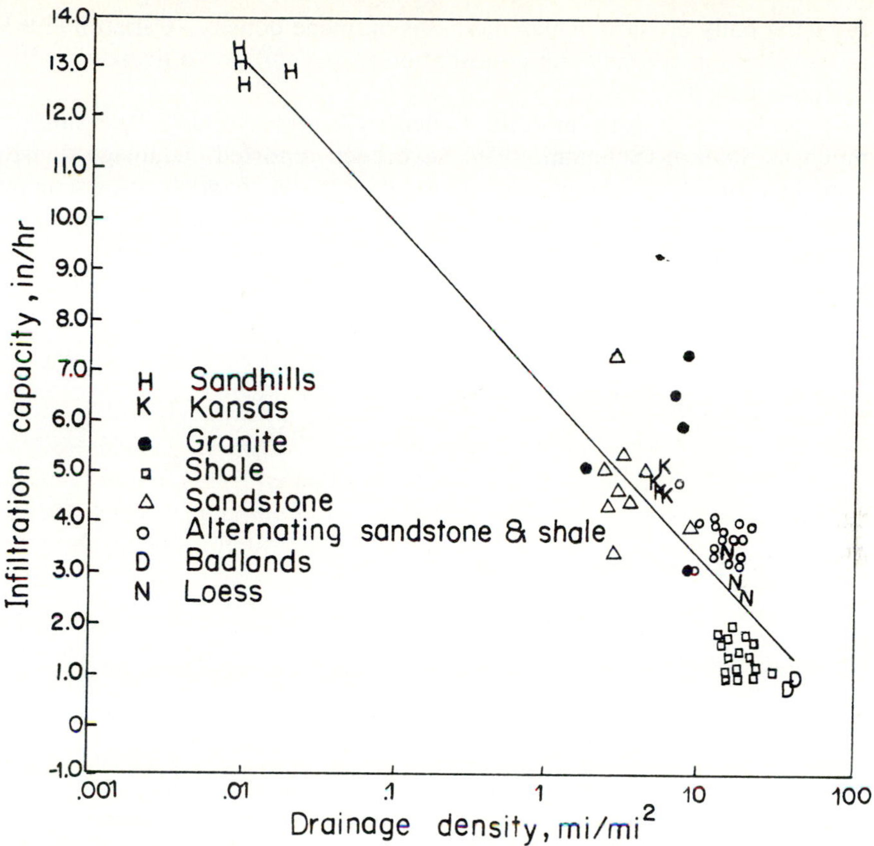

Figure 5.14 Relation between drainage density and infiltration capacity (after Rogers, 1971).

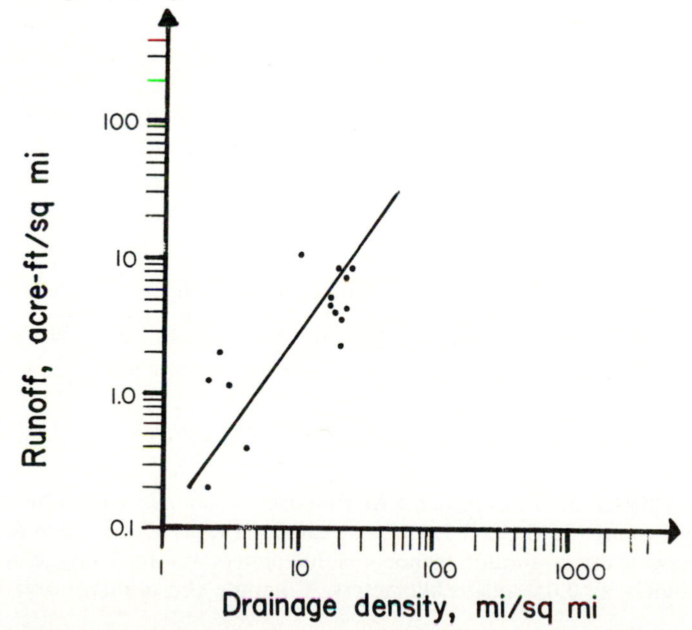

Figure 5.15 Relation between drainage density and runoff.

reducing gully erosion, which increases drainage density. Carlston (1963) observed a tendency for groundwater contribution to streamflow to decrease with increasing drainage density.

Numerical values for drainage density from less than 2/km (Smith, 1950) to as much as 800/km (Schumm, 1956) have been reported. Drainage density is determined by measuring the total length of all drainage channels using a map scale. The total length of all the channels is then divided by the drainage-basin area to obtain the drainage density.

EXAMPLE 5.8 Calculate drainage density of the Irvington drainage basin.

Solution For this basin, the total length of all the channels of all orders is $L = 399.9$ km, and $A = 83$ km^2. Therefore,

$$D_d = 399.9/83$$
$$= 4.82 \text{ km}^{-1}$$

5.3.14 Length of Overland Flow

As rain falls on a drainage-basin surface, it flows down slope toward a channel. The maximum length of this surface flow is called the length of overland flow. Of course, other shorter distances of overland flow can occur. The average length of overland flow is approximately one-half the average distance between stream channels. As a result, Horton (1945) recommended using one-half the reciprocal of the drainage density to determine the average of overland flow for the entire drainage basin. Because drainage density is a function of infiltration characteristics of the drainage basin, so, too, must be the length of overland flow.

Horton (1945) expressed the length of overland flow L_o as

$$L_o = \frac{1}{2D_d}\left(1 - \frac{S_c}{S_s}\right)^{-0.5} \tag{5.29}$$

where S_c and S_s are the average channel and surface slopes, respectively. Since S_c/S_s is much smaller than 1, Equation (5.29) can be approximated as

$$L_o = \frac{1}{2D_d} \tag{5.30}$$

EXAMPLE 5.9 Calculate overland flow length L_o for the Irvington drainage basin.

Solution From Example 5.8, D_d for this basin is 4.82 km^{-1}. Therefore, using Equation (5.29),

$$L_o = \frac{1}{2D_d} = \frac{1}{9.64} = 0.1 \text{ km}$$

EXAMPLE 5.10 Consider a fourth-order watershed with a bifurcation ratio as 4, a stream-length ratio as 2.5, and a stream-area ratio as 4.5. Also given are the average length of the first-order channels as 200 meters and the average area of the first-order channels as 0.05 square kilometers. Compute the drainage area of the fourth-order channel. What is the drainage density of the watershed? Compute the total number of channels this watershed has. How many channels are there in a unit area?

Solution

$$R_b = 4,\ w = 4,\ R_L = 2.5,\ \bar{L}_1 = 200 \text{ m},\ R_a = 4.5,\ \bar{A}_1 = 0.05 \text{ km}^2$$

Using Equation (5.17), with $w = 4$ and $r = R_L/R_b$,

$$A_4 = (50{,}000 \text{ m}^2)(4.5)^3 \frac{(2.5/4.0)^4 - 1}{(2.5/4.0) - 1}$$

$$= 1.03 \times 10^7 \text{ m}^2 = 10.296 \text{ km}^2$$

Using Equation (5.15a),

$$L_4 = 200 \text{ m } (4.0)^3(2.5)^0 + 200 \text{ m } (4.0)^2(2.5)^1$$

$$+ 200 \text{ m } (4.0)^1(2.5)^2 + 200 \text{ m } (4.0)^0(2.5)^3 = 28{,}925 \text{ m}$$

From Equation (5.25),

$$D_d = \frac{28{,}925 \text{ m}}{1.03 \times 10^7 \text{ m}^2} = \frac{0.0028 \text{ m}}{\text{m}^2}$$

D_d is also given by Equation (5.27):

$$D_d = \frac{(200)(4)^3}{1.03 \times 10^7 \text{ m}} \left[\frac{(2.5/4)^4 - 1}{(2.5/4) - 1}\right]$$

$$= \frac{0.0028 \text{ m}}{\text{m}^2}$$

The number of streams N_w in the area is given by Equation5.10a):

$$N_1 = R_b^3 = (4)^3 = 64$$

$$N_2 = R_b^2 = (4)^2 = 16$$

$$N_3 = R_b^1 = 4$$

$$N_4 = 1$$

The total number of channels = 85

The stream frequency is given by Equation (5.28):

$$F = (0.694)(0.0028 \text{ m/m}^2) = \frac{5.4 \times 10^{-6}}{\text{m}^2}$$

5.3.15 Channel-Length Frequency Distribution (CLFD)

When a rain storm falls uniformly over the entire drainage basin, runoff draining into each channel is carried to the basin outlet. Because the first-order channels drain the greatest areal portion of the drainage basin, they together account for the greatest amount of runoff. Further, the first-order channels nearest to the outlet (or stream gage) will deliver their flow to the outlet before the first-order channels farthest from the outlet. It is reasonable to assume that the effect of average flow conditions throughout the drainage basin results in similar flow velocities. Thus, the time of arrival of flow from each of the first-order channels is nearly proportional to its travel distance. It is possible to determine which channels deliver flow to the outlet at the same time by measuring the distance from the outlet to the head of each first-order channel. By dividing the longest flow distance into reasonably small intervals and plotting the number of first-order channels contained in each interval against the distance interval from the outlet, the result is a histogram called the channel-length frequency distribution (CLFD). The interval containing the greatest frequency of first-order channels will deliver the most flow to the outlet (allowance must be made for hydraulic routing effects). This condition is illustrated in Figure 5.16 for Marsh Creek, Pennsylvania (Rogers, 1972). Because of the hydraulic routing effects, the comparison between the CLFD and the hydrograph is not exact but is similar. This example shows that first-order channels do account for the shape of the hydrograph. Therefore, the density of drainage and the arrangement of the drainage are important in determining the hydrograph shape. The reason for the gap in the histogram for Marsh Creek, Pennsylvania, shown in Figure 5.16, is fewer drainage channels in this part of the basin caused by higher infiltration capacity of a sandy soil.

EXAMPLE 5.11 Construct the channel-length frequency histogram using the first-order channels for the Irvington drainage basin.

Solution The distance from the head of each first-order channel to the basin outlet is measured. Values of this distance are arranged in classes with an interval of 0.5 km, and the number of first-order channels is counted for each class, as shown in Table 5.6. The channel length frequency histogram is then plotted as shown in Figure 5.17.

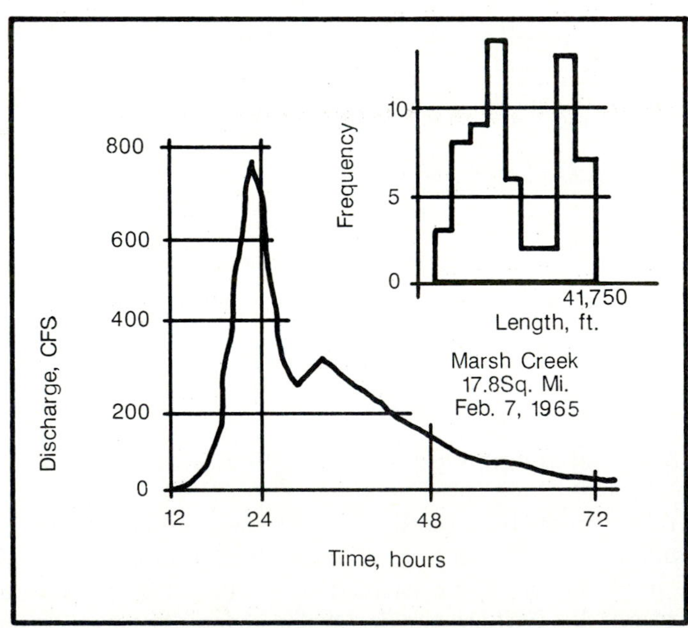

Figure 5.16 Hydrograph and channel-length frequency distribution (CLFD) for the Marsh Creek drainage basin, Pennsylvania.

TABLE 5.6 CHANNEL-LENGTH FREQUENCY DISTRIBUTION FOR THE IRVINGTON DRAINAGE BASIN

No.	Distance (km)	Number of first-order channels	Frequency (%)
1	0.5–1.0	2	0.20
2	1.0–1.5	10	1.02
3	1.5–2.0	15	1.53
4	2.0–2.5	14	1.43
5	2.5–3.0	20	2.04
6	3.0–3.5	33	3.36
7	3.5–4.0	38	3.87
8	4.0–4.5	24	2.45
9	4.5–5.0	10	1.02
10	5.0–5.5	23	2.34
11	5.5–6.0	30	3.06
12	6.0–6.5	33	3.36
13	6.5–7.0	33	3.36
14	7.0–7.5	50	5.10
15	7.5–8.0	28	2.85
16	8.0–8.5	47	4.79
17	8.5–9.0	42	4.28
18	9.0–9.5	36	3.67
19	9.5–10.0	40	4.08
20	10.0–10.5	32	3.26
21	10.5–11.0	26	2.65
22	11.0–11.5	31	3.16
23	11.5–12.0	40	4.08
24	12.0–12.5	48	4.89
25	12.5–13.0	35	3.57
26	13.0–13.5	42	4.28
27	13.5–14.0	51	5.20
28	14.0–14.5	43	4.38
29	14.5–15.0	35	3.57
30	15.0–15.5	25	2.55
31	15.5–16.0	23	2.34
32	16.0–16.5	21	2.14
33	16.5–17.0	1	0.10
Total		981	100.00

5.4 SUBSURFACE ENVIRONMENT

The subsurface environment of a drainage basin determines much of the character of the drainage basin itself. This subsurface environment includes the nature and structural properties of the bedrocks, in other words, the geology of the drainage basin. The nature and thickness of the soil derived from the geology have a profound effect on the drainage-basin character.

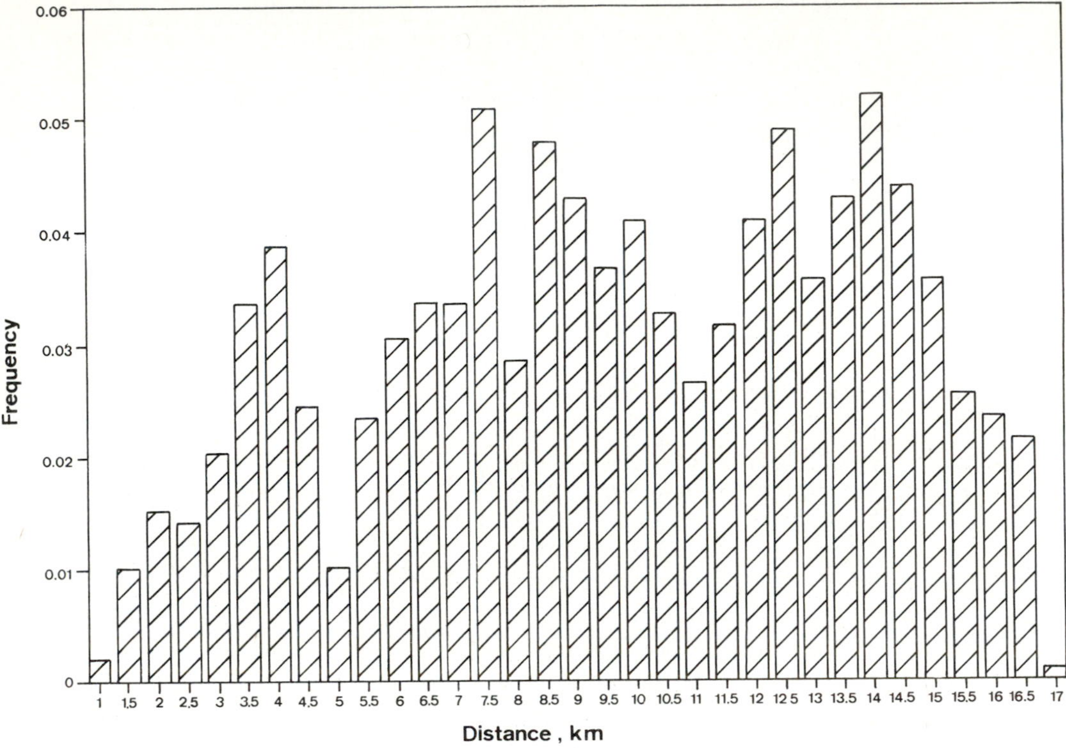

Figure 5.17 Channel-length frequency distribution for the Irvington drainage basin.

5.4.1 Geologic Elements Affecting Drainage-Basin Characteristics

Geology underlying a drainage basin affects the runoff characteristics of a drainage basin in two ways. (1) The rocks underlying a drainage basin often compose the parent material from which the soil is derived. This in situ soil then reflects the hydrologic character of the parent rocks. Fine-grain soils are derived from shales, silts, limestones, and some igneous rocks. Such soils have low infiltration capacities (Figure 5.14). On the other hand, soils derived from sandstones, conglomerates, and granites are coarse in nature and have high infiltration capacities. (2) Not all soils are in situ but are transported to the basin from elsewhere. Some of these soils are derived from windblown loess and sand, glacial till, and several forms of water-transported soil materials. An in-depth knowledge of geology is important in thoroughly understanding hydrology. The effects of geology on a drainage basin can be determined by making measurements and observations previously and subsequently described.

5.4.2 Soil and Its Spatial Variability

The nature of sedimentary rocks is such that the kinds of such rocks can vary spatially in areal distribution in almost any manner conceivable. Igneous rocks tend to vary somewhat less than sedimentary rocks. For these reasons, soils derived from these parent materials will also vary widely or perhaps vary but little. When transported soil material is included with the foregoing geology, the potential degree of variations in areal soil distribution can be visualized. The importance of soil distribution over a drainage basin is appreciated when one realizes that a drainage basin might be composed of uniform soil over the entire basin area. In this case, the entire drainage basin will have nearly uniform basin characteristics. On the other hand, if a drainage basin is covered with areas of different soils, then soil and hydrologic characteristics will vary throughout the basin. This condition will mean that drainage density, channel slope, length of overland flow, and channel profile will be different. These variations in drainage-basin characteristics caused by soil variations can result in large differences in hydrologic response to storm events. Inspection of soil variability throughout the watershed is an important preliminary step in analyzing basin characteristics.

Areal distribution is not the only important characteristic of soils. Soil thickness and character are also of great hydrologic importance. Soils are divided into four convenient divisions called A, B, C, and D zones. For hydrologic purposes, the A zone is the zone penetrated by plant roots. The D zone is the uppermost zone of parent rock material and represents rocks that are only little weathered. The B and C zones represent transition zones between the true plant-growing soil and the parent material. Engineering characteristics of soils along with maps showing their areal distribution and other descriptive information may be found in publications of the U.S. Department of Agriculture Soil Conservation Service, or SCS as it is commonly known. The SCS maintains an office for each county in each state of the United States. For most of the counties, the SDS has completed soil maps and a reference manual on engineering characteristics of soils found within that county. Some of the information published in this manual includes permeability, moisture-holding capacity, soil thickness, and a descriptive soil profile. All these data are important for hydrologic analyses.

5.4.3 Hydrogeological Properties of Soils

The foregoing discussion indicates that soil properties influence the movement of water. Because soils are made of aggregates of individual grains of material, they are classified as a porous material or a porous medium. Porous media have physical characteristics that affect the amount of water they can hold, that is, porosity, and the rate that water can move through the material, that is, hydraulic conductivity.

Porosity is defined as the ratio of the amount of water that can be contained in the interstitial spaces of a volume of rock or soil to the total volume of that rock or soil. Porosity is usually expressed as a percentage by multiplying this ratio by 100. Porosity is independent of grain size. For this reason, fine-grain soils might have porosities equal to or higher than coarse-grained soils. Porosity is determined mostly by grain assortment, shape, cementation, and fracturing. Values of porosity might reasonably range from 15 to 25%.

Hydraulic conductivity is a relatively recent term to replace the term permeability. For this reason, one should consider these two terms interchangeable when reading older literature. Hydraulic conductivity is defined as the amount of water that will flow through an area of material (1 ft^2 or 1 m^2) under a unit hydraulic gradient. Principally because of capillarity (adhesive and cohesive forces), hydraulic conductivity is mostly determined by grain size. Grain assortment, grain shape, cementation, and fracturing are also factors in determining hydraulic conductivity. The importance of hydraulic conductivity to hydrology is in its close association to infiltration and soil-moisture movement.

5.4.4 Groundwater

Many drainage basins supply groundwater flow to streams either on a continuous basis or on an intermittent basis. Thus, groundwater hydrology is closely related to surface-water hydrology. When the water table is high from long periods of rainfall and infiltration, the soil cannot contain as much moisture as it can when the water table is lower. Hence, greater runoff occurs during storm events. The behavior of the groundwater and the position of the water table is largely determined by the characteristics of the aquifer. Consequently, a knowledge of groundwater is important in understanding surface-water hydrology. In some drainage basins, inflow to the basin and outflow from the basin result from groundwater conditions determined by geological structure. This condition is complex and is not readily recognized without knowledge of geology in addition to groundwater and hydrology.

5.5 CLASSIFICATION OF STREAMS

Streams are classified according to the influence of baseflow on their discharge or the location of the water table. These streams are (1) perennial, (2) intermittent, and (3) ephemeral. Classification of streams does not consider the total stream length, but rather any reach of a stream. It is quite obvious that a first-order tributary of the Missouri River in South Dakota might be much different from the lower Mississippi River. For this reason, classification is intended to look at any reach along a stream that is of interest.

A perennial stream is a stream that flows the year around or at least at all times except during severe drought. This flow is maintained by groundwater flow (baseflow) to the stream. The bottom of the channel is below the groundwater table and, as a result, groundwater moves to the channel. As long as the channel remains below the groundwater surface, flow will continue. In Pennsylvania, for example, where about 4 inches of precipitation each month is sufficient to maintain a high water table, perennial streams are common. Rocky Mountain perennial streams are maintained by nearly continuous snowmelt. A perennial stream gains discharge from baseflow and it is therefore called a gaining stream.

An intermittent stream is a stream whose channel bed is intermittently below and above the groundwater table. When the water table is below the stream bed, the channel is dry; when the water table is above the stream bed, the stream flows water. The water table rises and lowers in response to recharge due to precipitation. An intermittent stream gains flow from baseflow and loses flow when the water table is below the channel bed. This stream is alternately a gaining stream and a losing stream.

An ephemeral stream is one that flows only during storm runoff events. The bed of an ephemeral stream is always above the water table. Because of this condition, water in the stream infiltrates from the stream bed during flow events and recharges the groundwater below. An ephemeral stream loses water as it flows. An example of this condition occurs in Walnut Gulch, Arizona, at the Agricultural Research Service test facility of the U.S. Department of Agriculture. More than 20,000 cfs flow was measured, yet this flow completely infiltrated within a few miles flow distance. If an ephemeral stream flows on porous material, it is always a losing stream. Sometimes an ephemeral stream occurs on impermeable material such as shale, clay, or some igneous rocks. Under these conditions, there is no groundwater table and the stream flows in response to storm events and carries its runoff downstream.

Any stream that is perennial, intermittent, or ephemeral throughout its length is called continuous; when alternating reaches are variously classified, the stream is called an interrupted stream. A large stream may change from perennial to interrupted to ephemeral as it moves from the basin outlet to its source.

From Figure 5.18, a stream is effluent or gaining when its bed intersects the groundwater table and it is fed by seepage from the groundwater reservoir. If the stream bed is above the groundwater table, then stream loses water to the groundwater reservoir and is called influent, or losing. A stream neither loses nor gains water because its stream bed is relatively impermeable. Such a stream is called an insulated stream. An influent or insulated stream, separated from the groundwater reservoir by an unsaturated zone, is called a perched stream. Because the groundwater table changes from time to time, this classification of stream is transitory.

5.6 FLOOD PLAINS

Flood plains are typically low areas adjacent to a river, ocean, or other body of water, and are subject to flooding. The amount of land inundated by a flood depends on the flood's magnitude. The floodplain of a river is the area of the valley floor adjacent to the incised channel. As water rises in the river over the bankful stage, this area gets flooded. There is, however, some difficulty in precisely defining the bankful stage and the floodplain. It is well documented that the floodplain is subject to frequent flooding. Floodplains in the eastern and central United States are flooded, on the average, twice in 3 years. Approximately 7% of the nation's land area is subject to flooding by the 100-year flood; this area is almost as big as Texas. Consequently, development of floodplains must be carefully regulated. Floodplains are built up from deposition of fine sediment. The sediment is deposited during

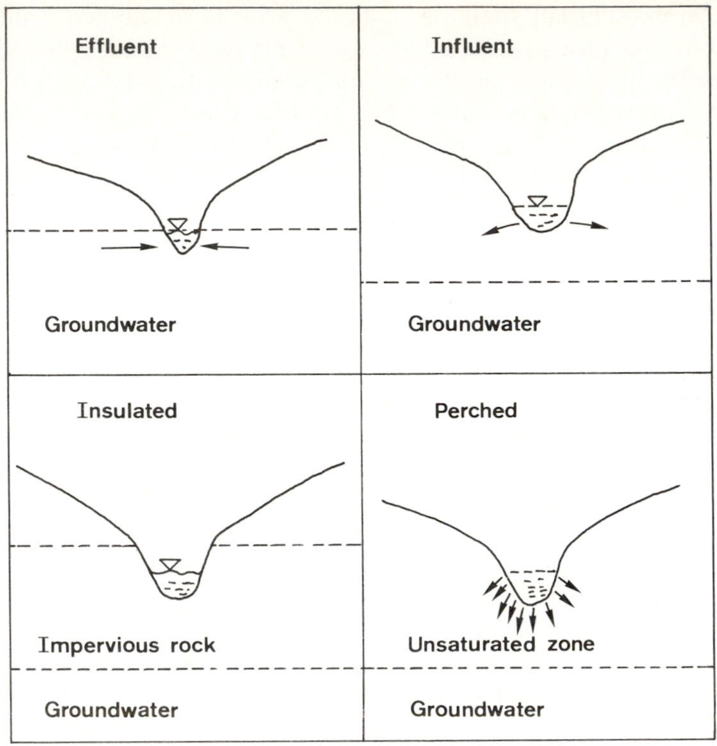

Figure 5.18 Classification of streams based on location of the groundwater table.

flooding. As the river meanders across its valley bottom, it first erodes one valley side and then the other. A natural levee forms along the banks of the incised channel caused by the deposition of coarse sediment when the streamflow inundates the floodplain.

On the basis of the capacity at peak flow, the floodplain land can be classified into floodway and pondage areas. Floodway, or flowage land, is the principal flow-carrying part of the natural cross-section of the stream. Any encroachment upon the floodway will increase flood heights. Pondage area is the land on which water is stored as dead water during flooding. These dead areas are not a part of a flow-carrying area, and are often referred to as flood fringes. If buildings are put in these areas, there would be no significant increase in flood heights there or above the construction sites.

The boundary between floodway and pondage lands is not a fixed one, but changes with the magnitude of the flood. The principal use of floodplain characteristics is in establishing regulations for land use for protection of the health, safety, and general welfare of the community. Examples of their use include defining encroachment lines, developing zoning ordinances, regulating urban development or subdivisions, and establishing building codes.

Beaches and small river valleys are easily recognizable as floodplains. Less obvious floodplains occur in dry washes and on alluvial fans in arid areas of the

United States around prairie potholes, in areas subject to high groundwater levels, and in low-lying areas where water may accumulate. Floodplains are characterized in a manner similar to floods. For example, the land flooded by the 50-year flood is called the 50-year floodplain.

EXERCISES

5.1. Take a topographic map of an area having streams. Delineate the drainage boundary of this basin area.

5.2. For the area in Exercise 5.1, move the basin outlet to another point upstream, and then delineate the drainage boundary.

5.3. Delineate the drainage areas of three major streams within the drainage basin in Exercise 5.1.

5.4. Select a drainage basin of at least third order in your area. Order the streams according to the Horton–Strahler ordering method. List the number of stream segments against each order.

5.5. Compute the area ratio, the bifurcation ratio, the length ratio, and the slope ratio for the watershed in Exercise 5.4.

5.6. Compute lengths of streams of all orders for the basin in Exercise 5.4. Also compute the average lengths of streams of given orders.

5.7. Compute the drainage density of the basin in Exercise 5.4.

5.8. Compute drainage areas of streams of all orders for the basin in Exercise 5.4. Also compute the average drainage areas of streams of given orders.

5.9. Determine the centroid of the drainage basin in Exercise 5.4.

5.10. Determine the slopes of streams of all orders for the basin in Exercise 5.4. Also compute the average slopes of streams of given orders.

5.11. Verify if the laws of basin areas, stream numbers, stream lengths, and stream slopes hold for the drainage basin in Exercise 5.4.

5.12. Construct the channel-length frequency distribution for the first-order streams of the drainage basin in Exercise 5.4.

5.13. Plot the watershed shape factors (given in Table 5.1) against area for various values of watershed length L. In the case of circularity ratio and compactness ratio, plot for various values of perimeter p_r. Show the calculations in tabular form. What do you conclude from this figure?

CHAPTER 6
Weather and Precipitation

Precipitation is the input to the hydrologic budget. Without precipitation there can be no runoff. The amount, distribution, and intensity of precipitation vary greatly in both space and time. This variability depends upon the general movement of the atmosphere and its characteristics, commonly called weather. Several important physical elements associated with weather are briefly treated in this chapter. For a more complete discussion, the reader should refer to such standard books on meteorology such as Landsberg and Jacobs (1951), Sutton (1953), Geiger (1957), Landsberg (1958), Petterssen (1958), Johnson (1960), and Budyko (1974), among others. Precipitation is a link between atmosphere and earth, and serves as a basis for association between hydrology and meteorology.

6.1 CHARACTERISTICS OF THE ATMOSPHERE

Air is a mixture of gases. It is not a chemical compound but rather a mixture of chemical elements. Clean air consists of gases only, whereas natural air contains solid and liquid mixtures of foreign matter. Natural air also contains water vapor. Dry air contains gases, as shown in Table 6.1. Except for small variations in carbon dioxide, the percentages of gases remain constant up to a height of about 15 miles (24 km). Above this height, chemical processes maintain a variable composition in the atmosphere.

Carbon dioxide is generated by natural processes on earth and is carried into the atmosphere, where it is mixed with other components of the air. The amount of carbon dioxide generated is balanced by its dissolution in precipitation, which is then

TABLE 6.1 COMPOSITION OF DRY AIR

	Gases	Percentage
N	Nitrogen	78.09
O	Oxygen	20.95
A	Argon	0.93
CO_2	Carbon dioxide	0.03
	Traces of other gases	0.003

carried to the oceans, where it is precipitated as lime. This process maintains a relatively constant percentage of carbon dioxide in the atmosphere.

Industrial processes over the last century have spilled a lot of carbon dioxide into the atmosphere. Although carbon dioxide is colorless, odorless, and harmless by itself, it slows the passage of heat. Having extra carbon dioxide means that light can come in and warm the earth, but heat cannot go back as easily as it once did. As a result, the earth should warm up because of the extra carbon dioxide. It is called the greenhouse effect because of the trapping of heat by the atmosphere, somewhat like the trapping of warmth by a greenhouse glass cover. There is evidence that burning of fuels on earth has increased the amount of carbon dioxide in the atmosphere by about 10% since the turn of the century. This added carbon dioxide is believed by many to represent the cause of increased retention of infrared rays, resulting in increased atmospheric temperatures due to the greenhouse effect. If this greenhouse effect does indeed develop along with increased carbon dioxide, the result could be a substantial change in the climate on the earth as we know it now. The climatic change will be accompanied by a multitude of hydrologic consequences.

The equivalent molecular weight of air is 28.95 and the molecular weight of water is 18. Thus, the specific gravity of water vapor is 0.622 that of dry air. Because water vapor is lighter than air, it would rise above the atmosphere unless condensation occurred. The water vapor content of natural air varies from near 0% in cold dry arctic air to about 4% in warm moist tropical air. The water vapor content of the atmosphere at any given time and location above the earth represents the potential precipitation at that place at that time. The amount of water vapor in the atmosphere varies with atmospheric conditions in space and time. The moisture and carbon dioxide present in the atmosphere intercept part of the long-wave (infrared) radiation from the earth. This results in heating of the atmosphere.

A small concentration of ozone, O_3, occurs in the atmosphere 12 to 31 miles (20–50 km) above the earth. This ozone protects the earth from harmful ultraviolet radiation from the sun. The atmosphere is most dense next to the earth and gradually decreases in density with altitude. About 50% of the atmosphere is below 3.5 miles (5.5 km) altitude, and about 75% is below 7 miles (11 km).

The atmosphere is not a uniform layer over the surface of the earth. Because it is a dynamic mass, the atmosphere is moving, circulating, eddying, and rising and falling in response to thermal changes. For this reason, the weight of the atmosphere varies with time and space over the earth. Units of atmospheric variables are shown in Table 6.2.

TABLE 6.2 ATMOSPHERIC UNITS

Variable	Units
Pressure	1 millibar (mb) = 1000 dynes/cm^2 = 0.0143 lb/in.^{2} or 0.029 in. Hg
Pressure (sea level)	1000 mb + 20–40 mb (14.52 psi) (29.53 in. Hg)
Temperature	Celsius (C) $\{[\frac{5}{9}(F - 32)]\}$
Mean sea-level air pressure	1013 millibars

6.2 ATMOSPHERIC WATER VAPOR AND ITS INDICES

Water vapor is a regular constituent of the atmosphere. By comparison with the amounts of other gases present in the atmosphere, the amount of water vapor may be relatively small. Nevertheless, the importance of water vapor cannot be overemphasized. This atmospheric water is the source of precipitation received on the land surface, which is the input to the hydrologic cycle. This atmospheric moisture exercises profound influence on temperature. It absorbs and reradiates terrestrial radiation and thus stabilizes the earth's temperature. The moisture content of the air is one of the principal determinants of evaporation. In turn, evaporation and transpiration are the sources of the atmospheric moisture, and evaporation from oceans is the principal source. Thus, it is important to understand the atmospheric water and its hydrologic consequences. Under most conditions of practical interest, water vapor behaves like any other gas and obeys the gas laws.

The amount of water in the atmosphere at any time varies from place to place throughout the world. Although the amount of water in the atmosphere at any instant is relatively small, large amounts of water pass through the atmosphere annually because water vapor is continuously precipitated and resupplied to the atmosphere at some place on the earth. The amount of precipitable water in the atmosphere at any time can range from near zero to several inches. The mean annual precipitation for the entire earth has been estimated at about 34 in. (86 cm) per year. Most of this precipitation falls on the water surface of the earth and is not evenly distributed throughout the world. Based on a global mass balance, the amount of precipitation must equal the amount of evaporation.

6.2.1 Vaporization, or Evaporation

The atmospheric vapor content is formed by evaporation and transpiration. Vaporization is the process by which water is converted to vapors; it results in the removal of heat energy from the water being vaporized. The rate of vaporization increases with increasing temperature, because rising temperature increases the kinetic energy of water molecules and decreases surface tension. When a solid such as ice or snow is directly transformed to water vapor or vice versa, the process is called sublimation.

6.2.2 Condensation

It is the reverse of vaporization. Condensation is the process by which water vapor is changed to the liquid or solid state; it adds heat energy to the water that condenses. Condensation and vaporization always occur simultaneously in a space that is in contact with water. In an unsaturated space, vaporization exceeds condensation and there will be net evaporation. Under the same air and water temperatures in a saturated space, condensation and vaporization equalize each other. If the space is supersaturated, condensation normally occurs with the introduction of suitable nuclei.

6.2.3 Latent Heat of Vaporization or Condensation

The latent heat of vaporization is defined as the amount of heat required to convert a unit mass of water to vapor without change in temperature. The latent heat of condensation is the amount of heat released when vapor is converted to a unit mass of water without change in temperature. Thus, the two latent heats are equivalent. The usual unit of measurement is calories per gram (cal/g). The latent heat of vaporization of water H_v changes with temperature and can be estimated accurately up to 40°C (104°F) as

$$H_v = 597.3 - 0.564T \tag{6.1}$$

where H_v is in cal/g, and T is the temperature in degrees Celsius.

6.2.4 Latent Heat of Fusion

The latent heat of fusion is the amount of heat required to convert 1 gram of ice to water at the same temperature. Conversely, this is also the amount of heat released when converting 1 gram of water at 0°C (32°F) to ice at the same temperature. The latent heat of fusion is 79.7 cal/g or approximately 80 cal/g.

6.2.5 Latent Heat of Sublimation

The latent heat of sublimation is defined as the amount of heat required to directly convert 1 gram of ice to vapor without change in temperature. Of course, when water vapor is directly converted to ice, an equivalent amount of heat is released. The latent heat of sublimation is the sum of the latent heat of fusion and the latent heat of vaporization. At 0°C (32°F), it is approximately 677 cal/g.

EXAMPLE 6.1 Compute the heat of vaporization in cal/g for water at (a) 10°C, (b) 15°C, (c) 25°C, (d) 70°F, (e) 80°F, and (f) 90°F.

Solution Equation (6.1) is used to compute the heat of vaporization.

(a) $H_v = 597.3 - 0.564(10) = 591.66$ cal/g

(b) $H_v = 597.3 - 0.564(15) = 588.84$ cal/g

(c) $H_v = 597.3 - 0.564(25) = 583.2$ cal/g

(d) $70°\text{F} = (70 - 32)\,\frac{5}{9}°\text{C} = 21.11°\text{C}$

$H_v = 597.3 - 0.564(21.11) = 585.39$ cal/g

(e) 80°F = (80 − 32) $\frac{5}{9}$°C = 26.67°C
$H_v = 597.3 - 0.564(26.67) = 582.26$ cal/g
(g) 90°F = (90 − 32) $\frac{5}{9}$°C = 32.22°C
$H_v = 597.3 - 0.564(32.22) = 579.13$ cal/g

EXAMPLE 6.2 How many calories are required to evaporate 5 gallons (U.S.) of water at 80°F? Compute the amount of ice (pounds) at 10°F that this heat would melt. The specific heat of ice may be taken as 0.5.

Solution

$$80°\text{F} = (80 - 32)\,\tfrac{5}{9}°\text{C} = 26.67°\text{C}$$

$$1 \text{ gallon of water} = 3785 \text{ cm}^3 \text{ or } 3785 \text{ g}$$

$$5 \text{ gallons of water} = 18{,}925 \text{ g}$$

Equation (6.1) is used to compute calories per gram.

$$H_v = 597.3 - 0.564(26.67) = 582.26 \text{ cal/g}$$

$$\text{Calories to evaporate 5 gallons of water} = 582.26 \times 18{,}925 \text{ cal} = 11.0193 \times 10^6 \text{ cal}$$

$$10°\text{F} = (10 - 32)\,\tfrac{5}{9}°\text{C} = -12.22°\text{C}$$

$$\text{One pound of ice} = 454 \text{ g}$$

$$\text{Calories required to heat 1 g of ice to } 0°\text{C} = 0.5 \times 12.22 = 6.11 \text{ cal}$$

$$\text{Latent heat of fusion} = 79.7 \text{ cal/g}$$

$$\text{Calories required to melt 1 g of ice} = 6.11 \text{ cal} + 79.7 \text{ cal} = 85.81 \text{ cal}$$

$$\text{Number of grams of ice melted} = \frac{11.0193 \times 10^6}{85.81} = 12.8415 \times 10^4$$

$$\text{Number of pounds of ice melted} = \frac{11.0193 \times 10^6}{85.81\ (454)} = 282.85$$

EXAMPLE 6.3 Compute the heat in cal/ft^2 required to melt a 6-in.-thick layer of ice at 10°F with a specific gravity of 0.9. Also compute the heat required to evaporate the resulting meltwater without raising its temperature above 32°F.

Solution

1 ft^3 of water is 62.4 lb

1 ft^3 of ice with specific gravity of 0.9 is 0.9(62.4) = 56.16 lb
= 56.16 × 454 = 25,496.64 g

0.5 ft^3 of ice is 12,748.32 g.

The calories required to heat the ice from 10°F to 32°F are computed as follows. 10°F = −12.22°C. Therefore, the calories required to heat ice from − 12.22°C to 0°C = 0.5(12.22) = 6.11 cal.

The calories required to heat the 6-in. layer of ice from 10°F to 32°F = 12,748.32(6.11) = 77,892.235 cal.

The calories required for melting the 32°F ice to 32°F water = 12,748.32(79.7) = 1,016,041.1 cal.

Therefore, the total calories required = 1,093,933.3 cal/ft^2.

The calories required to evaporate the resulting meltwater at 32°F = 12,748.32(597.3) = 7,614,571.5 cal/ft^2

6.2.6 Vapor Pressure

Each gas in a mixture of gases exerts a partial pressure. The partial vapor pressure is defined as the pressure that water vapor would exert if other gases were absent. The standard notation for vapor pressure is *e*, which is the difference between the pressure of the moist air and that of dry air usually expressed in millibars. The amount of vapor in the air depends on the temperature. It is a maximum at high temperatures and a minimum at low temperatures.

The amount of water in the atmosphere is related to temperature, as seen from Figure 6.1. The higher the temperature, the more the water vapor air can contain and transport. Because the specific gravity of water vapor is 0.622 of that of dry air, transfer of water vapor in the atmosphere is readily possible. As this water vapor is transported, any reduction in temperature must be accompanied by the reduction in water vapor content in accordance with Figure 6.1. This reduction in water vapor content is accomplished by condensation and precipitation. When air is cooled to −30°C, no water vapor can be carried by the air.

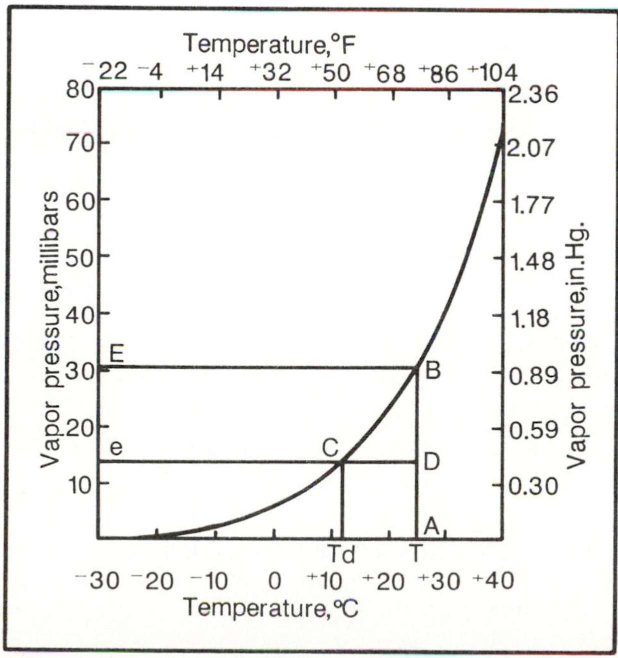

Figure 6.1 Relation between air temperature and vapor pressure.

The minimum amount of water vapor that can exist in a space depends on the temperature and does not depend on the existence of other gases. This space becomes saturated when it contains the maximum amount of water vapor at a given temperature. The saturation vapor pressure e_s is the pressure exerted by the vapor in a saturated space. Corresponding to the given temperature, this is the maximum vapor pressure. Vapor pressure over water is used most often.

The vapor pressure e in millibars can be computed empirically from psychrometric equation as

$$e = e_s - 10^{-4} \times 3.67p(T_a - T_w)\left(1 + \frac{T_w - 32}{1571}\right) \tag{6.2}$$

where p is the atmospheric pressure (mb), T_a is the dry-bulb temperature (°F), T_w is the wet-bulb temperature (°F), and e_s is the saturation vapor pressure (mb) corresponding to T_w.

The vapor pressure can also be expressed as

$$e = e_s - \frac{pc_p}{0.622H_v}(T_a - T_w) = e_s - \gamma(T_a - T_w) \tag{6.3}$$

where c_p is the specific heat of air at constant pressure, γ is the psychrometric constant, and temperatures are in °C. If p = 1013 mb at 1 atmosphere, c_p = 0.2396 cal g^{-1} °C^{-1}, and H_v = 590 cal g^{-1} at 10°C, $pc_p/0.622H_v$ = 0.662 mb °C^{-1} (0.496 mm Hg °C^{-1}, or 0.276 mm Hg °F^{-1}).

The saturation vapor pressure of moist air over water at pressure p and temperature T can be expressed as

$$e_s = \frac{r_s}{0.622 + r_s}p \tag{6.4}$$

where r_s is the saturation mixing ratio.

The saturation vapor pressure can be obtained from psychrometric tables or approximated in the range -50°C to 55°C (Bosen 1960) as

$$e_s \cong 33.8639[(7.38 \times 10^{-3}T + 0.8072)^8 - 1.9 \times 10^{-5}\,|1.8T + 48| + 13.16 \times 10^{-4}] \tag{6.5}$$

where e_s is in millibars, and T is the temperature in degrees Celsius.

The saturated vapor pressure over water can be converted to that over ice for the temperature range 0 to -50°C (32 to -58°F) (Bosen, 1961) as

$$\frac{e_s\text{ (ice)}}{e_s\text{ (water)}} \cong 1 + 9.72 \times 10^{-3}T + 4.2 \times 10^{-5}T^2 \tag{6.6}$$

where T is in degrees Celsius.

Saturation deficit or vapor pressure deficit of the air s_d is defined as

$$s_d = e_s - e \tag{6.7}$$

6.2.7 Density of Moist Air

The density of moist air ρ_a is defined as the mass of water vapor and the mass of dry air per unit volume of the moisture. If p_a is the pressure of the moisture, then $p_a - e$ is the partial pressure of the dry air alone.

The specific gravity of water vapor is 0.622 of that of dry air at the same temperature and pressure. The density of water vapor ρ_w in grams per cubic centimeter can be defined as

$$\rho_w = 0.622 \frac{e}{RT} \tag{6.8}$$

where T is the absolute temperature (°C), e is the vapor pressure (mb), and R is dry-gas constant $= 2.87 \times 10^{-3}$.

The density of dry air ρ_d in g/cm³ can be expressed as

$$\rho_d = \frac{p_d}{RT} \tag{6.9}$$

where p_d is the pressure in mb.

Because $\rho_a = \rho_w + \rho_d$, Equations (6.8) and (6.9) can be added to yield an expression for ρ_a as

$$\rho_a = 0.622 \frac{e}{RT} + \frac{p_d}{RT} = \frac{1}{RT}(0.622e + p_d) \tag{6.10}$$

By replacing p_d by $p_a - e$ in Equation (6.7),

$$\rho_a = \frac{1}{RT}(0.622e + P_a - e) = \frac{p_a}{RT}\left(1 - 0.378 \frac{e}{p_a}\right) \tag{6.11}$$

Because e/p_a is a positive quantity, Equation (6.11) shows that moist air is lighter than dry air.

EXAMPLE 6.4 Consider 1 m³ of dry air. What will be the weight in kilograms if the air temperature and pressure are 5°C and 1000 mb, respectively?

Solution

Equation (6.9) is used to compute the weight of dry air. $R = 2.87 \times 10^3$.

$$\rho_d = \frac{1000}{2.87 \times 10^3 \times (273 + 5)} \text{ g/cm}^3 = 0.001253 \text{ g/cm}^3 = 12.53 \text{ kg/m}^3$$

EXAMPLE 6.5 Compute the density of dry air at 20°C and at a pressure of 1000 mb, and the density of moist air with relative humidity of 50% at the same temperature and pressure.

Solution

(a) Equation (6.9) is used to compute the density of dry air:

$$\rho_d = \frac{1000}{2.87 \times 10^3 \times (273 + 20)} = 1.1892 \times 10^{-3} \text{ g/cm}^3 = 1.1892 \text{ kg/m}^3$$

(b) Equation (6.11) is used to compute the density of moist air. The partial vapor pressure is obtained by using Equation (6.5):

$$\begin{aligned} e_s &= 33.8639\,[(7.38 \times 10^{-3} \times 20 + 0.8072)^8 \\ &\quad - 1.9 \times 10^{-5}\,|1.8 \times 20 + 48| + 13.16 \times 10^{-4}] \\ &= 33.8639[(0.9548)^8 - 10^{-3} \times 1.596 + 13.16 \times 10^{-4}] \\ &= 33.8639\,[0.6907 - 0.001596 + 0.001316] = 23.38 \text{ mb} \end{aligned}$$

Therefore,

$$e = 0.5 \times 23.38 = 11.69 \text{ mb}$$

From Equation (6.5),

$$\begin{aligned} \rho_a &= \frac{1000}{2.87 \times 10^3 \times (20 + 273)}\left(1 - 0.378\,\frac{11.69}{1000}\right) \\ &= 1.18918 \times 10^{-3}\,(0.996) = 1.8834 \times 10^{-3} \text{ g/cm}^3 = 1.8834 \text{ kg/m}^3 \end{aligned}$$

6.2.8 Dew-Point Temperature

Dew-point temperature is the temperature to which air must be cooled at constant pressure in order for it to reach saturation. At the dew point, any further cooling will cause the water vapor to condense and precipitation will occur. At the dew-point temperature, the saturation vapor pressure e_s equals the existing vapor pressure e with no change in water content.

At this temperature, the saturation vapor pressure $e_s(T_d)$ is

$$e_s(T_d) = \frac{r_s}{0.622 + r_s}\,p \tag{6.12}$$

Following Bosen (personnel communication), the dew-point temperature can be estimated within 0.3°C (0.5°F) in the temperature range −40 to 50°C (−40 to 122°F) as

$$\begin{aligned} T - T_d \cong{}& (14.55 + 0.114T)(1 - \text{RH}) + [(2.5 + 0.007T)(1 - \text{RH})]^3 \\ &+ (15.9 + 0.117T)(1 - \text{RH})^{14} \end{aligned} \tag{6.13}$$

where T is in °C, and RH is the relative humidity expressed as a decimal fraction.

6.2.9 Wet-Bulb Temperature

A moist evaporating surface attains a certain temperature when the radiation energy balance is zero. This temperature is called the wet-bulb temperature. This is the equilibrium temperature of a thermometer covered with a cloth that has been wetted with pure water in moving air. This thermometer will be cooled until heat drawn from the air equals the gain of latent heat for evaporation.

6.2.10 Humidity

Humidity is a general term used to indicate moisture in the atmosphere. Vapor pressure and humidity are intertwined. Absolute humidity (g/m³) is the mass of water vapor contained in a unit volume of space, and can be determined from the equation of state for an ideal gas. It is the same as the density of water vapor. The relation between absolute humidity ρ_w and vapor pressure e can be shown to be

$$e = \rho_w \frac{R}{m_w} T \tag{6.14}$$

where R is the universal gas constant, m_w is the molecular weight of water vapor, and T is the absolute temperature. ρ_w can be written as

$$\rho_w = 217 \frac{e}{T} \tag{6.15}$$

where T is the absolute temperature in °C, and e in mb. Equation (6.15) is equivalent to Equation (6.8).

The specific humidity q (g/g) is the mass of water vapor contained within a unit mass of moist air, and can be calculated as

$$q = \frac{0.622\, e}{p - 0.378e} \tag{6.16}$$

where q is in g/g, e is in millibars, and p is the total pressure of the moist air in millibars. Because p is much larger than e (on the order of 100 times), Equation (6.12) is simply written as

$$q = 0.622 \frac{e}{p} \tag{6.17}$$

Analogous to specific humidity is the mixing ratio r, defined as the mass of water vapor per unit mass of perfectly dry air in humid mixture. It is approximately equal to q given by Equation (6.17).

The relative humidity RH is the gravimetric ratio of water vapor in unit space at a given temperature to the maximum water vapor that the moist space can hold at that temperature. For practical purposes, it is the ratio of the water vapor to that which would exist under saturated conditions, or it is the ratio of the actual vapor pressure to the saturation vapor pressure, and can be expressed as

$$\mathrm{RH} = \frac{e}{e_s} \tag{6.18}$$

where e_s is the saturated vapor pressure. Since most of the atmosphere is unsaturated, it is useful to estimate the degree of saturation. Saturation deficit is one of the important parameters used to compute evaporation.

Relative humidity is a function of temperature because the vapor pressure that determines relative humidity is a function of temperature. Relative humidity is not a direct measure of the moisture in the air. It is possible to have 100% relative humidity at 0°C and 100% relative humidity at 30°C. Even though the relative humidity is 100% in both cases, the amount of water vapor in the atmosphere is far greater at 30°C than at 0°C, as seen from Figure 6.1.

Bosen (1958) has given a simple formula for computing RH in percent directly from air temperature T (°C) and dew-point temperature T_d (°C) as

$$\mathrm{RH} = \left(\frac{112 \ -0.1T + T_d}{112 + 0.9T}\right)^8 \tag{6.19}$$

Equation (6.19) yields RH to within 0.6% in the temperature range −25 to 45°C (−13 to 113°F). If T is °F, Equation (6.19) becomes

$$\mathrm{RH} = \left(\frac{96 - 0.056T + 0.56T_d}{96 + 0.5T}\right)^8 \tag{6.21}$$

or

$$\mathrm{RH} = \left(\frac{173 - 0.1T + T_d}{173 + 0.9T}\right)^8 \tag{6.22}$$

Equation (6.22) is obtained by multiplying the numerator and denominator within parentheses of Equation (6.21) by 1.8.

EXAMPLE 6.6 The dry-bulb temperature is given as 25°C and the wet-bulb temperature as 15°C. Compute the dew-point temperature, relative humidity, saturation vapor pressure, and actual vapor pressure.

Solution Equations (6.2) and (6.3) or (6.5) are utilized to compute partial vapor pressure and saturation vapor pressure. Assume atmospheric pressure as 1013 mb.

$$\begin{aligned} e_s &= 33.8639[(7.38 \times 10^{-3} \times 25 + 0.8072)^8 \\ &\quad - 1.9 \times 10^{-5} \,|1.8 \times 25 + 48| + 13.16 \times 10^{-4}] \\ &= 33.8639[(0.9917)^8 - 0.001767 + 0.001316] \\ &= 33.8639[0.935497 - 0.001767 + 0.001316] = 33.8639(0.935046) \\ &= 31.6643 \text{ mb} \end{aligned}$$

Equation (6.3) is used to compute partial vapor pressure e:

$$e = 31.6643 - 0.662(25 - 15) = 31.6643 - 6.62 = 25.04 \text{ mb}$$

Also, Equation (6.5) is used to compute e. 25°C = 77°F, and 15°C = 59°F.

$$\begin{aligned} e &= 31.6643 - 10^{-4} \times 3.67 \times 1013 \times (77 - 59) \times \left(1 + \frac{59 - 32}{1571}\right) \\ &= 31.6643 - 6.6919(1.01719) = 24.8574 \text{ mb} \end{aligned}$$

Both equations give comparable values of partial vapor pressure.

$$\text{Relative humidity, RH} = \frac{25.04}{31.6643} = 0.7908 \text{ or } 79.08\%$$

Equation (6.13) is used to compute the dew-point temperature T_d:

$$\begin{aligned} T - T_d &= (14.55 + 0.114 \times 25)(1 - 0.79) + [(2.5 \\ &\quad + 0.007 \times 25)(1 - 0.79)]^3 + (15.9 + 0.117 \times 25)(1 - 0.79)^{14} \\ &= 3.654 + (0.56175)^3 + 18.825(0.21)^{14} \\ &= 3.654 + 0.17727 + 18.825 \times 10^{-10} \\ &\cong 3.8313 \end{aligned}$$

Therefore,

$$T_d = T - 3.8313 = 25 - 3.8313 = 21.1687°\text{C}$$

Equation (6.19) can also be used to calculate T_d

$$0.79 = \left(\frac{112 - 0.1 \times 25 + T_d}{112 + 0.9 \times 25}\right)^8$$

$$T_d = (112 + 22.5)(0.79)^{0.125} + 2.5 - 112$$

$$= 134.5 \times 0.97 + 2.5 - 112 = 21.095°\text{C}$$

which is comparable to the previous value of T_d.

EXAMPLE 6.7 Compute the relative humidity if the air temperature and the dew-point temperature are, respectively, 30°C and 20°C.

Solution Equation (6.19) is used to compute RH:

$$\text{RH} = \left(\frac{112 - 0.1 \times 30 + 20}{112 + 0.9 \times 30}\right)^8 = \left(\frac{129}{139}\right)^8 = (0.928)^8 = 0.5503 = 55.03\%$$

EXAMPLE 6.8 Compute the dew-point temperature if the air temperature and relative humidity are, respectively, 35°C and 80%.

Solution Equation (6.19) is used to compute T_d:

$$(0.8)^{0.125} = \frac{112 - 0.1 \times 35 + T_d}{112 + 0.9 \times 35}$$

$$0.973 = \frac{108.5 + T_d}{143.5}$$

$$T_d = 139.63 - 108.5 = 31.13°\text{C}$$

Equation (6.13) can also be used to compute T_d:

$$T - T_d = (14.55 + 0.114 \times 35)(0.2) + [(2.5 + 0.007 \times 35)(0.2)]^3$$

$$+ (15.9 + 0.117 \times 35)(0.2)^{14}$$

$$= 3.708 + (0.549)^3 + (19.995)(0.2)^{14}$$

$$\cong 3.873$$

$$T_d = 35 - 3.873 = 31.13°\text{C}$$

Both equations give the same answers.

6.3 TEMPERATURE VARIATION WITH ALTITUDE

The climatological data published by the U.S. National Weather Service contain normal and mean temperature values. The normal temperature is the average value for a particular time scale computed over a specific 30-year period, say, 1951 to 1980. The time scale can be daily, weekly, monthly, seasonal, or annual. The mean temperature is the same as the average temperature. If hourly temperature observations are available for any day, then the average daily temperature for that day is the sum of hourly temperature values divided by 24 hours. In the United States, the mean daily temperature is taken as the average of the daily maximum and daily minimum temperatures. The daily range in temperature is the difference between the maximum and minimum temperatures observed that day. The mean monthly temperature is taken as the average of the mean monthly maximum and minimum temperatures. The mean annual temperature is taken as the average of the monthly means of the year.

Changes in temperature in the atmosphere take place from dynamic processes. For example, a Chinook wind is a wind that blows from a higher elevation to a lower elevation. In this process, compression of air occurs and heating results. The temperature change in this process is adiabatic. By the same token, an upslope wind will cause cooling as the air expands. The upslope condition is a common cause of precipitation in areas of considerable topographic relief because moist air is elevated and adiabatic cooling occurs.

Temperature variations occur with the time of day. These diurnal temperature variations can be large in desert continental areas and small in oceanic areas. Such diurnal temperatures are cyclic and are added to annual cyclic temperature changes.

In an adiabatic process, there is no heat exchange between the working system and its environment. In such a process, gas is compressed or expanded without giving up or receiving heat. Boyle's law states that for a constant temperature, the product of gas density and volume is a constant, whereas Charles' law states that temperature is proportional to volume if the gas density is a constant. The application of these gas laws in the atmosphere refers to the usual composition of air and some water vapor. The latent heat of condensation is approximately 539 calories per gram of water condensed. Thus, for every gram of water condensed from the atmosphere, 539 calories of heat is released. Thus, condensation is a great source of heat energy to the atmosphere. As long as the water vapor does not condense, no latent heat of condensation is released and the energy content of the air does not change. This condition is called the dry adiabatic process. As water begins condensing, latent heat is released and the energy content of the air volume increases. This energy was not added from the outside. This process is called the moist adiabatic process.

Lapse rate is defined as the decrease in temperature with an increase in elevation. The average lapse rate in the atmosphere is approximately 3.8°F/1000 ft (2.1°C/328 m, or 0.7°C/100 m). This value is often rounded to 3.5°F/1000 ft (2°C/328 m). The dry adiabatic lapse rate is 5.4°F/1000 ft (3°C/328 m), which is the rate of temperature change of unsaturated air due to expansion or compression occurring during ascent or descent of air. (During ascent, pressure decreases and it increases during descent). These lapse-rate conditions exist in normal atmospheric conditions. The lapse rate shows that any air that is elevated is cooled. If moist air is elevated sufficiently, condensation will occur. If the saturated air is elevated adiabatically, its temperature will drop and its water vapor will condense. The condensation of water vapor releases the latent heat of vaporization, which, in turn, reduces the cooling rate of the rising air. This is why the saturated-adiabatic lapse rate is lower than the dry-adiabatic lapse rate. Clearly, the higher the vapor content, the lower the lapse rate, and, hence, the higher the air temperature. Above freezing, the saturated-adiabatic lapse rate is about half of the dry-adiabatic rate. When the moisture is precipitated during the air ascent, the temperature decreases at the pseudoadiabatic lapse rate. Because the heat is taken away by the falling precipitation, the process of cooling is pseudoadiabatic, but the rate is nearly the same as the saturated-adiabatic lapse rate.

A temperature inversion occurs in the atmosphere when cold dense air is wedged under lighter warm air and occurs when a cold front advances. Under these conditions, temperature will increase with elevation from the ground surface when the elevated warm air is encountered. At the cold air–warm air contact, rising smoke will no longer rise but will drift horizontally along the contact of the air masses. Temperatures will follow the normal lapse-rate decrease as elevation increases above the temperature inversion.

EXAMPLE 6.9 Consider a parcel of air at 25°C at an elevation of 1000 m above the mean sea level. While moving, this air encounters a mountain range 3000 m high. The air rises, moves over the mountains, and then descends to an elevation of 1500 m. If a descent of 1500 m produces saturation and condensation, and if the average pseudoadiabatic lapse rate is about half of the dry-adiabatic lapse rate, what will be the temperature of air?

Solution The air reaches saturation at an elevation of 1000 + 1500 = 2500 m. The air cools at the rate of 0.7°C/100 m. The drop in the temperature during a climb of 1500 m = 0.7 × (1500/100) = 10.5°C. The drop in the temperature during a climb of 500 m (from 2500 to 3000 m) = (0.7/2) × (500/100) = 1.75°C. The air temperature (°C) at an elevation of 3000 m = 25 − 10.5 − 1.75 = 12.75°C. The air descends to an elevation of 1500 m. The temperature rise = 0.7 × (1500/100) = 10.5°C. The final air temperature = 12.75 + 10.5 = 23.25°C.

6.4 AIR CIRCULATION

The energy from the sun is the driving force for the hydrologic cycle. Atmospheric and surface heat depends on the amount of incoming solar radiation. Incoming solar radiation, termed insolation, amounts to an average value of 1.94 cal/cm^2/min at the upper edge of the earth's atmosphere. This value of energy is termed the solar constant. The actual value of the solar constant at any point on the earth is determined by (a) the rate at which energy is radiated from the sun; (b) the distance the earth is from the sun, which ranges from 91–94 million miles (146–151 million km); (c) the inclination of the radiation surface to the incoming radiation; and (d) the amount of radiation absorbed by the atmosphere.

The average value of energy from radiation available to the earth–atmosphere system is about 0.30 cal/cm^2/min at the earth's surface. The distribution of this energy varies greatly with latitude and season. It varies locally with varying atmospheric conditions.

The total mass of the atmosphere has been computed at a value equivalent to 32.8 ft (10 m) of water evenly distributed over the surface of the earth. If all the incoming solar radiation reaching the earth–atmosphere system were used to heat the atmosphere, its temperature would rise 1.5°C (3°F) per day. This temperature rise does not happen, so the earth–atmosphere system as a whole must lose as much heat as it gains. This heat loss is brought about by long-wave (infrared) radiation from the earth. The incoming short-wave (ultraviolet) radiation is converted to heat when it encounters the earth or moisture and carbon dioxide in the atmosphere. A small amount of heat is transmitted by conduction between the earth and the atmosphere; and some heat is transmitted by convection. The heat capacity and conduction characteristics of the earth surface have an important influence on the climate.

Table 6.3, adapted from Petterssen (1964), shows the heat-transfer and heat-capacity characteristics for some common earth substances. The significance of this table is indicated by comparing the heat capacity of the various materials. Water stores far more heat than any of the other substances. Because there is a far greater volume of water on the earth's surface than the other substances, a far greater amount of heat is stored in the earth's water than elsewhere. The water on the surface of the earth contains currents, which are equivalent to stirred water in Table 6.3. The conductive capacity of stirred water is far higher than for any of the other substances. Thus, water stores more heat than the other common substances and conducts that heat faster. For this reason, those parts of the world adjacent to large bodies of water will have temperatures influenced by the heat released from the water body. This effect will usually be a higher and more even temperature. Inland land masses, away from the influence of heat released by water, contain surface substances that store little heat. These substances cannot supply heat to influence air temperatures and the result is large variations in diurnal and seasonal temperatures. The heat capacity and conductive capacity of the substances also account for the reason that land surfaces heat rapidly and water surfaces heat slowly. Land surfaces store little heat and reach their capacity rapidly, whereas a water body can store more heat and therefore takes longer to reach its capacity, as shown in Table 6.4.

TABLE 6.3 CONDUCTION PROPERTIES OF AIR AND TYPICAL SURFACE SUBSTANCES, CGS-CAL UNITS

Substance	Heat capacity per unit volume, [a]$c\rho$	Surface-stirred air ratio
Snow, fresh	0.03	0.02
Snow, aged	0.022	0.12
Sand, dry	0.3	0.11
Sand, wet	0.4	0.4
Sandy clay[b]	0.6	0.37
Soil, organic	0.57	0.4
Soil, wet	0.7	0.38
Ice	0.45	0.5
Water, still	1.00	0.39
Water, stirred	1.00	70
Air, stirred	0.0003	

[a]c = specific heat, and ρ = density.

[b]15% water.

TABLE 6.4 PENETRATION OF DIURNAL AND ANNUAL HEATING

Surface substance	Penetration (ft)	
	Diurnal	Annual
Sand, dry	0.6	11
Customary soils	1.6	30
Ice	1.8	34
Water, still	0.6	12
Water, stirred[a]	100	Variable
Air, stirred	5000	Tropopause

[a]The annual penetration depends greatly upon the stability, which usually increases downward.

An illustration of the effect of heat capacity of various substances on the surface of the earth can be seen by observing the range in temperatures at various locations on the earth given in Table 6.5. The 120°F range in temperature at Verkhoyansk is caused by the small heat storage available on the earth surface in a continental environment at that location. Reykjavik, at about the same latitude, has only an 18°F range in temperature because of the heat from the oceanic environment. In the tropics, Singapore has only a 2°F range in temperature because of the oceanic environment. Comparison of Minneapolis in a continental environment with Seattle in an ocean environment shows that Seattle has a 47° F smaller range in temperatures. It is apparent that continental land masses provide an environment for greater temperature ranges than does an oceanic environment. Also, the oceanic environment is warmer than the continental environment.

The foregoing discussion shows that energy from the sun provides heat to the atmosphere and to the surface of the earth. As the air is heated, it becomes less dense and starts to rise. As the warm air rises, it is replaced by colder air moving laterally to take the place of the warm air. This process is called thermal circulation, a comprehensive discussion of which is given by Rossby (1941). A classical example of thermal circulation exists at the equator, where incoming solar radiation is a maximum. Heated air rises at the equator and is replaced at the earth's surface by air converging toward the equator from the poles. This phenomenon gives rise to the term equatorial convergence, which is characterized by cumulus clouds forming as

TABLE 6.5. ANNUAL VARIATIONS IN TEMPERATURES AT SELECTED LOCATIONS

Location	Latitude	Temperature range
Verkhoyansk, Siberia	68 deg N	−50°C (−60°F) to 15°C (60°F)
Reykjavik, Iceland	63 deg N	0°C (32°F) to 11°C (52°F)
Singapore	1 deg N	25°C (78°F) to 26°C (80°F)
Minneapolis, Minnesota	43 deg N	−11°C (12°F) to 23°C (73°F)
Seattle, Washington	47 deg N	4°C (40°F) to 18°C (64°F)

the moisture-laden air rises to cooler temperatures. The presence of these converging cumulus clouds presents a striking view from an aircraft flying high over the equator. Their presence is also obvious in satellite photographs.

By assuming the earth to be stationary, Figure 6.2 shows the heated air rising at the equator, and this heated air is replaced by colder air moving laterally from the north pole. This movement sets up circulation in the manner shown. Because the earth is a rotating body, the effect of the coriolis force on the atmosphere must be considered. As the circulating air mass moves northward, the rotating earth increases its velocity, and the result is shown in Figure 6.3.

This generalized illustration of air circulation shows the surface winds and the upper air circulation. A detailed explanation of this figure is beyond the scope of this text. It is, however, important to visualize the complex circulation of the atmosphere in the context of later explanations. These explanations as well as others must necessarily be general for the purpose of this text.

The combination of air circulation above the earth and the rotation of the earth results in a general west-to-east movement of the atmosphere at upper levels. A feature of the upper air circulation is the jet stream, which moves in a west-to-east direction at velocities that often exceed 100 mph (160 km/h). The jet stream is believed to be the steering mechanism for weather pressure systems, although its true relation to such systems is not well understood.

EXAMPLE 6.10 Colorado Springs, Colorado, lies at an elevation of 6000 ft mean sea level (msl) at the foot of the Rocky Mountain Front Range. On a particular day, the temperature at Colorado Springs is 55°F and the dew point is 42°F. A deep low-pressure system moves into New Mexico and causes an east wind to blow across western Kansas and eastern Colorado. In western Kansas at an elevation 3000 ft, the temperature is 50°F. What would be the expected weather resulting from this condition at Colorado Springs?

Solution The east wind will blow the 50°F air up-slope to an elevation that is 3000 ft higher than in Kansas. The lapse rate of 3.5°F/1000 ft will cool the air 10.5°F, which will

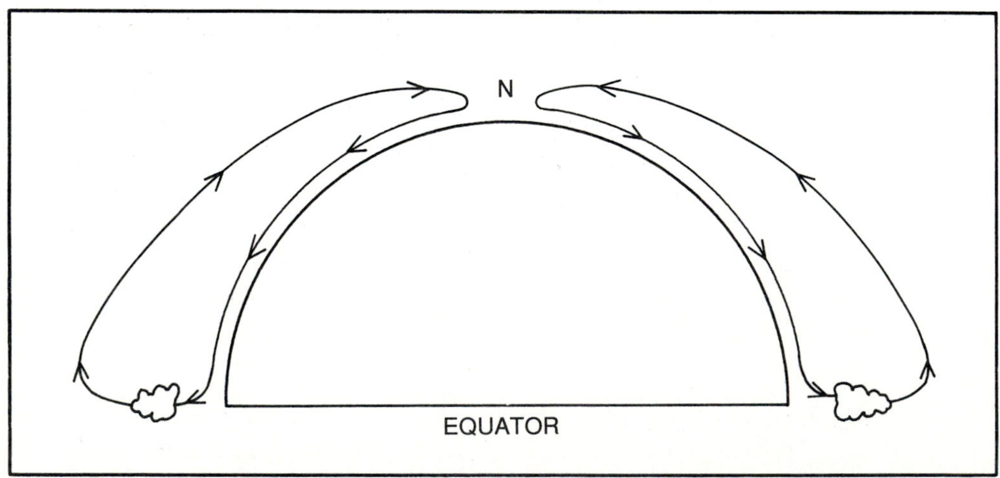

Figure 6.2 Theoretical air circulation on a stationary earth.

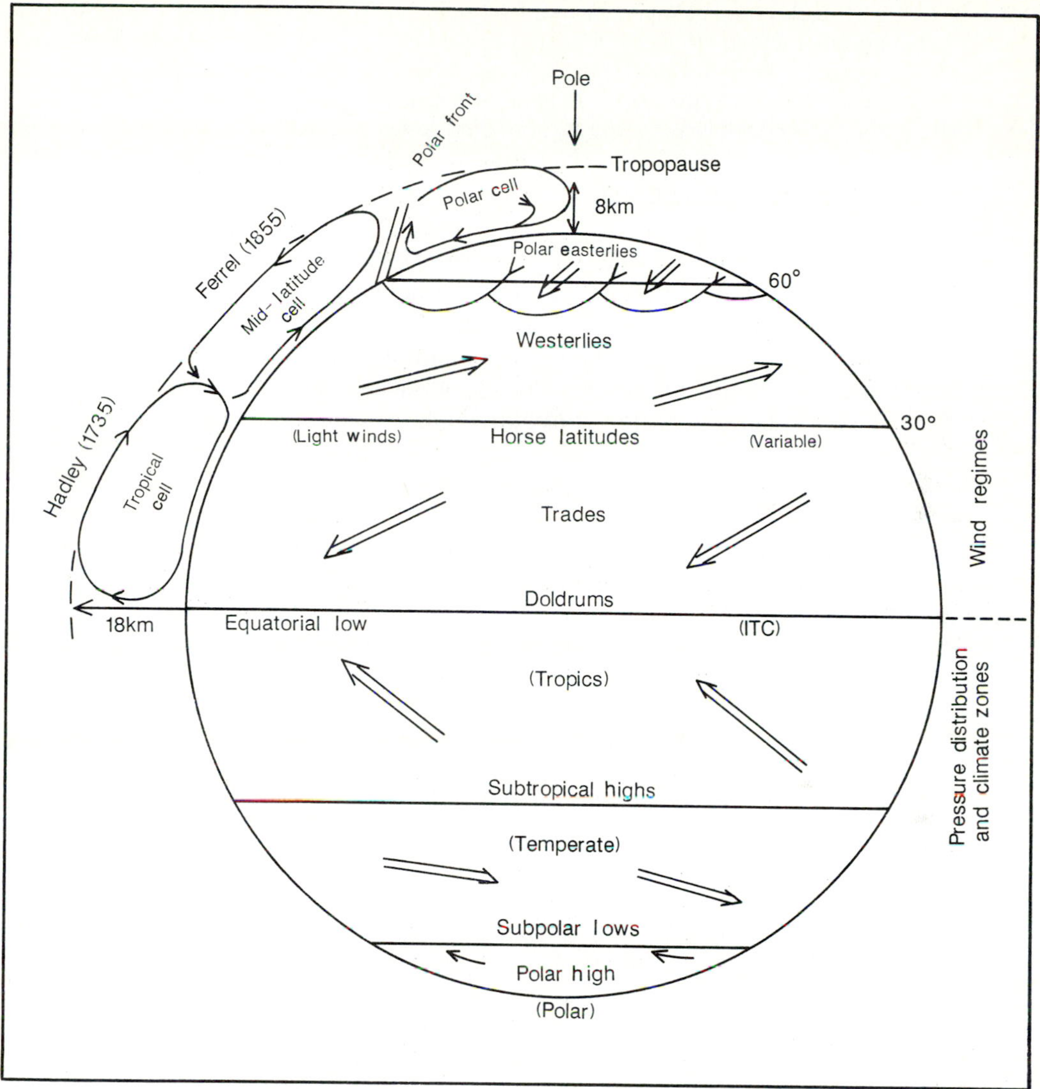

Figure 6.3 Typical air circulation in the Northern Hemisphere.

cool the air at Colorado Springs below the dew point of 42°F and rain will occur. At higher elevations, the temperature will be lowered even more, so that it is possible that snowfall would occur as well.

EXAMPLE 6.11 Much of the Missouri River basin in eastern South Dakota at an elevation of 1500 ft msl is covered with snow at a temperature of 29°F. A large, strong high-pressure system builds over western Kansas, causing west winds to blow over South Dakota and Wyoming. The temperature at a 10,000-ft station in Wyoming is 17°F. What hydrologic effect would be expected at the Missouri River in South Dakota?

Solution The elevation difference is 10,000 − 1500 = 8500 ft. 8500 ft × 3.5°F lapse rate/1,000 is 29.75°F, which represents the increase in air temperature as it moves from a 10,000-ft elevation in Wyoming to a 1500-ft elevation in South Dakota. Therefore, the temperature will rise to nearly 59°F at the Missouri River because of this Chinook effect and rapid snowmelt would occur. This rapid snowmelt will increase the runoff into the Missouri River and affect the operation of the mainstream dams located there.

6.5 WEATHER SYSTEMS AND FRONTS

The atmosphere is not distributed uniformly over the earth. The weight of the atmosphere is measured using a barometer. The normal atmosphere exerts a pressure of 1000 millibars (29.92 in. Hg). If the barometric pressure is higher at a point on the earth, it means that a greater thickness of atmosphere has accumulated at that point. If the barometric pressure is lower, it means that a thinner than normal amount of atmosphere is located above that site.

6.5.1 Cyclone, or Low-Pressure System

The circulation of the atmosphere results in air masses that rotate. Those that rotate counterclockwise in the northern hemisphere are similar in form to the vortex of water draining through a hole from a container. The counterclockwise rotation results in lower barometric pressure in the center of the rotating air mass. Such a circulation in the atmosphere is called a cyclone or low (-pressure system). Simply put, a cyclone is a large area of circling winds moving about an area of relatively low atmospheric pressure. These winds move counterclockwise around the center north of the equator and clockwise in the southern hemisphere. The term low refers to the lower than normal barometric pressure in the central part of the cyclone. Figure 6.4 illustrates a synoptic map presentation of a low pressure system.

Air circulation around a low-pressure air mass is counterclockwise in the northern hemisphere (clockwise in the southern hemisphere). The velocity of the air circulation is determined by the pressure gradient in the low. The greater and

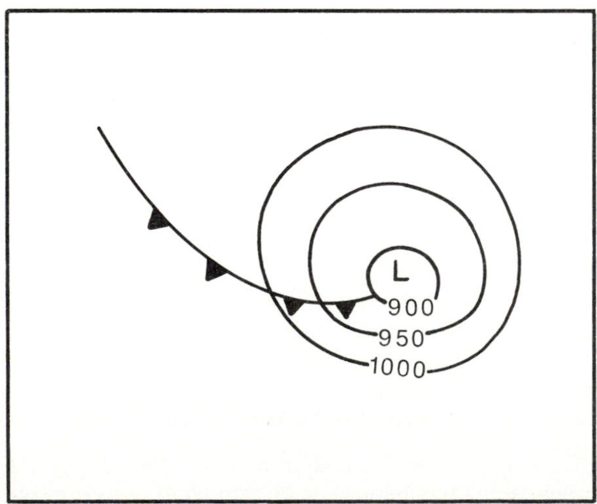

Figure 6.4 Cyclone, or low-pressure cell.

steeper the gradient, the higher the wind velocity. While the circulation is counterclockwise, the wind also converges to the center of the low-pressure system. Upon reaching the center of the low, the air is driven vertically to colder altitudes, where condensation can occur. Low-pressure systems are storm systems. Lows draw moisture to them and concentrate that moisture by elevating it in their circulation systems.

Low-pressure systems occur in the tropics, where they usually move from east to west. Those lows occurring in the temperate zone, extratropical, usually move west to east. A typical low is large in area and can be up to 1500 miles (2500 km) in diameter, although it can also be smaller. Wind strength in the low depends on the pressure gradient. The rate of movement of the low-pressure system is usually about 30 mph (50 km/h), but can be less or greater. A tropical cyclone may or may not be a hurricane, depending on its intensity. If below hurricane force but still a storm of above 39 mph, it is properly termed a "tropical storm." Below that, it is called a tropical depression.

6.5.2 Anticyclone, or High-Pressure System

An air mass that rotates clockwise results in an accumulation of air to a greater thickness than normal and exhibits higher than normal barometric pressure in its center. This air mass is called an anticyclone, or high-pressure system. The high refers to the higher than normal barometric pressure, as shown in Figure 6.5.

A higher-pressure system, or anticyclone, is characterized by clockwise circulation. There are two types of high-pressure systems: (a) a cold-core high and (b) a warm-core high. The cold-core high is the common high that advances, within the general west-to-east movement of the atmosphere, southward from the arctic behind a low in the northern hemisphere. Circulation around this high causes air to diverge around it near the surface of the earth. The intensity of this high decreases with altitude. In upper levels, the cold-core high is delineated by a pressure ridge in an

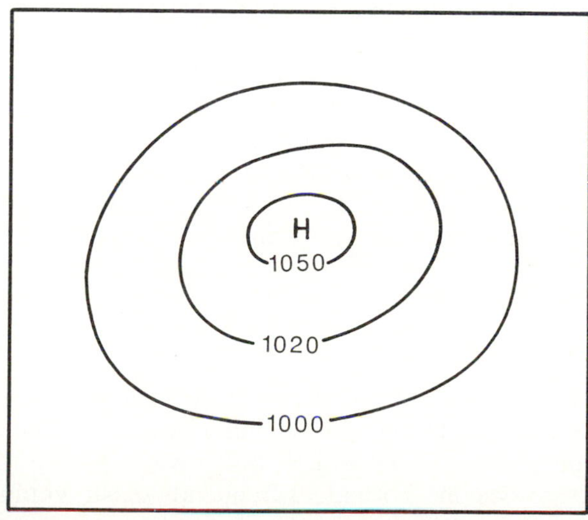

Figure 6.5 Anticyclone, or high-pressure cell.

easterly flow. The high pressure in a cold-core high is caused by the denser cold air rather than a thicker air mass. The warm-core high is not as common as the cold-core high. A characteristic of the warm-core high is that it extends to a greater altitude, where it blocks the normal west-to-east flow of air. This is the "blocking" high that blocks or slows down the normal west-to-east flow of the atmosphere and its pressure systems. A typical example is the high pressure over Bermuda that dominates the summer weather over the southern and eastern United States.

6.5.3 Hurricane, or Typhoon

A hurricane, or typhoon, is a large tropical cyclone, or low, that is typically small in area, up to 300 miles (500 km) in diameter. Hurricanes originate in the tropics and are characterized by counterclockwise (in the northern hemisphere), circular, slightly centripetal winds rotating about a calm area of low atmospheric pressure called the "eye," usually about 10 to 20 miles in diameter. These storms are characterized by high winds (at least 75 miles per hour) caused by a steep pressure gradient. Hurricanes, or typhoons, are accompanied by high, intense rainfall amounts, possibly tornadoes, lightning, and thunder. Because these systems are fed by the energy from a warm-water environment, they dissipate when they move over land or cold water.

The hurricanes that occur along the coast of the Gulf of Mexico usually originate in the southern section of the North Atlantic Ocean in the vicinity of the Cape Verde Islands, in the western Gulf of Mexico, or in the Caribbean Sea. Other locations where hurricanes occur are the western coast of Mexico; Northern Pacific Ocean areas such as the China Sea, Japan, and the Philippines; in the Bay of Bengal; and in the Arabian Sea. In the Orient, Hurricanes are known as "typhoons" or "baguios." "Willie-Willie," as they are called, occur east of Australia in the South Pacific and the islands of the Coral Sea. The other southern hemisphere location where hurricanes occur is the South Indian Ocean between Madagascar and northwest Australia, where they are referred to by their generic name of "cycles."

Two theories concerning formation of hurricanes are well accepted. First, they originate as a super thunderstorm with a tremendous updraft of superheated tropical air, subsequently being continuously fed by energy from warm tropical seas, and are given their circular motion by the prevailing winds controlled by the earth's rotation. The second theory is that the hurricanes are formed by the convergence of the seasonally energy-rich prevailing winds that converge, combine their forces, and then draw their strength from the warm seas. Regardless of their parentage, they are essentially of the same structure but differ widely in size, intensity, speed and direction of travel, life span, tornado occurrences, precipitation, and other characteristics.

6.5.4 Tornado

Tornadoes, or "twisters," are localized phenomena of centripetal winds of extremely high and destructive velocity (up to perhaps 500 miles per hour). These are very small low-pressure systems. These systems often occur within the southeast quadrant of an active low-pressure system. Tornadoes frequently occur within hurricanes as they do in nontropical cyclones. The familiar "dust devils" of the arid southwest are small, benign tornadoes of low velocity. Strong updraft and downdraft winds are common and result in shear forces that can be destructive to aircrafts. The diameter of a tornado is usually only a few hundred feet. It moves at about 30 mph (50 km/h), and its duration is short, but some have been known to be on the ground for several hours.

6.5.5 Cold Front

A cold front is the boundary between advancing cold air and retreating warm air. As the front advances, the dense cold air wedges itself under the warm air and lifts the warm air as the cold air advances, as shown in Figure 6.6. The elevated warm moist air is cooled and condensation results. Because a low-pressure system converges air to itself, it is always associated with a cold front. As a cold front passes, the wind shifts counterclockwise in the northern hemisphere and pressure rises because of the cold heavier air. The passage of a cold front is accompanied by a decrease in temperature and a decrease in relative humidity.

6.5.6 Warm Front

A warm front is the boundary of advancing warm air with retreating cold air. Warm air moves faster than the cold air. Being less dense, the warm air overrides the overtaken cold air. This overriding causes the warm air to be elevated, cooled, and condensation occurs, as shown in Figure 6.7. There is a clockwise wind shift and a

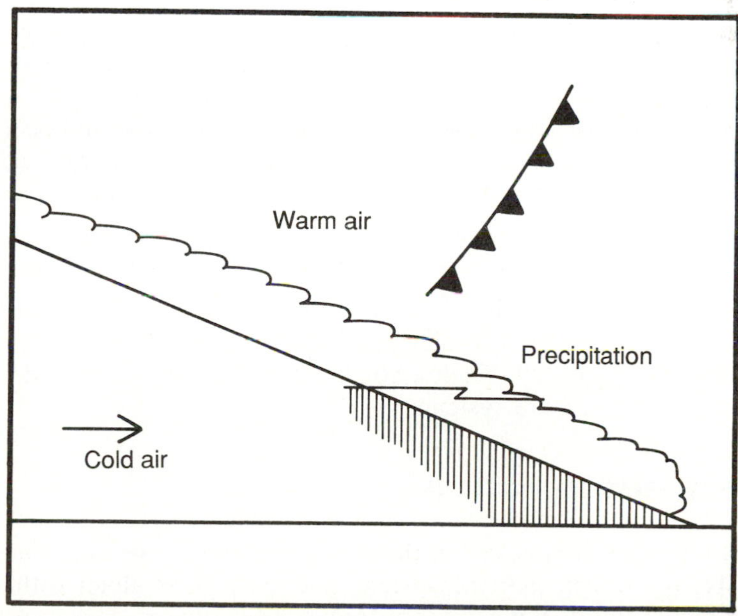

Figure 6.6 A cold front.

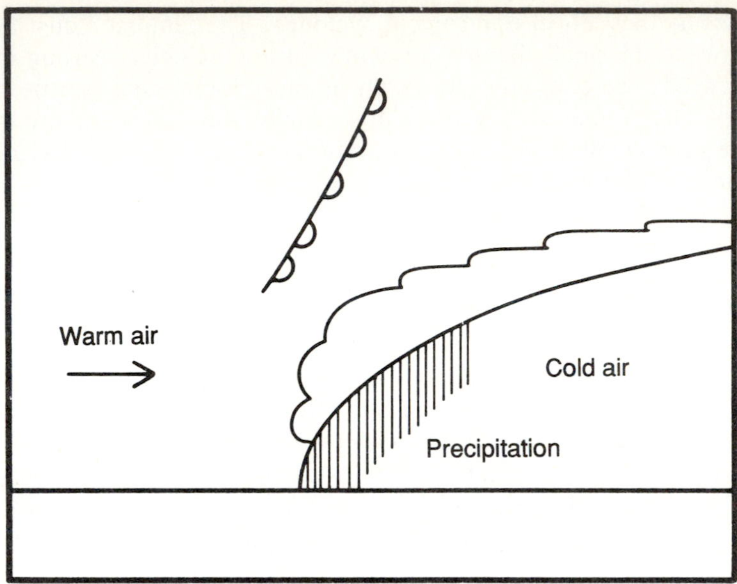

Figure 6.7 A warm front.

decrease in barometric pressure across the front. A warm front is accompanied by a temperature rise across the front. After the frontal passage, temperatures might be nearly steady. Steady, light precipitation is characteristic of warm fronts.

6.5.7 Stationary Front

A stationary front is either a cold front or a warm front that has ceased to move, as shown in Figure 6.8. This condition occurs when a blocking high stops the movement of the atmosphere. Such a condition can exist for a short time or for several weeks. The temperature gradient still exists across the stationary front and can continue unless enough mixing of the air occurs to break down the front. Often, the stationary front resumes movement and the warm or cold front resumes its nature.

6.5.8 Occluded Front

Occasionally, an advancing cold front will overtake another cold front. The warm air between the two fronts is elevated and this condition results in an occluded front, as shown in Figure 6.9. Such a condition is favorable for the development of cyclones, or low-pressure systems.

6.6 WEATHER-PRESSURE PATTERNS

The atmosphere contains all of the weather systems together. The dynamics of the atmosphere is such that the systems move together along with the atmosphere. Pressure systems do not necessarily move at the same rate, but, in general, do move in a west-to-east direction in the northern hemisphere.

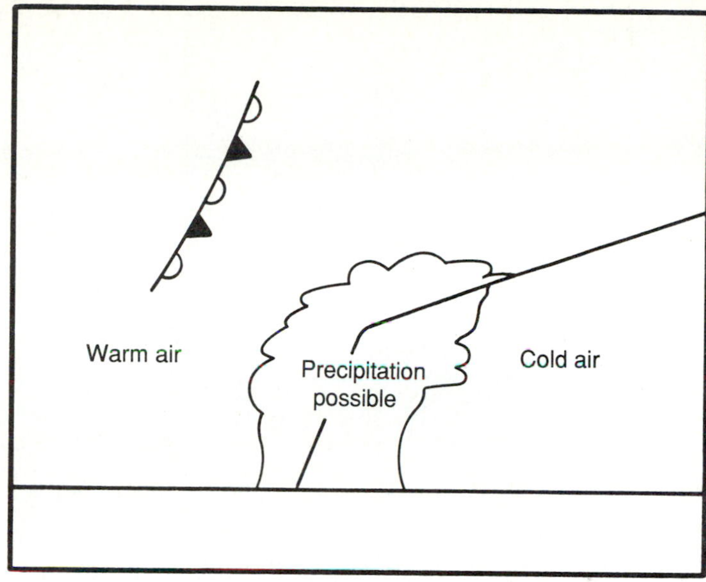

Figure 6.8 A stationary front.

As noted earlier, the heat capacity of continental land masses is small and that the range in temperatures is high. The result of this loss of heat is that cold, high-pressure air builds up during the winter months. As the highs build, some of the cold air breaks off and moves south, invading the warmer air at lower latitudes. An example of this can be seen on the synoptic map in Figure 6.10. This map of the world for January shows highs over Canada, Siberia, and North Africa. This is a typical condition during the winter months. By comparing this map with the average synoptic map for July in Figure 6.11, it can be seen that the indicated pressure systems are reversed. Where highs formerly existed in January, lows exist in July. By the same token, the lows that existed in the warm Gulf of Alaska and over Iceland in January become highs in July.

The weather patterns originate and move from these average synoptic relations. It is this knowledge of the location and movement of the pressure systems that

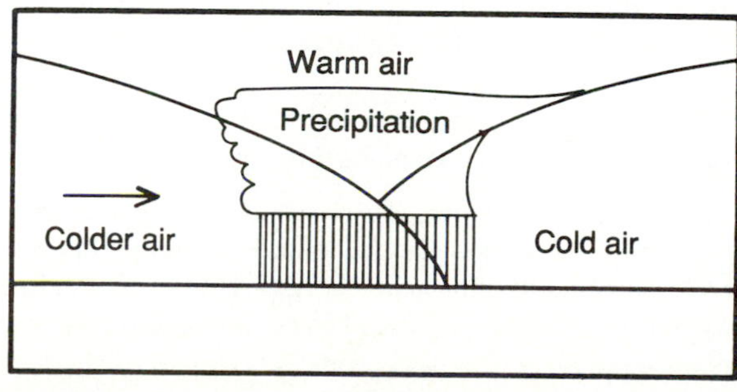

Figure 6.9 An occluded front.

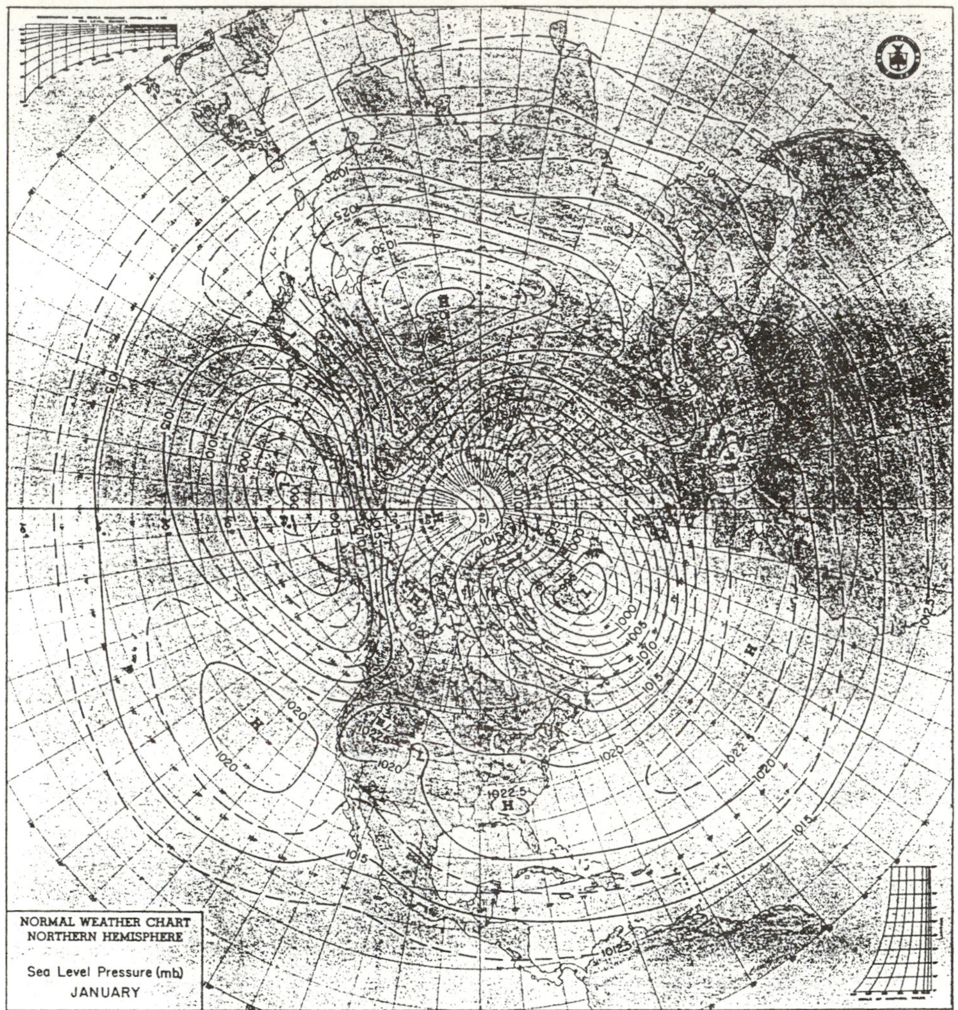

Figure 6.10 January mean surface pressure.

provide the key to understanding precipitation characteristics of an area. The path that each system takes might be different from time to time. The average path is reasonably constant. The weather that occurs at any point on the earth depends on its relation to the location and path of the weather systems.

EXAMPLE 6.12 Relatively moist air is moving over the high plains during the summer, resulting in some atmospheric instability. Explain this condition.

Solution This is a condition that results in the formation of strong convective storms. Increased heating as the day develops is often sufficient to cause thermal updrafts that result in condensation and release of additional heat energy. This cycle continues to form convective storms that can result in intense rainfall over small areas. Such storms

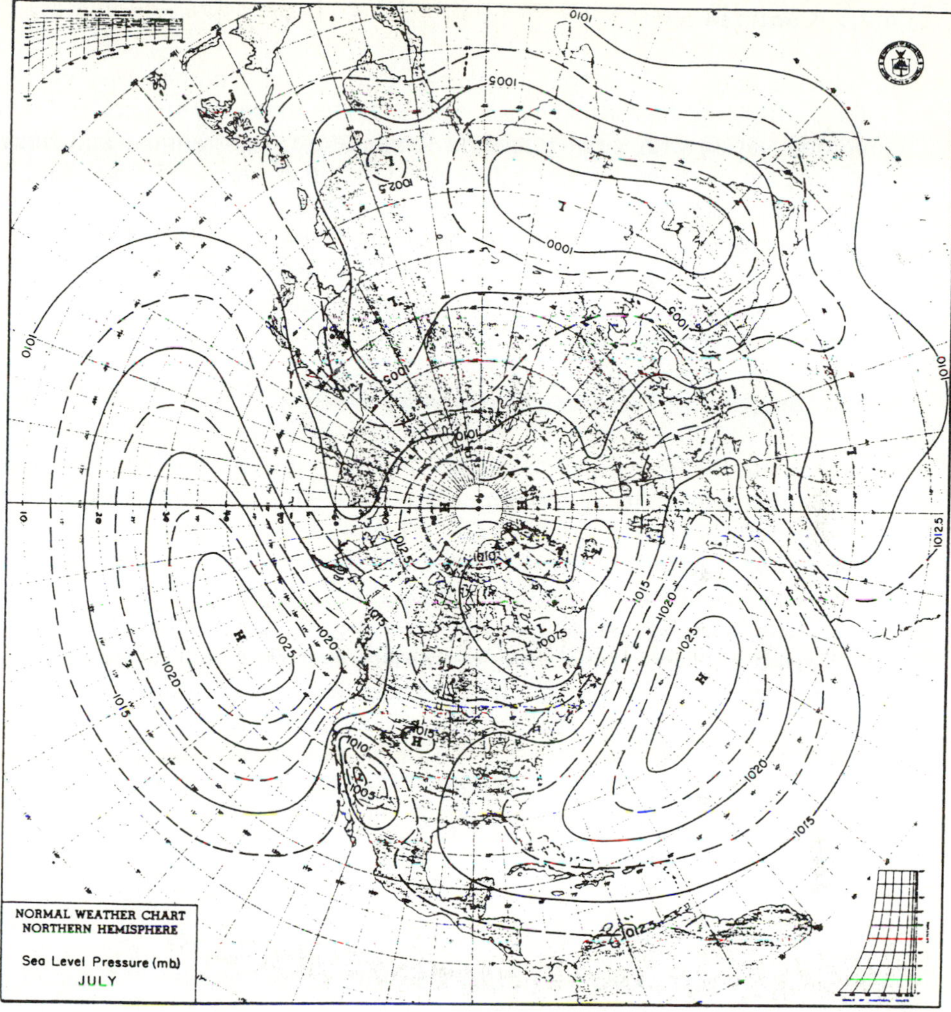

Figure 6.11 July mean surface pressure.

cause large floods on small basins, and at their upper limit can cause large floods on relatively large basins.

6.7 CLIMATIC CONTROLS

The climate at any point on earth depends upon a multitude of factors known as climatic controls. Climatic controls are not intended to be tools for short-term weather forecasting, but rather guidelines for estimating the type and nature of climate to be expected. This information is useful. The climatic controls are (1) continentality, (2) direction of prevailing storm movements, (3) topography, (4) altitude, (5) orientation of topography, and (6) location with respect to the jet stream.

6.7.1 Continentality

Continentality is the effect of the heat capacity of continental masses on climate. As noted earlier, the heat capacity of typical land surface substances is low and this condition varies with large ranges in temperature, both annual and diurnal. The effect of continentality is further dependent on the distance the land area is from the ocean. Heat from the oceanic environment will obviously affect adjacent land masses, as shown in Table 6.3. There is no simple relation between the distance from the ocean and continentality, but it is clear that temperature ranges are greater over continental masses than elsewhere, and the farther from the ocean, the greater will be the temperature range.

6.7.2 Direction of Prevailing Storm Movements

This refers to the typical path of the low-pressure systems moving from the ocean to any location of interest. A low-pressure system will pick up moisture over the ocean and transport that moisture over land. If the ocean water over which the low moves is warm, high amounts of moisture will be picked up in the system. An example is a low moving out of the Gulf of Alaska on to the northwest coast of the United States. If the ocean water over which the low moves is cold, little moisture will be picked up and transported. An example is a low moving eastward from the Pacific Ocean onto the coast of northern Chile. In this instance, moisture is actually lost over the cold oceanic water and the result is that the Atacama Desert in northern Chile is one of the driest spots on the earth. Any factor, such as topography, that influences the moisture-carrying capacity of a low that lies in the path of the low from the ocean to the continental location of interest is also a factor related to storm movement. These factors are other climatic controls.

6.7.3 Topography

Topography refers to the nature and elevation of land masses. Mountains, valleys, plateaus, and all variations of these features are part of the topography. A low-pressure system laden with moisture that is forced to move from sea level up over 10,000-ft (3280-m) mountains will encounter cooler temperatures at this altitude, which will cause the low to lose much of its moisture. This condition deprives other locations in the path of the low of precipitation if there is no other source of moisture to the low. Other characteristics of topography of a given area might also affect the transport of moisture by low-pressure systems.

6.7.4 Orientation of Topography

This factor has a profound effect on the movement of moisture-laden low-pressure systems across topography. An example of the effect of such conditions is the movement of moisture-laden low-pressure systems from the Pacific eastward across the United States. This low must cross several north–south oriented ranges of mountains. First is the Coastal range; next is the Cascade range, including the southern extension, the Sierra Nevada range; next is the Wasatch range, Rocky Mountain range, and finally the Front range. Each of these ranges of mountains reduces the moisture in the low such that if there has been no replacement, the plains east of the Front range would receive little if any moisture from this low. If these mountain ranges were oriented east and west, the effect would cause less moisture to be lost during the transit of the low because the low would not be subject to elevation changes of such magnitude.

6.7.5 Altitude

The amount and type of precipitation as well as temperature is related to altitude. It is well known that temperature decreases with an increase in altitude. This relation also affects the amount of moisture available for condensation. A low-pressure system can hold more moisture at sea level for a given temperature. For this reason, more moisture is available for precipitation under favorable circumstances. At high elevations where temperatures are lower, less moisture is available for precipitation. It can be said in general that the higher the altitude, the less the precipitation. Exceptions to this statement exist, but those exceptions have special conditions associated with them.

6.7.6 Location with Respect to the Jet Stream

The upper air jet stream is believed to be the steering mechanism for low-pressure systems. It is true that the general direction of weather systems in the temperate Northern Hemisphere is from west to east, however, variations in this direction occur. The path of the jet stream is generally farther north in the summer and moves south in the winter. Frequent abrupt changes in the general west–east direction occur such that the jet stream will make a southward dip and loop back north before returning to an eastward direction. These loops are usually accompanied by low-pressure storm systems. The climate of an area is affected by its location with respect to the jet stream and its associated effect.

6.8 MECHANISMS OF PRECIPITATION

Precipitation is defined as liquid or solid water that reaches the surface of the earth. This definition rules out virga, the evaporation of condensed water particles in the atmosphere before they reach the ground. Precipitation includes rain, snow, hail, sleet, dew, fog, drizzle, etc., and their variations. Condensation of water vapor in the atmosphere occurs when the relative humidity reaches 100% and a nucleus is present upon which precipitation may condense. The condensation nuclei might be foreign particles, or, under supercooled conditions, ice crystals. Foreign nuclei might consist of colloidal dust particles, carbon dioxide, salt, silver iodide, quartz, or any particle small enough to be present in the atmosphere. It has been shown that under pure conditions, water vapor can be cooled below its dew point without condensation occurring.

6.8.1 Bergeron's Theory of Precipitation

Bergeron (1933) postulated that water droplets can exist in clouds at subfreezing temperatures down to −40°C (−42°F). In the precipitation environment, solid freezing nuclei composed of clay minerals, organic and common salt, and other nuclei exist as freezing ice nuclei elements. When ice elements form in an undercooled cloud, an imbalance is created because the equilibrium vapor pressure over the water drops is higher than over the ice drops. This imbalance creates a vapor-pressure gradient from the water drops to the ice crystals and water tends to evaporate on one hand, and condense on the ice drops on the other hand. As a result, some cloud elements grow and other elements shrink. Uneven growth of drops causes some drops to fall faster than other drops, and this condition favors further growth through collision and coalescence while the drops are falling.

6.8.2 Coalescence Theory of Precipitation

This theory assumes that droplets form on nuclei when the dew point is reached. The weight of condensation causes the droplets to fall. The droplets collide with other droplets and they coalesce. The growth of the droplets increases their weight and the rate of fall is increased, whereby more collisions with other droplets occur and more growth of the droplets takes place. As the droplets grow, their fall velocity increases until the droplets reach a diameter of about 7 mm (0.28 in.). At this size, the fall velocity, about 10 m/s (30 fps), causes the droplets to flatten and break up into smaller droplets, whereupon the process of growth is repeated. In convective updrafts, droplets can fall and grow, break up, and rise several times.

6.8.3 Mechanisms of Cooling

Precipitation occurs as a result of cooling of water-vapor-laden air to its dew point or below. Condensation of atmospheric vapor can occur from (a) dynamic, or adiabatic, cooling, (b) moisture in two air masses of different temperatures, (c) contact cooling, and (d) radiational cooling.

Dynamic, or adiabatic, cooling occurs when air circulation carries air to higher elevations. The air might be elevated by thermal updrafts or by circulating into a low-pressure system. Any dynamics of the atmosphere that causes air to rise will encounter decreasing temperatures in accordance with the lapse rate and adiabatic cooling will occur.

Frontal conditions provide an atmospheric environment favorable for mixing of two air masses of different temperatures. An advancing cold front results in mixing conditions along the contact of the cold and warm air. Because the cold front is usually associated with a low-pressure system, there is a wind shift across the front and mixing of air occurs. Similar conditions exist along and across a warm front.

Contact cooling occurs when warm moist air is blown across a cold surface. An example of this type of cooling occurs in the Great Lakes area. During the early winter months before Lake Erie has lost the heat from its summer accumulation, northwest winds blow warm moist air across the lake to the New York and Pennsylvania shores. These shore land areas are cold and cool the warm moist air and precipitation results. These areas are noted for high amounts of precipitation from this condition.

Radiational cooling occurs when clear nights provide a favorable environment for the radiation of heat (infrared radiation) from the earth. During the day, insolation warms the atmosphere and the earth. The incoming ultraviolet rays are converted to heat in the atmosphere and on the earth. Infrared rays do not pass through clouds but are reradiated back to the earth when the sky is overcast. During clear nights, infrared rays are readily radiated into the atmosphere, and as a result, the temperature is lowered as this heat is low. This condition is a common cause of dew or frost forming. During the daytime, the relative humidity might increase as temperature increases. As the temperature falls, the relative humidity increases further until the dew point is reached.

6.9 TYPES OF PRECIPITATION

Precipitation is classified by the type of the mechanism that produces it. Each type of precipitation has characteristics that have important hydrological consequences. Three types of precipitation are distinguished: (a) convective, (b) cyclonic, and (c) orographic. It is important to understand the nature and characteristics of these types of precipitation.

6.9.1 Convective Precipitation

Convection is the process of heat transfer from one place to another by the actual movement of hot fluids (gases or liquids). This type of precipitation is caused by thermal uplifting of moisture-laden air. Convective storms are commonly known as thunderstorms. The mechanism that forms a convective storm begins with the sun heating moist air. A simple example of thermal heating is the sun heating a bare tilled field in the summertime. The absorption of the sun's rays is converted to infrared (heat) rays that heat the air and cause the air to rise. The elevated warm air is replaced by air moving laterally along the earth's surface, which, in turn, is heated and then rises and expands. This process is repeated over and over until the heating causes the air to reach an elevation where the temperature drops below the dew point and the condensation results. When condensation occurs, the latent heat released by condensation (539 cal/g) causes further heating to the upper air and results in a greater thermal uplift. Because the latent heat released per gram of water condensed is high, condensation adds high energy to the system. As moist air moves into the convective system, additional energy is added by the heat of condensation and the whole system continues to build and feed on itself. This buildup of energy and condensation will continue as long as a moisture supply is available or until the moisture load accumulated in the convective system becomes so great that the system releases moisture faster than it receives moisture. Convective storms can build up to tremendous heights. It is common for such storms to build to 30,000 ft (10,000 m); and reports of convective storm tops exceeding 75,000 ft (25,000 m) are known.

The tremendous energy within a convective storm creates updrafts and local winds exceeding 100 mph (166 km/h). These winds and updrafts and downdrafts have been known to tear aircraft apart. The modern term for some of these high winds is wind shear. A strong updraft movement carries raindrops to high elevations, where they might freeze and fall as hail. Hail might be caught again in an updraft and rise and fall many times, each time accumulating moisture that freezes and adds to the size of the hail stone. It is common to see hail stones that, when broken open, show many layers of accumulation. It is important to visualize the tremendous energy created within a convective storm, because the precipitation that results from a convective storm is compatible with this energy.

The energy within a convective storm is so great that tremendous amounts of water are stored within such a storm. When this water is released as rain, the rainfall intensity is high. It is possible for such rain to fall at rates of 15–20 in./h (38–51 cm/h) for short periods of time. Rainfall amounts of 6–12 in. (15–30 cm) falling in an hour from a convective storm are not uncommon. The maximum velocity that falling rain can attain is about 18 mph (30 km/h). When large drops, falling at this rate, break up into smaller drops, electrical energy is released as lightning.

The size of a convective storm is usually quite small. Such storms can be as large as 1000 mi^2 (1667 km^2), but usually range from 3–4 mi^2 (5–8 km^2) up to 600 mi^2 (1000 km^2). The small size of convective storms relates significantly to the flood-producing storms for small drainage basins. Convective storms can cover small drainage basins, but cannot completely cover large drainage basins.

The duration of a convective storm usually averages about 6 hours. The rate of movement of these storms is about equal to the average rate of movement of the atmosphere, which is about 30 mph (50 km/h). To summarize the characteristics of a convective storm, it produces intense and heavy rainfall, covers a small area, has a short duration, and moves at about 30 mph (50 km/h).

Convective storms occur most commonly in the tropics and rarely in the arctic regions. Approximately 200 convective storms occur yearly in any given area of the tropics, about 60 per year occur in the Gulf States, 15–20 per year along the Canadian border, and about 1 per year occurs in the arctic. The Pacific coast will have about 2 per year.

6.9.2 Cyclonic Precipitation

Cyclonic precipitation occurs in a low-pressure system and its associated cold front. This type of precipitation occurs as a consequence of warm moist air being drawn into the low-pressure system, where it is elevated aloft and circulated over cold air behind the front. Adiabatic cooling causes condensation along with mixing of the warm moist air with cold air. The amount of moisture drawn into the low-pressure system depends on the intensity of the low and the available source of moisture. A typical cyclonic precipitation system might be a low moving eastward across the plains and drawing moisture up from the Gulf of Mexico.

Because a low-pressure system is large and the circulation of the adiabatically cooled and mixed air is over a large area, the precipitation is distributed over a large area. A cyclonic storm can have a diameter of up to 1500 miles (2500 km), but it is usually smaller. Precipitation will occur from most of the low and along the front for such a distance that has a moisture supply from the low. The intensity of rainfall from such a condition is light to medium. Intense, convective storms can and do occur in the southeast quadrant of the low-pressure area. Such storms are called imbedded thunderstorms by forecasters. Cyclonic systems move with the atmosphere at about 30 mph (50 km/h). Because such storms are large in area, the duration of rainfall can be long for any location over which the storm moves. This type of storm accounts for most precipitation. In summary, cyclonic storms are large, move at a rate of about 30 mph (50 km/h), have low- to medium-intensity rainfall, and are of long duration.

6.9.3 Orographic Precipitation

Orographic precipitation occurs as a result of mechanical lifting of warm moist air. Movement of warm moist air onto a mountainous land surface, such as the northwest coast of the United States, causes the air to be mechanically lifted up the mountain slope, where it is adiabatically cooled. Any topographic condition where moist air is driven upslope will result in adiabatic cooling and is orographic precipitation. This upslope condition is common when winds blow moisture-laden air from east to west toward the Rocky Mountains upslope from elevations around 1000 ft (300 m) elevation up to elevations higher than 10,000 ft (3000 m). Another example of orographic precipitation is found in the mountain ranges extending along the Pacific coast states. A narrow band of high annual rainfall exists parallel to the ranges on their seaward side. On the lee side of the coastal mountains, annual precipitation is much lower. This is vividly seen on Washington's Olympia Peninsula. Other examples are common throughout the world.

The characteristic of orographic precipitation is that the precipitation must be stationary because it occurs at an elevation along the topography where the dew point occurs. The system driving the moist air on the topographic feature is usually a low-pressure system. For this reason, the duration of such precipitation can be long. Such precipitation is of medium to high intensity. The size of orographic precipitation might not cover a particularly large area. Its dimensional length can be great, but, generally, the width of such precipitation is fairly low. In summary, orographic precipitation is characterized by medium- to high-intensity rainfall, is stationary, and medium to long duration, and its dimensions are usually long and relatively narrow.

6.10 EXTREMES OF PRECIPITATION

In the United States, the wettest place is Mt. Waialeale on the Hawaiian island, Kauai, with a long-term average annual rainfall of 460 in. (1168 cm). Greenland Ranch in Death Valley, California, is considered to be the driest place, with the average annual rainfall of 1.78 in. (4.5 cm). In the world, the wettest place is still Mt. Waialeale, although Cherrapunji, India, with a little less annual average rainfall, holds the record for periods from 15 days to 2 years. At this place, during the month of July 1861, more than 366 in. (930 cm) of rainfall was recorded, which is about 80%

TABLE 6.6 WORLD'S GREATEST OBSERVED POINT RAINFALLS

Duration	Depth in.	Depth mm	Location	Date
1 min	1.50	38	Barot, Guadeloupe	Nov. 26, 1970
8 min	4.96	126	Fussen, Bavaria	May 25, 1920
15 min	7.80	198	Plumb Point, Jamaica	May 12, 1916
20 min	8.10	206	Curtea-de-Arges, Roumainia	July 7, 1889
42 min	12.00	305	Holt, Mo.	June 22, 1947
2 h, 10 min	19.00	483	Rockport, W. Va.	July 18, 1889
2 h, 45 min	22.00	559	D'Hanis, Tex. (17 mi NNW)	May 31, 1935
4 h, 30 min	30.8+	782+	Smethport, Pa.	July 18, 1942
9 h	42.79	1,987	Belouve, Reunion	Feb. 28, 1964
12 h	52.76	1,340	Belouve, Reunion	Feb. 28–29, 1964
18 h, 30 min	66.49	1,689	Belouve, Reunion	Feb. 28–29, 1964
24 h	73.62	1,870	Cilaos, Reunion	Mar. 15–16, 1952
2 days	98.42	2,500	Cilaos, Reunion	Mar. 15–17, 1952
3 days	127.56	3,240	Cilaos, Reunion	Mar. 15–18, 1952
4 days	146.50	3,721	Cherrapunji, India	Sept. 12–15, 1974
5 days	151.73	3,854	Cilaos, Reunion	Mar. 13–18, 1952
6 days	159.65	4,055	Cilaos, Reunion	Mar. 13–19, 1952
7 days	161.81	4,110	Cilaos, Reunion	Mar. 12–19, 1952
8 days	162.59	4,130	Cilaos, Reunion	Mar. 11–19, 1952
15 days	188.88	4,798	Cherrapunji, India	June 24–July 8, 1931
31 days	366.14	9,300	Cherrapunji, India	July 1861
2 months	502.63	12,767	Cherrapunji, India	June–July 1861
3 months	644.44	16,369	Cherrapunji, India	May–July 1861
4 months	737.70	18,738	Cherrapunji, India	Apr.–July 1861
5 months	803.62	20,412	Cherrapunji, India	Apr.–Aug. 1861
6 months	884.03	22,454	Cherrapunji, India	Apr.–Sept. 1861
11 months	905.12	22,990	Cherrapunji, India	Jan.–Nov. 1861
1 year	1,041.78	26,461	Cherrapunji, India	Aug. 1860–July 1861
2 years	1,605.05	40,768	Cherrapunji, India	1860–1861

of the normal annual precipitation. The driest place is at Calama in the Atacama Desert of northern Chile, where rain has never been recorded.

Table 6.6 shows the world's greatest point rainfalls. Most of the short-duration rainfalls are convective, whereas the long-duration rainfalls are orographic.

EXAMPLE 6.13 A southwest trade wind blows almost continuously across the Hawaiian Islands. What effect will this trade wind have on precipitation there?

Solution The trade winds carry evaporation from the Pacific Ocean across the Hawaiian Islands. Locations, such as on the island of Kauai, where elevations are high enough to result in lapse-rate temperature reductions below the dew point result in rainfall. On the highest elevations of Kauai, rainfall occurs almost continuously as a result of this orographic effect.

EXAMPLE 6.14 A large low-pressure system crosses the southern United States, from west to east. What hydrologic effect will result from this system?

Solution The counterclockwise circulation will draw moisture from the Gulf of Mexico and circulate it into the system, where it will be elevated and cooled below the dew point. Some moisture will be circulated northward into cooler air so that precipitation will occur over a large part of the low-pressure system. Light to medium precipitation will cover large drainage basins.

If the temperature variation within the system is large, severe convective storms will develop generally within the southwest quadrant of the system.

EXERCISES

6.1. List 10 cities in the United States receiving some of the highest annual precipitation. Give precipitation amounts.

6.2. List 10 cities in the United States that are some of the most windy. Give wind speeds in kilometers per hour.

6.3. List 10 of the hottest as well as of the coldest cities in the United States. Give temperatures in °C.

6.4. List 10 of the most humid as well as of the least humid cities in the United States. Give values of relative humidity.

6.5. List the type of precipitation received in your area for different seasons. Explain briefly why one type of precipitation is predominant in a given season?

6.6. Give the latent heat of vaporization in calories per gram for water at different temperatures, and graph the variation of this heat.

6.7. How does the weight of dry air vary with temperature for a constant pressure? Graph this variation.

6.8. How many calories are required to evaporate 1 liter of water 10°C? How many kilograms of ice would the same amount of heat melt? Take the specific heat of ice as 0.5.

6.9. Which 10 cities in the United States experience some of the largest temperature fluctuations? List the seasons against these fluctuations.

6.10. Which 10 cities in the United States are some of the most tornado-prone? Can you explain why?

6.11. Compute the heat of vaporization of water at 5°C, 20°C, 30°C, 60°F, 75°F, 85°F, and 95°F?

6.12. Compute the number of calories required to evaporate 10 gallons (U.S.) of water at 10°C. How many kilograms of ice can be melted by these calories? Assume the specific heat of ice as 0.5.

6.13. How many calories are required per square meter to melt a 30-cm-thick layer of ice at −5°C with a specific gravity of 0.8? How much heat will be required to evaporate the above meltwater without raising the temperature above 0°C?

6.14. What will be the weight of 5 m^3 of dry air if the air temperature and pressure are 10°C and 1000 mb, respectively?

6.15. What will be the density of dry air at 10°C and 1000 mb, and the density of moist air with a relative humidity of 50% at the same temperature and pressure?

6.16. What will be the dew-point temperature, relative humidity, the saturation vapor pressure, and actual vapor pressure if the dry-bulb temperature and the wet-bulb temperature are 30°C and 20°C, respectively?

6.17. What will be the relative humidity if the air temperature and dewpoint temperature are 25°C and 15°C, respectively?

6.18. What will be the dew-point temperature if the air temperature and the relative humidity are 25°C and 65%, respectively?

6.19. Consider a parcel of air at 20°C at an elevation of 500 m above the mean sea level. The air rises to an elevation of 4000 m and then descends to the original elevation. If a descent of 2000 m produces saturation and condensation and if the saturated adiabatic lapse rate is about half that of the dry-adiabatic lapse rate, what will be the air temperature?

6.20. Assume you are the reservoir operations officer for the Missouri River dams in the following situation: One foot of packed snow containing 2 inches of water covers the Missouri basin in North and South Dakota; the reservoirs are filled to the top of the conservation pool. Temperatures are at 30°F. Weather conditions predict strong west winds to blow from the Rocky Mountains across the two Dakotas. What would you do and why?

6.21. You are operating the Ft. Randall reservoir (elevation 1000 ft msl). Two feet of snow is on the ground containing 3 inches of moisture. The daytime temperature is 28°F. The weather report predicts west winds. You notice that the Harney Peak (elevation 5000 ft) in the Black Hills is reporting 28°F with west winds. What would you do and why?

6.22. What change in weather and hydrologic conditions would you expect from the following circumstances: A layer of snow covers all of Nebraska, temperatures are near 0°F, and a cold north wind is blowing. Shortly, the wind changes to the west and you note from reports that a west wind is blowing briskly across Nebraska, Wyoming, and eastern Colorado. Explain your answer making use of meteorological principles.

6.23. Warm saturated air crossing over an area contains initially 2 inches of water. While crossing a 20-mile area, it rises and cools to 10°F. Compute (a) the approximate amount of moisture condensed during crossing over the area; and (b) the average rainfall rate if half of the condensed moisture falls as rain and if the air is traveling at 20 miles per hour and if traveling at 60 miles per hour.

6.24. The following data are available for mean annual precipitation versus elevation:

MEAN ANNUAL PRECIPITATION (IN.)	ELEVATION (FT)
(WEST SIDE OF MOUNTAINS)	
26	300
30	1100
41	1930
44	2340
(EAST SIDE OF MOUNTAINS)	
24	7920
20	5630
3.7	4350

Determine the relation between the mean annual precipitation and the elevation.

CHAPTER 7
Measurement and Analysis of Precipitation Data

Precipitation constitutes the principal input to the hydrologic budget. Due to its variability, it is measured in space and time. The measured precipitation is useful in a variety of hydrologic applications. This chapter discusses measurement of precipitation and methods of analysis of measured precipitation before its use. It also indirectly shows the association of hydrology with geography and/or climatology.

7.1 PRECIPITATION VARIABILITY

Rainfall varies greatly in both time and space. The variability can be visualized by analyzing rainfall records of different gaging stations. Rainfall varies during the rainfall event, from one event to another, and from one time period to another (say, one hour, one day, one week, one month, etc.). Figure 7.1 shows variation of rainfall intensity at two gaging stations for the event of May 15, 1960, on a small watershed W-1 (area 3.26 mi^2; 8.35 km^2) near Hastings, Nebraska, where the gages are about 1477.3 m apart. Clearly, there is pronounced variability in rainfall from 0.33 to 9.78 cm/h on rain gage C-31-R, even though the duration of the event is only about 2 hours. When rainfall is plotted for another event, as shown in Figure 7.2, the two events have nothing in common. For this event, the rainfall intensity varies from 0.33 to 4.88 cm/h on rain gage C-31-R. Thus, each rainfall event is unique.

In a similar vein, rainfall is found to vary greatly from one day to another, from one month to another, and one year to another. Figures 7.3 and 7.4 plot daily rainfall for the month of June 1982 for Springhill and Cotton Valley in northwestern Louisiana. These two stations are about 16 km apart. Mean daily rainfall over a period of 15 years (1966–1982) is graphed for these stations in Figures 7.5 and 7.6. Likewise,

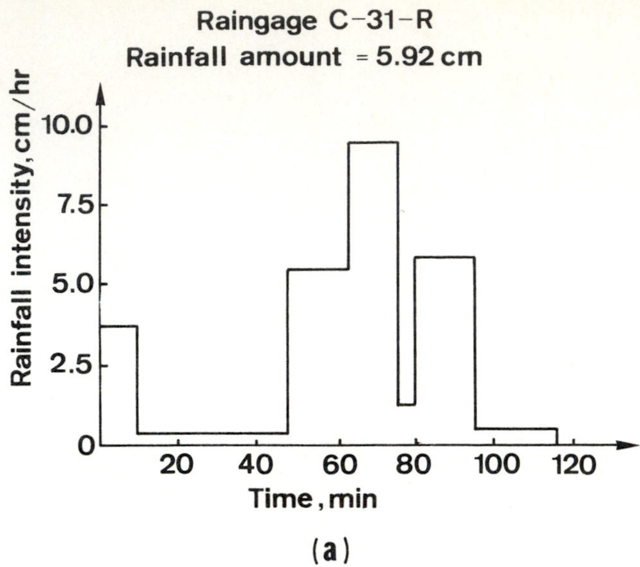

(a)

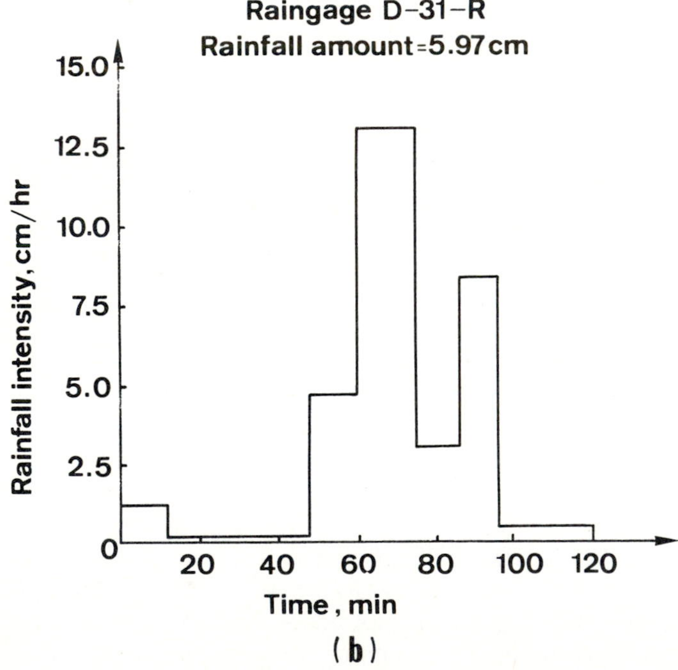

(b)

Figure 7.1 Rainfall hyetographs for the event of May 15, 1960, on watershed W-8, near Hastings, Nebraska. The drainage area is 8.35 km^2. The two rain gages are approximately 1477.3 m apart.

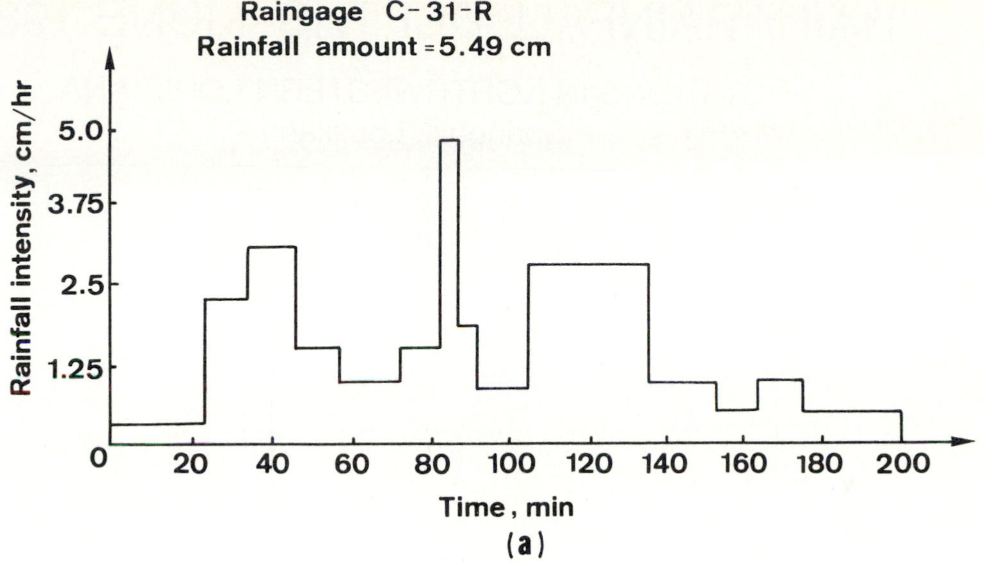

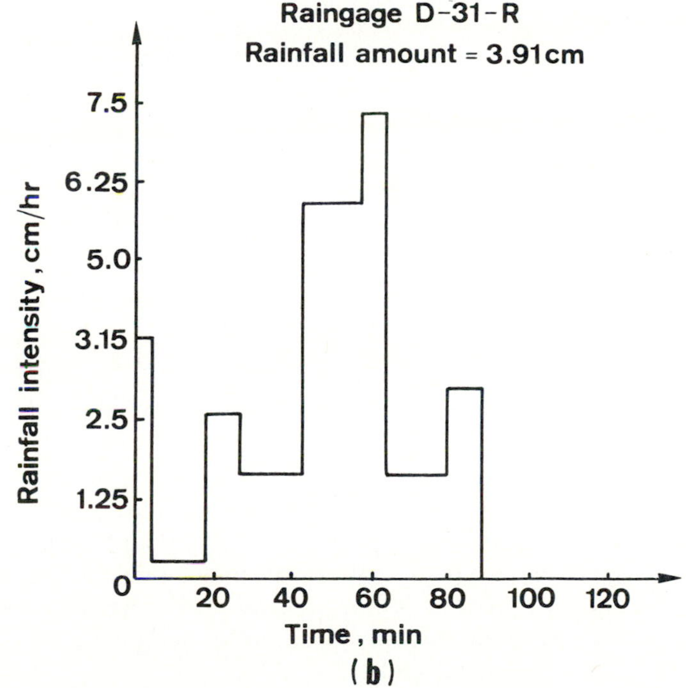

Figure 7.2 Rainfall hyetographs for the event of June 14–15, 1961, on watershed W-8, near Hastings, Nebraska. The drainage area is 8.35 km^2. The two rain gages are approximately 1477.3 m apart.

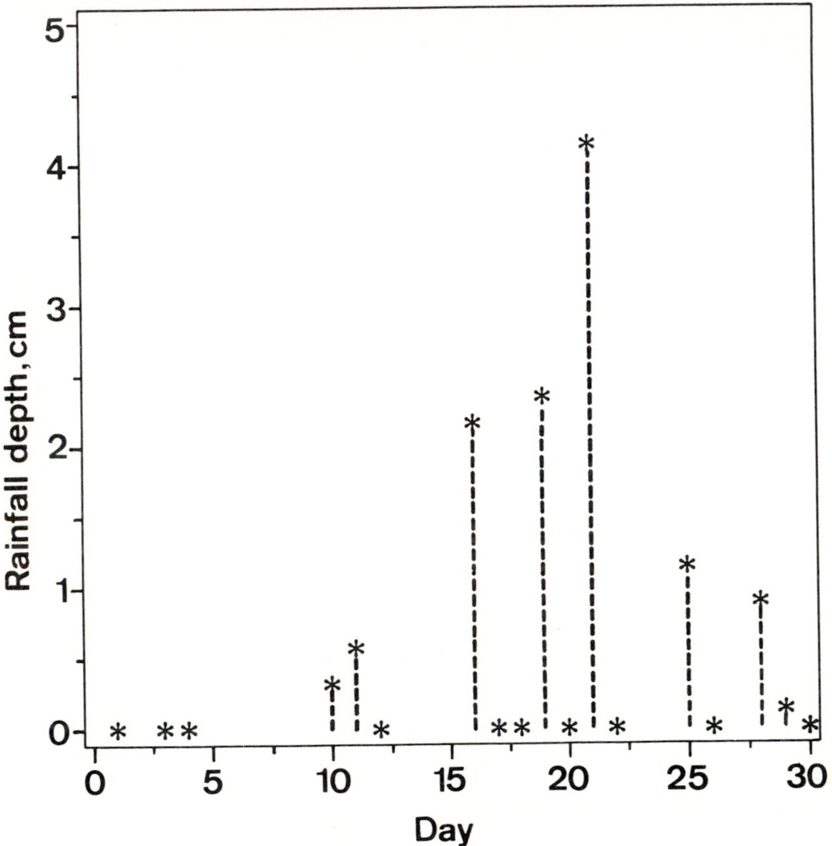

Figure 7.3 Amount of daily rainfall for June 1982 observed at Springhill in northwestern Louisiana.

monthly rainfall is plotted for the year 1982 for these two stations in Figures 7.7 and 7.8. Mean monthly rainfall is shown in Figures 7.9 and 7.10. These figures clearly show that rainfall varies greatly not only from one time interval to another, but also from one station to another. On a national scale, the mean annual precipitation varies considerably from place to place, as shown in Figure 7.11. Similar spatial variations occur on a global scale.

These variations show that although there is a network of rain gages, the maximum rainfall amount may not occur over one of the gages. Quite often, the maximum point rainfall may not be recorded by an official rain gage. Variations in space can also occur because of topographic effects. The assumption that the maximum rainfall recorded at a gage is, indeed, the maximum rainfall that occurred in a storm may not be justified.

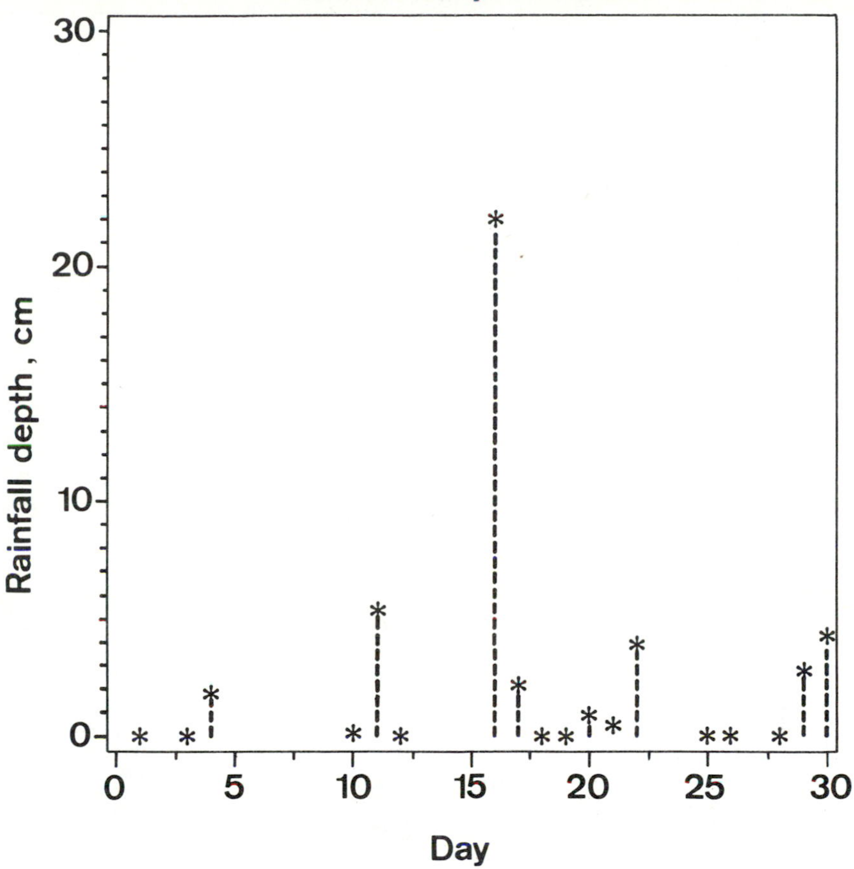

Figure 7.4 Amount of daily rainfall for June 1982 observed at Cotton Valley in northwestern Louisiana.

7.1.1 Seasonal Variation of Precipitation

It is known that seasonal variations in precipitation exist. Because of its varied terrain and position in relation to sources of maritime air, the United States exhibits distinct regional patterns, and seasonal distribution varies widely across the nation. In the state of Arizona, seasonal precipitation is bimodal and its modes occur mainly in July and December. In Nebraska, seasonal precipitation is unimodal and its mode occurs in the spring and summer months, whereas in Pennsylvania, precipitation is nearly evenly distributed over each month throughout the year. Pacific coast and western mountain states get their main precipitation in winter. Along the west coast, most of the precipitation occurs in November through March, with maximum

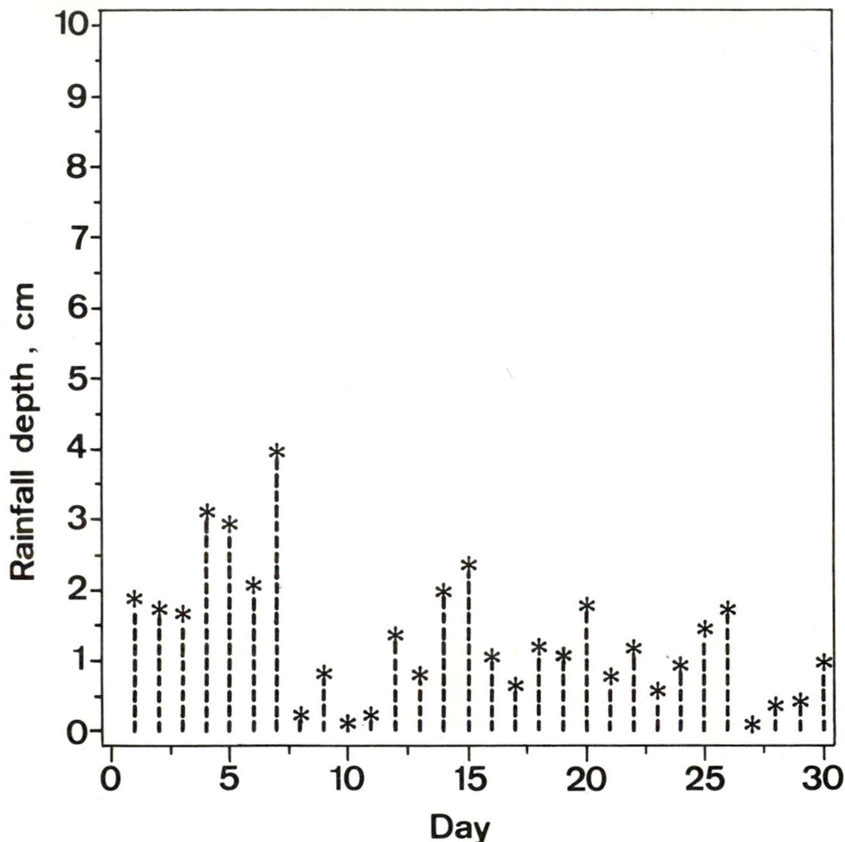

Figure 7.5 Mean daily amount of rainfall for the month of June for Springhill in northwestern Louisiana.

monthly values of 4 to 6 in. (10 to 15 cm). Minimum values of less than 1 in. (2.5 cm) occur in July and August. States in the Great Plains and upper Mississippi River basin show a more even distribution, but with a tendency toward maximum precipitation in the summer. This is because of the prevalence of thunderstorms during the summer. Most states east of the Mississippi River exhibit fairly uniform precipitation the year round, especially in the Northeast and northern Midwest. This is not, however, true of Florida, where winter rainfall is lowest. In the hurricane season, September rainfall is over 8 in. (20 cm). Along longitude 100°N, maximum monthly precipitation is less than 2 in. (5 cm), and maximum monthly rainfall occurs in May through June.

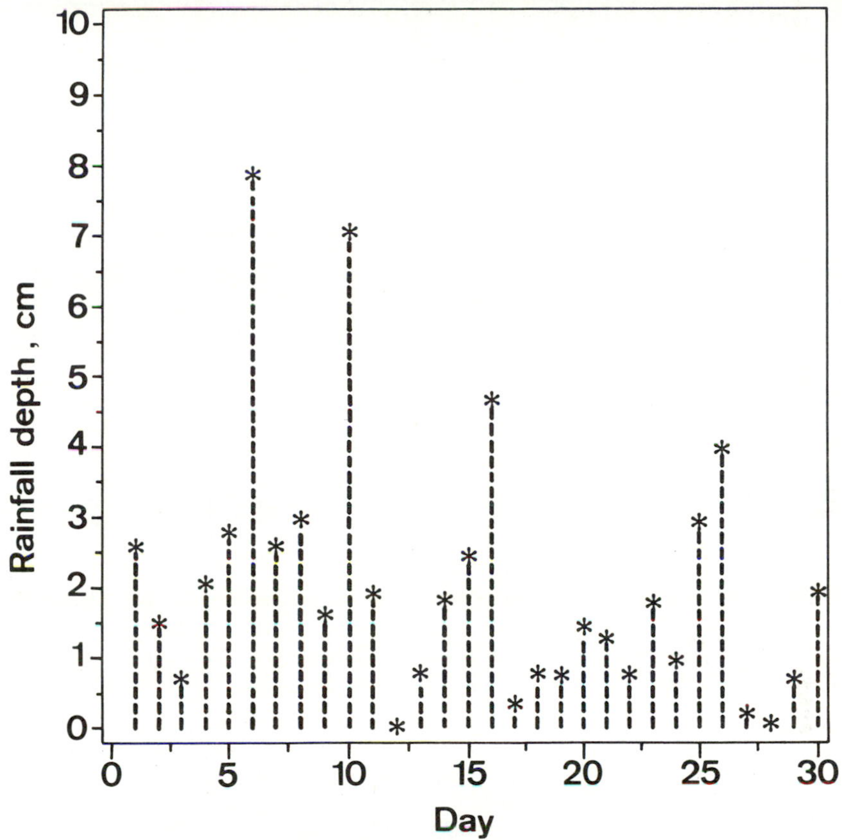

Figure 7.6 Mean daily amount of rainfall for the month of June for Cotton Valley in northwestern Louisiana.

Although the tropics (from about 10°N to 10°S) do not really have seasons, precipitation varies with the sun's position. The months of September, October, March, and April, when the sun is almost directly overhead, receive heavy rains. This position of sun is conducive to supplying abundant heat energy to the atmosphere and ocean. Other parts of the world have more definite seasons around the year and widely varying precipitation patterns. Many parts of the world have one well-defined rainy season, which may occur in summer or in winter, depending upon the general circulation, prevailing weather fronts and cyclonic storms, frequency of summer thunderstorms, and so on. Southern Africa has its rainy season mainly during the summer. This same is true of northern plains in India. Northern Austra-

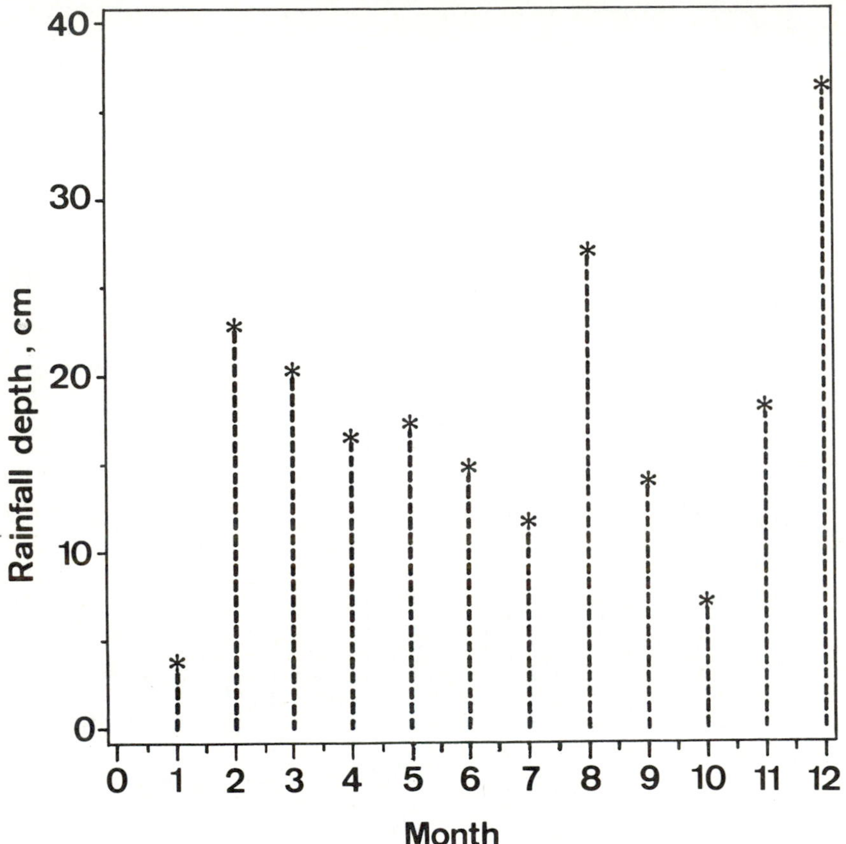

Figure 7.7 Monthly amount of rainfall for 1982 observed at Springhill in northwestern Louisiana.

lia has a rainy season in the summer, whereas southern Australia has one in winter. Egypt, portions of northern Africa, and Kharga have no rainy season.

The importance of these observations to hydrologists is that there are variations in precipitation in time and space and these variations must be taken into account in hydrologic analysis.

7.1.2 Effect of Urbanization on Precipitation

The effect of urban areas on precipitation has been studied by Changnon et al. (1976), Huff and Vogel (1978), amongst others. They noted that the heat from urban areas added to the energy in a convective storm caused the cell to store more energy and

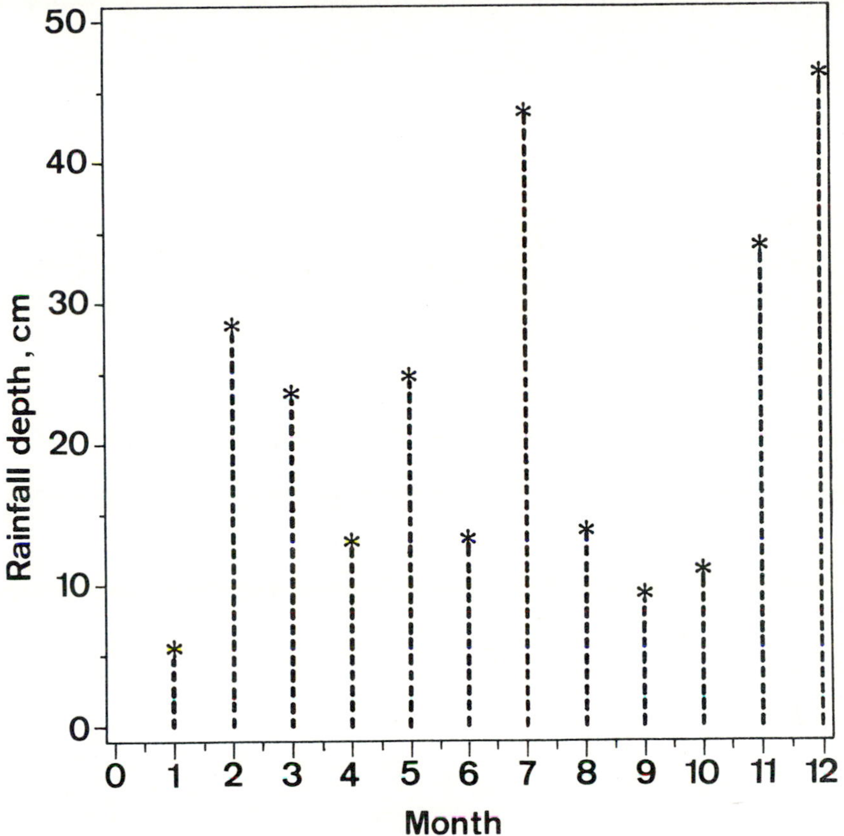

Figure 7.8 Monthly amount of rainfall for 1982 observed at Cotton Valley in northwestern Louisiana.

therefore enable it to store and carry more precipitation than before urbanization. The effect of this condition caused a rain deficit near the urban area and a rain increase further down in the direction of storm movement. This so-called La Porte effect (named from an occurrence at La Porte, Indiana) disturbs the spatial distribution of rainfall in urban areas. It is generally believed that the climate in La Porte has changed primarily due to the large industrial complex in the Chicago area (Masters, 1974). A 30 to 40% increase in precipitation has been found over 40 years. Cities on the downwind sides of the Great Lakes have somewhat higher seasonal precipitation than those upwind. These are, however, minor factors in the overall pattern of continental precipitation.

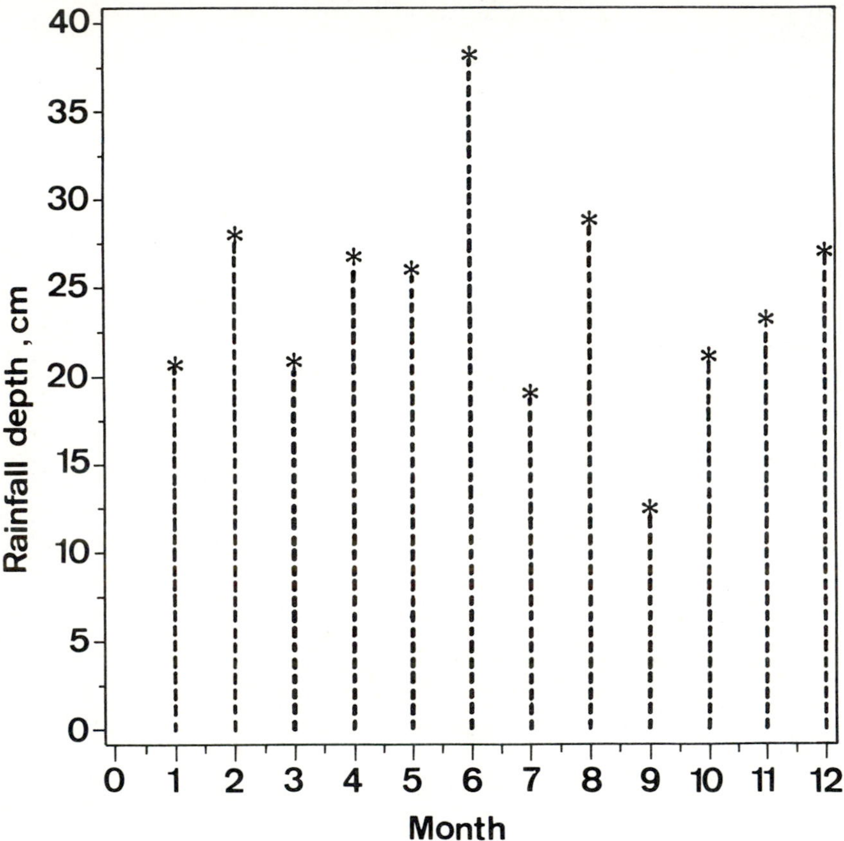

Figure 7.9 Mean monthly amount of rainfall for Springhill in northwestern Louisiana.

7.1.3 Geographic Distribution of Precipitation

The geographic distribution of precipitation depends upon latitude, orographic factors, and the distance the air mass moves away from the source of moisture. Rainfall tends to be heaviest near the equator and tends to diminish as the air flows toward higher latitudes. This is due to abundant warm, moist air that drifts continually from tropical seas and flows north and south. Although there is complex distribution of precipitation over the United States, there is a perceptible decrease in rainfall going north from the Gulf of Mexico. This decrease clearly shows the effect of air moving away from its source of moisture. Over North America, the major source of precipitation is maritime from the Pacific and Atlantic oceans and from the Gulf of Mexico,

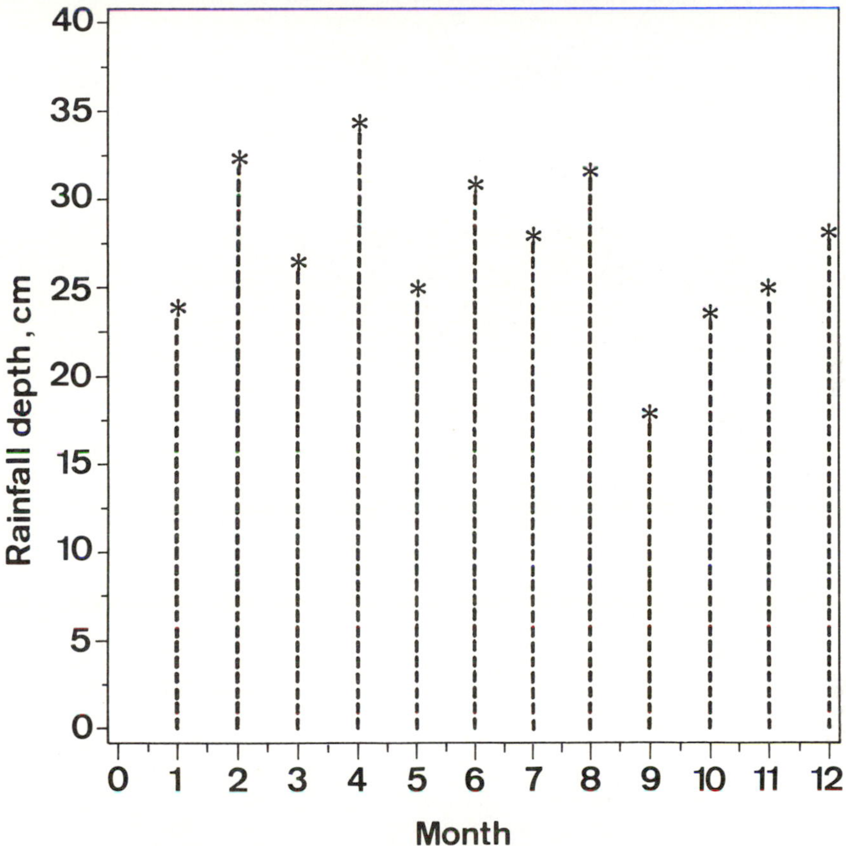

Figure 7.10 Mean monthly amount of rainfall for Cotton Valley in northwestern Louisiana.

even though evaporation from the land is significant. The Great Lakes contribute some moisture.

Annual precipitation (average over 30 years) in the United States varies from less than 5 in. (12.7 cm) in certain parts of California, Nevada, Arizona, and New Mexico to over 140 in. (356 cm) in western Washington. Average annual precipitation ranges from less than 20 in. (51 cm) along the eastern foothills of the Rockies to nearly 80 in. (203 cm) in the southeastern mountains. Moist air blowing out of the Pacific Ocean encounters the mountains eastward and it results in orographic lifting and cooling with large amounts of precipitation. Much of the precipitation in the central region is derived from the moisture picked up in the Gulf of Mexico.

It is interesting to note that two places, Kharga and Cherrapunji, at almost exactly the same latitude, are among the driest and wettest places, respectively, on

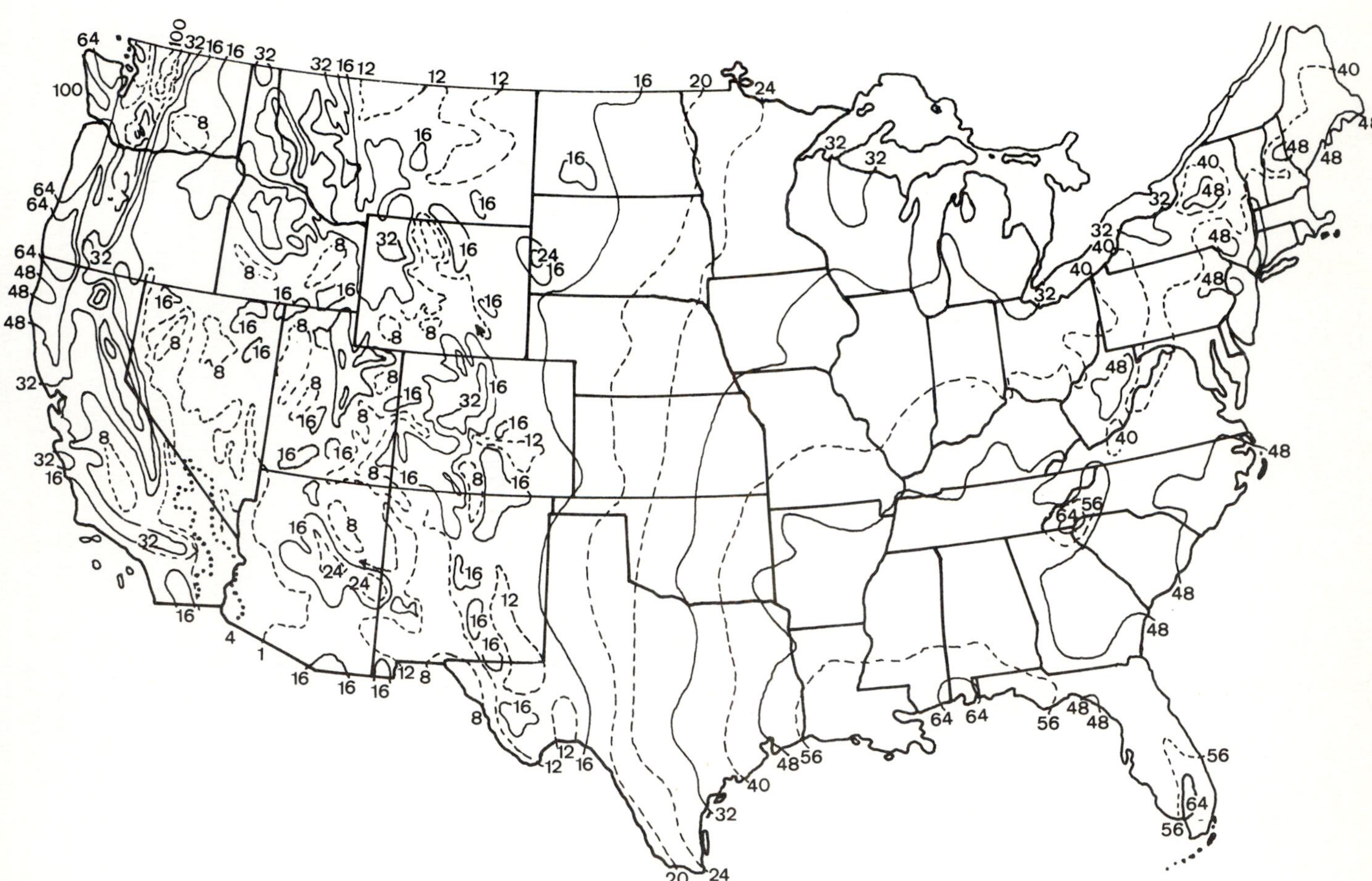

Figure 7.11 Mean annual precipitation (inches) in the United States.

the earth. This extreme difference in precipitation of these two places is caused by orographic factors. Along 25°N latitude in northeastern India is located Cherrapunji—a town that holds the world record for the most rainfall in a 12-month period—1041 in. (2644 cm) in 1860–1861. Its long-term average is 430 in. (1092 cm) per year. The town of Cherrapunji is located on the edge of a plateau that provides the first orographic barrier to the moisture-laden winds of a monsoon blowing north out of the Bay of Bengal. On the other hand, along the same latitude in the Western Desert of Egypt, the region around Kharga receives virtually no rainfall at all year after year. Its long-term average is probably 1 mm (0.04 in.) a year. Kharga lies in a large depression near the eastern side of the great Sahara Desert, more than 2000 miles (3200 km) from the Atlantic Ocean and about 500 miles (800 km) south of the Mediterranean Sea. By the time winds from the ocean reach Kharga, they have lost their moisture on the way and become as dry as the sands they blow across.

7.1.4 Long-Term Variation of Precipitation

Wet periods may occur in which wet years follow wet years. Droughts may also occur in which dry years follow dry years. There is some evidence of cyclical variations in precipitation (Shaw, 1942; Mitchell, 1964). However, this variation does not follow a regular pattern. To illustrate this point, annual precipitation at Bakersfield, California, for 1874 to 1983 is plotted in Figure 7.12, with the mean annual precipitation and the 5-year moving average precipitation. There appears to be a predictable pattern of fluctuation of the 5-year mean around the long-term mean, but the way annual precipitation fluctuates around the mean is not predictable. It is to be noted that one of the driest years (1976–1977) preceded the wettest years (1977–1978) in more than 100 years of record. The moving 5-year mean eliminates sharp fluctuations in the annual values. The 5-year mean values exhibit an alternat-

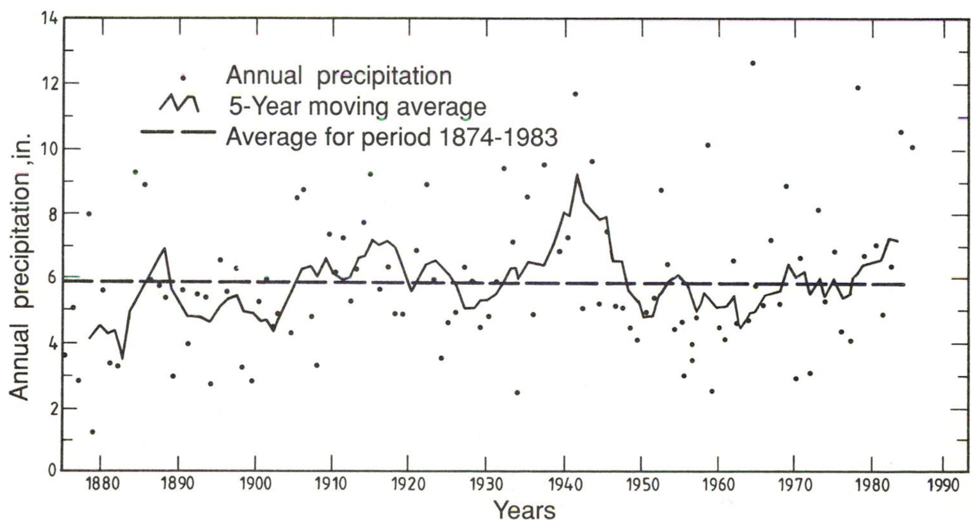

Figure 7.12 Annual precipitation at Bakersfield, California, for 1874 to 1983.

ing pattern of increasing and decreasing precipitation. Historic trends of annual precipitation may be an important factor in planning water management programs. Records of tree ring growth rates for periods of hundreds of years have been used to provide an indication of precipitation trends over a long period of time. Studies of annual tree rings have shown periods of reduced growth rate, which, in turn, indicate periods of deficient precipitation.

7.2 METEOROLOGIC HOMOGENEITY

The probability of a storm occurring within a given area is the same if all the meteorological conditions within that area are similar. Such an area is termed meteorologically homogeneous. The duration and intensity of storms within such an area might be different, but the total precipitation of such storms will be about the same. This does not mean that all areas with the same mean annual rainfall are meteorologically homogeneous. Different meteorological conditions might result in the same mean annual rainfall; but to be meteorologically homogeneous, an area must have the same mean annual rainfall and have the same annual precipitation pattern.

The factors that determine mean annual rainfall and meteorological homogeneity are (a) the distance from the ocean (continentality); (b) the direction of prevailing winds, that is, the direction of storm movement; (c) the mean annual temperature; (d) altitude; and (e) topography. These factors are more or less interdependent because of their relation to climatic influences.

As noted in the previous chapter, the distance from the ocean determines the continentality of an area. Seasonal variations in temperature increase with the distance from the ocean. These temperature variations also affect precipitation and the type of precipitation. The direction of prevailing winds is determined by the direction of the storm movements. Thus, the direction of prevailing winds also indicates the similarity of an area to the influence of the storm systems. The mean annual temperature of an area is also an indicator of the climatic influence, and is a function of latitude, distance from the ocean, altitude, and topographic influence. The effect of altitude and topography on precipitation is the same as on climate, and depends on the relation of these factors to the other factors. Climate and precipitation are the result of the interaction of all these variables and that the net effect is more or less unique. If an area is meteorologically homogeneous, the probability of a storm of a given magnitude and duration occurring is the same at any point within that area. Rainfall records within a meteorologically homogeneous area can be added together and treated as one record to generate a longer rainfall record.

7.3 MEASUREMENT OF RAINFALL

Because precipitation is the input to the hydrologic budget, it is essential to know how much precipitation has occurred during a specified period of time. The advent of probability concepts in hydrology has established the need for long, continuous records of precipitation. Perhaps the oldest and longest records of rainfall amounts go back to Egyptian measurements along the Nile River about a thousand years ago. Today, at least in the United States, there are many records exceeding 100 years. The National Weather Service operates thousands of rain gages throughout the nation. Many other rain gages are operated and maintained by other agencies and organizations.

There are several instruments in use for measuring precipitation. These are the standard nonrecording rain gage, tipping bucket recording rain gage, weighing recording rain gage, float recording rain gage, and nonrecording standpipe gage for snow.

7.3.1 Standard Nonrecording Rain Gage

The standard nonrecording rain gage, shown in Figure 7.13, was developed by the National Weather Service (then the U.S. Weather Bureau) for official rainfall observations. This gage consists of a metal can 8 in. (20 cm) in diameter and 24 in. (60 cm) high. The upper part of this container consists of a removable funnel that conducts the captured rain into another removable inner container. The inner container is

Figure 7.13 Standard nonrecording rain gage.

2.53 in. (6.42 cm) in diameter, equal to 0.1 of the 8-in. (20-cm) receiver area plus an amount equal to the displacement of the necessary graduated measuring stick. The inner container holds 2 in. (5 cm) of rainfall; the excess rainfall overflows the inner container and is held in the outer can. This gage must be operated by an attending observer. The observer monitors the gage at regular intervals, usually daily. The observing procedure is to remove the receiving funnel, insert the graduated scale into the inner tube, and read the amount of water contained therein. If the inner tube has overflowed, the observer inserts the measuring stick and displaces its volume. He then pours out the water in the inner container and then pours the overflow water from the outer can into the inner container and measures that depth. Continuing this procedure permits measuring the total depth of rainfall from any rain storm.

7.3.2 Tipping Bucket Recording Rain Gage

The tipping bucket recording rain gage was designed to preserve a record of the amount of rainfall occurring over any given time. This gage provides a continuous record of the rainfall depth and its intensity. There are several makes of tipping bucket rain gages, but the principle of each is the same. The Stevens tipping bucket rain gage, as shown in Figure 7.14, consists of an 8-in. (20-cm) diameter collector that directs the rain into a tipping bucket that empties when 0.01 in. (0.025 cm) of rain fills it. An electrical pulse is generated on each tilt and transmitted to a time-driven recorder. It is then possible to read the number of tilts occurring over any interval of time. Older models recorded the tilts using a pen that marked each tilt on a clock-driven recorder drum. During intense rainfall, the pen marks were so close together that it often was not possible to distinguish them. Modern digital and electronic recorders have eliminated this problem. Tilting bucket recorders are not the preferred recorders because the inertia of the tilting bucket prevents measurement of all rainfall.

7.3.3 Weighing-Type Recording Rain Gage

The weighing-type recording rain gage operates by continuously recording the weight of the accumulated rainfall. The data are recorded on tape or transmitted to a remote data gathering point. Older weighing gages record accumulated weight by pen and clock-driven drum. This type of gage can be used to record snowfall as well as rainfall.

7.3.4 Float Recording Rain Gage

Float recording gages operate by catching the accumulated rainfall in a tube that contains a float. The rise of the float with increasing amounts of rainfall is recorded with time. There are several designs of gages in which the float mechanism floats directly on water accumulated in the receiver or on a column of other fluid that is displaced by accumulating rainfall. The accumulated rainfall is recorded in a manner similar to the gages described before.

Figure 7.14 Stevens tipping bucket recording rain gage.

7.3.5 Standpipe Gage

Standpipe gages, also known as storage gages, are used to record snow precipitation, particularly at remote locations. These instruments consist of a 12-in. (30.5-cm) galvanized thin-wall tube made in 5-ft (1.6-m) sections. The standpipe is erected high enough to extend above the anticipated depth of snow accumulation. To prevent snow from accumulating on the lip of the tube and to minimize the effects of wind, an Alter shield is constructed around the top of the tube. An Alter shield consists of a ring mounted around the top of the standpipe upon which is hung strips of metal hinged to the ring. This shield reduces the effect of the wind and tends to keep snow from accumulating on the lip of the standpipe. Within the standpipe is antifreeze or a calcium chloride solution to melt the snow so that its water equivalent can be measured. To prevent evaporation of the snowmelt, a thin layer of oil is located on top of the fluid column.

7.3.6 Measurement by Radar

Radar measurements are used to estimate precipitation intensity. Radar can detect any type of hydrometeors in the atmosphere. The reflection of the hydrometeors is determined by electromagnetic energy of the radar pulse and is termed echo. The brightness of the echo is a measure of the precipitation intensity. The strength of reflected radar pulses is a function of the number and size of the raindrops. By using this established relation, it is possible to delineate light, medium, intense, and very intense rainfall. The distance from the radar site to the precipitation area is measured by the time between emission of the radar pulse and receipt of the echo. The radar antenna has to be oriented in the direction of reflection. Thus, the precipitation field can be located from the measurements of distance and direction. It is not possible to accurately determine the volume of rain that falls from radar alone. There can be significant interference from trees and buildings in measurement of rainfall by radars. Wind drift of particles, storm type, distance to the storm, and precipitation type also affect radar measurements. Use of a radar, in conjunction with rain-gage data, provides useful estimates for areas not covered by rain gages. Radar can be used to determine the areal extent of rainfall with reasonable accuracy.

7.3.7 Other Rain Gages

There are many kinds of inexpensive rain gages made of glass and plastic that are useful in observing rainfall for hydrologic purposes. These gages are not nearly as accurate as the more expensive and sophisticated gages. Nevertheless, such inexpensive gages are useful and desirable for many hydrologic computations. Official rain gages are often not located where rainfall information is desired and any reasonably accurate measured rainfall data are, therefore, useful. For this reason, measured observations by individuals who maintain inexpensive rain gages should be sought. Furthermore, when rainfall data are desired after the occurrence of a storm, it is useful to make a "bucket" survey of the storm area. A bucket survey is a search within the area of interest after a storm for containers, for example, pans, cans, buckets, tanks, or any container that could intercept rainfall. When making a bucket survey, it is necessary to assess the reliability of the observed container. One needs to know if the container was empty before the storm; if it has a tapered or irregular shape, can the volume of water in the container be adjusted mathematically to compute the rainfall; is the container located in an unobstructed area so that it has any chance to represent the true rainfall? If such data can be established to be reasonably reliable, then they can be useful. Field observations of this type should always be included in hydrologic analyses, if possible.

A bucket survey made by the U.S. Army Corps of Engineers (1967) is shown in Table 7.1 together with the available official National Weather Service rain gages for the storm of September 6–7, 1965, over the Papillion Creek drainage basin near and adjacent to Omaha, Nebraska. The observations from this bucket survey and the official gages were used to prepare the isohyetal map for the 400 mi^2 (1036 km^2) Papillion Creek drainage basin, shown in Figure 7.15, of which the Irvington drainage basin is a part.

TABLE 7.1 NEBRASKA RAINFALL BUCKET SURVEY, STORM OF SEPTEMBER 6–7, 1965 (AFTER U.S. ARMY CORPS OF ENGINEERS, 1967)

County	Township	Range	Section	Rainfall amounts (in.) night of September 6–7	Type of tank	Accuracy	Remarks
Douglas	15N	11E	5	8.0	Tank	Poor	
	16N	10E	11	5.0+	Plastic	Poor	Gage overflowed
	16N	11E	22	7.9	[a]	Good	
	16N	12E	28	5.0+	[a]	Poor	Gage overflowed
	16N	12E	29	4.5	[a]	Poor	
	16N	12E	32	5.5	Bucket	Poor	
Sarpy	13N	10E	24	4.0	[a]	Good	
	13N	11E	1	4.1	[a]	Good	Most intense rain, midnight to 2:30 A.M. on September 7
	13N	11E	23	3.5	[a]	Fair	
	13N	12E	10	4.5	Plastic	Fair	
	13N	13E	5	4.5	Plastic	Good	Most intense rain, 2:00 A.M. to 5:00 A.M. on September 7
	14N	10E	26	4.6	[a]	Good	
	14N	11E	10	4.5	[a]	Good	
	14N	12E	9	3.5	[a]	Fair	
	14N	13E	30	3.5	[a]	Fair	Most intense rain, 2:00 A.M. to 5:00 A.M. on September 7
Washington	17N	10E	20	4.8	[a]	Good	
	17N	11E	13	5.0	[a]	Good	
	17N	11E	16	8.0	Bucket	Poor	
	17N	11E	17	3.25	[a]	Poor	
	17N	11E	20	4.0	[a]	Poor	
	17N	11E	30	6.0	Can	Poor	
	18N	10E	28	5.7	Bucket	Poor	
	18N	11E	7	4.0	[a]	Poor	
	18N	11E	25	5.0	[a]	Good	
	18N	11E	11	4.8	—	—	
Gretna 3NE	14N	12E	2C	5.73		Good	National Weather Gage
Omaha, Eppley	15N	13E	$E\frac{1}{2}$ 1	6.45		Good	National Weather Gage
Omaha, North	16N	12E	E line 11	7.82		Good	National Weather Gage
Bennington 3E	16N	12E	20	8.9		Good	National Weather Gage

[a] Small glass rain gages, dimensions, approximately 3/4 in. × $4\frac{1}{2}$ in. or $5\frac{1}{2}$ in.

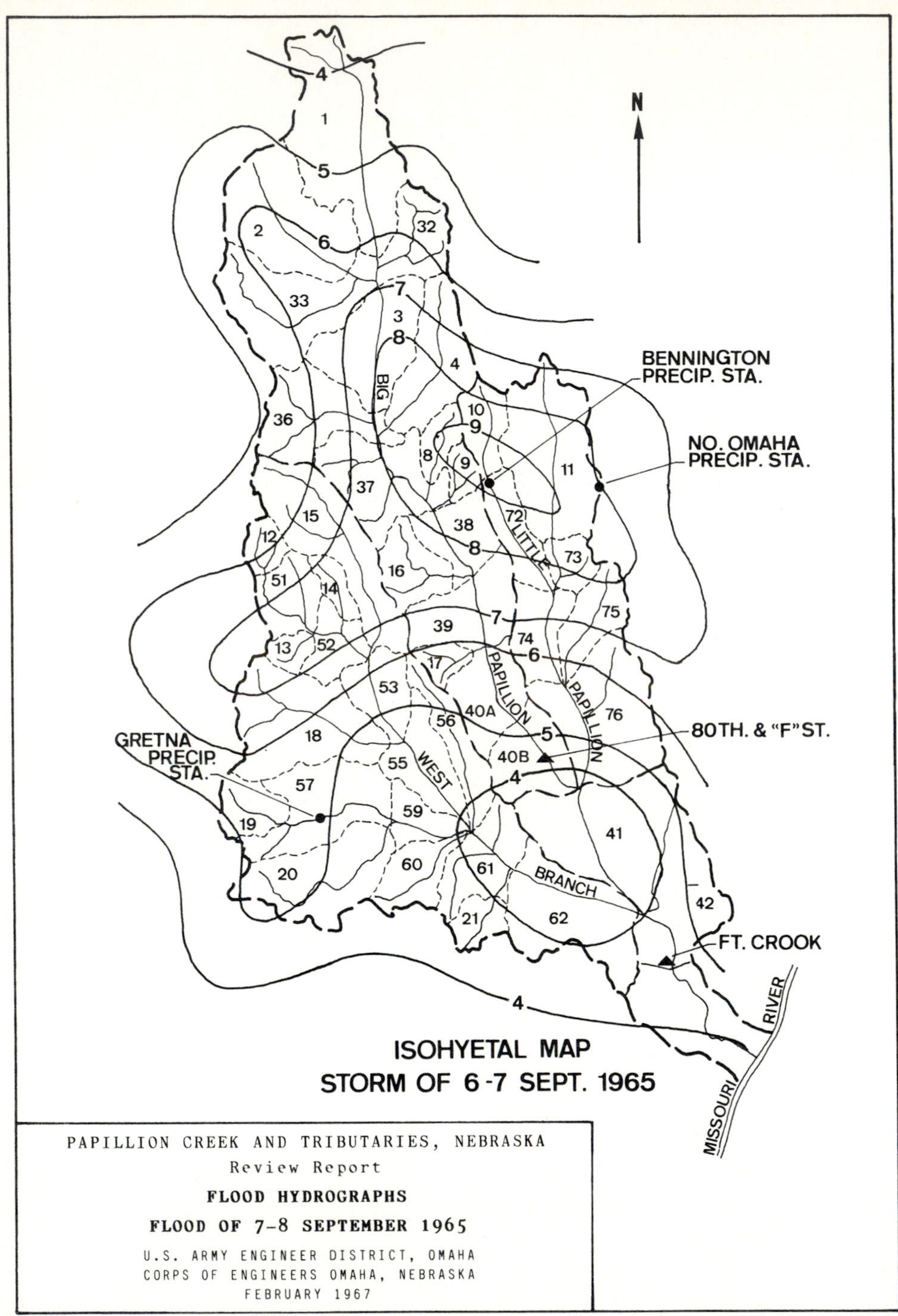

Figure 7.15 Isohyetal map for Papillion Creek and tributaries, Nebraska.

It is important to note the gage type, accuracy, and remarks on a bucket survey in addition to the other data. There are inherent problems in recording rainfall in a small glass tube gage that has flat shoulders on the glass at the opening. These shoulders tend to cause splattering and often record less rainfall than what actually falls. This and the location of the gage largely determine its accuracy. When gages overflow, the minimum rainfall is recorded at that point. Such values, even if rated poor, are still useful for construction of an isohyetal map. These values can be useful as a check on contouring. Their recorded values must reasonably fit the isohyetal contours.

Other comments can also be useful for hydrologic purposes and should be included if such observations are available. It is important to realize that important and useful information can be obtained beyond official data from independent investigations. The isohyetal map, shown in Figure 7.15, for Papillion Creek is a case in point. From this map, it is possible to determine the mean rainfall on the Irvington drainage basin, which was 7.5 in. for the storm of September 6–7, 1965. This mean precipitation will be useful later to compare with the measured runoff. It is interesting to note that the person who prepared Figure 7.15 drew a 9-in. (22.86-cm) isohyetal contour around the National Weather Service Bennington gage even though it showed only 8.9 in. (22.6 cm) of rainfall. Such action indicates some knowledge that this contour was justified.

7.4 DESIGN OF RAIN-GAGE NETWORKS

There is large variation in rainfall within a given storm (Linsley and Kohler, 1951; Causey, 1953). Ranges of precipitation from 2 in. (5.1 cm) to 8.9 in. (22.6 cm) in a single convective storm over a drainage basin (139 mi^2, or 360 km^2) have been observed. It is reasonable to believe that greater variations can exist. The important conclusion from such observations is that rain does not fall uniformly over a drainage basin. Therefore, it is necessary to provide a network of gages that adequately sample rainfall variability within the storm as well as over the drainage basin. Rainfall variability depends on topography, wind, direction of storm movement, as well as the type of storm. The location and spacing of gages depend on the type of precipitation to be measured as well as the use of precipitation measurements.

7.4.1 Number of Rain Gages

The National Weather Service maintains a network of about 3500 recording and 11,000 nonrecording gages of various types at about 13,000 stations, or one station per 230 mi^2 (600 km^2) on an average. However, this average spacing does not hold in considerable areas in the west and in Alaska. Fewer gages are located in remote areas and relatively many more in densely populated areas. The heavily urban areas of California have, on the average, one gage per 2 mi^2 (5 km^2). In the United Kingdom and Hawaii, there are about five to six gages per 100 mi^2 (256 km^2), whereas in Alaska, there is less than one gage per 200 mi^2 (512 km^2). These gages are a primary source of precipitation data and have been designed to serve a broad range of applications. The design of a rain-gage network for a drainage basin naturally depends upon the use for which the rainfall data are intended.

7.4.1.1 Mean areal rainfall criterion

If the objective is to determine the mean areal rainfall, then it is important to map spatial variability of rainfall. Because so many factors affect the accuracy of determining mean areal rainfall, there is no single answer to rain-gage location. Generally, the denser the gage network, the more accurate the representation. Gages are not evenly spaced; high-variability areas have more gages and relatively uniform rainfall areas have fewer gages. In addition, cost of installation, maintenance of a network, as well as its accessibility to the observer, are also important considerations.

In general, sampling errors of rainfall amount tend to increase with increasing mean areal rainfall, and decrease with increasing network density, duration of rainfall, and areal extent. Accordingly, larger average errors are produced by a particular network for storm rainfall than for monthly, seasonal, or yearly precipitation. Average errors are usually greater for summer than for winter rainfall, for the former is much more spatially variable. Indeed, for the same level of accuracy, network density may be two to three times higher for summer rainfall than for winter rainfall. Eagleson (1967) found that two properly located rain gages would be adequate for long-term watershed mean precipitation.

The adequacy of an existing rain-gage network of a watershed is assessed statistically. The optimum number of rain gages corresponding to an assigned percentage of error in estimation of mean areal rainfall can be obtained as

$$N = \left(\frac{C_v}{\varepsilon}\right)^2 \tag{7.1}$$

where N is the optimum number of rain gages, C_v is the coefficient of variation of the rainfall values of the gages, and ε is the assigned percentage of error in estimation of mean areal rainfall. If there are m rain gages in the watershed recording $P_1, P_2, \ldots, P_m$ values of rainfall for a fixed time interval, then C_v is computed as

$$C_v = \frac{100S}{\bar{P}} \tag{7.2}$$

in which $\bar{P}$ is mean rainfall defined as

$$\bar{P} = \frac{1}{m}\sum_{i=1}^{m} P_i \tag{7.3}$$

and S is the standard deviation of rainfall computed as

$$S = \frac{1}{m-1}\left[\sum_{i=1}^{m} P_i^2 - \frac{\left(\sum_{i=1}^{m} P_i\right)^2}{m}\right]^{0.5} \tag{7.4}$$

In Equation (7.1), the value of ε is usually taken as 10%. When ε is reduced, a greater number of rain gages will be needed.

7.4.1.2 General-purpose criterion

The World Meteorological Organization (1969) has recommended the minimum precipitation-network densities for general hydrometeorological purposes as follows:

1. One station per 600 to 900 km^2 (230 to 350 mi^2) of area in flat regions of temperate, mediterranean, and tropical zones.
2. One station per 100 to 250 km^2 (40 to 100 mi^2) of area in mountainous regions of temperate, mediterranean, and tropical zones.
3. One station per 25 km^2 (10 mi^2) of area in small mountainous lands with irregular precipitation.
4. One station per 1500 to 10,000 km^2 (600 to 4000 mi^2) of area in arid and polar zones.

If the objective is to estimate streamflow (hydrograph, volume, peak, etc.), then the precipitation-network density might be different (Johanson, 1971). Simple guidelines for specific hydrologic purposes have not yet been developed.

7.4.2 Location of Gages

The location of a rain gage is very important to the accuracy of its measurements (Brown and Peck, 1962; Weiss, 1963). The location of rain gages has been studied by Horton (1924), Alavarez and Henry (1970), and Huff (1970), amongst others. A gage should not be located on a windswept area, but should be fairly well protected from wind currents from all directions. The best site is on level ground with trees or bushes providing the protection. As a rough guide, to minimize the impact on its catch, a gage should not be located closer to an object than one and one-half times the height of the object. Or the objects serving as a windbreak should subtend angles of 20 to 30° from the gage orifice, but the angle should never exceed 45°. These objects should surround the gage fairly well. Clearing in a coniferous forest would serve as an ideal site for locating the rain gage.

7.4.3 Errors in Measurement

Errors in rainfall measurement result from three sources: instrumental defects, improper siting (or location) of the gage, and human errors. Each recording-type gage has inherent errors caused by mechanical parts of the instrument. Most instrumental errors cause a loss of some precipitation and result in a low measurement. It takes approximately 0.01 in. (0.002 mm) of rain to wet the collecting funnel and measuring cylinder surfaces if the gage is dry before the cylinder begins to collect measurable amounts of water. Over a year, this could amount to a loss of as much as 1 in. (25 mm) of measurable rainfall. Similarly, evaporation in a nonrecording gage, being read only once in 24 hours, could cause a small loss of measurable water over a year. The improper location of the rain gage can tend to either overcatch or undercatch rainfall. The observer must attend to the rain gage regularly to ensure that it works properly. The largest error in all gages is the effect of wind on the entrance of rain or snow into the instrument. Wind updrafts, currents, and eddys divert falling precipitation so that it is not possible to be sure that an accurate sample has been collected. Such errors are greater for light rain than for heavy rainfall. These errors are particularly serious for snow. According to Larson and Peck (1974), these errors in gage catch of rainfall can be as high as 20% for wind speeds of 20 mph (32 km/h) at orifice height.

EXAMPLE 7.1 A watershed has a network of five rain gages. Annual rainfall recorded by these gages is given for a year as

Rain gage	1	2	3	4	5
Annual rainfall (cm)	50	82	73	64	105

Calculate the optimum number of rain gages for this watershed for a 10% error in estimation of mean areal rainfall.

Solution For this data, $m = 5$, $\Sigma P_i = 374$, $\bar{P} = 74.8$ cm, $\Sigma P_i^2 = 29{,}674$, $(\Sigma P_i)^2 = 139{,}876$, $(1/5)(\Sigma P_i)^2 = 27{,}975.2$, and $S = (1698.8/4)^{0.5} = (424.7)^{0.5} = 20.61$.

$$C_v = \frac{100 \times 20.61}{74.8} = 27.55\%$$

$$N = \left(\frac{27.55}{10}\right)^2 = 7.59 \cong 8 \text{ rain gages}$$

The optimum number of rain gages for this watershed is 8. Hence, three more rain gages are needed.

7.5 CONSISTENCY OF RAINFALL RECORDS

Many hydrologic analyses require a long-term record of rainfall data. It was noted previously that the gage should not be located closer to a potential obstruction than one and one-half times the height of that obstruction. In the course of nature and progress of time, trees grow and buildings are built and rain gages must be moved. If a rain gage is moved or is affected by an obstruction, the rainfall it records may be different than before and the consistency of its record might be changed. Change of observational procedure may also affect the consistency of a rainfall record. If the record is not consistent, then the record must be corrected in order to make it consistent with the previous rainfall record if that gage is to be used for analysis. It is common to find changes in location of National Weather Service rain gages. Therefore, it is necessary to test the consistency of the rainfall record when using the record in hydrologic analysis.

Inconsistency in a rainfall record may be identified by graphical or statistical methods (Singh and Birsoy, 1975a; Buishand, 1982), including double-mass analysis, the von Neumann ratio test, cumulative deviations, likelihood ratio test, and run test. The double-mass analysis is the most popular of all in hydrology, and is discussed here.

Double-mass analysis tests the consistency of the rainfall record at a given station by comparing its accumulated annual record with that of the accumulated annual, or seasonal, mean values of several other nearby stations (Kohler, 1949). Figure 7.16 shows a double-mass analysis for the rain gage at Blair, Nebraska, and

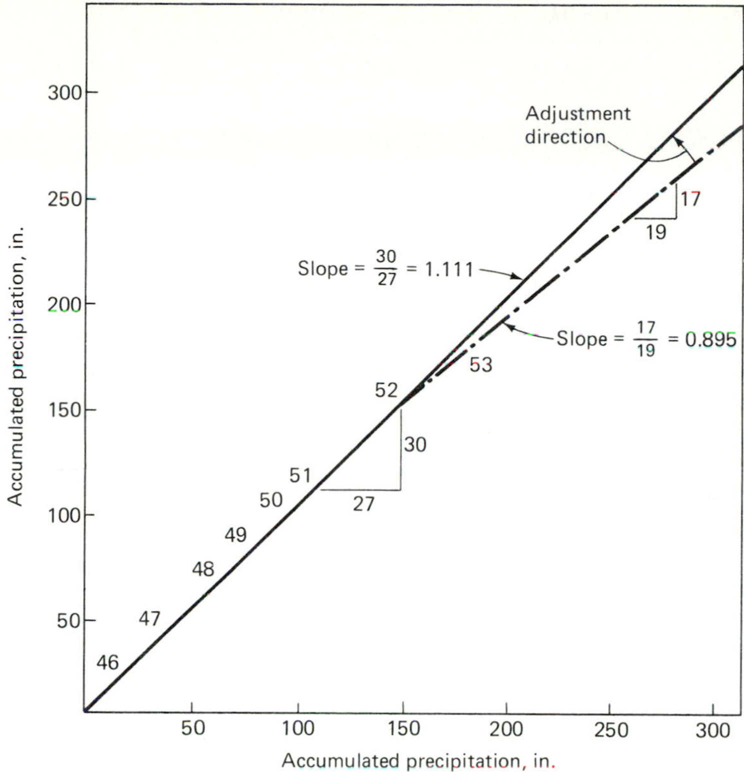

Figure 7.16 Double-mass analysis for checking consistency of rainfall at Blair, nebraska. On the x-axis is the eight-station accumulated mean precipitation and on the y-axis is the accumulated precipitation at Blair.

eight other stations. A change in the slope of the relationship occurred in 1952 when the gage was moved to another location. This change in slope indicates that the gage was moved or some other occurrence caused the gage to receive lesser amounts of rainfall. To make the present record consistent with the record prior to 1952, it will be necessary to adjust the record by the ratio of the slopes, or 1.11/0.895. When using the double-mass technique, it should be remembered that the data will have some scatter and an indicated change in slope should be confirmed by other evidence unless the change in slope is substantial (greater than 10%).

EXAMPLE 7.2 The average annual precipitation at station X and the average annual precipitation at 25 surrounding stations are given for 36 years in Table 7.2. Determine the consistency of record at station X. In what year is a change in regime indicated?

Solution We first compute the cumulative annual precipitation ΣP for station X as well as for 25 stations, as given in Table 7.3. Then we draw a double-mass curve using these cumulative values, as shown in Figure 7.17. It is seen that there is a change in the slope in the years 1954–1955. The slope of the line up to the year of break (1955–1956) is 1.187 and beyond 0.833. The difference in the two slopes is $1.187 - 0.833 = 0.354$, which equals $(0.354/1.187) \times 100 = 29.8\%$. Since the difference is greater than 10%, the data are inconsistent and require correction.

TABLE 7.2 ANNUAL PRECIPITATION IN INCHES

Year	Station X	25-Station Ave.	Year	Station X	25-Station Ave.
1976	18.8	26.4	1958	22.4	36.1
1975	18.5	22.86	1957	17.3	23.4
1974	31.0	38.6	1956	28.2	33.3
1973	29.5	29.7	1955	21.8	23.6
1972	20.8	28.4	1954	24.6	25.1
1971	28.7	35.1	1953	28.4	28.4
1970	18.3	23.6	1952	48.3	36.1
1969	30.5	37.1	1951	32.0	28.2
1968	22.9	23.4	1950	27.4	27.2
1967	21.6	29.0	1949	32.3	27.43
1966	22.35	28.2	1948	43.7	30.2
1965	20.3	24.64	1947	38.9	35.1
1964	28.5	26.4	1946	30.5	22.9
1963	29.5	33.3	1945	32.0	31.2
1962	20.6	23.1	1944	32.8	28.2
1961	27.0	23.4	1943	30.7	31.5
1960	24.1	23.1	1942	30.2	27.9
1959	28.5	31.2	1941	41.4	34.3

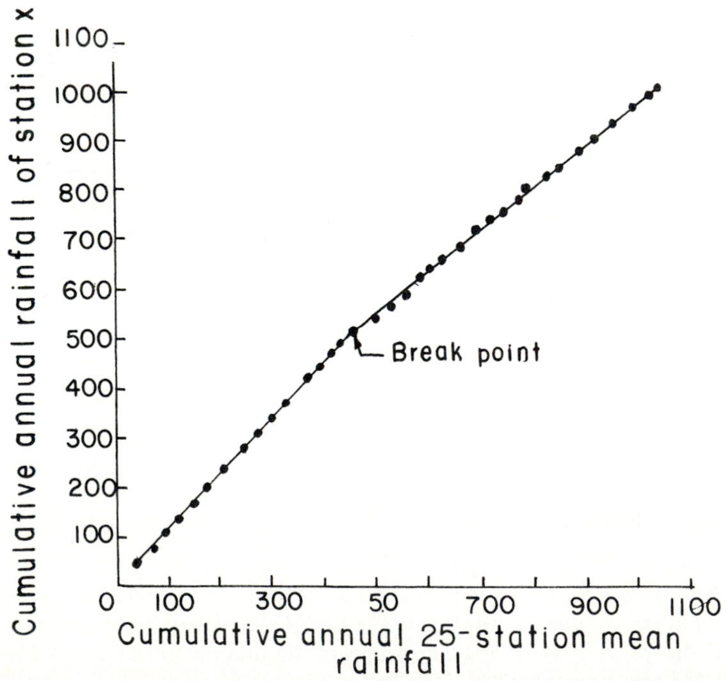

Figure 7.17 Double-mass curve for testing the homogeneity of the precipitation (inches) record at station X.

TABLE 7.3 CUMULATIVE ANNUAL PRECIPITATION

	Annual precipitation (in.)		Cumulative annual precipitation (in.)	
Year	Station X	25-Station Ave.	Station X	25-Station Ave.
1941	41.4	34.3	41.4	34.3
1942	30.2	27.9	71.6	62.2
1943	30.7	31.5	102.3	93.7
1944	32.8	28.2	135.1	121.9
1945	32.0	31.2	167.1	153.1
1946	30.5	22.9	197.6	176.0
1947	38.9	35.1	236.5	211.1
1948	43.7	30.2	280.2	241.3
1949	32.3	27.43	312.5	268.73
1950	27.4	27.2	339.9	295.93
1951	32.0	28.2	371.9	324.13
1952	48.3	36.1	420.2	360.23
1953	28.4	28.4	448.6	388.63
1954	24.6	25.1	473.2	413.73
1955	21.8	23.6	495.0	437.33
1956	28.2	33.3	523.2	470.63
1957	17.3	23.4	540.5	494.03
1958	22.4	36.1	562.9	530.13
1959	28.5	31.2	591.4	561.33
1960	24.1	23.1	615.5	584.43
1961	27.0	23.4	642.5	607.83
1962	20.6	23.1	663.1	630.93
1963	28.5	33.3	692.6	664.23
1964	28.5	26.4	721.1	690.63
1965	20.3	24.84	741.4	715.27
1966	22.35	28.2	763.75	743.47
1967	21.6	29.0	785.35	772.47
1968	22.9	23.4	808.25	795.87
1969	30.5	37.1	838.75	832.97
1970	18.3	23.6	857.05	856.57
1971	28.7	35.1	885.75	891.67
1972	20.8	28.4	906.55	920.07
1973	29.5	28.7	936.05	949.77
1974	31.0	38.6	967.05	988.37
1975	18.5	22.86	985.55	1011.23
1976	18.8	26.4	1004.35	1037.63

7.6 FILLING IN MISSING RECORDS

Many rain-gage records are incomplete. Breaks may vary in length from one or two days to several years. It is often necessary to estimate the missing data in order to utilize partial records, especially in data-sparse areas. Several methods are available for estimating missing data (Paulhus and Kohler, 1952; Aron and Rachford, 1974; Tung, 1983; Tabios and Salas, 1985), including the arithmetic average method, the normal ratio method, the inverse distance method, the modified inverse distance method, linear programming, the isohyetal method, the Lagrange method, interpolation methods, the Kriging method, and the total rainfall correlation method. Tabios and Salas (1985) made a comparative analysis of the methods of Thiessen polygons, Lagrange inverse distance, classical interpolation, multiquadratic interpolation, optimal interpolation, and Kriging. The Kriging, optimal interpolation, and multiquadratic interpolation methods were comparable and superior to other methods. However, the inverse distance and the Thiessen polygon methods gave fairly satisfactory results.

The problem of filling missing data or point rainfall estimation at an ungaged location involves transmittal of the rainfall amounts observed at nearby index gages to a gage with missing data or to an ungaged location, and can be formulated as

$$P_x = \sum_{i=1}^{N} a_i P_i, \qquad \text{where} \sum_{i=1}^{N} a_i = 1 \tag{7.5}$$

in which a_i is the weighting factor or factor of contribution of the ith gage, P_i is the rainfall at the ith gage, N is the number of index gages, and P_x is the rainfall to be estimated for gage x. Different methods of point rainfall estimation differ in the ways of estimating the weighting factors a_i, $i = 1, 2, \ldots, N$.

7.6.1 Arithmetic Average Method

The arithmetic average method averages the values of the precipitation of the surrounding gages and applies this value to the missing gage provided the normal annual precipitation of the surrounding gages is within 10% of the missing gage. For this method, $a_i = l/N$, $i = 1, 2, \ldots, N$.

7.6.2 Normal Ratio Method

If any of the surrounding gages has a normal annual precipitation exceeding 10%, then the normal ratio method is used, which weighs the effect of each surrounding precipitation gage. The normal ratio method is one of the simplest and most popular methods. This procedure is based on selecting m (m is usually 3) stations that are near and approximately evenly spaced around the station with the missing record. The missing data are estimated as

$$\begin{aligned} P_x &= \frac{1}{m}\left(\frac{N_x}{N_1} \cdot P_1 + \frac{N_x}{N_2} \cdot P_2 + \frac{N_x}{N_3} \cdot P_3 + \cdots + \frac{N_x}{N_m} P_m\right) \\ &= \sum_{i=1}^{m} \left(\frac{N_x}{mN_i}\right) P_i \end{aligned} \tag{7.6}$$

where P is the precipitation, N is the normal annual precipitation, x denotes the missing gage, and the subscripts $i = 1, 2, 3, \ldots, m$ denote the surrounding gages. Thus, the weight a_i assigned to the ith station for this method is $a_i = N_x/mN_i$. This method is useful where missing gage data are needed.

7.6.3 Inverse Distance Method

The inverse distance method involves computing weights of the surrounding rain gages on the basis of their distances from the gage with missing data (designate this gage as A). These distances are computed by establishing a set of axes running through this gage A; its location is (x_0, y_0), as shown in Figure 7.18. The squared distance D_i^2 between A and another gage i is $D_i^2 = (x_i - x_0)^2 + (y_i - y_0)^2$, $i = 1, 2, \ldots$; the location of the ith gage is (x_i, y_i). The weight of the ith station is

$$a_i = \frac{1/D_i^2}{\sum_{i=1}^{N} 1/D_i^2}$$

Normally, N is taken to be no more than 5.

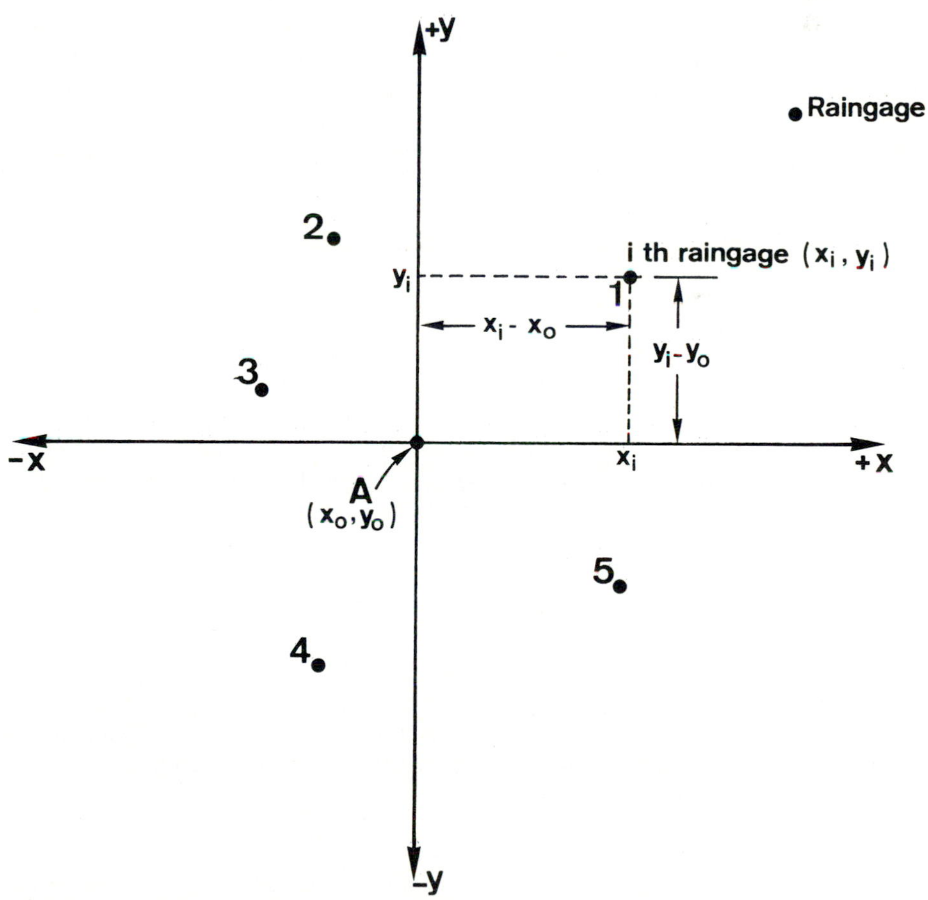

Figure 7.18 Establishing axes at the station with missing records for filling in the missing records.

EXAMPLE 7.3 Monthly precipitation in Washington, Baltimore, and Annapolis was observed to be 11.5 cm, 9.0 cm, and 12.4 cm, respectively. Precipitation for the same month could not be observed at Frederick. The normal annual precipitation values for Frederick, Washington, Baltimore, and Annapolis are 102, 114, 95, and 122 cm, respectively. Estimate the storm precipitation for Frederick. All these cities are within a 50-km radius.

Solution Let P_{ij} be the precipitation for the ith period ($i = m$ for monthly, and $i = A$ for annual) for the jth station (j = A for Annapolis, B for Baltimore, F for Frederick, and W for Washington).

$$P_{m\text{F}} = \frac{P_{\text{AF}}}{N}\left(\frac{P_{m\text{W}}}{P_{\text{AW}}} + \frac{P_{m\text{B}}}{P_{\text{AB}}} + \frac{P_{m\text{A}}}{P_{\text{AA}}}\right)$$

$$= \frac{102}{3}\left(\frac{11.5}{114} + \frac{9.0}{95} + \frac{12.4}{122}\right)$$

$$= 10.106 \cong 10.1 \text{ cm}$$

EXAMPLE 7.4 Table 7.4 lists the annual precipitation in inches for a part of the Tana River watershed in East Africa. (1) Estimate the five-station mean for the years of missing record. (2) Perform the consistency analysis of the precipitation values for Fort Hall, and adjust them for the period of record if necessary.

Solution For jth year, the precipitation at Fort Hall, P_{F}, is related to the five-station mean precipitation, P_{S}, as

$$\frac{P_{\text{F}}(j)}{\bar{P}_{\text{F}}} = \frac{P_{\text{S}}(j)}{\bar{P}_{\text{S}}}$$

Therefore,

$$P_{\text{S}}(j) = \frac{\bar{P}_{\text{S}}}{\bar{P}_{\text{F}}} P_{\text{F}}(j)$$

The computations are shown as follows. The cumulative precipitation for 26 years from 1937 to 1962 inclusive at Fort Hall is 1223, and that for five-station mean values is 1129.5. Therefore,

$$\bar{P}_{\text{F}} = 1223/26 = 47.038 \text{ in.}$$

$$\bar{P}_{\text{S}} = 1129.5/26 = 43.44 \text{ in.}$$

Using these values, the five-station mean annual precipitation is computed as given in Table 7.5.

For testing homogeneity of the precipitation record at For Hall, the cumulative precipitation is computed, as shown in Table 7.6. These are then plotted to construct a double-mass curve, as shown in Figure 7.19. From the straightness of the line, it is clear that the record is homogeneous.

EXAMPLE 7.5 Determine the rainfall amount, using the inverse distance method, at station A in Figure 7.18, if the rainfall at five surrounding stations is given as: station 1, P = 2.5 cm; station 2, P = 3.4 cm; station 3, P = 1.5 cm; station 4, P = 2.2 cm; and station 5, P = 1.8 cm. Station numbers are shown in Figure 7.17.

TABLE 7.4 ANNUAL PRECIPITATION IN INCHES FOR A PART OF THE TANA RIVER WATERSHED, EAST AFRICA

Year	Fort Hall	Five-station mean
1917	59.2	
1918	35.9	
1919	38.8	
1920	46.7	
1921	27.8	
1922	44.4	
1923	51.7	
1924	27.5	
1925	33.6	
1926	52.3	
1927	29.4	
1928	41.6	
1929	45.3	
1930	58.0	
1931	52.9	
1932	56.5	
1933	34.9	
1934	38.2	
1935	42.3	
1936	46.0	
1937	70.2	54.5
1938	39.7	40.0
1939	25.7	26.6
1940	48.1	43.2
1941	50.1	44.8
1942	49.3	41.7
1943	49.2	37.0
1944	27.7	32.1
1945	35.1	32.1
1946	47.1	44.5
1947	55.1	53.5
1948	47.3	43.0
1949	30.3	25.5
1950	51.3	44.6
1951	69.3	64.6
1952	37.7	34.9
1953	39.3	40.4
1954	52.3	44.5
1955	34.2	37.9
1956	53.0	44.7
1957	52.7	52.7
1958	59.0	50.8
1959	41.8	41.3
1960	36.9	35.2
1961	70.5	77.0
1962	50.1	42.4

City	Annual precipitation P_A (cm)	Monthly precipitation P_m (cm)
Washington (W)	114	11.5
Baltimore (B)	95	9.0
Annapolis (A)	122	12.4
Frederick (F)	102	?

TABLE 7.5 COMPUTING FIVE-STATION MEAN PRECIPITATION (INCHES) FOR THE YEARS OF MISSING RECORDS

Year	Fort Hall precipitation	Computed 5-station mean precipitation
1917	59.2	54.67
1918	35.9	33.15
1919	38.8	35.83
1920	46.7	43.13
1921	27.8	25.67
1922	44.4	41.00
1923	51.7	47.75
1924	27.5	25.40
1925	33.6	31.03
1926	52.3	48.30
1927	29.4	27.15
1928	41.6	38.42
1929	45.3	41.83
1930	58.0	53.56
1931	52.9	48.85
1932	56.5	52.18
1933	34.9	32.23
1934	38.2	35.28
1935	42.3	39.06
1936	46.0	42.48
Σ Year = 20	Σ Precipitation = 863 in.	Σ Precipitation = 796.97 in.

TABLE 7.6 CUMULATIVE PRECIPITATION (INCHES) FOR DOUBLE-MASS CURVE

Year	Cumulative precipitation at Fort Hall	Cumulative precipitation of 5-station mean values
1962	50.1	42.40
1961	120.6	119.40
1960	157.5	154.60
1959	199.3	195.90
1958	258.3	246.90
1957	311.0	299.40
1956	364.0	344.10
1955	398.2	382.00
1954	450.5	426.50
1953	489.8	466.90
1952	527.5	501.80
1951	596.8	566.40
1950	648.1	611.00
1949	678.4	636.50
1948	725.7	679.50
1947	780.8	733.00

TABLE 7.6 (*Continued*)

Year	Cumulative precipitation at Fort Hall	Cumulative precipitation of 5-station mean values
1946	827.9	777.50
1945	863.0	809.60
1944	890.7	841.70
1943	939.9	878.70
1942	989.2	920.40
1941	1039.3	985.20
1940	1037.4	1008.40
1939	1113.1	1035.00
1938	1152.8	1075.00
1937	1223.0	1129.50
1936	1269.0	1171.98
1935	1311.3	1211.04
1934	1349.5	1246.32
1933	1384.4	1278.55
1932	1440.9	1330.73
1931	1493.8	1379.58
1930	1551.8	1433.14
1929	1597.1	1474.97
1928	1638.7	1513.39
1927	1668.1	1540.54
1926	1720.4	1588.84
1925	1754.0	1619.87
1924	1781.5	1645.27
1923	1833.2	1693.02
1922	1877.6	1734.02
1921	1905.4	1759.69
1920	1951.1	1802.82
1919	1989.9	1838.65
1918	2025.8	1871.80
1917	2085.0	1926.47

Solution First, x- and y-axes are drawn through station A, and distances $[(x_i - x_0), (y_i - y_0)]$ are measured for all the surrounding stations. Calculations for obtaining rainfall at station A are shown in Table 7.7.

TABLE 7.7 ESTIMATION OF RAINFALL AT STATION A BY THE INVERSE DISTANCE METHOD[a]

Station number	Rainfall (cm)	$x_i - x_0$	$y_i - y_0$	D^2	D^{-2}	a_i	$P_i a_i$
1	2.5	1.2	0.9	2.25	0.44	0.12	0.3
2	3.4	0.5	1.1	1.46	0.68	0.19	0.65
3	1.5	0.8	0.3	0.73	1.37	0.38	0.57
4	2.2	0.5	1.2	1.69	0.59	0.16	0.35
5	1.8	1.1	0.8	1.85	0.54	0.15	0.27

[a] $\Sigma D_i^{-2} = 3.62$; $P_A = \Sigma P_i a_i = 2.14$ cm.

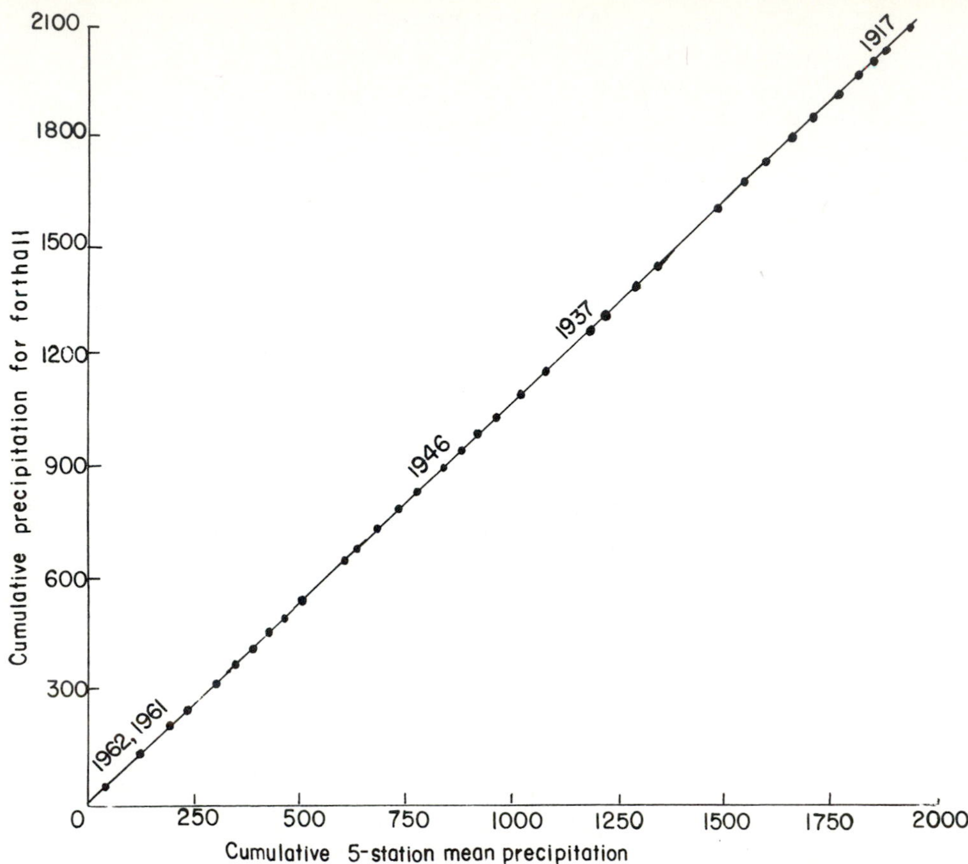

Figure 7.19 Double-mass curve for testing the homogeneity of the precipitation record at Fort Hall in the Tana River watershed, east Africa.

7.7 EXTENSION OF A POINT-RAINFALL RECORD

Point-rainfall data, or rainfall data recorded by a rain gage, are used in a variety of applications. When used in frequency analysis or for construction of rainfall intensity–duration–frequency curves, the rainfall record may not be of sufficient length. It may then be necessary to extend the point-rainfall record. One simple and popular method to accomplish the extension of the record is the station–year method.

The station-year (SY) method combines the records of a number of rain gages (including the rain gage whose record is to be extended) into a single composite record. These records may correspond to the same or different periods. The combined record is of the length equal to the sum of the lengths of individual records. For example, consider 10 rain gages each with 15 years of record. By combining these records, a composite record of $10 \times 15 = 150$ years is produced. Thus, any of these 10 rain gages can be assumed to have an equivalent record of 150 years by virtue of the SY method, and this composite record can then be used for intended analysis. The composite record is presumed to have the same characteristics as a 150-year record would have if it were recorded at one of the gages.

The SY method must satisfy the following restrictions before it can be applied: (1) The individual rain-gage records must be independent. This means that the rain gages are located sufficiently far apart and that these gages do not duplicate recording of the same rainfall values at the same time. (2) The area in which the gages are located is meteorologically homogeneous. This means that rainfall at all of the gages is generated by the same types of storms and sources. One rough indicator of the homogeneity is that the annual rainfall of all the gages is approximately the same. (3) The rainfall frequency distribution is about the same for all the gages, at least in the long run.

7.8 MEAN AREAL PRECIPITATION

Many factors affect spatial distribution of rain falling on the ground. Rain-gage networks must be designed to optimally sample this rainfall distribution. A method of estimating mean areal rainfall must be able to represent this distribution in a reasonable manner. In order to accomplish a reasonably reliable estimate of mean precipitation over a drainage basin, one should search for all rain-gage information that can be found as for a bucket survey.

Although there are several methods to estimate mean areal rainfall, no method accurately represents its distribution. The methods of estimating mean areal rainfall reported in hydrologic literature include the arithmetic mean, the grouped aspect weighted mean, the Thiessen polygon method, the individual area altitude weighted mean, the triangular weighted mean, the Myers method, the isohyetal method, trend surface analysis, the reciprocal distance squared method, the two-axis method, the double Fourier series method, the modified polygon method, the finite-element method, the analysis of variance, and the Kriging method. Comprehensive discussions of most of these methods can be found in Whitmore et al. (1961), Singh and Birsoy (1975a, 1975b), and Singh and Chowdhury (1986).

The mean areal rainfall P can simply be expressed as a weighted average of rainfall values of surrounding gages:

$$\bar{P} = \sum_{i=1}^{N} a_i P_i \tag{7.7}$$

in which N is the number of gages, a_i is the weight assigned to the ith gage, and P_i is the rainfall observed by the ith gage. The weights must satisfy

$$\sum_{i=1}^{N} a_i = 1 \tag{7.8}$$

and

$$0 \leq a_i \leq 1, i = 1, 2, \ldots, N \tag{7.9}$$

The methods of estimating $\bar{P}$ differ in the choice of weights only. In some cases, the weights depend only on the distance between gages. In other cases, the weights are optimized on the basis of correlation functions. Three of the simplest and most popular methods are discussed here.

7.8.1 Arithmetic Method

The arithmetic method is simply the arithmetic average of the readings of all the rain gages within the drainage area. In other words, all the weights in Equation (7.8) are the same, $a_i = 1/N$, $i = 1, 2, \ldots, N$. The accuracy of this method depends on the density and location of the gages. This method assumes that each gage represents the average rainfall that falls all around the gage. This assumption can be erroneous because of topographic and wind influences. Rain gages outside the boundary of the watershed should not be used for arithmetic averaging.

7.8.2 Thiessen Polygon Method

The Thiessen method attempts to define the area represented by each gage in order to weigh the effects of nonuniform rainfall distribution. The rain-gage stations are plotted on a map and each gage is connected to each adjoining gage by a line. Perpendicular bisectors of these connecting lines are drawn to form a polygon around each gage. The area within each polygon is assumed to be the area represented by each gage. This area is determined by a planimeter and is expressed as a fraction of the total area. This percent area of a gage defines its weighting factor a_i, $i = 1, 2, \ldots, N$, in Equation (7.8); $a_i = A_i/A$, where A_i is the polygon area of the ith gage, and A is the total drainage area. The weighted average rainfall for the drainage basin is computed by multiplying the rainfall at each station by its assigned area within the drainage basin, represented by the polygon area divided by the total drainage area, and totaling the values. The Thiessen method is believed to be more accurate than the arithmetic average. An advantage of the Thiessen method is that once the polygons have been defined and their areas measured, it is a simple procedure to compute the mean rainfall for other events. If the rain-gage network is changed, then new polygons must be constructed and their areas measured. A disadvantage of this method is that no allowance is made for topographic influences on rainfall distribution. The location of the rain gages is the sole information needed.

7.8.3 Isohyetal Method

The isohyetal method is generally considered to be the most accurate method for computing the mean rainfall over a drainage basin. This method is accomplished by plotting the rainfall values on a map of the rain-gage locations. Contours of equal rainfall amounts (isohyets) are then drawn. The choice of contour interval depends on the variation of rainfall amounts and is a matter of judgment. The accuracy of the computations depends on the adequacy of the rainfall isohyets. The mean rainfall is determined by planimetering the area between adjoining contours within the drainage basin, multiplying this area by the mean rainfall between contours, and then tabulating. These tabulated values are summed and divided by the total area of the drainage basin to obtain the mean rainfall for the drainage basin. Thus, the weights

a_i's are $a_i = A_i/A$, $i = 1, 2, \ldots$, where A_i is the area between two isohyets, and A is the watershed area. The advantage of the isohyetal method is that it permits consideration of topographic effects and is, therefore, more accurate in mountainous areas. A disadvantage of this method is that it requires planimeter measurements for each storm event.

EXAMPLE 7.6 Consider a rectangular area whose (x, y) coordinates are (0,0) (4,0), (0,4), and (4,4). The area is 4 km wide and 4 km long (one unit coordinate represents 1 km). This area has four rain gages. The locations of the gages and the rainfall amounts measured by them are as follows:

RAIN GAGE	GAGE COORDINATES (x,y) (KM, KM)	RAINFALL AMOUNT (CM)
A	(1,1)	5
B	(1,3)	10
C	(3,3)	8
D	(3,1)	12

Determine the mean rainfall for this area by the Thiessen polygon method, and compare it with that obtained by the arithmetic mean method.

Solution The mean areal rainfall $\bar{P}$ by the arithmetic method = (5 + 10 + 8 + 12)/4 = 8.75 cm.

To determine $\bar{P}$ by the Thiessen polygon method, first, the Thiessen polygons are constructed around each station, as shown in Figure 7.20. Stations are connected by straight lines (dashed lines in the figure) and perpendicular bisectors (solid lines in the figure) are then drawn on these connecting lines. The bisectors form polygons around

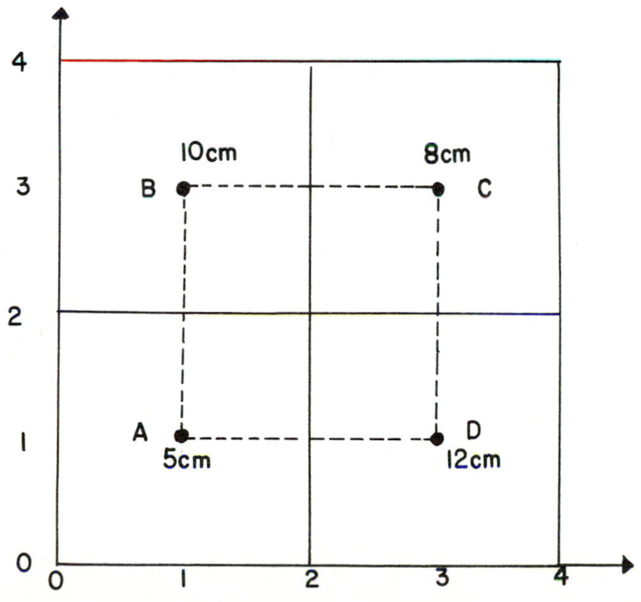

Figure 7.20 Construction of Thiessen polygons for a rectangular area.

each station. The area of the polygons are then planimetered, and the value of $\bar{P}$ is then obtained as

RAINFALL AMOUNT (CM)	THIESSEN POLYGON AREA (KM^2)	THIESSEN WEIGHTS (POLYGON AREA/ TOTAL AREA)
5	4	4/16 = 0.25
10	4	0.25
8	4	0.25
8	4	0.25
	Total area = 16 km^2	

Therefore,

$$\bar{P} = 0.25 \times 5 + 0.25 \times 10 + 8 \times 0.25 + 12 \times 0.25 = 8.75 \text{ cm}$$

Both methods yield the same value for this example.

EXAMPLE 7.7 Rainfall values observed at different points on a watershed (5.41 km^2) are shown in Figure 7.21. Compute the mean watershed rainfall by the arithmetic mean.

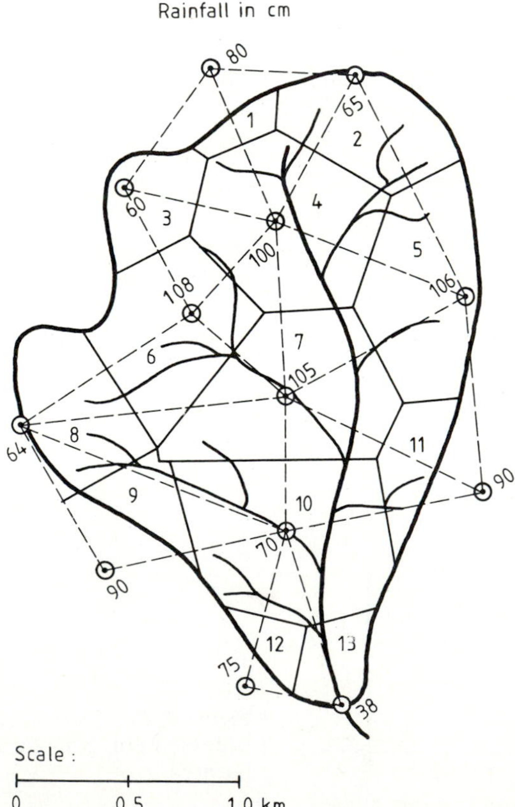

Figure 7.21 The Thiessen polygon method for computing the mean areal rainfall.

Solution

$$\bar{P} = \frac{1}{9}(64 + 60 + 65 + 108 + 100 + 106 + 105 + 70 + 38)$$

$$= 716/9 = 79.56 \text{ cm}$$

EXAMPLE 7.8 Compute the mean rainfall for the data in Example 7.7 using the Thiessen polygon method.

Solution The Thiessen polygons are constructed as shown in Figure 7.21. Calculations for this method are shown in Table 7.8.

TABLE 7.8 MEAN AREAL PRECIPITATION BY THE THIESSEN POLYGON METHOD[a]

Area number (1)	Observed precipitation (cm) (2)	Area (mi^2) (3)	Percent total area (4)	Weighted precipitation (5) = (2) × (4)/100
1	80	0.06	1.1	0.88
2	65	0.30	5.5	3.58
3	60	0.24	4.4	2.64
4	100	0.88	16.3	16.27
5	106	0.69	12.8	13.78
6	108	0.68	12.6	14.04
7	105	0.79	14.6	15.33
8	64	0.51	9.4	6.03
9	90	0.09	1.7	1.50
10	70	0.91	16.8	11.77
11	90	0.15	2.8	2.5
12	75	0.03	0.6	0.42
13	38	0.08	1.48	0.56
Total		5.41	100	89.43

[a] Area is 5.41 km^2; $\bar{P}$ = 89.43 cm.

EXAMPLE 7.9 Compute the mean rainfall for the data in Example 7.7 using the isohyetal method.

Solution Isohyets are drawn as shown in Figure 7.22. Calculations for this method are shown in Table 7.9.

7.8.4 Areal Rainfall from Point Rainfall

Tables have been developed that permit estimating areal rainfall by converting from point rainfall. These tables have been prepared by Huff and Neill (1957) but should be used cautiously because they can be in error substantially, depending on the type of storm causing the rainfall and the location of that gage within the storm. This issue will be addressed in more detail in Chapter 26.

Rainfall in cm

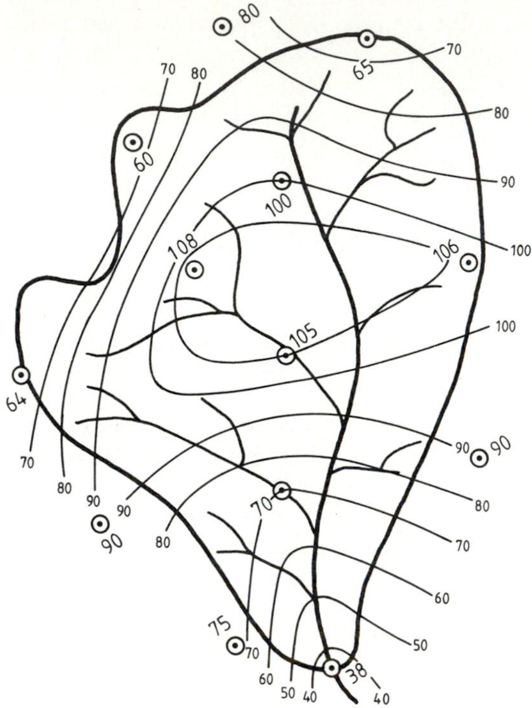

Scale :

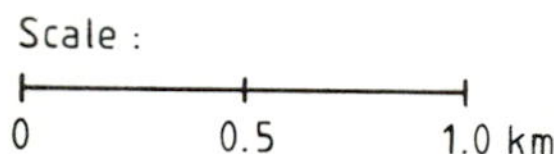

Figure 7.22 The isohetal method for computing the mean areal rainfall.

TABLE 7.9 MEAN AREAL PRECIPITATION BY THE ISOHYETAL METHOD[a]

Isohyet (cm) (1)	Areal enclosed (km^2) (2)	Net area (3)	Average precipitation (cm) (4)	Precipitation volume (km^2-cm) (5) = (3) × (4)
105	0.79	0.79	106.5	84.14
100	1.52	0.73	107.9	78.77
90	2.57	1.05	95	99.75
80	3.47	0.90	85	76.5
70	4.5	1.03	75	77.25
60	5.18	0.68	65	44.2
40	5.39	0.04	45	1.8
<40	5.41	0.02	39	0.78
	5.41			463.19

[a] $\bar{P} = 463.19/5.41 = 85.6$ cm.

7.9 GRAPHICAL REPRESENTATION OF RAINFALL DATA

Point precipitation data may be in the form of daily totals from nonrecording gages or on recorder charts in the form of mass curves or time-accumulation depth charts from recording gages. Rainfall intensities and their durations, as well as storm duration, can be obtained for each storm from time-accumulation depth charts. The durations for computing rainfall intensities can be selected depending upon the use of the data. For some hydrologic projects, hourly rainfall data may be sufficient.

Rainfall data are usually presented in two common graphical ways. A mass curve, shown in Figure 7.23, is a plot of the accumulated rainfall for the duration of the storm. This mass curve is plotted from data provided by recording rain gages. The storm commenced just prior to 2:00 P.M. and ended at about midnight. The most intense portion of the rainfall occurred between 4:30 and 7:30 P.M. Rainfall intensi-

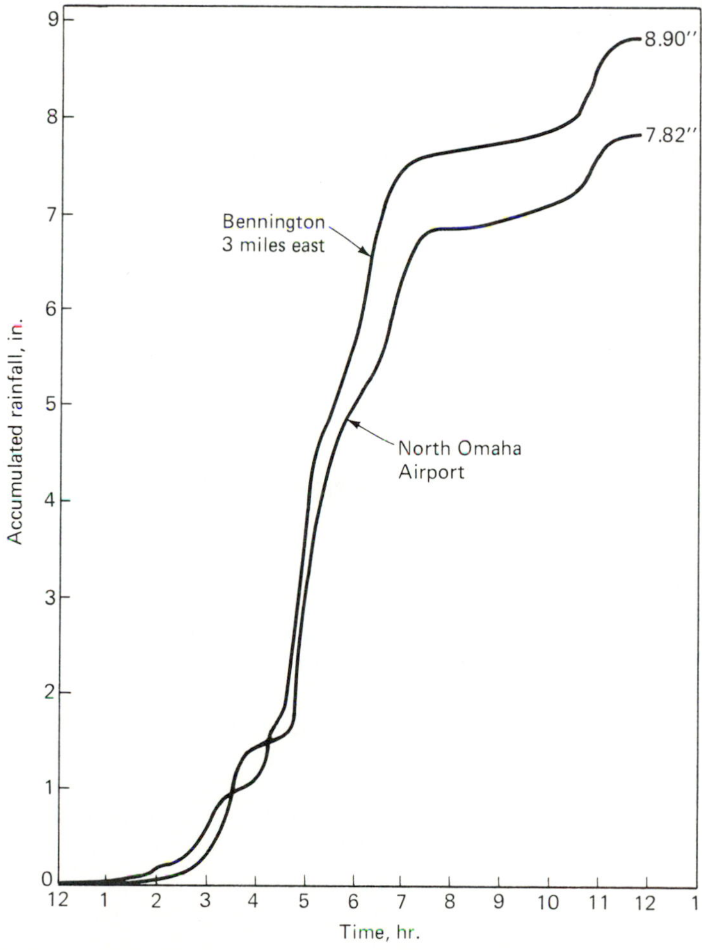

Figure 7.23 Rainfall mass curve for the storm of September 6–7, 1965, on the Irvington drainage basin, near Omaha, Nebraska.

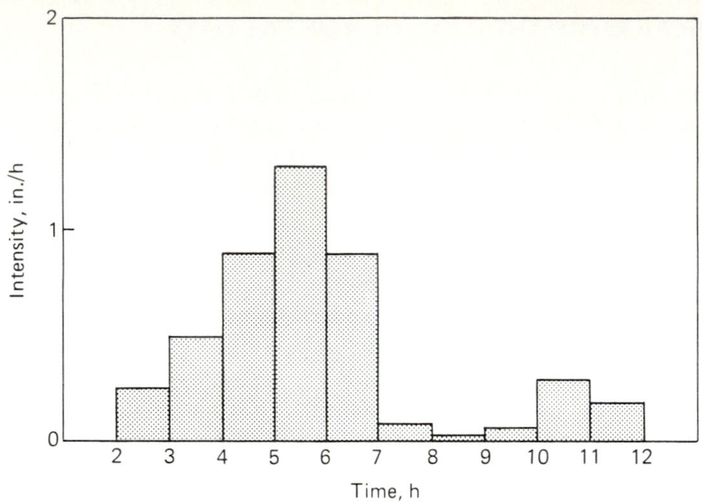

Figure 7.24 Hyteograph for the Bennington, Nebraska, gage for September 6–7, 1965.

ties can be computed for any period of the rainfall event. This mass curve allows for analysis of rainfall intensities and amounts for any portion of the storm. A hyetograph of the Bennington gage for the storm of September 6–7, 1965, is shown in Figure 7.24. Intensities for 1-hour periods were computed and plotted as a histogram against time. This graphic presentation can be made for any time periods within the storm duration. Such a hyetograph is useful in identifying the intensity, periods, and duration of the most intense rainfall and for use with infiltration data.

EXERCISES

7.1. Plot a bar diagram of monthly rainfall for your area for 1986. Also plot a bar diagram of the mean monthly rainfall. Comment on the temporal variability of rainfall in this area.

7.2. Estimate the mean areal rainfall for the area shown in Figure 7.25, using the arithmetic mean, the Thiessen polygon, and the isohyetal methods. Compute weighting factors for each method.

7.3. Estimate the mean areal rainfall using the isohyetal method. The area is 3000 hectares, and areas within isohyetal lines area are as follows:

ISOHYETAL INTERVAL (CM)	ENCLOSED AREA (HA)
0–1	150
1–2	900
2–3	600
3–4	450
4–5	900

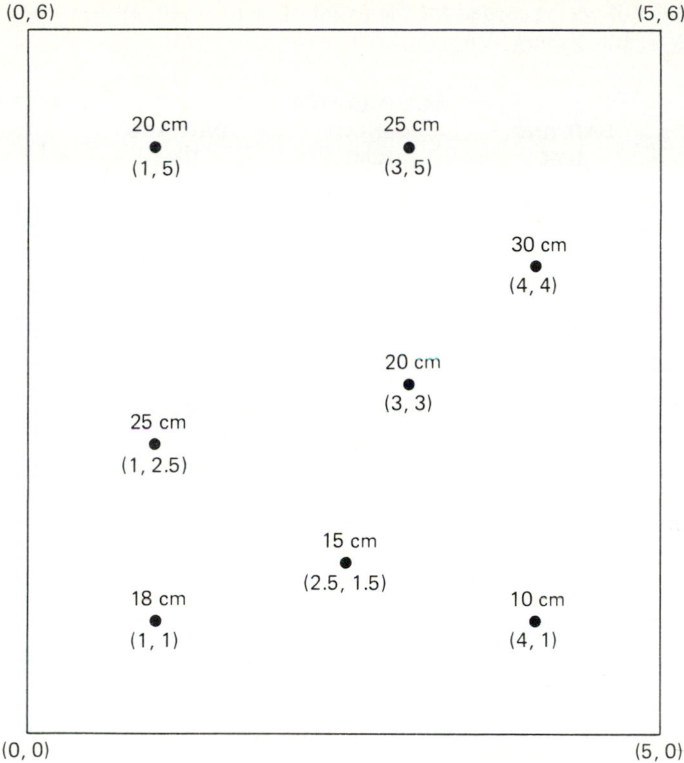

Figure 7.25 Observed rainfall in an area.

Assume that within each enclosed area a rain gage is located whose measured rainfall is the average of the isohyetal interval. Estimate the mean areal rainfall by the arithmetic mean method, and compare this estimate with the one obtained from the isohyetal method.

7.4. Remove the rain gage located at the point (3,3) in Exercise 7.2, and then construct the Thiessen polygons and estimate the weighting factors. What is the mean areal rainfall now? Comment on the effect of removing the rain gage on the Thiessen polygons.

7.5. Assume that rainfall is not known at the gage located at the point (3,3) in Exercise 7.2. Compute the rainfall at this point using the normal ratio method. The normal precipitations of the three neighboring gaging stations are as follows:

STATION COORDINATES	NORMAL ANNUAL PRECIPITATION (CM)	PRECIPITATION (CM)
(1,2.5)	28	25
(4,1)	15	10
(3,5)	30	25
(3,3)	25	?

7.6. Use the inverse distance method to solve Exercise 7.5, and compare the result with that of the normal ratio method.

7.7. Rainfall was recorded for the event of October 10, 1959, for Powells Creek watershed W-1, Blacksburg, Virginia.

DATE AND TIME	ACCUMULATED RAINFALL (CM)	DATE AND TIME	ACCUMULATED RAINFALL (CM)
10-8-59			
4:38 P.M.	0	5:20 P.M.	3.81
4:40 P.M.	0.076	5:25 P.M.	3.886
4:44 P.M.	0.152	5:40 P.M.	4.039
4:50 P.M.	0.406	5:50 P.M.	4.166
4:55 P.M.	0.584	6:00 P.M.	4.242
5:00 P.M.	1.219	6:20 P.M.	4.496
5:04 P.M.	2.210	6:48 P.M.	4.572
5.07 P.M.	2.845	6:50 P.M.	4.674
5.11 P.M.	3.607	7:00 P.M.	4.750
5:13 P.M.	3.734	7:10 P.M.	4.801

Construct the rainfall hyetograph (rainfall intensity versus time) for this event. What are the maximum and average rainfall intensities for this record?

7.8. A watershed has five rain gages. The annual rainfall recorded by these gages for 1986 is as follows:

Rain Gage	1	2	3	4	5
Annual Rainfall (cm)	80.5	100.5	150.3	90.7	120.2

Calculate the optimum number of rain gages for the watershed if a 10% error in the mean areal rainfall is acceptable.

7.9. Test the consistency of a record of 25 years of annual rainfall of a rain gage designated as A. Mean annual rainfall for five neighboring stations located in a meteorologically homogeneous region is available. This mean and rainfall data for rain gage A are as follows:

YEAR	1956	'57	'58	'59	'60	'61	'62	'63	'64
Rain Gage A (cm)	142	115	142	130	155	134	157	115	128
5-Rain-Gage Mean (cm)	114	106	117	118	129	124	122	94	102

YEAR	'65	'66	'67	'68	'69	'70	'71	'72	'73
Rain Gage A (cm)	157	113	126	116	106	76	118	114	112
5-Rain-Gage Mean (cm)	154	125	131	124	114	92	108	130	108

YEAR	'74	'75	'76	'77	'78	'79	'80
Rain Gage A (cm)	104	110	104	130	125	140	145
5-Rain-Gage Mean (cm)	114	104	117	129	140	133	163

Indicate the year in which the change in rainfall regime occurred. Adjust the annual rainfall of rain gage A, and compute the difference between the recorded and the adjusted annual rainfall for this gage.

7.10. Plot Equation (7.1) for various values of C_v and ε, and discuss the plot.

7.11. Figure 7.26 shows a 1-km^2 basin having three rain gages. Estimate the mean areal rainfall by the Thiessen method for the storm whose rainfall depths are given on the map. The areas of the polygons around gages A, B, and C, respectively, are 0.2, 0.4, and 0.4 km^2. Compare it with the arithmetic mean rainfall.

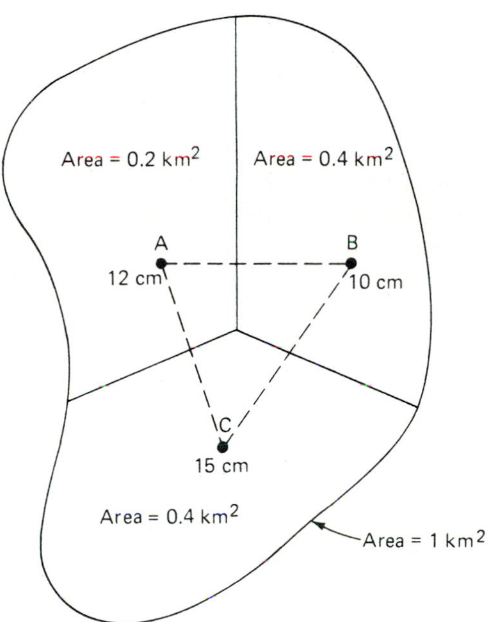

Figure 7.26 A watershed showing rain-gage locations.

7.12. Double-mass analysis is used for testing the consistency of precipitation data and for adjusting, interpolating, or extending such data. The following precipitation data are given:

YEAR	STATION A	MEAN OF 8 BASE STATIONS	YEAR	STATION A	MEAN OF 8 BASE STATIONS
1946	15.2	16.1	1956	19.9	16.4
1947	22.6	23.8	1957	28.2	24.6
1948	18.3	19.2	1958	16.9	15.4
1949	19.1	21.7	1959	22.7	18.8
1950	17.9	19.3	1960	21.3	18.2
1951	16.2	18.3			
1952	27.9	29.8			
1953	23.1	21.8			
1954	12.4	14.1			
1955	18.3	17.5			

In checking the historical record of Station A, it was noted that the station was moved during the latter part of 1950.

(a) Test the consistency and the general quality of the data for Station A by double-mass analysis.

(b) Adjust the data as required.

7.13. **(a)** How could you extend short-term rainfall records?

(b) What criteria would you use to substantiate the procedure?

7.14. **(a)** Explain how you estimate missing data for a rain gage in a rain-gage network.

(b) How would you check the consistency of the data for a gage?

7.15. If you wish to extend rainfall records in order to estimate return periods of rainfall, what two principal cautions should you exercise?

7.16. For what purpose would you use double-mass analysis?

7.17. **(a)** How can existing meteorological records be extended?

(b) What are the two criteria that must be satisfied prior to extending meteorological records?

7.18. Describe the relationship between

(a) area and rainfall depth

(b) area and rainfall intensity

(c) area and rainfall duration

7.19. If you were performing a rainfall study, how would you determine if a rain gage had a consistent recording history?

7.20. What are the methods of determining average precipitation over an area? Describe the merit of each.

7.21. In a given river basin, station A was not operating during the occurrence of a given storm, while stations B, C, and D, which were adjacent to station A, registered, respectively, precipitations totaling 12.3, 14.8, and 11.9 cm. Given the mean annual precipitations at the four stations, A, B, C, and D, as 1290, 1510, 1680, and 1375 mm, respectively, estimate the missing storm precipitation at A.

7.22. Fill in the blank spaces for obtaining the average basin precipitation by the Thiessen polygon method.

Observed precipitation	Area (km^2)	% of total area	Weighted precipitation	Average basin precipitation
A	B	C	$D = A \times C/100$	$E =$ Sum of $(A \times B/626)$
2.69	108	—		
1.54	35	—		
2.82	18	—		
1.92	110	—		
1.46	107	—		
0.65	71	—		
2.98	83	—		
4.50	66	—		
5.00	12	—		
1.95	9	—		
1.75	7	—		
	626	100		

7.23. In the semiarid Paraiba State in northeast Brazil, the recorded annual precipitations at given rain gage stations for the years 1973–1980 are as given:

	Mean annual precipitation at the stations (mm)				
Year	Sume A	Monteiro B	Jatoba C	Umburana D	Gangorra E
1973	—	576.5	629.9	441.8	596.6
1974	1210.1	1083.1	1009.0	951.3	1011.0
1975	950.0	752.9	904.6	603.7	706.5
1976	690.0	674.9	586.3	622.5	573.0
1977	860.0	557.1	695.8	803.2	798.4
1978	598.8	522.8	753.6	850.9	714.2
1979	503.4	460.8	625.1	403.4	479.8
1980	348.0	475.7	519.5	301.2	361.2

Find the missing annual precipitation for the year 1973 at Station A (Sume) and check the consistency of precipitation of all stations by double-mass curve analysis.

Suggest any corrections to be applied. Show the corrected readings for the station in question. Use the graphical methods for Stations A and B and A and C and the ratio method for Stations D and E.

7.24. Tabulated are the data derived from a drainage basin of 17,500 hectares, giving the areas covered within each of the isohyetal lines. Determine the average depth of precipitation for the storm within the basin by the isohyetal method.

No.	1	2	3	4	5	6	7
Interval of Isohyets (cm)	0–2	2–4	4–6	6–8	8–10	10–12	12–14
Enclosed Area (hectares × 1000)	5.3	4.4	3.2	2.6	2.3	1.9	1.4

7.25. The monthly precipitation depths (mm) for the year 1964 of rain gages at Cabaceiras, Boa Vista, Pocinhos, and Sao Joao do Cariri are given in the table. Estimate the missing values at Boa Vista and Pocinhos using the averages given in the last column for the years 1935–1963.

Months	J	F	M	A	M	J	J	A	S	O	N	D	Ave.
Cabaceiras	67.0	65.4	105.7	126.6	184.8	176.0	50.0	0.0	0.0	0.0	0.0	0.0	252.0
Boa Vista	71.5	132.5	72.9	??	76.2	86.8	100.1	8.5	22.5	1.7	0.0	0.0	410.0
Pocinhos	114.3	??	66.4	69.2	51.6	112.1	97.1	29.2	26.0	2.1	1.9	1.7	353.0
Sao Joao do Cariri	43.0	53.0	80.0	122.2	73.8	71.3	0.0	0.0	0.0	0.0	0.0	0.0	375.0

7.26. For a river basin in Sume Paraiba, northeast Brazil, the Thiessen polygon network is as given in Figure 7.27. Using the data of mean annual precipitation at 29 stations in a year given below, calculate the mean basin precipitation.

Station	Mean annual precipitation (mm)	Net area (km²)	Station	Mean annual precipitation (mm)	Net area (km²)
2	975.0	5.58	33	573.0	2.19
3	892.0	6.59	34	198.0	5.81
4	585.0	3.89	35	375.0	8.99
5	687.0	4.35	36	597.0	0.81
6	170.0	9.19	37	927.0	1.08
7	240.0	8.22	38	925.0	3.39
8	547.0	3.50	39	913.0	6.74
9	555.0	2.36	40	696.0	1.97
23	270.0	9.73	41	572.0	2.60
24	865.0	6.03	43	634.0	0.50
28	962.0	3.68	44	585.0	2.42
29	613.0	1.79	45	712.0	2.46
30	785.0	3.23	46	701.0	1.03
31	632.0	2.98	67	612.0	—
32	315.0	4.60	48	597.0	—

Note: Stations 43, 47, and 48 have a total representative area of 2.87 km².

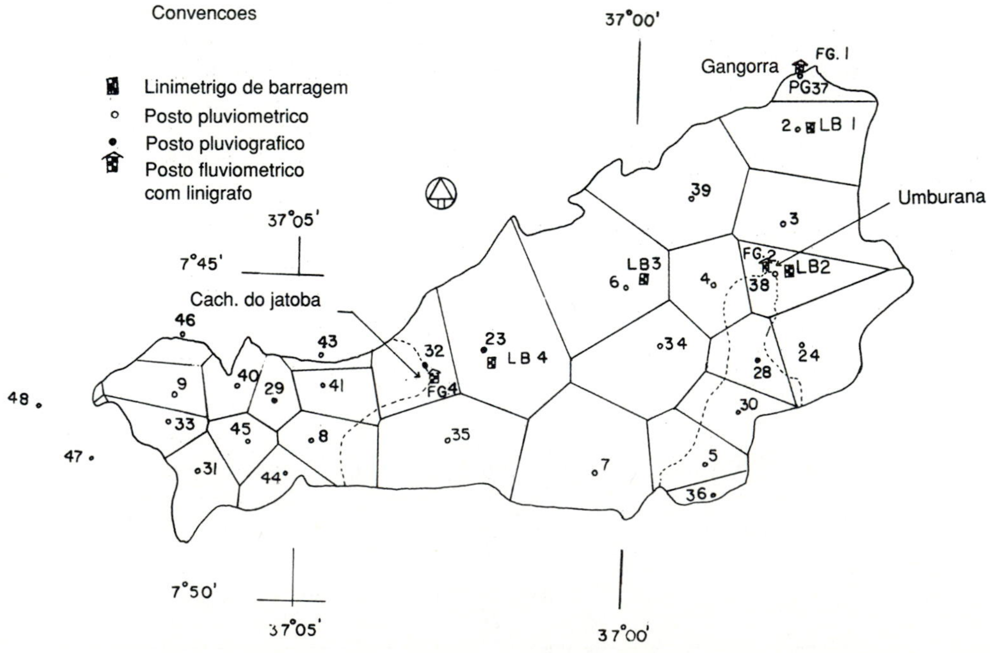

Figure 7.27 A watershed in Sume Paraiba, northeastern Brazil.

7.27. Observed precipitation for a storm that occurred on March 12, 1987, on the representative basin of Sume, Pb, Brazil, shown in Figure 7.27, at the 28 rain gage stations is as shown. The missing numbers of rain gages are those that do not pertain to the hydrological basin in question. The total area of the basin is 137.4 km^2.

Station Number	2	3	4	5	6	7	8	9	23	24	28	29	30	31	32
Observed Storm Precip. (mm)	25	31	12	35	62	17	81	23	24	31	28	35	47	11	27

Station Number	33	34	35	36	37	38	39	40	41	43	44	45	46	(47	(48) outside the basin
Observed Storm Precip. (mm)	196	49	27	33	29	55	31	47	54	27	83	42	55	(62	78)

(The effective area for stations 33, 47, and 48 = 2.8703 km^2).

(a) Construct the storm isohyetal map.

(b) Plot the area–depth curve for the above storm.

(c) Estimate the average depth of precipitation over the basin from the isohyetal map.

(d) Find the depth of precipitation over the basin by the Thiessen polygon method.

(e) Get the correction factor to be applied to the Thiessen polygon station weights.

7.28. Rainfall values observed at different points on the Atterbury watershed in Arizona are shown in Figure 7.28. Compute the mean areal rainfall by the arithmetic mean method.

7.29. Compute the mean areal rainfall for the data in Exercise 7.28 by the Thiessen polygon method.

7.30. Compute the mean rainfall for the data in Exercise 7.28 by the isohyetal method.

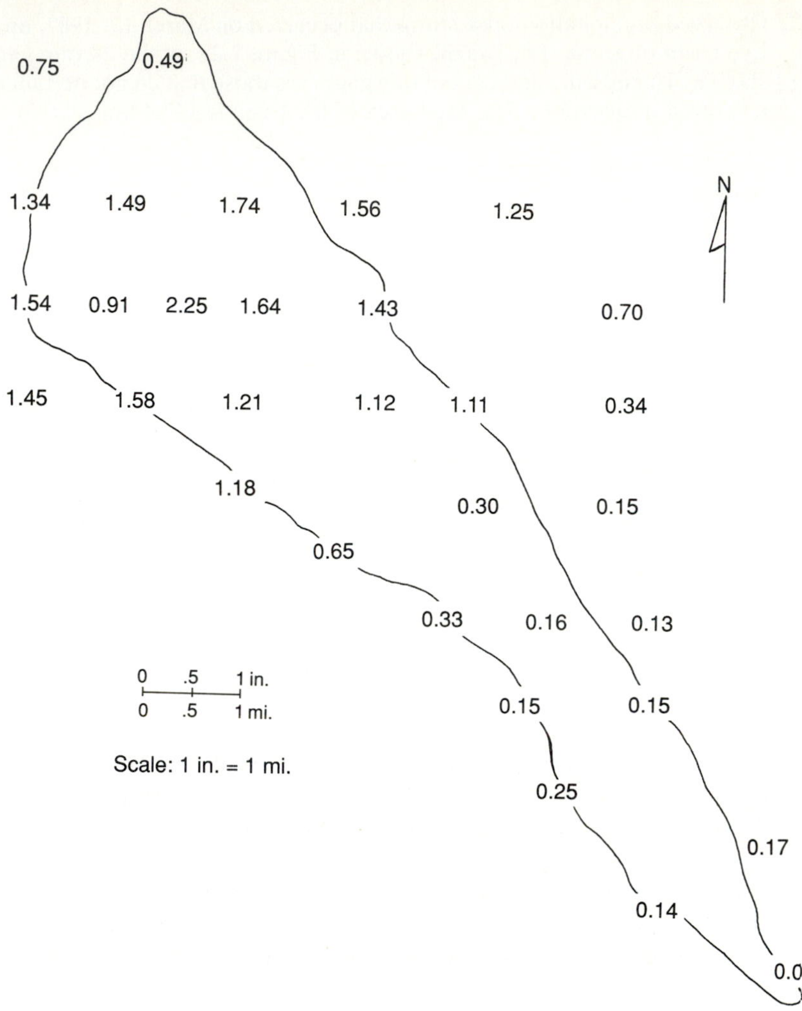

Figure 7.28 Rainfall for the Atterbury, Arizona, watershed.

PART 3
Subsurface Flow

CHAPTER 8
Infiltration and Soil Moisture

Infiltration is a vital natural process. Runoff, one of the most important components of the hydrologic cycle, is determined by subtracting from precipitation the abstractions that occur in the drainage basin. The most important abstraction affecting runoff is usually the portion of precipitation that is lost to infiltration. Infiltration supplies most of the water to plants and many animals. Were it not for infiltration, streams would cease to flow soon after a rain, frequent floods would devastate the river valleys, wells would go dry, springs would stop, farm soil would erode, crops would not grow, and droughts would be more frequent. This chapter presents basic concepts of infiltration and soil moisture, and the methods of determining them. It also shows the association of hydrology with soil science.

8.1 INFILTRATION PROCESS

When rainfall or melted snow contacts the soil surface, part of that water finds its way downward into the soil through its pore openings. Soil is, in general, an unconsolidated aggregate of mineral and rock fragments that range in size from tiny clay and silt particles to sand and sometimes even to pebbles and boulders. It is a complex system of organic as well as inorganic components. In soil science, soil is confined to the upper few feet near the surface—usually to a maximum depth of 5 ft (1.5 m). This is the plant-root zone. In hydrology, soil is all of the unconsolidated granular material, including organic material, overlying hard bedrock, and could thus range from 1 foot to the thousands of feet. The pores through which water moves may range from almost microscopic size to large, well-defined pathways between soil grains. Sometimes cracks or holes in the soil provide large channels for water to move rapidly through the unsaturated zone. Because soil surface undergoes almost constant change, both physical and chemical, pores between grains grow or diminish in size, affecting infiltration.

The water is driven into the porous soil by the force of gravity and by the capillary attraction of the soil pores. First, the water wets the soil grains by forming a hydrogen bond with mineral surfaces. The adhesive forces attracting two unlike molecules of water and soil material also act in concert with the capillary (surface tension) forces. As more water enters the soil, the result of these forces is to cause the water to penetrate openings in the soil surface and migrate downward, with numerous menisci forming at the interfaces between air and water. With more water entering into the soil, the menisci are destroyed. Then the force of gravity takes over and moves the water down toward the water table. When the supply of water to the soil (e.g., rain) stops, the water percolates (or moves down) by gravity until the menisci begin to reappear, and the forces of surface tension begin to dominate again, holding water in the pores. Figure 8.1 illustrates this phenomenon. This process is called infiltration. Laymen often refer to this process as "soaking in"; percolation is another term often used to describe this process.

The rate at which water enters the soil at its surface is defined as infiltration rate. The infiltration rate as a function of time defines the infiltration curve. This rate is determined by a multitude of factors. Under most conditions, the greatest rate of infiltration occurs when water first contacts the soil surface, as shown in Figure 8.2. Because the capillary forces are the strongest when the soil is the driest, the rate of infiltration is most rapid near the beginning of precipitation. As water penetrates the soil to some depth, the resistance to the forces acting on the water increases and the rate of infiltration decreases; as a result, a fairly constant rate of infiltration is attained after some time has passed. When the availability of water is not a limiting factor, the maximum rate of infiltration at any time is called the infiltration capacity of the soil (Horton, 1933). The infiltration capacity asymptotically approaches a constant value after a certain period of time has passed, and is called the constant infiltration capacity; other names meaning the same are the steady-state infiltration capacity, the time-invariant infiltration capacity, the minimum infiltration capacity, the ultimate infiltration capacity, etc. The time interval used to define the constant infiltration capacity is usually selected as some time after the beginning of measurement when the infiltration rate has reached a near constant rate. Because all soils are not alike, and because some soils would never attain a near constant rate of infiltration, it is common to define infiltration capacity for some specified time period such as 30 minutes, 1 hour, or any other time. Permanently installed infiltrometers, such as the double-ring infiltrometer, measure infiltration over long periods of time. These measurements would not necessarily be comparable to shorter infiltration-capacity measurements. The purpose of this time specification is to permit a constant relation between infiltration observations and to permit others to understand the basis under which the measurements were conducted if such measurements were to be replicated.

Soils with different characteristics have different infiltration capacities. The infiltration capacity of one soil might be very low such that the infiltration curve after a period of time might be nearly asymptotic to the abscissa; another soil might have a steeper infiltration curve at the same time. Because drier soils have a more rapid

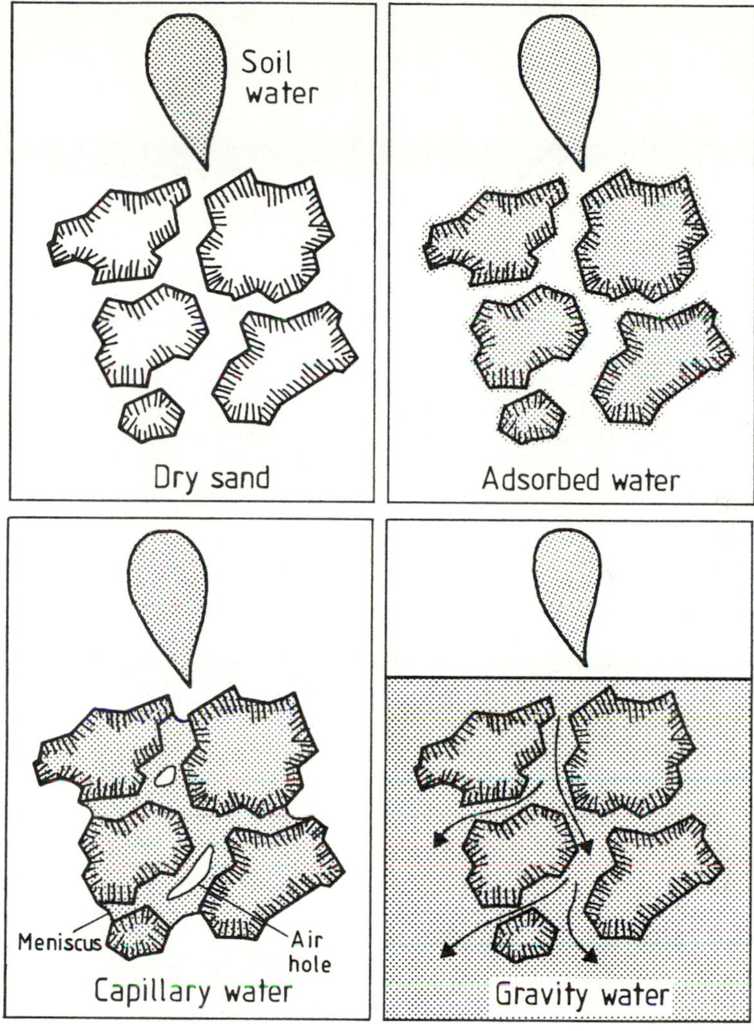

Figure 8.1 Water-holding characteristics of sandy soil.

initial infiltration rate, the moisture content in the soil at the beginning of an infiltration measurement will have an important effect on the shape of the infiltration curve for a given soil. Soils with a high moisture content will have lower initial infiltration rates than will drier soils.

The amount of runoff from a given area depends on the rainfall intensity and the infiltration rate, both in units of inches per hour (or centimeters per hour). If the infiltration rate exceeds the rainfall intensity, no runoff will occur; on the other hand, if the rainfall intensity exceeds the infiltration rate, then runoff will occur. In water-balance studies, the total infiltration occurring during a storm event is of primary interest. The portion of rainfall that infiltrates the soil will first replace any soil-moisture deficit; when the soil-moisture deficit is replaced, the remaining infiltration will continue migrating downward until it joins the groundwater table.

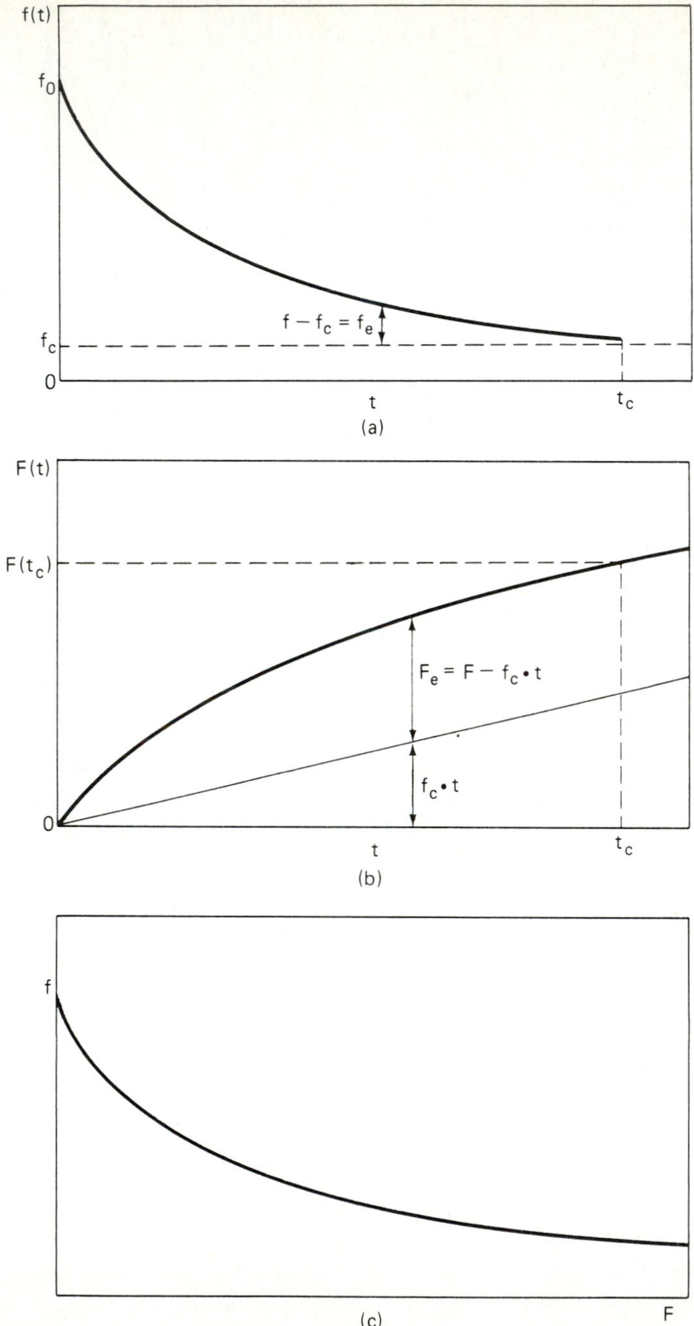

Figure 8.2 (a) A typical infiltration-capacity curve. (b) A cumulative infiltration curve. (c) Infiltration-capacity rate versus cumulative infiltration. f = infiltration-capacity rate, t = time, F = cumulative infiltration, f_c = constant infiltration-capacity rate, f_0 = excess infiltration-capacity rate, t_c = time to constant infiltration-capacity rate, $F(t_c)$ = amount of infiltration up to time t_c, and F_e = amount of excess infiltration.

There is a close relation between the runoff hydrograph and infiltration characteristics of the drainage basin. This relation can be illustrated by two extreme examples. The Nebraska Sandhills are among the best ranch lands because they contain huge amounts of groundwater. These Sandhills have a measured infiltration capacity of about 15 in./h (38 cm/h). This extremely high infiltration capacity will infiltrate the most intense rainfalls without permitting any runoff because storms producing rainfall greater than 15 in./h are not known or are rare in this region and are not likely to occur. As a result, there are no surface channels in this area. Streams in these Sandhills are fed by baseflow from the water table and storm hydrographs are mainly the result of increased baseflow and surface accumulation of precipitation near and on the stream itself. Consequently, the hydrographs reflect small discharge increases from large storms, but the discharge increase is of long duration.

Opposite conditions exist in the Bad Lands of South Dakota. The soils of these Bad Lands have very low infiltration capacities, on the order of 0.05 in./h (0.13 cm/h) or less. As a result, very little precipitation infiltrates the soil and the remainder must run off. This relation causes a discharge hydrograph of high magnitude and of short duration. The ability of a drainage basin to absorb, retain, or shed water that falls on it as rain or snow provides a key to the character of the resulting hydrograph. Horton (1933) first recognized this fact and suggested the theory of infiltration capacity.

8.2 ZONES OF SUBSURFACE WATER

The water existing in the soil is called subsurface water and is of two kinds: soil water and groundwater, as shown in Figure 8.3. The zone in which soil water occurs is called the unsaturated zone, and the zone in which groundwater occurs is called the saturated zone. The two zones are fundamentally different because of the way water occurs in them. In the groundwater zone, all the soil or rock pores are completely filled by water. The top of this zone is termed the water table. The soil-water zone occurs above the water table all the way up to the soil surface. In this zone, some soil pores contain water, some are partially filled, and some are essentially empty. The space not occupied by water is filled by air. The unsaturated zone acts as a storage reservoir, containing water for vegetative growth, and as a conduit for water moving down to recharge the groundwater. It provides pathways for water moving up toward the ground surface and the atmosphere.

8.3 SOIL WATER

Soil water is subsurface water in the unsaturated zone between the land surface and the water table. Infiltration supplies water to this zone. The water in this zone can remain in storage, move downward by gravity to the water table and the groundwater, or move upward through evaporation and transpiration. Thus, the unsaturated zone acts as a storage reservoir or as a conduit for soil water in transit. Some porous materials contain many small openings that are not connected to other openings. Such porous media may store a certain amount of water; they are incapable of

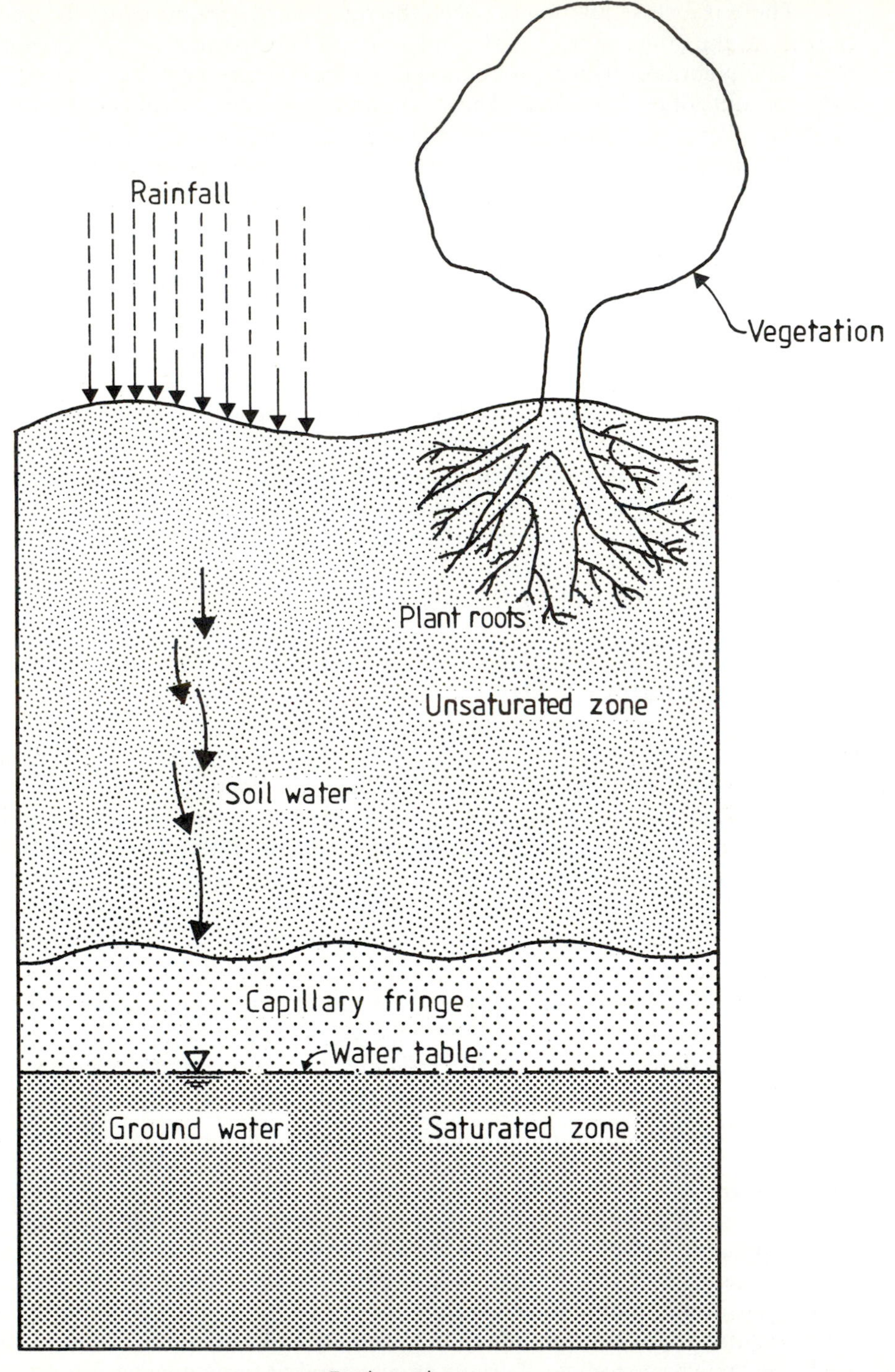

Figure 8.3 Zones of subsurface water.

transmitting water. Soil, of course, is capable of not only storing water, but also transmitting it.

The maximum amount of water that soil can hold in permanent storage is called field capacity. This is the capillary water held by surface tension after gravity drainage has stopped, and can be defined as the volume of capillary water per unit volume of soil. The soil water in storage is depleted by the plant-root zone. The depletion occurs due to osmotic-pressure differences between fluid in the plant roots and water in soil storage. Water flows from soil to plant roots as long as the suction force in the roots exceeds the capillary force in the soil pores. When the capillary force exceeds the roots' suction force, the plants can no longer extract soil moisture, and the residual amount of soil water is called the permanent wilting point. Thus, soil water in storage in the root zone can vary from field capacity to the wilting point.

Analogous to infiltration, forces associated with the action of hydrogen bonding in water, termed capillary forces, and gravity forces are responsible for the downward movement of water in the unsaturated zone. The water infiltration into the soil may follow one of the following paths during its movement in the unsaturated zone: (1) It may move toward the plant roots. (2) It may move toward the soil surface, where it will be transpired or evaporated and returned to the atmosphere. (3) It may move down by gravity to join the saturated zone. During its movement, the water will flow from regions of higher potential (energy) to regions of lower potential.

The potential controlling the flow of soil water is principally made up of two components: capillary potential and gravitational potential. The gravitational potential energy of an object depends upon its weight and its position in the earth's gravitational field, or its distance from the earth's center. Capillary potential is more complicated. Capillary forces result from cohesive and adhesive molecular forces acting within the body of a liquid. Cohesive forces cause surface tension, which acts at the interface between two liquids or between a liquid surface and a gas such as air. All liquids exhibit surface tension, but water manifests a much greater surface tension than other liquids except mercury. The reason that water has a higher surface tension is that cohesive forces attract water molecules together strongly wherever the water is in contact with air or other gases. Adhesive forces result from the attraction between water molecules and solid surface. An example is absorbed water, which is a thin film on a solid surface that is held strongly against gravity. The cause of these cohesive and adhesive forces is hydrogen bonding. The capillarity depends upon molecular forces of adhesion with a solid surface and surface-tension forces within the liquid. It raises water in a tube against the action of gravity. In soil with interconnected granular pores or tubes, capillarity may move water many feet above an open water surface such as a water table. However, capillarity always operates in the earth's gravitational field, and the capillary rise of water in a soil pore will continue only until the gravitational force balances it. Thus, the capillary potential energy depends on molecular forces within a liquid and on pore diameter. The capillary potential gradient will usually run from larger pores to smaller ones. The direction of the gradient can be up, down, or horizontal, with water moving from a wetter soil to a drier one. Soil water moves due to the total potential energy gradient (gravitational plus capillary).

8.4 DEFINITIONS AND NOTATIONS

***Infiltration*:** The process of the entry of water into a soil through the soil surface is defined as infiltration.

***Percolation*:** The process of water movement within the soil profile in a vertical direction is called percolation. It is evident that infiltration precedes percolation.

***Infiltration Capacity,* f_p:** This is the maximum rate at which soil can absorb water through its surface and has the dimensions of L/T. A distinction should be recognized between f and f_p; $0 \leq f \leq f_p$.

***Infiltration Rate,* f:** Infiltration rate is the rate at which water enters into the soil surface. It is expressed as volume per unit area per unit time, and has the dimensions of L/T. If there is no limit to the availability of water, then the infiltration rate is the same as infiltration capacity, or $f = f_p$. Under this condition, referring to Figure 8.2, initially (when $t = 0$), the infiltration rate is the highest and is denoted by f_0, and approaches a constant value denoted by f_c. The tendency to approach f_c is an approximation that continues to be debatable. The term f_c is also called the ultimate, or constant, rate of infiltration capacity. The time required to attain f_c is denoted by t_c. The difference between f and f_c, denoted as $f - f_c$, is called the rate of excess infiltration, f_e.

***Cumulative Infiltration,* F:** This denotes the volume of infiltration from the beginning of rainfall, t_0, to time t. It is also called the infiltration volume or accumulated infiltration and has the dimension of L. Cumulative infiltration can be represented by

$$F(t) = \int_0^t f(w)\, dw \tag{8.1}$$

or

$$f(t) = \frac{dF}{dt} \tag{8.2}$$

The volume of infiltration at any time is the area under the curve in Figure 8.2(a) up to that time. This volume is plotted as a function of time in Figure 8.2(b). Figure 8.2(c) shows the variation of f with F. Clearly, the rate of infiltration depends on how much water has already infiltrated.

When t is very large, F approaches a constant value, which is called the ultimate volume of infiltration, F_c. Therefore, the volume of potential infiltration is $F_p = F_c - F$. The difference between F and $f_c t$, or $F - f_c t$, is called the volume of excess infiltration, F_e. Likewise, the final volume of excess infiltration is $F_{ce} = F_c - f_c t$, and the volume of potential excess infiltration is $F_{pe} = F_{ce} - F_e$.

***Average Infiltration Rate,* f_a:** The average infiltration capacity rate (L/T) can be defined in two ways. First,

$$f_a = \frac{f_{p_i} + f_{p_{i+1}}}{2} \tag{8.3}$$

where subscript i denotes the point in time where f_p is to be computed. The choice of the two points in time, t_i and t_{i+1}, is to be determined for a given soil. Second,

$$f_a = \frac{F(T)}{T} \tag{8.4}$$

where T is the time period that has to be determined for each soil.

***Capillary Potential*:** This is the hydraulic head caused by capillary forces and has the dimension of L. It is variously known as capillary pressure, pressure head, pore pressure, moisture tension, moisture suction, and negative pressure. Capillary potential is always less than atmospheric pressure.

***Capillary Suction,* S:** This is the capillary potential with an opposite sign. A positive suction will represent a negative hydraulic head. It has the dimension of L.

***Hydraulic Conductivity,* K:** This is defined as the volume rate of flow of water through a unit area of soil under a unit gradient. This term is dependent on available moisture and has the dimensions of L/T.

***Saturated Conductivity,* K_s:** This represents the soil conductivity when the soil is saturated and has the dimensions of L/T.

***Relative Conductivity,* k_r:** This is the ratio of hydraulic conductivity for a given moisture content to the saturated conductivity and is dimensionless. This term is especially useful for two-phase flow where both air and water are present in the soil.

8.5 FACTORS AFFECTING INFILTRATION

Because soil characteristics and climatic conditions that influence the growth of vegetation vary widely, numerous factors are expected to affect infiltration. A comprehensive list of these factors is given by Gray et al. (1970), as shown in Figure 8.4. Some of these factors are discussed here.

1. Gravity causes water to enter the soil. The depth of surface-water detention on the soil affects the gravitational force acting on the soil surface. The greater the depth of water, the greater is the hydrostatic head or the gravitational force. The force of gravity must overcome all forces resisting the downward movement of water such as adhesion, and viscous forces.

2. The effect of existing soil moisture on infiltration is illustrated by the effect of wet and dry conditions on the movement of water through porous material. Dry soil creates a strong capillary potential in the capillary size pores between soil grains. This strong capillary attraction initially exerts a force that acts to supplement the force of gravity. The capillary force is strongest just under the dry ground surface. The strength of the capillary force is inversely proportional to the size of the pore openings; the force is small for large pore openings and large for small pore openings. As the surface of the soil becomes saturated with water, the capillary potential is satisfied and tends to resist movement of water through capillary-size

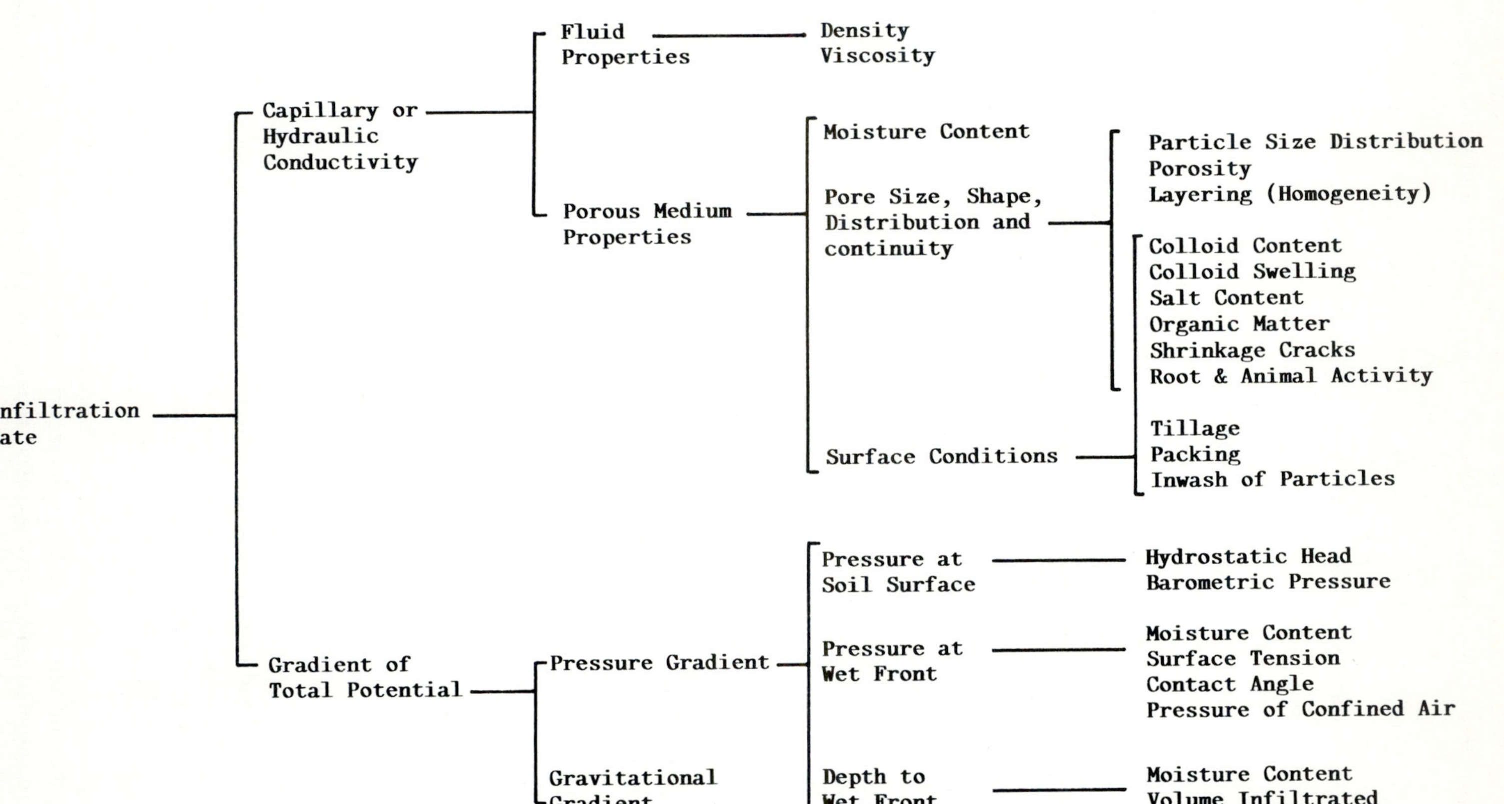

Figure 8.4 Factors affecting the infiltration rate (after Gray et al., 1970).

openings and decrease infiltration rate. Wetness in the soil thus creates resistance to infiltration. This resistance is consistent with Figure 8.2, where the infiltration rate decreases as the soil becomes wetter from more infiltration. This resistance to infiltration is highest for small saturated pore openings and small for large pore openings.

Many soils contain clay minerals such as illite and montmorillonite (bentonite) that swell when they become wet. Swelling of these clay minerals reduces the capacity of such soils to infiltrate water by further plugging the pore openings. Where such clay is not present, infiltration rates are directly proportional to the median grain diameter in semiarid regions where the influence of vegetation is minimal, as shown in Figure 8.5 (Rogers, 1971).

3. Soils are compacted by the falling rain and, as a result, infiltration is reduced. A soil loosened in a newly plowed field will infiltrate rainfall rapidly, but as the rain continues, the soil is compacted and the infiltration rate is reduced. Experiments have shown that soils protected from compaction by a layer of burlap infiltrate water more rapidly than unprotected soils (Duley, 1939).

4. A factor related to compaction is the inwash of fine material into the soil, which has the effect of reducing the size of pore openings. Fine airborne particles are continuously transported by wind and deposited over the land surface. As rain falls on the soil covered by these fine particles, they are carried into the intersticial spaces between soil grains. This condition results in obstructing the pore openings and, therefore, reducing the infiltration rate. Any alteration of the soil surface by placing fill on a soil surface will affect infiltration of the soil. If the fill is composed of finer material than the original soil material, infiltration will be reduced.

5. Soils are subject to compaction by man and animals. Unsurfaced roadways, animal trails, heavily pastured areas and feedlots, and other areas subject to compaction by heavy machinery or other use reduce infiltration rate. The net effect of

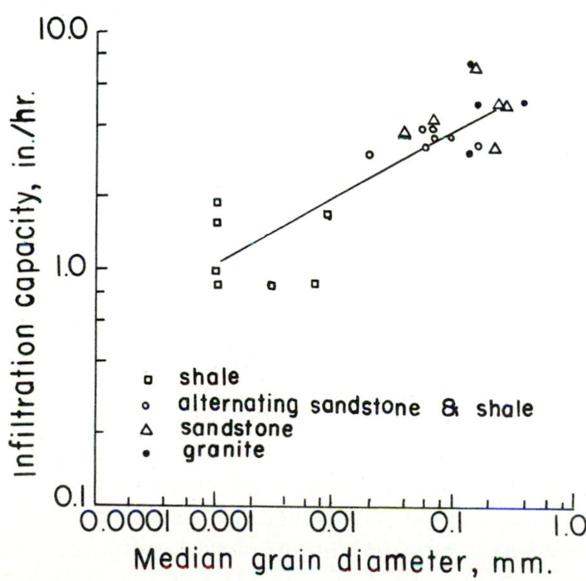

Figure 8.5 Infiltration capacity versus the median grain diameter.

reduced infiltration from such areas can be large depending on the intensity and areal extent of the compaction. Large areas of northern and eastern United States have been covered by glaciers. The weight of the ice has compacted soil material (till) in some areas to a degree that it is nearly impermeable. In Pennsylvania, for example, much of the state has developed less than a foot of soil since the last (Wisconsin) glacial advance. The till underlying this thin soil is nearly impervious. As a result, when the thin soil is saturated with water, no further infiltration occurs and all additional rainfall runs off.

Man has altered the surface of the earth for his use and as a result he has changed its infiltration characteristics. The most radical changes made by man are the impervious structures placed on the earth surface such as paved roads, parking lots, and buildings. All such structures have reduced the infiltration rate at these locations to near zero. As a result, much more runoff occurs from such areas. More widespread but less severe is the agricultural development of the natural earth surface. The net effect of agricultural development is the decreased infiltration rate of the soil thus causing greater runoff. Removal of native vegetation and exposing bare soil material to compaction by raindrops have significantly increased runoff from such areas. The result has been to enlarge and deepen existing drainages receiving runoff from such agricultural land. This is clearly visible over large areas of agricultural land in Nebraska, Iowa, and elsewhere. Both urbanization and agricultural development decrease infiltration and increase runoff.

6. Natural processes, such as burrowing by animals and insects, provide additional openings for water to penetrate the soil. Because of the very large number of insects on the earth, their effect is probably the larger of the two, although both are significant. The total effect of all burrowing is important because these burrows are larger than the pore openings and provide greater access to the soil for surface water.

As plants grow, their roots penetrate the soil. The depth of soil penetration by roots can be between a few inches to almost 100 ft (30.5 m). A reasonable estimate for average root penetration might be about 30 in. (76 cm). The roots of annual plants die and decay each year and some roots of perennial plants also die and decay. The root openings left from the decay of such roots provide tubular channel access for infiltrating water and increase the rate of infiltration. The action of roots during growth and development also loosen the soil and aid infiltration. Measurements made in such vegetation show very high infiltration in fine-grained soils that would otherwise have very low infiltration.

In climates subject to alternating freezing and thawing, the soil is loosened. The frost heaving results in considerable loosening of soil particles. As a result, infiltration is increased. Sun cracking and checking occurs in soils in dry climates when drought occurs. The degree and magnitude of such cracking are related to the clay content of the soil and the degree of dehydration. Those soils high in clay material produce larger and more extensive cracks than soils low in clay content. Cracks can be as large as several inches across and several feet deep; usually, they are smaller. The net effect of such cracking is to greatly enhance infiltration until such cracks are filled by swelling of the clay with increased moisture content.

Some soils contain soluble material that is dissolved by infiltrating water. This soluble material is usually carried to greater depths, where it can be redeposited.

The soil layer from which the soluble material was removed usually develops increased infiltration rates as a result of this solution process.

7. Vegetative cover provides protection for the soil from compaction by rain. Soils with a mature corn crop have greater infiltration than with a newly planted corn crop. Some of this increased infiltration is due to increased root openings and some is caused by protection the mature foliage provides the soil surface from compaction by raindrops. In any case, heavy native vegetation has many root openings to increase and assist infiltration. Dense vegetation also provides resistance to lateral flow of water through the vegetation, and increases the depth of flow, increasing the opportunity for water to infiltrate. Heavy vegetation provides a layer of decaying vegetative matter that promotes bacterial, insect, and animal growth. The accumulation of litter in forests provides a dense layer of vegetative matter that reduces lateral-flow velocities and increases infiltration. The removal of vegetation from a drainage area increases the amount of runoff. This increase comes about from reduced infiltration. Row crops have greater runoff than from hay crops because of the greater capacity of rows to carry water along with lower infiltration rates between rows.

8. Temperature is believed to influence infiltration. The viscosity of water roughly doubles for each 40°F (22°C) decrease in temperature. For this reason, the infiltration of water must dccrcasc as temperature decreases (Musgrave, 1955), as shown in Figure 8.6. Frozen soil is believed to reduce infiltration rates because frozen water contained in the pore openings between the grains obstructs infiltration. Water expands when frozen and, therefore, loosens soil grains, so an argument can be made that frozen soil must not reduce infiltration.

Horton (1940) has shown, as in Figure 8.7, that infiltration rates change seasonally. Infiltration capacity is highest during the summer months and lowest during the

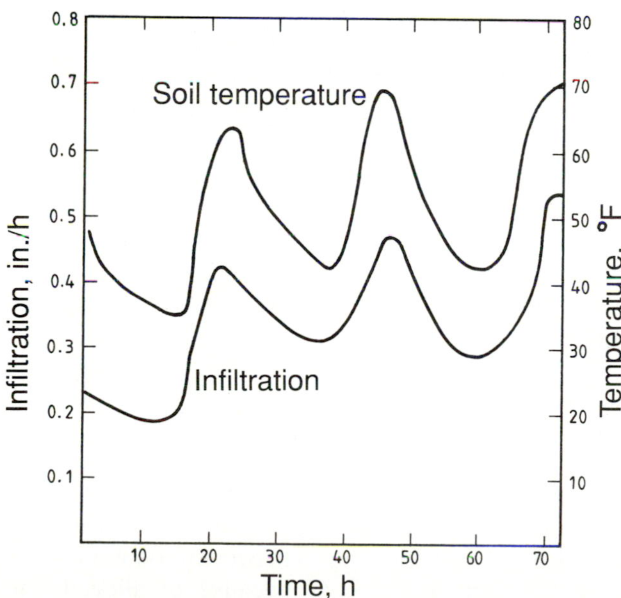

Figure 8.6 Correlation between soil temperature and the rate of infiltration, where infiltration is proportional to water viscosity (after Musgrave, 1955).

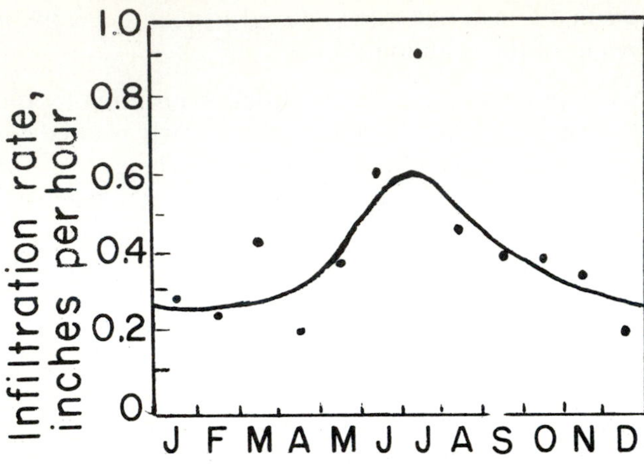

Figure 8.7 Seasonal variation of the apparent minimum infiltration capacity, North Concho River Basin (after Horton, 1940).

winter months. This relation does not necessarily confirm the effect of frozen ground. The seasonal influence might be caused by the removal, or at least the diminished effect, of vegetation. Another cause might be the effect of increased water viscosity caused by decreased winter temperatures.

9. Infiltration measurements are influenced by air entrapped in the soil by an advancing curtain of infiltrating water. The air entrapment reduces infiltration, but its effect is temporary, because the compressed air below the advancing curtain eventually breaks through the water curtain and infiltration continues in a normal manner. This effect does not appear to be important in practical hydrology.

It is clear that many factors that control infiltration can interact in many ways. Some are interrelated and some are not. Median grain diameter of the soil is the dominant factor in semiarid regions, whereas vegetation becomes the dominant influence on infiltration in humid regions. There are so many possible combinations of these factors that determine infiltration that no way has yet been developed to measure and quantify their various influences. However, a qualitative estimate of infiltration characteristics can be obtained merely by on-site observation of the drainage basin.

8.6 MEASUREMENT OF INFILTRATION

Three methods of determining infiltration are discussed; (1) flooding infiltrometers, (2) rainfall simulators, and (3) hydrograph analysis.

8.6.1 Flooding Infiltrometers

Flooding infiltrometers consist of a metal ring driven into the ground surface such that the inner part of the ring can be filled with water. The rate at which this water infiltrates the ground can be measured using a hook gage or other measuring instruments. Early infiltrometers consisted of a single ring. Theoretical objections were voiced regarding the single-ring infiltrometer concept because of alleged lateral movement of water in the soil under the ring. This lateral movement was thought to cause too high an infiltration reading because some of the infiltrated water penetrated laterally under the ring.

Because of this objection, a double-ring infiltrometer was developed, as shown in Figure 8.8. This infiltrometer consists of two concentric rings driven into the ground surface. Both rings are filled with water. The outer ring is believed to supply the water that might migrate laterally and at the same time saturate the soil next to the inner ring. Water in the inner ring is then thought to be able to move vertically into the soil. A hook gage is used to measure the rate of infiltration from the inner ring. Because of the general acceptance of the double-ring theory, the double-ring infiltrometer is considered to be the standard for infiltration measurements. In spite of its acceptance, the double-ring infiltrometer does not duplicate rainfall conditions and it must have a substantial depth of water in the ring in order to provide sufficient depth for measuring instruments. Another objection to the double-ring infiltrometer is that it is too cumbersome to install and operate and takes too long to make a measurement. For this reason, a portable infiltrometer was developed.

A single-ring portable infiltrometer, as shown in Figure 8.9, provides an inexpensive, simple way to make infiltration measurements (Rogers, 1970). This infiltrometer has been used on gaged drainage basins and found to be sufficiently accurate for practical use. It can measure the infiltration rate initially and at selected time intervals. Thus, infiltration curves can be prepared for different soil-moisture conditions. For comparison with the rainfall simulator of the Agricultural Research Ser-

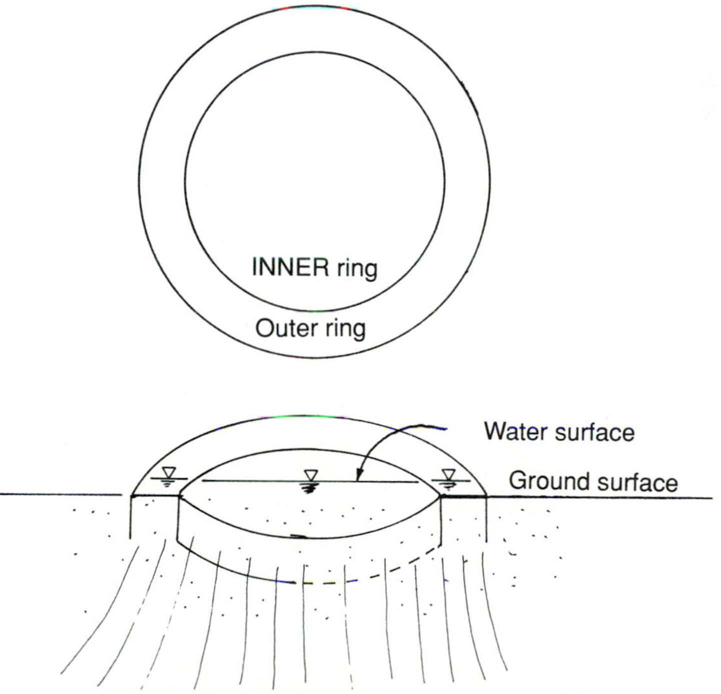

Figure 8.8 Double-ring floating infiltrometer.

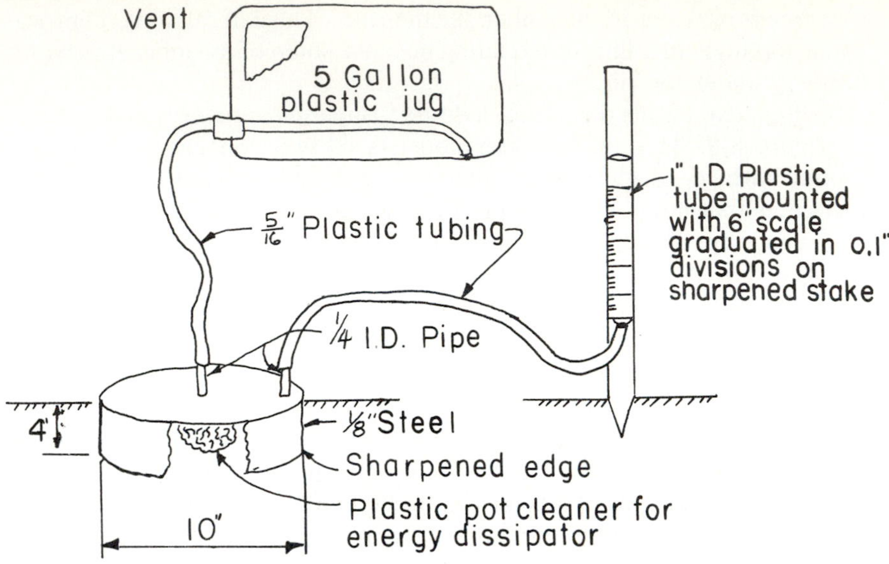

Figure 8.9 Single-ring portable infiltrometer.

vice of the U.S. Department of Agriculture, different rates of infiltration were obtained with the two infiltrometers, and it was found to provide consistent measurements that could be applied to practical problems. A measurement using this infiltrometer can be completed in 45 minutes. This portable infiltrometer is adaptable to most conditions. Soil surfaces covered with heavy vegetation, thick forest litter, or bedrock are unsuitable for any ring infiltrometer, and infiltration measurement must be made using other methods.

The installation and operation of the portable infiltrometer is quite simple. The ring is located at a point where a measurement is desired. The plastic pot cleaner is placed under the center-pipe connection to act as an energy dissipator for the injected water and the ring is then pounded evenly into the ground to a depth about 0.5 in. (1.25 cm) above the ground surface. If the instrument is not evenly driven into the ground, leakage can occur under and around the ring. The soil immediately outside the ring should be tamped to provide an adequate seal. This tamping will not affect the soil inside the ring. The gage tube is then driven into the ground at a convenient distance from the ring. The gage tube is driven sufficiently deep so that a flexible tube connection can be made between the gage and the outer pipe on the ring. The center pipe on the ring is connected to the outlet of the water container. The water container should be of at least 5-gallon (20-liter) capacity and should be arranged so that air can be blown under pressure into the water container by the operator.

A stopwatch and an elapsed-time watch are required to make an infiltration measurement. Initial (zero) elapsed time begins when water is injected into the infiltrometer. The operator blows water into the water container by pressure. The pressure forces the water to flow rapidly into the infiltrometer ring, which, in turn, forces the enclosed air to evacuate through the gage tube. When injected water rises in the tube thus displacing all air, the water-inlet hose is blocked by pinching and the time required for the water in the gage tube to fall 1 inch is measured by use of a stopwatch. With a little practice, measurements can be made as soon as 1 minute after zero elapsed time and at selected time intervals thereafter. In most cases, a reasonable infiltration-capacity measurement can be made after 30 minutes. Because the area relation between the gage tube and the infiltrometer ring is 1 to 100, it is easy to compute the infiltration rate by timing the 1-inch fall in the gage tube.

The infiltrometer measures infiltration at one location on the drainage basin. To obtain a representative infiltration value for the entire drainage basin, it is necessary to measure infiltration at a sufficient number of random locations. The resulting infiltration observations may be plotted on semilogarithm paper and the total infiltration for any period of rainfall can be calculated.

EXAMPLE 8.1 For the Irvington drainage basin, 24 random measurements using the infiltrometer of Figure 8.6 have been made at 10, 20, and 30 minutes, as given in Table 8.1. Construct the mean infiltration curve for the basin.

Solution An average infiltration is calculated at each time interval using the 24 random observations. The average values corresponding to different times are shown in Table 8.1, which are then plotted in Figure 8.10. Since the observations are for only three times, this figure involves extrapolation.

TABLE 8.1 RANDOM INFILTRATION MEASUREMENTS IN THE IRVINGTON DRAINAGE BASIN

Observation	Time (min.)	Time (s) per 0.01 in.	Infiltration (in./h)	Observation	Time (min.)	Time (s) per 0.01 in.	Infiltration (in./h)
1	10	2.9	12.1	2	10	14.5	2.48
	20	3.9	9.23		20	13.8	2.61
	30	5.0	7.20		30	20.0	1.8
3	10	4.4	8.18	4	10	8.7	4.14
	20	5.0	7.20		20	9.5	3.79
	30	5.2	6.92		30	14.0	2.57
5	10	4.5	8.0	6	10	14.9	2.42
	20	7.0	5.14		20	15.5	2.32
	30	8.4	4.29		30	18.2	1.98
7	10	7.0	5.14	8	10	24.6	1.46
	20	7.1	5.07		20	27.0	1.33
	30	8.7	4.14		30	27.5	1.3
9	10	7.6	4.74	10	10	14.3	2.52
	20	13.8	2.61		20	16.2	2.22
	30	15.7	2.29		30	21.1	1.71
11	10	4.1	8.78	12	10	37.8	1.1
	20	4.2	8.57		20	34.1	1.06
	30	4.4	8.18		30	43.3	0.83

TABLE 8.1 (*Continued*)

Observation	Time (min.)	Time (s) per 0.01 in.	Infiltration (in./h)	Observation	Time (min.)	Time (s) per 0.01 in.	Infiltration (in./h)
13	10	10.8	3.33	14	10	20.5	1.76
	20	12.4	2.90		20	33.6	1.07
	30	16.1	2.24		30	34.7	1.04
15	10	18.3	1.97	16	10	51.4	0.7
	20	19.5	1.85		20	55.7	0.65
	30	21.2	1.71		30	59.8	0.60
17	10	36.4	0.99	18	10	14.8	2.43
	20	40.8	0.88		20	17.1	2.11
	30	52.6	0.68		30	22.0	1.64
19	10	6.3	5.71	20	10	9.3	3.87
	20	12.4	2.90		20	10.9	3.30
	30	13.9	2.59		30	13.1	2.75
21	10	13.8	2.61	22	10	41.2	0.87
	20	17.7	2.03		20	55.6	0.65
	30	19.2	1.88		30	65.4	0.57
23	5	7.4	4.86	24	10	8.9	4.04
	20	10.4	3.46		20	9.5	3.79
	30	14.4	2.50		30	11.2	3.21

Mean of 24 Observations

Time (min.)	10	20	30
Infiltration (in./h)	3.93	3.20	2.69

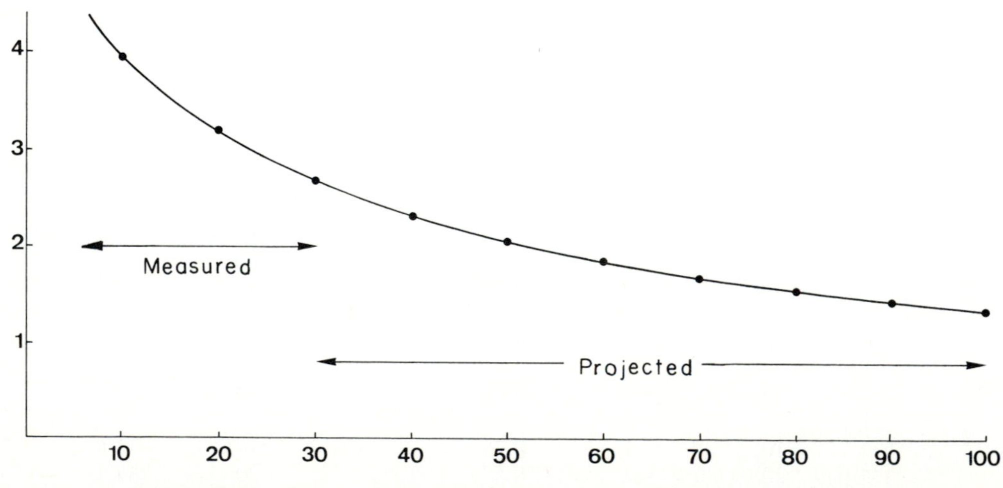

Figure 8.10 Mean infiltration curve for the Irvington drainage basin.

8.6.2 Rainfall Simulators

Critics of flooding infiltrometers object to the necessary ponding of water and the accompanying hydrostatic head that such infiltrometers place on the infiltrating water. A further objection is the absence of the effect of raindrop impact on soil. In order to cope with these objections, rainfall-simulator infiltrometers were developed. These infiltrometers apply water by sprinklers to a small tract of land confined by a metal ringlike barrier at a rate sufficient to exceed the infiltration capacity of the soil. The amount of water applied is measured and the excess is collected and measured. The most common rainfall simulator infiltrometers are (a) the Rocky Mountain, (b) the Modified North Fork, and (c) the Modified Type F. The main differences distinguishing each are the size of the plot of ground to which the water is applied and the method of measuring the applied water and runoff water.

A plot of land, situated so that surface water can drain from it, is confined by some boundary barrier so that runoff from this plot can be collected and measured. Sprinklers and gages are located on the plot so that measured amounts of water can be applied to the plot. Infiltration is measured by applying sprinkled water at a given rate and measuring the runoff over a given time. Infiltration capacity for this time is then calculated. A major objection to the rainfall simulators is that they are cumbersome and difficult to move and operate. The Agricultural Research Service of the U.S. Department of Agriculture developed a truck-mounted rainfall simulator that could be readily moved and set up (Rauzi et al., 1968). The objection to this instrument is its high cost. Because it is truck-mounted, this infiltrometer requires a large expenditure to obtain and operate.

8.6.3 Hydrograph Analysis

It is possible to obtain a reasonable estimate of infiltration of a drainage basin by analyzing the runoff hydrograph and measured rainfall. A network of rain gages is used to obtain a reasonable estimate of mean rainfall over the drainage basin for a given runoff event. The amount of runoff from such a storm is measured by means of a stream gage. The difference between the amount of water falling on the drainage basin and the amount of water that runs off is the basis for estimating infiltration of that basin. The hydrologic budget requires accounting for evapotranspiration, depression storage, and interception. Evapotranspiration is negligible for short-duration storms and very small for long-duration storms. Depression storage and interception losses can be estimated, as will be discussed in Chapter 10.

Infiltration estimates made by hydrograph analysis take into account variations in infiltration that might exist throughout the drainage basin. Such estimates are truly drainage-basin infiltration estimates. On the other hand, infiltration measurements made by an infiltrometer are only for one point within the drainage basin. Since a drainage basin does not have uniform infiltration throughout its area, infiltration measurements must be made throughout the basin in such a manner as to determine the mean value of infiltration for the basin. A network of infiltration observations determined by statistical methods is usually satisfactory and will agree well with infiltration determined by hydrograph analysis.

8.7 EMPIRICAL EVALUATION OF AVERAGE INFILTRATION

Two methods that estimate average infiltration capacity or constant infiltration capacity are presented here.

8.7.1 Correlation with Related Variables

Drainage-basin infiltration can be related to such variables as medium-grain diameter, drainage density, runoff volume, and sediment yield (Horton, 1939). Medium-grain diameter is determined by sieve analysis of random soil samples. This relation is valid only for semiarid and arid lands where the effect of vegetation is minimal. The relation in Figure 8.5 shows that constant infiltration capacity increases as the median grain size increases. Fine-grain material has a lower infiltration capacity than does coarse-grain material. Because obtaining representative soil samples requires about the same time and effort as actually making an infiltration measurement, it is more practical to measure infiltration unless the required soil samples already exist.

Drainage density is inversely related to constant infiltration capacity, as shown in Figure 8.11. This relation is particularly practical because of the ease in determining drainage density. The rationale for this relationship is based on the fact that water that infiltrates the ground cannot run off over the surface to form channels. Small amounts of surface runoff require and form few channels. That is why many channels are formed where infiltration rates are low and few channels are formed where infiltration rates are high.

There has been no definitive study sufficient to establish the relationship between infiltration capacity and drainage density. If this limited study done in the

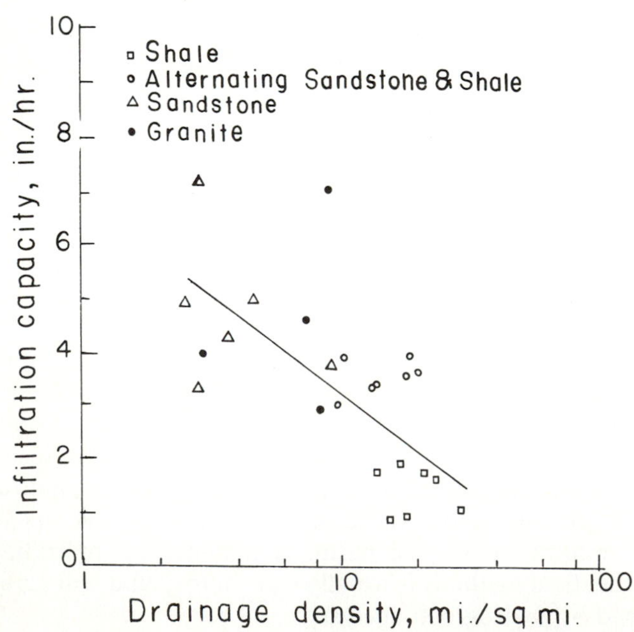

Figure 8.11 Relation between drainage density and infiltration capacity.

Cheyenne River basin of eastern Wyoming, as shown by Figure 8.11, is a reasonably representative relation, then the measured drainage density of 4.8 km/km^2 for the Irvington drainage basin translates to a mean infiltration capacity of approximately 4+ inches per hour. This value is also consistent with that obtained by hydrograph analysis, which is to be discussed in Example 8.5.

A consequence of the rationale is apparent from the relation between runoff and infiltration capacity, as shown in Figure 8.12. This inverse relation shows that the greater the runoff volume, the lower the infiltration capacity. Precipitation that does not infiltrate runs off. Greater runoff volumes result from lower infiltration capacities. Where appropriate data exist, this relation can be used to estimate infiltration capacity.

The inverse relation between sediment yield and infiltration capacity, shown in Figure 8.13, derives from the relation between runoff volume and infiltration capacity. Runoff volume and sediment yield are closely related to each other, and both, therefore, must be related to infiltration capacity.

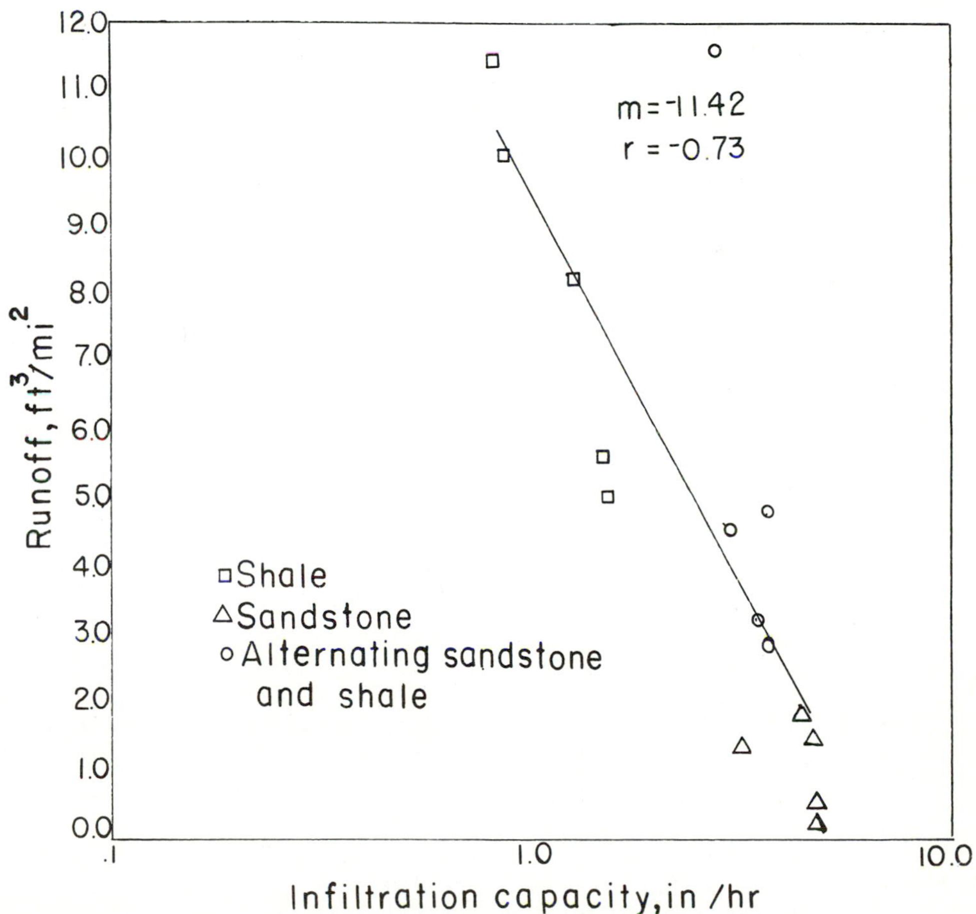

Figure 8.12 Relation between runoff volume and infiltration capacity.

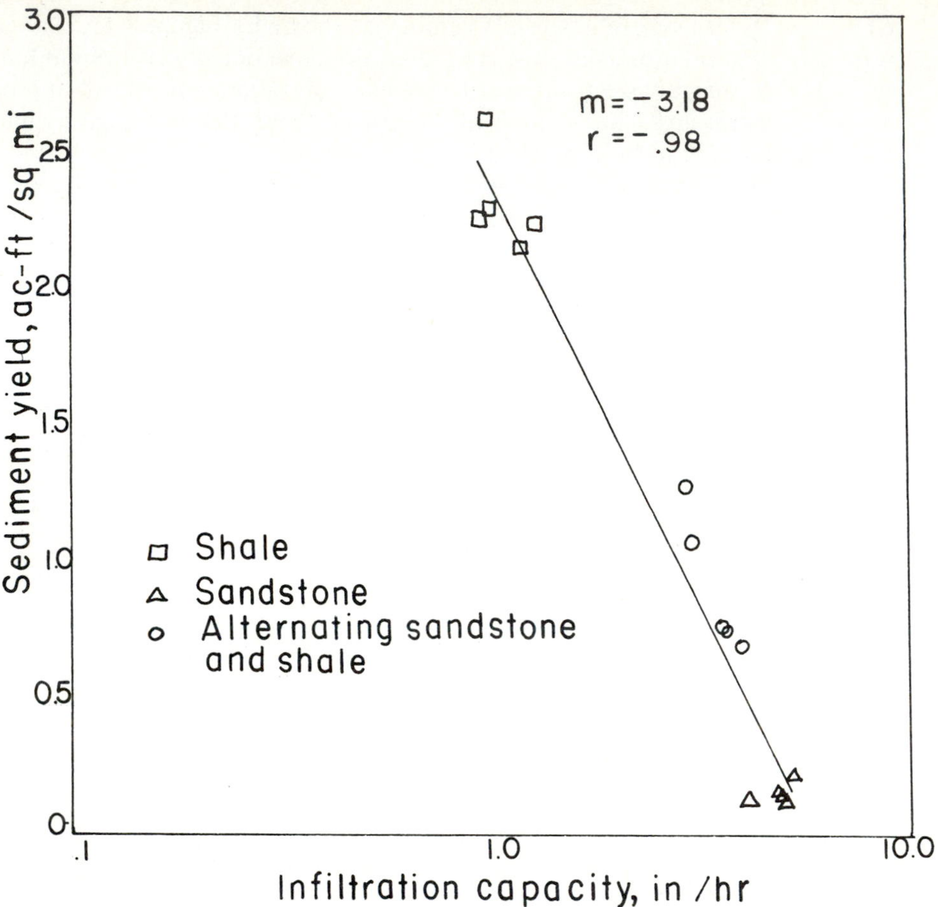

Figure 8.13 Relation between sediment yield and infiltration capacity.

8.7.2 Indices

ϕ Index. The ϕ index is the mean infiltration rate occurring for the duration of the storm. This mean infiltration rate, as shown in Figure 8.14, is the rainfall rate above which the rainfall volume is equal to the runoff volume. The interception and depression storage losses are excluded in the ϕ index.

The average rainfall intensity replacing depleted soil moisture and the average infiltration rate are equal at the ϕ index. Despite exclusion of depression storage and interception, the ϕ index is a useful tool for many hydrologic studies. A particularly useful application is the study of a single drainage basin based on long records such as is done by Natural Resources Districts. For such studies, interception and depression storage losses are nearly a constant loss and their inclusion in infiltration does not materially affect the prediction of runoff. When studying many runoff events, the mean infiltration rate can be determined by analyzing the many hydrographs. The resulting ϕ index is probably a reasonable average loss value for some purposes.

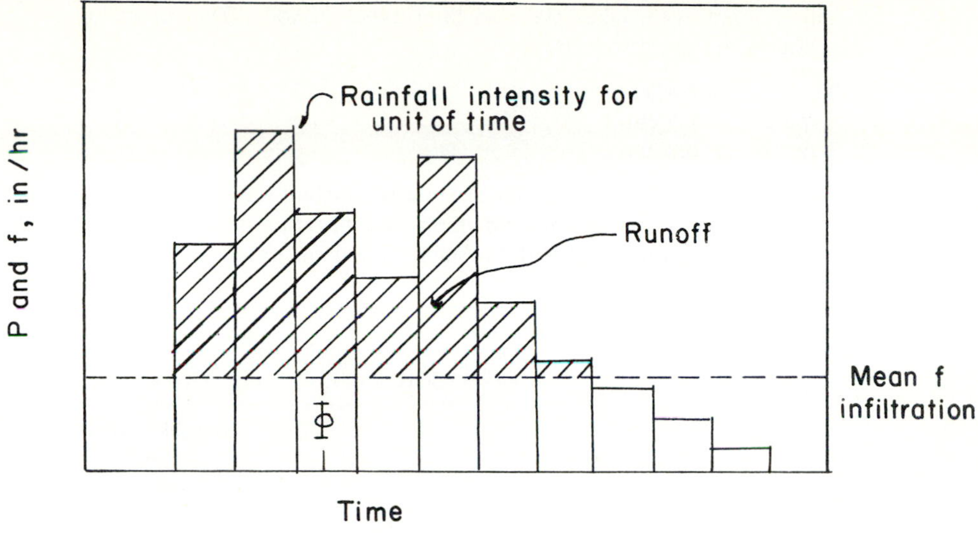

Figure 8.14 ϕ index.

W Index. The W index refines the ϕ index, by including interception and depression storage, and is represented by

$$W = \frac{F}{T} = \frac{P - V_Q - S}{T} \tag{8.5}$$

where W is the average infiltration rate, P is the total storm rainfall corresponding to T, V_Q is the total storm runoff, S is the total volume of interception and depression storage (or effective surface retention), and T is the total time during which rainfall intensity exceeds the infiltration capacity. When $S = 0$, the W index reduces to the ϕ index. Or the W index essentially equals the ϕ index minus the average rate of retention due to interception and storage. For very wet conditions, this equality will approximately hold.

The difficulty with the W index relates to determining the values of depression storage and interception. Interception varies with season and depression storage varies with rainfall amount. Furthermore, much of the depression storage becomes infiltration or evaporation.

EXAMPLE 8.2 A rainfall event is given as follows:

TIME		INTENSITY, *P*
(H)	(MIN)	(CM/H)
00	10	2
00	20	4
00	30	3
00	40	2
00	50	1
00	60	0

This rainfall event takes place on three types of soils. The runoff volumes/unit area produced on these soils are

Type 1 0.8333 cm
Type 2 1.5 cm
Type 3 0.0 cm

Assume that the infiltration-capacity rate is constant for each soil and that there are no other abstractions. Compute and graph the infiltration rate during the rainfall event for each soil, and compare it with the infiltration-capacity rate. Also compute the rainfall excess intensity.

Solution

SOIL TYPE 1

Time (min)	p (cm/h)	f (cm/h)	Rainfall excess intensity (cm/h)
0–10	2	1.5	0.5
10–20	4	1.5	2.5
20–30	3	1.5	1.5
30–40	2	1.5	0.5
40–50	1	1	0
50–60	0	0	0

Infiltration-capacity rate = 1.5 cm/h.

Figure 8.15 shows the infiltration-capacity rate.

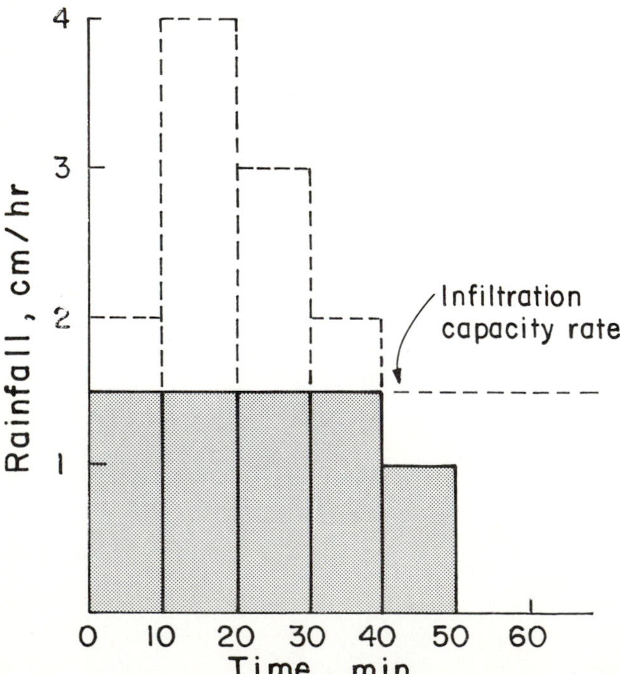

Figure 8.15 Infiltration-capacity rate for soil Type 1.

SOIL TYPE 2

Time (min)	p (cm/h)	f (cm/h)	Rainfall excess intensity (cm/h)
0–10	2	0.6	1.4
10–20	4	0.6	3.4
20–30	3	0.6	2.4
30–40	2	0.6	1.4
40–50	1	0.6	0.4
50–60	0	0	0

Infiltration-capacity rate = 0.6 cm/h.

Figure 8.16 shows the infiltration-capacity rate.

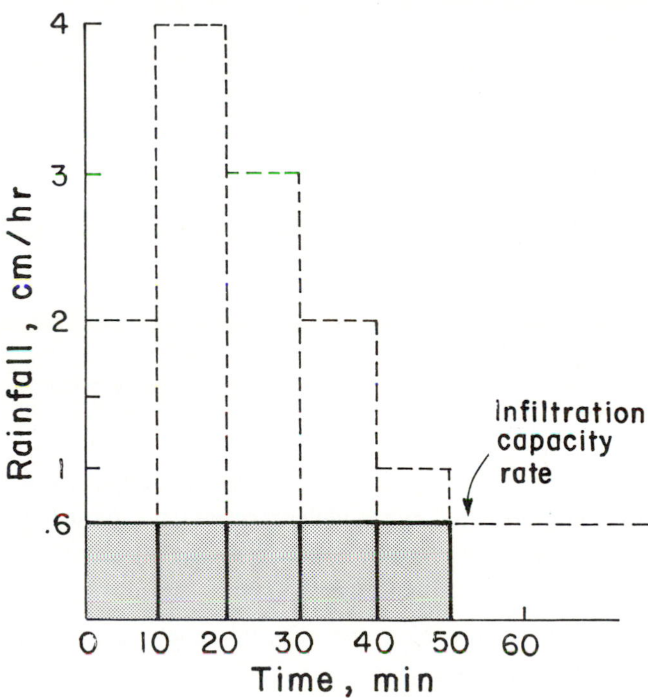

Figure 8.16 Infiltration-capacity rate for soil Type 2.

SOIL TYPE 3

Time (min)	p (cm/h)	f (cm/h)	Rainfall excess intensity (cm/h)
0–10	2	2	0
10–20	4	4	0
20–30	3	3	0
30–40	2	2	0
40–50	1	1	0
50–60	0	0	0

Infiltration-capacity rate ≤ 4 cm/h.

Figure 8.17 shows the infiltration-capacity rate.

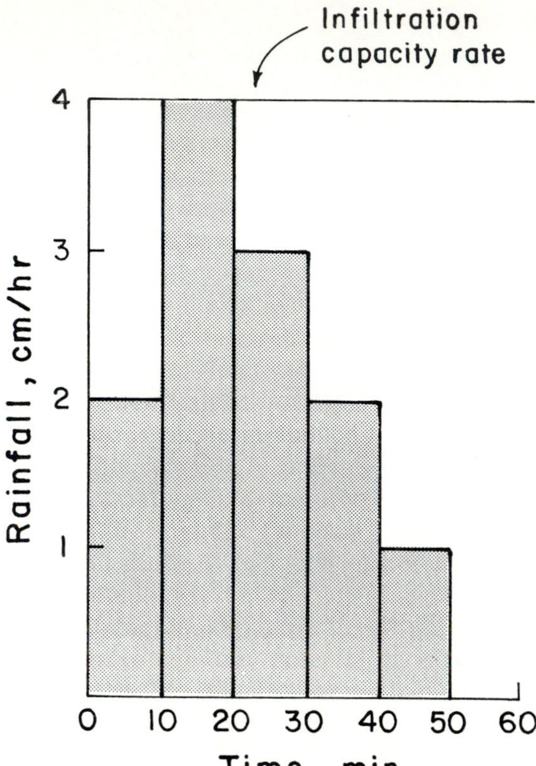

Figure 8.17 Infiltration-capacity rate for soil Type 3.

EXAMPLE 8.3 Compute and graph the accumulated mass infiltration in time for each soil in Example 8.2. Then compute and graph the rainfall excess in time. Assume the depression storage model as

$$Q = (p - f)[1 - \exp(-kp_e)$$

where $k = 0.5/\text{cm}$, Q is the overland flow rate, and p_e is the rainfall excess rate. Compute and graph the overland flow rate, Q.

Solution First, we compute the accumulated infiltration for each soil type.

Time	Infiltration mass (cm)			Accumulated infiltration mass (cm)		
(min)	Type 1	Type 2	Type 3	Type 1	Type 2	Type 3
0				0	0	0
10	0.25	0.1	0.33	0.25	0.1	0.33
20	0.25	0.1	0.667	0.50	0.2	1.00
30	0.25	0.1	0.500	0.75	0.3	1.500
40	0.25	0.1	0.333	1.00	0.4	1.833
50	0.1667	0.1	0.1667	1.1667	0.5	2.000
60	0	0	0	1.1667	0.5	2.000

Accumulated infiltration as a funciton of time is shown in Figure 8.18.

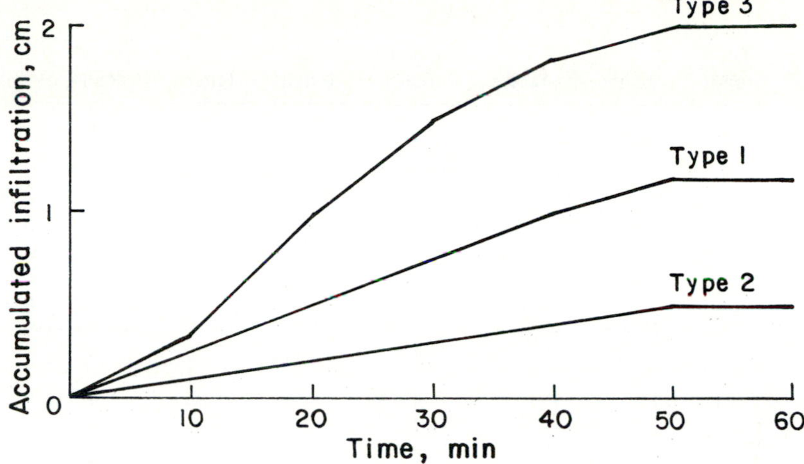

Figure 8.18 Accumulated infiltration as a function of time for three soil types.

Next, the rainfall excess is computed.

Time (min)	Rainfall excess (cm)			Total excess rain (cm)		
	Type 1	Type 2	Type 3	Type 1	Type 2	Type 3
0				0	0	0
10	0.0833	0.2333	0	0.0833	0.2333	0
20	0.4167	0.5067	0	0.5	0.8	0
30	0.25	0.4	0	0.75	1.2	0
40	0.0833	0.2333	0	0.8333	1.4333	0
50	0	0.0667	0	0.8333	1.5	0
60	0	0	0	0.8333	1.5	0

The total rainfall excess is plotted against time in Figure 8.19.

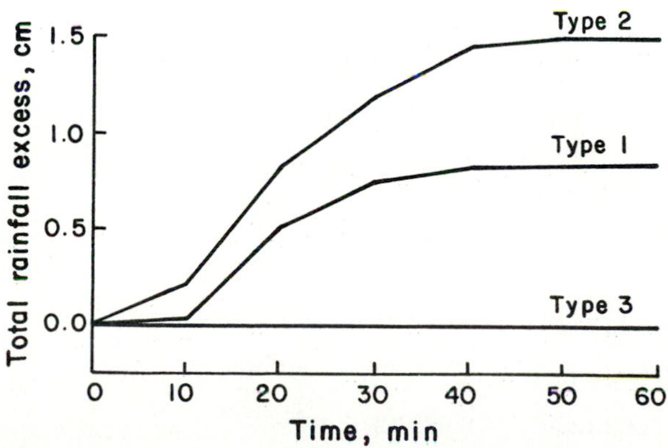

Figure 8.19 Total rainfall excess as a function of time for three soil types.

Next, the overland discharge rate is computed.

Time (min)	P (cm)	F (cm)	$P - F$ (cm)	p (cm/h)	f (cm/h)	$p - f$ (cm/h)	$\dfrac{Q}{p - f}$	Q (cm/h)
				Soil Type 1				
10	0.3333	0.25	0.0833	2	1.5	0.5	0.0408	0.0204
20	1	0.5	0.5	4	1.5	2.5	0.211	0.553
30	1.5	0.75	0.75	3	1.5	1.5	0.313	0.470
40	1.8333	1.0	0.8333	2	1.5	0.5	0.340	0.170
50	2	1.1667	0.8333	1	1	0	0.340	0
60	2	1.1667	0.8333	0	0	0	0.340	0
				Soil Type 2				
10	0.3333	0.1	0.2333	2	0.6	1.4	0.110	0.154
20	1	0.2	0.8	4	0.6	3.4	0.330	1.122
30	1.5	0.3	1.2	3	0.6	2.4	0.452	1.085
40	1.8333	0.4	1.4333	2	0.6	1.4	0.512	0.717
50	2	0.5	1.5	1	0.6	0.4	0.528	0.211
60	2	0.5	1.5	0	0	0	0.528	0

For soil type 3, there is no overland flow.

The overland flow is plotted against time in Figure 8.20.

EXAMPLE 8.4 Determine ϕ index and W index for watershed SW-12 Riesel (Waco), Texas, with an area of 2.97 acres. A rainfall storm occurred on December 15, 1967, on this watershed and produced 1.159 cm of runoff. Assume the effective surface retention as 20% of the rainfall amount. The rainfall observations are as follows:

Time (min)	Intensity (cm/h)	Amount (cm)	Time (min)	Intensity (cm/h)	Amount (cm)
0–25	0.305	0.127	150–215	0.00	0.00
25–45	0.686	0.229	215–335	0.406	0.533
45–55	0.305	0.051	335–355	1.067	0.356
55–115	1.295	0.432	355–415	0.016	0.025
115–123	0.584	0.076	415–435	0.686	0.229
123–131	2.108	0.279	435–725	0.305	0.864
131–140	0.330	0.051	725–805	0.025	0.051
140–150	0.762	0.127			

Solution

$$\text{Total rainfall } P = 3.429 \text{ cm} \cong 3.43 \text{ cm}$$

$$\text{Total runoff } V_Q = 1.159 \text{ cm} \cong 1.16 \text{ cm}$$

$$\text{Total time } T = 8.0833 \text{ h}$$

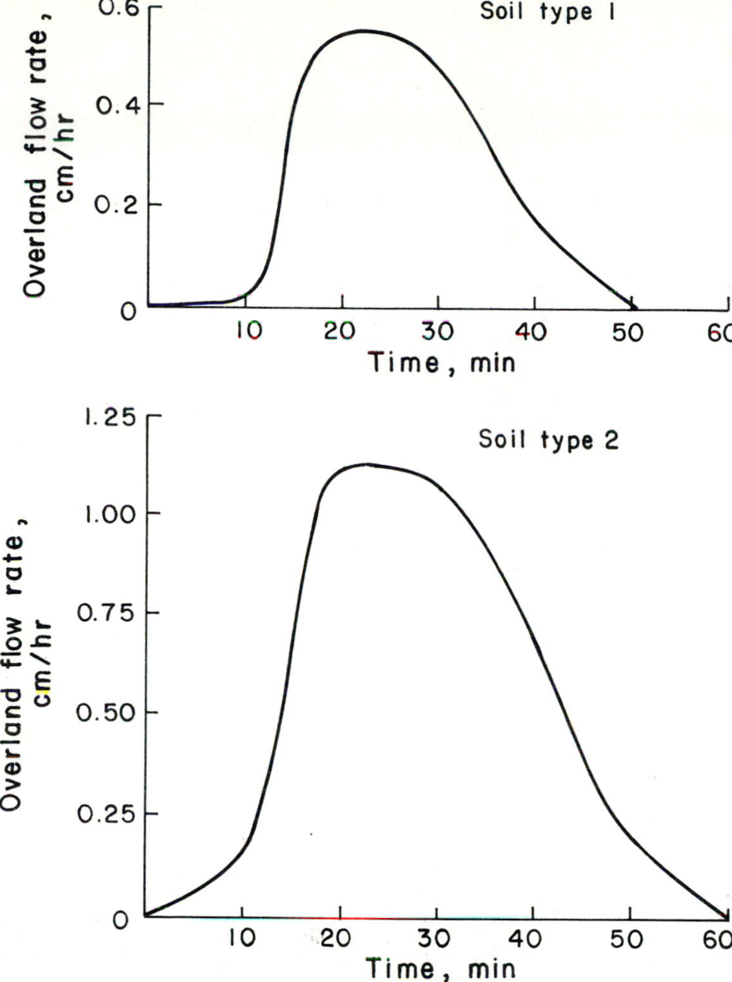

Figure 8.20 Overland flow rate as a function of time for two soil types.

Three guesses of ϕ are made.

$$\Phi_{\text{initial}} = 0.28 \text{ cm}$$

$$\Phi_{\text{second}} = 0.328 \text{ cm}$$

$$\Phi_{\text{third}} = 0.354 \text{ cm}$$

These are obtained as follows:

$$\Phi_{\text{initial}} = \frac{3.43 - 1.16}{8.083} = 0.28 \text{ cm/h}$$

$$\Phi_{\text{second}} = \frac{(3.43 - 0.025 - 0.051) - 1.16}{8.083 - 1.41} = \frac{2.19}{6.67} = 0.3283 \text{ cm/h}$$

$$\Phi_{\text{third}} = \frac{(3.354 - 0.864 - 0.051 - 0.051 - 0.127) - 1.16}{6.67 - (3.42 + 0.15)} = 0.3548 \text{ cm/h}$$

We check on Φ for those intensities that exceed Φ.

p (cm/h) (1)	Φ (cm/h) (2)	T (h) (3)	Sum (cm) [(1) − (2)] × (3)
0.686	0.354	0.333	0.110
1.295	0.354	0.330	0.343
0.584	0.354	0.133	0.031
2.108	0.354	0.133	0.233
0.762	0.354	0.166	0.067
0.406	0.354	1.333	0.069
1.067	0.354	0.333	0.237
0.686	0.354	0.333	0.110
	Sum = $\Sigma(p - Q) \times T$ = 1.16 cm runoff		

This shows that $\phi = 0.354$ cm/h is acceptable.

The W index is equivalent to the ϕ index minus the rate of retention by interception and storage depression.

$$W = \phi - \text{rate of retention}$$

$$W = 0.3544 - \left(\frac{0.2}{5.09}\right)(3.43 \text{ cm}) = 0.221 \text{ cm/h}$$

$$W \text{ index} = 0.221 \text{ cm/h}$$

EXAMPLE 8.5 The rainfall over the Irvington drainage basin for the storm of September 6–7, 1965, and the corresponding measured runoff from this storm follow. Compute the total infiltration, and compare with the observed infiltration given in Example 8.1.

Solution The rainfall storm of September 6–7, 1965, at Little Papillion Creek of the Irvington drainage basin (32 mi^2) is as follows:

Portion	Drainage area: Square miles	Acres	Average rainfall (in.)
A	2.329	1,490.5	5.5
B	3.934	2,517.9	6.5
C	6.724	4,303.2	7.5
D	9.281	5,939.6	7.5
E	8.10	5,181.8	8.45
F	1.51	968.4	7.75
Total	31.878	20,401.4	

$$\text{Total rainfall (a-in.)} = 5.5 \times 1{,}490.5 + 6.5 \times 2517.9 + 7.5 \times 4303.2 + 7.5 \times 5939.6 + 8.45 \times 5181.8 + 7.75 \times 968.4$$

$$= 8{,}197.75 + 16{,}366.35 + 32{,}274.00 + 44{,}547.00 + 43{,}786.21 + 7{,}505.10$$

$$= 152{,}676.41 \text{ acre-inches}$$

$$\text{Average rainfall} = \frac{152{,}676.41}{20{,}401.4} = 7.48 \text{ in.}$$

$$= \frac{7.48}{12} \times 20{,}401.4 \times 43{,}560 = 553{,}946{,}973.4 \text{ ft}^3$$

The streamflow hydrograph for the rainfall storm of September 6–7, 1965, from Little Papillion Creek is as follows:

Time Sept. 7	Average discharge		Time Sept. 8	Average discharge	
	cfs	ft^3		cfs	ft^3
1200–100	450	1,620,000	2400–100	135.5	487,800
100–200	1552.5	5,589,000	100–200	139	500,400
200–300	2965	10,674,000	200–300	184.5	664,200
300–400	4502	16,207,200	300–400	300	1,080,000
400–500	5852.5	21,069,000	400–500	360	1,296,000
500–600	5925	21,330,000	500–600	231	831,600
600–700	4675	16,830,000	600–700	202.5	729,000
700–800	3185	11,466,000	700–800	169	608,400
800–900	1875	6,750,000	800–900	141	507,600
900–1000	1042.5	3,753,000	900–1000	123	442,800
1000–1100	742.5	2,673,000	1000–1100	111.5	401,400
1100–1200	673.5	2,424,600	1100–1200	94.5	340,200
1200–1300	617	2,221,200	1200–1300	87	313,200
1300–1400	548.5	1,974,600	1300–1400	76	273,600
1400–1500	414.5	1,492,200	1400–1500	68	244,800
1500–1600	286.5	1,031,400	1500–1600	62	223,200
1600–1700	226.5	815,400	1600–1700	57	205,200
1700–1800	196	705,600	1700–1800	52	187,200
1800–1900	173	622,800	1800–1900	48	172,800
1900–2000	157	565,200	1900–2000	44.5	160,200
2000–2100	147.5	531,000	2000–2100	42.5	153,000
2100–2200	140.5	505,800	2100–2200	41	147,600
2200–2300	136	489,600	2200–2300	39	140,400
2300–2400	132	475,200	2300–2400	37.5	135,000
		131,815,800 ft^3			10,245,600 ft^3

$$\text{Total runoff} = 131{,}815{,}800 + 10{,}245{,}600 = 142{,}061{,}400 \text{ ft}^3$$

$$\text{Percent runoff} = \frac{\text{Total runoff (ft}^3)}{\text{Total rainfall (ft}^3)} = \frac{142{,}061{,}400}{553{,}946{,}973.4} = 0.2564$$

$$= 25.64\% \text{ runoff}$$

$$\text{Average rainfall excess on the basin} = 7.48 \text{ in.} \times \frac{25.64}{100}$$

$$= 1.92 \text{ in. runoff}$$

Therefore,

$$\text{Infiltration} = 7.48 - 1.92 = 5.56 \text{ in.}$$

Thus, 5.56 in. of rainfall was lost on the Irvington drainage basin. From infiltration measurements, the infiltration is

First hour: 2.54 in.
Second hour: 2.25 in.
Interception: 0.25 in.

Therefore,

$$\text{Infiltration} = 5.04 \text{ in.}$$

Because infiltration is the greatest loss on this basin, this high value is expected. This 5.56-in. loss is related to the ϕ index.

8.8 ANALYTICAL MODELS OF INFILTRATION

The infiltration curve, shown in Figure 8.2, is typical of drainage basins. This curve is called the f curve for its standard notation for infiltration. Many simple analytical models have been proposed to determine this curve for hydrological analyses. There appears to be renewed interest in two conceptual models: the Green–Ampt (GA) model and the Philip two-term (PTT) model. Fok (1975) and Singh and Yu (1990) have shown that these two models are related to one another.

Several of the infiltration models can be derived by representing the soil matrix as a conceptual system. The behavior of this system is assumed to mimic the infiltration process. The system can be represented by a single element or a network of elements arranged in series and/or parallel. What takes place inside each element is not considered, but only what goes in and what comes out of the element. Thus, spatial variability of soil-moisture movement in the vertical direction is not considered explicitly. Following Dooge (1973), each element, as shown in Figure 8.21, known as an absorber, can be linear or nonlinear. Inflow to the absorber can be considered as the infiltration rate, outflow from it as the constant infiltration rate, and its storage as the potential infiltration. It is hypothesized that the infiltration rate f is related to the actual or potential infiltration volume (F_p). This hypothesis was first invoked by Overton (1964) in derivation of some well-known infiltration models including the Kostiakov, Horton, Holtan, Overton, GA, and PTT models. Thus, the concept of treating infiltration by a systems approach appears to provide not only a unifying framework for these various models, but also to establish connections between them.

For deriving analytical models, consider a column of soil matrix with a unit area of vertical infiltration as shown in Figure 8.21. The initial storage space available in the soil column is denoted as S_0. If the soil is initially dry, then S_0 equals the effective porosity multiplied by the total volume of the column. At time $t = 0$, a thin

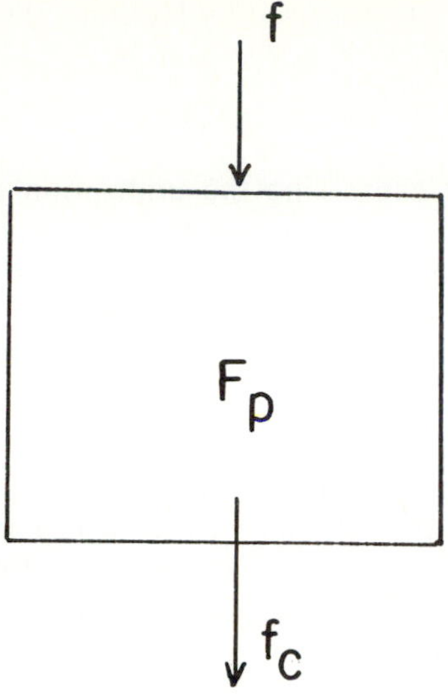

Figure 8.21 A conceptual element representing infiltration.

layer of water starts to cover the upper surface of the column. At any time t, the infiltration rate is denoted as $f(t)$, the seepage rate $f_s(t)$, and the potential water storage space as $S(t)$, which is the same as F_p.

The spatially lumped continuity equation for the soil column can be written as

$$\frac{dS(t)}{dt} = f_s(t) - f(t) \tag{8.6}$$

The amount of water stored in the soil column, $W(t)$, from $t = 0$ to any time t is

$$W(t) = S_0 - S(t) = \int_0^t [f(t) - f_s(t)]\, dt \tag{8.7}$$

But $W(t)$ is the same as $F(t)$, so

$$F(t) = S_0 - S(t) = S_0 - F_p \tag{8.8}$$

Overton (1964) expressed Equation (8.8) as

$$bS_0 = F_p + F \tag{8.9}$$

where b is a percentage factor based on vegetation whose value varies from 0.3 for weeds to 1 for bluegrass. Equation (8.6) contains three unknowns: S, f, and f_s. Therefore, two additional relationships amongst these unknowns are required. The function $f_s(t)$ may be taken as

$$f_s(t) = f_c \tag{8.10}$$

where f_c is constant. Singh and Yu (1990) proposed a generalized relationship as

$$f(t) = f_s(t) + \frac{a[S(t)]^n}{[S_0 - S(t)]^m} = f_s(t) + \frac{a[F_p(t)]^n}{[F(t)]^m} \tag{8.11}$$

where a, m, and n are positive real constants for a given soil–vegetation–land use complex. According to Equation (8.11), $f(t) \to \infty$ at $t = 0$, and $f(t) \to f_s(t)$ at $t = \infty$.

A special case of Equation (8.11) was used by Holtan (1961) and Overton (1964) that can be expressed as

$$f - f_c = aF_p^n \tag{8.12}$$

This is an empirical relation expressing the infiltration rate as a function of the potential volume of infiltration. Here S_0 is taken to equal the difference between the total pore space and soil moisture in about 50-cm (20-in.) depth of soil stratum. The constant a is usually between 0.25 and 0.8 (for f in inches per hour) and depends on antecedent moisture content.

Thus, the governing equations for the infiltration system are Equations (8.6), (8.10), and (8.11). A number of infiltration models, including those mentioned before, can be derived using variants of Equation (8.11) (Singh and Yu, 1990), as will be shown in the ensuing discussion. The direct use of Equations (8.6) and (8.7) as an infiltration model circumvents the time-dependency problem by introducing soil moisture as the dependent variable. This form of model is especially useful when the water-supply rate only intermittently exceeds the infiltration-capacity rate—the rate at which the water drains from the soil-moisture zone.

8.8.1 Overton Model

Overton (1964) employed the hypothesis that infiltration rate is a function of potential infiltration volume or soil moisture. In other words, $m = 0$ and $f_s(t) = f_c$ in Equation (8.11). Accordingly, Equations (8.9) and (8.12) can be combined and F_p is eliminated so that

$$\frac{dF}{dt} = a(bS_0 - F)^n + f_c \tag{8.13}$$

The initial condition is when $t = 0$, $F = 0$. An analytical solution of Equation (8.13) does not seem trackable for an arbitrary value of n. However, if n is an integer, such as -1, 1, 2, etc., then this equation can be solved. For $n = 2$, the Overton model is obtained:

$$\frac{dF}{dt} = a(bS_0 - F)^2 + f_c \tag{8.14}$$

Its solution is

$$F = bS_0 - d \tan [\, \tan^{-1} (bS_0/d) - Jt] \tag{8.15}$$

where $d = (f_c/a)^{0.5}$, and $J = (af_c)^{0.5}$. The time t_c to f_c is

$$t_c = (1/J) \tan^{-1} (bS_0/d) \tag{8.16}$$

Equation (8.15) can be written more compactly as

$$F = bS_0 - d \tan [J(t_c - t)] \tag{8.17}$$

By differentiating Equation (8.17), f is determined as

$$f = f_c \sec^2 [J(t_c - t)] \tag{8.18}$$

Based on field data, Overton (1964) found $n = 2$ to yield satisfactory results. However, the value of a was noted to vary from storm to storm for this special case.

The Overton model has four parameters: a, b, S_0, and n. Parameter b can be obtained from knowledge of the vegetative cover. Overton (1964) determined b as 1.0 for bluegrass, 0.7 for crabgrass and alfalfa, 0.45 for lespedeza and timothy, 0.35 for alfalfa, and 0.30 for weeds. The value of S_0 is obtained from the difference at the beginning between the total pore space and the soil moisture in the soil column (say, 50 cm deep). The volume of pore space is the product of porosity and the length of the soil column. The value of f_c is approximated by the hydraulic conductivity of the last soil horizon. The parameters to be estimated experimentally are a and n. If observations on f and F as functions of time are available for a given soil, then the value of a and n can be calculated as follows.

Equation (8.12) can be expressed by using Equation (8.9) as

$$f - f_c = a(bS_0 - F)^n \tag{8.19}$$

or

$$\log (f - f_c) = \log a + n \log (bS_0 - F) \tag{8.20}$$

For various times, the values of $f - f_c$ and $bS_0 - F$ are computed and plotted on log–log paper. A straight line is fitted through the points. The slope and intercept of the straight line provide the values of a and n.

In order to express f and F as functions of time, Equation (8.13) is expressed as

$$\frac{dF}{f_c + a(bS_0 - F)^n} = dt \tag{8.21}$$

On integrating,

$$\int_0^F \frac{dF}{f_c + a(bS_0 - F)^n} = t \tag{8.22}$$

Equation (8.22) can be integrated numerically.

Likewise, Equation (8.13) can be recast as

$$\left(\frac{f - f_c}{a}\right)^{1/n} = bS_0 - F = bS_0 - \int_0^t f\,dt$$

or

$$\int_0^t f\,dt + \left(\frac{f - f_c}{a}\right)^{1/n} = bS_0 \tag{8.23}$$

Again, Equation (8.23) can be solved numerically to yield f as a function of t. Alternatively, the value of F at a specified time obtained from Equation (8.22) can be inserted in Equation (8.19) to obtain f at that time.

8.8.2 Holtan Model

Holtan (1961) used Equation (8.12) with $n = 1.387$ and $a = 0.62$ (for f in inches per hour) to estimate f. The value of a is plot-averaged, and it will vary from one watershed to the other, but will be the same for all storms occurring on the given watershed. The value of a is usually in the range of 0.2 and 0.8. This model appears to be well suited for inclusion in a model of watershed hydrology because it relates f to the soil-moisture content. With a slight modification, this model has been used by Holtan in his watershed hydrology model (Holtan, et al., 1975).

$$f = \mathrm{GI} \times bS_0^{1.387} + f_c \tag{8.24}$$

where GI is a growth index of crop (vegetation) in percentage of maturity. As before, b is the vegetation parameter determined at plant maturity as the percentage of the ground surface area occupied by plant stems or root crowns.

Singh and Yu (1990) derived a general solution of the Holtan model by using the governing equations presented before. For the Holtan model, Equation (8.11) becomes

$$f(t) = f_c + a[S(t)]^n \tag{8.25}$$

Substitution of Equation (8.25) in Equation (8.6) yields

$$\frac{\mathrm{d}S(t)}{\mathrm{d}t} = -a[S(t)]^n \tag{8.26}$$

Its solution is

$$S(t) = [S_0^{1-n} - (1-n)at]^{1/(1-n)} \tag{8.27}$$

By inserting Equation (8.27) into Equation (8.25),

$$f(t) = f_c + a[S_0^{1-n} - (1-n)at]^{n/(1-n)} \tag{8.28}$$

which is the general solution of the Holtan model.

8.8.3 Huggins–Monke (HM) Model

Huggins and Monke (1966) proposed a model similar to Equation (8.12), which can be expressed as

$$f - f_c = a\left(\frac{S_0 - F}{\eta}\right)^n \tag{8.29}$$

where η is the total porosity of soil overlying the impeding stratum, a is a parameter, and n is an exponent. Parameters a and n can be determined from infiltration measurements following the procedure described for the Overton model. By following the procedure used to derive Equation (8.28), the general solution of the HM model can be expressed as

$$f(t) = f_c + a_*[S_0^{1-n} - (1-n)a_*t]^{n/(1-n)} \tag{8.30}$$

where $a_* = a/\eta^n$.

8.8.4 Kostiakov Model

If $n = -1$ and $f_c = 0$, then Equation (8.12), with F_p replaced by F, becomes

$$f = aF^{-1} \quad \text{or} \quad \frac{\mathrm{d}F}{\mathrm{d}t} = aF^{-1} \tag{8.31}$$

By solving Equation (8.31) with the initial condition $F = 0$ at $t = 0$,

$$F = (2at)^{0.5} \tag{8.32}$$

or

$$f = a_* t^{-0.5}, \qquad a_* = (2a)^{0.5}/2 \tag{8.33}$$

This is a special case of the Kostiakov model (Kostiakov, 1932). In general, the Kostiakov model is expressed as

$$F = at^b \tag{8.34}$$

or

$$f = (ab)t^{b-1} \tag{8.35}$$

where $a > 0$ and $0 < b < 1$ are parameters. Clearly, as $t \to 0, f \to \infty$ and as $t \to \infty, f \to 0$. In reality, f does not attain these values. This model cannot predict the start of runoff after the occurrence of rainfall. Because of its simplicity, this model is frequently used in agricultural irrigation studies. The parameters a and b can be estimated by plotting on log–log paper the infiltration rate f or accumulated infiltration F against time and fitting a straight line:

$$\log F = \log a + b \log t \tag{8.36}$$

or

$$\log f = \log (ab) + (b - 1) \log t \tag{8.37}$$

8.8.5 Green–Ampt (GA) Model

A simple model, based on Darcy's law, was proposed by Green and Ampt (1911) for infiltration into uniform soil with uniform initial moisture content due to a pool of negligible depth of water. The water is assumed to infiltrate into the soil. The infiltrated water defines a sharply wetting front separating the wetted and unwetted zones, as shown in Figure 8.22. The GA model results from application of Darcy's law to the wetted zone as

$$f = K \frac{L + S_c}{L} \tag{8.38a}$$

where K is the hydraulic conductivity of the wetted zone, L is the distance from the ground surface to the wetting front, and S_c is the capillary suction at the wetting front. If η is porosity, or fillable pore space, then $F = \eta L$, and Equation (8.38a) can be expressed as

$$f = K \frac{F + \eta S_c}{F} = K\left(1 + \frac{\eta S_c}{F}\right) \tag{8.38b}$$

or

$$\frac{\mathrm{d}F}{\mathrm{d}t} = K \frac{F + \eta S_c}{F}$$

which, when integrated with $F = 0$ at $t = 0$, yields

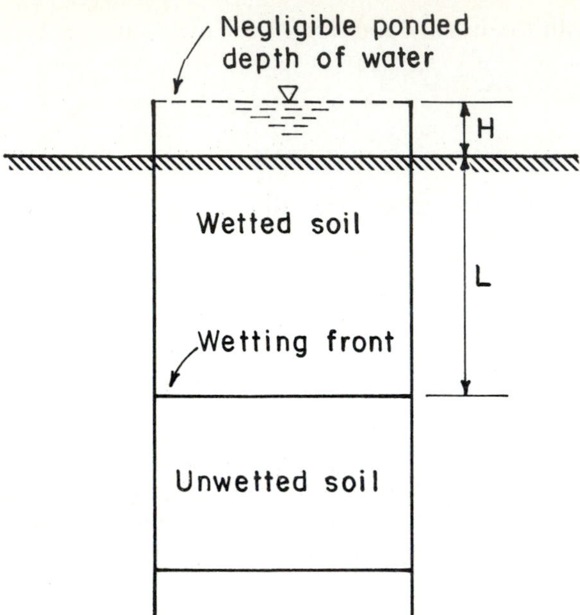

Figure 8.22 Schematic diagram of the Green–Ampt model.

$$t = \frac{F}{K} - \frac{\eta S_c}{K} \ln \frac{\eta S_c + F}{\eta S_c} \tag{8.39}$$

which is the GA model. The quantity $\eta S_c = S_f$ is defined as the storage suction factor and has a narrow range between 0 and 10 cm (Morel-Seytoux, 1981).

The GA model can also be derived by assuming $n = 1$ and replacing aF_p by aF^{-1} in Equation (8.12). Therefore.

$$f - f_c = aF^{-1} \quad \text{or} \quad \frac{dF}{dt} - f_c = aF^{-1} \tag{8.40}$$

By solving Equation (8.40) with $F = 0$ and $t = 0$,

$$t = (1/f_c)[F - (a/f_c) \ln (1 + Ff_c/a)] \tag{8.41}$$

Equation (8.41) is of the same form as Equation (8.39). These two equations become equivalent if $f_c = K$ and $a/f_c = \eta S_c = S_f$. Equation (8.41) is also analogous to the model of Philip (1954).

There has been a resurgence of interest in the GA model, for its parameters K, η, and S_c are physically meaningful and measurable quantities. It is more convenient from the standpoint of watershed modeling because it relates the volume of infiltrated water to the time from the beginning of infiltration.

If infiltration observations are available, then the GA parameters can be estimated by empirical fitting. By referring to Equation (8.38b),

$$f = K \frac{a}{F} \qquad a = K\eta S_c \tag{8.42}$$

The values of $1/F$ are then computed, and f is plotted against $1/F$ on arithmetic paper. By fitting a straight line through the data points, the values of K and a are obtained.

Brooks and Corey (1964), based on laboratory tests, found S_c to be a logarithmic function of effective saturation S_e, which is the ratio of the available moisture content $\theta - \theta_r$ to the maximum available moisture content $\eta - \theta_r$,

$$S_e = \frac{\theta - \theta_r}{\eta - \theta_r} \tag{8.43a}$$

where $\eta - \theta_r$ is termed the effective porosity, η_e; θ is the moisture content; and θ_r is the residual moisture content after it has been thoroughly derained. If $\theta_r \leq \theta \leq \eta$, S_e has the range $0 \leq S_e \leq 1.0$. The Brook–Corey relation can be expressed as

$$S_e = \left(\frac{S_b}{S_c}\right)^{\lambda} \tag{8.43b}$$

where S_b and λ are constants to be obtained experimentally.

According to Bouwer (1966), the effective hydraulic conductivity for an unsaturated flow is about half the corresponding value for saturated flow. Rawls et al. (1983), using the Brooks–Corey equation, analyzed about 5000 soil samples across the United States and determined average values of the GA parameters for different soil classes, as shown in Table 8.2. The ranges for η and η_e are not large, but the range of S_c can be large for a given soil.

TABLE 8.2 GREEN–AMPT INFILTRATION PARAMETERS FOR VARIOUS SOIL CLASSES (AFTER RAWLS ET AL., 1983)[a]

Soil class	Porosity, η	Effective porosity, θ_e	Wetting front soil suction head, ψ (cm)	Hydraulic conductivity, K (cm/h)
Sand	0.437 (0.374–0.500)	0.417 (0.354–0.480)	4.95 (0.97–25.36)	11.78
Loamy Sand	0.437 (0.363–0.506)	0.401 (0.329–0.473)	6.13 (1.35–27.94)	2.99
Sandy Loam	0.453 (0.351–0.555)	0.412 (0.283–0.541)	11.01 (2.67–45.47)	1.09
Loam	0.463 (0.375–0.551)	0.434 (0.334–0.534)	8.89 (1.33–59.38)	0.34
Silt Loam	0.501 (0.420–0.582)	0.486 (0.394–0.578)	16.68 (2.92–95.39)	0.65
Sandy Clay Loam	0.398 (0.332–0.464)	0.330 (0.235–0.425)	21.85 (4.42–108.0)	0.15
Clay Loam	0.464 (0.409–0.519)	0.309 (0.279–0.501)	20.88 (4.79–91.10)	0.10
Silty Clay Loam	0.471 (0.418–0.524)	0.432 (0.347–0.517)	27.30 (5.67–131.50)	0.10
Sandy Clay	0.430 (0.370–0.490)	0.321 (0.207–0.435)	23.90 (4.08–140.2)	0.06
Silty Clay	0.479 (0.425–0.533)	0.423 (0.334–0.512)	29.22 (6.13–139.4)	0.05
Clay	0.475 (0.427–0.523)	0.385 (0.269–0.501)	31.63 (6.39–156.6)	0.03

[a] The numbers in parentheses below each parameter are one standard deviation around the parameter value given.

8.8.6 Horton Model

The Horton model (Horton, 1939, 1940) is one of the popular infiltration models. Horton recognized that the infiltration capacity decreased with time until it approached a more-or-less constant value. He hypothesized that infiltration is similar to the exhaustion process according to which the rate of performing work is proportional to the amount of work remaining to be performed. In the case of infiltration, the work remaining to be performed at any time t is equal to that of changing the infiltration capacity f to its ultimate constant value f_c. The rate of performing work is $\mathrm{d}f/\mathrm{d}t$. The amount of work remaining to be performed is $f - f_c$. Since f decreases with time,

$$\frac{\mathrm{d}f}{\mathrm{d}t} = -k(f - f_c) \tag{8.44}$$

where k is a proportionality factor dependent on soil type and initial moisture content. The initial condition can be taken as $f = f_0$ (initial infiltration rate) at $t = 0$. Equation (8.44) can be integrated to yield

$$f = f_c + (f_0 - f_c) \exp(-kt) \tag{8.45}$$

This is the Horton model. It can also be expressed in terms of F as a function of t. Equation (8.45) becomes

$$\frac{\mathrm{d}F}{\mathrm{d}t} = f_c + (f_0 - f_c) \exp(-kt) \tag{8.46}$$

Upon integrating with the condition $F = 0$ at $t = 0$,

$$F = f_c t + \frac{1}{K}(f_0 - f_c)\,[1 - \exp(-kt)] \tag{8.47}$$

The Horton model can also be derived from Equation (8.12) by taking $n = 1$, f_c replaced by f_0, and F_p replaced by $-(F - f_c t)$:

$$f - f_c = -aF_e = -a(F - f_c t) \tag{8.48}$$

Solving Equation (8.48) with the same initial conditions yields

$$F = f_c t + \frac{1}{a}(f_0 - f_c)[1 - \exp(-at)] \tag{8.49}$$

Differentiating Equation (8.49),

$$f = f_c + (f_0 - f_c) \exp(-at) \tag{8.50}$$

Equation (8.49) is equivalent to Equation (8.34), and Equation (8.50) to Equation (8.45), with $a = k$.

The Horton model is simple in form and fits well to experimental data. The principal weakness of the model is in determination of reliable values of its parameters f_0, f_c, and k. As a rule of thumb, the ratio of f_0/f_c is of the order of 5. The parameter f_c can be interpreted to be the hydraulic conductivity at natural saturation. The simplest way to estimate parameters is by curve fitting on field observations of infiltration. The variation of f with t can be plotted on arithmetic paper. The flattening of the curve toward the extreme right suggests the value of f_c. Then any two points can be selected and their values inserted in Equation (8.45). This gives two equations with two unknowns, f_0 and k, which can then be solved. Parameters f_0, f_c, and a can be estimated in two other ways.

First, take the logarithm of Equation (8.50):

$$\ln (f - f_c) = \ln (f_0 - f_c) - at \tag{8.51}$$

Equation (8.51) represents a straight line when plotted on semilog paper whose slope $-a$ and intercept $\ln (f_0 - f_c)$ can readily be determined. For given infiltration data, f_c is taken as the lowest value of f when it tends to become constant. The value of $f - f_c$ at $t = 0$ is $f_0 - f_c$.

Second, a least squares method can be used to directly estimate parameters of Equation (8.49), which is cast in the form

$$F = a_0 + a_1 t - a_0 \exp(-at), \qquad a_0 = (f_0 - f_c)/a, \qquad a_1 = f_c \tag{8.52}$$

Equation (8.52) can be fitted to the experimental data. During the last five minutes or so (assuming that the experiment is continued until the infiltration becomes constant), the exponential term becomes small. Therefore,

$$F = a_0 + a_1 t \tag{8.53}$$

and a_0 and a_1 can be estimated using the data for this time interval. To determine a, one can write

$$-at = \ln [-(F - a_0 - a_1 t)/a_0] \tag{8.54}$$

By applying linear regression to the remainder of the data excluding those for $F < a_0 - a_1 t$, parameter a can be estimated.

8.8.7 Philip Two-Term Model

For a uniform soil with a uniform soil-moisture content and excess water-supply rate at the surface, Philip (1957) found a solution to the flow equation in the form of an infinite series. Because of rapid convergence, the first two terms of the series are considered sufficient and constitute the Philip two-term (PTT) model,

$$F = st^{0.5} + At \tag{8.55}$$

or

$$f = \tfrac{1}{2} st^{-0.5} + A \tag{8.56}$$

in which s is called sorptivity, a function of initial and surface water contents of the soil and soil-water diffusivity; and A is a parameter depending upon soil properties. Singh and Yu (1990) have shown that under certain assumptions, the PTT model and the GA model are equivalent.

The PTT model can also be derived from Equation (8,12), in which if $n = -1$ and F_p is replaced by $F - f_c t$, then

$$f - f_c = aF_e = a(F - f_c t)^{-1} \tag{8.57}$$

Solving Equation (8.57) with the same initial conditions, $F = 0$ at $t = 0$, yields

$$F = (2at)^{0.5} + f_c t \tag{8.58}$$

Upon differentiation,

$$f = (a/2)^{0.5} t^{-0.5} + f_c \tag{8.59}$$

which is analogous to Equation (8.56), with $A = f_c$ and $(2a)^{0.5} = s$.

Although parameters s and A have precise physical meaning, it is difficult to determine them in the field. Philip has shown that A is approximately $K/3$. By comparison with the GA model (Youngs, 1968), A is approximately $2K/2$. These parameters can be obtained from empirical fitting, if infiltration observations are available. The observed values of f are plotted against the values of $t^{-0.5}$ on arithmetic paper and a straight line is fitted through the data points. According to Equation (8.56), the intercept of this line is A and the slope is $s/2$.

Smiles and Knight (1976), however, suggested that infiltration data should be graphed with $Ft^{-0.5}$ against $t^{0.5}$. The resulting curve becomes linear quickly. The two parameters s and A emerge as the intercept and slope of the line according to Equation (8.55).

8.8.8 SCS Model

The Soil Conservation Service (1972), SCS, developed a procedure, often called the curve number method, for estimating runoff from small agricultural watersheds. Details of this method are presented in Chapter 14. This method, referred to here as the SCS model, has recently been adapted to estimate infiltration (Hjelmfelt, 1979). The SCS model is based on three assumptions. First, there is a maximum amount of water S that a watershed can hold through depression and soil storage. This is analogous to F_p. Second, the ratio of actual storage F to potential storage S is equal to the ratio of runoff (V_Q) to rainfall (P) minus initial abstraction (I_a):

$$\frac{F}{S} = \frac{V_Q}{P - I_a} \tag{8.60}$$

Third, I_a and S are linearly related as

$$I_a = aS \tag{8.61}$$

where a is constant. In the SCS method, a was taken originally as 0.2, which seems satisfactory for large storms, but leads to underestimation of runoff from small to medium storms. For such storms, $a = 0.1$ may be more appropriate.

From the water budget,

$$F = P - I_a - V_Q \tag{8.62}$$

By combining Equations (8.60) to (8.62) and eliminating V_Q,

$$F = P - I_a - \frac{(P - aS)^2}{P + (1 - a)S} \tag{8.63}$$

This is the SCS model for accumulated infiltration as a function of P, S, and I_a. By differentiating Equation (8.63),

$$f = p(1 - b)^2 \tag{8.64}$$

in which $p = dP/dt$ = rainfall intensity and

$$b = \frac{P - aS}{P + (1 - a)S} \tag{8.65}$$

It is seen from Equation (8.64) that as $P \to \infty$, $f \to 0$, which may not be true. The SCS model relates f to P, p and S, where S is the only unknown parameter if $a = 0.1$ or 0.2. The parameter S is related to the curve number derived by the Soil Conservation Service (1972) for soil–vegetation–land use complexes as shown in Chapter 14. The SCS model yields satisfactory values of infiltration from small agricultural watersheds and can be used to quantify the effect of changing land use.

EXAMPLE 8.6 In an infiltration experiment on Columbia soil, Mein and Larson (1971) reported infiltration data, as given in Table 8.3, for a rainfall rate of 5.56×10^{-3} cm/s and an initial moisture content of 0.2 (volume/volume). The porosity is 0.518 and the saturated hydraulic conductivity is 1.39×10^3 cm/s. Calibrate the Kostiakov model for this experimental data and assess its accuracy.

Solution The Kostiakov model has two parameters, a and b, and can be expressed as

$$\log F = \log a + b \log t$$

which is an equation of a straight line in the logarithmic domain. Thus, a and b can be determined using a least squares method. For these experimental data, $a = 0.0193$ and $b = 0.77$, and

$$F = 0.0193t^{0.77}$$

where F is in cm and t in s. For f in cm/h and t in s,

$$f = 53.5t^{-0.23}$$

TABLE 8.3 INFILTRATION MEASUREMENTS DUE TO MEIN AND LARSON (1971)

Time (s)	F (cm)	f (cm/h)	Time (s)	F (cm)	f (cm/h)
472.2	2.62	20.02	831.2	4.28	14.46
489.7	2.72	19.54	883.3	4.48	14.03
507.5	2.81	19.21	951.7	4.74	13.63
531.5	2.94	18.75	1008.5	4.95	13.18
559.8	3.08	18.32	1081.5	5.21	12.97
582.3	3.19	17.60	1142.5	5.43	12.63
609.4	3.32	17.00	1230.5	5.72	12.15
644.1	4.48	16.70	1357.2	6.14	11.79
682.9	3.66	16.42	1450.6	6.43	11.29
714.7	3.80	15.74	1547.3	6.73	11.09
751.1	3.95	15.43	1686.6	7.146	10.75
783.9	4.09	14.93	1793.2	7.45	10.40

Some sample computed values of f and F are

TIME (s)	F (cm)	f (cm/h)	TIME (s)	F (cm)	f (cm/h)
472.2	2.21	12.98	1008.5	2.97	11.9
531.5	2.42	12.63	1230.5	4.62	10.41
644.1	2.81	12.08	1547.3	5.51	9.88
783.9	3.27	11.55	1793.2	6.17	9.55

For larger times, the model fits better. The errors in fit are under 20% for F and under 35% for f.

EXAMPLE 8.7 Calibrate the Horton model using the data in Example 8.6, and assess its accuracy.

Solution The least squares method is applied to estimate the Horton parameters f_0, f_c, and a. During the last five minutes of the experiment, the exponential term became small. Let that time be t_c = 1547.30 s. This gives f_c = 0.00295 cm/s, f_0 = 0.0112 cm/s, and

$$F = 2.2 + 0.00295t, \qquad t > 1547.3 \text{ s}$$

For $t > 1547.3$ s,

$$\ln[-F(t) + a_0 + a_1 t)] = \ln a_0 - at$$

where $a = 0.0038$ is obtained by fitting. Therefore,

$$F = 2.2[1 - \exp(-0.0038t)] + 0.00295t, \qquad t < 1547.3 \text{ s}$$

Some sample values of calculated f and F are as follows;

TIME (s)	F (cm)	f (cm/h)	TIME (s)	F (cm)	f (cm/h)
472.2	3.26	24.39	1008.5	5.16	12.05
531.5	3.41	15.18	1230.5	5.85	11.19
644.1	4.94	13.75	1793.2	7.54	10.81

For larger times, the model fits better. The errors in fit are under 25% for F and f.

EXAMPLE 8.8 Calibrate the Philip two-term model using the data in Example 8.6, and assess its accuracy.

Solution The Philip two-term model is

$$F = (2at)^{0.5} + f_c t$$

$$f = \left(\frac{a}{2}\right)^{0.5} + f_c$$

By fitting

$$(2a)^{0.5} = 0.05 \qquad \text{and} \qquad a = 0.0013$$

Therefore,

$$F = 0.05t^{0.5} + 0.00298t$$

$$f = 0.0255t^{-0.5} + 0.00298$$

Some sample values of calculated f and F are

TIME (s)	F (cm)	f (cm/h)
472.2	2.49	14.95
531.5	2.74	14.71
644.1	3.19	14.35
783.9	3.74	14.01
1008.5	4.59	13.62
1230.5	5.42	13.35
1547.3	6.58	13.06
1793.2	7.47	12.9

The PTT model fits better with errors of about 5% in F and 25% in f. High error in f is at the beginning and declines quickly.

EXAMPLE 8.9 For experimental data given in Table 8.4, fit the following infiltration models: (a) Kostiakov model, (b) Horton model, (c) Holtan model, (d) Green–Ampt model, and (e) Philip two-term model. Show graphically the model fit to experimental

TABLE 8.4 INFILTRATION AND RUNOFF RATES

Time from start of rain (min)	Accumulated runoff (in.)	Runoff rate (in./h)	Accumulated infiltration (in.)	Infiltration rate (in./h)
3	0.000	0.000	0.336	6.730
5	0.072	3.124	0.440	3.605
10	0.425	3.292	0.696	3.438
15	0.721	3.628	0.960	3.101
20	1.035	3.953	1.207	2.776
25	1.373	4.130	1.430	2.600
30	1.720	4.163	1.645	2.567
35	2.064	4.164	1.862	2.565
40	2.415	4.242	2.071	2.487
45	2.765	4.236	2.282	2.494
50	3.112	4.219	2.496	2.510
55	3.479	4.346	2.690	2.384
60	3.851	4.455	2.879	2.275
65	4.213	4.427	3.077	2.303
70	4.585	4.393	3.267	2.336
75	4.952	4.407	3.460	2.323
80	5.323	4.490	3.650	2.240
85	5.682	4.349	3.852	2.381
90	6.055	4.353	4.040	2.377
95	6.443	4.521	4.212	2.208
100	6.809	4.448	4.408	2.282
105	7.185	4.654	4.593	2.264
110	7.552	4.414	4.787	2.316
115	7.932	4.534	4.967	2.195
120	8.307	4.562	5.153	2.168
125	8.669	4.404	5.352	2.325
130	9.060	4.645	5.521	2.085
135	9.443	4.761	5.700	1.969
140	9.804	4.627	5.899	2.102

data. Also plot the relative error in model fit against time. Relative error = (observed value − computed value)/observed value.

Solution

(a) *Horton model.* The value of f_c is taken as 1.95 in./h. Then ln $(f - f_c)$ is computed for each value of t. The values of ln $(f - f_c)$ are plotted against t on graph paper and a straight line is fitted through the points. The slope of the straight line is then computed. The value of the slope is found to be $a = 0.0184$, and the intercept ln $(f_0 - f_c) = 0.4233$. Therefore, $f_0 - f_c = 1.527$ and $f_0 = 3.477$ in./h. The Horton model becomes

$$f = 1.95 + 1.527 \exp(-0.0184t)$$

where t is in min and f in in./h. By integrating this equation,

$$F = 0.0325t + 1.38[1 - \exp(-0.0184t)]$$

The values of f and F are then computed for given times using these equations. Relative errors between observed f_{obs} and computed f_{model} values of f are calculated and the same is done for F. Some sample calculated values are as follows:

TIME (min)	f_{obs} (in./h)	f_{model} (in./h)	ERROR, E	F_{obs} (in.)	F_{model} (in.)	ERROR, E
3	6.730	3.395	0.496	0.336	0.172	0.488
15	3.101	3.109	−0.002	0.960	0.821	0.145
30	2.567	2.829	−0.102	1.645	1.562	0.083
60	2.275	2.456	−0.080	2.879	2.874	0.002
90	2.377	2.242	0.057	4.040	4.044	−0.001
120	2.168	2.118	0.023	5.153	5.131	0.004
140	2.102	2.066	0.017	5.899	5.828	0.012

A plot of computed and observed values is shown in Figure 8.23. The relative error between observed and computed f is graphed in Figure 8.24. The coefficient of determination between f_{obs} and f_{model} is found to be 0.827. The model fits the data quite well for $t \geq 5$ min.

(b) *Kostiakov model.* The values of log f are computed for different values of log t and are then plotted on log–log paper. A straight line is fitted through the points, which yields.

$$b - 1 = -0.214 \qquad ab = 5.84$$

Therefore,

$$f = 5.84t^{-0.214}$$

and

$$F = 7.43t^{0.786}$$

where t is in min and f in in./h. By using these equations, f and F are calculated and then relative errors between observed and computed values are computed. A sample of calculated values is as follows.

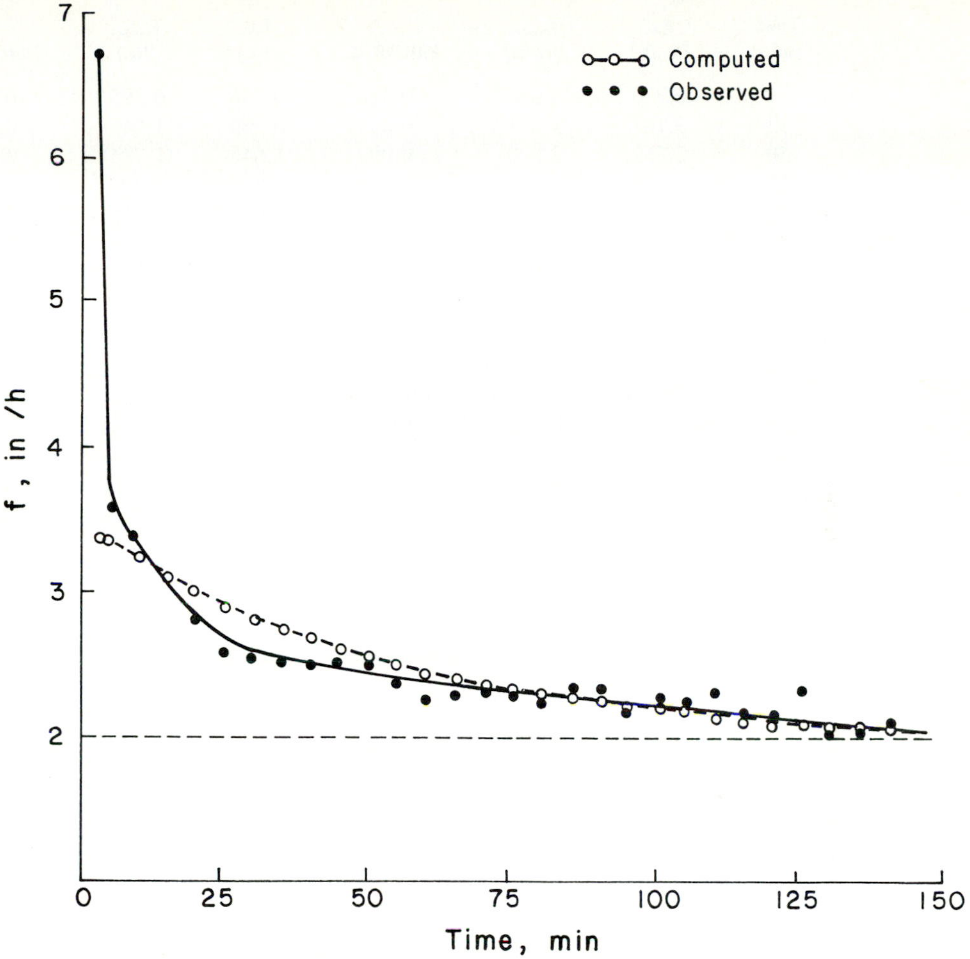

Figure 8.23 A comparison of the observed and computed infiltration rates for the Horton model.

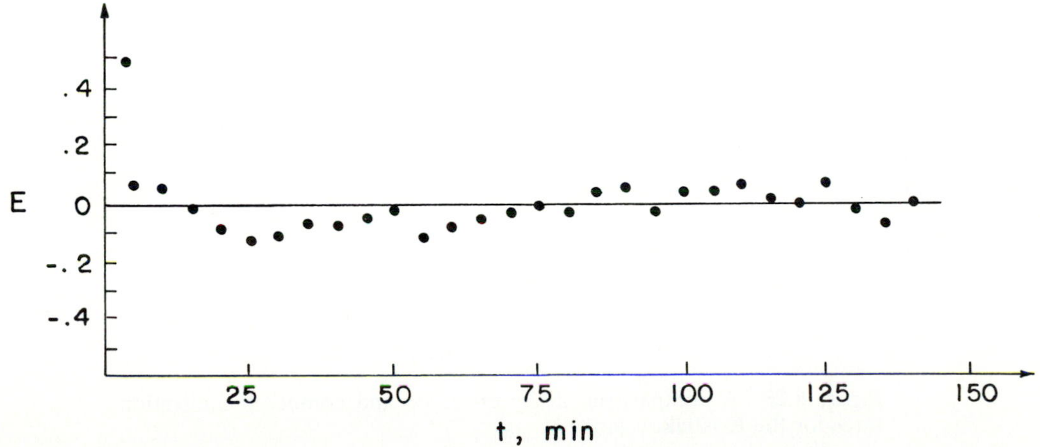

Figure 8.24 Relative error in the infiltration rate computed by the Horton model.

TIME (min)	f_{obs} (in./h)	f_{model} (in./h)	ERROR, *E*	F_{obs} (in.)	F_{model} (in.)	ERROR, *E*
3	6.730	4.616	0.314	0.336	0.295	0.125
15	3.101	3.271	−0.055	0.960	1.041	−0.084
30	2.567	2.820	−0.099	1.645	1.794	−0.091
60	2.275	2.432	−0.069	2.879	3.094	−0.075
90	2.377	2.229	0.062	4.040	4.255	−0.052
120	2.168	2.096	0.033	5.153	5.334	−0.035
140	2.102	2.028	0.035	5.899	6.021	−0.021

A plot of observed and computed f is shown in Figure 8.25. The values of E are graphed in Figure 8.26. The coefficient of determination between f_{obs} and f_{model} is found to be 0.912. The model fits the data quite well, especially for $t \geq 5$ min.

(c) *Holtan model.* For the Holtan model, f_c is taken as 1.95 in./h, $F_p = S_0 - F$, where S_0 is the available soil-moisture storage. A porosity of 30% is assumed. So $S_0 = 0.3L$, L = depth of soil column = 36 in., and gives $S_0 = 0.3 \times 36 = 10.8$ in. The values of $\ln (f - f_c)$ and $\ln F_p$ are computed and plotted on log–log paper and a straight line is fitted through the data. This produces

$$n = 3.415 \qquad a = 0.00045$$

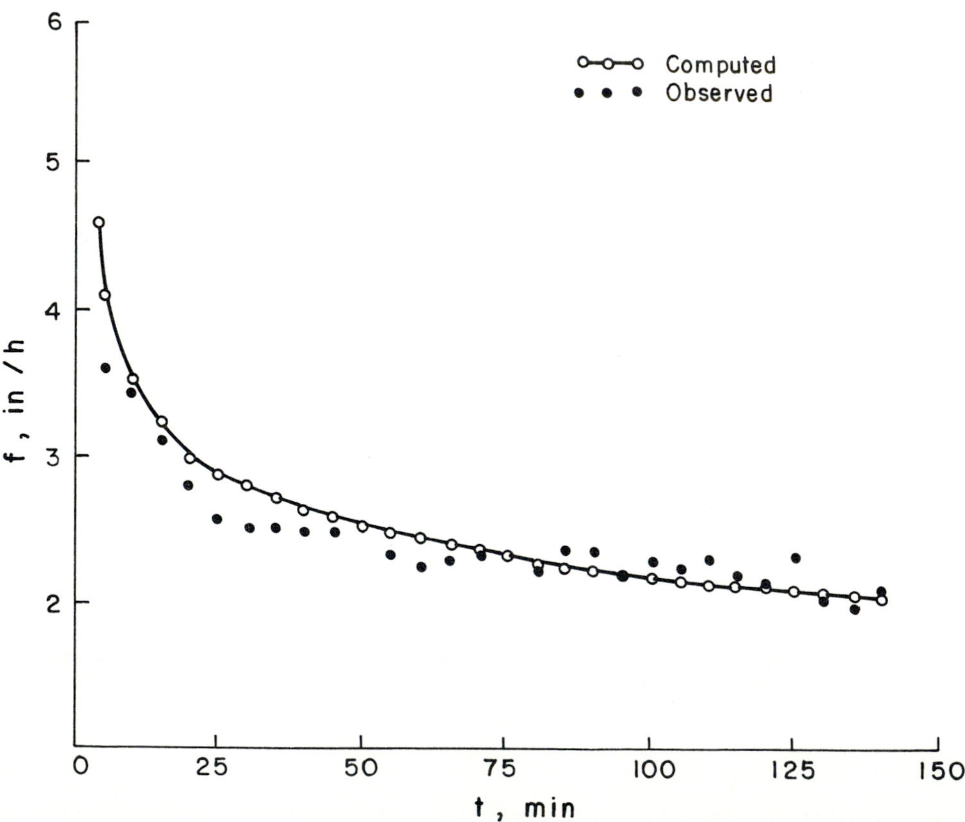

Figure 8.25 A comparison of the observed and computed infiltration rates for the Kostiakov model.

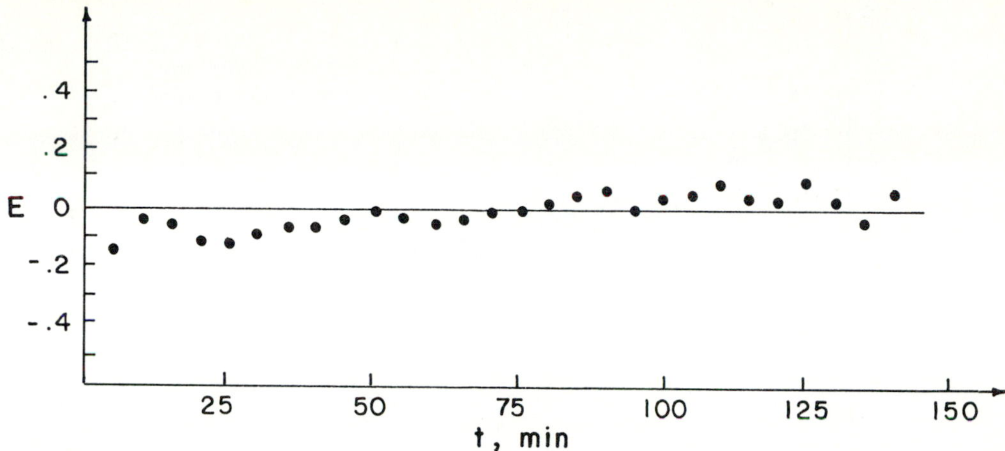

Figure 8.26 Relative error in the infiltration rate computed by the Kostiakov model.

Therefore,

$$f = 1.95 + 0.00045(10.8 - F)^{3.415}$$

or

$$\frac{dF}{dt} = 1.95 + 0.00045(10.8 - F)^{3.415}$$

$$\frac{dF}{1.95 + 0.00045(10.8 - F)^{3.415}} = dt$$

This is solved numerically. Substituting F from this solution into the expression for f before, we get f as a direct function of t. The values of f and F are then calculated and then their corresponding errors. A sample of calculated values is as follows:

TIME (min)	f_{obs} (in./h)	f_{model} (in./h)	ERROR, E	F_{obs} (in.)	F_{model} (in.)	ERROR, E
3	6.73	3.316	0.507	0.336	0.17	0.494
15	3.101	3.057	0.044	0.960	0.90	0.061
30	2.567	2.816	−0.097	1.645	1.60	0.027
60	2.275	2.478	−0.089	2.879	2.90	−0.007
90	2.377	2.257	−0.050	4.040	4.05	−0.003
120	2.168	2.116	0.024	5.153	5.15	−0.001
140	2.102	2.052	0.024	5.899	5.85	0.008

The coefficient of determination between f_{obs} and f_{model} is found to be 0.828, suggesting a good model fit. Figure 8.27 plots f_{obs} and f_{model}, and Figure 8.28 graphs E versus time.

(d) *Green–Ampt model.* This model can be expressed as

$$f = K + \frac{aK}{F}, \qquad a \text{ and } K \text{ are constants}$$

First, values of $1/F$ are obtained. Then f is plotted against $1/F$ on arithmetic paper and a straight line is fitted through the data points. This fit yields

$$K = 1.887, \qquad aK = 1.462, \qquad a = 0.775$$

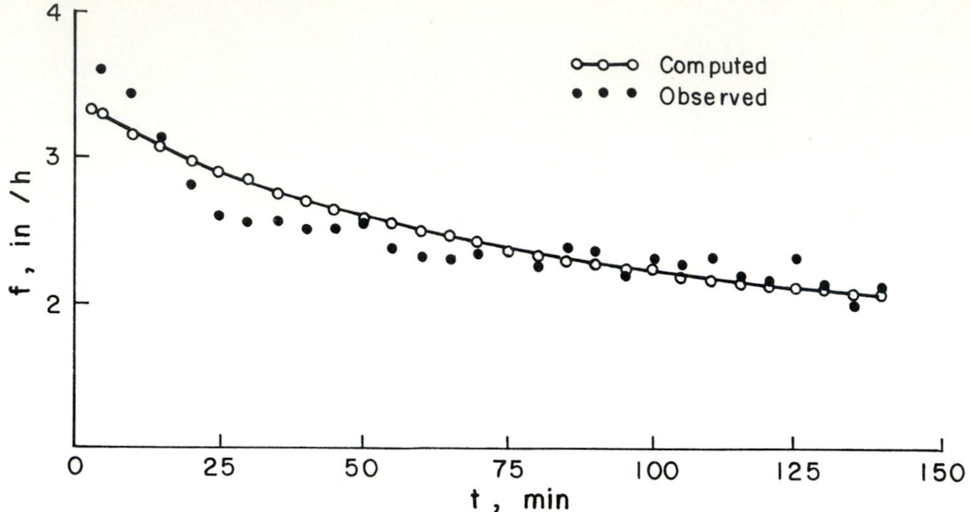

Figure 8.27 A comparison of the observed and computed infiltration rates for the Holtan model.

Therefore,

$$f = 1.887 + \frac{1.462}{F}$$

and

$$\frac{dF}{dt} = 1.887 + \frac{1.462}{F} = \frac{1.887F + 1.462}{F}$$

or

$$\frac{F}{1.887F + 1.462}\, dF = dt$$

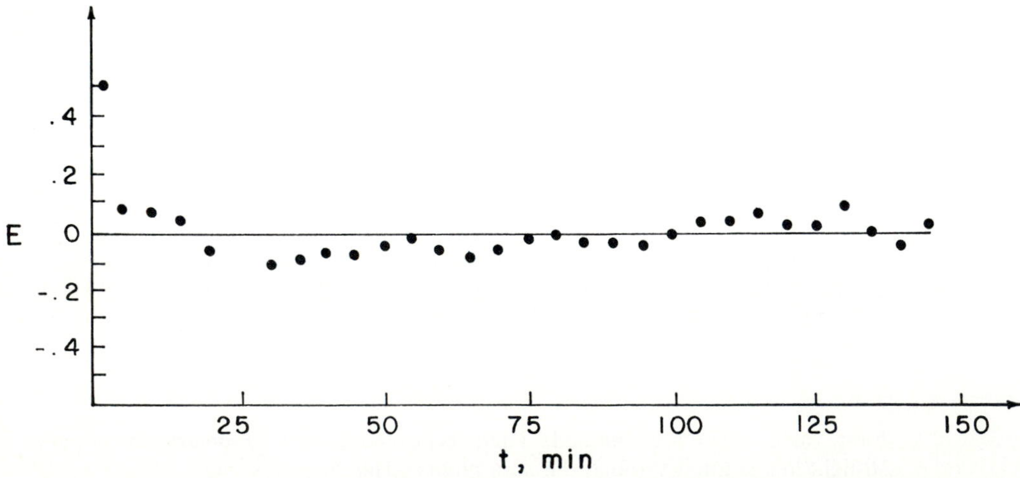

Figure 8.28 Relative error in the infiltration rate computed by the Holtan model.

This results in

$$t = \frac{F}{K} + \frac{a}{K} \ln \frac{a + F}{a}, \qquad F = 0 \text{ at } t = 0$$

$$= \frac{F}{1.887} + 0.411 \ln \frac{0.775 + F}{0.775}$$

By using these equations, values of f and F are calculated for different values of t, and then the corresponding errors are obtained. A sample of calculated values is as follows:

TIME (min)	f_{obs} (in./h)	f_{model} (in./h)	ERROR, E	F_{obs} (in.)	F_{model} (in.)	ERROR, E
3	6.730	6.239	0.073	0.336	0.25	0.256
15	3.101	3.410	−0.100	0.960	1.20	−0.222
30	2.567	2.776	−0.081	1.645	1.90	−0.155
60	2.275	2.395	−0.053	2.879	3.15	−0.094
90	2.377	2.249	0.054	4.040	4.25	−0.052
120	2.168	2.171	−0.001	5.153	5.35	−0.382
140	2.102	2.135	−0.016	5.899	6.05	−0.274

The coefficient of determination between f_{obs} and f_{model} is found to be 0.933, suggesting a good model fit. Figure 8.29 shows a comparison of observed and calculated f, and Figure 8.30 shows errors in calculated f.

(e) *Philip two-term model.* The values of $t^{-0.5}$ are calculated for observed times and then f is plotted against $t^{-0.5}$ on arithmetic paper. A straight line is fitted through the points, which produces

$$f_c = 1.433 \quad \text{and} \quad (a/2)^{0.5} = 7.207$$

Therefore,

$$f = 7.207t^{-0.5} + 1.433$$

and

$$F = 1.433 + 14.414t^{0.5}$$

where t is in min and f in in./h. Values of f and F are calculated using these equations and then their corresponding errors. A sample of observed and calculated values is as follows:

TIME (min)	f_{obs} (in./h)	f_{model} (in./h)	ERROR, E	F_{obs} (in.)	F_{model} (in.)	ERROR, E
3	6.730	5.593	0.169	0.339	0.487	−0.310
15	3.101	3.294	−0.062	0.960	1.288	−0.342
30	2.567	2.749	−0.071	1.645	2.032	−0.235
60	2.275	2.363	−0.039	2.879	3.293	−0.144
90	2.377	2.193	0.077	4.040	4.428	−0.100
120	2.168	2.091	0.036	5.153	5.500	−0.067
140	2.102	2.042	0.029	5.899	6.186	−0.048

The coefficient of determiantion between f_{obs} and f_{model} is found to be 0.929, indicating a good model fit. Figure 8.31 compares observed and calcualted f, and Figure 8.32 shows E as a function of time.

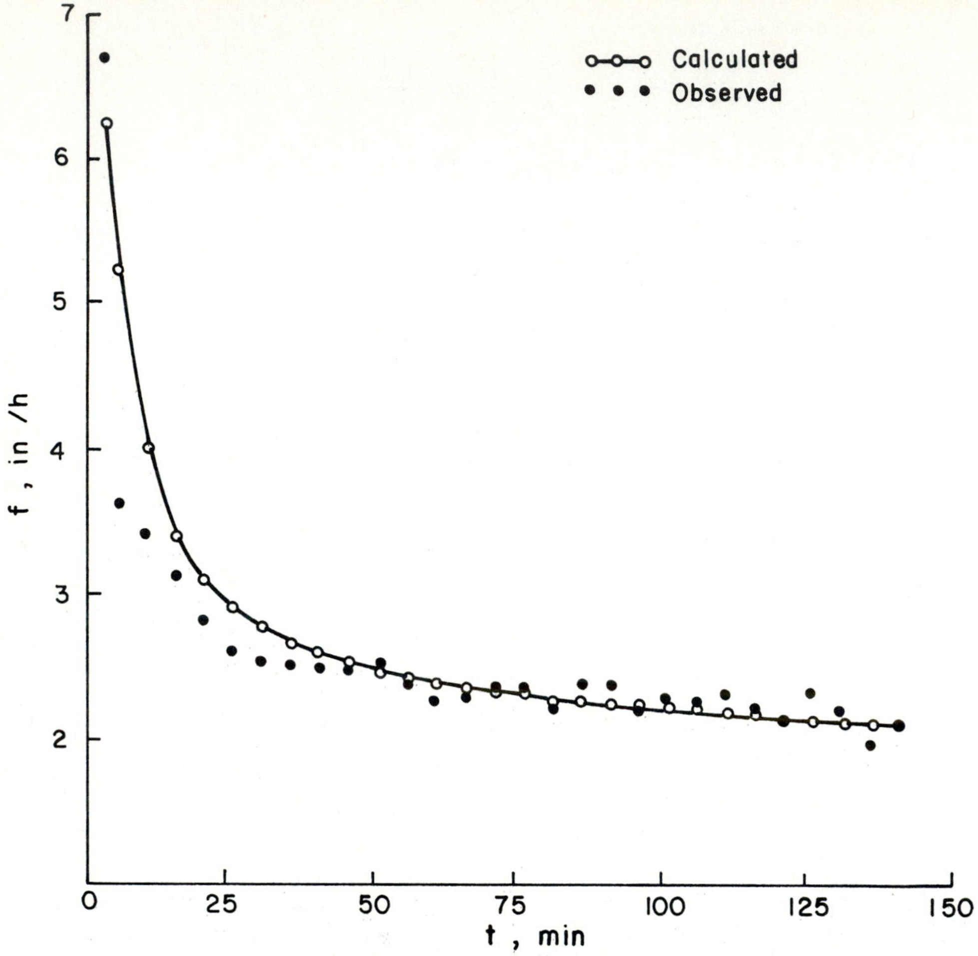

Figure 8.29 A comparison of the observed and computed infiltration rates for the Green–Ampt model.

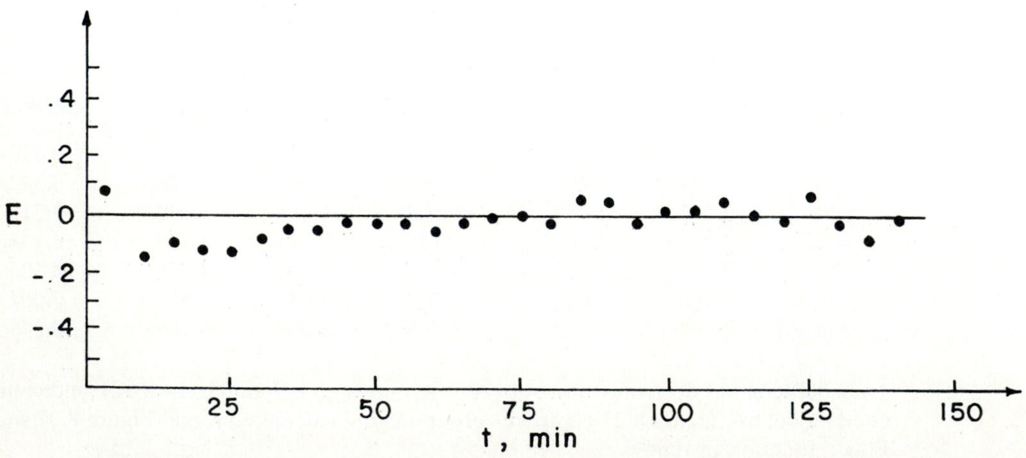

Figure 8.30 Relative error in the infiltration rate computed by the Green–Ampt model.

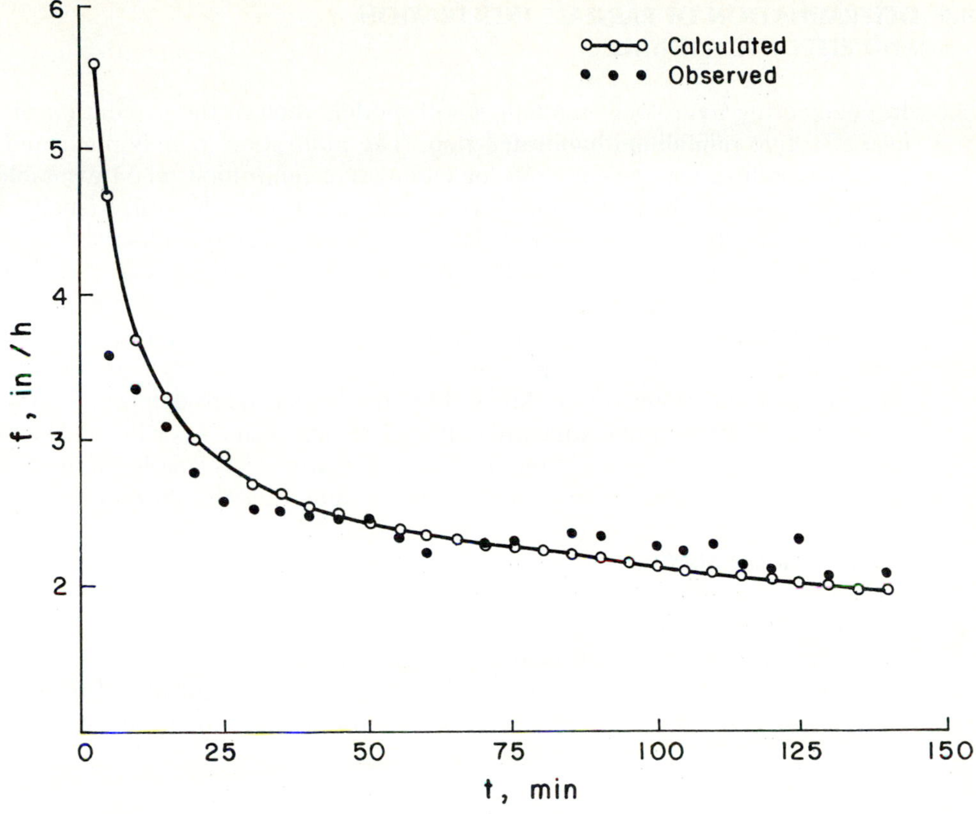

Figure 8.31 A comparison of the observed and computed infiltration rates for the Philip two-term model.

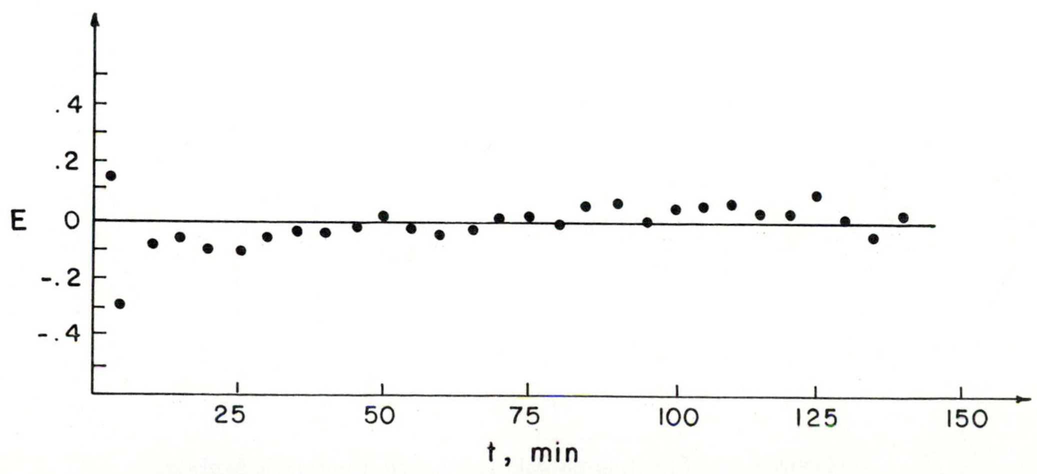

Figure 8.32 Relative error in the infiltration rate computed by the Philip Two-term model.

8.9 DETERMINATION OF RAINFALL INFILTRATION AND EFFECTIVE RAINFALL

In engineering hydrology, rainfall–runoff models require the separation of rainfall into effective rainfall and infiltrated rain. The infiltration models presented earlier prescribe infiltration capacity rate or cumulative infiltration when available water equals or exceeds the infiltration demand (or capacity) of the soil. Usually, rainfall occurs erratically, sometimes exceeding infiltration capacity and sometimes not and sometimes occurring intermittently. The infiltration models, therefore, must be adapted to take account of erratic, intermittent characteristics of rainfall. Only limited work has been reported on determination of rainfall infiltration under such conditions. The work by Bauer (1974), Morel-Seytoux (1981), and Peschke and Kutilek (1982) are noteworthy. Most of the methods used to determine rainfall infiltration assume that rainfall and streamflow data are available or the total amount of infiltrated water for a given rainfall event is known. Two simple methods are discussed here to present the concepts of determining rainfall infiltration.

8.9.1 ϕ Index Method

Let us consider a rainfall event of 11 cm (4.3 in.), as shown in Figure 8.33. The direct runoff resulting from this event is 3 cm (1.2 in.). Assume that other losses such as interception, depression storage, evaporation, etc., are negligible. The ϕ index determines the line above which the total amount of rainfall equals the direct runoff of 3

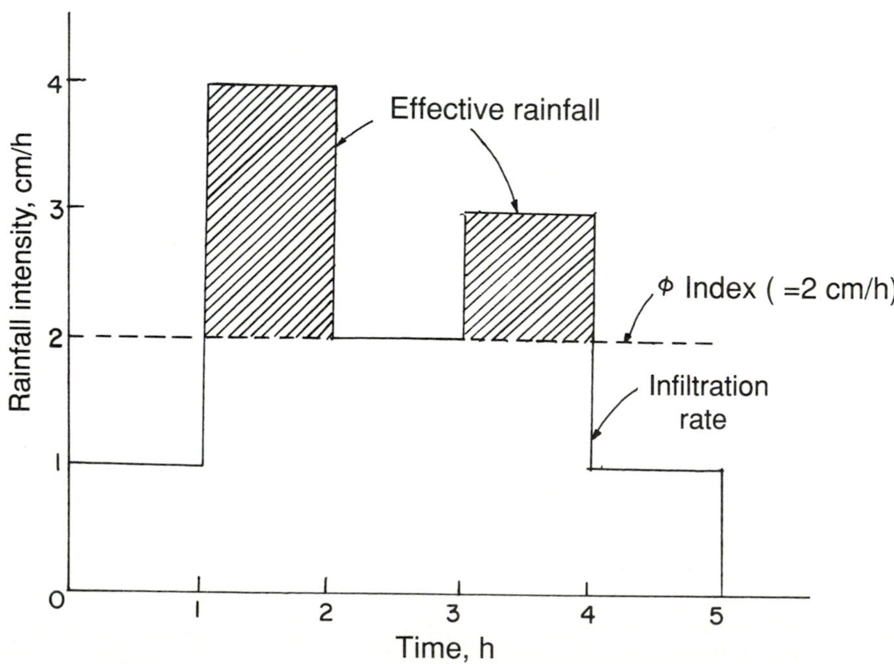

Figure 8.33 Effective rainfall determined using the ϕ method.

cm. This can be done by trial and error or algebraically. Its calculation can be facilitated by noting that

$$I_e \le \phi \le I_m \tag{8.66}$$

in which I_e is the total rainfall minus the total direct runoff divided by the rainfall duration, and I_m is the maximum rainfall intensity. From Figure 8.33,

$$1.8 \le \phi \le 4 \text{ cm/h} \tag{8.67}$$

A simple calculation shows that $\phi = 2$ cm/h. The effective rainfall and rainfall infiltration are shown in that figure. Below the ϕ-index line, the accumulated rainfall is 8 cm (3.1 in.), and above the line, it is 3 cm (1.2 in.), equal to the amount of direct runoff. It is to be noted that rainfall infiltration is not constant and is less than or equal to ϕ, although the infiltration capacity is constant at ϕ.

8.9.2 Variable Infiltration Method

Peschke and Kutilek (1982) presented a general procedure for determining infiltration caused by time-varying rainfall using the Green–Ampt and Kostiakov models. Due to the variability of rainfall, infiltration can occur at three rates during the course of an event: (1) It can occur at the infiltration-capacity rate. (2) It can occur at a rate equal to rainfall intensity. (3) It can occur at a rate less than rainfall intensity. These rates, of course, do not follow a particular sequence, but depend upon accumulated infiltration and rainfall rate. In other words, saturated and unsaturated phases can alternate. Under these conditions, two cases can be distinguished: (1) When there is only one ponding time, and (2) when ponding occurs more than once. Of course, the first case is a special case of the latter. The basic assumption is that for a given initial moisture, the potential infiltration rate at any time is uniquely determined by the cumulative infiltration up to that time. The first case is discussed using the Horton model as an example.

CASE 1: ONE PONDING

When direct runoff is produced by a rainfall event, the rainfall infiltration can be divided into two parts: (1) The unsaturated phase without runoff at $t < t_p$: when soil moisture at the surface is less than saturation and the hydraulic head on the surface is less than 0, the infiltration rate $f(t) = I(t)$, where t is time, the t_p is the ponding time when $I(t)$ exceeds the infiltration-capacity rate, $f_p(t)$, and when the effective rainfall starts. (2) The saturated phase at $t \ge t_p$ when direct runoff is produced; the soil surface is saturated and the hydraulic head equals or exceeds 0, $f(t) \le I(t)$ for $t \ge t_p$ and at $t = t_p$, $f(t) = I(t)$. Therefore,

$$f(t) = \min\,[I(t), g(F)] \tag{8.68}$$

where $g(F)$ is some function of F and specifies the rate of infiltration. The ponding time t_p can be obtained for variable rainfall $I(t)$ as

$$\int_0^{t_p} I(w)\,dw = \int_0^{t_p} f(w)\,dw = \int_0^{t_s} f_p(w)\,dw \tag{8.69}$$

$$I(t_p) = f(t_p) = f_p(t_s) \tag{8.70}$$

where t_s refers to the time to infiltrate $\int_0^{t_p} I(w)\,dw$ = the depth of water infiltrated due to $f_p(t)$.

Verma (1982) used these assumptions to modify the Horton model for the case when once the ponding occurred, then $f_p(t)$ would be satisfied. It has somewhat limited application, but may be helpful to understand the more realistic cases. By combining Equations (8.49) and (8.50),

$$f_p(t) = (f_0 - f_c) - a[F(t) - f_c t] + f_c \tag{8.71}$$

With $f(t) < f_p(t)$, $F(t)$ is the amount of rainfall absorbed in time t_s due to f_p. From Figure 8.34, Equation (8.50) can be written for t_s as

$$I(t_p) = f_c + (f_0 - f_c) \exp(-at_s)$$

giving

$$t_s = -\frac{1}{a} \ln \frac{I(t_p) - f_0}{f_0 - f_c} \tag{8.72}$$

By combining Equation (8.71) for $t = t_s$ and $F(t) = \int_0^{t_p} I(w)\,dw$ with Equation (8.72),

$$I(t_p) = f_0 - a \int_0^{t_p} I(w)\,dw = f_c \ln [(t_p - f_c)/(f_0 - f_c)] \tag{8.73}$$

which can be solved numerically to yield t_p. If $I(t) = I$ = constant,

$$t_p = \frac{1}{aI} \{f_0 - I + f_c \ln [(f_0 - f_c)/(I - f_c)]\} \tag{8.74}$$

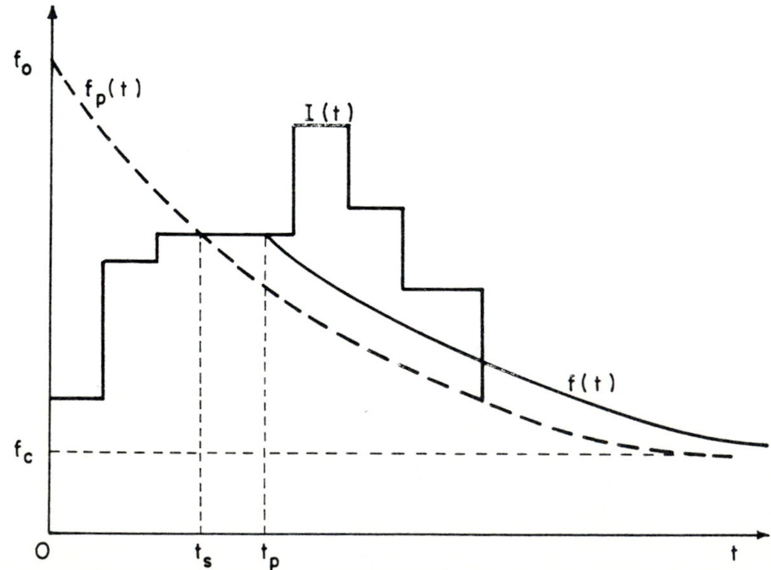

Figure 8.34 Rainfall infiltration using the Horton model with one ponding.

For rainfall infiltration, the Horton infiltration curve is shifted by t_s along the t-axis. After ponding, this shifted curve represents the rainfall infiltration. Therefore,

$$f(t) = f_c + (f_0 - f_c) \exp [-a(t - t_s)] \tag{8.75}$$

and also

$$f(t_p) = f_c + (f_0 - f_c) \exp [-a(t_p - t_s)] \tag{8.76}$$

from which

$$[f(t) - f_c]/[f(t_p) - f_c] = \exp [-a(t - t_s)] \tag{8.77}$$

By inserting $f(t_p)$ and $I(t_p)$ from Equation (8.72) and using $Q(t) = I(t) - f(t)$,

$$\begin{aligned} Q(t) &= I(t) - f_c \\ &= \left\{f_0 - f_c - a \int_0^{t_p} I(w)\, dw - f_c \ln [I(t_p) - f_c] \Big/ [f_0 - f_c] \exp [-a(t - t_p)\right\} \end{aligned} \tag{8.78}$$

or

$$\begin{aligned} Q(t) &= I(t) - f_c \\ &= [f_0 - f_c - a \int_0^{t_p} I(w)\, dw + af_c t_s] \exp [-a(t - t_p)] \end{aligned} \tag{8.79}$$

where $Q(t)$ is the runoff rate at $t \geq t_p$. Until $t = t_p$, $f(t) = I(t)$ and $Q(t) = 0$. Equation (8.79) can be simplified further. The term within the brackets equals $I(t_p) - f_c$ from Equation (8.73). For $I(t) = I$,

$$Q(t) = (I - f_c) \{1 - \exp [-a(t - t_p)]\}, \qquad t \geq t_p \tag{8.80}$$

8.10 RELATION BETWEEN PRECIPITATION, INFILTRATION, AND RUNOFF

If all the precipitation infiltrates, then no runoff can occur. If the precipitation exceeds the losses, runoff will increase by a similar amount. This concept is illustrated in Figure 8.35. This figure is simplified in order to present the relation desired.

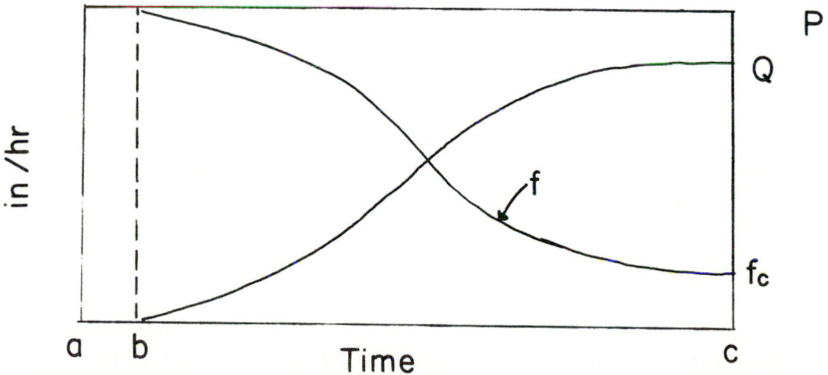

Figure 8.35 Relation between precipitation intensity (P), infiltration rate and runoff (Q).

Precipitation is occurring uniformly over a drainage area at a constant rate commencing at time *a*. Interception, depression storage, and infiltration losses use up precipitation as fast as it falls. Once the initial values of these losses are satisfied at time *b*, infiltration remains the only loss if there is no more depression storage. Infiltration loss diminishes with time, and at time *b*, runoff commences. As time increases, infiltration decreases further and is accompanied by an increase in runoff. A nearly constant infiltration capacity is reached at time *c*. At this time, runoff becomes a maximum and nearly constant, reflecting the character of infiltration capacity. This illustration only grossly reflects the processes that occur in a drainage basin. In nature, there is a seldom continuous, uniform rainfall or uniform infiltration.

8.11 SOIL MOISTURE

Soil moisture is the term applied to the water held in the soil by molecular attraction. The forces acting to retain water in the soil are adhesive and cohesive forces. These forces act against the force of gravity and against evaporation and transpiration. Thus, the amount of moisture in the soil at any given time is determined by the strength and duration of the forces operating on the moisture, and the amount of moisture initially present.

The degree of resistance to movement of water in soil material is called capillary tension, or capillary potential; it is a measure of the force required to remove moisture from the soil. The strength of capillary potential for a given soil is determined by the amount of soil moisture, grain size, and grain assortment. The units of capillary potential are conveniently expressed in terms of an equivalent depth of water, but is always less than atmospheric pressure. Capillary potential is measured by use of a tensiometer, an instrument that is made by sealing a tube in a porous cup, filling with water, and attaching a pressure gage on the top of the tube with all air excluded. When the cup is buried in soil, water moves by osmosis through the cup to and from the soil. This movement exerts a pressure on the tensiometer, which is indicated by the gage. The tensiometer is used in irrigation and will measure capillary potential down to about 0.85 atmosphere for this purpose. If measurements for greater tensions are required, they can be determined by use of a centrifuge. Other methods used to measure capillary potential include porous sorption blocks, electrical resistance, neutron scattering, and gravimetric analysis.

Field capacity is the moisture content of a soil that has been saturated with water and allowed to drain for 24 hours. This term, presumably, represents the maximum moisture-holding capacity of a soil. It is assumed that after 24 hours, no more moisture is moving through the soil under the force of gravity. Whether or not this is accurate, field capacity is a useful term because it represents a reference point for comparing moisture-holding capacities of soils. For example, a good medium-textured soil will hold, at field capacity, more than 2 in. (5 cm) of water per foot of soil, a clay soil will hold about 1.25 in. (3.2 cm) of water per foot, and sand will hold about 1.75 in. (4.5 cm) per foot. These values are important because they represent the potential soil-moisture replacement values that must be supplied by infiltration in order to bring moisture-depleted soil up to field capacity. At field capacity, the capillary potential is approximately equal to a pressure of 0.36 atmosphere.

Permanent wilting point is the term applied to the soil moisture content at which plants permanently wilt. At this point, plants will not recover even though moisture is added to the soil. The permanent wilting point is approximately equivalent to 0.15 atmosphere.

Antecedent moisture condition is used to describe the state of the soil in terms of the amount of moisture contained by that soil prior to the occurrence of rainfall. The importance of this term is shown by considering rainfall occurring on a dry drainage basin. As shown in Figure 8.36, during the first 30 minutes of the storm, the rainfall intensity is less than the infiltration rate and no runoff occurs. After 30 minutes, the rainfall intensity exceeds the infiltration rate and the hatched portion of the rainfall hyetograph above the infiltration curve becomes runoff. The portion of rainfall below the infiltration curve infiltrates the soil. Runoff ceases at 100 minutes even though rainfall continues for 10 minutes longer because rainfall intensity is less than infiltration capacity. If this storm hyetograph occurred on a wetter drainage basin, the hyetograph would need to be moved to the right until the beginning of the rainfall coincided with the initial infiltration rate of the wetter drainage basin. Under these conditions, more runoff would occur. Prediction of runoff, using the relation between infiltration curve and precipitation, would be very difficult because the moisture conditions on the drainage basin could not be predicted prior to each storm event. This is one of the main problems in predicting runoff because the soil-moisture content at any given time is almost never known. Moreover, the average soil moisture is probably never known. Estimates of the antecedent moisture conditions

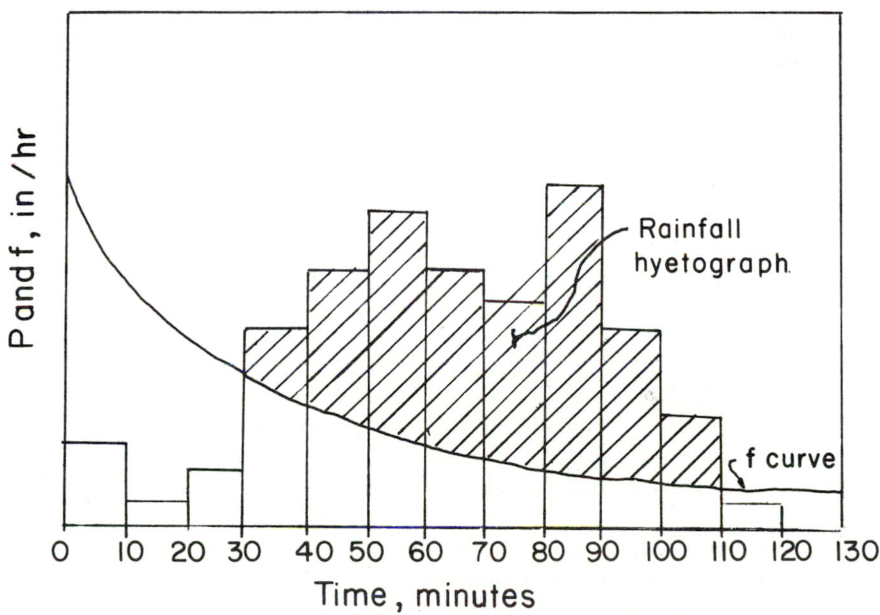

Figure 8.36 Infiltration curve and rainfall hyetograph.

are usually made by relating the time of a rainfall event to the time and amount of the last rainfall event. This relation may be reasonable because the depletion of soil moisture is related to time.

The antecedent moisture may be assumed to decrease in accordance with a mathematical decay function of the form

$$\mathrm{AMI}_t = \mathrm{AMI}_0 k^t \tag{8.81}$$

in which AMI_t is the antecedent moisture index at any time t, AMI_0 is the initial antecedent moisture index following precipitation at time t, and k is a recession constant. The values of k range between 0.85 and 0.98 with a value of 0.90 often used provided t is in units of days. Equation (8.81) shows that the moisture in the soil depletes as an exponential decay function.

8.12 SOIL-MOISTURE LOSSES

Moisture is depleted from the soil by evaporation and transpiration by plants. Evaporation depletes the soil surface. Continued evaporation might dessicate soils to a depth of 1 ft (30.5 cm) or more. Lesser depths of dessication are more common. During dry periods in the high plains, dessicated soil of 6 to 8 in. (15 to 20 cm) have been observed.

The amount and depth of moisture depletion by plant roots depends on the type of plant. Xerophytes, such as cacti, have shallow roots and remove moisture only in the surface portion of the soil. Phreatophytes, such as alfalfa, Palo Verde, and Salt Cedar trees, extend root systems in search of soil moisture down to depths of 100 ft (33 m). Alfalfa is a common crop and its root system more commonly reaches a depth of some 4 ft (1.3 m). Except for trees and shrubs, a depth of about 30 in. (76 cm) is a reasonable average penetration for most vegetation. The potential loss of soil moisture by transpiration is substantial. When evaporation losses are added to such losses, it is clear that substantial rainfall is required to replace such moisture losses. Dry soil conditions cause very high infiltration intake when rainfall occurs. High infiltration intake, in turn, results in delays in runoff occurring from a storm event and results in less runoff.

8.13 APPLICATIONS

Infiltration supplies water to the soil. It replenishes soil water and recharges groundwater. It is also used for disposal of wastes and water harvesting.

8.13.1 Groundwater Recharge

In addition to recharge by natural infiltration, infiltration ponds are constructed and operated for recharging the underlying groundwater reservoir. These infiltration ponds are extensively used in the San Joaquin Valley in California. Water that cannot be stored on the surface can be used for groundwater recharge. Floodwater generated by a big storm or water released from a dam may be that excess water. Infiltration ponds also help reduce flood crests by storing some of the floodwaters.

8.13.2 Irrigation

Different methods of irrigation are border, furrow, basin, sprinkler, and drip. The purpose of irrigation is to supply water to plant roots as uniformly as possible. This purpose is accomplished through infiltration. The uniformity of irrigation is achieved by control of infiltration. Fertilizers are extensively used to boost agricultural production. Uniformity of fertilizer application is directly linked to uniformity of irrigation. In areas with a limited water supply, drip irrigation may be applied, which conserves water. It applies water to the soil around individual plants and trees. Water wets the surface and is moved down by capillary suction and gravity into the soil for plant roots to extract and eventually transpire.

8.13.3 Management of Disposal Fields

Land disposal of treated sewage effluent and liquid waste from vegetable and animal processing plants requires prediction of infiltration rates for the disposal fields. Management of these fields requires maintaining optimum infiltration rates. Liquid wastes generally contain suspended solids in varying amounts that tend to remain on the ground, but fine, particulate matter may enter into the soil. As a result, clogging occurs on the surface and results in decreased infiltration rates, undermining the land-treatment systems for disposal of liquid waste.

8.13.4 Water Harvesting

Many oil-rich countries in the Middle East and Africa have very low annual precipitation (4 in./year, or 10 cm/year, in some cases). In these regions, a very large fraction of rainfall becomes infiltration. The infiltrated water is evaporated for the most part and does not reach the water table. Application of crude oil as a waterproofing and soil-binding agent has been made in Saudi Arabia and other countries to decrease infiltration in collection systems designed to conserve rainwater.

EXERCISES

8.1. Discuss the movement of water in the dry homogeneous soil under the following conditions: (a) dry homogeneous soil, (b) a clay layer in the soil, (c) a sand layer in the soil, (d) fine soil overlying a coarse sand or gravel subsoil, (e) a layer of coarse soil aggregates in the soil, (f) when deep vertical channels are cut in the soil and filled with organic matter, (g) straw plowed under and left in a layer in the soil, (h) straw mixed with soil, (i) three soils—sand, loam, and clay—side by side, and (j) hilly or rolling terrain.

8.2. For alapaha loamy sand in the Georgia coastal plain, infiltration and runoff were observed for a rainfall intensity of 6.73 in./h, as given in Table 8.3. Initial soil moisture for the 0- to 12-inch depth was 2.3 in. and final soil moisture for this depth was 3.10 in.

Initial soil moisture for the 12- to 36-in. depth was 5.18 in. and the final soil moisture for this depth was 6.01 in. Ninety percent of the surface was covered with weeds and 10% was bare. These data have been used in Example 8.9. Plot the infiltration rate and the accumulated infiltration as a function of time. Also plot the infiltration rate versus the accumulated infiltration.

8.3. Fit the Horton model to the infiltration data of Exercise 8.2. Provide detailed calculations. Discuss the accuracy of fit. Give values of the Horton model parameters.

8.4. Rainfall intensity I for a storm is given as follows:

TIME (h)	I (cm/h)	TIME (h)	I (cm/h)	TIME (h)	I (cm/h)
0	2	3	8	5	1
1	10	4	2	6	0
2	2				

The amount of runoff resulting from the storm is 15 cm. Compute the ϕ index and the rainfall infiltration using the ϕ index.

8.5. Compute the rainfall infiltration for Exercise 8.4 using the Horton model. Use the Horton parameters from Exercise 8.3.

8.6. Fit the Kostiakov model to the infiltration data of Exercise 8.2. Give detailed calculations. Discuss the accuracy of fit. Give values of the Horton model parameters.

8.7. Fit the Overton model to the infiltration data of Exercise 8.2. Give detailed calculations. Discuss the accuracy of fit. Give values of the Overton model parameters.

8.8. Compare the accuracies of the Horton, Kostiakov, and Overton models, and discuss their respective representations of the infiltration data in Exercise 8.2.

8.9. Fit the Green–Ampt model to the infiltration data of Exercise 8.2. Give detailed calculations. Discuss the accuracy of fit. Give values of the Green–Ampt model parameters.

8.10. Fit the Philip two-term model to the infiltration data of Exercise 8.2. Give detailed calculations. Discuss the accuracy of fit. Give values of the Philip two-term model parameters.

8.11. The following rainfall event took place on a given watershed:

TIME		RAINFALL INTENSITY (cm/h)
HOUR	MINUTE	
00	10	5
00	20	10
00	30	0

The amount of runoff produced by this rainfall is 0.5 cm. Assume the infiltration capacity to be constant. Determine the actual rate of infiltration and the capacity rate of infiltration for this watershed. Other abstractions can be assumed to be negligible.

8.12. A rainfall event is given as follows:

TIME (h)	INTENSITY (cm/h)
0	1
1	4
2	2
3	0

The amount of runoff generated by this event is 3 cm. Compute infiltration as a function of time. Assume that the infiltration-capacity rate is constant, and that there are no other abstractions. Compute the amount of effective rainfall and its intensity as a function of time.

8.13. Consider the Holtan model

$$f - f_c = aF_p^n$$

where

$$\frac{dF_p}{dt} = -(f - f_c), \qquad \text{with } t = 0, f = f_0$$

Solve this differential equation for $n = 1$, assuming the rate of rainfall is greater than the infiltration capacity. What well-known infiltration equation is this equivalent to?

8.14. Consider a model of the form:

$$f - f_c = \frac{a}{S_2 + S_3}$$

where

$$\frac{d}{dt}(S_2 + S_3) = f - f_c$$

with initial conditions $S_2 + S_3 = 0$ at $t = 0$. S_2 and S_3 represent storages of two soil elements connected in parallel. What well-known infiltration equation is this?

8.15. An infiltration test was conducted in the rice field of an irrigated plot in the DNOCS area northeast of Brazil that resulted in the following:

No.	1	2	3	4	5	6	7	8	9
Time elapsed, min	$\frac{1}{2}$	1	2	5	10	15	20	40	60
Depth of penetration of moisture, cm	2	2.9	3.4	3.6	3.75	3.9	3.95	4.0	4.0

Estimate the infiltration capacity of the soil.

8.16. A clayey loam has exhibited the following values of infiltration capacity:

Infiltration capacity, in./h	0.075	0.0557	0.045	0.0353	0.03	0.025	0.0153
Time, min	0	2	4	6	8	10	15
Infiltration capacity, in./h	0.013	0.0145	0.0135	0.013	0.0128		
Time, min	20	30	40	50	60		

Determine the value of k of the Horton model.

8.17. Compute the ϕ index for the following rainfall–runoff storm:

Rainfall (time in hour–minute and intensity in cm/h):

10	26	3.00	10	31	10.50	10	34	8.00	10	38	11.00
10	41	8.50	10	46	7.50	10	48	4.50	10	52	4.00
11	0	1.50	11	15	0.50	11	30	0.0			

Runoff (time in hour–minute and rate in cm/h):

13	30	0.003	13	31	0.015	13	34	0.050	13	37	0,080
13	40	0.100	13	45	0.120	14	00	0.150	14	37	0.120
14	47	0.100	15	10	0.080	15	35	0.050	16	05	0.040
17	21	0.030	18	0	0.000						

8.18. Compute the W index for the data in Exercise 8.17. Assume that the average rate of retention by interception and depression storage is 0.10 cm/h.

8.19. Compute ϕ index for the following rainfall–runoff storm:

Rainfall (time in hour–minute and intensity in cm/h):

00	10	1.4	00	35	0.3	00	55	2.1	01	25	0.9
01	45	0.3	02	00	0.0						

Runoff (time in hour–minute and rate in cm/h):

00	10	0.0	01	20	0.7	02	30	0.0

8.20. Compute the W index for the data in Exercise 8.19. Assume that the average rate of retention by interception and depression storage is 0.10 cm/h.

8.21. The soil-water content on a dry-weight basis of a soil profile and the bulk densities are given as follows. Calculate the soil-water content for each layer on a volume basis and the total soil water depth in cm.

DEPTH (cm)	DENSITY (BULK)	WATER CONTENT
0–30	1.3	15
30–60	1.4	20
60–90	1.5	22

8.22. Given a rainfall pattern:

TIME (h)	RAINFALL INTENSITY (cm/h)
0	5
1	10
2	0

and the volume of resulting runoff of 12 cm, determine the ϕ index. Remember that rainfall is represented by a bar diagram.

8.23. For the Horton model, the initial infiltration rate (f at $t = 0$) is equal to 10 cm/h, f_c, the final infiltration rate (f at a very large value of t) is equal to 1 cm/h, and k, the time constant, is equal to 2/h. Determine the volume of water infiltrated up to 5 hours and that between 10 and 15 hours.

8.24. For the Horton model, the infiltration rate at the beginning of rainfall is 10 cm/h and decreases to 1 cm/h after 10 hours. A total of 70 cm of water infiltrated during the 10-hour period. Compute the value of k of the Horton model.

8.25. Consider the following rainfall storm:

TIME (h)	RAINFALL INTENSITY (cm/h)
0	2
5	4
10	0

Evaporation, interception, and depression all amount to 3 cm during the storm period. These can be assumed to be uniform throughout the storm. The runoff from this storm is 12 cm. Determine the infiltration rate by the ϕ-index method.

8.26. A storm produces the following event:

TIME (h)	INTENSITY (cm/h)
0	4
1	2
2	3
3	1
4	0

The amount of surface runoff produced by this storm is 2 cm. Determine the ϕ index for computing the rate of infiltration.

8.27. The following rainfall event takes place on a given area:

TIME		INTENSITY
h	min	(cm/h)
00	10	5
00	20	10
0	30	0

The amount of runoff produced by this rainfall is 2 cm. Assume the infiltration-capacity rate to be constant. Determine the actual rate of infiltration and the capacity rate of infiltration for this area. Assume that there are no other abstractions.

8.28. If the soil-water content on a dry-weight basis of a soil profile and the bulk densities are as given, calculate the soil-water content for each layer on a volume basis and the total soil-water depth in cm.

DEPTH (cm)	DENSITY (BULK)	WATER CONTENT
0–30	1.3	15
30–60	1.4	20
60–90	1.5	22

8.29. An ephemeral drainage basin has a drainage area of 91.53 mi^2. The accompanying hydrologic data were obtained for a storm on June 23. The antecedent moisture conditions are considered to be normal. Infiltration measurements have been made and are believed to be representative of the entire drainage basin.

(a) Plot the infiltration area on semilog paper.

(b) Making use of the plotted infiltration data, determine the duration of effective rainfall and the average hourly infiltration losses.

(c) Compute runoff volume in inches.

	DISCHARGE, cfs		
TIME	**JUNE 23**	**JUNE 24**	**JUNE 25**
0200		3370	450
0400		3800	300
0600		3350	150
0800		3080	80
1000		2960	0
1200		3050	
1400		2300	
1600		2050	
1800	0	1400	
2000	550	1070	
2200	1400	850	
2400	2350	600	

The drainage area is 91.53 mi^2.

TIME	**RAINFALL ON JUNE 23**
1650	0.01
1700	0.03
1720	0.60
1740	1.00
1800	1.40
1820	0.84
1840	0.66
1900	0.49
1920	0.01

The average of 46 infiltration measurements yielded the following information:

TIME DURATION (min)	**INFILTRATION RATE (in./h)**
10	4.20
20	3.28
30	2.75

CHAPTER 9
Groundwater and Baseflow

The discussion in Chapter 2 indicates that groundwater stored beneath the land surface is probably at least 200 times the volume of annual runoff from the world's rivers. This enormous body of water plays a crucial role in the operation of the hydrologic cycle. It sustains streamflow during precipitation-free periods and constitutes the major source of water supply in many localities for municipalities, agriculture, and industries. Management for stormwater and disposal of chemical waste are also accomplished by the use of groundwater technology. Groundwater has become so important in recent years that a new branch of hydrology, called groundwater hydrology, has evolved. This chapter presents some elementary aspects of groundwater and its contribution to streamflow as baseflow. It also shows the relation between hydrology and geology.

9.1 DEFINITIONS AND NOTATIONS

9.1.1 Aquifers

A geologic formation, a part of a formation, or a group of formations that yields significant quantities of water is defined as an aquifer (Hantush, 1964).

A geologic formation that contains no interconnected pores and, therefore, can neither absorb nor transmit water is termed an aquifuge. Solid granite is an example.

A geologic formation that is porous and contains water but is not capable of transmitting water in significant quantities is termed an aquiclude. Clay or shale is an example.

A geologic formation whose hydraulic conductivity is too small to permit development of wells or springs but may be sufficiently large to influence the hydraulics of aquifers adjacent to it is termed an aquitard.

9.1.2 Types of Aquifers

Different geologic formations give rise to different types of aquifers. Hantush (1964) has given a generalized sectional view combining different types of aquifers, as shown in Figure 9.1.

CONFINED AQUIFERS

In these aquifers, groundwater is confined under pressure greater than atmospheric by overlying impervious or semipervious strata. These are also called pressure aquifers. The water level in a well penetrating a confined aquifer will rise above the bottom of the upper confining layer, but may or may not reach the land surface. When the water level reaches the land surface, the aquifer is artesian and the well is a flowing one. Rises and falls of water in artesian wells result primarily from changes in pressure rather than changes in storage volume.

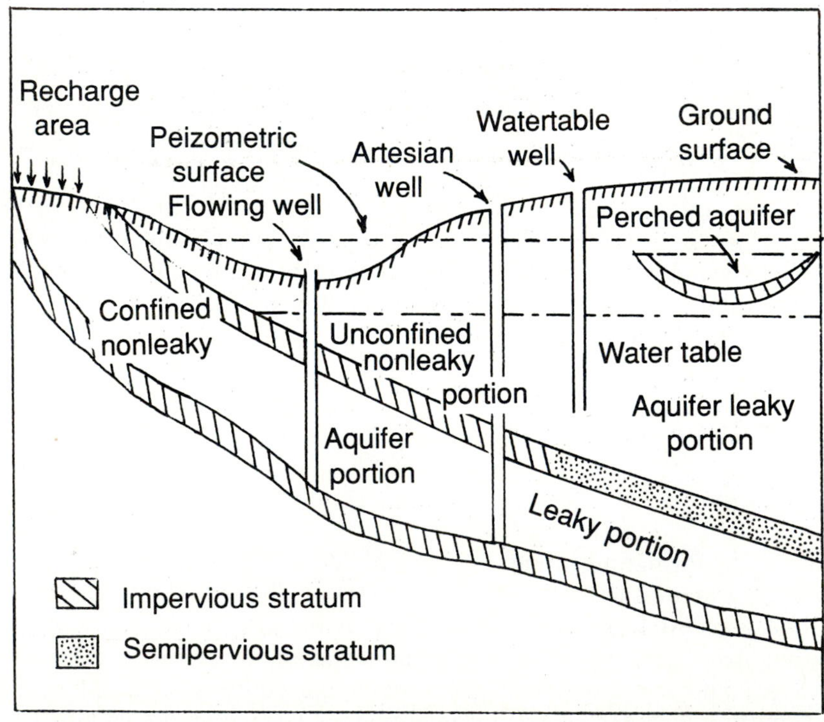

Figure 9.1 Different types of aquifers (adapted from Hantush, 1964).

UNCONFINED AQUIFERS

These are also called free, phreatic, or water-table aquifers. In these aquifers, the upper surface of the zone of saturation is under atmospheric pressure and is constituted by the water table. The water level in a well penetrating an unconfined aquifer does not rise above the water table. Rises and falls in the water table result primarily from changes in the volume of water in storage in the aquifer. The water table changes in form and slope, depending upon areas of recharge and discharge, pumping from wells, aquifers, and aquifer properties. When the piezometric surface falls below the bottom of the upper confining stratum, the confined aquifer becomes an unconfined one.

PERCHED AQUIFERS

These are special cases of unconfined aquifers. A perched aquifer occurs wherever a relatively small, impervious or semipervious stratum supports a groundwater body that is above the main water table. If the base of the supporting stratum penetrates the main groundwater body, then this aquifer is called a semiperched aquifer. Clay lenses in sedimentary deposits often possess perched water bodies overlying them. These water bodies yield only small quantities of water. The water table of a perched aquifer is called a perched water table.

LEAKY AQUIFERS

These aquifers, whether confined or unconfined, lose or gain water through adjacent semipervious layers. For example, a confined aquifer that has at least one semipervious confining stratum is a leaky confined aquifer. Similarly, an unconfined aquifer resting on a semipervious bed is an example of a leaky unconfined aquifer.

9.1.3 Piezometric Surface

The piezometric, or potentiometric, surface of a confined aquifer is an imaginary surface coinciding with the hydrostatic pressure level of the water in the aquifer. The elevation of this surface at a given point is defined by the water level in a well penetrating a confined aquifer at that point. This surface plays a key role in the identification, analysis, and synthesis of flow in confined aquifers. If this surface falls below the top of the aquifer at any point, then the piezometric surface and water table coincide and the aquifer becomes unconfined at that point.

9.1.4 Water Table

The upper surface of the zone of saturation under atmospheric pressure is called the water table. This is also known as the phreatic surface.

9.1.5 Groundwater Basin

A groundwater basin can be defined as a physiographic unit containing one or more than one aquifer. These aquifers are connected and interrelated. In a given physiographic unit, the drainage and groundwater basins may have more or less the same configuration or entirely a different one. For example, in limestone and sand hill areas, the two basins have different configurations, whereas in a valley between mountain ranges, the two closely coincide with one another.

9.1.6 Recharge Area

A region supplying water to a groundwater basin is called a recharge area.

9.1.7 Porosity

Porosity is the ratio of the volume of the interstices to the total volume of a given rock or soil, and can be expressed as a decimal fraction or as a percentage. Grain size and shape, arrangement of grains, degree of compaction and cementation, and degree of assortment are some of the factors affecting porosity. Quantitatively,

$$\eta = \frac{v_i}{V} = \frac{v_w}{V} = \frac{V - v_m}{V} = 1 - \frac{v_m}{V} \tag{9.1}$$

where η is porosity as a decimal fraction, V is the total volume, v_i is the volume of interstices, v_w is the volume of water in a saturated sample, and v_m is the volume of mineral particles. η can also be expressed (Lohman, 1972) as

$$\eta = \frac{\rho_m - \rho_d}{\rho_m} = 1 - \frac{\rho_d}{\rho_m} \tag{9.2}$$

where ρ_d is the density of a dry sample (bulk density), and ρ_m is the mean density of mineral particles (grain density).

Hantush (1964) distinguishes between porosity and effective porosity. The latter represents a portion of the pore space in a saturated permeable material in which the flow of water takes place. Not all the pore space of a material filled with water is accessible to flow, since the voids of this material are partly filled with water held in place by molecular and surface-tension forces.

9.1.8 Void Ratio

The void ratio, η_c, of a soil is the ratio of the volume of its interstices to the volume of its mineral particles:

$$\eta_c = \frac{v_i}{v_m} = \frac{v_w}{v_m} = \frac{\eta}{1 - \eta} \tag{9.3}$$

9.1.9 Permeability

The permeability of a porous medium is a measure of its ability to transmit fluid under a hydropotential gradient. It is considered to be approximately proportional to the square of the mean grain diameter,

$$k \cong Cd^2 \tag{9.4}$$

where k (L^2) is the intrinsic (or specific) permeability, d is the mean grain diameter, and C is a dimensionless constant called the shape factor depending upon porosity range, distribution of particle size, shape of grains, etc. Permeability is a property of the medium and independent of the properties of the fluid.

9.1.10 Hydraulic Conductivity

A porous medium has a hydraulic conductivity, K (L/T), of unit length per unit time if it will transmit in unit time a unit volume of groundwater through a unit cross-sectional area normal to the direction of flow under a unit hydraulic gradient. Mathematically, from Darcy's law,

$$K = -\frac{q}{\mathrm{d}h/\mathrm{d}L} \tag{9.5a}$$

K (L/T) is also simply called the conductivity of the aquifer, transmission constant, or the coefficient of permeability. k and K are related to each other as

$$K = \frac{\gamma k}{\mu} \tag{9.5b}$$

Thus, K depends not only upon the permeability of the aquifer, but also upon the specific or unit weight γ and dynamic viscosity μ of the fluid. The dependence of K on k, γ, and μ makes it vary not only from aquifer to aquifer and from liquid to liquid, but also from direction to direction and from temperature to temperature.

K may vary with the direction of flow even in homogeneous media, in which case it is called anisotropic conductivity. This occurs commonly in unconsolidated sedimentary deposits where the horizontal component of K may be two- to tenfold or even larger than the vertical component. The temperature does not vary appreciably in many aquifers. Thus, K is frequently assumed to be constant with temperature and direction of flow.

9.1.11 isotropy and Homogeneity

If all the properties of the aquifer are constant or independent of its location, then the aquifer is homogeneous; otherwise, the aquifer is nonhomogeneous, or heterogeneous. An aquifer is isotropic if its properties are independent of the direction. For example, the hydraulic conductivity must be the same in the x, y, and z directions; otherwise, the aquifer is anisotropic. In other words, homogeneity depends on the location, whereas isotropy depends upon the direction. A homogeneous aquifer may be isotropic or anisotropic, but a heterogeneous aquifer must necessarily be anisotropic.

9.1.12 Transmissivity

The transmissivity, T (L^2T^{-1}), of an aquifer is the rate at which water is transmitted through its unit width under a unit hydraulic gradient, and is also called the coefficient of transmissivity, or transmissibility. It characterizes the ability of the aquifer to transmit water. In aquifers having approximately uniform thickness, T is expressed as the product of the thickness b and the average value of the hydraulic conductivities in a vertical section of the aquifer:

$$T = bK \tag{9.6}$$

9.1.13 Specific Yield

The specific yield represents the water yielded from water-bearing material. More precisely, it is the ratio of the volume of water that the material, after being saturated, will yield by gravity to its own volume (Meinzer, 1923),

$$S_y = \frac{v_g}{V} \tag{9.7}$$

where v_g is the volume of water drained by gravity. It is also defined as

$$S_y = \eta - S_r \tag{9.8}$$

where S_r is the specific retention. The specific yield is also known as the effective porosity.

9.1.14 Apparent Specific Yield

The apparent specific yield, s_{ya} (dimensionless), is defined as the ratio of the volume of water added to or removed directly from the saturated aquifer to the resulting change in the volume of aquifer below the water table. The effects of such factors as air entrapped near the water table, water-table position, and the rate of change of water-table elevation are included in S_{ya}. As an approximation, it is considered constant. Although there may be little relation between S_y and S_{ya}, $S_{ya} \leq S_y$.

9.1.15 Specific Retention

The specific retention, S_r (dimensionless), of a soil is defined as the ratio of the volume of water that the soil, after being saturated, will retain against the force of gravity to its own volume:

$$S_r = \frac{v_r}{V} = \eta - S_y \tag{9.9}$$

where v_r is the volume of water retained mostly by molecular attraction and surface tension against gravity. S_r is also known as field capacity or water-holding capacity.

9.1.16 Specific Storage

The specific storage, S_s (L^{-1}), of an aquifer can be defined as the volume of water that a unit volume of the aquifer under a unit decline in the average head releases from storage due to expansion of water and compression of the aquifer. Hantush (1964) showed that

$$S_s = \gamma\eta\beta\left(1 + \frac{\sigma}{\eta\beta}\right) \tag{9.10}$$

where σ is the vertical compressibility of the solid skeleton of the material, or the reciprocal of its modulus of elasticity, and β is the compressibility of water, or the reciprocal of its bulk modulus of elasticity. In Equation (9.10), the factor $\gamma\eta\beta$ represents the fraction of storage derived from the expansion of water and the product $\gamma\sigma$ represents the fraction derived from the compression of the aquifer. The specific storage is regarded as a constant. It is a property of the aquifer material, its contained water, and the overburden stress.

9.1.17 Storage Coefficient

The storage coefficient or storativity, S_c (dimensionless), of an aquifer can be defined as the volume of water that a vertical column of the aquifer of unit cross-sectional area releases from or takes into storage as the average head within this column changes a unit distance.

For unconfined aquifers, S_c is virtually equal to S_y, as most of the water is released from storage by gravity drainage and only a small part comes from compression of the aquifer and expansion of water. S_c ranges from 0.1 to 0.3 and averages 0.2 for most unconfined aquifers. In confined aquifers, the water released from or taken into storage is entirely due to compressibility of the aquifer and of water. S_c can be expressed as

$$S_c = bS_s \tag{9.11}$$

where b is the aquifer thickness. For most confined aquifers, S_c ranges from about 10^{-5} to 10^{-3}, and is about 10^{-6} per foot of thickness.

EXAMPLE 9.1 In an area of 120.0 ha, the water table dropped by 5.0 m. If the porosity is 28.0% and specific retention is 9.0%, determine the specific yield of the aquifer and change in groundwater storage.

Solution

$$\text{Porosity} = S_y + S_r$$

$$28\% = S_y + 9\%$$

Therefore, $S_y = 19\%$, or 0.19.

Change in groundwater storage

$$= \text{area of aquifer} \times \text{drop in groundwater level} \times S_y$$

$$= 120.0 \times 5.0 \times 0.19 = 114 \text{ ha-m}$$

$$= 114 \times 10^4 \text{ m}^3$$

EXAMPLE 9.2 A phreatic aquifer extends over an area of 12.0 km^2. The water table was initially at 20.0 m below the ground level. After irrigation with a depth of 30.0 cm of water, the water table rose to a depth of 19.2 m below ground level. When 8×10^6 m^3 of water was pumped out, the groundwater table (g.w.t.) dropped to 21.5 m below ground level. Determine the specific yield of the aquifer and the deficit in soil moisture before irrigation.

Solution

$$\text{Volume of water pumped out} = \text{area of aquifer} \times \text{drop in g.w.t.} \times S_y$$

$$8 \times 10^6 = 12.0 \times 10^6 \times (21.5 - 19.2) \times S_y$$

Therefore, $S_y = 0.2899 = 29.0\%$.

Volume of irrigation of water recharging the aquifer

$$= \text{area of aquifer} \times \text{rise of g.w.t.} \times S_y$$

Consider an area of 1.0 m^2 of aquifer:

$$1 \times y = 1 \times 0.8 \times 0.29$$

Therefore, the recharge volume (depth) $y = 0.232$ m.

Soil-moisture deficiency (below field capacity) before irrigation

$$= 0.30 - 0.232 = 0.068 \text{ m}$$

$$= 68 \text{ mm}$$

EXAMPLE 9.3 A phreatic aquifer, extending over an area of 220.0 km^2, has a storativity of 0.15. Estimate the amount of water lost from storage if the water level falls 0.16 m during a drought.

Solution

Volume of water lost

$$= \text{storativity} \times \text{area} \times \text{drop in water level}$$

$$= 0.15 \times 220 \times 10^6 \times 0.16$$

$$= 5.28 \times 10^6 \text{ m}^3$$

9.2 AREAS OF NATURAL RECHARGE AND DISCHARGE

Natural recharge of groundwater reservoirs occurs from infiltration of precipitation into the soil and influent streams. Natural discharge may occur due to evapotranspiration by phreatophytes, springs, and baseflow to streams (effluent). Areas where recharge and/or discharge will occur depend upon geologic conditions. For example, there is more recharge in humid regions than in arid regions. In the conterminous United States, the average annual recharge to groundwater is about 3 in. (7.6 cm), which is about 10% of the average annual precipitation. This amounts to about 400 billion gallons per day (1.5×10^9 m^3/day) of natural recharge. This is equivalent to the amount of natural discharge, most of which goes to streams. Nearly 35 to 40% of the total water discharged by streams each year in the United States is the groundwater contribution. In arid Australia, groundwater recharge amounts to about 1% of annual rainfall and evapotranspiration is about 87% of the annual rainfall.

In a particular area, the depth of the water table below the surface is a good index of the effectiveness of recharge. The water table is quite shallow (only a few feet below the surface) in humid regions, is much deeper in arid regions, and is very deep (more than 1000 ft, or 300 m) in desert basins. Groundwater moves from areas of recharge to areas of discharge.

9.3 GEOLOGY OF AQUIFERS

Geological conditions determine the extent and hydrologic nature of the water-bearing formations—the aquifers. Much of the information about aquifers is derived from the wells drilled into them. Water-bearing formations could be unconsolidated sand and gravel, volcanic rocks, sandstone, or limestone.

9.3.1 Unconsolidated Sediments

An unconsolidated sediment is a loose, granular deposit of natural earth materials whose particles are not cemented together. Examples of unconsolidated sediment include most soils, sands at the beach, sand and gravel deposits of stream channels, and glacial deposits. Any of these formations may become aquifers, but alluvium sediments are the most important. Alluvium is sediment consisting of a mixture of sand, silt, gravel, and clay. The sediment is deposited by running water. The deposit may accumulate to form a more-or-less random pattern of coarse-grained channel deposits and fine-grained floodplain deposits. It is of restricted areal extent and generally occurs locally. Coarse-grained deposits are the most productive aquifers. Most alluvial aquifers are made up of relatively young, unconsolidated sediments.

9.3.2 Consolidated Sediments

When originally loose grains are compacted and cemented, they become sedimentary rocks. Sand changes to sandstone, silt changes to siltstone, and clay changes to shale. The consolidated sediments are areally extensive and sometimes form major landscape features. Erosion of these sediments has formed the Grand Canyon in the southwestern United States. Of all the sedimentary rocks, sandstone is the most important water-bearing formation. It accounts for nearly 25% of all sedimentary rock formations. Depending upon the degree of compaction and amount of cementation, its porosity may vary from 2 to about 30%. By comparison with alluvium, sandstones are more uniform in their hydrologic properties.

9.3.3 Limestone

Limey deposits change to limestone or sedimentary rock through compaction due to burial beneath younger sediments. Limestone is mainly composed of calcium carbonate resulting from either natural waters or organic processes such as lime secretion by animals. Limestone is often a hard, dense rock having low porosity. Sometimes high temperatures and pressures transform limestone to marble, often classified as metamorphic rock. Its major constituent, calcium carbonate, is soluble in water. When water, which almost always has carbon dioxide in solution, seeps through cracks of a limestone formation, it causes a certain amount of calcium carbonate to dissolve, and enlarges the cracks or openings. Over a long period of time (say, a few hundred years), this process results in limestone caverns. Limestones can be porous and permeable enough to become good aquifers.

9.3.4 Volcanic Rocks

Volcanic rocks occur widely and are distinguished by fracture and porous zones. The basalt lavas are the most widespread volcanic rocks. These lavas make up the Hawaiian islands, the Columbia plateau in the northwestern United States, and the Daccan plateau in India. When the lava cools, it shrinks and fractures develop. As a result, outcrops of basalt exhibit columnar appearance. The vertical fractures allow rainwater to penetrate and seep down to porous zones between the basalt flows. The porous zones may develop when the porous sediments deposited on the top of the older flows between eruptive periods. Volcanic rocks make good aquifers, for fractures and porous zones provide both conduits for groundwater transmission and reservoirs for water storage.

9.3.5 Crystalline Rocks

Crystalline rocks include igneous and metamorphic rocks. Examples of crystalline rocks are granite (the most common igneous rock) and slate (a common metamorphic rock). These rocks are almost always uniformly hard and dense, and have very low porosities (1 to 3%) and permeabilities. Hence, they do not make good aquifers. Groundwater exists in and is transmitted through interconnected fracture systems. The groundwater storage is usually limited.

9.4 GROUNDWATER FLOW EQUATIONS

The governing equations of groundwater flow are (1) the continuity equation or the law of conservation of mass, and (2) Darcy's law or a storage-discharge relation. These equations can be expressed in lumped or distributed form, as well as in one or more than one dimension.

9.4.1 Water-Balance Equation

Consider a stream-aquifer system as shown in Figure 9.2. The water-balance equation states that during a time interval Δt, the inflow Q_{I} to the groundwater system or aquifer equals the outflow Q_0 from the system plus the rate of change of water storage in the system. Inflow to a groundwater system may consist of natural recharge, drainage, artificial recharge, irrigation return flow, streamflow, and deep percolation. Outflow from a groundwater system may include pumping, artificial drainage, springs, subsurface outflow, and phreatophyte transpiration. Symbolically,

$$Q_{\mathrm{I}} - Q_0 = S_c \frac{\mathrm{d}V}{\mathrm{d}t} \tag{9.12}$$

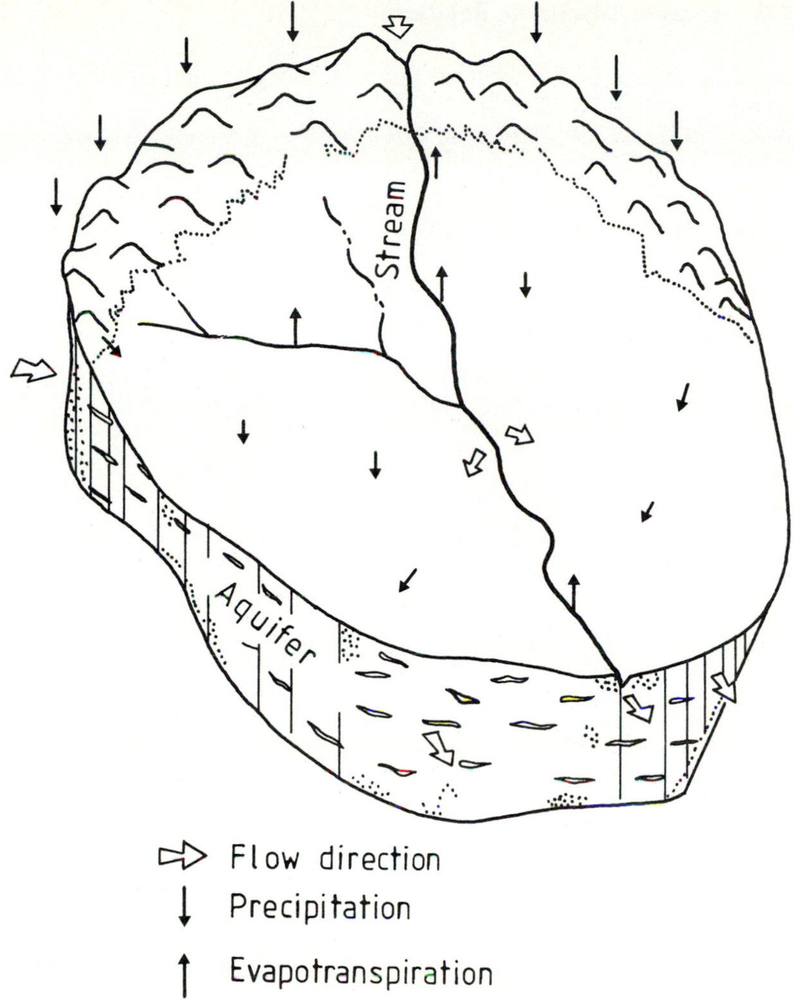

Figure 9.2 Schematic representation of a natural stream–aquifer system.

The storage V can be expressed as

$$V = Ah \tag{9.13}$$

where A is the area of the aquifer, and h is the mean water level in the aquifer. Differentiating Equation (9.13),

$$dV = A\,dh \tag{9.14}$$

Substituting Equation (9.14) into Equation (9.12),

$$S_c \frac{dh}{dt} = \frac{1}{A}(Q_1 - Q_0) \tag{9.15}$$

which is the spatially lumped continuity equation.

9.4.2 Storage–Discharge Relation

A familiar storage–discharge relation is the one that applies to a linear storage element expressed as

$$Q = aV \tag{9.16}$$

where Q is the discharge (L^3/T), and a is a storage constant (T^{-1}). By inserting Equation (9.13) into Equation (9.16),

$$Q = aAh = a_*h \qquad a_* = aA \tag{9.17}$$

9.4.3 Darcy's Law

The one-dimensional form of Darcy's law can be written as

$$Q = qA = -KA\frac{dh}{dx} \tag{9.18}$$

where h is the piezometric head (L) and is equal to $Z + p/\rho g$, p is the pressure ($ML^{-1}T^{-2}$), $\gamma = \rho g$ is the specific weight ($ML^{-2}T^{-2}$), x is the space coordinate in the direction of flow (L), q is the specific discharge (Q/A) or velocity (LT^{-1}), and Z is the static head above a given datum (L).

EXAMPLE 9.4 Calculate the discharge and the seepage velocities for water flowing through a pipe filled with sand with a hydraulic conductivity of 1.5×10^{-6} m/s and an effective porosity of 0.2. The hydraulic gradient is 0.01 and the cross-sectional area of the pipe is 150.0 cm².

Solution

$$Q = A \times V = AK\,dh/dx$$

$$= 150 \times 1.5 \times 10^{-4} \times 0.01 = 2.25 \times 10^{-4}\ \text{cm}^3/\text{s}$$

$$\text{Seepage velocity, } V_s = \frac{Q}{\eta_e A} = \frac{2.25 \times 10^{-4}}{0.2 \times 150} = 7.5 \times 10^{-6}\ \text{cm/s}$$

9.4.4 Continuity Equation

Consider a control element of volume $dx\,dy\,dz$ below the water table, as shown in Figure 9.3. For this element, the net flow in the x direction can be rewritten as

$$\text{Mass flux out} - \text{mass flux in} = \frac{\partial}{\partial x}(\rho Q_x)\,\Delta x$$

where Q_x is the volumetric discharge equal to the product of the Darcy velocity q_x and the area $dy\,dz$ normal to the x-axis.

$$Q_x = q_x\,dy\,dz$$

Likewise in the y and z directions,

$$Q_y = q_y\,dx\,dz$$

$$Q_z = q_z\,dy\,dx$$

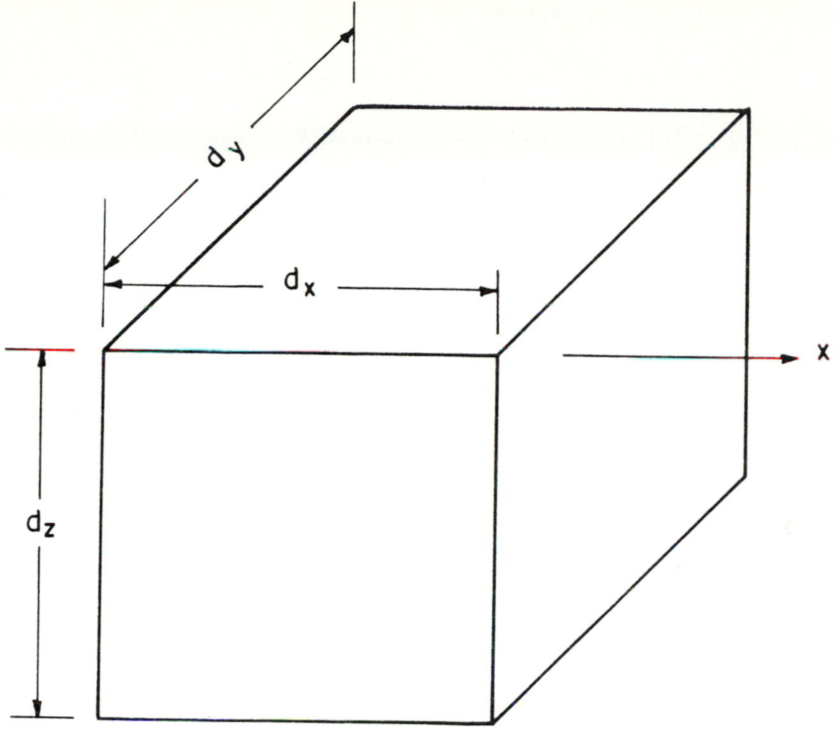

Figure 9.3 Volume element.

By considering the flow in and out in all directions,

$$\begin{aligned}&\text{Mass flux out} - \text{mass flux in}\\&= \frac{\partial}{\partial x}(\rho Q_x)\,\mathrm{d}x + \frac{\partial}{\partial y}(\rho Q_y)\,\mathrm{d}y + \frac{\partial}{\partial z}(\rho Q_z)\,\mathrm{d}z\\&= \left[\frac{\partial}{\partial x}(\rho q_x) + \frac{\partial}{\partial y}(\rho q_y) + \frac{\partial}{\partial z}(\rho q_z)\right]\mathrm{d}x\,\mathrm{d}y\,\mathrm{d}z\end{aligned}$$

By the law of the conservation of mass, the difference between the total flux in and the total flux out must equal the rate of change of mass M inside the control element:

$$\left[\frac{\partial}{\partial x}(\rho q_x) + \frac{\partial}{\partial y}(\rho q_y) + \frac{\partial}{\partial z}(\rho q_z)\right]\mathrm{d}x\,\mathrm{d}y\,\mathrm{d}z = -\frac{\partial M}{\partial t} \tag{9.19}$$

The negative sign on the right side is to make the net outflow positive when flow is from storage.

9.4.5 Differential Equations for Confined Aquifers

Several flow equations can be derived by combining Equations (9.18) and (9.19) and then specializing in the resulting equation. We discuss some of those that are useful for building simple models of groundwater flow in these aquifers. When Equation

(9.18) is used in Equation (9.19),

$$\frac{\partial}{\partial x}\left(K_x \frac{\partial h}{\partial x}\right) + \frac{\partial}{\partial y}\left(K_y \frac{\partial h}{\partial y}\right) + \frac{\partial}{\partial z}\left(K_z \frac{\partial h}{\partial z}\right) = \frac{1}{\rho \, dx \, dy \, dz} \frac{\partial M}{\partial t} \tag{9.20}$$

Equation (9.20) is based on the assumption that water is incompressible. Its right side can be written as

$$\frac{1}{\rho \, dx \, dy \, dz} \frac{\partial M}{\partial t} = S_s \frac{\partial h}{\partial t}$$

Therefore,

$$\frac{\partial}{\partial x}\left(K_x \frac{\partial h}{\partial x}\right) + \frac{\partial}{\partial y}\left(K_y \frac{\partial h}{\partial y}\right) + \frac{\partial}{\partial z}\left(K_z \frac{\partial h}{\partial z}\right) = S_s \frac{\partial h}{\partial t} \tag{9.21}$$

Equation (9.21) specifies the space–time distribution of the piezometric head in nonhomogeneous, anisotropic, confined aquifers.

If the aquifer is homogeneous but anisotropic, then Equation (9.21) reduces to

$$K_x \frac{\partial^2 h}{\partial x^2} + K_y \frac{\partial^2 h}{\partial y^2} + K_z \frac{\partial^2 h}{\partial z^2} = S_s \frac{\partial h}{\partial t} \tag{9.22}$$

Equation (9.22) can be further simplified by assuming the aquifer to be homogeneous and isotropic, that is, $K_x = K_y = K_z = K$,

$$\frac{\partial^2 h}{\partial x^2} + \frac{\partial^2 h}{\partial y^2} + \frac{\partial^2 h}{\partial z^2} = \frac{S_s}{K} \frac{\partial h}{\partial t} \tag{9.23}$$

Equation (9.23) is a linear parabolic partial differential equation that is found in many branches of applied science.

Further simplification of Equation (9.23) can be made by assuming the aquifer to be of constant thickness b and two-dimensional flow. Therefore,

$$\frac{\partial^2 h}{\partial x^2} + \frac{\partial^2 h}{\partial y^2} = \frac{S_c}{T} \frac{\partial h}{\partial t} \tag{9.24}$$

where $T = bK$, and $S_c = bS_s$. In Equation (9.24), the piezometric head does not vary with elevation.

Equation (9.24) can be expressed in terms of drawdown s, where

$$s = h_0 - h \tag{9.25}$$

in which h_0 is a reference value of the piezometric head, usually assumed as the initial, or stationary, value. By substituting Equation (9.25) into Equation (9.24),

$$\frac{\partial^2 s}{\partial x^2} + \frac{\partial^2 s}{\partial y^2} = \frac{S_c}{T} \frac{\partial s}{\partial t} \tag{9.26}$$

Another simplification of Equation (9.23) occurs when the right side vanishes, implying a steady-state condition:

$$\frac{\partial^2 h}{\partial x^2} + \frac{\partial^2 h}{\partial y^2} + \frac{\partial^2 h}{\partial x^2} = 0 \tag{9.27}$$

Equation (9.27) is the famous Laplace equation.

9.4.6 Differential Equations for Unconfined Aquifers

The governing equations of flow in unconfined aquifers can be derived from Equations (9.18) and (9.19). Their derivations are facilitated by noting the difference between confined and unconfined aquifers. In unconfined aquifers, water is derived principally by drainage of pores; water expansion and rock compaction also contribute to release of water but to a relatively negligible extent.

Consider an aquifer whose water table is initially horizontal. A pumping well is used to draw water from the aquifer. Pumping causes the water table to go down, forming a cone of depression, as shown in Figure 9.4. Most of the pumped water is derived by the dewatering of the aquifer within the cone of depression. The water coming from the aquifer beyond this cone is negligibly small. Therefore, the change in aquifer storage can be satisfactorily considered by determining the change in the volume of the cone of depression and multiplying by the apparent specific yield.

A flow equation for unconfined aquifers can be obtained from Equation (9.19) with $S_c = 0$:

$$\frac{\partial}{\partial x}\left(K_x \frac{\partial h}{\partial x}\right) + \frac{\partial}{\partial y}\left(K_y \frac{\partial h}{\partial y}\right) + \frac{\partial}{\partial z}\left(K_z \frac{\partial h}{\partial z}\right) = 0 \tag{9.28}$$

If the aquifer is homogeneous and isotropic, Equation (9.28) reduces to

$$\frac{\partial^2 h}{\partial x^2} + \frac{\partial^2 h}{\partial y^2} + \frac{\partial^2 h}{\partial z^2} = 0 \tag{9.29}$$

which is the same as Equation (9.27)—the famous Laplace equation. It may be emphasized that the right side of Equation (9.29) is 0, because $S_c = 0$ and not because the flow is steady. Equation (9.29) is applicable to both steady and transient cases. This equation can be used to compute the location of the water table in time and

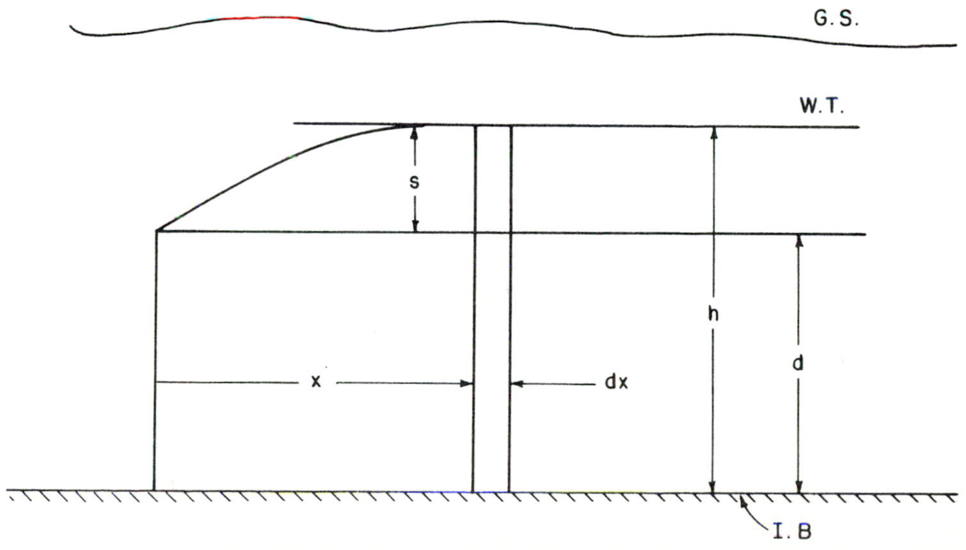

Figure 9.4 Dupuit–Forchheimer flow.

space. However, in transient cases, the flow domain entails a moving boundary constituted by the time-varying water table, which is not known a priori. Thus, the solution of Equation (9.29) becomes difficult. The change in water-table position with time accounts for the change in storage.

9.4.6.1 Dupuit–Forchheimer approximation

This approximation states that the flow is horizontal and uniform everywhere in a vertical plane, and the velocity of flow is proportional to the tangent of the hydraulic gradient instead of the sine, as defined by Equation (9.18). These assumptions render the formulation less rigorous but mathematically and practically advantageous.

Consider an element penetrating the full depth of the aquifer, as shown in Figure 9.4 for flow in the x direction only. The water table is sloping above the horizontal impermeable boundary. For unidirectional flow, the discharge per unit width q at any vertical section can be given by

$$q = -Kh \frac{\mathrm{d}h}{\mathrm{d}x}$$

In Equation (9.30), h represents both the thickness of the flow and the piezometric head at the water table. $dh/\mathrm{d}x$ is the tangent of the angle the water table makes with the horizontal. It is implied that the pressure-head distribution is hydrostatic or the piezometric head along any vertical is constant. Equation (9.30) is valid where the water-table slope is small.

9.4.6.2 Boussinesq equation

The Dupuit–Forchheimer approximation forms the basis of the Boussinesq equation. A volume balance can be made for the derivation of this equation.

The net outflow rate must equal the negative rate of change of storage, which can be determined from the definition of apparent specific yield S_{ya} corresponding to a change $\mathrm{d}h$ of the water table. This leads to

$$\frac{\partial}{\partial x}\left(Kh \frac{\partial h}{\partial x}\right) + \frac{\partial}{\partial y}\left(Kh \frac{\partial h}{\partial y}\right) = S_{ya} \frac{\partial h}{\partial t} \tag{9.31}$$

For a homogeneous unconfined aquifer,

$$\frac{\partial}{\partial x}\left(h \frac{\partial h}{\partial x}\right) + \frac{\partial}{\partial y}\left(h \frac{\partial h}{\partial y}\right) = \frac{S_{ya}}{K} \frac{\partial h}{\partial t} \tag{9.32}$$

Equation (9.32) is the nonlinear Boussinesq equation (Boussinesq, 1904).

9.4.6.3 Laplace formulation with radial symmetry

Consider radial flow, as shown in Figure 9.5. Consider a ring-shaped element whose cross-section is $\mathrm{d}r\,\mathrm{d}y$ and radius r. The flow of groundwater in this element can be written as

$$\frac{\partial}{\partial y}\left(2\pi r K\,\mathrm{d}r \frac{\partial h}{\partial y}\right)\mathrm{d}y + \frac{\partial}{\partial r}\left(2\pi r K\,\mathrm{d}y \frac{\partial h}{\partial r}\right)\mathrm{d}r$$

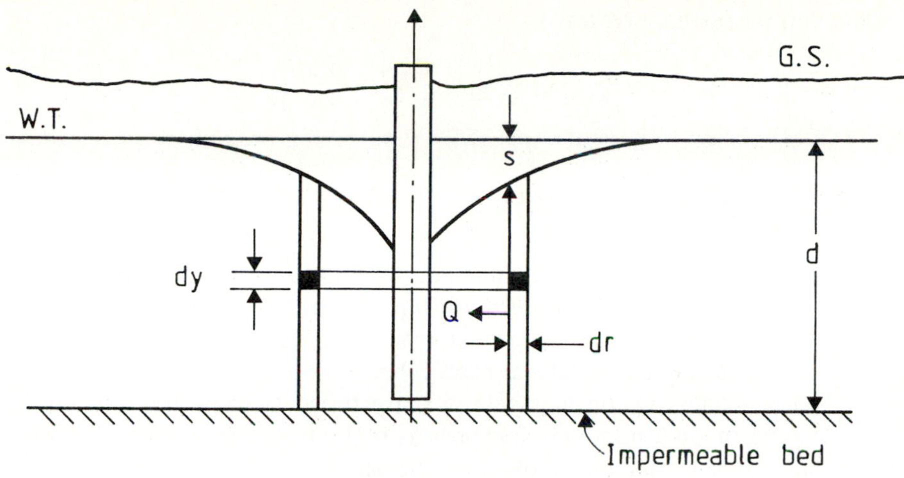

Figure 9.5 Radial flow.

which must be equated to zero by virtue of continuity. Here h represents the departure of the pressure from the hydrostatic. Therefore,

$$\frac{\partial^2 h}{\partial r^2} + \frac{1}{r}\frac{\partial h}{\partial r} + \frac{\partial^2 h}{\partial y^2} = 0 \tag{9.33}$$

This is the Laplacian formulation with radial symmetry.

9.4.6.4 Dupuit–Forchheimer formulation with radial symmetry

The flow Q through the cylindrical surface of radius r and height $d - s$ as shown in Figure 9.5 can be expressed as

$$Q = -2\pi r K(d - s)\frac{\partial s}{\partial r} \tag{9.34}$$

The rate of change of flow through the element of thickness $\mathrm{d}r$ is

$$\frac{\partial Q}{\partial t}\,\mathrm{d}r = -2\pi K\frac{\partial}{\partial r}\left[r(d - s)\frac{\partial s}{\partial r}\right]\mathrm{d}r \tag{9.35}$$

The rate of change of storage in the element is

$$-\frac{\partial V_w}{\partial t} = -2\pi r S_{ya}\,\mathrm{d}r\frac{\partial s}{\partial t} \tag{9.36}$$

By equating Equation (9.35) to Equation (9.36),

$$\frac{\partial}{\partial r}\left[r(d - s)\frac{\partial s}{\partial r}\right] = \frac{S_{ya}}{K}r\frac{\partial s}{\partial t} \tag{9.37}$$

Equation (9.37) can be linearized by assuming s to be too small as compared with d:

$$\frac{\partial}{\partial r}\left(r\frac{\partial s}{\partial r}\right) = \frac{S_{ya}}{\mathrm{dK}}r\frac{\partial s}{\partial t} \tag{9.38}$$

This can be further written as

$$\frac{\partial^2 s}{\partial r^2} + \frac{1}{r}\frac{\partial s}{\partial r} = \frac{S_{ya}}{\mathrm{d}K}\frac{\partial s}{\partial t} \tag{9.39}$$

9.5 STREAM-AQUIFER INTERACTION

Interaction between the water-table aquifer and the stream can be classified in three general categories: (1) The water-table aquifer contributes baseflow to the stream, (2) the stream contributes recharge to the water-table aquifer, and (3) the stream and aquifer alternately contribute to each other.

The streams in the first category are those streams that always flow on the water table or on a relatively impervious bed where the water table adjacent to the stream is at a higher elevation than the stream surface. Under these conditions, the adjacent water table always contributes baseflow to the stream if such water is available. Examples of these conditions occur respectively in the sand hills of northern Nebraska and most of Pennsylvania. The geology and climate of these two areas are very different but combine to produce similar baseflow results. The sand hills of Nebraska contain huge groundwater storage. Precipitation is on the order of 50 cm (20 in.) per year. High infiltration characteristics of the sand dunes result in rather high recharge. The streams flow on the water table in the valleys and the hills contain large supplies of water for baseflow supply. The mountains of Pennsylvania contain much smaller reservoirs for groundwater storage. Baseflow is maintained in spite of this small storage by a rather uniform precipitation that averages about 10 cm (4 in.) per month.

The streams in the second category are those in which the stream bottom is always above the water table. This condition exists in ephemeral streams. An example of this condition is many of the desert streams in Arizona such as Walnut Gulch near Tombstone, where the porous stream-channel bottom permits high seepage loss from the channel to the water table below. Ephemeral streams are quite common in arid and semiarid areas.

The streams in the third category are those streams in which the stream channel and the water table are alternately higher and lower. An example of this condition is the Platte River of Nebraska and Colorado. During the winter and spring months after the irrigation season, accumulated precipitation builds the water table along the broad valley to an elevation higher than the stream surface. During this time, the baseflow contribution is generally at a maximum. As the summer irrigation season approaches, diversions for irrigation and heavy withdrawals from thousands of wells in the valley cause a reduction in the water table to a level below the stream. The stream then tries to supply the water table and does so until it becomes dry. This condition continues until the end of irrigation season in the fall. Surface runoff in the Platte River continues to supply the water table until the postirrigation precipitation raises the water table to an elevation equal to the stream elevation. From this point on, the water table continues to rise and begins to supply the Platte River. This condition recurs on an annual cycle with the water table and the Platte River alternately supplying each other.

An aquifer and a stream are interconnected, as shown in Figure 9.2. Two simple models are discussed here.

9.5.1 Storage Models

These models are the simplest of all and are based on Equation (9.15). By rewriting this equation,

$$S_c \frac{\mathrm{d}h}{\mathrm{d}t} = q_r \tag{9.40}$$

where q_r is the net inflow to the system:

$$q_r = (Q_\mathrm{I} - Q_0)A = q_\mathrm{I} - q_0 \tag{9.41}$$

The stream-aquifer flow may be approximated using Equation (9.17) as

$$q = a(H - h) \tag{9.42}$$

in which q is the stream-aquifer flow per unit area, H is the mean water level of the stream, and a is the subsurface outflow constant. Q_0 in Equation (9.41) can be decomposed into two parts: one coming out of the aquifer and the other consisting of the remaining outflows such as pumping, evaporation, etc. Then by using Equation (9.42), Equation (9.40) becomes

$$S_c \frac{\mathrm{d}h}{\mathrm{d}t} + ah = I \tag{9.43}$$

where $I = aH + q_r$, with q_r excluding the stream-aquifer interaction. Equation (9.43) describes in a lumped manner the stream-aquifer interaction. This equation can be solved using the Laplace transform. Let the initial condition be $h(t_0) = h_0$. If $t_0 = 0$, the solution is

$$h(t) = h_0 \exp\,(-at/S_c) + \frac{1}{S_c} \int_0^t I(\tau) \exp\left[\frac{-a(t-\tau)}{S_c}\right] \mathrm{d}\tau \tag{9.44}$$

The parameter a can be defined from Equation (9.42) as the stream-aquifer flow per unit aquifer area due to a unit difference of mean head between the aquifer and stream, which may be either influent or effluent:

$$Q_s = qA = aA(H - h) \tag{9.45}$$

where Q_s is stream-aquifer discharge (L^3/T); and a is a constant and accounts for stream-bed characteristics, aquifer geometry, transmissivity, recharge, and withdrawal. Its dimension is 1/T. It is related to the time that the aquifer takes to respond to an input.

9.5.2 Steady-State Flow Model

We consider a stream connected to an unconfined aquifer subject to natural recharge I, as shown in Figure 9.6. For steady-state flow, the one-dimensional governing equation under the Dupuit approximation (Bear, 1972) is derived from Equation (9.29) as

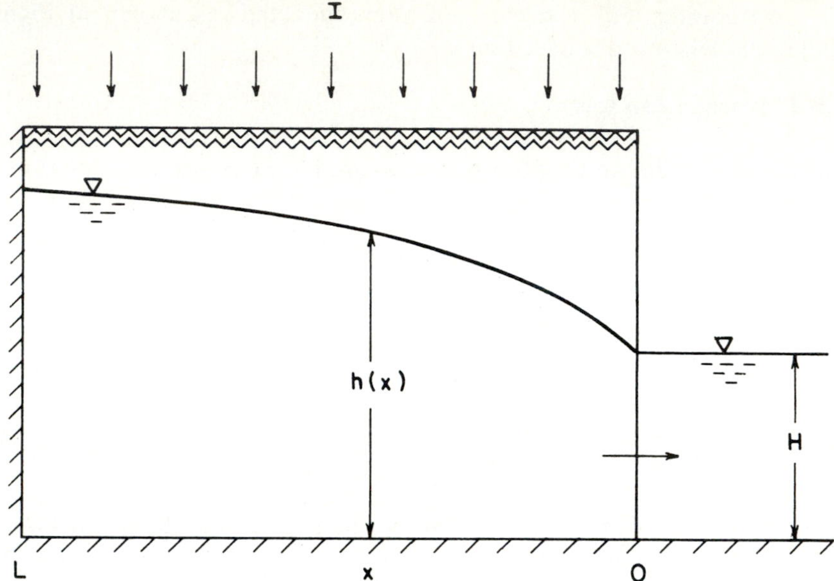

Figure 9.6 Stream–aquifer interaction in an unconfined aquifer under stream-effluent conditions.

$$\frac{d}{dx}\left(Kh\frac{dh}{dx}\right) + I = 0 \tag{9.46}$$

The initial and boundary conditions can be specified as

$$h(0) = H \tag{9.47}$$

$$\frac{dh}{dx} = 0 \qquad \text{at } x = L \tag{9.48}$$

Equation (9.46) can be solved by direct integration:

$$Kh\frac{dh}{dx} = I(L - x) \tag{9.49}$$

Integrating again:

$$h^2 - H^2 = I\,x\,(2L - x)/K \tag{9.50}$$

Solving for h:

$$h = H\left[1 + \frac{I\,x\,(2L - x)}{KH^2}\right]^{0.5} \tag{9.51}$$

9.5.3 Unsteady-State Flow Model

The linearized governing equation based on the Dupuit approximation can be written as

$$d^2 \frac{\partial^2 h}{\partial x^2} + \frac{I}{S_c} = \frac{\partial h}{\partial t} \tag{9.52}$$

where $d^2 = T/S_c$ is the hydraulic diffusivity of the aquifer. Let I be specified as

$$\begin{aligned} I &= I_0, \qquad t < 0 \\ I &= 0, \qquad t > 0 \end{aligned} \tag{9.53}$$

The following initial and boundary conditions can be used:

$$h - H = I_0 x \frac{2L - x}{2T}, \qquad t < 0 \tag{9.54}$$

$$h - H = 0, \qquad x = 0 \tag{9.55}$$

$$\frac{\partial(h - H)}{\partial x} = 0, \qquad x = L \tag{9.56}$$

Equation (9.52) can be solved using standard techniques such as separation of variables. The solution is

$$h - H = \sum_{m=1}^{\infty} A_m \exp\left[-\left(\frac{m\pi d}{2L}\right)^2 t\right] \sin\frac{m\pi x}{2L} \tag{9.57}$$

where

$$A_m = (1 - \cos m\pi \frac{8I_0 L}{Tm^3\pi^3} \tag{9.58}$$

The average head in the aquifer can be expressed as

$$\bar{h} = \frac{1}{2L}\int_0^{2L} h \, \mathrm{d}x \tag{9.59}$$

By inserting Equation (9.57) into Equation (9.59) and solving,

$$\bar{h} - H = \sum_{m=1}^{\infty} A_m \exp\left[-\left(\frac{\pi m}{2L}\right)^2 \frac{T}{S_c} t\right] \frac{1 - \cos m\pi x}{m} \tag{9.60}$$

Note that

$$q = \frac{T}{L}\frac{\partial \bar{h}}{\partial x}\bigg|_{x=0} \tag{9.61}$$

Therefore,

$$q = \sum_{m=1}^{\infty} A_m \exp\left[-\left(\frac{m\pi}{2L}\right)^2 \frac{Tt}{S_c}\right] \frac{Tm\pi}{2L^2} \tag{9.62}$$

It can be shown that the terms of Equation (9.62) corresponding to $m > 1$ are much smaller than the term for $m = 1$ (Kraijenhoff Van de Leur, 1958; Venetis, 1969). By retaining the first term only,

$$q = A_1 \exp\left(-\frac{\pi^2 Tt}{4S_cL^2}\right)\frac{T\pi}{2L^2} \tag{9.63}$$

We can also write

$$\bar{h} - H = A_1 \frac{2}{\pi} \exp\left(-\frac{\pi^2 T}{4S_cL^2}t\right) \tag{9.64}$$

9.6 WELL HYDRAULICS

When a well is pumped, water levels in its neighborhood are lowered. The amount of this lowering at a given point defines the drawdown at that point. The drawdown evidently varies in space and time. At a fixed point in time, the variation of drawdown with distance from the well represents the drawdown curve. This curve has a conic appearance in three dimensions and is called the cone of depression. The extent of this depression represents the area of influence of the well. Determination of the drawdown at and away from the well under various conditions of pumpage and recharge is of fundamental significance. Simple models are considered here to determine the drawdown and its spatial variation.

9.6.1 Steady radial flow to a well in a confined aquifer

We consider radial flow to a well fully penetrating a homogeneous and isotropic confined aquifer, as shown in Figure 9.7. The flow to the well is two-dimensional.

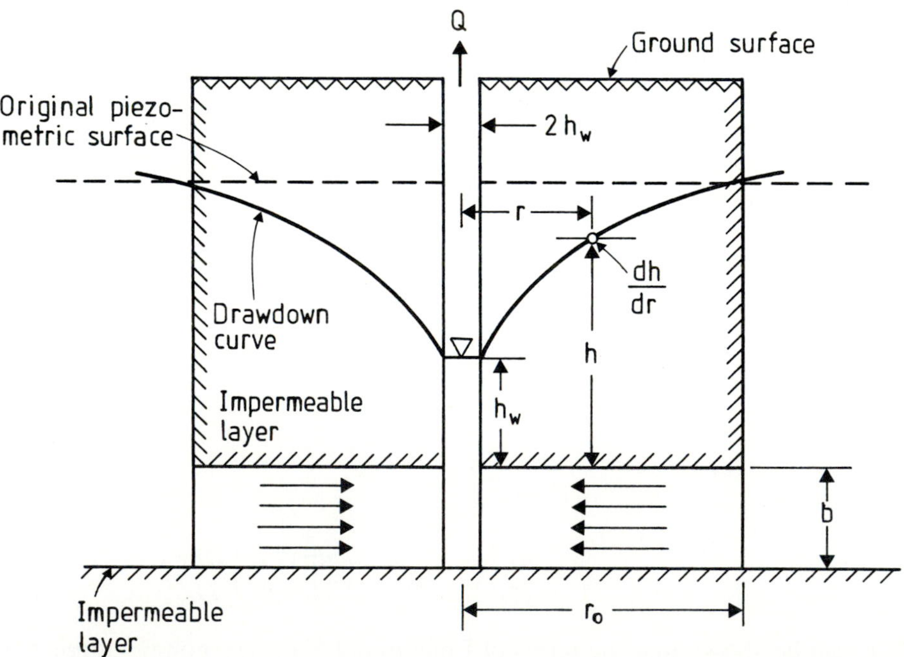

Figure 9.7 Steady radial flow to a well penetrating a confined aquifer on an island.

Under the Dupuit assumption, the flow equation relating drawdown to well discharge Q can be derived. Using polar coordinates with the well as the origin.

$$v = K\frac{dh}{dr} \tag{9.65}$$

and

$$A = 2\pi rb \tag{9.66}$$

Therefore,

$$Q = Av = 2\pi rbK\frac{dh}{dr} \tag{9.67}$$

Equation (9.67) can be solved using the boundary conditions, which can be specified in two ways: (1) The confined aquifer is finite as on an island, as shown in Figure 9.7. Then, at the well, $h = h_w$ when $r = r_w$ and $h = h_0$ when $r = r_0$. (2) The confined aquifer is extensive, as shown in Figure 9.8. Then, at the well, $h = h_w$ when $r = r_w$, and $h = h_0$ when $r = \infty$.

By using the finite confined aquifer conditions, the solution to Equation (9.67) for constant Q can be written as

$$Q = 2\pi Kb\frac{h_0 - h_w}{\ln(r_0/r_w)} \tag{9.68}$$

For the extensive confined aquifer conditions, the solution to Equation (9.67) is

$$Q = 2\pi Kb\frac{h - h_w}{\ln(r/r_w)} \tag{9.69}$$

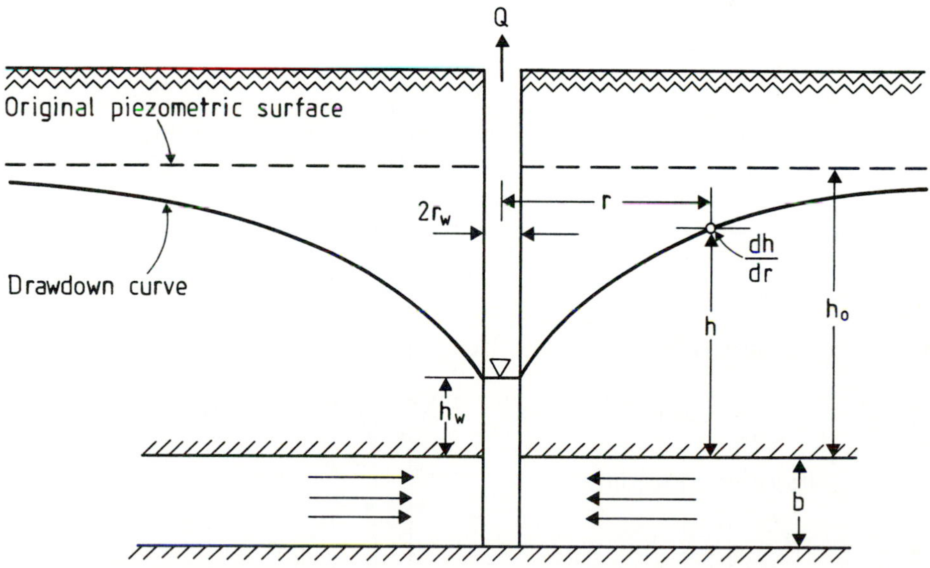

Figure 9.8 Radial flow to a well penetrating an extensive confined aquifer.

This is known as the equilibrium, or Thiem, equation. It is clear from Equation (9.69) that h increases as r increases. From the boundary condition, the maximum h is h_0, the initial uniform head. This then suggests that steady radial flow does not exist in an extensive aquifer. For practical purposes, h can be assumed to approach h_0 at a radius of influence $r = r_0$ as in Equation (9.68). To relate the variation in h with r, Q can be eliminated by equating Equations (9.68) and (9.69):

$$h - h_w = (h_0 - h_w) \frac{\ln (r/r_w)}{\ln (r_0/r_w)} \tag{9.70}$$

This shows a linear variation of head with the logarithm of the distance, irrespective of the rate of discharge.

Equation (9.69) has K as an unknown parameter. This can be determined by observing h at two different points for a constant Q as

$$K = \frac{Q}{2\pi b(h_2 - h_1)} \ln \frac{r_2}{r_1} \tag{9.71}$$

To apply this equation, the steady-state condition must approximately be attained. This means that pumping must be carried out with a constant Q for a sufficiently long time. The observation wells must be located close to the pumping well so that their drawdowns are readily observable.

9.6.2 Steady Radial Flow to a Well in an Unconfined Aquifer

We consider radial flow to a well penetrating fully a homogeneous isotropic unconfined aquifer as shown in Figure 9.9. Under the Dupuit approximation, Equation (9.67) applies with b replaced by h:

$$Q = Av = 2\pi rKh \frac{dh}{dr} \tag{9.72}$$

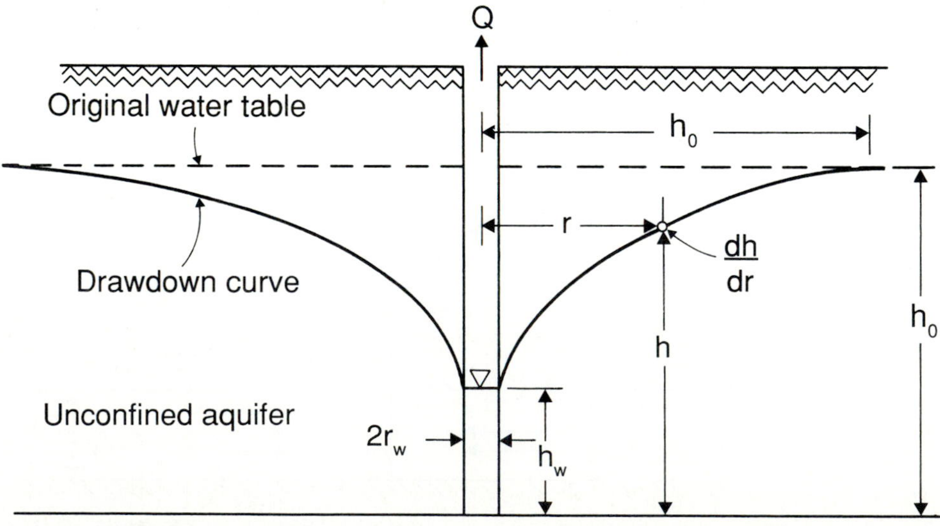

Figure 9.9 Radial flow to a well penetrating an unconfined aquifer.

The boundary conditions are $h = h_w$ at $r = r_w$ and $h = h_0$ at $r = r_0$. Thus, the solution of Equation (9.72) is

$$Q = \pi k \frac{h_0^2 - h_w^2}{\ln (r_0/r_w)} \tag{9.73}$$

This equation does not accurately describe the drawdown curve near the well due to large vertical flow components, although Q is estimated reasonably accurately for specified h. r_0 is normally taken between 150 and 350 m.

9.6.3 Unsteady Radial Flow to a Well with Constant Discharge

We consider the case where the groundwater flow is unsteady, that is, the piezometric head changes with time. Mathematically, the analysis of unsteady flow in confined aquifers is similar to that in unconfined aquifers. This is because the unconfined flows are, for the most part, treated by the linearized Boussinesq equation, which is identical in appearance to the differential equation for two-dimensional confined flows. One must, however, note that the assumptions leading to the linearized Boussinesq equation are considerably more restrictive than those for confined flow.

When a well fully penetrating an aquifer of infinite areal extent is pumped at a constant rate, water is continuously withdrawn from the aquifer and the cone of depression extends radially outward from the well. The distribution of drawdown at any distance from the pumping well can be given by the nonequilibrium Equation (9.24), which can be written in polar coordinates as

$$\frac{\partial^2 h}{\partial r^2} + \frac{1}{r}\frac{\partial h}{\partial r} = \frac{S_c}{T}\frac{\partial h}{\partial t} \tag{9.74}$$

In term of drawdown $s = h - h_0$, Equation (9.74) can be written as

$$\frac{\partial^2 s}{\partial r^2} + \frac{1}{r}\frac{\partial s}{\partial r} = \frac{S_c}{T}\frac{\partial h}{\partial t} \tag{9.75}$$

The initial and boundary conditions can be written as

$$s(r, 0) = 0 \tag{9.76}$$

$$s(\infty, t) = 0 \tag{9.77}$$

$$\lim_{r \to 0} r \frac{\partial s}{\partial r} = -\frac{Q}{2\pi T} \tag{9.78}$$

Equation (9.76) is an initial condition. Equation (9.77) is a boundary condition. Equation (9.78) is also a boundary condition and states that the well is a line sink. These conditions together imply that the discharge changes abruptly from 0 to Q at $t = 0$. Equations (9.74) to (9.78) are normally satisfied in confined aquifers with constant thickness. Their application to unconfined aquifers requires that vertical components of flow are negligible and that changes in aquifer storage due to water expansion and aquifer compression are vanishingly small in relation to gravity drainage of pores as the water table falls on account of pumping. These conditions are not satisfied in the immediate vicinity of a well for quite a while after an abrupt change of discharge.

Equation (9.75) can be solved by using the Boltzmann variable,

$$u = \frac{r^2}{4Dt} \tag{9.79}$$

where D is diffusivity (equal to T/S_c for confined flow and equal to T/S_{ya} for unconfined flow). Equations (9.75) and (9.78) can be written as

$$\frac{d^2 s}{du^2} + \left(\frac{1}{u} + 1\right)\frac{ds}{du} = 0 \tag{9.80}$$

$$\lim_{r \to 0} u \frac{ds}{au} = -\frac{Q}{4\pi T} \tag{9.81}$$

Let $w = ds/du$. Then Equation (9.80) can be reduced to

$$\frac{dw}{du} + \left(\frac{1}{u} + 1\right)w = 0 \tag{9.82}$$

Its integrating factor is

$$\exp\left[\int\left(\frac{1}{u} + 1\right) du\right] = u \exp(u) \tag{9.83}$$

The solution of Equation (9.83) is

$$wu \exp(u) = c$$

or

$$\frac{ds}{du} = \frac{C}{u}\exp(-u) \tag{9.84}$$

where C is a constant of integration. By using Equation (9.81), $C = -Q/4\pi T$. By integrating Equation (9.84) with $s = 0$ at $t = 0$,

$$s = \frac{Q}{4\pi T}\int_u^\infty \frac{1}{x}\exp(-x)\, dx \tag{9.85}$$

where u is defined by Equation (9.79). In Equation (9.85), the integral is the exponential integral or the Theis well function (Theis, 1935), and is often given the symbol $W(u)$. Therefore,

$$s = \frac{Q}{4\pi T} W(u) \qquad \text{or} \qquad T = \frac{Q}{4\pi s} W(u) \tag{9.86}$$

and S_c may be determined by rewriting Equation (9.79) as

$$S_c = \frac{4Tu}{r^2/t} \tag{9.87}$$

The values of the well function for a practical range of u are available in several standard textbooks on groundwater hydrology (Todd, 1980; Walton, 1970). Theis devised a graphical method of superposition to obtain solution of Equations (9.86) and (9.87).

The well function can be expanded in an infinite series as

$$W(u) = -0.5772 - \ln u + u - \frac{u^2}{2 \times 2!} + \frac{u^3}{3 \times 3!} - \frac{u^4}{4 \times 4!} + \cdots \tag{9.88}$$

For small values of u, $u \leq 0.01$ (for a large time at a given distance). $W(u)$ may be approximated by the first two terms (Cooper and Jacob, 1946; Jacob, 1949) in Equation (9.88). Thus,

$$\begin{aligned} s &= \frac{Q}{4\pi t}(-0.5772 - \ln u), \qquad u \leq 0.01 \\ &= \frac{Q}{4\pi T} \ln \frac{2.25DT}{r^2} \end{aligned} \tag{9.89}$$

This is commonly known as the Jacob method. By writing this equation in terms of x and y coordinates,

$$x^2 + y^2 = 2.25Dt \exp\left(-\frac{4\pi Ts}{Q}\right) \tag{9.90}$$

It is clear that contours of s are circles centered at the well.

Equation (9.89) can be written as

$$s = \frac{Q}{2\pi T} \ln \frac{1.5(tD)^{0.5}}{r} \tag{9.91}$$

If we compare Equation (9.91) with Equation (9.71), a radius of influence R can be defined by taking $s = 0$:

$$R = 1.5(tD)^{0.5} \tag{9.92}$$

For very large times, the difference in drawdowns at two different points can be obtained from Equation (9.92) as

$$s_2 - s_1 = \frac{Q}{2\pi T} \ln \frac{r_1}{r_2} \tag{9.93}$$

which is the steady-state equation.

EXAMPLE 9.5 A well is penetrating an aquifer with a hydraulic conductivity of 12.0 m/day and storativity of 0.00735. The aquifer is 28.0 m thick and is pumped at a rate of 2800.0 m^3/day. Estimate the drawdown after 1 week of pumping at a distance 16.0 m from the well.

Solution

$$T = k \times b = 12.0 \times 28.0 = 336 \text{ m}^2/\text{day}$$

$$u = \frac{r^2 S_c}{4Tt} = \frac{(16.0)^2 \times 0.00735}{4 \times 336 \times 7} = 0.0002$$

From the table of w(u) and u, u = 0.0002 and $w(u)$ = 7.94. Therefore,

$$s = h_0 - h = \frac{Q}{4\pi T} w(u) = \frac{2800 \times 7.94}{4\pi \times 336} = 5.27 \text{ m}$$

Therefore, the drawdown is 5.27 m at a distance 16.0 m from the well after 1 week of pumping.

EXAMPLE 9.6 A fully penetrating well is pumped at a rate of 1500 m^3/day from an aquifer whose S_c and T values are 4×10^{-4} and 208.8 m^2/day, respectively. Find the drawdown at a distance 3.0 m from the production well after 1 hour of pumping and at a distance 495.0 m after 2 days of pumping.

Solution At $r = 3.0$ m and $t = 1.0$ h,

$$u = \frac{r^2 S_c}{4Tt} = \frac{9 \times 4 \times 10^{-4}}{4 \times 208.8 \times 1/24} = 1.03 \times 10^{-4}$$

From the table of $W(u)$ and u, $W(u) = 8.62$.

$$s = \frac{Q}{r\pi T} W(u)$$

Therefore,

$$s = \frac{1500 \times 8.62}{4\pi \times 208.8} = 4.93 \text{ m}$$

and the drawdown is 4.93 m at a distance 3.0 m after 1 h. At $r = 495.0$ m and $t = 2$ days,

$$u = \frac{(495)^2 \times 4\ 10^{-4}}{4 \times 208.8 \times 4} = 5.86 \times 10^{-2}$$

From the table of $W(u)$ and u, $W(u) = 2.314$. Therefore,

$$s = \frac{1500 \times 2.314}{4\pi \times 208.8} = 1.32 \text{ m}$$

and the drawdown is 1.32 m at a distance 495.0 m from the well after 2 days of pumping.

EXAMPLE 9.7 Find the approximate values of the radius of influence in the preceding example after 1 hour and 2 days of continuous pumping.

Solution Assume that the drawdown at the radius of influence, s_{ri}, is 2.5 cm, which is practically zero.

$$W(u) = \frac{4\pi T s_{ri}}{Q} = \frac{4\pi \times 208.8 \times 0.025}{1500} = 0.0437$$

From tables of well function, we get $u = 2.1$. Therefore,

$$r = \left(\frac{4Ttu}{S_c}\right)^{0.5}$$

When $t = 1.0$ h,

$$r = \left(\frac{4 \times 208.8 \times 2.1}{4 \times 10^{-4} \times 24}\right)^{0.5} = 427.43 \text{ m}$$

When $t = 2.0$ days,

$$r = \left(\frac{4 \times 208.8 \times 2 \times 2.1}{4 \times 10^{-4}}\right)^{0.5} = 2961.35 \text{ m}$$

EXAMPLE 9.8 A well penetrating a confined aquifer is pumped at a uniform rate of 96,000 ft^3/day. An observation well is located 200 ft away. Drawdowns measured in the observation well during the period of pumping (t) are given in Table 9.1. Compute the values of S_c and T using the Theis method.

Solution Values of r^2/t are computed and listed in Table 9.1. Values of s and r^2/t are plotted on logarithmic paper, as shown in Figure 9.10. On another logarithmic plot are values of $W(u)$ and u obtained from any groundwater hydrology (i.e., Glover, 1977) and a curve is drawn through the points, as shown in Figure 9.11. The two sheets are superposed and moved with axes parallel until the observational points coincide with the curve, as shown in Figure 9.11. This produces a convenient match point whose values are $W(u) = 1.0$, $u = 10^{-1}$, $s = 0.56$ ft, and $r^2/t = 2.75 \times 10^7$ ft/day. Then from Equation (9.86),

$$T = \frac{(96{,}000 \text{ ft}^2\text{/day}) \ (1.0)}{(4\pi(0.56 \text{ ft})} = 13{,}700 \text{ ft}^2\text{/day}$$

TABLE 9.1 DRAWDOWN OF THE WATER LEVEL IN AN OBSERVATION WELL (r = 200 ft) (AFTER LOHMAN, 1972)

Time from start of pumping, t (min)	Observed drawdown, s (ft)	r^2/t (ft^2/day)
1.0	0.66	5.76×10^7
1.5	0.87	3.84×10^7
2.0	0.99	2.88×10^7
2.5	1.11	2.30×10^7
3	1.21	1.92×10^7
4	1.36	1.44×10^7
5	1.49	1.15×10^7
6	1.59	9.6×10^6
8	1.75	7.2×10^6
10	1.86	5.76×10^6
12	1.97	4.80×10^6
14	2.08	4.10×10^6
18	2.20	3.2×10^6
24	2.36	2.40×10^6
30	2.49	1.92×10^6
40	2.65	1.44×10^6
50	2.78	1.15×10^6
60	2.88	9.60×10^5
80	3.04	7.20×10^5
100	3.16	5.76×10^5
120	3.28	4.80×10^5
150	3.42	3.84×10^5
180	3.51	3.20×10^5
210	3.61	2.74×10^5
240	3.67	2.50×10^5

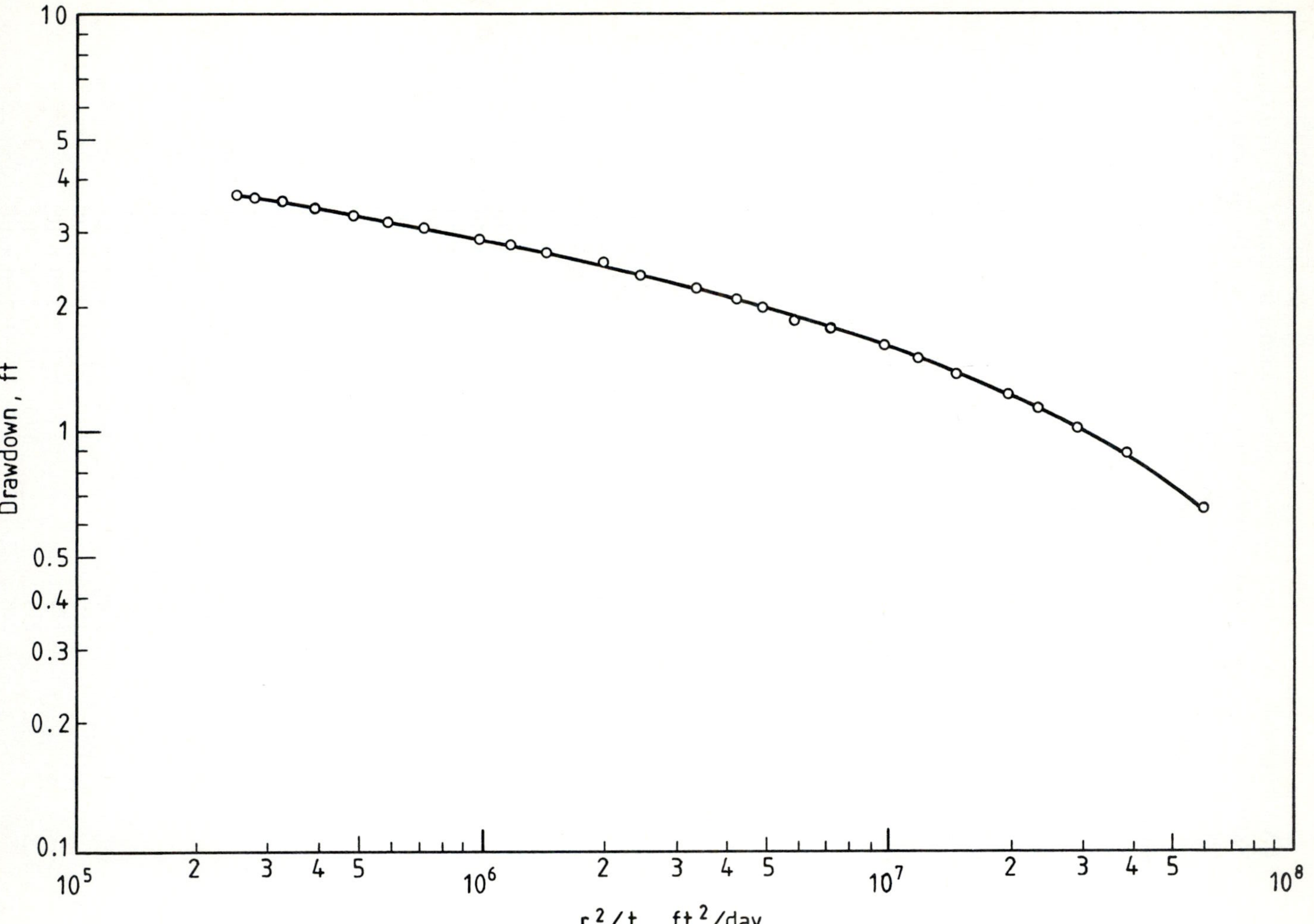

Figure 9.10 A plot of drawdown s versus r^s/t.

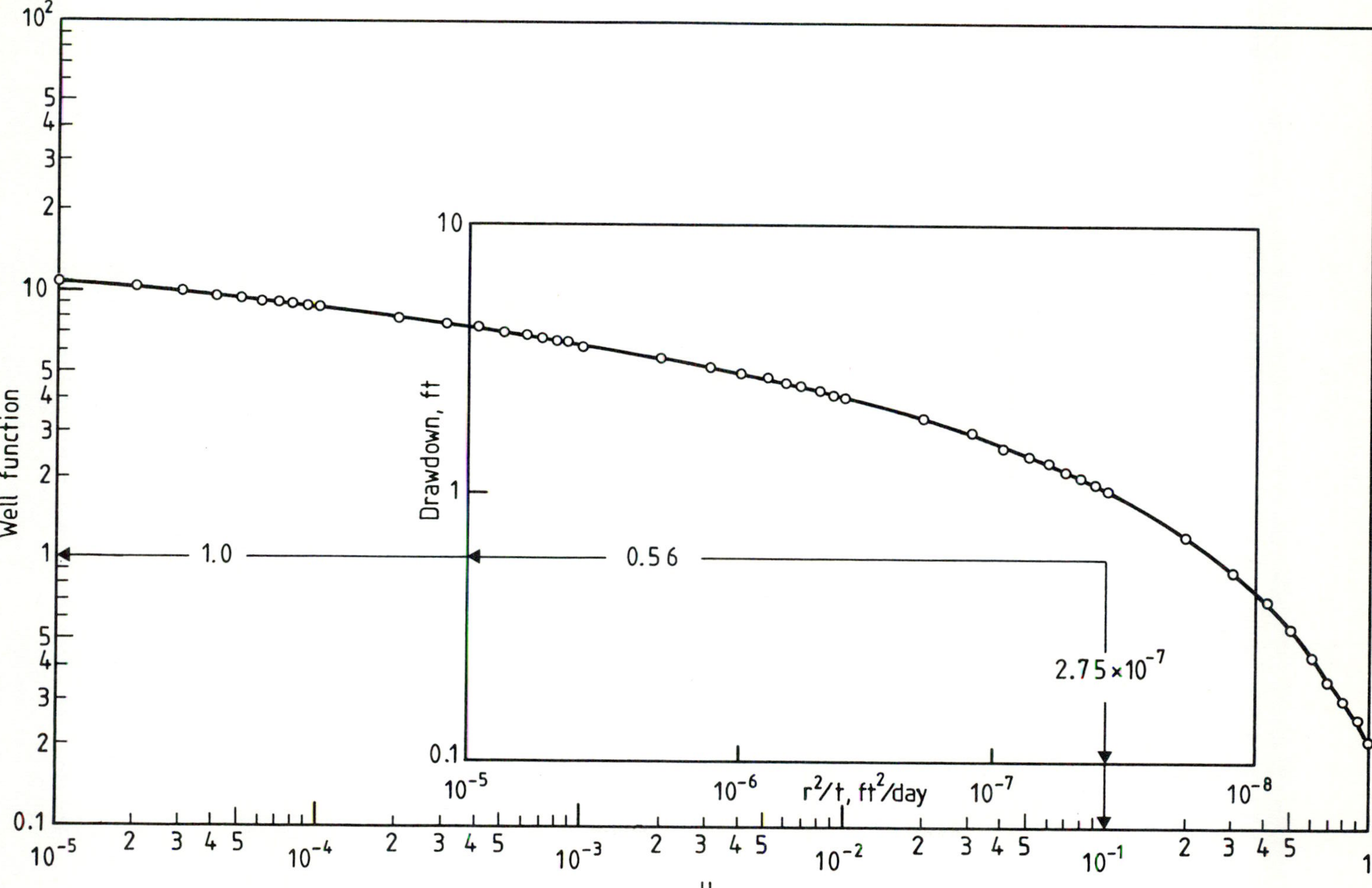

Figure 9.11 Theis method of superposition of solution of the nonequilibrium equation.

From Equation (9.87),

$$S_c = \frac{(4)(13{,}700 \text{ ft}^2/\text{day})(10^{-1})}{2.75 \times 10^7 \text{ ft}^2/\text{day}} = 2 \times 10^{-4}$$

EXAMPLE 9.9 Compute T and S_c for the pumping test data of Table 9.1 using the Cooper–Jacob approximation.

Solution Values of drawdown s are plotted against the logarithm of time t and a straight line [following Equation (9.90)] is fitted, as shown in Figure 9.12. When $s = 0$ and $t = t_0$,

$$0 = \frac{2.30}{4\pi T} \log \frac{2.5Dt_0}{r^2}$$

yielding $2.25\ Dt/r^2 = 1$, and, therefore, $S_c = 2.25\,Tt_0/r^2$. The value of T can be obtained by taking the difference of two values s_2 and s_1 of s corresponding to their respective times t_2 and t_1 from Equation (9.90):

$$s_2 - s_1 = \frac{2.30}{4\pi T} [\log (t_2/t_1)]$$

If t_2/t_1 is taken to be 10, then $T = 2.3Q/4\pi\Delta s$, $\Delta s = s_2 - s_1$. Therefore, using Figure 9.12,

$$T = \frac{2.3 \times 96{,}000}{4\pi \times 1.30} = 13{,}525 \text{ ft}^2/\text{day}$$

and then

$$S_c = \frac{2.25 \times 13{,}525 \times 0.35}{(200.0)^2 \times 24 \times 60} = 1.85 \times 10^{-2}$$

EXAMPLE 9.10 A farm has two wells A and B located 1000 meters apart. Both wells fully penetrate a saturated thickness of 25 meters. The hydraulic conductivity is 0.001 m/s and the storage coefficient is 0.25. If well A is pumped at the rate of 0.20 m^3/s for a period of 120 days, how much decline will it cause to the water table of well B?

Solution Aquifer transmissivity = hydraulic conductivity × saturated thickness

$$T = K \times b = 0.001 \times 86{,}400 \times 25$$

$$= 2160 \text{ m}^2/\text{day}$$

$$u = \frac{r^2 S_c}{4Tt} = \frac{(1000)^2 \times 0.25}{4 \times 2160 \times 120} = 0.241$$

From the table of $w(u)$ and u, $w(u) = 1.09$.

$$s = \frac{Q}{4\pi \text{T}} w(u)$$

$$= \frac{0.2 \times 86{,}400}{4\pi \times 2160} \times 1.09 = 0.695 \text{ m}$$

Therefore, the water table at well B will decline 69.5 cm after 120 days of pumping well A.

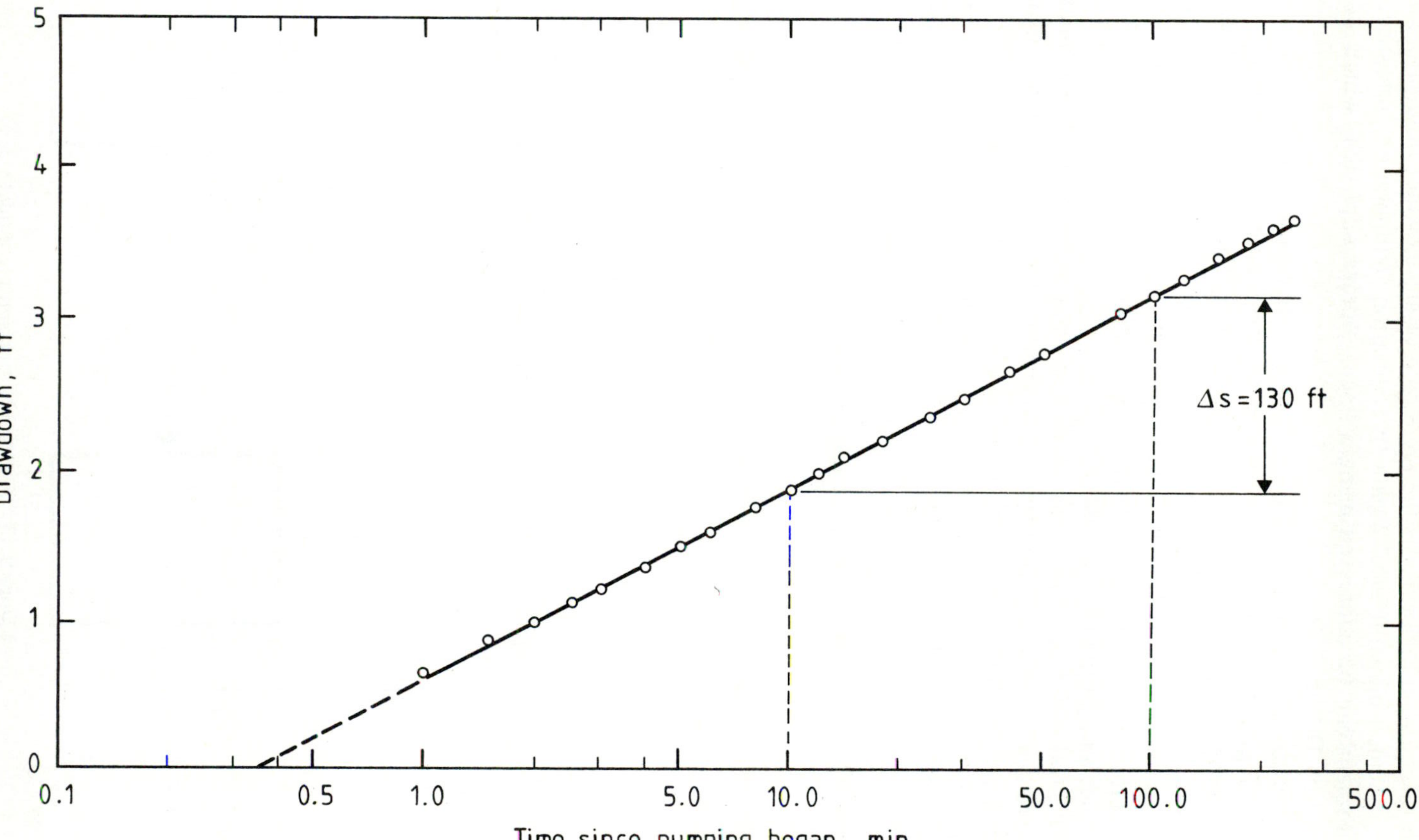

Figure 9.12 Cooper-Jacob method for solution of the nonequilibrium equation.

9.7 RECHARGE AND GROUNDWATER RUNOFF

A major problem in hydrology is the prediction of groundwater runoff from a basin due to specified climatological inputs (precipitation, temperature, pressure, etc.). This problem can be addressed using a linear storage approach, which was first proposed by Dooge (1960).

A subsurface basin may consist of one or more groundwater reservoirs, as shown in Figure 9.13. Each reservoir can be represented by one or more linear storage elements. The storage elements are combined in a manner that best represents the basin. Five types of reservoirs can be distinguished because their actions depend on the restrictions imposed on their inflow or outflow: (1) the soil-moisture zone, (2) the shallow zone (outflow ≥ 0), (3) the shallow zone (outflow ≤ 0), (4) the deep zone, and (5) the mixed zone.

The soil-moisture zone is essentially the upper root zone, which quickly responds to rainfall and may contribute to interflow. Most of the evapotranspiration takes place from this zone. The recharge of ground water reservoirs is determined by the soil-moisture status of this zone. It will be positive when the soil moisture is at field capacity and zero when the soil moisture is below the field capacity.

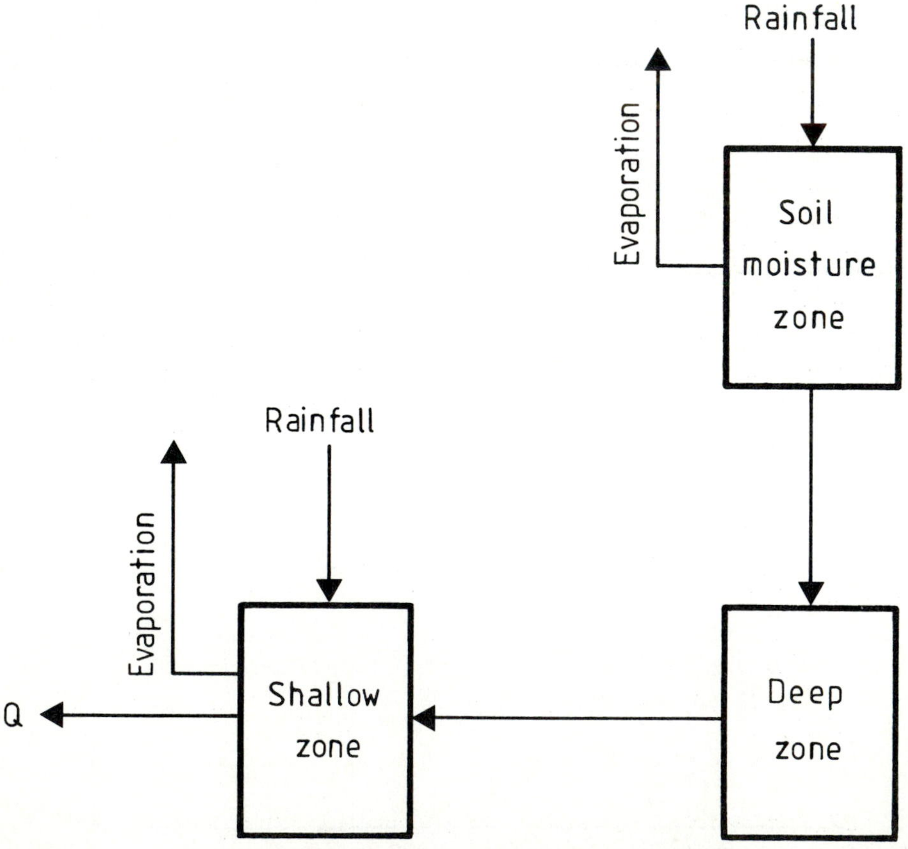

Figure 9.13 A groundwater routing model.

In the shallow zone, the water-table reservoir is close to the surface. Therefore, there may be direct recharge of groundwater by precipitation and direct abstraction from the groundwater (negative recharge) by evapotranspiration. If the evapotranspiration exceeds precipitation during a given time interval, then the net recharge of this reservoir will be negative; otherwise, zero or positive. This zone may also result from a drainage system in an irrigated area, and responds relatively fast. The shallow zone can be subdivided further into two zones, one where negative outflow can occur and the other where negative outflow cannot occur. Negative outflow can occur during floods. When there is a flood, there is a sudden rise in river level. Bank storage increases and the water is caused to flow from the river to the groundwater reservoir.

The deep zone has a water table well below the surface. Therefore, recharge by precipitation and abstraction by evapotranspiration will not take place in this zone but to the soil lying between the surface and the water table. The recharge will always be equal to or greater than zero, depending upon the soil-moisture status above the water table and would sensibly be constant throughout its period.

A groundwater reservoir can also be of mixed type. It may act like a shallow zone until its storage falls below a certain critical level. Then, it will act like a deep zone. All these zones can exist in a given groundwater basin. Each zone can be represented by a single storage element or a number of elements arranged in series or any other fashion. Thus, the key to developing a relationship between recharge and groundwater is to determine the response of a single storage element due to a specified pattern of recharge. The recharge, measured or computed, is normally available in terms of volume. Its intensity is usually constant for a deep zone but time-varying for a shallow zone during a given period of time. Since the approach is linear, it may suffice to consider a single storage element subject to uniform recharge.

9.7.1 Single Storage Element

The storage S (or V) of this element at any time t can be specified by Equation (9.16) recast as

$$S(t) = kQ(t) \tag{9.94}$$

where k is the delay time. The volume balance for the element subject to recharge rate I can be expressed as

$$\begin{gathered}\frac{\mathrm{d}S(t)}{\mathrm{d}t} = I(t) - Q(t), \qquad S(0) = 0 \\ I(t) = I, \qquad 0 \le t \le T_0, \qquad I(t) = 0, \qquad t \ge T_0\end{gathered} \tag{9.95}$$

The solution of Equations (9.94) and (9.95) can be written as

$$\begin{gathered}Q = \mathrm{I}[1 - \exp(-t/k)], \qquad 0 < t \le T_0 \\ Q = \mathrm{I}[1 - \exp(-T_0 k)] \exp[-(t - T_0)/k]\end{gathered} \tag{9.96}$$

$$= I[\exp(T_0/k) - 1] \exp(-t/k), \qquad T_0 \le t < \infty \tag{9.97}$$

The volume outflow V_m during any period, mT_0 and $(m + 1)T_0$, $m = 0, 1, \ldots,$ can be computed using the previous equations. If $m = 0$,

$$V_0 = \int_0^{T_0} Q \, dt = \int_0^{T_0} I[1 - \exp(-t/k)] \, dt$$
$$= I[T_0 + k \exp(-T_0/k) - k] \tag{9.98}$$

Let $R_0 = IT_0$. Then Equation (9.98) can be written as

$$\frac{V_0}{R_0} = 1 - \frac{k}{T_0}\left[1 - \exp\left(-\frac{T_0}{k}\right)\right] \tag{9.99}$$

For any time interval after T_0, the volume of outflow can be computed by integrating Equation (9.97):

$$V_m = \int_{mT_0}^{(m+1)T_0} Q \, dt$$
$$= \int_{mT_0}^{(m+1)T_0} I[\exp(T_0/k) - 1] \exp(-t/k) \, dt \tag{9.100}$$
$$= Ik[\exp(T_0/k) - 1]\{\exp(-mT_0/k) - \exp[-(m+1)T_0/k]\}$$

Therefore,

$$\frac{V_m}{R_0} = \frac{k}{T_0}\left[\exp\left(\frac{T_0}{k}\right) - 1\right]^2 \exp\left[-(m+1)\frac{T_0}{k}\right] \tag{9.101}$$

Equations (9.99) to (9.101) can be used to compute V_m, $m = 0, 1, 2, \ldots$, due to R_0 if k is known.

We are, however, more interested in determining the outflow during a given period due to recharge in that period and preceding ones. This is equivalent to determining the contributions to a given outflow V_n of a number of past recharges, $R_n, R_{n-1}, R_{n-2}, \ldots$. Thus,

$$V_n = R_n\left\{1 - \frac{k}{T_0}\left[1 - \exp\left(\frac{-T_0}{k}\right)\right]\right\}$$
$$+ R_{n-1}\frac{k}{T_0}[\exp(T_0k) - 1]^2 \exp\left(\frac{-2T_0}{k}\right)$$
$$+ R_{n-2}\frac{k}{T_0}\left[\exp\left(\frac{T_0}{k}\right) - 1\right]^2 \exp\left(\frac{-3T_0}{k}\right) + \ldots \tag{9.102}$$
$$= \sum_{j=0}^{\infty} a_j R_{m-j}$$

where

$$a_0 = 1 - \frac{k}{T_0}\left[1 - \exp\left(-\frac{T_0}{k}\right)\right] \tag{9.103}$$

$$a_j = \frac{k}{T_0}\left[\exp\left(\frac{T_0}{k}\right) - 1\right]^2 \exp\left[-\frac{(j+1)T_0}{k}\right], \quad j \geq 1 \tag{9.104}$$

Equation (9.102) was also derived in the context of the rainfall–runoff relationship by Singh and Birsoy (1977).

From a practical standpoint, however, it may be more convenient to employ a form of coefficient routing in terms of the recharge in the present period and the recharge and the outflow in the immediately preceding period. This can be expressed as

$$V_n = C_0 R_n + C_1 R_{n-1} + C_2 V_{n-1} \tag{9.105}$$

in which

$$C_0 = 1 - \frac{k}{T_0}\left[1 - \exp\left(-\frac{T_0}{k}\right)\right] \tag{9.106}$$

$$C_1 = \frac{k}{T_0}\left[\left(1 - \exp\left(-\frac{T_0}{k}\right)\right) - \exp\left(-\frac{T_0}{k}\right)\right] \tag{9.107}$$

$$C_s = \exp\left(-T_0/k\right) \tag{9.108}$$

Coefficients C_0, C_1, and C_2 depend only on the value of T_0/k.

9.7.2 Storage Element with Negative Recharge

Negative recharge can only occur with shallow zones. Two cases can be distinguished. First, negative recharge and negative outflow occur concurrently. This case can be handled by the previous procedure. Second, negative recharge occurs but negative outflow cannot. This case cannot be handled by the previous procedure as such, and requires special calculations. In this case, negative recharge will deplete the storage until outflow and storage will fall to zero and will remain so until precipitation again exceeds evapotranspiration. During the period of zero outflow, the storage will remain zero and the actual evapotranspiration, of necessity, must be equal to the amount of precipitation. On the other hand, if potential evapotranspiration continues, then the water level will continue to fall, creating a deficit that must be satisfied before outflow can occur again. The storage in these two cases may fall to zero during a time interval rather than at its end. To check this, the storage at the start and the end of the time period must be computed. Determination of zero outflow is not sufficient because positive outflow can occur part of the time and so can negative outflow.

EXAMPLE 9.11 A watershed is represented by two linear elements connected in series: (1) a deep water-table element representing the upper portion, and (2) a shallow water-table element representing the lower portion, which receives outflow from the first element. The storage delay time k is taken as 1 month for both elements. The area represented by each element is considered as equal. Compute the groundwater contribution to streamflow if the recharge due to precipitation is known for each elements as:

Month	Deep water-table element recharge due to precipitation (cm)	Shallow water-table element recharge due to precipitation (cm)
January	6.8	6.5
February	0.5	0.5
March	8.0	8.0
April	3.2	3.2

Assume the initial recharge to and initial outflow from these elements to be zero.

Solution Using Equations (9.106) to (9.108) in which, with $k = 1$ month and $T_0 = 1$ month,

$$C_0 = e^{-1}, \quad C_1 = 1 - 2e^{-1}, \quad C_2 = e^{-1}$$

Month	Deep water-table recharge (cm)	Deep water contribution to shallow water (cm)	Shallow water-table recharge (cm)	Total shallow water recharge (cm)	Groundwater contribution (cm)
January	6.8	2.5	6.5	9.0	3.31
February	0.5	4.56	0.5	4.06	5.46
March	8.0	4.75	8.0	12.75	8.04
April	3.2	5.04	3.2	8.24	9.36

9.8 BASEFLOW

Streamflow may be comprised of surface flow, interflow, bank flow, baseflow, or a combination of these. The baseflow is normally defined as that portion of streamflow that is derived from groundwater storage or other delayed sources. It represents withdrawal of water from groundwater storage. Depending upon their interest, hydrologists have called it by various names. Groundwater flow, groundwater runoff, seepage flow, sustained flow, low flow, and percolation flow are some of those names.

Baseflow is defined as the groundwater contribution to streamflow. Most baseflow originates from the water table. The water table is defined as the water-saturated soil or sediment whose surface is in communication with the atmosphere through the pore openings between sediment grains and is therefore under atmospheric pressure. A water table can exist in limestone cavities, fractures in volcanic rock or other igneous rock, and fractures in sedimentary or metamorphic rocks. Most water-table aquifers exist in granular material such as sand or gravel. The water table must be supported by an underlying impervious or nearly impervious layer of rock.

Perennial streams depend on baseflow (groundwater contribution) for discharge between runoff events. The amount and duration of baseflow occurring in a stream is dependent on the amount of precipitation, the geologic conditions permitting infiltrated water to be stored underground, and the hydrogeologic controls governing groundwater flow to the stream. The geologic factors that determine the amount of baseflow storage also govern the amount of runoff from a drainage basin. Infiltrated precipitation first satisfies the soil-moisture deficiency and when this deficiency is satisfied, infiltrated water contributes to groundwater supplies.

Many communities are dependent on streamflow for water supplies. During periods of drought, streamflow is dependent on groundwater contribution. Therefore, knowledge of baseflow is important for predicting future water supplies from streams. During wet periods when groundwater supplies are high, there might be little subsurface storage space for infiltrating precipitation. During such wet periods, surface runoff and, as a result, flow potential are increased. Many cities have not adequately planned for these conditions and, as a result, severity of flooding has increased with time.

9.9. BASEFLOW AND AQUIFERS

Some baseflow originates from buried rocks that are under lithostatic pressure and such groundwater is forced upward into a stream. Aquifers providing such water are called artesian aquifers. Baseflow originating from artesian aquifers is not common. Artesian aquifers are not fed directly by infiltration and are not influenced by infiltration.

Groundwater moves slowly. Lateral movement of groundwater is slower than vertical movement from infiltration because the value of the hydraulic gradient is smaller for lateral movement. For this reason, groundwater "mounds" along streams and becomes elevated above the stream channel, thus providing a gradient that directs baseflow to the stream. If geologic and soil conditions are agreeable, the water table will continue to elevate during wet periods, and this increased elevation will increase the hydraulic gradient and increase the baseflow contribution to the stream. During dry periods, the opposite will occur and the baseflow contribution will decrease.

Most of the water-table aquifers in the world are comprised of granular material such as sand and gravel. These aquifers constitute the reservoirs from which most irrigation and municipal groundwater is pumped. The behavior of such aquifers can be reasonably predicted using groundwater equations such as Darcy's law and similar equations. Therefore, it is possible to determine the amount and rate of baseflow contribution.

A typical condition associated with a water-table aquifer is illustrated in Figure 9.14. The dashed line shows the water table in a sand aquifer, D. Precipitation has infiltrated the ground surface and caused mounding to occur under the topographic

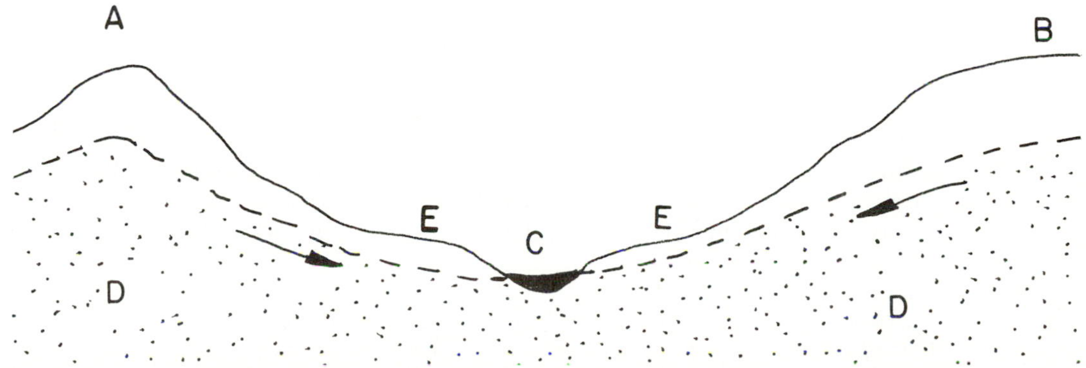

Figure 9.14 Baseflow supply by a water table in a granular aquifer.

hills where slow lateral movement causes water to accumulate. The stream, C, is fed by baseflow moving laterally from the mounded areas, A and B. The gradient of the water table results in the water table being closer to the surface at E than at A and B. Thus, some surface runoff from storm events occurs at E, where the small groundwater storage is soon filled and the remaining storm water must be drained over the surface to the stream channel. The supply from baseflow to the channel will continue as long as the water table at A and B are at a higher elevation than the stream at C.

Other water-table aquifers such as fractured rocks and solution cavity rock openings also contribute baseflow to streams. These aquifers may supply substantial quantities of baseflow to streams, particularly in mountainous areas, and are important contributors to streamflow. However, their behavior is more complicated, and the amount of their baseflow contribution is less easy to predict.

Because water-table aquifers supply baseflow, the movement of such groundwater is controlled by Darcy's law (Darcy, 1856). The rate of flow entering the stream is determined by Equation (9.18). It can be seen that the volumetric rate of flow entering the stream is determined by the aquifer hydraulic conductivity K, the hydraulic gradient, and the area of aquifer supplying the baseflow.

The hydraulic gradient is indicated in Figure 9.14 by the dashed line, which represents the water-table surface. The water table mounds under the dunes and migrates laterally to the stream. If no additional infiltration recharges this water-table aquifer, then the hydraulic gradient I decreases as water moves from higher elevations to the stream. Consequently, less water will be discharged to the stream as drainage continues and I decreases in value. This process is called baseflow recession. The rate at which baseflow decreases is a measure of the amount of baseflow available to a stream. The rate of change of the discharge integrated over time determines the amount of baseflow in the supply reservoir for the stream.

9.10 BASEFLOW CONTRIBUTION TO STREAMFLOW

Baseflow contribution to streamflow varies widely according to the geologic nature of the water-table reservoir. A measure of the amount of water available from a water-table aquifer is the coefficient of storage, or specific yield. The coefficient of storage is defined as the amount of water released from, or taken into, storage per unit change in head, per unit surface area of the aquifer normal to that change in head. The value of the coefficient of storage can range from 0.01 to 0.3. These coefficients mean that for every cubic meter (or ft^3) of saturated aquifer above the elevation of the stream surface, 0.01 m^3 (or ft^3) to 0.3 m^3 (or ft^3) of water will find its way to the stream channel as baseflow. The value of the coefficient of storage is dependent on geology, which for a drainage area might be fairly uniform or might vary widely. A sound knowledge of geology is therefore important for prediction of baseflow contribution to streams.

EXAMPLE 9.12 A portion of the Sandhills of Nebraska, through which a stream is flowing, has a hydraulic conductivity K of 27 cubic feet per day per square foot. Surveyed water-table measurements indicate a hydraulic gradient to the stream of 0.009. Determine the baseflow per square foot delivered to the stream each day.

Solution By using equation (9.18),

$$Q = (27)\ (0.009)\ (1) = 0.24\ \text{ft}^3/\text{day}$$

9.11 BASEFLOW AND GEOLOGY

A relation between geology and baseflow is expected, based on the foregoing discussion. Geology determines groundwater-reservoir storage characteristics. Basic geology of an area determines the bedrock characteristics, which, in turn, determine the character of the soils that result from bedrock weathering. Geologic processes such as glaciation, wind and water erosion, and sorting modify the weathered products of geology to produce water reservoirs of various characteristics.

In many areas of the earth, bedrock geology is folded, faulted, fractured, and exposed at the surface of the earth. These rocks develop storage characteristics that directly affect the baseflow supply from them. The magnitude of this storage and the baseflow supply characteristics vary depending on the nature of the rocks and the type and magnitude of the tectonic changes that have occurred.

9.12 BASEFLOW RECESSION

Streamflow recession represents withdrawal of water from storage with no inflow. It constitutes the falling limb of a hydrograph to the right of the point of contraflexure. The baseflow recession is the lower part of the falling limb and is represented by a curve expressing the relation between baseflow and time. Since baseflow is principally groundwater effluent, the baseflow recession curve represents withdrawal mainly from groundwater storage. The bank storage may also contribute to baseflow depending upon the position of the water table, river-bank material, soil characteristics near the river bank, and the rise and fall of the stream.

Baseflow recession has many practical applications (Hall, 1968, 1982): (1) low flow forecasting (Kunkle, 1962; White, 1977) for (a) controlling pollution through streamflow regulation, (b) evaluating the effect of agricultural practices, (c) locating suitable areas for induced infiltration, and (d) determining the increase in baseflow due to infiltration; (2) controlling withdrawal of groundwater for irrigation during low flow periods; (3) making water-supply estimates and forecasts (Dahl, 1985); (4) determining storage requirements for maintenance of adequate flow for waste dilution; (5) comparing drainage basins and geology with the aquifers at their maximum capacity and performance (Knisel, 1963); (6) simulating total runoff from physical and climatological data; (7) synthesis of daily streamflows through the techniques of operational hydrology; and (8) determining groundwater recharge (Meyboom, 1961).

Baseflow can most easily be evaluated by use of a stream gage. It has been shown in the previous chapter that the runoff hydrograph recession records the withdrawal of surface runoff from a drainage basin. When all surface runoff has departed from the drainage basin, the flow in the stream consists of water supplied by baseflow. Baseflow is a maximum immediately after a storm event because precipitation has just supplied recharge to the water table and conditions are favorable for maximum groundwater flow. As water is withdrawn from the water-table aquifer in the form of baseflow, groundwater supplies are diminished and the amount of baseflow decreases. This process continues even with decreasing baseflow until all groundwater available for baseflow is exhausted unless additional recharge occurs from another storm event.

The maximum supply that the baseflow reservoir may hold can be determined by analyzing the runoff hydrograph for a storm of sufficient magnitude to completely saturate the supply reservoir. This condition might or might not occur. If the baseflow reservoir is small, such as in many parts of Pennsylvania, such a storm event is likely to have occurred. For a large baseflow reservoir, such as in the Sandhills of Nebraska, such a storm event is not likely to occur.

Several methods have been suggested for representing baseflow recession (Toebes and Strang, 1964; Hall, 1968; Toebes et al., 1969). We will discuss some of these methods here.

9.12.1 Simple Exponential

The most widely used equations of baseflow recession are perhaps of exponential type of the following forms (Barnes, 1939; Mitchell, 1948; Laurenson, 1961; Toebes and Strang, 1964; Hall, 1968; Toebes et al., 1969):

$$Q = Q_0 \exp(-t/a) \tag{9.109}$$

$$Q = Q_0 k^t \tag{9.110}$$

$$Q = Q_0 (10)^{-t/b} \tag{9.111}$$

where Q_0 is flow at a selected time, or at $t = 0$; Q is flow at t unit times later (a unit time is often taken as 1 day), a is a constant, k is a constant, b is a constant, and t is time.

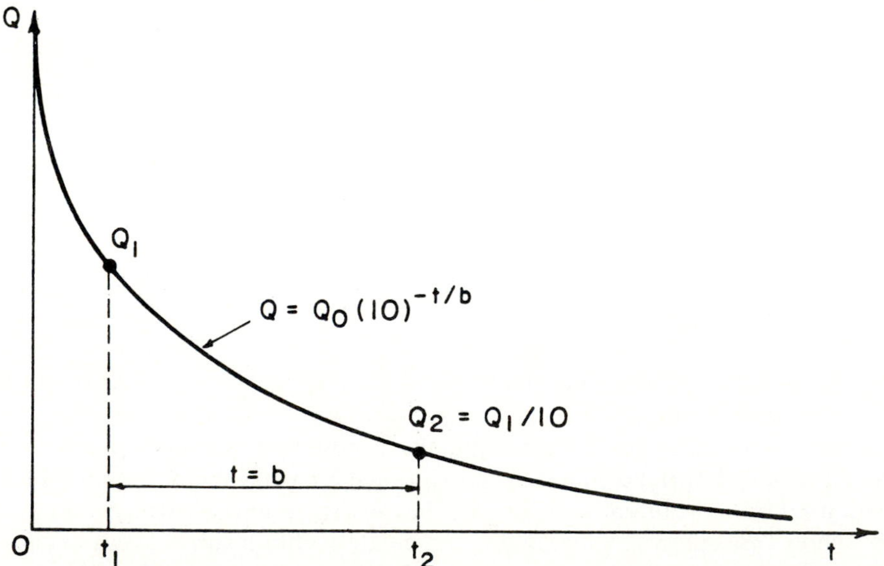

Figure 9.15 Baseflow recession. Determination of storage delay time b.

Equations (9.109) to (9.111) are obtained from linear solutions to the diffusion equation (Hall, 1968). Horton (1933), Barnes (1939), Laurenson (1961), and others showed that baseflow recession could be empirically fitted by Equation (9.109). Since Equations (9.110) and (9.111) are alternative forms of Equation (9.109),

$$k = \exp(-1/a) \tag{9.112}$$

$$k = (10)^{-1/b} \tag{9.113}$$

Parameter k is known as the recession constant, or the depletion factor, and will always be less than unity (normally greater than 0.9). Although Equation (9.110) implies that k is dimensionless, its value depends on the unit of time selected. In Equation (9.112), "1" in the exponential implies that same one unit of time. This is also true of Equation (9.113).

Constants a and b are known as storage delay factors, and both have the dimension of time. For example, constant b denotes the time required for the flow to decrease in the amount by a factor of 10 or one log cycle, as shown in Figure 9.15. Similarly, constant a denotes the time required for the flow to decrease by a factor equal to e, or one natural log cycle, as shown in Figure 9.16.

EXAMPLE 9.13 The baseflow values observed 25 days apart from an experimental watershed are 0.057 m^3/s and 0.0057 m^3/s. Compute the recession constant k.

Solution At $t = 0$, $Q = 0.057$ m^3/s. At $t = 25$ days, $Q = 0.0057$ m^3/s.

By substituting these values in Equation (9.110)

$$0.0057 = 0.057k^{25}$$

$$\log k = \frac{\log 0.0057 - \log 0.057}{25}$$

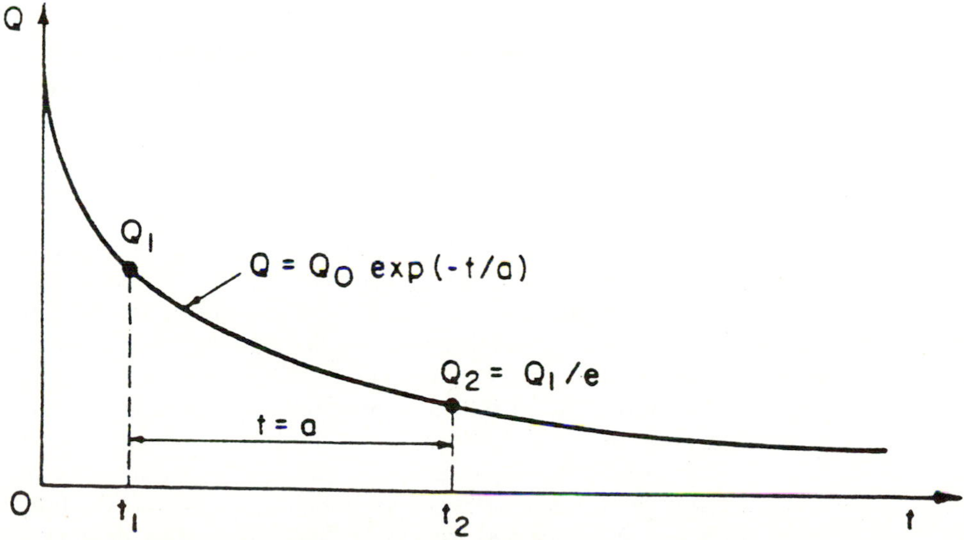

Figure 9.16 Baseflow recession. Determination of storage delay time a.

Therefore,

$$k = 0.912 \qquad \text{if } t \text{ is in days}$$

and

$$Q_t = Q_0(0.912)^t$$

9.12.2 Determination of Recession Constants

The recession constant k or parameters a and b can be determined in a number of ways. We describe some of them here.

GRAPHICAL METHOD

Constants a and b can be determined by constructing Figures 9.15 and 9.16, and choosing any two values of discharge that satisfy the condition specified there. One problem of this method is that these constants vary with the choice of discharges. Therefore, it may be appropriate to obtain a number of values of a and b, and then obtain an average. Alternatively, by taking logarithm of Equations (9.109 to 9.111),

$$\ln Q = \ln Q_0 - \frac{t}{a} \tag{9.114}$$

$$\ln Q = \ln Q_0 + t \ln k \tag{9.115}$$

$$\log Q = \log Q_0 - \frac{t}{b} \tag{9.116}$$

Baseflow discharge Q can be plotted against time on semilog paper using either of Equations (9.114) to (9.116). By fitting a straight line to the plotted data, the recession parameters can be determined. This method is more commonly used (Barnes, 1939, 1940; Langbein, 1940; Laurenson, 1961; Toebes, et al., 1969). The difficulty, of course, lies in obtaining a good straight-line fit.

On the other hand, one can also plot Q_{t-1} versus Q_t on a simple graph. For example, if Equation (9.110) were valid, this plot will result in a straight line passing through the origin. Its slope will specify k. Similar interpretations can be advanced for Equation (9.110) and (9.111). For example, Equation (9.109) can be recast as

$$Q_t = Q_{t-1} \exp(-T/a) \tag{9.117}$$

where Q_t is the discharge at time t; Q_{t-1} is the discharge at time one unit of time earlier, $t - 1$; and T is the time interval between Q_{t-1} and Q_t (usually denoted by 1). Therefore, a plot of Q_{t-1} versus Q_t should give a straight line having the slope of exp $(-T/a)$. Obviously, the slope depends upon the choice of the time interval, which is commonly taken as one day. One can do likewise for Equation (9.111).

Anderson and Burt (1980), based on laboratory experimentation, have argued that graphical techniques may falsely interpret the factors controlling streamflow recession. Some graphical techniques produce breaks of slope interpreted as indicating a change in flow characteristics, even though no change of the flow process has actually occurred. Nevertheless, recession flow graphs may be very useful for prediction of low flows.

EXAMPLE 9.14 Baseflow for the Amite River at Denham Springs, Louisiana, is given as follows:

Date	Time (h)	Discharge (m^3/s)	Date	Time (h)	Discharge (m^3/s)
4–24–79	12 P.M.	130	4–27–79	6 A.M.	98
	6 P.M.	125		12 P.M.	96
	12 A.M.	122		6 P.M.	92
				12 A.M.	90
4–25–79	6 A.M.	120			
	12 P.M.	118	4–28–79	6 A.M.	90
	6 P.M.	115		12 P.M.	88
	12 A.M.	110		6 P.M.	84
				12 A.M.	82
4–26–79	6 A.M.	108			
	12 P.M.	104	4–29–79	6 A.M.	85
	6 P.M.	102		12 P.M.	76
	12 A.M.	100			

Plot the baseflow discharge as a function of time and determine the depletion constant k graphically.

Solution The data are rearranged as in what follows. The discharge is then plotted on semilog paper, as shown in Figure 9.17. By using Equation (9.110),

$$
\begin{aligned}
k &= \left(\frac{Q}{Q_0}\right)^{1/t} \\
&= \left(\frac{82}{110}\right)^{1/(34/3)} \\
&= \left(\frac{82}{110}\right)^{3/34} \\
&= 0.9744
\end{aligned}
$$

Date	Time (h)	Unit of time	Discharge (m^3/s)	Date	Time (h)	Unit of time	Discharge (m^3/s)
4–24–79	12 P.M.	0	130	4–27–79	6 A.M.	11	98
	6 P.M.	1	125		12 P.M.	12	96
	12 A.M.	2	122		6 P.M.	13	92
					12 A.M.	14	90
4–25–79	6 A.M.	3	120				
	12 P.M.	4	118	4–28–79	6 A.M.	15	90
	6 P.M.	5	115		12 P.M.	16	88
	12 A.M.	6	110		6 P.M.	17	84
					12 A.M.	18	82
4–26–79	6 A.M.	7	108				
	12 P.M.	8	104	4–29–79	6 A.M.	19	85
	6 P.M.	9	102		12 P.M.	20	76
	12 A.M.	10	100				

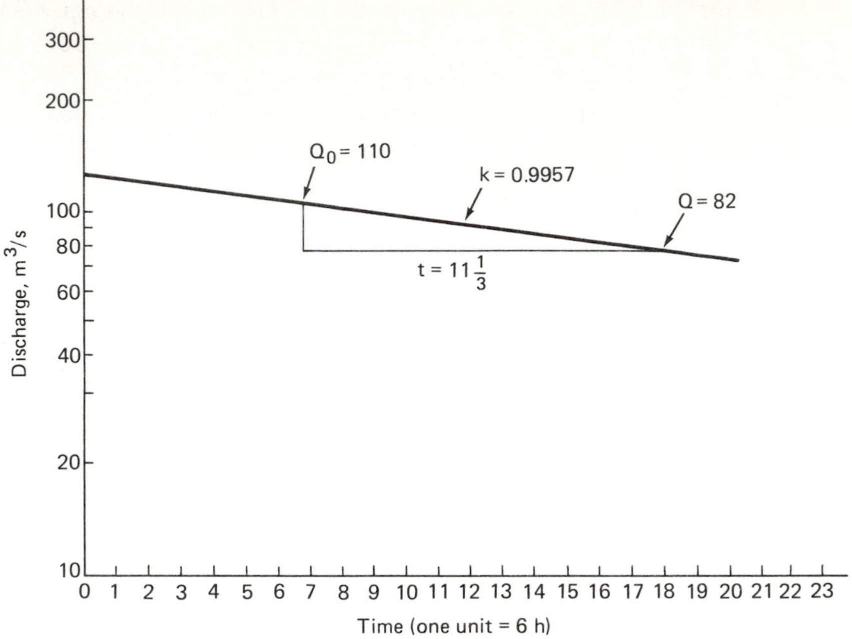

Figure 9.17 Determination of recession constant k.

LEAST SQUARES METHOD

James and Thompson (1970) determined the recession constants by using the least squares method. Any one of Equations (9.109) to (9.111) can be used with this method. For illustration, Equation (9.110), recast as

$$Q_t = kQ_{t-1} \tag{9.118}$$

is used. The recession constant k can be estimated by minimizing the sum of squares of differences between the observed and computed values of discharge:

$$R = \sum_{t=1}^{N} (Q_t - kQ_{t-1})^2 \tag{9.119}$$

where N is the number of discharge values, and Q_t corresponds to the observed values. The least squares estimate of k can be obtained by setting the derivatives of R with respect to k equal to zero and solving for the value of k. This yields

$$k = \frac{\sum_{t=1}^{N} Q_t Q_{t-1}}{\sum_{t=1}^{N} Q_{t-1}^2} \tag{9.120}$$

It is clearly seen that when $N = 1$, $k = Q_1/Q_0$, which is what Equation (9.118) gives.

Equation (9.120) provides a convenient method for estimating k. However, the baseflow data should be free of errors as far as possible. James and Thompson (1970) noted that k would be more reliable if the flow sequence was longer and contained larger flows. An average of results obtained from a series of sequences would be desirable.

EXAMPLE 9.15 For the data of Example 9.14, compute the recession constant k using the least squares method.

Solution From Equation (9.110),

$$\ln Q = \ln Q + t \ln k$$

Date	Time	Unit of Time	Discharge (m^3/s)	$\ln Q$
4–24–79	12 P.M.	0	130	4.8675
	6 P.M.	1	125	4.8283
	12 A.M.	2	122	4.8040
4–25–79	6 A.M.	3	120	4.7875
	12 P.M.	4	118	4.7707
	6 P.M.	5	115	4.7449
	12 A.M.	6	110	4.7005
4–26–79	6 A.M.	7	108	4.6821
	12 P.M.	8	104	4.6444
	6 P.M.	9	102	4.6250
	12 A.M.	10	100	4.6052
4–27–79	6 A.M.	11	98	4.5850
	12 P.M.	12	96	4.5643
	6 P.M.	13	92	4.5218
	12 A.M.	14	90	4.4998
4–28–79	6 A.M.	15	90	4.4998
	12 P.M.	16	88	4.4773
	6 P.M.	17	84	4.4308
	12 A.M.	18	82	4.4067
4–29–79	6 A.M.	19	85	4.4427
	12 P.M.	20	76	4.3307

By using the least squares method,

$$\text{Correlation coefficient} = -0.9939$$

$$\text{slope} = \ln k = -0.0246$$

$$k = 0.9757$$

METHOD OF MOMENTS

The recession constants can be estimated by the method of moments. Since there is only one parameter, to consider only the first moment about the origin will suffice. Again, any one of Equations (9.109) to (9.111) can be used. For illustrative purposes, we consider Equation (9.109). Let $M_1(Q)$ be the first moment of Q about the origin. Then

$$M_1(Q) = \int_0^\infty tQ_0 \exp(-t/a)\, dt \Big/ \int_0^\infty Q_0 \exp(-t/a)\, dt \qquad (9.121)$$

$$= a$$

Thus, the recession constant a for a given baseflow sequence is equal to the first moment of that sequence about its origin. This first moment is obtained from the data. As in the least squares method, it may be desirable to have an error-free and sufficiently long baseflow record.

EXAMPLE 9.16 Compute the recession constant k for the data of Example 9.14 using the method of moments.

Solution One unit of time is considered as 6 hours. Calculations for obtaining the first moment follow:

Date	Time	Unit of time	Discharge, Q (m^3/s)	Time to centroid (h)	Subarea (m^3/s/unit of time)
4–24–79	12 P.M.	0	130	0.5	127.5
	6 P.M.	1	125	1.5	123.5
	12 A.M.	2	122	2.5	121
4–25–79	6 A.M.	3	120	3.5	119
	12 P.M.	4	118	4.5	116.5
	6 P.M.	5	115	5.5	112.5
	12 A.M.	6	110	6.5	109
4–26–79	6 A.M.	7	108	7.5	106
	12 P.M.	8	104	8.5	103
	6 P.M.	9	102	9.5	101
	12 A.M.	10	100	10.5	99
4–27–79	6 A.M.	11	98	11.5	97
	12 P.M.	12	96	12.5	94
	6 P.M.	13	92	13.5	91
	12 A.M.	14	90	14.5	90
4–28–79	6 A.M.	15	90	15.5	89
	12 P.M.	16	88	16.5	86
	6 P.M.	17	84	17.5	83
	12 A.M.	18	82	18.5	83.5
4–29–79	6 A.M.	19	85	19.5	80.5
	12 P.M.	20	76		

Area = 2032 (m^3/s − unit of time)

$$M_1 = \frac{1}{2032\ (\text{m}^3/\text{s} - \text{unit of time})} \times [18{,}678.5\ \text{m}^3/\text{s} - (\text{unit of time})^2]$$

$$= 9.192175 \text{ unit of time}$$

From Equation (9.109),

$$M_1(Q) = a, \text{ the storage delay factor}$$

From Equations (9.109) and (9.110),

$$k = \exp(-1/a), \text{ where "1" is one unit of time}$$

Therefore,

$$k = \exp\left(\frac{-1}{9.192175}\right) = 0.8969$$

9.12.3 More on Exponential Recession

Equation (9.109) represents a first-order process or an exhaustion phenomenon expressed by

$$\frac{ds}{dt} = -Q \qquad Q(0) = Q_0 \tag{9.122}$$

and

$$S = aQ \tag{9.123}$$

where s is storage. The solution of Equations (9.122) and (9.123) is Equation (9.109). It can be shown that if discharges are replaced by volumes in Equation (9.109), its form does not change. The volume of flow V during the time from $t - T$ to t is given by

$$\begin{aligned} V &= \int_{t-T}^{t} Q\, dt = \int_{t-T}^{t} Q_0 \exp(-t/a)\, dt \\ &= aQ_0 [\exp(T/a) - 1] \exp(-t/a) \end{aligned} \tag{9.124}$$

For a fixed time interval T such as 1 day,

$$V = V_0 \exp(-t/a) \tag{9.125}$$

in which

$$V_0 = aQ_0 [\exp(T/a) - 1] \tag{9.126}$$

It then follows that the recession constant a can be estimated from Equation (9.125) using any one of the previous procedures. Indeed, it may be preferable to use volume in place of discharge because the former is less sensitive to errors than the latter. Shirmohammadi et al. (1984) used volumes for partitioning streamflow data in surface and subsurface flows.

If Equation (9.110) is utilized to express V, then

$$V = V_0 k^t \tag{9.127}$$

where

$$V_0 = \frac{Q_0}{\ln k}(1 - k^{-T}) \tag{9.128}$$

If Equation (9.111) is utilized to estimate V, then

$$V = V_0 (10)^{-t/b} \tag{9.129}$$

where

$$V_0 = \frac{bQ_0}{\ln 10}(10^{T/b} - 1) \tag{9.130}$$

EXAMPLE 9.17 Compute the value of k using Equation (9.127) for the data of Example 9.14.

Solution From Equation (9.127),

$$\ln V = \ln V_0 + t \ln k$$

By using the least squares method,

$$\text{Correlation coefficient} = -0.9939$$

$$\text{Slope} = \ln k = -0.0247$$

$$k = 0.9757$$

The calculations follow:

Date	Time	Hour	Discharge, Q (m^3/s)	Volume, V ($m^3/s - h$)	$\ln V$
4–24–79	12 P.M.	0	130	780	6.6593
	6 P.M.	1	125	750	6.6201
	12 A.M.	2	122	732	6.5958
4–25–79	6 A.M.	3	120	720	6.5793
	12 P.M.	4	118	708	6.5624
	6 P.M.	5	115	690	6.5367
	12 A.M.	6	110	660	6.4922
4–26–79	6 A.M.	7	108	648	6.4739
	12 P.M.	8	104	624	6.4362
	6 P.M.	9	102	612	6.4167
	12 A.M.	10	100	600	6.3969
4–27–79	6 A.M.	11	98	588	6.3767
	12 P.M.	12	96	576	6.3561
	6 P.M.	13	92	552	6.3136
	12 A.M.	14	90	540	6.2916
4–28–79	6 A.M.	15	90	540	6.2916
	12 P.M.	16	88	528	6.2691
	6 P.M.	17	84	504	6.2226
	12 A.M.	18	82	492	6.1985
4–29–79	6 A.M.	19	85	510	6.2344
	12 P.M.	20	76	456	6.1225

9.12.4 Double Exponential

When baseflow is plotted on semilog paper, the resulting curve is found to be nonlinear. This curve can perhaps be better represented by a double exponential of the form

$$Q = Q_0 \exp(-mt^n) \tag{9.131}$$

where $k = \exp(-m)$ is the recession constant, and n is a constant. Equation (9.131) was first obtained by Horton (1933, 1935), and has been used by many workers (Yevjevich, 1963; Toebes and Strang, 1964; Toebes et al., 1969).

Parameters m and n can be obtained either graphically or using the method of least squares. By taking the log twice of both sides of Equation (9.131),

$$\ln (Q_0/Q) = mt^n \tag{9.132}$$

$$\ln [\ln (Q_0/Q)] = \ln m + n \ln t \tag{9.133a}$$

or

$$\log [\log (Q_0/Q)] = \log m + n \log t - 0.369222 \tag{9.133b}$$

Equation (9.133) represents a straight line.

9.12.5 Hyperbola

Another equation for the case where a stream is located on a horizontal impermeable lower boundary with an initial curvilinear water table and zero water-level elevation in the stream is of the form

$$Q = Q_0/(1 + ct)^2 \tag{9.134}$$

where c is a constant. This equation has been frequently used in Europe for spring discharge (Hall, 1968; Toebes et al., 1969). In those conditions where groundwater storage is at its maximum after cessation of surface runoff or if no further precipitation or snowmelt occurred until the streamflow ceased Equation (9.134) will hold.

Parameter c can be estimated as

$$c = \frac{1}{t}\left[\left(\frac{Q_0}{Q}\right)^{0.5} - 1\right] \tag{9.135}$$

However, the value of c in this manner will depend upon the choice of Q. A better estimate of c can be obtained by the method of least squares.

Equation (9.134) can be derived by using Equation (9.122) and a nonlinear storage–discharge relation:

$$Q = aS^n \quad n \neq 1 \tag{9.136}$$

with the initial condition $Q(0) = Q_0$. This then yields

$$Q = Q_0(1 + ct)^{n/(1-n)} \tag{9.137}$$

where $c = (1 - a)a^{1/n}/Q_0^{n/(1-n)}$, a constant. Clearly, when $n = 2$, Equation (9.134) results.

9.12.6 Ice-Melt Hyperbola

For snow and ice-melt conditions, the baseflow recession can be adequately represented (Toebes et al., 1969) by

$$Q = at^{-n} + b \tag{9.138}$$

where a, b, and n are constants. As time increases, Q asymptotically approaches a constant value b. This may typify baseflow recession where permanent snow and ice are present.

Constants a, b, and n can be determined either graphically or by using the method of least squares. By taking the log of Equation (9.138),

$$\log (Q - b) = \log a - n \log t \tag{9.139}$$

Thus, a plot of $Q - b$ versus t will result in a straight line on log–log paper, provided b is known or has been chosen correctly. By studying streamflow data, an approximate value of b can be obtained as an initial approximation. Later, a more accurate value of b can be obtained.

If $b = 0$, then

$$Q = at^{-n} \tag{9.140}$$

Equation (9.140) was used by Tjomsland et al. (1978) to study recession characteristics of small Norwegian rivers. By integrating it from the time element t to ∞,

$$S = \int_t^\infty Q \, dt = \int_t^\infty = at^{-n} \, dt = \frac{a}{n-1} t^{1-n} \tag{9.141}$$

where S represents the total storage of water available for runoff at time t. By substituting Equation (9.140),

$$S = \frac{a^{1/n}}{n-1} Q^{(n-1)/n} \tag{9.142}$$

Equation (9.142) was used to represent the recession curve.

9.12.7 Ice-Melt Exponential

Another equation for baseflow recession in watersheds with permanent ice and snow is of the form (Toebes et al., 1969)

$$Q = a + (Q_0 - a)k^t \tag{9.143}$$

where a and k are constants. This is similar to Equation (9.138), where Q asymptotically approaches a constant value for large values of t, which is largely supplied by melting of permanent ice and snow fields. Unlike Equation (9.138), the recession curve here is relatively flat for small values of t.

Parameters a and k can be determined by plotting Equation (9.143) on semilogarithmic paper:

$$\log (Q - A) = \log (Q_0 - a) + t \log k \tag{9.144}$$

For an appropriate value of a, a plot of $Q - a$ versus t on a semilog paper will produce a straight line. The value of a can be obtained from streamflow data.

EXAMPLE 9.18 A perennial stream discharges 1000 cfs immediately after all surface runoff has passed a certain point. Estimate the discharge at this same point 10 days later assuming no rainfall occurs during this 10-day interval.

Solution From Equation (9.110),

$$Q = 1000(0.9)^{10} = 348.7 \text{ cfs}$$

9.13 MASTER BASEFLOW RECESSION CURVE

The master baseflow recession curve is a composite recession curve based on the largest baseflow discharge values. This curve can be constructed in several ways (Toebes et al., 1969): (1) by experimentation, (2) by the strip method, or the tabulation method, and (3) by the correlation method. Jones and McGilchrist (1978) found that a master recession curve did not give an adequate description of recession flows for three Australian rivers.

9.13.1 Experimentation

This method involves making observations of discharge at a number of time intervals encompassing the entire dry-weather period. The discharge values are then plotted against time on semilog paper, and a best-fit straight line is drawn through the plotted points. The resulting curve is a recession curve. This curve is useful for prediction purposes and for developing correlations with rainfall.

9.13.2 Strip Method, or Tabulation Method

This requires that streamflow records are available. Recessions during periods of dry weather are extracted from the records. Surface flow should be eliminated as far as possible. Discharge values are plotted against time on semilog paper. This is repeated for a number of dry periods. Recessions within the same discharge range will have sensibly equal slopes. Segments of discharge are combined by plotting individual recessions on tracing paper. These are then superimposed and adjusted horizontally until the main parts overlap. A mean curve through the overlapping parts results in the master recession curve.

9.13.3 Correlation Method

This involves plotting the discharge for the initial day of each segment on log–log paper, starting with the largest discharge, against discharge T time units later (where $T = \frac{1}{2}$, 1, 2, 5, etc.). The plotting is continued for as many segments as possible until a good correlation is established for drawing a line. The points above the line represent surface flow and those below it represent slow recession.

EXAMPLE 9.19 Six hydrographs observed in 1964 from Puketurua experimental watershed in New Zealand are available. For purposes of simplicity, only baseflow portions of these hydrographs are given here and properly aligned in Table 9.2. Construct a master baseflow recession curve. Discharge is observed at 12-hour intervals.

TABLE 9.2 BASEFLOW FROM SIX SELECTED HYDROGRAPHS IN 1964 FROM THE PUKETURUA EXPERIMENTAL WATERSHED IN NEW ZEALAND

Baseflow discharge (m^3/s)						
1 August	2 Sept. 6	3 July 19	4 June 5	5 Nov. 3	6 Dec. 24	7 Ave. discharge
12:00 hours						
0.069						0.069
0.057						0.057
0.054						0.054
0.053						0.053
0.046						0.046
0.044	00:00 hours					0.044
0.043	0.043	00:00 hours				0.043
0.040	0.040	0.042				0.041
0.038	0.038	0.038				0.038
0.037	0.036	0.036				0.036
0.034	0.032	0.034				0.034
0.032	0.031	0.032				0.032
0.031	0.030	0.031				0.031
0.029	0.029	0.029				0.029
	0.028	0.028				0.028
	0.027	0.026				0.027
	0.023	0.025	00:00 hours			0.024
	0.022		0.023			0.023
	0.021		0.022			0.021
	0.020		0.021			0.020
	0.019		0.018			0.019
	0.018		0.017			0.018
	0.017		0.017			0.017
	0.016		0.016			0.016
	0.015		0.015	12:00 hours		0.015
	0.015		0.014	0.015	00:00 hours	0.014
			0.013	0.015	0.014	0.013
			0.012	0.013	0.013	0.012
			0.011	0.012	0.011	0.011
				0.011	0.010	0.010
				0.010	0.009	0.009
				0.009	0.007	0.007
				0.008		0.006

Solution Since the columns expressing recession are already adjusted vertically until the discharges approximately agree horizontally, the discharges are averaged horizontally in column 7 of Table 9.2, and these mean discharges constitute the master baseflow recession curve.

9.14 BASEFLOW AND PHYSIOGRAPHIC CHARACTERISTICS

Limited attempts have been made to predict the master recession curve from watershed characteristics. Tjomsland et al. (1978) considered for 14 Norwegian rivers (area 0.6 to 29.3 km^2) 28 hydrologic and geologic parameters for establishing correlations with baseflow, and found only the drainage density, the percentage lake area, and the weighted lake-inflow-area index to be significant. The multiple regression equation explained the more than 85% variance in storage available for runoff. This finding is consistent with the study conducted by Carlston (1963) who, using the Jacob water-table model (Jacob, 1943, 1944), showed that mean minimum monthly baseflow varied inversely with the square of the drainage density for 15 watersheds in central and eastern United States. In a related study, Weisman (1977) found significant relation with the recession constant k decreasing and increasing pan evaporation for three watersheds in New Zealand. For low values of average daily pan evaporation, the recession constant approached a constant value.

EXAMPLE 9.20 Carlston (1963) has provided data on area, drainage density, and average minimum monthly flow (baseflow) per unit area, as shown in Table 9.3. Plot baseflow against drainage density on log–log paper, and determine the relation between them. Do likewise for the mean annual flood and drainage density. Then, determine the relation between baseflow and mean annual flood.

Solution Figure 9.18 plots baseflow versus drainage density, and Figure 9.19 does likewise for the mean annual flood.

TABLE 9.3 MONTHLY BASEFLOW, MEAN ANNUAL FLOOD, AND MEAN ANNUAL PRECIPITATION FOR 15 WATERSHEDS (AFTER CARLSTON, 1963)

Watershed no.	Area (mi^2)	Drainage density	Average minimum monthly flow (cfs/mi^2)	Mean annual flood (cfs/mi^2)	Mean annual precipitation (in.)
1	5	7.6	0.13	62	38
2	16.9	7.0	0.29	44	38
3	18.1	9.5	0.26	77	49
4	16.7	4.2	0.61	32	42
5	6.0	4.7	0.35	23	45
6	5.1	3.0	1.22	19	45
7	11.3	3.6	0.80	14	40
8	55.7	6.3	0.27	44	44
9	20.9	8.0	0.10	86	43
10	14.7	8.0	0.36	102	46
11	11.4	6.0	0.84	75	45
12	31.9	5.0	0.83	38	47
13	5.5	5.0	0.91	44	49
14	10.9	7.4	1.86	72	62
15	11.7	8.0	2.45	67	63

EXERCISES

9.1. During a period of drought, the static water level in an aquifer has dropped to an elevation of 620.4 ft msl. The drought is broken by a prolonged period of rain in which the measured infiltration was 3.4 inches. The storage coefficient is 0.25. At what mean sea level elevation (msl) will the static water level be located after the rainfall?

9.2. A 25-cm well 100.0 m deep is proposed in an aquifer having transmissivity of 2000 m^2/

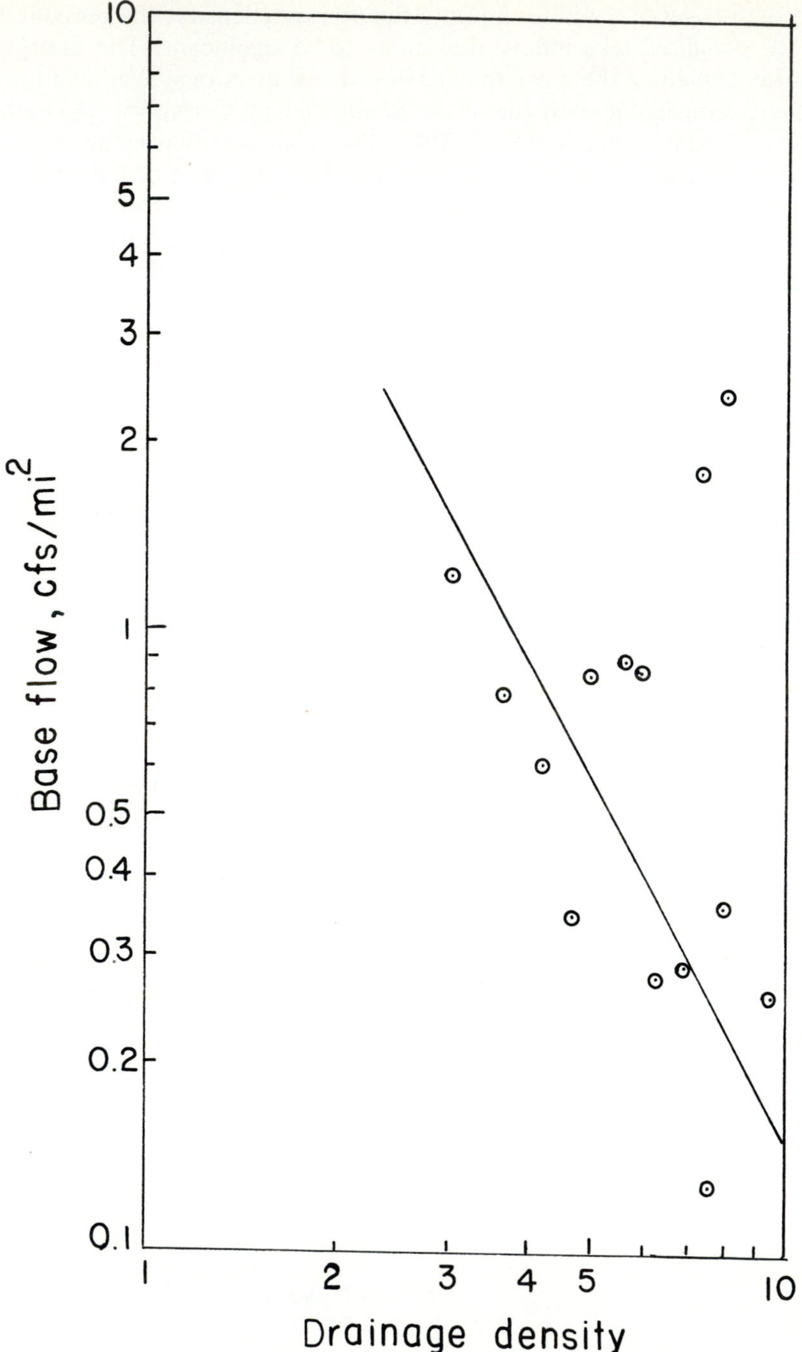

Figure 9.18 Relationship between baseflow and drainage density for some watersheds in the central and eastern United States.

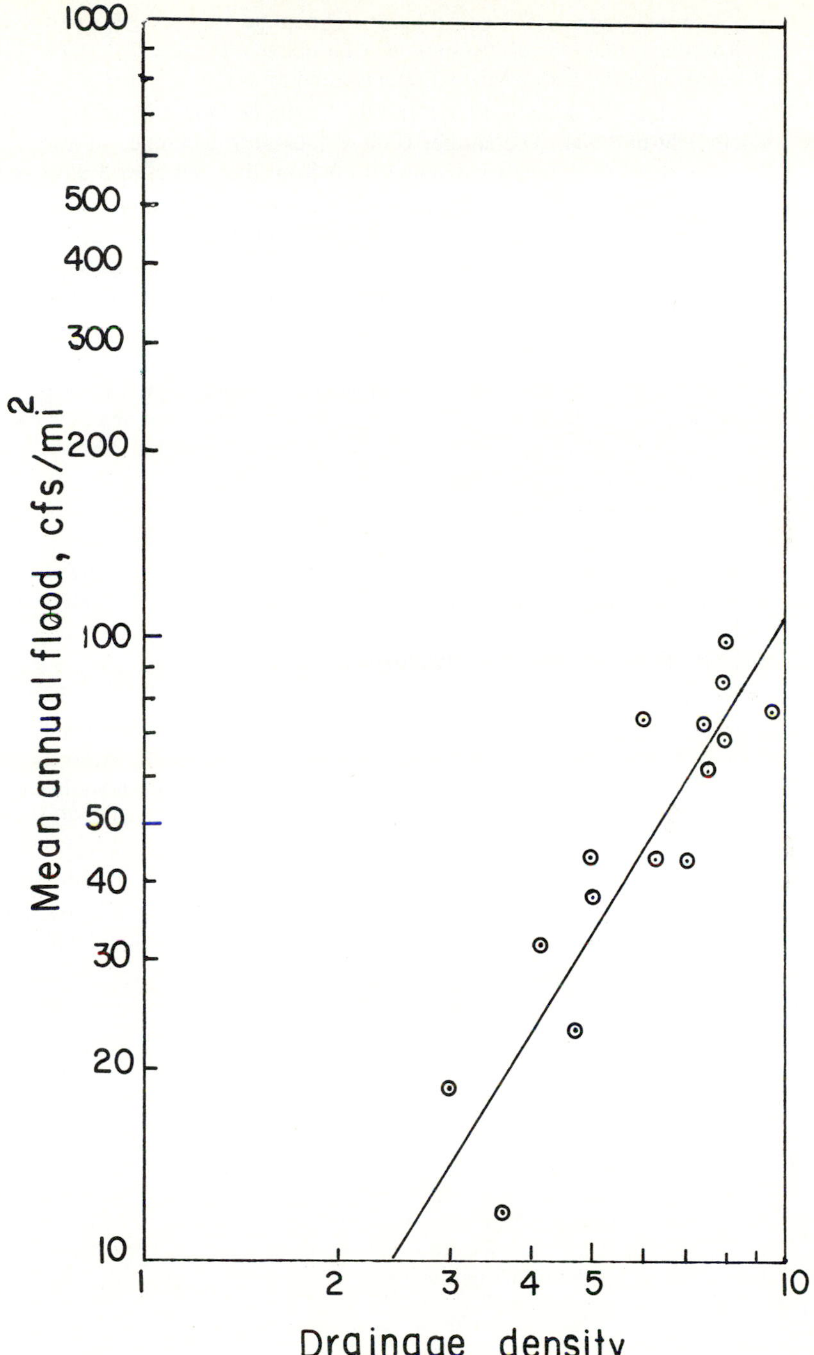

Figure 9.19 Relationship between the mean annual flood and drainage density for some watersheds in the central and eastern United States.

day and a storage coefficient of 0.004. The static water level is expected to be 20.0 m below the ground level. Assuming a pumping rate of 0.035 m^3/s, what will be the drawdown in the well after (a) 200 days and after (b) 2 years?

9.3. A well is located in an aquifer with a hydraulic conductivity of 50 ft/day and a storage coefficient of 0.004. The aquifer is 80 ft thick and is pumped at a rate of 4500 ft^3/day. What is the drawdown at a distance 15 ft from the well after 3 days of pumping?

9.4. A 20-cm well fully penetrates an artesian aquifer 20.0 m thick. After 12 hours of pumping at the rate of 20.0 liter/s, the drawdown in the well is 2.5 m, and after 2 days, the drawdown is 2.75 m. Determine the transmissivity and storage coefficient of the aquifer. What is the permeability of the aquifer medium? After what time will the drawdown be 3.6 m?

9.5. A pumping test was conducted on a fully penetrating well in an artesian aquifer. The pumping rate was 0.05 m^3/s. The aquifer has a storage coefficient of 7.0×10^{-4} and a transmissivity of 1800 m^2/day. The observation well is at a distance of 25.0 m from the pumping well. Determine the drawdown in the observation well at 5, 50, and 500 min from the time of starting the pump.

9.6. A 16-inch-diameter well has been constructed in a water-table aquifer. A 14-inch pump capable of continuous discharge at 100 gpm has been installed to test the well. Observation wells have been located 40 feet and 90 feet from the pumped well. Pumping continuously for 24 hours resulted in drawdowns in the observation well of 3.0 and 2.0 ft. At the end of 36 hours, the drawdowns were 3.3 and 2.3 ft. At this time, the pumping level is 13 ft. Determine the entrance losses.

9.7. How far has the cone of depression expanded in Exercise 9.6 after 36 hours of pumping?

9.8. A test well is constructed with 16-inch-diameter casing in a water-table aquifer. Observation wells are located at 50 ft and 100 ft from the test well. After 24 hours of continuous pumping of 1000 gpm, the drawdowns in the observation wells were 3.13 and 2.15 ft. After another 12 hours, the drawdowns were 3.42 and 2.43 ft. Determine the coefficient of storage.

9.9. A nearby farmstead has a supply well located 300 ft from the pumping well in Exercise 9.8. The pump suction in the farmstead well is located 10 ft below the original static water level. How many days would elapse before the farmstead well would go dry as a consequence of continuous pumping of the other well?

9.10. A pumping test has determined the aquifer transmissivity to be 2000 ft^2/day and the storage coefficient to be 0.0002. Your client requires an average water supply of 500,000 gallons per day. You plan a 12-inch-diameter well. Assume that 200 days of continuous withdrawal from storage is probable each year. If the entrance losses to the well total 2 ft, at what depth would you set the pump?

9.11. A rural home receives its water supply from a shallow well in a water-table aquifer. The transmissivity of the aquifer is 200,000 gpd/ft and the coefficient of storage is 0.30. A farmer constructs an irrigation well 500 ft away from the residence well and plans to pump 1500 gpm for the irrigation season. How long can the irrigation well be pumped before it affects the residence well?

9.12. Baseflow is given as

Time (days)	0	1	2	3	4	5
Discharge (m^3/s)	100	80	60	40	20	10

Plot the baseflow recession and fit Equations (9.109) to (9.111). Which equation better represents the recession curve?

9.13. Reconstruct the baseflow discharges using the fitted equations in Exercise 9.12. Compare these discharges with observed ones and compute relative errors in computed values.

9.14. Flow sequences for eight storms for the Cumberland River at Williamsburg, Kentucky, are given in Table 9.4. Separate the baseflow and then compute the recession constant for the baseflow of each storm.

9.15. Construct the master baseflow recession curve for the Cumberland River at Williamsburg, Kentucky, using the data of Table 9.4.

9.16. For a stream, the baseflow recession constant is 0.92. Initially, the baseflow discharge is 10 m^3/s. Compute the baseflow discharge on the tenth day.

9.17. The baseflow discharge for a stream is 100 m^3/s on a given day. The recession constant for this stream is 0.95. What was the baseflow 5 days earlier?

9.18. Consider flow sequence 2 starting on January 30, 1954, given in Table 9.4. Separate the baseflow and compute the baseflow volume for each time interval. Plot the baseflow volume as a function of time.

9.19. Estimate the recession constant k using the volumes obtained in Exercise 9.18.

9.20. Estimate the recession constant k using the method of moments for flow sequence 2 given in Table 9.4.

9.21. Estimate the recession constant k using the least squares method for flow sequence 2 given in Table 9.4.

9.22. The streamflow recession observation indicated that the flow on June 10, 1983, was 1000 m^3/s. The recession constant was found to be 0.9 with a time interval of 24 hours. Compute the flow in this stream on June 15, 1983.

9.23. The groundwater discharge at zero time is 60 m^3/s and 45 m^3/s 5 hours later. Determine the recession constant. The discharge observations are available at 1-hour intervals.

TABLE 9.4 FLOW SEQUENCES OBTAINED FROM DAILY FLOWS (CFS) FOR THE CUMBERLAND RIVER AT WILLIAMSBURG, KENTUCKY, FOR WATER YEAR 1954 (AFTER JAMES AND THOMPSON, 1970)[a]

Day	Seq. 1 Dec. 17	Seq. 2 Jan. 30	Seq. 3 Mar. 16	Seq. 4 Mar. 28	Seq. 5 May 21	Seq. 6 June 4	Seq. 7 June 19	Seq. 8 July 26
0	800	3580	4540	5580	2450	902	2000	420
1	566	2750	3380	3700	1970	824	1260	296
2	372	2130	2650	3150	1560	674	824	219
3	280	1740	2260	2690	1300	542	596	178
4	222	1500	2110	2390	1100	440	460	154
5	212	1450	2030	2080	934	380	372	142
6	216	1360	1840	1820	818	332	312	118
7	208	1210	1720	1640	722	304	288	95
8	202	1070		1530	638	276	264	90
9	190	927		1470	560		222	
10	169	842			500		193	
11	160	788			460		172	
12		746					151	
13		698					136	
14		626					127	
15		548					130	
16		506						

[a] Dates given are initial dates.

PART 4
Above-Surface Flow and Hydrologic Abstractions

CHAPTER 10
Evaporation and Transpiration

Evaporation and transpiration are abstractions in the hydrologic-budget equation. The importance of these abstractions depends upon the time scale of the hydrologic problem. These abstractions are small for a runoff event and can be neglected. The bulk of evaporation and transpiration takes place during the time between runoff events, which is usually long. Hence, these abstractions are most important during this time interval. The combined effect of evaporation and transpiration is called evapotranspiration and is usually designated by the notation ET. Over large land areas in temperate zones, about two-thirds of the annual precipitation is evapotranspired, and the remaining one-third runs off in streams to the oceans. In arid regions, evapotranspiration may be even more significant, returning up to 90% or more of the annual precipitation to the atmosphere.

This chapter discusses some aspects of evaporation and transpiration and their measurement and evaluation. Evaporation also links hydrology to atmospheric science, and, through transpiration, to agricultural sciences.

10.1 EVAPORATION PROCESS

Evaporation is the process whereby liquid water is converted to water vapor by the transfer of water molecules to the atmosphere. The net result is that visible water is converted to invisible water vapor. This process is accompanied by a transfer of energy. Clearly, the molecules that are especially energetic will be able to break all bonds with their neighbors and move up into the air. The latent heat of vaporization is 539 cal/gm at 1 atmosphere pressure (at 100°C). Latent heat of vaporization is the heat required, in calories, to convert 1 gram of water at 1 atmosphere (atm) pressure to vapor. Thus, for practical purposes, it requires 539 cal of heat to evaporate 1 gram of water. One calorie (cal) is the amount of heat required to raise 1 gram of water from, say, 14.5°C to 15.5°C. The amount of heat energy required to raise the temperature of water to its boiling point is less than that required to convert the liquid water to a vapor. The amount of energy required to vaporize water is the sum of the energy required to bring the water temperature to 100°C plus the latent heat of vaporization. The thermal energy of latent heat is merely transformed kinetic energy from fast moving molecules. This removes energy and, hence, heat from the liquid. Therefore, evaporation tends to reduce the temperature of the liquid phase.

When the water molecule becomes a gas molecule, its behavior is different. Each molecule in a gas is more isolated and much farther from neighboring molecules than is the case in a liquid. Hence, hydrogen bonding properties, which are so important in the liquid phase, are weak or absent altogether in the vapor phase. The vapor molecules are far apart and not affected by hydrogen bonding. The water vapor is only about 60% as heavy as other atmospheric gases, and it tends to rise in the air above a water surface. As an example, in relatively humid air on a typical summer day, the vapor molecules above the surface of a water body (say, a lake) will be more than 40 times as far from their neighboring molecules as those in the water below. Clearly, the average spacing between molecules increases during evaporation, and water undergoes enormous expansion. At ordinary temperatures, 1 gram of liquid water occupies about 1 milliliter. At 25°C (77°F), 1 gram of saturated water vapor occupies approximately 42,000 milliliters. This amounts to an expansion of 42,000 times. Except for its high thermal energy content, the water vapor shares many of the properties of other common gases. It can compress or expand, and its molecules exert a pressure of their own called partial pressure.

Water also evaporates directly from the solid state, and this process is called sublimation. A significant amount of water vapor enters the atmosphere by sublimation from snow and ice. This contribution of water vapor to the atmosphere, however, is far less than evaporation from the liquid water contained in the oceans, lakes, and rivers on the earth.

When water is warmed, the water molecules become more active and move from the liquid water to the atmosphere. As more water molecules move into the atmosphere, the density of the water molecules near the water surface increases. Collisions between molecules emerging from the water with those already in the air cause some water molecules to return to the water. As a result, there is a constant transfer of molecules to and from the water, but the transfer from the water is dominant. Other molecules are caused to move farther into the air by molecular action. After a period of time, there is nearly the same number of molecules returning to the water surface as leaving the water surface in a thin film next to the water. Within this thin film, as shown in Figure 10.1, the air is nearly saturated with water

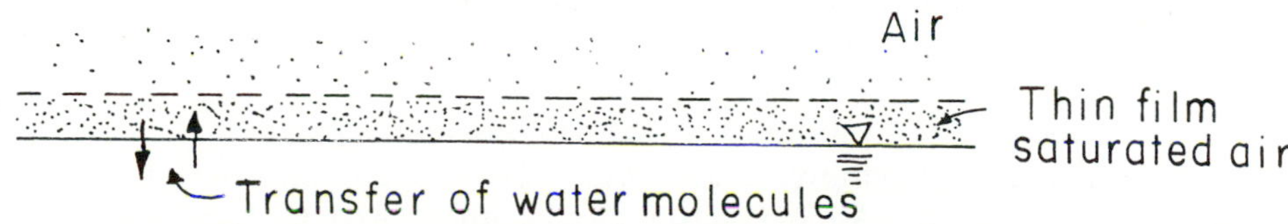

Figure 10.1 Thin-film concept of evaporation.

vapor and little further increase in water vapor can occur between the water and the saturated film next to the water. Water-vapor molecules can still move from this saturated film to the air so that there can still be a very small amount of transfer of water molecules from the water to the saturated film. Additional room for more molecules is made as some water-vapor molecules escape from the thin film and move upward. As a result, there is a nearly saturated thin layer of water vapor next to the liquid. Above this thin saturated layer, the density of water-vapor molecules decreases with elevation.

The rationale for this thin-layer concept can be illustrated with Dalton's law:

$$E = C(e_s - e_d) \tag{10.1}$$

where E is the evaporation rate (cm/day); e_s is the saturated vapor pressure at the temperature of the water surface; e_d is the saturated vapor pressure of air at the dewpoint, T_d, and is equivalent to the vapor pressure at the air temperature, T_a; and C is a coefficient depending on barometric pressure, wind, and other factors. The quantity $e_s - e_d$ is the saturation deficit. Figure 10.2 defines the notation for temperature, vapor pressure, and wind velocity that will be used throughout the chapter. In this figure, T_s is the temperature of the water surface, T_a is the temperature of air at height z, T_1 is the air temperature at height z_1, T_2 is the air temperature at height z_2, e_s is the saturated vapor pressure of air at the water surface, e_d is the vapor pressure of air at height z, e_1 is the vapor pressure of air at height z_1, e_2 is the vapor pressure of air at height z_2, u is the horizontal wind velocity at height z, w is the vertical wind velocity at height z, u_1 is the horizontal wind velocity at height z_1, u_2 is the horizontal wind velocity at height z_2, and q is the specific humidity at height z.

As the vapor pressure of the air becomes saturated and equals the vapor pressure of the water, evaporation becomes zero. By the same token, evaporation is a maximum when the vapor pressure of the air is low, thus producing the maximum gradient between the vapor pressure of the air and that of water. The difference in the vapor-pressure gradient between the water surface and the air varies with the height above the water surface. Thus, the rate of evaporation can be represented as

$$E = -K\frac{de}{dz} \tag{10.2}$$

where e is the vapor pressure, z is the elevation above the water surface, and K is a transfer coefficient.

The amount of water vapor that can be held in the air is a function of temperature. Figure 10.3 shows the relation between the saturated vapor pressure and temperature. It can be seen that the saturated vapor pressure is nearly zero at −30°C (25.6°F) and rises exponentially to a near maximum at about +45°C (113°F). Thus, warm air can hold far more water vapor than cold air. This relation provides the basis for understanding not only evapotranspiration, but also precipitation.

Heat energy, essential for evaporation, is derived generally from solar radiation. The relative humidity of air and the wind velocity across the water surface also influence the rate of evaporation. Near the equator, where the sea and winds are warm, the rate of evaporation is quite high. Tropical climates are typically humid, as in the East Indies, the Amazon basin, and in central Africa. Near the poles, there is little evaporation, for the cold winds cannot hold much moisture and weak sunshine

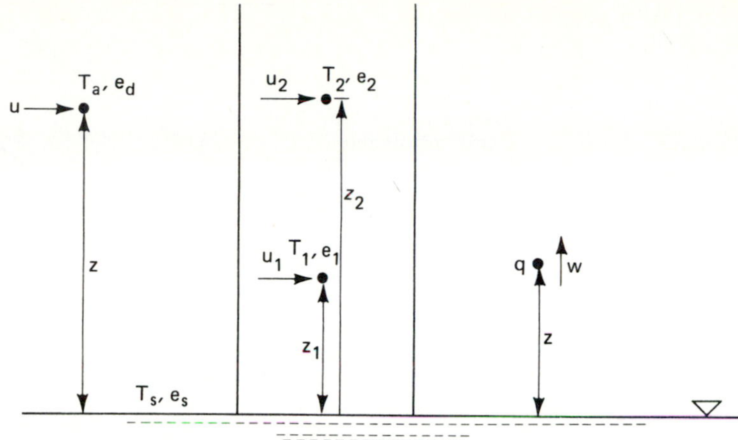

Figure 10.2 Notation for temperature, vapor pressure, and wind velocity.

transmits little heat. Artic climates are typically dry: Alaska's North Slope, despite the spongy muskeg in summer, has an arid climate.

10.2 FACTORS AFFECTING EVAPORATION

The most important factors affecting evaporation are (a) vapor-pressure difference (gradient), (b) temperature, (c) barometric pressure, (d) humidity, (e) wind, (f) water quality, and (g) water depth and soil type.

10.2.1 Vapor Pressure and Humidity

Vapor pressure and humidity are interdependent, as discussed in Chapter 6. According to Dalton's law, the vapor-pressure gradient or the difference between the saturated vapor pressure and the vapor pressure of the air (e_a) determines the rate of evaporation. The greater this difference, the greater the vapor-pressure gradient, that is, when $e_s - e_a$ is large, evaporation is large. When $e_s - e_a$ is small or when $e_s = e_a$, evaporation is small or zero.

The relation between the vapor-pressure gradient and evaporation is complicated by the temperature of the air and the water. If the water has lower temperature than the air, evaporation will be low because the vapor pressure of cold water is low, as shown in Figure 10.3. To facilitate computation, the saturation vapor pressures of water for various values of mean air temperature are listed in Table 10.1. Also listed are the slopes of the curves relating saturation vapor pressure and air temperature. When air is warmer than water, evaporation continues until the vapor pressure of air equals that of water, and then no further evaporation occurs. If air is colder than water and the vapor pressure of the water and vapor pressure of air are equal, the air becomes supersaturated and fog develops.

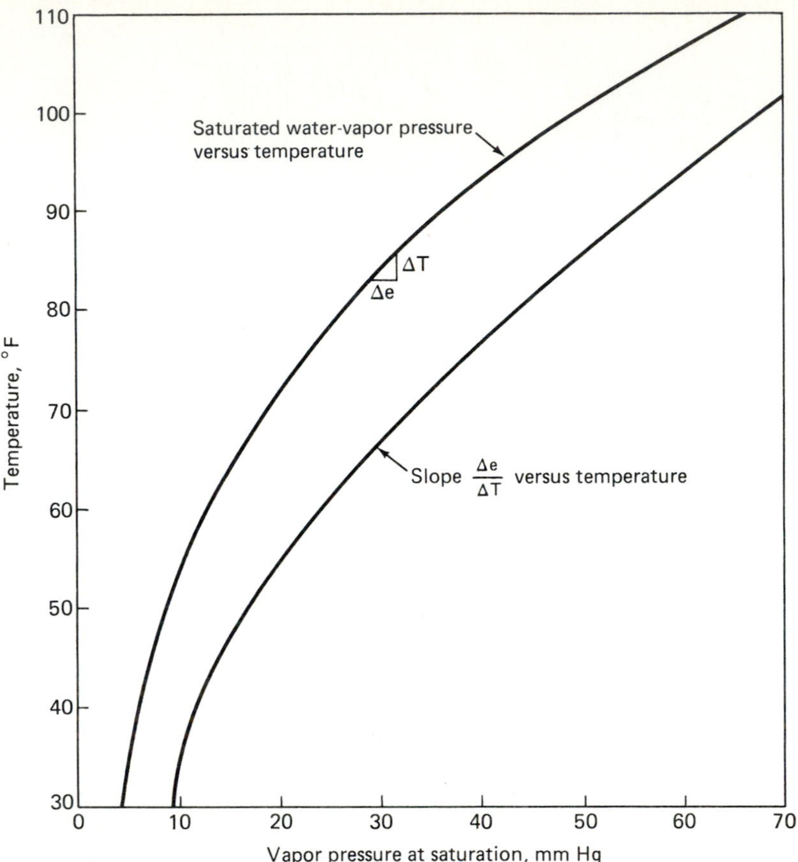

Figure 10.3 The relationship between temperature (=T) and saturation vapor pressure (=e).

10.2.2 Temperature

The movement of water molecules increases with temperature. The kinetic energy of water molecules increases as temperature rises and this increased energy permits molecules to escape from liquid water to the air more rapidly than otherwise. For this reason, the warmer the water, the more active are the molecules and the greater is the transfer of molecules from the water to the air. Experiments have shown that evaporation increases as water is warmed (Fortier, 1907). These experiments have been confirmed by maps showing decreased evaporation at higher latitudes (Horton, 1943a).

Of course, temperature and other factors are interrelated, and it is not possible to isolate the effect of each factor on evaporation. A demonstration of this interrelation is shown in Figure 10.4. This figure shows that the mean rate of evaporation from a U.S. Weather Bureau Class A pan at Davis, California, is the same for months of different temperatures. Evaporation increases from the colder months of

TABLE 10.1 SATURATION VAPOR PRESSURE OF WATER

Temperature		Saturation vapor pressure, e_s		Slope
°C	°F	mb	mm of Hg	(mm Hg/°F)
0	32	6.11	4.58	0.30
5.0	41.0	8.72	6.54	0.45
7.5	45.5	10.37	7.78	0.54
10.0	50.0	12.28	9.21	0.60
12.5	54.5	14.49	10.87	0.71
15.0	59.0	17.05	12.79	0.80
17.5	63.5	20.00	15.00	0.95
20.0	68.0	23.38	17.54	1.05
22.5	72.5	27.25	20.44	1.24
25.0	77.0	31.67	23.76	1.40
27.5	81.5	36.71	27.54	1.61
30.0	86.0	42.42	31.82	1.85
32.5	90.5	48.89	36.68	2.07
35.0	95.0	57.07	42.81	2.35
37.5	99.5	64.46	48.36	2.62
40.0	104.0	73.14	55.32	2.95
45.0	113.0	94.91	71.20	3.66

December and January to the warmest month of July. It is not clear why the evaporation is as high during the first half of the year, when the temperature is lower, as it is during the second half of the year, when the temperature is higher. It is possible that the vapor pressure of the air, wind velocity, atmospheric pressure, and other factors are sufficiently different to account for the same evaporation at different

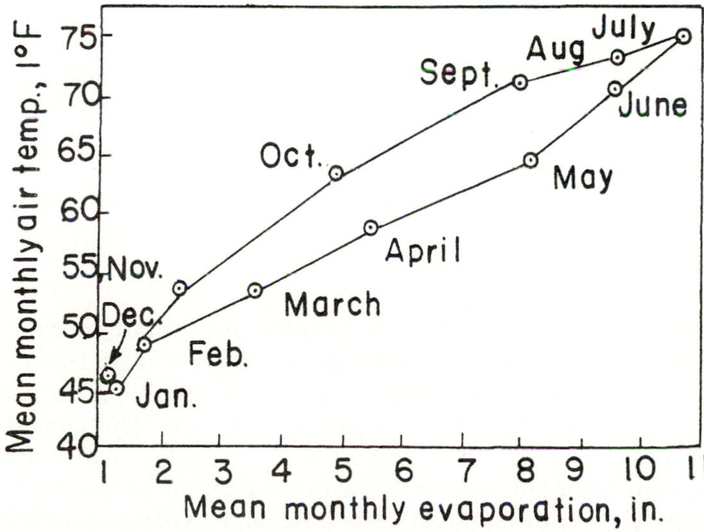

Figure 10.4 Evaporation from the U.S. Weather Bureau Class A pan at Davis, California.

monthly temperatures. This illustrates the complexity of evaporation affected by interrelated factors.

EXAMPLE 10.1 Select 10 days with a wide range of temperatures for the city of Baton Rouge, Louisiana. Show the pan evaporation for those days.

Solution The data on temperature and pan evaporation are taken from *Climatological Data,* U.S. Dept. of Commerce, National Oceanic and Atmospheric Administration Environmental Data Service.

DAY	min. °F	max. °F	PAN EVAPORATION (in.)
February 10, 1971	37	75	0.24
March, 3, 1971	44	82	0.11
April 6, 1971	40	79	0.32
May 4, 1971	53	85	0.27
June 14, 1971	66	101	0.29
July 12, 1971	72	102	0.33
August 30, 1971	69	99	0.23
September 20, 1971	64	91	0.10
October 12, 1971	47	79	0.13
November 10, 1971	41	76	0.00

Due to the effect of other factors such as relative humidity, vapor pressure, and wind speed, pan evaporation is not directly related to the temperature.

10.2.3 Atmospheric Pressure

Because fewer air molecules are present in less dense air (lower atmospheric pressure), there is less likelihood of escaping water molecules colliding with air molecules. Consequently, evaporation is greater at lower atmospheric pressures than at higher atmospheric pressures. Atmospheric pressure decreases with altitude and, therefore, evaporation is higher at higher altitudes. This effect is somewhat offset by decreasing temperatures with increasing altitude, and, as a consequence, the relation between altitude and evaporation is somewhat obscured. It has been shown that evaporation does increase in lakes and in evaporation pans with increase in elevation (Meyer, 1944; California Department of Public Works, 1947). Other factors affecting evaporation also exist under conditions of changing atmospheric pressure.

10.2.4 Wind

It was shown in Figure 10.1 that a thin film of saturated vapor would exist over a water surface under certain conditions. If this film were left undisturbed, it would act as an insolating buffer between the water surface and the unsaturated air above and evaporation would then slow down to a low rate. This condition seldom exists in nature because wind tends to remove the saturated film and reexpose the water surface to unsaturated air such that a vapor-pressure gradient exists providing favorable conditions for evaporation. The relation between a water surface and wind drag on that surface is complex. Evaporation is believed to increase with wind velocity until the vapor-pressure gradient reaches some nearly constant relation. At this point, a further increase in wind velocity will not increase evaporation appreciably. The effect of increased wind velocity on evaporation is believed to be related to the size of the water body. Wind removes water vapor from small bodies of water rather quickly, whereas on large bodies of water, the effect of wind requires a longer time to accomplish the same effect.

10.2.5 Water Quality

The vapor pressure of water is reduced when solids are dissolved in water. Pure water has a higher vapor pressure than salt water. Because there is a wide range of water purity in nature, it would not be possible to predict the effect of dissolved minerals in any given body of water on evaporation. The effect is probably small. A salt content of 1% slows the rate of evaporation by about 1%. Since the ocean has generally a salt content of a little over 3%, sea-water evaporation is lower than the freshwater rate by about 3%.

10.2.6 Water Depth and Soil Type

Temperature rises increase evaporation rates. Shallow-water bodies are more rapidly heated than are deep-water bodies and are, therefore, evaporated more rapidly. Precipitation that is ponded on a drainage basin and intercepted precipitation are evaporated rapidly after rainfall when insolation provides renewed heating. The effect of this heating is increased because dark soils and other dark objects absorb incoming radiation and convert it to heat more effectively than lighter-colored soils and objects. For this reason, soils are an important factor affecting evaporation. Solar radiation penetrates shallow water bodies and is absorbed by the soil forming the pond bottoms. This absorption results in rapid heating of the water body. This same process operates even more effectively on rainfall that is intercepted by soils and by other objects.

This discussion on evaporation is based on the assumption that a free water surface from which evaporation can occur exists. This condition exists on the oceans, lakes, and other permanent bodies of water and perennial streams. Such a condition also exists on abstract surfaces that are wet with rain, such as soil, vegetation, buildings; artificial surfaces, such as parking lots and roads; and animal life. Evaporation also occurs beneath the soil surface as heat energy penetrates and vaporizes the moisture contained on and between the soil grains. It is possible for evaporation to penetrate the soil to a depth of more than 1 foot if enough time elapses between rainfall events.

10.3 MEAN ANNUAL EVAPORATION

The National Weather Service maintains stations throughout the United States that measure evaporation rates. These data are published in the Monthly Climatological Summary for each state. Figure 10.5 is a map of the average annual lake evaporation for the United States (Kohler et al., 1959). The greatest annual evaporation, about 86 in. (218 cm), occurs in the southwestern part of the contiguous United States.

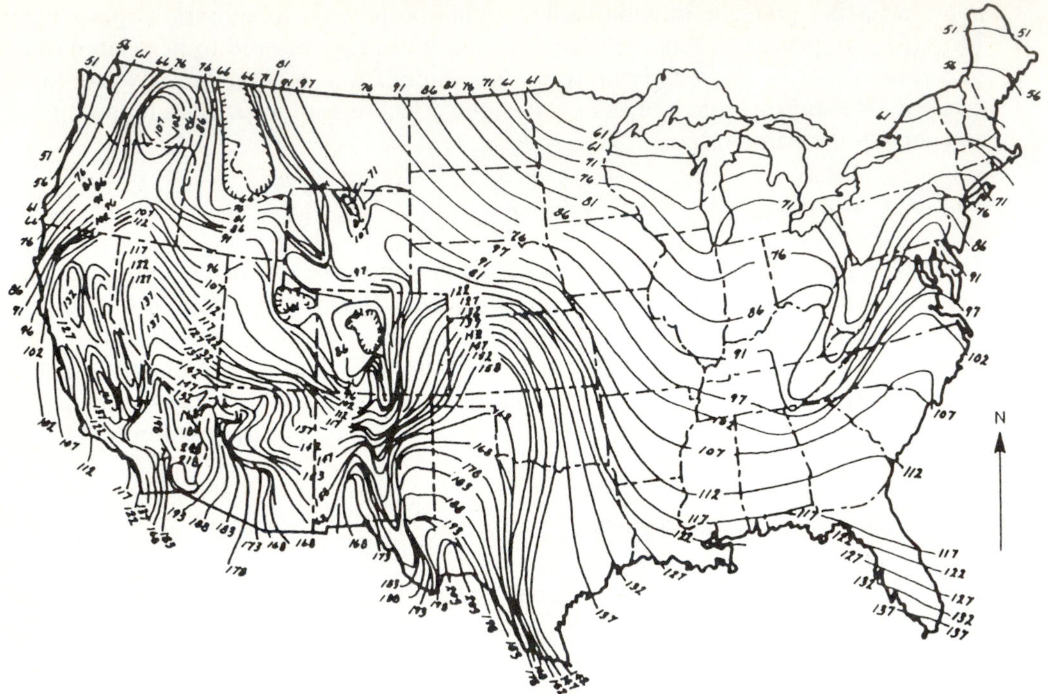

Figure 10.5 Average annual lake evaporation in the United States (after Kohler et al., 1959).

Lake Mead on the Colorado River would have about this amount of evaporation. This amount of evaporation represents a tremendous loss of water that can otherwise be used for beneficial purposes. The lowest evaporation rate, 20 in. (51 cm), on this map are in Maine and Washington. Topography, latitude, and climate, together with the evaporation-controlling factors, account for the wide variation in evaporation rates.

10.4 LAKE EFFECT

Water absorbs and stores heat energy. A larger water body stores greater energy than a smaller one. Lakes and ponds have varying effects on evaporation, depending on the depth of the water body. A deep lake, such as the Great Lakes, enters the spring season after the ice has melted with a temperature at the surface of 32°F (0°C). As the air temperature rises with the onset of warm weather, the water-surface temperature also rises as the water body absorbs insolation. The weight of water increases with a rise in temperature up to 39.2°F (4°C)—the temperature of maximum density of water—and sinks to the bottom, displacing colder but lighter water. This turnover continues until all the water in the deep lake is at 39.2°F (4°C). From this point in the process on, the surface water continues to warm and becomes less dense, and remains on the surface. Any mixing of this warmer surface water and colder water from below occurs as a result of wave action, lake currents, and conduction. The lake is a heat sink during this time in which it stores heat and little evaporation occurs. With the air temperature cooling in the fall, this process is reversed in the Great Lakes region. During the fall and winter, the lake becomes a heat source and encourages evaporation as the vapor-pressure gradient is increased. The vapor pressure of the warm lake is much higher than the cold air, thus providing favorable conditions for evaporation. The warm lake vapor provides moisture that results in large snowfall amounts as northwest winds drive the moisture onto cold land adjacent to the southeast boundary of the lake.

Shallow lakes and ponds are heated by the sun to the bottom, raising the water temperature quickly, and increasing evaporation with rising temperature. Such water bodies, including ponding on a drainage basin, are rapidly evaporated because they have little heat storage. The rate of evaporation is largely determined by the depth of the water body. Evaporation rates are higher for shallower water bodies.

10.5 APPLICATION OF EVAPORATION IN HYDROLOGY

Evaporation is small during a runoff event, for the vapor-pressure gradient is low. This is because when rainfall occurs, the vapor pressure of the air is high and the relative humidity is high. The amount of water vapor contained in the atmosphere is high relative to the amount of moisture it can hold at the prevailing temperature. As a result, little evaporation occurs. Even if the rainfall event continues over several days, significant evaporation cannot occur because of the low vapor-pressure gradient. In general, a runoff event from a convective storm lasts only a few hours, and this short time, coupled with a low vapor-pressure gradient, would result in very small, even negligible, evaporation. Although a cyclonic storm has a longer duration, evaporation occurring over its duration and the duration of runoff is still small and can be neglected.

Evaporation becomes important when long-term storage for water supply is the primary purpose of the study. The average annual lake evaporation in Figure 10.5 shows that central Kansas can expect an evaporation of about 55 in. (140 cm) of water. This amounts to an average of about 1 in. (2.54 cm) per week. Evaporation from storage reservoirs, therefore, can be large and represents a direct loss for the project purpose. Table 10.2 shows loss of water due to evaporation from some reservoirs in the United States. Evaporation from soils in the drainage basin is important because this antecedent moisture deficit must be satisfied before any significant runoff will occur. On the other hand, irrigation of agricultural lands is scheduled, on the basis of the moisture deficit, to alleviate plant stress. The rate of evaporation is related to the soil type as well as the rate of replenishment of moisture in the soil from rainfall or water delivery.

By suppressing evaporation, loss of usable water can be prevented. Oil and chemicals (cetyl alcohol) have been used to reduce the rate of evaporation from water surfaces. Less than 10 grams of this chemical can make a film covering about 43,000 ft^2 (4000 m^2) of water surface. If the film can remain unbroken over the entire surface of a reservoir, as much as 70% reduction in evaporation can be effected. In water-scarce areas, chemical films can be used to suppress evaporation. In some

TABLE 10.2 LOSS OF WATER DUE TO EVAPORATION FROM SOME RESERVOIRS IN THE UNITED STATES[a]

Reservoir	Surface area		Annual evaporation		Water loss		Number of people ($\times 10^6$) to be supplied water at 500 liters/day		
	Acres	Hectares	cm	in.	$\times 10^6$ liters	$\times 10^6$ gallons	Daily	Monthly	Yearly
Elephant Butte (Rio Grande River), New Mexico	13,760	5,568.67	190.5	75	106,083.16	28,023.99	212.17	7.07	0.589
Lake Mead (Colorado River), Nevada	124,200	50,263.74	190.5	75	957,524.25	252,949.18	1915.05	63.83	5.32
Oahe (Missouri River), South Dakota	375,680	152,037.7	113.8	44.8	1,729,300	45,653.5	3459.0	115.3	9.61
Garrison (Missouri River), North Dakota	390,400	157,994.9	77.1	30.34	1,217,000	321,288	2434	81.1	6.76
Fort Peck (Missouri River), Montana	244,480	98,341.1	87.12	34.3	861,600	227,462.4	1723	57.4	4.79

F. D. Roosevelt Grand Coulee (Columbia River), Washington	83,200	33,671.0	123.2	48.5	414,600	109,454.4	829.2	27.6	2.3
Kentucky (Tennessee River), Kentucky	260,480	105,416.3	74.9	29.5	789,500	208,428	1579	52.64	4.39
Lake Texoma (Denison) at Red River, Oklahoma	142,720	57,758.8	101.9	40.12	588,300	155,311.2	1177	39.22	3.27
Bull Shoals (White River), Arkansas	71,040	28,749.9	79.2	31.2	227,740	60,123.4	455.5	15.18	1.27
Shashta (Sacramento River), California	29,440	11,914.4	133.6	52.6	159,100	42,002.4	318.2	10.61	0.88
Clark Hill (Savannah River), Georgia	78,720	31,858	130.3	51.3	414,900	109,533.6	829.9	27.66	2.30

[a] 1 month = 30 days.

areas (e.g., South West Africa), sand storage dams have also been used to save runoff water that would otherwise evaporate.

Knowledge of the evaporative process can be used beneficially to dispose of unwanted waste and to cool water heated during industrial processes. Evaporation ponds can serve as an efficient means to dispose of waste. If the pond is shallow, the rate of evaporation is generally high. Of course, ponds must be properly lined to prevent the waste water from percolating down to the underlying groundwater. In the San Joaquin Valley in California, evaporation ponds are commonly used to dispose of oil-field waste water as well as agricultural waste water.

Evaporative cooling is widely used in industry to cool water. Many industrial sites, such as steel mills, power stations, oil refineries, natural gas compressor stations, etc., discharge large quantities of waste heat, and, therefore, require cooling systems as part of their operation. Cooling towers, spray ponds, and similar techniques that use evaporation for cooling are commonly employed.

10.6 MEASUREMENT OF EVAPORATION

Evaporation is usually measured by evaporation pans and atmometers. Evaporation pans are most widely used for measuring evaporation. The use of an evaporation pan is as follows. Water is placed in a relatively small, shallow pan and the water level is measured. Water levels in the pan are measured using a hook gage or a similar device by which the levels can be measured accurately. Water is added daily to keep the levels within a limited range. The change in the water level measures the evaporation, which is then converted to the equivalent value for evaporation from a lake. The rate of evaporation from a pan is not the same as from a lake, and, for this reason, a conversion coefficient, called the pan coefficient, must be determined to make the two values compatible. Evaporation from the pans is affected by the size, depth, and location of the pan. Evaporation observations are easy to make using an evaporation pan.

10.6.1 The U.S. Weather Bureau Class A Land Pan

The Class A pan, as shown in Figure 10.6, is 4 ft (1.22 m) in diameter and 10 in. (25.4 cm) deep. The pan is located so that its bottom is raised 6 in. (15.4 cm) above the ground surface. The pan is filled with water to a level at least 2 in. (5.1 cm) below the top and not more than 3 in. (7.6 cm) below the top of the pan. The decrease in the water level in the pan within a specified time is the amount of evaporation. The conversion coefficient to be applied to this pan evaporation is 0.7. It actually ranges from 0.6 to 0.8. The disadvantages of this pan are that its coefficient is less than unity and the data obtained using this pan vary in time and space. Its advantages are that (a) more data are available from using this pan, (b) the coefficient is quite stable, (c) it is easily accessible, (d) it has a reasonable cost, and (e) it is located above much blowing dirt and snow.

The U.S. Weather Bureau Class A land pan is generally recommended for evaporation studies. It is the most widely used and is quite simple and easy to install. Its wide use provides a large amount of comparable data. Pan measurements

Figure 10.6 U.S. Weather Bureau Class A land pan.

are satisfactory because they approximate lake evaporation reasonably close and have the advantage of less cost and labor. One wishing to use evaporation data can obtain the necessary information from pan records kept by the National Weather Service, from evaporation maps, or from supplemental pan observations.

EXAMPLE 10.2 Tabulate the mean monthly Class A pan evaporation values for Omaha, Nebraska, for the period May through October. These evaporation values will apply to the Irvington drainage basin.

Solution The pan evaporation values are given in Fornsworth and Thompson (1982).

MEAN MONTHLY CLASS A PAN EVAPORATION (in).	MONTH	NO. OF YEARS	COEFFICIENT OF VARIATION (%)
7.8	May	20	14
8.82	June	22	10
8.7	July	22	10
7.94	August	22	7
5.75	September	23	13
4.81	October	17	13

This record began June 1958 and continued to September 1979.

10.6.2 The U.S. Bureau of Plant Industry Sunken Pan

This sunken pan is 6 ft (1.8 m) in diameter and 2 ft (0.6 m) deep. It is buried in the ground to a depth within 4 in. (10 cm) of its top. The water level is to be kept no more than 0.5 in. (1.25 cm) above or below the ground level. The conversion coefficient for this pan is 0.95. It is slightly smaller in the spring and slightly larger in the fall. Because this coefficient is near unity, it is considered the best index. It is little used, however, because of ifs difficulty in installation and because it tends to gather blowing dust and snow.

10.6.3 The Colorado Sunken Pan

This pan, developed by the Colorado Experiment Station, is 3 ft (0.9 m) square and has a depth that can range from 18 in. (46 cm) to 36 in. (0.9 m). It is buried in the ground to within 4 in. (10 cm) of the top. The water level in the pan is to be maintained within 1 in. (2.5 cm) of the ground level. The coefficient for this pan ranges from 0.75 to 0.86; the mean value of 0.78 is used.

10.6.4 The U.S. Geological Survey (USGS) Floating Pan

The USGS floating pan is based on the theory that evaporation from a floating pan on a body of water would be nearly the same as from the surrounding water. This pan is 3 ft (0.9 m) square by 18 in. (46 cm) deep It is supported by gimbals in the center of a drum float raft that has dimensions of 14 ft (4.27 m) by 16 ft (4.9 m). The pan is submerged in the water so that the sides project 3 in. (7.5 cm) above the water surface. The water level in the pan is kept at nearly the same level as the surrounding water. The coefficient for this pan ranges from 0.78 to 0.82. The recommended coefficient is 0.80.

10.6.5 Atmometers

An atmometer is an instrument for measuring evaporation. There are several kinds of atmometers, but they are, in general, composed of a porous bulb that draws water from a container. As water is evaporated from the bulb surface, more water is drawn from the container. The reduction of water in the container is correlated with evaporation by establishing proper coefficients, and then the measurement of evaporation is obtained.

EXAMPLE 10.3 Compute the daily evaporation from a Class A pan if the daily rainfall and the amount of water added to bring the water level in the pan to the fixed point are as follows:

Day	1	2	3	4	5	6
Rainfall (cm)	0	0.5	0.1	0	0	0.4
Water added (cm)	1.5	1.7	0.5	1.2	0.7	1.3

Solution The daily pan evaporation equals the amount of water required to bring its water level to the fixed point plus the water contributed by rainfall. For each day, this amount is shown as follows and varies from day to day depending upon atmospheric conditions.

Day	1	2	3	4	5	6
Evaporation (cm)	1.5	2.2	0.6	1.2	0.7	1.7

EXAMPLE 10.4 If a lake has a 500-hectare surface area, compute the daily lake evaporation for the data in Example 10.3. Assume the pan coefficient is 0.8. Compute the 6-day evaporation and loss of water in kilograms or tons.

Solution The daily lake evaporation is obtained by multiplying the daily pan evaporation by the pan coefficient:

Day	1	2	3	4	5	6
Pan evaporation (cm)	1.5	2.2	0.6	1.2	0.7	1.7
Lake evaporation (cm)	1.2	1.76	0.48	0.96	0.56	1.36

The 6-day evaporation is the sum of the daily values for 6 days = 6.32 cm. The 6-day loss of water = (6.32 m/100) × 500 × 10,000 m^2 = 316,000 m^3. Since 1 m^3 = 1000 liters, the 6-day loss = 316 × 10^6 liters. The density of water is approximately 1000 kg/m^3. Therefore, the 6-day loss of water in kilograms is 316 × 10^6 kg. One metric ton has 1000 kg. Therefore, the 6-day loss of water in tons is 316 × 10^3 metric tons.

EXAMPLE 10.5 Compute the mean daily evaporation loss in hectare-meters for the month of July from a stream reach 100 km long and 50 m wide, on the average. The mean daily evaporation measured by a Class A pan for July is 0.6 cm. Assume the pan coefficient as 0.8.

Solution

$$\text{Mean daily evaporation} = 0.6 \text{ cm} \times 0.8 = 0.48 \text{ cm}$$

This evaporation takes place from the entire stream reach whose area is 100,00 m × 50 m = 5 × 10^7 m^2, or 500 ha. Therefore,

$$\text{Daily evaporation} = \frac{500 \text{ ha} \times 0.48 \text{ m}}{100} = 2.4 \text{ ha-m}$$

This loss of water due to evaporation must be considered in managing water of this stream reach.

EXAMPLE 10.6 It is desired to estimate evaporation from a lake on June 24, 1986. A Class A pan is located near the lake. On this day, the rainfall amount is 0.5 cm and the amount of water added to restore the water level to the value at the beginning of that day is 0.7 cm. Assume the pan coefficient as 0.8.

Solution

$$\text{Class A evaporation} = 0.5 \text{ cm} + 0.7 \text{ cm} = 1.2 \text{ cm}$$

$$\text{Evaporation from the lake} = 1.2 \text{ cm} \times 0.8 = 0.96 \text{ cm}$$

Thus, on June 24, the water level declined by 0.96 cm.

10.7 DETERMINATION OF EVAPORATION FROM WATER SURFACES

Evaporation from water surfaces can be determined by (1) the water budget, (2) the energy budget, (3) mass transfer methods, (4) combination methods, and (5) evaporation formulas.

10.7.1 Water Budget

The water-budget equation for estimating evaporation (Horton, 1943b) can be written as

$$E = I + P - O - O_s + \Delta S \tag{10.3}$$

where E is the evaporation, I is the inflow, P is the precipitation, O is the outflow, O_s is the seepage, and ΔS is the change in storage. Seepage, O_s, cannot be measured or evaluated directly and accurately, and the extent to which this quantity is accurate will affect the true value of evaporation. Inflow, outflow, precipitation, and change in storage can be measured reasonably accurately. This equation has been used to measure evaporation from reservoirs such as Lake Hefner in Oklahoma and Elephant Butte in New Mexico (Harbeck et al., 1954). The water-budget method of determining long-term evaporation can be used as a standard for comparing other methods. This method is not perfect, but it is satisfactory for practical purposes.

EXAMPLE 10.7 Estimate the evaporation for a month for a lake of 500 hectare surface area. The mean discharge from the lake is estimated to be 1.00 m^3/s. The monthly rainfall is about 10 cm. A stream flows with an average discharge of 2.00 m^3/s into the lake. The water level in the lake dropped about 5 cm in the month. The seepage losses are negligible.

Solution

$$\text{Monthly inflow} = I = 2 \times 3600 \times 24 \times 30 \text{ m}^3 = 51.84 \times 10^5 \text{ m}^3$$

$$\text{Monthly outflow} = O = 1 \times 3600 \times 24 \times 30 \text{ m}^3 = 25.92 \times 10^5 \text{ m}^3$$

$$\text{Monthly rainfall} = P = \frac{10}{100} \times 500 \times 10{,}000 \text{ m}^3 = 5 \times 10^5 \text{ m}^3$$

$$\text{Change in storage} = \Delta S = \frac{5}{100} \times 500 \times 10{,}000 \text{ m}^3 = 2.5 \times 10^5 \text{ m}^3$$

By applying Equation (10.3) for an estimation of monthly evaporation,

$$E = (51.84 + 5.0 - 25.92 + 2.5) \times 10^5 \text{ m}^3 = 33.42 \times 10^5 \text{ m}^3$$

$$= \frac{33.42 \times 10^5}{500 \times 10^4} \times 100 \text{ cm} = 66.84 \text{ cm}$$

10.7.2 Energy Budget

The energy-budget method for determining evaporation is similar to the water-budget method except that the energy budget deals with the conservation of energy rather than water (Richardson, 1931; Cummings, 1935). The energy available for evaporation is obtained by considering the incoming energy, the outgoing energy, and the energy stored in the water body for a given time interval (Harbeck et al., 1954).

The energy balance for a water body, shown in Figure 10.7, can be expressed as

$$H_s = H_n + H_{ai} - H_{ao} - H_g - H_e - H_w \tag{10.4}$$

where H_s is the increase in the energy storage in the water body, H_n is the net radiation and equals $H_c - H_r - H_b$, H_c is the incoming solar radiation, H_r is the reflected radiation, H_b is the back radiation (long-wave) from the water body, H_{ai} is

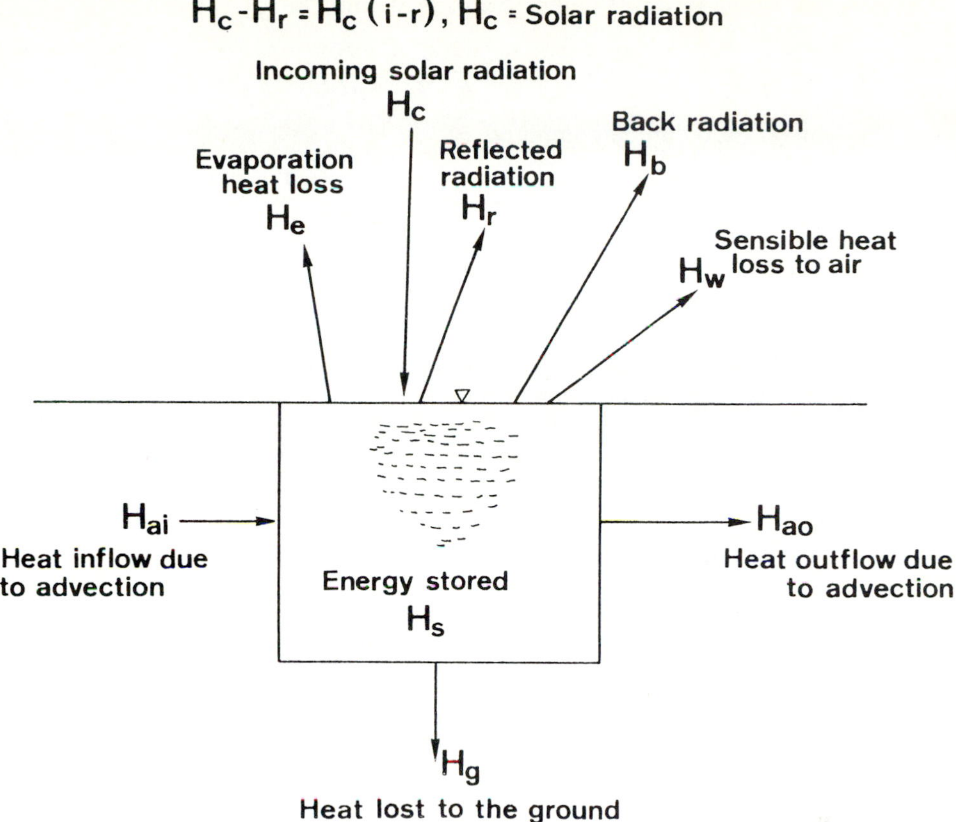

Figure 10.7 Energy budget for a water body.

the heat inflow into the water body due to advection, H_{ao} is the heat outflow from the water body due to advection, H_g is the heat loss from the water body to the ground by conduction, H_e is the latent heat flux due to evaporation (negative) or condensation (positive) and equals ρEL, ρ is the density of water (cm^3/g), L is the latent heat of vaporization (cal/g), E is the rate of evaporation (cm/day), and H_w is the sensible heat loss from the water body to the atmosphere. These energy terms are expressed in cal/cm^2 per day, or langleys.

The quantity $H_c - H_r$ represents the incoming solar radiation to the water body with a reflection coefficient, or albedo, of r $(0 \leq r \leq 1)$, and can be expressed as $H_c - H_r = H_c(1 - r)$, where $H_r = rH_e$, and H_c is the solar radiation. The quantity $H_{ai} - H_{ao} = H_a$ represents the net advected energy. Thus, Equation (10.4) can be rewritten as

$$H_s = H_n - H_a - H_g - H_e - H_w \tag{10.5}$$

For a soil column, $H_a = 0$. For short periods of time, H_a and H_s can be negligibly small. All energy terms except H_w can be measured or evaluated indirectly. The sensible heat term H_w, which is not amenable to easy measurement, is estimated using the Bowen ratio (Bowen, 1926), B, defined as

$$B = \frac{H_w}{H_e} = \frac{H_w}{\rho LE} \tag{10.6}$$

The Bowen ratio can be computed as

$$B = 0.61 \frac{p}{1000} \frac{T_o - T_a}{e_o - e_a} \tag{10.7}$$

where p is the atmospheric pressure (millibars), T_a is the temperature of the air (°C), T_o is the water-surface temperature (°C), e_a is the vapor pressure of the air (millibars), and e_o is the saturation pressure (millibars) corresponding to T_o. Bowen (1926) found 0.58 and 0.66 as the limiting values for B, and recommended a value of 0.61 under normal atmospheric conditions.

From Equations (10.5) and (10.6), E (cm/day) can be expressed as

$$E = \frac{H_n - H_a - H_g - H_s}{\rho L(1 + B)} \tag{10.8}$$

The term H_a can be expressed as

$$H_a = \frac{c_p H_e (T_e - T_b)}{L} \tag{10.9}$$

where c_p is the specific heat of water (cal/g °C), T_e is the temperature of evaporated water (°C), and T_b is the temperature at an arbitrary datum usually taken as 0°C.

The net radiation H_n can be computed as

$$H_n = R_s(1 - r) - R_L \tag{10.10}$$

where R_s is the solar (sun and sky) radiation (short-wave) received at the earth's surface (cal/cm^2-day); $(1 - r)R_s$ is the net short-wave radiation (cal/cm^2-day); r is the short-wave reflectance, or albedo, which equals 0.05 to 0.15 for water; and R_L is the net outgoing thermal radiation (cal/cm^2-day).

The quantity R_s can be computed as

$$R_s = I_0\left(a + b\frac{n}{N}\right) \tag{10.11}$$

where I_0 is the solar radiation received at the earth's outer surface or the earth's atmosphere (cal/cm^2-day), n is the actual number of hours of bright sunshine, N is the possible maximum number of sunshine hours, a is a constant depending upon the region and has been reported to vary from 0.18 to 0.34, and b is a constant and has been reported to vary from 0.52 to 0.56. Criddle (1958), and Jensen (1973) have tabulated values of I_0 for various latitudes (°N) and months of the year, as shown in Table 10.3. The values of N are listed in Table 10.4.

Penman (1948) suggested an expression for R_s in mm of water/day as

$$R_s = \left(0.18 + 0.55\frac{n}{N}\right)I_0 \tag{10.12}$$

TABLE 10.3 MEAN MONTHLY SOLAR RADIATION INCIDENT AT THE EARTH'S OUTER SPACE (EXTRATERRESTRIAL RADIATION), I_0, IN MM OF EVAPORABLE WATER/DAY, IN NORTHERN HEMISPHERE WITH L = 560 CAL/G (AFTER CRIDDLE, 1958)

	North latitude (°N)									
Month	90°	80°	70°	60°	50°	40°	30°	20°	10°	0°
January	—	—	—	1.3	3.6	6.0	8.5	10.8	12.8	14.5
February	—	—	1.1	3.5	5.9	8.3	10.5	12.3	13.9	15.0
March	—	1.8	4.3	6.8	9.1	11.0	12.7	13.9	14.8	15.2
April	7.9	7.8	9.1	11.1	12.7	13.9	14.8	15.2	15.2	14.7
May	14.9	14.6	13.6	14.6	15.4	15.9	16.0	15.7	15.0	13.9
June	18.1	17.8	17.0	16.5	16.7	16.7	16.5	15.8	14.8	13.4
July	16.8	16.5	15.8	15.7	16.1	16.3	16.2	15.7	14.8	13.5
August	11.2	10.6	11.4	12.7	13.9	14.8	15.3	15.3	15.0	14.2
September	2.6	4.0	6.8	8.5	10.5	12.2	13.5	14.4	14.9	14.9
October	—	0.2	2.4	4.7	7.1	9.3	11.3	12.9	14.1	15.0
November	—	—	0.1	1.9	4.3	6.7	9.1	11.2	13.1	14.6
December	—	—	—	0.9	3.0	5.5	7.9	10.3	12.4	14.3

Another expression for R_s in mm of water/day has been reported by Black et al. (1954) as

$$R_s = \left(0.23 + 0.48\,\frac{n}{N}\right)I_0 \tag{10.13}$$

More accurately, however, R_s can be estimated from measured values of I_0 for a specific location and associated values of a, b, n, and N.

The quantity R_L can be estimated as

$$R_L = R_0\left(a_1 + \frac{R_s}{R_{s0}}\,b_1\right) \tag{10.14}$$

where R_0 is the net outgoing long-wave radiation on a clear day (cal/cm^2-day), R_{s0} is the solar radiation expected under cloudless or clear sky conditions (cal/cm^2-day),

TABLE 10.4 MEAN MONTHLY VALUES OF POSSIBLE SUNSHINE HOURS, N

Latitude north (°N)	Month											
	Jan.	Feb.	Mar.	Apr.	May	June	July	Aug.	Sept.	Oct.	Nov.	Dec.
0°	12.1	12.1	12.1	12.1	12.1	12.1	12.1	12.1	12.1	12.1	12.1	12.1
10°	11.6	11.8	12.1	12.4	12.6	12.7	12.6	12.4	12.9	11.9	11.7	11.5
20°	11.1	11.5	12.0	12.6	13.1	13.3	13.2	12.8	12.3	11.7	11.2	10.9
30°	10.4	11.1	12.0	12.9	13.7	14.1	13.9	13.2	12.4	11.5	10.6	10.2
40°	9.6	10.7	11.9	13.2	14.4	15.0	14.7	13.8	12.5	11.2	10.0	9.4
50°	8.6	10.1	11.8	13.8	15.4	16.4	16.0	14.5	12.7	10.8	9.1	8.1

and a_1, b_1 are constants. Penman (1948) has shown that

$$R_L = \left(0.1 + 0.9\,\frac{n}{N}\right)R_0 \tag{10.15}$$

The quantity R_0 is expressed as

$$R_0 = \sigma T^4(\varepsilon - 1) \tag{10.16}$$

where σ is the Stefan–Boltzmann constant = 11.71×10^{-8} cal/cm^2-°K^4-day or 2.0×10^{-9} mm of water/day; T is the temperature of the water surface (°K); and ε is the emissivity, defined as the ratio of the long-wave radiation down from the atmosphere to that upward from the earth, and can be expressed as

$$\varepsilon = a_2 + b_2(e_a)^{0.5} \tag{10.17}$$

where e_a is the mean vapor pressure in the air (millibars) usually at a height of 2 meters, and a_2, b_2 are constants. The value of a_2 has been reported to vary from 0.43 to 0.68 and that of b_2 between 0.029 and 0.082. The value of ε varies from 0.906 to 0.985 for water surfaces.

By substituting Equation (10.16) into Equation (10.15),

$$R_L = \sigma T^4(\varepsilon - 1)\left(0.1 + 0.9\,\frac{n}{N}\right) \tag{10.18}$$

Penman (1948) expressed R_L in mm of water as

$$R_L = \sigma T^4[0.56 - 0.092(e_a)^{0.5}]\left[0.1 + 0.9\,\frac{n}{N}\right] \tag{10.19}$$

The energy-budget method aids in understanding the evaporation process. This method also requires an approximate water budget, because inflow, outflow, and storage of water represent energy values that must be considered along with their respective temperatures. With improving instrumentation, the energy method has the potential to be used on a broad scale in the future.

EXAMPLE 10.6 Estimate the solar radiation incident at the earth's surface for the month of June for a place at latitude 50°N. The mean observed number of sunshine hours is 12.

Solution The value of N from Table 10.4 is 16.4 hours. The value of I_0 from Table 10.3 is 16.7 mm of water/day. Therefore, by using Equation (10.12),

$$R_s = \left(0.18 + 0.55\,\frac{12}{16.4}\right)16.7 = 10.73 \text{ mm of water/day}$$

If Equation (10.13) is used, then

$$R_s = \left(0.23 + 0.48\,\frac{12}{16.4}\right)16.7 = 10.71 \text{ mm of water/day}$$

EXAMPLE 10.9 Estimate the long-wave radiation for a day in June for a place at 50°N latitude. The mean observed number of sunshine hours is 12. The mean air temperature is 30°C. Take $\varepsilon = 0.95$.

Solution

$$\sigma = 1.17 \times 10^{-7}\ \text{cal/cm}^2/\text{K}^4/\text{day}$$

$$T = 30 + 273 = 303°\text{K}$$

$$n = 12\ \text{h}$$

$$N = 16.4\ \text{h}$$

From Equation (10.18),

$$R_L = 1.17 \times 10^{-7} \times (0.95 - 1) \times (303)^4 \times \left(0.1 + 0.9\,\frac{12}{16.4}\right)$$

$$= -37.40\ \text{cal/cm}^2/\text{day}$$

EXAMPLE 10.10 Compute the emissivity if the temperature is 30°C, the saturation vapor pressure is 80 mb, and the relative humidity is 20%. Take $a_2 = 0.52$ and $b_2 = 0.065$.

Solution

$$\text{Vapor pressure, } e_a = \text{saturation pressure} \times \text{relative humidity}$$

$$= 80 \times 0.2 = 16\ \text{mb}$$

Therefore, by using Equation (10.17),

$$\varepsilon = 0.52 + 0.065(16)^{0.5} = 0.52 + 0.26 = 0.78$$

10.7.3 Mass-Transfer Methods

The simplest mass-transfer method employs Equation (6.1)—Dalton's law—with C expressed as a function of wind speed:

$$E = f(u)(e_s - e_d) \tag{10.20}$$

where E is the evaporation rate, $f(u)$ is a function of wind speed u, and $e_s - e_d$ is the saturation vapor-pressure deficit. Thus, evaporation is related to wind speed and is proportional to the vapor-pressure deficit, which is the difference between the saturated vapor pressure at the water-surface temperature and the vapor pressure of air at some height above the water surface. Several empirical expressions for $f(u)$ have been proposed, some simple and some fairly complex (Veihmeyer, 1964; Jensen, 1973).

A commonly used expression for $f(u)$ is due to Meyer (1944):

$$f(u) = C\left(1 + \frac{u_s}{16}\right) \tag{10.21}$$

where C is a constant that is equal to 0.36 for daily data on an ordinary lake and 0.5 for wet surfaces, and u_s is the wind speed (km/h) at an elevation of 8 m. The values of e_d and e_s in Equation (10.20) are in mb, and E in cm/day. When Equations (10.20) and (10.21) are combined, this becomes Meyer's method.

Based on the Lake Hefner studies in Oklahoma (U.S. Geological Survey, 1954),

$$f(u) = Nu \tag{10.22}$$

where u is the wind speed (m/s) at 2 m above the water surface, and N is a coefficient defined as

$$N = 0.0291A^{-0.05} \tag{10.23}$$

where A is the surface area of the water surface in m^2. Errors in computed E can be as high as 30% if Equation (10.23) is used. If A is less than 4×10^6 m^2, errors may be even larger; it is not advisable to use this equation.

EXAMPLE 10.11 Using Meyer's equation, estimate the lake evaporation for the month of June. The mean air temperature is 30°C, the water temperature is 25°C, the wind speed at 8 m above the ground is 15 km/h, and the relative humidity is 70%.

Solution The saturated vapor pressure e_s for 25°C from Table 10.1 is 23.76 mm Hg. The saturated vapor pressure at 30°C = 31.82 mm Hg. Therefore, vapor pressure = 31.82 × 0.7 = 22.274 mm Hg. From Equation (10.21),

$$f(u) = 0.36 \times \left(1 + \frac{15}{16}\right) = 0.698$$

From Equation (10.20),

$$E = 0.698 \times (23.76 - 22.274) = 1.037 \text{ cm/day}$$
$$= 1.037 \times 30 = 31.11 \text{ cm/month}$$

EXAMPLE 10.12 Estimate the daily lake evaporation using the Lake Hefner equation if the lake area is 500 ha. The wind velocity is 10 km/h and the vapor pressure and temperature are the same as in Example 10.10.

Solution From Equation (10.23),

$$N = 0.0291 \times (500 \times 10^4)^{-0.05} = 0.0291 \times 0.462 = 0.01346$$

From Equation (10.22),

$$f(u) = 0.01346 \times 10 \times \frac{1000}{3600} = 0.0374$$

From Equation (10.20),

$$E = 0.0374 \times (23.76 - 22.274) = 0.056 \text{ cm/day}$$
$$= 0.056 \times 30 = 1.667 \text{ cm/month}$$

10.7.4 Combination Methods

The combination methods combine the energy-budget and mass-transfer methods. The most popular combination method for computing evaporation from free water surfaces is that developed by Penman (1948). This method combines fundamental physical principles and empirical concepts based on standard meteorological observations. The physical principles are the energy-balance equation and the mass-transfer (or aerodynamic) equation.

In a simplified form, the energy terms H_s, H_a, and H_g can be neglected in the energy balance. Equation (10.4) can be written using a more common notation, with $H = H_n$, $E = H_e$, and $G = H_w$, as

$$H = E + G \tag{10.24}$$

where H is the net radiation energy available at the earth's surface, E is the energy used for evaporating water, and G is the energy used for heating the air. The value of G can be defined also by Dalton's law or by an aerodynamic equation such as

$$G = \gamma f_1(u)(T_s - T_a) \tag{10.25}$$

where γ is the hygrometric constant (0.27 mm of mercury/°F) to make units consistent, $f_1(u)$ is a function of wind speed, T_s is the temperature of the water surface, and T_a is the air temperature.

In practice, T_a can be easily measured, but T_s is difficult to measure. However, if it is assumed that the temperature difference, $T_s - T_a$, is small, then the slope of the curve of saturated vapor pressure against temperature can be written as

$$\Delta = \frac{de}{dT} = \frac{e_s - e_d}{T_s - T_d} \cong \frac{e_a - e_d}{T_a - T_d} \tag{10.26}$$

if the gradient is small, where e_d is the saturation vapor pressure at a small height d from the water surface, where the air temperature is T_a, and T_d is the dewpoint temperature. Equation (10.25) can be written as

$$G = \gamma f_1(u)[(T_s - T_d) - (T_a - T_d)] \tag{10.27}$$

By using Equation (9.26) in Equation (9.27),

$$\begin{aligned} G &= \gamma f_1(u)\left(\frac{e_s - e_d}{\Delta} - \frac{e_a - e_d}{\Delta}\right) \\ &= \frac{\gamma}{\Delta} f_1(u)(e_s - e_d) - \frac{\gamma}{\Delta} f_1(u)(e_a - e_d) \end{aligned} \tag{10.28}$$

Penman assumed $f_1(u) = f(u)$. There is experimental evidence to support this assumption. Therefore, Equation (10.28), with G replaced by $H - E$, becomes

$$H - E = \frac{\gamma}{\Delta} f(u)(e_s - e_d) - \frac{\gamma}{\Delta} f(u)(e_a - e_d) \tag{10.29}$$

From Equation 10.20,

$$H - E = \frac{\gamma}{\Delta} E - \frac{\gamma}{\Delta} E_a \tag{10.30}$$

where E_a expressed as

$$E_a = f(u)(e_a - e_d) \tag{10.31}$$

represents drying power of air and is highly correlated with the evaporation given by the mass-transfer method. Solving for E,

$$E = \frac{\Delta H + \gamma E_a}{\Delta + \gamma} = \frac{\Delta}{\Delta + \gamma} H + \frac{\gamma}{\Delta + \gamma} E_a \tag{10.32}$$

This is the Penman equation for free water surface evaporation. In order to apply this equation, H and E_a must be determined. The drying capacity of the air, E_a, in mm/day was experimentally determined by Penman as

$$E_a = (e_a - e_d)(0.175 + 0.0022u_2) \tag{10.33}$$

where e_a and e_d are in mm of mercury, and u_2 is the average wind velocity in km/h at a height of 2 m. u_2 can be computed from observations at any height as

$$u_2 = u_h\left(\frac{2.0}{h}\right)^{0.143} \tag{10.34}$$

u_h is the observed wind speed (km/h) at a height of h meters.

The Penman method can be modified to provide evapotranspiration (ET) estimates for crops as

$$\text{ET} = cE \tag{10.35}$$

where c is a crop efficient ranging in value from 0.6 in winter to 0.8 in summer at latitudes around 50°N and approaching 0.7 toward the equator in humid regions; and a somewhat smaller value than 0.7 in semiarid regions.

EXAMPLE 10.13 Calculate the free water surface evaporation in June using the Penman method from an area near Fortworth, Texas, whose latitude is approximately 33°N. The available data include air temperature = 30°C, wind speed at 2-m height = 10 km/h, relative humidity = 60%, mean observed sunshine hours = 12, and reflection coefficient = 0.05.

Solution First, we compute R_s using Equation (10.12). Corresponding to the latitude 33°N, I_0 from Table 10.3 is 16.56 mm of evaporable water, and N from Table 10.4 is 14.37 h. From Equation (10.12),

$$R_s = \left(0.18 + 0.55\,\frac{12}{14.37}\right) \times 16.56 = 10.587 \text{ mm water/day}$$

Vapor pressure at 30°C is obtained from Table 10.1. $e_a = 31.82 \times 0.60 = 110.092$ mm of Hg. $T = 273 + 30 = 303$°K. $\sigma = 2.01 \times 10^{-9}$ mm of water/day. By using Equation (10.19),

$$R_L = 2.01 \times 10^{-9} \times (303)^4[0.56 - 0.092(110.092)^{0.5}]\left(0.1 + 0.9\,\frac{12}{14.37}\right)$$

$$= 16.942(0.158)(0.855) = 2.281 \text{ mm water/day}$$

From Equation (10.10),

$$H = 10.587(1 - 0.05) - 2.281 = 7.777 \text{ mm water/day}$$

From Equation (10.33),

$$E_a = (0.175 + 0.0022 \times 10)(31.82 - 110.092) = 2.507 \text{ mm water/day}$$

The value of Δ from Table 10.1 is 1.85, and γ is 0.410. Therefore, by using Equation (10.32),

$$E = \frac{1.85}{1.85 + 0.49} \times 7.777 + \frac{0.49}{1.85 + 0.49} \times 2.507$$

$$= 6.148 + 0.525 = 6.673 \text{ mm water/day}$$

10.8 TRANSPIRATION

Transpiration is the process by which plants utilize water for their metabolism and growth. Plants remove water from the soil through their root system and transpire this water to the atmosphere as vapor through stomata in their leaves. Transpiration can be considered as similar to evaporation except that the surface from which the water escapes is not a free water surface, but rather through plant leaves. The number of stomata on plant leaves varies according to plant species. The density of stomata can vary from 50,000 to nearly 800,000 per in.^{2}, or per 6.45 cm^2, with many more on the lower leaf surface than on the upper. These stomata actively transpire water vapor to the atmosphere during daylight, but close after darkness begins, whereupon transpiration ceases.

Transpiration is affected by physiological and environmental factors. Stomata tend to open and close in response to environmental conditions such as light and dark, heat and cold, and so on. Their function is vital to plant metabolism because stomata allow carbon dioxide to enter the plant in the process of photosynthesis. Important physiological factors include (a) density and behavior of stomata, (b) extent and character of protective coverings, (c) leaf structure, and (d) plant diseases. All these factors are influenced by the types of plants and the density of those plants. Leaves that are shaded by higher leaves transpire less water than those exposed to direct sun rays. With adequate soil moisture, the greater the plant density, the greater the amount of moisture lost by transpiration. Plant diseases reduce transpiration and are often more common and severe in plants that are densely grouped. Only about 1% of the water entering the roots is incorporated into plant tissue. Most of the water passes from root to leaf to atmosphere. The driving force is the total potential-energy gradient from soil water to leaf surface. Water is extracted by plant roots from the surrounding soil by the osmotic suction within the roots.

Environmental factors that affect transpiration are essentially the same as for evaporation, but can be considered a bit differently. For practical purposes, vapor-pressure gradient, temperature, solar radiation, wind, and available soil moisture are the most important factors affecting transpiration.

The difference in vapor pressure between the space immediately inside the leaf and the outside air determines the vapor-pressure gradient and is a measure of the energy required to move the water from the leaf to the air. This vapor pressure is affected by the morphological features of the leaf, the position of the leaf, and the relation of the leaf to neighboring leaves. Plants transpire little, if any, when the vapor-pressure gradient is low as during a rainfall event. On the other hand, plants transpire rapidly during times in sunshine when warm, dry air surrounds them and plenty of soil moisture is available.

The rate of transpiration is doubled for approximately each 18°F rise in temperature. This relation is consistent with the change in vapor pressure of air. Because a leaf is dark colored, it absorbs incoming solar radiation effectively and its temperature becomes higher than the surrounding air. This condition exists even though evaporation cooling from transpiration lowers the temperature of the leaf. Experiments have shown that it is impossible to lower the temperature of a leaf below that of the surrounding air when the leaf is in direct sunlight. If the temperature of a leaf and that of the air are raised equally, there will be an equal rise in their vapor pressures.

Solar radiation is the primary source of energy. The solar constant is 1.94 cal/cm^2/min at the edge of the earth's atmosphere with about 0.30 cal/cm^2/min reaching the earth's surface. Absorption of this energy by a leaf raises its temperature and its aqueous vapor pressure. Thus, transpiration increases with increasing insolation. A good correlation exists between insolation and transpiration. The more insolation that is reflected back into the atmosphere, the less insolation is available for heating vegetation. Different crops and different soils reflect different amounts of insolation and result in different amounts of transpiration.

Transpiration can be altered by spraying leaves with chemicals. It has been found that spraying leaves with Bordeaux mixture, a common insecticide made from copper sulfate and lime, increases transpiration. Spraying with lampblack increases transpiration further, whereas spraying with hexadeconal reduces transpiration. These sprays probably exercise their effect on transpiration by increasing or decreasing the amount of insolation absorbed.

Wind usually increases transpiration by removing the film of moisture-laden air next to the leaf and consequently increasing the vapor-pressure gradient. Gentle winds have been found to be more effective in increasing transpiration than strong winds.

Transpiration is greatly affected by the amount of soil moisture present. As the plant uses moisture from the soil, the capillary forces holding moisture in the soil become stronger and it is more difficult for the plant roots to remove moisture. This is especially true when the soil moisture approaches the permanent wilting point. At this time, all transpiration ceases and the plant dies. Transpiration measurements are related to moisture conditions at field capacity. This standard set of conditions assumes optimum moisture available for plant growth and provides a basis for comparing plant transpiration. It would be very difficult to measure the conditions of stress on a plant growing under lesser moisture conditions.

10.9 MEASUREMENT OF TRANSPIRATION

Methods for measuring transpiration are related to the size and nature of the plants. Small plants are placed in a small closed container, such as a plastic bag, and the amount of moisture transpired in this closed container is measured. A drying agent is usually placed in the container to absorb the transpired water vapor. This moisture can be measured by weighing the drying agent before and after the test. Allowances are made for variations in humidity during the measurement.

A phytometer is a large vessel, filled with soil, in which plants are rooted. The soil surface is sealed to prevent evaporation, so the only moisture escape is by transpiration. The lost moisture can be determined by weighing the plant and container before and after the test. This method yields good results as long as natural environmental conditions are maintained.

A potometer operates in a manner similar to a phytometer except that it uses water instead of soil. There is disagreement among scientists whether this duplicates natural growing conditions.

10.10 EVAPORTRANSPIRATION

Evapotranspiration (ET) and consumptive use include both the transpiration by vegetation and evaporation from water surfaces, soil, snow, ice, and vegetation. Consumptive use differs from evapotranspiration only in that it includes the water used to make plant tissue. For all practical purposes, these two terms are synonymous. This means the water used by plants for building plant tissue and the water intercepted by plants. ET and consumptive use convert water to a form, water vapor, from which it cannot be used again. This use is in contrast to water impounded for electrical generation, where water is only borrowed, and when it is used, it is returned to the stream for reuse. The same can be said for water taken for domestic use, where most of the water is returned to the system in liquid form for reuse. Some water consumed by animals is converted to vapor in the metabolic process and, therefore, is not available for further use. The units of ET and consumptive use are in. (cm) of depth for a specified period.

10.11 MEASUREMENT OF EVAPOTRANSPIRATION

Consumptive use can be measured by (a) tanks and lysimeters, (b) using field plots, and (c) studies of groundwater fluctuations.

A tank is a watertight container that is set into the ground with its rim nearly flush with the ground surface. The size of the tank varies, but can be up to 30 ft (10 m) square and 10 ft (3.3 m) deep. The size should be sufficient to simulate natural growing conditions for the type of plants being studied. The tank is mounted on a scale to assist in necessary moisture measurements. Each plant requires certain moisture conditions for optimum growth and these conditions are maintained during consumptive-use measurements. ET is determined by measuring the quantity of water necessary to maintain constant, optimum moisture conditions in the tank.

A lysimeter is essentially a tank with a pervious bottom. The bottom arrangement is such that excess soil moisture will drain through the soil, which can be collected and measured. This condition offers an advantage over a tank because it prevents accumulation of water at the bottom to cause unnatural growing conditions for the plants. The consumptive use is the difference between the amount of water applied to the lysimeter and the amount draining out along with an adjustment for moisture content.

Specially designed field plots are used to determine ET under field conditions. These plots are designed so that surface runoff water from the plot can be collected and measured. Deep percolation is captured by underground drain tiles. To determine ET, water input in the form of precipitation or irrigation is measured. Water losses as runoff or deep percolation are subtracted from the water input and a correction is made for soil moisture.

A variation of the field-plot method is the water-balance method with inflow–outflow measurements from large drainage areas. This method requires measurements of water entering the area and the amount of water leaving the area during a known time interval. These values must be adjusted for the change in groundwater storage. Difficulty in accounting for all these variables makes this method less accurate than other methods.

Sometimes conditions exist where plants obtain moisture from the water-table capillary fringe or the zone of saturation. The height or elevation of the water table is influenced by the amount of water used by the plants. These conditions result in fluctuations in the water table reflecting withdrawals by plant transpiration. The equation for consumptive use under these conditions is

$$U = S(24a + b) \tag{10.36}$$

where U is the consumptive use in in. (cm) per day, S is the specific yield of the soil, a is the rate of rise of the capillary fringe in in./h (cm/h) from midnight to 4 A.M., and b is the net change in in. (cm) in saturated water elevation during the day.

10.12 DETERMINATION OF EVAPOTRANSPIRATION

Evapotranspiration equations have been developed to predict consumptive use for different crops and different conditions. These are empirical relations that have been useful to replace the difficult and expensive measurements. Full discussions of these equations can be found in Veihmeyer (1964), Gray et al. (1970), and Jensen (1973), amongst others.

10.12.1 Blaney–Criddle Method

Blaney and Criddle (1962) developed an empirical relation between evapotranspiration, mean air temperature, and mean percentage of daytime hours. This relation has been used extensively, especially in the western United States. The underlying assumption of this procedure is that the heating of the air and evaporation share the heat budget in a fixed proportion. As a result, ET varies directly with the sum of the products of mean monthly air temperature and monthly percentage of daytime hours with an actively growing crop with sufficient soil moisture as

$$\text{ET} = kF \qquad F = \frac{(1.8T + 32)p}{310.37} \tag{10.37a}$$

where ET is the monthly evapotranspiration or consumptive use in cm, k is an empirically derived seasonal consumptive-use coefficient applicable to a particular crop (depending on the units of ET), F is the monthly consumptive-use factor, T is the mean monthly air temperature in °C, and p is the mean monthly percentage of annual daytime hours. The values of monthly consumptive-use factor k for a sample of crops are given in Table 9.5 (Blaney and Criddle, 1962; Criddle, 1958).

TABLE 10.5 THE BLANEY–CRIDDLE MONTHLY CONSUMPTIVE USE FACTOR k

Crops and location	Jan.	Feb.	Mar.	Apr.	May	Jun.	Jul.	Aug.	Sep.	Oct.	Nov.	Dec.
Alfalfa:												
Mesa, Arizona	0.35	0.55	0.75	0.90	1.05	1.15	1.15	1.10	1.00	0.85	0.65	0.45
Los Angeles, California	0.35	0.45	0.60	0.70	0.85	0.95	1.00	1.00	0.95	0.80	0.55	0.36
Davis, California				0.70	0.80	0.90	1.10	1.00	0.80	0.70		
Logan, Utah				0.55	0.80	0.95	1.00	0.95	0.80	0.50		
Corn:												
Mandon, North Dakota					0.50	0.65	0.75	0.80	0.70			
Cotton:												
Phoenix, Arizona				0.20	0.40	0.60	0.90	1.00	0.95	0.75		
Bakersfield, California					0.30	0.45	0.90	1.00	1.00	0.75		
Weslaco, Texas			0.20	0.45	0.70	0.85	0.85	0.80	0.55			
Grapefruit:												
Phoenix, Arizona	0.50	0.50	0.60	0.65	0.70	0.75	0.75	0.75	0.75	0.70	0.60	0.50
Oranges:												
Los Angeles, California	0.30	0.35	0.40	0.45	0.50	0.55	0.55	0.55	0.50	0.50	0.45	0.30
Potatoes:												
Davis, California				0.45	0.80	0.95	0.90					
Logan, Utah						0.40	0.65	0.85	0.80			
North Dakota					0.45	0.75	0.90	0.80	0.40			
Grain, small:												
Wheat at Phoenix, Arizona	0.20	0.40	0.80	1.10	0.60							
Oats at Scottsbluff, Nebraska					0.50	0.90	0.85					
Sorghum:												
Phoenix, Arizona						0.40	1.00	0.85	0.70			
Great Plains Field Station, Texas						0.30	0.75	1.10	0.85	0.50		

Equation (10.37a) is also expressed as

$$\text{ET} = kF \qquad F = \frac{T_p}{100} \tag{10.37b}$$

where ET is in inches, T is in °F, and p is the monthly daytime hours given as a percentage of the year.

Equations (10.37a) and (10.37b) are for monthly estimates, and can be written for seasonal estimates of ET in in. (cm) as

$$\text{ET} = \Sigma kF = K\Sigma F \tag{10.38}$$

where K is an average seasonal crop consumptive-use coefficient. Both meteorological and crop effects are included in the Blaney–Criddle method. By accounting for rainfall in the month or season, one can compute the crop irrigation water requirement for a given irrigation efficiency.

10.12.2 Thornthwaite Method

Thornthwaite (1948) derived an equation to be used for limited water conditions. This equation produces monthly estimates of ET for the east central United States, using assumptions similar to those of the Blaney–Criddle method, and can be written as

$$\text{ET} = cT^a \tag{10.39}$$

where ET is in cm, c is a coefficient, T is the mean monthly temperature in °C, and a is an exponent. Both a and c depend on the location, and a can be estimated as

$$a = (67.5 \times 10^{-8})I^3 - (77.1 \times 10^{-6})I^2 + 0.0179I + 0.492 \tag{10.40}$$

where I is the heat index expressed as

$$I = \sum_{j=1}^{12} (T_j/5)^{1.51} \tag{10.41}$$

where T_j is the mean temperature of the jth month. The heat index is an integral element of Thornthwaite's classification of climates. The value of ET is modified by a factor to account for the number of daylight hours and the number of days in a month. Thornthwaite has tabulated values of this factor corresponding to various degrees and months of the year.

Assuming each month has 30 days and each day has 12 hours of sunshine, then Equation (10.39) reduces to

$$\text{ET} = 1.62(10T/I)^a \tag{10.42}$$

By taking its logarithm,

$$\log \text{ET} = \log 1.62 + a(\log 10 + \log T - \log I) \tag{10.43}$$

Obviously, ET $= 1.62$ when $I = 10T$ for $10T = \log I$. It has been shown by Thornthwaite that all lines obeying this equation have a common point of convergence at $T = 26.5$°C and E $= 1.35$ cm.

10.12.3 Jensen–Haise Method

By analyzing 3000 observations of ET determined by a soil-sampling procedure for a 35-year period, Jensen and Haise (1963) developed the following relation:

$$\text{ET} = C_T(T - T_x)R_s \tag{10.44}$$

where R_s is the solar radiation in ly/day, C_T is a temperature constant = 0.014, T_x is the intercept of the temperature axis = 26.4 if T is in °F; and $C_T = 0.025$ and $T_x = -3$ for T in °C. These coefficients are constant for a particular area. Jensen (1966) defined C_T as

$$C_T = \frac{1}{C_1 + C_2 C_H} \qquad C_H = \frac{50 \text{ mb}}{e_2 - e_1} \tag{10.45}$$

in which e_2 and e_1 are, respectively, the saturation vapor pressure at the mean maximum and mean minimum temperatures for the warmest month of the year, and $C_2 = 13°\text{F}$, or 7.6°C. Jensen et al. (1970) defined C_1 as

$$C_1 = 68°\text{F} - (3.6°\text{F} \times \text{elevation in ft/1000 ft}) \tag{10.46}$$

with

$$\begin{aligned} T_x = {} & 27.5°\text{F} - 0.25(e_2 - e_1)\ °\text{F/mb} \\ & - (\text{elevation in ft/1000})\ °\text{F} \end{aligned} \tag{10.47}$$

For temperatures in °C:

$$C_1 = 38 - (2°\text{C} \times \text{elevation in m/305}) \tag{10.48}$$

and

$$T_x = -2.5 - 0.14(e_2 - e_1)\ °\text{C/mb} - (\text{elevation in m/550}) \tag{10.49}$$

EXAMPLE 10.14 Estimate the monthly consumptive use of a corn crop grown in Nebraska for the month of June if the average monthly June temperature is 25°C, the average daytime hours in percent of the year is 11, and the mean monthly consumptive-use factor is 0.7.

Solution From Equation (10.37),

$$F = \frac{(1.8 \times 25 + 32) \times 11}{310.37} = 21.514$$

Therefore,

$$\text{ET} = 21.514 \times 0.7 = 15.06 \text{ cm}$$

EXAMPLE 10.15 Estimate the potential evapotranspiration for the data in Example 10.12. Use $c = 1.62$ in the Thornthwaite method.

Solution From Equation (10.41),

$$I = \sum_{j=1}^{12} \left(\frac{25}{5}\right)^{1.51} = 12 \times 5^{1.51} = 136.341$$

From Equation (10.40),

$$a = 67.5 \times 10^{-8}(136.341)^3 - 77.1 \times 10^{-6}(136.341)^2$$
$$+ 0.0179 \times 136.341 + 0.492$$
$$= 1.711 - 1.433 + 2.441 + 0.492 = 3.21$$

$$\text{ET} = 1.62\left(10 \times \frac{25}{136.341}\right)^{3.21} = 11.34 \text{ cm}$$

EXAMPLE 10.16 Compute the potential evapotranspiration using the Jensen–Haise method if the following data are available: the mean air temperature = 30°C, the latitude = 33°N, R_s = 800 ly/day, the mean maximum temperature of the warmest month = 35°C, the mean minimum temperature of the warmest month = 20°C, and the elevation = 100 m.

Solution From Table 10.1, T_2 = 35°C, e_2 = 56.2 mb, T_1 = 20°C, and e_1 = 23.4 mb. From Equation (10.45),

$$C_H = \frac{50}{56.2 - 23.4} = 1.524$$

$$C_1 = 38 - \left(2 \times \frac{100}{305}\right) = 37.34$$

$$C_T = \frac{1}{37.34 + 7.6 \times 1.524} = 0.0204$$

From Equation (10.49),

$$T_x = -2.5 - 0.14(56.2 - 23.4) - \frac{100}{550} = -7.274$$

From Equation (10.44),

$$\text{ET} = 0.0204 \times [30 - (-7.274)] \times 800 \text{ ly/day}$$
$$= 608.31 \text{ ly/day} = 10.45 \text{ mm/day}$$

10.12.4 Turc Method

Turc (1961) analyzed data collected from 254 watersheds located in virtually all parts of the world and related evaporation to rainfall and temperature as

$$E = \frac{P}{[0.9 + (P/I_T)^2]^{0.5}} \tag{10.50}$$

where E is the annual evaporation or ET in mm, P is the annual precipitation in mm, $I_T = 300 + 25T + 0.05T^3$, and T is the mean temperature in °C. Turc (1961) also suggested another equation incorporating the effect of soil-moisture variability on ET as

$$E = \frac{P + E_{10} + K}{\left[I + \left(\frac{P + E}{I_T} + \frac{K}{2I_T}\right)\right]^{0.5}} \tag{10.51}$$

in which E is the evaporation in a 10-day period in mm, E_{10} is the estimated evaporation (in a 10-day period) from bare soil assuming no precipitation and is not greater than 10 mm, K is a crop factor expressed as

$$K = 25(cM/G)^{0.5} \tag{10.52}$$

where $100M$ is the final yield of dry matter in kg/ha, $10G$ is the length of growing season in days, and c is a crop coefficient. I_T is the evaporation capacity of the air:

$$I_T = (T + 2)R_s^{0.5}/16 \tag{10.53}$$

where T is the mean air temperature over the 10-day period in °C, and R_s is incoming solar radiation energy in cal/cm²/day.

When soil moisture is not a limiting factor and $I_T > 10$, Turc proposed

$$E = \frac{P + E_{10} + 70}{\left[1 + \left(\frac{P + E}{I_T} + \frac{70}{2I_T}\right)\right]^{0.5}} \tag{10.54}$$

where E is in mm for 10 days.

Under general climatic conditions of Western Europe, Turc (1961) computed ET in mm/day for 10-day periods as

$$\text{ET} = 0.013 \frac{T}{T + 15}(R_s + 50) \tag{10.55}$$

for relative humidity (RH) > 50%, and

$$\text{ET} = 0.013 \frac{T}{T + 15}(R_s + 50)\left(1 + \frac{50 - \text{RH}}{70}\right) \tag{10.56}$$

for RH < 50%, when T is the average temperature in °C, and R_s is in ly/day.

EXAMPLE 10.17 Compute the mean daily evapotranspiration for London, Ontario, Canada, for July 1964 by the Penman method. Compare these values with the monthly means computed by the Thornthwaite method and the Meyer method. In the Penman method, assume that humidity and wind are consistent throughout the month.

Solution

Penman Method

1. Calculate I_0, the mean extraterrestrial radiation, for the London area in mm of water per day. This can be obtained from the tables providing the values of I_0 at varying latitudes for each month of the year. The latitudinal location of the London area is 43°N. At 40°N, I_0 for July = 940.6 cal/cm²/day. At 60°N, I_0 for July = 891.9 cal/cm²/day. Hence, at 43°N, I_0 = 932.8 cal/cm²/day. The approximate heat of vaporization = 590 cal/g. Therefore, I_0 in mm of water/day = (932.8 × 10)/590 = 15.8 mm/day.

2. Determine n/N. Tables are available for the maximum possible hours of sunshine and the average monthly values at various latitudes for each month of the year. N at 40°N for July = 14.7 h. N at 45°N for July = 15.3 h. At 43°N, N = 15 h approximately. The values of n, the actual hours of sunshine, are obtained for the month of July for the London area from weather reports. These data are given in Table 10.6. Compute the ratio n/N for every day of July.

3. Obtain the daily temperature values for July from weather records. Determine the corresponding values of saturation vapor pressure from Figure 10.2 directly. Be-

TABLE 10.6 AVERAGE DAILY TEMPERATURE AND *n/N* FOR JULY, 1964, LONDON, ONTARIO

Date	Temp. (max. °F)	Temp. (min. °F)	Average temp. (°F)	No. of hours of bright sun	n/N
1	89	67	78	10.70	0.713
2	76	65	70.5	0.00	0.000
3	84	59	61.5	5.30	0.353
4	73	54	63.5	12.70	0.847
5	78	51	64.5	14.80	0.987
6	79	50	64.5	11.70	0.780
7	82	59	70.5	6.30	0.420
8	74	58	66.0	0.00	0.000
9	83	61	72.0	7.00	0.467
10	84	57	70.5	14.00	0.933
11	86	53	69.5	10.20	0.680
12	64	59	61.5	9.60	0.640
13	71	57	64.0	3.30	0.220
14	69	56	62.5	2.60	0.173
15	76	56	66.0	13.10	0.873
16	82	53	67.5	11.40	0.760
17	94	58	81.0	13.20	0.880
18	97	69	83.0	9.60	0.640
19	88	69	78.5	12.70	0.857
20	88	65	76.5	10.00	0.667
21	88	68	78.0	10.80	0.747
22	87	62	74.5	11.20	0.567
23	86	60	73.0	8.50	0.567
24	86	65	75.5	8.80	0.587
25	87	68	77.5	7.70	0.513
26	86	61	73.5	13.20	0.880
27	88	58	73.0	12.40	0.827
28	91	66	78.5	6.50	0.433
29	74	57	65.5	12.00	0.800
30	74	50	62.0	13.50	0.900
31	74	41	57.5	13.10	0.873

cause the relative humidity is also known from the records, determine the actual vapor pressure (e_a). Values of e_a are given in Table 10.7.

4. Determine the value of $\Delta = (de/dT)|_{T_a}$ from Figure 10.2.

Average monthly temperature for July, 1964, for London:

HOURS	MEAN TEMPERATURE
01	64.5°F
07	65.1°F
13	78.6°F
19	74.9°F

Average (monthly) temperature = 70.78°F

Mean monthly relative humidity:

HOURS	R.H.
01	85%
07	84%
13	59%
19	64%

TABLE 10.7 TEMPERATURE AND VAPOR PRESSURE, JULY, 1964

Date	Mean temp. (°F)	Mean temp. (°C)	Absolute °K temp. (°C abs.)	Saturation vap. pres. (in. Hg)	Saturation vap. pres. (mm Hg)	Actual vap. pres. (mm Hg)
1	78	25.5	298.5	0.966	24.6	18.0
2	70.5	21.3	294.3	0.751	19.2	14.05
3	61.5	16.4	289.4	0.531	13.5	9.80
4	63.5	17.5	290.5	0.590	15.0	11.00
5	64.5	18.0	291.0	0.611	15.6	11.40
6	64.5	18.0	291.0	0.611	15.6	11.40
7	70.5	21.3	294.3	0.751	19.1	14.0
8	66.0	18.85	291.85	0.644	16.4	12.0
9	72.0	22.20	295.20	0.791	20.2	15.0
10	70.5	21.40	294.40	0.751	19.1	14.0
11	69.5	20.70	293.70	0.726	18.45	13.5
12	61.5	16.40	289.40	0.531	13.5	9.9
13	64.0	17.80	290.80	0.600	15.30	11.2
14	62.5	17.00	290.00	0.569	14.30	10.48
15	66.0	18.85	291.85	0.645	16.40	12.00
16	67.5	19.70	292.70	0.678	17.30	12.65
17	81.0	27.2	300.20	1.060	27.0	19.80
18	83.0	27.83	300.83	1.138	28.90	21.20
19	78.5	25.80	298.80	0.983	25.00	18.30
20	76.5	24.60	297.60	0.920	23.50	17.20
21	78.0	25.50	298.50	0.966	24.50	18.50
22	74.5	23.50	296.50	0.860	21.80	16.00
23	73.0	22.70	295.70	0.818	20.7	15.20
24	75.5	24.20	297.20	0.889	22.6	16.60
25	77.5	25.20	298.20	0.952	24.3	17.80
26	73.5	23.00	296.00	0.832	21.2	15.50
27	73.0	22.70	295.70	0.818	20.8	15.25
28	78.5	25.80	298.80	0.982	24.9	18.20
29	65.5	18.60	291.60	0.633	16.1	11.80
30	62.0	22.30	295.30	0.559	14.0	10.25
31	59.5	15.30	288.30	0.476	12.10	8.85

Average relative humidity = 73%

July 1, 1964

$r = 0.05$; $I_0 = 15.8$ mm/day; $n/N = 0.715$; $\sigma = 2.01 \times 10^{-9}$ mm/day-°K^4; $T_a = 298.5$°K; and $e_a = 18$ mm Hg. By substituting these values in Equation (10.10) for H,

$$\begin{aligned} H &= 15.8(1 - 0.05)(0.18 + 0.55 \times 0.715) \\ &\quad - (2.01 \times 10^{-9} \times 298.5^4)[0.56 - 0.09(18)^{0.5}] \\ &\quad \times (0.10 + 0.9 \times 0.715) \\ &= 15(0.18 + 0.393) - [2.01 \times 10^{-1} \times (2.985)^4] \\ &\quad \times (0.56 - 0.092 \times 3.16 \times 1.34)(0.10 + 0.645) \\ &= 15 \times 0.573 - (0.201 \times 75.6)(0.56 - 0.39)(0.745) \\ &= 8.6 - 1.92 = 6.68 \text{ mm/day} \end{aligned}$$

TABLE 10.8 EVAPORATION AND EVAPOTRANSPIRATION RATES FOR JULY, 1964

Date	H (mm/day)	Evaporation, E (mm/day)	Evapotranspiration (mm/day)
1	6.68	6.19	4.95
2	2.37	2.75	2.10
3	4.05	3.82	3.06
4	6.56	5.12	4.10
5	7.23	5.73	5.38
6	6.22	5.12	4.10
7	4.59	4.26	3.41
8	2.35	2.57	2.06
9	4.97	4.55	3.64
10	7.34	6.21	4.96
11	5.90	5.17	4.14
12	5.41	4.37	3.44
13	4.28	3.80	3.04
14	3.00	2.86	2.29
15	7.43	6.03	4.82
16	6.33	5.40	4.32
17	7.84	7.45	5.97
18	6.43	6.16	4.93
19	7.66	6.55	5.24
20	6.25	5.75	4.60
21	8.19	7.20	5.76
22	6.67	5.97	4.78
23	5.47	5.55	4.44
24	5.67	5.25	4.10
25	5.26	5.08	4.06
26	6.87	6.08	4.86
27	6.45	5.80	4.44
28	4.55	4.60	3.68
29	6.33	5.21	4.17
30	6.44	5.06	4.05
31	6.34	4.94	3.95

Wind velocity at 2 meters = 8.2 mph = 196.8 miles/day

$$e_s = 24.6 \text{ mm Hg/day}$$

$$E_a = 0.35(24.6 - 18)[(0.5 + 196.8)/100] = 4.55 \text{ mm/day}$$

$$\Delta = 0.82 \text{ mm Hg/°F}$$

Therefore,

$$E = \frac{0.82 \times 6.68 + 0.27 \times 4.55}{0.82 + 0.27} = \frac{6.73}{1.09} = 6.19 \text{ mm/day}$$

$$\text{Estimated evapotranspiration} = 6.19 \times 0.8 = 4.952 \text{ mm/day}$$
$$= 5.86 \text{ in./month}$$

where the crop coefficient is 0.8. Similarly, evapotranspiration can be compared for other days. Its values are given in Table 10.8.

The crop factor, C, is 0.8 (for July). The cumulative evaporation for the entire month = 154.89 mm ≅ 155 mm = 15.5 cm = 6.1 in. The estimated evapotranspiration = 124.0 millimeters = 12.4 cm = 4.88 in.

Meyer's Method

$$V_2 = 8.2 \text{ mph}; h = 25 \text{ ft}; V_h = ?$$

Thus,

$$V_h = \frac{8.2 \times \log 25}{\log 6.6} = \frac{8.2 \times 1.3929}{0.8195} = 14 \text{ mph}$$

Similarly, the temperature is to be corrected. The mean temperature for July is 70.8°F. In an adiabatic atmosphere, the temperature gradient is 3°F/1000 ft (elevation). Therefore, 25-ft elevation = 0.021°F. Hence, the temperature at 25-ft elevation = 70.8°F. The corresponding saturation vapor pressure = 0.76 in. Hg. The actual air pressure = 0.7615 × 0.73 = 0.56 in. Hg. Assume that the saturation air pressure is close to the vapor pressure of the water surface. Hence,

$$\begin{aligned} E &= 0.36(0.7615 - 0.56)[1 + 14/10)] \\ &= 0.36(0.2015)(2.4) = 0.1735 \text{ in./day} \\ &= 4.4 \text{ mm/day} \\ &= 135 \text{ mm/month (July) (considering 31 days in a month)} \\ &= 13.5 \text{ cm/month} \\ E &= 5.4 \text{ in./month} \end{aligned}$$

Thornthwaite Method

The exponent a can be evaluated in terms of annual heat index, I, as $a = (67.5\ 10^{-8})I^3 - (77.1 \times 10^{-6})I^2 + 0.01791I + 0.492$. I is as follows:

$$I = \sum_{n=1}^{12} \left(\frac{T_n}{5}\right)^{1.51}$$

If each month has 12 h of sunshine each day and 30 days a month, the basic equation reduces to

$$E = 1.62[10T/I]^a$$

or

$$\log E = \log 1.62 + a[\log 10T - \log I]$$

When $\log 10T = \log I$, $E = 1.52$, and $I = 10T$. $C = 1.37$ at 50°N. The average temperature for the month of July = 27°C. The evaporation = 14 cm = 5.55 in.

EXERCISES

10.1. The National Weather Service of the U.S. Department of Commerce maintains records of pan evaporation for various parts of the United States. List the monthly pan evaporation for the period 1970–1975 for Washington, D.C., and Los Angeles, California. Discuss why the differences in evaporation values exist.

10.2. Tabulate Class A pan coefficients for several parts of the United States corresponding to various seasons.

10.3. The atmospheric pressure = 1013 mb, the air temperature = 25°C, and the wet-bulb temperature = 18°C. Calculate the vapor pressure in mb, the relative humidity in %, the vapor-pressure deficit in mb, and the dewpoint temperature in °C.

10.4. Compute the mean daily evapotranspiration for New Orleans, Louisiana, for August 1980 by the Penman method. Assume that humidity and wind are consistent throughout the month. Then compute the monthly evapotranspiration.

10.5. Compute the monthly evapotranspiration using the Thornthwaite method for August 1980 for New Orleans, Louisiana, and compare this value with that obtained in Exercise 10.4.

10.6. Calculate the daily potential evapotranspiration by the Penman method from an area having the following characteristics: latitude = 30°N, elevation = 300 m above mean sea level, mean monthly temperature = 15°C, mean relative humidity = 70%, mean observed sunshine hours = 10, wind velocity at 2-m height = 50 km/day, and the ground surface is covered with green crop.

10.7. Compute the daily evaporation from a water body located in the area of Exercise 10.6.

10.8. The annual streamflow of a 4000-mi^2 watershed was 3000 cfs. The average precipitation in a 1-year period was 25 inches. Give a rough estimate of the combined amount of water evaporated and transpired.

10.9. Compute the daily and total water loss from Class A evaporation if the amount of water added or subtracted to bring the pan water level to the fixed point and rainfall on successive days are as follows:

Day	1	2	3	4	5
Rainfall (in.)	0	0.65	0.12	0	0.01
Water added to pan (in.)	0.25	−0.55	0.07	0.28	0.10

10.10. The pan coefficient for Class A pan evaporation is 0.7 for Exercise 10.9. What is the evaporation during the 5-day period for the lake having a surface area of 250 acres.

10.11. Determine the daily evaporation from a lake for which the following mean values were obtained: air temperature = 88°F, water temperature = 62°F, wind speed = 20 mph, and relative humidity = 35%.

10.12. When the mean monthly temperature is 75°F, and the average hours of daytime in percent of the year is 10.3, determine the monthly consumptive use of an alfalfa crop. Take the monthly consumptive-use coefficient of 0.84.

10.13. The annual evaporation from a lake, with a surface area of 1600 hectares, is 3 meters. Determine the average daily evaporation rate in hectare-centimeters per day during the year.

10.14. Thirty centimeters of water evaporated from a 200-hectare vertical-walled reservoir during 24 hours. Storm water was added to the reservoir at a constant rate of 30 m^3/s during this period. Determine the volume of hectare-centimeters of water released during the day (through the bottom of the reservoir) if the water level in the reservoir was the same at the beginning and end of the day.

10.15. Over a period of 12 months, 550 mm of rainfall occurred over a given catchment having 15,000 km^2 of area, and 450 m^3 of streamflow was recorded by a weir installed at the base of the basin. Account for the losses, including the evaporation and transpiration, and describe as to how the water budget can be made over the basin.

10.16. On a saturation vapor pressure versus temperature chart, sketch the air, dewpoint, and wet-bulb temperatures, and the vapor-pressure deficit.

10.17. A watershed has an area of 15,000 km^2. It received an annual rainfall of 50 cm. The rate of flow in the river draining the area was observed to be 600 m^3/s. Estimate the combined amounts of evaporation and transpiration that took place from the watershed during the year of record. You can assume that the rate of change in storage was negligible during the year.

10.18. Calculate the amount of evaporation expected from a lake per week under the following average daily conditions:

Temperature of water = 78°F

Wind milage = 1680 miles/week 4 m above the water surface

Temperature of air = 90°F 2 m above the water surface

CHAPTER 11 Interception and Depression Storage

This chapter presents an elementary discussion of the processes of interception and depression storage, and their evaluation. Through interception, hydrology is related to forest science, and through depression storage to earth science.

11.1 INTERCEPTION

Interception is defined as the precipitation water retained on the drainage basin through its adherence to abstract objects such as leaves or other vegetation, buildings, animals, or any such objects above the surface of the ground.

When a raindrop strikes a leaf or other abstract objects, the drop spreads over the leaf as a thin film because of the adhesive forces acting on the drop. Further rainfall will act in a similar manner until a layer of water sufficiently thick has accumulated for the gravitational forces to become greater than the adhesive forces. At this point, some water will run off the leaf and a near constant quantity of water is retained on the leaf. This near equilibrium condition continues as additional rainfall accumulates, and the interception loss becomes constant if other factors remain unchanged.

Because the adhesive forces are caused by the attraction of water to an unlike object, the strength of the adhesive forces vary with the nature and composition of the intercepting object and the thickness of the intercepted water layer. Hence, some objects intercept more water than others, resulting in different amounts of interception between drainage basins or even subareas on the same drainage basin.

The intercepted water is retained on its object and is eventually evaporated. Thus, interception is a loss of water that might otherwise become part of runoff. Water that is intercepted and adheres to the soil surface is sometimes not considered to be interception, even though it fits the definition. This water is in fact interception. It cannot become a part of the soil moisture and be available to plants, nor can it become runoff. This water intercepted by the soil surface becomes evaporation in the same manner as any intercepted water.

A certain amount of intercepted water never contributes to soil moisture, groundwater, or streamflow. Most of the rainstorms producing 5 mm or less of rainfall in a forest watershed are evaporated from the crowns or litter. For bigger storms, the interception loss will be about the same, depending upon the vegetation density and the period between storms. For a given storm, this loss may not seem much, but becomes significant on an annual basis. If there are 80 storms occurring in a year, then yearly interception would amount to 160 mm at the rate of 2 mm per storm. In the eastern United States, 50 to 120 storms occur every year. Hence, 160 mm of loss is not insignificant.

Discussion of interception is facilitated by defining the following terms.

Interception storage is defined as the amount of water or snow held by vegetation or other objects (buildings, etc.) at any given time.

Crown interception loss (I_c) is the amount of water evaporated from water or sublimated directly from snow intercepted by the crowns of vegetation.

Throughfall (P_t) is that part of precipitation that falls or drips through the crown of the vegetation.

Stemflow (P_s) is that portion of intercepted water that accumulates and runs down stems.

Forest floor interception loss (I_f) is the amount of water that is intercepted by and evaporated from the forest floor before infiltrating into the soil.

Total interception loss (I_t) is the amount of water evaporated from rain or sublimated from snow that is intercepted by vegetation and other objects. Therefore, $I_t = I_c + I_f$.

11.2 FACTORS AFFECTING INTERCEPTION

Three principal factors determine the amount of water to be intercepted on a drainage basin. These factors are (a) storm characteristics; (b) plant species, age, density, and the condition of vegetation or other objects; and (c) the season of the year.

11.2.1 Storm Characteristics

The number and spacing of precipitation events, intensity and amount of precipitation, and wind speed determine the availability of water for interception. Fine water droplets, such as occur in fog or mist, accumulate on abstract objects and vegetation to a greater thickness than does water from large heavy droplets falling in an intense rainstorm. Such large droplets contain sufficient kinetic energy that results in leaf vibration removing some intercepted water that otherwise would remain on a leaf. Convective storms result in less interception than do gentle, light frontal storms. By the same token, storms accompanied by high winds tend to remove intercepted water through turbulent air movement. Horton (1919) developed a relation between the amount of precipitation and the percent retained as interception, as shown in Figure 11.1. The percentage of interception is large for a small rainfall and levels off to a constant value for larger storms. The relation does not show any relation to other storm characteristics or their variability. For large rainfall amounts, the net quantity of intercepted water can exceed 0.25 in. (0.62 cm). It is not uncommon to have interception values greater than this amount when conditions are favorable.

11.2.2 Vegetation Characteristics

The species, age, density, and condition of vegetation or of other abstract objects determine the adhesive effect of these objects. It is reasonable to believe that different plant species and kinds have different characteristics such as leaf size, number of

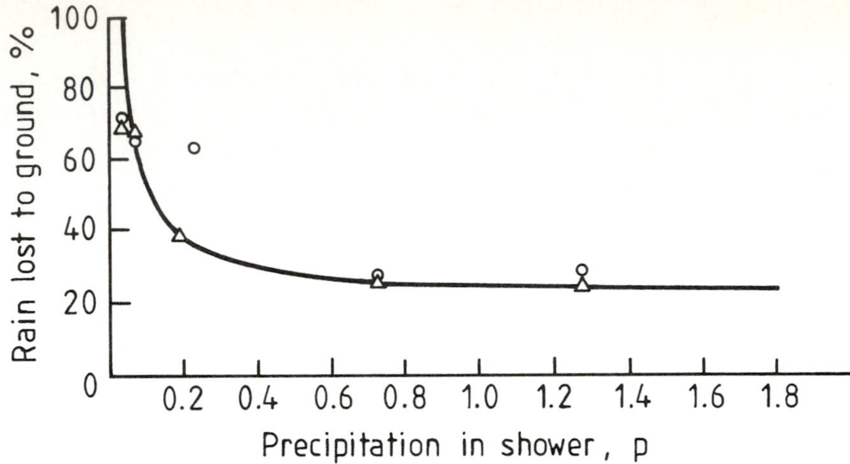

Figure 11.1 Total percentage of precipitation in a shower intercepted by various trees in 1917–1918. The curve is drawn with respect to triangular points. Circles represent the mean of all the interceptometers, including some uncorrected ones for trunk water and peripheral pans. Triangles represent the mean of interceptometers, excluding peripheral pans or those uncorrected for trunk water (after Horton, 1919).

leaves, and molecular attraction. The density and health of the plants, and the condition of plant growth further affect the opportunity for interception. More plants with larger and more dense leaf arrangement result in greater leaf area for interception. Dry plants and other abstract objects will tend to soak up intercepted rainfall until the plant or object has reached its capacity to hold water. Grasses, crops, and shrubs often intercept as much rainfall as forests. Table 11.1 summarizes observations made on crops during growing seasons and on a variety of crops (Viessman et al., 1989).

TABLE 11.1 OBSERVED INTERCEPTION BY VARIOUS CROPS AND GRASSES

Vegetation	Interception (%)	Remarks
Crops		
Alfalfa	36	
Corn	16	
Soybeans	15	
Oats	7	
Grasses		
Natural grasses	14 to 19	
Blue grass	17	Prior to harvest
Mixed species	26	
Buffalo grass	31	
Bindweed	17	
Little bluestem	50 to 60	Water applied at the rate of 0.5 in. in 0.5h
Big bluestem	57	
Tall panic grass	57	

11.2.3 Season of the Year

Interception is affected by the season of the year. Plants just emerging from the ground or tree leaves just emerging from their bud intercept less water than plants and leaves that are mature. Plants that have matured and died and are shedding or have shed leaves have lesser and varying ability to intercept rainfall or snowfall. For this reason, plant interception is less in spring and winter than in late summer.

11.3 ESTIMATION OF INTERCEPTION

Interception loss can be measured by noting the definition of total interception loss, $I_t = I_c + I_f$. The term I_f can be measured by collecting and weighing samples of the forest floor. The first sample is taken shortly after rainfall and the average loss of weight in later samples is interpreted as the amount of intercepted water evaporated. Under mature pine, I_f may average out to be about 5% of the annual precipitation, and under mature hardwoods, it may be about 3%.

The term I_c is obtained from the gross precipitation (P_g) and P_t and P_s by noting that $I_c = P_g - P_t - P_s$. P_t is measured by rain gages under the canopy. P_g is measured by rain gages located in nearby openings. P_s is measured by collars around stems. P_s is usually 1 to 2% of P_g, except in beech forests, where it may be 5 to 8%. P_s is hydrologically negligible in spruce and fir.

Many studies have been made to determine the amount of water intercepted by different species of vegetation. These studies show that a dense, even aged stand of conifers intercepts one-fourth to more than one-third of the precipitation that falls annually (Trimble, 1959). The percentage of a given rainfall that can be intercepted (Figure 11.1) ranges from 100% down to something less than 25%. The percentage of annual precipitation intercepted by vegetation is believed to be on the order of 10 to 25%. This is a relatively significant amount of water, and its loss represents a significant deficit in potential annual runoff.

Equations have been developed to estimate the amount of interception from a drainage basin. These equations have limited use because it is not possible to measure several of the variables. For example, interception equations require the interception storage retained on the foliage against wind and gravity, the ratio of the surface area of intercepting leaves to the horizontal projection of this area, and the water evaporated per hour during the interception period. Such data are usually not available.

Horton (1919) derived a series of empirical formulas for estimating interception per storm by various types of vegetal cover. His model relates the total interception loss I for a storm to the storage capacity of the vegetation and the rate of evaporation:

$$I = S + kED \tag{11.1}$$

where S is the storage capacity of the vegetation for the projected area of the canopy (its value varies from 0.25 to 1.25 mm), k is the ratio of the vegetal surface area to its projected area, E is the rate of evaporation from the vegetal surface, and D is the duration of rainfall. If interception is assumed to increase exponentially from zero to some value, then a variant of Equation (11.1) can be expressed as

$$I = (S + ED)[1 - \exp(-cP)] \tag{11.2}$$

where P is the amount of rainfall; and c, a constant, is $1/(S + ED)$. Another equation for expressing interception loss is of the form

$$I = S\left[1 - \exp\left(-\frac{P}{S}\right)\right] + kED \tag{11.3}$$

If $K = kED/P$ is assumed constant, Equation (11.3) becomes

$$I = S\left[1 - \exp\left(-\frac{P}{S}\right)\right] + KP \tag{11.4a}$$

For large values of P, Equation (11.4a) becomes

$$I = S + KP \tag{11.4b}$$

which is the same as Equation (11.1). Horton (1919) derived the values of S and K as $S = 0.015$ in. (0.381 mm) and $K = 0.23$ for ash trees, and $S = 0.03$ in. (0.762 mm) and $K = 0.22$ for oak trees.

It is difficult to quantify the variables of these equations. Hence, such mathematical models have little practical utility. Coniferous trees intercept more water than deciduous trees. Dense grass has nearly the same interception amount as fully grown trees. This amount may be as much as 20% of the annual rainfall.

EXAMPLE 11.1 Estimate the amount of interception by vegetation from a storm of 5-cm rainfall. The value of $K = 0.25$ and $S = 1$ mm.

Solution From Equation (11.4b),

$$I = \frac{1}{10} + 0.25(5)$$

$$= 0.1 + 1.25 = 1.35 \text{ cm}$$

Helvey (1971) presented equations for the estimation of the annual I_c (cm) from annual precipitation (P) and the number of storms (n) in temperate regions:

$$I_c = aP + bn \tag{11.5}$$

where a and b are constants, depending upon the type of the forest. For example, $a = 0.1$ and $b = 0.1$ for pines; $a = 0.21$ and $b = 0.13$ for spruce, fir, and hemlock; $a = 0.06$ and $b = 0.1$ for deciduous hardwoods in summer; and $a = 0.03$ and $b = 0.05$ for deciduous hardwoods in winter. Equation (11.5) is based on the assumption that the forest is mature. For thinned stands, I_c is reduced in proportion to the reduction in basal area, crown cover, or biomass.

EXAMPLE 11.2 Calculate the annual interception loss in cm from a mature deciduous forest for the following data:

SEASON	P (cm)	NO. OF STORMS (n)
Summer	60	30
Winter	55	25

Solution For the summer season, $a = 0.06$ and $b = 0.1$, and for the winter season, $a = 0.03$ and $b = 0.05$. Therefore, for the summer:

$$I_c = 0.06(60) + 0.1(30) = 3.6 + 3.0 = 6.6 \text{ cm}$$

and for the winter:

$$I_c = 0.03(55) + 0.05(25) = 1.65 + 1.25 = 2.90 \text{ cm}$$

Total $I_c = 6.60 + 2.90 = 9.5$ cm, which is about 8.26% of the annual precipitation. An additional 3% (or 3.45 cm) may be added for I_f, giving a total loss I_t of 11.26% of P, or 12.95 cm.

For practical purposes, a general estimate based on observed runoff conditions may be required. Interception is closely related to depression storage, and the two can be combined as a single loss. This loss, often referred to as initial loss, can be reasonably estimated on a gaged drainage basin where infiltration measurements exist.

11.4 DEPRESSION STORAGE

Depression storage, or ponding, is that water on a drainage basin that drains into closed depressions and never reaches the outlet of the basin. This water becomes trapped in ponds; some eventually evaporates and the remainder infiltrates into the ground. Depression storage occurs on most drainage basins. It is most common and serious in glaciated areas where large, isolated depressions occur. A careful inspection of drainage basins is required in order to identify their presence. Even though it might be difficult to quantify depression storage in some circumstances, the relative magnitude of its effect on runoff can be estimated.

11.5 FACTORS AFFECTING DEPRESSION STORAGE

Depression storage has taken on a relatively new importance in recent times. To save moisture and reduce erosion, many farmers have terraced sloping farm land. Terraces are berms built up nearly on contour with the topography. The height and spacing of these terraces determine the amount of runoff that they intercept. If the terraces were exactly on level contour, they would impound intercepted runoff water and permit little or no drainage to occur. The terraces are usually built on a slight grade in order to permit retention of runoff water long enough so as to increase infiltration, reduce erosion, and yet drain away slowly to the natural drainage system in the basin, thus making room for the next rainfall event. These conservation practices increase depression storage on the drainage basin. Their presence can be easily identified on the basin by inspection.

Terraces are a popular agricultural land treatment developed by the Soil Conservation Service (1971) of the U.S. Department of Agriculture to preserve soil moisture and to prevent erosion in sloping farm land. Typical terrace designs are shown in Figure 11.2. By examining this figure, it is possible to construct terraces of different sizes in order to intercept and retain different amounts of runoff water. The spacing of terraces on the land slope is determined by the climate, land slope, and the soil type. Where rainfall amounts are greater, terraces should be spaced closer together than in areas where rainfall is less.

Land terraces are effective in intercepting runoff and perform their intended function well. Therefore, where terraces exist, ponding behind them provides substantial storage and reduces the amount of runoff that contributes to the peak discharge from a storm event. Terraces, which are on a grade from the contours, intercept surface runoff and then permit that runoff water to drain slowly along the terrace grade to the natural drainage system. This runoff water does not add to the peak of the hydrograph, but is delayed in its runoff so that it lags behind the normal hydrograph of runoff. In this instance, the total volume of water draining from an area might be nearly the same as before terraces were constructed, but the peak flow will be decreased. In practice, even the runoff volume will be less because the delay in the runoff time as water flows along the terrace provides greater opportunity for infiltration to occur. Greater infiltration reduces runoff volume and enhances soil moisture. Terracing intercepts natural stream channels of lower orders. The effect of terracing is to reduce the frequency of such channels so that the discharge from them is reduced.

11.6 ESTIMATION OF DEPRESSION STORAGE

It is possible to identify and measure depression storage areas that are obvious from topographic maps. These areas should be deducted from the total drainage area of the drainage basin for future analysis. However, depression storage areas are not always obvious and for this reason cannot be easily identified short of a detailed

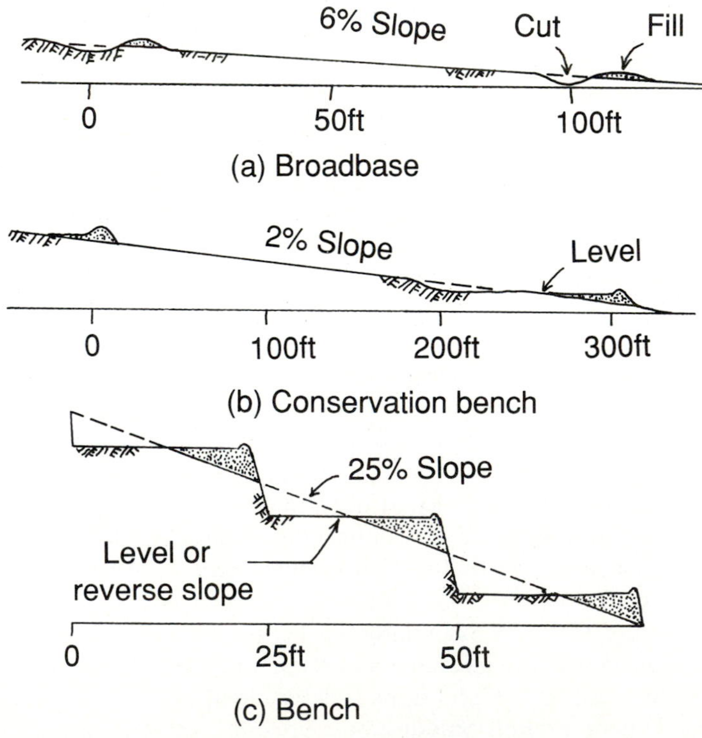

Figure 11.2 Typical terraces.

topographic survey of the drainage area. These areas lead to depression storage, which can be determined along with interception losses from measurements of stream flow, precipitation, and infiltration using the hydrologic budget. The amount of infiltration can be determined by measuring the mean infiltration capacity on the drainage basin and plotting those points as a straight line on semilogarithmic paper. From this infiltration data, the amount of infiltration can be determined over the duration of the storm event. Streamflow and precipitation are measured. From these measurements, combined interception and depression storage can be computed using the water budget if evapotranspiration loss is neglected. Experience has shown that the combined effect of interception and depression storage can often range from 0.5 in. (1.25 cm) to 2+ in. (5+ cm). This is a significant amount of water. Interception and depression storage are not necessarily evenly distributed over the drainage-basin area.

Linsley et al. (1949) expressed the amount of water stored at a given time by surficial depressions as

$$V = S_d\,[1 - \exp(-kP_e)] \tag{11.6}$$

where V is the amount of water stored at time t, S_d is the maximum depression storage capacity, P_e is the precipitation excess (gross precipitation minus evaporation, interception and infiltration) at time t, and k is a constant.

Equation (11.6) is based on the hypothesis that

$$\frac{dV}{dP_e} = k(S_d - V) \tag{11.7}$$

This hypothesis also enables an estimation of the constant k. From Equations (11.6) and (11.7),

$$\frac{dV}{dP_e} = kS_d \exp(-kP_e) \tag{11.8}$$

Where $P_e = 0$, dV/dP_e equals 1, for all the water essentially fills the depressions. Therefore,

$$k = \frac{1}{S_d} \tag{11.9}$$

The value of S_d may be obtained from topographic maps or field measurements.

If the value of P_e is very large, then V approaches the value of S_d. If P_e is negligible, then the value of V is also negligible. Thus, from Equation (11.6), it is clear that $0 \leq V \leq S_d$.

EXAMPLE 11.3 The mean rainfall for the storm of September 6–7, 1965, was 19 cm for the Irvington drainage basin. Streamflow measurements showed runoff to be 4.83 cm. The amount of infiltration from the infiltration curve is about 11.18 cm. Estimate interception and depression storage. The basin has healthy crops and terracing. Assume evapotranspiration is negligible for this storm.

Solution By applying the hydrologic budget,

Interception and depression storage = rainfall − runoff − infiltration

= 19 − 4.83 − 11.18 = 2.99 cm

EXAMPLE 11.4 Estimate the amount of depression storage for a storm occurring on a watershed with the area of 10 km^2 and depression-storage capacity of 100,000 m^3. The precipitation excess for the storm is 10 cm.

Solution

$$S_d = 100{,}000 \text{ m}^3$$

$$= \frac{100{,}000 \times 100}{10 \times 1000 \times 1000} \text{ cm} = 1 \text{ cm}$$

$$P_e = 10 \text{ cm}$$

$$k = \frac{1}{S_d} = \frac{1}{1} = 1 \text{ cm}$$

Therefore, by using Equation (11.6),

$$V = 1\,[1 - \exp(-1 \times 10)]$$

$$\cong 1 \text{ cm}$$

This shows that the entire depression-storage capacity is filled during the storm.

EXERCISES

11.1. The values of interception as computed by Horton's equation for 1-inch storms for various crops can be noted as follows:

CROP	HEIGHT (ft)	INTERCEPTION (in.)
Corn	6	0.03
Cotton	4	0.33
Tobacco	4	0.07
Small grains	3	0.16
Meadow grass	1	0.08
Alfalfa	1	0.11

Compute the interception loss as a percentage of precipitation. Is there a relation between interception loss and crop height?

11.2. Consider a mostly residential urban area. For a 20-min rainfall storm, the total precipitation amount may be (a) 0.13 cm, (b) 0.25 cm, (c) 1.0 cm, and (d) 2.0 cm. For which storm will runoff be most affected by the estimation of the interception loss and why?

11.3. Consider a forested watershed. For two storms having the same water equivalence, precipitation occurs as rainfall for one storm and as snowfall for the other storm. In which case will the interception loss be greater?

11.4. Select a watershed in your area. Demarcate the areas of depression on its topographic map. Determine the surface areas of depression of this watershed.

11.5. Does urbanization increase depression storage? If yes, why and how much? Discuss your answer.

11.6. What precipitation excess would cause the 2.5 m^3 of water stored in a depression having 10 times the storage capacity?

PART 5
Streamflow Measurement and Hydrograph Analysis

CHAPTER 12
Streamflow Measurement

Most hydrologic analyses involve runoff from a drainage area, and hence its measurement is of vital importance. As discussed in the previous chapters, it is difficult to accurately measure abstractions (loss variables): infiltration, interception, evaporation, transpiration, and depression storage. In contrast, streamflow can be measured accurately. Measurements of the loss variables are used for other purposes in agricultural sciences, earth sciences, atmospheric sciences, etc. Streamflow data are collected primarily for hydrologic studies.

If precipitation and runoff can be accurately measured, it is then possible to estimate the total loss on a drainage basin. This information can help predict runoff from similar drainage basins that have no gages. Streamflow measurements are used to develop physical or statistical relations between other variables and runoff volume or peak discharge. These relations form the basis for many calculations to predict streamflow characteristics of ungaged basins.

Streamflow is measured in units of discharge (m^3/s or cfs). Direct measurement of discharge is an expensive, time-consuming procedure. Two steps are employed, therefore, to obtain streamflow measurements. First, the discharge of a specified stream is related to the water-surface elevation or stage using a series of careful measurements, and a stage-discharge relationship, popularly called a rating curve, is established for that stream. Second, the water stage is observed routinely, and the corresponding discharge is obtained from the rating curve. The measurement of stage is easy and relatively inexpensive.

Measurements of streamflow can be continuous or intermittent. Intermittent measurements are for the gages that are manually read on some periodic basis, or might consist of recording only crest flows over some specified period of time.

Continuous records are most useful, although intermittent records can be very valuable where they are the only data available.

There are four principal methods of measuring streamflow discharge: (1) weir stations, (2) control meter stations, (3) power plants, and (4) velocity-area stations. A complete discussion of these methods can be found in books on hydrometry, which is the science and practice of water measurement. This chapter deals with only the salient aspects of streamflow measurement. The reader is referred to Kolupaila (1960), Boyer (1964), Grover and Harrington (1966), Bos (1976), Ackers et al. (1978), Herschy (1978), Rantz and others (1983a, 1983b), amongst others. This chapter also shows application of hydraulics and, hence, the association of hydrology with hydraulics.

12.1 MEASUREMENT OF STAGE

The term stage, river stage or water stage, refers to the water-surface elevation, above an arbitrary datum, at a point along a stream. The datum can be the mean sea level (msl) and is often slightly below the point of zero flow in the stream.

12.1.1 Nonrecording Stream Gages

Commonly used nonrecording stream gages are staff gages and wire gages, which are manually operated. A staff gage, as shown in Figure 12.1, is the simplest way to measure the river stage. It may be mounted vertically or at an angle from the vertical. The staff is rigidly attached to a permanent structure such as a bridge pier,

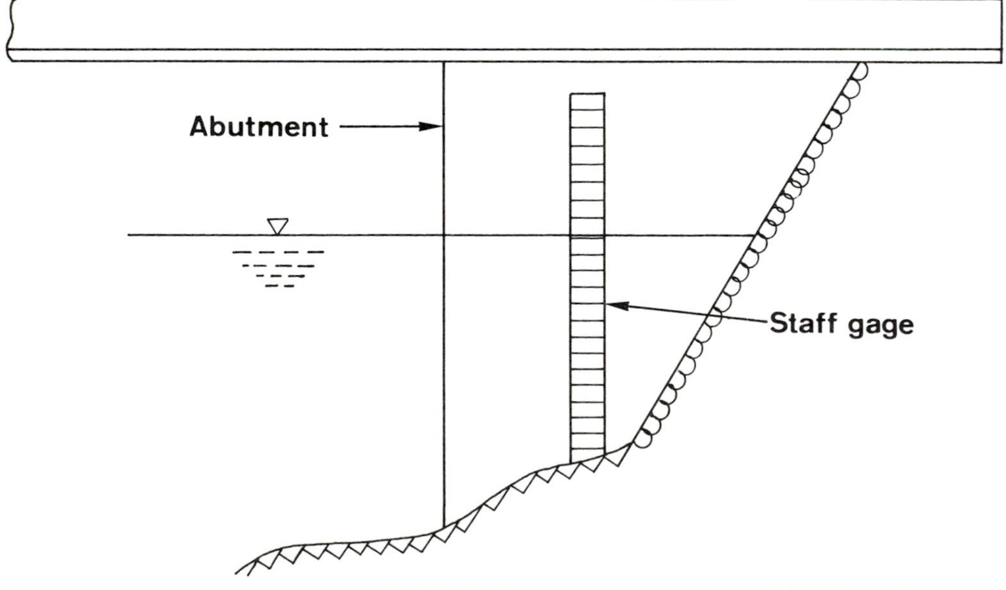

Figure 12.1 Staff gage.

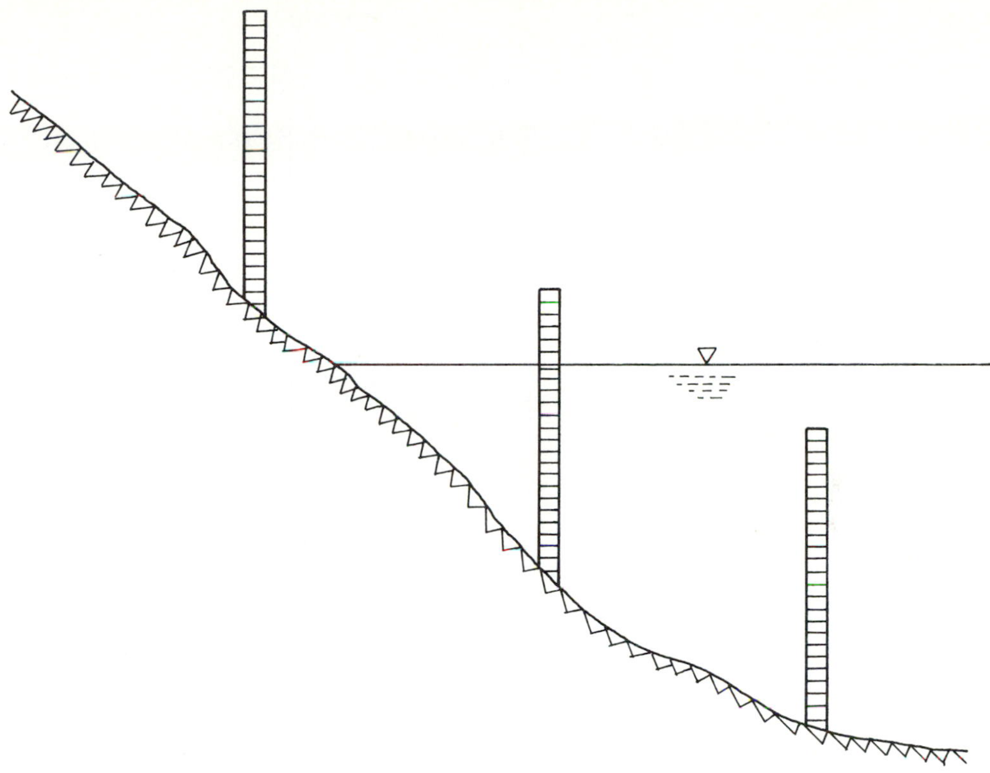

Figure 12.2 Sectional staff gage.

wall, abutment, etc. The gage indicates water-surface elevation on a staff that is graduated with clear and accurate markings in tenths of a foot or in centimeters. A portion of the scale is immersed in the water at all times. Sometimes, a single gage is not adequate for all stages; a sectional staff gage, as shown in Figure 12.2, is used. Sectional gages are installed to provide overlap between various gages with their readings corresponding to the same datum.

A wire gage measures the water-surface elevation from above such as from a bridge or other overhead structure. A weight is lowered from the structure until it reaches the water surface. The gage has a drum with a circumference equal to 1 ft or 1 m of wire. The number of revolutions of the drum is measured by a mechanical counter, which, in turn, measures the length of the wire transmitted to reach the water surface. The operating range of a wire-weight gage is about 25 m (75 ft).

12.1.2 Crest-Stage Gage

A special application of a staff-gage installation is the crest gage. The crest gage is designed to obtain a measurement of the peak discharge in a channel reach during a flood event. A crest gage consists of an ordinary staff gage of sufficient width and selected length to fit into a 2-inch galvanized pipe, as shown in Figure 12.3. This galvanized pipe is fitted with threaded pipe caps on either end in which are drilled

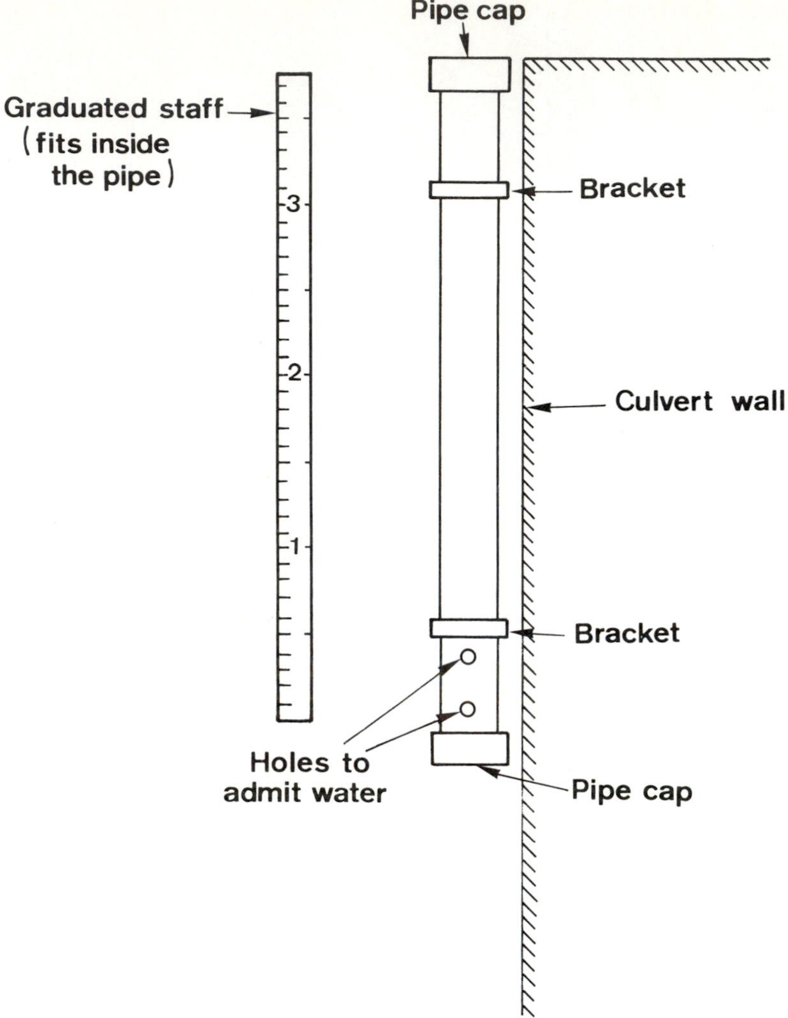

Figure 12.3 USGS crest-staff gage.

several holes of approximately 0.25 inch in diameter. The pipe is mounted vertically in the stream channel with its bottom at the stream datum. The staff gage is inserted at the top of the pipe along with about a capful of ground cork and the top ventilated pipe cap is replaced.

During a flood event, water enters the lower cap perforations and rises in the pipe, carrying the ground cork with it. At the highest stage, the cork adheres to the wetter staff, and as the water recedes, a visual record of the highest stage is indicated by the cork adhering to the staff. The operator only needs to remove the staff, read the high stage, and wipe the staff clean for additional use. Additional ground cork may be added, as necessary, to replace that which is lost.

12.1.3 Recording Stream Gages

Recording stream gages are instruments that continuously record the water stage at a given location along the stream. Two types are in general use: (1) the float type and (2) the bubble gage or manometer–servo water-level sensor.

The float-type recorder is shown in Figure 12.4. The recorder is located in a stilling well that consists of a vertically mounted culvert or other similar structure that may be from 1 to 3 feet in diameter, as shown in Figure 12.5. The culvert is installed on the bank of the channel to a depth at least equal to the lowest level of the channel bottom. A 2-inch open-ended galvanized pipe is run horizontally from the bottom or near the bottom of the deepest part of the channel and fastened to the culvert. The top of this pipe is usually used as the stream-gage datum because it marks the lowest stage of streamflow or nearly so.

The recorder is mounted in a weatherproof housing at the top of the culvert and stilling well. Water rises in the stilling well through the 2-inch galvanized pipe to a level equal to the water elevation in the channel. The purpose of the stilling well is to dampen the water-surface fluctuations so that the float records changes in water elevation, but does not reflect wave action or other interference. Water-level changes are recorded on punched tape at selected time intervals, often every 15 minutes. This instrument can be adapted to remote monitoring by telephone or radio.

Older recorders, from which many records exist, used a pen that marked stage elevations on a chart mounted on a clock-driven drum. These instruments required the chart to be changed periodically.

The bubble gage, or manometer–servo water-level sensor, is shown in Figure 12.6. This instrument eliminates the need for a stilling well, but requires battery power and a 116-ft^3 dry nitrogen cylinder for operating up to 6 months. Pressure corresponding to the water head in the channel is imparted to the recorder through a tube in the bottom of the channel. This tube is supplied with nitrogen gas pressure equal to the water head by a servo motor that automatically adjusts for changes in water head. This instrument is attached to a digital recorder, similar to the float-type recorder, and can be remotely monitored by telephone or radio.

The bubble gage may be preferable to the float-type recording gage for the following reasons. The stilling well, which is expensive, is not needed. Large changes (up to 30 m) in water-surface elevation can be measured. The inlet is less likely to be blocked because of the gas pressure. The recorder assembly can be located far away from the sensing point.

12.1.4 River-Stage Data

The stage data are presented chronologically in time as a time series. Their plot, as shown in Figure 12.7, is called a stage hydrograph. The primary use of this data is the determination of discharge. Other uses of this data are in flood-insurance studies, design of flood-protection works, flood warning and evacuation, urban development, flood-damage assessment, water diversion, navigation, etc. Long-term stage

Figure 12.4 Stevens float-type digital recording gage. Some models record the water level, precipitation, temperature, pressure, conductivity, and other variables. Encoding modules add to the versatility of this instrument.

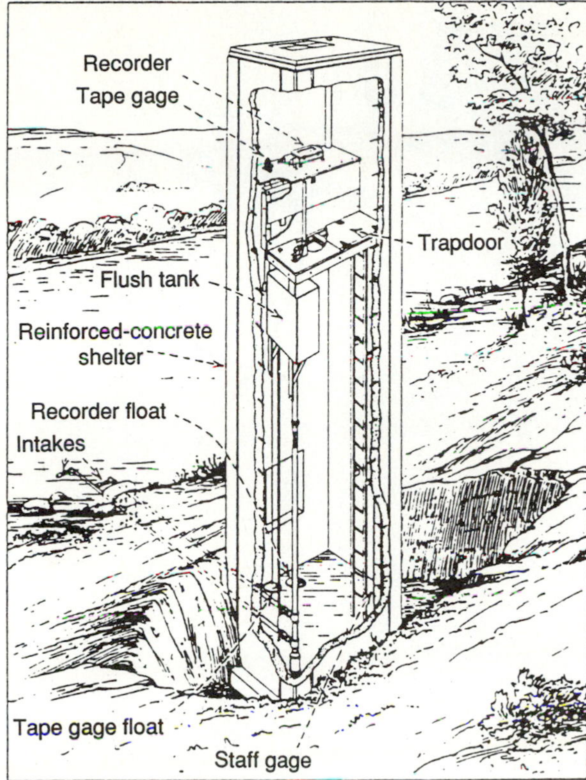

Figure 12.5 Stilling well for the float-type recorder.

data are needed to estimate peak river stages for application in the design of hydraulic structures such as bridges, culverts, weirs, etc.

12.1.5 Selection of Gage Site

The location of a stream gage depends upon the purpose for which the river-stage data are collected. If the primary use of data is flood warning or navigation, the gage to be located should be most easily accessible. If the primary use is to obtain a record of discharge, then its location requires understanding of open-channel hydraulics, especially channel controls, and involves greater care and judgment. These controls impact the relation between stage and discharge, which will be discussed later in the chapter.

12.2 MEASUREMENT OF VELOCITY

12.2.1 Velocity Distribution

Velocity distribution in a channel is not uniform over the width and depth of the channel. Velocity is greatest in the deepest part of the channel and is zero along the boundary of flow, as shown in Figure 12.8. The greatest velocity occurs just under

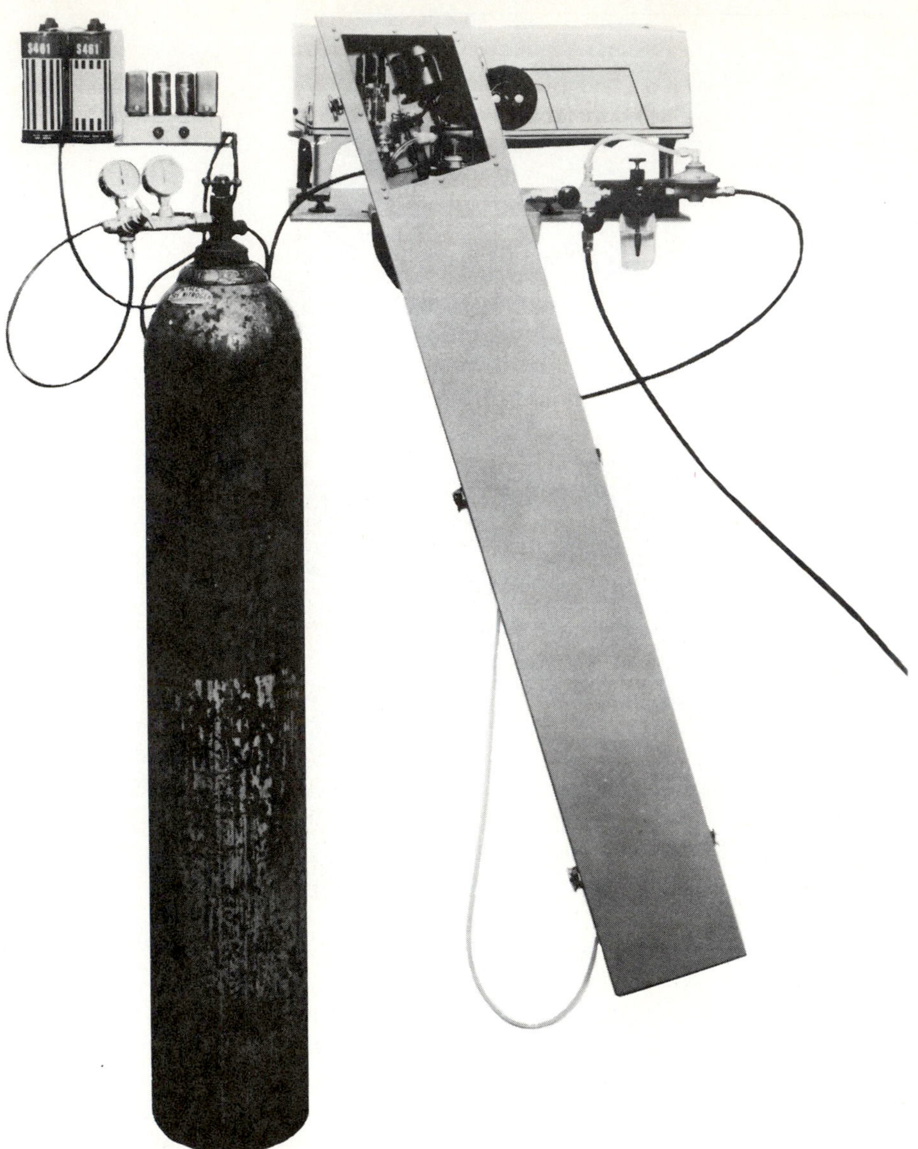

Figure 12.6 Stevens manometer-servo water-level recorder.

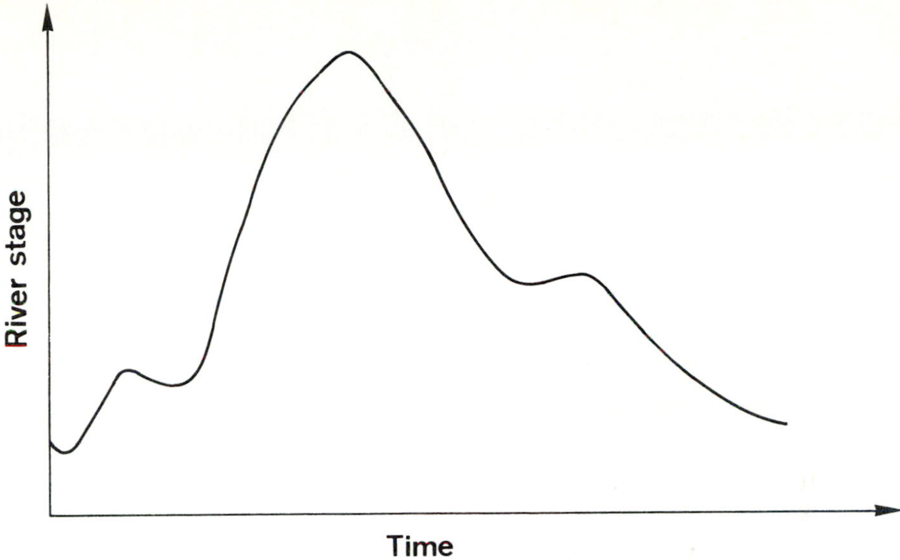

Figure 12.7 River-stage hydrograph.

the water surface, in the deepest part of the channel of a straight reach. Eddys and bends might cause the highest velocity to be located somewhere on either side of the channel, but, in general, the highest velocity will be above the deepest part of the channel. The velocity distribution in a stream across a vertical section is approximately logarithmic in nature, as shown in Figure 12.9. For computing the average vertical velocity, a large number of points will be needed to define this distribution. The average velocity in the vertical section is determined as

$$\bar{v} = \frac{1}{d} \int_{s=0}^{s=d} v(s)\ \mathrm{d}s \qquad (12.1)$$

where v varies with the depth, $0 \leq s \leq \mathrm{d}$. The integral defines the area encompassed by the velocity-distribution curve. d is the depth of flow at a given section, and s is

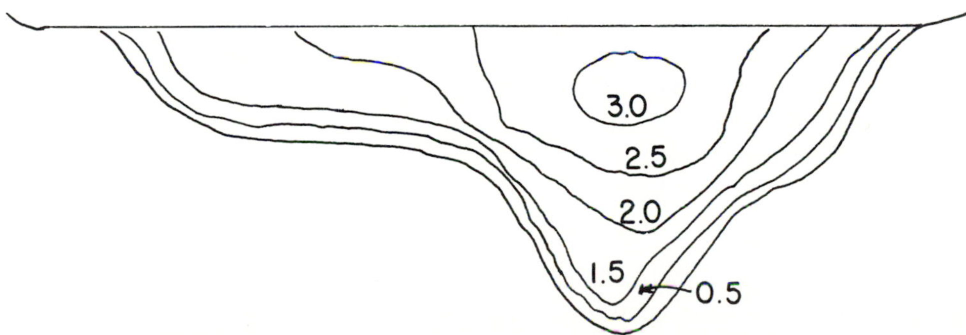

Figure 12.8 Velocity distribution in a nonuniform channel.

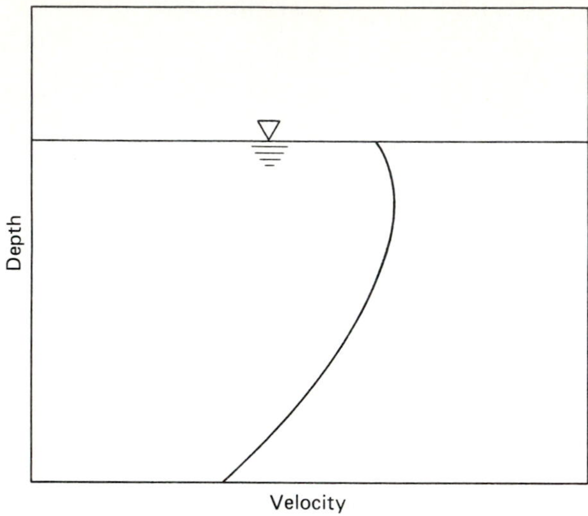

Figure 12.9 Vertical distribution of velocity in a stream channel.

the depth at which velocity is measured or known. Thus, $s = 0$ defines the stream bed, and $s = d$ defines the water surface.

1. The average velocity $\bar{v}$ in shallow streams with depths of flow not exceeding 3 m is taken as the velocity measured at 0.6 times the depth of flow, $v_{0.6}$, below the water surface,

$$\bar{v} = v_{0.6} \tag{12.2}$$

   This procedure needs a single-point measurement.

2. The average velocity $\bar{v}$ in moderately deep streams is computed as

$$\bar{v} = \frac{v_{0.8} + v_{0.2}}{2} \tag{12.3}$$

   where $v_{0.2}$ is the velocity measured at 0.2 times the depth of flow below the water surface, and $v_{0.8}$ is the velocity measured at 0.8 times the depth of flow below the water surface.

3. The average velocity in rivers having flood flows is obtained from a single measurement as

$$\bar{v} = cv_{0.5} \tag{12.4}$$

   where $v_{0.5}$ is the surface velocity measured within a depth of 0.5 m below the water surface; and c is a reduction factor, which is usually between 0.85 and 0.95, and is obtained from measurements taken at lower stages.

12.2.2 Estimating the Mean Channel Velocity of Floats

It is possible to estimate the mean velocity of flow in a channel, when necessary, by timing the movement of a float along a measured channel reach.

$$\bar{v} = \frac{L}{T} \tag{12.5}$$

where L is the distance traveled in meters, and T is the time of travel in seconds. This method must be used in a straight reach of channel on a windless day so that the float can be maintained in the center of the channel. Because velocity varies across the width and depth of flow in the stream, coefficients must be applied to convert surface-float velocity to mean channel velocity. These coefficients, shown in Table 12.1, have been developed from empirical relations. Somewhat better results are obtained when using a rod float 1 to 2 inches in diameter. The rod must be weighted on one end such that it floats vertically in the water and should be of such a length that the immersed length is 0.9 times the depth of water. Under these conditions, the rod tends to integrate the variations in vertical velocity distribution and tends to move at approximately the mean velocity. Table 12.2 gives coefficients for conversion of rod-float velocity to mean channel velocity.

12.2.3 Current Meters

The current meter is the most commonly used instrument to measure velocity of flow in streams. Several types of current meters are available. Common amongst them is the Price current meter.

As shown in Figure 12.10, a current meter is an instrument equipped with a propeller or cup-equipped rotating wheel that is driven by the water current. It is calibrated to register flow velocity relative to the number of wheel rotations. These rotations may be counted by means of an audible signal transmitted to earphones for

TABLE 12.1 COEFFICIENTS FOR SURFACE-FLOAT VELOCITIES

Average depth of reach (ft)	Coefficient
1	0.66
2	0.68
3	0.70
4	0.72
5	0.74
6	0.76
9	0.77
12	0.78
15	0.79
20 and over	0.80

TABLE 12.2 COEFFICIENTS FOR ROD-FLOAT VELOCITY MEASUREMENTS

Ratio of length of submerged rod to depth of channel	Ratio of mean velocity to float velocity
0.90	1.00
0.75	0.95
0.50	0.92

every fifth rotation and also may be had to register every rotation. For a current meter, the relation between rotations per second, N, and flow velocity, v (m/s), is of the form

$$v = a + bN \tag{12.6}$$

where a and b are constants of the meter and depend upon the size of the meter. Equation (12.6) is called the rating equation of a current meter. Constant a is equivalent to the velocity required to overcome mechanical friction. For a standard 12.5-cm-diameter Price meter (cup type), $a = 0.65$ and $b = 0.03$. For smaller 5-cm-diameter cup pigmy meters, $a = 0.3$ and $b = 0.003$. Operation of the current

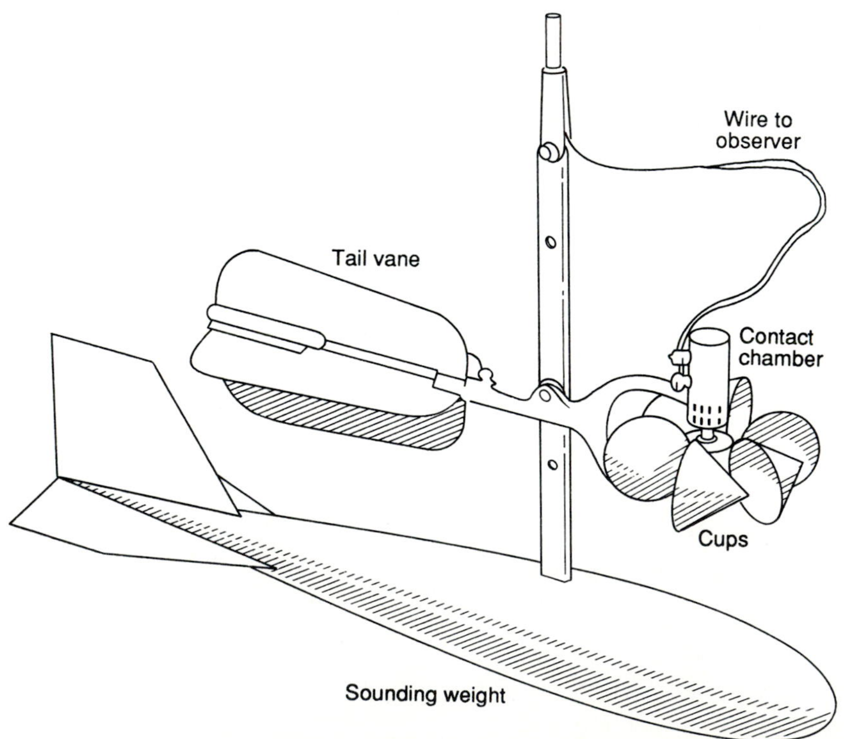

Figure 12.10 Price current meter.

meter is over a timed period. Calibration charts are then used to convert the recorded rotations to velocity. Recently developed current meters register velocity that is read directly on a meter.

An advantage of the Price current meter is that the cup-shaped meter wheel measures water flow accurately without being pointed directly in the line of flow. Propeller-type wheels must be pointed directly into the line of flow to register accurately. This condition is sometimes difficult to accomplish. A current meter containing a heavy finned weight is used to assist in keeping the instrument aligned with the direction of flow. The weight is also helpful in sounding the channel for depth as the instrument is suspended from above.

EXAMPLE 12.1 Compute the flow velocity in a channel using a 12.5-cm-diameter cup-type Price meter. The meter is kept at 0.6 times the depth of flow below the water surface. The recorded measurements are as follows:

DEPTH (m) (1)	REVOLUTIONS (N/s) (2)	FLOW VELOCITY (m/s) (3)
0	0	0
0.5	30	1.55
1.0	60	2.45
1.5	70	2.75
2.0	90	3.35
2.5	100	3.65
3.0	120	4.25

For the Price current meter used, $a = 0.65$ and $b = 0.03$. The velocities are computed using Equation (12.6), and are indicated in column (3).

12.3 MEASUREMENT OF DISCHARGE

12.3.1 Velocity–Area Station Method

Velocity–area stations have stream gages that measure velocities and cross-sectional areas of flow to obtain discharges. The basic equation used to obtain the discharge from velocity–area stations is

$$Q = VA \tag{12.7}$$

where Q is the discharge in cubic meters per second (cms) or cubic feet per second (cfs), V is the mean velocity of flow through the cross-sectional area of the channel in meters per second (mps) or feet per second (fps), and A is the cross-sectional area of flow in square meters (m^2) or square feet (ft^2). Equation (12.7) can be used as a continuity equation because the mass flow into a point must equal the mass flow out from that point. To determine the discharge, both the mean velocity and cross-sectional area of flow must be measured.

In order to conduct stream-discharge measurements, the stream channel must be subdivided into several smaller widths. The vertical dashed lines in Figure 12.11

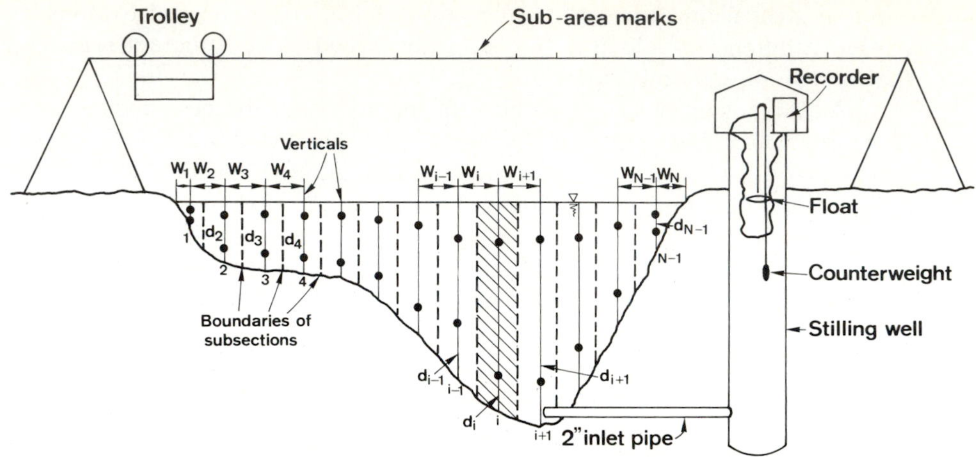

Figure 12.11 Schematic sketch of a velocity–area station.

represent subdivisions of the stream width. These subwidths are permanently marked on the bridge railing and are used each time velocity measurements are made. The number of width subdivisions depends on the actual width of the stream. Generally, the minimum number of subdivisions is 10, and for wide streams, many more are required. These subdivisions are not necessarily equally spaced. Accuracy of mean velocity and discharge measurements is nearly proportional to the number of these vertical subdivisions. Resistance to flow varies with depth of flow. Because channel depth in natural streams varies across the channel, the greater the number of subdivisions, the greater the accuracy. The actual number of subdivisions used is determined by experience and knowledge of open-channel hydraulics. The following guidelines can be used to select the number of subdivisions: (1) The subwidth should not be greater than 1/15 to 1/20 of the stream width. (2) The difference in velocities of adjacent subdivisions should not exceed 20%. (3) The discharge in each subdivision should be less than 10% of the total discharge.

The average velocity in each of the subwidth divisions is measured. To do this, the gager goes to the center of each subwidth station. The mean depth of flow of this subdivision is measured using sounding rods, sounding weights, or an echo-depth recorder (an electroacoustic instrument). A current meter is used to measure velocity for this subdivision. Velocity is measured from observations at points 0.2 and 0.8 of the total depth. These two velocity measurements are averaged to obtain the mean velocity in that subwidth. In subwidths where water is too shallow to reasonably operate the current meter at these depths, a single measurement is made at 0.6 of the total depth to obtain the mean velocity. The discharge for the subdivision is computed by taking the product of the mean depth, subarea width, and mean velocity. In other words, Equation (12.7) is applied to the subdivision. This gaging procedure is followed for each subdivision. Total discharge for the stream is determined by summing the discharges for all subwidths. The mean stream velocity can be computed by dividing the total discharge by the total cross-sections area. The velocity–area method when using the current-meter method is also referred to as the standard current-meter method.

A velocity–area station must be located in a reach of the channel that provides stable banks and a reasonably stable bottom. This condition usually produces a stable stage (water-surface elevation). The stream cross-section should be well-defined and should remain practically unaltered. The station should be easily accessible at all times. The reach of the channel should be reasonably straight so that unwanted eddying does not interfere with velocity measurement. Such stations are often located at bridges, where the bridge abutments provide stable bank conditions. A bridge also provides a stable operating platform for stream-gage measurements. Velocity–area stations not located at bridges require a cable-car arrangement similar to that shown in Figure 12.11. The stream gager uses this equipment to accomplish the same procedure as if he were on a bridge but with less convenience.

Where water-flow depths are sufficiently shallow, measurements are accomplished by wading the stream. Measurements made by wading are accomplished by means of a graduated meter rod attached to the current meter. Subwidths are delineated by a tape or by a measured line strung across the channel at the point of measurement.

The total discharge Q can be analytically computed using the velocity–area station method. From Figure 12.11, the stream width is divided into N subwidths, W_i, $i = 1, 2, \ldots, N$. Because the discharge is calculated using the method of midsections, the total cross-sectional area can be divided into $N - 1$ segments. This means there are $N - 1$ verticals. Therefore,

$$Q = \sum_{i=1}^{N-1} q_i \tag{12.8}$$

where q_i is the discharge of the ith segment and is obtained as

$$q_i = a_i v_i \tag{12.9}$$

with a_i is the area of the ith segment = (depth at the ith vertical) × ($\frac{1}{2}$ width to the left + $\frac{1}{2}$ width to the right) expressed as

$$a_i = d_i \frac{W_i + W_{i+1}}{2}, \qquad i = 2, 3, \ldots, N - 2 \tag{12.10}$$

and v_i is the average velocity at the ith vertical, and d_i is the depth of the ith segment. For the first segment $i = 1$, and the last segment $i = N - 1$, the areas are represented by triangles as

$$a_1 = \bar{W}_1 d_i \qquad a_{N-1} = \bar{W}_{n-1} d_{N-1} \tag{12.11}$$

where

$$\bar{W}_1 = \frac{(W_1 + W_2/2)^2}{2W_1}$$

and

$$\bar{W}_{N-1} = \frac{(W_N + W_{N-1}/2)^2}{2W_N}$$

EXAMPLE 12.2 The data for a stream gaging station are given as follows:

Distance from left water edge (m)	0	1	3	5	7	9	10
Depth (m)	0	1.0	2.0	3.0	2.0	1.0	0
Average velocity (m/s)	—	0.20	0.3	0.5	0.3	0.2	0

Compute the discharge in the stream.

Solution First, the average width and then the cross-sectional area of flow of each subsection are computed using Equation (12.10). The average width for the first section, W_1, is

$$\bar{W}_1 = \frac{(1 + 2/2)^2}{2 \times 1} = 2.0 \text{ m}$$

The average width of the last section, W_{N-1}, is

$$\bar{W}_{N-1} = \frac{(1 + 2/2)^2}{2 \times 1} = 2.0 \text{ m}$$

The average width of the rest of the sections, $\bar{W}_i$, is

$$\bar{W}_i = \frac{(2 + 2)}{2} = 2.0 \text{ m}$$

The cross-sectional flow area of each subsection is computed from Equation (12.10), that is,

$$a_1 = d_1\bar{W}_1 = 1.0 \times 2.0 = 2.0 \text{ m}^2$$

$$a_2 = d_2\bar{W}_2 = 2.0 \times 2.0 = 4.0 \text{ m}^2$$

and so on. For each subsection, the discharge is computed using Equation (12.9), that is,

$$q_1 = a_1 v_1 = 2.0 \times 0.20 = 0.4 \text{ m}^3/\text{s}$$

$$q_2 = a_2 v_2 = 4.0 \times 0.3 = 1.2 \text{ m}^3/\text{s}$$

and so on. The calculations are shown in tabular form as follows:

Distance from left water edge (m)	Average width, $\bar{W}$ (m)	Depth of flow, d (m)	Subsection area of flow (m^2)	Average velocity, v (m/s)	Subsection discharge (m^3/s)
0	0.0	0	0	0	0
1	2.0	1.0	2.0	0.2	0.4
3	2.0	2.0	4.0	0.3	1.2
5	2.0	3.0	6.0	0.5	3.0
7	2.0	2.0	4.0	0.3	1.2
9	2.0	1.0	2.0	0.2	0.4
10	0.0	0.0	0.0	0.0	0.0

The total discharge is calculated using Equation (12.8) or summing the subsection discharges.

$$Q = 6.2 \text{ m}^3/\text{s}$$

12.3.2 Moving-Boat Method

For large rivers such as the Mississippi River, the standard current-meter method is not a convenient method, especially during flooding. Not only is it too time consuming to take velocity measurements, but it is virtually impossible to keep the boat stable on a rapidly moving water surface. Under these conditions, a moving-boat method is very useful. A special propeller-type current meter is towed to a boat. This meter is free to move about a vertical axis at a velocity v_m perpendicular to the direction of flow. If the flow velocity is v_f, the meter aligns itself in the direction of the resultant velocity v_r, which makes an angle θ with the direction of the boat. The meter itself registers the velocity v_r. From Figure 12.12, if v_m is normal to v_f, then

$$v_m = v_r \cos \theta \qquad \text{and} \qquad v_f = v_r \sin \theta \tag{12.12}$$

The width w between the two verticals is

$$w = v_m T \tag{12.13}$$

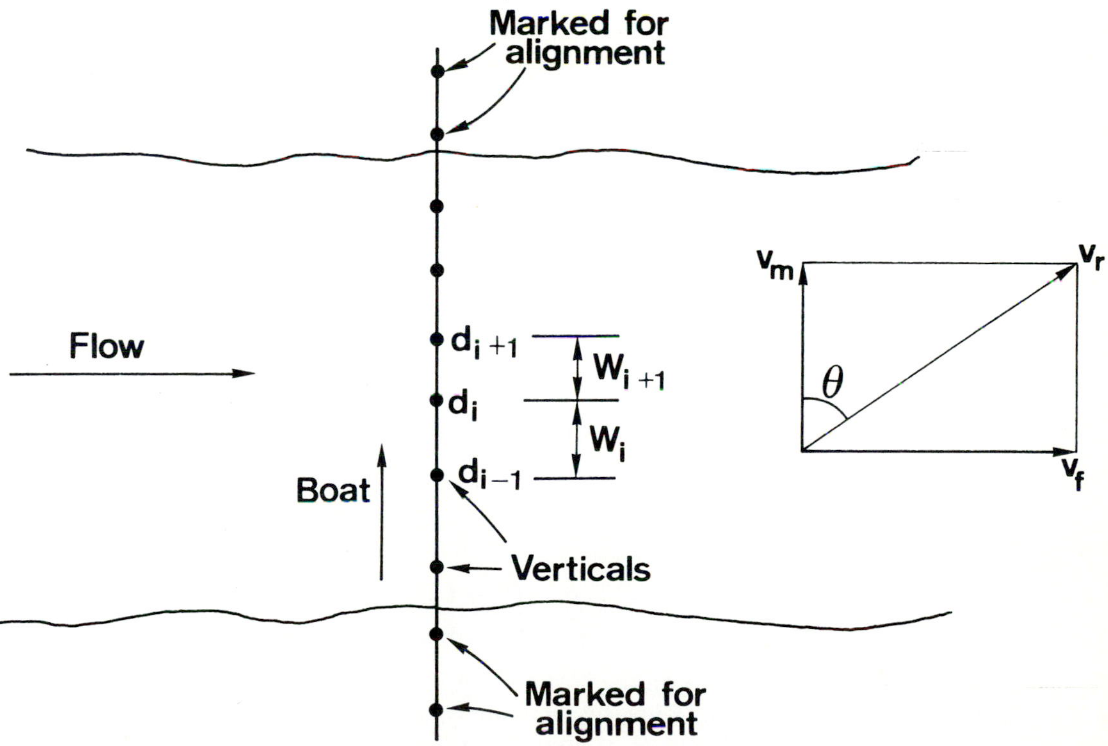

Figure 12.12 Moving-boat method.

where T is the time of travel from one vertical to the other. The depths at the two verticals, i and $i + 1$, are d_i and d_{i+1}, respectively. The discharge in the subarea between the two verticals then is

$$q_i = \frac{d_i + d_{i+1}}{2} w_{i+1} v_f \tag{12.14}$$

where v_f is the average velocity in the vertical. By substituting Equations (12.12) and (12.13) into Equation (12.14),

$$q_i = \frac{d_i + d_{i+1}}{2} (v_r^2 \sin \theta \cos \theta) T_i \tag{12.15}$$

The discharges of subareas are summed up to obtain the total stream discharge as in Equation (12.8).

The boat is moved from one bank of the river to the other along a cross-sectional line. This line is defined by permanent landmarks where the reach has islands, shoals, bars, etc. When the boat is in sufficient water, the instruments are commissioned. The current meter is immersed at a depth of 0.5 m from the water surface to measure surface velocities and the angle of the meter to the boat, θ. The echo-depth recorder measures the depths at the vertical sections. A large number of observations are taken during the back-and-forth journey of the boat to obtain good average readings.

12.3.3 Chemical Gaging

Chemical gaging is also referred to as the dilution method, and is especially useful in very small streams, mountain streams strewn with boulders, etc. This method employes the conservation of mass of the tracer to be used. Common salt, fluorescent dyes, and radioactive materials are the main tracers used. A tracer should be able to mix freely with flow; should not react with sediment, channel boundaries, or vegetation; and should not evaporate. The tracer of specified concentration C_1 is injected into the stream at a constant rate Q_c at a defined location. Samples are taken at a downstream point where the concentration gradually rises to a constant value C_2. Let C_0 be the initial concentration of tracer in the stream. At steady state, the mass conservation for the tracer between the two locations can be expressed as

$$C_0 Q + C_1 Q_c = (Q + Q_c) C_2$$

which yields

$$Q = \frac{C_1 - C_2}{C_2 - C_0} Q_c \tag{12.16}$$

Equation (12.16) is based on the assumption of steady flow, which does not always exist in natural streams. Thus, by knowing Q_c, C_i ($i = 0, 1, 2$), the stream discharge can be measured. Because Q_c is constant, this method of chemical gaging is called the constant-rate injection method, or plateau gaging.

12.3.4 Ultrasonic Method

In this method, ultrasonic signals are used to measure flow velocity. Two transducers are installed at the same elevation from the bed on both sides of the stream. The path connecting the transducers makes an angle with the direction of flow. The transducers receive as well as transmit ultrasonic signals. The time taken by a signal from one transducer to the other is recorded. The average velocity of flow along the path is then obtained by knowing the path length and its angle with the flow direction. This average velocity v_s corresponds to the height h above the bed and is not equal to the average velocity v_f for the entire cross-section of flow. However, v_s can be related to v_f, and a relation between v_s/v_f and h can be established for a given stream cross-section. In this way, v_f is obtained, and then the stream discharge. If this estimation involves a single-path signal, it is referred to as single-path gaging. For a given depth of flow, multiple single paths can be used to obtain v_s for different values of h. The values of v_s are then averaged to yield v_f. This estimation is called multipath gaging.

Ultrasonic equipment is commercially available and easy to use. This method of gaging is suitable for automatic recording of data. It is accurate, efficient, and can accommodate rapid changes in the magnitude and direction of flow. However, unstable cross-sections, large loads of sediment in suspension, steep temperature changes, changing weed growth, air entrapment, changes in salinity, etc., can reduce accuracy of this method.

12.3.5 Indirect Methods

These methods measure flow depths at specified locations, and translate these depths into discharges using depth–discharge relations applicable to those locations. Flow-measuring structures and slope–area methods exemplify indirect methods.

FLOW-MEASURING STRUCTURES

Different types of flow-measuring structures are in use. Broad-crested weirs, flumes made of concrete, masonry or metal sheets, and V-notches are common examples of such structures. When a flow-measuring structure is installed in a stream, it produces a unique control section in the flow. The discharge Q through the structure is related to the water-surface elevation H adjacent to or within the structure as

$$Q = f(H) \tag{12.17}$$

where f is some function. For example, for a weir, Equation (12.17) becomes

$$Q = CBH^x \tag{12.18}$$

in which B is the width of the weir crest, C is the discharge coefficient, and x is an exponent. Both C and x are specific for a weir, and are obtained by calibration. The value of x is 1.5 for broad-crested weirs and 2.5 for triangular weirs. The value of C takes into account the channel geometry, the friction loss due to the weir, horizontal and vertical contractions of flow, the form of the weir, etc. Tables of coefficient C are given by Brater and King (1976).

Equation (12.18) is applicable for free flows in which the downstream conditions do not affect the flow at the structure. For conditions deviating from free flows, such as submerged or drowned flows, a reduction factor is applied to the free-flow discharge. The weir should be located in a long, straight, uniform stretch of the stream free of rocks, islands, bars, or other obstructions. It should be placed such that water approaches perpendicular to the crest. The measurement of head must be made at a point upstream of the crest not less than 4 to 6 times the head over the weir.

SLOPE–AREA METHOD

This method is based on the principle of energy conservation. A stream reach is selected, as shown in Figure 12.13. From Bernouli's equation applied to the ends of the reach (sections 1 and 2),

$$z_1 + y_1 + \frac{v_1^2}{2g} = z_2 + y_2 + \frac{v_2^2}{2g} + h_L \tag{12.19}$$

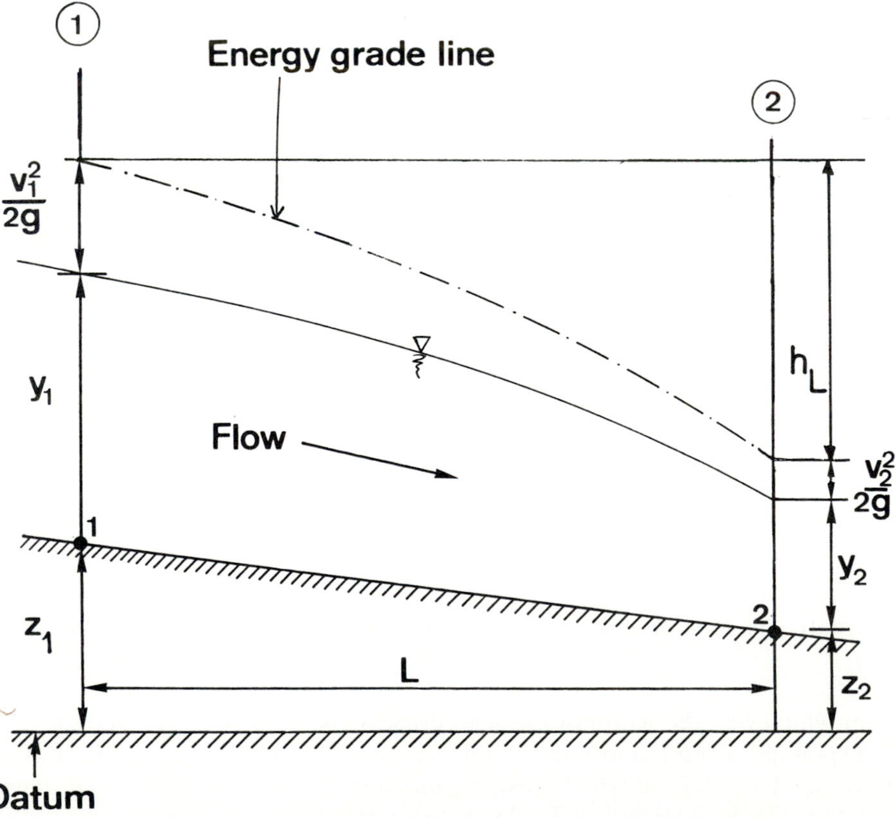

Figure 12.13 Channel reach for the slope–area method.

where z is the elevation above the chosen datum, y is the depth of flow, v is the velocity of flow, h_L is the head loss in the reach, and subscripts 1 and 2 refer to the sections. If the water-surface elevation is h, then $h = z + y$. The quantity h_L includes head loss due to friction, h_f, and head loss due to eddys, h_e. Equation (12.19) can then be expressed as

$$h_1 + \frac{v_1^2}{2g} = h_2 + \frac{v_2^2}{2g} + h_e + h_f$$

or

$$h_f = (h_1 - h_2) + \left(\frac{v_1^2}{2g} - \frac{v_2^2}{2g}\right) - h_e \tag{12.20}$$

The head loss due to friction can be used to express the energy slope S_f of Manning's equation for uniform flow:

$$S_f = \frac{h_f}{L} = \frac{Q^2}{K^2} \tag{12.21}$$

where K is the channel conveyance expressed (in MKS units) as

$$K = \frac{1}{n} AR^{2/3} \tag{12.22}$$

in which R is the hydraulic radius (A/P, P = wetted perimeter), and n is Manning's roughness factor. In order to account for nonuniformity of flow, an average energy slope $\bar{S}_f$ is obtained from the average conveyance. The conveyance factor K is estimated as

$$K = (K_1 K_2)^{0.5}, \qquad K_1 = \frac{1}{n_1} A_1 R_1^{2/3}, \qquad K_2 = \frac{1}{n_2} A_2 R_2^{2/3} \tag{12.23}$$

The head loss due to eddys is estimated as

$$h_e = K_e \left|\frac{v_1^2}{2g} - \frac{v_2^2}{2g}\right| \tag{12.24}$$

where K_e is the eddy-loss coefficient and depends upon the nature of flow and the channel conveyance. Table 12.3 gives values of K_e.

By knowing the water-surface elevations, channel cross-sections, and Manning's n at the two sections, the discharge from the stream reach can be computed. This method is very powerful but involves a trial-and-error solution. To initiate the

TABLE 12.3 VALUES OF EDDY-LOSS COEFFICIENT

Cross-section of channel reach	Values of K_e	
	Expansion	Contraction
Uniform	0	0
Gradual transition	0.3	0.1
Abrupt transition	0.8	0.6

solution, the velocity at the two sections can be assumed to be the same. Then h_f is computed. This is used to compute S_f and then Q. Because A_1 and A_2 are known from before, v_1 and v_2 are now computed, and then h_e is computed. After this, a refined value of h_f is calculated, and other calculations are reiterated, until the discharge values of two successive iterations are within an acceptable range.

SLOPE-AREA METHOD FOR FLOODS

Frequently, it is necessary to estimate the discharge that occurred in an ungaged stream from high-water marks left by a flood event. The high-water marks may include stranded vegetation (grass, straw, etc.), silt lines on river banks, trace of erosion on the banks, and silt or straw lines on buildings, trees, bridges, etc. Newspaper accounts, records in archives, and interviews with people familiar with flood may provide valuable information. High-water marks enable reconstructing the water-surface elevation and the slope of the water surface. The value of the slope is determined from high-water marks located upstream and downstream. The slope–area method is then used to estimate the flood discharge. The value of Manning's roughness coefficient can be assumed based on the type and condition of the stream. This method is expected to produce approximate values of discharge.

The stream reach selected should be straight and uniform, and the high-water marks are of good quality. If the stream is contracting or expanding, then a contracting section should be preferred. As an approximation, the reach length should be greater than 75 times the mean depth. The recorded fall in the water-surface elevation based on high-water marks should be greater than 0.15 m.

EXAMPLE 12.3 Estimate the flood discharge through a 5-m wide rectangular channel for the following data. The depth of water is 2 m and 1.8 m at two sections 500 m apart. The drop in water-surface elevation is 0.25 m. Manning's roughness coefficient is 0.025. Assume eddy loss to be zero.

The channel conveyance at both sections is computed using Equation (12.22).

$$A_1 = 5 \times 2 = 10 \text{ m}^2, \quad P_1 = 5 + 2 + 2 = 9 \text{ m}, \quad R_1 = \frac{10}{9} = 1.11 \text{ m}$$

$$A_2 = 5 \times 1.8 = 9 \text{ m}^2, \quad P_2 = 5 + 1.8 + 1.8 = 8.6 \text{ m}, \quad R_2 = \frac{9}{8.6} = 1.05 \text{ m}$$

$$K_1 = \frac{1}{0.025} \times 10 \times (1.11)^{2/3} = 428.82$$

$$K_2 = \frac{1}{0.025} \times 9 \times (1.05)^{2/3} = 371.90$$

The average conveyance K is computed from Equation (12.23):

$$K = (428.82 \times 371.90)^{0.5} = 399.35$$

To start the computation, $h_f = 0.25$ m is assumed. Eddy loss $h_e = 0$. Therefore, $S_f = 0.25/500 = 0.0005$. This yields, from $Q = K(S_f)^{0.5}$, $Q = 399.35 \times (0.0005)^{0.5} = 8.9297$ m^3/s.

Now the velocities at sections 1 and 2 are computed so Equation (12.20) can be used to compute h_f.

$$h_f = 0.250 + \frac{8.9297 \times 8.9297}{10 \times 10 \times 19.62} - \frac{8.9297 \times 8.9297}{9 \times 9 \times 19.62}$$

$$= 0.250 + 0.040642 - 0.050175 = 0.0404$$

This new value of h_f is now used to compute S_f and the results of the calculations are as shown in tabular form:

Trial	h_f (m)	S_f	Q (m³/s)	$v_1^2/2g$	$v_2^2/2g$	h_f from equation (12.20) (m)
1	0.25	5×10^{-4}	8.9297	0.0406	0.0502	0.2404
2	0.2404	5×10^{-4}	8.9297	0.0406	0.0402	0.2404

The discharge in the channel is 8.929 m³/s.

12.4 STAGE-DISCHARGE RELATION

A stage-discharge relation for a velocity–area station or gaging section is obtained by plotting measured stage on the ordinate and the measured discharge on the abscissa, as shown in Figure 12.14. This relation is also called the rating curve. It represents the integrated effect of a wide range of channel and flow parameters. The combined effect of these parameters is designated as control. If the rating curve for a gaging station does not change with time, the control is called permanent; otherwise, it is called shifting control. The stage-discharge relation is of fundamental importance in the acquisition of discharge measurements. All direct discharge-measuring methods require construction of this relationship before measured stages can be translated into discharges. Kennedy (1984) has given a good discussion of discharge ratings at gaging stations used by the U.S. Geological Survey.

12.4.1 Simple Rating Curve

When measured values of stage and discharge are plotted on arithmetic paper, the result is an approximate parabolic curve, as shown in Figure 12.14. This curve can be expressed as

$$Q = a(h - b)^c \tag{12.25}$$

in which b is a constant representing the gage reading for zero discharge, and a and c are rating curve constants. When the data are plotted on logarithmic paper, the plot is a straight line, as shown in Figure 12.15. Equation (12.25) becomes

$$\log Q = \log a + c \log (h - b) \tag{12.26}$$

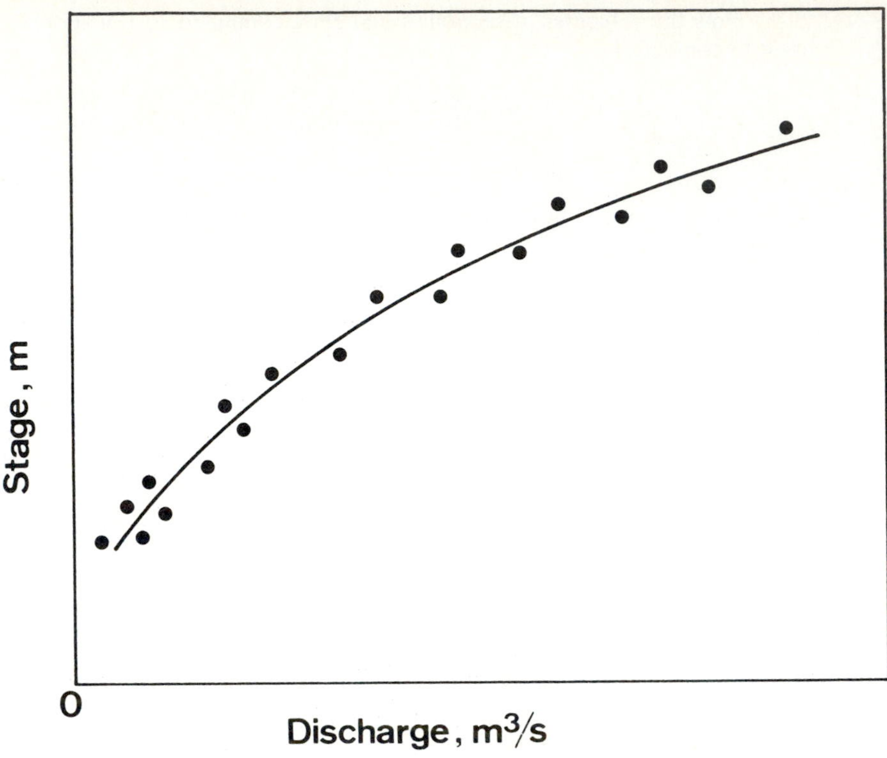

Figure 12.14 Stage-discharge relationship.

The best values of constants a and c can be obtained using the least squares method. However, constant b must be found beforehand, and this can be estimated in several ways. A trial-and-error method can be used to yield b, which then gives the best-fit curve. Another way is to extrapolate the rating curve corresponding to $Q = 0$ and then plot log Q versus log $(h - b)$. If the plot is a straight line, then the value of b obtained by extrapolation is acceptable. Otherwise, another value in the neighborhood of the previous value of b is selected and the procedure is repeated. A graphical method has been reported by Running (cited in Wisler and Brater, 1949, p. 404).

Another method of computing b is analytical. From a smooth curve of Q versus h, three values of discharge, Q_1, Q_2, and Q_3, are selected such that $Q_1/Q_2 = Q_2/Q_3$. The corresponding values of stage are h_1, h_2, and h_3. Then, by using Equation (12.25),

$$\frac{(h_1 - b)^c}{(h_2 - b)^c} = \frac{(h_2 - b)^c}{(h_3 - b)^c}$$

or

$$\frac{h_1 - b}{h_2 - b} = \frac{h_2 - b}{h_3 - b}$$

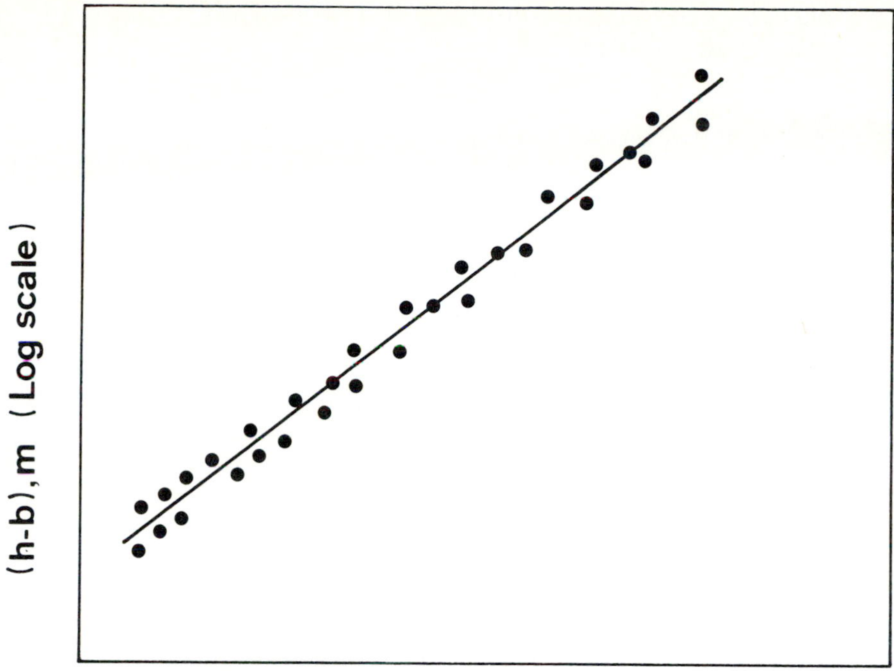

Figure 12.15 Stage-discharge relationship on logarithmic paper.

from which the value of b is derived as

$$b = \frac{h_1 h_3 - h_2^2}{h_1 + h_3 - 2h_2} \tag{12.27}$$

Alternatively, all three parameters, a, b, and c, can be obtained by optimization.

The simple rating curve is generally satisfactory for a majority of streams where rapid fluctuations of stage are not experienced at the gaging section. The adequacy of the curve is measured by the scatter of data around the fitted curve. When there is a permanent control, the rating curve is essentially permanent. If the rating curve is made using a range of stages from low to high, it can be used to interpolate the discharge for any stage of flow between the measured stages without measuring that flow. It is important to check the stability of the curve by periodic discharge measurements, and to extend it with each new observed high stage. Changes in channel shape, due to scouring or sedimentation, can change the effect of control and thereby change the rating curve.

For some gaging stations, there may be two or more controls each for a particular range of stage. The rating curve for such a station is discontinuous; the point of discontinuity corresponds to the stage reflecting the change in control. An example is when submergence of a weir control starts when the tailwater level below the control rises above the lowest point of the control. Even under such conditions, the simple rating curve may be satisfactory if the control is permanent, free of backwater, and the stream slope is steep.

12.4.2 Shifting Control

When the control of a gaging station changes, its rating curve changes. The change may result from (1) scour or deposition, (2) varying backwater, (3) rapidly changing flow, and (4) changes in flow caused by dredging, channel encroachment and weed growth, etc. For the shifting control due to cases (1) and (4), frequent current-meter gaging is needed. The discharge is then estimated by noting the difference between the stage at the time of a discharge measurement and the stage obtained from the rating curve corresponding to the same discharge. This difference serves as a correction and is applied to all stages before reading the rating curve. If this correction varies from one measurement to another, it can be assumed to vary linearly in time. The effect of backwater and unsteady flow is amenable to analytical treatment.

12.4.3 Constant-Fall Rating Curve

The backwater may develop as a result of an obstruction downstream or high stages in an intersection stream, and may be variable in time. Under the conditions of shifting control due to backwater effects, a given stage will indicate different dis-

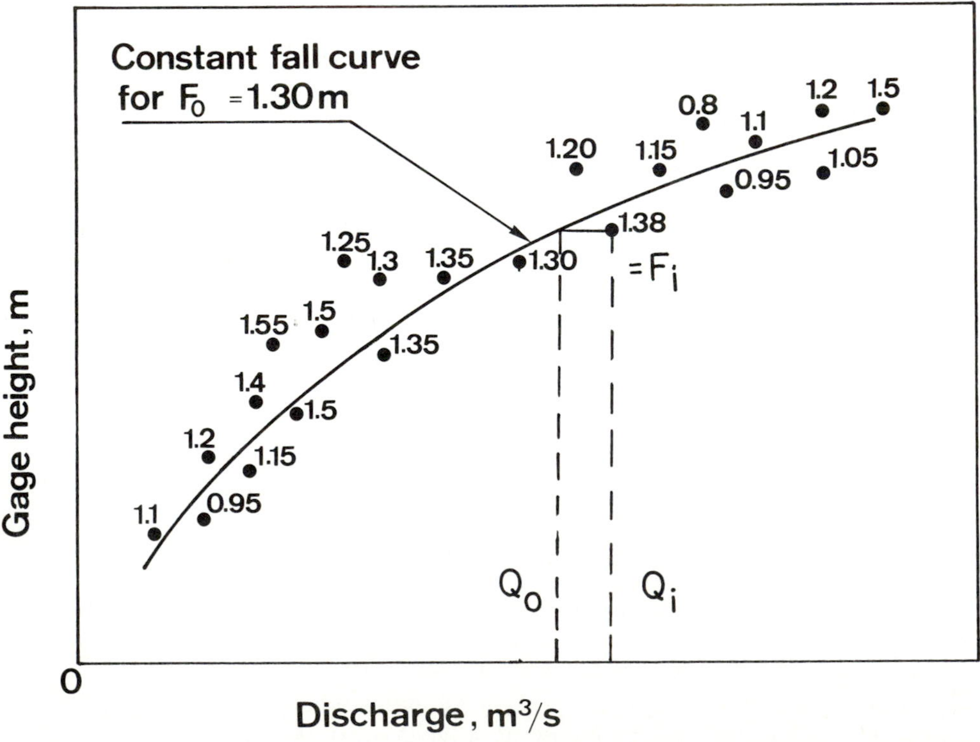

Figure 12.16 Constant-fall rating curve.

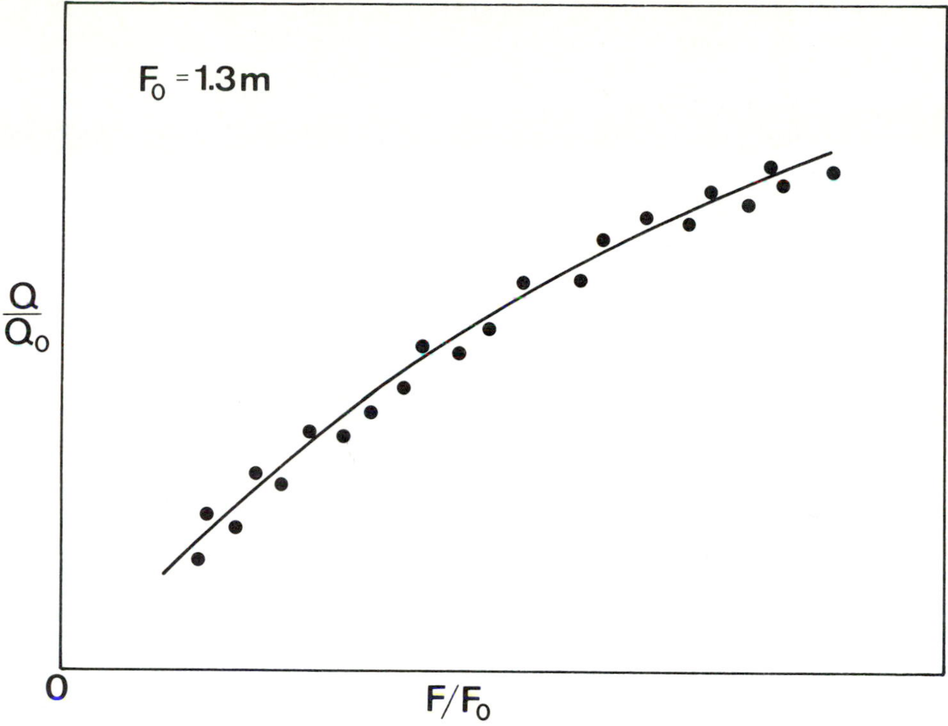

Figure 12.17 Adjustment curve for constant-fall rating curve.

charge values. This results from differences in water-surface slope at the control. The discharge is estimated by establishing another gage some distance downstream of the main gaging station. This other gage is called the secondary, or auxiliary, gage. Measurements are made at both gages. The difference between stages of these gages indicates the fall (F) or slope ($S = F/L$) of the water surface in the reach. The gages should be so far apart as to produce at least 30 cm (1 ft) of F. The discharge then is a function of stage h (at the main gage) and fall F, and is expressed as

$$\frac{Q}{Q_0} = \left(\frac{S}{S_0}\right)^k = \left(\frac{F}{F_0}\right)^m \tag{12.28}$$

where Q_0 is the normalized discharge at the stage when $F = F_0$ or $S = S_0$, and k and m are exponents. k has a value of about 0.5. If the water-surface profile between the two gages is not a straight line, then m need not be 0.5 and must be determined empirically. This method is called the slope-stage-discharge rating-curve method.

This method works as follows. Measurements are made at the main gage as well as at the auxiliary gage. The value of F is computed for each discharge measurement. If the variation in observed F values is not high, an average value of F is computed and this average value is taken as F_0. All observed values of stage and discharge for values of $F \cong F_0$ are plotted as a simple rating curve, as shown in Figure 12.16. This is the Q_0 versus h curve and is called the constant-fall curve. For each measurement with $F \neq F_0$, values of Q/Q_0 and F/F_0 are calculated and plotted, as shown in Figure 12.17. The plot of Q/Q_0 versus F/F_0 is called the auxiliary curve, or adjustment curve. If this curve plots as a straight line on logarithmic paper, its slope defines m in Equation (12.28). The constant-fall curve and auxiliary curve provide the required stage-discharge information. The discharge can be computed for a given stage by first computing the ratio F/F_0 and then obtaining the ratio Q/Q_0 from the auxiliary curve. Corresponding to the given stage, a value of Q_0 is obtained from the constant fall curve. This value of Q_0 is multiplied by Q/Q_0 to get Q, that is, $Q = (Q/Q_0) \times Q_0$.

12.4.4 Normal-Fall Rating Curve

Under some conditions, F varies widely and its variation is related to stage. Then a normal fall, F_n, is defined as a function of stage, which replaces F_0 in Equation (12.28), as shown in Figure 12.18. This is a second auxiliary curve. Correspondingly, normal flow Q_n replaces Q_0. Analogous to a constant-fall rating, a normal-fall rating is employed, as shown in Figure 12.19. In the same vein, the auxiliary curve is obtained by plotting Q/Q_n versus F/F_n, as shown in Figure 12.20. For a given stage, the value of actual fall F is computed first. Then a value of F_n is obtained from the second auxiliary curve, and the ratio F/F_n is computed. Against this value of F/F_n, the value of Q/Q_n is obtained from the first auxiliary curve. The value of Q_n is read from the rating curve, which then is multiplied by Q/Q_n to yield the actual value of discharge, that is, $Q = (Q/Q_n) \times Q_n$.

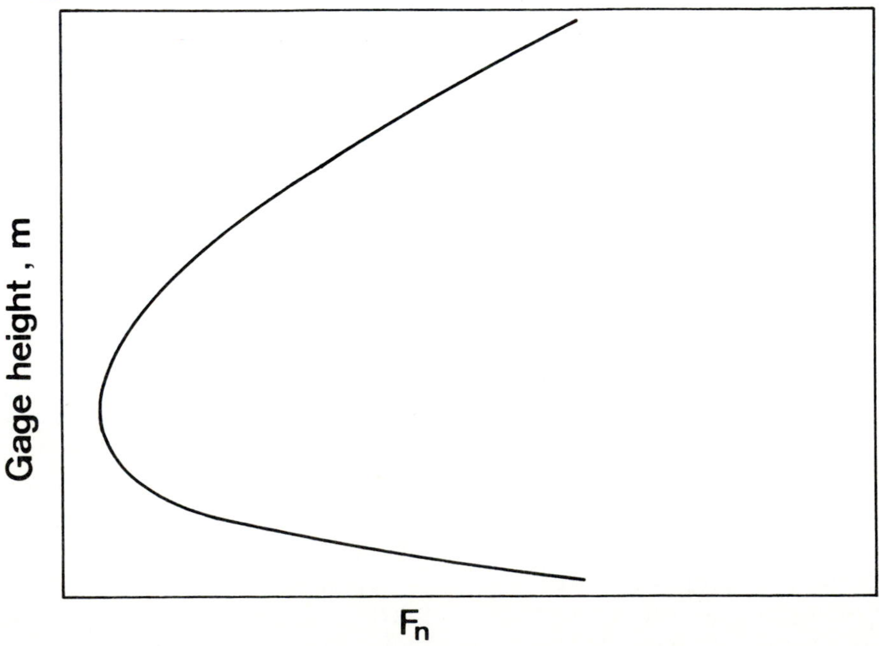

Figure 12.18 Normal-fall curve.

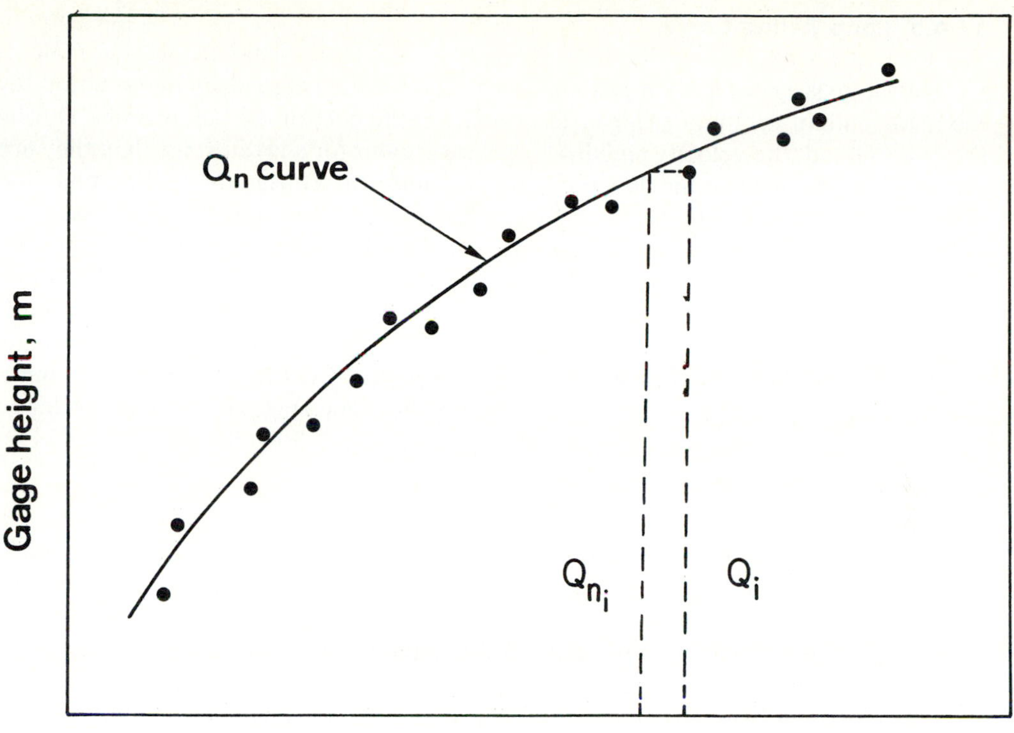

Figure 12.19 Q_n curve for a normal-fall rating curve.

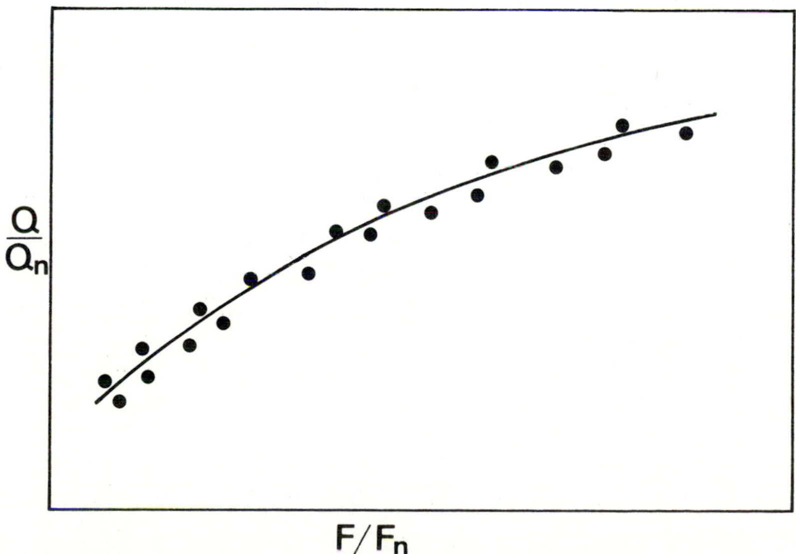

Figure 12.20 Auxiliary curve for a normal-fall rating curve.

12.4.5 Loop Rating Curve

During the passage of a flood past a gaging section, the stage-discharge relation for the rising limb of the hydrograph is different from the one for the falling limb. During the rise of flood, the velocity and discharge are greater for a given stage than they are for the same stage when the flow is steady and uniform. During the falling stage, the reverse is true. Thus, the stage-discharge relation for unsteady flows is a loop, as shown in Figure 12.21. It may be seen that for the same stage, there is greater discharge through the stream reach during the rising stage than it is during the falling stage. This loop may be plotted from stage and discharge measurements of a flood. This curve can be used as an approximation for other floods of about the same magnitude and duration. For floods with multiple peaks, the loop may be complicated. Thus, a better procedure is to relate for the same stage the normal discharge Q_N under steady uniform flow to the measured unsteady-flow discharge Q_M (Chow, 1959) as

$$\frac{Q_M}{Q_N} = \left(1 + \frac{1}{V_c S_0}\frac{dh}{dt}\right)^{0.5} \tag{12.29}$$

where V_c is the velocity of the flood wave or wave celerity; S_0 is the channel slope, the slope of the water surface for uniform flow, and dh/dt is the rate of change of the stage. The value of V_c is approximately equal to $1.4V$, where V is the average velocity obtained from Manning's equation for the given stage, and S_0 in Equation

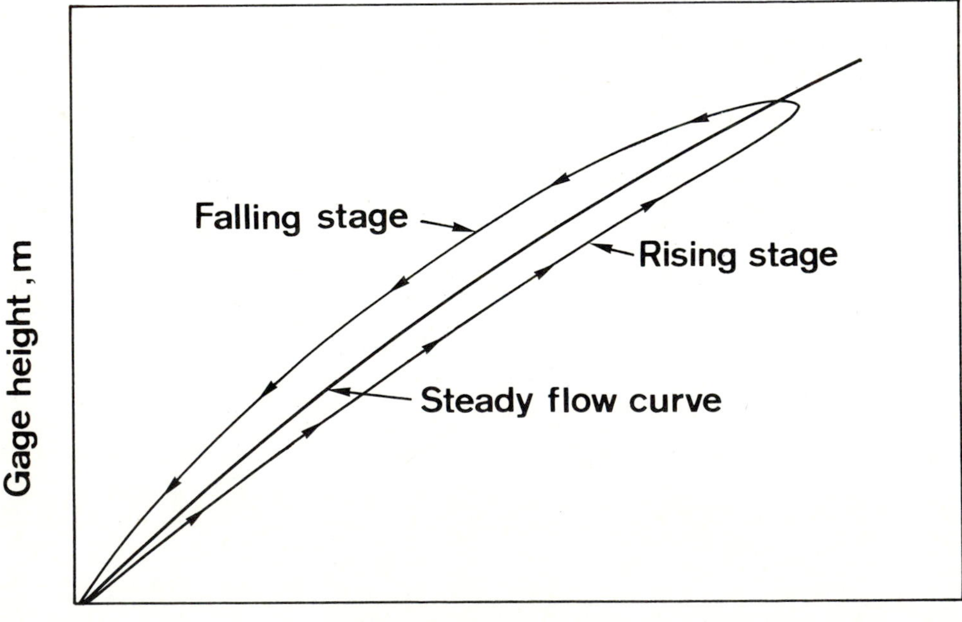

Figure 12.21 Stage-discharge relationship for unsteady flow.

(12.29) is replaced by the energy slope S_f. From measurements, dh/dt and $1/V_c S_0$ are computed. The term $1/V_c S_0$ is plotted against the observed stage. The actual discharge for an observed stage is determined by first estimating dh/dt for that stage, reading $1/V_c S_0$ from the plot, and then using Equation (12.29).

Another method is to construct the rating curve for those observations for which flow is uniform (water surface slope = stream-bed slope). By assuming V_c to be constant, a plot of Q_M/Q_N versus the slope of the flood wave expressed as a rate of change of stage (dh/dt) is prepared, where Q_N is the discharge corresponding to the measured stage when $S_f = S_0$. This curve serves as a correction curve. Thus, for a given stage, the discharge is first obtained from the rating curve and dh/dt is computed for which the actual discharge is then obtained from the correction curve.

12.5 EXTENSION OF RATING CURVES

The rating curve constructed for gaging site does not, in general, cover the full range of gage heights likely to be encountered in the future. This may be especially true for extremes, both low and high. The curve has the highest curvature in the range of low stages, where measurements may be scarce. The design of hydraulic structures requires computation of a stage for a design discharge using other means. Thus, it may be necessary to extend the rating curve beyond the range of values used in its construction. Extension of a rating curve is risky and should be carefully checked with the result obtained from another method. Before extrapolation, the site, control, and physical factors should all be carefully examined. A completely satisfactory method for extrapolation does not exist. Three simple methods are described in what follows.

12.5.1 Simple Rating Curve

When the rating curve is described by Equation (12.25), it plots as a straight line on logarithmic paper. The straight line can then be easily extended. Of course, this involves determination of constant b. The simple rating curve, however, cannot accommodate any significant change in the hydraulic geometry of the stream at high flows.

12.5.2 Conveyance Method

This method is based on the conveyance relation for nonuniform flow:

$$Q = K(S_f)^{0.5} \tag{12.30}$$

where all the terms are as defined before. The conveyance K can be expressed either using Chezy's or Manning's equation. With Chezy's relation,

$$K = CAR^{0.5}, \qquad K = K_* C, \qquad K_* = AR^{0.5} \tag{12.31}$$

and with Manning's relation for Q in m³/s,

$$K = \frac{1}{n} AR^{2/3}, \qquad K = \frac{K_*}{n}, \qquad K_* = AR^{2/3} \tag{12.32}$$

where C is Chezy's roughness coefficient, n is Manning's roughness factor, R is the hydraulic radius, and A is the cross-sectional area of flow.

Both A and R are computed from a stage, and thus K is computed from a stage if an appropriate roughness factor is known. A plot of K_* versus stage is prepared, as shown in Figure 12.22, using field observations, and a best-fit curve is obtained. This curve is extended beyond the observed range. For an observed Q, K_* is obtained from this curve. Then by using this value of K_* and observed Q, the value of $S_f^{0.5}/n$ or $S_f^{0.5}C$ is obtained from Equations (12.30) and(12.31) or (12.32). If desired, a plot of stage versus $S_f^{0.5}/n$ or $S_f^{0.5}C$ can be constructed, as shown in Figure 12.23. These two curves are utilized to compute the actual discharge for any observed stage. For example, K_* and $S_f^{0.5}/n$ or $S_f^{0.5}C$ are first obtained corresponding to the specified stage. Then, Equation (12.30) is used to compute the actual discharge. Sometimes, a single value of $S_f^{0.5}/n$ or $S_f^{0.5}C$ is obtained for the maximum observed

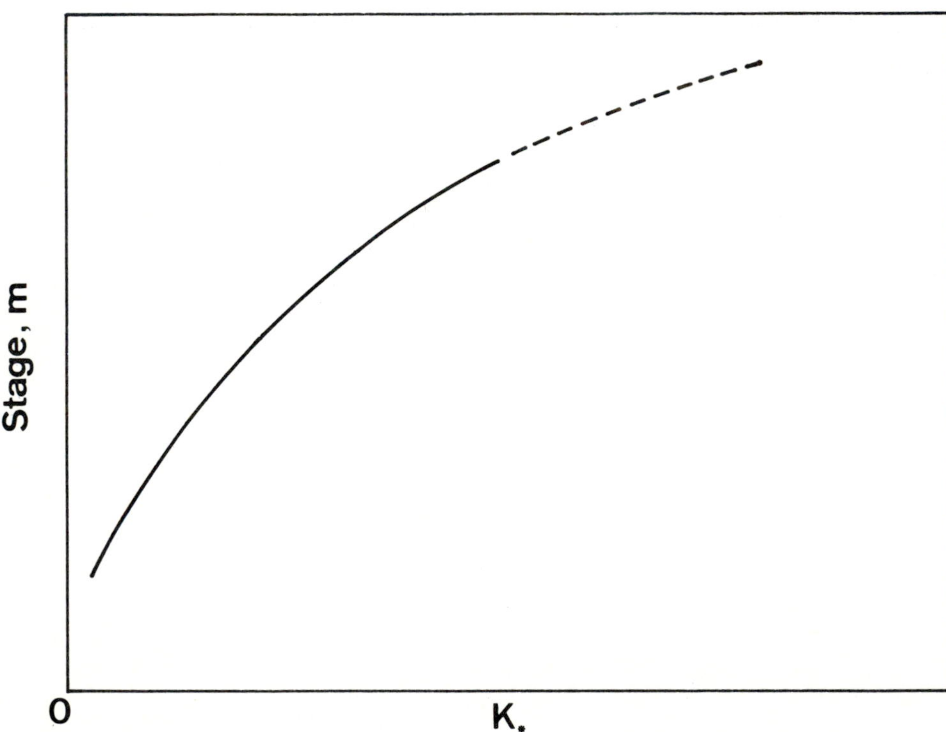

Figure 12.22 K_* versus stage for the conveyance method of rating curve extension.

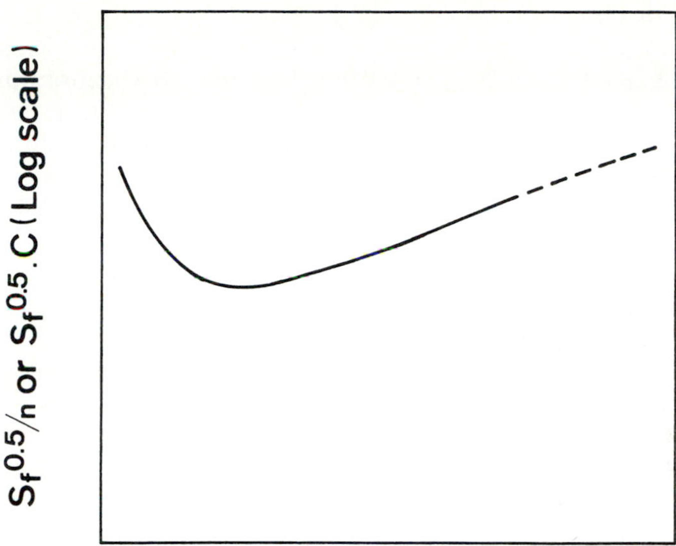

Figure 12.23 $S_f^{0.5}/n$ or $S_f^{0.5}C$ versus stage for the conveyance method of rating curve extension.

discharge, and this value is taken for all discharges beyond the observed range. In this way, a plot of stage versus $S_f^{0.5}/n$ or $S_f^{0.5}C$ is not needed.

12.5.3 Mean-Depth Method

This method is based on Chezy's formula:

$$Q = AC(RS_f)^{0.5} \tag{12.33}$$

If $C(S_f)^{0.5}$ is assumed constant for the gaging station, and R can be approximated by the mean depth, then

$$Q = ABD^{0.5} \qquad B = CS_f^{0.5} \tag{12.34}$$

Since both A and D are functions of stage, a plot of Q versus $AD^{0.5}$ for all measurements is obtained. This plot is approximately a straight line that can be easily extended. Thus, the discharge can be found by computing $AD^{0.5}$ for any desired stage.

12.6 RELATION BETWEEN THE RATING CURVE AND THE RUNOFF HYDROGRAPH

During runoff from a storm event, water drains from all channels on the drainage basin that have been covered by continuous effective precipitation. For illustrative purposes, assume that uniform effective rainfall covers the entire drainage basin and results in a runoff hydrograph. The stage of this hydrograph is continuously recorded at the velocity–area recording gage for which a valid rating curve exists. Clearly, the drainage channels nearest the basin-outlet gage deliver water to the gage before channels farther from the gage. The amount of water draining from these few nearby channels is relatively small and they, therefore, cause water in the channel at the gage to flow at a relatively low stage. The stage of flow increases with time as more and more channels deliver water to the gage in addition to those already providing runoff at that point. The maximum stage is reached (a) when all drainage channels are delivering water to the gage or (b) when rainfall ceases. When rainfall ceases, stage at the gage falls as surface water already on the basin drains away.

The result of this process is that the recording gage provides a continuous record of the stage of flow as runoff begins from a storm event until the flow reaches its highest discharge and on to the time at which the stage lowers to its level before the storm began. The summation of the discharges over this period of time represents the total volume of runoff from the drainage basin for this event.

It is apparent, then, that the rating curve is simply a plot of discharge (m^3/s or ft^3/s) versus runoff volume at any instant. The runoff volume is represented by the stage (depth) in meters or feet for a channel cross-sectional area in square meters or square feet. The result is that the rating curve is a plot of volume per unit time versus volume. This engineering relation is useful for hydrologic purposes.

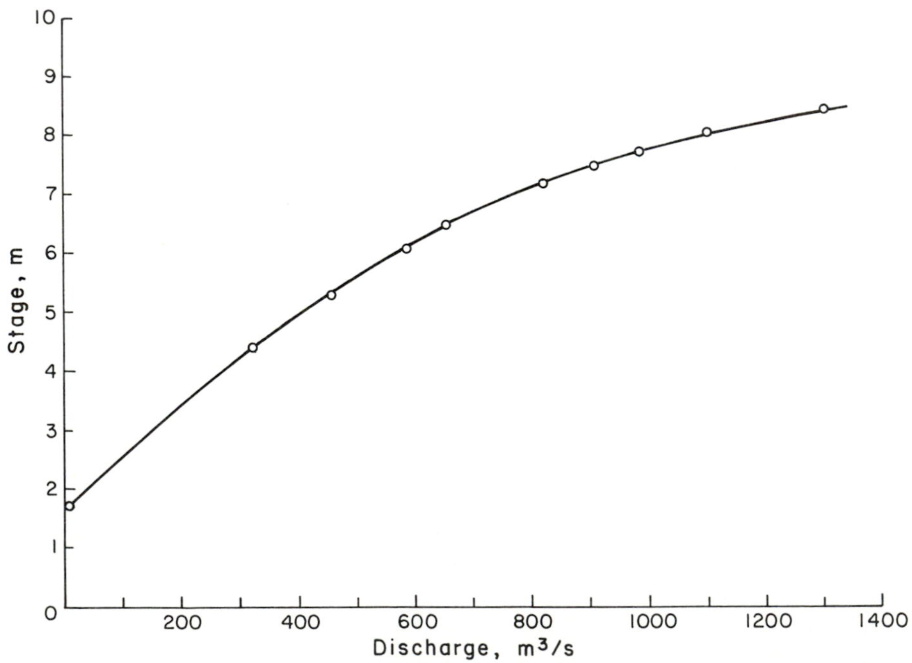

Figure 12.24 Stage-discharge relationship.

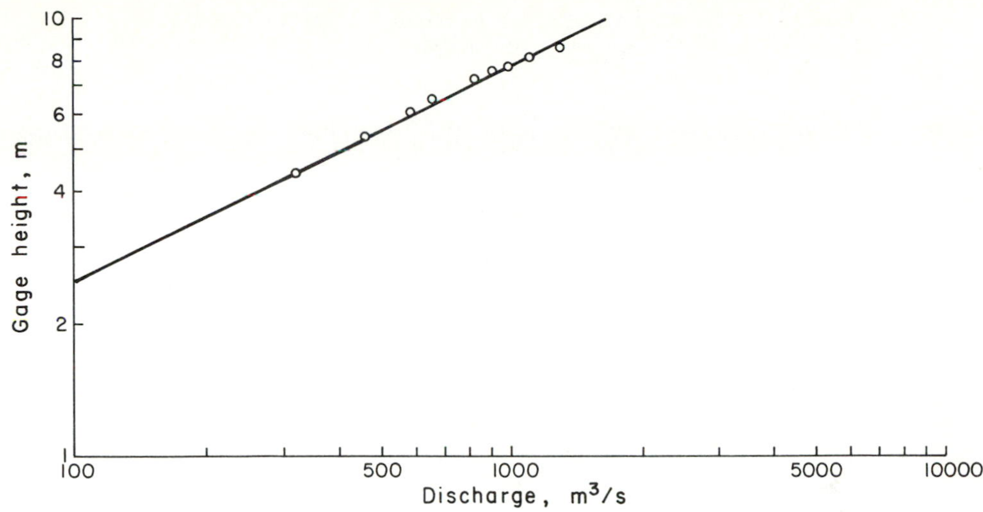

Figure 12.25 Stage-discharge relationship on lag-lag paper.

EXAMPLE 12.4 The following data are available for the Leaf River at Hattiesburg, Mississippi, for 11–5–69 and 11–17–69.

GAGE HEIGHT (m)	DISCHARGE (m^3/s)
4.42	319.79
5.33	455.63
6.1	582.98
6.46	650.9
7.19	820.7
7.5	905.6
7.74	984.84
8.05	1098.04
8.47	1296.14

Plot the rating curve on arithmetic as well as log–log paper. From the arithmetic plot, determine the stage at zero discharge.

Solution A plot of gage height versus discharge is made on arithmetic paper, as shown in Figure 12.24. From this figure, the stage at zero discharge is found to be 1.62 m. The gage height against discharge is plotted on log–log paper, as shown in Figure 12.25.

EXAMPLE 12.5 The following table gives the discharge, the base stage, and the stage at an auxiliary gage 610 meters downstream. Develop a slope-stage-discharge relationship from these data. Compute the average error of the rating using the tabulated data. What is the estimated discharge for the base and auxiliary stages of 7.6 and 7.4 meters, respectively?

BASE STAGE (m)	DISCHARGE (m^3/s)	AUXILIARY STAGE (m)
4.27	69.92	3.96
7.25	837.68	7.09
5.39	599.96	5.06
7.50	2419.67	7.18
6.21	798.06	5.96
5.18	209.42	5.00
5.68	962.2	5.33
8.05	1556.5	7.83
6.77	2099.86	6.4
4.94	270.27	4.66
6.43	1231.05	6.14
7.80	2377.2	7.5
7.07	2646.05	6.69

Solution First, the value of fall is computed from the difference between the base stage and the auxiliary stage for each value of the base stage. Then an average value of fall is obtained as $0.29 \cong 0.3$. The value of F_0 is chosen as 0.3 m. The base stage is thereafter plotted against the observed discharge, and a curve is drawn for the fall = 0.3 m, as shown in Figure 12.26. From this curve, the base discharge Q_0 is read against each base

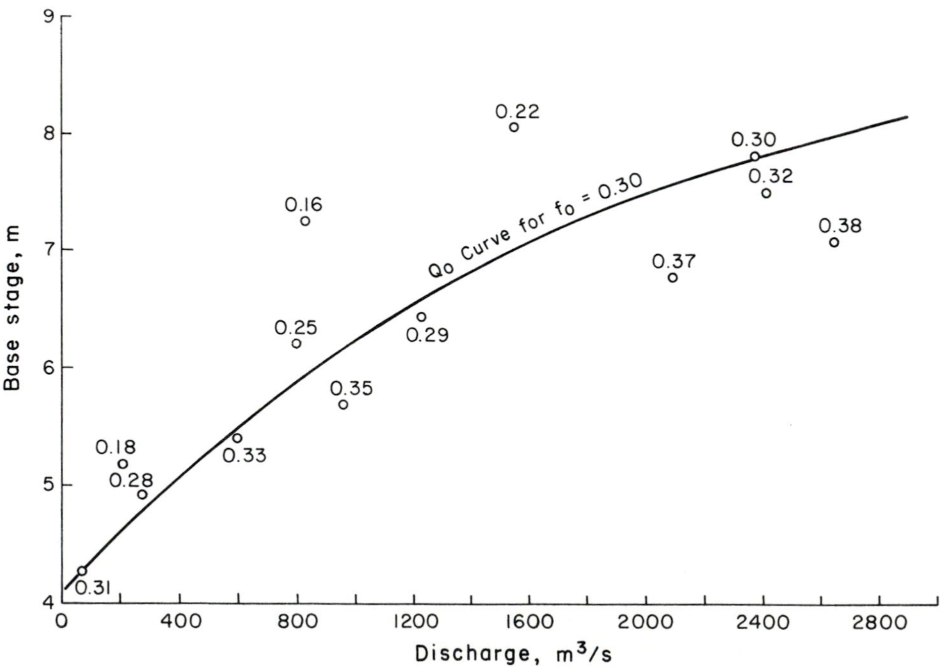

Figure 12.26 Stage versus discharge and rating curve for the fall of 0.3 m.

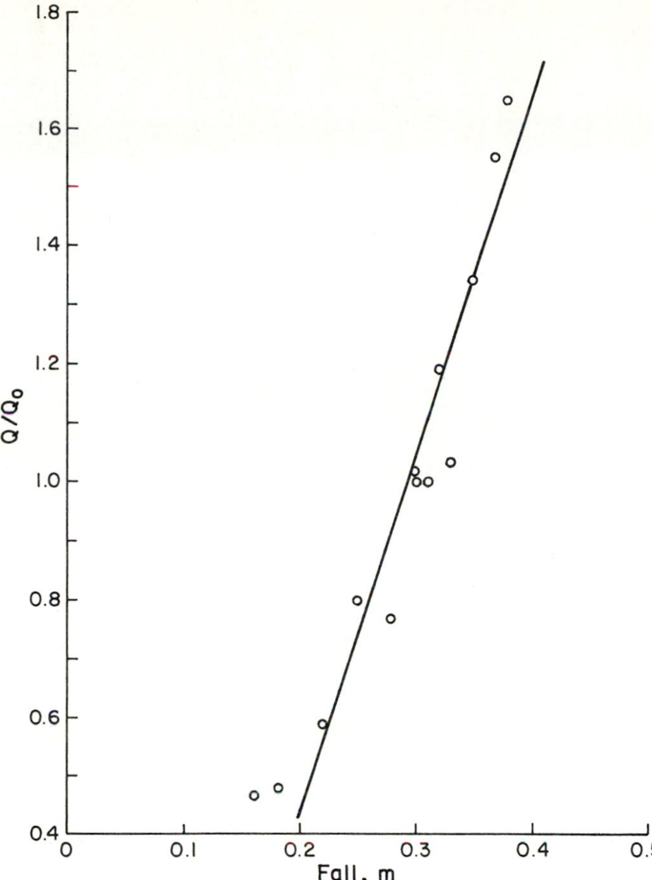

Figure 12.27 Q/Q_0 versus fall.

stage. The value of Q/Q_0 is obtained for each value. The observed fall is plotted against Q/Q_0 on arithmetic paper and a curve is drawn through the points as shown in Figure 12.27. In this plot, a straight line appears to represent the points well. This is the fall correction curve. For each value of the observed fall, Q/Q_0 is read from the fitted straight line. By multiplying this ratio by the base discharge, the computed discharge is obtained. Results of the calculations are shown in Table 12.4. The difference between the observed and computed discharges is calculated to yield the error given in column (9) of the table. The average error is obtained by summing the absolute values (ignoring algebraic values) and dividing by the number of values, 879.49/13 = 67.65 m^3/s.

EXAMPLE 12.6 The following data are given for a station rating curve. Extend the curve and estimate the flow at a stage of 4.42 meters by both the logarithmic and $A\sqrt{D}$ methods.

TABLE 12.4 RESULTS OF THE CALCULATIONS FOR EXAMPLE 12.5

Base stage (m) (1)	Discharge, Q (m^3/s) (2)	Auxiliary stage (m) (3)	Fall (m) (4)	Base discharge Q_0 (m^3/s) (5)	Q/Q_0 (6)	Fall correction from the curve (7)	Computed discharge (m^3/s) (8)	Error (m^3/s) (9)
4.27	69.96	3.96	0.31	70.00	1.00	1.1	70.0	−0.08
7.25	837.68	7.09	0.16	1780.00	0.47	0.4	712.0	125.68
5.39	599.96	5.06	0.33	580.00	1.03	1.22	707.6	−107.64
7.50	2419.67	7.18	0.32	2040.00	1.19	1.16	2366.4	53.27
6.21	798.06	5.96	0.25	1000.00	0.80	0.74	740.0	58.06
5.18	209.42	5.00	0.18	450.00	0.47	0.40	180.0	29.42
5.68	962.20	5.33	0.35	720.00	1.34	1.34	964.8	−2.6
8.05	1556.50	7.83	0.22	2650.00	0.59	0.56	1484.0	72.5
6.77	2099.86	6.40	0.73	1350.00	1.56	1.46	1971.0	128.86
4.94	270.27	4.66	0.28	350.00	0.77	0.92	322.0	−79.73
6.43	1231.05	6.14	0.29	1130.00	1.09	0.98	1107.4	123.65
7.80	2337.20	7.50	0.30	2380.00	1.00	1.04	2475.2	−98.0
7.07	2646.05	6.69	0.38	1600.00	1.65	1.52	2432.0	214.05

STAGE (m)	AREA (m^2)	DEPTH (m)	DISCHARGE (m^3/s)
0.52	24.43	0.46	28.87
1.06	111.48	0.64	138.67
1.30	166.29	0.98	217.91
1.71	221.10	1.40	302.81
2.04	304.71	1.58	427.33
2.38	367.88	1.73	537.7
2.81	464.5	1.86	707.5
4.42	761.78	2.74	
0.76	62.61	0.55	76.41
1.22	145.85	0.85	186.78
1.55	199.74	1.19	267.44
1.82	270.34	1.49	370.73
2.08	317.72	1.65	455.63
2.67	447.78	1.83	682.03
3.04	487.73	1.98	772.59

Solution First, the stage is graphed against discharge on log–log paper, as shown in Figure 12.28. From this graph, the discharge corresponding to the stage = 4.42 m is 1600 m^3/s.

Second, the quantity $A\sqrt{D}$ is computed corresponding to each value of discharge:

DISCHARGE (m^3/s)	$A\sqrt{D}$	DISCHARGE (m^3/s)	$A\sqrt{D}$
28.87	16.57	76.41	46.43
138.67	89.57	186.78	134.47
217.91	164.62	267.44	217.89
302.81	261.61	370.73	329.99
427.33	383.01	455.63	408.12
523.70	483.87	682.03	605.75
707.50	633.49	722.50	686.30

The discharge is then plotted against $A\sqrt{D}$ on arithmetic paper, as shown in Figure 12.29. For the stage = 4.42 m, $A\sqrt{D}$ = 1260.97, which produces the discharge of 1415 m^3/s from the figure.

EXAMPLE 12.7 Fit an equation of the form:

$$Q = C(y - y_0)^n$$

for the rating curve to the data of Example 12.4, where Q is the discharge, y is the stage, y_0 is the stage at $Q = 0$, and C and n are constants. What are the units of C and n?

Solution The equation is first transformed logarithmically as

$$\log Q = \log C + n \log (y - y_0)$$

Then parameters C and n are determined using the method of least squares, which produces

$$\Sigma \log Q \log (y - y_0) = \log C \, \Sigma \log (y - y_0) + n \, \Sigma [\log (y - y_0)]^2$$

$$\Sigma \log Q = n \log C + n \, \Sigma \log (y - y_0)$$

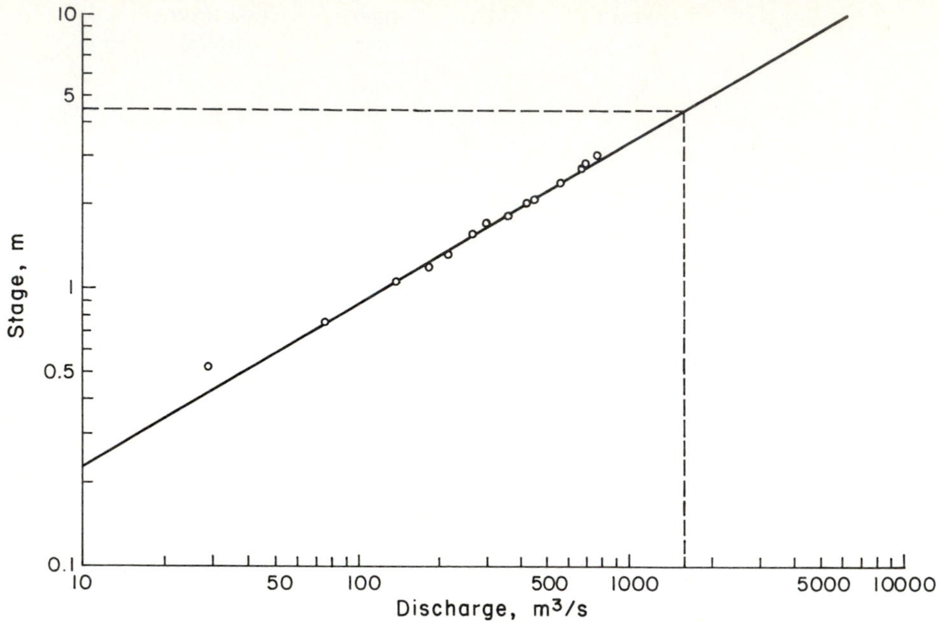

Figure 12.28 Stage-discharge curve.

The value of $y_0 = 1.62\ m$ is given in Example 12.4. The results of the calculations follow:

N	Q	$y - y_0$	$\log Q$	$\log (y - y_0)$	$\log Q \log (y - y_0)$	$[\log (y - y_0)]^2$
1	319.79	2.80	2.50	0.45	1.13	0.20
2	455.63	3.71	2.66	0.57	1.52	0.32
3	582.98	4.48	2.77	0.65	1.80	0.42
4	650.9	4.84	2.81	0.68	1.91	0.47
5	820.7	5.57	2.91	0.75	2.19	0.56
6	905.6	5.88	2.96	0.77	2.28	0.59
7	984.84	6.12	2.99	0.79	2.36	0.62
8	1098.04	6.43	3.04	0.81	2.46	0.65
9	1296.14	6.85	3.11	0.84	2.61	0.70
Σ	7114.62	46.68	25.75	6.31	18.26	4.53

$$18.26 = \log C(6.31) + n(4.53)$$

$$25.75 = \log C(9) + n(6.31)$$

or

$$-25.44 = -8.79 \log C - 6.31n$$

$$25.75 = 9 \log C + 6.31n$$

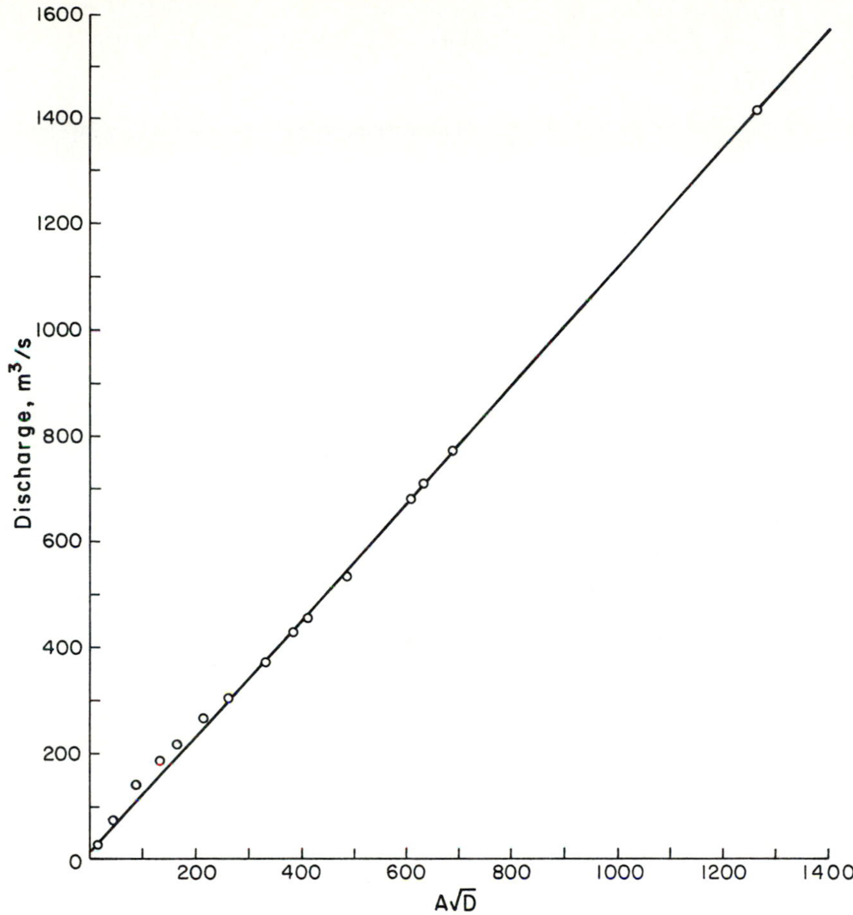

Figure 12.29 Discharge versus $AD^{0.5}$.

Summing the two equations,

$$0.31 = 0.21 \log C$$

or

$$\log C = 1.48$$

or

$$C = 29.94$$

$$25.75 = 9(1.48) + 6.31n$$

$$n = 1.97$$

$$Q = 29.94(y - y_0)^{1.97}$$

Exponent n has no units. Parameter C has dimensions of L^{3-n}/T or units of $m^{3-1.97}/s = m^{1.03}/s$.

EXAMPLE 12.8 Estimate the discharge at a flood stage of 8.47 m both graphically and by use of the equation in Example 12.7. Use the Leaf River data of Example 12.4.

Solution

$$\text{Flood stage} = 8.47 \text{ m}$$

$$Q = 29.94(8.47 - 1.62)^{1.97}$$

$$= 1326.06 \text{ m}^3/\text{s}$$

Graphically, from Figure 12.24, at the stage of 8.47 m, $Q = 1296.14 \text{ m}^3/\text{s}$.

EXAMPLE 12.9 Estimate the depth at a flow of 10.33 m^3/s both graphically and by use of the equation in Example 12.7.

Solution

$$\text{Discharge} = 10.33 \text{ m}^3/\text{s}$$

$$10.33 = 29.94(y - 1.62)^{1.97}$$

$$\ln 10.33 = \ln 29.94 + 1.97 \ln (y - 1.62)$$

$$\ln (y - 1.62) = -0.54$$

$$y - 1.62 = e^{-0.54}$$

$$y = 1.62 + 0.58$$

$$y = 2.20 \text{ m}$$

Graphically, from Figure 12.24, at the discharge of 10.33 m^3/s, the depth is 1.75 m.

12.7 HYDRAULIC GEOMETRY

The hydraulic geometry of a stream represents the relation between the stream width, depth, and velocity of flow of the stream to its discharge (Leopold and Maddock, 1953). This relation means that for any given discharge up to the mean annual discharge on any natural stream, there is a unique relation between that discharge and the stream width, depth, and mean velocity of flow. There is a different relation for discharges exceeding the mean annual discharge. The hydraulic geometry of streams demonstrates that an equilibrium condition exists between the streams draining a watershed and the water supplied to those streams. This relation can be quantified by rewriting Equation (12.7) as

$$Q = wdv \tag{12.35}$$

where w is the width of the water surface in the stream in feet or meters at a given cross-section, d is the mean depth of the water in the stream in feet or meters at the cross-section, and v is the mean velocity of flow in ft/s or m/s at that cross-section. The plotted relations in Figure 12.30 show that the width, depth, and velocity are related to discharge according to the following equations:

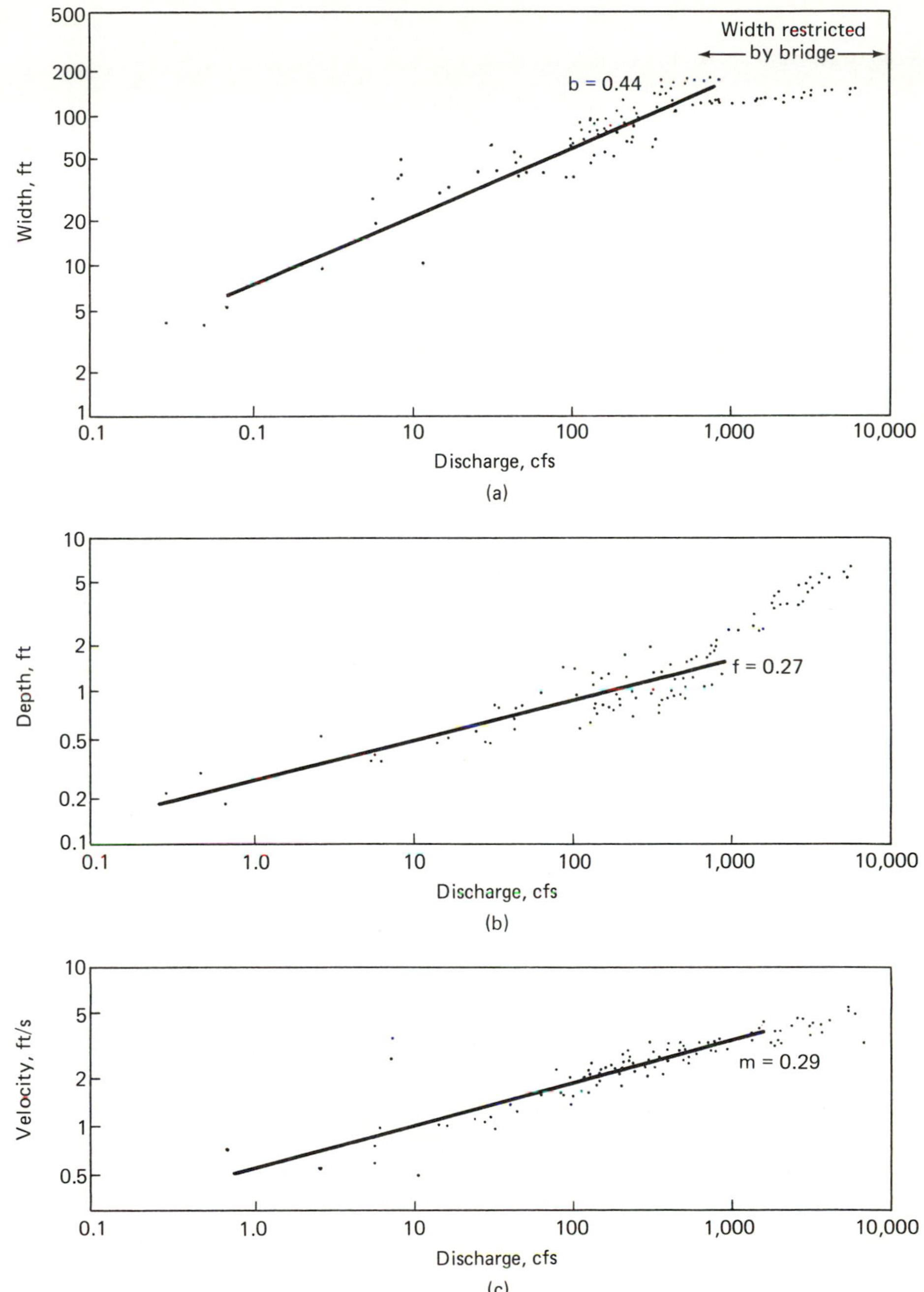

Figure 12.30 Hydraulic geometry for the Power River at Arvada, Wyoming.

$$w = aQ^b \tag{12.36}$$

$$d = cQ^f \tag{12.37}$$

$$v = kQ^m \tag{12.38}$$

By substituting these relations in Equation (12.35),

$$Q = aQ^b \times cQ^f \times kQ^m \tag{12.39}$$

or

$$Q = ackQ^{b+f+m} \tag{12.40}$$

from which

$$ack = 1 \tag{12.41}$$

and

$$b + f + m = 1 \tag{12.42}$$

These equations verify the mathematical relation between the geometry of any stream channel and its discharge. The relations are illustrated in Figure 12.30. The slopes of the curves in Figure 12.30 are believed to be determined by the nature of the soil or rock material in which the stream channel is incised. Slopes b, f, and m determine the ordinate intercepts for each set of curves for every stream.

The hydraulic geometry of a stream remains constant for any stream as long as its drainage environment is not changed. Urbanization, agricultural development, or any other factors that change the drainage environment will change the hydraulic geometry of a natural stream. If such changes are made, then a new relation is established between discharge and the width, depth, and velocity of a channel. When development, whether urban or agricultural, is taking place on the drainage basin, the hydraulic geometry resumes a state of flux until static conditions are reestablished.

In Figure 12.30 the slopes for the width, depth, and velocity are $b = 0.44$, $f = 0.27$, and $m = 0.29$, respectively, and their sum is equal to unity in accordance with Equation (12.42). The product of intercepts a, c, and k is also unity.

The hydraulic geometry of streams is a complex and detailed subject involving sediment transport and other factors that are beyond the scope of this text. The geometry characteristics of the stream itself yields information related to hydraulic geometry that assists in predicting the expected runoff from a drainage basin. One of the important relations developed from hydraulic geometry is that the bank full discharge of a natural stream has a return period of about 2.33 years. This return period represents the mean annual discharge, which is discussed in Chapter 22.

EXAMPLE 12.10 The following data are available on the cross-section of a natural stream. Construct curves showing the relationships between the depth y and the section elements A, R, D, and Z. Determine the geometric elements for $y = 1.2$ m from the curves.

DISTANCE FROM A REFERENCE POINT NEAR LEFT BANK (m)	STAGE (m)
−1.5	1.67
−1.2	1.38
−0.6	1.20
0	0.57
0.3	0.24
0.6	0.06
0.9	0.09
1.5	0.06
2.1	−0.03
2.7	−0.03
3.3	−0.12
3.9	−0.03
4.5	0.21
5.1	0.78
5.7	0.96
Bank right: 6.0	1.23

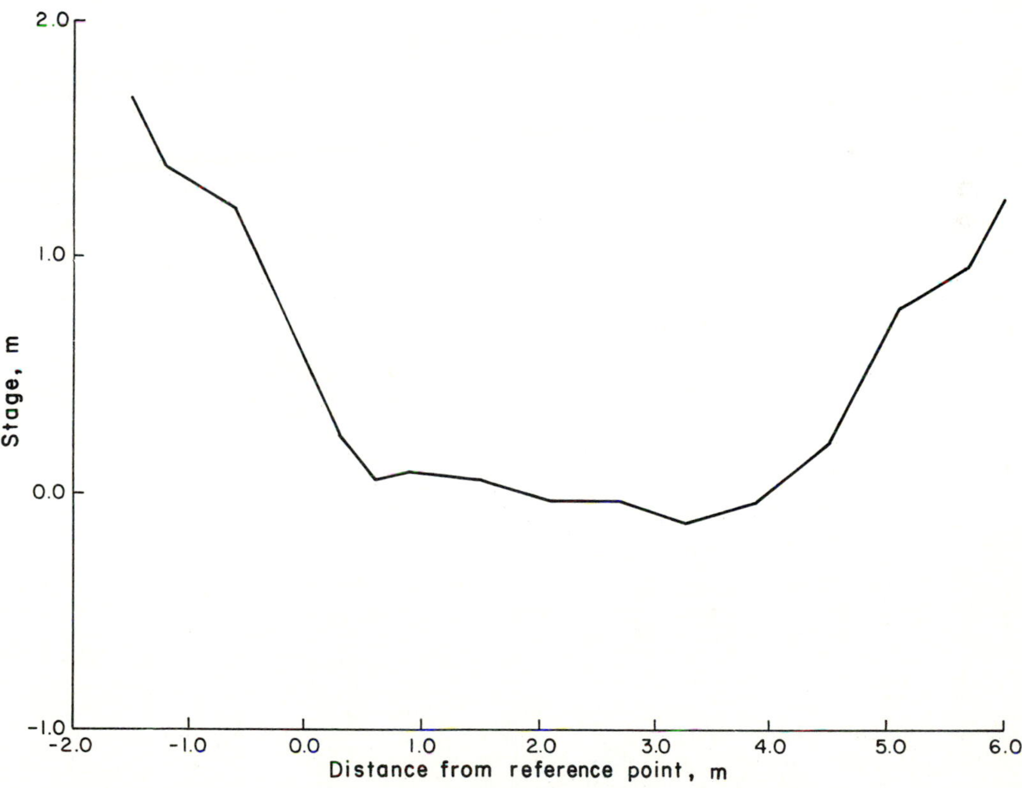

Figure 12.31 Cross-Section of a natural stream.

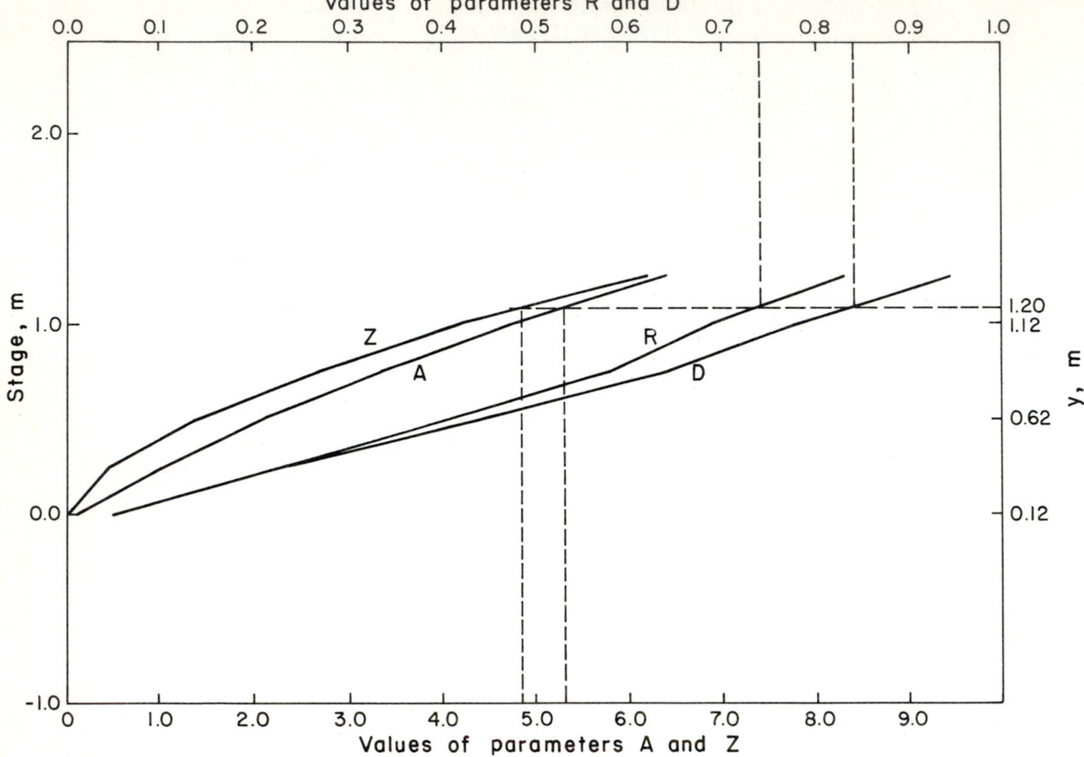

Figure 12.32 Stage versus hydraulic elements.

Solution First, the stage is plotted against the distance from the reference point on arithmetic paper, as shown in Figure 12.31. This plot produces the cross-section of the stream. By using the formulas

$$R = \frac{A}{P}; \quad D = \frac{A}{T}; \quad Z = A\sqrt{D}; \quad Z' = AR^{2/3}$$

the following quantities are obtained:

Stage height (m)	A (m^2)	P (m)	T (m)	R (m)	D (m)	Z	Z'
0.00	0.11	2.12	2.1	0.05	0.05	0.02	0.01
0.25	0.99	4.39	4.25	0.23	0.23	0.47	0.37
0.50	2.11	5.11	4.75	0.41	0.44	1.40	1.16
0.75	3.36	5.79	5.25	0.58	0.64	2.69	2.33
1.00	4.78	6.88	6.15	0.69	0.78	4.22	3.73
1.25	6.38	7.67	6.8	0.83	0.94	6.19	5.63

The values of A, R, D, and Z versus the stage height and y are plotted on the graph paper as shown in Figure 12.32. From this figure, the following values are obtained for $y = 1.2$ m:

$$A = 5.3 \text{ m}^2$$

$$R = 0.74 \text{ m}$$

$$D = 0.84 \text{ m}$$

$$Z = 4.85 \text{ m}^{5/2}$$

12.8 STREAMFLOW RECORDS IN THE UNITED STATES

Streamflow records derived from stream gages provide valuable information for hydrologic analysis. The U.S. Geological Survey in cooperation with the state geological surveys publishes an annual summary of stream-gage information for each gage maintained in each state. The information contained in these summaries is illustrated for the Loup River near Genoa, Nebraska (U.S. Geological Survey, 1979) in Figure 12.33. The water year is defined as the period from October 1 to September 30 of the following year. In this example, the water year is from October 1, 1978 to September 30, 1979.

12.8.1 Interpretation of Streamflow Records

A discussion of the data illustrated in Figure 12.33 is helpful in interpreting the data presented here. Stream gages are identified according to major drainage areas and subdrainage areas within the major drainage system. The gage number, 06793000, indicates that the Loup River is located within the Missouri River drainage basin identified by the digits 06. Other major drainage areas are similarly identified. For example, streams draining to the North Atlantic are identified by numbers beginning with 01. The Loup River gage near Genoa, Nebraska, is further identified by the number 7930. A decimal is assumed to be located ahead of the last two digits, 00, to provide for additional gages that might be installed at some future date. If such a condition occurred, the gage number 7930 would then carry an added decimal number.

The notation in parentheses below the gage number and gage name indicates that this gage is one of the national stream-quality accounting network stations. The water-quality records for this station are published adjacent to the stream gage records each year.

Location: The location of each gage is described in terms of its latitude and longitude, section, township, and range, county, and a complete description using roads and landmarks. The terms right and left bank are a standard description used to identify channel banks and are reckoned by facing downstream. In this manner of

PLATTE RIVER BASIN

06793000 LOUP RIVER NEAR GENOA, NE
(National stream-quality accounting network station)

LOCATION.--Lat 41°25'05", long 97°43'25", in SW1/4NE1/4 sec.25, T.17 N., R.4 W., Nance County, Hydrologic Unit 10210009, on right bank 12 ft (4 m) downstream from bridge on State Highway 39, 2 mi (3 km) south of Genoa, 3 mi (5 km) upstream from Beaver Creek, and 6 mi (10 km) downstream from diversion dam of Loup River Public Power District.

DRAINAGE AREA.--14,400 mi² (37,300 km²), approximately, of which about 5,650 mi² (14,600 km²) contributes directly to surface runoff.

WATER-DISCHARGE RECORDS

PERIOD OF RECORD.--August 1928 to June 1932, October 1943 to current year (October 1953 to April 1955, monthly discharge only).

REVISED RECORDS.--WDR NE-74: Drainage area.

GAGE.--Water-stage recorder. Datum of gage is 1,540.13 ft (469.432 m) National Geodetic Vertical Datum of 1929, Aug. 17, 1928, to June 30, 1932, nonrecording gage at present site at datum 1.49 ft (0.454 m) higher. Oct. 1, 1943 to Sept. 16, 1974 (Apr. 26 to Dec. 22, 1949, wire-weight gage only) at present site and datum. Sept. 17, 1974 to Nov. 21, 1977 at site 300 ft (90 m) upstream at present datum.

REMARKS.--Records fair except those for winter period, which are poor. Natural flow of stream affected by power developments, ground-water withdrawals and diversions for irrigation, and return flow from irrigated areas. Records do not include flow of Loup River power canal (station 06792500) which diverts at point 6 mi (10 km) upstream and returns to Platte River below mouth of Loup River; diversion began Dec. 2, 1936.

EXTREMES FOR PERIOD OF RECORD.--Maximum discharge, 129,000 ft³/s (3,650 m³/s) Aug. 13, 1966, gage height, 13.93 ft (4.246 m), from rating curve extended above 42,000 ft³/s (1,190 m³/s) on basis of indirect measurement of peak flow; no flow at times during 1956, 1959, 1961, 1963, 1970, 1973, 1974, 1975.

EXTREMES FOR CURRENT YEAR.--Maximum discharge, 12,300 ft³/s (348 m³/s) Mar. 22, gage height, 8.20 ft (2.499 m), maximum gage height, 10.87 ft (3.313 m), Mar. 15, backwater from ice; minimum daily discharge, 2.8 ft³/s (0.079 m³/s) Nov. 6.

DISCHARGE, IN CUBIC FEET PER SECOND, WATER YEAR OCTOBER 1978 TO SEPTEMBER 1979
MEAN VALUES

DAY	OCT	NOV	DEC	JAN	FEB	MAR	APR	MAY	JUN	JUL	AUG	SEP
1	12	4.1	1850	2150	54	560	660	42	14	124	12	18
2	11	3.8	2000	1950	40	700	1110	59	48	65	12	17
3	11	3.6	450	1800	38	1300	493	80	20	65	11	16
4	11	3.5	300	1500	40	1800	382	45	19	38	12	16
5	11	2.9	500	1000	46	2500	186	35	19	47	11	15
6	9.9	2.8	780	380	80	3700	220	32	18	129	11	16
7	9.9	2.9	880	150	170	3000	161	30	18	194	11	14
8	9.6	4.3	840	70	160	2800	118	30	20	30	11	14
9	9.4	14	820	40	135	4000	119	36	25	18	11	14
10	8.5	7.2	820	20	120	4400	118	1380	23	16	11	14
11	7.6	5.5	800	50	130	5400	289	1090	22	18	11	13
12	6.4	5.8	800	60	135	6400	1080	218	21	18	11	15
13	6.1	11	820	100	145	6300	1300	65	21	19	11	14
14	6.0	14	840	1000	150	6200	882	31	20	16	13	14
15	5.9	16	840	500	120	7800	374	24	20	22	14	14
16	5.8	16	860	70	100	9000	165	22	21	16	14	15
17	5.9	19	880	62	90	9600	156	20	22	14	15	15
18	5.8	9.9	900	52	88	8000	130	20	22	15	15	14
19	5.7	150	920	44	92	7000	95	50	22	14	15	14
20	5.8	900	1000	70	100	5240	569	30	20	14	18	15
21	5.6	820	1100	45	54	3360	1550	18	19	14	18	14
22	7.0	500	1200	40	56	9930	403	16	19	13	18	14
23	6.9	540	1250	58	58	8910	187	15	23	12	17	13
24	7.8	700	1300	52	60	3810	66	15	22	13	16	13
25	5.9	1400	1400	48	200	2000	57	14	21	13	17	12
26	6.0	2100	1500	44	300	1230	52	14	21	44	18	13
27	5.6	2000	1550	47	400	1170	50	14	34	13	19	14
28	4.8	1900	1700	60	1000	930	45	14	67	13	20	14
29	4.2	1800	1800	70	---	735	43	14	69	13	19	15
30	4.2	1750	1900	50	---	1000	40	16	200	12	19	15
31	3.8	---	2000	60	---	940	---	14	---	11	18	---
TOTAL	226.1	14706.3	34600	11642	4161	129715	11100	3503	930	1063	449	434
MEAN	7.29	490	1116	376	149	4184	370	113	31.0	34.3	14.5	14.5
MAX	12	2100	2000	2150	1000	9930	1550	1380	200	194	20	18
MIN	3.8	2.8	300	20	38	560	40	14	14	11	11	12
AC-FT	448	29170	68630	23090	8250	257300	22020	6950	1840	2110	891	861

CAL YR 1978 TOTAL 276418.6 MEAN 757 MAX 30000 MIN 2.8 AC-FT 548300
WTR YR 1979 TOTAL 212529.4 MEAN 582 MAX 9930 MIN 2.8 AC-FT 421600

Figure 12.33 Sample page from the U.S. Geological Survey, Water Resources Data.

downstream facing, the left bank is on the left side of the channel and the right bank is on the right side of the channel.

Drainage Area: The total area within the drainage-basin boundary is shown as the drainage-basin area. This area is subject to change because the measurement might be refined at a later date. The term "approximately" indicates there is some uncertainty as to the exact drainage area because some areas within the drainage-basin boundary might not contribute to surface runoff. The area contributing directly to surface runoff is shown to be less than the total drainage area. In this example, the high infiltration of the sand hills, within which this stream is located, prevents surface runoff from a large portion of the drainage basin. Isolated or closed drainage areas within a basin from other causes are also subtracted from the total drainage area. The total effective drainage area based upon the best information available is then published under this heading. It should be recognized that the drainage areas shown are subject to change as better information becomes available. These changes in drainage area are published as they become available and are referenced appropriately.

Period of Record: The dates show the period during which a gage was maintained at this location. In this example, there was a period between June 1932 and October 1943 when the gage was not maintained. For this reason, no records are available while the gage was not operated. Later, monthly discharges only were recorded. This information allows a knowledge of what records are available and the extent of those records.

Revised Records: This section shows any changes in the records kept for this gage. In this instance, the drainage area was revised in 1974 and published in the report for that year. The actual change made was a reduction in the effective drainage area. Any other revisions made during the operation of the gage are also listed under this heading.

Gage: A complete description of the gage and its history is presented in this section. The datum of the gage from which all water stage elevations are measured is shown. If the gage is moved, it is indicated and the dates of such action are noted. The type of gage, recording, crest, or whatever is indicated and the dates of its service are indicated. Any changes in the type of gage or its method of operation are identified.

Remarks: This section describes the accuracy of records at the gage and the factors that affect its accuracy. This information is presented so the user can take the described factors into account for the intended purposes. It is possible that water is diverted upstream for some purpose and might or might not be returned to the stream at some point. If such is the case, the condition is described under this heading. The quality of streamflow data depends on the likely error, as indicated in Table 12.5.

TABLE 12.5 ACCURACY OF STREAMFLOW DATA, PROBABLE ERRORS IN PERCENT

Record quality	Individual measurements	Published records
Excellent	<2	< 5
Good	<5	<10
Fair	<8	<15
Poor	>8	>15

Extremes for Period of Record: The highest and lowest flows recorded during the period of the gage operation are indicated along with the method of determining those discharges. In this instance, the discharge of 129,000 cfs was not actually measured, but was determined by extending the rating curve to the recorded gage height. Such high discharges are usually not possible to measure because of the danger to personnel and lack of sufficient equipment. The fact that this discharge was not measured indicates that it could be in error and should be used accordingly. Whenever a discharge is extrapolated by extending the rating curve beyond the highest measurement, serious error could result. Such a discharge measurement should be verified by use of high-water marks and equal conveyance calculations where possible. It must be clearly understood that the use of discharges that have a high probability of error can result in serious error and should always be so identified in any hydrologic analysis. It should be remembered that the data presented are for information only. The U.S. Geological Survey does not make recommendations for their use, but simply provides the information they have available.

Extremes for Current Year: The highest and lowest discharges for the year are identified along with their gage heights and any comments affecting their accuracy. Sometimes additional high discharges are noted if their characteristics seem significant.

Daily Mean Discharge Values: This table lists the mean discharge values for each day during the water year. These values are computed from the recording gage. The variation in the mean daily discharges in this example is probably due, to a large extent, to the diversion of flow upstream through the Loup River power canal. The Loup River drains a large portion of the Nebraska Sandhills and is supplied mainly by baseflow from groundwater in the Sandhills. For this reason, the mean daily discharges should have little variation. Mean daily discharge figures dampen peak discharge values and one would not need to obtain the original gage hydrographs to identify peak values for a period of interest if such data were needed. A summary of the flow data is tabulated at the bottom of the table and is sufficiently clear as to need no further explanation.

12.8.2 Adjustment of Flow Records

Streamflow records might need to be adjusted for several reasons such as water diversion upstream, controlling the effect of regulation works, to synthesize a long period of record from several shorter periods, or to adjust a time trend. These techniques are left to more advanced texts.

12.9 STREAMFLOW NETWORK

Streamflow data are needed for a multitude of uses, and may be needed for future applications not known at present. How much data should be collected, and where should they be collected? To put it another way, how many gaging stations should be used, and where should they be located? These questions are answered by the design of a streamflow network. If there were no financial limitations, these questions would be moot. However, this is not the real world. Therefore, to a great extent, the use to which streamflow data are put determines the number of gaging stations.

Three types of gaging stations are generally recognized: basic data stations, operational stations, and special stations. The purpose of the basic data stations is to collect data for future use, the nature of which may not be known. These stations may be discontinued after sufficient data have been collected. The operational stations are installed to collect data for specific purposes such as water supply, water allocation, streamflow forecasting, reservoir operation, project operation, etc. These stations are terminated when their intended purpose is served. The objective of special stations is to meet a specific data need that may arise in such specific cases as research, project investigation, special studies, litigation cases, etc. Their operation is terminated when the specific need is fulfilled. Different streams are subject to different degrees of human interference (e.g., urbanization, deforestation, etc.). Some gaging stations should be designated as benchmark stations and these stations should be maintained permanently on streams that are relatively unaffected by human activities (Langbein, 1968).

12.9.1 Number of Gaging Stations

Streamflow data may be needed at any point on the stream. These data can be produced by a combination of data collection and data generation techniques. The number of sites to be selected for measurement depends upon the cost of collection and operation, the value of the data, watershed size, development, objective of data collection, accuracy, hydrologic characteristics, etc. Some of these factors are interrelated. For example, large watersheds involve costlier projects and, therefore, larger data and high accuracy are needed. Thus, watersheds and streams may be classified as major and minor. An approximate area of 1300 km^2 can be taken to divide the watersheds.

TABLE 12.6 VARIABILITY AND ACCURACY OF STATISTICAL CHARACTERISTICS OF STREAMFLOW (AFTER WORLD METEOROLOGICAL ORGANIZATION, 1965)

	Standard error in percent for indicated length of record (year)			
Streamflow characteristics	5	10	25	50
Mean annual flow	44	30	20	14
Standard deviation of annual flow	35	25	15	11
Mean monthly flow	75	50	32	22
Standard deviation of monthly flow	35	25	15	11
50-year peak flow	87	58	36	24
50-year 7-day high flow	93	62	38	27
2-year 7-day low flow	78	53	32	23
20-year 7-day low flow	110	73	43	30

Every major stream should be gaged at or near its mouth. Likewise, a number of its tributaries should also be gaged. Naturally, gaging depends upon the existing and likely development in the watershed. Linsley et al. (1975) report that the minimum drainage area for streamflow gaging might range from 26 km^2 (10 mi^2) in a well-developed region to 260 km^2 (100 mi^2) or more in developing areas. Thus, the density of the population (the number of people per unit area) is also a factor in determining the number of gaging stations (Langbein, 1968). According to the World Meteorological Organization (1965), the major streams are gaged as follows: The first gaging station is selected at the most upstream location where the drainage area is about 1300 km^2. The second station is located at a point in the downstream direction where the drainage area is approximately doubled. The area is doubled to account for areas drained by tributaries entering the main stream. This method of spacing the gaging stations permits interpolation of intermediate flows within the range of accuracy given in Table 12.6. The accuracy in this table refers to the statistical characteristics of streamflow, rather than to the basic data such as individual discharge measurements, and this accuracy is limited by the length of the record.

12.9.2 Location of Gaging Stations

The cost and ease of installation, the cost of operation, and the accuracy of measurements depend upon the location of gaging stations. The following guidelines are helpful in selection of a site: (1) The site should be readily accessible throughout the year. (2) The site should satisfy the requirements of depth and velocity measurements in terms of hydraulic characteristics of the streams. (3) The site should be away from the influence of hydraulic structures such as bridges, dams, etc. In unavoidable circumstances, the site must be located upstream of the structure. (4) If a site has to be located upstream or downstream of a confluence, it should be away at least three times the maximum width of the channel or 0.8 km, whichever is greater. (5) The stream reach should be stable and fairly straight and uniform both upstream

TABLE 12.7 DESIRABLE SUBWIDTHS FOR DIFFERENT TYPES OF CHANNELS

Channel characteristics	Minimum number of observations—verticals	Width segments (m)
Channel with width of waterway less than 15 m or with river bed changing abruptly	15	1 to 1.5
Channels with width of waterway between 15 and 90 m	15	3 to 6
Channels with width of waterway between 90 to 180 m	15	6 to 15
Channels with width greater than 180 m	25	—

and downstream of the site for at least four times the normal width of the stream during floods or 0.8 km, whichever is less.

12.9.3 Demarcation of Gaging Sites

The gaging site should be aligned perpendicular to the direction of flow. The stream cross-section at the gaging site can be demarcated using masonry or concrete pillars from one side of the stream to the other. Each pillar can be identified by fixing a flag at its center. The pillars can be spaced about 30 m apart.

Once the gaging site is demarcated, a decision has to be made about the number of segments along the section at which the depth and velocity measurements are to be made. The number of segments depends upon the width of the stream and other hydraulic characteristics. Of course, greater accuracy is achieved with a larger number, but at a higher cost. General guidelines for deciding the number of segments are given in Table 12.7. As a general limitation, no more than 10% of the total discharge should pass through any segment, and the variation of discharge from one segment to the other should be no more than 10%.

EXERCISES

12.1. A drainage basin has an area of 50 km^2. The average depth of water covering the basin during a rainfall event is 10 cm. Compute the volume of water in cubic meters. How many people can be supplied water if the per-capita water demand is 150 liters per day.

12.2. A drainage basin has an area of 1000 km^2. The flow measured at the basin outlet for a period of 8 days follows. Compute the 8-day mean flow rate. Compute the total discharge in cubic meters and hectare-meters.

Day	1	2	3	4	5	6	7	8
Flow (m^3/s)	10	50	80	70	60	50	30	20

12.3. A town has a population of 10,000. If the per-capita water demand is 150 liters per day, for how many days will the total discharge of Exercise 12.2 be sufficient to meet the water-supply demand?

12.4. An area is to be irrigated with 1 cm of water depth. How many hectares can be irrigated with the total discharge of Exercise 12.2?

12.5. Compute the stream discharge if the stream gaging station data are as follows:

DISTANCE FROM LEFT WATER EDGE (m)	DEPTH, d (m)	VELOCITY (m/s) AT $0.2d$	VELOCITY (m/s) AT $0.8d$
0	0.0	0.0	0.0
1	1.25	0.5	0.3
2	2.0	0.75	0.5
3	3.0	1.0	0.6
4	2.5	0.85	0.5
5	2.0	0.80	0.5
6	1.5	0.70	0.4
7	0.0	0.0	0.0

12.6. Compute the discharge through a triangular stream with side slope 3 horizontal to 1 vertical. The stream reach is 500 m long. The data during a flood at the two ends are as follows:

SECTION	BED ELEVATION (m)	WATER-SURFACE ELEVATION (m)
Upstream	100	101.0
Downstream	99	100.5

Assume Manning's roughness coefficient to be 0.025.

12.7. The following measurements are available for the main gage and the auxiliary gage, which is downstream:

MAIN GAGE (m)	AUXILIARY GAGE (m)	DISCHARGE (m^3/s)
125.0	124.50	250
125.0	123.50	480

If the main gage reading is 125.0 and the auxiliary gage reading is 124.0, what is the discharge in the stream?

12.8. Measurements taken during a flood at a gaging station indicate that the water-surface elevation increased at a rate of 10.0 cm/h. The normal discharge for the river stage obtained from a steady-flow rating curve is 150 m^3/s. The slope of the river bed is 3×10^{-4}. Assuming the velocity of the flood wave to be 2.5 cm/s, compute the river discharge.

12.9. Compute the velocity of flow if the current meter has the number of revolutions as: 0, 80, 90, 95, 100, 120, 130, 125, 115, 95, 85, and 70. The coefficients a and b, respectively, are 0.45 and 0.03.

12.10. A 400 g/l solution of sodium dichromate is used for gaging a stream. The solution is injected at a constant rate of 10 l/s. At a downstream section, the equilibrium concentration is measured as 10 parts per million. Estimate the discharge in the stream.

12.11. Compute the discharge of the Pearl River near Bogalusa, Louisiana, using the velocity–area station method. Pertinent data for discharge measurement as collected by the U.S. Geological Survey are given in Table 12.8.

12.12. The following streamflow observations (1982–1984) were made by the U.S. Geological Survey at Bogue Chitto River near Bush, Louisiana. Construct the rating curve for this river.

GAGE HEIGHT (ft)	DISCHARGE (cfs)
3.04	573
6.84	1,840
8.03	3,270
10.60	8,760
18.80	81,900
21.15	130,000
6.06	1,370
8.54	3,090
4.46	940
4.64	1,080
3.87	783
4.07	840
6.91	2,350
8.54	3,860
7.42	280
6.87	2,470
6.15	2,050
5.04	1,560
3.42	682

TABLE 12.8 STREAMFLOW MEASUREMENTS OF THE PEARL RIVER NEAR BOGALUSA, LOUISIANA[a]

	BASE GAGE READINGS		
TIME	RECORDER	INSIDE	OUTSIDE
0825	7.68	7.67	7.72
1120 start	7.68	7.67	7.72
1250 end	7.68	7.67	7.72
1330	7.68	7.67	7.72
Correct M.G.H.	7.68	7.67	7.72

TABLE 12.8 *(Continued)*

Angle coeffi-cient	Dist. from initial point	Width	Depth	Obser-vation depth	Revolu-tions	Time (s)	Velocity At point	Velocity Mean in vertical
RWE	8	16					11	20
	40	21	10.3	0.2	5	41	0.28	0.18
				0.8	2	60	0.09	
	50	10	10.5	0.2	9	46	0.45	0.44
				0.8	8	42	0.43	
	60	10	9.9	0.2	11	43	0.58	0.40
				0.8	4	41	0.23	
	70	10	10.4	0.2	14	41	0.76	0.64
				0.8	10	43	0.53	
	80	10	10.9	0.2	14	43	0.74	0.65
				0.8	10	41	0.56	
	90	10	11.2	0.2	16	41	0.87	0.70
				0.8	10	43	0.53	
	100	10	11.2	0.2	15	41	0.82	0.76
				0.8	13	42	0.70	
	110	10	11.1	0.2	19	40	1.05	0.91
				0.8	14	41	0.77	
	120	10	10.6	0.2	19	41	1.02	0.91
				0.8	15	42	0.80	
	130	10	10.0	0.2	21	42	1.11	1.06
				0.8	19	42	1.00	
	140	10	10.0	0.2	20	42	1.06	1.00
				0.8	17	41	0.93	
	150	10	9.4	0.2	20	42	1.07	0.96
				0.8	16	42	0.85	
	160	10	9.4	0.2	21	42	1.11	0.98
				0.8	16	42	0.85	
	170	10	9.0	0.2	22	40	1.21	1.00
				0.8	15	42	0.80	
	180	10	9.0	0.2	21	40	1.15	1.04
				0.8	17	40	0.94	
	190	10	8.8	0.2	21	41	1.06	0.94
				0.8	15	42	0.81	
	200	10	8.1	0.2	19	40	1.06	0.95
				0.8	16	42	0.84	
	210	10	7.6	0.2	19	40	1.04	0.94
				0.8	16	42	0.85	
	220	10.0	8.5	0.2	18	42	0.95	0.82
				0.8	13	42	0.70	

TABLE 12.8 *(Continued)*

Angle coefficient	Dist. from initial point	Width	Depth	Observation depth	Revolutions	Time (s)	Velocity At point	Velocity Mean in vertical
	230	10	8.2	0.2	16	42	0.86	0.74
				0.8	11	41	0.61	
	240	10	7.4	0.2	16	42	0.86	0.76
				0.8	12	41	0.66	
	250	10	6.7	0.2	15	40	0.84	0.77
				0.8	13	42	0.70	
	260	10	6.3	0.2	14	41	0.76	0.76
				0.8	14	41	0.77	
	270	10	7.0	0.2	13	42	0.69	0.64
				0.8	11	42	0.60	
	280	10	7.9	0.2	13	43	0.68	0.68
				0.8	12	40	0.67	
	290	10	6.4	0.2	11	42	0.59	0.61
				0.8	12	43	0.63	
	300	11.5	5.0	0.2	7	44	0.36	0.36
				0.8	7	44	0.37	
LWE	313	6.5						

[a] Date 6–16–88. Width = 305 ft, area = 2476 ft^2, no. of x-sections = 29, velocity = 7.5 ft/s, gage height = 7.67 ft, discharge = 1853 cfs, boat downstream side of bridge 2 miles below gage.

12.13. Fit a curve to the streamflow data of Exercise 12.12. Determine parameters of the curve. Plot the curve on both rectangular paper and logarithmic paper.

12.14. Develop the constant fall rating curve for streamflow data collected by the U.S. Geological Survey for Bayou Courtableau at Washington, Louisiana. Table 12.9 gives such data. Estimate the error of the rating curve.

12.15. Develop the normal fall rating curve using the data in Table 12.10.

12.16. Establish a discharge-rating curve for the following data available for a hydrograph on September 6–7, 1985, at Little Papillion Creek, Irvington, Nebraska.

Time	September 7 Gage height (ft)	September 7 Discharge (cfs)	September 8 Gage height (ft)	September 8 Discharge (cfs)
1:00 A.M.	8.10	900	4.75	130
2:00	12.20	2205	4.95	148
3:00	15.60	3725	5.60	221

Time	September 7		September 8	
	Gage height (ft)	Discharge (cfs)	Gage height (ft)	Discharge (cfs)
4:00	18.90	5280	6.60	379
5:00	26.20	6425	6.40	341
6:00	19.20	5425	5.60	221
7:00	16.00	3925	5.30	184
8:00	12.70	2445	5.00	152
9:00	9.70	1305	4.75	130
10:00	8.20	780	4.60	116
11:00	8.00	705	4.50	107
12:00 N	7.75	642	4.35	94
1:00 P.M.	7.55	592	4.20	80
2:00	7.20	505	4.10	72
3:00	6.30	324	4.00	64
4:00	5.80	249	3.95	60
5:00	5.50	208	3.85	54
6:00	5.30	184	3.80	50
7:00	5.10	162	3.75	46
8:00	5.00	152	3.70	43
9:00	4.90	143	3.68	42
10:00	4.85	138	3.64	40
11:00	4.80	134	3.62	38
12:00 M	4.75	130	3.60	37

TABLE 12.9 STREAMFLOW MEASUREMENTS OF BAYOU COURTABLEAU AT WASHINGTON, LOUISIANA

Base gage height (ft)	Auxiliary gage height (ft)	Fall (ft)	Discharge (cfs)
19.0	19.91	0.91	1210
19.04	20.25	1.21	1410
18.83	20.03	1.2	1340
19.55	20.75	1.2	1440
21.04	22.45	1.41	1950
17.82	18.61	0.79	904
19.04	20.21	1.17	1310
18.54	19.47	0.93	1070
19.24	20.23	0.99	1280
17.9	18.98	1.08	1190
19.32	20.33	1.01	1310
19.9	21.16	1.26	1600
20.13	21.25	1.12	1530
18.22	19.57	1.35	1290
17.29	18.32	1.03	998
17.88	19.17	1.29	1240
19.09	20.03	0.94	1280
18.32	19.66	1.34	1340
19.32	20.41	1.09	1330

TABLE 12.10 STREAMFLOW MEASUREMENTS OF BAYOU COURTABLEAU AT WASHINGTON, LOUISIANA

Gage height (ft)	Auxiliary gage height (ft)	Fall (ft)	Discharge (cfs)
17.12	17.41	0.29	458.0
18.14	18.78	0.64	833.0
18.59	18.98	0.39	670.0
24.37	26.89	2.52	3440.0
19.12	19.77	0.65	970.0
14.86	14.90	0.04	98.6
18.88	19.37	0.49	839.0
19.00	19.91	0.91	1210.0
13.67	13.93	0.26	210.0
12.40	12.65	0.25	165.0
12.65	12.90	0.25	157.0
22.84	25.17	2.33	3040.0
19.04	20.25	1.21	1410.0
18.83	20.03	1.20	1340.0
19.55	20.75	1.20	1440.0
18.64	19.06	0.42	710.0
18.06	18.35	0.29	551.0
17.51	17.57	0.06	211.0
14.18	14.43	0.25	220.0
17.08	17.20	0.12	299.0

12.17. The tabulated data that follow are measured discharges and stages at an auxiliary station 750 m downstream of the principal gage. Establish a stage-discharge relationship. Also estimate the discharge corresponding to a stage of 7.5 m in the river and a fall of 0.8 m.

Measured discharge in m^3/s ($\times$ 1000)	Stage of the river (m)	
	Principal gage	Auxiliary gage
3.2	1.791	1.640
5.45	2.195	1.990
8.15	2.645	2.432
13.85	3.215	3.068
16.4	3.618	3.431
19.45	3.912	3.723
22.15	4.415	3.942
24.62	4.331	4.126
26.45	4.432	4.241
30.22	4.637	4.458
31.62	4.715	4.521
34.55	4.856	4.614

12.18. Do the following data indicate a simple, normal, or constant fall rating curve? Show all computations to arrive at your conclusion.

Base stage (ft)	Discharge (cfs)	Auxiliary stage (ft)
14.02	2,400	13.00
23.80	29,600	23.25
17.70	21,200	16.60
24.60	85,500	23.55
20.40	28,200	19.55
17.00	7,400	16.40
18.65	34,000	17.50
26.40	55,000	25.70
22.20	74,200	21.00
16.20	9,550	15.30
21.10	43,500	20.13
25.60	84,000	24.60
23.20	93,500	21.95

12.19. From the following data, determine if a constant-fall or normal-fall rating curve is required.

Gage height (ft)	Discharge (cfs)	Fall (ft)
3.2	20	2.2
4.0	350	1.0
9.0	8000	3.5
6.0	1700	2.8
3.6	150	1.2

12.20. Calculate the flow rate Q at the gaging station from the field data taken.

Distance from initial point (ft)	Depth (ft)	Observation depth (ft)	Velocity (ft/s)
0			
3	3.3	0.2	2.55
		0.8	1.96
5	3.4	0.2	2.44
		0.8	1.84
7	3.4	0.2	2.50
		0.8	2.05
9	3.6	0.2	2.55
		0.8	2.05
12	3.5	0.2	2.44
		0.8	1.96
15	0	0	0

12.21. Compute the streamflow for the following measurement data:

Distance (ft)	0	2	4	6	8	10	12	14	16	18	20	22
Depth, d (ft)	0	1.0	4.3	7.2	8.5	7.4	5.6	4.7	3.5	2.1	1.4	0
Velocity (ft/s)												
$0.2d$	0	1.4	1.0	2.6	2.9	2.7	2.5	2.3	2.1	1.8	1.5	0
$0.8d$	0	0.7	1.2	1.8	2.0	1.9	1.7	1.5	1.3	1.1	1.0	0

12.22. The following data are measured discharges, stages, and stages at an auxiliary station 2000 ft downstream from the main gage. Develop a slope-stage-discharge relation from these data. Determine the exponent m by plotting Q_a/Q_0 vs. F/F_0 on logarithmic paper. The slope of the line is the exponent m. Find the discharge for a stage of 24.95 ft and a fall of 1.43 ft. Note that it is convenient to take $F_0 = 1$ ft.

Measured Q (cfs)	Stage (ft)	
	Main gage	Auxiliary gage
93,900	11.71	10.81
80,000	9.73	8.73
242,000	16.11	15.10
27,000	6.00	4.99
553,000	24.95	23.97
417,200	21.02	19.99
560,000	21.75	20.19
421,000	23.51	22.77
241,000	14.48	13.02
112,700	10.75	9.53
365,000	18.30	17.00
702,000	27.22	26.07
570,000	26.45	25.58
297,600	21.60	20.98
129,800	13.72	13.02

12.23. Plot the rating curve on arithmetic and on log–log paper for the data that follow. From the arithmetic plot, determine the stage at $Q = 0$. This rating table is for the Leaf River at Hattiesburg, Mississippi (October 1, 1968).

Gage height (ft)	Discharge (cfs)	Difference (cfs)
14.5	11,300	140
		160
17.5	16,100	160
		180
20	20,600	180
		200
21.2	23,000	200
		250
23.6	29,000	250
		300

Gage height (ft)	Discharge (cfs)	Difference (cfs)
24.6	32,000	300
		350
25.4	34,800	350
		400
26.4	38,800	400
		500
27.8	45,800	500
		600

12.24. Determine Equation (12.25) for the rating curve. What are the units of a and c?

12.25. Estimate the discharge at a flood stage of 27.80 ft, both graphically and by the use of Equation (12.25) in Exercise 12.24.

12.26. Estimate the depth at a flow of 365 cfs, both graphically and by the use of Equation (12.25) in Exercise 12.24.

CHAPTER 13
Streamflow Hydrograph

A streamflow hydrograph at any point on a stream is a graph of the time distribution of water discharge at that point. The graph is plotted with discharge on the ordinate and time on the abscissa. The units of time can be in minutes, hours, or in days to accommodate the duration of the flow hydrograph. A recording stream gage defines a continuous hydrograph representing all discharges that occur during the life of the gage. It is more usual in hydrology to consider a hydrograph for a certain storm event. The hydrograph is a continuous curve, whereas a rainfall hyetograph is discrete. A similar graph is made relating depth or stage to time, and is called a stage hydrograph. When streamflow velocity is related to time, the graph is called a velocity hydrograph. The term hydrograph usually refers to the time distribution of water discharge.

A hydrograph for a given storm reflects the influence of all the physical characteristics of the drainage basin and, to some extent, also reflects the characteristics of the storm causing the hydrograph. A hydrograph can be considered a thumbprint of the drainage basin. The actual shape of a hydrograph is determined by the rate at which water is transmitted from the various parts of the drainage basin to the gage. Most of this water is carried by the channels, but some water flows overland directly to the gage. Clearly, the number of channels delivering water to the gage at the same time reflects hydrograph shape. When a streamflow hydrograph is plotted using short time intervals, the influence of tributary channels on the shape of the hydrograph is apparent.

No two drainage basins produce identical hydrographs for the same storm. Hydrographs from similar drainage basins may be similar, but not the same. Similarly, no two storms produce identical hydrographs from the same basin. Figure 13.1 shows two similar basins, watersheds P-2 and P-3, Riesel (Waco), Texas. Each

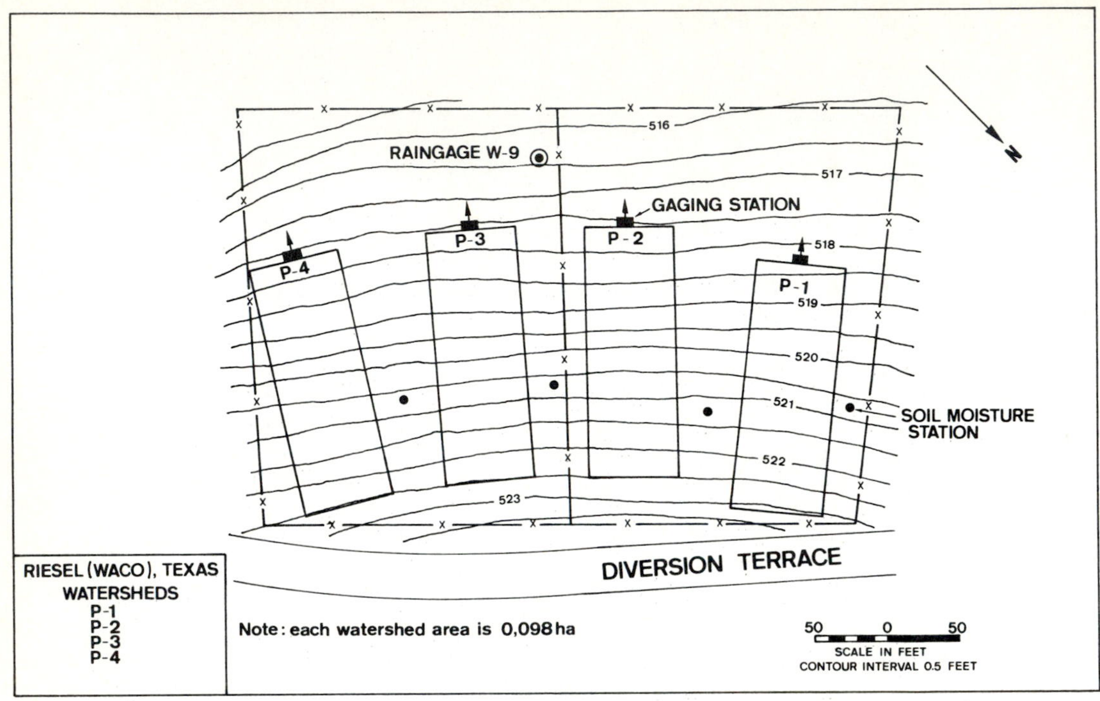

Figure 13.1 Two similar watersheds, Riesel (Waco), Texas.

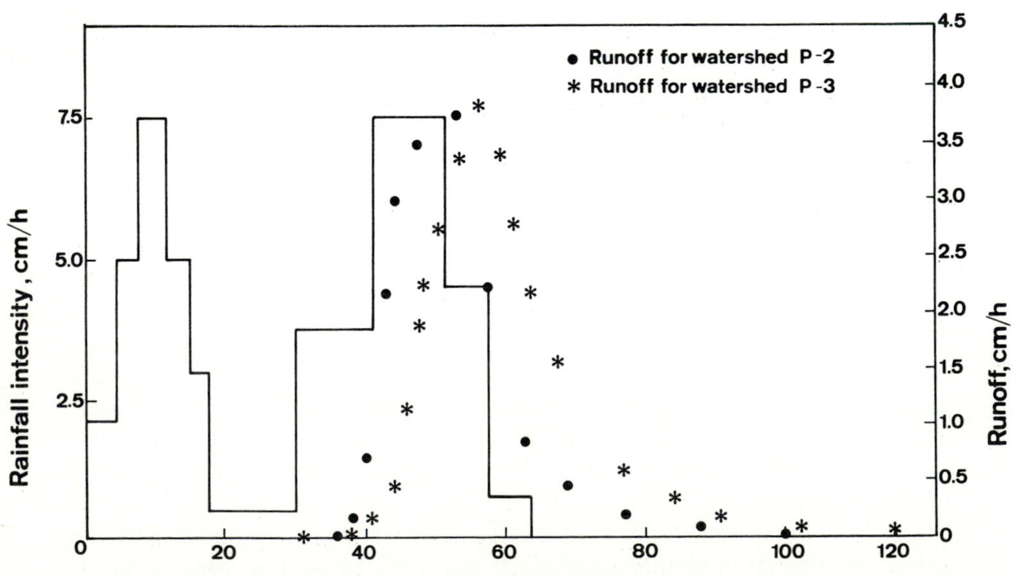

Figure 13.2 Rainfall from watersheds P-2 and P-3, Riesel (Waco), Texas, for the rainfall event of June 25, 1961.

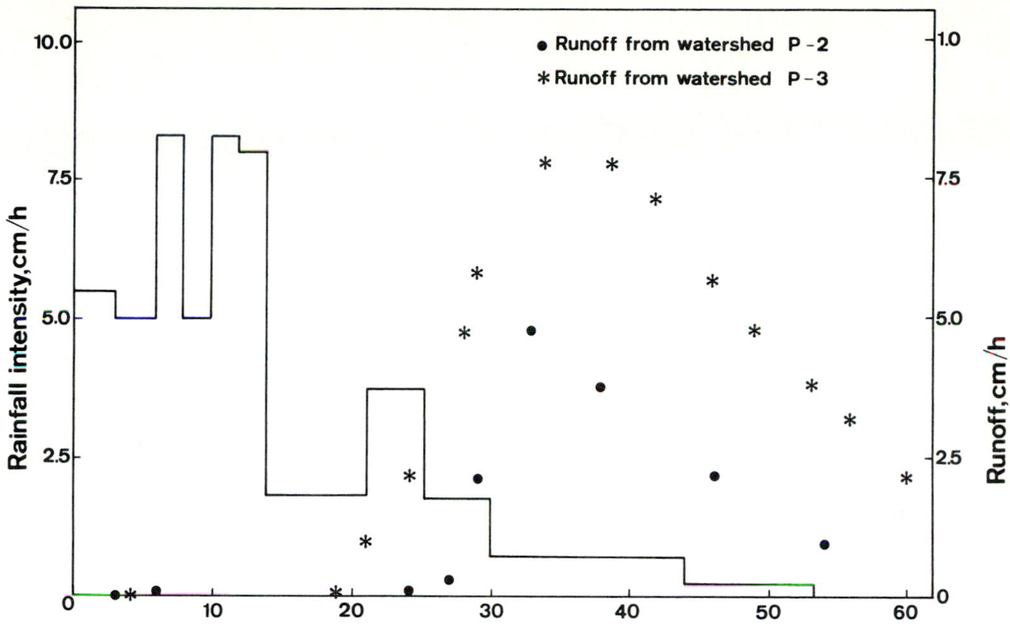

Figure 13.3 Runoff from watersheds P-2 and P-3, Riesel (Waco), Texas, for the rainfall event of July 16–17, 1961.

watershed is 0.098 ha in area. Two rainfall events are considered for these watersheds. Runoff hydrographs from the watersheds for rainfall event of June 25, 1961, are shown in Figure 13.2. The rainfall amount for the event was 3.76 cm. Although both watersheds produced similar hydrographs, the hydrographs are not the same. The runoff from watershed P-2 was 1.22 cm and from watershed P-3 was 1.32 cm. The hydrograph peaks were 4.24 cm/h and 3.89 cm/h for watersheds P-2 and P-3, respectively. For another rainfall event, runoff hydrographs from these two watersheds are shown in Figure 13.3. The rainfall amount was 2.51 cm, and runoff amounts were 0.15 cm and 0.38 cm, respectively, for watersheds P-2 and P-3. Although hydrographs are similar in shape, they are quite different. The hydrograph peaks were 0.48 cm/h and 0.79 cm/h, respectively, for watersheds P-2 and P-3.

13.1 COMPONENTS OF STREAMFLOW

A streamflow hydrograph of a given storm is a hydrograph of total runoff. Depending upon their sources, the components of runoff are shown in Figure 13.4. These components are (a) direct runoff and (b) baseflow. The direct runoff is divided into surface runoff and quick interflow, whereas the baseflow is divided into delayed interflow and groundwater runoff. The division into quick and delayed interflows is essentially arbitrary.

Total runoff corresponds to a given storm event. The volume of total runoff is determined by including in the streamflow hydrograph all runoff between the baseflow discharge occurring prior to the storm up to the same baseflow discharge after

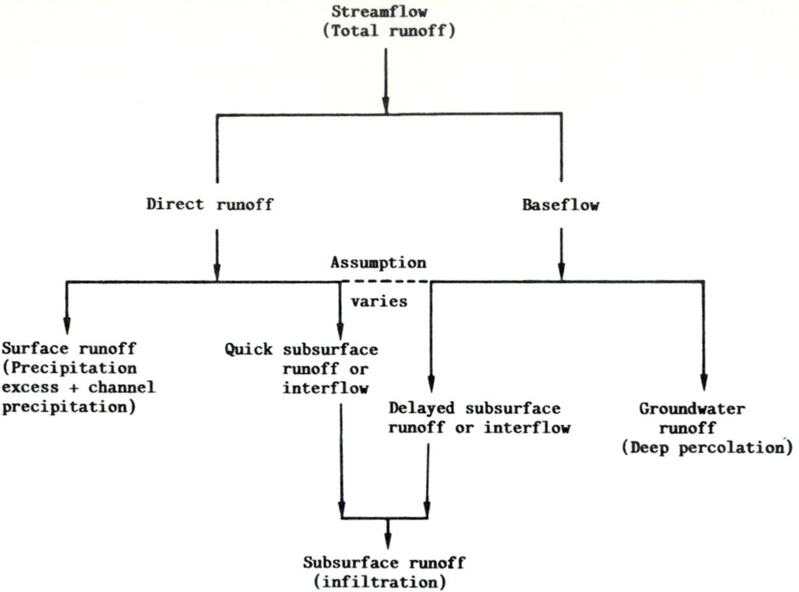

Figure 13.4 Components of streamflow or total runoff, with their sources indicated in parentheses.

the storm. Sometimes the hydrograph is cut off at a greater baseflow discharge to consider the volume of total runoff. This volume is then calculated by summing the discharges over the specified time base. Such volume determination is arbitrary and subject to a certain amount of uncertainty.

13.1.1 Surface Runoff

Surface runoff, or overland flow, as shown in Figure 13.5, is that runoff water that travels over the ground surface to a drainage channel. The channel to which surface runoff flows may be of any order. Any drainage depression, no matter how small, that directs the flow of water is a channel. This microdrainage is usually not included in stream ordering for practical reasons, but could be included if desired. Most surface runoff flows to first-order channels because they collectively drain the greatest area of the drainage basin. Surface runoff also includes that precipitation that falls directly on water flowing in the channel, whereas overland flow does not include this water. Although surface runoff or overland flow is present, it does not necessarily occur as sheet flow from all parts of the drainage basin. Sheet flow usually occurs from an impervious surface such as a paved parking lot, but can only occur on a natural drainage basin when rainfall intensity uniformly exceeds the infiltration capacity. This condition does not frequently happen. Variations in the distribution of soil type and of rainfall over a drainage basin usually result in limited sheet flow.

Surface runoff is believed to be the principal contributor to the peak discharge from a storm event. Because this water runs off over the surface to the channel, it is believed to be the first to reach the channel and, therefore, forms the rising limb and peak of the hydrograph.

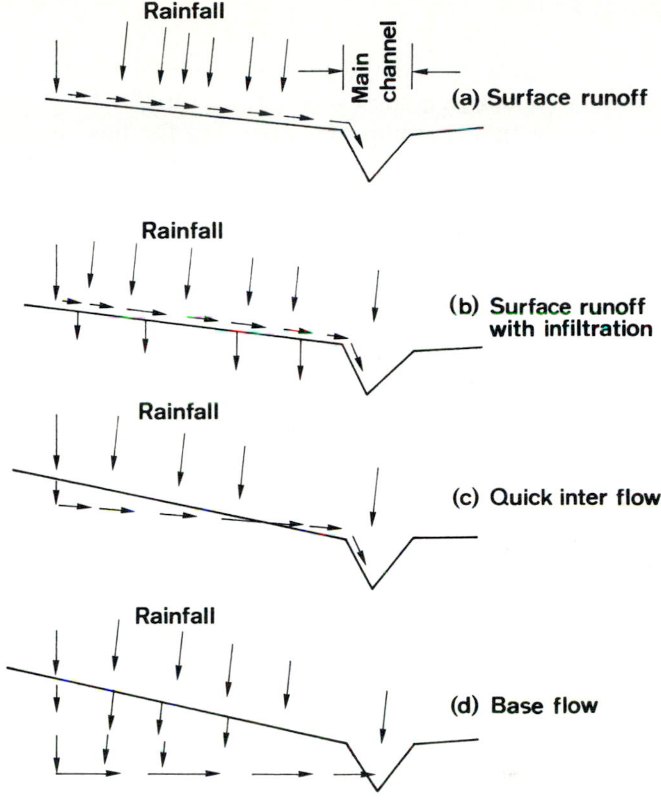

Figure 13.5 Runoff components.

13.1.2 Interflow

Interflow, also called subsurface storm flow, is that surface water that infiltrates the surface layer and moves laterally beneath the surface to a channel, as shown in Figure 13.5. Interflow can occur on forest floors, where the leaves, needles, and other debris cover the ground. Precipitation infiltrates this surface material and moves laterally through the forest litter over the ground surface. Interflow might occur in shallow soils filled and loosened by tree roots, rock debris covering the ground surface, or surface soils loosened by any cause. During interflow, the movement of water is subject to greater flow resistance than surface runoff. As a result, interflow does not move as rapidly as surface runoff and is, therefore, delayed in reaching the channel and the watershed outlet until after surface runoff has passed. Accordingly, interflow does not add to the peak discharge, but reaches the gage or outlet after the peak discharge has passed. Interflow might occur at any time in the course of water movement over the surface or interflow can emerge and continue its flow over the surface. Chow (1964) has distinguished interflow into quick interflow and delayed interflow, with the former merging with surface runoff and the latter with baseflow. This division, of course, is arbitrary.

13.1.3 Direct Runoff

Direct runoff is usually considered to be the sum of surface runoff and interflow. Direct runoff is frequently equated with surface runoff. These two flow components move more rapidly than groundwater flow and for this reason are often lumped together for hydrologic purposes. Such lumping is reasonable for certain purposes because it is logical to believe that some interflow near the gage or outlet will arrive at that point before surface runoff from farther up the basin.

13.1.4 Baseflow

Baseflow, or groundwater flow, is the flow component contributed to the channel by groundwater. Groundwater occurs from surface-water infiltration to the water table and then moving laterally to the channel through the aquifer, as shown in Figure 13.5. Such water moves much more slowly than direct runoff and, for this reason, does not contribute to the peak discharge for a given storm. Flow in a perennial stream prior to a storm is from baseflow. During a storm event, the baseflow is augmented by infiltration. For this reason, baseflow is separated from direct runoff. This separation is an uncertain task and, as a result, it is possible to separate some interflow along with the baseflow or perhaps include some baseflow with interflow. Drainage basins with highly permeable, thick soils usually have a high groundwater-flow component and relatively small direct-flow component, whereas basins with heavy-clay, low-infiltration soils have a small or zero groundwater component and a high direct-runoff component. A portion of the groundwater-flow component occurs from water infiltrating the banks of the channel during high-water flows. When the channel water recedes, this portion of the groundwater stored in the channel banks moves back to the channel as groundwater flow.

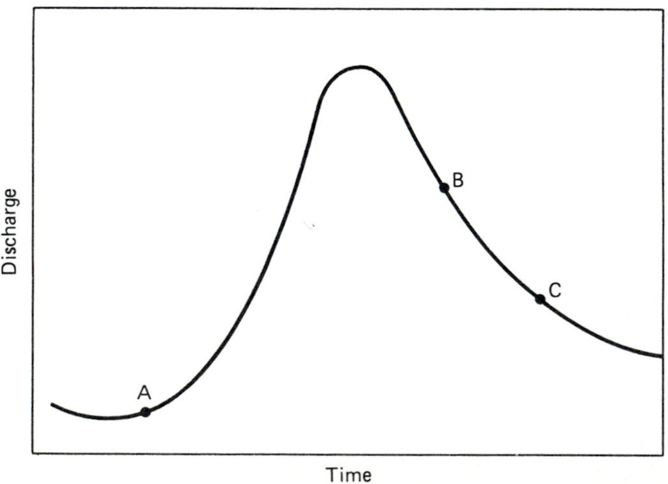

Figure 13.6 A typical streamflow hydrograph.

13.1.5 Delineation of Runoff Components

A streamflow hydrograph is shown in Figure 13.6, with points illustrating the three flow components. By common definition, point A marks the beginning of surface runoff, which is believed to end at the change in slope shown as B; point B is considered to be the beginning of interflow, which ends at point C; point C marks the beginning of groundwater flow, which continues beyond the end of the hydrograph. Of course, this division of hydrograph-flow components is subjective and has no quantitative basis. There is no way to verify the source of water during most of the hydrograph and certainly the separation boundaries cannot be verified.

13.2 FACTORS AFFECTING HYDROGRAPH CHARACTERISTICS

Several factors affect a streamflow hydrograph during a runoff event on a drainage basin. These factors are (a) drainage characteristics, (b) rainfall characteristics, and (c) soil and land use.

13.2.1 Drainage Characteristics

The drainage characteristics that effect streamflow derive principally from the parent geology of the drainage basin. The effect of geology is transmitted through the characteristics of the drainage basin to the hydrograph of runoff from that basin. Geology determines the perviousness of the drainage basin. As a result, drainage basins with low infiltration produce high-density drainage channels and high runoff with a high peak discharge and a relatively steep rising limb. An adjacent drainage basin with high infiltration will produce low runoff with a lower peak discharge and a more gentle rising limb and a longer recession. Geology also determines the drainage pattern. A treelike drainage pattern is called a dendritic drainage and is most common. However, geology might cause the pattern to be in a rectangular form or in a trellislike form. Other forms of drainage are possible and are usually related to the geology of the drainage basin.

The pattern of drainage, in conjunction with other geologic factors, has a pronounced effect on hydrograph characteristics. The greatest effect on the shape of the hydrograph occurs when the geology and soil vary within the drainage basin. The resulting channel-length frequency distribution may cause water to be delivered to the gage in surges such that the recession or even the rising limb is quite irregular. Another geologic condition that markedly affects the hydrograph rise occurs when porous and permeable limestone underlies a drainage basin. Such a condition often results in "sinks," where channel flow disappears underground into the limestone at such sinks. Flow might or might not rise again further downstream. If it rises again, the flow is usually diminished and the hydrograph reflects the characteristics of a storage reservoir represented by a large underground storage cavern. Similar conditions occur in volcanic rocks. Some glacial deposits have a marked effect on hydrograph.

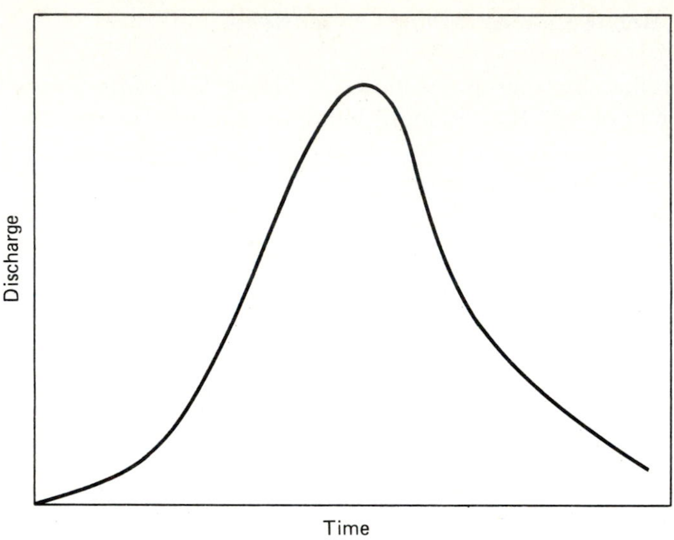

Figure 13.7 Hydrograph from storm falling on the lower end of a drainage basin.

13.2.2 Rainfall Distribution

A storm falling on the lower end of the basin near the outlet will have a higher peak discharge with a steeper rising limb, as shown in Figure 13.7, than a storm producing the same volume of water falling on the upper end of the drainage basin, as shown in Figure 13.8. The same volume of water draining from a storm covering the entire drainage basin will have hydrograph characteristics between the hydrographs from upper and lower storms. Further differences in hydrograph characteristics occur when runoff producing storms move laterally up or down the drainage basin or are

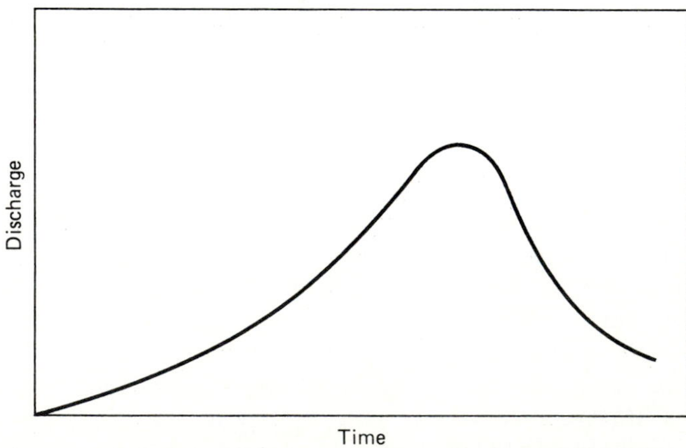

Figure 13.8 Hydrograph from the storm falling on the upper end of a drainage basin.

stationary over the basin. Other important rainfall characteristics affecting the runoff hydrograph are rainfall depth, intensity, duration, and time distribution.

13.2.3 Soil and Land Use

Soil and land use by man modify the hydrograph characteristics. Most changes in land use, except dams and diversions, increase the amount of runoff from a given drainage basin. Urbanization, farming, forest removal, cutting grass, and building roads or any other structure increase runoff. The modified hydrograph reflects the influence of these changes. Dams and diversions decrease runoff and modify the hydrograph to reflect their influence.

13.3 ELEMENTS OF THE HYDROGRAPH

The elements of the hydrograph are shown in Figure 13.9.

13.3.1 Rising Limb

As surface runoff reaches the gage, the water begins to rise in the channel. With continuing elapse of time, more and more surface runoff reaches the gage and the water in the channel continues to rise until it reaches a maximum discharge, recorded as the maximum gage height, and after this stage, the water begins to recede. The rising portion of the hydrograph indicated by the rising stage is called the rising limb. The rising limb graphically represents increasing discharge over time as the limb rises and discharge increases. The simplified hydrograph shown in Figure 13.9 indicates a smooth increase in discharge. In practice, it is more likely that frequent stage observations will show a fluctuating discharge, which presents a

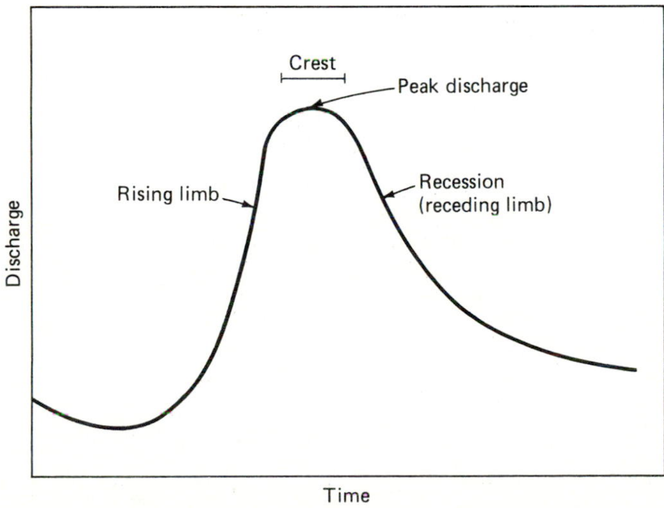

Figure 13.9 Elements of a hydrograph.

rather ragged rising limb. This fluctuating rising limb reflects the delivery of water to the gage by groups of channels with approximately equal travel time.

13.3.2 Crest

The time interval of the greatest discharge at the peak of the hydrograph is called the crest. The crest might be of a short time interval represented by a sharp peak or of a fairly long interval represented by a flat peak. The crest is not necessarily composed of equal discharges, but rather represents a subjective zone of nearly equal highest discharges. The greatest discharge within the crest is the peak discharge, which is of primary interest in hydrologic design. Some hydrographs have several well-defined peaks reflecting channel-length frequency distribution, but the peak discharge is always the highest discharge within the hydrograph.

13.3.3 Recession Limb

The portion of the hydrograph after the peak is known as the receding limb, falling limb, or recession curve. The receding limb represents decreasing discharge as water is withdrawn from the drainage-basin storage after rainfall ceases. The slope of the receding limb indicates the rate at which water is drained from the basin. The lower part of the recession, which has a much lower slope, is believed to represent groundwater contribution because the water is withdrawn much more slowly than the other components.

13.3.4 Streamflow Recession

Barnes (1940) has shown that the streamflow recession can be expressed as

$$Q_t = Q_{t-1}K_r \tag{13.1}$$

or

$$Q_t = Q_0 K_r^t \tag{13.2}$$

where Q_0 is the initial discharge at any time, Q_t is the discharge at time interval t later, and K_r is the recession or depletion constant dependent upon the units of time and is less than unity. The time interval is usually taken as 24 hours, or one day, but a smaller value may be desirable for smaller drainage basins. The number "1" in Equation (13.1) reflects the time interval and not the numerical value of 1.

During recession, there is no inflow to the drainage basin. Therefore, the rate of change of the drainage-basin storage S_t depends entirely on the streamflow discharge:

$$\frac{dS_t}{dt} = -Q_t \tag{13.3}$$

After a long time, the basin is completely depleted of water. This condition can be expressed as

$$S_t = 0, \qquad t \to \infty \tag{13.4}$$

Subject to this condition, an expression for S_t in terms of Q_t can be obtained by inserting Equation (13.2) into Equation (13.3) and then solving:

$$\frac{dS_t}{dt} = -Q_0 K_r^t$$

which yields

$$S_t = \frac{Q_t}{\ln K_r} \tag{13.5}$$

From Equation (13.5), the storage remaining at anytime t in the basin can be computed if Q_t and K_r are known.

The storage of water in the basin is comprised of surface storage (both surface detention and channel storage), interflow storage, and groundwater storage. Clearly, then, the recession constant K_r will be different for these three types of storage. In order to account for these storages, K_r can be considered to be made up of the recession constant for surface storage, K_{rs}; the recession constant for interflow, K_{ri}; and the recession constant for groundwater storage, K_{rb}. Equation (13.2) plots as a straight line on semilogarithmic paper, with K_r defining the slope of the line. When streamflow recession is plotted, it is not a straight line, but a curve with gradually decreasing slope or increasing value of K_r. This, of course, is due to the different storages contributing to streamflow. When the time interval is assumed in days, typical values of the recession constants of these storage coefficients are $K_{rs} = 0.1$ to 0.5, $K_{ri} = 0.5$ to 0.85, and $K_{rb} = 0.85$ to 0.99. If interflow is not significant, K_{ri} can be assumed to be zero.

Equation (13.2) can be written for recession of each component of streamflow, with K_r replaced by the component's recession constant. Thus, streamflow recession can be approximated by three straight lines representing the three components on semilogarithmic paper (Barnes, 1940). In reality, the transition from one line to the next is not always sharp and it becomes difficult to identify the points marking the change in slope of the line. The three straight lines can be constructed as follows. Plot the streamflow recession on semilogarithmic paper. The last portion of the curve represents the recession of groundwater runoff. This portion will plot as a straight line and is projected backward. The slope of this line is the value of K_{rb}. The difference between the total streamflow recession and this straight line is plotted, which gives the recession of interflow and surface runoff, and the procedure is repeated. In this manner, recessions of surface flow and interflow, along with their respective constants K_{rs} and K_{ri}, can be established.

EXAMPLE 13.1 The streamflow recession observation indicated that the flow on June 10, 1983, was 1000 m^3/s. The recession constant was found to be 0.9 with the time interval of 24 hours. Compute the flow in this stream on June 15, 1983.

Solution Equation (13.2) is used to compute the flow on June 15, 1983. $Q_0 = 1000$ m^3/s, $K_r = 0.9$, and $t = 5$.

$$Q_t = 1000 \times (0.9)^5 = 1000 \times 0.59 = 590 \text{ m}^3/\text{s}$$

The flow on June 15, 1983, is 590 m^3/s.

EXAMPLE 13.2 If the initial discharge is given as 180 m^3/s and the depletion constant is 0.9, what will be the discharge at the end of the third day?

Solution

$$Q_0 = 180 \text{ m}^3/\text{s}$$
$$K_r = 0.9$$
$$t = 3$$

Therefore, using Equation (13.2),

$$Q_3 = 180 \times (0.9)^3 = 180 \times 0.729 = 131.22 \text{ m}^3/\text{s}$$

13.4 HYDROGRAPH TIME CHARACTERISTICS

The shape, and therefore the characteristics, of a hydrograph can be measured in terms of time, as shown in Figure 13.10. A detailed discussion of hydrograph time characteristics is given in Singh (1988).

13.4.1 Time to Peak

The time to peak is the time elapsed from the beginning of the rising limb to the peak discharge. This value is largely determined by the drainage-basin characteristics such as travel distance, drainage density, channel slope, channel roughness, and soil-infiltration characteristics. The time to peak is altered somewhat by the distribu-

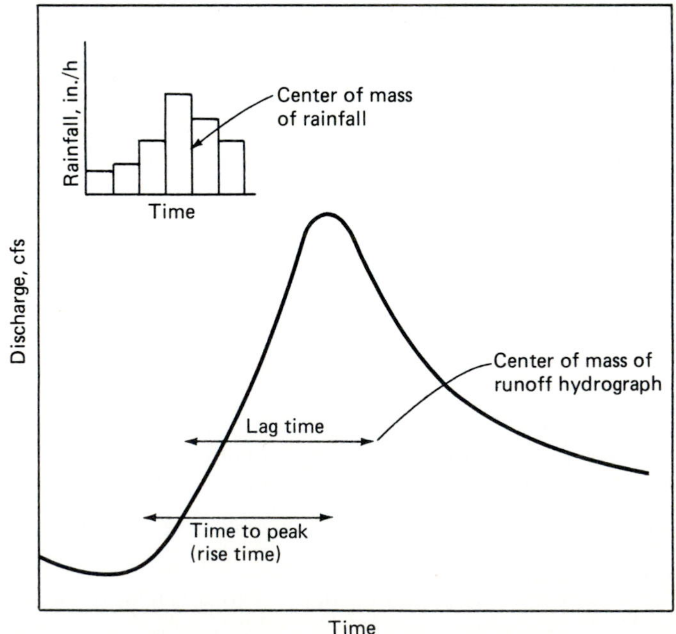

Figure 13.10 Hydrograph response time.

tion of rainfall over the drainage basin. For example, the hydrograph of a storm falling on the upper part of a drainage basin has a longer time to peak than for a storm falling on the lower end of the drainage basin. For a given amount of runoff, a longer time to peak has a lower peak discharge than a shorter time to peak. The time interval represented by the time to peak varies according to the duration of the rainfall event.

13.4.2 Time of Concentration

The time of concentration is the time required for a drop of water falling on the most remote part of the drainage basin to reach the basin outlet or gage. It includes the time required for all portions of the drainage basin to contribute runoff to the hydrograph and this time then represents the maximum discharge that can occur from a given storm intensity over the drainage basin. This definition assumes that all initial losses are satisfied and that uniform rainfall continues over the entire drainage basin for a period at least equal to the time of concentration. Except for very small drainage basins, it is rare for the time of concentration to be attained under natural conditions. By assuming a uniform rainfall over the entire drainage basin, the discharge increases as water from progressively farther distances arrives at the gage. Because the storm is continuing, this runoff is added to drainage nearer to the gage and the discharge rises. The hydrograph continues to rise as time elapses and rainfall continues until drainage from the most remote point on the basin arrives at the gage. At this time, the discharge becomes a constant, as shown in Figure 13.11, because all drainage areas within the basin are contributing to the discharge and no additional drainage area can be added to the runoff. If the rainfall continued at the same uniform rate, the hydrograph peak would become flat at its maximum discharge and would continue so until rainfall intensity changed or rainfall ceased. The time of concentration is a unique condition for the time to peak. It is important to distinguish between the time to peak and the time of concentration. In practice, the hydrograph peak is sharply defined and the storm duration is less than the time of concentration.

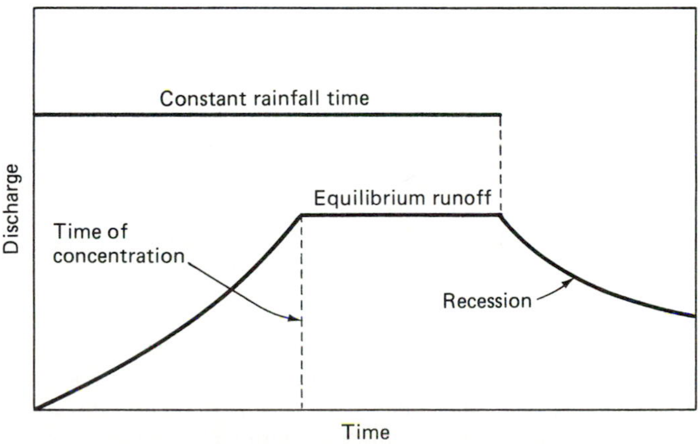

Figure 13.11 Time of concentration.

The time of concentration can be determined by many formulas. One of the most commonly used formulas is by Kirpich (1940):

$$t_c = 0.0078\ (L/S^{0.5})^{0.77} \tag{13.6}$$

where t_c is the time of concentration in minutes, L is the length of travel in feet from the most remote point on the drainage basin along the drainage channel to the basin outlet, and S is the slope in feet per foot determined by the difference in elevation of the most remote point and that of the outlet divided by L. The equation assumes uniform rainfall occurring on the drainage basin.

The Soil Conservation Service (1972) uses Equation (13.6) in the form

$$t_c = L^{1.15}/7700H^{0.38} \tag{13.7a}$$

where t_c is the time of concentration in hours, and H is the difference in elevation in feet between the most distant point on the drainage basin and the outlet. Equation (13.7a) can be converted to Equation (13.6) as

$$t_c(\text{min}) = \frac{60L^{0.77}}{7700(H/L)^{0.38}} \tag{13.7b}$$

where L is in ft.

The time of concentration is determined by the drainage-basin characteristics of length and slope. The distance of travel reflects the drainage-basin area. The channel slope is explicitly considered and the channel roughness is included through the constants.

EXAMPLE 13.3 Compute the time of concentration for a watershed for which the distance from the most remote portion to the outlet is 1.25 miles and the slope is 0.01.

Solution Equation (13.6) is used to compute t_c. $L = 1.25 \times 5280 = 6600$ ft. $S = 0.01$. Therefore,

$$t_c = 0.0078 \times \left(\frac{6600}{\sqrt{0.01}}\right)^{0.77} = 0.0078 \times \left(\frac{6600}{0.1}\right)^{0.77}$$

$$= 0.0078 \times (66{,}000)^{0.77} = 0.0078 \times 5141.0 = 40.1 \text{ min}$$

EXAMPLE 13.4 A 139 mi^2 drainage basin of Big Papillion Creek is located above Center Street in Omaha, Nebraska. The lower 21.7 mi^2 is heavily urbanized and the remainder is mostly agricultural. A channel-length frequency distribution analysis was prepared that showed the greatest discharge from a uniform storm would come from the urbanized area and a secondary peak would come from the agricultural area upstream. A question arose as to whether the subarea drainage within the urbanized area would have coincident peaks to confirm that such peaks could be added together to determine the total discharge. A uniform rainfall of 6-h duration was used as a basis for design.

Solution The time of concentration for the urbanized area is determined using Equation (13.6) at a point just above the Center Street bridge. The most remote part within the urbanized area has a time of concentration of 2.7 h. This calculation shows that the entire urban area would be delivering its maximum discharge for a period of at least 3.3 h, so that the peaks of all the subbasins within the drainage area would be coincident and can be added. Actual observations of a subsequent storm supported by photographs confirmed that the highest peak came from the urban area and a secondary and lower peak arrived from the agricultural land upstream.

13.4.3 Lag Time

Lag time is defined as the elapsed time between the center of mass of effective rainfall and the center of mass of the direct runoff hydrograph, as shown in Figure 13.10. Because of the difficulty in determining the center of mass of the direct-runoff hydrograph, lag time is also defined as the elapsed time between the center of mass of the effective rainfall and the peak of the direct-runoff hydrograph. The definition assumes uniform effective rainfall over the entire drainage basin. This assumption is not always met, but represents an average concept.

Many attempts have been made to empirically estimate lag time from drainage-basin characteristics (Panu and Singh, 1981; Singh and Aminian, 1986). Area, length, and main-channel slope have been identified to be the most important characteristics. The simplest equation to estimate lag time t_L is

$$t_L = cA^b \tag{13.8}$$

where c is a constant, b is an exponent, and A is the drainage area. For t_L in hours and A in mi^2, exponent b has been reported to vary between 0.23 and 0.94, with a value between 0.3 and 0.4 most frequently used. The value of c varies widely, depends upon units of measurement, and has to be determined for each watershed.

The Soil Conservation Service (1972) developed an equation using the curve-number method to estimate watershed lag time t_L (from the center of mass of the effective rainfall to the time of peak runoff) that can be expressed as

$$t_L = \frac{L^{0.8}(S_p + 1)^{0.7}}{1900\,S^{0.7}} \tag{13.9}$$

where t_L is in hours, L is the hydraulic length of the watershed in feet, S is the average watershed land slope in percent, and S_P is the potential watershed storage in inches = 1000/CN − 10, CN = hydrologic soil–vegetative cover complex number. Equation (13.9) was developed for less than 2000 acres (810 ha, or 8.1 km^2). The curve-number method will be discussed in more detail in the next chapter.

13.4.4 Law of Basin Lag

In a study of four nested drainage basins near Sydney in New South Wales, Australia, Boyd (1978) discovered that the basin lag followed a law similar to the law of basin areas. This law can be written as

$$t_L^w = t_L^1 R_L^{w-1} \tag{13.10}$$

where t_L^w is the lag time of the drainage basin of order w, t_L^1 is the lag time of the first-order basin, and R_L is basin-lag ratio that was found to be 1.737 for the watersheds used.

13.4.5 Relation between Time of Concentration and Lag Time

The Soil Conservation Service (1975) suggested that

$$t_c = 1.67t_L \tag{13.11a}$$

where t_L is defined with the peak discharge of direct runoff. When the other definition of t_L based on centroids is used, then

$$t_c = 1.42t_L \tag{13.11b}$$

McCuen et al. (1984) found Equations (13.11a) and (13.11b) to agree well with data from 39 urban watersheds. These equations can only be valid for conditions of peak discharge when the time of concentration is reached.

EXAMPLE 13.5 Plot lag time versus order of the watershed for the following data adapted from Boyd (1978).

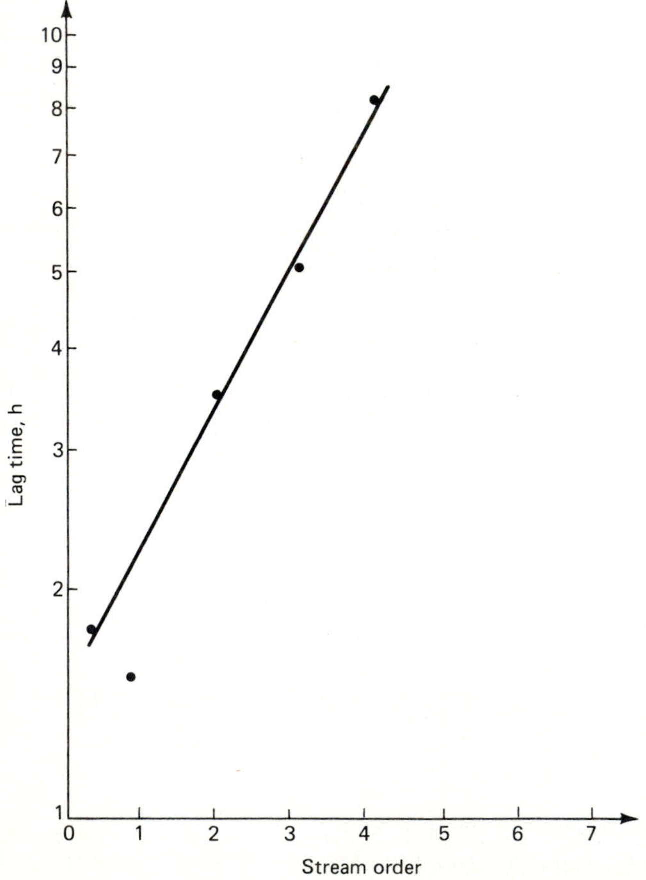

Figure 13.12 Lag time versus watershed order.

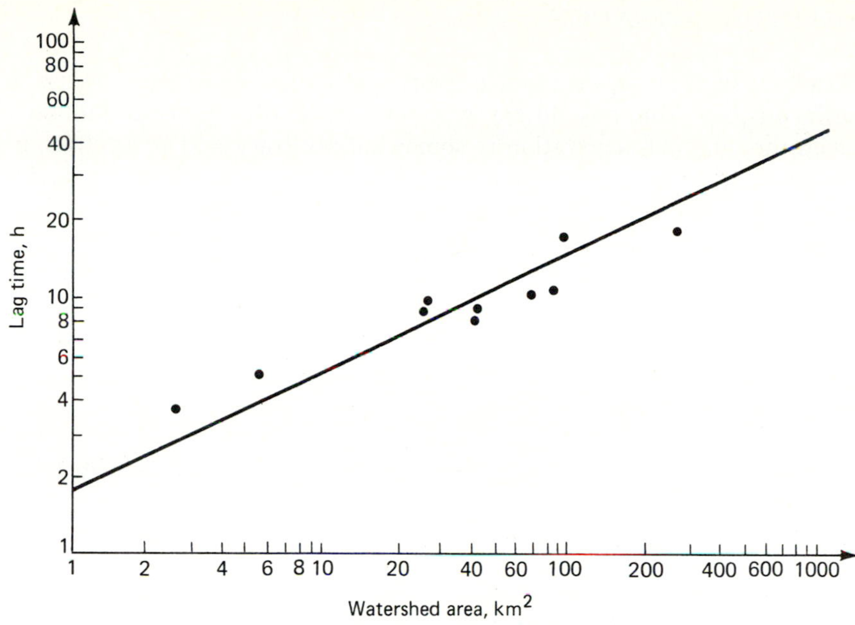

Figure 13.13 Lag time versus watershed area.

BASIN	HORTON–STRAHLER ORDER	AREA (km^2)	MEAN LAG TIME (h)
Blicks		251.0	18.00
South	4	89.6	17.12
Bobo		80.3	10.95
Blicks		69.9	11.58
Hacking	4	39.8	8.43
Eastern	3	24.6	9.12
Deep		25.4	9.7
First		14.2	9.1
Cawleys	3	5.41	5.08
Research	1	0.39	1.50
Kelly	2	2.54	3.61

Also plot lag time versus watershed area and find an empirical relation expressing the plot.

Solution The lag time is plotted against watershed order, as shown in Figure 13.12, and against watershed area, as shown in Figure 13.13. The relation between lag time and stream order is linear on semilog paper, whereas the lag time plots as a straight line against watershed area on log–log paper.

13.5 BASEFLOW SEPARATION

Baseflow separation, also called hydrograph analysis, is the process of separating surface runoff from baseflow or groundwater runoff on the streamflow hydrograph. Even though such separation is somewhat arbitrary and subjective, it is useful in hydrograph analysis. Several techniques have been developed to perform baseflow separation. Some of the techniques separate the hydrograph into direct runoff and groundwater runoff, and some into surface runoff, interflow, and baseflow. Two of the two-component separation methods are described: (a) the area method and (b) the subjective method.

13.5.1 Area Method

The area method of baseflow separation is based upon a nonlinear relation between time and area (Linsley et al., 1958):

$$N = bA^{0.2} \tag{13.12}$$

where A is the drainage-basin area in either square miles or in square kilometers; b is a coefficient that has a value of unity for A in square miles and 0.8 when A is in square kilometers; N is the time in days from the hydrograph peak, marking the beginning of groundwater flow, as shown in Figure 13.14. This equation is not suitable for smaller watersheds and should be checked for a number of streamflow hydrographs. It generally gives a longer time base. For example, if $A = 1000$ km^2, then, using Equation (13.12), $N = 0.8 \times (1000)^{0.2} = 3.18$ days, that is, if rainfall occurs for 6 hours, its effect will be felt for more than 3 days. Hence, it is better to use a subjective method by observing the hydrograph.

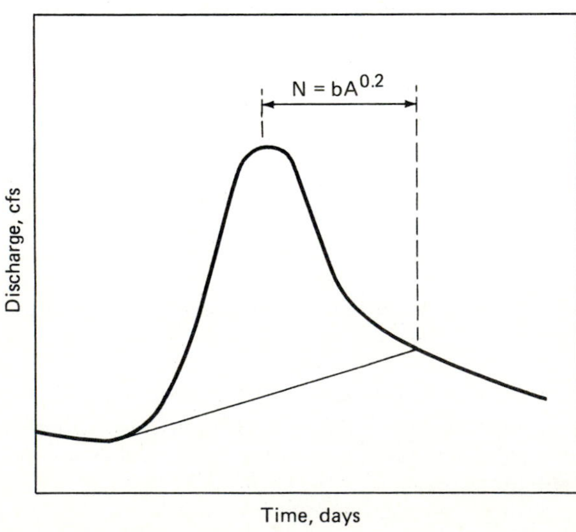

Figure 13.14 Baseflow separation based upon area.

13.5.2 Subjective Method

There are several subjective methods of baseflow separation. One simple method is to visually inspect the hydrograph and choose a discharge on the recession that appears to represent the beginning of baseflow based upon inspection of previously recorded baseflow hydrographs. A baseflow hydrograph has a slow recession because the rate of movement of groundwater is slow. The slope of the recession is low, as shown in Figure 13.15, and is similar for all preceding recessions. The separation is made by drawing a straight line from the selected point on the hydrograph recession to the point on the hydrograph where the rise commences or where the peak is. This linear separation may be theoretically objectionable because channel banks become saturated as water rises in the channel and this same bank storage water drains back to the channel as the stage recedes. This process is not linear and, hence, linear separation is not correct. Because the actual contribution of baseflow to the hydrograph is not known, many hydrologists object to any baseflow separation.

For either of these two baseflow separation methods, it is convenient to draw a separation line directly from the chosen groundwater discharge on the receding limb to the point under the hydrograph peak. Although this linear separation probably does not represent the true boundary between direct runoff and groundwater runoff, the error may be acceptable providing the groundwater runoff on the receding limb is reasonably delineated.

13.5.3 Three-Component Separation

The three-component separation involves separating surface runoff, interflow, and groundwater flow. A method, developed by Barnes (1940), is illustrated in Figure 13.16. This method is based on Equation (13.2), and involves determining recession

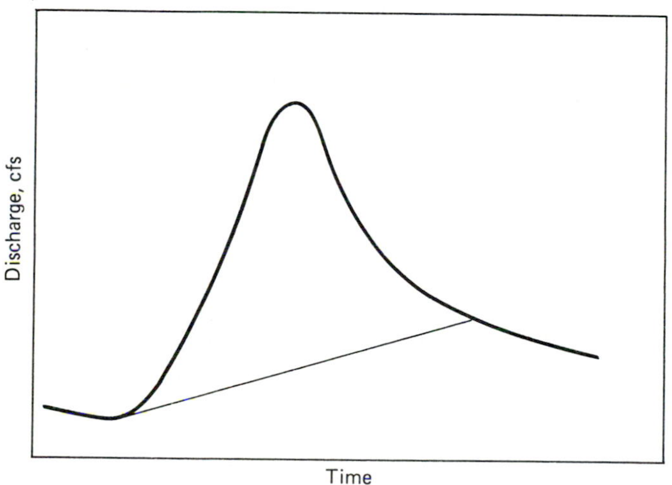

Figure 13.15 Subjective baseflow separation.

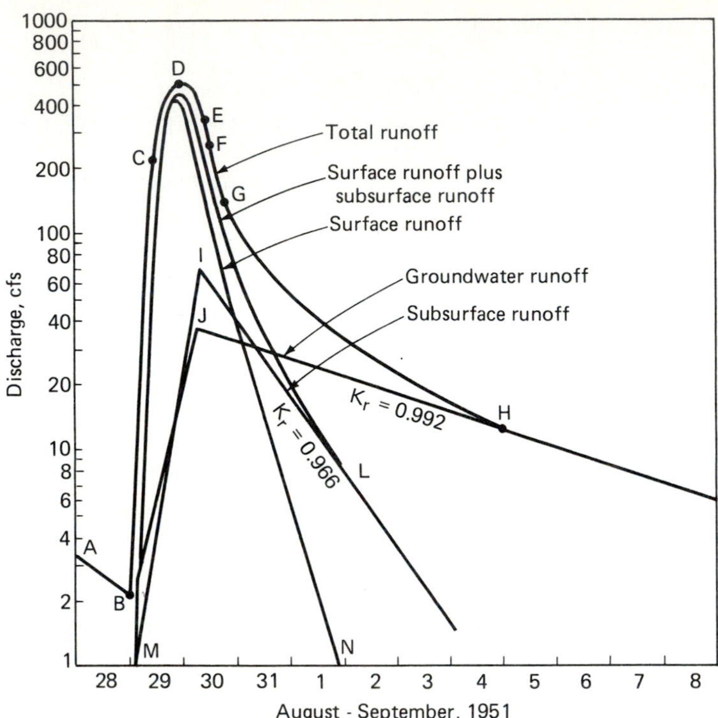

Figure 13.16 Three-component hydrograph separation.

of surface flow, interflow, and baseflow, as explained before. First, streamflow recession is plotted on semilogarithmic paper. For example, in Figure 13.16, the groundwater recession plots approximately as a straight line, with $K_r = 0.992$. By extending this straight line under the hydrograph to the point directly under the point of inflection E and to B on line AB, points B and J are connected arbitrarily by a straight line. The area under the hydrograph above BJH is considered to be direct flow and that area below BJH is considered to be groundwater flow. The direct runoff is replotted and a straight line IL with $K_r = 0.966$ is fitted and extended to point I directly under inflection point E and to the beginning point M. The line MIL divides the replotted hydrograph into surface runoff on top and interflow below.

Application of this method requires considerable smoothing of the runoff hydrograph. Contributions of direct runoff from various parts of the drainage basin often produce numerous humps in the hydrograph recession and preclude computations for surface flow and interflow. Despite these problems, groundwater flow from these irregular recessions can usually be accomplished using this method.

EXAMPLE 13.6 The mean daily discharge (m^3/s) hydrograph of streamflow from a watershed is shown as follows:

DATE	MEAN DAILY DISCHARGE (m^3/s)	DATE	MEAN DAILY DISCHARGE (m^3/s)
1	278	11	179
2	265	12	167
3	5350	13	157
4	8150	14	147
5	6580	15	139
6	1540	16	131
7	505	17	123
8	280	18	117
9	219	19	111
10	195	20	105
		21	100

Find the values of the recession coefficients of surface runoff, interflow, and baseflow.

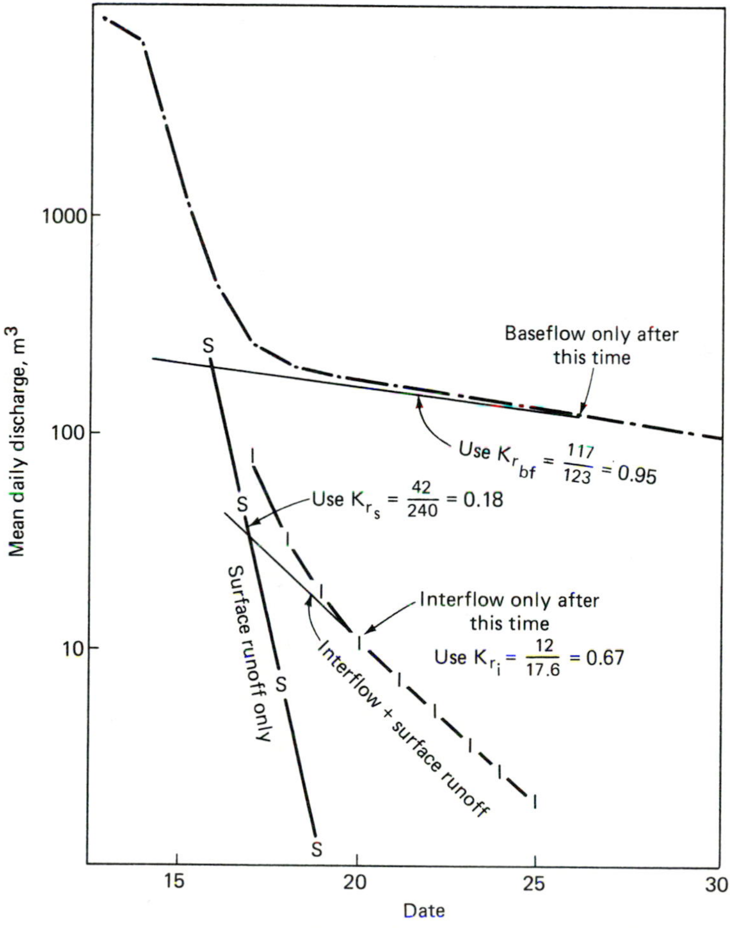

Figure 13.17 Three-components of separation hydrograph recession.

RECESSION CONSTANTS

Date (1)	Recorded hydrograph (m^3/s) (2)	K_{r_b} (3)	Baseflow (m^3/s) (4)	Runoff (surface runoff + interflow) (m^3/s) (5)	K_{r_i} (6)	Computed interflow (m^3/s) (7)	Surface runoff (m^3/s) (8)	K_{r_s} (9)
21	100		100					
20	105	0.95	105					
19	111	0.95	111					
18	117	0.95	117					
17	123[a]	0.95	123	0				
16	131		129	2		2		
15	139		136	3	0.67	3		
14	147		143	4	0.75	4		
13	157		151	6	0.67	6		
12	167		159	8	0.75	8		
11	179		167	12[b]	0.67	12[b]	0	
10	195		176	19		18	1	
9	219		185	34		27	7	0.14
8	280		195	85		50	45	0.16
7	505		205	300		60	240	0.19
6	1540		215	1325		89	1236	0.19
5	6580		226	6354		132	6222	0.20

[a] Semilog graph shows a departure from straight-line base flow before the 17th; therefore, the baseflow was computed using $K_{r_b} = 0.95$.

[b] Semilog graph shows a departure from the straight-line interflow before the 11th; therefore, the interflow was computed using $K_{r_i} = 0.67$.

Solution The recession hydrograph is plotted on semilog paper and then separated into three components, as shown in Figure 13.17. The slopes of straight lines yield the recession coefficients. The results of the computations appear on the preceding page.

EXAMPLE 13.7 Perform hydrograph separation for the data in Example 13.6, and compute and graph interflow, baseflow, and surface runoff. Use the depletion equation.

Solution Following the method of three-component hydrograph separation, baseflow, interflow, and surface runoff are obtained. The results of the calculations follow. The hydrographs of surface runoff, interflow, and baseflow are shown in Figure 13.18.

Date	Runoff (m^3/s)	Baseflow (m^3/s)	Direct runoff (m^3/s)	Interflow (m^3/s)	Surface runoff (m^3/s)
1	278	278	0		
2	265	265	0	0	0
3	5350	252	5098	99	4999
4	8150	238	7912	197	7715
5	6580	226	6354	132	6222
6	1540	215	1325	89	1236
7	505	205	300	60	240
8	280	195	85	40	45
9	219	185	34	27	7
10	195	176	19	18	1
11	179	167	12	12	0
12	167	159	8	8	
13	157	151	6	6	
14	147	143	4	4	
15	139	136	3	3	
16	131	129	2	2	
17	123	123	0		
18	117	117			
19	111	111			
20	105	105			
21	100	100			

$K_{rbf} = 0.95$ $\quad K_{r_i} = 0.67$

EXAMPLE 13.8 Assume that the streamflow hydrograph in Example 13.6 is to be separated into direct runoff and baseflow. Compute the recession constants of these two components, and then compute and graph hydrographs of these components.

Solution The hydrograph is separated into baseflow and direct runoff by plotting on semilog paper, as shown in Figure 13.19. The results of the calculations for this separation follow. Hydrographs of baseflow, direct runoff, and total runoff are shown in Figure 13.20.

Date	Runoff (m^3/s)	K_{r_b}	Baseflow (m^3/s)	Direct runoff (m^3/s)	Date	Runoff (m^3/s)	K_{r_b}	Baseflow (m^3/s)	Direct runoff (m^3/s)
1	278		278	0	12	167	0.94	167	0
2	265		265	0	13	157	0.94	157	
3	5350		271	5079	14	147	0.95	147	
4	8150		276	7874	15	139	0.94	139	
5	6580		259	6321	16	131	0.94	131	
6	1540		243	1297	17	123	0.95	123	
7	505		228	277	18	117	0.95	117	
8	280		214	66	19	111	0.95	111	
9	219		201	18	20	105	0.95	105	
10	195		189	6	21	100		10	
11	179		178	1					

$$K_{r_d} = \left(\frac{Q_t}{Q_0}\right)^{1/t} = \left(\frac{18}{6321}\right)^{1/4} = 0.23$$

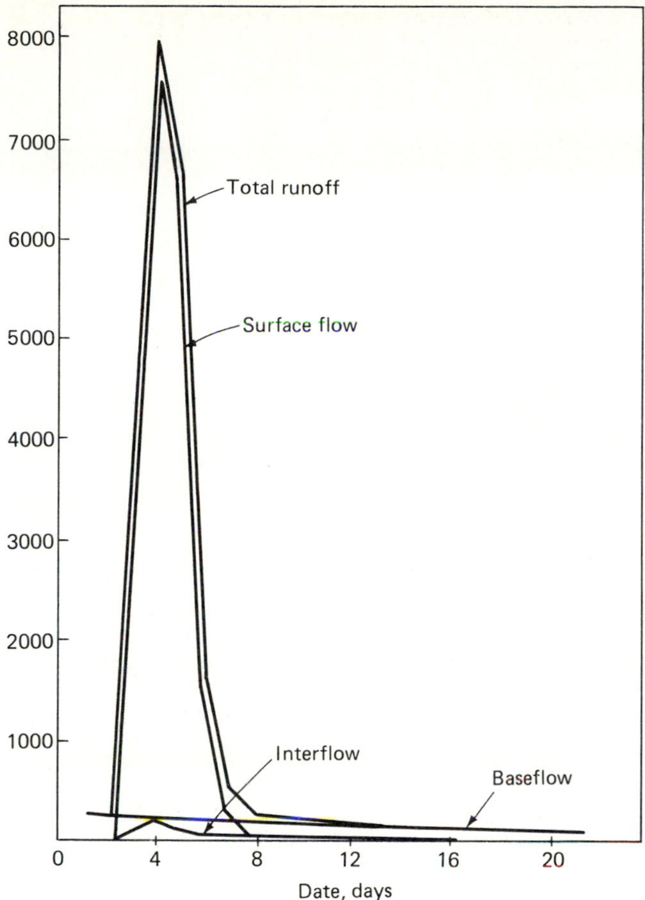

Figure 13.18 Total runoff, surface flow, interflow, and baseflow hydrographs.

13.6 COMPLEX HYDROGRAPHS

Complex hydrographs are those hydrographs that result from storms with two or more rainfall events. It is the nature of rainfall to vary in time and space. The variation in rainfall can include complete cessation and resumption of rainfall one or more times during a given period of time. The hydrograph resulting from such a storm has two or more peaks separated approximately in time by the time interval between rainfalls, as shown in Figure 13.21. Such a complex hydrograph is simplified using baseflow recession, and then analyzed using standard procedures. If a storm covers the entire drainage basin with nearly uniform rainfall, the hydrograph recession will nearly be the same for all storms of a given rainfall amount. The recession for such storms can be used to extend the interrupted recession of hydrograph A in Figure 13.21 to produce a usable simple hydrograph for hydrologic analysis. Separation of baseflow can be accomplished from this simplified hydrograph using one of the methods described earlier. Simplification of a complex hydrograph requires that hydrograph A has an initial recession that can be clearly identified and extended.

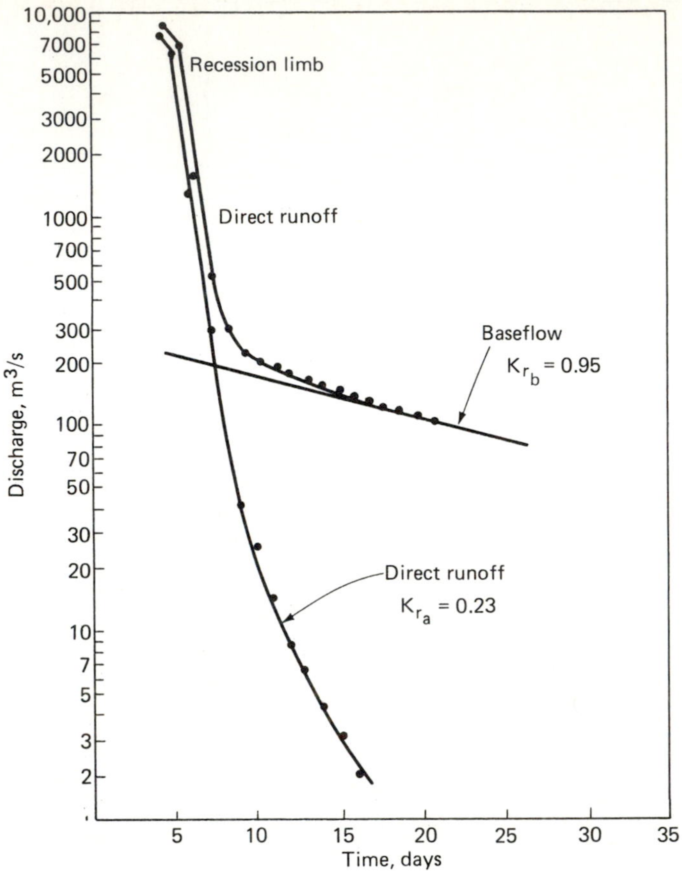

Figure 13.19 Two-component hydrograph separation.

13.7 EFFECTIVE RAINFALL

The effective rainfall is that portion of rainfall that contributes to direct runoff. Similarly, rainfall excess constitutes that portion of rainfall that contributes to surface runoff. The difference between effective rainfall and rainfall excess is that the former includes the latter plus some abstractions. However, the terms effective rainfall and rainfall excess are freely interchanged in hydrologic usage. Thus, a rainfall storm is considered to be composed of two portions: one that contributes to runoff and the other that contributes to abstractions, including interception, evaporation, transpiration, depression and detention storage, and infiltration. Because the effective rainfall entirely becomes direct runoff, its volume must equal the volume of direct runoff.

For a given rainfall–runoff event, the volume of direct runoff is determined by separating the baseflow. This yields the volume of effective rainfall and, in turn, the volume of rainfall that is used up by abstractions. The distribution of this volume in time (determination of loss function) and then its subtraction from the rainfall hyetograph yields effective rainfall. This is illustrated in Figure 13.22. However, due to

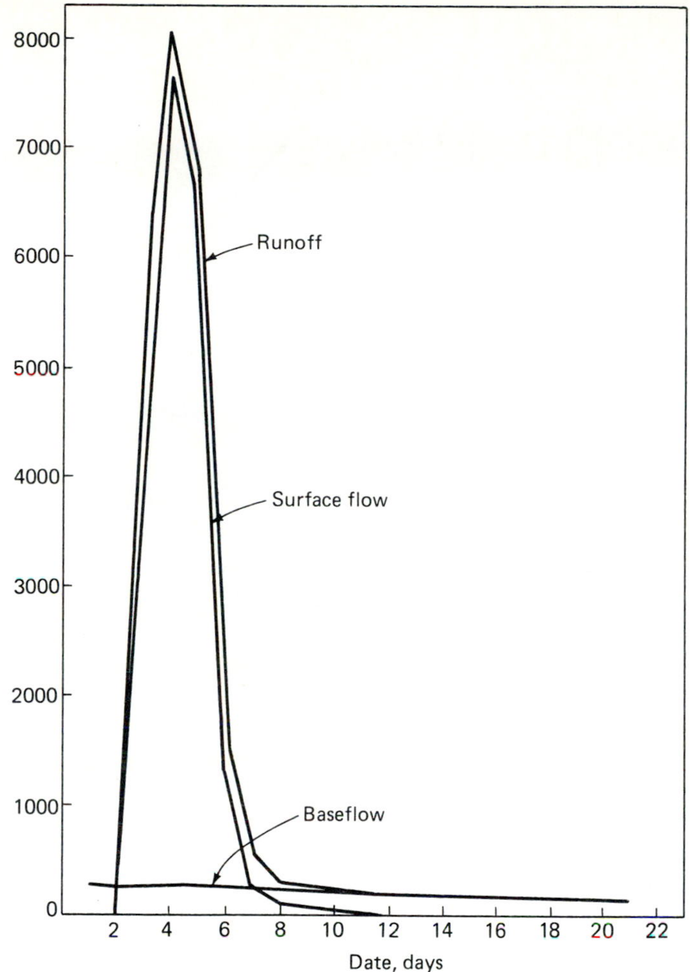

Figure 13.20 Total runoff, direct runoff, and baseflow hydrographs.

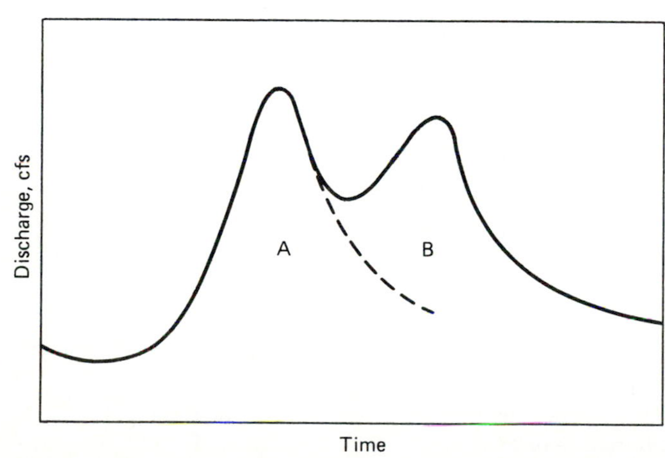

Figure 13.21 Simplifying a complex hydrograph.

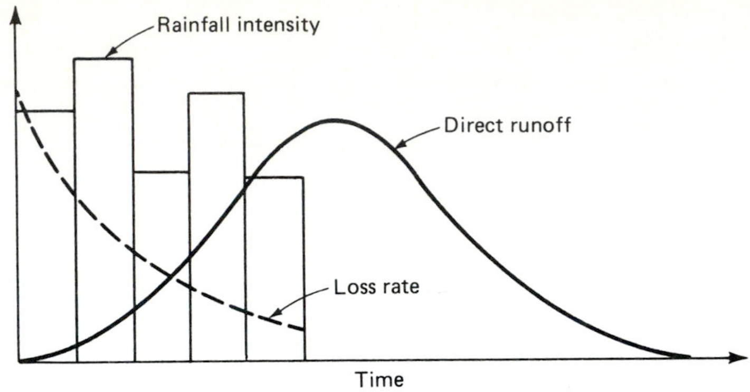

Figure 13.22 Determination of effective rainfall from rainfall–runoff data.

extreme space–time heterogeneity in antecedent conditions, extrapolation of this loss function for any other rainfall event is extremely difficult.

Let us briefly pause for a moment. If we neglect all other abstractions than infiltration for simplicity, it is immediately seen that during and after a rainfall episode, runoff and infiltration occur simultaneously. Indeed, infiltration continues to occur as long as surface runoff occurs or there is water over the ground. By looking at durations of rainfall and runoff, it is usually true that the duration of surface runoff is much longer than the duration of rainfall. But infiltration is allowed to occur only during the period of rainfall. The question is: What happens to the infiltration during the period of surface runoff in excess of that of rainfall? One might argue that infiltration may be very small after the cessation of rainfall and hence neglected. This may not be true. Besides, from a physical standpoint, in order to compensate for infiltration accounted for after the period of rainfall, infiltration has to be allowed to occur at a rate higher than that at which it actually occurs. This is inevitable to ensure volume continuity of rainfall excess and surface runoff. Therefore, the concept of rainfall excess or effective rainfall is an artificial one. So also can be said about direct runoff. In nature, rainfall occurs as a continuum. And so does runoff.

EXERCISES

13.1. Streamflow in a river was receding and was observed as 1500 m^3/s on May 5, 1988. The recession constant for this river was found to be 0.8 with the time interval of 24 hours. What was the flow in the river on May 10, 1988?

13.2. Streamflow in a river was receding and was observed as 200 m^3/s on a given day. Three days later the streamflow was observed as 150 m^3/s. What is the recession constant for this river?

13.3. Streamflow in a river was receding and was observed as 180 m^3/s. If the recession constant with the time interval of 24 hours is found to be 0.9 for this river, what was the streamflow in this river 3 days ago?

13.4. Plot Equation (13.12) on graph paper, with drainage area in square kilometers. How good is this relation for different watershed sizes? Take the value of b as 0.8.

13.5. Compute the time of concentration for a watershed that has a slope of 0.05, and the main-channel length from the upper extremity to the outlet is 2 kilometers.

13.6. Separate baseflow and direct runoff components for the following streamflow hydrograph:

Time	Runoff (m^3/s)	Time	Runoff (m^3/s)	Time	Runoff (m^3/s)
8:30 A.M.	50.0	12:30 noon	44.5	4:30 P.M.	44.5
8:30 P.M.	60.0	12:30 midnight	123.0	4:30 A.M.	160.0
6:30 A.M.	165.0	8:30 A.M.	160.0	12:30 noon	133.0
4:30 P.M.	101.8	8:30 A.M.	78.5	12:30 midnight	60.0
4:30 A.M.	31.7	12:30 noon	24.5	4:30 P.M.	16.0
6:30 P.M.	0.0				

The drainage area of the basin is 65 square kilometers.

13.7. An average daily flow hydrograph is listed. Determine the recession constant both analytically and graphically.

Time (days)	0	1	2	3	4	5	6	7	8	9
Discharge (m^3/s)	0.20	0.14	0.30	0.39	0.24	0.12	0.083	0.05	0.13	0

13.8. Tabulated are data for a flood in a river draining an area of 1471 square miles. Separate the groundwater flow, and compute the volumes of groundwater flow and direct runoff.

Date (1943)	Hour	Flow (×1000 cfs)	Date (1943)	Hour	Flow (×1000 cfs)
4–18	2400	1.5	4–22	1200	7.0
4–19	0600	1.5		2400	5.7
	1200	1.6	4–23	1200	4.7
	1800	2.0		2400	3.9
	2400	3.1	4–24	1200	3.2
4–20	0600	5.2		2400	2.8
	1000	7.4	4–25	1200	2.4
	1200	9.5		2400	2.2
	1400	12.0	4–26	1200	2.0
	1800	12.45		2400	1.9
	2400	12.1	4–27	2400	1.7
4–21	0600	11.1	4–28	2400	1.5
	1200	10.2	4–29	2400	1.4
	1800	3.3			
	2400	8.5			

13.9. Tabulated are the total hourly discharge rates at some cross-section of a stream. The drainage area above the section is 0.4 ha. Plot the hydrograph on arithmetic paper and label the rising limb (concentration curve), the crest segment, and the recession limb. Determine the hour of cessation of direct runoff using a semilog plot for discharge versus time. Use the baseflow portion of your semilog plot to determine the groundwater recession constant. Carefully construct and label baseflow separation curves on the discharge hydrograph using two different methods.

Time (hour)	Discharge (m^3/s)	Time (hour)	Discharge (m^3/s)
0	2.89	8	5.94
1	2.83	9	4.25
2	2.77	10	2.97
3	6.23	11	2.12
4	14.49	12	1.70
5	17.83	13	1.53
6	13.02	14	1.37
7	9.34	15	1.23

13.10. Tabulated are the ordinates of a hydrograph at 24-hour intervals. Assuming the recession constant for baseflow to be 0.9, separate the baseflow from direct runoff by more than one method. Compute the volume of each case.

Time (day)	Flow (m^3/s)	Time (day)	Flow (m^3/s)
1	6.2	8	91.4
2	970.7	9	78.8
3	707.5	10	67.6
4	396.2	11	58.3
5	253.6	12	50.1
6	162.4	13	43.0
7	121.7	14	37.4

13.11. Perform the hydrograph separation by two or more methods for watershed W-3, North Danvible, Vermont. Compare them. Area = 2067 acres (3.23 mi^2). Event of July 16–17, 1967.

RAINFALL			RUNOFF		
Time of of day	Intensity (in./h)	Accumulated amount (in.)	Time of of day	Rate (cfs)	Accumulated amount (in.)
0450	0.1846	0.00	0301	1.938	0.0000
0542	0.1333	0.16	0529	2.105	0.0024
0618	0.0286	0.24	0616	2.751	0.0033
0700	0.0496	0.26	0759	6.169	0.0070
0901	0.0000	0.36	0850	9.150	0.0107
1001	0.1154	0.36	0929	9.504	0.0121
1027	0.0960	0.41	1114	9.504	0.0209
1052	0.2483	0.45	1414	16.757	0.0398
1121	0.1385	0.57	1459	16.757	0.0458
1134	0.0383	0.60	1800	10.588	0.0656
1221	0.0356	0.63	2400	4.981	0.0880
1402	0.0000	0.69	0600	3.339	0.1008
			1200	3.481	0.1115
			1459	3.356	0.1164
			1914	2.855	0.1227
			2129	2.751	0.1257
			2400	2.751	0.1290

What are the recession coefficients?

13.12. Perform the hydrograph separation for a complex hydrograph given as follows for watershed 100 Chickasha, Oklahoma. Estimate the recession coefficients. Area = 2,339,800 acres (3656 mi^2).

RUNOFF EVENT OF APRIL 12–18, 1967

Date	Time of day	Flow (cfs)	Date	Time of day	Flow (cfs)
4–12	0200	42.0		1630	2024.8
	0218	52.0		1718	2104.7
	0235	57.9		1842	2235.3
	0254	89.1		1954	2322.0
	0312	94.3		2230	2437.9
	0330	83.0		2400	2463.9
	0348	79.4	4–14	0036	2474.6
	0400	87.5		0148	2450.8
	0424	108.5		0412	2390.6
	0512	122.1		0554	2316.5
	0542	356.1		0724	2155.2
	0612	532.9		0842	2131.8
	0618	590.8		1036	1916.5
	0718	643.1		1218	1774.4
	0830	750.2		1312	1644.8
	0848	771.8		1418	1559.0
	1018	905.2		1430	1415.4
	1112	1064.7		1530	1364.4
	1206	1190.6		1706	1190.1
	1318	1399.1		1830	1062.7
	1454	1538.8		1854	1095.5
	1554	1596.9		1942	948.8
	1718	1698.4		2112	854.0
	1842	1789.3		2230	813.1
	2012	1912.0		2400	736.7
	2124	1941.5	4–15	0348	594.2
	2236	1990.7		0648	523.9
	2324	1994.2		0848	502.1
	2400	1933.5		1424	504.4
4–13	0054	1922.9		2054	474.7
	0142	1868.5		2400	446.5
	0212	1915.4	4–16	1200	409.5
	0348	1739.3	4–17	1200	350.3
	0600	1634.3	4–18	1200	305.1
	0742	1603.4			
	0854	1626.4			
	0954	1650.0			
	1106	1680.7			
	1230	1767.8			
	1424	1903.1			

Note: To convert runoff in cfs to in./h, multiply by 4.237×10^{-7}.

13.13. The streamflow hydrograph of the Amite River at Denham Springs observed during April 16–May 3, 1979 is tabulated. Separate the hydrograph into three components of surface runoff, interflow, and baseflow. Tabulate and plot them. Compute their recession constants. Compute the volumes of surface runoff and interflow.

Date	Time	Flow (m^3/s)	Date	Time	Flow (m^3/s)
4–21	6 A.M.	32	4–28	6 A.M.	145
	12 P.M.	127		12 P.M.	124
	6 P.M.	263		6 P.M.	106
	12 A.M.	425		12 A.M.	93
4–22	6 A.M.	555	4–29	6 A.M.	83
	12 P.M.	759		12 P.M.	78
	6 P.M.	1034		6 P.M.	75
	12 A.M.	1294		12 A.M.	74
4–23	6 A.M.	1478	4–30	6 A.M.	71
	12 P.M.	1617		12 P.M.	67
	6 P.M.	1727		6 P.M.	64
	12 A.M.	1798		12 A.M.	61
4–24	6 A.M.	1886	5–1	6 A.M.	59
	12 P.M.	1940		12 P.M.	57
	6 P.M.	1903		6 P.M.	56
	12 A.M.	1778		12 A.M.	54
4–25	6 A.M.	1597	5–2	6 A.M.	53
	12 P.M.	1419		12 P.M.	52
	6 P.M.	1240		6 P.M.	51
	12 A.M.	1059		12 A.M.	50
4–26	6 A.M.	—	5–3	6 A.M.	49
	12 P.M.	719		12 P.M.	48
	6 P.M.	589		6 P.M.	47
	12 A.M.	464		12 A.M.	46
4–27	6 A.M.	374			
	12 P.M.	281			
	6 P.M.	207			
	12 A.M.	159			

13.14. Separate the hydrograph of Exercise 13.13 into two components of direct runoff and baseflow. Tabulate and plot them. Compute the groundwater recession constants.

13.15. The average daily streamflows resulting from a heavy storm on a basin of 1347 square miles are tabulated. Compute the total flow volume in second-foot-days, acre-feet, inches/miles, and millions of gallons.

DAY	**1**	**2**	**3**	**4**	**5**	**6**
Mean daily flow (cfs)	3,200	10,450	6,580	3,230	1,560	670

13.16. The following runoff data are available from the Cypress Creek gaging station near Janice, Mississippi:

DAY	HOUR	DISCHARGE (cfs)
1	0600	3
	1800	5
2	0600	67
	1800	100
3	0600	150
	1800	182
4	0600	140
	1800	109

(a) Calculate the runoff in SFD.
(b) If the drainage area is 52 square miles, determine the runoff in inches over the drainage area.

13.17. Given are the area average daily discharges for the Tombigbee River at Columbus. If the drainage area is 4490 square miles, determine:
(a) total volume of runoff in Second-foot-days (SFD)
(b) mean daily discharge in cfs
(c) unit daily discharge in cfs/mi and
(d) inches of runoff over the drainage area

DAY	DISCHARGE (cfs)
1	8,880
2	7,190
3	6,070
4	8,290
5	13,100
6	14,100
7	13,100
8	10,500
9	6,670
10	6,630

13.18. Given are the average streamflows for 6-hour intervals. If the drainage area is 617 mi^2, calculate the runoff in
(a) SFD
(b) acre-ft
(c) inches over the drainage area

TIME (h)	FLOW (cfs)
30	1,743
36	2,075
42	2,408
48	2,740
54	5,380
60	8,020
66	10,660
72	13,300
78	11,425
84	9,550
90	7,675
96	5,800

PART 6
Precipitation-Runoff Relation

CHAPTER 14
Estimation of Surface-Runoff Volume

Runoff is a general term used to indicate the accumulation of precipitation excess. Of greater interest is either runoff volume from a storm event or for some other time period or peak discharge for the event. Sometimes, time to the peak discharge or the duration of the hydrograph might be of special interest.

Runoff volume V_Q is the total volume of runoff water occurring over a period of time, and is expressed as

$$V_Q = \int_0^t Q(t)\,\mathrm{d}t \tag{14.1}$$

where $Q(t)$ is the discharge at time t. One time period of special interest is the volume of runoff over the duration of the hydrograph. This value of the volume enables a comparison to be made between the volume of precipitation falling on a drainage basin and the volume of runoff occurring as a result of that precipitation. Baseflow and low-flow volumes also might be desired. Interest in runoff volume is often associated with the water supply for municipal or industrial use, farm irrigation, or similar applications.

An estimate of runoff volume from a drainage basin involves precipitation, infiltration, evaporation, transpiration, interception, and depression storage, each of which is complex and can interact with the other variables to either enhance or reduce runoff. These variables are variously distributed within a drainage basin. The manner in which the variables interact in time and space makes a direct determination of runoff very difficult. Therefore, we estimate runoff using methods that reflect the combined effect of the variables on an individual drainage basin. Because no two drainage basins are exactly alike, no two solutions can be exactly alike. The general approach to the solution, however, can be alike.

14.1 SCS CURVE NUMBER (SCS-CN) METHOD

The Soil Conservation Service (SCS) was established to provide technical assistance to farmers, ranchers, and other rural residents through engineers located in each county. Uniform recommended techniques have been established and promulgated through the *SCS' National Engineering Handbook.* This service has been very satisfactory for the purpose for which it was established. One of the techniques is the SCS curve-number (SCS-CN) method for estimating runoff volume (Soil Conservation Service, 1969).

The fundamental hypotheses of the SCS-CN method are as follows: (1) Runoff starts after an initial abstraction I_a has been satisfied. This abstraction consists principally of interception, surface storage, and infiltration. (2) The ratio of actual retention of rainfall to the potential maximum retention S is equal to the ratio of direct runoff to rainfall minus initial abstraction. Mathematically,

$$\frac{P - I_a - V_Q}{S} = \frac{V_Q}{P - I_a} \tag{14.2a}$$

This can be written as

$$V_Q = \frac{(P - I_a)^2}{(P - I_a) + S} \tag{14.2b}$$

where V_Q is the runoff volume uniformly distributed over the drainage basin, P is the mean precipitation over the drainage basin, and S is the retention of water by the drainage basin. The quantity I_a can be expressed as a function of S. The Soil Conservation Service expressed $I_a = 0.2S$. Physically, this means that for a given storm, 20% of the potential maximum retention is the initial abstraction before runoff begins. Presumably, $0.8S$ represents other retention losses, including interception, infiltration, evapotranspiration, and depression storage. Therefore,

$$V_Q = \frac{(P - 0.2S)^2}{P + 0.8S} \tag{14.3a}$$

or

$$V_Q = P - S\left(1.2 - \frac{S}{P + 0.8S}\right) \tag{14.3b}$$

Evidently this is a one-parameter model containing S as the parameter and is illustrated graphically in Figure 14.1.

Equation (14.3b) is a form of the hydrologic budget,

$$V_Q = P - L \tag{14.4a}$$

in which L accounts for losses expressed as

$$L = S\left(1.2 - \frac{S}{P + 0.8S}\right) \tag{14.4b}$$

In a limiting case, as $P \to \infty$, $L \to 1.2S$ and not S.

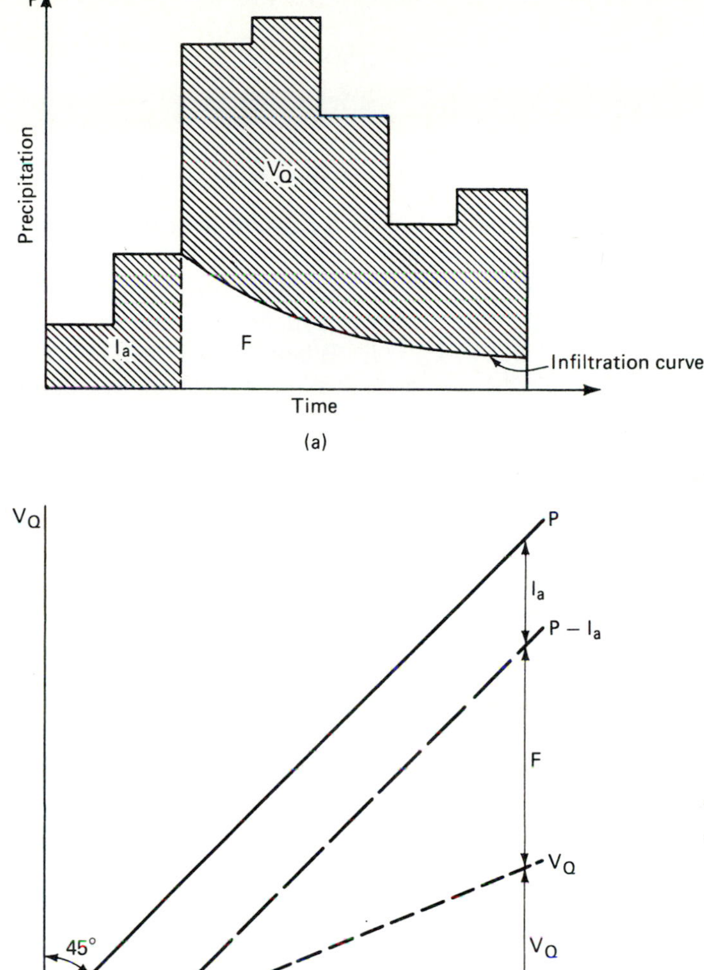

Figure 14.1 (a) SCS relation between precipitation, runoff, and retention. (b) A mass-curve representation of the SCS relation between precipitation, runoff, and retention.

14.1.1 Estimation of *S*

The parameter S depends upon characteristics of the soil–vegetation–land use (SVL) complex and antecedent soil-moisture conditions in a watershed. For each SVL complex, there is a lower limit and an upper limit of S. The Soil Conservation

Service expressed S as a function of what was termed as curve number as

$$CN = \frac{1000}{S + 10} \tag{14.5}$$

or

$$S = 1000/CN - 10 \tag{14.6}$$

where CN is the curve number; it is a relative measure of retention of water by a given SVL complex and takes on values from 0 to 100. This number is derived from the character of the soil; vegetation, including crops; and the land use of that soil, as well as intensity of use. The unit of S is inches. Obviously, when CN equals 100, S becomes zero. This leads to $V_Q = P$. When $S \to \infty$, $CN \to 0$. This yields $V_Q = 0$ for all P when $S = \infty$ and $CN = 0$. Substitution of Equation (14.6) into Equation (14.3a) yields

$$V_Q = \frac{\left(P - \frac{200}{CN} + 2\right)^2}{P + \frac{800}{CN} - 8} \tag{14.7}$$

In this equation, CN is the only parameter to be determined.

14.1.2 Determination of Curve Number

The CN value is determined from (a) soil type and (b) antecedent moisture conditions. As shown in Table 14.1, soils have been classified in four groups. A complete listing of all soils in the United States along with their group classifications can be obtained from the Soil Conservation Service (1969). A short description is given here.

CLASSIFICATION OF SOILS

Group A. Soils in this group have a low-runoff potential (high-infiltration rates) even when thoroughly wetted. They consist of deep, well to excessively well-drained sands or gravels. These soils have a high rate of water transmission.

TABLE 14.1 SOIL GROUP CLASSIFICATION

Group	Soil characteristics	Minimum infiltration rate (in./h)
A	Deep sand, deep loess, and aggregated silts	0.3–0.45
B	Shallow loess and sandy loam	0.15–0.30
C	Clay loams, shallow sandy loam, soils in organic content, and soils usually high in clay	0.05–0.15
D	Soils that swell upon wetting, heavy plastic clays, and certain saline soils	0–0.05

Group B. Soils in this group have moderate infiltration rates when thoroughly wetted and consist chiefly of moderately deep to deep, well-drained to moderately well-drained soils with moderately fine to moderately coarse textures. These soils have a moderate rate of water transmission.

Group C. Soils have slow infiltration rates when thoroughly wetted and consist chiefly of soils with a layer that impedes the downward movement of water, or soils with moderately fine to fine texture. These soils have a slow rate of water transmission.

Group D. Soils have a high-runoff potential (very slow infiltration rates) when thoroughly wetted. These soils consist chiefly of clay soils with high swelling potential, soils with a permanent high-water table, soils with a clay pan or clay layer near the surface, and shallow soils over nearly impervious material. These soils have a very slow rate of water transmission.

ANTECEDENT MOISTURE CONDITION

Antecedent moisture condition (AMC) refers to the water content present in the soil at a given time. The AMC value is intended to reflect the effect of infiltration on both the volume and rate of runoff, according to the infiltration curve of Figure 14.1. The Soil Conservation Service developed three antecedent soil-moisture conditions and labeled them as I, II, and III. These AMCs correspond to the following soil conditions:

AMC I: Soils are dry but not to the wilting point; satisfactory cultivation has taken place.
AMC II: Average conditions.
AMC III: Heavy rainfall, or light rainfall and low temperatures have occurred within the last 5 days; saturated soil.

Table 14.2 provides seasonal rainfall limits for the three antecedent soil-moisture conditions.

TABLE 14.2 ANTECEDENT MOISTURE CONDITIONS FOR DETERMINING THE VALUE OF CN

Antecedent moisture condition (AMC)	Total rain in previous 5 days	
	Dormant season	Growing season
I	Less than 0.50 in. (1.27 cm)	Less than 1.4 in. (3.5 cm)
II	0.50 to 1.1 in. (1.27 to 3.25 cm)	1.4 to 2.1 in. (3.5 to 5.25 cm)
III	More than 1.1 in. (3.25 cm)	Over 2.1 in. (5.25 cm)

SELECTION OF CN

The procedure for selecting the actual value of CN for any given application begins in Table 14.3. The value of CN is shown for AMC condition II and for a variety of land uses, soil treatment, or farming practices. The hydrologic condition refers to the state of the vegetation growth. For example, a poor condition refers to pasture heavily grazed with sparse vegetation, a fair condition is for pasture moderately grazed with between half and three-fourths of the basin under plant cover, and a good condition refers to pasture lightly grazed with more than three-fourths of the basin under plant cover. For a more complete explanation, see Soil Conservation Service (1969). Upon selecting the applicable crop cover, treatment, and hydrologic condition, the value of CN is found under the appropriate soil group for AMC II. The value of CN is lower for soils with high infiltration than for soils with low infiltration.

Table 14.4 provides conversion from AMC II to AMC I and AMC III. After determining the CN value for AMC II, Table 14.4 is entered at this value in column 1. Values for AMC I, AMC III, S, and P are found in adjoining columns. These data can then be applied to determine runoff volume V_Q. It may be noted that estimation of V_Q involves precipitation and the abstractions—interception, infiltration, evapotranspiration, and depression storage. It is only by considering all these variables that a reasonable solution can be obtained.

14.1.3 Initial Abstraction I_a

An exact determination of I_a is very difficult. However, for practical purposes, I_a can be related to S. Based on analysis of data from a large number of small watersheds, the Soil Conservation Service (1969) found I_a to be roughly equal to $0.2S$. It can also be estimated by relating to the antecedent soil-moisture index, as done by Hamon (1963), and Singh and Dickinson (1975) amongst others.

14.1.4 Graphical Solution

Equations (14.3a) and (14.6) can be used to estimate the volume of surface runoff if the amount of precipitation and the curve number are known. The Soil Conservation Service (1969) developed a graphical solution, as shown in Figure 14.2, of these equations. Either Equations (14.3a) and (14.6) or their graphical solution can be used for estimation of surface-runoff volume.

14.1.5 Additional Remarks

The SCS-CN method is satisfactory for the purposes it was developed but has severe deficiencies for other purposes and should be used with caution. The deficiencies of the SCS-CN method are as follows: (1) The basin is covered with a soil group that has uniform hydrologic characteristics throughout the basin area. (2) Rainfall is uniform and is distributed uniformly over the basin area. For example, if S is in millimeters, then Equation (14.6) can be expressed as

$$S = \frac{25{,}400}{\text{CN}} - 254$$

TABLE 14.3 RUNOFF CURVE NUMBERS FOR HYDROLOGIC SOIL-COVER COMPLEXES (AFTER SOIL CONVERSATION SERVICE, 1969)[a]

Cover			Hydrologic soil group			
Land Use	Treatment or Practice	Hydrologic condition	A	B	C	D
Fallow	Straight row	—	77	86	91	94
Row crops	Straight row	Poor	72	81	88	91
	Straight row	Good	67	78	85	89
	Contoured	Poor	70	79	84	88
	Contoured	Good	65	75	82	86
	Contoured & terraced	Poor	66	74	80	82
	Contoured & terraced	Good	62	71	78	81
Small grain	Straight row	Poor	65	76	84	88
	Straight row	Good	63	75	83	87
	Contoured	Poor	63	74	82	85
	Contoured	Good	61	73	81	84
	Contoured & terraced	Poor	61	72	79	82
	Contoured & terraced	Good	59	70	78	81
Closed-seeded Legumes[b] or rotation meadow	Straight row	Poor	66	77	85	89
	Straight row	Good	58	72	81	85
	Contoured	Poor	64	75	83	85
	Contoured	Good	55	69	78	83
	Contoured & terraced	Poor	63	73	80	83
	Contoured & terraced	Good	51	67	76	80
Pasture or range		Poor	68	79	86	89
		Fair	49	69	79	84
		Good	39	61	74	80
	Contoured	Poor	47	67	81	88
	Contoured	Fair	25	59	75	83
	Contoured	Good	6	35	70	79
Meadow		Good	30	58	71	78
Woods		Poor	45	66	77	83
		Fair	36	60	73	79
		Good	25	55	70	77
Farmsteads		—	59	74	82	86
Road (dirt)[c]		—	72	82	87	89
(hard surface)[c]		—	74	84	90	92

[a] Antecedent moisture condition II, and $I_a = 0.2S$.

[b] Close drilled or broadcast.

[c] Including the right of way.

TABLE 14.4 CURVE NUMBERS (CN) AND CONSTANTS FOR THE CASE $I_a = 0.2S$ (AFTER SOIL CONSERVATION SERVICE, 1969)

CN for condition	CN for conditions		S values[a]	Curve[a] starts where	CN for condition	CN for conditions		S values[a]	Curve[a] starts where
II	I	III	(in.)	$P =$ (in.)	II	I	III	(in.)	$P =$ (in.)
100	100	100	0.000	0.00	60	40	78	6.67	1.33
99	97	100	0.101	0.02	59	39	77	6.95	1.39
98	94	99	0.204	0.04	58	38	76	7.24	1.45
97	91	99	0.309	0.06	57	37	75	7.54	1.51
96	89	99	0.417	0.08	56	36	75	7.86	1.57
95	87	98	0.526	0.11	55	35	74	8.18	1.64
94	85	98	0.638	0.13	54	34	73	8.52	1.70
93	83	98	0.753	0.15	53	33	72	8.87	1.77
92	81	97	0.870	0.17	52	32	71	9.23	1.85
91	80	97	0.989	0.20	51	31	70	9.61	1.92
90	78	96	1.11	0.22	50	31	70	10.0	2.00
89	76	96	1.24	0.25	49	30	69	10.4	2.08
88	75	95	1.36	0.27	48	29	68	10.8	2.16
87	73	95	1.49	0.30	47	28	67	11.3	2.26
86	72	94	1.63	0.33	46	27	66	11.7	2.34
85	70	94	1.76	0.35	45	26	65	12.2	2.44
84	68	93	1.90	0.38	44	25	64	12.7	2.54

83	67	93	2.05	0.41	43	25	63	13.2	2.64
82	66	92	2.20	0.44	42	24	62	13.8	2.76
81	64	92	2.34	0.47	41	23	61	14.4	2.88
80	63	91	2.50	0.50	40	22	60	15.0	3.00
79	62	91	2.66	0.63	39	21	59	15.6	3.12
78	60	90	2.82	0.56	38	21	58	16.3	3.26
77	59	89	2.99	0.60	37	20	57	17.0	3.40
76	58	89	3.16	0.63	36	19	56	17.8	3.56
75	57	88	3.33	0.67	35	18	55	18.6	3.72
74	55	88	3.51	0.70	34	18	54	19.4	3.88
73	54	87	3.70	0.74	33	17	53	20.3	4.06
72	53	86	3.89	0.78	32	16	52	21.2	4.24
71	52	86	4.08	0.82	31	16	51	22.2	4.44
70	51	85	4.28	0.86	30	15	50	23.3	4.66
69	50	84	4.49	0.90					
68	48	84	4.70	0.94	25	12	43	30.0	6.00
67	47	83	4.92	0.98	20	9	37	40.0	8.00
66	46	82	5.15	1.03	15	6	30	56.7	11.34
65	45	82	5.38	1.08	10	4	22	90.0	18.00
64	44	81	5.62	1.12	5	2	13	190.0	38.00
63	43	80	5.87	1.17	0	0	0	∞	∞
62	42	79	6.13	1.23					
61	41	78	6.39	1.28					

[a] For CN in column 1.

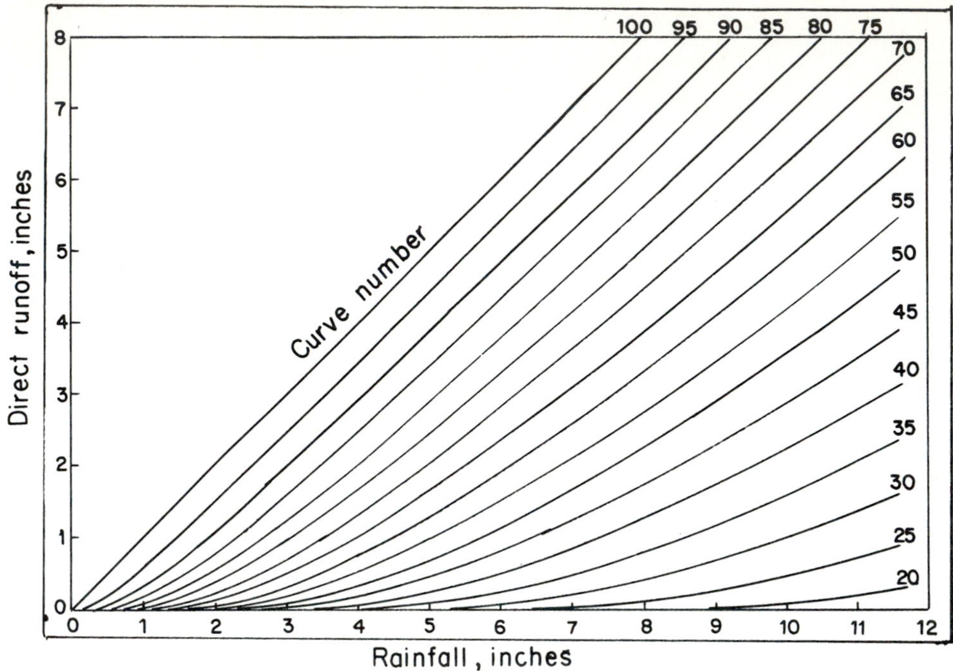

Figure 14.2 Volume of direct runoff as a function of rainfall and curve number.

(3) All other hydrologic characteristics are uniform. Most drainage basins do not satisfy these assumptions and, as a result, the SCS-CN method overpredicts by a large magnitude. This overpredicting is not a serious deficiency when used for farmers, ranchers, and other rural residents, but is deficient for most other applications.

Several features of Equation (14.6) may be noted (Hawkins, 1978; Smith, 1978): (1) Equation (14.3) is valid only for $P \geq 0.2S$; otherwise, $V_Q = 0$. (2) The SCS-CN method is formulated in inches. Thus, the 10 and 1000 in Equations (14.5) and (14.6) must carry inch dimensions, although conversions can be made to the metric system. (3) CN has no physical meaning. It is a convenient transformation of S to establish a 0 to 100 scale. Since this is the only parameter in Equation (14.7), it must reflect the effect of soil, vegetation, and antecedent conditions on storm runoff. (4) The method does not consider the effect of variations in rainfall intensity and its duration. (5) The method does not properly predict I_a for shorter, more intense storms because I_a is assumed constant. (6) The method cannot be extended to properly predict infiltration within a storm. (7) The method assumes a maximum depth of infiltration S after which all rainfall becomes runoff. The choice of an S used to approximate runoff for short storms can produce poor results for extended storms.

EXAMPLE 14.1 Estimate the runoff volume from an area receiving 5 inches of rainfall. The antecedent soil-moisture condition II is assumed valid. The soil–vegetation–land use complex of the area is

LAND USE/CONDITION	AREA (%)	SOIL GROUP
Residential ($\frac{1}{2}$-acre lots)	20	C
Wooded: poor cover	20	C
Meadow: good condition	30	D
Open space: good condition	30	D

Solution Table 14.3 provides values of CN for the four land uses, respectively: 80, 77, 78, and 80. A weighted CN, therefore, is

$$\text{CN} = 0.20 \times 80 + 0.2 \times 77 + 0.3 \times 78 + 0.3 \times 80 = 78.8 \cong 79$$

The value of S from Equation (14.6) is

$$S = \frac{1000}{79} - 10 = 2.66 \text{ in.}$$

The value of V_Q from Equation (14.3a) is

$$V = \frac{(5 - 0.2 \times 2.66)^2}{5 + 0.8 \times 2.66} = \frac{(4.468)^2}{7.128} = 2.80 \text{ in.}$$

Therefore, the runoff volume corresponding to 5 in. of rainfall is 2.80 in.

EXAMPLE 14.2 A watershed is half wooded (good condition) and half residential ($\frac{1}{4}$-acre lots). Each part of the watershed has 50% soil group B and 50% soil group C. Determine the runoff volume if the rainfall is 5 in.

Solution First, the value of CN is obtained for each land use–soil group, and then a weighted CN is obtained. For each land use, there are two soil groups:

LAND USE	SOIL GROUP	AREA	CN
Wooded	B	0.5(0.5) = 0.25	55
	C	0.5(0.5) = 0.25	70
Residential	B	0.5(0.5) = 0.25	75
	C	0.5(0.5) = 0.25	83

The weighted CN is

$$\text{CN} = 0.25(55 + 70 + 75 + 83) = 70.75$$

The value of S from Equation (14.6) is

$$S = \frac{1000}{70.75} - 10 = 4.13 \text{ in.}$$

The value of V_Q from Equation (14.3a) is

$$V_Q = \frac{(5 - 0.2 \times 4.13)^2}{5 + 0.8 \times 4.13} = \frac{(4.17)^2}{8.30} = 2.10$$

The resulting runoff volume is 2.10 in.

14.2 WATER-BALANCE METHODS

These methods attempt to estimate various components of the water budget. The complexity of the methods and significance of their components are greatly influenced by the period for which runoff is to be determined. Normally, the larger the time period, the simpler the method. For example, the yearly water balance is much simpler than the daily water balance. Conceptually, the philosophy remains similar.

The central idea of a water-balance model is one of soil-moisture accounting considering moisture-holding and moisture-transmitting characteristics of the soil and underlying strata. The soil profile is usually divided into two (sometimes more) zones: an upper zone and a lower zone. The depths of these zones may vary from one watershed to another. The moisture-holding capacities of these zones can be specified from knowledge of soil characteristics. Fundamental to determination of soil moisture is simulation of wetting and drying phases of the soil profile. The wetting of zones depends on rainfall and occurs in a sequential order from the upper zone to the lower zone, each filling to capacity before discharging to the lower zone. When rain occurs, the moisture content of the upper zone begins to fill. The amount of water in excess of the maximum capacity of the zone is percolated to the lower zone.

The soil moisture is depleted by evapotranspiration. The moisture of the upper zone is readily available for evapotranspiration at a potential rate. The water available in the lower zone takes place at a rate less than the potential evapotranspiration. Evapotranspiration occurs at a potential rate as long as enough water is available. It is reduced by the amount of moisture available in the lower zone to its moisture-holding capacity.

Rainfall is partitioned into infiltration, abstractions, and surface runoff. When rainfall intensity exceeds the infiltration rate and other abstractions, the excess amount is treated as surface runoff. All infiltrated water is stored in the upper zone until its capacity is filled. Thereafter, any additional infiltrated water automatically goes to the lower zone. The entire rainfall minus steady infiltration is assumed to be runoff after both zones are filled to their capacity.

Deep seepage can be estimated as approximately equal to steady-state infiltration after both zones are saturated. Interflow can be considered as a portion of this seepage. Once these component variables are determined, they are substituted in the water-balance equation to determine the value of runoff for the period under consideration. Figure 14.3 shows a schematic of a typical water-balance model.

EXAMPLE 14.3 Compute the daily soil-moisture budget for the month of May 1964 based on the two-zone depletion procedure. Initial conditions on April 29 are:

Upper zone moisture (UZM) = 1 in.

Lower zone moisture (LZM) = 3 in.

Maximum upper zone storage (UZS) = 1 in.

Maximum lower zone storage (LZS) = 3 in.

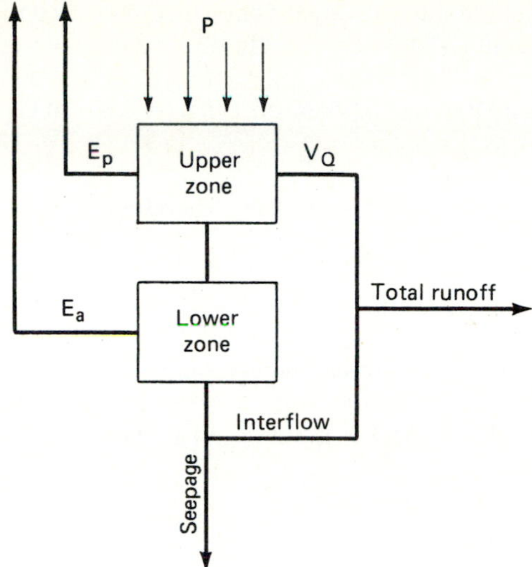

Figure 14.3 A typical water-balance model.

The potential evapotranspiration rate E_P is 0.06 in./day for April 30, 0.14 in./day for May 1–15, and 0.15 in./day for May 16–31. The precipitation in May is as follows:

MAY	PRECIPITATION (in.)	MAY	PRECIPITATION (in.)
1	0.0	14	0.02
5	0.0	16	0.15
7	0.01	24	0.24
8	0.49	26	0.26
9	0.01	28	0.05
13	0.33	29	0.45

Solution The two-zone depletion procedure works as follows: Evapotranspiration occurs at the potential rate until available moisture storage is depleted. Evapotranspiration from the lower zone occurs when the upper-zone moisture storage is depleted, at a rate proportional to the available moisture. Therefore, the water budget can be expressed for the two zones as

$$\text{UZM} = \text{UZMO} + P - E$$

where UZMO is the initial upper-zone moisture at the start of computation, UZM is the upper-zone moisture at the end of the day, P is the precipitation, and E is the evapotranspiration, potential (E_P) or actual (E_a). When UZM > 0,

$$\text{LZM} = \text{LZMO}$$

where LZMO is the initial lower-zone moisture. When UZM = 0,

$$\text{LZM} = \text{LZMO} - \frac{\text{LZMO}}{\text{LZS}} E_a$$

Soil-moisture recharge occurs first in the upper zone when $P > E_P$ until UZM = UZS, and then in the lower zone until LZM = LZS. Additional water may run off the surface.

April 30

$P = 0.0$, $E_P = 0.06$, and UZMO = 1. Therefore, UZM = 1.0 − 0.06 = 0.94 in., and LZM = 3.00 in.

May 1

$E_P = 0.14$ in., $P = 0.0$ in., and UZMO = 0.94 in. Therefore, UZM = 0.94 + 0.0 − 0.14 = 0.80 in., and LZM = 3.0 in.

May 2

$E_P = 0.14$ in., $P = 0.0$ in., and UZMO = 0.80 in. Therefore, UZM = 0.8 + 0 − 0.14 = 0.66 in., and LZM = 3.0.

In a like manner, the soil moisture can be computed for each day of the month.

May 31

$P = 0.0$ in., $E_P = 0.15$ in., and UZMO = 0.0 in., and LZMO = 1.581 in. Therefore, LZM = 1.581 − (1.581 × 0.15)/3 = 1.502 in.

The results are tabulated in Table 14.5.

14.3 ANNUAL AND SEASONAL RUNOFF

It is sometimes useful to estimate the annual or seasonal runoff expected from a drainage basin. This information might be desired for long-term water-supply purposes. The best source of this information is from gaged data obtained from long-term measured information. Such gaged data is seldom available for the watershed of interest, so it frequently becomes necessary to determine annual or seasonal runoff by transporting data from other gaged watersheds.

Streamflow records, published by the U.S. Geological Survey, contain summarized daily mean flows as well as a monthly tabulation of such flows. These data can be transposed to a drainage basin of interest, providing the drainage basins have a substantial degree of similarity. The drainage basins should be reasonably near to the same size because the size of a drainage basin is often related to the amount of baseflow from an area. Large basins usually have higher baseflow in humid areas, but in arid or semiarid areas, there are usually greater channel losses for large basins than for small watersheds. The drainage basins should contain similar soil, which would reflect similar infiltration and other abstractions. These basins should be located within a meteorologically homogeneous area so that precipitation is approximately the same. The type of ground cover should also be nearly the same because there is a great difference between runoff from row crop lands and that from grass or forest lands.

The monthly distribution of runoff reflects, to a great extent, the monthly distribution of rainfall. For example, the Safford watershed in Arizona produces nearly 84% of its runoff in July and August. In Illinois, Wisconsin, and Virginia, the runoff is more evenly distributed throughout the year.

Large variation in annual runoff can occur from one part of the country to the other (Langbein and others, 1949), as shown in Table 14.6. The tabulation of data in this table demonstrates that annual runoff can range from less than 1 in. (2.5 cm) to more than 24 in. (63.5 cm). This large variation in annual runoff emphasizes the care that must be observed in transposing runoff data. This and similar data can be used

TABLE 14.5 DAILY MOISTURE BUDGET FOR MAY 1964

Date	Precipitation (in.)	Potential evapotranspiration (in.)	UZM (in.)	LZM (in.)	Remarks[a]
1	0.0	0.14	0.80	3.0	
2	0.0	0.14	0.66	3.0	
3	0.0	0.14	0.52	3.0	
4	0.0	0.14	0.38	3.0	
5	0.0	0.14	0.24	3.0	
6	0.0	0.14	0.10	3.0	
7	0.01	0.14	0.0	2.97	$E_a < E_p$
8	0.49	0.14	0.35	2.97	$E_a = E_p$
9	0.01	0.14	0.22	2.97	
10	0.0	0.14	0.08	2.97	
11	0.0	0.14	0.0	2.9106	$E_a < E_p$
12	0.0	0.14	0.0	2.7746	$E_a < E_p$
13	0.33	0.14	0.19	2.7746	$E_a = E_p$
14	0.02	0.14	0.07	2.7746	$E_a = E_p$
15	0.00	0.14	0.00	2.7098	$E_a < E_p$
16	0.15	0.15	0.00	2.7098	$E_a = E_p$
17	0.00	0.15	0.00	2.5744	$E_a < E_p$
18	0.00	0.15	0.00	2.4459	$E_a < E_p$
19	0.00	0.15	0.00	2.3259	$E_a < E_p$
20	0.00	0.15	0.00	2.2094	$E_a < E_p$
21	0.00	0.15	0.00	2.0986	$E_a < E_p$
22	0.00	0.15	0.00	1.9936	$E_a < E_p$
23	0.00	0.15	0.00	1.8936	$E_a < E_p$
24	0.24	0.15	0.09	1.8936	$E_a = E_p$
25	0.00	0.15	0.00	1.8556	$E_a < E_p$
26	0.26	0.15	0.11	1.8556	$E_a = E_p$
27	0.00	0.15	0.00	1.8310	$E_a < E_p$
28	0.00	0.15	0.00	1.7410	$E_a < E_p$
29	0.00	0.15	0.00	1.6640	$E_a < E_p$
30	0.00	0.15	0.00	1.581	$E_a < E_p$
31	0.00	0.15	0.00	1.502	$E_a < E_p$

[a] E_a = actual evapotranspiration, E_P = potential evapotranspiration.

for preliminary estimates of annual runoff from an ungaged basin. However, such transposed data can be in substantial error, and gaged data should be obtained for design purposes.

The cost of installing and monitoring a short-term stream gage is justified in most cases because the need for accurate data in any worthwhile hydrologic investigation cannot be underestimated. If a project is worth doing, it justifies accurate data. The actual expense of a stream gage is relatively small in terms of the initial cost. Long-term monitoring can be expensive. In order to minimize the cost of obtaining annual and seasonal runoff data, a short-term stream gage can be installed. The record from this stream gage can be related to a nearby established stream gage. If a usable correlation between the monthly values of runoff for the two gages can be established, then the short-term gage can be abandoned and the relationship can be used for future prediction.

TABLE 14.6 MEAN ANNUAL PRECIPITATION, TEMPERATURE, AND RUNOFF FOR SELECTED DRAINAGE BASINS

Stream	Drainage area (mi^2)	Period	Mean annual precipitation (in.)	Mean annual temperature (°F)	Weighted mean temperature (°F)	Mean annual runoff (in.)
Mexican Springs Wash at Mexican Springs, NM	32.7	1937–1941	15	47.5		0.4
Cannonball River at Breien, ND	4,066	1921–1945	15.6	42.1	53.5	.61
Churchill River at Island Falls, Sask.	71,000	1929–1943	16	30	40.2	4.1
S. Fork Palouse near Pullman, WA	81.1	1934–1940	19.6	47.4	40.1	2.8
Stream A, Wagonwheel Gap, CO	0.347	1911–1926	21.1	34.0	36.5	6.1
Saline River at Rescott, KS	2,820	1920–1936	22.1	54.8	63.1	.76
Cajon Creek near Keenbrook, CA	40.9	1931–1943	22.8	56.1	48.1	4.5
Elkhorn River at Waterloo, NE	6,900	1921–1945	23.0	48.7	57.7	1.7
Deep Creek near Hesperia, CA	137	1905–1915	27.1	51.5	39	10
Strawberry Creek near San Bernardino, CA	8.6	1921–1941	30.9	57.1	49.2	8

Washita River near Durwood, OK	7,310	1921–1945	31.2	60.8	65	3.2
Kings River at Piedra, CA	1,694	1895–1940	31.4	44	36	18.8
Ralston Creek near Iowa City, IA	3.01	1925–1935	33	50.2	58.1	6.7
Miami River at Dayton, OH	2,513	1924–1942	37.0	51.0	53.6	11.5
Neuse River near Clayton, NC	1,140	1921–1945	45.4	60.3	62.3	13.9
Middle Westfield River at Goss Heights, MA	52.6	1922–1934	45.6	46.8	46	25.9
West River at Newfane, VT	308	1920–1924; 1929–1933	46.5	42.3	42.3	25
Kissimmee River near Okeechobee, FL	3,260	1931–1942	50	72.5	76.1	7.3
Little River near Horatio, AL	2,690	1931–1944	50.7	61.3	61.2	17.3
Elk River at Queen Shoals, WV	1,145	1921–1945	51.8	52.0	53.3	24.0
Average of ten comparable basins in southeastern Alabama	—	1938–1947	59.3	66	66.2	22.7
Amite River near Denham Springs, LA	1,330	1939–1947	59.4	67	67.1	19.5

The simplest linear model for annual runoff takes the form

$$V_Q = aP - b \tag{14.8}$$

where V_Q is the annual runoff, P is the annual watershed mean precipitation, and a and b are constants estimated by linear regression analysis. The difference between annual precipitation and annual runoff is principally comprised of evapotranspiration E, and one can write

$$V_Q = P - E \tag{14.9}$$

Therefore, by equating Equation (14.8) and (14.9),

$$E = b + P(1 - a) \tag{14.10}$$

From Equations (14.8) and (14.10), it is seen that runoff and evaporation increase linearly with precipitation. This assumption is reasonable in temperate and subhumid regions, where precipitation is moderate and well-distributed temporally. The key to determining annual runoff is the determination of annual evapotranspiration. Ayers (1962) suggested that annual evaporation is approximately half the annual precipitation for Equation (14.8). On the other hand, Equation (14.10) expresses a linear relation between annual evaporation and annual precipitation. Budyko (1948) reported that

$$E = P[1 - \exp(-E_p/P)] \tag{14.11}$$

where E_p is the annual potential evapotranspiration. Some well-known empirical equations for estimating annual evapotranspiration were presented in Chapter 9.

EXAMPLE 14.4 Values of monthly rainfall and the corresponding monthly runoff for a watershed are available. Develop an empirical equation relating runoff to rainfall.

Month	P (cm)	V_Q (cm)	Month	P (cm)	V_Q (cm)	Month	P (cm)	V_Q (cm)
1	4	0.20	11	50	10.5	21	14	3.2
2	40	9	12	35	7.5	22	16	1.5
3	30	5	13	12	1.5	23	26	5.5
4	25	4.5	14	18	2.0	24	32	6.5
5	20	2.5	15	8	1.0	25	34	8.0
6	15	2.0	16	24	4.5	26	44	8.5
7	10	1.0	17	28	5.5	27	36	7.0
8	5	0.5	18	38	8.5	28	46	8.0
9	50	14	19	48	10.0	29	2	0.25
10	45	10.5	20	42	9.5	30	17	3.5

Solution A linear relation between V_Q and P was derived using the least squares method. This relation was also derived for the log-transformed values of V_Q and P. The results are summarized as follows:

Month	P (cm) (x)	V_Q (cm) (y)	log P	log V_Q
1	4	0.20	0.60206	−0.69897
2	40	9	1.60206	0.95424
3	30	5	1.47712	0.69897
4	25	4.5	1.39794	0.65321
5	20	2.5	1.30103	0.39794
6	15	2.0	1.17609	0.30103
7	10	1.0	1.00000	0.00000
8	5	0.5	0.69897	−0.30103
9	50	14	1.69897	1.14613
10	45	10.5	1.65322	1.02119
11	50	10.5	1.69897	1.02119
12	35	7.5	1.54407	0.87506
13	12	1.5	1.07918	0.17609
14	18	2.0	1.25527	0.30103
15	8	1.0	0.90309	0.00000
16	24	4.5	1.38021	0.65321
17	28	5.5	1.44716	0.74035
18	38	8.5	1.57978	0.92949
19	48	10.0	1.68124	1.00000
20	42	9.5	1.62325	0.97772
21	14	3.2	1.14613	0.50515
22	16	1.5	1.20412	0.17609
23	26	5.5	1.41497	0.74036
24	32	6.5	1.50515	0.81291
25	34	8.0	1.53148	0.90309
26	44	8.5	1.64345	0.92942
27	36	7.0	1.55630	0.84510
28	46	8.0	1.66276	0.90309
29	2	0.25	0.30103	−0.60206
30	17	3.5	1.23045	0.54407
Sum	814	161.65	39.9955	16.60502

$$V_Q = a + bP$$

$$a = -1.2240$$

$$b = 0.2437$$

correlation coefficient, $r = 0.9678$

$$r^2 = 0.9365$$

$$V_Q = -1.2240 + 0.2437P$$

$$\log V_Q = a + b \log P$$

$$a = -1.2548$$

$$b = 1.3564$$

$$r = 0.9764$$

$$r^2 = 0.9533$$

$$\log V_Q = -1.2548 + 1.3564 \log P$$

$$V_Q = 10^{-1.2548} P^{1.3564}$$

$$= 0.0556P$$

The linear relation is as good as a log–linear relation.

14.4 FLOW-MASS CURVE

The flow-mass curve is a plot of the cumulative runoff amount, or flow volume, against time. In other words, it is a graphical representation of Equation (14.1), where V_Q varies with time, and the lower limit of the integral denotes the start of time or of the curve and can be taken other than zero. Thus, on the ordinate is V_Q and on the abscissa is time t. Because $Q(t)$ represents the runoff hydrograph, $V_Q(t)$ is an integral curve, or summation curve, of the hydrograph. Rippl (1882) was probably the first to use the flow-mass curve, and, hence, it is also known as Rippl's diagram or mass curve. The time for the mass curve is in chronological order, and can be in days, weeks, months, or any other unit. The ordinate is in million m^3, ha-m, m^3/s-day, ft^3, acre-ft, etc.

A typical mass curve is shown in Figure 14.4. At any point in time, the curve gives the cumulative volume of flow up to that time. Similarly, the slope of the curve at any point provides the rate of flow at that time by virtue of $Q = dV_Q/dt$. If any two points on the mass curve are connected by a straight line, then the slope of this line represents the average flow rate between those points in time. The practical implication of this is that if a reservoir had adequate storage, then the average rate of flow

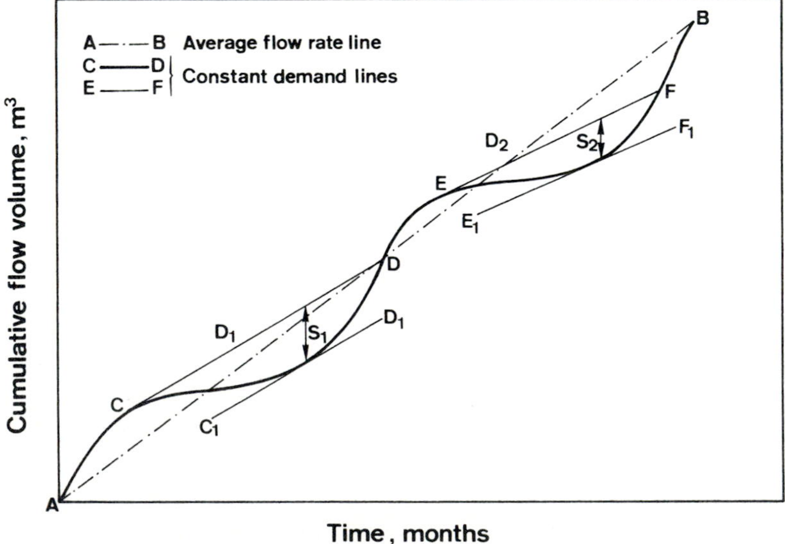

Figure 14.4 Flow mass curve. The reservoir is full at C and E. S_1 and S_2 are cumulative deficiencies.

could be maintained during those times. This argument can be extended further. If the starting and terminating points of the mass curve are connected by a straight line, then the slope of this line represents the average rate of flow over the whole period. If enough storage were provided, then the reservoir would be able to supply at this average rate for the period of the curve. The straight line can be called supply line. Thus, the mass curve can be used for design of storage reservoirs.

14.4.1 Computation of Storage Amount

Consider a reservoir is full initially or at the start of a dry season. At any time, the amount of water withdrawn from the reservoir is the difference between the cumulative supply and the cumulative demand from the beginning up to that time. The maximum of this amount, whenever it occurs, is the storage S to be provided. Thus, we can express

$$S = \text{maximum of } (\Sigma V_D - \Sigma V_S) \tag{14.12}$$

where V_D is the cumulative demand, and V_S is the cumulative supply. Note that the difference between the cumulative demand and the cumulative supply varies with time, and defines the cumulative deficiency. When the mass curves of supply and demand are plotted, the maximum difference in their ordinates gives the maximum cumulative deficiency, which is the storage amount S for that dry season. The values of S are calculated for different dry periods and the largest of these values is the minimum of storage required for the reservoir. The value of S can be obtained either graphically using the mass curve or arithmetically. From Figure 14.4, line CD represents the constant-demand line; its slope is D_1, which is the demand rate. Line EF represents another demand line with demand rate of D_2. Vertical distances between these lines and their corresponding tangents specify the values of storage such as S_1 and S_2 in the figure. The largest value of S_1, S_2, . . . , is the minimum required storage.

EXAMPLE 14.5 Compute the minimum reservoir storage required to maintain a demand of 50 m^3/s. Monthly river flows to the reservoir are available as follows:

MONTH	JAN	FEB	MAR	APR	MAY	JUN	JUL	AUG	SEP	OCT	NOV	DEC
Mean monthly flow (m^3/s)	94	82	45	20	26	43	90	110	86	70	53	40

To simplify plotting of the mass curve, assume each month has 30 days.

Solution First, monthly flow volumes are computed for each month. For a month, the volume is the mean flow during the month times the number of days in the month. Values of monthly flow volumes are shown in column (4) of Table 14.8. These volumes are then accumulated as shown in column (5) of this table. Cumulative flow volumes are plotted against time as 1 month to obtain the flow mass curve, as shown in Figure 14.5. For simplicity of plotting, all months are assumed to be of equal length of 30 days. A demand line having a slope of 50 m^3/s is drawn tangential to the hump at the beginning of

TABLE 14.8 MONTHLY FLOWS FOR CONSTRUCTION OF FLOW MASS CURVE

Month (1)	No. of days (2)	Mean flow (m^3/s) (3)	Monthly flow volume (m^3/s-day) (2) × (3) (4)	Cumulative flow volume (m^3/s-day) (5)
January	31	94	2,914	2,914
February	28	82	2,296	5,210
March	31	45	1,395	6,605
April	30	20	600	7,205
May	31	26	806	8,011
June	30	43	1,290	9,301
July	31	90	2,790	12,091
August	31	110	3,410	15,501
September	30	86	2,580	18,081
October	31	70	2,170	20,251
November	30	53	1,590	21,841
December	31	40	1,240	23,081

the flow mass curve. This line is denoted as AB in Figure 14.5. In order to compute the required storage, a line parallel to the demand line is drawn tangentially to the valley bottom of the mass curve. This line is denoted as A_1B_1 in the figure. The vertical distance between these two parallel lines gives the minimum storage S required to meet the constant demand rate of 50 m^3/s. This is found to be S = 2300 m^3/s.

EXAMPLE 14.6 Solve Example 14.5 arithmetically, and explain if there is any discrepancy between the value of storage obtained here and that in the previous example.

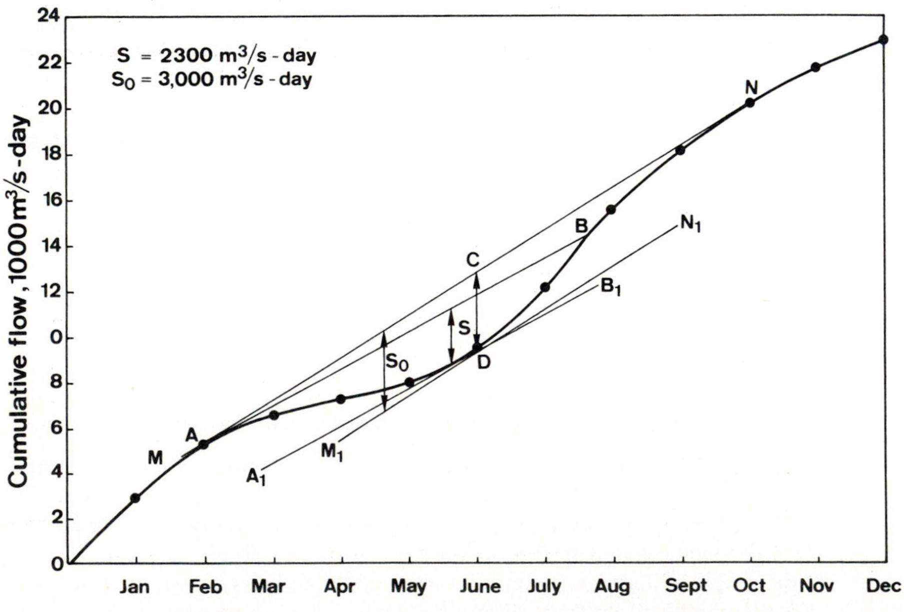

Figure 14.5 Mass curve of monthly flow volumes.

Solution For each month, flow volume and demand volume are computed. Then the difference between these values is obtained. If the difference is positive, that is, the flow volume is greater than the demand volume, there is no deficiency and there is an excess flow volume that is noted. On the other hand, if the difference is negative, then there is deficiency and it is noted. By performing these calculations for each month, cumulative excess demand and cumulative excess flow are computed. These calculations are shown in Table 14.9. It is seen that the maximum cumulative excess demand is in the month of June and equals -2164 m^3/s-day. This is the value of the storage required to meet the demand of 50 m^3/s.

This value of storage differs from the one obtained in Example 14.5. This difference is attributed to the assumption of flow variability from one month to the next. In the arithmetic calculation, it has been tacitly assumed that flow varies linearly from one month to the other, whereas in the graphical solution, the mass curve is curvilinear. As a result, the methods will produce the same result only coincidently.

It may also be noted that column (8) of Table 14.9 indicates if the reservoir is going to be refilled. In this example, the reservoir will be refilled in the month of August. Thereafter, the flow in excess of 2164 m^3/s-day will be spillover.

14.4.2 Computation of Maintainable Demand

When a reservoir is designed for a specific storage, a question arises: What is the maximum firm demand that can be maintained by the reservoir? This question can be solved by constructing the flow mass curve. Valleys and ridges are identified on the mass curve, as shown in Figure 14.6. Then tangents are drawn from the ridges, and each tangent crosses the next valley at a particular slope. The tangential line from the ridge to the crossing of valley defines the demand that can be sustained by the given storage over that period of the valley (or dry period). From the figure, the demand line over the first valley is A_1B_1, over the second valley is A_2B_2, and so on for other valleys. The slope of each demand line gives the demand rate. The slope of the first demand line is D_1, of the second is D_2, and of the third is D_3. The minimum of these demand rates represents the maximum firm demand that can be sustained by the given storage. This analysis is based on the assumption that the mass curve has ridges and valleys. In other words, the demand line intersects the mass curve, thus ensuring that the reservoir is refilled; otherwise, there would be insufficient inflow to the reservoir. By drawing tangents at the valleys parallel to the demand lines, the amount of storage is obtained. The vertical distance between two successive tangents at the ridges yields the amount of excess water that will be spilled over, or wasted.

EXAMPLE 14.7 Compute the maximum uniform demand rate that can be maintained by the reservoir with storage of 3000 m^3/s-day. Use the data of Example 14.5.

Solution The solution of this problem is obtained by trial and error. A vertical line CD of 3000^3/s-day is drawn at the valley bottom of the mass curve, as shown in Figure 14.5. Then a line is drawn that is tangential to the hump of the mass curve and passes through point C. This line is MN in the figure. Parallel to this line is drawn a tangent at the valley bottom of the mass curve, which is M_1N_1, and the vertical distance is measured. If this vertical distance is of the same magnitude as 3000 m^3/s-day, then the initial guess is right. Otherwise, the procedure is repeated until the right location of C

TABLE 14.9 COMPUTATION OF STORAGE ARITHMETICALLY

Month (1)	Mean monthly flow (m^3/s) (2)	Flow volume (m^3/s-day) (3)	Demand rate (m^3/s) (4)	Demand volume (m^3/s-day) (5)	Deficiency (m^3/s-day) (3) − (5) (6)	Cumulative excess demand (m^3/s-day) (7)	Cumulative excess flow volume (m^3/s-day) (8)
January	94	2914	50	1550	+1364	—	1364
February	82	2296	50	1400	+896	—	2260
March	45	1395	50	1550	−155	−155	—
April	20	600	50	1500	−900	−1055	—
May	26	806	50	1550	−899	−1954	—
June	43	1290	50	1550	−210	−2164	—
July	90	2790	50	1550	+1450	—	1450
August	110	3410	50	1550	+1860	—	3310
September	86	2580	50	1500	+1380	—	4690
October	70	2170	50	1550	+620	—	5310
November	53	1590	50	1500	+90	—	5400
December	40	1240	50	1550	−310	−310	—

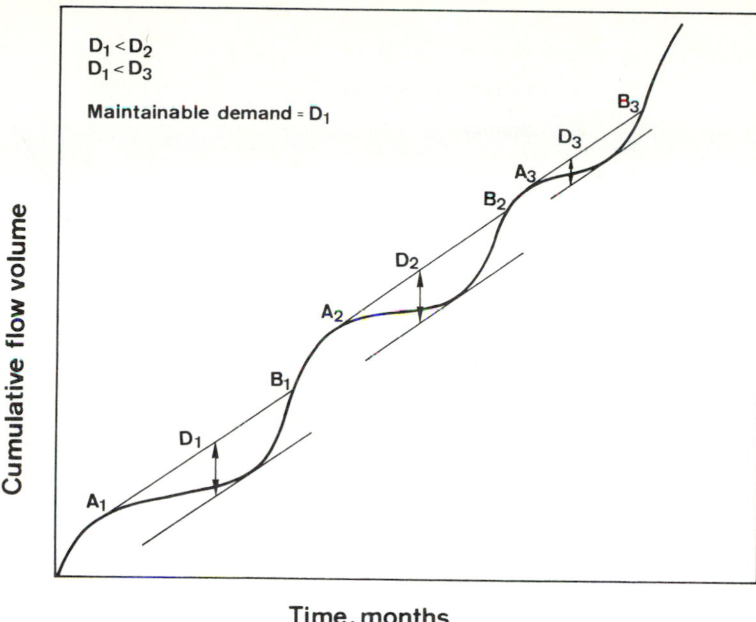

Figure 14.6 Flow mass curve for computation of maintainable demand.

is found. The slope of line MN gives the maximum demand, which, in our case, is 63 m^3/s.

14.4.3 Consideration of Variable Demand

In practice, the demand varies with time, for the needs of irrigation, water supply, recreation, power generation, waste disposal, etc., vary during the year. This variable demand needs to be taken into account for designing the storage reservoirs. The flow mass curve again can be used. A mass curve of demand is prepared and superposed on the flow mass curve with proper matching of time, as shown in Figure 14.7. Sometimes, the demand mass curve is also referred to as the variable use curve. For superposition of two curves, the matching of time is essential, because during each time interval, say, a month, the inflow is different and so is demand. In addition to actual societal demand, provision must be made for such natural demands as evaporation, seepage, leakage, etc. These demands are also variable in time. The natural demands can either be deducted from the inflow to the reservoir or added to the societal demand. The needed storage is represented by the maximum vertical distance between the flow mass curve and the demand mass curve, assuming the reservoir is full at the first intersection of the two curves, say, at point A. Figure 14.7 shows this case of variable demand.

EXAMPLE 14.8 Compute the amount of storage needed to meet the demands varying from month to month, as given in Table 14.10. The reservoir area is 10 km^2. For converting rainfall to flow to reservoir, a runoff coefficient of about 0.6 can be assumed. Prior commitments are for 10 cm per unit area for each month.

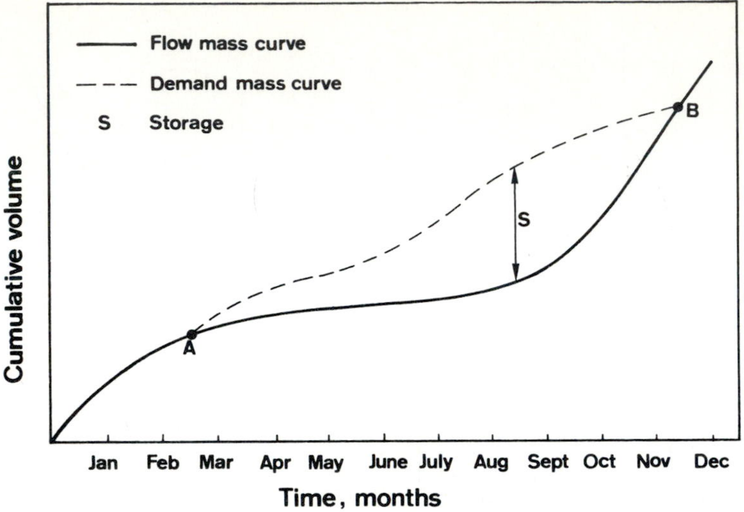

Figure 14.7 Flow and demand mass curves.

Solution Calculations similar to Table 14.9 are carried out for this example. These essentially involve arithmetic operation of budgeting for each month, and are shown in Table 14.11. The required storage for this example is 97 cm. Thus,

$$S = 97 \text{ cm} = \frac{97}{100} \times 10 \times 1000 \times 1000 = 9.7 \times 10^6 \text{ m}^3$$

TABLE 14.10 FLOW, RAINFALL, AND DEMAND DATA

Month	Mean flow[a] (cm)	Societal[a] demand (cm)	Monthly evaporation (cm)	Other monthly losses (cm)	Monthly rainfall (cm)
January	70	20	5	1	10
February	50	25	8	2	8
March	40	28	10	2	6
April	30	32	12	1	5
May	10	25	15	2	4
June	20	30	16	2	3
July	300	50	16	1	15
August	350	40	15	2	20
September	250	30	13	1	15
October	100	20	10	2	12
November	80	10	8	1	10
December	70	15	5	1	8

[a] This quantity is on a per-unit reservoir-area basis. To convert it to rate, this is to be multiplied by the area and divided by the number of days in the month, with appropriate accounting of units.

TABLE 14.11 COMPUTATION OF STORAGE FOR VARIABLE DEMAND

			Demand								
Month (1)	Inflow (cm) (2)	Rainfall (cm) (3)	Societal (cm) (4)	Prior commitments (cm) (5)	Evaporation (cm) (6)	Losses (cm) (7)	Total demand (4) + (5) + (6) + (7) (cm) (8)	Total inflow (2) + (3) (cm) (9)	Difference Between flow and demand (9) − (8) (cm) (10)	Cumulative excess demand (cm) (11)	Cumulative excess flow (cm) (12)
January	70	10	20	10	5	1	36	80	+44	—	44
February	50	8	25	10	8	2	45	58	+13	—	57
March	40	6	28	10	10	2	50	46	−4	−4	—
April	30	5	32	10	12	1	55	35	−20	−24	—
May	10	4	25	10	15	2	52	14	−38	−62	—
June	20	3	30	10	16	2	58	23	−35	−97	—
July	300	15	50	10	16	1	77	315	+238	—	238
August	350	20	40	10	15	2	67	370	+303	—	541
September	250	15	30	10	13	1	54	265	+211	—	752
October	100	12	20	10	10	2	42	112	+70	—	822
November	80	10	10	10	8	1	29	90	+61	—	883
December	70	8	15	10	6	1	32	78	+46	—	929

EXERCISES

14.1. Compute the runoff volume for 3 inches of rainfall for a watershed that is fallow with a straight row and the soil is group A. How different is the runoff volume when the soil group is changed to B, C, or D?

14.2. Compute the runoff volume from a watershed that is 40% residential ($\frac{1}{4}$-acre lots), 10% streets and roads paved with curbs and storm sewers, 10% commercial and business, 20% industrial, and 20% open spaces. The hydrologic soil group is B and the rainfall is 5 inches.

14.3. Compute the runoff volume from a watershed that is $\frac{1}{2}$ wooded, $\frac{1}{4}$ residential ($\frac{1}{2}$-acre lots), and $\frac{1}{4}$ pasture with good condition. The watershed has 60% soil group B and 40% soil group C. The rainfall amount is 5 inches.

14.4. Compute the daily moisture budget for the month of July 1980 for a watershed near New Orleans, Louisiana, based on the two-zone moisture-depletion procedure. Initial conditions on June 30, 1980, are assumed as: upper-zone moisture = 5 cm, maximum upper-zone storage = 5 cm, lower-zone moisture = 15 cm, and maximum lower-zone moisture storage = 15 cm. Obtain values of precipitation and potential evapotranspiration from ''Climatological Records'' of the National Weather Service.

14.5. Outline an approach for estimating the runoff volume from an ungaged drainage basin.

14.6. Select a watershed in your area for which monthly rainfall and monthly runoff data are available for at least 30 months. Fit a straight line between the monthly runoff and the monthly rainfall. How good is this relation? Transform the data logarithmically and then fit the straight line. How good is this relation? Is there any improvement due to the log transformation?

14.7. Select a watershed and fit a straight line between the monthly runoff and the monthly rainfall. How good is this relation? Also fit a relation between the annual runoff and the annual rainfall. Is this relation better than the one for monthly data in Exercise 14.6? If the answer is yes, what is the probable cause?

14.8. Verify Equation (14.10) for a watershed in your area.

14.9. Verify Equation (14.11) for a watershed in your area.

14.10. Compute graphically the minimum reservoir storage required to maintain a steady demand of 25 m^3/s. The monthly river flows to the reservoir are available as follows:

MONTH	JAN	FEB	MAR	APR	MAY	JUN	JUL	AUG	SEP	OCT	NOV	DEC
Mean monthly flow (m^3/s)	50	45	35	40	15	10	15	20	50	45	55	60

14.11. Solve Exercise 14.10 arithmetically, and explain if there is any discrepancy between the value of storage obtained here and that obtained graphically.

14.12. Compute the maximum uniform demand rate that can be maintained by the reservoir with storage of 2500 m^3/s-day. Use the data of Exercise 14.10.

14.13. Compute the amount of storage needed to meet the demands varying from month to month as given in Table 14.10. The reservoir area is 15 km^2. For converting rainfall to flow to the reservoir, a runoff coefficient of about 0.5 can be assumed. Prior commitments are for 10 cm per unit area for each month. Consider the monthly loss of 1 cm for each month in Table 14.10.

14.14. The average annual precipitation over a medium-sized watershed A of Afagodas da Ingazeira near Compina Grande, Paraiba, Brazil, and the annual runoff during the corresponding water years are as follows:

YEAR	ANNUAL PRECIPITATION (mm)	ANNUAL RUNOFF (mm)
1973	634.7	425.3
1974	1033.8	646.1
1975	751.0	446.8
1976	470.0	289.1
1977	723.0	455.5
1978	770.6	470.1
1979	473.3	295.8
1980	519.5	309.1

Estimate the annual runoff from an adjoining watershed B that is hydrologically similar to watershed A when the annual precipitation during the year is 650 mm. Project the value of runoff if the 5-year moving means of annual precipitation and runoff were used.

14.15. In a watershed of 1350 square miles, there is a daily flow of 1600 cfs. Determine the total runoff volume in second-foot-days, acre-feet, inches, millions of imperial gallons (1 gallon = 10 lb of water), hectare-centimeters, centimeters, and millions of liters.

14.16. In a watershed, there exists a linear relation between rainfall and runoff. Establish this relation for the following data:

RAINFALL (cm)	RUNOFF (cm)
3	1.5
10	3.5

Determine the amount of runoff from this watershed if the rainfall is 7 cm.

14.17. A 50-in. rainfall fell during a monsoon season on three 10-square mile watersheds. The three watersheds, classified as good, average, and bad, have runoff coefficients, respectively, of 48.8%, 36.6%, and 24.4%. Determine the runoff yield of each of these watersheds in inches and cubic feet.

14.18. On a small watershed, 1800 hectares in area, 15 cm of rain fell. The amount of infiltration due to this rainfall was 6 cm. Assume that all other abstractions were negligible. Compute the amount of runoff. If the runoff coefficient for this watershed is 0.5, what would be the amount of infiltration?

14.19. Hydrologic records of the Lake of the Woods drainage area in Kenora, Ontario, from 1927 to 1947 show that the average annual precipitation was 30.2 inches; the yearly evaporation from the water area is 21.5 inches; the mean yearly runoff is 6.7 inches; the total drainage area is 27,170 square miles, and the water area is 17% of the total area. Determine the following:

(a) the average yearly runoff from the water area
(b) the average yearly runoff from the land area

14.20. The data given is for a lake having an average area of 10 acres. Determine the total outflow from the lake for the 3-month period if the level of the lake is the same at the beginning and end of the period. Use a pan coefficient of 0.75.

PERIOD	PAN EVAPORATION (in.)	PRECIPITATION (in.)	INFLOW (in. over lake)
Oct	3.2	2.7	0.8
Nov	1.7	2.4	1.1
Dec	1.0	3.1	1.3

CHAPTER 15
Unit Hydrograph Method

15.1 UNIT HYDROGRAPH THEORY

In 1932, Sherman developed the concept of unit hydrograph (UH) for determining the surface or direct runoff hydrograph (DRH) from the effective rainfall hyetograph (ERH), which has since been widely used in applied hydrology. The unit hydrograph (UH), or simply unit graph, of a watershed can be defined as the DRH resulting from one unit (1 in. or 1 cm) of effective rainfall (ER) occurring uniformly over the watershed at a uniform rate during a unit period of time. This unit period is not necessarily equal to unity; it can be any finite duration up to the time of concentration. This unit period of ER is the period for which the UH is determined. As soon as this period changes, so does the UH for a specified watershed. Thus, there can be as many UHs for a watershed as many periods of effective rainfall. We often use 1-hour UH, 6-hour UH, 12-hour UH, or 1-day UH. Here 1 hour, 6 hour, 12 hour, or 1 day is not the duration for which the UH occurs, but it is the duration of the ER for which the UH is derived.

Since the UH is applicable for direct or surface runoff only, separation of baseflow from the total runoff hydrograph is required. In order to obtain the DRH, several assumptions must be observed in applying the UH method. It is not possible to fit these assumptions perfectly, but they must be reasonably satisfied before the UH method can be used. Application of the UH, without regard to the limiting assumptions, may result in erroneous results.

The basic postulates of the UH concept are perhaps best described by Johnstone and Cross (1949):

1. For a given drainage basin, the duration of surface runoff is essentially constant for all uniform-intensity storms of the same length, regardless of differences in the total volume of surface runoff.
2. For a given drainage basin, if two uniform-intensity storms of the same length produce different total volumes of surface runoff, then the rates of surface runoff at corresponding times t, after the beginning of two storms, are in the same proportion to each other as the total volumes of surface runoff.
3. The time distribution of surface runoff from a given storm period is independent of concurrent runoff from antecedent storm periods.

To these postulates the following can be added:

4. The ER is uniformly distributed within its duration or specified period of time. This means that the rainfall intensity is uniform throughout the drainage basin during the time rain falls. This steady rainfall intensity must occur even if the unit time is 1 hour or 2 hours or whatever.
5. The ER is uniformly distributed throughout the drainage-basin area. This means that uniform runoff occurs from all parts of the drainage basin and that the rainfall event that causes the runoff is distributed in such a manner that the ER is, indeed, uniformly distributed over the entire drainage basin. Because ER is defined as that rainfall that produces surface runoff, this limitation means that the hydrologic losses must be uniform over the entire drainage basin.

The definition of the UH together with these postulates constitutes what is now called the unit hydrograph theory.

15.2 MATHEMATICAL REPRESENTATION

The unit hydrograph theory assumes that the watershed is linear and time-invariant. That is, the DRH is derived from the ERH by a linear operation. To avoid ambiguity in interpretation of the UH, it must be emphasized that it corresponds to a particular duration of ER and has volume of one unit (i.e., 1 in. or 1 cm). Because ER is assumed to occur uniformly during this period, its duration defines its intensity. This duration is referred to as the unit storm duration and may change from one application to another. Thus, for a given watershed, there can be a large number of unit hydrographs, each corresponding to a specific duration of ER.

The mathematical representation of the UH theory is facilitated by proper interpretation of its postulates indicated before. Postulates 1 and 2 together make up the principle of proportionality. In other words, if the duration of ER is fixed but its volume changes, then the duration of DR does not change, but its ordinates do in proportion to the volume of ER. This is illustrated in Figure 15.1.

Mathematically, this postulate can be expressed as follows. Let the ERH of D-hour duration be denoted as

$$\begin{aligned} I(t) &= I, \qquad 0 \le t < D \\ &= 0, \qquad t \ge D \end{aligned} \tag{15.1}$$

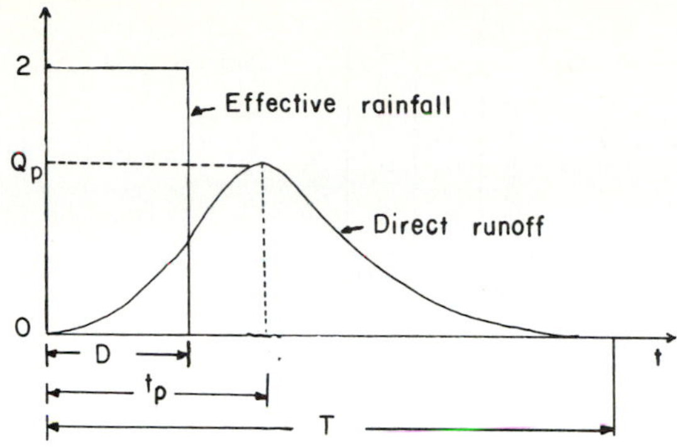

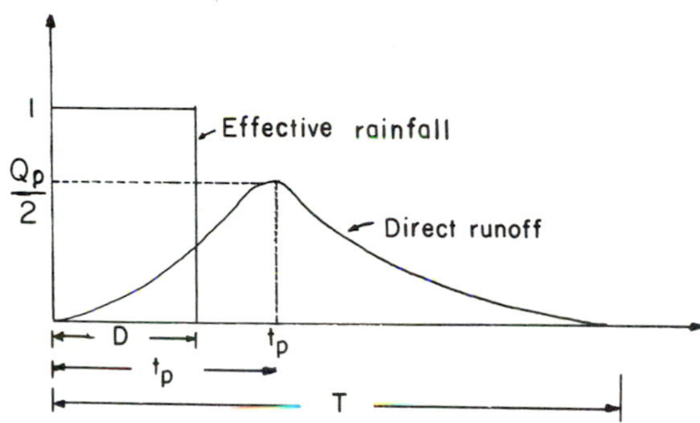

Figure 15.1 Application of the principle of proportionality to the hydrograph calculation.

with I as a constant ER intensity (L/T), the D-hour UH as $h(D,t)$, and the resulting DRH as $Q(t)$. Then

$$Q(t) = h(D,t)I(t)D = h(D,t)ID \tag{15.2}$$

The quantity ID denotes the volume of ER. This equation states that DRH is proportional to the ER volume. When $ID = 1$, the DRH and UH are numerically the same.

The third postulate is the principle of superposition, which allows the decomposition of a complex ERH into rectangular blocks or pulses and then superimposing on one another the hydrographs of these rectangular pulses, each of steady intensity, to obtain the total DRH. As illustrated in Figure 15.2(a), the ERH can be broken down into three rectangular pulses. The DRH due to each pulse is determined. The hydrographs of DR due to these pulses are added to obtain the DRH due to the complex ERH. This can be generalized to any ERH regardless of its complexity.

For expressing the principle of superposition mathematically, we refer to Figure 15.2(b). Let each pulse be of duration D hours, the intensity of pulse a by I_1 and its DRH by $Q_1(t)$, the intensity of pulse b by I_2 and its DRH by $Q_2(t)$, and the intensity of pulse c by I_3 and its DRH by $Q_3(t)$. If the D-hour UH is $h(D,t)$, then the DRH is

$$Q(t) = Q_1(t) + Q_2(t) + Q_3(t) \tag{15.3}$$

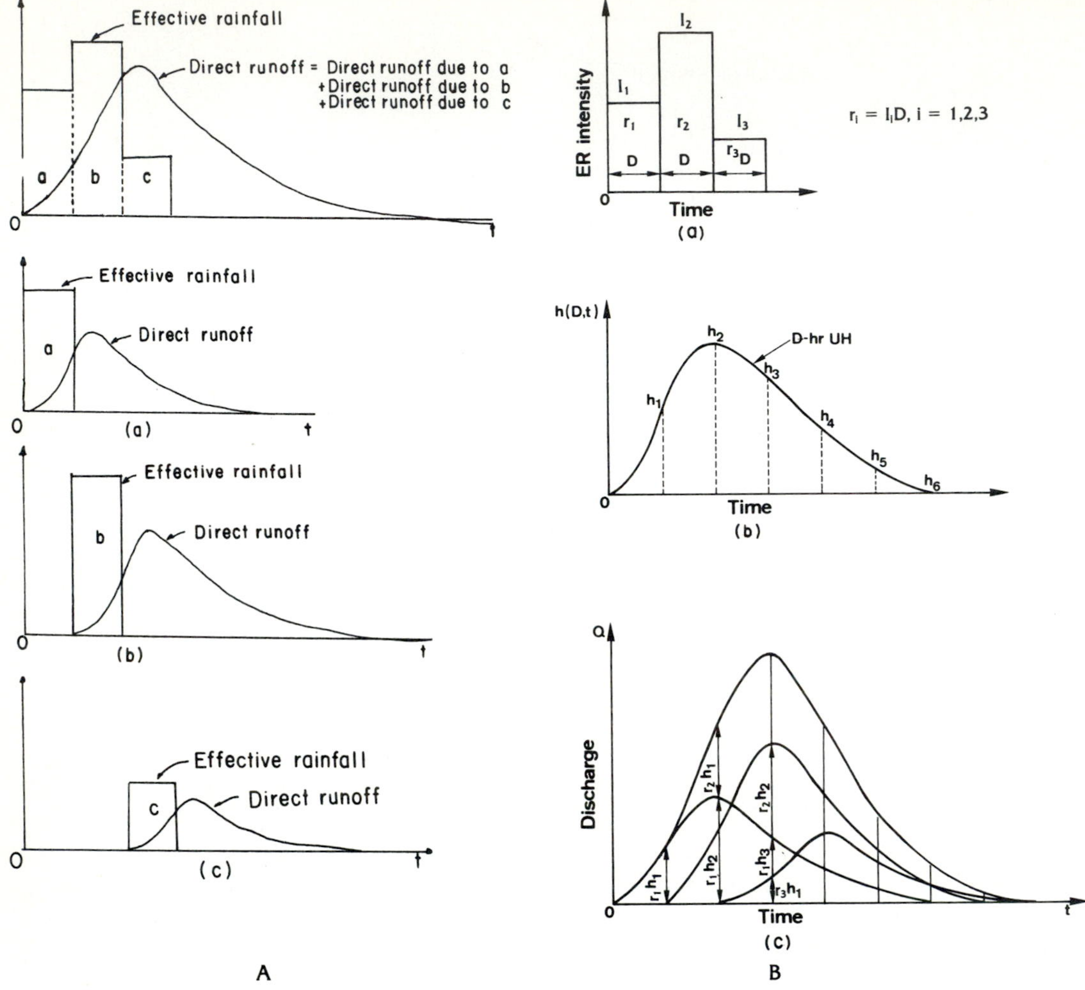

Figure 15.2 (A) Application of the principle of superposition to the hydrograph calculation. (B) Application of the principles of superposition and proportionality to hydrograph calculation.

where $Q_2(t)$ does not start until $t = D$, and $Q_3(t)$ does not start until $t = 2D$. Therefore, to account for this time lag, it is more appropriate to write

$$Q(t) = Q_1(t) + Q_2(t - D) + Q_3(t - 2D) \tag{15.4}$$

The individual DRHs can be expressed as

$$Q_1(t) = h(D,t)I_1D$$

$$Q_2(t - D) = h(D,t - D)I_2D, \qquad h(D,t - D) = 0, \qquad t \leq D$$

$$Q_3(t - 2D) = h(D,t - 2D)I_3D, \qquad h(D,t - 2D) = 0, \qquad t \leq 2D$$

Hence, as shown in Figure 15.2(b),

$$\begin{aligned} Q(t) &= h_1(D,t)I_1D + h_2(D,t-D)I_2D + h_3(D,t-2D)I_3D \\ &= \sum_{j=1}^{3} h(D,t-(j-1)D)I_jD \end{aligned} \tag{15.5}$$

Equation (15.5) can be generalized for the ERH having any number of pulses, each of duration D. If there are n pulses, then Equation (15.5) becomes

$$Q(t) = \sum_{j=1}^{n} h(D,t-(j-1)D)I_jD \tag{15.6}$$

Since the DRH at any time t is the result of the ERH pulses occurring up to that time but not subsequent to it, Equation (15.6) must be modified as

$$Q(t) = \sum_{j=1}^{t} h(D,t-(j-1)D)I_jD \tag{15.7}$$

For purposes of simplicity, time t can be taken as a discrete variable taking values at D time intervals.

15.3 LIMITATIONS OF THE UNIT HYDROGRAPH THEORY

Johnstone and Cross (1949) commented on the propositions of the unit hydrograph theory:

> All these propositions are empirical. It is not possible to prove them mathematically. In fact, it is a rather simple matter to demonstrate by rational hydraulic analysis that not a single one of them is mathematically accurate. Fortunately, nature is not aware of them.

Nevertheless, many investigators support the view that, at least as a practical tool, the UH theory is useful. This theory is particularly adequate in the range of floods experienced on natural watersheds. In addition, this theory is amenable to linear mathematics, whose methods are simple and best understood. Therefore, it will continue to enjoy its popularity until workable nonlinear methods are developed that are accurate without being unduly complex and costly.

15.3.1 Space Invariance of Effective Rainfall

The ER of a specified duration seldom occurs uniformly over the watershed of a reasonable size. Therefore, the spatial invariance of ER is only an assumption, not a reality. The spatial variation generally becomes more and more pronounced as the size of the watershed increases. Rainfall storms that produce intense rainfall usually do not extend over large areas. A fair semblance of a uniform spatial distribution of rainfall may seldom be obtained on watersheds exceeding 5000 km^2—an upper limit conventionally observed in the application of the UH method. In addition, storm movement, upstream, downstream, or across the watershed may result in nonuniform areal distribution of rainfall.

The nonuniform areal distribution of rainfall can cause variation in hydrograph shape (Dickinson and Ayers, 1965). For example, if high-intensity rainfall occurs over the area near the mouth, then the resulting hydrograph would usually experience a rapid rise, a sharp peak and rapid recession. On the other hand, if this same rainfall occurred over the upstream portion of the watershed, then the resulting hydrograph would usually experience a slow rise, lower and broader peak, and slow recession. However, errors in the UH due to nonuniformity in areal distribution of rainfall can be minimized by decomposing the watershed into subwatersheds, each being small to ensure approximately uniform areal distribution, deriving the UH for each subwatershed, and then appropriately combining these UHs to develop the UH for the entire watershed.

15.3.2 Time Invariance of Effective Rainfall

The ER usually does not occur uniformly, even for as short a duration as 5 minutes. However, the effect of temporal variation of rainfall intensity on the UH shape depends principally on the watershed size. For example, rainfall bursts lasting only a few minutes may produce well-defined peaks in the hydrograph from small watersheds (say, a few hectares in area), but may cause little change in the hydrograph shape from large watersheds (say, several hundred square kilometers in area). Only those changes that last for several hours will cause a distinguishable change in the hydrograph shape on large watersheds (Dickinson and Ayers, 1965).

The first two postulates together show that uniform runoff must occur from all portions of the drainage basin. This condition rarely happens in natural watersheds. For this reason, much consideration must be given to any departure from these limitations, especially on large drainage basins where variations in rainfall intensity in time and space are more likely to occur. Rainfall from cyclonic storms is likely to cover large areas nearly uniformly, whereas convective storms cover only small areas uniformly. A factor that complicates the uniform runoff requirement is the spatial variation of infiltration characteristics on larger drainage basins. Even where soil associations are shown to be of the same group, considerable differences in infiltration rates can exist. Such variations in infiltration may destroy the concept of uniform runoff and results in nonlinear conditions. Thus, the UH theory should be used with caution. Unfortunately, the UH procedure has been used indiscriminately, without satisfying the limiting assumptions. This practice presents the impression that such use is valid and leads to further indiscriminate use.

15.3.3 Validity of Linearity Hypothesis

The principles of proportionality and superposition define linearity of the UH theory. The theory assumes that the ordinates of the DRH are directly proportional to the volume for a given unit storm, thus permiting addition or subtraction of unit hydrographs to obtain the hydrograph for a storm of longer or shorter duration from the data obtained for a given storm. It also follows that the DRH of any unit duration may be made up by addition of successive hydrographs of shorter durations. This

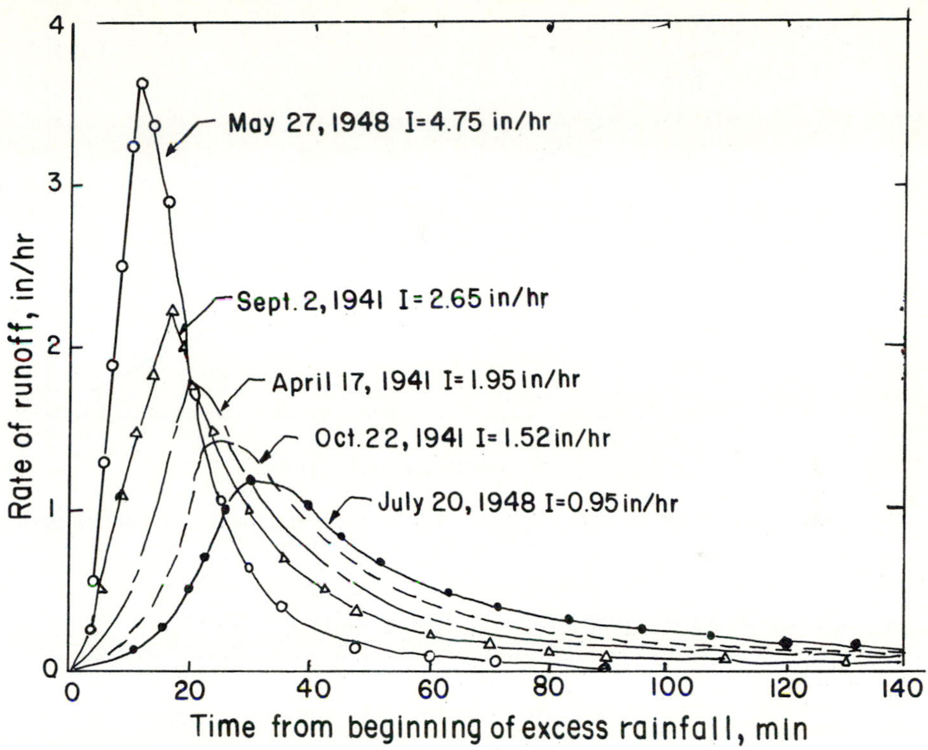

Figure 15.3 Unit hydrographs of watershed W-1 for different rainfall intensities (after Minshall, 1960).

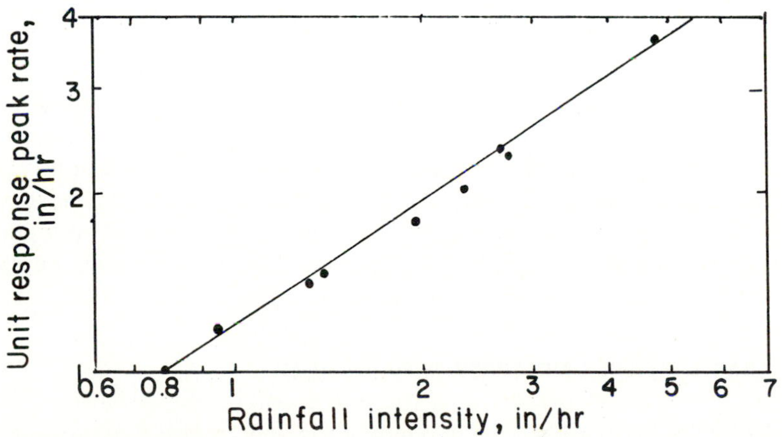

Figure 15.4 Relationship between the unit-hydrograph peak rate and rainfall intensity (after Minshall, 1960).

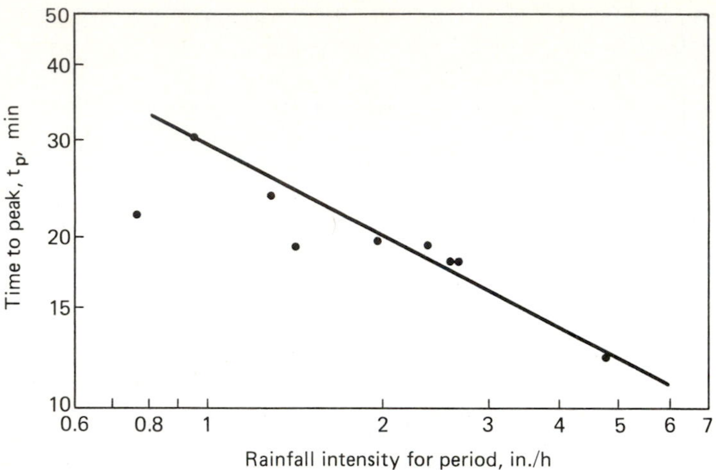

Figure 15.5 Relationship between the time to the unit-hydrograph peak rate and rainfall intensity (after Minshall, 1960).

permits addition or subtraction of hydrographs of various durations to obtain a hydrograph of the desired duration even though such a duration might not have been observed.

All watersheds in nature are nonlinear; some are more nonlinear and some less. They are linear only by assumption. If the hydrographs from the storms of the same duration are compared, it is commonly found that their ordinates are not in proportion to their volumes and that their time bases are not the same. The peaks of the UHs for small rainfall events are usually lower than those for larger ones. Also, the length of the recession depends on the hydrograph peak. Moreover, the watershed linearity requires a linear relationship between storage and discharge, and that the velocity of flow at every point must be constant for all discharges. These conditions are too stringent and are almost never valid.

Perhaps the first conclusive evidence invalidating the linearity hypothesis appeared in a study by Minshall (1960). Since then several investigators have probed the question of linearity, and developed nonlinear models for the rainfall–runoff relationship. Based on his analysis of small watersheds in the Midwestern United States, Minshall (1960) found that one UH could not adequately define the shape of the hydrograph derived from a storm of unit duration. Consequently, different UHs, as shown in Figure 15.3, would be required to represent the watershed runoff response if the rainfall intensity varied. His investigation showed a good relationship between rainfall intensity and the UH peak and the time from the beginning of the ER to the runoff peak, as shown in Figures 15.4 and 15.5.

15.4 INSTANTANEOUS UNIT HYDROGRAPH (IUH)

The difficulty arising from the dependence of the UH on the duration D of the ER is circumvented by letting D be diminished indefinitely. The UH so obtained is called the instantaneous unit hydrograph (IUH). Thus, the IUH, $h(0,t) = h(t)$, is a hypothetical UH due to the ER whose duration tends to zero as a limit, but whose volume remains unity (say, 1 cm). It is then evident that the IUH is independent of the duration of ER. Mathematically,

$$h(t) = h(0,t) = \lim_{D \to 0} h(D,t)$$

and

$$\delta(t) = \lim_{D \to 0} I(t,D)D \tag{15.8}$$

where $\delta(t)$ is the Dirac Delta function defined as

$$\begin{aligned} &\int_{-\infty}^{\infty} \delta(t)\, \mathrm{d}t = 1 \\ &\delta(t) = 0, \qquad t \neq 0 \end{aligned} \tag{15.9}$$

Physically, this function can be thought of as a spike of infinitesimally small thickness and infinitely large height such that the area under the spike is 1. Therefore, Equation (15.7) can be written as

$$Q(t) = \int_0^t h(t - \tau)I(\tau)\, \mathrm{d}\tau, \qquad h(\tau) = 0 \qquad \text{for } \tau < 0 \tag{15.10}$$

Because the mathematical operations are linear, Equation (15.10) can be expressed in an alternative form as

$$Q(t) = \int_0^t h(\tau)I(t - \tau)\, \mathrm{d}\tau \tag{15.11}$$

with

$$h(t) \geq 0 \qquad \text{for any } t \geq 0$$

$$\lim_{t \to \infty} h(t) = 0$$

$$\int_0^{\infty} h(s)\, \mathrm{d}s = 1$$

Experimental evidence shows that

$$\int_0^{\infty} (t - \bar{t})^3\, h(t)\, \mathrm{d}t > 0 \tag{15.12}$$

where $\bar{t}$ is the centroid of the IUH located at the t-axis:

$$\bar{t} = \int_0^{\infty} th(t)\, \mathrm{d}t \tag{15.13}$$

Equation (15.12) states that the IUH must be skewed to the right, that is, its peak should be to the left of the centroid with a long tail extending to the right.

15.5 S-HYDROGRAPH (SH)

If the duration of ER is indefinitely long and its intensity is one unit per unit of time (say, 1 cm/h), then the hydrograph so obtained is termed the summation hydrograph, S-hydrograph, or simply SH. This hydrograph assumes a deformed S-shape and its ordinates ultimately approach the rate of ER in the limit or at the time of equilibrium. If the ERH is broken down into pulses each of 1-hour duration and 1 cm/h intensity, then the SH will be obtained by superposing the DRHs due to each of these pulses. Since each pulse satisfies what is required for 1-hour UH, the DRH of this pulse and the UH will be numerically identical. For this ERH, the SH, $U(t)$, is obtained by simply replacing I_jD by 1 in Equation (15.7) as

$$U(t) = \sum_{j=1}^{t} h(t,t-(j-1)) = \sum_{j=1}^{t} Q_j(t-(j-1)) = \sum_{j=1}^{t} Q(t-(j-1)) \quad (15.14)$$

where $Q_j(t-j) = Q(t-j)$ represents the DRH due to the jth pulse (either more or less). However, if the intensity of each pulse of duration D hours is other than unity, then the resulting DRH will not have a unit amount. Therefore,

$$U(t) = \frac{1}{D}\sum_{j=1}^{t} Q(D,t-(j-1)D) = \sum_{j=1}^{t} h(D,t-(j-1)D) \quad (15.15)$$

Clearly, if $D \to 0$, $h(D,t-(j-1)D) \to h(t)$, the lower limit for summation will approach the origin, and

$$U(t) = \int_0^t h(s)\,\mathrm{d}s \quad (15.16)$$

15.6 RELATION BETWEEN IUH, UH AND SH

If the definitions of the IUH, UH, and SH are closely scrutinized, connections between them become apparent. It is clear from the definitions of the UH and IUH that

$$h(D,t) = \frac{1}{D}\int_{t-D}^{t} h(s)\,\mathrm{d}s \quad (15.17)$$

UH and SH are related by

$$h(D,t) = \frac{1}{D}[U(t) - U(t-D)] \quad (15.18)$$

As $D \to 0$,

$$h(t) = \frac{\mathrm{d}U(t)}{\mathrm{d}t} \quad (15.19)$$

It may be didactic to summarize that the three hydrographs—IUH, UH, and SH—are ascribed to three distinct characteristics of effective rainfall—volume, duration, and intensity—as shown in Table 15.1. Furthermore, it may be noted that the

TABLE 15.1 EFFECTIVE RAINFALL CHARACTERISTICS FOR HYDROGRAPHS

	Effective rainfall characteristics		
Hydrograph	Volume	Duration	Intensity
IUH	Unity	Zero	Indefinite
UH	Unity	Finite	1/Duration
SH	Indefinite	Indefinite	Unity

time to the peak of $h(D,t)$ can be determined by differentiating Equation (15.18) with respect to t, and equating the derivative to zero:

$$h(t) - h(t - D) = 0 \tag{15.20}$$

This says that the peak of $h(D,t)$ occurs at a time when the ordinate of the IUH is equal to the ordinate at D time units earlier. The SH attains its maximum value at a time equal to D hours less than the time base of the initial $h(D,t)$.

15.7 DIMENSIONAL CONSIDERATIONS

The dimensions of the UH can be expressed in two ways. If the UH is derived from a DRH due to the ER of specified duration whose depth is not unity, then the dimensions of the UH ordinates will be equal to the dimension of the runoff divided by the dimensions of the depth of ER. Runoff, however, can be expressed either in terms of discharge having volumetric rate units or specific discharge having volumetric rate per unit area units. Therefore, the dimensions of the UH ordinates are either the dimensions of discharge per unit depth or the dimensions of discharge per unit area per unit depth. Let L denote the dimension of length and T the dimension of time. Then the two forms of the dimensions of the UH ordinates are

(a) $\dfrac{L^3/T}{L} = \dfrac{L^2}{T}$

(b) $\dfrac{L^3/T}{L^2L} = \dfrac{1}{T}$

As pointed out by Diskin (1979), it is erroneous to assign the dimension of discharge (either L^3/T or L/T) to the UH ordinates. Proper understanding of the dimension of the UH ordinates is particularly useful in synthetic UH computations.

15.8 CHANGING THE DURATION OF THE UNIT HYDROGRAPH

Once a UH is obtained from a storm of duration D, the UH of duration equal to any integral multiple of D can be easily obtained by using the principle of superposition. Figure 15.6 shows the UH of duration D repeated three times. A simultaneous addition of ordinates using the principle of superposition will result in a runoff

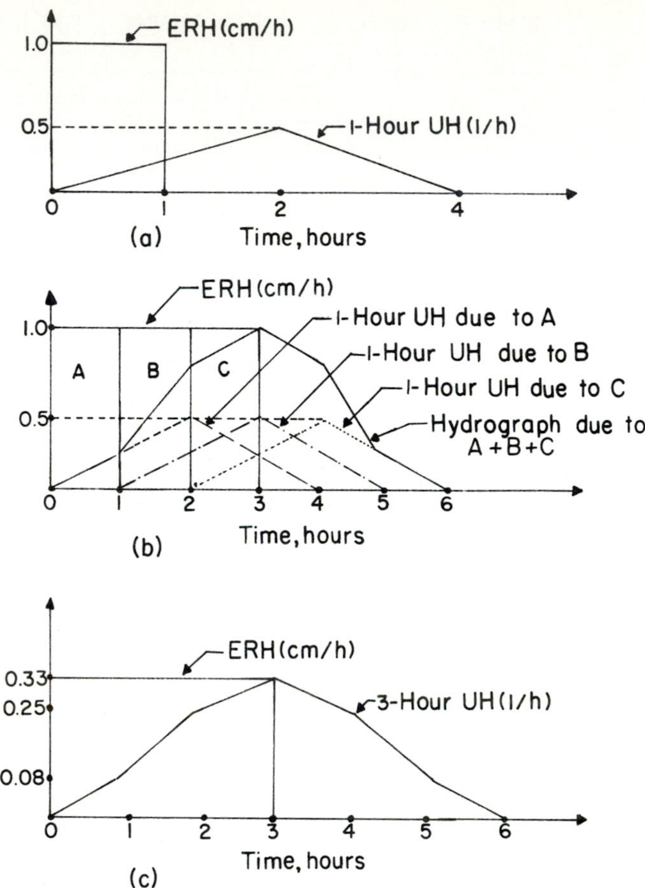

Figure 15.6 Derivation of a 3-hour UH from a 1-hour UH.

hydrograph having three units of volume as a result of the storm having the duration of three units of time ($3D$). Dividing the ordinates of this hydrograph by 3 will give rise to the $3D$ UH, that is,

$$h(D_*,t) = \tfrac{1}{3}[h(D,t) + h(D,t - D) + h(D,t - 2D)] \tag{15.21}$$

where $D_* = 3D$. Equation (15.21) can be generalized as

$$h(D_*,t) = \frac{1}{n}\sum_{j=1}^{n} h(D,t - (j - 1)D), \qquad D_* = nD \tag{15.22}$$

This method will, however, not work if the UH is required for a duration other than an integral multiple of the given duration. The solution in such cases can be obtained by resorting to the SH. This is referred to as the SH method, and involves the following steps, as shown in Figure 15.7:

1. Construct the SH using the given UH. This SH is assumed due to a continuous ER occurring at a constant rate of 1 unit depth/unit time (say, 1 cm/h). This follows:

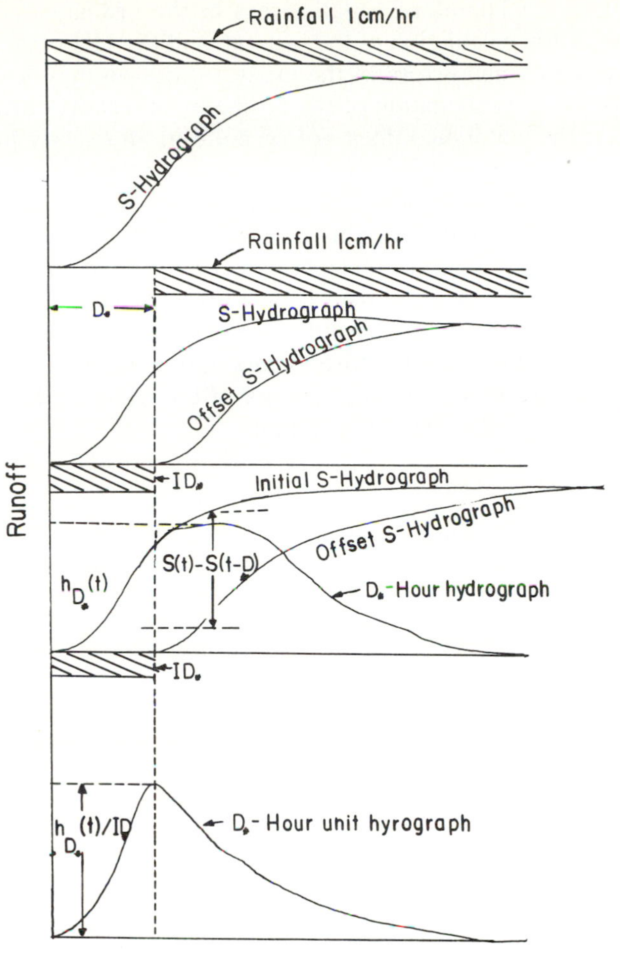

Figure 15.7 Derivation of the unit hydrograph from the S-hydrograph.

$$U(t) = D[h(D,t) + h(D,t - D) + h(D,t - 2D) + \cdots] \cdots \quad (15.23a)$$

$$= D \sum_{j=1}^{t} h(D,t - (j - 1)D) \quad (15.23b)$$

2. Advance or offset the position of the SH by a period equal to the desired duration of D_* hours. Call this SH as an offset SH, expressed by $U(t - D_*)$.
3. Compute the difference between the ordinates of these two SHs, that is, $U(t) - U(t - D_*)$.
4. Divide the differences in the ordinates by D_*. This gives the desired UH of duration D_*:

$$h(D_*,t) = [U(t) - U(t - D_*)]/D_* \quad (15.24)$$

A word of caution is appropriate to mention here. If possible, the derivation of a short-period UH from one of longer duration should be avoided. From Equation (15.24), it is seen that the process of deriving a greater-duration UH from a shorter one entails averaging of ordinates. Therefore, small errors in the ordinates of the short-duration UH are ironed out in calculation of the ordinates of the UH of greater duration. On the other hand, any small errors in the greater duration may lead to disproportionately large errors in the calculation of the ordinates of a short-duration UH. This would indicate that the watershed does not respond in an exactly linear manner.

Frequently, the SH constructed from a given UH becomes wobbly when it approaches unity. If the UH of a desired duration is calculated from this SH, its ordinates may be rapidly fluctuating, sometimes becoming even negative. This may be because the UH we started with is not the true UH. Further, the SH, strictly speaking, may not be continuously known. This would also correspond to the duration of the UH from which it was derived. Therefore, errors in the UH will result in oscillations in the SH. This may again indicate nonlinearity of the watershed. Measurement and data errors may also cause these oscillations.

EXAMPLE 15.1 A rainfall–runoff event is selected to derive a 2-hour unit hydrograph for a watershed. The corresponding ERH and DRH have been computed as

$$I(t) = 5 \text{ cm/h}, \qquad 0 \le t \le 2$$
$$= 0, \qquad t \ge 2$$
$$Q(t) = 2.5t, \qquad 0 \le t \le 2$$
$$= 10 - 2.5t, \qquad 2 \le t \le 4$$

where t is measured in hours. Determine the 2-hour UH. Determine the S-hydrograph for the watershed. Compute the 3-hour and 4-hour UHs.

Solution Since $I(t)$ satisfies the assumptions required by the UH theory, the 2-hour UH can be obtained by simply dividing $Q(t)$ by its volume, which is 10 cm. Therefore,

$$h(2,t) = 0.25t, \qquad 0 \le t \le 2$$
$$= 1 - 0.25t, \qquad 2 \le t \le 4$$

To determine $U(t)$, we first need to determine $Q(t)$ as a result of $I(t)$, whose intensity is 1 cm/h. The duration of $I(t)$ is, of course, 2 hours. This is given as

$$Q(2,t) = 0.5t, \qquad 0 \le t \le 2$$
$$= 2 - 0.5t, \qquad 2 \le t \le 4$$

Numerically, $Q(2,t)$ is the same as $h(2,t)$. Then the S-hydrograph is given as

$$U(t) = 0.5t, \qquad 0 \le t \le 2$$
$$= 1, \qquad t \ge 2$$

By using $U(t)$, we can compute the 3-hour UH. To this end, we compute the DRH due to the 3-hour ERH, which can be expressed as

$$Q(3,t) = U(t) - U(t - 3)$$

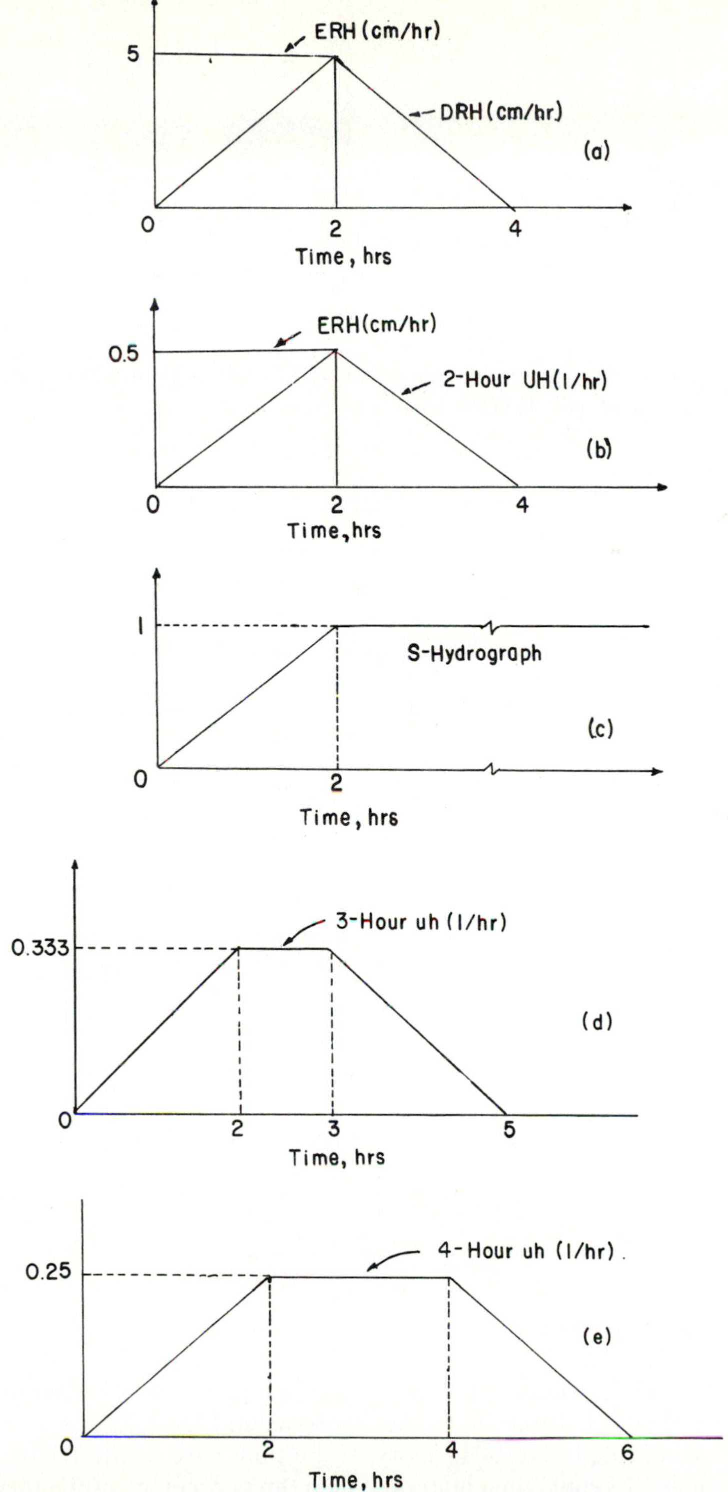

Figure 15.8 (a) Effective rainfall and direct runoff, (b) 2-hour UH, (c) S-hydrograph, (d) 3-hour UH, and (e) 4-hour UH.

Therefore,

$$Q(3,t) = 0.5t, \qquad 0 \le t \le 2$$
$$= 1 \qquad 2 \le t \le 3$$
$$= 2.5 - 0.5t, \qquad 3 \le t \le 5$$

Therefore, the 3-hour UH is

$$h(3,t) = t/6, \qquad 0 \le t \le 2$$
$$= 1/3, \qquad 2 \le t \le 3$$
$$= (5/6) - (t/6), \qquad 3 \le t \le 5$$

Likewise, we can compute the 4-hour UH from the SH or directly from the 2-hour UH. In any event, this is given by

$$h(4,t) = 0.125t, \qquad 0 \le t \le 2$$
$$= 0.25 \qquad 2 \le t \le 4$$
$$= 0.75 - 0.125t, \qquad 4 \le t \le 6$$

These hydrographs are sketched in Figure 15.8.

EXAMPLE 15.2 A 1-hour unit hydrograph is given as a triangle whose base is 3 hours, peak is $\frac{2}{3}$ cm/h, and the times of rise and recession, respectively, are 1 h and 2 h. Compute a 2-hour unit hydrograph using the S-hydrograph method.

Solution First, the SH is constructed as shown in column (3) in the table that follows. Then this SH is offset by 2 h to obtain the 2-h SH. By taking the differences between ordinates of these two SHs, the 2-h hydrograph is obtained, which when divided by 2 cm produces the 2-h UH as shown in the table.

Time (1)	1-h UH (1/h) (2)	SH (cm/h) (3)	2-h Offset SH (cm/h) (4)	2-h Hydrograph (cm/h) (5)	2-h UH (1/h) (6)
0	0	0	0	0	0
1	2/3	2/3	0	2/3	1/3
2	1/3	1	0	1	2/3
3	0	1	2/3	1/3	1/6
4	0	1	1	0	0
5	0	1	1	0	0
6	0	1	1	0	0
7	0	1	1	0	0

15.9 DISTRIBUTION GRAPH

Bernard (1935) expressed the UH concept in terms of the distribution graph (DG). A DG is a UH of direct-runoff volume, as shown in Figure 15.9, and the area under it is 100%. According to the UH theory, if the time base of the UH is divided into any given number of equal time intervals, then the percentage of the total volume of flow

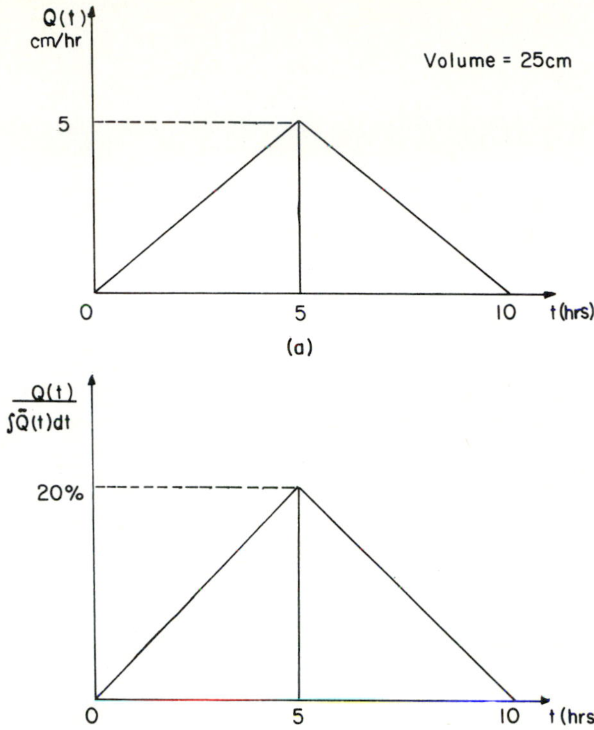

Figure 15.9 (a) A direct-runoff hydrograph and (b) a distribution graph.

occurring during a given time interval will approximately be the same irrespective of the amount of total runoff. The DG will normally be more stable than the UH for a given watershed. Its shape may vary from one watershed to another, reflecting the differences in their runoff characteristics. Therefore, the DG may be preferable to the UH for comparing hydrograph characteristics from areas of different sizes.

15.10 USES OF THE UNIT HYDROGRAPH

The unit hydrograph of a specified duration has many uses in environmental and water-resource development. It may not be necessary to provide an exhaustive catalog of these uses here, but it will suffice to mention a couple of illustrative uses.

The unit hydrograph can be used to determine the watershed response due to a given rainfall event if an estimate can be made of the abstractions not contributing to the runoff. This has immediate application in flood forecasting and warning. With knowledge of the time distribution of runoff response, both the flood peak and its time of occurrence are known. For the maximum possible rainfall, the maximum possible flood can be determined. Thus, the UH can serve as an indicator of the flood-producing characteristics of a given watershed.

Another important application is in the determination of the effect of flood-protection works on watershed response. This can be done by comparing the UH of a stated duration derived prior to the protection works with the UH of the same

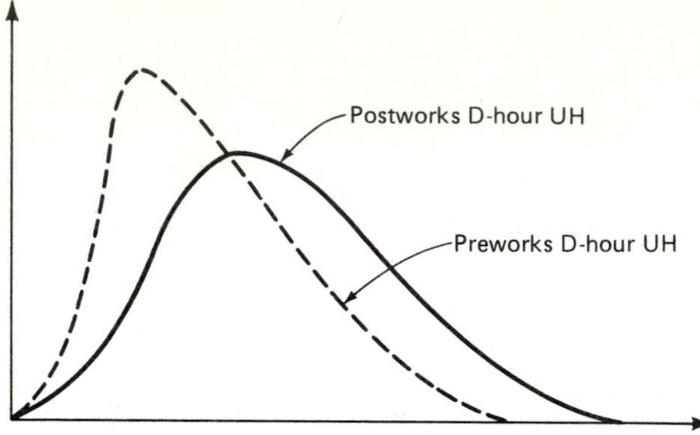

Figure 15.10 Comparison of pre- and postworks unit hydrographs of a stated duration.

duration derived after the completion of the works, as shown in Figure 15.10. This comparison helps in the proper management of water resources.

Still another application is in determining soil erosion from upland watersheds. If sediment concentration for a given watershed is known, the unit sediment graph for a specified duration can be constructed (Rendon-Herrero, 1978; Williams, 1978; Singh et al., 1982). This, in turn, can be utilized to estimate erosion due to a specified rainfall on the watershed.

15.11 DURATION OF THE UNIT HYDROGRAPH

Choosing the duration of the UH is determined more by the availability of data than by the problem and the watershed under consideration. If only daily records of rainfall and streamflow are available, there is little option in choosing the duration. However, if rainfall and streamflow are available at shorter intervals of time, say, $\frac{1}{2}$ h, 1 h, or so, then a convenient duration can be selected not to exceed the least value of (1) the basin lag, (2) the time to peak, and (3) the time of concentration. A recommended duration is about one-fourth of the basin lag. In general, a duration of 12 h is used for watersheds larger than 1200 km^2. These are based on heuristics and practical experience. In a particular watershed, significant departures may be found. Hence, caution is recommended for using these guidelines.

15.12 DERIVATION OF UH AND IUH FROM MEASURED DATA

There are several methods to derive the UH that are also applicable for deriving the IUH. When the duration of the UH is very short, then the UH and IUH are practically indistinguishable from each other. In practice, the question arises: How short should this duration be? The answer to this question depends on the shape of the particular UH. Approximately, this can be determined as follows. Assume that the calculated UH of duration D is the IUH. Use Equation (15.17) to calculate the corresponding UH of duration D. If this UH does not significantly differ from the original UH, the assumption is justified. If the difference is significant, the dissimilarity may suggest the probable shape of the IUH, which may be sketched and tested as before. However, we must bear in mind that the watersheds are usually nonlinear and time-variant. Some of the difference will be due to these attributes. Therefore, if the difference is within a few percent, further corrections may not be warranted.

15.12.1 Single-Event Method

The single-event method requires choosing a single, isolated, short-duration, high-intensity, flood-producing storm that is reasonably uniform throughout its duration and over the watershed. The duration of the storm should be between 10 to 30% of basin lag or time to peak. The advantages of using a short-duration storm are as follows: (1) The variations in the time distribution of intensity of ER are much less. (2) Determination of ER is much easier. (3) It is more reliable to derive the UH of a storm of longer duration than from one of shorter duration. However, such short-duration, flood-producing storms are not a common occurrence. The selected storm is then analyzed. By computing abstractions, the ERH is obtained and its duration noted. This ERH should be approximately uniform.

The measured runoff hydrograph is analyzed next. (1) The baseflow is separated from the total runoff hydrograph using one of the methods discussed in Chapter 12 to obtain the direct-runoff hydrograph, which is then tabulated and plotted, as shown in Figure 15.11. (2) The volume of runoff under the DRH is then computed and converted to the equivalent depth (in centimeters or inches) distributed uniformly over the entire drainage basin. This can be done by planimetering or by counting squares on a graph sheet. (3) The ordinates of the hydrograph are divided by the runoff volume (in inches or centimeters) to obtain the UH, as shown in Figure 15.11, which is for the duration of ER. For the following example, the ordinates of the UH for a 1-in. (2.54-cm) runoff are twice those for the DRH of a 0.5-in. (1.27-cm)

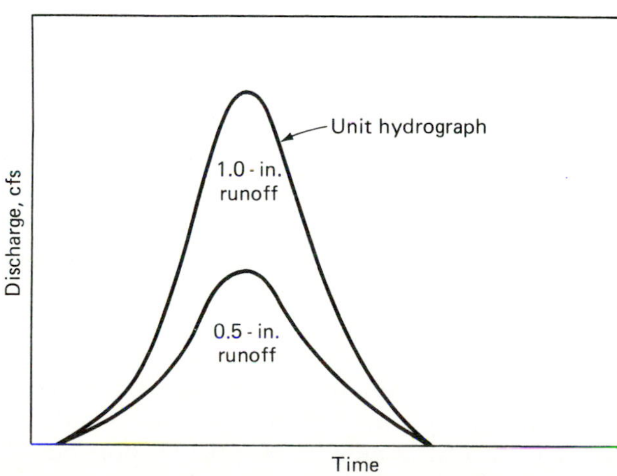

Figure 15.11 Unit hydrograph from a surface-runoff hydrograph of 0.5 in.

TABLE 15.2 RAINFALL, RUNOFF, AND LOSS DATA FOR THE STORM OF MAY 30–31, 1948, ON BUCKHORN CREEK NEAR MASONVILLE, COLORADO[a]

Time (1)	Rainfall (in.) (2)	Loss (in.) (3)	Effective rainfall (in.) (4)	Total runoff (cfs) (5)	Baseflow (cfs) (6)	Direct runoff (cfs) (7)
7:00–7:20 P.M.	0.34	0.34	0.0	40	40	0
–7:40	0.43	0.43	0.0	50	50	0
–8:00	1.52	1.26	0.26	120	50	70
–8:20	0.70	0.70	0.0	1021	55	966
–8:40	0.37	0.37	0.0	2450	55	2395
–9:00	0.11	0.11	0.0	4050	60	3990
–9:20	0.04	0.04	0.0	4150	60	4090
–9:40	0.00	0.00	0.0	2860	65	2795
–10:00	0.04	0.04	0.0	1790	65	1725
–10:20	0.00	0.00	0.0	1150	70	1080
–10:40	0.00	0.00	0.0	810	70	740
–11:00	0.00	0.00	0.0	600	75	525
–11:20	0.00	0.00	0.0	490	75	415
–11:40	0.00	0.00	0.0	410	80	330
–12:00	0.00	0.00	0.0	340	80	260
–12:20	0.00	0.00	0.0	280	85	195
–12:40	0.00	0.00	0.0	240	85	155
–01:00	0.00	0.00	0.0	210	90	120
–01:20	0.00	0.00	0.0	180	90	90
–01:40	0.00	0.00	0.0	160	95	65
–02:00	0.00	0.00	0.0	140	95	45
–02:20	0.00	0.00	0.0	120	100	20
–02:40	0.00	0.00	0.0	110	100	10
–03:00	0.00	0.00	0.0	100	100	0

[a] Drainage area = 39.9 mi^2.

runoff volume. If all the limiting assumptions are approximately satisfied, the resulting UH is valid.

EXAMPLE 15.3 The data obtained from a storm that occurred May 30–31, 1948, on Buckhorn Creek near Masonville, Colorado, are given in Table 15.2. Compute the unit hydrograph for this storm. What is the period of this unit hydrograph?

Solution When losses are subtracted from rainfall, the ERH is obtained as shown in columns (4) of Table 15.2. The baseflow separation produces the DRH, which is given as column (7). This DRH, however, corresponds to the ER volume of 0.26 in. Therefore, to obtain the UH, the DRH ordinates are divided by 0.26 in., as shown in column (4) of Table 15.3. The period of the UH is 20 min.

15.12.2 Representative UH Method

Because it is often difficult to be sure that the limiting assumptions of the UH are satisfied, it is a prudent practice to prepare several UHs for the same unit duration and derive a representative UH. To that end, a number of isolated short-duration,

TABLE 15.3 COMPUTATION OF THE UH FOR EXAMPLE 15.2

Time (1)	Effective rainfall (in.) (2)	Direct runoff (cfs) (3)	UH (cfs/in.) (4) = (3)/0.26
7:00–7:20 P.M.	0.0	0	
–7:40	0.0	0	0
–8:00	0.26	70	269
–8:20	0.00	966	3,715
–8.40		2395	9,211
–9:00		3990	15,346
–9:20		4090	15,730
–9:40		2795	10,750
–10:00		1725	6,634
–10:20		1080	4,154
–10:40		740	2,846
–11:00		525	2,019
–11:20		415	1,596
–11:40		330	1,269
–12:00		260	1,000
–12:20		195	750
–12:40		155	596
–01:00		120	462
–01:20		90	346
–01:40		65	250
–02:00		45	173
–02:20		20	77
–02:40		10	38
–03:00		0	0

high-intensity, flood-producing storms are selected. These storms should be of approximately the same duration, with ±10% variation. Other qualifications to be satisfied by the selected storms are as follows: (1) Storms are isolated and occur individually. (2) Storms are approximately uniform during their duration and over the watershed. By definition, the ER of each storm should be uniform. It is implied that these storms occur over the entire watershed, not a portion thereof. This limits the application of the UH theory to watersheds less than 5000 km^2. (3) The storms should be flood-producing storms. In other words, the ER of each storm is high. A range of ER values of 1.0 to 5.0 cm is sometimes suggested. (4) The duration of storm rainfall should be approximately $\frac{1}{5}$ to $\frac{1}{3}$ of the basin lag. This is also approximately the duration of the UH. This implies that the duration of the UH is related to the size of the watershed, for the basin lag is related to the size. For watersheds 250 km^2 or larger, a duration of 6 hours is generally satisfactory. (5) A suitable number of storms is approximately 5. However, if more storms are available that satisfy the previous qualifications, then they all should be used.

For each storm, the measured rainfall hyetograph and runoff hydrograph are analyzed following the same method as in the single-event method discussed before. The UH is then derived for each storm. The UHs thus derived are plotted on the graph. If the UHs are not of the same duration, they should be changed, using the S-hydrograph method, to the same duration. If the limiting assumptions are satisfied, all of the resulting UHs will be identical or nearly so. If the limiting assumptions are not satisfied, the resulting UHs will be different. Such different UHs are often averaged to obtain a mean or representative UH. To derive the mean UH, the averages of peak ordinates and their times are computed first. A mean UH of best fit, judged visually, is drawn through the average peak for an assumed base length. If the area under this UH is different from unity, an adjustment is made through an adjustment of the peak value.

This averaging should be valid providing the differences in the hydrographs are small. However, if the derived UH peaks are higher for small measured volumes than those for larger measured volumes, this indicates that the drainage basin is nonlinear and should not be used for the UH procedure. Barnes (1959) discovered that unit hydrographs prepared from small measured volumes did not convert to reasonable unit hydrographs. As a result, he recommended that hydrographs of at least 0.5 in. (1.25 cm) be used to prepare a UH. This observation is consistent with drainage-basin nonlinearity and is probably an indication of such a condition. The effect of limiting the usable hydrograph volumes to near 1-in. (or 1-cm) runoff volume reduces the variation in the several derived UHs that are averaged to obtain a representative UH for use. The UH of a similarly adjusted hydrograph for any predicted runoff volume may be added to an estimated baseflow to obtain a total hydrograph for design purposes.

EXAMPLE 15.4 Compute the peak of a 4-h UH if the following data are available. The effective rainfall of 4-h duration has produced a flood-hydrograph peak of 400 m^3/s. This flood hydrograph has a baseflow of 40 m^3/s. The average watershed rainfall is 4 cm, and average loss rate is 0.4 cm/h.

Solution

$$\text{Duration of the effective rainfall} = 4 \text{ h}$$

$$\text{Amount of mean areal rainfall} = 4 \text{ cm}$$

$$\text{Amount of loss} = 0.4 \times 4 = 1.6 \text{ cm}$$

$$\text{Amount of effective rainfall} = 4 - 1.6 = 2.4 \text{ cm}$$

$$\text{Flood-hydrograph peak} = 400 \text{ m}^3/\text{s}$$

$$\text{Baseflow} = 40 \text{ m}^3/\text{s}$$

$$\text{Peak of the direct-runoff hydrograph} = 360 \text{ m}^3/\text{s}$$

$$\text{Peak of 4-h UH} = 360/2.4 = 150 \text{ m}^3/\text{s-cm}$$

EXAMPLE 15.5 The following are the data for a 4-h unit hydrograph for a basin of 32.5 km^2. Construct the S-hydrograph and derive the 2-h and 6-h unit hydrographs.

Time (h)	Flow (m^3/s-cm)	Time (h)	Flow (m^3/s-cm)
0	0	12	52
1	9.5	13	43
2	59.5	14	33
3	105	15	26
4	143	16	19
5	167	17	14
6	145	18	10
7	124	19	5
8	107	20	2.5
9	90	21	0
10	76		
11	64		

TABLE 15.4 RESULTS OF THE CALCULATIONS FOR THE S-HYDROGRAPH

Time (h)	UH (m^3/s-cm)	Lagged UHs (m^3/s-cm)				S-curve (m^3/s)
(1)	(2)	(3)	(4)	(5)	(6)	(2) + · · · + (6)
0	0					0
1	9.5					9.5
2	59.5					59.5
3	105					105
4	143	0				143
5	167	9.5				176.5
6	145	59.5				204.5
7	124	105				229
8	107	143	0			250
9	90	167	9.5			266.5
10	76	145	59.5			280.5
11	64	124	105			293
12	52	107	143	0		302
13	43	90	167	9.5		309.5
14	33	76	145	59.5		313.5
15	26	64	124	105		319
16	19	52	107	143	0	321
17	14	43	90	167	9.5	323.5
18	10	33	76	145	59.5	323.5
19	5	26	64	124	105	324
20	2.5	19	52	107	143	323.5
21	0	14	43	90	167	
22		10	33	76	145	
23		5	26	64	124	
24		2.5	19	52	107	
25		0	14	43	90	
26			10	33	76	
27			5	26	64	
28			2.5	19	52	
29			0	14	43	
30				10	33	

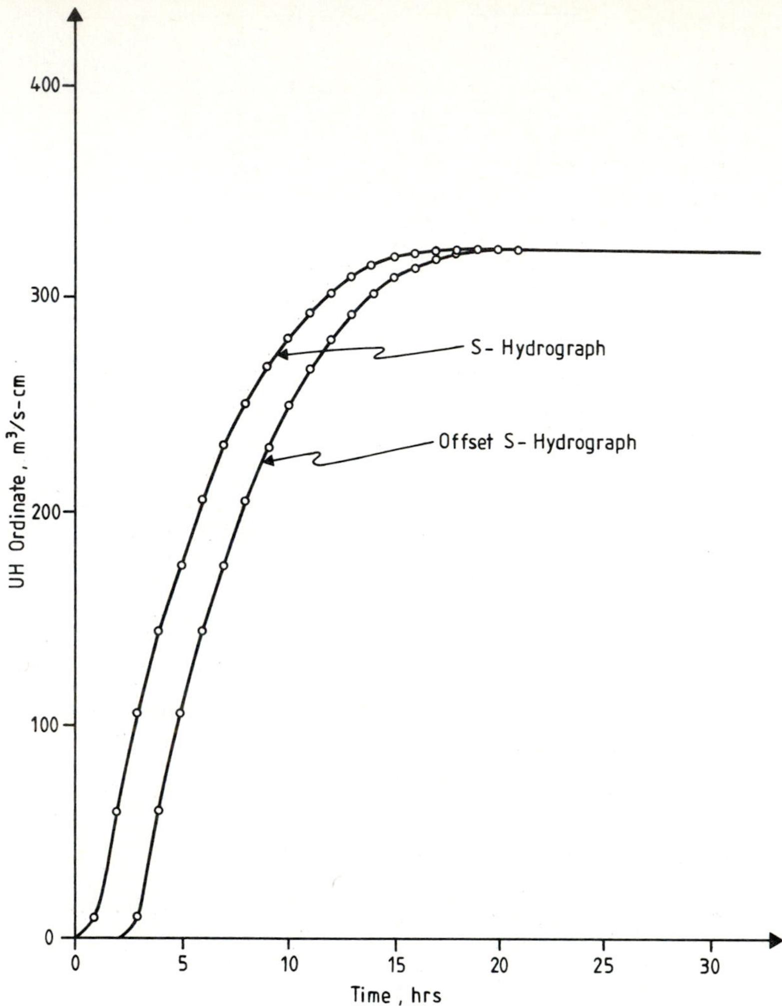

Figure 15.12 Construction of a S-hydrograph from a 4-hour UH.

Solution The S-hydrograph is constructed by adding a string of the unit hydrographs each lagged by 4 hours from one another. The results of the calculations are shown in Table 15.4. The S-hydrograph so obtained is shown in Figure 15.12.

In order to derive the 2-hour and 6-hour unit hydrographs, the S-hydrograph is offset by 2 hours, as shown in Figure 15.12. The difference between ordinates of the S-curve and the offset S-curve is noted. This difference is then multiplied by 2 to obtain the 2-hour unit hydrograph. The results of the calculations are shown in Table 15.5. The 6-hour unit hydrograph is obtained in a similar fashion, and the results of the calculations are shown in Table 15.5. The 2-h and 6-h unit hydrographs are shown in Figure 15.13.

TABLE 15.5 CALCULATIONS FOR 2-HOUR AND 6-HOUR UNIT HYDROGRAPHS

Time (h) (1)	S curve (m^3/s) (2)	2-h lagged S curve (m^3/s) (3)	(2) − (3) = (4)	2-h UH (m^3/s-cm) (2) × (4) = (5)	6-h lagged S curve (m^3/s) (6)	(2) − (6) = (7)	6-h UH (m^3/s-cm) (7)/1.5 = (8)
0	0		0	0		0	0
1	9.5		9.5	19		9.5	6.3
2	59.5	0	59.5	119		59.5	39.7
3	105	9.5	95.5	191		105	70
4	143	59.5	83.5	167		143	95.3
5	176.5	105	71.5	143		176.5	117.7
6	204.5	143	61.5	123	0	204.5	136.3
7	229	176.5	52.5	105	9.5	219.5	146.3
8	250	204.5	45.5	91	59.5	190.5	127
9	266.5	299	37.5	75	105	161.5	108
10	280.5	250	30.5	61	143	137.5	92
11	293	266.5	26.5	53	176.5	116.5	78
12	302	280.5	21.5	43	204.5	97.5	65
13	309.5	293	16.5	33	229	80.5	54
14	313.5	302	11.5	23	250	63.5	42
15	319.0	309.5	9.5	19	266.5	52.5	35
16	321	313.5	7.5	15	280.5	40.5	27
17	323.5	319.0	4.5	9	293	30.5	20
18	323.5	321	2.5	5	302	21.5	14
19	324.0	323.5	0.5	1	309.5	14.5	10
20	323.5	323.5	0	0	313.5	10	7
21	323.5				319.0	4.5	3
22	323.5				321	2.5	2
23	323.5				323.5	0	0
24	323.5				323.5		
25	323.5				324.0		
26	323.5				323.5		
27	323.5				323.5		

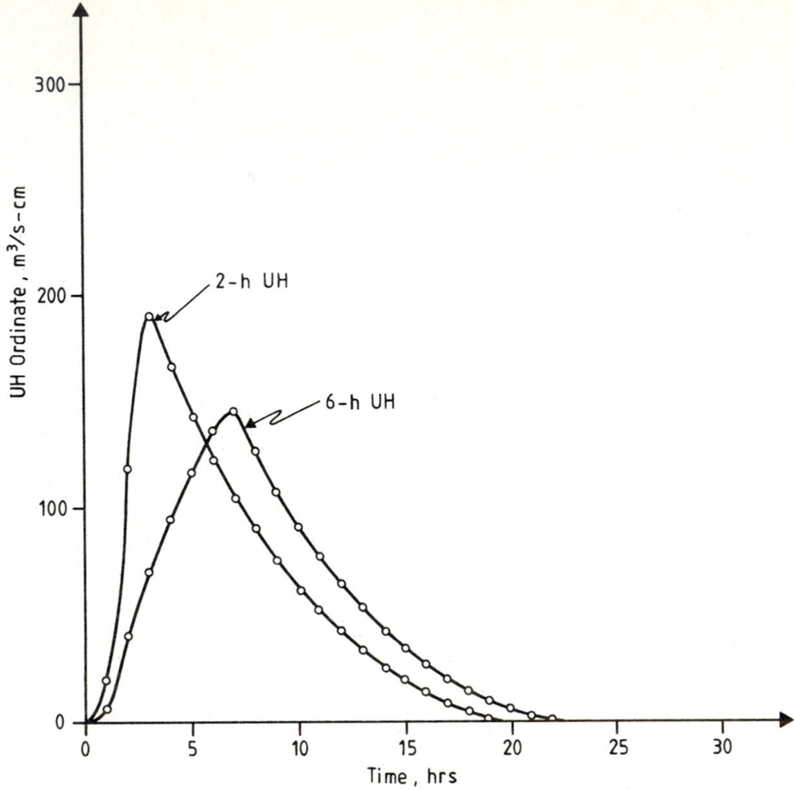

Figure 15.13 Two-hour and 6-hour unit hydrographs.

15.12.3 UH from Complex Storms

Often simple, suitable isolated storms are not available. It may then be necessary to use complex storms for derivation of the UH. If the storm pattern is such that it can be separated into isolated periods of rainfall, then the UHs can be derived for all the periods using hydrograph separation. These can then be combined in a suitable manner to obtain a single average UH. However, a complex storm often consists of consecutive periods of varying rainfall amounts and the effects of these amounts cannot be isolated. Under these conditions, two methods of deriving the UH are discussed here.

COLLINS' METHOD

The Collins method (Collins, 1939) is a trial-and-error procedure. The basic steps of the procedure are as follows:

1. Determine the duration for which the UH is to be derived. This period should usually be equal to or less than one-fourth of the period of rise of the DRH.

2. Determine the time base of the UH, which is equal to the time from the end of the ER to the end of the DRH plus the duration of the UH.
3. Assume a UH with these qualifications. The easiest way is to take the UH as a rectangular pulse over its time base. Another way is to assume it as a triangle, preferably with its rise over one-third and recession over two-thirds of the time base.
4. Apply the assumed UH to all effective rainfall amounts except the largest. Compute the DRH.
5. Subtract this DRH from the observed DRH. Adjust the residuals to the UH terms so they have a unit volume. This yields the first trial UH.
6. Determine the initial UH for the second trial, and start a second round of adjustment by keeping, of course, the unit volume. This can be done by averaging the initial and first trial UHs as

$$\bar{h} = \frac{h^0 + h^1 F}{1 + F} \tag{15.25}$$

where h^0 is the ordinate of the initial UH, h^1 is the ordinate of the first trial UH, $\bar{h}$ is the ordinate of the initial UH for the second trial, and

$$F = \frac{\Sigma(Q - \Sigma I h^0)}{\Sigma\Sigma I h^0} \tag{15.26}$$

where I is the effective rainfall amount, and Q is the actual observed discharge.
7. Repeat this procedure until the differences between the UHs of successive trials are within acceptable limits. The procedure is illustrated in the following example.

EXAMPLE 15.6 A rainfall–runoff record is given in Table 15.6 for watershed C, Riesel (Waco), Texas. Its drainage area is 2.343 km^2. Derive a UH for this multiperiod storm by the Collins method.

Solution

1. The DRH is given so baseflow separation is not needed. Calculate the amount of direct runoff:

$$\text{Depth of runoff} = \frac{\text{volume}}{\text{watershed area}} = 3.36 \text{ cm}$$

2. Calculate the infiltration rate by trial and error, knowing that the total effective rainfall must equal 3.36 cm. Infiltration is computed using the Philip two-term infiltration equation for which the parameters are $A = 0.254$ cm/h and $S = 0.522$ cm/h$^{0.5}$. Rainfall infiltration and effective rainfall are shown in Table 15.7. This gives a total infiltration of 0.86.
3. Rainfall–runoff data are given at varying time intervals. Therefore, a time interval of 20 minutes is selected. Effective rainfall–direct runoff data are then interpolated at this time interval, as given in Table 15.8. The direct-runoff ordinates are expressed in cm/h.

TABLE 15.6 DATA FOR THE RAINFALL–RUNOFF EVENT OF JUNE 10, 1941, ON WATERSHED C, RIESEL (WACO), TEXAS

Rainfall

Time (min)	Intensity (cm/h)	Time (min)	Intensity (cm/h)	Time (min)	Intensity (cm/h)	Time (min)	Intensity (cm/h)
0.0	3.05	4.0	1.52	6.0	3.56	9.0	1.52
11.0	1.91	15.0	3.81	19.0	3.66	24.0	5.33
28.0	5.08	31.0	10.16	34.0	7.62	42.0	4.57
45.0	2.03	48.0	3.56	51.0	2.03	54.0	1.52
58.0	0.00	59.0	0.76	63.0	0.00	66.0	1.52
68.0	0.00	71.0	0.76	75.0	0.00	79.0	0.76
81.0	0.38	85.0	0.00				

Volume of rainfall = 4.22 cm

Direct runoff

Time (min)	Discharge (cm/h)	Time (min)	Discharge (cm/h)	Time (min)	Discharge (cm/h)	Time (min)	Discharge (cm/h)
0.0	0.00	7.0	0.001	15.0	0.001	23.0	0.003
28.0	0.007	30.0	0.015	33.0	0.041	35.0	0.070
38.0	0.135	42.0	0.234	45.0	0.353	50.0	0.518
57.0	0.787	62.0	1.123	67.0	1.763	73.0	2.123
75.0	2.192	81.0	2.240	85.0	2.192	90.0	2.101
100.0	1.849	115.0	1.448	135.0	1.049	157.0	0.726
187.0	0.419	207.0	0.295	237.0	0.160	262.0	0.103
293.0	0.064	337.0	0.039	415.0	0.020	495.0	0.012

Volume of observed direct runoff = 3.36 cm

TABLE 15.7 INFILTRATION AND EFFECTIVE RAINFALL FOR THE RAINFALL–RUNOFF EVENT OF JUNE 10, 1941, ON WATERSHED C, RIESEL (WACO), TEXAS

Rain infiltration

Time (min)	Infiltration rate (cm/h)	Time (min)	Infiltration rate (cm/h)	Time (min)	Infiltration rate (cm/h)	Time (min)	Infiltration rate (cm/h)
0.0	1.27	4.0	1.27	6.0	1.08	9.0	0.93
11.0	0.86	15.0	0.77	19.0	0.72	24.0	0.67
28.0	0.64	31.0	0.62	34.0	0.60	42.0	0.57
45.0	0.56	48.0	0.55	51.0	0.54	54.0	0.53
58.0	0.52	59.0	0.52	63.0	0.51	66.0	0.50
68.0	0.50	71.0	0.49	75.0	0.49	79.0	0.48
81.0	0.48	85.0	0.47				

Effective rainfall hyetograph

Time (min)	ER (cm/h)	Time (min)	ER (cm/h)	Time (min)	ER (cm/h)	Time (min)	ER (cm/h)
0.0	1.78	4.0	0.35	6.0	2.55	9.0	0.63
11.0	1.09	15.0	3.06	19.0	2.97	24.0	4.68
28.0	4.45	31.0	9.55	34.0	7.04	42.0	4.01
45.0	1.48	48.0	3.01	51.0	1.50	54.0	1.0
58.0	0.0	59.0	0.25	63.0	0.0	66.0	1.02
68.0	0.0	71.0	0.27	75.0	0.0	79.0	0.28
81.0	0.0	85.0	0.0				

4. Calculate the volume of runoff for the UH.

$$\text{Watershed area} = 2.343 \text{ km}^2 = 2.343 \times 10{,}000 \text{ m}^2$$
$$= 23{,}430 \text{ m}^2$$

   For 1 cm of runoff, volume $= 23{,}430 \times 1/100$ m^3 $= 234.3$ m^3.

5. Assume an initial hydrograph h^0 of the given volume. For the first estimate, a rectangular unit hydrograph is assumed. By choosing the duration of flow for 7 hours, or 420 minutes, the ordinate is 1/7 h = 0.1428/h. Thus, 21 ordinates are available at the interval of 20 min, or $\frac{1}{3}$ hour.
6. Multiply the assumed UH by the first ER amount, and lag the first product by the unit period, as shown in column (4) of Table 15.9.
7. Repeat step 6 for each ER amount, as shown in columns (4) to (7).
8. Sum the products of ER and h^0 for each unit duration, as shown in column (8) of Table 15.9.
9. Tabulate the actual discharge, as shown in column (9).
10. Subtract the sum obtained in step 8 from the actual discharge, as shown in column (10).
11. Proceed to calculate the first trial UH values, as shown in column (11). To calculate this step, we work only on the time periods where the maximum I has effect, that is, neglect the calculation for the time units 1 and 2, as the differ-

TABLE 15.8 EFFECTIVE RAINFALL AND DIRECT RUNOFF CORRESPONDING TO A 20-MIN TIME INTERVAL

Time unit (20 min, or $\frac{1}{3}$ h)	Effective rainfall (cm)	Direct runoff (cm/h)
1	0.562	0.000
2	1.968	0.070
3	0.526	0.613
4	0.083	1.886
5	0.000	2.023
6		1.424
7		1.032
8		0.873
9		0.558
10		0.289
11		0.222
12		0.126
13		0.098
14		0.078
15		0.046
16		0.033
17		0.029
18		0.021
19		0.014
20		0.012
21		0.010
22		0.009
23		0.007
24		0.006
25		0.000

ences between columns (8) and (9) are not caused by the maximum value of I. These differences vanish as the UH approaches its correct value.

12. Determine the value of F using Equation (15.26). From Table 15.9, $\Sigma(Q - \Sigma Ih^0) = 6.068$, and $\Sigma\Sigma Ih^0 = 3.334$. Therefore, $F = 1.82$.
13. Take the average of the initial and first trial UHs using Equation (15.25). Therefore,

$$\bar{h}_1 = \frac{h^0 + 1.82h^1}{1 + 1.82}$$

Sum the ordinates $\bar{h}_1$ and then divide each ordinate $\bar{h}_1$ by this sum to ensure the unit volume of the UH. The results of the calculations are shown in Table 15.10. This produces the initial UH for the second trial.

14. Repeat this process until differences between the final UHs of successive trials are acceptable. The results of the calculations are shown in Table 15.11. The derived UH and DRH are shown in Table 15.12.

TABLE 15.9 CALCULATION OF THE FIRST TRIAL UNIT HYDROGRAPH BY THE COLLINS METHOD

Time unit (h) (1)	Effective rainfall (cm) (2)	Assumed initial UH, h^0 (1/h) (3)	Ih^0 I_1h^0 (4)	I_2h^0 (5)[a]	I_3h^0 (6)	I_4h^0 (7)	ΣIh^0 (8)	Actual hydrograph, Q (cm/h) (9)	$Q - \Sigma Ih^0$ (cm/h) (10)	$\frac{Q - \Sigma Ih^0}{I_{max}} = h^1$ (11)[b]
1	0.562	0.143	0	0	0	0	0	0	0	0
2	1.968	0.143	0.080	0	0	0	0.080	0.070	0	0
3	0.526	0.143	0.080	0	0	0	0.080	0.613	0.533	0.270
4	0.083	0.143	0.080	0	0.075	0	0.155	1.886	1.730	0.879
5	0.000	0.143	0.080	0	0.075	0.012	0.167	2.023	1.855	0.943
6		0.143	0.080	0	0.075	0.012	0.167	1.424	1.257	0.639
7		0.143	0.080	0	0.075	0.012	0.167	1.032	0.864	0.439
8		0.143	0.080	0	0.075	0.012	0.167	0.873	0.706	0.35
9		0.143	0.080	0	0.075	0.012	0.167	0.558	0.391	0.199
10		0.143	0.080	0	0.075	0.012	0.167	0.289	0.122	0.062
11		0.143	0.080	0	0.075	0.012	0.167	0.222	0.055	0.028
12		0.143	0.080	0	0.075	0.012	0.167	0.126	−0.041	−0.021
13		0.143	0.080	0	0.075	0.012	0.167	0.098	−0.07	−0.035
14		0.143	0.080	0	0.075	0.012	0.167	0.078	−0.089	−0.045
15		0.143	0.080	0	0.075	0.012	0.167	0.046	−0.121	−0.062
16		0.143	0.080	0	0.075	0.012	0.167	0.033	−0.134	−0.068
17		0.143	0.080	0	0.075	0.012	0.167	0.029	−0.139	−0.071
18		0.143	0.080	0	0.075	0.012	0.167	0.021	−0.146	−0.071
19		0.143	0.080	0	0.075	0.012	0.167	0.014	−0.154	−0.078
20		0.143	0.080	0	0.075	0.012	0.167	0.012	−0.155	−0.079
21		0.143	0.080	0	0.075	0.012	0.167	0.011	−0.157	−0.080
22		0.000	0.080	0	0.075	0.012	0.167	0.009	−0.158	−0.080
23		0	0	0	0.075	0.012	0.087	0.007	−0.079	−0.040
24		0	0	0	0.075	0.012	0.087	0.006		
25		0	0	0	0	0.012	0.087	0		

[a] Does not include the highest effective rainfall period.
[b] Column (11) lists the first trial UH values (called h^1).

TABLE 15.10 DERIVATION OF THE INITIAL UH FOR THE SECOND TRIAL (F = 1.82)

h^0 (1)	$F \times h^1$ (2)	$\bar{h}_1$ (3)	Initial UH ordinates for the second trial[a] (4)
0.143	0.492	0.225	0.221
0.143	1.600	0.618	0.607
0.143	1.716	0.659	0.648
0.143	1.162	0.463	0.455
0.143	0.799	0.332	0.328
0.143	0.653	0.282	0.277
0.143	0.362	0.179	0.176
0.143	0.112	0.090	0.089
0.143	0.050	0.069	0.067
0.143	−0.038	0.037	0.036
0.143	−0.064	0.028	0.027
0.143	−0.082	0.021	0.021
0.143	0.112	0.011	0.011
0.143	−0.124	0.007	0.007
0.143	−0.128	0.005	0.005
0.143	−0.135	0.003	0.003
0.143	−0.142	0.000	0.000
0.143	−0.144	0.000	0.000
0.143	−0.145	0.000	0.000
0.143	−0.146	0.000	0.000
0.143	−0.074	0.000	0.000

[a] Adjusted to the required unit-hydrograph volume.

TABLE 15.11 RESULTS OF SUBSEQUENT TRIALS OF UH CALCULATIONS[a]

Second trial	Third trial	Fourth trial	Fifth trial	· · ·	Tenth trial
0.166	0.166	0.136	0.121		0.104
0.607	0.667	0.693	0.706		0.723
0.648	0.689	0.701	0.706		0.706
0.455	0.434	0.424	0.420		0.417
0.328	0.304	0.298	0.297		0.296
0.277	0.279	0.287	0.291		0.296
0.176	0.170	0.177	0.172		0.172
0.089	0.075	0.071	0.069		0.065
0.067	0.069	0.072	0.075		0.078
0.036	0.035	0.034	0.033		0.032
0.027	0.029	0.030	0.025		0.030
0.021	0.025	0.026	0.012		0.027
0.011	0.013	0.013	0.010		0.012
0.007	0.010	0.010	0.009		0.010
0.005	0.009	0.009	0.007		0.009
0.003	0.007	0.007	0.004		0.007
0.000	0.004	0.004	0.004		0.004
0.000	0.004	0.004	0.004		0.004
0.000	0.003	0.004	0.004		0.004
0.000	0.000	0.000	0.001		0.003
0.000	0.011	0.007	0.004		0.003

[a] For convergence, the error between two consecutive UH ordinates was set to be 0.000001.

TABLE 15.12 FINAL UH AND THE RESULTING DRH BY THE COLLINS METHOD

Time (h) (1)	UH (1/h) (2)	Observed DRH (cm/h) (3)	Computed DRH (cm/h) (4)	Error in calculated DRH $\frac{(3)-(4)}{(3)}$
0.333	0.104	0.00	0.00	—
0.667	0.723	0.07	0.058	0.171
1.000	0.706	0.613	0.611	0.003
1.333	0.417	1.886	1.875	0.006
1.667	0.296	2.023	2.013	0.005
2.000	0.296	1.424	1.418	0.004
2.333	0.172	1.032	1.027	0.005
2.667	0.065	0.873	0.869	0.005
3.000	0.078	0.558	0.556	0.003
3.333	0.032	0.289	0.288	0.004
3.667	0.030	0.222	0.221	0.005
4.000	0.027	0.126	0.126	0.0
4.333	0.012	0.098	0.097	0.010
4.667	0.010	0.078	0.078	0.0
5.000	0.009	0.046	0.046	0.0
5.333	0.007	0.033	0.033	0.0
5.667	0.004	0.029	0.028	0.034
6.000	0.004	0.021	0.021	0.0
6.333	0.003	0.014	0.014	0.0
6.667	0.003	0.012	0.012	0.0
7.000	0.003	0.010	0.010	0.0
7.333	0.0	0.009	0.009	0.0
7.667	0.0	0.007	0.007	0.0
8.000	0.0	0.006	0.002	0.666
8.333	0.0	0.0	0.0	—

LEAST SQUARES METHOD

Equation (15.7) states that the DRH can be constructed by multiplying the ER of each time period and the appropriate UH ordinate. This means that the DRH is the sum of the properly lagged hydrographs. The problem is one of deriving the UH, given the DRH and ERH. This then is the inverse of the derivation of the DRH through the use of the UH. We want to derive the UH such that the difference between the DRH computed with the derived UH and the observed DRH is minimum. This minimum is expressed as

$$E = \min \sum_{i=1}^{N} [Q_0(i) - Q_c(i)]^2 \tag{15.27}$$

where E is the error, $Q_0(i)$ is the ith observed DR ordinate, $Q_c(i)$ is the computed DR ordinate, and N is the number of ordinates. The least value of E is the least sum of squares of errors, and corresponds to the best UH. This is called the least squares method.

For purposes of illustration, consider an ERH composed of three pulses, as shown in Figure 15.6(a), each of duration D and of intensity I_0, I_1, and I_2, respectively. The D-hour UH is given in Figure 15.6(b), and its ordinates at the D-hour interval are $h_0, h_1, h_2, \ldots$. The DRH ordinates at the D-hour interval are Q_0, Q_1, $Q_2, \ldots$. From Equation (15.7), we can write

$$\begin{aligned} Q_0 &= r_0 h_0 \\ Q_1 &= r_1 h_0 + r_0 h_1 \\ Q_2 &= r_2 h_0 + r_1 h_1 + r_0 h_2 \\ Q_3 &= r_3 h_0 + r_2 h_1 + r_1 h_2 + r_0 h_3 \\ Q_4 &= r_4 h_0 + r_3 h_1 + r_2 h_2 + r_1 h_3 + r_0 h_4 \end{aligned} \tag{15.28}$$

where $r_j = I_j D, j = 0, 1, 2, 3, \ldots . I_j = 0, i \geq 3$, and $r_j = 0, j \geq 3$. Equation (15.28) can be expressed using matrix notation as

$$Q = RH \tag{15.29}$$

where

$$Q = \begin{bmatrix} Q_0 \\ Q_1 \\ Q_2 \\ \vdots \\ Q_N \end{bmatrix}_{N \times 1}, R = \begin{bmatrix} r_0 & 0 & 0 & 0 \\ r_1 & r_0 & 0 & 0 \\ r_2 & r_1 & r_0 & 0 \\ \vdots & \vdots & \vdots & \vdots \\ 0 & 0 & r_{m-1} & r_m \end{bmatrix}_{N \times m}; H = \begin{bmatrix} h_0 \\ h_1 \\ h_2 \\ \vdots \\ h_m \end{bmatrix}_{m \times 1}$$

The least squares solution of H is obtained by multiplying Equation (15.29) by the transpose of R, R^T, and then multiplying by the inverse of the product of $[R^T R]^{-1}$ as

$$R^T Q = R^T R H$$

$$[R^T R]^{-1} R^T Q = [R^T R]^{-1} [R^T R] H$$

This yields

$$H = [R^T R]^{-1} R^T Q \tag{15.30}$$

Computation of H can be easily performed on a pocket calculator or a small microcomputer. This is a powerful method and usually produces a reasonable UH.

EXERCISES

15.1. The ordinates of a 4-hour UH (due to a 1 cm ER) of a watershed are given. Compute the ordinates of the DRH due to effective rainfall of 5 cm occurring uniformly for 4 hours.

Time (h)	0	2	4	8	10	12	14	16	18	20	22	24	25
UH Ordinate (m^3/cm-s)	0	10	25	50	65	70	65	55	30	20	10	5	0

15.2. The ordinates of a one-hour UH (due to 1 cm ER) are given. Construct an S-hydrograph. Derive a 3-hour UH.

Time (h)	UH ordinate (m^3/cm-s)	Time (h)	UH ordinate (m^3/cm-s)
0	0	6.0	1.42
0.75	16.42	6.75	1.13
1.5	31.15	7.5	0.85
2.25	27.18	8.25	0.57
3.0	15.0	9.0	0.5
3.75	7.36	9.75	0.42
4.5	3.96	10.50	0.14
5.25	2.27	11.25	0

15.3. Consider the following rainfall–runoff storm.

EFFECTIVE RAINFALL

Time (h-min)	10:00	11:00
Intensity (cm/h)	15.0	0.00

DIRECT RUNOFF

Time (h-min)	10:00	11:30	12:00
Intensity (cm/h)	0.0	10.0	0.0

Determinate the UH.

15.4. Construct the S-hydrograph using the UH derived in Exercise 15.3.

15.5. Derive the UH for 2-h, 4-h, and 6-h durations from the UH derived in Exercise 15.3.

15.6. For a storm of 3-hour duration on a watershed of 312,000 hectares, the following hydrograph has been observed:

	Discharge (m^3/s)		
Hour	Day 1	Day 2	Day 3
3 A.M.	17.2	132.0	48.5
6 A.M.	15.8	115.0	43.0
9 A.M.	175.0	100.0	37.0
Noon	270.0	88.0	32.0
3 P.M.	230.0	78.0	26.0
6 P.M.	200.0	68.0	23.0
9 P.M.	172.0	60.0	26.0
Midnight	150.0	54.0	17.0

Assume a constant baseflow of 17.0 m^3/s. Derive the unit hydrograph for this storm.

15.7. Construct the S-hydrograph using the UH derived in Exercise 15.6. Then derive the 6-hour UH.

15.8. Three UHs derived from separate storms on a small watershed are given. All are believed to have resulted from 4-hour rainstorms. The drainage area is 9.7 km^2.

Time (h)	UH (m^3/s-cm)		
	Storm 1	Storm 2	Storm 3
0	0	0	0
1	3.11	0.71	0.45
2	10.33	3.54	1.64
3	14.15	10.13	4.91
4	11.04	13.16	9.54
5	8.77	11.46	12.45
6	6.65	8.63	11.32
7	4.95	6.23	8.07
8	3.68	4.81	6.08
9	2.69	3.68	4.67
10	1.34	2.55	3.45
11	1.13	1.71	2.55
12	0.62	0.99	1.71
13	0.28	0.57	0.99
14	0.14	0.23	0.45
15	0.0	0.0	0.0

Compute a representative UH for this watershed.

15.9. An S-hydrograph is given such that at time $t = 0$, its ordinate is 1 cm/h and it remains so for an indefinite period of time. Determine a 2-hour unit hydrograph. Using this unit hydrograph, determine a 4-hour unit hydrograph.

15.10. A 2-hour unit hydrograph is given by a triangle whose base is 4 hours and height at the center of the base is 0.5/hour. Determine the surface-runoff hydrograph for a rainfall event whose effective rainfall is given as a rectangle of 1 cm/h height and 4-h base.

15.11. The direct-runoff hydrograph due to an effective rainfall event is given by a triangle such that its base is 8 hours and its height at the midpoint of the base is 1 cm/h. The duration and intensity of the effective rainfall are 4 hours and 1 cm/h, respectively. Derive and sketch a 4-hour unit hydrograph.

15.12. A 2-hour unit hydrograph is defined by a triangle whose base is 4 hours and height at the midpoint of the base is 0.5/hour. Compute a 4-hour unit hydrograph.

15.13. Construct an S-hydrograph if a 2-hour unit hydrograph is given as in Exercise 15.12.

15.14. Construct a 3-hour unit hydrograph if a 2-hour unit hydrograph is given as in Exercise 15.12.

15.15. A 2-hour unit hydrograph is given by a triangle whose base is 4 hours and height at the midpoint of the base is 0.5/hour. Compute the direct runoff due to a 4-hour effective rainfall whose intensity is 2 cm/h. Compute the peak discharge.

15.16. Derive a 3-hour unit hydrograph from a 2-hour unit hydrograph as given in Exercise 15.15. If a 3-hour effective rainfall occurs at a uniform rate of 1 cm/h, what should be the direct runoff? Give its peak discharge.

15.17. The effective-rainfall and direct-runoff hydrographs for an event are given by a rectangular pulse and a symmetric triangle:

$$\begin{aligned} ER &= 2 \text{ cm/h}, && 0 \le t \le 2 \text{ h} \\ &= 0 \text{ cm/h}, && t \ge 2 \text{ h} \\ DR &= 0 \text{ cm/h}, && t = 0 \text{ h} \\ &= 4/3 \text{ cm/h}, && t = 3 \text{ h} \\ &= 0 \text{ cm/h}, && t = 6 \text{ h} \end{aligned}$$

Compute the unit hydrograph for this event. What period unit hydrograph is this?

15.18. An S-hydrograph is given as

$$\begin{aligned} U(t) &= 0 \text{ cm/h} && t = 0 \text{ h} \\ &= 1 \text{ cm/h}, && t \ge 4 \text{ h} \end{aligned}$$

Compute a 3-hour unit hydrograph.

15.19. The effective rainfall and direct runoff for an event are

$$\begin{aligned} ER &= 1 \text{ cm/h}, && 0 \le t < 1,\ t \text{ in hours} \\ &= 0 \text{ cm/h}, && t \ge 1 \\ DR &= 0 \text{ cm/h}, && t = 0 \text{ h} \\ &= 0.5 \text{ cm/h}, && t = 2 \text{ h} \\ &= 0 \text{ cm/h}, && t = 4 \text{ h} \end{aligned}$$

Determine the S-hydrograph, and determine a 2-hour unit hydrograph as a rectangular pulse and a symmetric triangle.

15.20. A 1-hour unit hydrograph is given by a rectangle whose base is 4 hours and height is 0.25/hour. Construct an S-hydrograph using this UH.

15.21. Determine the surface runoff from a basin due to an effective rainfall event whose intensity is 2 cm/h and duration is 3 hours. A 1-hour UH for the basin is as in Exercise 15.20.

15.22. A 2-hour unit hydrograph is given by a rectangle whose base is 4 hours and height is 0.25 cm/hour. Derive a 4-hour unit hydrograph using the given 2-hour unit hydrograph.

15.23. Construct and draw an S-hydrograph using the 2-hour unit hydrograph of Exercise 15.22.

15.24. An S-hydrograph is given such that at time $t = 0$, its ordinate is 1 cm/h and it remains so for an indefinite period of time. Determine a 2-hour unit hydrograph. Using this unit hydrograph, determine a 4-hour unit hydrograph.

15.25. A 1-hour unit hydrograph is given as a triangle whose base is 3 hours, peak is $\frac{2}{3}$ cm/h, and the times of rise and recession, respectively, are 1 h and 2 h. Compute a 2-h unit hydrograph using the S-hydrograph method.

15.26. A rainfall event fell on a watershed whose surface was dry. Its hyetograph is given as follows:

TIME (h)	INTENSITY (cm/h)
0	2
1	4
2	2
3	0

The volume of direct runoff resulting from this event is 3 cm. Compute the rate of infiltration by the ϕ index method.

The area of this watershed is 10 km^2. The direct-runoff hydrograph is represented by a triangle with a base of 6 hours and a peak occurring at 3 hours (or the midpoint of the base). Compute the peak discharge in m^3/s.

15.27. Rainfall and runoff data are given for a watershed 5220 ha in area.

Rainfall			Runoff		
Day	Hour	Intensity (mm/h)	Day	Hour	Rate (m^3/s)
1	20	0.5	1	8	0.56
	21	0.5		12	0.61
	22	0.5		16	0.61
	23	2.7		20	0.59
	24	1.0		24	0.66
2	1	5.2	2	4	3.35
	2	1.7		8	4.75
	3	1.3		12	4.25
	4	1.1		16	2.40
	5	1.3		20	1.25
			3	4	0.94
				8	0.89
				12	0.84
				16	0.79
				20	0.74
				24	0.70
			4	4	0.68
				8	0.65
				12	0.64
				16	0.62
				20	0.59
				24	0.59

The maximum runoff was recorded as 4.82 m^3/s at hour 9 on the second day.

(a) Calculate a 6-hour unit hydrograph.

(b) Calculate a 12-hour unit hydrograph, where the total rainfall occurred in two periods of 6 hours with different intensities.

(c) Calculate the runoff hydrograph for the following complex storm:

6 hours of rainfall with an intensity of 2 mm/h
6 hours of rainfall with an intensity of 7 mm/h
6 hours of rainfall with an intensity of 11 mm/h

Determine the maximum discharge.

15.28. From the storm hydrograph data given, separate the baseflow and direct-runoff components using a standard method and then construct the unit hydrograph for the watershed in question. The drainage area of the basin is 65 km^2.

Time (h)	Runoff (m^3/s)	Time (h)	Runoff (m^3/s)
8:30 A.M.	50.0	4:30 P.M.	101.8
12:30 Noon	44.5	8:30 P.M.	78.5
4:30 P.M.	44.0	12:30 Midnight	60.0
8:30 P.M.	60.0	4:30 A.M.	43.2
12:30 Midnight	123.0	8:30 P.M.	31.7
4:30 A.M.	160.0	12:30 Noon	24.5
6:30 A.M.	165.0	4:30 P.M.	16.0
8:30 A.M.	160.0		
12:30 Noon	133.0		

15.29. Three 6-hour storm hydrographs that occurred on the same watershed with an area of 175 km^2 are given. Construct the mean unit hydrograph.

Time (h)	Hydrograph, Q (m^3/s)		
	(1)	(2)	(3)
1.5	165	120	90
3.0	1080	720	510
4.5	2250	1500	1200
6	2305	2070	1990
7.5	1560	3210	2880
9	970	2070	2340
12	600	1290	1470
15	390	750	920
18	210	335	480
21	60	69	180
24	420	330	270
27	1650	1080	810
30	2310	1870	1750
33	2050	3570	2400
36	1260	2610	2790
39	780	1630	1800
42	485	995	1145
45	300	540	695
48	120	185	285
51	—	35	45

15.30. Develop an S-hydrograph using the 4-hour unit hydrograph given.

Time (h)	UH (m^3/s-m)	Time (h)	UH (m^3/s-m)
0	0	20	120
4	180	24	35
8	560	28	8
12	540	32	0
16	260		

Derive a 2-hour unit hydrograph.

15.31. Derive the unit hydrograph from the given flood hydrograph caused by a 6-hour storm of 7.5-cm depth.

Time (h)	Runoff (m^3/s)	Time (h)	Runoff (m^3/s)
8:30 A.M.	50	12:30 Noon	163
12:30 Noon	45	4:30 P.M.	135
4:30 P.M.	44.5	8:30 P.M.	110
8:30 P.M.	60	12:30 Midnight	70
12:30 Midnight	160	4:30 A.M.	45
2:30 A.M.	240	8:30 A.M.	32.5
4:30 A.M.	263	10:30 A.M.	27.5
7:30 A.M.	240	12:30 Noon	20
8:30 A.M.	225	4:30 P.M.	0

Determine the drainage area of the watershed.

15.32. A 6-hour unit hydrograph is given for a watershed as follows:

Time (h)	UH (ft^3/s-in.)
0	0
6	830
12	1023
18	455
24	212
36	99
42	27

Determine the peak outflow from the watershed corresponding to 5 in. of rainfall. The runoff coefficient for the watershed is 0.4; the baseflow is 250 ft^3/s and is constant throughout.

15.33. A 6-hour unit hydrograph is given. Develop an S-curve.

Time (h)	UH (m^3/s-cm)	Time (h)	UH (m^3/s-cm)
0	0	30	120
6	180	36	35
12	560	42	8
18	540	48	0
24	260		

Compute a 12-hour unit hydrograph.

15.34. Observed flows due to a storm of 3-hour duration on a drainage area of 122 square miles are given. Derive the unit hydrograph. Assume a constant baseflow equal to 600 cfs.

Hour	Discharge (cfs) Day 1	Day 2	Day 3
3 A.M.	600	4600	1700
6 A.M.	650	4400	1500
9 A.M.	6000	3500	1300
Noon	9500	3100	1100
3 P.M.	8000	2700	900
6 P.M.	7000	2400	800
9 P.M.	6100	2100	700
Midnight	5300	1900	600

15.35. A 12-hour unit hydrograph is given for a watershed. Construct an S-curve for the watershed. Derive a 6-hour unit hydrograph.

Time (h)	12-h UH (m^3/s-cm)	Time (h)	12-h UH (m^3/s-cm)
0	0	54	3350
6	3600	60	2600
12	8600	66	2100
18	9300	72	1700
24	8500	78	1400
30	7200	84	1100
36	6000	90	800
42	5150	96	400
48	4100	102	0

15.36. The following data are available for a flood hydrograph from a watershed of 1250 km^2 generated by a 6-hour effective rainfall of 6.68 mm. Determine the unit hydrograph for the watershed.

Date	Time (h)	Discharge (m^3/s)	Date	Time (h)	Discharge (m^3/s)
3–21–89	0300	240	3–22–89	0300	560
	0600	0		0600	400
	0900	550		0900	230
	1200	1650		1200	100
	1500	2100		1500	35
	1800	1760		1800	0
	2100	1170			
	2400	765			

15.37. Given is a unit hydrograph of a watershed as a result of a 4-hour rainstorm. Find the peak flow resulting from four successive 4-hour periods of rainfall producing 0.9, 2.2, 3.5, and 2.0 cm of runoff, respectively. Ignore baseflow.

Time (h)	UH (m^3/s-cm)	Time (h)	UH (m^3/s-cm)
0	0	8	6.09
1	0.45	9	4.67
2	1.64	10	3.45
3	4.90	11	2.54
4	9.54	12	1.70
5	12.46	13	1.0
6	11.33	14	0.45
7	8.07	15	0

15.38. The following 24 hourly observations at the mouth of a watershed are given. The rise in discharge is due to a storm of an intensity of 3 cm/h for a 10-hour duration. The watershed area is 100 km^2.

Time (h)	Discharge (m^3/s)	Time (day)	Discharge (m^3/s)
1	66.25	8	91
2	971	9	78
3	708	10	68
4	396	11	58
5	254	12	50
6	162	13	43
7	122	14	37

(a) Plot the data on semilogarithmic paper and determine the recession constants for surface runoff, interflow, and groundwater flow.
(b) Separate the baseflow from the direct runoff.
(c) Calculate the volume of water in each of the three parts of the hydrograph.
(d) Plot the unit hydrograph from the given data.

15.39. A 1-hour unit hydrograph is given for a rainstorm of 2.54 cm (1 in.) per $\frac{3}{4}$ hour:

Time interval ($\frac{3}{4}$ h)	UH ordinates	
	(ft^3/s-in.)	(m^3/s-2.54 cm)
0.00	0	0
0.75	580	16.42
1.50	1100	31.15
2.25	960	27.18

Time interval ($\frac{3}{4}$ h)	UH ordinates (ft^3/s-in.)	UH ordinates (m^3/s-2.54 cm)
3.00	530	15.00
3.75	260	7.36
4.50	140	3.96
5.25	80	2.27
6.00	50	1.42
6.75	40	1.13
7.50	30	0.85
8.25	20	0.57
9.00	20	0.57
9.75	15	0.42
10.50	5	0.14
11.25	0	0.0
12.00	0	0.0

(a) Compute the unit hydrograph due to a rainfall storm of 1 cm/h.
(b) Construct an S-hydrograph due to this 1-cm/h rainstorm.
(c) Construct a 6-hour unit hydrograph.

15.40. Given are the daily flows for the Leaf River at Hattiesburg, Mississippi, for the dates shown. The drainage area is 1760 mi^2.
(a) Determine the surface runoff (SRO) in inches, using the straight-line, arbitrary, and recession methods for separating SRO from baseflow.
(b) Develop a unit hydrograph from the flow data. (Use the straight-line technique.)
(c) Develop an S curve by assuming a reasonable effective value of rainfall duration, t_R.
(d) Construct unit hydrographs for $t_R' = 0.75t_R$, $3t_R$.

	Q (cfs)		Q (cfs)
April 28	916	May 8	2500
29	890	9	1660
30	842	10	1330
May 1	806	11	1150
2	2030	12	1030
3	5880	13	936
4	7450	14	878
5	7890	15	836
6	7120	16	824
7	4650	17	1130

15.41. Determine the surface runoff (SRO) in inches over the drainage area (A_d) for the following total flow hydrograph. Use the straight-line technique for separating baseflow from SRO. A_d is 1.4 mi^2.

Day	Time (h)	Flow (cfs)
1	12:00 M	100
	6:00	100
	12:00 N	300
	6:00	200
2	12:00 M	100
	6:00	100

Determine the ordinates of the unit hydrograph.

15.42. Given the simple unit hydrograph and the storm sequence, calculate the peak runoff discharge and the time to peak. Neglect groundwater flow. The duration of rainfall is 1 h.

Time (h)	UH ordinates (cfs/in.)	Accumulative rainfall (in.)
0	0	0
1	100	0.5
2	50	1.6
3	25	1.9
4	0	1.9

15.43. A storm during the middle of August 1974 produced the following storm hydrograph on the Tibbee Creek near Tibbee, Mississippi, where the drainage area is 928 mi^2.

Date	Q (cfs)	Date	Q (cfs)	Date	Q (cfs)
Aug. 9	75	Aug. 14	955	Aug. 19	85
10	67	15	498	20	58
11	63	16	296	21	45
12	54	17	175	22	36
13	393	18	120		

In later August and early September 1974, the following storm hydrograph occurred on the same drainage area.

Aug. 27	19	Sept. 5	826	Sept. 13	75
28	18	6	474	14	61
29	18	7	264	15	49
30	42	8	214	16	40
31	248	9	174	17	32
Sept. 1	590	10	140	18	26
2	1040	11	114	19	21
3	1320	12	92	20	17
4	1330				

(a) Calculate the inches of surface runoff from each storm. Use the straight-line separation technique.
(b) From which storm would you develop a unit hydrograph for this point on the creek? State your reasons for this decision and back your reasons with any available data.
(c) Calculate the groundwater recession coefficient for each storm. Explain any differences between the coefficients for each storm.

15.44. A hydrograph of runoff from a 214-mi^2 drainage area is given. The time of rainfall is approximately 24 hours.

Date	Discharge (cfs)	Date	Discharge (cfs)
Jan. 28	140	Feb. 5	2600
29	134	6	2040
30	132	7	1270
31	217	8	938
Feb. 1	976	9	588
2	2360	10	330
3	2620	11	224
4	2840	12	190
		13	157
		14	151

(a) Determine the total surface runoff using an arbitrary method to separate baseflow from surface runoff.
(b) Determine the time of concentration for the hydrograph in part (a).
(c) Calculate the ordinates of a unit hydrograph for a rainfall duration of 48 hours.

15.45. The following 24 hour unit hydrograph was developed for a drainage basin of 300 sq. mi.

Time (days)	UH (cfs/in.)	Time (days)	UH (cfs/in.)
0	0	7	460
1	270	8	350
2	1000	9	250
3	2080	10	150
4	1550	11	75
5	850	12	30
6	620	13	0

Assuming 100% runoff, what would be the resulting hydrograph for the following 3-day storm?

Date	24-h precipitation (in.)
July 18	0.5
19	2.0
20	1.0

Assuming that the ratio of runoff to precipitation is 0.6, what would be the magnitude and date of the peak discharge?

15.46. Daily flow data from the Chuquatonchee Creek near West Point, Mississippi, is given for a drainage area (A_d) of 514 mi^2.

Date	Flow (cfs)	Date	Flow (cfs)
8/10/74	4.5	8/18/74	47.0
8/11/74	4.0	8/19/74	15.0
8/12/74	2.6	8/20/74	9.0
8/13/74	69.0	8/21/74	6.8
8/14/74	984.0	8/22/74	5.0
8/15/74	873.0	8/23/74	3.5
8/16/74	496.0	8/24/74	2.7
8/17/74	190.0	8/25/74	2.2

(a) Determine the daily baseflow.
(b) Compute the volume of direct surface runoff in SFD and in inches over A_d.
(c) Is it advisable to use this total flow hydrograph to develop a unit hydrograph for this area? Why?

15.47. A typical 4-hour storm produces 2.0 in. of runoff from the Big Bear Creek basin, with the flow in the stream as follows:

Time (h)	Flow (cfs)	Time (h)	Flow (cfs)
0	0	12	250
2	90	—	—
4	300	16	100
6	500	—	—
8	420	20	0
—	—		

Estimate as accurately as possible the peak flow and the time of occurrence in a flood created by an 8-hour storm that produces 1.0 in. of runoff during the first 4 hours and 1.5 in. of runoff during the second 4 hours. Assume that baseflow is negligible.

CHAPTER 16

Empirical Synthesis of the Unit Hydrograph

Frequently, measured rainfall and streamflow data are not available for the watersheds under study. This is especially true for small watersheds and remote areas. Unit hydrographs (UHs) are derived for such watersheds from empirical equations relating salient UH characteristics to watershed characteristics. These empirical equations are applicable to specific regions for which they are derived, and are not universal. These should be applied to the watersheds having similar characteristics. The UHs, derived from such equations, are called synthetic UHs. Three of the popular methods are discussed in what follows.

16.1 THE SNYDER METHOD

Snyder (1938) developed a model that involves construction of a synthetic UH for a drainage basin without gaged data. This UH is derived from measured drainage-basin characteristics and is based solely upon such measurements. Snyder (1938) was perhaps the first to have established a set of formulas relating the physical geometry of the watershed to three basic parameters of the UH. These formulas were based on a study of 20 watersheds located mainly in the Appalachian Highlands of the eastern United States, which varied in size from 10 to 10,000 mi^2 (25.6 to 25,600 km^2).

The basic parameter that Snyder defined is t_p, the time of lag to peak in hours taken as the time from the center of mass of the effective rainfall of unit duration to the peak of the UH, or, simply, the watershed lag, as shown in Figure 16.1. He derived all other characteristic parameters of the UH in terms of t_p. According to the Snyder method,

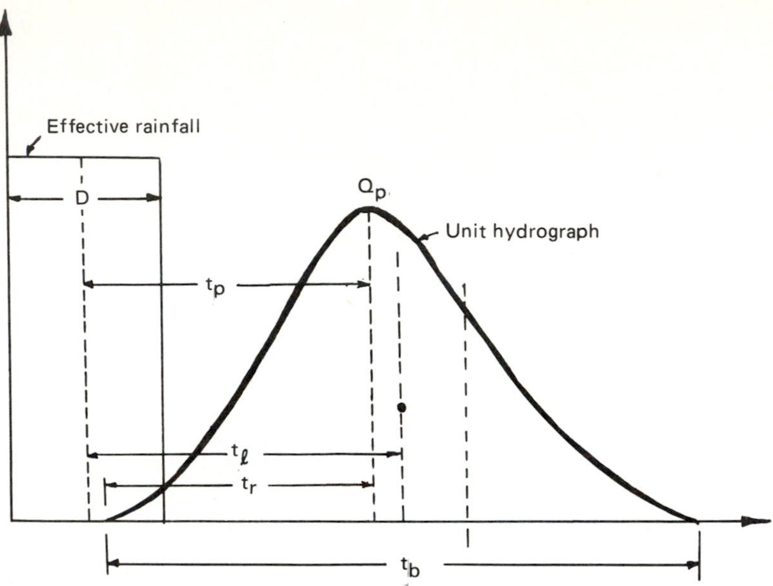

Figure 16.1 Schematic for the Snyder method.

$$t_p = C_t(LL_c)^{0.3} \tag{16.1}$$

where L is the length of the main stream in miles from the outlet to divide; L_c is the distance in miles from the outlet to a point on the stream nearest the center of area of the watershed; and C_t is a constant varying normally in the range from 1.8 to 2.2, with some indication of lower values for watersheds with steeper slopes.

In a study of 27 watersheds in Pennsylvania, Miller et al. (1983) found C_t to vary from 1.01 to 4.33. This much of variation was also observed by Bull (1968) for 25 watersheds in the lower Missouri River basin in Kansas and Missouri. Hudlow and Clark (1969) found its range from 0.40 to 2.26 for 13 watersheds (0.5 to 75 mi^2, or 1.28 to 192 km^2) in central Texas. Linsley (1943) noted its range from 0.3 to 0.7 for the northwestern states in the United States. Cordery (1968) found C_t to vary from 0.4 to 2.4 for 12 watersheds (0.021 to 248 mi^2, or 0.054 to 634.9 km^2) in eastern New South Wales, Australia. The quantity LL_c is a measure of the size and shape of the watershed. It is thus seen that watershed slopes were not considered.

The unit duration of the effective rainfall (ER) was defined by

$$D = t_p/5.5 \tag{16.2}$$

Due to this ER, the peak h_p of the resulting UH was given by

$$h_p = C_p/t_p \tag{16.3}$$

where C_p is a constant ranging from 0.56 to 0.69, and h_p has the dimensions (1/T). If h_p is desired in the dimensions of L^2/T, then Equation (16.3) must be multiplied by watershed area A and an appropriate conversion factor, for example,

$$h_p = \frac{2.78 C_p A}{t_p} \tag{16.4}$$

where A is the drainage area in km². From Equations (16.4) and (16.2), the peak ordinate is proportional to the average ordinate of the effective rainfall expressed as

$$\frac{1 \text{ cm} \times \text{watershed area}}{\text{duration of effective rainfall}}$$

The conversion factor of 2.78 becomes 640 if A is in square miles and h is in inches/hour. Miller et al. (1983) found C_p to vary from 0.23 to 0.67 for 27 watersheds in Pennsylvania. Cordery (1968) found its range as 0.4 to 1.1 for 12 watersheds in eastern new South Wales, Australia. Hudlow and Clark (1969) found its variability from 0.31 to 1.22 for watersheds in central Texas. For the Sacramento and lower San Joaquin rivers in California, Linsley (1943) noted its variability from 0.35 to 0.59.

The time base t_b in days of the UH was defined as

$$t_b = 3 + 3(t_p/24) \tag{16.5}$$

The constants in Equation (16.5) are fixed by the procedure of hydrograph separation. By determining t_p, h_p, and t_b, the UH can be constructed such that its volume is equal to one unit. Obviously, many UHs can be sketched that will satisfy this requirement.

As an aid to sketching a reasonable UH, the U.S. Army Corps of Engineers (1940) developed a relation between h_p and the width of the UH at values of 50% (W_{50}) and 75% (W_{75}) of h_p, as shown in Figure 16.2. These time widths can be evaluated from the following equations:

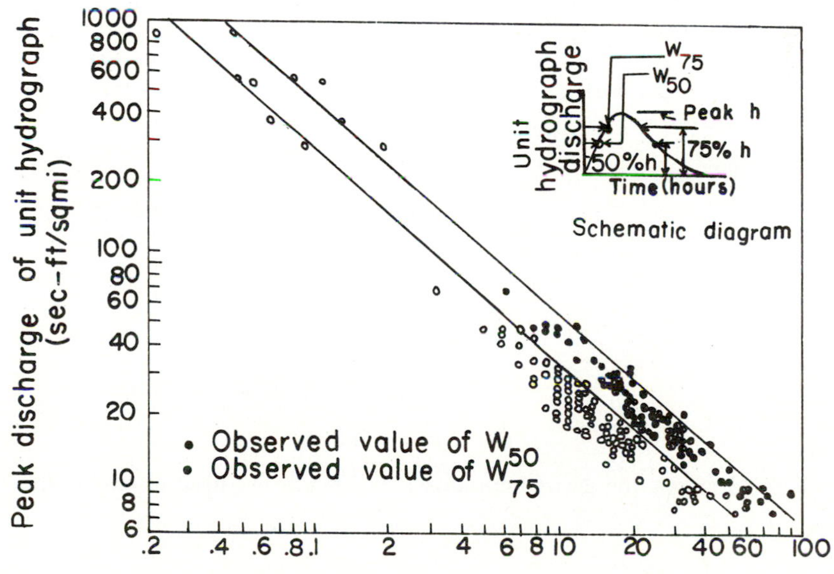

Figure 16.2 Unit-hydrograph width at 50 and 75% of peak value (after U.S. Army Corps of Engineers, 1940).

$$W_{50} = \frac{830}{q_p^{1.1}} \tag{16.6}$$

$$W_{75} = \frac{470}{q_p^{1.1}} \tag{16.7}$$

where q_p is the UH peak h_p in square feet per second per square mile. The Corps of Engineers have suggested, as a guide for shaping the UH, that these widths should be positioned such that one-third of the width is placed to the left and two-thirds to the right of the peak. However, Hudlow and Clark (1969) found in their study that an allocation of four-tenths of the width to the left and six-tenths of the width to the right produced optimum results. On the other hand, Viessman et al. (1977) stated that as a general rule of thumb, the time widths W_{50} and W_{75} should be proportioned on each side of the peak in a ratio of 1 : 2, with the larger proportion on the right of the peak. Equation (16.5) is reasonable for large watersheds, but is known to produce exceptionally large t_b for smaller watersheds. As a general rule of thumb, t_b can be taken as three to five times the time to peak for sketching a UH. Some workers (Hudlow and Clark, 1969) have attempted to fit a mathematical function to the UH.

If the duration of ER, say, t_R, is different from D defined before, a modified lag time t_p^* can be obtained from

$$t_p^* = t_p + \frac{t_R - D}{4} \tag{16.8}$$

Using this t_p^*, h_p and t_b can be determined from Equations (16.3) and (16.4), and then the UH can be constructed.

EXAMPLE 16.1 Derive a 3-hour unit hydrograph by the Snyder method for a watershed of 54 km² area. It has a main stream that is 10 km long. The distance measured from the watershed outlet to a point on the stream nearest to the centroid of the watershed is 3.75 km. Take C_t as 2.0 and C_p as 0.65.

Solution

$$\text{Watershed area, } A = 54 \text{ km}^2 = 20.86 \text{ mi}^2$$

$$\text{Length of main stream, } L = 10 \text{ km} = 6.215 \text{ mi}$$

$$L_c = 3.75 \text{ km} = 2.331 \text{ mi}$$

Therefore,

$$t_p = 2(6.215 \times 2.331)^{0.3} = 4.46 \text{ h}$$

$$D = t_p/5.5 = 0.81 \text{ h}$$

Because the desired duration t_R is 3 hours, we need to modify the value of t_p:

$$t_p = 4.46 + \frac{3 - 0.81}{24} = 5 \text{ h}$$

Therefore,

$$h_p = \frac{0.65}{5} = 0.13/\text{h}$$

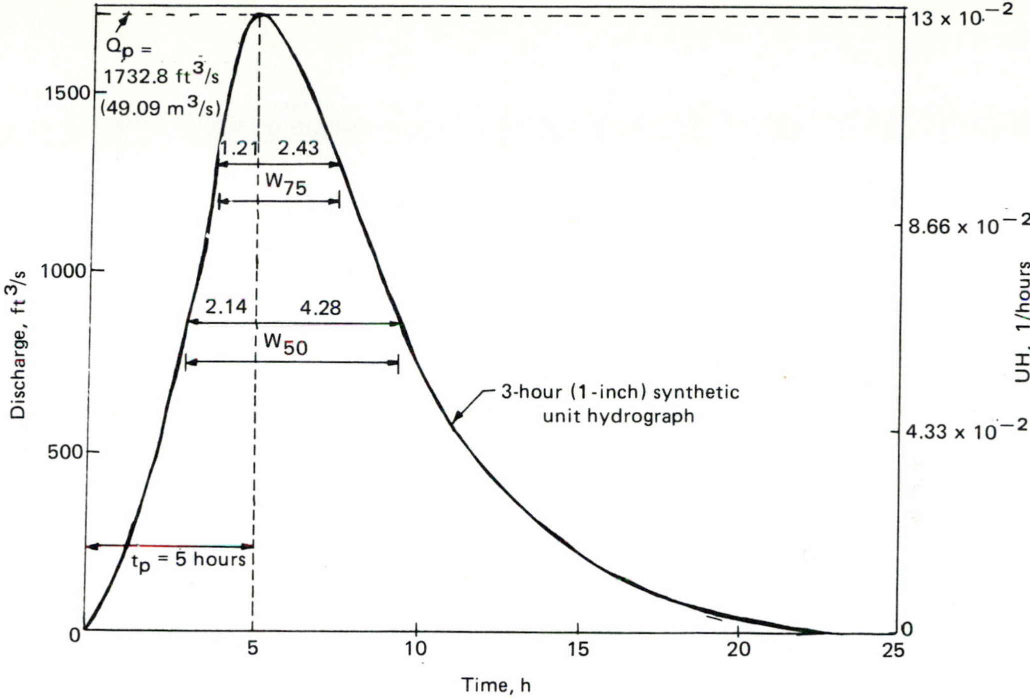

Figure 16.3 Three-hour unit hydrograph from the Snyder method.

The peak ordinate in terms of discharge:

$$q_p = h_p = 640\ c_p/t_p \text{ ft}^3\text{/s/mi}^2$$

$$= 0.65\ \frac{640}{5} = 83.075 \text{ ft}^3\text{/s/mi}^2$$

$$Q_p = q_p A = 83.075 \times 20.86 = 1732.8 \text{ ft}^3\text{/s} = 49.09 \text{ m}^3\text{/s}$$

The widths of the UH at 50% and 75% of peak discharge from Figure 16.2, respectively, are

$$W_{50} = \frac{830}{(83.075)^{1.1}} = 6.42 \text{ h}$$

$$W_{75} = \frac{470}{(83.075)^{1.1}} = 3.64 \text{ h}$$

W_{50} and W_{75} are partitioned about $t = t_p$ such that one-third is toward the rising limb and two-thirds is toward the recession limb. The 3-hour UH is plotted in Figure 16.3. The UH is drawn such that the area under the curve is 1 inch.

16.2 THE SCS DIMENSIONLESS HYDROGRAPH METHOD

The method of hydrograph synthesis employed by the Soil Conservation Service (1971) of the U.S. Department of Agriculture uses an average dimensionless hydrograph derived from an analysis of a large number of natural UHs for watersheds

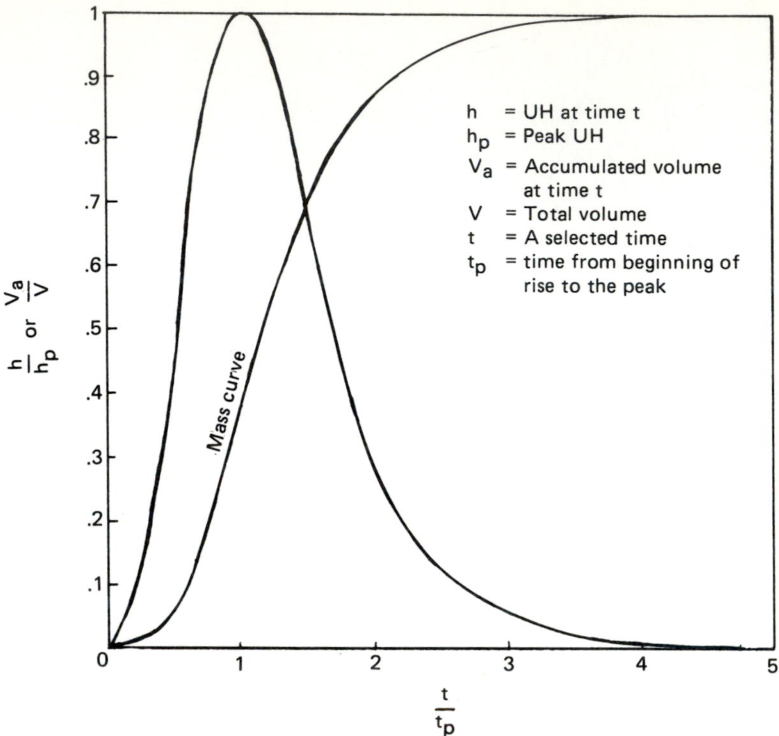

Figure 16.4 Dimensionless unit hydrograph and mass curve (after Soil Conservation Service, 1971).

varying widely in size and geographical locations. This dimensionless UH, as shown in Figure 16.4, has its ordinate expressed as h/h_p and its abscissa as t/t_p, where h is the UH ordinate expressed as the discharge at any time t, h_p is the peak discharge, and t_p is the time from the beginning of rise to the peak. Its ordinates can also be expressed as V_a/V, where V_a is the accumulated value of h at time t, and V is the total volume. This UH has a point of inflection approximately 1.7 times t_p and t_p is approximately $0.2t_b$, the base time. The dimensionless UH, also given in Table 16.1, has approximately 37.5% of the total volume on the rising side, which can be represented by one unit of time and one unit of discharge.

16.3 THE SCS TRIANGULAR HYDROGRAPH METHOD

The SCS method suggests that the dimensionless UH can also be represented by an equivalent triangular hydrograph, as shown in Figure 16.5, having the same units of time and discharge, and having, in turn, the same percentage of volume on the rising side of the triangle. This enables defining t_b, the base of the triangle, in relation to t_p. If one unit of time t_p equals 0.375 of volume, then

$$t_b = 1/0.375 = 2.67 \text{ units of time}$$

TABLE 16.1 RATIOS FOR DIMENSIONLESS UNIT HYDROGRAPH AND MASS CURVE (AFTER SOIL CONSERVATION SERVICE, 1971)

Time ratios (t/t_p)	Discharge ratios (h/h_p)	Mass-curve ratios (V_a/V)
0	0.000	0.000
0.1	0.030	0.001
0.2	0.100	0.006
0.3	0.190	0.012
0.4	0.310	0.035
0.5	0.470	0.065
0.6	0.660	0.107
0.7	0.820	0.163
0.8	0.930	0.228
0.9	0.990	0.300
1.0	1.000	0.375
1.1	0.990	0.450
1.2	0.930	0.522
1.3	0.860	0.589
1.4	0.780	0.650
1.5	0.680	0.700
1.6	0.560	0.751
1.7	0.460	0.790
1.8	0.390	0.822
1.9	0.330	0.849
2.0	0.280	0.871
2.2	0.207	0.908
2.4	0.147	0.934
2.6	0.107	0.953
2.8	0.077	0.967
3.0	0.055	0.977
3.2	0.040	0.984
3.4	0.029	0.989
3.6	0.021	0.993
3.8	0.015	0.995
4.0	0.011	0.997
4.5	0.005	0.999
5.0	0.000	1.000

Chu and Lytle (1972) found for 23 watersheds (0.18 to 750 mi^2, or 0.46 to 1920 km^2) in South Dakota that

$$t_b = 0.1021 \left(\frac{L}{S^{0.5}}\right)^{0.968} \tag{16.9}$$

where L is the main-channel length in miles, and S is the main-channel slope in ft/ft. The correlation coefficient was approximately 0.982.

For simplicity one may use

$$t_b = 0.1 \frac{L}{S^{0.5}} \tag{16.10}$$

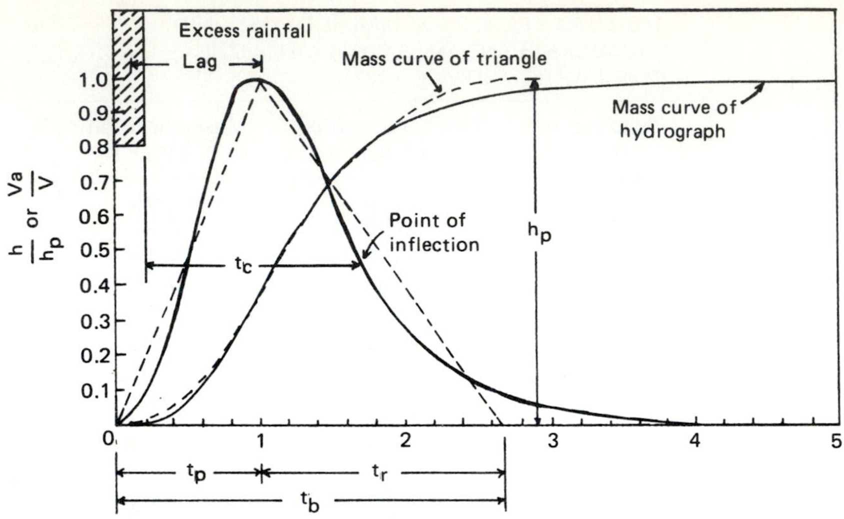

Figure 16.5 Dimensionless curvilinear unit hydrograph and equivalent triangular hydrograph (after Soil Conservation Service, 1971).

Therefore, the time of recession t_r can be computed as

$$t_r = t_b - t_p = 1.67 \text{ units of time} = 1.67 t_p$$

Consequently, the volume under the triangular unit hydrograph V is

$$V = \frac{h_p}{2}(t_p + t_r) \tag{16.11}$$

Hence,

$$h_p = \frac{KV}{t_p} \tag{16.12}$$

where

$$K = \frac{2}{1 + (t_r/t_p)} = 0.75$$

Therefore,

$$h_p = 0.75 V/t_p \tag{16.13}$$

If h_p is to be expressed in volumetric units (L^2/T), then the watershed area and appropriate conversion factor can be multiplied by h_p.

From Figure 16.5, it is clear that

$$t_p = \frac{D}{2} + t_L \tag{16.14}$$

where t_L is the lag time between the center of the ER and the peak discharge, and D is the duration of the ER. Furthermore,

$$t_L = 0.6t_c \tag{16.15}$$

where t_c is the time of concentration. By substituting Equations (16.14) and (16.15) into Equation (16.13),

$$h_p = \frac{1.5V}{D + 1.2t_c} \tag{16.16}$$

Thus, the SCS model reduces to a one-parameter model with t_c as the parameter. Parameter t_c can be computed from watershed characteristics.

The Soil Conservation Service defined t_c in two ways: (1) the time of travel from the divide to the outlet, and (2) the time from the end of the ER to the point of inflection on the recession limb of the UH. The first definition was used to compute t_c in the SCS method. However, the point of inflection, or t_c, as mentioned before, is roughly 1.7 times t_p:

$$t_c = 1.7t_p \tag{16.17}$$

By using Equations (16.14), (16.15), and (16.17),

$$t_c + D = 1.7t_p$$

$$1.2t_c + 2D = 2t_p$$

Solving these equations,

$$D = 0.133t_c \tag{16.18}$$

Thus, the UH is completely specified by one parameter. The above relationships can be used to construct either the triangular UH or the curvilinear UH. The SCS model is one of the popular models for synthesizing the UH for small watersheds of less than 500 mi^2. Morgan and Johnson (1962) in Illinois found that errors in prediction of peak discharge ranged up to nearly 200% on 12 watersheds (10.1 to 101 mi^2, or 25.86 to 258.6 km^2).

EXAMPLE 16.2 Derive a 3-hour unit hydrograph for the watershed of Example 16.1 using the SCS method and compare it with that of the Snyder method.

Solution

$$t_p = \frac{D}{2} + t_L = \tfrac{3}{2} + 4.46 = 5.96 \text{ h}$$

$$t_r = 1.67 \times 5.96 = 9.95 \text{ h}$$

$$t_p = 2.67 \times 5.96 = 15.91 \text{ h}$$

$$h_p = 0.75/5.96 = 0.125/\text{h}$$

The UH is sketched in Figure 16.6. A comparison with the results of the Snyder method shows;

METHOD	h_p (1/h)	t_p (h)	t_b (h)
SCS	0.125	5.96	15.91
Snyder	0.13	5.0	25.0

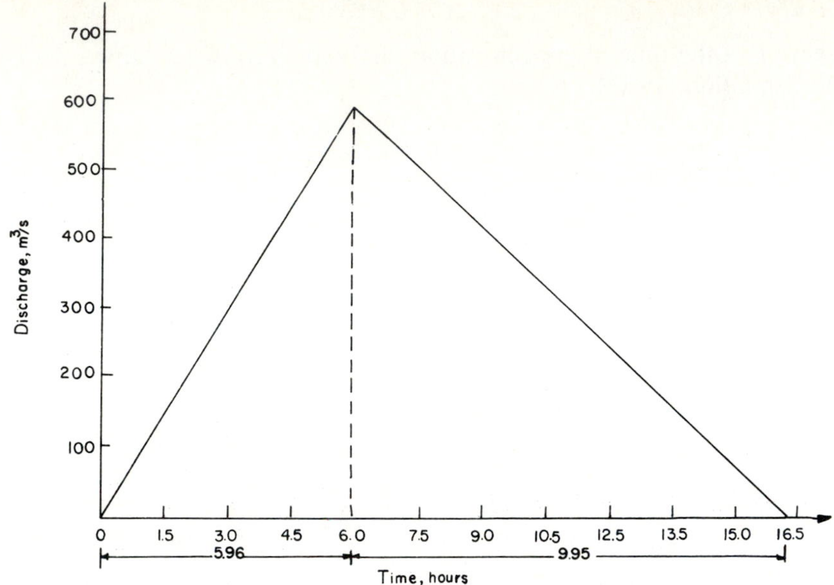

Figure 16.6 Three-hour unit hydrograph from the SCS method.

EXAMPLE 16.3 Derive a synthetic unit hydrograph by the Snyder method for a watershed having the following characteristics:

$$\text{Area} = 10 \text{ mi}^2$$

$$L = 5 \text{ mi}$$

$$L_c = 2 \text{ mi}$$

Take $C_p = 0.6$ and $C_t = 2.0$.

Solution

1. Lag time:

$$t_p = C_t(LL_c)^{0.3} = 2.0 \times (5 \times 2)^{0.3}$$
$$= 4 \text{ h}$$

2. Duration of ER:

$$t_r = t_p/5.5 = 4/5.5$$
$$= 0.73 \text{ h}$$

3. Peak Q_p:

$$Q = \frac{640 C_p A}{t_p} = \frac{640 \times 0.6 \times 10}{4}$$
$$= 960 \text{ cfs}$$

4. Time base:

$$t_b = 4\left(\frac{t_r}{2} + t_p\right) = 4 \times \left(\frac{0.13}{2} + 4\right)$$

$$= 17.5 \text{ h}$$

5. Hydrograph construction: For

$$\frac{960}{10} = 96 \text{ cfs/mi}^2, \; W_{50} = 5.4 \text{ h}$$

and

$$W_{75} = 3.1 \text{ h}$$

The W_{50} and W_{75} are apportioned in a ratio of 1 : 2. The peak ordinate is at 4 h. Therefore, the width of the UH at 50% of the peak is 1.8 h to the left (or 0.73/2 + (4.0 − 1.8) = 2.56 h from the origin) and 3.6 h at 75% peak or (0.365 + 4 + 3.6 = 7.96 h from the origin) to the right of the peak. Thus, coordinates values:

$$(2.56, 480), (7.96, 480)$$

$$(3.33, 720), (6.43, 720)$$

EXAMPLE 16.4 Compute the 4-hour UH hydrograph for the watershed in Example 16.3 using the SCS method.

Solution

1. $t_1 = 4$ h

2. $t_p = \dfrac{D}{2} + t_1$

$$= \frac{4}{2} + 4 = 6 \text{ h}$$

3. $Q_p = \dfrac{484A}{t_p}$

$$= \frac{484 \times 10}{6} = 807 \text{ cfs (at } t = 6 \text{ h)}$$

(a) The time base: $T = 5t_p = 30$ h.
(b) $t/t_p = 0.5$ and $q/q_p = 0.45$; thus, $t = 3$ h and $q = 363$ cfs.
(c) $t/t_p = 2$ and $q/q_p = 0.28$; thus, $t = 12$ h and $q = 226$ cfs.
(d) $t/t_p = 3$ and $q/q_p = 0.07$; thus, $t = 18$ h and $q = 56$ cfs.

16.4 TIME–AREA METHODS

The time–area methods were developed in recognition of the importance of the time distribution of rainfall on runoff in the hydrologic design of storage and regulation works. There are many time–area methods that have appeared in the literature; many of them differ only in the method of presentation. Dooge (1973) has given an excellent survey of these methods. The central idea in these methods is a time contour or an isochrone. An isochrone is a contour joining those points in the watershed that are separated from the outlet by the same travel time. The isochrones cannot cross one another, cannot close, and can only originate or terminate on the watershed boundary. In general, the isochrones have discontinuities at the ridge lines because there the travel time depends on which of the two alternative flow paths the water takes. The isochrones are closely spaced in the flatter, lower portions of the watershed and are more widely spaced in the steeper, upper portions thereof.

The time–area (TA) diagram, therefore, indicates the distribution of travel times of different parts of the watershed. When the area of that part of the watershed whose time of travel is less than or equal to a given value, say, τ, is plotted against the value τ, the time–area diagram is obtained, as shown in Figure 16.7. In other

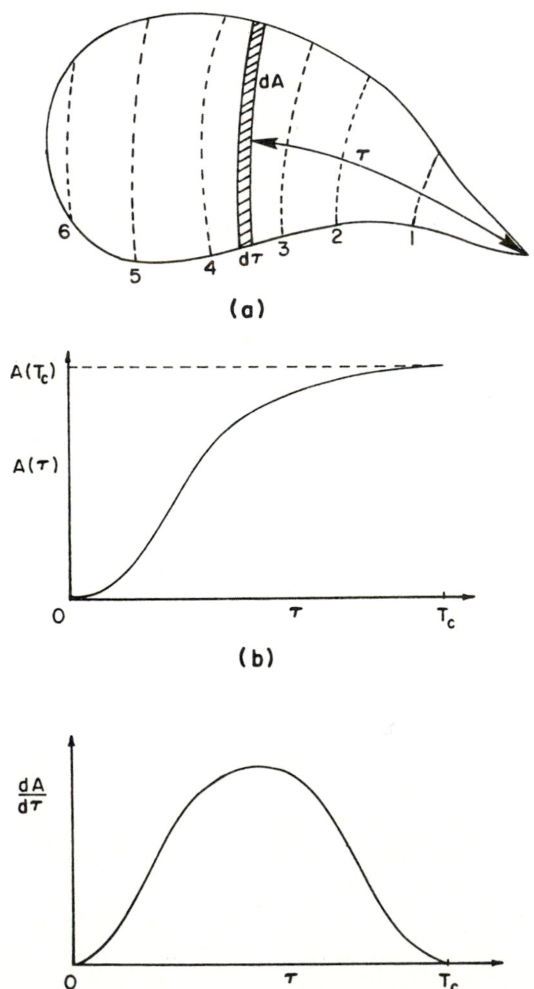

Figure 16.7 (a) Isochrones of travel time, (b) time-area diagram, and (c) time-area-concentration curve.

words, this is the graph of the area enclosed by an isochrone against time. This figure shows for any value of τ, the area that will contribute to the maximum discharge at the outlet due to rainfall of duration equal to τ. It is obvious that $0 \le \tau \le T_c$ and $0 \le A(\tau) \le A(T_c)$.

Often, it is more convenient to use the time–area concentration (TAC) curve, as shown in Figure 16.7(c), which is the derivative of the time–area diagram. Evidently, its base length will correspond to t_c. In hydrologic literature, frequently the TA diagram, the TAC curve, the TA curve, and the TAC diagram are interchangeably used.

16.4.1 Travel Time

The definition of travel time employed in this discussion is the one given by Laurenson (1964). Each point in the watershed can be associated with a particular time period representing the time difference between the occurrence of an element of the ER at that point and the realization of the effect of this element at the outlet. This time period is the travel time of that point. This definition can be made more specific by considering the time of travel of a point to be equal to the time difference between the occurrence of the elemental ER at the point and the center of mass of the resulting DR at the outlet. Therefore, this implies travel of an effect rather than a drop of water. This effect is transmitted by both wave movement and translation of water. This travel time can also be termed as storage delay time.

This definition is at variance with the one that regards it as the time it would take a drop of water to travel from the point to the outlet. As emphasized by Laurenson (1964), it is unrealistic to talk in terms of drops of water. On the ground and in storms, water does not exist as a collection of independent drops, but as an amorphous mass.

16.4.2 Construction of TA and TAC Diagrams

We can construct TA and TAC diagrams in many different ways. We present the dimensionless time–area diagram method.

This method of constructing TA and TAC diagrams is due to Laurenson (1964). The fundamental assumption made here is that the travel time for any element of area is approximately proportional to $\Sigma(L/S^{1/2})$, where L and S are, respectively, the length and slope of any reach of the flow path, and the summation is performed along the flow path from the point in question to the watershed outlet.

Furthermore, in the dimensionless time–area diagram, the abscissa is taken as relative travel time and hence is dimensionless. The relative travel time is the proportion of the maximum travel time, $\tau/\tau_{\max}$. Therefore, the TA diagram can be first proposed without knowing the travel time for any point in the watershed. When the travel time is determined for any point, then the abscissa can be made dimensional. The mechanics of constructing TA and TAC diagrams are explained in the following steps:

1. Draw the elevation contours on the topographic map of the watershed under study.

2. Mark a large number of points on the contours. Distribute these points uniformly over the watershed.
3. Tabulate for each point the distances between adjacent contours along the flow path to the watershed outlet.
4. Raise these individual distances to the 3/2 power because the travel time is assumed proportional to $L/S^{0.5}$ or $L^{3/2}/E^{1/2}$, where E is the contour (elevation) interval. Since E is constant, the travel time is proportional to $L^{3/2}$.

   It may be noted that the watershed outlet may not be on a contour, and a correction would have to be made for the lowest channel segment. This correction amounts to multiplying its length by $(E/E_1)^{1/2}$, where E_1 is the elevation drop of this length.
5. Sum the lengths raised to the 3/2 power for each point.
6. Reduce the sums obtained in step 5 proportionately to give a delay time of unity for the extreme upstream portion of the watershed.
7. Mark the relative delay time obtained in step 6 on the points in the map. Draw the isochrones of relative travel times, $\tau_j, j = 1, 2, \ldots$.
8. Compute the area between the isochrones. This can be done by planimetering. These areas are denoted by $a_j, j = 1, 2, \ldots$.
9. Plot these areas, $a_j, j = 1, 2, \ldots$, against their times of travel $\tau_j, j = 1, 2, \ldots$. This figure gives the TAC histogram. If desired, a continuous curve can be fitted to the histogram.
10. Plot the cumulative areas, $\sum_{j=1} a_j, j = 1, 2, \ldots$, against τ_j, which gives the TA diagram.

16.4.3 Mathematical Representation

From the previous discussion, it is clear that a time–area method is analogous to the UH method and is linear. Therefore, we may want to determine its IUH. First, consider an elemental area dA within a watershed, as shown in Figure 16.8. Let $t - \tau$ be its time of travel. Then the ER at time τ falling on this area will contribute to the discharge at the outlet at time τ. Let this contribution be d$Q(t)$. Therefore,

$$dQ(t) = I(\tau)\, dA(t - \tau) \tag{16.19}$$

By dividing the multiplying the right side by dτ,

$$dQ(t) = \frac{dA(t - \tau)}{d\tau} I(\tau)\, d\tau \tag{16.20}$$

By integrating and using the initial condition $Q(0) = 0$, Equation (16.20) becomes

$$Q(t) = \int_0^t \frac{dA(t - \tau)}{d\tau} I(\tau)\, d\tau \tag{16.21}$$

By comparing Equation (16.21) with Equation (15.10), it is evident that the IUH of the time area method is

$$h(t - \tau) = \frac{dA(t - \tau)}{d\tau} \tag{16.22}$$

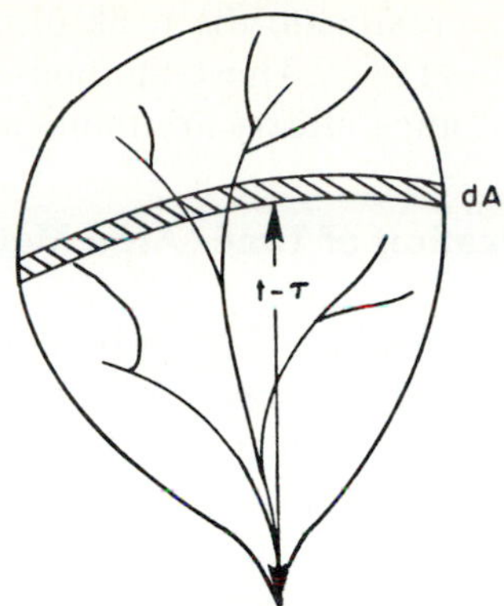

Figure 16.8 Watershed area.

or, simply,

$$h(t) = \frac{dA(t)}{d\tau} \tag{16.23}$$

Therefore, the IUH of the time–area method is the TAC curve. This is the fundamental assumption in all time–area methods.

Similarly, the UH and SH can be determined from the time–area method. The D-time unit UH, $h_D(t)$, can be written as

$$h_D(t) = \frac{1}{D}[U(t) - U(t - D)] \tag{16.24}$$

Then

$$U(t) = \int_0^t h(x)\, dx = \int_0^t \frac{dA}{d\tau}\, d\tau = A(t) \tag{16.25}$$

Similarly,

$$U(t - D) = A(t - D) \tag{16.26}$$

Therefore,

$$A(t) - A(t - D) = Dh_D(t) \tag{16.27}$$

The quantity D is a finite number to be chosen arbitrarily. Equation (16.27) has an important implication in hydrology. This says that the TA diagram can be constructed from a D-hour UH. This relationship was utilized by Mathur (1974) in representing a natural watershed by a series of linear channels.

16.4.4 Assumptions and Limitations

The fundamental assumption in the time–area methods is one of translation. They allow for the delay experienced by water in reaching the watershed outlet. They also allow for spatially nonuniform effective rainfall. However, they do not allow for the storage effects that are primarily responsible to produce attenuation in peak flow and to increase the time base of the hydrograph. It is, therefore, not surprising that the time–area methods tend to overestimate the peak discharge. This is evident from the results reported by Mathur (1974). These methods may yield acceptable results for small watersheds where storage effects are minimal.

16.4.5 Procedure for Application of Time–Area Methods

If a watershed is divided into a number of strips by drawing isochrones and if to the ER occurring on each strip the time of travel from the strip to the outlet is applied, the resulting outflow from the whole watershed at its outlet will specify the DR hydrograph. In other words, the rate of runoff from any strip is equal to the intensity of the ER on that strip, and this runoff will arrive at the watershed outlet by the time of concentration of the strip. It may be noted that the time of concentration of each strip is assumed independent of the rate of runoff from the strip and from the surrounding strips. Nash (1958) perhaps explained it best by an example:

> The runoff at any time t is equal to the area enclosed by the 1-hour contour multiplied by the mean intensity of effective rainfall of the hour previous to t plus the additional area enclosed by the 2-hour contour multiplied by the mean intensity of effective rainfall during the period from 1 to 2 hours before t and so on.

For purposes of applying the time–area method, let us consider a watershed, as shown in Figure 16.9(a). Determine its T_c and divide it into a number of time intervals of equal length. Draw the isochrones to divide the watershed into the same number of strips as the number of intervals. Determine the area of these strips and plot the time–area histogram, as shown in Figure 16.9(b).

Consider the ERH for which the DRH is desired. Replot this hyetograph such that the time interval is the same as for the TAC diagram. This is shown in Figure 16.9(c). By applying the time–area concept, the DRH can be mathematically expressed as

$$Q_i = \sum_{j=1}^{i} a_j I_{i-j} \tag{16.28}$$

where a_i is the area enclosed by the ith isochrone, and I_i is the intensity of effective rainfall in the ith interval. Equation (16.28) is a discrete convolution of I and a.

EXAMPLE 16.5 Consider an effective rainfall hyetograph as

TIME (min)	INTENSITY (cm/h)
0	5
5	4
10	5
15	4
20	2
25	0

and the time–area histogram for a specified watershed as

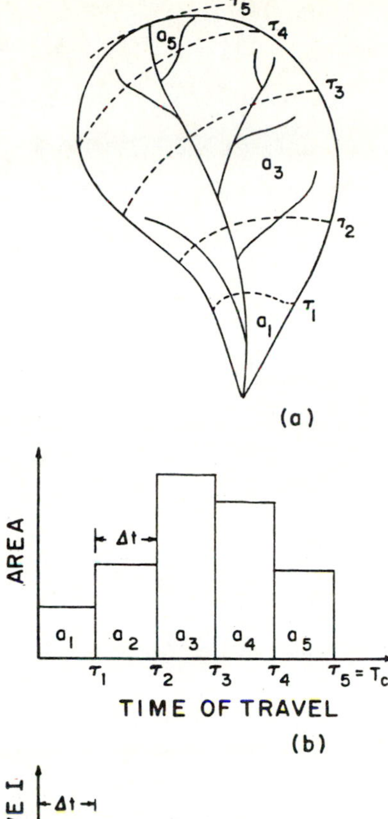

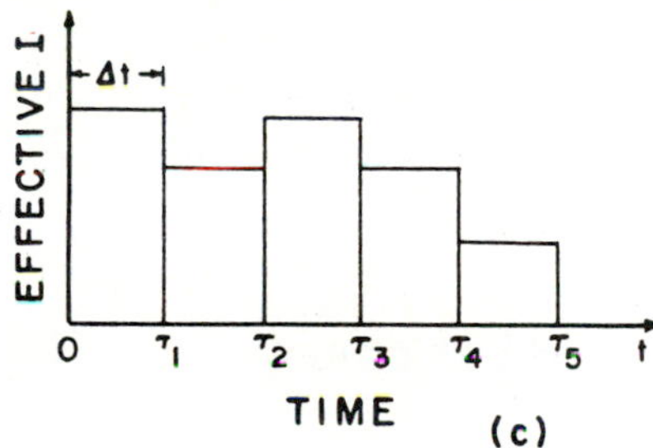

EFFECTIVE I

Δt

0 τ1 τ2 τ3 τ4 τ5 t

TIME (c)

Figure 16.9 (a) Watershed decomposition by isochrones, (b) time-area histogram, and (c) effective rainfall hyetograph.

TIME (min)	AREA (ha)
5	1
10	2
15	4
20	3
25	2

Compute the runoff hydrograph due to the effective rainfall.

Solution The watershed area is 12 ha. We compute the runoff hydrograph ordinates using Equation (16.28).

$Q_0 = a_0 I_0 = 0$

$Q_1 = a_1 I_0 = 1 \times 5 = 5$

$Q_2 = a_1I_1 + a_2I_0 = 1 \times 4 + 2 \times 5 = 14$

$Q_3 = a_1I_2 + a_2I_1 + a_3I_0 = 1 \times 5 + 2 \times 4 + 4 \times 5 = 33$

$Q_4 = a_1I_3 + a_2I_2 + a_3I_1 + a_4I_0 = 1 \times 4 + 2 \times 5 + 4 \times 4 + 3 \times 5 = 45$

$Q_5 = a_1I_4 + a_2I_3 + a_3I_2 + a_4I_1 + a_5I_0 = 1 \times 2 + 2 \times 4 + 4 \times 4 + 3 \times 4 + 2 \times 5 = 52$

$Q_6 = a_2I_4 + a_3I_3 + a_4I_2 + a_5I_1 = 2 \times 2 + 4 \times 4 + 3 \times 5 + 2 \times 4 = 43$

$Q_7 = a_3I_4 + a_4I_3 + a_5I_2 = 4 \times 2 + 3 \times 5 + 2 \times 4 = 31$

$Q_8 = a_4I_4 + a_5I_3 = 3 \times 2 + 2 \times 4 = 14$

$Q_9 = a_5I_4 = 2 \times 4 = 8$

$Q_{10} = 0$

By dividing Q_i, $i = 1, 2, \ldots, 10$, by the watershed area, the runoff is obtained in cm/h:

TIME (min)	RUNOFF (cm/h)
0	0
5	0.42
10	1.17
15	2.75
20	3.75
25	4.33
30	3.58
35	2.58
40	1.17
45	0.67
50	0

EXERCISES

16.1. Derive the unit hydrograph based on Snyder's method for a watershed where A = 20.5 m^2, L = 10 mi, L_c = 5.8 mi, and c_t = 2.0.

16.2. Use the SCS method to derive the unit hydrograph for the watershed in Exercise 16.1. The mean channel slope is 0.2%.

16.3. The stream length for a basin is L = 6.56 mi, and the length to the centroid is L_c = 3.5 mi. Using an average value of C_t as 2.0 and an average C_p of 0.60, derive Snyder's unit hydrograph for this basin. Use a W_{50} = 9 h and W_{75} = 5 h. Convert Snyder's UH to a 2-hour duration before plotting.

16.4. It is desired to construct a highway embankment just downstream of an area having the following characteristics; drainage area = 2.5 mi^2, length = 3.2 mi, slope = 2.1%, land use: medium density residential ($\frac{1}{4}$-acre lots), soils: 50% D and 50% C group. Derive a 1-h UH for this area by the SCS method. Derive a hydrograph for this area assuming it will have 2.0 in. of runoff. Use 1 cfs per square mile for baseflow.

16.5. Consider a drainage area of 2.5 mi^2 with main stream length = 2.6 mi, and length to centroid = 1.4 mi. Using Snyder's coefficients C_t = 1.95 and C_p = 0.62, determine the basin lag time and time of concentration. Determine the peak of Snyder's unit hydrograph, the 50% width, the 75% width, and the duration.

16.6. Select a proper duration for the drainage area of Exercise 16.5. Then derive the UH for this duration. Convert the UH of Exercise 16.5 to a 3-h duration.

16.7. Derive a synthetic UH by Snyder's method for a watershed having the following characteristics: area = 15 mi^2, L = 7 mi, and L_c = 2 mi. Take C_t = 2.0 and C_p = 0.6.

16.8. Compute a 4-h UH for the watershed in Exercise 16.7 using the SCS method, and compare with the UH derived in Exercise 16.7.

16.9. Consider a rectangular watershed 100 m wide and 1000 m long. Its average slope is 0.01. Compute its time of concentration and lag time. Construct the time–area curve and time–area–concentration curve. If the effective rainfall is a rectangular pulse of 5 cm/h for a period of 2 hours, compute the DRH for this watershed.

16.10. Consider a triangular watershed 200 m wide at the upstream boundary and 1000 m long. The watershed is symmetric about its length of flow and has a slope of 0.01 in the direction of flow. Construct the TA and TAC curves. Compute the DRH for an effective rainfall pulse of 5 cm/h for a period of 2 hours. Compare this DRH with the one of Exercise 16.9.

16.11. A basin has an area of 360 km^2. L = 25.75 km and L_c = 9.66 km. Derive a synthetic unit hydrograph for this basin.

16.12. For a drainage basin whose area approximates 100 hectares, develop a 4-h unit hydrograph using the methods of Snyder and SCS. The following information regarding the basin topography is available: L = 4 kilometers and L_c = 1.5 kilometers.

16.13. For a drainage basin whose area approximates 60 hectares, develop a 6-h unit hydrograph using the methods of Snyder and SCS. The following information regarding the basin topography is available: L = 2 kilometers and L_c = 1 kilometer.

16.14. The following information is available on a small natural watershed:

Area = 0.9 square miles

Length of the main water course = 1.20 miles

Distance from the outlet to a point on the stream nearest the centroid of the basin = 0.90 miles

Coefficient C_t = 0.10

Coefficient C_p = 0.6

(a) Use Snyder's method to derive the unit hydrograph.
(b) Derive the unit hydrograph using the SCS method.
(c) Compare the results of the two methods.

16.15. Determine the peak of Snyder's unit hydrograph, the 50% width, the 75% width, and the duration. Convert this unit hydrograph to a 3-h duration UH (just convert the lag and peak). The watershed has a drainage area = 2.5 mi^2, main stream length = 2.6 mi, and length to centroid = 1.4 mi. Take C_t = 1.95 and C_p = 0.62.

CHAPTER 17
Conceptual Models of the Unit Hydrograph

The unit hydrograph (UH) serves as a link between the direct-runoff hydrograph (DRH) and the effective rainfall hyetograph (ERH). This link can be established using the so-called conceptual elements. These elements are linear channels and linear reservoirs. These elements have certain properties that characterize watershed response to rainfall and, therefore, can be utilized to derive the instantaneous unit hydrograph (IUH) of the watershed.

17.1 WATERSHED RESPONSE

Runoff from a watershed consists of two phases: (1) the overland flow phase, or the land phase, and (2) the channel phase. Inflow at any point in the watershed travels a certain distance to reach the point of measurement. During this travel, it undergoes changes caused by overland and channel storage. The transformation undergone by the inflow is due to (1) translation and (2) a storage effect consisting of (a) time lag and (b) shape modification.

Let $f_1(t)$ be the inflow rate, and $f_2(t)$ the outflow from a channel, as shown in Figure 17.1; $f_2(t)$ is an exact reproduction of $f_1(t)$, shifted by a time interval C.

$$f_2(t) = f_1(t - C), \qquad t \geq C$$
$$f_2(t) = 0, \qquad t < C \tag{17.1}$$

This is an example of pure translation, or we can write

$$f_1(t) = f_2(t + C), \qquad t \geq 0 \tag{17.2}$$

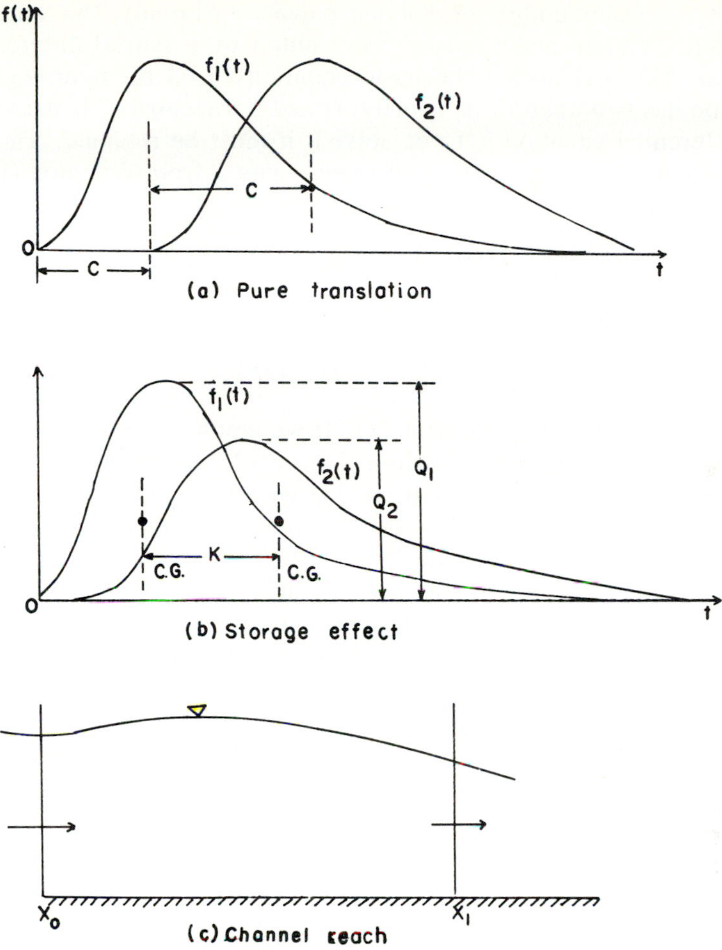

Figure 17.1 (a) Pure translation, (b) storage effect, and (c) channel reach.

Figure 17.1 also gives an example of the storage effect. The inflow is given by $f_1(t)$ and the outflow by $f_2(t)$. The center of the area of inflow is shifted by k hours, which is a measure of lag time. The reduction of peak from Q_1 to Q_2 may be taken as a measure of the change in shape.

17.1.1 Continuity Equation

The spatially lumped form of the continuity equation for a watershed can be written as

$$I(t) = Q(t) + \frac{dS(t)}{dt} \tag{17.3}$$

where I is the rate of inflow, Q is the rate of outflow, and S is the storage. These quantities are all functions of time t. Equation (17.3) is also referred to as an equation of the water budget or volume balance. Although the watershed is truly a distributed system and must be represented by a partial differential equation(s), Equation (17.3) is one of the basic equations used for hydrologic analysis. This equation has two unknowns, namely, Q and S. Because Q is the variable for which the differential equation is to be solved, it must be retained. Then we must have another equation that can be used to eliminate S from Equation (17.3).

17.1.2 Storage Characteristics

Perhaps the simplest storage–discharge relationship used in hydrology is

$$S = kQ \tag{17.4}$$

where k is the storage parameter (T). If we graph S versus Q for a watershed, the plot is normally a loop, as shown in Figure 17.2. Equation (17.4) is a straight-line approximation of the loop. Both are, however, single-valued, whereas the actual relationship between Q and S is double-valued. A plot of Q versus time and S versus time for actual rainfall storms will show that for watersheds that are too small, the peak of storage that more or less occurs at or near the end of the effective rainfall precedes the peak of direct runoff by a significant amount of time. However, Equation (17.4) yields that S and Q peak concurrently. Substituting Equation (17.4) into Equation (17.3) yields

$$I = Q + k\frac{dQ}{dt} \tag{17.5}$$

The peak of Q occurs mostly after I has ceased to exist. If $I = 0$, Equation (17.5) becomes

$$-Q = k\frac{dQ}{dt} \tag{17.6}$$

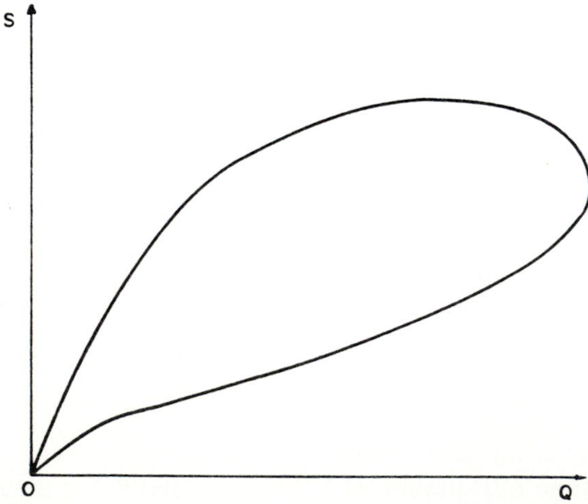

Figure 17.2 A typical storage–discharge relation.

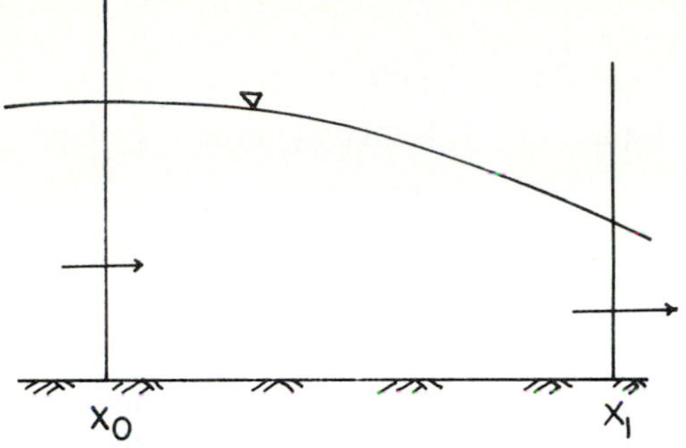

Figure 17.3 A channel reach.

When $Q = Q_p$, $dQ/dt = 0$, where Q_p is the peak of Q. Then Equation (17.6) gives $Q = 0$ at $Q = Q_p$, a patently absurd result. This indicates that only the recession side of the runoff hydrograph, preferably after the inflection point, can be approximated by Equation (17.5). This may not be appropriate for the rising limb of the hydrograph. Besides Q, there are obviously other variables on which S depends.

Storage in a channel reach consists of prism storage and wedge storage. Under steady-state conditions, only prism storage exists. Wedge storage is associated with unsteady flow. Wedge storage is positive for the rising hydrograph and negative for the falling hydrograph. A plot of Q versus S for steady flow conditions in a channel will be a single-valued curve and the loop is caused by the superposition of wedge storage over prism storage. The nature of the loop indicates that the effects of unsteady flow are considerable and must be taken into consideration. Even in channel reaches, it has been found that the storage is a function not only of inflow, but of outflow also. It is all the more so when the storage on the entire watershed is considered. It then appears that storage on the watershed is a function of inflow, outflow, and their time derivatives. Nevertheless, for purposes of simplicity, Equation (17.4) is frequently employed and has been found to yield reasonable results.

17.2 LINEAR CHANNEL

The equation of continuity in one-dimensional form for a channel with no lateral inflow can be written as

$$\frac{\partial Q}{\partial x} + \frac{\partial A}{\partial t} = 0 \tag{17.7}$$

and the properties of the channel can be assumed to be

$$A = C(x)Q \tag{17.8}$$

where A is the cross-sectional area of flow, Q is the rate of outflow, and C is a constant at a section of the channel varying with distance. Combining Equations (17.7) and (17.8) leads to

$$\frac{\partial Q}{\partial x} + C(x)\frac{\partial Q}{\partial t} = 0 \tag{17.9}$$

Referring to Figure 17.3, let the inflow at x_0 be denoted by $Q(x_0,t)$ and let

$$Q(x_0,t) = f_0(t) \tag{17.10}$$

Given $Q(x_0,t)$, the solution of Equation (17.9) gives Q at any x. Let

$$Q(x_1,t) = f_1(t) \tag{17.11}$$

It can be shown by solving Equation (17.9) that

$$Q(x_1,t) = f_1(t) = f_0(t - T) \tag{17.12}$$

where T is the time of travel from x_0 to x_1. The outflow at x_1 is merely the inflow at x_0 delayed by time T. Equation (17.8) defines a linear channel whose effect on inflow is pure translation, that is, it shifts the function $Q(x_0,t)$ by a time interval equal to T, where

$$T = C(x_1 - x_0) \tag{17.13}$$

Thus,

$$Q(x_1,t) = Q(x_0,t - T), \qquad t \geq T \tag{17.14}$$

The translation time of a linear channel is also referred to as delay time.

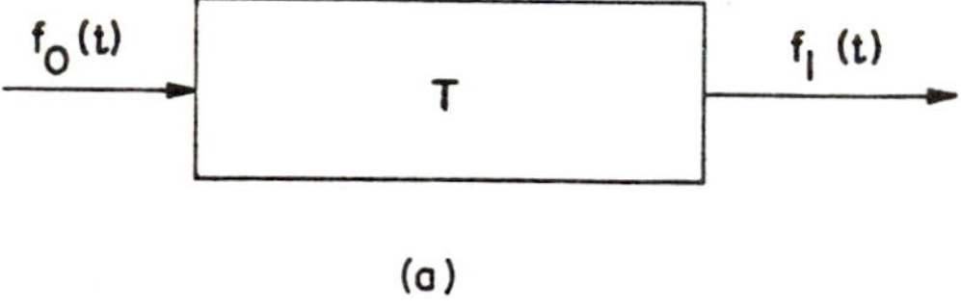

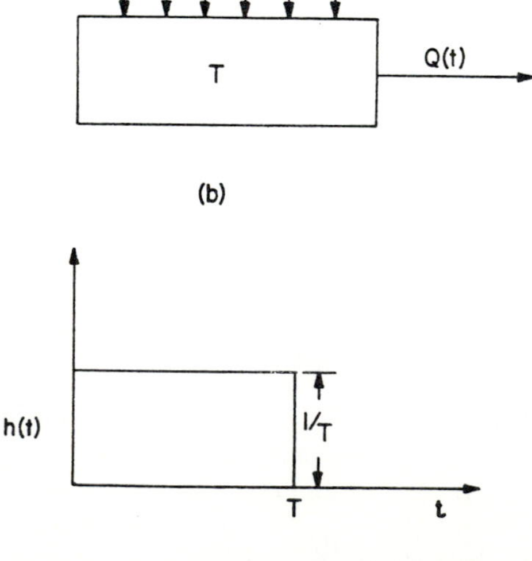

Figure 17.4 A linear channel: (a) pure translation, (b) lateral inflow, and (c) IUH.

The linear channel is analogous to the idea of a linear reservoir, to be discussed in the ensuing section, where the storage–outflow curve is a straight line. The channel unaffected by backwater has a definite rating curve at every point of the channel. For a linear channel, the rating curve at every point specifies a linear relationship between discharge and area. This implies that the velocity is constant for all discharges, but may vary from point to point along the reach.

17.2.1 Instantaneous Unit Hydrograph

If the inflow, represented by the delta function $\delta(t)$, is taking place only at the upstream end of the channel, then referring to Figure 17.4, its IUH, $h(t)$, is

$$h(t) = \delta(t - T) \tag{17.15}$$

where T is the translation time. If, however, the input is considered to occur uniformly along its length, then the IUH is

$$h(t) = \frac{1}{T}[u_1(t) - u_1(t - T)] \tag{17.16}$$

where $u_1(t)$ signifies the unit step function defined as

$$\begin{aligned} u_1(t) &= 0, \qquad t < 0 \\ &= 1, \qquad t \geq 0 \end{aligned}$$

The form of the IUH is shown in Figure 17.4(c).

The IUH in Equation (17.16) has only one parameter, T, which can be estimated by the first moment, M_1, as

$$M_1 = \frac{T}{2} \tag{17.17}$$

17.2.2 Direct-Runoff Hydrograph

The DRH of a linear channel can be determined for any arbitrary ERH. Since an ERH is composed of a number of pulses, it can be expressed by using the unit step function $u_1(t)$ as

$$\begin{aligned} I(t) = I_0 u_1(t) &+ \sum_{j=1}^{m} (I_j - I_{j-1}) u_1(t - jD) \\ &+ I_m u_1[t - (m + 1)D] \end{aligned} \tag{17.18}$$

where D is the time interval, m is the number of pulses, and I_j, $j = 0, 1, 2, \ldots, m$, are ER intensities, as shown in Figure 17.5. If we let

$$w_0 = I_0, \qquad w_j = I_j - I_{j-1}, \qquad j = 1, 2, \ldots, m$$

then

$$I(t) = \sum_{j=0}^{m} w_j u_1(t - jD) \tag{17.19}$$

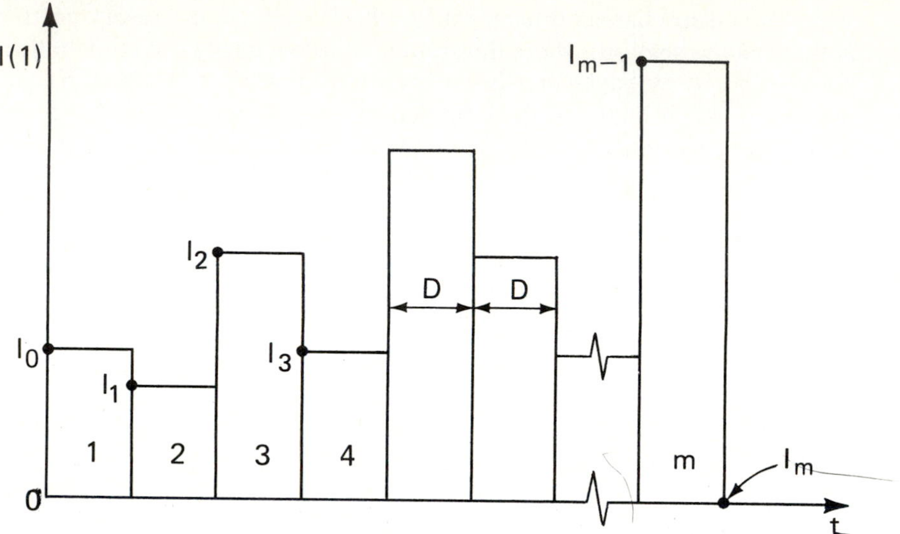

Figure 17.5 An effective rainfall hyetograph (ERH) composed of m pulses, each of duration of D time units.

The DRH of a linear channel with its IUH given by Equation (17.15) is

$$\begin{aligned} Q(t) &= \int_0^t I(\tau)\,\delta(t - T - \tau)\,d\tau = I(t - T) \\ &= \sum_{j=0}^{m} w_j u_1(t - jD - T) \end{aligned} \tag{17.20}$$

On the other hand, if the IUH of a channel is prescribed by Equation (17.16), then the DRH is

$$\begin{aligned} Q(t) &= \frac{1}{T}[u_1(t) - u_1(t - T)] \int_0^t I(\tau)\,d\tau \\ &= \frac{1}{T}\int_{jD}^{t} [u_1(t) - u_1(t - T)] \sum_{j=0}^{m} w_j u_1(t - jD)\,d\tau \end{aligned}$$

Since the ERH and IUH are of finite durations, the limits of the integral should be properly considered.

$$\begin{aligned} Q(t) = \frac{1}{T}\sum_{j=0}^{m} w_j[u_1(t - jD)(t - jD) \\ - u_1(t - jD - T)(t - jD - T)] \end{aligned} \tag{17.21}$$

These mathematical operations can be conveniently done using the Laplace transform.

EXAMPLE 17.1 Consider a linear channel having a translation time of 6 hours. An ERH is given by Equation (17.18) with $I_0 = 5$ cm/h, $I_1 = 2$ cm/h, $I_2 = 1$ cm/h, and $I_3 = 2$ cm/h, $I_4 = 0$ cm/h, $D = 0.5$ h, and $m = 4$. Compute the DRH with the IUH given by Equation (17.16).

Solution The ERH can be expressed by Equation (17.19) in which

$$w_0 = 5, \quad w_1 = -3, \quad w_2 = -1, \quad w_3 = 1, \quad \text{and} \quad w_4 = -2$$

Therefore,

$$I(t) = 5u_1(t) - 3u_1(t - 0.5) - u_1(t - 1) + u_1(t - 1.5) - 2u_1(t - 2)$$

Its Laplace transform is

$$I(s) = \frac{1}{s}[5 - 3 \exp(-0.5s) - \exp(-s) + \exp(-1.5s) - 2 \exp(-2s)]$$

We are given

$$h(t) = \frac{1}{6}[u_1(t) - u_1(t - 6)]$$

Its Laplace transform is

$$h(s) = \frac{1}{6s}[1 - \exp(-6s)]$$

Thus, the DRH is given by

$$\begin{aligned}
Q(s) &= \frac{1}{6s}[1 - \exp(-6s)][5 - 3 \exp(-0.5s) - \exp(-s) \\
&\quad + \exp(-1.5s) - 2 \exp(-2s)]\frac{1}{s} \\
&= \frac{1}{6s^2}[5 - 3 \exp(-0.5s) - \exp(-s) + \exp(-1.5s) \\
&\quad - 2 \exp(-2s) - 5 \exp(-6s) + 3 \exp(-6.5s) \\
&\quad + \exp(-7s) - \exp(-7.5s) + 2 \exp(-8s)]
\end{aligned}$$

By inverting the Laplace transform,

$$\begin{aligned}
Q(t) &= \frac{1}{6}[5t - 3u_1(t - 0.5)(t - 0.5) - u_1(t - 1)(t - 1) \\
&\quad + u_1(t - 1.5)(t - 1.5) - 2u_1(t - 2)(t - 2) \\
&\quad - 5u_1(t - 6)(t - 6) + 3u_1(t - 6.5)(t - 6.5) \\
&\quad + u_1(t - 7)(t - 7) - u_1(t - 7.5)(t - 7.5) \\
&\quad + 2u_1(t - 8)(t - 8)
\end{aligned}$$

The results of the calculations are as follows:

TIME (h)	DISCHARGE (cm/h)	TIME (h)	DISCHARGE (cm/h)
0	0	4.5	0.833
0.5	0.417	5.0	0.833
1.0	0.583	5.5	0.833
1.5	0.667	6.0	0.833
2.0	0.833	6.5	0.417
2.5	0.833	7.0	0.250
3.0	0.833	7.5	0.167
3.5	0.833	8.0	0.000
4.0	0.833		

17.3 LINEAR RESERVOIR

The continuity equation for a reservoir is given by Equation (17.5). A linear reservoir is defined as the storage–discharge relation given by Equation (17.4). Equations (17.3) and (17.4), when combined, yield

$$I - Q = k\frac{dQ}{dt} \tag{17.22}$$

By utilizing the operator notation, $D = d/dt$,

$$Q = \frac{1}{1 + kD} I(t)$$

Mathematically, this is equivalent to

$$Q = e^{-t/k} \int e^{t/k} I \, dt \tag{17.23}$$

The operator $1/(1 + kD)$ represents its effect on the inflow. Let

$$H = \frac{1}{1 + kD}$$

The operator H represents the effect of a linear reservoir with a proportionality coefficient k and is often referred to as the storage operator.

17.3.1 Instantaneous Unit Hydrograph

The IUH of the linear reservoir obtained by replacing $I(t)$ by $\delta(t)$ in Equation (17.23) is

$$h(t) = \frac{e^{-t/k}}{k} \tag{17.24}$$

It may be noted that the IUH can also be derived rather conveniently from the S-hydrograph, $U(t)$, of the linear reservoir, which is obtained from Equation (17.23) with $I = 1$:

$$U(t) = 1 - e^{-t/k} \tag{17.25}$$

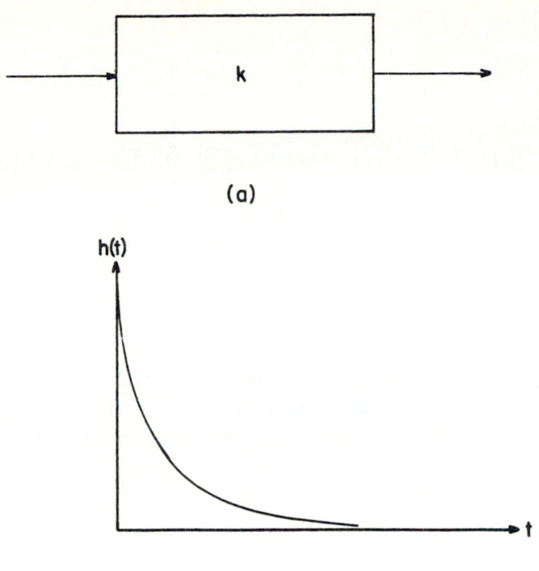

Figure 17.6 A linear reservoir: (a) lag time k and (b) IUH.

By differentiating with respect to t,

$$h(t) = \frac{\mathrm{d}U(t)}{\mathrm{d}t} = \frac{e^{-t/k}}{k}$$

Its appearance is shown in Figure 17.6. If the inflow were a unit pulse of duration D, the UH $h(D;t)$, as shown in Figure 17.7, and the direct runoff $Q(t)$ (ignoring the

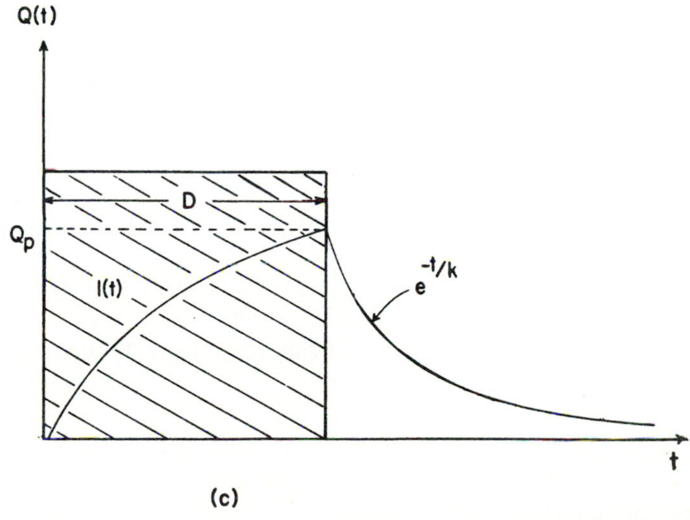

Figure 17.7 A linear reservoir: Hydrograph due to a pulse of D-hour duration.

dimensions of $h(D;t)$ and $Q(t)$) would be

$$Q(t) = h(D;t) = I[1 - e^{-t/k}], \qquad t \leq D \tag{17.26a}$$

and

$$Q(t) = h(D;t) = Q_p e^{-(t-D)/k}, \qquad t \geq D \tag{17.26b}$$

where Q_p represents the hydrograph peak and is given by

$$Q_p = I[1 - e^{-D/k}] \tag{17.26c}$$

Here $I = 1/D$. We note that the hydrograph peak will occur at the end of the duration of the unit pulse.

As $t \to \infty$, $Q(t) = I(t)$. This implies an equilibrium condition: outflow becoming equal to inflow. As $t \to 0$, $Q = 0$. As $I(t)$ terminates at $t = D$, the recession starts immediately. For an instantaneous inflow that fills the reservoir of storage S in $t = 0$, $Q_p = S/k$. The equation of outflow is simply

$$Q(t) = \frac{S}{k} e^{-t/k} \tag{17.27}$$

For a unit inflow, $S = 1$ and $I(t) = \delta(t)$. Consequently, $Q(t) \to h(t)$, the IUH given by Equation (17.27) is the same as given by Equation (17.24).

17.3.2 Properties of the Impulse Response

An impulse form of the ER has a finite volume V or S. Equation (17.27) prescribes the response of the linear reservoir due to such an impulse. At $t = 0$, $Q(t) = Q_0$. Therefore,

$$Q_0 = \frac{S}{k} \tag{17.28}$$

The initial state Q_0 is directly proportional to the impulse ER volume S and inversely proportional to parameter k. Equation (17.27) can then be written as

$$Q(t) = Q_0 \exp(-t/k) \tag{17.29}$$

The volume of the impulse response is

$$V = \int_0^\infty Q(t)\, dt = \int_0^\infty Q_0 \exp(-t/k)\, dt = Q_0 k \tag{17.30}$$

We can determine the lag time of this response. To this end, we compute the first moment about the Q-axis:

$$M_1^Q = \int_0^\infty t \cdot Q(t) \cdot dt = \int_0^\infty t \cdot Q_0 \exp(-t/k)\, dt = Q_0 k^2 \tag{17.31}$$

Therefore, the lag time T is

$$T = \frac{M_1^Q}{V} = \frac{Q_0 k^2}{Q_0 k} = k \tag{17.32}$$

Thus, parameter k of the linear reservoir is identical to the reservoir lag time, or the first moment of the impulse response.

Likewise, we can compute the mean discharge. We compute the first moment about the t axis:

$$M_1^t = \int_0^\infty \frac{[Q(t)]^2}{2} \, dt = \frac{Q_0^2}{2} \int_0^\infty \exp(-2t/k) \, dt = \frac{Q_0^2 k}{4} \tag{17.33}$$

Therefore, the mean discharge Q_m is

$$Q_m = \frac{M_1^t}{V} = \frac{Q_0^2 k}{4} \cdot \frac{1}{Q_0 k} = \frac{Q_0}{4} \tag{17.34}$$

17.3.3 Parameter Estimation

Parameter k has the dimension of time and is equal to the average lag time imposed on the inflow by the reservoir. This is referred to as lag time and equals the difference in time between the centers of areas of inflow and outflow. Evidently, for a particular event on a gaged watershed, k can be estimated from records of inflow and outflow provided certain qualifications are met. For example, for a rainfall–runoff event, these qualifications are as follows: (1) The event is relatively isolated in time. (2) The event is fairly uniformly distributed over the watershed. (3) The event has a single well-defined peak. The value of k would be constant if the UH theory holds. However, its variation with rainfall and watershed characteristics has been reported by various investigators. Parameter k can be estimated by computing the first moment of the IUH in Equation (17.24) about its origin, which is specified as

$$M_1 = k \tag{17.35}$$

17.3.4 Direct-Runoff Hydrograph

The DRH of a linear reservoir for an arbitrary ERH can be determined by either direct convolution or using Laplace transform:

$$Q(t) = \int_0^t \frac{1}{k} \exp[-(t - \tau)/k] I(\tau) \, d\tau$$

By inserting Equation (17.19) for $I(t)$,

$$\begin{aligned} Q(t) &= \sum_{j=0}^{m} w_j u_1(t - jD) \frac{1}{k} \int_{jD}^{t} \exp[-(t - \tau)/k] \, d\tau \\ &= \sum_{j=0}^{m} w_j u_1(t - jD)\{1 - \exp[-(t - jD)/k]\} \end{aligned} \tag{17.36}$$

This same expression can also be obtained by using the Laplace transform. Note that

$$h(s) = \frac{1}{1 + ks}$$

$$I(s) = \sum_{j=0}^{m} w_j \exp(-jDs)/s$$

Therefore,

$$Q(s) = \sum_{j=0}^{m} w_j \exp(-jDs)/[s(1 + ks)]$$

Upon inversion,

$$Q(t) = \sum_{j=0}^{m} w_j u_1(t - jD)\{1 - \exp[-(t - jD)/k]\}$$

which is the same as Equation (17.36).

EXAMPLE 17.2 Consider a linear reservoir with $k = 6$ hours. Compute the DRH for the ERH of Example 17.1.

Solution

$$I(t) = 5u_1(t) - 3u_1(t - 0.5) - u_1(t - 1)$$
$$+ u_1(t - 1.5) - 2u_1(t - 2)$$

Its Laplace transform is

$$I(s) = \frac{1}{s}[5 - 3\exp(-0.5s) - \exp(-s)$$
$$+ \exp(-1.5s) - 2\exp(-2s)]$$

The IUH is

$$h(t) = \frac{1}{k}\exp(-t/k)$$

Let $a = 1/k$. The Laplace transform of $h(t)$ is

$$h(s) = a\left(\frac{1}{s + a}\right)$$

Therefore,

$$Q(s) = h(s)I(s)$$
$$= a\left(\frac{1}{s + a}\right)[5 - 3\exp(-0.5s) - \exp(-s)$$
$$+ \exp(-1.5s) - 2\exp(-2s)]\frac{1}{s}$$
$$= \frac{1}{s(s + a)}[5 - 3\exp(-0.5s) - \exp(-s)$$
$$+ \exp(-1.5s) - 2\exp(-2s)]$$

The quantity

$$\frac{1}{s(s + a)} = \frac{1}{sa} - \frac{1}{a(s + a)}$$

Therefore,

$$Q(s) = 5\left(\frac{1}{s} - \frac{1}{s+a}\right) - 3\exp(-0.5s)\left(\frac{1}{s} - \frac{1}{s+a}\right)$$
$$- \exp(-s)\left(\frac{1}{s} - \frac{1}{s+a}\right) + \exp(-1.5s)\left(\frac{1}{s} - \frac{1}{s+a}\right)$$
$$- 2\exp(-2s)\left(\frac{1}{s} - \frac{1}{s+a}\right)$$

On inverting,

$$Q(t) = 5u_1(t)[1 - \exp(-t/k)] - 3u_1(t - 0.5)$$
$$\{1 - \exp[-(t - 0.5)/k]\} - u_1(t - 1)$$
$$\{1 - \exp[-(t - 1)/k]\} + u_1(t - 1.5)$$
$$\{1 - \exp[-(t - 1.5)/k]\}$$
$$- 2u_1(t - 2)\{1 - \exp[-(t - 2)/k]\}$$

The results of the calculations are as follows:

TIME (h)	DISCHARGE (cm/h)	TIME (h)	DISCHARGE (cm/h)
0	0	5.5	0.38
0.5	0.4	6.0	0.35
1.0	0.53	6.5	0.32
1.5	0.57	7.0	0.30
2.0	0.68	7.5	0.27
2.5	0.63	8.0	0.25
3.0	0.58	9.0	0.21
3.5	0.53	12.0	0.13
4.0	0.49	16.0	0.07
4.5	0.45	20.0	0.03
5.0	0.41		

Linear channels and linear reservoirs, in conjunction with time–area–concentration (TAC) curves, have been variously combined to derive the IUH of a watershed (Dooge, 1959). Most of the conceptual models employ some arrangement of these components, as shown in Figure 17.8. We discuss some of these models here.

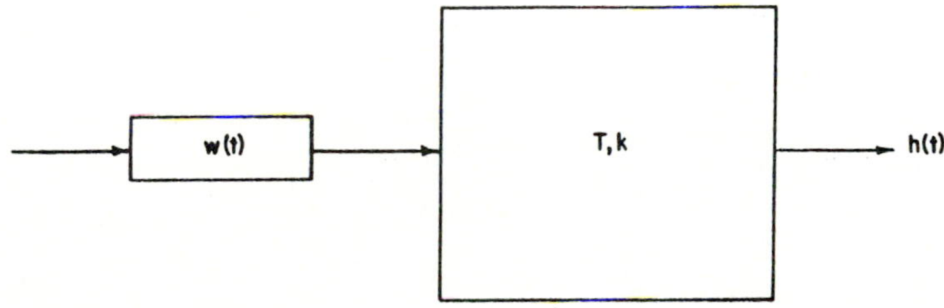

Figure 17.8 A model routing the effective rainfall through a TAC curve ($w(t)$) and an arrangement of linear channels (T) and linear reservoirs (k). This structure represents most of the conceptual models discussed in this chapter.

17.4 CASCADE OF LINEAR CHANNELS

Mathur (1972a, 1972b) proposed to conceptually represent a natural watershed by a set of linear channels arranged in series. This involves folding a two-dimensional area into a unidimensional translatory scheme. To illustrate, let us assume that the travel times from various portions of the watershed to the outlet are known. In other words, its TAC diagram is known. From Figure 17.9, the travel time from portion 1 to the outlet is marked as T. For a uniform translation time ΔT, the tributary B can be decomposed into three reaches, representing the areas ΔA, $\Delta A'$, and $\Delta A''$. Similarly, the main channel and other tributaries can be divided into several reaches, each having ΔT translation time. Then subareas of tributary reaches can be superimposed on the subareas of the reaches of main channel having the same translation time. This is shown in Figure 17.10. These superimposed areas can be construed as subwatersheds or mean delay time areas.

It is thus seen that a linear channel, as used here, functionally represents a specific portion of the watershed area. Its position is determined by the topographic characteristics. Therefore, the order in which these linear channels are arranged is also fixed.

The inflow to a linear channel can come from its upstream channel and its subwatershed area. Distributed runoff from any of the intermediate subwatersheds is assumed to enter the upper bound of its corresponding linear channel, along with the outflow translated down to it from its previous reaches. Translated inflow contributed by previous reaches is defined as primary inflow and distributed runoff from its corresponding subwatershed area reaching the upper bound is defined as secondary inflow.

Mathematically, referring to Figure 17.11, if we let,

$$dQ(t) = I(\tau)\, dA$$

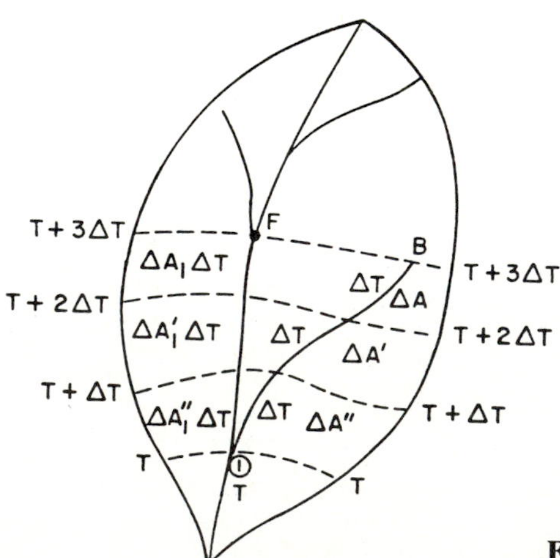

Figure 17.9 Delineation of subwatershed areas.

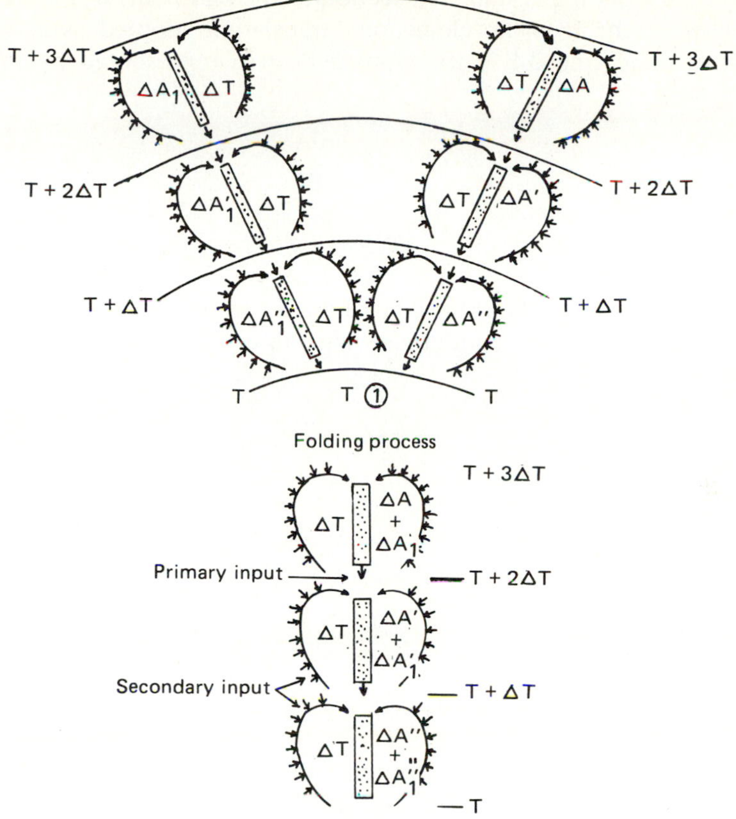

Figure 17.10 Equivalent linear-channel cascade (after Mathur, 1972a).

then

$$Q(t) = \int_0^t I(\tau) \frac{dA}{d\tau} \, d\tau \tag{17.37}$$

Equation (17.37) relates three parameters of the folding process: t refers to the total

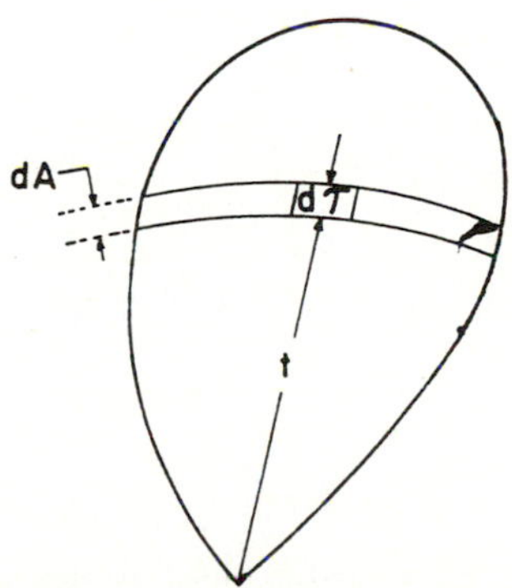

Figure 17.11 Watershed area.

translation from the confluence section to the end point of the influence area for a tributary reach; $d\tau$ is the elementary translation adopted as a basic unit for each linear channel; and dA is the equivalent area represented by the channel reach. Therefore,

$$h(0,t-\tau) = \frac{dA}{d\tau} \tag{17.38}$$

$$Q(t) = \int_0^t h(0,t-\tau)I(\tau)\,d\tau \tag{17.39}$$

Note that Equations (17.38) and (17.39) refer to runoff from the same subwatershed for the same prevailing conditions. We can express $h(0,t)$ in terms of the S-hydrograph, $U(t)$, as

$$U(t) = \int_0^t h(0,s)\,ds \tag{17.40}$$

Then

$$U(t-\tau) = \int_0^{t-\tau} h(0,t-\tau)\,d\tau$$

and

$$U(t-T-\tau) = \int_0^{t-T-\tau} h(0,t-T-\tau)\,d\tau$$

Because the UH ordinates are proportional to differences in ordinates at two positions on $U(t)$ separated by the UH duration T,

$$h(T,t-\tau) = \frac{1}{T}[U(t-\tau) - U(t-T-\tau)]$$

We can choose T as small as we wish. By substituting $U(t)$ in terms of $h(0,t)$ from Equation (17.40),

$$h(T,t-\tau) = \frac{1}{T}\left[\int_0^{t-\tau} h(0,t-\tau)\,d\tau - \int_0^{t-T-\tau} h(0,t-T-\tau)\,d\tau\right]$$

By substituting Equation (17.38) to replace $h(0,t)$,

$$h(T,t-\tau) = \frac{1}{T}\left[\int_0^{t-\tau} \frac{dA}{d\tau}\,d\tau - \int_0^{t-T-\tau} \frac{dA}{d\tau}\,d\tau\right]$$

Therefore,

$$Th(T,t-\tau) = A_{t-\tau} - A_{t-T-\tau} \tag{17.41}$$

If $T = 1$ hour,

$$h(1,\,t) = A_t - A_{t-1} \tag{17.42}$$

Equation (17.42) shows that the subwatershed areas represented by linear channels are directly given by the ordinates of the UH.

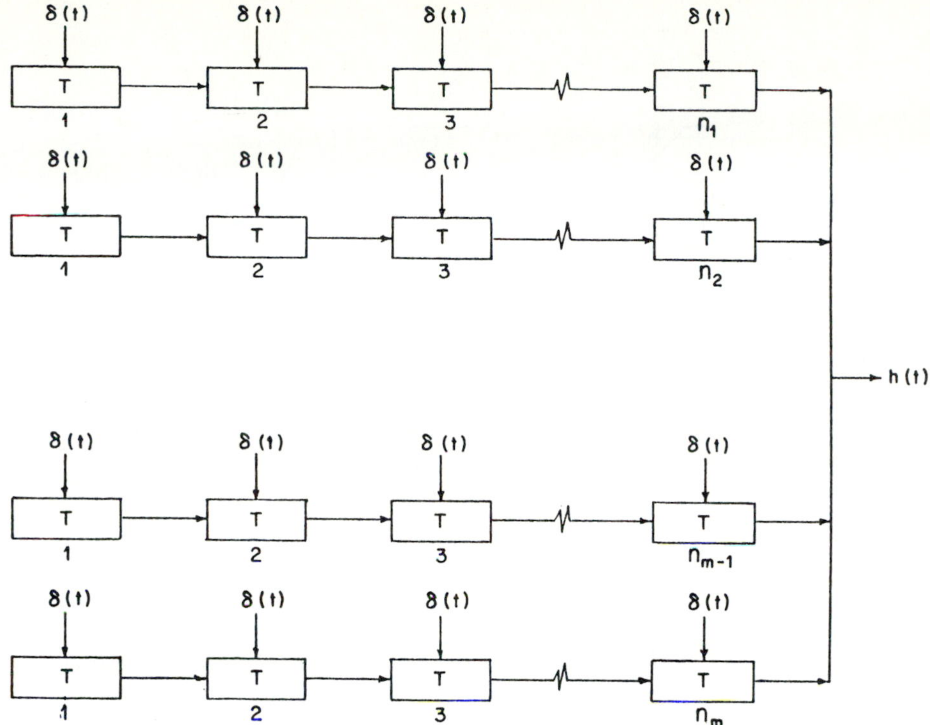

Figure 17.12 Cascade of linear channels arranged in parallel. Linear channels may or may not have the same translation time. Each channel represents a particular part of the drainage area.

It is seen that the TAC curve is a response of linear channels arranged in series and parallel, as shown in Figure 17.12. Each channel represents a particular portion of the watershed and has a particular area to drain. If storage effects are neglected, then the two-dimensional flow pattern in the watershed can be simulated by such an arrangement of linear channels. Thus, any TAC curve can be represented by a particular arrangement of linear channels and vice versa.

17.5 NASH MODEL

In a series of papers, Nash (1957, 1958, 1959, 1960) developed a model, as shown in Figure 17.13, based on a cascade of equal linear reservoirs for derivation of the IUH from a natural watershed. Kalinin and Milyukov (1958) independently derived an identical model for routing of flows in open channels. This is one of the most popular and frequently used models in applied hydrology. We derive it using the Laplace transform.

When the instantaneous unit effective rainfall, specified by $\delta(t)$, is fed into the first reservoir, then Equation (17.22) becomes

$$h + k\frac{dh}{dt} = \delta(t)$$

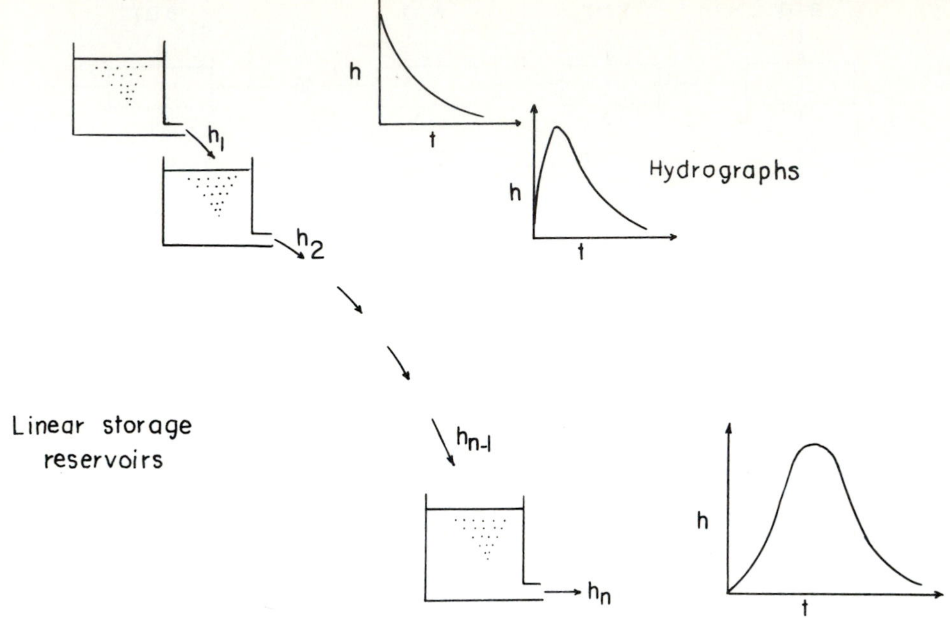

Figure 17.13 Cascade of equal linear reservoirs or the Nash model.

Taking the Laplace transform,

$$L\left[h + k\frac{\mathrm{d}h}{\mathrm{d}t}\right] = L[\delta(t)]$$

$$h(s) + ks\,h(s) = 1$$

$$h(s) = \frac{1}{1 + ks} = \frac{1}{k\left(\frac{1}{k} + s\right)}$$

The outflow from the first reservoir due to an instantaneous inflow is

$$h_1(s) = \frac{1}{k\left(\frac{1}{k} + s\right)}$$

For the second reservoir,

$$h_2 + k\frac{\mathrm{d}h_2}{\mathrm{d}t} = h_1$$

Then

$$L\left[h_2 + k\frac{\mathrm{d}h_2}{\mathrm{d}t}\right] = L[h_1]$$

$$h_2(s)(1 + ks) = \frac{1}{k\left(\frac{1}{k} + s\right)}$$

Thus

$$h_2(s) = \frac{1}{k^2\left(\frac{1}{k} + s\right)^2}$$

Similarly, for the nth reservoir, we can write

$$h_n(s) = \frac{1}{k^n\left(\frac{1}{k} + s\right)^n}$$

By taking the inverse of the Laplace transform,

$$h_n(t) = \frac{1}{k}\left(\frac{t}{k}\right)^{n-1} \frac{e^{-t/k}}{\Gamma(n)} \tag{17.43}$$

which is the Nash model.

17.5.1 Some Characteristics of the Nash Model

Equation (17.43) shows that the outflow is expressed by a single term identical to the gamma probability density function. Hence, we can write

$$h(t) = \frac{1}{k}p\left(\frac{t}{k}; n - 1\right) \tag{17.44}$$

where p(. . .) is the gamma probability density function, which is extensively tabulated in statistical books.

The time to peak t_p can be obtained by differentiating Equation (17.43) with respect to t and putting $\mathrm{d}h/\mathrm{d}t = 0$,

$$t_p = k(n - 1) \tag{17.45}$$

$$h_p = \frac{1}{k(n - 1)!}\left[(n - 1)^{n-1}e^{-(n-1)}\right] \tag{17.46a}$$

$$= \frac{1}{k}\,p(n - 1; n - 1) \tag{17.46b}$$

Figure 17.14 shows a plot of h_p versus n. It can be seen that as n increases, routing a storage of the same lag time has a comparatively less peak-reducing effect.

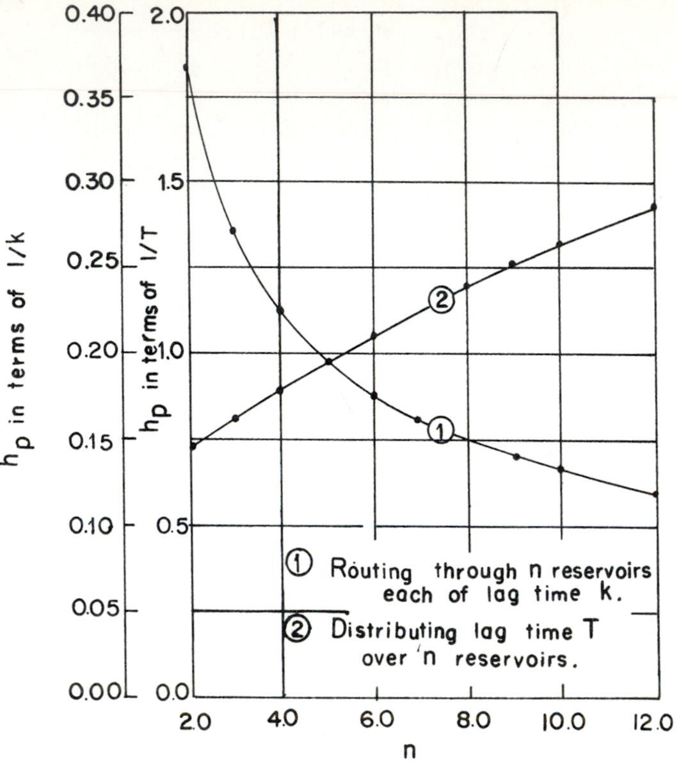

Figure 17.14 Routing $\delta(t)$ of the effective rainfall through a cascade of a linear reservoir.

Another important characteristic of storage is observed when a constant lag time T is assumed. We can determine T as

$$T = \frac{\int_0^\infty ht \, dt}{\int_0^\infty h \, dt} \tag{17.47}$$

where T denotes the time between the centers of the areas of inflow and outflow. This lag time is distributed over a number of reservoirs, each of lag time k. That is,

$$T = nk \tag{17.48}$$

Each reservoir has a lag time $k = T/n$. Although T remains constant, n is increased, that is, the total lag time is distributed over more and more reservoirs. The ultimate lag effect remains T irrespective of the value of n. We can then study the effect on the outflow hydrograph of varying n. Thus, the value of h_p for various n can be used as a measure of change in shape. Figure 17.14 gives a plot of h_p versus n for $n \geq 2$. From this figure and Equation (17.46), it can be seen that as n increases, the peak increases. That is, the peak-reduction effect reduces as T is distributed over more and more reservoirs, which means that the shape-modification effect decreases as the storage gets distributed. Because the lag time remains the same, in the extreme case of distributed storage, the storage effect tends to become one of translation only. In other words, as $n \to \infty$, the outflow is

$$h_n = \delta\,[(n - 1)K] \tag{17.49}$$

This indicates that if a finite total delay time is divided up among a large number of equal linear reservoirs, the effect on an instantaneous inflow is equivalent to translation by an amount equal to the total delay time. Thus, we can perceive the reservoir action as being related to concentrated storage and translatory action as related to completely distributed storage. We can show this result mathematically. Routing an inflow $f_1(t)$ through reservoirs of lag time T/n to obtain the final outflow $f_2(t)$ can be written as

$$f_2(t) = \frac{1}{\left(1 + \dfrac{T}{n}D\right)^n} f_1(t) \tag{17.50}$$

Taking the limits,

$$\lim_{n\to\infty} \frac{1}{\left(1 + \dfrac{T}{n}D\right)^n} = e^{-TD} \tag{17.51}$$

$$\begin{aligned} f_2(t) &= e^{-TD} f_1(t) \\ &= \sum_{j=0}^{\infty} (-1)^j \frac{(TD)^j}{j!} f_1(t) \end{aligned} \tag{17.52}$$

The right side of Equation (17.52) is equal to the expansion of $f_1(t - T)$ in a Taylor series about t. Therefore,

$$f_2(t) = f_1(t - T) \tag{17.53}$$

Hence,

$$\lim_{n\to\infty} \frac{1}{\left(1 + \dfrac{T}{n}D\right)^n} = H(T) \tag{17.54}$$

which equals the delay operator. The Nash model has the advantage that the S-hydrograph is given by

$$
\begin{aligned}
U(t) &= \int_0^t h(0, t)\, \mathrm{d}t \\
&= \frac{1}{k}\int_0^t \left(\frac{t}{k}\right)^{n-1} \frac{e^{-t/k}}{(n-1)!}\, \mathrm{d}t \\
&= \int_0^m \frac{m^{n-1}e^{-m}}{(n-1)!}\, \mathrm{d}m \qquad (17.55) \\
&= \frac{\int_0^m m^{n-1}e^{-m}\, \mathrm{d}m}{(n-1)!} \\
&= G\left(\frac{t}{k}; n-1\right)
\end{aligned}
$$

where $G(\cdot)$ is the ratio of the incomplete to the complete gamma functions. Thus, the ordinates of the finite-period unit hydrograph can be found from the differences between two tabulated values:

$$
\begin{aligned}
h(D, t) &= \frac{U(t) - U(t-D)}{D} \\
&= \frac{1}{D}\left[G\left(\frac{t}{k}; n-1\right) - G\left(\frac{t-D}{k}; n-1\right)\right]
\end{aligned}
\qquad (17.56)
$$

17.5.2 Parameter Estimation

The Nash model has two parameters, n and k, that can be estimated using the method of moments. The first two moments of the IUH expressed by Equation (17.43) can be written as

$$M_1 = nk \qquad (17.57a)$$

$$M_2 = nk^2(n+1) \qquad (17.57b)$$

Thus, n and k can be determined from the first two moments of the IUH.

17.5.3 Dimensionless IUH

Dimensionless forms of the IUH derived from the Nash model have been used for synthesis of direct runoff. Frequently, the normalizing quantities have been taken as h_p and t_p or T. The dimensionless IUH becomes

$$\frac{h(t)}{h_p} = \left[\frac{t}{t_p} \exp\left(1 - \frac{t}{t_p}\right)\right]^{n-1} \qquad (17.58)$$

Equation (17.58) contains only one parameter, n. Under appropriate circumstances, it can be used for regionalization. By computing n and k for one watershed, the

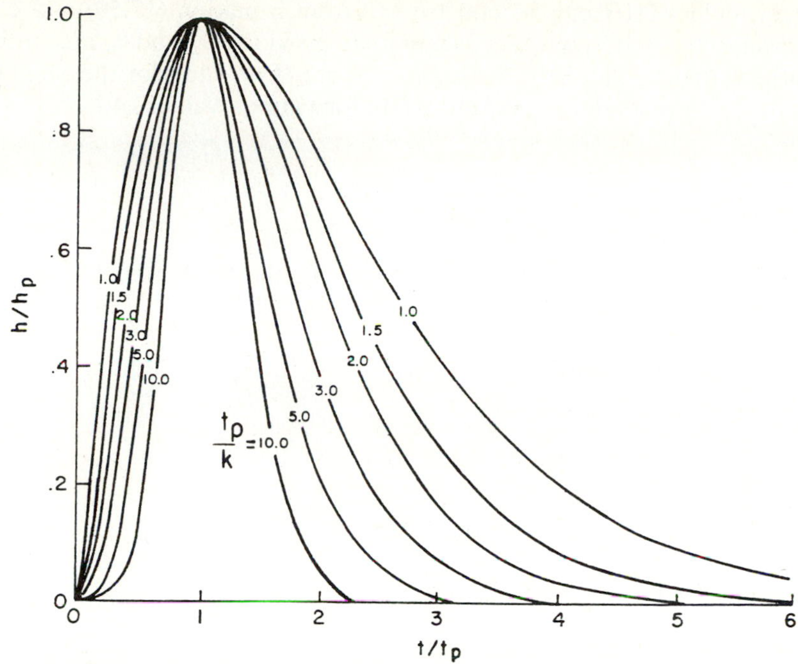

Figure 17.15 Dimensionless IUHs by the Nash model for various values of t_p/k.

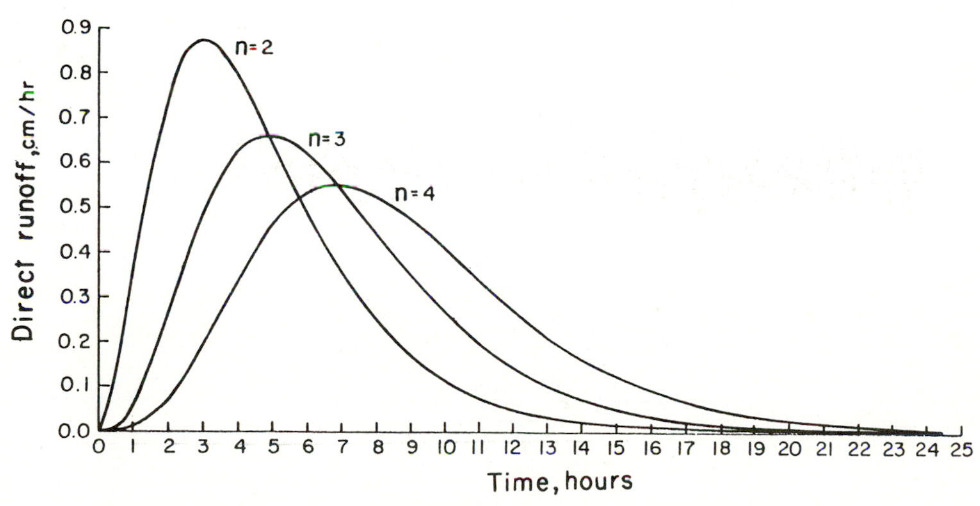

Figure 17.16 The direct-runoff hydrograph by the Nash model for the ERH in Example 17.3. $k = 2$ hours.

dimensionless IUH can be constructed from Equation (17.59) and can be used to determine the IUH for other watersheds provided h_p and t_p are known. The Soil Conservation Service (1972) used $n - 1 = t_p/k \cong 4$ to plot the dimensionless UH. Figure 17.15 plots dimensionless IUHs for various values of t_p/k.

17.5.4 DRH

The DRH for any $I(t)$ is now derived by using the Nash model. We use the Laplace transform here:

$$h(s) = \frac{1}{(1 + ks)^n}$$

Therefore,

$$Q(s) = \sum_{j=0}^{m} w_j \exp(-jDs) \frac{1}{s(1 + ks)^n}$$

On inverting,

$$Q(t) = \sum_{j=0}^{m} w_j u_1 (t - jD) - \sum_{j=0}^{m} w_j u_1 (t - jD) \sum_{i=0}^{n-1} \frac{1}{i!} [(t - jD)/k]^i \times \exp[-(t - jD)/k] \tag{17.59}$$

EXAMPLE 17.3 Compute the DRH by the Nash model for the ERH in Example 11.2. Take $k = 2$ hours and $n = 2$, 3, and 4.

Solution Figure 17.16 shows the direct-runoff hydrographs computed by the Nash model for various values of n.

17.5.5 Inclusion of Translation

The Nash model can be easily extended to include translation T by inserting a linear channel, as shown in Figure 17.17. The governing equation then becomes

$$h(t - T) = \frac{1}{(1 + kD)^n} \delta(t - nT) \tag{17.60}$$

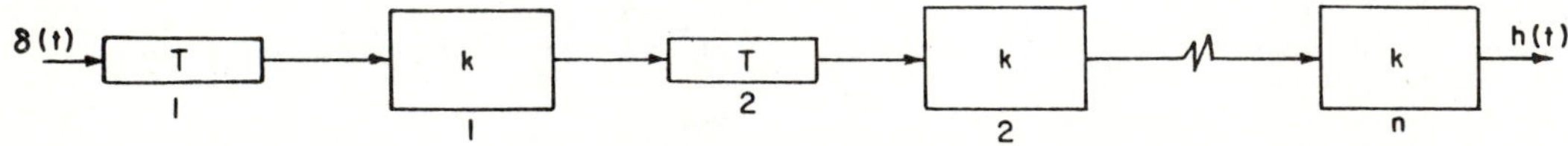

Figure 17.17 A cascade of equal linear reservoirs and channels arranged in series.

Let $t^* = t - nT$. Then

$$h(t^*) = \frac{1}{(1 + kD)^n}\, \delta(t^*)$$
$$= \frac{1}{k\Gamma(n)} \left(\frac{t^*}{k}\right)^{n-1} \exp(-t^*/k) \tag{17.61}$$

Its first two moments can be written as

$$M_1 = nT + nk \tag{17.62a}$$

$$M_2 = (nT + nk)^2 + nk^2 \tag{17.62b}$$

The DRH for any $I(t)$ by this model can be derived directly from Equation (17.59) by introducing the translation time:

$$Q(t) = \sum_{j=0}^{m} w_j u_1(t - jD) - \sum_{j=0}^{m} w_j u_1(t - jD) \sum_{i=0}^{n-1} \frac{1}{i!} \left(\frac{t - nT - jD}{k}\right)^i \times \exp\left[-\frac{(t - nT - jD)}{k}\right] \tag{17.63}$$

EXERCISES

17.1. A watershed has an area of 1 km^2. Compute the DRH due to an ER occurring at a uniform rate of 2 cm/h for a period of 4 h. The linear reservoir has a k of 3 h. Repeat the calculations for ER durations of 6 h and 8 h.

17.2. Generate the IUH using the Nash model for various values of n, keeping k constant. Then discuss the effect of n on the IUH.

17.3. Generate the IUH using the Nash model for various values of k, keeping n constant. Then discuss the effect of k on the IUH.

17.4. The inflow–outflow data observed on the reach from Burrendong to Wellington of the Macquarie River in Australia for a major flood that occurred in March 1926 are given:

DATE	TIME (h)	INFLOW (10^3 m^3/s)	OUTFLOW (10^3 m^3/s)
24	12	0.18	0.12
	15	0.32	0.20
	18	0.70	0.40
	21	1.70	0.80
	24	3.22	1.36
25	03	4.56	2.00
	06	5.66	2.64
	09	4.40	3.24
	12	3.12	3.80

DATE	TIME (h)	INFLOW (10^3 m³/s)	OUTFLOW (10^3 m³/s)
	15	3.94	3.94
	18	2.16	3.62
	21	1.80	3.00
	24	1.46	2.52
26	03	1.18	2.14
	06	0.98	1.76
	09	0.80	1.44
	12	0.66	1.18
	15	0.48	0.74
	18	0.34	0.50
	21	0.32	0.30
	24	0.34	0.50
27	12	0.32	0.30
	24	0.14	0.18
28	24	0.06	0.09

Compute the storage for the reach as a function of time. Plot inflow, outflow, and storage against time.

17.5. Plot storage against the outflow of Exercise 17.4 and explain the significance of the plot.

17.6. Represent the plot of Exercise 17.5 by a straight line passing through the origin. Discuss if this representation is adequate.

17.7. Suppose the river reach of Exercise 17.4 is represented by a linear channel with a translation time of 10 hours. What will be the outflow from this reach if the inflow hydrograph is as given in Exercise 17.4.

17.8. Compute the first moment about the origin of inflow and that of outflow of Exercise 17.4. Take the difference of these two first moments. What does this difference represent hydrologically?

17.9. Consider the inflow hydrograph of Exercise 17.4. Taking the pulse duration of 6 h, represent the hydrograph by an equivalent histogram, and express its equation similar to Equation (17.19). Be sure to compute the values of w_j, $j = 0, 1, 2, \ldots$.

17.10. A linear channel has a translation time of 5 hours, and its IUH is given by Equation (17.16). Compute the DRH for this channel if the ERH is given by Equation (17.18), with $I_0 = 2$ cm/h, $I_1 = 5$ cm/h, $I_2 = 2$ cm/h, $I_3 = 1$ cm/h, and $I_4 = 0$ cm/h, $D = 0.5$ h, and $m = 4$.

17.11. A linear reservoir has a value of storage parameter $k = 5$ h. Compute the DRH for the ERH given in Exercise 17.10.

17.12. Represent the TAC curve of a watershed by a rectangle. The time of concentration of the watershed is 6 hours, and its area is 10 km². Compute the DRH from this watershed for the ERH of Exercise 17.11.

CHAPTER 18
Estimation of Peak Discharge

The estimation of peak discharge is required for many hydrologic applications, including design of hydraulic structures, drainage design, flood-prevention works, etc. Of the many methods available to predict peak discharge, two methods are discussed here, namely, (a) the rational method and (b) the drainage-basin peak-discharge rating-curve method.

18.1 RATIONAL METHOD

The rational method, also known as the rational formula, is an empirical relation that is credited, in the United States, to Kuichling (1889). Prior applications of the rational formula occurred in England, but such records are somewhat obscure. The rational formula is

$$Q_p = CIA \tag{18.1}$$

where Q_p is the peak discharge in cfs, C is a dimensionless runoff coefficient whose value depends on hydrologic characteristics of the drainage area, I is the rainfall intensity in in./h for a duration equal to or greater than the time of concentration of the drainage basin, and A is the area of the drainage basin in acres. This formula was termed "rational" because the units of the quantities are approximately consistent. Examination of this formula does not obviously confirm this consistency because the units are cfs = inch × acre/hour. The consistency of the units occurs because a flow of 1 cfs produces a volume of water approximately equal to 1 acre-inch each hour or approximately 2 acre-feet per day.

Equation (18.1) yields the maximum flood discharge if I is the maximum intensity of rainfall. However, the term "maximum" is not clearly defined yet and is subjective. A concept of frequency is thus introduced. We say that if the intensity of rainfall could be as high as 10 cm/h once in 100 years, then the corresponding Q_p will also have a frequency of once in 100 years. If the intensity of rainfall is 5 cm/h once in 20 years, then the corresponding Q_p has a frequency of once in 20 years.

The runoff coefficient C is a dimensionless decimal value that estimates the decimal portion of rainfall that becomes runoff. For example, areas paved with concrete, such as streets, are estimated to produce 0.80 to 0.95 of the rainfall as runoff. Lawn areas composed of flat sandy soil might produce only 0.05 to 0.10 of the rainfall as runoff. Other typical urban drainage areas and their runoff coefficients are shown in Table 18.1.

The second variable in the rational formula, I, has two aspects that must be dealt with: (1) rainfall intensity and (2) time of concentration. The rainfall intensity used is for a duration equal to the time of concentration, which can be calculated

TABLE 18.1 VALUES OF RUNOFF COEFFICIENT C (AFTER CHOW, 1962)

Type of drainage area	Runoff coefficient, C
Lawns:	
Sandy soil, flat, 2%	0.05–0.10
Sandy soil, average, 2–7%	0.10–0.15
Sandy soil, steep, 7%	0.15–0.20
Heavy soil, flat, 2%	0.13–0.17
Heavy soil, average, 2–7%	0.18–0.22
Heavy soil, steep, 7%	0.25–0.35
Business:	
Downtown areas	0.70–0.95
Neighborhood areas	0.50–0.70
Residential:	
Single-family areas	0.30–0.50
Multiunits, detached	0.40–0.60
Multiunits, attached	0.60–0.75
Suburban	0.25–0.40
Apartment dwelling areas	0.50–0.70
Industrial	
Light areas	0.50–0.80
Heavy areas	0.60–0.90
Parks, cemeteries	0.10–0.25
Playgrounds	0.20–0.35
Railroad yard areas	0.20–0.40
Unimproved areas	0.10–0.30
Streets:	
Asphaltic	0.70–0.95
Concrete	0.80–0.95
Brick	0.70–0.85
Drives and walks	0.75–0.85
Roofs	0.75–0.95

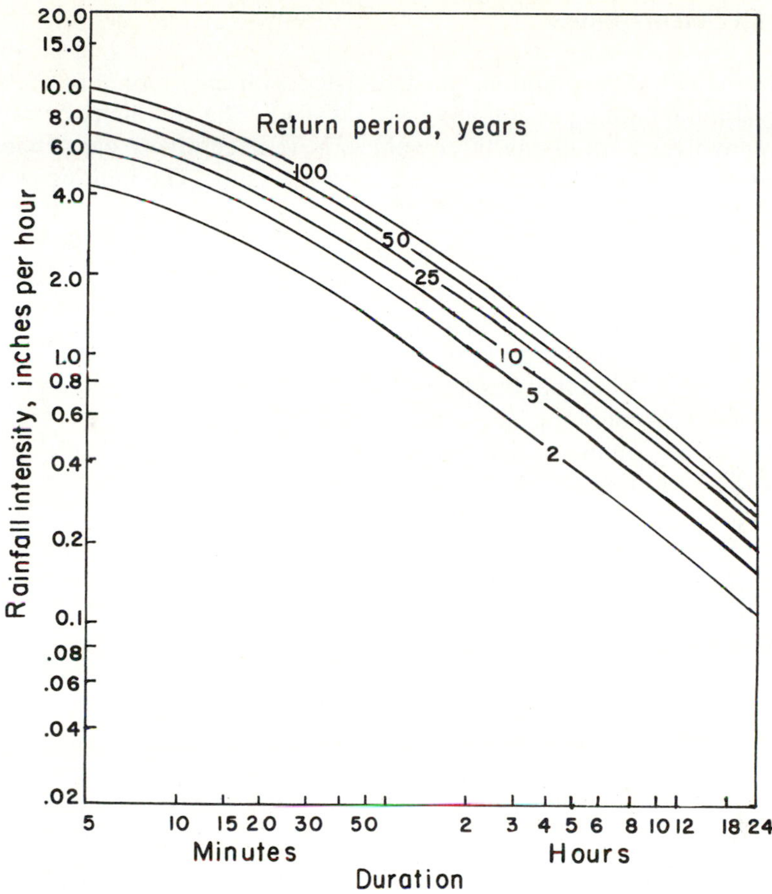

Figure 18.1 Rainfall–intensity–duration frequency curves for Omaha, Nebraska (National Weather Service).

using one of the methods discussed in Chapter 12. Many municipalities have devised graphs utilizing channel length and slope for determining time of concentration to be used for all hydrological analyses within their jurisdictions. When the time of concentration is determined, it is then possible to determine the design rainfall for the applicable return period.

The National Weather Service has developed rainfall intensity–duration–frequency curves for many areas in the United States. Figure 18.1 is such a curve for Omaha, Nebraska, that estimates the rainfall intensity in inches per hour for durations up to 24 hours and having a return period of 2, 5, 10, 25, 50, and 100 years. Entering this curve for a duration of 10 minutes (equal to the time of concentration) results in a rainfall intensity of about 8.4 inches per hour for a storm with a 100-year return period.

The final variable, A, in the rational formula requires determining the drainage basin area, A, in acres. This is usually done by a planimeter on a topographic, cadastral or other map.

18.1.1 Assumptions

Several limiting assumptions are associated with the rational formula, which is valid only if these limiting assumptions are satisfied. The simplicity of the formula, which only involves multiplying three values, is so deceptively easy that the limiting assumptions are often ignored and large errors result. The limiting assumptions are as follows:

1. The peak rate of runoff is a maximum and is a direct function of the average rainfall intensity throughout the time of concentration. In other words, the computed discharge is the maximum that can occur for the selected rainfall intensity from that basin and that discharge occurs at the time of concentration and beyond.
2. The maximum discharge resulting from a rainfall intensity equal to or greater than the time of concentration is a simple, direct function of the rainfall. In other words, runoff is directly proportional to rainfall.
3. The frequency of the peak discharge is the same as the frequency of rainfall. This assumption is not strictly correct, but does not create significant problems for the limited purposes of this method.
4. The relation between peak discharge and the drainage area is the same as the relation between peak discharge, intensity, and duration of rainfall. This means that the drainage basin is considered as linear.
5. The coefficient of runoff, C, is the same for storms of various frequencies. This means that all the losses on the drainage basin are a constant. This assumption may be reasonable for a drainage basin covered with an impervious surface, but not for other drainage surfaces.
6. The coefficient of runoff is the same for all storms on the drainage basin regardless of antecedent moisture conditions. Variation in the runoff coefficient is often a function of drainage area. This implies that the drainage area for which this assumption is valid is necessarily small.
7. Rainfall remains constant over the entire watershed during the time of concentration. The significance of this assumption is that because of the spatial variability of rainfall, the drainage area for which the rational method will apply is limited.
8. Runoff occurs nearly uniformly from all parts of the watershed. This means that the runoff coefficient must be nearly the same over the entire drainage basin. This assumption is less likely to be valid as the drainage-basin size increases.

Clearly, these limitations restrict the drainage-basin size to which the rational formula can be applied. High-intensity storms, for which the rational method is intended, are small by their nature and are of short duration. Because the duration of such storms must be equal to the time of concentration, it follows that the applicable drainage area must be small in order to have a short time of concentration. Hydrologic characteristics vary with increasing drainage-basin area, and the runoff coefficient is not uniform over a drainage basin and particularly on a larger drainage basin. Small drainage basins are more likely to satisfy the uniformity assumptions of the rational formula. It is believed that these assumptions are valid for paved areas with gutters and sewers of fixed dimensions and hydraulic characteristics. Hence, the rational formula is generally accepted for sewer design for urban areas. The maximum area applicable to the rational formula is believed to be approximately 240 acres. Some investigators believe the maximum area should not be more than 100 acres in order to accommodate the rainfall-intensity and uniformity limitations. The rational formula is often used beyond its limitations and its use is often recommended by official guidelines of many municipalities. This use may result in overdesign and waste of resources.

EXAMPLE 18.1 Compute the time of concentration and peak discharge at the mouth of a watershed having the following characteristics: pasture land, fairly flat with an average slope of about 1%, tight clay soils, drainage area of 40 ha, distance from the farthest point to the outlet is 2000 m, and the difference in elevation between the farthest point and the outlet is 10 m. Consider a rainfall of 8.5 cm/h occurring uniformly for a period equal to the time of concentration.

Solution

The watershed gradient = (10/2000) × 100 = 0.5%

The time of concentration from Equation (12.7)

$$= 0.0078\ (2000 \times 3.281)^{0.77} \times \left(\frac{10}{2000}\right)^{-0.385}$$

$$= 0.0078 \times (6562)^{0.77} \times \frac{1}{(0.005)^{0.385}}$$

$$= 0.0078 \times 869.196 \times \frac{1}{0.13} = 0.0078 \times 869.196 \times 7.689$$

$$= 52.132 \text{ min} = 0.87 \text{ h}$$

The value of the runoff coefficient = 0.4

Therefore,

$$Q_p = 0.4 \times 8.5 = 3.4 \text{ cm/h}$$

$$= 0.4 \times \frac{8.5}{2.54} \times 40 \times 2.471 = 132.306 \text{ cfs}$$

EXAMPLE 18.2 Determine the peak discharge and time of concentration for Ward Creek at Capital Heights Avenue in Baton Rouge, Louisiana. The drainage area is 4.10 mi^2. The main channel length and slope are 16,210 feet and 7.2 feet/mile, respectively. The rainfall intensity is 1.73 inches/hour, occurring uniformly for the time of concentration. The runoff coefficient is 0.6. Also compute the time of concentration.

Solution

$$\text{The time of concentration} = 0.0078 \times 16{,}210^{0.77}/0.0014^{0.385}$$

$$= 172 \text{ min} \cong 3 \text{ h}$$

$$Q_p = 0.6 \times 1.73 = 1.038 \text{ in./h}$$

$$= \frac{1.038}{12} \times \frac{4.1 \times 5280 \times 5280}{43{,}560} = 2720 \text{ cfs}$$

18.2 DRAINAGE-BASIN PEAK-DISCHARGE RATING-CURVE METHOD

This method is also known as the volume-peak relation (Singh and Aminian, 1986) or the standardized peak-discharge distribution (Rogers, 1980; Rogers and Singh, 1986, 1988). It is simply a peak-discharge rating curve for a drainage basin in which peak discharge is plotted against runoff volume (in centimeters or inches uniformly distributed over the entire drainage basin) rather than stage. This method relies on measured discharge data from a stream gage or weir. In order to avoid excessive repetition of the rather long title "drainage-basin peak-discharge rating curve," henceforth it will be identified by either peak-discharge rating curve or drainage-basin rating curve. The peak-discharge rating curve is distinguished from the stream-gage rating curve; the data used to prepare the former are taken from the latter.

The rationale for the term "peak-discharge rating curve" is based on the fact that this peak-discharge–runoff volume relation is a transformation of the stream-gage rating curve. The stream-gage rating curve is a plot of the discharge versus stage (in meters or feet), whereas the drainage-basin rating curve is a plot of the peak discharge versus the distributed depth of the total hydrograph volume over the entire drainage basin.

Although the discharges for both the stream-gage rating curve and the drainage-basin rating curve are plotted against depth values (say, stage in feet and depth in inches), these depth values are, in fact, volumes. For the stream-gage case, the stage is a depth of water for a given cross-sectional area in the channel for a unit length that is a volume. For the drainage-basin rating curve, the depth is distributed uniformly over the entire drainage-basin area, the product of which is a volume. The reason for pointing out the relation between the stream-gage rating curve and the drainage-basin rating curve is to demonstrate that the peak-discharge rating curve is simply a transformation of the stream-gage rating curve in a manner that reflects characteristics of the drainage basin under study.

The development of the drainage-basin rating curve was motivated by the established and accepted UH method. Both methods plot peak discharge against runoff volume (in centimeters or inches). This relation is desirable because it is customary to relate the amount of rainfall (in centimeters or inches) to the runoff volume from a drainage basin (expressed in centimeters or inches) uniformly distributed over the entire drainage basin. This relation is logical and can be applied to compute design-flood events from runoff volumes from a drainage basin. One advantage of the drainage-basin rating curve is that it does not require baseflow separation, which is an uncertain procedure, at best.

One of the problems with the UH derived for various unit storms is that its plotted data have considerable scatter. As a result, use is made of averaging two or more of the UHs to determine the usable UH. The distribution of scattered points relating peak discharge to runoff volume is most frequently nonlinear and, therefore, would cause a great deal of scatter when plotted on arithmetic paper.

18.2.1 Relation to the Unit Hydrograph

The UH is a valid concept and has proven to be satisfactory for those conditions for which it is applicable. The drainage-basin rating curve has many similarities to the UH, but is deemed to be a considerable improvement for several reasons.

1. Qualitative observations have indicated that a given drainage basin produces baseflow contributions that are related to the storm event falling on it. This is particularly true for drainage basins with deep soils in the valley bottoms. At any rate, the baseflow produced is a characteristic of each drainage basin for each runoff event. Thus, there is little reason to separate baseflow, but rather the total hydrograph should be used for analysis. This concept has two advantages: (a) The uncertainty in separating out baseflow produces errors that can be avoided by not attempting baseflow separation. (b) A different baseflow must be added to the direct-runoff hydrograph when converting to a UH for some other unit duration; this results in yet additional errors. These accumulating errors are avoided using the drainage-basin rating curve.

2. The condition of uniform rainfall over the drainage basin is unnecessary. The drainage-basin rating curve is valid for any rainfall distribution and for any storm movement. This condition constitutes a significant advantage because all storm events can be used. This simplifies the collection of data for analysis.

3. The condition of uniform runoff over the entire drainage-basin area is unnecessary. Nonuniform runoff distribution causes drainage-basin nonlinearity. The peak-discharge rating curve method handles any nonlinear runoff condition as well as any linear runoff condition. For this reason, it can be applied to any drainage basin.

4. The peak-discharge rating curve is valid for any duration of rainfall. The UH method can also be used to construct a hydrograph for any duration of rainfall. The drainage-basin rating curve can determine the peak discharge and base time of the hydrograph for a given runoff volume. This statement must be qualified to exclude drainage basins where the duration of runoff exceeds the time of concentration. When the time of concentration has been reached, the peak no longer increases, but becomes constant while the runoff volume increases. This condition is rare for drainage basins other than very small ones.

5. The peak-discharge rating curve can be applied to drainage basins of any size. It has been used successfully on drainage basins exceeding 33,000 mi^2 (85,000 km^2). Because the UH method is bound by the limitations imposed by uniform rainfall and uniform runoff, the drainage-basin size for which these conditions could occur is very limited. Thus, the peak-discharge rating-curve method opens up the possibility for analysis of drainage basins not possible when using the UH method.

6. The peak-discharge rating curve is valid for snowmelt hydrographs. It has been used on drainage basins where hydrographs of both snowmelt and rainfall were mixed for the input data. Snowmelt hydrographs plotted well with rainfall hydrographs. Preliminary investigations using only snowmelt hydrographs in Quebec, Canada, showed reasonable results.

7. The peak-discharge rating curve is valid for very small runoff volumes as well as very large runoff volumes. This relation has been demonstrated using runoff volumes ranging from 0.02 in. (5 mm) to 9.36 in. (23.8 cm). If runoff from a drainage basin exceeds its time of concentration, the volume of runoff from that hydrograph will be larger than for a hydrograph at or less than the time of concentration. For this reason, hydrographs for a time exceeding the time of concentration cannot be used. This same caution would apply to the UH method. Fortunately, except for very small basins, most drainage basins never reach their time of concentration.

18.3 CONSTRUCTION OF A DRAINAGE-BASIN PEAK-DISCHARGE RATING CURVE

18.3.1 Data Requirements

The peak-discharge rating curve is constructed from records ordinarily kept for any stream gage. The essential information that is required is the following:

1. The stream-gage rating curve or rating table that is valid for the time period during which the recorded hydrographs were obtained.
2. Several recorded hydrographs.
3. The drainage area is needed in order to determine the distributed-runoff volumes.

The number of hydrographs used is less important than the quality of the hydrographs. The hydrographs should be simple hydrographs that are the result of a single rainfall event although complex hydrographs can be used and simplified if it is determined that those hydrographs were the result of multiple rainfall events. The hydrographs should not be complicated by upstream obstructions or downstream obstructions. Such obstructions are often caused by ice blockage in the stream; hence, the winter hydrographs should be examined carefully before they are used. Other obstructions can also affect the shape of the hydrograph, so that each hydrograph should be carefully examined to see if there has been significant distortion of the hydrograph caused by some external influence. Useful hydrographs for this method or any other method can be only those that result from the influence of drainage-basin characteristics and not influenced by some alien condition. The number of hydrographs chosen usually consists of the number that is reasonably available. The larger the number, the more reliable the rating curve. It is useful to choose hydrographs that cover the range from small to large if possible.

18.3.2 Preparation of Data

The hydrographs to be used to prepare a peak-discharge rating curve must be taken from the recorded data provided by the stream gage. These data, whether they be digital printouts of some type or recorded charts, will be in the form of stage height versus time. The hydrograph is indicated simply by a recorded increase or decrease in stage over time. These stages are converted to discharges by use of the stream-gage rating curve. The time intervals at which the stage is recorded by the gage can range from continuous as for a chart to predetermined intervals such as 10 or 15 minutes for digital equipment. The time interval used to convert the stage hydrograph to a discharge hydrograph should be short enough so that substantial errors are not introduced by leaving out part of the hydrograph. In other words, the time interval available on the stage hydrograph might be shorter than required to produce a reasonable discharge hydrograph so that it is not necessary to use all the detail available. Good judgment in converting the stage hydrograph to the discharge hydrograph avoids introducing significant errors. These steps are required for producing any measured discharge hydrograph.

Because the peak-discharge rating curve method requires the runoff volume under the hydrograph, it is necessary to define the hydrograph in a manner that is consistent so that all hydrographs will be comparable. Choosing the point of the beginning of the hydrograph is simply to choose the point at which the hydrograph begins to rise from the antecedent stream flow. This antecedent stream flow is usually a gentle recession so that identification of the point of rise is easy. When it is necessary to use a hydrograph with more complex rises, a judgment should be made relative to the volume of flow that is involved in the alternative decisions, which are discussed as follows.

Whereas choosing the point of rise on the hydrograph is usually easy, determining the end of the hydrograph is not. Because it is not yet really known as to how to determine the end of a hydrograph, it must be done on some consistent basis so that each hydrograph can be compared to other hydrographs on the same basis. The peak discharge can be used as a reference point, because the end of a hydrograph occurs at some time after the peak discharge. An empirical equation has been developed by Rogers and Zia (1982) that seems to provide the necessary consistency:

$$Q_t = Q_0 + Q_p^{0.6} \tag{18.2}$$

where Q_t is the discharge in cfs (0.0283 m^3/s) identifying the termination discharge of the hydrograph, Q_p is the peak discharge of the hydrograph in cfs (0.0283 m^3/s), and Q_0 is the baseflow discharge in cfs (0.0283 m^3/s) immediately prior to the hydrograph rise. This equation has no recommendation except that it is consistent. Any other choice for terminating the hydrograph that is consistent would work equally well.

Once the hydrograph is identified, the next step is to determine the volume of water, V, under this hydrograph. This is a simple procedure and is easily done from tabulated hydrograph discharges. This volume of water is then distributed uniformly over the entire drainage area to determine the value of V (in inches or centimeters). This procedure is repeated for all hydrographs.

18.3.3 Construction and Meaning of the Peak-Discharge Rating Curve

A linear plot on log–log paper of runoff volume V on the abscissa and peak discharge Q_p on the ordinate is made as shown in Figure 18.2. An equation for this plot can be determined using the least squares method and a measure of the fit can be determined. This equation takes the form

$$\log Q_p = \log b + m \log V \tag{18.3}$$

where b is the intercept when the log of V is zero, m is the slope of the linear relation, and all other notations are as previously stated.

18.3.4 Causes of Drainage-Basin Nonlinearity

The slope, m, in Equation (18.3) indicates the degree of linearity or nonlinearity of a drainage basin and is thus a critical element for the analysis of drainage basins. A drainage basin is defined to be linear when the peak discharge is directly proportional

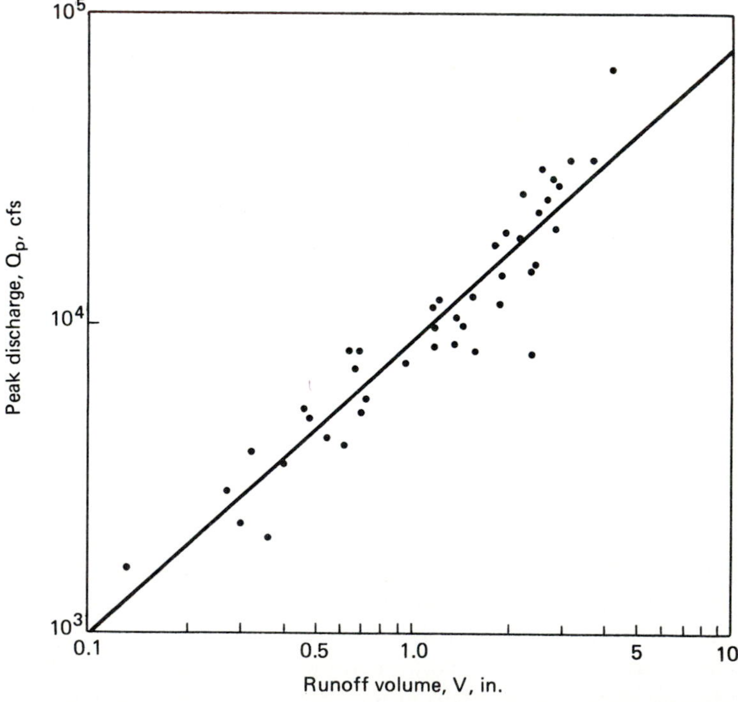

Figure 18.2 Peak discharge versus runoff volume for the Cheat River near Parsons, West Virginia (drainage area = 718 mi²). Log Q_p = 3.937 + 0.951 log V : r^2 = 0.919.

to the runoff volume. If m is equal to 1, the runoff volume is directly proportional to the peak discharge and the drainage basin is linear. If the value of m is less than 1, Equation (18.3) is nonlinear and the peak discharge is a power function of runoff volume in the form

$$Q_p = bV^m \tag{18.4}$$

Drainage-basin nonlinearity is caused by the nonuniform distribution of runoff losses. The principal loss to runoff is infiltration. Drainage basins that have nonuniform infiltration are nonlinear unless this loss is compensated for in some manner by other losses.

Studies have shown that a typical drainage basin, which is located on nearly uniform soils, has higher infiltration on and near the basin divides than in the valleys. This difference in infiltration comes about because typical soils contain a range of grain sizes. Erosion of the drainage basin results in fine, clay-size material transported from the higher elevations and deposited in the valleys at lower elevations. As shown in Chapter 8, fine-grain material has a lower infiltration capacity than does coarse-grain material, where vegetation is not the dominant factor. Removal of fine-grain material from soils at higher elevations and deposition of this material at lower elevations results in higher infiltration rates on the divide areas and lower infiltration rates in the valley areas. Infiltration differences of as much as 8.0 in./h (20.3 cm/h) in the divides to 0.04 in./h (9 mm/h) in the valleys have been measured. These and other drainage basins that have unevenly distributed infiltration caused by vegetation or geology are nonlinear. Such drainage basins produce nonuniform runoff from the watershed. Even though rainfall covers the entire drainage basin, the high-infiltration areas produce little or no runoff and high runoff occurs in the low-infiltration areas. Infiltration appears to be transitional on uniform soils from the divide to the valleys. For this reason, the change in runoff is also transitional.

The effect of nonuniform runoff from a drainage basin on the hydrograph is indicated by a change in the rise time. A change in rise time for hydrographs with the same volume of runoff affects the peak discharge. A simple experiment using a triangular hydrograph can verify this observation. Comparison of hydrographs between basins of different nonlinearity (slope m) for the same runoff volume shows different rise times. A definitive relation between these conditions has not yet been established. Other factors may also contribute to drainage-basin nonlinearity.

The value of m in Figure 18.2 is 0.95. This value is slightly less than that required for the basin to be linear, but close enough that a large error would not be introduced in determining discharges if it were assumed to be linear. The size of the Cheat River drainage basin, 718 mi^2 (1860 km^2), is far too large to satisfy the other uniformity requirements of the UH method.

The value of m can range from 1 to at least -0.83. Equation (18.3) shows the value of m to be positive; and it is positive for the great majority of drainage basins. Because the value of m indicates the degree of nonlinearity of a drainage basin, the smaller the value of m, the greater the degree of nonlinearity. The greater the degree of nonlinearity, the less will be the increase in peak discharge for increasing volumes of runoff. Negative values of m are an anomaly that is not yet well understood because they show that the discharge actually decreases with increasing runoff volume. At least one drainage basin, with a negative value of m, contains sinks and underground caverns within its drainage area. It is likely that such drainage basins that exhibit a negative value of m have a rather high degree of drainage-basin leakage. Such leakage would not be recorded by the gage and would become larger with increasing volume, consistent with the negative value of m.

It is not uncommon to have values of m as low as 0.4. If the value of m were assumed to be 1 for a basin that actually had a value of m equal to 0.4, then huge overdesign errors would result. The value of m cannot exceed 1 because this would require more runoff volume than was contained in the storm event that fell on the basin. The value of 1 for m can arise when all or part of the rainfall contributes to the runoff volume. If the losses are uniformly distributed over the entire drainage basin, then m will be equal to 1. In a like manner, if there are no losses on the drainage basin, such as an ideal impervious basin, then the value of m is also 1.

A special case of linearity arises when a large drainage basin, which obviously contains nonuniform infiltration, is linear with a value of m equal, or nearly equal, to 1. Such a drainage basin delivers water to the gage in such a manner as to present the mean effect of the varying losses to the gage. The statistical effect of these random conditions at the gage is linear.

Figure 18.2 shows that 92% of the variation in peak discharge Q_p can be explained by runoff volume V alone ($r^2 = 0.92$). Thus, this relation provides an accurate method of peak-discharge prediction for various runoff volumes. The runoff volumes range from 0.13 in. (33 mm) to 3.7 in. (9.4 cm). The prediction of peak discharge can be made for at least this range of runoff volumes and possibly beyond. There were 41 hydrographs where use was consistent for varying discharges and runoff volumes.

The data contained in Figure 18.2 were observed over a 10-year time span. Because of the wide variation in runoff volumes, there had to have been a great difference in the storms that caused these runoff volumes. It then follows that the nature, distribution, movement, duration, and type of storm that caused these hydrographs did not have a significant effect upon the peak discharge. More will be said about the distribution of rainfall later.

The antecedent moisture conditions, land treatment, vegetation, precipitation characteristics, and all other factors that affect runoff and peak discharge are implicit in Figure 18.2, because the runoff volume results after these effects have had their influence. It is not possible to evaluate these individual effects other than to lump them together in the difference between precipitation and runoff volume. Such practice is useful when peak discharge is the primary goal.

18.4 APPLICATION OF THE PEAK-DISCHARGE RATING CURVE

Because the peak-discharge rating curve deals both with peak discharge and runoff volume, it has several applications of interest. These applications deal with predicting peak discharge and with the effect of flood-retarding structures on downstream flood levels.

The strength of the peak-discharge rating-curve procedure is that it relies on real measured data and not upon synthetic data. The relation between peak discharge and runoff volume is so good that reasonably accurate results can be expected even when using few data points. It has already been shown that the curve is continuous between runoff volumes of as little as 0.02 in. (5 mm) up to more than 9 in. (23 cm). For this reason, installation of a gage on an ungaged drainage basin is justified because sufficient data can be obtained in a short time to provide a reasonable basis for decisions. Such observations can be extended as desired to increase accuracy. Sufficient data can usually be obtained within a few months, depending on rainfall distribution and amounts. Of course, accuracy will be increased with the increased number and range of observed runoff volumes.

Equation (18.3) can be used to predict peak discharge from an anticipated runoff volume for any drainage basin, for any rainfall distribution and duration, and for any antecedent moisture condition.

The peak-discharge rating-curve data can be extrapolated within reason in order to predict peak discharges at other locations along the same drainage system. If the location of interest is upstream of the gage, then it can usually be assumed that the drainage characteristics of the drainage basin are approximately the same and, therefore, the value of m for a peak-discharge rating curve at that point would be approximately the same. If the location of interest is downstream from the gage, then it must be ensured that the drainage characteristics of the added drainage area in the downstream direction are not significantly different from the characteristics of the drainage area above the gage. This decision is subjective and should be done carefully. Examination of urbanization, geology, and soils should indicate whether or not hydrologic characteristics of the added drainage area are significantly different.

Assuming that the added downstream drainage area is reasonably similar, then it can be assumed that the slope, m, of Equation (18.3) for the downstream point of interest would not be significantly different from the upstream gage if a gage were in fact present at the new location of interest. The actual volume of runoff from any drainage-basin area is the depth, V, in Equation (18.3), times the drainage-basin area. Therefore, the difference in runoff volume for the same depth of runoff between two drainage basins with like nonlinearity, m, is simply the ratio of the drainage areas.

EXAMPLE 18.3 Determine the peak discharge that will occur from 2 in. (5 cm) of runoff on the Little Papillion Creek drainage basin.

Solution As shown in Figure 18.3, the peak-discharge rating-curve equation established for this drainage basin is

$$\log Q_p = 3.6844 + 0.5236 \log V \qquad (r^2 = 0.88) \tag{18.5}$$

Substituting 2 in. for the runoff volume yields 6950 cfs. The measured runoff for 1.9 in. (4.8 cm) is 6425 cfs.

EXAMPLE 18.4 As indicated by Figure 18.4, dams were constructed on the Salt Creek drainage basin in Nebraska to provide flood reduction. Evaluate the effect of these flood-retarding structures.

Solution The peak-discharge rating curve in the form of Equation (18.3) was used to evaluate the effect of the dams on flood levels at the Lincoln gage number 06803500 in

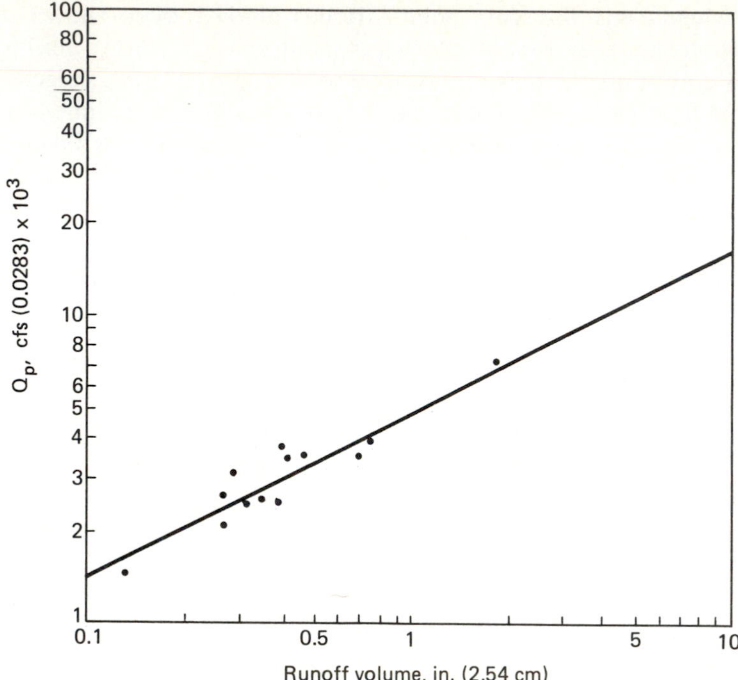

Figure 18.3 Peak discharge versus runoff volume for the Little Papillion Creek at Irvington, Nebraska (drainage area = 32 mi²).

Nebraska for a 2-in. (5-cm) runoff volume. The predam peak-discharge rating-curve equation is

$$\log Q_p = 4.1712 + 0.6897 \log V \qquad (r^2 = 0.86) \tag{18.6}$$

Substituting 2 in. for the runoff volume and solving yields Q_p = 23,923 cfs. The postdam peak-discharge-rating curve equation is

$$\log Q_p = 3.9197 + 0.6322 \log V \qquad (r^2 = 0.81) \tag{18.7}$$

Substituting 2 in. for the runoff volume and solving yields Q_p = 12,883 cfs.

There is, therefore, an 11,040-cfs reduction in flood levels at the Lincoln gage as a result of the flood-retarding structures. This is equivalent to a 46% reduction in the flood peak for this runoff volume. The value of *m* for the predam condition is 0.6897, which is highly nonlinear. The postdam value of *m* is 0.6322, which is more nonlinear. The dams were placed near the basin margins, which had the effect of increasing the infiltration losses above them and caused the drainage basin to become more nonlinear after the dams were built than before. This relation tends to confirm the concept that nonuniform infiltration causes nonlinearity.

EXAMPLE 18.5 Figure 18.5 shows the peak-discharge rating curve for the Susquehanna River at Wilkes Barre, Pennsylvania. The drainage area above this gage is 9960 mi² (25,800 km²) and the value of *m* is 0.9362. This drainage area is mountainous and contains numerous urbanized areas. It is desired to construct a bridge 115 miles downstream at Harrisburg, Pennsylvania. The drainage area above Harrisburg is 24,100 mi²

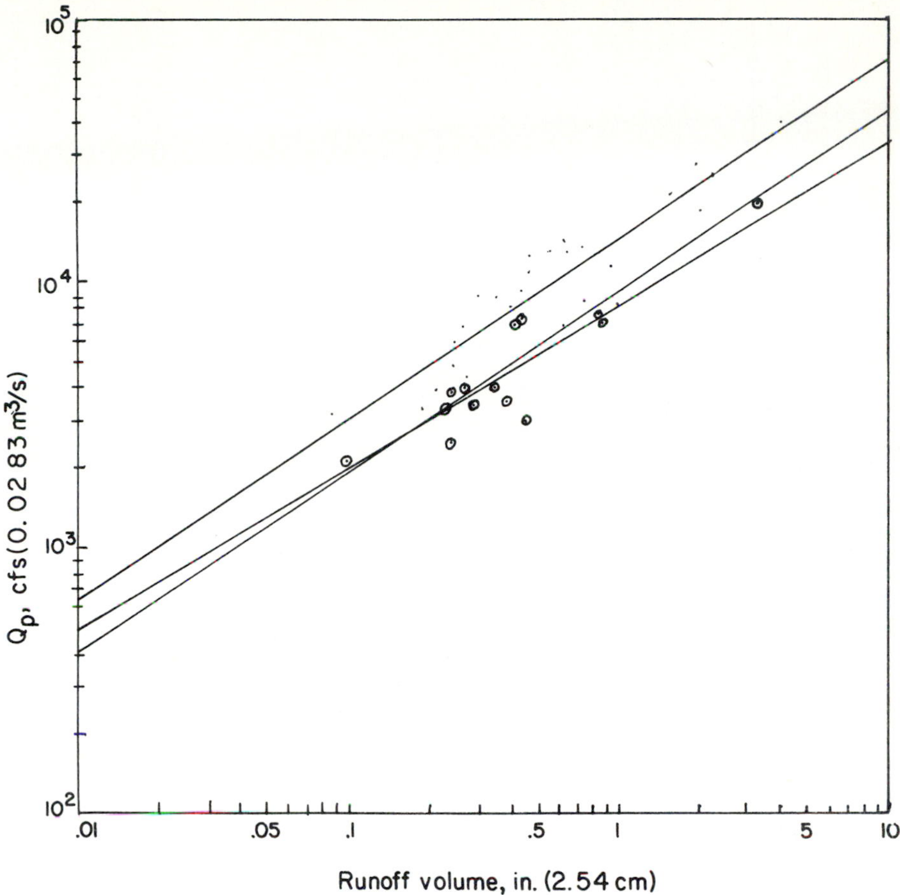

Figure 18.4 Peak discharge versus runoff volume for predam and postdam conditions at the Salt Creek at Lincoln, Nebraska (drainage area = 689 mi² for the predam condition; drainage area = 469.4 mi² for the postdam condition). Log Q_p = 4.1712 + 0.6897 log V, r^2 = 0.86 for the predam condition; log Q_p = 3.9197 + 0.6322 log V, r^2 = 0.81 for the postdam condition.

(62,400 km²) and the expected runoff volume is 6 in. Estimate the discharge for this runoff at Harrisburg.

Solution The general drainage characteristics between Wilkes Barre and Harrisburg are quite similar to those above Wilkes Barre, so that an assumption is made that the nonlinearity, with slope of 0.9362, for Wilkes Barre will be similar at Harrisburg. The drainage area above Harrisburg is 2.42 times that for Wilkes Barre. Thus, 6 in. of runoff at Harrisburg is equivalent to 6 × 2.42, or 14.52 in., at Wilkes Barre. The peak-discharge rating-curve equation for Wilkes Barre is

$$\log Q_p = 4.7086 + 0.9362 \log V \qquad (r^2 = 0.85) \tag{18.8}$$

Substituting 14.52 in for V yields 625,797 cfs (17,979 m³s).

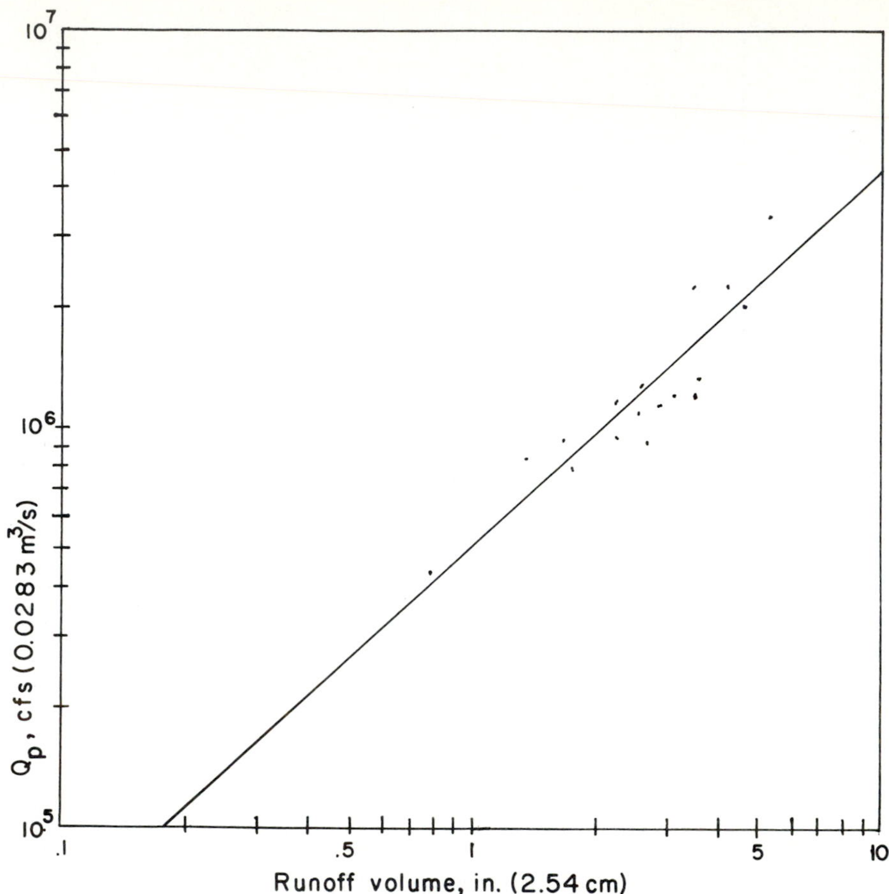

Figure 18.5 Peak discharge versus runoff volume for the Susquehanna River at Wilkes Barre, Pennsylvania (drainage area = 9960 mi²). Log $Q_p = 4.7086 + 0.9362 \log V$, $r^2 = 0.85$.

It turns out that there is a gage at Harrisburg and the peak-discharge rating curve, as shown in Figure 18.6, for this gage yields a discharge of 630,627 cfs (17,847 m^3s) for 6 in. of runoff. The result is very close to the actual value because the value of m for Harrisburg is 0.9679, indicating very similar drainage characteristics.

Application of the drainage-basin peak-discharge rating curve requires knowledge of hydrology and observance of the limitations imposed by external influences. This example should not lead one to believe that it is possible to extend the peak-discharge rating curve without reservation for design purposes. The Wilkes Barre to Harrisburg extrapolation is fortuitous. If there were no gage at Harrisburg to check the extrapolation, it would be prudent to install such a gage in order to determine final discharges for design purposes. The extrapolated value from the Wilkes Barre gage is sufficient for initial planning, until better data is obtained, but should not be used for final design involving important economic considerations. Extrapolation of the peak-discharge rating curve for final design could only be justified when the economic importance of the proposed project does not justify more accurate information. Many hydrologic failures

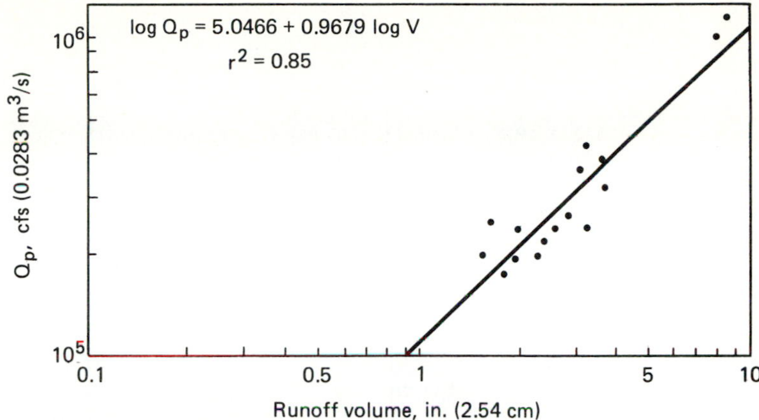

Figure 18.6 Peak discharge versus runoff volume for the Susquehanna River at Harrisburg, Pennsylvania (drainage area = 24,100 mi^2). Log $Q_p = 5.0466 + 0.9679 \log V$, $r^2 = 0.85$.

have resulted from misapplication of procedures when the limitations of the techniques have not been observed. Such methods do well when applied within their limitations. When these limitations are ignored, large errors result.

EXERCISES

18.1. Compute the peak discharge from a watershed of 2000 acres for a runoff coefficient of 0.8 and a rainfall intensity of 2.5 in./h.

18.2. Graph the rational equation and then use the graph to discuss its validity.

18.3. Determine the peak discharge for 5 inches of rainfall from the Little Papillion Creek drainage basin, Nebraska. Use Equation (18.5).

18.4. Rainfall amounts and the corresponding peak-discharge data are given in Table 18.2 for Hubbard Brook 3 in New Hampshire. The brook has a drainage area of 42 ha. Establish the peak-discharge rating curve for this watershed. How good is this relation?

18.5. Rainfall amounts and the corresponding peak-discharge data are given in Table 18.3 for watershed Charleston 78, South Carolina, with a drainage area of 4576.3 ha. Develop the volume–peak-discharge relation and determine the amount of variance that can be explained by this relation.

18.6. Rainfall amounts and the corresponding peak-discharge data are given in Table 18.4 for watershed Fernow 4 in West Virginia with a drainage area of 39 ha. Develop the volume–peak-discharge relation and determine the amount of variance that can be explained by this relation.

18.7. Compute the peak discharge by the rational method for a rainfall intensity of 5 cm/h on a drainage basin 50 hectares in area. The soil is sandy and the watershed slope is 7%. The runoff coefficient is 0.2.

18.8. A 100-acre area is undergoing urbanization. The runoff coefficient estimated for the area prior to urbanization is 0.25 and that after urbanization is 0.55. Determine the

TABLE 18.2 RAINFALL AND PEAK-DISCHARGE DATA FOR HUBBARD BROOK 3 IN NEW HAMPSHIRE

Rainflow (in.)	Peak flow ($ft^3/s/mi^2$)	Rainfall (in.)	Peak flow ($ft^3/s/mi^2$)
0.130	6.80	0.150	7.90
0.210	8.80	0.100	5.00
0.110	7.20	0.300	13.30
0.090	4.20	0.730	28.30
4.860	302.70	0.120	5.50
0.310	9.10	0.360	11.70
0.790	48.80	0.440	22.80
1.570	73.40	0.210	11.90
1.590	79.70	0.130	5.50
0.100	5.30	0.640	33.90
0.250	8.20	1.990	115.00
0.180	8.50	0.330	16.10
0.270	9.00	0.080	4.40
0.140	9.80	0.600	30.30
0.240	18.80	0.130	5.80
0.110	6.40	0.170	6.90
0.230	9.60	0.090	3.70
0.290	12.60	0.170	17.50
0.140	7.30	0.880	43.60
0.100	4.50	0.780	47.30
0.190	10.60	0.250	10.80
0.410	14.80	0.570	35.80
2.360	71.50	0.410	12.50
0.880	27.70	0.220	11.50
0.260	8.90	0.270	11.30
0.200	8.90	1.850	88.40
0.220	8.40	0.140	6.00
0.500	26.40	0.440	27.20
0.170	6.60	0.420	16.70
0.200	9.50	0.230	10.70
0.130	7.80	0.660	25.50
0.940	38.50	0.370	9.80
0.200	11.00	0.180	10.40
0.200	7.70	0.110	10.50
0.560	33.30	0.670	42.40
0.580	19.30	0.110	5.20
0.080	3.90	0.300	31.50
		1.050	115.20
		0.250	14.30
		0.240	10.50

TABLE 18.3 RAINFALL AND PEAK-DISCHARGE DATA FOR WATERSHED CHARLESTON 78 IN SOUTH CAROLINA

Rainflow (in.)	Peak flow ($ft^3/s/mi^2$)	Rainfall (in.)	Peak flow ($ft^3/s/mi^2$)
0.610	12.00	0.150	3.90
0.520	8.90	0.460	8.60
0.160	4.80	0.460	10.10
0.230	5.20	0.260	7.80
0.490	9.20	0.770	13.00
0.100	3.50	0.540	9.20
0.670	12.90	1.220	14.10
0.290	8.10	0.490	10.20
3.180	101.20	0.080	3.40
0.330	6.30	0.160	4.20
0.160	5.20	0.150	4.00
0.120	4.20	0.180	5.00
0.080	3.30	0.180	5.20
0.260	6.00	0.370	8.40
0.180	5.20	1.620	25.90

TABLE 18.4 RAINFALL AND PEAK-DISCHARGE DATA FOR WATERSHED FERNOW 4, WEST VIRGINIA

Rainflow (in.)	Peak flow ($ft^3/s/mi^2$)	Rainfall (in.)	Peak flow ($ft^3/s/mi^2$)
0.160	5.30	0.380	12.20
1.510	30.80	0.250	7.40
1.180	33.40	0.870	20.70
0.600	18.90	1.180	49.90
0.360	12.00	0.310	8.40
0.450	12.20	0.540	12.90
1.590	69.50	0.210	6.30
0.820	34.20	0.420	9.20
1.020	45.80	0.460	12.90
0.910	34.90	0.460	12.30
0.360	11.40	2.200	101.20
0.460	17.70	1.140	30.40
0.210	7.20	1.120	24.70
0.160	7.10	0.820	33.70
0.210	7.50	0.210	5.30
0.450	14.30	2.390	120.50
0.180	5.70	0.390	13.00
0.190	6.00	0.940	27.60
1.450	35.20	0.520	17.40
0.140	7.20	1.670	54.60
0.150	7.50	1.690	31.70
1.950	86.50	0.880	16.30
0.290	7.20	0.950	34.70
0.360	9.30		
0.600	13.90		
0.240	7.30		
0.480	10.80		

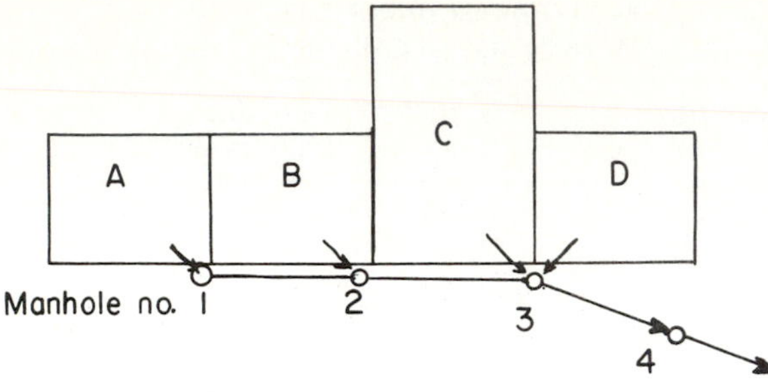

Figure 18.7 Four areas along with sewer lines. Area A = area B = area D = 10 acres; and area C = 30 acres.

peak discharge in both cases due to a rainfall intensity of 15 cm/h. How much discharge can be attributed to urbanization?

18.9. An urban area is composed of two parcels of land designated as A and B. Parcel A is 10 acres in area, 3000 ft long, has a slope of 0.05, with a runoff coefficient of 0.25, and drains into parcel B. Parcel B is 5 acres in area, 2000 ft long, has a slope of 0.1, with a runoff coefficient of 0.5. What is the peak discharge from this urban area for a rainfall with an intensity of 5 in./h occurring for a period of 1 hour?

18.10. Four areas, A, B, C, and D, as shown in Figure 18.7, are given along with the following basic data: The average coefficient of imperviousness is 0.8; the inlet time is 8 minutes for areas A, B, and D, and 25 minutes for area C. Assume that the time of flow between manholes is 5 minutes. The time intensity of the rainfall equation to be used is

$$I = \frac{105}{t + 15}$$

where I is the intensity of rainfall in units of inches per hour, and t is the duration of rainfall in minutes. Use the rational method to determine the flow in cfs units in each of the sewer lines between manholes. Tabulate your calculations and give a key to the column headings.

Note: The direction of flow from each area and between manholes is shown by the arrows on the diagram.

18.11. The time of concentration for a rectangular area is 30 minutes. The direction of overland flow is parallel to the longer sides of the rectangle. Should one expect a greater peak rate of runoff from this area from a storm of intensity 4 in./h of 10-minute duration or from a storm of intensity 1 in./h of 40-minute duration? Why?

CHAPTER 19
Snow and Snowmelt Runoff

Snow is a form of precipitation. Its texture changes with temperature. It can transform to ice or melt to become water. The water is stored on watersheds in the various forms of snow, which range from newly fallen crystalline snow to glacial ice. Release of water from the various forms of snow and ice results from increased temperatures, but the rate of melting is different for different forms of snow and ice. Light snow melts faster than old snow that has been altered to ice.

Snow is of considerable importance in water-resources planning and development in those areas where winter precipitation falls as snow. Such areas are found in both the Northern and Southern Hemispheres. In some of these areas, snow may be the dominant source of streamflow, and these areas may depend on snow for water supply. The Rocky Mountain region in the United States is an example of such an area. Snowmelt runoff causes flooding, especially when combined with heavy rainfall. Thus, design of flood-prevention works have to include the effect of snowmelt runoff. Snowmelt-runoff estimates are needed for forecasting seasonal water yields, river regulation and storage works, determination of design floods, planning flood-control programs, etc.

Because snow is a conveniently stored form of water, its usage can be delayed until it is needed for a particular use. Further storage of snowmelt water is frequently accomplished by construction of dams. Accumulated snow and ice withdraw large quantities of water from participating in evaporation and affect local climate because of the heat energy required to bring about any changes of state. Snow also plays an important role in the transportation, civil aviation, electrical transmission lines, etc.

This chapter briefly describes the general principles to help understand the role of snow hydrology in water-resources development. For a comprehensive discussion of snow hydrology, reference is made to the U.S. Army Corps of Engineers (1956) and Gray and Male (1981) amongst others.

19.1 CHARACTERISTICS OF SNOW

Snow is a solid form of water. Its formation results from crystallization of ice particles in the atmosphere during precipitation. The snow crystals grow while they are in the atmosphere and can reach large size, although smaller sizes are more common. Newly formed snow crystalizes in hexagonal shapes. These crystals alter in form when they accumulate on the ground by metamorphosis and over time increase in density. The crystalline structure of fully ripe snow is granular and similar to porous soil material. Movement of water through this snow is similar to the movement of water through porous media and can be described using the same physical laws.

Certain terms, commonly employed to describe snow characteristics, are defined first. The water equivalent (cm or in.) of a snow cover is the depth of water that has the same weight as the snow cover. Thus, snow can be measured or described in terms of depth of water. Water equivalent and density of snow are interrelated. The density of snow, ρ_s (g/cm^3 or %), is the fraction of snow volume occupied by its water equivalent. Snow density is considered to be the same as specific gravity, and is computed as the depth of the measured water equivalent of the snow cover divided by its depth. Many factors affect the snow density: heat exchange between the snow cover and the air, heat exchange between ground and snow cover, wind, temperature, percolation of melt water, depth of snow cover, etc.

Newly fallen snow has a density ranging from 0.01 to 0.15. Névé or firn has a density of about 0.5 and natural ice (glacial ice) has a density of about 0.91. For hydrologic purposes, fully ripe snow has a density of about 0.45 to 0.50. An average value of 0.1 g/cm^3 appears satisfactory for engineering applications. Seligman (1962) has suggested average densities for some snowpacks, as given in Table 19.1.

TABLE 19.1 SNOWPACK DENSITIES

Condition of snow	Density
Wild snow	0.01 to 0.03
Ordinary snow, immediately after falling in still air	0.05 to 0.065
Settling snow	0.07 to 0.19
Settled snow	0.2 to 0.3
Very slightly wind-toughened, immediately after falling	0.06 to 0.08
Average wind-toughened snow	0.28
Hard wind slab	0.35
New firn snow (partly consolidated into ice)	0.4 to 0.55
Advanced firn snow	0.55 to 0.65
Thawing firn snow	0.6 to 0.7

Quality of snow is related to the ice content of the snow cover. It is expressed as a decimal fraction or ratio of the weight of the ice content to the total weight of the snow cover. Generally, snow quality is about 0.95, but may drop down to 0.7 or so during melting. This reduction in snow quality depends on the melt rate, density of snow cover, infiltration capacity of the underlying land surface, etc.

Thermal quality of snow is the ratio of the amount of heat required to produce a given amount of water from snow to the amount of heat required to produce the same amount of water from pure ice at 0°C. At subfreezing temperatures, the thermal quality, C, may exceed 1. For ripe snowpacks, C is less than 1 and is close to 0.97. C can be expressed as

$$C = \frac{L_s}{L} + \frac{c_p T}{L} \tag{19.1}$$

where L is the latent heat of fusion of water (cal/g), T is the temperature of the snowpeak (°C), L_s is the latent heat of fusion of snow, and c_p is the specific heat (cal/g-°C) of snowpack.

Cold content (cal/cm²), H_c, of the snowpack is defined as the heat required per unit area to raise the snowpack temperature to 0°C. The water equivalent (cm), W_c, of the cold content is the equivalent amount of water needed to raise the snowpack temperature to 0°C. The terms H_c and W_c can be expressed as

$$H_c = \rho_s c_p dT \tag{19.2}$$

and

$$W_c = \frac{H_c}{\rho_w L} = \frac{\rho_s dT}{160} = \frac{W_0 T}{160} \tag{19.3}$$

where ρ_s is the snowpack density (g/cm³), ρ_w is the liquid-water density at °C = 1 g/cm³, d is the depth of snowpack (cm), T is the snowpack temperature below 0°C, and W_0 is the initial water equivalent (cm) of the snowpack for an assumed specific heat of ice of 0.5.

The latent heat of fusion of ice is 79.7 cal/g at 0°C. This combined with the latent heat of vaporization, 596 cal/g, brings the total energy required to evaporate water from ice to about 676 cal/g. For this reason, snow reduces the air temperature by absorbing a large amount of solar energy. The latent heat of fusion of snow depends on the presence of liquid water in the snowpack.

Snow cover undergoes structural changes. The change from loose, dry, low-density subfreezing form to coarse, granular, high-density form is called ripening. A ripe snow cover is considered primed when it is ready to melt and yield runoff.

Snow may reflect a large part of the solar radiation received at the snow surface, depending upon solar elevation, wavelength, temperature, depth of snow cover, and so on. The amount of radiation reflected as a fraction of the total radiation received is called the reflectivity, or albedo. Depending upon the condition of snow-cover surface and the height of sun, the value of albedo may change from 0.29 for very porous, dirty, saturated with water, snow to 0.86 for compact, dry, clean snow. Table 19.2 shows values of albedo for some snowpacks (Kondrat'ev, 1954).

TABLE 19.2 ALBEDO OF SOME SNOWPACKS

Condition of snowpack surface	Height of sun (°)	Albedo (%)
Compact dry clean	30.3	86
	29.7	88
	25.1	95
Clean, wet, fine grain	33.3	64
	34.5	63
	35.3	63
Wet, clean, granular	33.7	61
	32.0	62
Porous, very wet, greyish color	35.3	47
	36.3	46
	37.3	45
Very porous, grey, full of water,	32.8	43
sea ice visible	31.7	43
Very porous, light brown, saturated with water	29.7	31
Very porous, dirty, saturated with water, sea ice visible	37.3	29

Snow absorbs most incident long-wave radiation and acts like a black body with respect to long-wave radiation. Snow radiation, R, can be described by Stefan's law as

$$R = \sigma T^4 \tag{19.4}$$

where R is the total radiation in all wavelengths, T is the temperature in °Kelvin, and $\sigma = 0.826 \times 10^{-10}$ langleys per minute. Note that the maximum temperature of the snow surface is 32°F, the maximum radiation intensity from Equation (19.4) is 0.459 ly/min, or 27.5 ly/h (U.S. Army Corps of Engineers, 1956).

EXAMPLE 19.1 A snowpack is 2 m deep. A sample taken from the snowpack is 0.1 m^3 and weighs 20 kg. When this sample is mixed with 100 kg of water at 20°C, the resulting temperature is 15°C. Compute the density, water equivalent, and quality of the snowpack.

Solution

$$\text{Volume of water equivalent of the sample} = 20/1000 \text{ m}^3 = 0.02 \text{ m}^3$$

$$\text{Density} = \frac{\text{volume of water equivalent}}{\text{volume of snow}} = \frac{0.02}{0.1} = 0.2, \text{ or } 20\%$$

$$\text{Water equivalent} = \text{density} \times \text{depth} = 0.2 \times 2 \text{ m} = 0.4 \text{ m} = 40 \text{ cm}$$

For quality of snow, we need to determine the weight of the ice content of the sample, which is obtained by considering the heat requirements. The amount of heat supplied by water = 100,000 × (20 − 15) = 500,000 cal. The amount of heat required to melt the snow = 80 × ice content (g), where the latent of fusion is ≅ 80 cal/g. The amount of heat required to raise the temperature of the melted snow to 15°C = 20,000 × (15 − 0) = 300,000 cal. Now balancing the heat,

$$500{,}000 = 80 \times \text{ice content} + 300{,}000$$

Therefore,

$$\text{Ice content (g)} = \frac{500{,}000 - 300{,}000}{80} = \frac{200{,}000}{80} = 2500 \text{ g} = 2.5 \text{ kg}$$

Thus,

$$\text{Quality of snow} = \frac{2.5}{20} = 0.125, \text{ or } 12.5\%$$

19.2 POINT SNOW MEASUREMENT

By comparison with rainfall, snowfall over an area is more uniform, but its accumulation and retention on the ground is highly heterogeneous because of the buoyancy of snow. Because snow is a form of precipitation, its measurement at weather stations is made at the time of fall. Snow measurement consists of the actual depth of snow unaffected by wind and the water content of such a depth of snow. Because the density of newly fallen snow varies, so, too, does its water content. To determine the water content of new snow, a column of such snow is melted down and its equivalent depth of water is determined. The same water content can be determined by weighing the snow column. A rule of thumb often used to estimate water content is 10 centimeters (4 inches) of snow is equivalent to 1 centimeter (0.4 inch) of water. This assumes an average snow density of 0.10. Except for special applications, such as ski slopes, the water content of snow is more useful than actual snow depth. A measurement of a 24-h snowfall can be made using a ruler, and an estimate of water content made with the average density.

The water content of snow at the time of fall can be measured using (a) precipitation gages, (b) snow boards, and (c) snow stakes. For small snowfalls, the standard rain gage with the funnel and overflow tube removed is often satisfactory. The National Weather Service operates three types of snow gages (U.S. Weather Bureau, 1951): (1) The standard 8-inch cylindrical rain gage is 24, 36, or 42 inches in height. This gage contains an antifreeze solution to melt the snow, permitting up to 30-day intervals between actual measurements. (2) The Sacramento conical seasonal storage gage has capacities of 60, 100, 200, and 300 inches. This gage is monitored once or twice a year to gage the accumulated fluid. (3) The standpipe seasonal precipitation gage is composed of a 12-inch tube that converges to an 8-inch-diameter tube. Tube lengths are available to accommodate 250 and 475 inches of precipitation. This gage is monitored intermittently.

All seasonal precipitation gages contain antifreeze solutions so that the accumulated snow will be melted into an easily measured water depth. Location of these seasonal gages is selected so that the effect of wind will be minimized.

Snow boards, at least 16 in. square, are placed on the snow surface. The freshly fallen snow is identified and the water content is then determined. A number of snow boards can be used to obtain more representative data. These should be located where the effect of wind is minimum.

Snow stakes are used where a large accumulation of snow occurs. These stakes are graduated in feet and inches similar to a stadia rod so that the accumulated snow depth can be read directly. Larger stakes, called aerial snow-depth markers, are used for monitoring snow depth by use of low-flying aircraft.

19.3 SNOW SURVEYING

Snow sampling is accomplished by means of a kit of tools that includes a segmented tube that, when assembled, is driven into the snow to its total depth. This tube is designed so as to retrieve a continuous core of snow from which the depth and water content of the snow are determined by weighing.

Snow surveying is used to get an estimate of watershed precipitation. A series of observation stations that might consist of any or all of the seasonal gages, stakes, or snow-sampling stations is combined to form a snow course (Soil Conservation Service, 1972). This course is regularly monitored by personnel on snowshoes, skis, or by snow vehicles to obtain the desired data. Snow courses are designed and located so as to provide typical accumulations that can be used to establish correlations to predict runoff from a drainage area. Networks of snow courses are established to produce sufficient measurements that can then be used to develop correlations with measured runoff at stream gages for reasonably accurate and dependable prediction of snowmelt. It would not be practical to maintain a dense network of snow-measurement stations over a large area that is difficult to monitor on snowshoes, skis, or even on snow vehicles. For this reason, only sufficient stations are monitored to provide dependable correlations with runoff.

Other special snow-measuring devices are used to determine the water-equivalent content of snow. Pressure pillows filled with antifreeze and containing transducers to transmit changes in pressure as snow accumulates on the pillow are dropped from aircraft. These pressure pillows are monitored electronically. They are reasonably accurate until depths accumulate that permit bridging of snow over the pillow. Nuclear radiation snow gages (Warnick and Penton, 1971) have been developed that measure the water equivalent of a snowpack. Some measuring problems with these instruments occur as snow density changes.

Because of high winds and other weather factors, mountain areas are not evenly covered with snow. Some mountain areas might have no snow cover, whereas great depths occur on other areas. Success in forecasting snowmelt runoff depends to a great extent on consistency of snow cover from year to year. Satellite imagery (Schneider et al., 1976) is used to compare snow cover from time to time to assist in evaluating snow distribution.

19.4 AREAL SNOWFALL

Point measurements of snowfall can be utilized to estimate depth as well as water equivalent of snow cover over the watershed. Because the snow-measurement network is usually limited, the normal averaging method does not apply either. Owing to large orographic and topographic effects, the usual methods of obtaining mean areal rainfall such as the arithmetic mean or the Thiessen polygon method do not apply to snowfall. However, regional orographic effects may be constant from year to year and for storm to storm at least for small watersheds when compared with the areal extent of storms occurring in the region. This implies that a basic pattern of areal snow distribution exists in a region.

19.4.1 Areal Snow Depth

Two simple methods are used to estimate the depth of a snow cover: (a) the ratio method and (b) the isopercentile method.

According to the ratio method,

$$\frac{P_a}{P_0} = \frac{N_0}{N_a} \tag{19.5}$$

where P_0 is the observed snowfall depth at a point or group of stations, P_a is the watershed snowfall, N_0 is the normal annual snowfall for the observation station or stations, and N_a is the normal annual snowfall for the watershed. Equation (19.5) states that the ratio of station snowfall to watershed snowfall is constant. The value of N_a can be calculated by drawing mean annual isohyets and planimetering the areas between the isohyets.

In the isopercentile method, point snowfall, storm or annual, is computed as a percentage of the normal annual snowfall. The percentage values of the measurement stations are used to construct isopercentile lines. An isohyetal map of snowfall is also constructed, which reflects the effect of topographic variability. The isopercentile map indicates departure from the isohyetal pattern. By superimposing the isopercentile map on the isohyetal map, the areal snow depth can be computed.

19.4.2 Areal Water Equivalent

The water content of a snow cover determines its contribution to runoff. Various averaging methods can be used to estimate the areal water equivalent from snow measurements. Index methods can also be used to estimate watershed water equivalent. Precipitation, runoff, meteorological factors, etc., can be used to develop these index methods.

19.5 SNOWMELT

Snowmelting is a thermodynamic process. Its study should, therefore, consider the various factors influencing the transmission of heat to the snowpack. Ranked in order of their importance, these factors are (1) sensible heat conduction from moist air to the snow surface, (2) latent heat of condensation, (3) solar radiation, (4) heat transmitted by rain falling on the snowpack, and (5) heat conduction from the underlying ground. The first two factors occur through turbulent diffusion of warm moist air.

Heat is required to convert snow to water, as illustrated by the fact that 79.7 calories of heat are needed to convert 1 gram of snow at 0°C to water at the same temperature. This heat might be obtained by (a) incoming solar radiation from the sun, (b) the latent heat of fusion (79.7 cal/g) resulting from the condensation of water vapor on snow, or (c) heat by conduction from the atmosphere or lithosphere in contact with the snow. These factors might act separately or together to almost any degree, thus providing a complex set of melting conditions. Equations to be presented later involve these three conditions as basic elements.

Insolation results in 1.94 langleys per minute of energy (1.94 cal/cm^2/min) arriving at the outer edge of the atmosphere. Of this energy, about 0.30 langley reaches the earth's surface, where it is available to heat the earth's environment. The actual value of the heat available varies with latitude, season, and atmospheric conditions.

Albedo gives the amount of solar radiation that is reflected into space by snow. The value of albedo, in percent, is 80 to 95 for clean snow, 40–50 for dirty snow, 10–20 for green forests, and 15–25 for crop-covered cultivated fields. Thus, it is apparent that the amount of the insolation reaching the earth and converted to heat is affected by the albedo of the earth's environment. That insolation not reflected as albedo is converted to heat in some form.

Latent heat of condensation affects snowmelt as warm moist air is condensed on snow when the dewpoint is above 0°C. Under these conditions, each gram of condensed water vapors releases 596 cal of heat. This heat is sufficient to convert snow at 0°C to nearly 7 cm of water. The effect of wind on melting is important in this process. As water is condensed on the snow, the atmosphere is deprived of moisture. Wind transports additional moisture-laden air over the snow, where additional condensation occurs.

Additional melting is caused by conductive heat transfer as snow is in contact with warmer air, rainfall, or the rock or soil surface of the earth. The effectiveness of these factors depends on the temperature differences and the mass of the heat source. Generally, only a small amount of heat is conducted from the earth. Conductive heat from the atmosphere and from rainfall can be effective melters of snow when conditions are optimum.

19.5.1 Energy-Budget Method

This method involves an accounting for a given period of time of incoming energy, outgoing energy, and the change in energy storage for a snowpack. The net energy is then expressed as the heat equivalent of snowmelt. The energy balance of snowpack can be expressed as

$$H_m = H_r + H_a + H_e + H_g + H_p + H_q \tag{19.6}$$

where H_m is the heat equivalent of snowmelt, H_r is the net radiation $= H_s + H_l$, H_s is the net solar (short-wave) radiation, H_l is the long-wave (terrestrial) radiation, H_a is the heat transmitted by warm moist air, H_e is the latent heat of condensation, H_g is the heat transmitted by the underlying ground, H_p is the heat transmitted by rainfall, and H_q is the change of energy storage of the snowpack. The energy balance of the snowpack is formulated similar to that for evaporation from a water body, which is discussed in Chapter 10.

The terms H_a, H_g, H_p, and H_s are all usually positive; H_l is negative in the open (long-wave radiation loss); and H_e and H_q may be positive or negative. Thus, the amount of snowmelt, M, in centimeters of water for a snowpack of thermal quality, C, can be expressed as

$$M = \frac{H_m}{80C} \tag{19.7}$$

Condensation of water vapor has been shown to be one of the most important sources of heat for snowmelt. The equation for a 6-h snowmelt from condensation of water vapor (Wilson, 1941) is

$$M = av(e - 6.11) \tag{19.8}$$

where M is the snowmelt in cm of water (cm/6 h), a is an empirical constant = 0.005 (the value of $a = 0.012$ for M in in.), v is the wind velocity 4.5 m above the ground (km/h), and e is the vapor pressure of the air in millibars 1.25 m above the snow. The value 6.11 is the saturation vapor pressure (mb) over the ice at 0°C. In order for condensation to occur, the dewpoint temperature should be more than 0°C; otherwise, evaporation will occur.

Snowmelt occurring as a consequence of conductive and convective heat transfer from air to snow for a 6-h period is shown by Sverdrup (1934) to be

$$M = cvT_a \tag{19.9}$$

where T_a is the air temperature in °C, and c is the turbulence coefficient involving the latent heat of ice, air density, turbulence, etc. For wind velocities 4.5 m above the ground surface and temperature measurements 1.25 m above the ground surface, the value of c is 0.003 to 0.006.

Anderson and Crawford (1964) modified Equation (19.9) for hourly snowmelt as

$$M = \frac{cvT}{C} \tag{19.10}$$

where M is the snowmelt rate (cm/h), v is the wind velocity (km/h), T is the surface air temperature (°C), c is the turbulence coefficient = 0.001 for v at 4.5 m and T at 1.25 m, and C is the snow quality.

Radiation is the next important source of heat for snowmelt. Since snow radiates like a black body, the amount of radiation is given by Planck's law, and the total energy radiated is given by Stefan's law expressed as Equation (19.4). The long-wave radiation by a snowpack is influenced by temperature, vegetative cover, and cloud conditions. The hourly snowmelt rate (cm/h) due to short-wave radiation is expressed as

$$M = \frac{H_s}{80C} \tag{19.11}$$

where H_s is the net absorbed radiation in langleys. When $C = 1$, there is an exchange of long-wave radiation between the snowpack and its surroundings. If the net long-wave radiation is positive, then the hourly snowmelt is given as

$$M = \frac{H_l}{80C} \tag{19.12}$$

TABLE 19.3 VALUE OF 12-h SNOWMELT IN CLEAR WEATHER WITHIN 40° AND 48° LATITUDES (AFTER WILSON, 1941)

Month	M_0 (cm)
March	0.89
April	1.07
May	1.22
June	1.35

where H_l is the net long-wave radiation in langleys. If H_l is negative, there would be an equivalent heat loss from the snowpack.

Wilson (1941) expressed 12-h snowmelt, M, as

$$M = M_0 (1 - 0.75m) \tag{19.13}$$

where M (cm) is the snowmelt occurring in half a day (noon to midnight or midnight to noon), M_0 (cm) is the snowmelt occurring in half a day in clear weather, and m is the degree of cloudiness, with 0 for clear weather and 1.0 for a complete overcast. The values of M_0 within latitudes 40° and 48° were prescribed by Wilson, as shown in Table 19.3.

When it rains, the temperature is usually low. The hourly snowmelt (cm/h) due to rainfall can be expressed as

$$M = \frac{PT_w}{80C} \tag{19.14}$$

where P is the depth of precipitation in cm, and T_w (°C) is the wet-bulb temperature (assumed to be that of the ambient rain). It can be seen that 1 cm of rain at 10°C would melt only about 0.125 cm of water from snow with $C = 1$. For M in in., $80C$ will be replaced by 144.

Daily snowmelt, M (cm), due to rainfall can be expressed as

$$M = 0.0126T_aP \tag{19.15}$$

where P is the daily rainfall (cm), and T_a is the mean daily temperature of saturated air taken at a height of 3 m.

Conduction of heat from the underlying ground to the snowpack depends on the soil texture, soil-moisture content, and amount of ground cover offered by vegetation, forest litter, etc. The heat transfer through conduction is

$$H_g = k\frac{dT}{dh} \tag{19.16}$$

where k is the thermal conductivity of soil, and dT/dh is the temperature gradient perpendicular to the ground surface. Snowmelt due to this source is approximated at

0.05 cm/day during a snowmelt season. However, this does provide moisture to the ground, which when saturated is most conducive to snowmelt runoff.

19.5.2 Light's Equation

Light (1941) developed a theoretical melting formula utilizing theories of atmospheric turbulence. His formula is expressed as

$$M = \frac{\rho_a K_0^2}{80 \ln\left(\frac{a}{z_0}\right) \ln\left(\frac{b}{z_0}\right)} u\left[c_a T + (e - 6.11)\frac{423}{p}\right] \tag{19.17}$$

where M is the effective snowmelt (combined snowmelt and condensate), ρ_a is the air density, K_0 is von Karman's coefficient = 0.38, u is the wind velocity at the anemometer level, z_0 is the roughness parameter = 0.25, c_a is the specific heat of air at constant pressure = 0.24, T is the air temperature at hygothermograph level, e is the vapor pressure of air (mb), p is the atmospheric pressure (mb), a is the elevation (cm) of anemometer in mb at hygothermograph level, and b is the elevation (cm) of hygothermograph. All values are expressed in cgs units.

By adopting the reference elevation of instruments at a = 50 ft and b = 10 ft, Light (1941) expressed snowmelt in a simplified form as

$$M = u_m[0.00184(T_f - 32)10^{-0.0000156h} + 0.00578(e - 6.111)] \tag{19.18}$$

where M is the effective snowmelt in inches per 6 hours, u_m is the average wind velocity in miles per hour, T_f is the air temperature in °F, e is the vapor pressure in mb, and h is the station elevation above sea level in feet.

Equation (19.17) or (19.18) is for snowmelt at a point, that is, the inflow of melt water to the snow cover, and not for subsequent disposition of the melt-water. In order to apply this formula to actual watersheds, Light developed a procedure for determining areal melting rates, taking into account overall changes in the air mass, produced by melting snow and by surface characteristics of the watershed.

EXAMPLE 19.2 A snowpack has a thermal quality of 0.8. How much will be the snowmelt due to heat of 200 langleys?

Solution By applying Equation (19.7),

$$M = \frac{200}{80 \times 0.8} \text{ cm} = 3.125 \text{ cm}$$

because the density of water is 1 g/cm^3 at 0°C.

EXAMPLE 19.3 Consider a snowpeak at an elevation of 2000 m above mean sea level at an altitude of 45°N. The snowpack is subjected to the following: air temperature = 15°C, mean wind velocity = 10 km/h, relative humidity = 70%, average rainfall intensity = 0.1 cm/h for 6 hours, and wet-bulb temperature = 10°C. Determine the 12-h snowmelt due to each heat source, the total 12-h snowmelt, and the percentage contribution to the snowmelt from each heat source. Assume the sky is clear during the 12-h period in the month of March.

Solution First, the snowmelt due to each heat source is estimated.

Condensation. To apply Equation (19.8), the saturated vapor pressure, e_s, of the air is to be determined first. $e_s = 20$ mb.

$$RH = 70 = 100 \times \frac{e}{e_s}$$

This yields

$$e = 0.7 \times 20 = 14 \text{ mb}$$

Assuming $a = 0.005$, the 6-h snowmelt due to condensation is

$$M = 0.005 \times 10 \times (14 - 6.11) = 0.395 \text{ cm for 6 h}$$
$$= 0.79 \text{ cm for 12 h}$$

Convection. To apply Equation (19.9), c is assumed to be 0.005.

$$M = 0.005 \times 10 \times 15 = 0.75 \text{ cm for 6 h}$$
$$= 1.5 \text{ cm for 12 h}$$

Radiation. To apply Equation (19.13), $M_0 = 0.89$ cm from Table 19.3, and $m = 0$.

$$M = 0.89 \times (1 - 0.75 \times 0) = 0.89 \text{ cm}$$

Rainfall. Assume $C = 1$. Applying Equation (19.14),

$$M = \frac{0.1 \times 12 \times 10}{80} = 0.15 \text{ cm}$$

By adding values of snowmelt due to the various heat sources, the total 12-h snowmelt = 0.79 + 1.5 + 0.89 + 0.15 = 3.33 cm. Percentagewise, 12-h snowmelt is 23.72% due to condensation, 45.05% due to convection, 26.73% due to radiation, and 4.5% due to rainfall.

19.6 LIQUID-WATER CONTENT OF SNOWPACK

It is assumed that the snowpack is homogeneous, has a maximum water-holding capacity, and fills from top to bottom. The liquid-water content present in a snowpack is zero when the snowpack is below 0°C. For temperatures equal to or greater than zero, water may exist in one of three forms: hygroscopic, capillary, and gravitational. The amount of water present in a snowpack is its actual liquid-water content, W_a, and may be less than the maximum water-holding capacity, W_m, of the snowpack ($W_a \leq W_m$). The difference, W_d between W_m and W_a, $W_d = W_m - W_a$, is the liquid-water deficiency of the snowpack. When this deficiency is more than satisfied, runoff begins to occur. For a snowpack at 0°C,

$$P_d = 100(1 - C) \tag{19.19}$$

where P_d is the percent actual liquid-water content, by weight, in the snowpack.

The amount of water S_d (cm) required to satisfy the liquid-water deficiency of the snowpack can be expressed as

$$S_d = \frac{P_d(W_0 + W_c)}{100} \tag{19.20}$$

The source of water supply to the snowpack is rainfall. If the average rainfall intensity is r cm/h, then the volume of water stored per unit area is

$$S = t(r + m) \tag{19.21}$$

where m is the average melt rate (cm/h), and t is the time from the beginning of rainfall.

For a snowpack of temperature <0°C, the storage can be expressed in terms of the cold content and the liquid-water-holding capacity.

$$S = S_d + W_c = P_d\left(\frac{W_0 + W_c}{100} + \frac{W_0 T}{160}\right) \tag{19.22a}$$

If W_c is small by comparison, then

$$S = W_0\left(\frac{P_d}{100} + \frac{T}{160}\right) \tag{19.22b}$$

The time, t_c (h), required to supply the cold content, W_c, is

$$t_c = \frac{W_0 T}{160\,(r + m)} \tag{19.23}$$

Similarly, the time, t_d (h), required to satisfy the storage requirement of the snowpack is

$$t_d = \frac{S_d}{r + m} \tag{19.24}$$

Upon satisfaction of the liquid-water deficiency, the snowpack becomes ripe, and continued rain or snowmelt will move through the pack as gravitational water with a transmission rate of u (cm/h). This water draining through the pack under the action of gravity constitutes the transient storage, S_t (cm), expressed as

$$S_t = \frac{d(r + m)}{u} \tag{19.25}$$

The time t_r taken by water to pass through the snowpack depth, $d = W_0/\rho_s$, is

$$t_r = \frac{d}{u} = \frac{W_0}{\rho_s u} \tag{19.26}$$

Now the total volume of water stored by the pack prior to occurrence of runoff is

$$\begin{aligned} S &= W_c + S_d + S_t \\ &= W_0\left(\frac{T}{160} + \frac{P_d}{100} + \frac{r + m}{\rho_s u}\right) \end{aligned} \tag{19.27}$$

Thus, the total time elapsed before runoff commences is approximately

$$t = t_c + t_d + t_r$$
$$= W_0\left[\frac{T}{160\ (r + m)} + \frac{P_d}{100\ (r + m)} + \frac{1}{\rho_s u}\right] \qquad (19.28)$$

The value of t_r is usually small and may be neglected.

EXAMPLE 19.4 Consider a snowpack 2 m deep, with a density of 0.25 and a temperature of −5°C. The air temperature is 15°C, the wind velocity is 10 km/h, and the relative humidity is 50%. Compute the snowmelt due to condensation. Also compute the cold content of the snowpack. Assume an area of 1 m².

Solution To apply Equation (19.8), the vapor pressure is to be determined. From the previous example, $e_s = 20$, so $e = 0.5 \times 20 = 10$ mb. Assuming $a = 0.005$, $M = 0.005 \times 10 \times (10 - 6.11) = 0.195$ cm for 6 h = 0.39 cm for 12 h. To apply Equation (19.3), the term W_0 is computed first.

$$\text{Density} = \frac{\text{volume of equivalent water}}{\text{volume of snow}}$$

$$0.25 = \frac{\text{volume of equivalent water}}{2}$$

Therefore, the volume of equivalent water = 0.25 × 2 = 0.5 m. This gives $W_0 = 0.5$ m for an area of 1 m². $T = 5$°C.

$$W_c = \frac{W_0 T}{160} = \frac{0.5 \times 5}{160} = 0.0156 \text{ m} = 1.56 \text{ cm}$$

EXAMPLE 19.5 Consider a snow pack with the following characteristics:

DEPTH (m)	LAYER	TEMPERATURE (°C)	DENSITY (%)
0–1	Bottom	−5	35
1–2	Top	−10	30

The water-holding capacity is 3% and the melt rate is 0.5 cm/h for 6 hours. Compute the thermal quality of the snowpack, the cold content, the 3-h penetration of melt water, the time to warm the snowpack to 0°C, and the time to offset the liquid-water deficiency.

Solution The thermal quality of the snowpack, C, is computed from Equation (19.1). For the bottom layer,

$$C = \frac{80}{80} + \frac{0.5 \times 5}{80} = 1 + 0.031 = 1.031, \text{ or } 103.1\%$$

For the top layer,

$$C = \frac{80}{80} + \frac{0.5 \times 10}{80} = 1 + 0.063 = 1.063, \text{ or } 106.3\%$$

The average thermal quality of the snowpack = (1.031 + 1.063)/2 = 1.047, or 104.7%. Note that −5°C requires 2.5 cal to rise to 0°C, and −10°C requires 5 cal to rise to 0°C, assuming $c_p = 0.5$. Indeed, we can see how thermal quality changes with increasing temperature.

THERMAL QUALITY (%)	TEMPERATURE (°C)	FREE WATER (%)
101.8	−3	0
101.2	−2	0
100.6	−1	0
100	0	0
99	0	1
98	0	2
97	0	3
96	0	4

The cold content is computed from Equations (19.2) and (19.3). For the bottom layer,

$$H_c = 0.35 \times 0.5 \times 100 \times 5 = 87.5 \text{ cal/cm}^2$$

$$W_c = \frac{0.35 \times 100 \times 5}{160} = 1.094 \text{ cm}$$

For the top layer,

$$H_c = 0.3 \times 0.5 \times 100 \times 10 = 150 \text{ cal/cm}^2$$

$$W_c = \frac{150}{80} = 1.875 \text{ cm}$$

The total cold content of the snowpack = 237.5 cal/cm^2, or 2.969 cm. The 3-hour penetration of melt water, d_p, can be expressed as

$$d_p = \frac{t(r + m)}{\rho_s\left(\frac{T}{160} + \frac{W_m}{100}\right)}$$

For the top layer,

$$t = 3 \text{ h}, \quad r = 0 \text{ cm/h}, \quad m = 0.5 \text{ cm/h}, \quad \rho_s = 0.30, \quad T = 10°\text{C}, \quad W_m = 3\%$$

$$d_p = \frac{3 \times (0 + 0.5)}{0.30 \times \left(\frac{10}{160} + \frac{3}{100}\right)} = \frac{1.5}{0.30 \times (0.0625 + 0.03)} = \frac{1.5}{0.0278} = 53.96 \text{ cm}$$

The melt water does not reach the bottom layer during the 3-hour period. This computation ignores the density changes due to melting.

Note that during the 3 hours, the melt = 1.5 cm. This will produce 80 × 1.5 = 120 cal. The amount of heat released when 1.5 cm of melt is brought to −10°C is 1.5 × 0.5 × 10 = 7.5 cal. Thus, total melt energy = 120 + 7.5 = 127.5 cal. This heat will produce extra melt and, hence, the actual penetration will be deeper than 53.96 cm. Also the cold content in the melt layer does not have to be fulfilled by melt water, that is, it is supplied in the melt process.

The time to warm the snowpack to 0°C is given by Equation (19.23). For the top layer, $W_0 = 0.3 \times 100$ cm, $T = 10°$C, $r = 0$, and $m = 0.5$ cm/h. Therefore,

$$t_c = \frac{0.3 \times 100 \times 10}{160 \times (0 + 0.5)} = \frac{300}{80} = 3.75 \text{ h}$$

For the bottm layer, $W_0 = 0.35 \times 100$, $T = 5°C$, $r = 0$, and $m = 0.5$ cm/h. Therefore,

$$t_c = \frac{0.35 \times 100 \times 5}{160 \times 0.5} = \frac{175}{80} = 2.19 \text{ h}$$

The total time = 3.75 + 2.19 = 5.94 h. This does not account for the fact that 5.94 × 0.5 the amount of water is warmed from external energy; this would reduce the time required by the entire snowpack.

The time to offset the liquid-water deficiency of the snowpack is given by Equation (19.24), which can be recast as

$$t_d = \frac{P_d(W_0 + W_c)}{100(r + m)}$$

For the top layer, $P_d = 3$, $W_0 = 0.3 \times 100$, $W_c = 1.875$ cm, $r = 0$, and $m = 0.5$ cm/h. Therefore,

$$t_d = \frac{3 \times (30 + 1.875)}{100 \times 0.5} = 1.91 \text{ h}$$

For the bottom layer, $W_0 = 0.35 \times 100$, and $W_c = 1.094$ cm. Therefore,

$$t_d = \frac{3 \times (35 + 1.094)}{100 \times 0.5} = 2.17 \text{ h}$$

The total time = 1.91 + 2.17 = 4.08 h.

19.7 SNOWMELT RUNOFF

A multitude of methods have been developed to predict snowmelt runoff from a watershed. It is to be emphasized that snowmelt at a point involves a change of state requiring heat, whereas the production of snowmelt runoff from a watershed is a much more complicated process. A couple of simple methods are presented here.

19.7.1 Degree-Day Method

A degree day is a term that expresses a 1-degree temperature departure from a reference temperature. For snowmelt studies, the degree day is usually computed by subtracting the average of the daily maximum and daily minimum temperatures from 0°C or 32°F. Other methods of computing degree days have been used. The degree days are correlated with snowmelt to obtain a relation that can be used to predict snowmelt. The normal form of a degree-day equation is given as

$$Q = a(T_a - T_b) \tag{19.29}$$

where Q is the daily snowmelt runoff (cm), T_a is the mean daily air temperature (°C or °F), and T_b is the reference temperature (close to 32°F or 0°C), and a is a coefficient. The values of a and T_b are selected by trial and error. The U.S. Army Corps of Engineers (1960) have developed snowmelt equations for this purpose.

For open sites,

$$M = 0.06(T_{\text{mean}} - 24) \tag{19.30}$$

$$M = 0.04(T_{\text{max}} - 27) \tag{19.31}$$

For forest sites,

$$M = 0.05(T_{\text{mean}} - 32) \tag{19.32}$$

$$M = 0.04(T_{\text{max}} - 42) \tag{19.33}$$

These equations are said to apply for T_{mean} ranging from 34°F to 66°F and for T_{max} ranging from 44°F to 76°F.

The advantages of the degree-day method are its simplicity and readily available air-temperature data.

19.7.2 Regression Equations

These equations relate snowmelt runoff to measurable meteorological variables such as sunshine hours, temperature, wind velocity, relative humidity, radiation, rainfall, cloud cover, and so on. The U.S. Army Corps of Engineers (1956) developed a regression equation for the partly forested Boise River basin above Twin Springs, Idaho, expressed as

$$Q = 0.00605G + 0.1120(T_m - 25) \tag{19.34}$$

where Q is the daily snowmelt runoff (cm), G is an estimated value of the daily all-wave radiation (langleys) exchange in the open, and T_m is the daily maximum temperature (°C) at Boise. Equation (19.34) predicts the snowmelt runoff within 0.28 cm of the observed value about 67% of the time.

Pysklywec (1966), based on data from a small experimental plot in the North Nashwaakis watershed in New Brunswick, Canada, suggested the following regression equation:

$$\begin{aligned} Q = 0.615 &+ 0.0375n + 0.00607H_l + 0.00201(T - 36)u \\ &+ 0.0437(\text{RH})u + 0.007r(T - 32) \end{aligned} \tag{19.35}$$

where Q is the daily snowmelt runoff (in./day), n is the sunshine (h/day), H_l is the long-wave radiation (ly/day), T is the temperature (°F), u is the wind velocity (mph), RH is the relative humidity, and r is the rainfall (in./day).

EXAMPLE 19.6 An unforested mountain area covered with accumulated snowpack during late spring has a daily mean temperature of 47°F. How much snowmelt could be expected under these conditions?

Solution From Equation (19.30),

$$M = (0.06)(47 - 24) = 1.38 \text{ in.}$$

EXAMPLE 19.7 If the maximum temperature reached 66°F, how much snowmelt could be expected from the forested areas?

Solution From Equation (19.33),

$$M = (0.04)(66 - 42) = 0.96 \text{ in.}$$

EXAMPLE 19.8 A 30-mph wind is blowing 50 feet above the ground surface over an open snow-covered area and the temperature is 38°F. What amount of snowmelt can be expected?

Solution From Equation (19.9),

$$M = (0.004)(30)(38 - 32) = 0.72 \text{ in.}$$

EXAMPLE 19.9 A drainage basin is covered with a thick layer of snow upon which 1.5 inches of rain, at a temperature of 42°F, is falling. How much snowmelt will result from the effect of the rainfall?

Solution From Equation (19.14),

$$M = 1.5(42 - 32)/144 = 0.10 \text{ in.}$$

EXAMPLE 19.10 Warm moist air having a vapor pressure of 30.1 millibars 20 feet above the snow is blown north over snow-covered plains by a wind blowing 15 mph 50 feet above ground level. How much snowmelt can be expected?

Solution From Equation (19.8),

$$M = (0.012)(15)(30.1 - 6.11) = 4.32 \text{ in.}$$

It should be noted that 4.32 inches of snowmelt is approximately equivalent to 4.32 feet of newly fallen snow. For this reason, if less snow is available for melting, then less melt will occur.

19.7.3 Hydrograph Recession

Except for snowmelt on rocks or very shallow soils, virtually all of the snowmelt runoff reaches streams as subsurface and groundwater flow; overland flow is practically non-existent. It is argued that the runoff during the snowmelt season is sum of

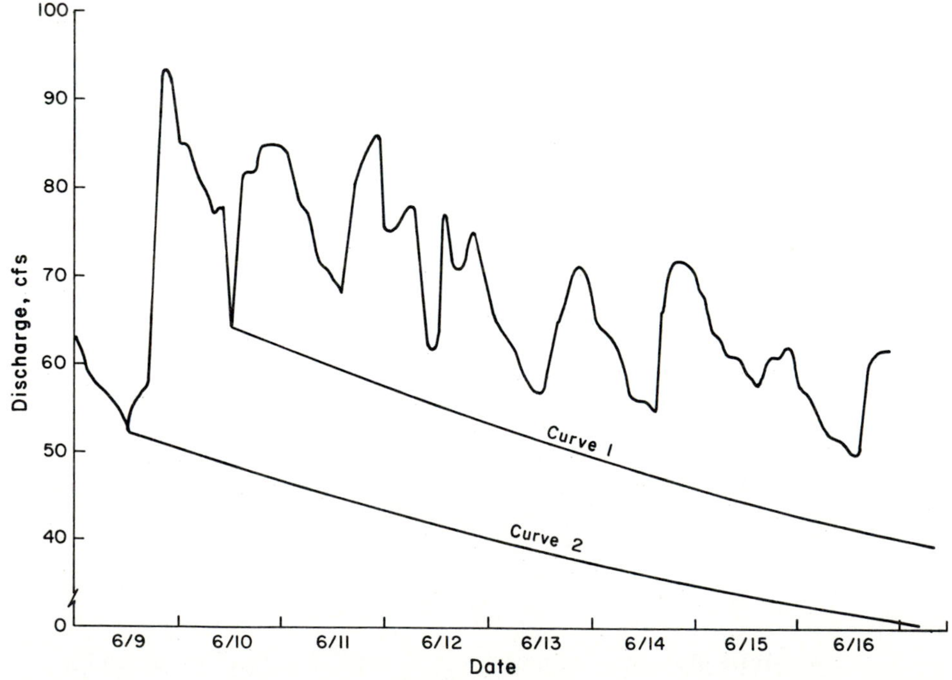

Figure 19.1 Separation of a snowmelt hydrograph.

the individual days' contributions in accordance with Darcy's law. Thus, the recession of the streamflow hydrograph at the end of the snowmelt season may be applied to each day's contribution. Garstka et al. (1958) developed a method of accounting for a day's contribution to the snowmelt hydrograph. The recession hydrograph can be expressed as an exponential decay function, as discussed in Chapter 13.

The procedure for evaluating the snowmelt runoff from the streamflow hydrograph and consequent daily snowmelt contribution can be described as follows: (1) The snowmelt hydrograph of the average daily discharge and the precipitation data are plotted to show the general picture. (2) A recession analysis is then performed, plotting runoff only for those days when no precipitation was recorded. This involves hydrograph separation. It is assumed that the first, second, and succeeding peaks are in correspondence with snowmelt days. The hydrograph separation or recession analysis is performed to determine a particular day's contribution. As shown in Figure 19.1, the area between curves 1 and 2 represents the snowmelt runoff attributed to day 1. (3) Determine the first day's volume for the melt runoff volume. (4) Compute the recession volume for the given day's melt contribution.

19.8 STREAMFLOW FORECASTING

Snow surveys can be utilized to forecast the probable total seasonal volume of flow of rivers in advance of the main spring and summer runoffs. The method of forecasting (Work, 1953) correlates the water content of the snowpack of a watershed each

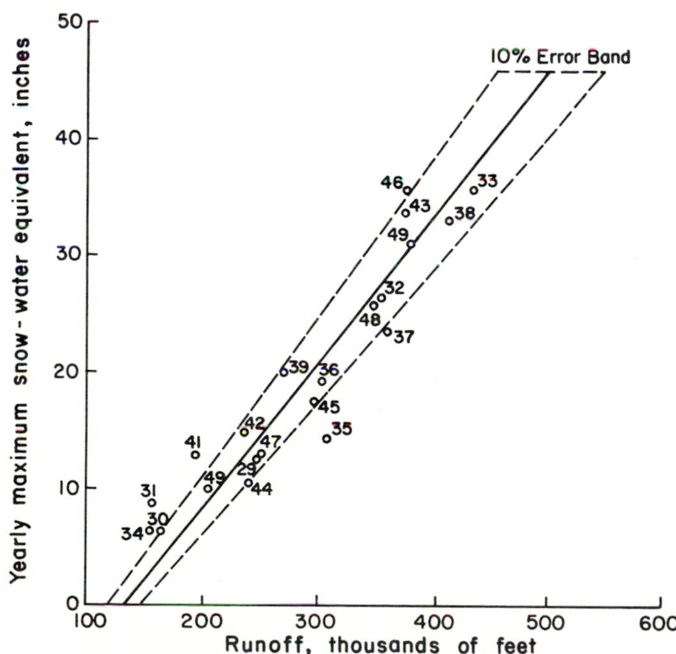

Figure 19.2 Relationship of snow cover on Diamond Lake, Oregon, snow course to April–September runoff volume of North Fork Rogue River above Prospect, Oregon (after Soil Conservation Service, 1953).

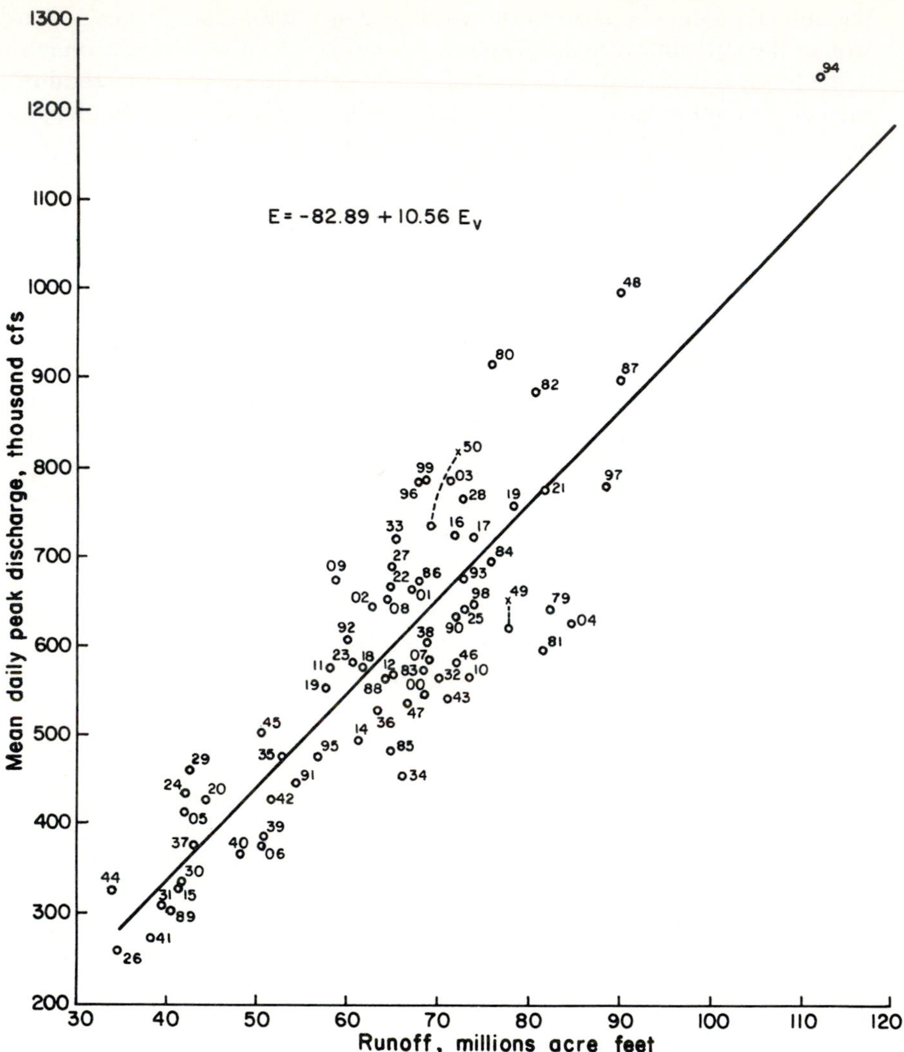

Figure 19.3 Relationship of peak discharge to April–June runoff volume for Columbia River at Dallas, Oregon (after Soil Conservation Service, 1953).

year with the resultant flow volume. The water content is used as an index, not a direct estimate, of the amount of water in the snowpack of the watershed. The relationship between the water content of the snowpack and the runoff is usually established for specific periods such as April–June and April–September or any other period of significance. This relationship is obtained by plotting on a graph the water content of the snowpack against the corresponding seasonal runoff for each year for which dependable data are available and fitting a line. Figure 19.2 shows this relationship for the North Fork of the Rogue River in Oregon. The yearly value of snow-water content used in the relationship in each instance is the maximum water content accumulated in the snowpack for the winter regardless of when the accumulation occurred. The resulting line of best fit becomes more reliable with availability and use of new data.

In addition to the total quantity of water, of vital importance is the manner in which the flow of water will take place in time. Work (1953) has reported that the peak flow of snow-fed streams is closely related to the total volume of flow during the melting season. Figure 19.3 shows the relation between peak discharge and runoff volume for the April–June period for the Columbia River at Dalles, Oregon. The correlation between runoff volume and peak discharge becomes more clear-cut as melting progresses and the period remaining of melting is shortened. Thus, the snow-water equivalent can be used to forecast peak discharge. However, abnormal weather changes may introduce significant errors in volume and subsequent discharge estimates.

EXERCISES

19.1. Contrast snow with rain as input into the hydrologic cycle.

19.2. Which components of the energy budget can be estimated and which are the most difficult for the practicing hydrologist?

19.3. Consider a 50-cm snowpack having a density of 0.4 and a uniform −10°C with a water-holding capacity for liquid water of 3% by weight. How long would it be before water would begin to be released from the snowpack with a rain of 1 cm/h at +10°C falling at a steady rate and accompanied by a net convective heat transfer to the pack of 20 cal per hour? Neglect any changes in liquid-water capacity resulting from the rain or metamorphism. Also neglect the slight decrease in depth resulting from convective melt. Assume there are no other net energy exchanges except from the rain and convection. How long would it take if the thermal quality of the pack was 97% instead of as described?

19.4. Consider a snowpack 1 m deep. A sample taken from the pack has a volume of 0.02 m^3 and a weight of 4 kg. When this sample is mixed with 50 kg of water at 30°C, the resulting temperature is 5°C. Compute the water equivalent, density, and quality of the snowpack.

19.5. Consider a snowpack of thermal quality of 0.95. If the amount of heat applied is 100 langleys, determine the snowmelt.

19.6. Consider a snowpack with a density of 0.3, a depth of 3 m, and a temperature of −10°C. The air temperature is 20°C and the relative humidity is 30%. Compute the vapor pressure and the cold content of the snowpeak for a surface area of 1 m^2.

19.7. What are the factors affecting snowmelt? Can they be ranked in order of their importance?

19.8. Determine a 12-h snowmelt in April due to condensation, convection, radiation, and rainfall for a snowpack at an elevation of 1000 m above sea level at an altitude of 45°N. The air temperature is 20°C, the mean wind velocity is 15 km/h, the relative humidity is 50%, the average rainfall intensity is 0.2 cm/h for 4 hours, and the wet-bulb temperature is 10°C. Assume the sky is cloudy during the 12-h period. What would be the total 12-h snowmelt and what would be percent contribution due to each heat source?

19.9. Consider a snowpack 3 m deep with the following characteristics:

DEPTH (m)	LAYER	TEMPERATURE (°C)	DENSITY (%)
0–1	Bottom	−5	40
1–2	Intermediate	−7	35
2–3	Top	−10	30

Compute the thermal quality of the snowpack.

19.10. Take the water-holding capacity as 5% of the snowpack in Exercise 19.9 and the melt rate as 0.5 cm/h for 6 hours. Compute the cold content, the 5-h penetration of melt water, the time to warm the snowpeak to 0°C, and the time to meet the liquid-water deficiency.

19.11. If the latent heat of fusion is 79.7 cal/g and the heat of vaporization of water is given by the equation $H = 597.3 - 0.564T$ (cal/g), where T is in °C, calculate the heat required to vaporize ice at 0°C. The specific heat of ice = 0.5 cal/g/°C. The specific heat of water = 1.0 cal/g/°C.

PART 7
Flow Routing

CHAPTER 20
Reservoir Flood Routing

The term ''flood routing'' represents a procedure to determine the outflow hydrograph at a downstream point of a river (or reservoir) given the inflow hydrograph at an upstream point thereof. Consider, for example, a river reach, as shown in Figure 20.1, for which the streamflow hydrograph $Q(t)$ is known at location A. The flood routing can be utilized to determine the streamflow hydrograph at location B. We can do likewise for location C, given the inflow hydrograph at location B. In this manner, the movement and shape of a flood wave in a river can be predicted. Figure 20.1 shows that the streamflow hydrograph experiences a decrease in its peak, an elongation in its peak time, and an increase in its time base while preserving its volume as the flood wave moves downstream. This is a manifestation of the effect that the channel exercises on the flood hydrograph.

The shape of the outflow hydrograph depends upon the channel geometry, bed slope, length of the channel reach, downstream control, initial channel flow, and the upstream inflow hydrograph. For a channel having a large valley storage capacity, the reduction in peak discharge is large. If the channel is steeply sloping, less reduction in peak discharge is permitted, for there will be less valley storage. Longer channels provide greater valley storage and, consequently, greater reduction in peak discharge. If the channel is partially full prior to the arrival of a flood wave, such that there is less storage space available to store flood waters, then a less reduction of the flood peak will occur. The amount of reduction in peak discharge is higher for sharp crested inflow hydrographs than for moderately rising inflow hydrographs. In a similar manner, the effect of reservoir geometry can be analyzed.

20.1 GOVERNING EQUATIONS

The propagation of a flood wave in a channel reach or a reservoir is a gradually varied unsteady flow. The laws governing the behavior of such flows are the conservation of mass and momentum derived from mathematical physics. These laws are

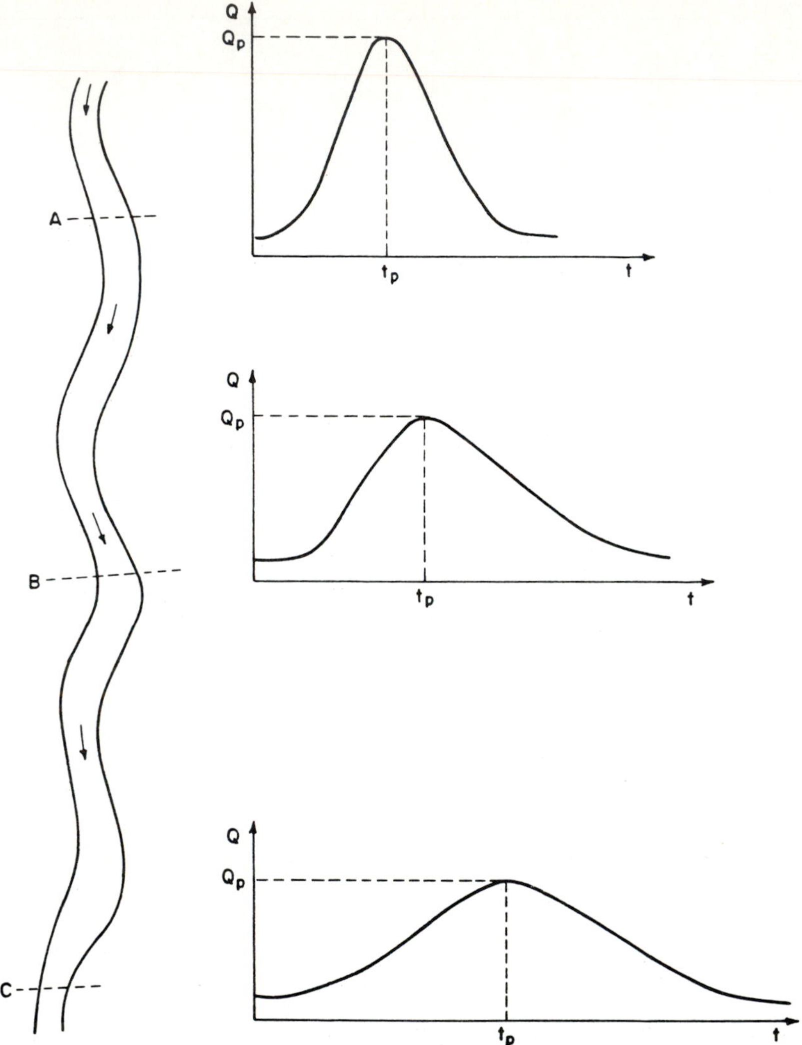

Figure 20.1 Streamflow hydrographs at locations A, B, and C along a river representing the movement of a flood wave. Q_p represents the hydrograph peak and t_p its time.

expressed, respectively, as continuity and momentum equations. For a wide rectangular channel with no lateral inflow, these equations can be written per unit width as

$$\frac{\partial h}{\partial t} + \frac{\partial Q}{\partial x} = 0 \tag{20.1}$$

$$\frac{\partial u}{\partial t} + u\,\frac{\partial u}{\partial x} + g\,\frac{\partial h}{\partial x} = g(S_0 - S_f) \tag{20.2}$$

where h is the depth of flow, Q is the discharge $= uh$, u is the mean velocity of flow, S_0 is the channel bed slope, S_f is the friction slope, g is the acceleration due to

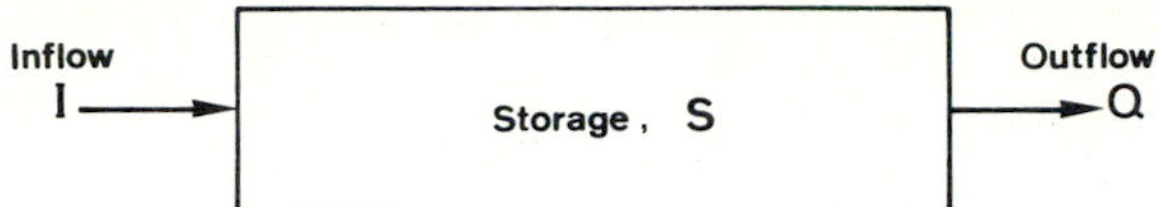

Figure 20.2 Black-box representation of a channel reach.

gravity, x is the space coordinate in the direction of flow, and t is time. Equations (20.1) and (20.2) are often referred to as the St. Venant equations. Derivation of these equations can be found in any standard textbook on open-channel hydraulics (see, e.g., Chow, 1959).

Equation (20.1) is an expression of mass balance with $\partial h/\partial t$ indicating the rate of change in storage and $\partial Q/\partial x$ the difference between inflow and outflow, or simply net outflow. If the channel reach is considered as a black box, as shown in Figure 20.2, where we are only interested in determining the outflow at the downstream end due to the inflow at the upstream end and ignore the variability of flow along the channel, then Equation (20.1) can be integrated in space and written in spatially lumped form as

$$\frac{dS}{dt} = I - Q \tag{20.3}$$

where S is the storage or volume of water within the reach, I is the inflow rate, Q is the outflow rate, and dS/dt is the rate of change of storage. Equation (20.3) is also referred to as the volume balance or storage equation. For a small interval of time, Δt, Equation (20.3) can be expressed as

$$\Delta S = \bar{I}\,\Delta t - \bar{Q}\,\Delta t \tag{20.4}$$

where $\bar{I}$ is the average inflow rate during Δt, $\bar{Q}$ is the average outflow rate during Δt, and ΔS is the change of storage. Equation (20.4) can be expressed alternatively as

$$S_2 - S_1 = \left(\frac{I_1 + I_2}{2}\right)\Delta t - \left(\frac{Q_1 + Q_2}{2}\right)\Delta t \tag{20.5}$$

in which the subscripts 1 and 2, respectively, denote the beginning and end of the time interval $\Delta t = t_2 - t_1$, $\bar{I} = (I_1 + I_2)/2$, $\bar{Q} = (Q_1 + Q_2)/2$, and $\Delta S = S_2 - S_1$. Implied in Equation (20.5) is that the variation of I and Q is linear during the small time interval Δt. To reemphasize, Δt must be sufficiently small for Equation (20.5) to be valid, and smaller than the time of travel of the flood wave through the reach.

Equation (20.2) is also called the equation of motion. There are four principal forces acting upon the flood wave: inertial, pressure, gravity, and friction. The terms $\partial u/\partial t$ and $u\,\partial u/\partial x$ are local and convective acceleration terms and can be combined as du/dt, and arise due to inertial forces. $\partial h/\partial x$ is the slope of the water-surface profile and arises due to pressure. The gravity and friction forces give rise, respectively, to S_0 and S_f. The relative magnitudes of the various terms in Equation (20.2) depend upon the channel reach under consideration. Sometimes all the terms are important, and at other times only some of the terms are important. In most cases of hydrologic interest, gravity and friction terms are dominant, followed by the pressure term, and inertial terms are of least importance. As a result, Equation

(20.2) is used in simplified form. Two of the most commonly used simplifications are

$$S_0 = S_f \tag{20.6}$$

and

$$\frac{\partial h}{\partial x} = S_0 - S_f \tag{20.7}$$

Equation (20.6) is popularly known as the kinematic-wave approximation, and Equation (20.7) as the diffusion-wave approximation.

Thus, the governing equations for flood routing are Equation (20.1) and either Equation (20.2) or its simplified form such as Equation (20.6) or (20.7). In many hydrologic problems, consideration of momentum is essentially ignored and instead an empirical equation relating S to Q and/or I, in conjunction with Equation (20.3), is used.

Although flood routing is normally discussed in relation to routing a flood wave from upstream to downstream in a stream or a reservoir or routing lateral inflow in a channel, its scope is much broader indeed. The routing can be reversed, that is, given the flood hydrograph at a downstream location, say, location C in Figure 20.1, the routing can be performed to determine the flood hydrograph upstream at location B. This operation can be termed as upstream routing. Also, the routing can be done with mass curves or merely peak rates or peak stages of runoff. It may include determining inflow or outflow hydrographs, mass curves or peak rates of runoff in reservoirs, farm ponds, lakes, swamps, and hydraulic structures. On the other hand, the routing can be extended to include low flows.

Flood routing is used in the design of detention storage, channels and many types of hydraulic structures, operation of control structures, evaluation of the effect of a water-control structure on flood flows, short-term forecasting of floods, determining the unit hydrographs at various points, deriving synthetic unit hydrographs, etc.

20.2 METHODS OF FLOOD ROUTING

Several methods have been developed for routing floods in reservoirs and rivers. Reviews of some of these methods have been presented by Weinmann and Laurenson (1979), Dooge (1980), and Fread (1982), Singh (1988), amongst others. Broadly speaking, these methods can be classified as (1) hydraulic and (2) hydrologic. The hydraulic methods employ Equation (20.1) and either Equation (20.2) or its simplified form. The hydrologic methods use Equation (20.3). Weinmann and Laurenson (1979) have pointed out that this classification is pertinent for well-defined regular channels, but may be fallacious for natural channels whose complex physical properties may defy exact mathematical representation of their hydraulics and necessarily lead to some degree of simplification.

Each method possesses distinctive features that make it suitable for particular applications. The hydrologic methods are, in general, simpler, but may not be satisfactory in problems other than those determining the progress of a flood wave down a long river. For example, when a flood passes through a junction, it produces backwater; when it is regulated by a dam, it produces surges. The backwater effect and the effect of surges cannot be accurately evaluated by a hydrologic method. The hydraulic methods are generally more accurate and versatile. However, their high demands on computing technology as well as on quantity and quality of input data may restrict their efficiency in practical applications. Only the hydrologic methods are considered in this chapter.

20.3 RESERVOIR FLOOD ROUTING

When a flood wave passes through a full reservoir, it experiences attenuation of its peak, elongation of its peak time, and lengthening of its time base. This is the effect of a reservoir on a flood hydrograph. To better grasp the modifying effect of a reservoir on the traveling flood wave, we consider an inflow hydrograph and the resulting outflow hydrograph, as shown in Figure 20.3. The inflow hydrograph is at the upstream of the reservoir and is not influenced by the reservoir. The outflow hydrograph is the hydrograph with the reservoir or that the reservoir produces by modifying the inflow hydrograph. It is seen that the peak of the outflow hydrograph is smaller than the peak of the inflow hydrograph, the peak of the outflow hydrograph occurs after the peak of the inflow hydrograph, the time base of the outflow hydrograph is longer than the time base of the inflow hydrograph, and the volume of flow is the same for both the inflow and outflow hydrographs. The reduction in the peak outflow is called attenuation. The time difference between the peak of inflow and that of outflow is called lag time, or lag. The attenuation and lag are two important parameters of flood routing. These effects are seen by noting that during the rise of the outflow hydrograph or from point A to point C, inflow is greater than outflow, and the area between the two hydrographs indicates the build up of storage in the reservoir. During the recession of the outflow hydrograph, outflow is greater than inflow, and the area between the two hydrographs indicates depletion of storage. When the outflow hydrograph has receded and the reservoir has returned to the

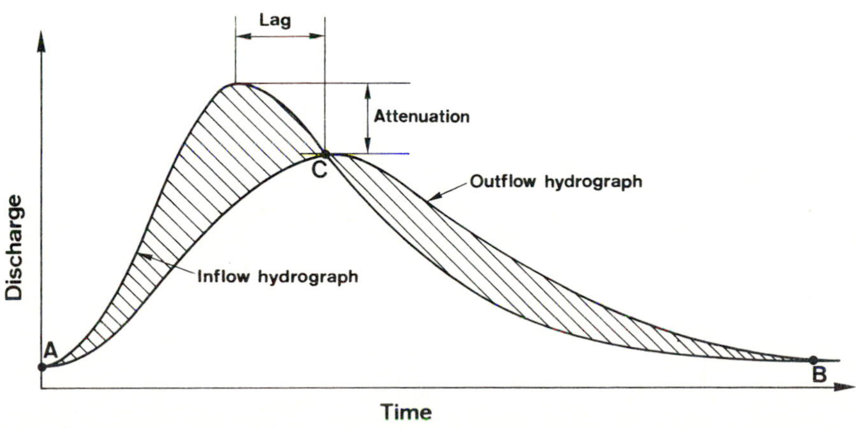

Figure 20.3 Routing of an inflow hydrograph through a reservoir.

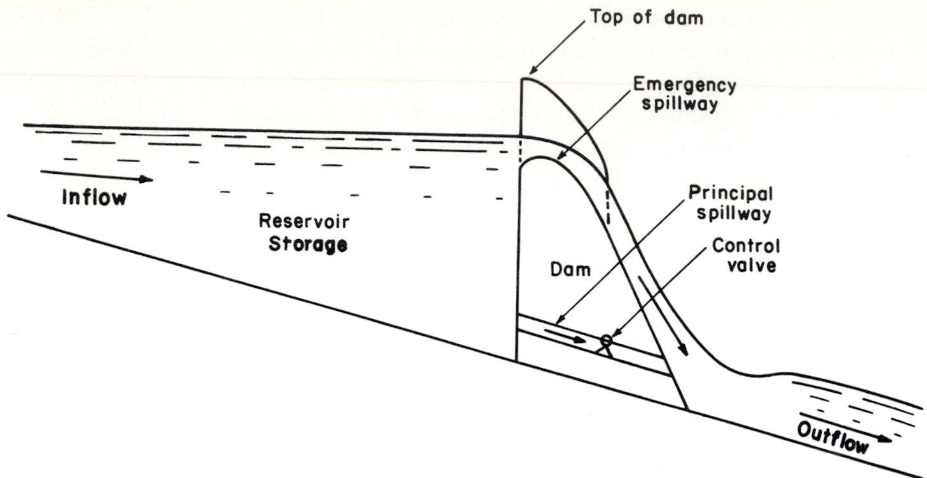

Figure 20.4 Dam, reservoir, and associated spillway works.

initial level, the build up of storage that occurred during the rise of the outflow hydrograph is completely offset by the depletion of storage that occurred during the recession of the outflow hydrograph. From the figure, the shaded area during time AC is the same as the shaded area during time CB. This also implies that the storage is maximum at point C—the intersection of inflow and outflow hydrographs. If the outflow from a reservoir is uncontrolled, as, for example, in the case of a freely operating spillway, the peak of the outflow hydrograph occurs at point C. This is logical because the storage, or hydraulic head, over the spillway is maximum at this point.

The water stored in the reservoir may discharge in a controlled or uncontrolled manner. The water release as pipe flow through turbines or outlet works, called principal spillways, represents an example of controlled conditions, whereas the flow over an emergency spillway represents uncontrolled conditions, as shown in Figure 20.4.

Consider a reservoir having a spillway for outflow. When a flood passes through this reservoir, the floodwaters are discharged through the spillway. The rate of outflow, Q, depends only on the reservoir water-surface elevation, h. The amount of water stored, S, also depends on h. Thus, Q and S are connected through the

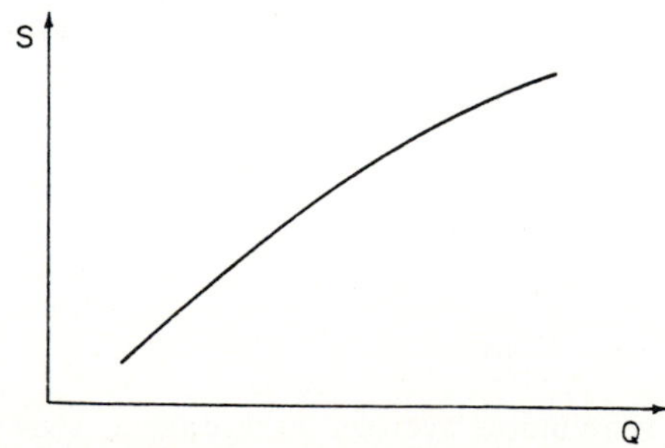

Figure 20.5 A typical storage–outflow discharge relation.

water-surface elevation, and this is the fundamental characteristic of a reservoir. A typical storage–outflow relation is shown in Figure 20.5. This relation, in conjunction with Equation (20.3), is utilized in most reservoir routing methods. This characteristic is seen to hold in many other situations not involving a dam. A few examples include (1) water ponding upstream of a highway culvert, (2) channel reaches with swamps, (3) river passing through a lake, and (4) channels where the inflow hydrograph rises and falls so slowly that nearly steady flow occurs. When a flood wave propagates through the reservoir, the water-surface elevation changes with time. This means that Q and S also change with time. Thus, for a given $I(t)$, determination of $h(t)$, $S(t)$, and, consequently, $Q(t)$ is the reservoir flood routing. If the spillway is uncontrolled, discharge Q goes freely over the spillway, depending upon the depth of flow over the spillway and the spillway geometry. The depth of flow over the spillway depends upon the depth of water in the reservoir or storage. Therefore,

$$S = S(y)$$

$$Q = Q(y)$$

where y represents the water-surface elevation. The outflow from the spillway can be computed using the general weir formula:

$$Q = CLh^{3/2} \tag{20.8}$$

where C is the weir coefficient, L is the length of the spillway crest, and h is the depth of water above the spillway crest or stage.

This discussion indicates that the data required for reservoir flood routing include storage versus water-surface elevation, outflow versus elevation, the inflow hydrograph, and initial values of storage and outflow. Of the several methods available for flood routing through reservoirs, two commonly used methods are presented here. Flood routing through a reservoir may represent the simplest of all routing problems.

20.3.1 Storage Indication Method

This is one of the most commonly used methods of reservoir flood routing, and is also referred to as the modified Puls method. This method employs Equation (20.5), wherein Q may incorporate controlled Q_c as well as uncontrolled discharge Q_s:

$$Q = Q_c + Q_s \tag{20.9}$$

By separating the known quantities from the unknown ones and rearranging, Equation (20.5) yields

$$(I_1 + I_2) - (Q_{c_1} + Q_{c_2}) + \left(\frac{2S_1}{\Delta t} - Q_{s_1}\right) = \frac{2S_2}{\Delta t} + Q_{s_2} \tag{20.10}$$

The left side is known and the right side is unknown. The inflow hydrograph is known. Q_c is the controlled discharge and may pass through turbines, outlet works, or over the spillway; it is known. The remaining terms on the left side, S_1 and Q_1, are assumed known at the start of routing, and are determined for each subsequent step. The right side of Equation (20.10) can be written as

$$\frac{2S}{\Delta t} + Q = f(y)$$

where y is the water-surface elevation. In order to utilize Equation (20.10), elevation–storage and elevation–discharge relationships must be specified. For simplicity, let us assume that Q_c is negligible. Therefore, Q can be taken to imply Q_s.

Equation (20.10) is then rewritten as

$$(I_1 + I_2) + \left(\frac{2S_1}{\Delta t} - Q_1\right) = \left(\frac{2S_2}{\Delta t} + Q_2\right) \tag{20.11}$$

The elevation-discharge relationship of Equation (20.8) is utilized to determine the outflow, given the depth of water over the spillway. From topographic information, y corresponding to h can be obtained. Then the curve for either y versus Q or h versus Q can be constructed, as shown in Figure 20.6.

To construct the elevation–storage curve, we again utilize the topographic map. The depth of water above the spillway h multiplied by the reservoir surface area can yield the (temporary) storage of the reservoir. The reservoir surface area is known from the topographic map. Therefore, the temporary storage S for any h is approximately

$$S = \frac{h}{2}(A_1 + A_2) \tag{20.12}$$

where A_1 denotes the surface area of the reservoir when $h = 0$, and A_2 is the surface area of the reservoir when the depth of flow is h. Then curves for either y versus S or h versus S can be constructed, as shown in Figure 20.7. It may be noted that elevation–discharge and elevation–storage curves of Figures 20.6 and 20.7 can be easily combined to form a storage–discharge curve. Thus, the storage in a reservoir depends only on the outflow, as shown in Figure 20.5.

The next step in routing by Equation (20.10) is selection of the routing time interval Δt. Its value should be such that neither is it too large nor too small. If it is too long and exceeds the travel time through the reservoir, then the crest segment of outflow containing the peak discharge could pass through the reservoir in between

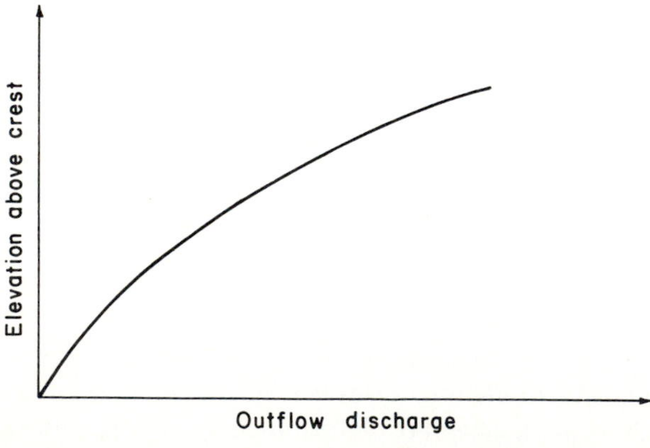

Figure 20.6 Elevation above the spillway crest versus reservoir outflow.

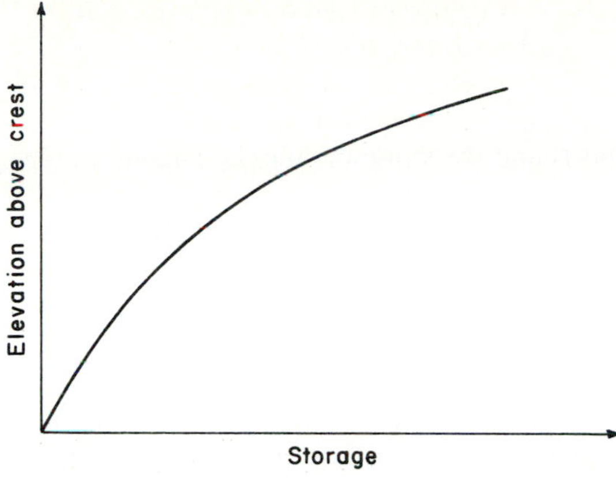

Figure 20.7 Elevation above the spillway crest versus storage in the reservoir.

time intervals and, therefore, not be computed. On the other hand, if Δt is too short, then it takes longer to perform flood routing. Further, Δt is assumed so short that I is approximately linear during it. Linsley et al. (1975) recommended Δt to be one-third to one-half the travel time through the reservoir. As a thumb rule, Δt is approximately 20 to 40% of the time of rise of the inflow hydrograph.

It may be noted that subtracting $2Q$ from $[(2S/\Delta t) + Q]$ yields $[(2S/\Delta t) - Q]$, which appears on the left-hand side of Equation (20.11). Before routing is performed, the curves of $(2S/\Delta t) \pm Q$ versus Q are thus constructed. This can be accomplished as follows. Select an initial outflow discharge Q. Estimate the corresponding elevation using the elevation–discharge curve of Figure 20.6. Determine the corresponding storage from the elevation–storage curve of Figure 20.7. Then determine $(2S/\Delta t) + Q$ and plot against Q, as shown in Figure 20.8.

The routing by Equation (20.10) can now be performed as follows:

1. At the beginning of routing, inflow (I_1 and I_2), storage S_1, and outflow Q_1 are known from initial reservoir conditions. Compute the left side of Equation (20.11), which equals $(2S_2/\Delta t) + Q_2$.

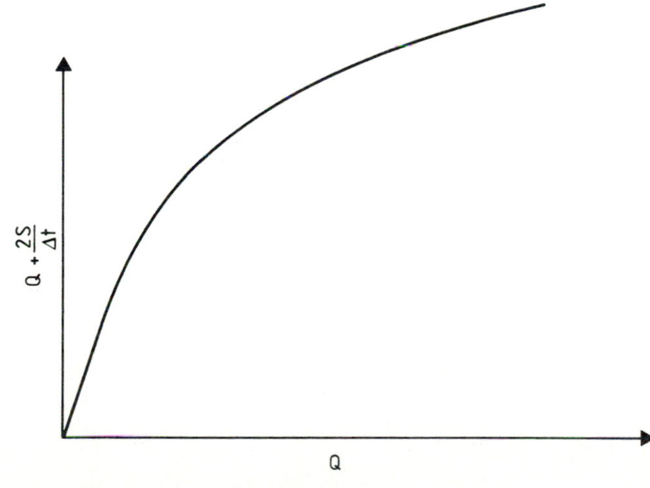

Figure 20.8 Discharge versus $(2S/\Delta t) + Q$ for reservoir routing.

2. Determine Q_2 for $(2S_2/\Delta t) + Q_2$ specified in step 1 using Figure 20.8. This value of Q_2 is the instantaneous outflow at the end of Δt.
3. Determine $(2S_2/\Delta t) - Q_2$ by subtracting twice the value of Q_2 of step 2. This gives the value of $(2S_1/\Delta t) - Q_1$ for the next time interval.
4. Add the two inflows I_1 and I_2 and the value in step 3. This gives $(2S_2/\Delta t) + Q_2$ for the next time interval.
5. Determine Q_2 against $(2S_2/\Delta t) + Q_2$ of step 4 using Figure 20.8.
6. Determine $(2S_2/\Delta t) - Q_2$. This gives the value of $2S_1/\Delta t - Q_1$ for the next time interval.
7. Go back to step 4 and repeat the procedure.

This method has two shortcomings. First, the assumption that the outflow begins at the same time as inflow does imply that the inflow passes through the reservoir instantaneously regardless of its length. However, this weakness is not serious if the ratio T_t/T_m is less than or equal to $\frac{1}{2}$, where T_m denotes time to peak of the inflow hydrograph, and T_t is the travel time defined as L/u with L being the length of the reach and u the average steady-state velocity. Second, it is difficult to choose an appropriate Δt. It is seen that negative outflow occurs during recession whenever $\Delta t > 2S_2/Q_2$ or $Q_2/2 > S_2/\Delta t$. This further means that the rising hydrograph is distorted. This problem can be circumvented by plotting Figure 20.8 on log–log paper and comparing the plot with the line of equal values. If the plotted values lie above the line of equal values, Figure 20.8 must be abandoned and a new value of Δt selected. Furthermore, negative outflow can usually be avoided by taking Δt less than T_t.

EXAMPLE 20.1 The outflow Q and storage S for an emergency spillway are linearly related as $Q = S/10{,}000$, where Q is in cubic meters per second and S is in cubic meters. Determine the outflow rate from the reservoir at the end of 2 hours if the inflow rate is 500 cubic meters per second at that time. At the beginning, assume inflow, outflow, and storage to be all zero.

Solution From the storage indication Equation (20.11), with initial inflow, outflow, and storage as zero,

$$I_2 = \frac{2S_2}{\Delta t} + Q_2$$

$\Delta t = 2$ h $= 7200$ s, $I_2 = 500$ m^3/s, and $S_2 = 10{,}000Q_2$. Therefore,

$$500 \times 7200 = 2 \times 10{,}000 \times Q_2 + 7200 \times Q_2 = 27{,}000Q_2$$

Therefore,

$$Q_2 = \frac{3{,}600{,}000}{27{,}200} = 132.35 \text{ m}^3\text{/s}$$

EXAMPLE 20.2 The storage and discharge characteristics for a reservoir are given in Table 20.1. Consider the routing period as 6 hours. The dam has an unregulated opening 5 m high with its elevation at 125 m and a spillway-crest elevation of 355 m. Discharges for pool elevations above 355 m include both unregulated and spillway discharges. Route the inflow hydrograph given in Table 20.2 by the storage indication method. Determine the attenuation and lag.

TABLE 20.1 STORAGE AND DISCHARGE CHARACTERISTICS OF A DAM

Elevation (m)	Discharge (m^3/s)	Storage ($\times 10^8$ m^3)
125	0.02	0.210
130	0.13	0.900
140	0.45	1.59
160	2.70	2.40
180	7.50	3.03
200	15.50	3.54
220	30.00	3.97
240	50.0	4.54
260	75.0	4.75
280	105.0	5.10
300	142.0	5.40
320	185.0	5.70
340	230.0	6.00
350	256.0	6.15
355	268.0	6.19
360	280.0	6.30
365	293.0	6.39
370	306.0	6.43
375	321.0	6.49
390	360.0	6.70
400	392.0	6.85
420	456.0	7.15

TABLE 20.2 AN INFLOW HYDROGRAPH

Time (h)	Discharge (m^3/s)	Time (h)	Discharge (m^3/s)
00	20	114	360
06	60	120	330
12	220	126	280
18	350	132	200
24	380	138	160
30	430	144	130
36	480	150	110
42	530	156	90
48	610	162	60
54	670	168	50
60	730	174	30
66	780	180	22
72	870	186	21
78	770	192	20
84	700	198	19
90	670	204	18
96	630	210	17
102	550	216	17
108	490	222	17

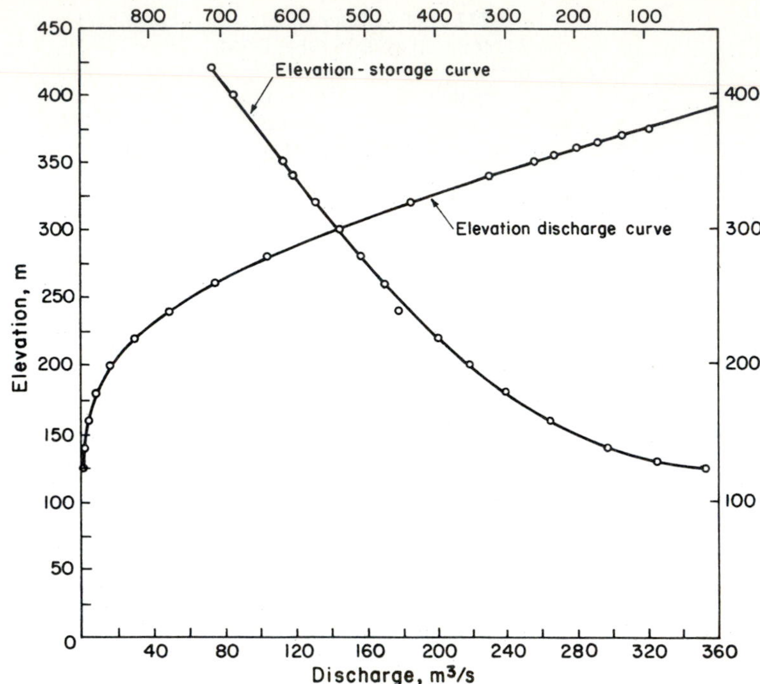

Figure 20.9 Reservoir elevation–discharge–storage characteristics.

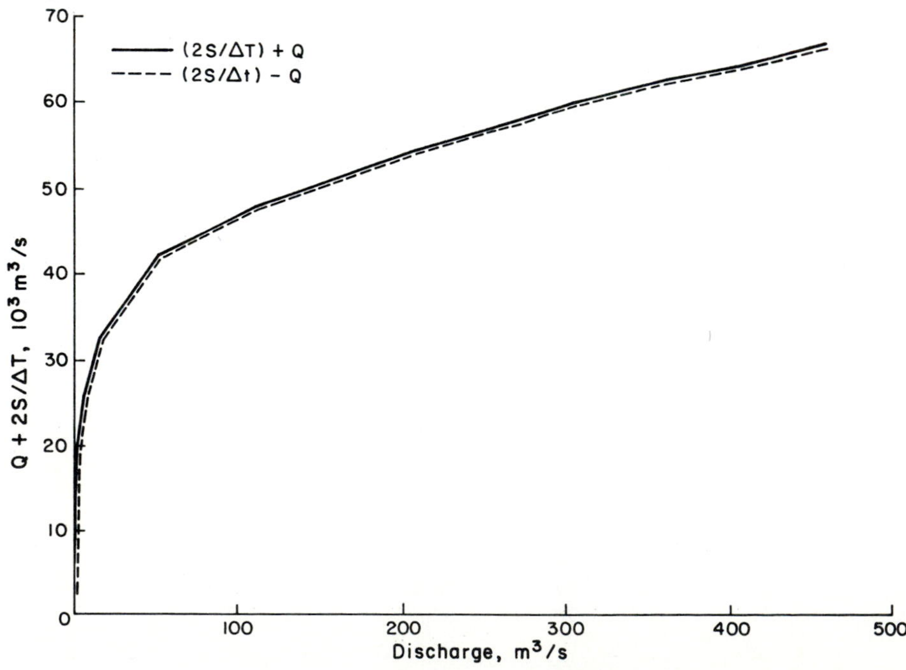

Figure 20.10 $(2S/\Delta t) \pm Q$ versus Q curves.

TABLE 20.3 FLOOD ROUTING BY THE STORAGE INDICATION METHOD

Time	Routing period[a]	Inflow, I (m^3/s)	$I_1 + I_2$	Total outflow, Q (m^3/s)	$(2S/\Delta t) - Q$	$(2S/\Delta t) + Q$	Reservoir elevation (m)
0	1	20	80	268.00	57,093.34	57,173.34	355.00
6	2	60	280	255.50	56,662.35	56,942.35	349.81
12	3	220	570	251.25	56,439.84	57,009.84	348.17
18	4	350	730	252.49	56,504.85	57,234.85	348.65
24	5	380	810	256.96	56,720.93	57,530.93	350.40
30	6	430	910	265.24	57,000.43	57,910.43	353.85
36	7	480	1010	271.43	57,367.57	58,377.57	356.43
42	8	530	1140	277.12	57,823.31	58,963.31	358.80
48	9	610	1280	285.37	58,392.55	59,672.55	362.07
54	10	670	1400	299.43	59,073.68	60,473.68	367.47
60	11	730	1510	321.27	59,831.14	61,341.14	375.10
66	12	780	1650	338.32	60,665.58	62,314.48	381.66
72	13	870	1640	357.46	61,599.55	63,239.55	389.02
78	14	770	1470	377.93	62,483.69	63,953.69	395.60
84	15	700	1370	394.01	63,165.67	64,535.67	400.63
90	16	670	1300	407.12	63,721.44	65,021.44	404.72
96	17	630	1180	418.06	64,185.32	65,365.32	408.14
102	18	550	1040	425.80	64,513.71	65,553.69	410.56
108	19	490	850	430.04	64,693.60	65,543.56	411.89
114	20	360	690	429.81	64,683.93	65,373.93	411.82
120	21	330	610	425.99	64,521.94	65,131.94	410.62
126	22	280	480	420.54	64,290.84	64.770.84	408.92
132	23	200	360	412.41	63,946.01	64,306.01	406.38
138	24	160	290	401.94	63,502.12	63,792.12	403.11
144	25	130	240	390.37	63.011.37	63,251.37	399.49
150	26	110	200	378.19	62,494.98	62,694.98	395.68
156	27	90	150	365.66	61,963.64	62,113.64	391.77
162	28	60	110	353.51	61,406.61	61,516.61	387.50
168	29	50	80	341.77	60,833.06	60,913.06	382.99
174	30	30	52	329.91	60,253.25	60,305.25	378.43
180	31	22	43	316.93	59,671.39	59,714.39	373.64
186	32	21	41	300.70	59,112.99	59,153.99	367.96
192	33	20	39	288.30	58,577.38	58,616.38	363.19
198	34	19	37	280.04	58,056.28	58,093.28	360.02
204	35	18	35	273.66	57,545.95	57,580.95	357.36
210	36	17	34	266.65	57,047.66	57,081.66	354.44
216	37	17	34	253.81	56,574.03	56,608.03	349.16
222	38	17		245.11			345.81

[a] Routing period (Δt) = 6 hours.

Solution The reservoir elevation–discharge–storage characteristics are plotted in Figure 20.9. The quantities $(2S/\Delta t) - Q$ and $(2S/\Delta t) + Q$ are computed and plotted in Figure 20.10. Clearly, the two curves are very close to one another because Q is much smaller than $(2S/\Delta t)$. The inflow hydrograph is routed following the steps outlined before, and calculations are given in Table 20.3. The inflow and outflow hydrographs are shown in Figure 20.11. Thus,

Reduction in peak flow = 440 m^3/s

The lag between inflow and outflow peaks = 39.5 h

The outflow peak = 430 m^3/s

EXAMPLE 20.3 The elevation–storage and elevation–discharge data are given for a small reservoir as follows:

ELEVATION (ft)	STORAGE (acre-ft)	DISCHARGE (cfs)
862	0	0
865	40	0
870	200	0
875	500	0
880	1000	0
882	1200	100
884	1630	230
886	2270	394
888	3150	600

Assuming the pool elevation to be 875 ft at midnight on the first, find the maximum pool elevation and peak outflow rate for the following inflow hydrograph:

DATE	HOUR	INFLOW (cfs)	DATE	HOUR	INFLOW (cfs)
1	Midnight	40	7	Noon	245
2	Noon	35		Midnight	192
	Midnight	37	8	Noon	144
3	Noon	125		Midnight	118
	Midnight	340	9	Noon	95
4	Noon	575		Midnight	80
	Midnight	722	10	Noon	67
5	Noon	740		Midnight	56
	Midnight	673	11	Noon	50
6	Noon	456		Midnight	42
	Midnight	320			

Solution First, we construct the curve $(2S/\Delta t) + Q$ versus Q. To that end, the elevation–storage curve is constructed, as shown in Figure 20.12, and the elevation–discharge is plotted, as shown in Figure 20.13. The value of Δt is 12 h. The relation between $(2S/\Delta t) + Q$ and Q is plotted in Figure 20.14. Calculations for these plots are shown in Table 20.4. The reservoir routing is then performed using Equation (20.11).

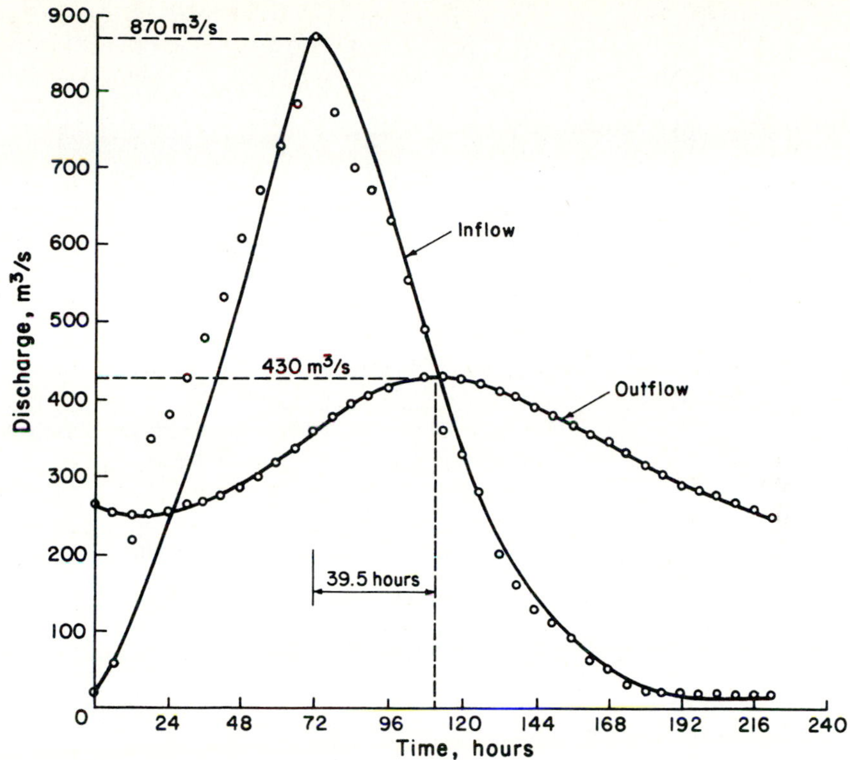

Figure 20.11 Inflow hydrograph and outflow hydrograph obtained by the storage indication method of flood routing.

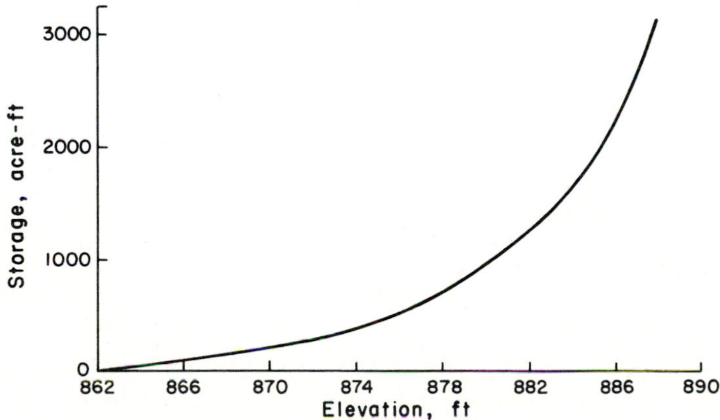

Figure 20.12 Storage versus elevation.

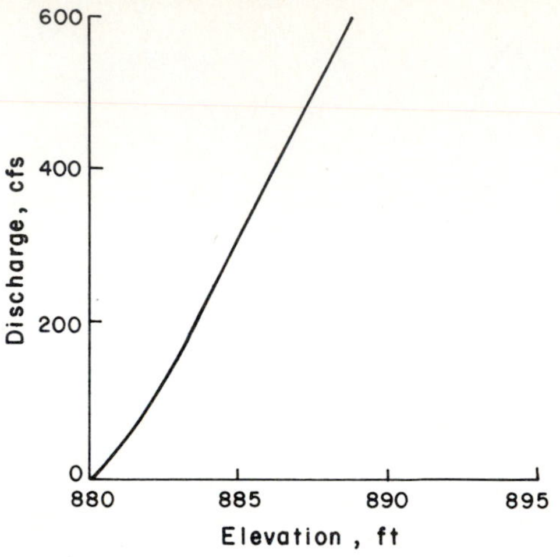

Figure 20.13 Discharge versus elevation.

Routing calculations are presented in Table 20.5, which yield:

$$\text{Maximum elevation} = 887 \text{ ft}$$

$$\text{Peak outflow rate} = 460 \text{ cfs}$$

EXAMPLE 20.4 A small reservoir has an area of 500 acres at the spillway level, and the banks are essentially vertical for several feet above the spillway level. The spillway is 20 ft long and has a coefficient of 3.75. Considering the inflow hydrograph of Example 20.3 as the inflow to the reservoir, compute the maximum pool level and the maximum discharge to be expected if the reservoir is initially at the spillway level at midnight on the first.

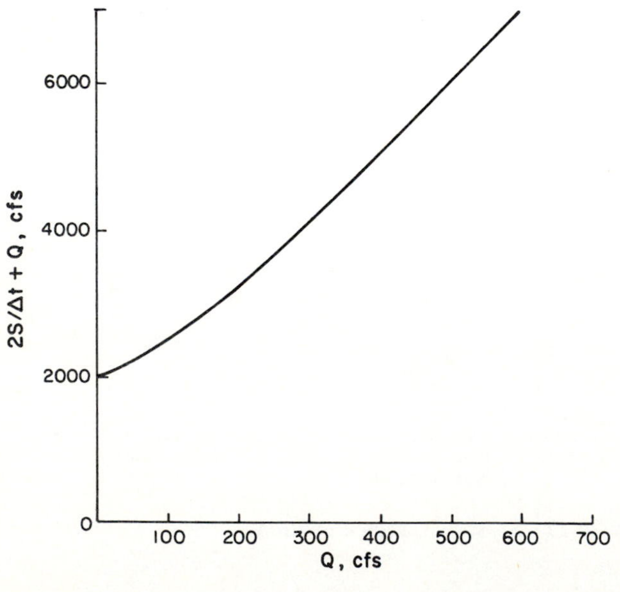

Figure 20.14 $(2S/\Delta t) + Q$ versus Q.

TABLE 20.4 COMPUTATION OF $(2S/\Delta t) + Q$ VERSUS Q FOR EXAMPLE 20.3

h (ft)	S (ft^3)	Q (cfs)	$(2S/\Delta t) + Q$ (cfs)
862	0	0	0
865	1,742,400	0	80.7
870	8,712,000	0	403.3
875	2,178,000	0	1008.3
880	4,356,000	0	2016.7
882	53,143,200	100	2560.3
884	71,002,800	230	3517.2
886	98,881,200	394	4971.8
888	13.721,400	600	6952.5

TABLE 20.5 FLOW ROUTING BY THE STORAGE INDICATION METHOD FOR EXAMPLE 20.3

Date	Time	I_i (cfs)	$I_i + I_{i+1}$	$(2S_i/\Delta t) - Q_i$	$(2S_{i+1}/\Delta t) + Q_{i+1}$	Q_{i+1}
1	Midnight	40	75	1008		0
2	Noon	35	72	1083	1083	0
	Midnight	37	162	1155	1155	0
3	Noon	125	465	1317	1317	0
	Midnight	340	915	1782	1782	0
4	Noon	575	1297	2457	2697	120
	Midnight	722	1462	3234	3754	260
5	Noon	740	1413	3976	4696	360
	Midnight	673	1129	4509	5389	440
6	Noon	456	776	4718	5638	460
	Midnight	320	565	4604	5494	445
7	Noon	245	437	4349	5169	410
	Midnight	192	336	4036	4786	375
8	Noon	144	262	3722	4372	325
	Midnight	118	213	3414	3984	285
9	Noon	95	175	3137	3627	245
	Midnight	80	147	2892	3312	210
10	Noon	67	123	2699	3039	170
	Midnight	56	106	2542	2822	140
11	Noon	50	92	2408	2648	120
	Midnight	42			2500	90

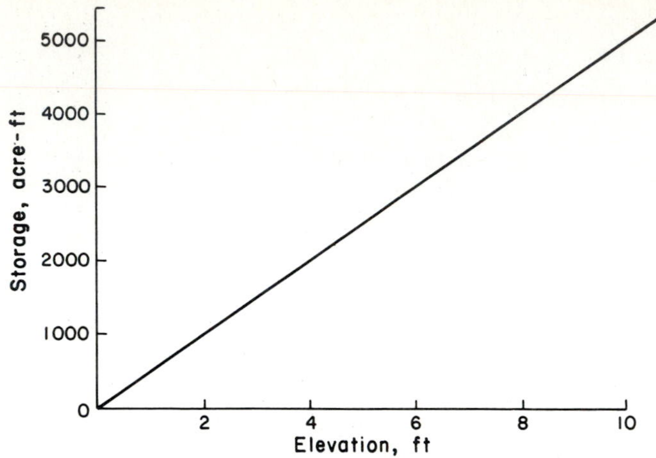

Figure 20.15 Storage–elevation relation.

Solution We compute storage for different values of elevation, which are as follows

ELEVATION ABOVE SPILLWAY (ft)	STORAGE (acre-ft)
0	0
1	500
2	1000
3	1500
4	2000
5	2500

The elevation–storage curve is plotted in Figure 20.15. Then, the discharge corresponding to the various values of elevation is determined, as follows:

ELEVATION ABOVE SPILLWAY (ft)	DISCHARGE (using $Q = CLh^x$)
0	0
1	75
2	212
3	390
4	600
5	838
6	1102
7	1189
8	1697
9	2025
10	2372

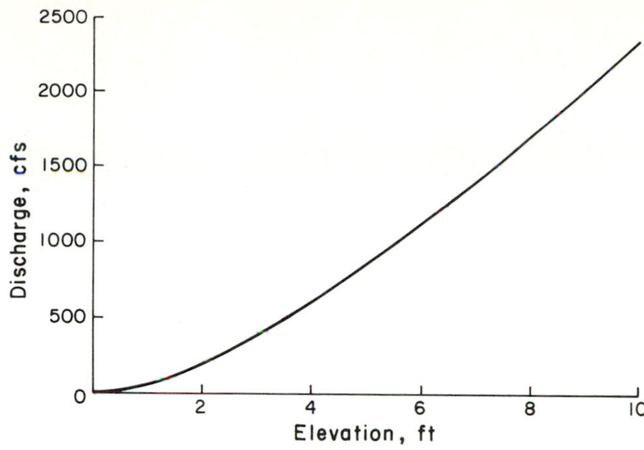

Figure 20.16 Discharge–elevation curve.

The elevation–discharge curve is plotted in Figure 20.16. The quantity $(2S/\Delta t) + Q$ is obtained for different values of Q, and calculations are given in Table 20.6. A plot of $(2S/\Delta t) + Q$ versus Q is shown in Figure 20.17. The results of the calculations for flow routing are shown in Table 20.7. These yield

$$\text{Maximum pool level} = 3.75 \text{ ft}$$

$$\text{Maximum discharge} = 545 \text{ cfs}$$

EXAMPLE 20.5 The elevation-discharge and elevation-area data for a small reservoir without spillway gates are as follows:

Elevation (ft)	0	1	2	3	4	5	6	7	8	9
Area (acres)	1000	1020	1040	1050	1060	1080	1100	1120	1140	1160
Outflow (cfs)	0	525	1490	2730	4200	5880	7660	9620	11800	14300

The inflow hydrograph of a flood is given in Table 20.8. Assuming the water level reaches the spillway crest (elevation) on April 18, 1943, compute the peak rate of

TABLE 20.6 COMPUTATION OF $(2S/\Delta t) + Q$ VERSUS Q FOR EXAMPLE 20.4

h (ft)	S (ft^3)	Q (cfs)	$(2S/\Delta t) + Q$ (cfs)
0.50	10,890,000	26.5	530.7
1.00	21,789,000	75.0	1083.0
1.50	32,670,000	137.8	1650.3
2.00	43,560,000	212.1	2228.8
2.50	54,450,000	296.5	2817.3
3.00	65,340,000	389.7	3414.7
3.50	76,230,000	491.1	4020.3
4.00	87,120,000	600.0	4633.3

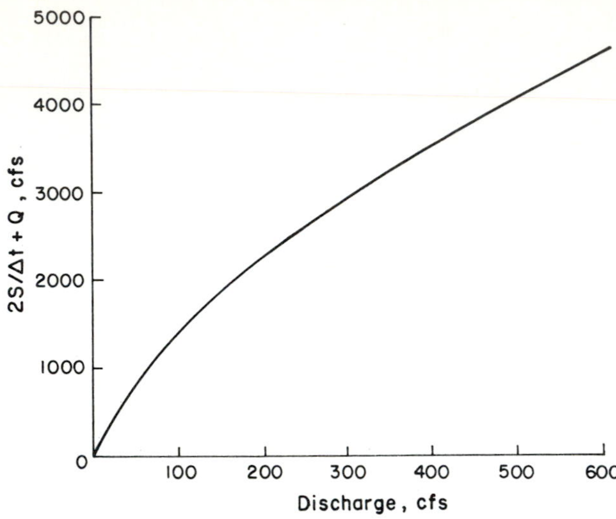

Figure 20.17 $(2S/\Delta t) + Q$ versus discharge.

outflow and the highest water level in the reservoir. Use a 6-h routing period. Plot the graph of computed outflow versus time. Determine the maximum outflow and its time of occurrence. Corresponding to the maximum outflow, compute the maximum elevation reached.

TABLE 20.7 FLOW ROUTING FOR EXAMPLE 20.4

Date	Time	I_i (cfs)	$I_i + I_{i+1}$	$(2S_i/\Delta t) - Q_i$	$(2S_{i+1}/\Delta t) + Q_{i+1}$	Q_{i+1}
1	Midnight	40	75	0		0
2	Noon	35	72	69	75	3
	Midnight	37	162	127	141	7
3	Noon	125	465	257	289	16
	Midnight	340	915	640	722	41
4	Noon	575	1297	1305	1555	125
	Midnight	722	1462	2122	2602	240
5	Noon	740	1413	2754	3584	415
	Midnight	673	1129	3117	4167	525
6	Noon	456	776	3156	4246	545
	Midnight	320	565	2827	3721	447
7	Noon	245	437	2652	3392	370
	Midnight	192	336	2429	3089	330
8	Noon	144	262	2205	2765	280
	Midnight	118	213	1997	2467	235
9	Noon	95	175	1820	2210	195
	Midnight	80	147	1655	1995	170
10	Noon	67	123	1512	1802	145
	Midnight	56	106	1395	1635	120
11	Noon	50	92	1287	1501	107
	Midnight	42			1379	92

TABLE 20.8 INFLOW HYDROGRAPH

Date	Hour	Flow (×100 cfs)
4–18–43	24	1.5
4–19–43	6	1.5
	12	1.6
	18	2.0
	24	3.1
4–20–43	6	5.2
	10	7.4
	12	9.6
	14	12.0
	18	12.45
	22	12.10
4–21–43	6	11.10
	12	10.20
	18	9.30
	24	8.5
4–22–43	12	7.0
	24	5.7
4–23–43	12	4.7
	24	3.9
4–24–43	12	3.2
	24	2.8
4–25–43	12	2.4
	24	2.2
4–26–43	12	2.0
	24	1.9
4–27–43	24	1.07
4–28–43	24	1.5
4–29–43	24	1.4

Solution We plot elevation versus area, as shown in Figure 20.18. Then we determine storage for different values of elevations as follows:

S.N.	Elevation (ft)	Storage (acre-ft)	Storage (ft^3)
1	1	10.6	$10.6 \times 43{,}560 = 0.462 \times 10^7$
2	2	40.0	1.74×10^7
3	3	64.6	28.2×10^7
4	4	99.2	4.35×10^7
5	5	189.0	8.24×10^7
6	6	299.6	13.5×10^7
7	7	428.2	18.3×10^7
8	8	577.8	25.2×10^7
9	9	747.4	32.6×10^7

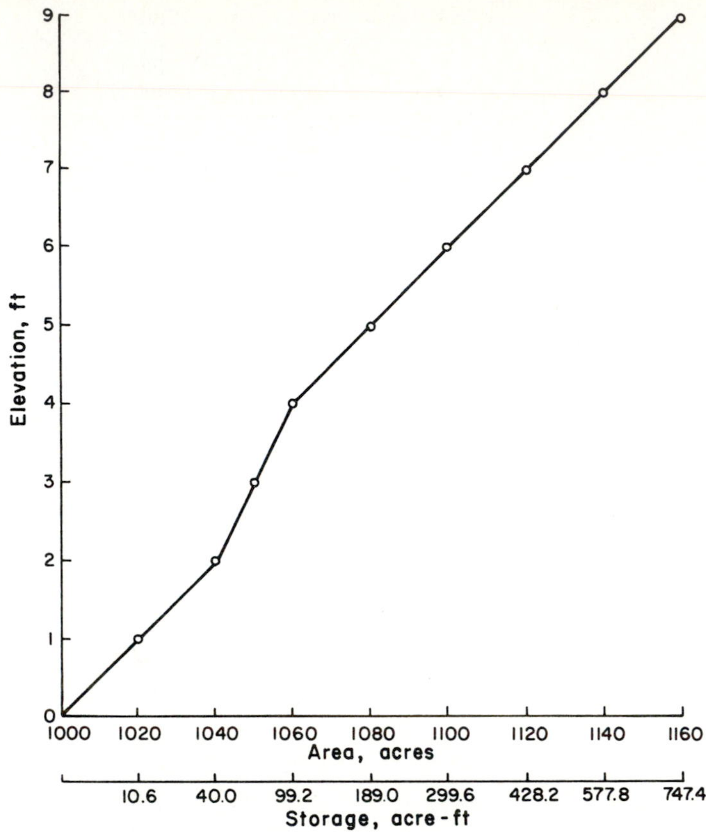

Figure 20.18 Storage versus elevation.

The discharge values and the computed storage values are each plotted against elevation, as shown in Figure 20.19. From this figure, we determine the values of discharge and storage for the same values of elevation. Figure 20.20 shows a plot of storage versus discharge. The computed values of storage and discharge are as follows:

S.N.	Elevation (ft)	Discharge (cfs)	Storage ($\times 10^7$ ft^3)
1	1	525	0.462
2	2	1490	1.72
3	3	2730	2.82
4	4	4200	4.35
5	5	5880	8.24
6	6	7660	13.5
7	7	9620	18.3
8	8	11800	25.2
9	9	14300	32.6

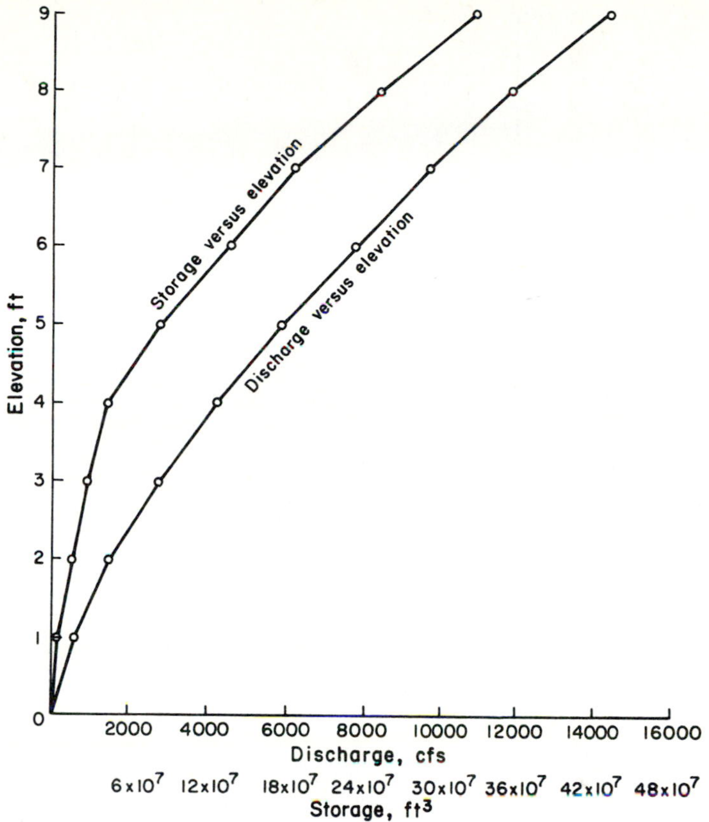

Figure 20.19 Discharge versus elevation and storage versus elevation.

Then we compute $(2S/\Delta t) + Q$ for various values of Q, as follows:

S.N.	Discharge (cfs)	Storage ($\times 10^7$ ft^3)	$2S/\Delta t$	$(2S/\Delta t) + Q$ (cfs)
1	2,000	1.6	$\dfrac{1.6 \times 2 \times 10^7}{3600 \times 6} = 1482$	3,482
2	4,000	3.6	3,333	7,333
3	6,000	8.7	8,052	14,052
4	8,000	14.1	13,072	21,072
5	10,000	19.2	17,708	27,708
6	12,000	25.5	23,620	35,620
7	14,000	31.6	29,220	43,220

Figure 20.21 shows $(2S/\Delta t) + Q$ versus Q. Flow routing is then performed, and the results of the routing calculations are shown in Table 20.9. The inflow and outflow hydrographs are plotted in Figure 20.22.

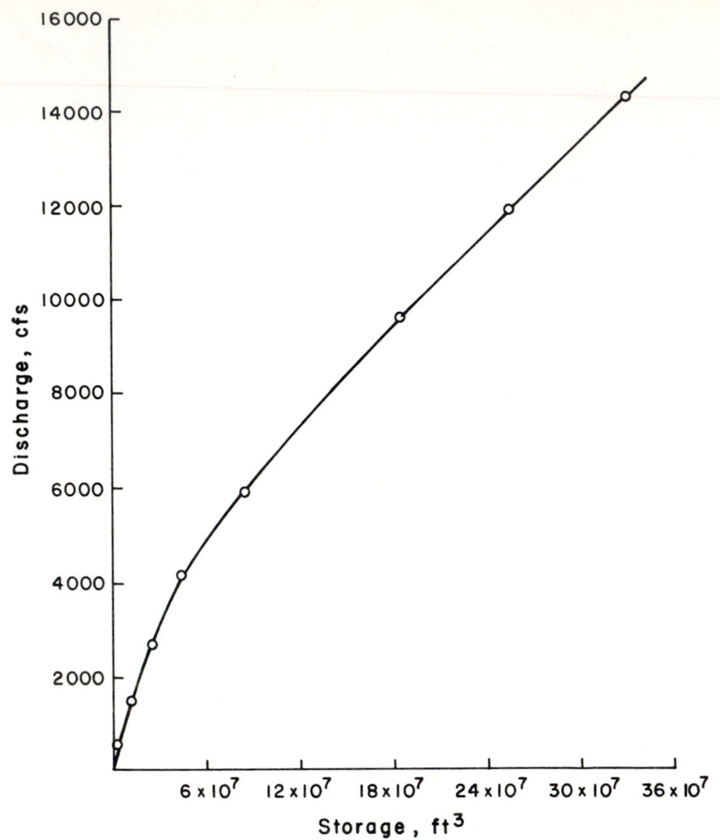

Figure 20.20 Storage–discharge relation.

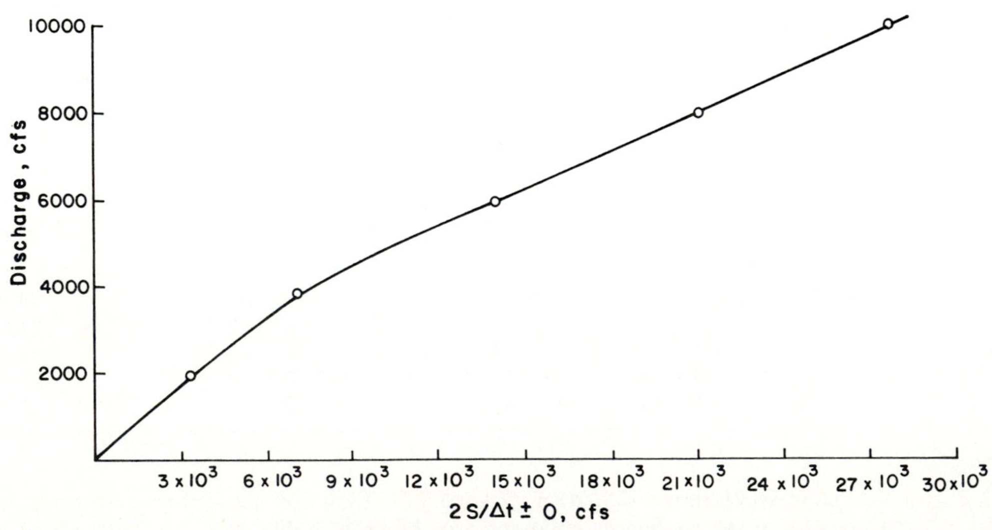

Figure 20.21 $(2S/\Delta t) \pm Q$ versus Q.

TABLE 20.9 FLOW ROUTING FOR EXAMPLE 20.5

Date	Hour	Inflow (cfs)	$(2S/\Delta t) - Q$ (cfs)	$(2S/\Delta t) + Q$ (cfs)	Q
4–18–43	24	1,500	−600	2,400	1,500
4–19–43	6	1,500	−600	2,400	1,500
	12	1,600	−600	2,500	1,550
	18	2,000	−550	3,000	1,775
	24	3,100	−650	4,550	2,600
4–20–43	6	5,200	−450	7,650	4,100
	10	7,400	1,250	12,150	5,450
	12	9,600	3,650	18,250	7,300
	14	1,200	6,250	24,250	9,000
	18	12,450	10,550	30,700	10,075
	24	12,100	11,300	35,100	11,900
4–21–43	6	11,100	11,000	34,500	11,750
	12	10,200	9,900	32,300	11,200
	18	4,300	8,400	29,400	10,500
	24	8,500	6,800	26,200	9,700
4–22–43	12	700	500	22,300	8,650
	24	5,700	3,300	17,700	7,200
4–23–43	12	4,700	1,800	13,700	5,950
	24	3,900	600	10,400	4,900
4–24–43	12	3,200	−300	7,700	4,000
	24	2,800	−600	5,700	3,650
4–25–43	12	2,400	−600	3,600	2,100
	24	2,200	−300	4,000	2,000
4–26–43	12	2,000	−300	3,600	1,950
	24	1,900	−300	3,600	1,950
4–27–43	24	1,700	−500	3,300	1,900
4–28–43	24	1,500	−500	2,700	1,600
4–29–43	24	1,400	−500	2,400	1,450

EXAMPLE 20.6 For a reservoir, elevation versus surface area data are given:

Area (acres)	520	450	400
Outflow (cfs)	156	163	150
Elevation (ft)	60	58	56

Determine the time required to lower the water table from an elevation of 60 ft to 56 ft. During the entire period, there is a steady flow of 100 cfs.

Solution A graph of elevation versus area is constructed, as shown in Figure 20.23, and storage for different elevations is determined as follows:

ELEVATION (ft)	STORAGE ($\times 10^6$ ft^3)
56	0
57	0.522
58	2.32
59	5.59
60	11.45

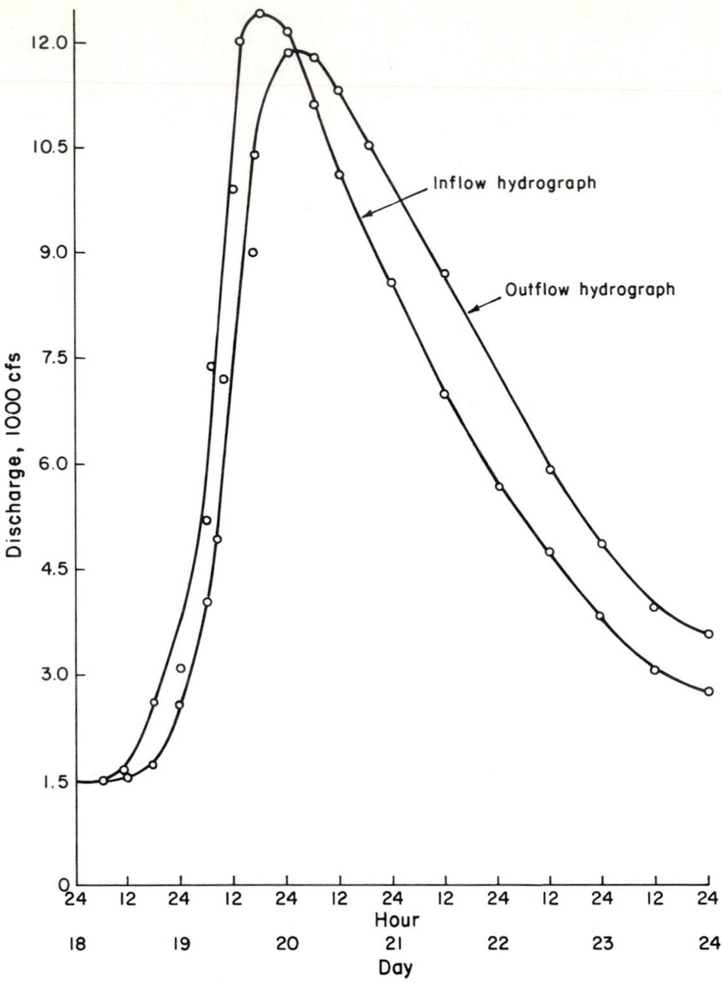

Figure 20.22 Discharge versus time.

By using the computed values of storage, a curve of storage versus elevation is constructed, as shown in Figure 20.24. Another graph of discharge versus elevation is plotted in Figure 20.25. Using these figures, we calculate $(2S/\Delta t) + Q$ as follows:

Elevation (ft)	$(2S/\Delta t) + Q$ (cfs)	$(2S/\Delta t) - Q$ (cfs)
56	150	−150
57	$\dfrac{2 \times 0.522 \times 10^6}{24 \times 60 \times 60} + 151.42 = 163.38$	$\dfrac{2 \times 0.522 \times 10^6}{2.4 \times 60 \times 60} - 151.42 = -139.4$

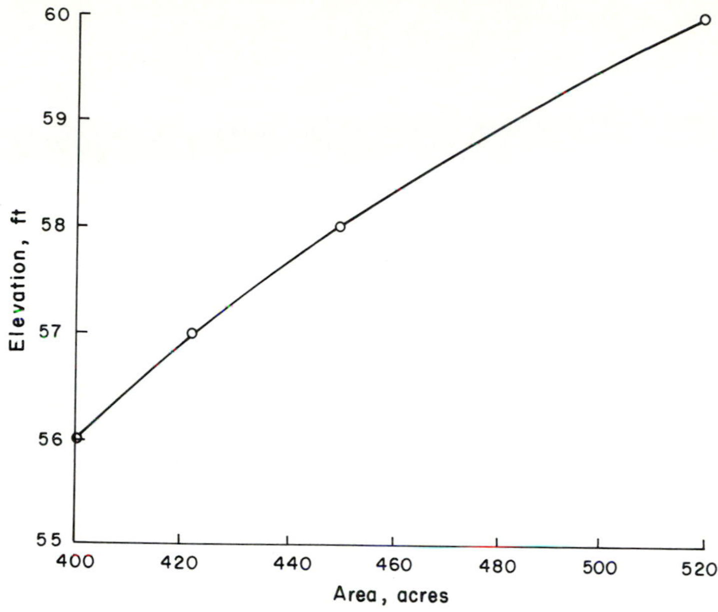

Figure 20.23 Elevation–area curve.

Elevation (ft)	$(2S/\Delta t) + Q$ (cfs)	$(2S/\Delta t) - Q$ (cfs)
58	$\dfrac{2 \times 2.326 \times 10^6}{24 \times 60 \times 60}$	$\dfrac{2 \times 2.326 \times 10^6}{24 \times 60 \times 60}$
	$+ 153 = 206.85$	$- 153 = -99.1$
59	$\dfrac{2 \times 5.59 \times 10^6}{24 \times 60 \times 60}$	$\dfrac{2 \times 5.59 \times 10^6}{24 \times 60 \times 60}$
	$+ 154.5 = 284$	$- 154.5 = -24.9$
60	$\dfrac{2 \times 11.45 \times 10^6}{24 \times 60 \times 60}$	$\dfrac{2 \times 11.45 \times 10^6}{24 \times 60 \times 60}$
	$+ 156 = 421.0$	$- 156 = 109$

Then $(2S/\Delta t) + Q$ is plotted against elevation, as shown in Figure 20.26. By routing the flow through the reservoir, we construct the time-elevation curve. The results of the routing calculations are as follows:

Days	Inflow (cfs)	$(2S/\Delta t) + Q$ (cfs)	$(2S/\Delta t) - Q$ (cfs)	Outflow (cfs)	Elevation (ft)
0	100	421	109	156	60
1	100	309	0	154.6	59.2
2	100	200	−110	155	57.8
3	100	10	—	—	—
4	100	—	—	—	—

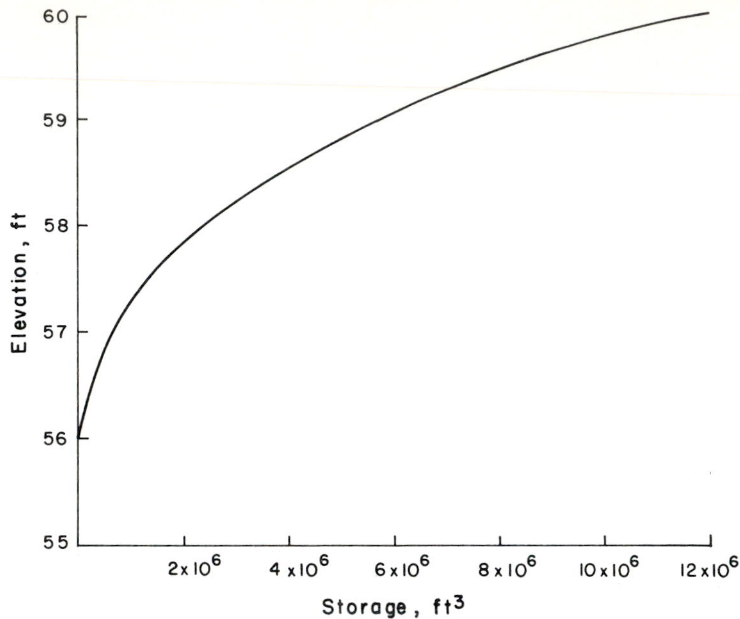

Figure 20.24 Storage–elevation curve.

When $(2S/\Delta t) + Q = 90$ cfs, the corresponding value of elevation is not available on the curve. It is therefore suggested to take $\Delta t = 12$ h for determination of a few more points. However, the time-elevation curve is plotted using the available readings, as shown in Figure 20.27. From this figure we get 2.8 days.

20.3.2 Goodrich Method

Goodrich (1931) proposed a method based on the graphical integration of Equation (20.3), which can finally be expressed as Equation (20.11). This method involves plotting discharge versus elevation, as well as $(2S/\Delta t) + Q$ versus elevation on the same graph. The flood routing by this method involves the following steps:

1. At the beginning, I_1, I_2, S_1, and Q_1 are known. The left side of Equation (20.11) is then computed. This gives $(2S_2/\Delta t) + Q_2$ at the end of the first time interval.
2. Corresponding to $(2S_2/\Delta t) + Q_2$ obtained in step 1, the value of elevation is obtained from the curve of $(2S/\Delta t) + Q$ versus elevation.
3. For the elevation in step 2, the value of Q_2 is obtained from the curve of Q versus elevation.
4. Twice the value of Q_2 of step 3 is subtracted from $(2S_2/\Delta t) + Q_2$ of step 1, and $(2S_2/\Delta t) - Q_2$ is obtained at the end of the first Δt.
5. The value of $(2S_2/\Delta t) - Q_2$ is taken as the initial value, and $(2S_1/\Delta t) + Q_1$ for the next time interval.
6. By adding I_1 and I_2 to $(2S_1/\Delta t) - Q_1$ of step 5, and left side of Equation (20.11) is obtained. Then by going back to step 2, the procedure is repeated.

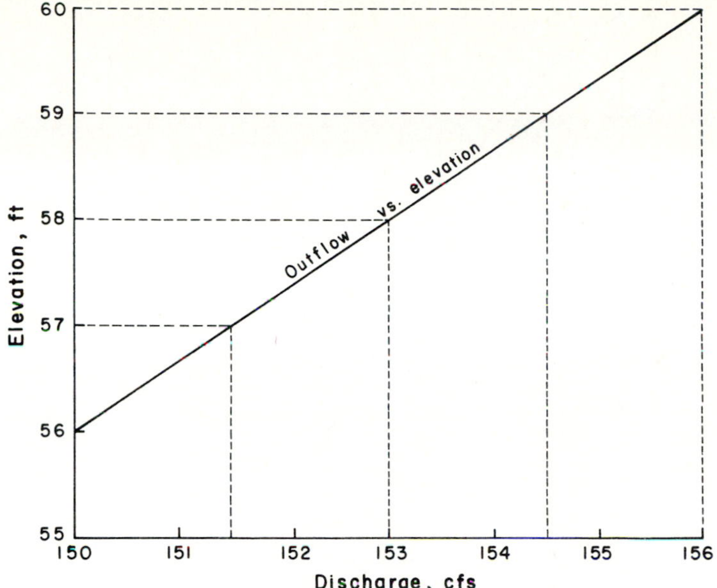

Figure 20.25 Discharge–elevation curve.

EXAMPLE 20.7 Route the flood hydrograph of Example 20.2 using the Goodrich method, and determine the attenuation and lag. Assume that initially the reservoir elevation is 355 m.

Solution The routing period is 6 h = 0.0216×10^6 s.

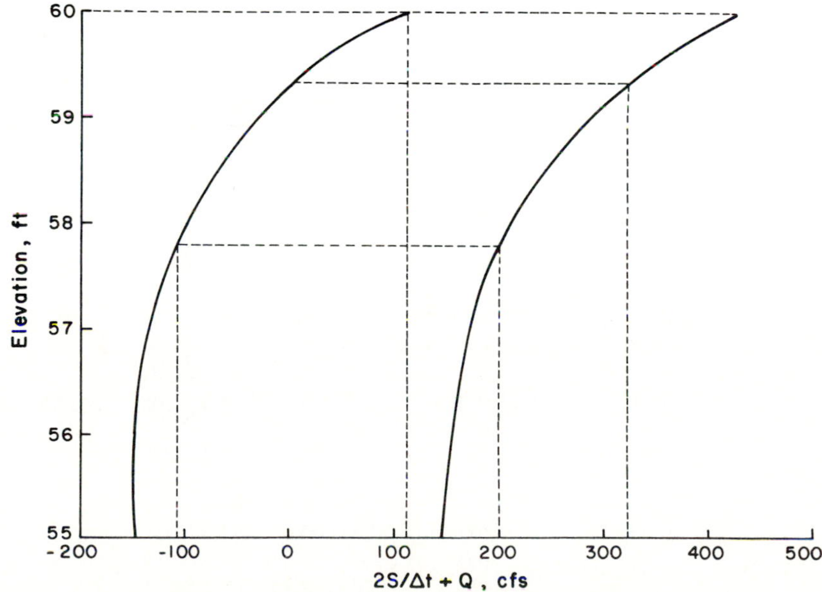

Figure 20.26 $(2S/\Delta t) \pm Q$ versus elevation.

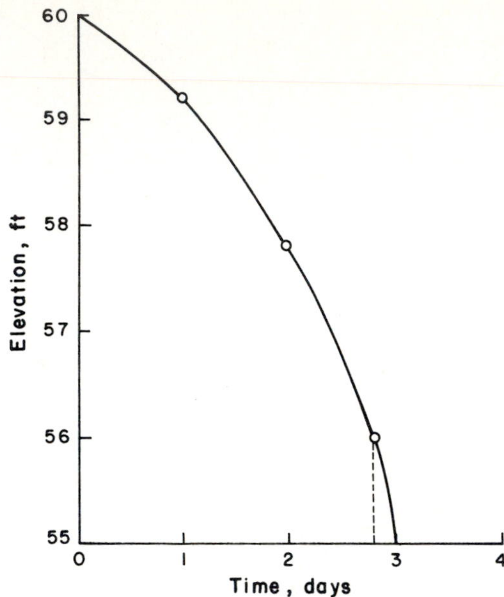

Figure 20.27 Elevation versus time.

TABLE 20.10 COMPUTATION FOR ELEVATION–DISCHARGE–STORAGE CURVES[a]

Elevation (m)	Outflow (m^3/s)	$\frac{2S}{\Delta t} + Q$ (m^3/s)
125	0.02	1,944.46
130	0.13	8,333.46
140	0.45	14,722.67
160	2.70	22,422.85
180	7.50	28,063.06
200	15.50	32,793.28
220	30.00	36,789.26
240	50.00	42,037.04
260	75.00	44,056.48
280	105.00	47,327.22
300	142.00	50,142.00
320	185.00	52,962.78
340	230.00	55,785.56
350	256.00	57,465.30
355	280.00	58,613.33
365	293.00	59,459.67
370	306.00	59,843.04
375	321.00	60,413.59
390	360.00	62,397.04
400	392.00	63,817.93
420	456.00	66,658.70

[a] Δt = 6 h = 21,600 s = 0.0216×10^6 s.

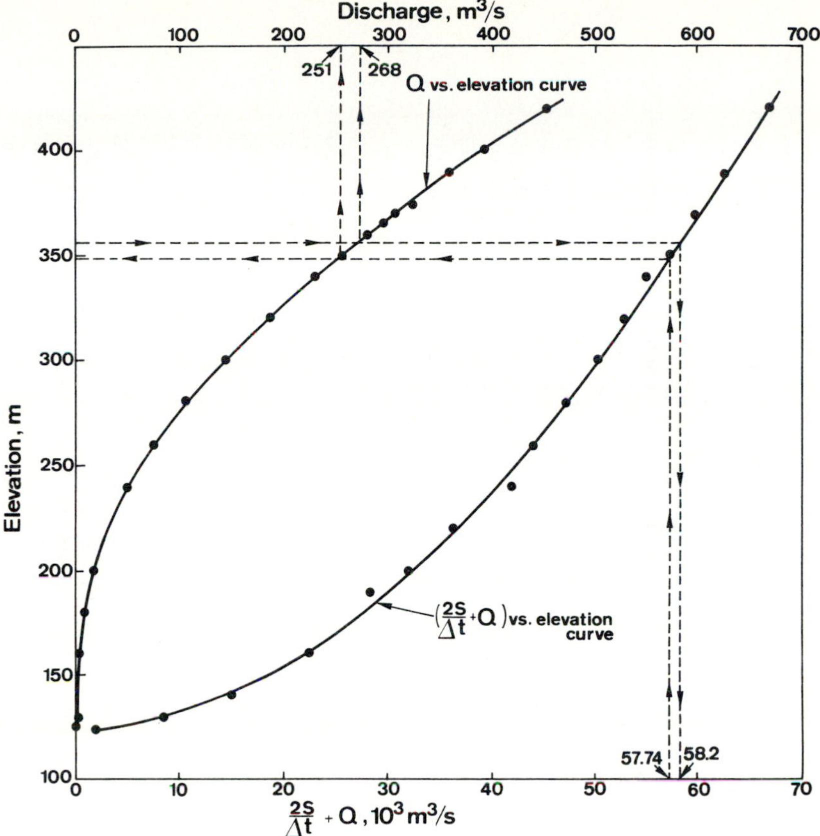

Figure 20.28 Elevation–discharge–storage curves for the Goodrich method of flood routing.

By using the given elevation–storage–discharge data, the quantity $(2S/\Delta t) + Q$ is computed for each elevation value, as shown in Table 20.10. Then a graph of Q versus elevation and of $(2S/\Delta t) + Q$ versus elevation is prepared, as shown in Figure 20.28.

At $t = 0$, the elevation = 355 m. From Figure 20.28 $(2S/\Delta t) + Q = 58.2 \times 10^3$ m³/s, and, corresponding to this value, $Q = 268$ m³/s. Thus, $(2S/\Delta t) - Q = 58.2 \times 10^3 - 2 \times 268 = 57.664 \times 10^3$ m³/s.

For the first $\Delta t = 6$ h, $I_1 = 20$ m³/s, $I_2 = 60$ m³/s, and $Q_1 = 268$ m³/s. This gives

$$I_1 + I_2 + \left(\frac{2S_1}{\Delta t} - Q_1\right) = 20 + 60 + 57{,}664 = 57{,}744 \text{ m}^3/\text{s} = \left(\frac{2S_2}{\Delta t} + Q_2\right)$$

Corresponding to this, the elevation from Figure 20.28 = 348 m, and $Q_2 = 251$ m³/s. For the next time increment,

$$\frac{2S_1}{\Delta t} - Q = 57{,}744 - 2 \times 251 = 57{,}242 \text{ m}^3/\text{s}$$

TABLE 20.11 FLOOD ROUTING BY THE GOODRICH METHOD

Time	Routing period[a]	Inflow, I (m^3/s)	$I_1 + I_2$	Total outflow, Q (m^3/s)	$(2S/\Delta t) - Q$	$(2S/\Delta t) + Q$	Reservoir elevation (m)
0	0	20	0	268.00		58,200.00	355.00
6	1	60	80	251.00	57,664.00	57,744.00	348.00
12	2	220	280	251.00	57,242.00	57,522.00	348.00
18	3	350	570	251.00	57,020.00	57,590.00	348.00
24	4	380	730	251.00	57,088.00	57,818.00	350.00
30	5	430	810	268.00	57,316.00	58,126.00	355.00
36	6	480	910	273.00	57,590.00	58,500.00	359.00
42	7	530	1010	275.00	58,964.00	57,954.00	360.00
48	8	610	1140	296.00	58,414.00	59,554.00	362.00
54	9	670	1280	297.00	58,962.00	60,242.00	362.50
60	10	730	1400	308.00	59,648.00	61,048.00	368.00
66	11	780	1510	333.00	60,432.00	61,942.00	375.00
72	12	870	1650	360.00	61,276.00	62,926.00	392.00
78	13	770	1640	370.00	62,206.00	63,846.00	395.00
84	14	700	1470	400.00	63,106.00	64,576.00	402.00
90	15	670	1370	400.00	63,776.00	65,146.00	402.00
96	16	630	1300	425.00	64,346.00	65,646.00	420.00
102	17	550	1180	425.00	64,796.00	65,976.00	420.00
108	18	490	1040	425.00	65,126.00	66,166.00	420.00
114	19	360	850	425.00	65,316.00	66,166.00	420.00
120	20	330	690	425.00	65,316.00	66,006.00	420.00
126	21	280	610	425.00	65,156.00	65,766.00	420.00
132	22	200	480	404.00	64,916.00	65,396.00	404.00
138	23	160	360	400.00	64,588.00	64,948.00	402.00
144	24	130	290	385.00	63.148.00	64,438.00	398.00
150	25	110	240	370.00	63,668.00	63,908.00	395.00
156	26	90	200	357.00	63,168.00	63,368.00	389.00
162	27	60	150	350.00	62.654.00	62,804.00	387.00
168	28	50	110	336.00	62,104.00	62,214.00	385.00
174	29	30	80	312.00	61,542.00	61,622.00	370.00
180	30	22	52	310.00	60,998.00	61,050.00	368.00
186	31	21	43	296.00	60,430.00	60,473.00	362.00
192	32	20	41	296.00	59,881.00	59,922.00	362.00
198	33	19	39	276.00	59,330.00	59,369.00	362.00
204	34	18	37	273.00	58,817.00	58,854.00	359.00
210	35	17	35	270.00	58,308.00	58,343.00	357.00
216	36	17	34	251.00	57,803.00	57,837.00	350.00
222	37	17	34	250.00	57,335.00	57,369.00	347.00

[a] Routing period (Δt) = 6 hours

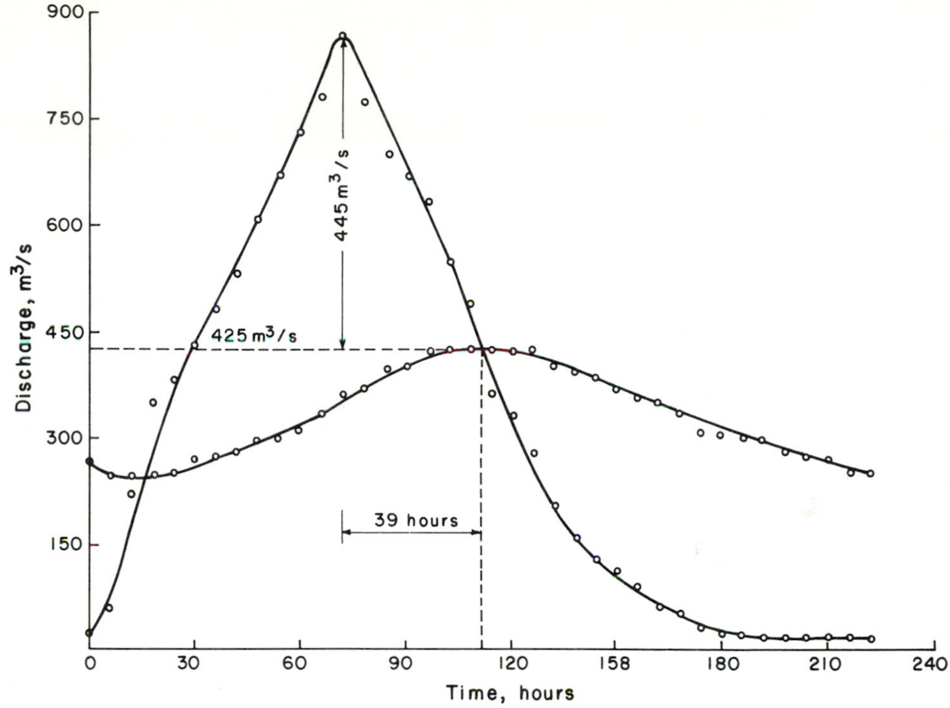

Figure 20.29 Inflow hydrograph and outflow hydrograph obtained by the Goodrich method of flood routing.

This procedure is repeated, and the results of the calculations are shown in Table 20.11. The calculated outflow hydrograph and the given inflow hydrograph are shown in Figure 20.29.

EXERCISES

20.1. How much is the flood wave modified in a reservoir having the following storage characteristics?

$(2S/\Delta t) + Q$ (cfs)	3000	4000	5000	6000	7000	8000
Outflow, Q (cfs)	950	1100	1280	1350	1430	1500

where S is the storage in ft^3 and Δt is the time interval of 6 hours.

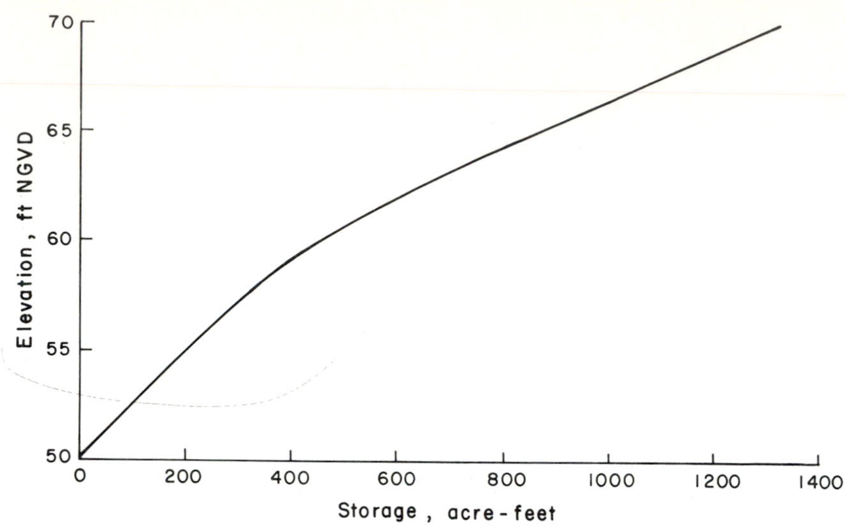

Figure 20.30 Storage–elevation curve.

20.2. The following hydrograph is to be used as a design hydrograph for a detention reservoir. The capacity curve of the site is as shown in Figure 20.30. Outflow is controlled by the ubiquitous 36-in. CMP with a downstream invert of 50.0 in. Using the orifice equation with a C value of 0.52, rate the pipe up to an elevation 65.0 NGVD. (This corresponds to a head of 15 feet.) Then route the inflow hydrograph through the reservoir to determine the maximum amount of storage used and, thus, the required height of the embankment. Use a routing interval of 1 hour. What is the maximum outflow from the pipe? Basin area is 8.26 mi^2.

TIME (h)	DISCHARGE (cfs)	TIME (h)	DISCHARGE (cfs)	TIME (h)	DISCHARGE (cfs)
1	5	2	13	3	37
4	64	5	131	6	388
7	744	8	723	9	649
10	624	11	582	12	366
13	302	14	332	15	296
16	258	17	178	18	131
19	97	20	80	21	68
22	57	23	54	24	48
25	44	26	40	27	36
28	33	29	32	30	31
31	29	32	28		

20.3. An embankment will be traversed by the well-know 36-in CMP. The $(2S/\Delta t) + Q$ versus Q curve for the site is as shown in Figure 20.31. The rating curve for the culvert is shown in Figure 20.32. Route the hydrograph of Exercise 20.2 through the site by storage indication and determine the minimum allowable embankment height.

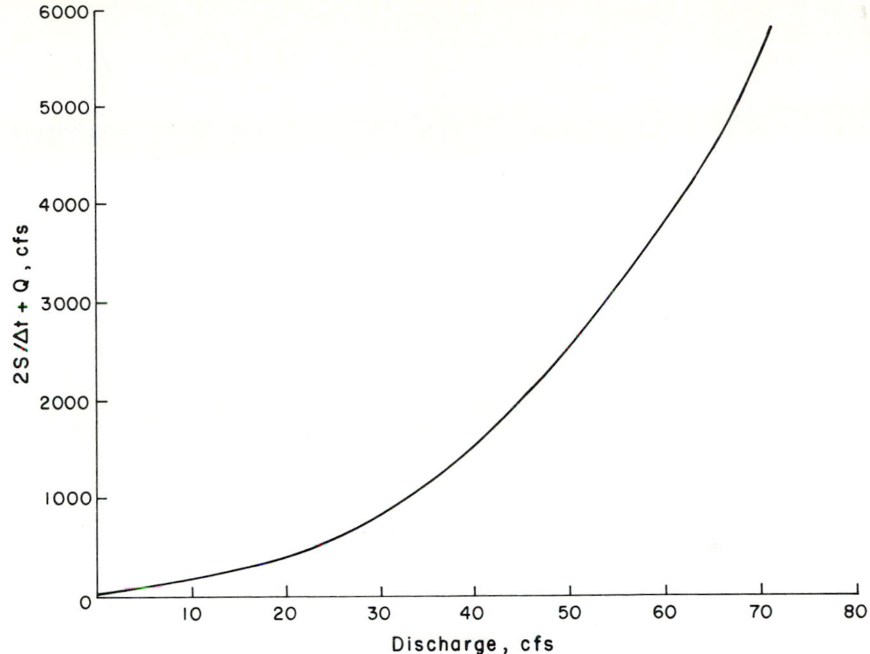

Figure 20.31 $(2S/\Delta t) + Q$ versus Q.

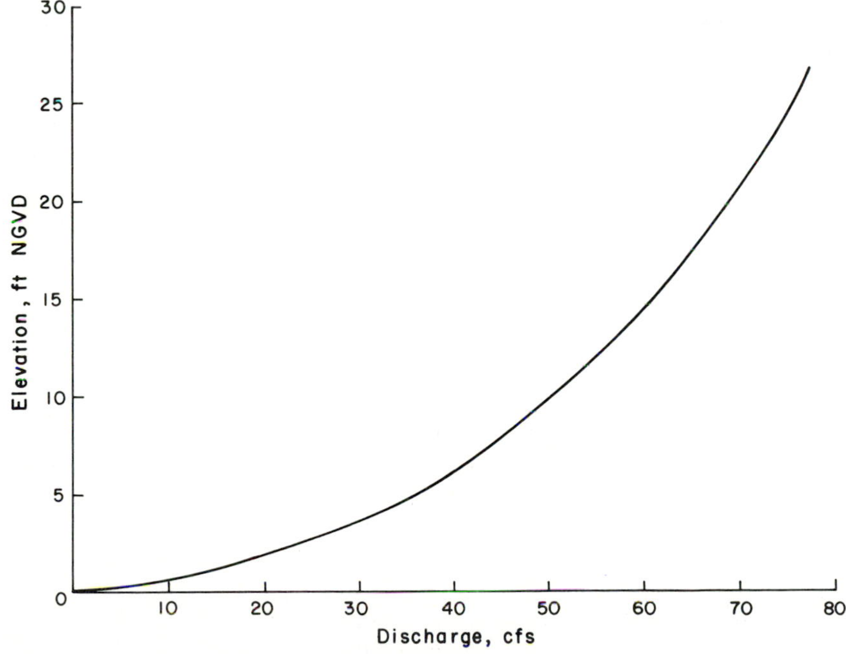

Figure 20.32 Discharge–elevation curve.

20.4. A survey of a potential reservoir site provided the following data:

ELEVATION (m)	AREA (ha)	ELEVATION (m)	AREA (ha)
510	0	515	12
511	1	516	16
512	5	517	18
513	8	518	21
514	10	519	25
		520	30

Draw a storage–elevation curve. Assume that the storage below inlet is zero. The inlet elevation is 511 m and there is no permanent pool.

20.5. A rectangular reservoir equipped with an outflow-controlling weir has the following characteristics:

$$S = 5\text{ h}$$

$$Q = 2\text{ h}$$

where S is the storage in m^3/s-day, Q is the discharge in m^3/s, and h is the water elevation in m. Route the inflow hydrograph given and determine the head on the weir as a function of time. The initial elevation is 0.25 m.

TIME (h)	DISCHARGE (m^3/s)	TIME (h)	DISCHARGE (m^3/s)
6:00 A.M.	30	1:00 P.M.	430
7:00	30	2:00	400
8:00	60	3:00	300
9:00	120	4:00	240
10:00	240	5:00	180
11:00	420	6:00	30
12:00	450		

20.6. Compute the maximum pool elevation and maximum discharge as a result of the inflow hydrograph of Exercise 20.5. Assume the initial pool elevation to be 500 m. The reservoir has the following storage–elevation and discharge–elevation data:

ELEVATION (m)	STORAGE (ha-m)	DISCHARGE (m^3/s)
480	0	0
485	2.5	0
490	15	0
495	35	0
500	60	0
501	70	10
502	80	30
503	100	60
504	110	80
505	135	120
506	150	140

ELEVATION (m)	STORAGE (ha-m)	DISCHARGE (m^3/s)
507	200	180
508	230	200
509	280	270
510	300	300

20.7. Consider the data of Exercise 20.6 as elevation in meters, storage in 10^5 m^3, and discharge in m^3/s. Determine the maximum pool level and maximum outflow rate for the following inflow hydrograph:

TIME (h)	INFLOW (m^3/s)	TIME (h)	INFLOW (m^3/s)	TIME (h)	INFLOW (m^3/s)
0	40	12	200	22	110
2	65	14	190	24	100
4	80	16	160	26	80
6	110	18	150	28	60
8	120	20	129	30	40
10	170				

20.8. A linear storage indication curve for a simple reservoir can be defined as

$$Q = \frac{Q}{2} + \frac{S}{\Delta t}$$

Assume $\Delta t = 3$ h, and the initial storage to be zero. Route the inflow hydrograph of Exercise 20.7.

20.9. A vertical walled reservoir has a surface area of 1000 ha. The spillway is 50 m long. The initial inflow and outflow are both taken as 100 m^3/s. Determine the reservoir storage for the following water-surface elevations above the spillway crest: h (m) = 0, 0.2, 0.4, 0.6, 0.8, 1.0, 1.2, 1.4, 1.6, 1.8, and 2.0. Determine the outflow rates for these values of h. Plot the storage–discharge curve as well as storage–indication curve. Determine the outflow rates over the spillway at the end of each 6-h interval for the inflow hydrograph: 100, 200, 500, 1000, 1500, 1200, 800, 400, 300, 200, 100, and 100 m^3/s.

20.10. A reservoir has the following characteristics:

$$S = 0.01h^2 \qquad \text{and} \qquad Q = 60h^{0.5}$$

where h is the depth of water, Q is the outflow in m^3/s, and S is in ha-m. Determine the maximum depth to which water will rise as a result of the inflow hydrograph of Exercise 20.5.

20.11. Consider a reservoir, initially dry, whose surface area is 1 km^2. It receives, from the beginning, a uniform inflow of 40 cm/h on a unit area basis for 5 hours. The outflow from the reservoir is observed as 0 at the beginning as well as 10 hours later. When the time, measured from the beginning, equals 5 hours, the outflow equals inflow. For simplicity, assume that the variation in outflow is linear during its rise as well as during its recession. Compute the storage in the reservoir and graph it against time. At what time will the storage achieve its peak? How long will it take for the reservoir to dry?

20.12. If $Q = 25H^{1.5}$ over the spillway and the storage–elevation relation is $S = 500H$, route the flow of March 16 through the reservoir. H = ft, Q = cfs, and S = acre-ft. Assume the reservoir is initially full.

Chuquatonchee Creek near Egypt, Mississippi
(Drainage area = 170 mi², or 440 km²)
March 16, 1972

DAY	FLOW	DAY	FLOW
13	328	19	497
14	331	20	373
15	2530	21	319
16	22900	22	265
17	5690	23	228
18	1330		

20.13. A simple reservoir has a linear storage indication curve defined by $Q/2 = S\ \Delta t$, where $\Delta t = 1.0$ h. If S at 8 A.M. is zero cfs-h, use the continuity equation to route the following hydrograph through the reservoir:

Time	8 A.M.	9 A.M.	10 A.M.	11 A.M.	12 Noon	1 P.M.
I (cfs)	0	200	400	200	0	0

20.14. The outflow rate (cfs) and storage (cfs-h) for an emergency spillway of a certain reservoir are linearly related by $Q = S/3$, where the number 3 has units of hour. Use this and the continuity equation to determine the peak outflow from the reservoir for the following inflow event:

Time	0	2	4	6	8
I (cfs)	0	400	600	200	0

CHAPTER 21
Channel-Flow Routing

Flood routing in a stream is somewhat more complicated than reservoir routing. In the case of a reservoir, the water level is assumed to be level at all times. Storage in the reservoir and outflow therefrom are assumed to be unique functions of the depth of water behind the dam. In other words, storage and outflow are uniquely related. However, in a stream, the water surface is not always parallel to its bed. In fact, the average free-surface slope is higher during the stream rise, but lower during the stream recession than the bed slope. Therefore, the relation between storage and outflow is more complicated in a channel than in a reservoir. Linsley et al. (1975) have shown that the storage in a channel is comprised of prism storage and wedge storage, as shown in Figure 21.1. Prism storage is defined by the volume of water that would exist if the flow were uniform at the downstream depth. Wedge storage is the volume of water between the actual water-surface profile and the top surface of the prism storage. If the depth of flow is constant at the downstream end of the channel reach, the prism storage would be constant. The wedge storage changes during propagation of a flood wave. It increases the flood volume during the rising stage and decreases it during the receding stage. The wedge storage can change from a positive value during the rising flood to a negative value during the receding flood. Thus, when storage is plotted against discharge, the resulting curve is a loop reflecting the rising and falling stages. Different methods, developed for streamflow routing, differ in the mechanics of accounting for the relationship between storage and discharge. Excellent discussions on many of these methods have been reported by Dooge (1980) and Fread (1982), amongst others.

Further complicating channel-flow routing are local inflows and seepage that may occur during the passage of a flood wave in a channel reach. In discussion of the routing methods, these will not be explicitly considered. There are, however,

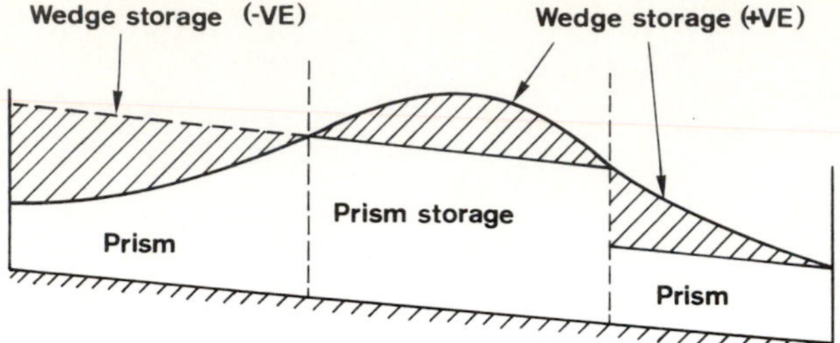

Figure 21.1 Prism and wedge storage in a channel reach.

approximate ways to incorporate their effect. For example, one common practice to include seepage is to subtract it from the routed outflow hydrograph. Similarly, local inflows can be considered by adding them to the outflow hydrograph.

21.1 CONVEX METHOD

This method is based on the assumption that the outflow at any time t depends upon the outflow Q and inflow I one routing time interval Δt earlier. Let c be some constant, $0 \leq c \leq 1$. We can write

$$Q_{t+\Delta t} = cI_t + (1 - c)Q_t \tag{21.1}$$

This assumption requires that for the rising hydrograph, if $I_t \geq Q_t$, then $I_t \geq Q_{t+\Delta t} \geq Q_t$, and for the recession hydrograph, if $I_t \leq Q_t$, then $I_t \leq Q_{t+\Delta t} \leq Q_t$. Therefore, if Δt is properly chosen, then $Q_{t+\Delta t}$ will fall somewhere on or between I_t and Q_t in magnitude, but not above or below them. It can be shown that the quantities I_t, Q_t, and $Q_{t+\Delta t}$ are members of a convex set and hence the name of this method.

Equation (21.1) can also be expressed as

$$c = \frac{Q_{t+\Delta t} - Q_t}{I_t - Q_t} \tag{21.2}$$

From Figure 21.2, it is seen that

$$\frac{Q_{t+\Delta t} - Q_t}{\Delta t} = \frac{I_t - Q_t}{k}$$

where k is some time parameter yet to be determined. Therefore,

$$\frac{\Delta t}{k} = \frac{Q_{t+\Delta t} - Q_t}{I_t - Q_t} \tag{21.3}$$

By comparing Equations (21.2) and (21.3),

$$c = \frac{\Delta t}{k} \qquad \text{or} \qquad \Delta t = ck \tag{21.4}$$

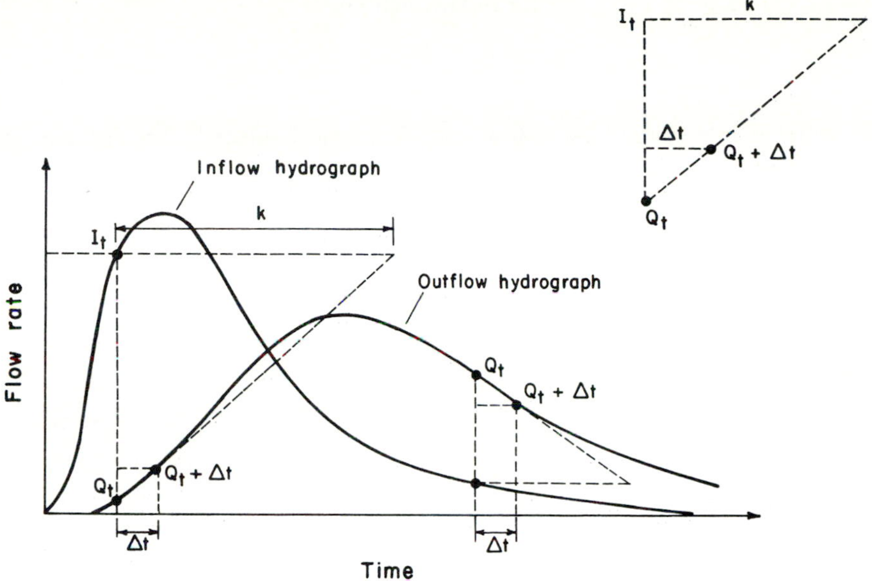

Figure 21.2 Construction of triangles for the convex method of flood routing.

The convex method is, therefore, a two-parameter method of flood routing in a channel. This method has the advantage that $I_{t+\Delta t}$ does not enter into computation of $Q_{t+\Delta t}$, and, therefore, can be used for forecasting. To illustrate, if the routing interval is 1 day and if the inflow and outflow are known today, then the outflow can be predicted for tomorrow without knowledge of tomorrow's inflow.

This method differs from other methods of streamflow routing in that the outflow begins one routing time interval after inflow begins, which is more realistic. In this method, the peak outflow does not intersect with the inflow hydrograph as in the case of reservoir routing methods. However, the maximum storage in the stream, as in most routing methods, occurs at the time when inflow equals outflow.

21.1.1 Estimation of Parameters

The parameter k has the same dimension as time. It has the same physical connotation as the time of travel and can be computed by

$$k = \frac{LA}{Q} = \frac{L}{u} \tag{21.5}$$

where L is the reach length, A is the average cross-sectional area, Q is the steady-state discharge, and u is the average steady-state velocity of Q. (This k is approximately equal to the k of the Muskingum method to be discussed later.)

Parameter c is the ratio of two velocities:

$$c = \frac{u}{u_a} \tag{21.6}$$

where u_a is a special wave function. The Soil Conservation Service (1964) has derived an empirical expression for c:

$$c = \frac{u}{1.7 + u} \tag{21.7}$$

Parameter c is approximately related to the weighting factor x of the Muskingum method:

$$c = 2x \tag{21.8}$$

A question arises: How can one select the steady-state velocity? In a natural channel, u varies with stage and from one season to another. This variation is not consistent with an increase or decrease in stage either. A flood wave is an unsteady flow phenomenon. Its velocity continuously varies with time. Although it may appear, therefore, that parameters c and k should change from one step of computation to another, mass conservation requires them to be constant. Furthermore, since one inflow hydrograph will always be different from another, these parameters will change likewise. One way to select u is to compute the velocity corresponding to a discharge equal to $\frac{1}{2}$ the peak inflow. Another way is to determine u for a discharge equal to $\frac{3}{4}$ of the peak inflow. If the inflow hydrograph has two or more peaks, then u should correspond to the discharge with the largest value of TQ defined as $TQ = \frac{3}{4}$ of the discharge multiplied by the duration of $\frac{3}{4}$ of the discharge. Whichever peak yields the highest TQ should be selected.

From Equation (21.4), it is clear that Δt can be determined from known values of c and k. The value of Δt computed in this manner may not be a convenient one for routing calculations. The Soil Conversation Service (1964) recommends selecting a desired value of Δt^* and then modifying c by

$$c^* = 1 - (1 - c)^{\Delta t^*/\Delta t} \tag{21.9}$$

where c^* is the modified c required with the use of the desired time interval Δt^*. A choice of Δt^* can be such that

$$\Delta t^* \leq \frac{t_p}{5} \tag{21.10}$$

where t_p is the time from the beginning of the rise to the peak of the inflow hydrograph. If the inflow hydrograph has more than one peak, then Δt^* should be chosen using t_p for the shortest of the rise periods of the peaks. Let us consider an example to illustrate this method.

EXAMPLE 21.1 Consider a channel reach receiving an inflow hydrograph defined by a triangle such that its base is 6 hours and its peak of 3000 m^3/s occurs at $t = 2$ hours. There is no tributary inflow. Assume that $c = 0.4$ and $\Delta t = 0.3$ hour. Route this inflow by the convex method.

Solution Equation (21.1) can be written as

$$Q_{t+0.3} = 0.4I_t + 0.6Q_t$$

which is used for routing the inflow hydrograph. The inflow hydrograph and the outflow hydrograph obtained by routing are graphed in Figure 21.3.

EXAMPLE 21.2 Consider a rectangular channel that is 5 meters wide and 1000 m long, with a bed slope of 0.001 and a Manning roughness coefficient of 0.03. The discharges occurring concurrently at the upstream and downstream ends of the channel, respectively, are 10 m^3/s and 6 m^3/s. What is the discharge downstream for an upstream flow of 10 m^3/s? Neglect evaporation, infiltration, seepage, and other losses.

Solution We need to determine c and Δt. Equation (21.7) can be used to determine c by knowing u. By using the Manning equation, the depth of flow y can be obtained for given inflow and outflow. Assume the friction slope to be equal to the bed slope.

$$10 = \frac{(0.001)^{1/2}}{0.03} \frac{(5y)^{5/3}}{(2y + 5)^{2/3}}$$

This yields $y = 1.83$ m and the wetted area $A = 9.15$ m^2. Therefore, $u = 10/9.15 = 1.093$ m/s. Also

$$6 = \frac{(0.001)^{1/2}}{0.03} \frac{(5y)^{5/3}}{(2y + 5)^{2/3}}$$

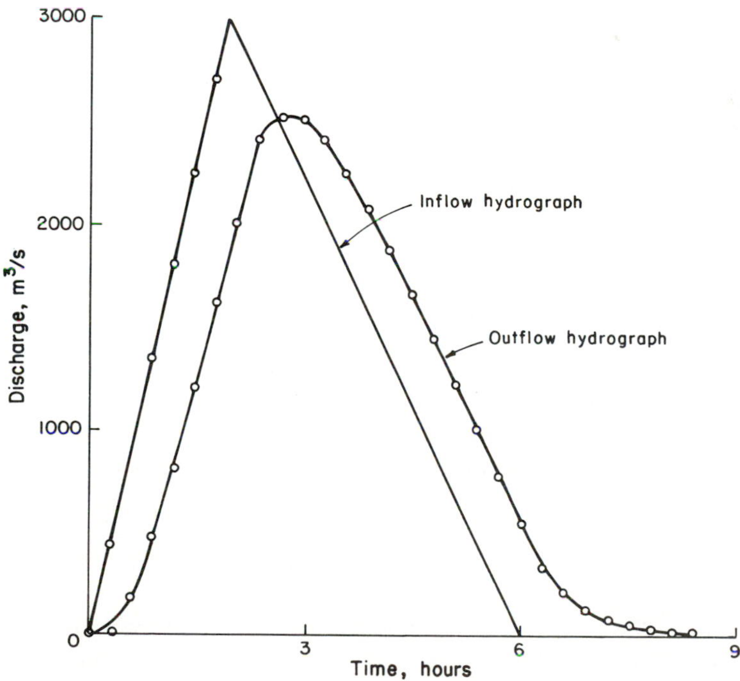

Figure 21.3 Inflow and outflow hydrographs obtained by the convex method.

This yields $y = 1.275$ m, $A = 6.375$ m^2, and $u = 0.941$ m/s. The average velocity $u = (1.093 + 0.941)/2 = 1.017$ m/s. Therefore, Equation (21.7) yields

$$c = \frac{1.017}{1.7 + 1.017} = 0.374$$

Equation (21.5) yields

$$k = \frac{1000}{1.017} = 983.28 \text{ s}$$

Now Equation (21.4) gives

$$\Delta t = ck = 0.374 \times 983.28 = 367.8 \text{ s} = 0.102 \text{ h}$$

Therefore,

$$\begin{aligned} Q_{t+\Delta t} &= 0.374 I_t + (1 - 0.374) Q_t \\ &= 0.374 \times 10 + (1 - 0.374) \times 6 \\ &= 7.5 \text{ m}^3/\text{s} \end{aligned}$$

This is the answer.

21.2 MUSKINGUM METHOD

The Muskingum method is one of the most popular methods of channel-flow routing. This method consists of a spatially lumped form of the continuity equation

$$\frac{dS}{dt} = I - Q \tag{21.11}$$

and a linear storage–discharge relation expressed as

$$S = k[xI + (1 - x)Q] \tag{21.12}$$

where S is the storage or volume of water within the reach, I is the inflow rate, Q is the outflow rate, and k and x are routing parameters. It is assumed that at $t = 0$, $I = Q$. Parameter k is considered as the average reach travel time and has the dimension of time, and is known as the storage time constant. Parameter x is the dimensionless coefficient used to weigh the relative effects of inflow and outflow on reach storage, and is known as a weighting factor. Theoretically, x can vary from 0 to 1.

Equations (21.11) and (21.12) can be combined and then expressed for a short time interval Δt as

$$\left(\frac{I_1 + I_2}{2}\right) \Delta t - \left(\frac{Q_1 + Q_2}{2}\right) \Delta t = k[x(I_2 - I_1) + (1 - x)(Q_2 - Q_1)] \tag{21.13}$$

where Δt is the routing period or discretization time interval. Equation (21.13) is rearranged and solved for Q_2 as

$$Q_2 = C_0 I_2 + C_1 I_1 + C_2 Q_1 \tag{21.14}$$

where

$$C_0 = (-kx + 0.5\,\Delta t)/C_3 \tag{21.15a}$$

$$C_1 = (kx + 0.5\,\Delta t)/C_3 \tag{21.15b}$$

$$C_2 = (k - kx - 0.5\,\Delta t)/C_3 \tag{21.15c}$$

and

$$C_3 = k - kx + 0.5\,\Delta t \tag{21.15d}$$

Coefficients C_0, C_1, and C_2 are such that

$$C_0 + C_1 + C_2 = 1 \tag{21.16}$$

Clearly, coefficients C_0, C_1, and C_2 are appropriate only for a specified Δt. To transform these coefficients to a different time step, x and k can be computed as

$$x = \frac{C_1 - C_0}{2(1 - C_0)} \tag{21.17}$$

$$k = \frac{\Delta t(1 - C_0)}{C_1 + C_2} \tag{21.18}$$

Equation (21.14) is known as the Muskingum routing equation. If x and k are known, then streamflow routing of any inflow hydrograph can be performed. If $x = 0.5$ and $k = \Delta t$, then, from Equations (21.14) and (21.15), it follows that $C_0 = 0$, $C_1 = 1$, and $C_2 = 0$. Therefore,

$$Q_2 = I_1 \tag{21.19}$$

This says that the inflow hydrograph is translated with a time lag of k. The validity of this conclusion has considerably been debated in hydrologic literature (Singh, 1988). Some consider it to be erroneous, whereas some state that it happens because a large value of routing time Δt is chosen and is equated with k, which roughly equals the travel time of the flood wave in the reach under consideration. This explanation is based on the observation that when Δt and k are not the same, C_0 and C_2 are not zero even for $x = 0.5$. Thus,

$$Q_2 \neq I_1 \tag{21.20}$$

The choice of Δt is more or less arbitrary, but must be such that it is much smaller than the travel time of the flood wave. Furthermore, Δt should be such that the linear variation of I and Q is valid. In general, the smaller the Δt, the more accurate should be the result. A heuristic argument is developed in what follows for a reasonable Δt with relation to k and x.

If $\Delta t = 2kx$, then $C_0 = 0$, $C_1 = 2x$, and $C_2 = 1 - 2x$. Therefore,

$$Q_2 = 2xI_1 + (1 - 2x)Q_1 \tag{21.21}$$

This equation parallels Equation (21.1) for the convex method of flow routing. In order for Q_2 to be positive at all times, x must be less than or equal to 0.5. At the same time, it is seen that

$$\Delta t \geq 2kx \tag{21.22}$$

These observations are useful in estimating k and x.

Although Equation (21.14) can be used to perform flood routing numerically, considerable insight is obtained by solving Equations (21.11) and (21.12) analytically, subject to specified initial conditions. Analytical solutions are derived in Appendix B.

21.2.1 Estimation of Parameters

The Muskingum parameters k and x can be easily determined graphically, as discussed here, or by the least squares method, as presented in Appendix C. The graphical method consists in choosing x such that the loop resulting from the plot of S versus $xI + (1 - x)Q$ becomes as close to a straight line as possible. The slope of the straight line fitted through the loop gives k. Obviously, this method, as practiced, results in a trial-and-error procedure. However, if we closely follow the steps involved in the graphical method and the result obtained therefrom, it becomes immediately clear that the least squares method is a numerical expression of the graphical method; hence, the two methods are equivalent. Therefore, the least squares method should constitute a natural replacement for the graphical method.

21.3 MUSKINGUM CREST-SEGMENT ROUTING

Equation (21.14) can be recast such that the right side contains only inflow terms. This suggests that the Muskingum method can also be used to route only a portion of the inflow hydrograph. To this end, we rewrite Equation (21.14) as

$$Q_j = C_0 I_j + C_1 I_{j-1} + C_2 Q_{j-1}, \qquad j = 1, 2, \ldots \tag{21.23}$$

in which 2 is replaced by j and 1 by $j - 1$. The outflow Q_{j-1} can be replaced by the substitution

$$Q_{j-1} = C_0 I_{j-1} + C_1 I_{j-2} + C_2 Q_{j-2} \tag{21.24}$$

resulting in

$$Q_j = C_0 I_j + (C_0 C_2 + C_1) I_{j-1} + C_1 C_2 I_{j-2} + C_2^2 Q_{j-2}$$

By repeated substitutions for the outflow term, Q_{j-2}, Q_{j-3}, Q_{j-4}, . . . , on the right side, Q_j can be expressed solely as a function of inflow as

$$Q_j = \sum_{i=1}^{j} k_i I_{j-i+1} \tag{21.25}$$

where

$$\begin{aligned} k_1 &= C_0 \\ k_2 &= C_0 C_2 + C_1 \\ k_i &= k_{i-1} C_2, \qquad i > 2 \end{aligned} \tag{21.26}$$

TABLE 21.1 COMPUTATION OF WEIGHTED DISCHARGE FOR VARIOUS VALUES OF WEIGHTING COEFFICIENT x

						Weighted discharge (m^3/s)			
Date	I (m^3/s)	Q (m^3/s)	$\bar{I}$ (m^3/s)	$\bar{Q}$ (m^3/s)	S (m^3/s-day)	$x = 0.1$	$x = 0.2$	$x = 0.3$	$x = 0.4$
1	2.63	2.41				2.43	2.45	2.48	2.50
2	3.88	2.89	3.26	2.65	0.61	2.99	3.09	3.19	3.29
3	5.89	3.99	4.89	3.44	2.06	4.18	4.37	4.56	4.75
4	9.06	5.81	7.48	4.9	4.64	6.14	6.46	6.79	7.11
5	12.52	8.21	10.80	7.01	8.43	8.64	9.07	9.50	9.93
6	15.46	10.76	14.0	9.49	12.94	11.23	11.70	12.17	12.64
7	17.84	13.31	16.7	12.64	17.6	13.76	14.22	14.67	15.12
8	19.20	15.26	18.52	14.29	21.83	15.65	16.05	16.44	16.84
9	19.52	16.74	19.36	16.0	25.2	17.02	17.30	17.57	17.85
10	19.60	17.75	19.56	17.25	27.5	17.93	18.12	18.30	18.49
11	19.37	18.35	19.48	18.05	28.93	18.45	18.58	18.66	18.76
12	19.00	18.69	19.18	18.52	29.6	18.72	18.75	18.78	18.81
13	18.60	18.8	18.8	18.74	29.66	18.78	18.76	18.74	18.72
14	18.07	18.69	18.33	18.74	29.25	18.63	18.57	18.50	18.44
15	17.24	18.41	17.65	18.55	28.35	18.29	18.18	18.05	17.94
16	16.34	17.98	16.8	18.20	26.95	17.82	17.65	17.49	17.32
17	15.12	17.27	15.73	17.63	25.03	17.06	16.84	16.63	16.41
18	13.71	16.42	14.42	16.85	22.62	16.15	15.88	15.61	15.34
19	12.06	15.29	12.89	15.86	19.65	14.97	14.97	14.32	14.00
20	10.36	13.82	11.12	14.56	16.3	13.47	13.13	12.78	12.44
21	8.44	12.18	9.4	13.0	12.7	11.81	11.43	11.06	10.68
22	6.65	10.33	7.55	11.26	9.0	9.96	9.57	9.23	8.86
23	5.18	8.5	5.92	9.42	5.5	8.17	7.84	7.50	7.17
24	3.88	6.6	4.53	7.55	2.48	6.33	6.06	5.78	5.51
25	2.92	5.04	3.4	5.82	0.06	4.83	4.62	4.40	4.19
26	2.29	3.74	2.61	4.39	−1.72	3.60	3.45	3.31	3.16
27	2.12	2.83	2.21	3.29	−2.8	2.76	2.69	2.62	2.55

Equation (21.25) can be viewed as a discrete convolution of I and k, with k being considered as the Muskingum IUH or UH. Thus, $k_1, k_2, \ldots,$ represent ordinates of the IUH.

EXAMPLE 21.3 The inflow–outflow data are given for a river reach:

TIME (days)	INFLOW (m^3/s)	OUTFLOW (m^3/s)	TIME (days)	INFLOW (m^3/s)	OUTFLOW (m^3/s)
1	2.63	2.41	15	17.24	18.41
2	3.88	2.89	16	16.34	17.98
3	5.89	3.99	17	15.12	17.27
4	9.06	5.81	18	13.71	16.42
5	12.52	8.21	19	12.06	15.29
6	15.46	10.76	20	10.36	13.82
7	17.84	13.31	21	8.44	12.18
8	19.20	15.26	22	6.65	10.33
9	19.52	16.74	23	5.18	8.50
10	19.60	17.75	24	3.88	6.60
11	19.37	18.35	25	2.92	5.04
12	19.00	18.69	26	2.29	3.74
13	18.60	18.80	27	2.12	2.83
14	18.07	18.69			

Compute the Muskingum parameters x and k. Also compute the Muskingum routing coefficients, C_0, C_1, and C_2.

Solution First, we compute for the various times the weighted observed discharge, $xI + (1 - x)Q$ for the selected values of $x = 0.1, 0.2, 0.3,$ and 0.4. We also compute the corresponding values of observed storage. Results of the calculations are shown in Table 21.1. The weighted discharge is plotted against storage, as shown in Figures 21.4 to 21.7. The plot has the appearance of a loop. The plot corresponding to $x = 0.2$ most closely approximates a straight line. Therefore, the weighting coefficient $x = 0.2$. From

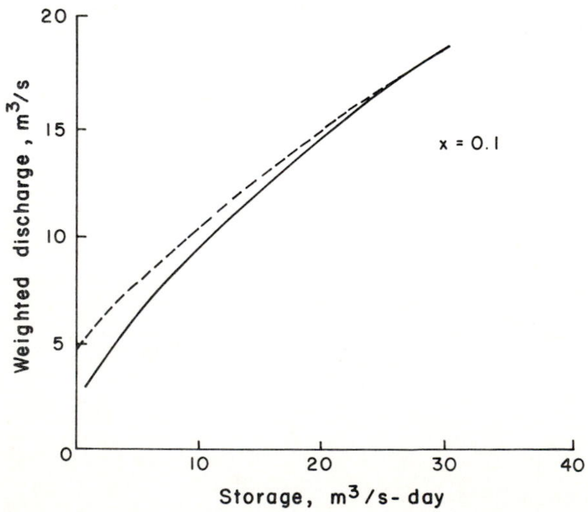

Figure 21.4 Weighted discharge versus storage for $x = 0.1$.

the slope of the straight line representing the loop for $x = 0.2$, $k = 1.82$ days. Therefore,

$$x = 0.2 \qquad k = 1.82 \text{ days}$$

$$C_0 = \frac{-kx + 0.5\,\Delta t}{k - kx + 0.5\,\Delta t} = 0.0695$$

$$C_1 = \frac{kx + 0.5\,\Delta t}{k - kx + 0.5\,\Delta t} = 0.4417$$

$$C_2 = \frac{k - kx - 0.5\,\Delta t}{k - kx + 0.5\,\Delta t} = 0.4888$$

EXAMPLE 21.4 Consider a river reach having the following characteristics: $x = 0.2$, $k = 2$ days, and the time interval = 1 day. Compute the Muskingum coefficients. Route the following flood hydrograph through the reach:

TIME (days)	DISCHARGE (m^3/s)	TIME (days)	DISCHARGE (m^3/s)
1	120	12	290
2	180	13	260
3	260	14	240
4	414	15	224
5	480	16	210
6	560	17	198
7	604	18	186
8	580	19	174
9	460	20	164
10	404	21	154
11	320	22	146

Plot the inflow and outflow hydrographs on the same sheet, and determine the peak attenuation and peak time delay for this flood.

Solution

$$C_0 = \frac{-kx + \Delta t/2}{k - kx + \Delta t/2}$$

$$= \frac{-\,2(0.2) + 0.5}{2 - 2(0.2) + 0.5} = 0.0476$$

$$C_1 = \frac{kx + \Delta t/2}{k - kx + \Delta t/2}$$

$$= \frac{2(0.2) + 0.5}{2 - 2(0.2) + 0.5} = 0.4286$$

$$C_2 = \frac{k - kx - \Delta t/2}{k - kx + \Delta t/2}$$

$$= \frac{2 - 2(0.2) - 0.5}{2 - 2(0.2) + 0.5} = 0.5238$$

$$Q_2 = C_0 I_2 + C_1 I_1 + C_2 Q_1$$

$$\Delta t = 1 \text{ day} > 2kx$$

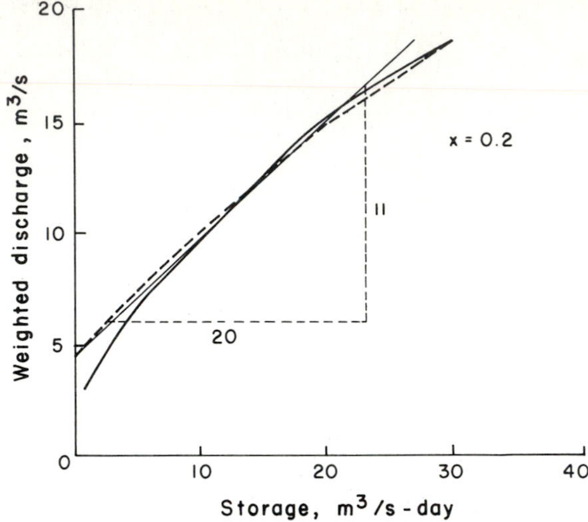

Figure 21.5 Weighted discharge versus storage for $x = 0.2$.

The results of the flow-routing calculations are presented in Table 21.2.

$$\text{Peak attenuation} = 604 - 551.1 = 52.9 \text{ m}^3/\text{s}$$

$$\text{Peak delay time} = 2 \text{ days}$$

The inflow and outflow hydrographs are shown in Figure 21.8.

EXAMPLE 21.5 Compute the outflow on the eighth day for the river reach in using the Muskingum crest-segment routing. The inflow hydrograph and the Muskingum parameters are given in Example 21.4.

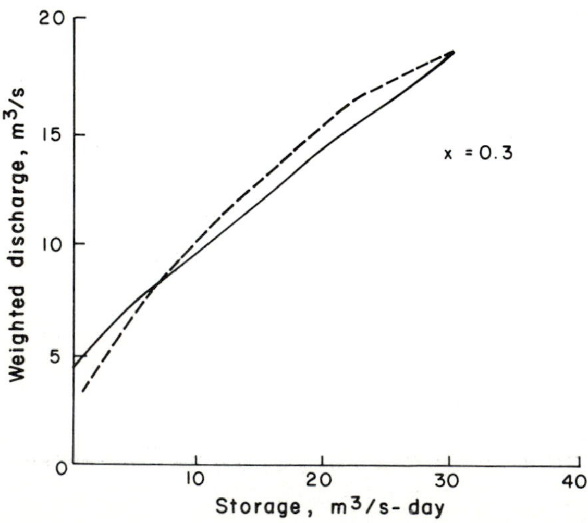

Figure 21.6 Weighted discharge versus storage for $x = 0.3$.

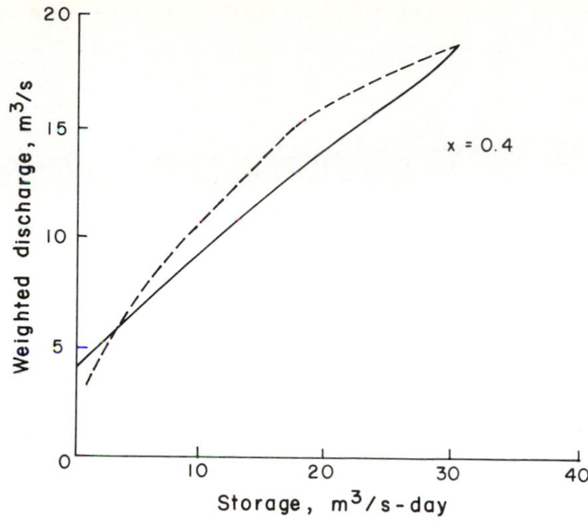

Figure 21.7 Weighted discharge versus storage for $x = 0.4$.

TABLE 21.2 FLOW ROUTING BY THE MUSKINGUM METHOD

Time (days)	Input (m^3/s)	C_0I_2 (m^3/s)	C_1I_1 (m^3/s)	C_2Q_1 (m^3/s)	Computed outflow (m^3/s)
1	120				120
2	180	8.57	51.43	62.86	122.9
3	260	12.38	77.15	64.35	153.9
4	414	19.70	111.44	80.60	211.7
5	480	22.85	177.44	110.91	311.2
6	560	26.70	205.73	163	395.4
7	604	28.75	240	207.13	475.9
8	580	27.60	258.9	249.27	535.8
9	460	21.90	248.6	280.63	551.1
10	404	19.23	197.16	288.68	505.1
11	320	15.23	173.15	264.56	452.9
12	290	13.80	137.15	237.25	388.2
13	260	12.38	124.3	203.34	340.0
14	240	11.40	111.44	178.1	300.9
15	224	10.7	102.86	157.63	271.2
16	210	10.0	96	142.05	248.1
17	198	9.4	90	130	229.4
18	186	8.85	84.86	120.12	213.8
19	174	8.28	79.72	112	200
20	164	7.8	74.6	104.76	187.2
21	154	7.33	70.3	98	175.6
22	146	6.95	66	92	165

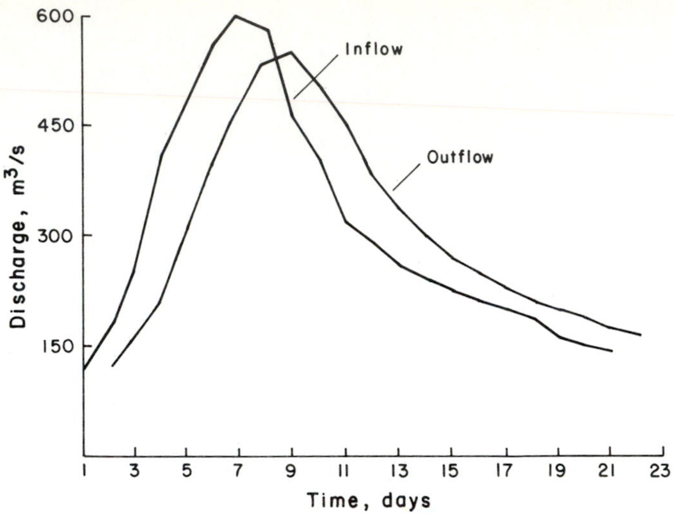

Figure 21.8 Inflow and outflow hydrographs.

Solution

$$C_0 = 0.0476$$
$$C_1 = 0.4286$$
$$C_2 = 0.5238$$
$$k_1 = C_0 = 0.0476$$
$$k_2 = C_0C_2 + C_1 = 0.4535$$
$$k_3 = k_2C_2 = 0.2375$$
$$k_4 = k_3C_2 = 0.1244$$
$$k_5 = k_4C_2 = 0.0652$$
$$k_6 = k_5C_2 = 0.0342$$
$$k_7 = k_6C_2 = 0.0179$$
$$k_8 = k_7C_2 = 0.0094$$

$$\begin{aligned} Q_8 &= k_1I_8 + k_2I_7 + k_3I_6 + k_4I_5 + k_5I_4 + k_6I_3 + k_7I_2 + k_8I_1 \\ &= (0.0476 \times 580) + (0.4535 \times 604) + (0.2375 \times 560) \\ &\quad + (0.1244 \times 480) + (0.0652 \times 414) + (0.0342 \times 260) \\ &\quad + (0.0179 \times 180) + (0.0094 \times 120) \\ &= 534.46 \text{ m}^3/\text{s} \end{aligned}$$

EXAMPLE 21.6 Given in Table 21.3 are inflow and outflow hydrographs for a section of river. Plot the inflow and outflow hydrographs and determine the best values of k and

TABLE 21.3 INFLOW AND OUTFLOW HYDROGRAPHS

Date	Hour	I (cfs)	Average inflow, $\bar{I}$ (cfs)	Outflow, Q (cfs)	Average outflow, $\bar{Q}$ (cfs)	$\Delta S/\Delta t = \bar{I} - \bar{Q}$ (cms)	$S/\Delta t$
9–23–66	6.0	1,000		1000			
			2,500		1150	2,500–1,150	1,350
	12.0	4,000		1300			
			6,775		1400	6,775–1,400	6,725
	18.0	9,550		1500			
			11,635		2300	11,635–2,300	16,060
	24.0	13,720		3100			
9–24–66	6.0	12,450		6020			
			11,325		6960	11,325–6,960	20,425
	12.0	10,200		7900			
			9,200		8350	9,200–8,350	21,275
	18.0	8,200		8800			
			7,350		8750	7,350–8,750	19,875
	24.0	6,500		8700			
9–25–66	6.0	5,500		8200			
			5,100		7850	5,100–7,850	17,125
	12.0	4,700		7500			
			4,400		7125	4,400–7,125	14,400
	18.0	4,100		6750			
			3,850		6450	3,850–6,450	11,800
	24.0	3,600		6150			
9–26–66	6.0	3,100		5800			
			2,900		5450	2,900–5,450	9,250
	12.0	2,700		5100			
			2,550		4800	2,550–4,800	7,000
	18.0	2,400		4500			
			2,250		4200	2,250–4,200	5,050
	24.0	2,100		3900			

x for the Muskingum routing method. Calculate the channel storage at the 6-h interval.

Solution First, the inflow and outflow hydrographs are plotted as shown in Figure 21.9. At the point of intersection of I and Q, we compute dI/dt and dQ/dt. The value of x is obtained as:

$$x = \frac{dQ/dt}{dI/dt + dQ/dt} = \frac{1}{1.5 + 1} = \frac{1}{2.5} = 0.4$$

Corresponding to this value of x, the value of $x\bar{I} + (1 - x)\bar{Q}$ is obtained as given in Table 21.4. Storage is then plotted against $x\bar{I} + (1 - x)\bar{Q}$, as shown in Figure 21.10. Therefore,

$$k = \frac{2.85 \times 2 \times 10^8}{5.7 \times 1000 \times 60 \times 60 \times 24} = 1.16 \text{ days}$$

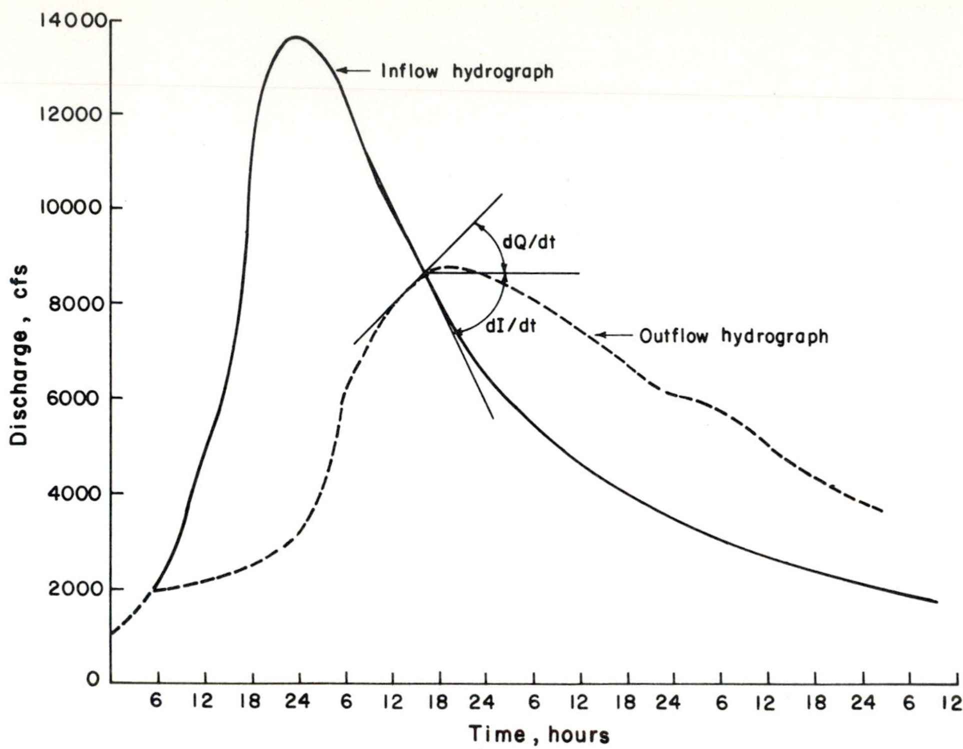

Figure 21.9 Inflow and outflow hydrographs.

TABLE 21.4 RESULT OF THE COMPUTATION OF $x\bar{I} + (1 - x)\bar{Q}$

S.N.	$\bar{I}$ (cfs)	$\bar{Q}$ (cfs)	$x\bar{I}$ (cfs)	$(1 - x)\bar{Q}$ (cfs)	$x\bar{I} + (1 - x)\bar{Q}$ (cfs)	S for 6 h (ft³)
1	2,500	1150	1000	690	1690	29,160,000
2	6,775	1140	2710	840	3550	145,260,000
3	11,635	2300	4654	1380	6034	346,896,000
4	11,325	6960	4530	4176	8706	441,180,000
5	9,200	8350	3680	5010	8690	459,540,000
6	7,350	8750	2840	5250	8090	429,300,000
7	5,100	7850	2040	4710	6750	369,900,000
8	4,400	7125	1760	4275	6035	311,040,000
9	3,850	6450	1540	3870	5410	254,880,000
10	2,900	5450	1160	3370	4530	199,800,000
11	2,550	4800	1020	2830	3900	151,200,000
12	2,250	4200	900	2520	3420	109,080,000

The Muskingum constants, C_0, C_1, and C_2, are

$$C_0 = -\frac{1.16 \times .4 - 0.5 \times 0.25}{1.16 - 1.16 \times 0.4 + 0.5 \times 0.25}$$

$$= -\frac{0.339}{1.285 - 0.464} = -\frac{0.339}{0.821} = -0.412$$

$$C_1 = \frac{1.16 \times 0.4 + 0.5 \times 0.25}{1.16 - 1.16 \times 0.4 + 0.5 \times 0.25}$$

$$= \frac{0.589}{0.821} = 0.717$$

$$C_2 = \frac{1.16 - 1.16 \times 0.4 - 0.5 \times 0.25}{1.16 - 1.16 \times 0.4 + 0.5 \times 0.25}$$

$$= \frac{0.571}{0.821} = 0.695$$

$$C_0 + C_1 + C_2 = 1$$

$$-0.412 + 0.717 + 0.695 = 1$$

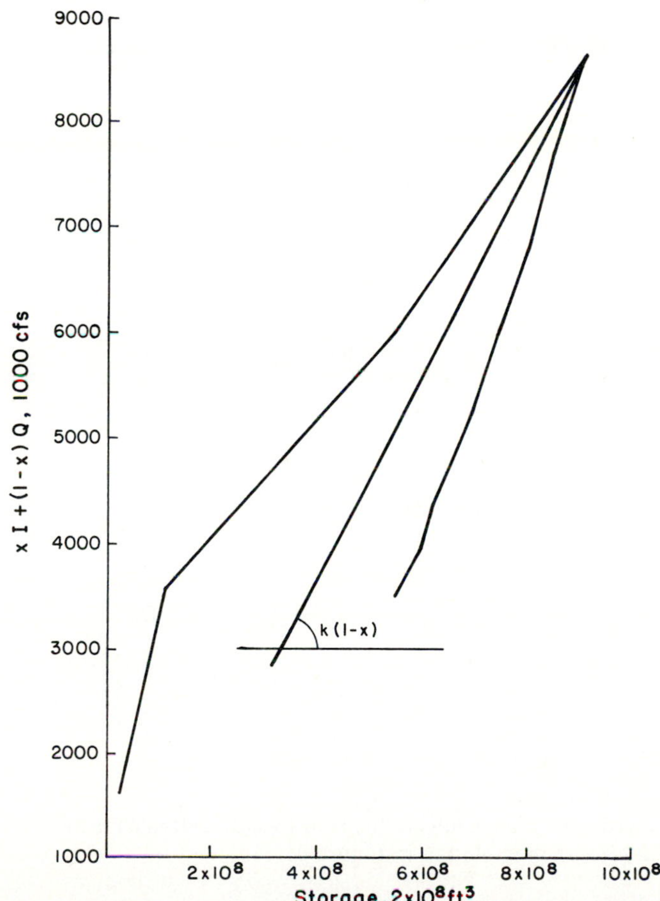

Figure 21.10 $x\bar{I} + (1 - x)\bar{Q}$ versus storage.

TABLE 21.5 INFLOW HYDROGRAPH

Date	Hours	Inflow (cfs)
4–18–43	24.00	1,500
4–19–43	6.00	1,500
	12.00	1,600
	18.00	2,000
	24.00	3,100
4–20–43	6.00	5,200
	10.00	7,400
	12.00	9,600
	14.00	12,000
	18.00	12,450
	24.00	12,100
4–21–43	6.00	11,100
	12.00	10,200
	18.00	9,300
	24.00	8,500
4–22–43	12.00	7,000
	24.00	5,700
4–23–43	12.00	7,000
	24.00	3,900
4–24–43	12.00	3,200
	24.00	2,800
4–25–43	12.00	2,400
	24.00	2,200
4–26–43	12.00	2,000
	24.00	1,900
4–27–43	24.00	1,700
4–28–43	24.00	1,800
4–29–43	24.00	1,400

EXAMPLE 21.7 Compute the outflow hydrograph using the Muskingum method if $k = 1.5$ days and $x = 0.15$. The inflow hydrograph is given in Table 21.5.

Solution We first determine the Muskingum constants.

$$C_0 = \frac{kx - 0.5\,\Delta t}{k - kx + 0.5\,\Delta t} = \frac{1.5 \times 0.15 - 0.5 \times 0.25}{1.5 - 1.5 \times 0.15 + 0.5 \times 0.25} = -0.072$$

$$C_1 = \frac{kx + 0.5\,\Delta t}{k - kx + 0.5\,\Delta t} = +\,0.25$$

$$C_3 = \frac{k - kx - 0.5\,\Delta t}{k - kx + 0.5\,\Delta t} = 0.822$$

The inflow hydrograph is then routed. The results of the routing calculations are shown in Table 21.6. The outflow hydrograph is plotted in Figure 21.11.

TABLE 21.6 FLOW ROUTING

Date	Days	Hours	Inflow	C_0I_2	C_1I_1	C_2Q_1	Q_2
4–18–43	0	24.00	1,500	—	—	—	1500
4–19–43	0.25	6.00	1,500	−108	375	1233	1500
	0.5	12.00	1,600	−115	375	1233	1493
	0.75	18.00	2,000	−144	400	1228	1484
	1.00	24.00	3,100	−223	500	1220	1497
4–20–43	1.25	6.00	5,200	−374	775	1230	1631
		10.00	7,400	−534	1300	1340	2106
	1.5	12.00	9,600	−691	1850	1735	2894
		14.00	12,000	−865	2400	2380	3915
	1.75	18.00	12,450	−895	3000	3220	5325
	2.0	24.00	12,100	−871	3112.5	4370	6611.5
4–21–43	2.25	6.00	11,100	−800	3025	5450	7675
	2.5	12.00	10,200	−735	2775	6300	8230
	2.75	18.00	9,300	−670	2550	6770	8650
	3	24.00	8,500	−612	2325	7100	8813
4–22–43	3.5	12.00	7,000	−506	2125	7250	8870
	4	24.00	5,700	−416	1750	7300	8634
4–23–43	4.5	12.00	7,000	−505	1425	7100	8620
	5	24.00	3,900	−280	1750	6700	8170
4–24–43	5.5	12.00	3,200	−230	975	6720	7365
	6.0	24.00	2,800	−200	800	6050	6650
4–25–43	6.5	12.00	2,400	−172.5	700	5470	5997.5
	7.0	24.00	2,200	−158.5	600	4920	5361.5
4–26–43	7.5	12.00	2,000	−144	550	4410	4916
	8.0	24.00	1,900	−136.5	500	4050	4413
4–27–43	9.0	24.00	1,700	−122.5	475	3610	3962
4–28–43	10.0	24.00	1,500	−108	425	3250	3567
4–29–43	11.0	24.00	1,400	−101	375	2930	3204

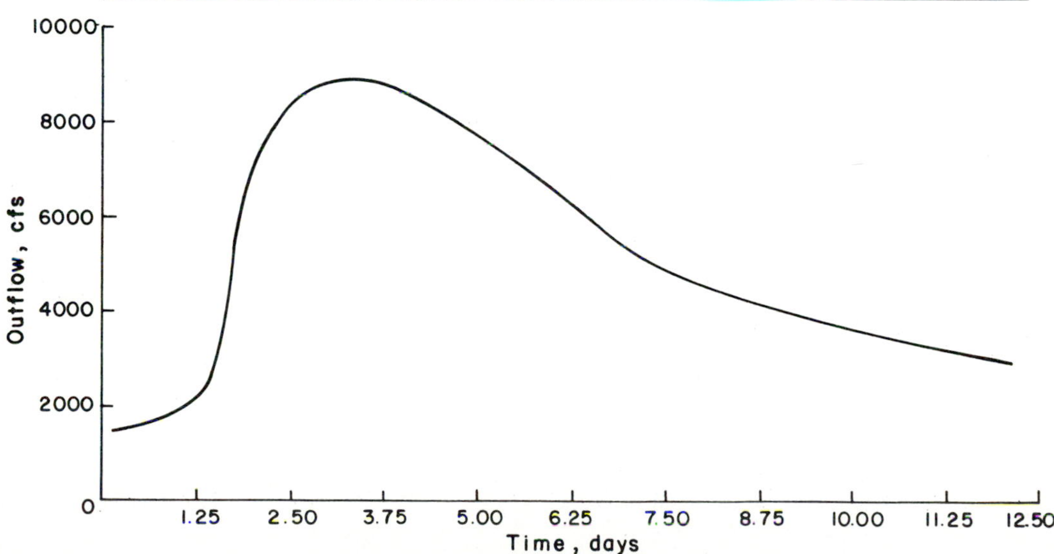

Figure 21.11 The outflow hydrograph.

EXERCISES

21.1. A flood wave has the following inflow hydrograph:

Time (h)	6	12	18	24	6	12
Inflow (cfs)	1495	1785	933	568	398	391

How is the flood wave modified in a natural channel if k, the storage constant of channel, is 0.75 day and the inflow affects storage by 25%.

21.2. Let S be related to outflow Q as

$$S = kQ$$

where $k = 1$ hour. The inflow in a stream is given by a rectangle whose base is 5 hours and height 100 m^3/s. Compute the outflow by routing this inflow.

21.3. If the Muskingum k value is 15 hours and x is 0.25, what would be a reasonable value of Δt for purposes of flood routing?

21.4. If $C_1 = 0.3$ and $C_2 = 0.3$, then determine C_3.

21.5. The Muskingum method of flood routing can be written as

$$Q_n = C_0 I_n + C_1 I_{n-1} + C_2 Q_{n-1}$$

Assume $C_0 = 0.0477$, $C_1 = 0.428$, and $C_2 = 0.524$. The following inflow is given:

DATE	INFLOW (cfs)
10	100
11	200
12	300
13	400
14	500
15	700
16	1000

Compute the outflow on the dates 14, 15, and 16 directly.

21.6. The following information is given on inflow and outflow hydrographs:

ROUTING PERIOD (days)	INFLOW (cfs)	OUTFLOW (cfs)
1	93	85
2	137	103
3	208	143
4	320	206
5	442	288
6	546	373
7	630	456
8	678	527
9	691	582
10	692	621

ROUTING PERIOD (days)	INFLOW (cfs)	OUTFLOW (cfs)
11	684	644
12	671	656
13	657	658
14	638	655
15	609	645
16	577	626
17	534	602
18	484	569
19	426	529
20	366	482
21	298	427
22	235	364
23	183	299
24	137	237
25	163	179
26	81	131

Compute parameters k and x of the Muskingum method of flood routing.

21.7. The following information is available on the inflow hydrograph:

ROUTING PERIOD (days)	INFLOW (cfs)
1	90
2	140
3	210
4	320
5	440
6	550
7	630
8	680
9	690
10	690
11	680
12	670
13	660
14	640
15	610
16	580
17	530
18	480
19	430
20	370
21	300
22	240
23	180

Using the values of x and k determined in Exercise 21.6, perform the flood routing by the Muskingum method.

21.8. Tabulated are inflow and outflow hydrographs for a river reach. Determine the storage in the reach and plot a curve showing storage at any instant as a function of simultaneous outflow (M = midnight; N = noon).

DATE	HOUR	INFLOW (m^3/s)	OUTFLOW (m^3/s)
1	M	1.05	1.05
2	N	3.54	1.47
	M	9.63	3.68
3	N	16.28	8.13
	M	20.45	13.37
4	N	20.96	17.67
	M	19.06	19.14
5	N	12.91	18.07
	M	9.06	16.25
6	N	6.94	11.16
	M	5.44	8.69
7	N	4.08	6.66
	M	3.34	5.10
8	N	2.69	4.02
	M	2.27	3.23
9	N	1.90	2.63
	M	1.59	2.18
10	N	1.42	1.81
	M	1.19	1.56

21.9. Determine parameters k and x of the Muskingum method of flood routing using data in Exercise 21.8. Use the slopes of Q and I for estimation of x.

21.10. Taking the parameters of Exercise 21.9, route the inflow hydrograph of Exercise 21.8 by the Muskingum method and compare the routing hydrograph with the outflow hydrograph of Exercise 21.8. How much is the error in the hydrograph peak and its time?

21.11. The inflows to a river reach are given for three days:

DATE	INFLOW (m^3/s)
1	500
2	800
3	1200

Compute the outflow from the reach on the third day using the Muskingum method. The Muskingum coefficients are $C_0 = 0.05$, $C_1 = 0.4$, and $C_2 = 0.55$. Assume the outflow on the first day is 500 m^3/s.

21.12. The following hydrographs apply to a certain reach of a river.

DATE	HOUR	INFLOW (cfs)	OUTFLOW (cfs)
May 10	1200	15	15
	1500	35	30
	1800	63	42
	2100	54	56
	2400	42	45
May 11	300	36	40

(a) Determine the maximum volume of storage in the reach.

(b) Find the peak and time of peak downstream if $k = 4$ h and $x = 0.2$ for a reach downstream of this outflow hydrograph. The initial baseflow in the entire stream is 15 cfs.

21.13. The inflow hydrograph to a river reach is as follows:

TIME (h)	FLOW (cfs)
0	1000
12	7500
24	7800
36	4700
48	2700

Route this flood down the river reach assuming $Q = 1000$ cfs at both upper and lower ends of the reach at time zero. Consider the case when $k = 18$ h and $x = 0.3$. Use a routing period of 12 hours.

21.14. Precipitation began at noon on June 14 and caused a flood hydrograph in a stream. As the hydrograph passed, the following measured streamflow data at cross-sections A and B were obtained.

TIME (JUNE 14–17)	SECTION A INFLOW (cfs)	SECTION B OUTFLOW (cfs)
6 A.M.	10	10
N	10	10
6 P.M.	30	13
M	70	26
6 A.M.	50	43
N	40	45
6 P.M.	30	41
M	20	35
6 A.M.	10	28
N	10	19
6 P.M.	10	15
M	10	13
6 A.M.	10	11
N	10	10

(a) Determine the Muskingum k if $x = 0.2$ for the river reach.

(b) Determine the hydrograph at section B if a different storm produced the following hydrograph at section A.

TIME	SECTION A INFLOW (cfs)
6 A.M.	100
N	100
6 P.M.	200
M	500
6 A.M.	600
N	400
6 P.M.	300
M	200
6 A.M.	100
N	100

PART 8
Watershed Simulation

CHAPTER 22
Erosion and Sediment Yield

Estimates of watershed sediment yield are required for solution of a number of problems. Design of dams and reservoirs; transport of pollutants; design of soil-conservation practices; design of stable channels; design of debris basins; depletion of reservoirs, lakes, and wetlands; determination of the effects of basin management; off-site damage evaluation; and cost evaluation of a water-resources project are some of the example problems. Sediment is a pollutant or a carrier of pollutants such as radioactive material, pesticides, and nutrients. Sediment has been called the greatest single pollutant of our nation's streams (Ritchie, 1972) and will have to be controlled if we are to attain the water-quality goals set forth in the Federal Water Pollution Control Act Amendments of 1972, PL 92-500. Section 208 of PL 92-500 has directed states to develop ''Area-wide Water Quality Management Plans'' to achieve these water-quality goals. These plans will have to include the control of sediment pollution from urban, agricultural, and forest activities.

22.1 DEFINITIONS AND SYMBOLS

22.1.1 Sediment Yield

American Society of Civil Engineers (1970) defines sediment yield as the total sediment outflow from a watershed or drainage basin, measurable at a point of reference and in a specified period of time.

22.1.2 Erosion

Ellison (1946) has defined soil erosion as a process of detachment and transportation of soil materials by erosive agents. For erosion by water, these agents are rainfall and runoff.

22.1.3 Soil Erodibility

Some soils erode more readily than others, even when everything else remains the same. This difference is caused by the properties of the soil itself. Or it can be ascribed to soil erodibility, which is the ease with which a soil erodes or the natural susceptibility of a soil to erosion.

22.1.4 Interrill Erosion

This involves a relatively uniform removal of soil from the land surface between the rills. The primary force for this erosion is the raindrop impact, whose erosive potential depends on raindrop size, distribution, fall velocities, and total mass of impact. Sheet erosion is approximately the same as interrill erosion. Sometimes, small rills, caused by localized concentration of flow, are included in sheet erosion.

22.1.5 Rill Erosion

This involves soil detachment principally by concentrated flow of water. It may cause intensive soil movement from a limited part of the land surface. Rills carry runoff from interrill areas as well as the rain falling directly on them. Rills may develop due to topographic variations, tillage operations, or irregularities on the surface, but are easily smoothed by agricultural practices.

22.1.6 Gully Erosion

Gullies often start as rills and enlarge until they become permanent topographic features. These may be so large as to prevent crossing by vehicles such as trucks or tractors, and may yield tremendous volumes of sediment. Gully erosion is the massive removal of soil by large concentrations of runoff.

22.1.7 Channel Erosion

Channel erosion includes gully erosion, streambank erosion, valley trenching, degradation, and floodplain scour. In many watersheds, these are more significant sources of sediment than sheet erosion.

22.1.8 Gross Erosion

Gross erosion is the summation of erosion from all sources within the watershed. It includes interrill and rill erosion from tilled cropland, meadows, pastures, woodlands, construction sites, industrial areas, abandoned areas, and surface-mined areas and mine wastes; gully erosion from all sources; and erosion from streambeds and streambanks, road and roadside erosion, and floodplain scour. The relative importance of each of these sources varies from one watershed to another.

22.1.9 Sediment Delivery Ratio

The ratio of sediment delivered at a given location in the stream system to the gross erosion from the drainage area above that location is the sediment delivery ratio for that drainage area.

22.1.10 Bed Load

The part of the sediment load composed of the relatively coarse material (coarser than the division grain size that moves along the streambed). The movement of bed load is executed by rolling, sliding along the bed, or saltation (bouncing along the bed or moving by the impact of bouncing particles) of bed particles by the action of the moving water at a rate related to stream discharge. The sediment may move a short distance during a given runoff event.

22.1.11 Suspended Load

It is the relatively fine part of the sediment load that is distributed throughout the flow cross-section and stays in suspension for appreciable lengths of time. The particle suspension is the result of vertical velocity fluctuations characteristic of turbulent flow.

22.1.12 Bed-Material Load

It is that part of the sediment load, moving along the bed or in suspension, that principally constitutes the bed. It may be transported as bed load or suspended load. It is composed of relatively coarse material and is generally the product of channel-type erosion.

22.1.13 Wash Load

It is that part of the sediment load that is not a significant part of the bed. The sediment is washed through the channel in nearly continuous suspension. It is normally assumed to move through the system in a single runoff event. It consists of the fine suspended particles and is usually uniformly distributed in a flow cross-section. The main source is sheet erosion, although channel erosion may also produce wash load. Road cuts and roadside ditches may constitute an additional source of wash load. Many workers substitute fine-sediment load for wash load. Woo et al. (1986) argue and recommend three criteria for distinguishing between these two types of loads. According to them, fine-sediment load is defined as the load of silts and clays, which have diameters smaller than 0.0625 mm. This load includes the electrochemically interacting clay particles that affect the fluid properties and settling velocity of larger particles.

22.1.14 Total Sediment Load

This consists of sediment moving as suspended load as well as that moving as bed load. Gyr (1983) has discussed criteria for separating the sediment transport of rivers into bed load, wash load, and suspended load.

22.2 SEDIMENT-YIELD PROCESS

The sediment-yield process involves detachment of soil particles (primary and aggregated) from the soil surface, transportation of these particles downslope, and their eventual deposition.

22.2.1 Factors Affecting Sediment Yield

The sediment-yield process may be considered to consist of two phases: (1) the upland phase and (2) the lowland stream or simply the channel phase (Bennett, 1974). The upland phase occurs on an upland area, which is an area within a watershed where runoff is predominantly overland flow (Foster and Meyer, 1975). As shown in Figure 22.1, an upland area includes both hillsides and bottomlands. The soil particles of major interest are in the silt and clay ranges. Rainfall characteristics play a major role in determining the sediment yield in the upland phase (Park et al., 1982). Major factors affecting the yield in this phase are (1) soil characteristics, (2) climate, (3) vegetation, (4) topography, and (5) human activities. The major soil properties include texture, structure, permeability, compactness, and infiltration capacity. Permeability and infiltration capacity influence runoff. The remaining factors determine the behavior of soil when subjected to raindrop impact and to the force of running water. Rainfall characteristics, temperature, and snow are the primary climatic factors. Rainfall intensity, amount, duration, and space–time distribution are perhaps the most significant factors in most watersheds, but temperature may also be significant where freeze and thaw cycles occur frequently. Vegetal cover, density, areal extent, and height are important characteristics (Hartley, 1984; Heede, 1984; Reid and Dunne, 1984; Loughran et al., 1986). These dissipate some of the raindrop energy, retard runoff water, influence permeability through root effects, and influence soil moisture through transpiration. The major topographic character-

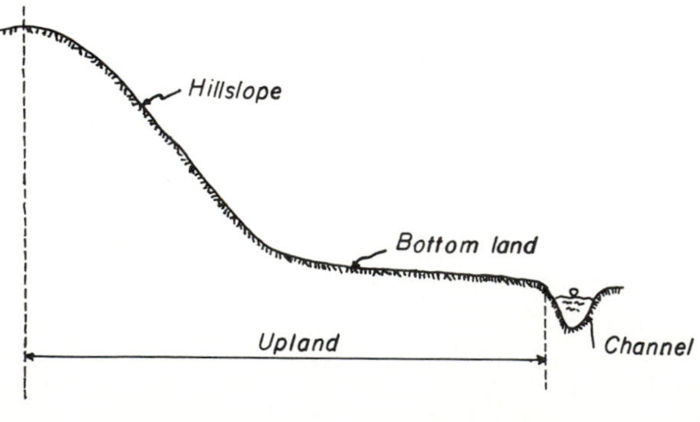

Figure 22.1 Definition sketch of an upland area.

istics are slope, slope length, slope configuration, and surficial features (Mutchler and Greer, 1980; Mutchler and McGregor, 1983). Land use, construction practices, and agricultural operations are some of the human influences on sediment yield.

The channel phase receives sediment from the upland phase. By a channel, we mean a well-defined watercourse flowing through a valley in which the material composing the valley has been deposited by the stream in the recent geologic past. Flows in the channels may be ephemeral or perennial. These channels are characterized by rapidly varying cross-sectional shape in the direction of flow and are usually meandering. Rainfall has little influence on the sediment yield. The velocity, depth of flow, channel slope, wash load, water temperature, mean fall diameter, grain-size distribution, hydraulic roughness, bed deformation, cross-sectional area, discharge, density, and viscosity of sediment—water mixture and seepage force on the channel bed—are some of the variables affecting sediment yield in the channel phase. Furthermore, some of these variables are interdependent; the details of their interdependence are not fully understood, but have been investigated for wide ranges of most of the important variables. The channel flow can usually transport all of the fine material (≤ 0.062 mm) supplied by the upland erosion.

22.2.2 Stages of Upland Erosion

Upland erosion by water may occur in three stages that follow one another in time and to some extent in space. The first stage is sheet erosion, which may be idealized as the removal by any means of a sheet of sediment of uniform thickness over an entire area. The second stage is rill erosion, which is the development of small channels of runoff concentration. The rills form due to natural areal variations in the resistance to erosion of the soil, and small variations in elevation and slope. These are easily obliterated by normal agricultural tillage practices. The third stage is gully erosion. The flows from rills concentrate in gullies, which are relatively permanent topographic features. The influence of rainfall on sediment detachment and transport in gullies is negligible. These stages are called the sources of erosion.

Meyer et al. (1975) divide, according to the source of eroded sediment, the upland erosion into two stages: (1) interrill erosion and (2) rill erosion. Runoff on erodible soil surfaces concentrates in many small, definable flow concentrations (Foster, 1971). Erosion occurring on the areas between the rills is interrill erosion. This classification is analogous to that of watershed runoff into overland flow and channel flow.

22.2.3 Mechanics of Upland Erosion

Based on the definition of soil erosion (Ellison, 1947a, 1947b, 1947c, 1947d, 1947e), sediment removal and transport are executed by four processes (Meyer and Wischmeier, 1969): (1) detachment by raindrop impact, (2) transport by raindrop splash, (3) detachment by runoff, and (4) transport by runoff. Although all of these processes do not occur on all source areas, each plays a part in the total erosion process. In each stage of upland erosion, a different mode of sediment removal and transport predominates. For example, the raindrop impact is usually the dominant factor influencing erosion rate on interrill areas (Meyer et al., 1975). Runoff splash accounts for only a small portion of total soil movement. Runoff is dominant in determining the rate of rill erosion. An excellent discussion of soil erosion and sedimentation by water is given by Foster and Meyer (1977).

22.3 SEDIMENT YIELD MODELING

A number of sediment yield models have been developed to address wide-ranging soil and water-resources problems. On the basis of soil and water problems to be solved, Williams (1981) classified the models into three broad categories: (1) erosion-control planning, (2) water-resources planning and design, and (3) water-quality modeling. The complexity of a model is usually dictated by the nature of the problem it is intended to solve.

Erosion-control planning for agricultural fields, construction sites, reclaimed mines, and forest management requires perhaps the simplest models. The only estimate needed in such applications is the average annual soil loss for various erosion-control systems. Annual time scale, small areas, low cost, short project span, and little risk of failure are some of the considerations allowing for model simplicity. On the other hand, sediment yield estimates required for designing structures ranging from temporary sediment basins at construction sites to large dams, and for evaluating the effects of hydraulic works on floodplain and channel degradation and deposition need to be fairly accurate and, hence, more complex models. Similarly, sediment yield models required to determine water quality depend upon the water-quality parameter to be modeled. For highly toxic chemicals, a short time step may be necessary to define changes in concentration during rainfall–runoff events. This means a more detailed model. For other chemicals such as nutrients, a longer time step is appropriate, and, hence, a simpler model.

22.3.1 Musgrave Equation

One of the earliest and most successful equations used in erosion-control planning is the Musgrave equation (Musgrave, 1947). It takes into account the effect on soil erosion of soil erodibility, vegetal cover, land slope, slope length, and rainfall intensity, and can be expressed as

$$A = CRS^{1.35}L^{0.35}P_{30}^{1.75} \tag{22.1}$$

where A is the long-term average soil loss from sheet and rill erosion in inches per year, C is the inherent erodibility of the soil in inches per year (soil-erodibility factor), R is a relative soil-cover factor (crop-management factor), S is the degree of the slope in percent, L is the length of the slope in feet, and P_{30} is 2-year, 30-minute rainfall amount in inches. The value of A can be converted to tons per acre by multiplying by an assumed volume weight of 150 tons per acre-inch. Based on field data from experimental plots and watersheds, Musgrave (1947) prepared a table of soil losses in different soils that provided a basis of comparison between a soil of similar physical characteristics with no erosion measurements to a soil of the same characteristics with erosion measurements. To that end, he employed a common value of S as 10%, L as 72 feet, and P_{30} as 1.25 in. When applying Equation (22.1), the given values of S, L, and P_{30} must be divided by their respective common values.

The Musgrave equation is the Corn Belt equation modified by inclusion of a rainfall factor. This has been widely used for estimating soil loss from watersheds in flood-abatement programs. Lloyd and Eley (1952) developed a graphical solution of Equation (22.1) that was used by the Soil Conservation Service and the Forest Service, both of the U.S. Department of Agriculture, in the northeastern states for prediction of sheet and rill erosion from agricultural and forest lands. Because the role of runoff in transport of sediment is well established, its absence in Equation (22.1) may lead to gross errors when applied to small time periods.

EXAMPLE 22.1 Compute the soil loss using Equation (22.1) for a wheat-producing area in the midwestern United States. The soil is a silt loam, which is generally similar in physical properties and erodibility to the Marshall silt loam. The land slope is 5%, the slope length is 150 feet, and the maximum 30-minute rainfall is 1.35 inches. The erodibility C of the soil is 0.32 inches per year. The wheat crop can be assumed to have a cover factor of 20% of that of continuous row crops.

Solution By using Equation (22.1),

$$A = 0.32 \times 0.2 \times (5/10)^{1.35} \times (150/72)^{0.35} \times (1.35/1.25)^{1.75}$$

$$= 0.037 \text{ inches/year}$$

$$= 5.55 \text{ tons/acre}$$

22.3.2 Universal Soil-Loss Equation (USLE)

The USLE, similar to but more versatile than the Musgrave equation, is perhaps the most widely used and accepted model for planning erosion-control practices in the United States. The principal application of the USLE has been to provide specific and reliable guides for selecting appropriate erosion-control practices for farmlands and construction areas. Other useful applications have been in determining upland erosion for reservoir sedimentation and stream loading, control of pollution from cropland, and alternative land use and treatment combinations. Although the USLE is designed to predict long-term average soil losses for specific conditions, it has also been applied to determine individual soil losses (Williams and Berndt, 1972; Williams, 1975a, 1975b).

Using 10,000 plot-years of data from various research facilities, mainly in the eastern half of the United States, Wischmeier and Smith developed the USLE in 1960. Following Wischmeier and Smith (1978), it can be written as

$$A = RKLSCP \tag{22.2}$$

where A is the soil loss per unit area, expressed in the units selected for K and for the period selected for R; in practice, the units are usually selected so that A is computed in tons per acre per year, but other units can also be selected; R is the rainfall and runoff factor—the number of rainfall–erosion index units, plus a factor for runoff from snowmelt or applied water where such runoff is significant; K is the erodibility factor—the soil-loss rate per erosion index unit for a specified soil as measured on a unit plot, which is defined as 72.6 feet in length of a uniform 9% slope continuously in clean-tilled fallow; L is the slope length factor—the ratio of soil loss from the field slope length to that from the 72.6-ft length under identical conditions; S is the slope steepness factor—the ratio of soil loss from the field slope to that from the 9% slope under identical conditions; C is the crop-management factor—the ratio of soil loss from an area with specified cover and management to that from an identical area in tilled continuous fallow; and P is the support practice factor—the ratio of soil loss with a support practice like contouring, strip cropping, or terracing to that with straight row farming up and down the slope.

22.3.2.1 Rainfall and runoff factor, *R*

The factor R must take into account the effect of raindrop impact, as well as the resulting amount and rate of runoff. Wischmeier (1959, 1962) showed that R used to estimate the average annual soil loss must include the cumulative effects of the many moderate-sized storms as well as the effects of the occasional severe storms, and derived a relation, called the rainfall erosion index, EI, to determine R. The local value of EI generally equals R. The effect of runoff from thaw, snowmelt, or irrigation is not included in EI here. A procedure for computing EI, where this type of runoff is significant, is given in Wischmeier and Smith (1978).

Based on 22-year station rainfall records, Wischmeier (1959) defined EI for a given storm as a product of total storm energy E in hundreds of foot-tons per acre, and the maximum 30-minute intensity I_{30} in inches per hour. It is a statistical interaction term combining total energy with peak intensity in a storm, and thus combining particle detachment with transport capacity. Based on the work of Gunn and Kinzer (1949), Wischmeier and Smith (1958) expressed

$$E = 916 + 331 \log I \tag{22.3}$$

where E is the kinematic energy in foot-tons per acre-inch, and I is the intensity in inches per hour. Based on Hudson (1971) and Carter et al. (1974), a limit of 3 inches/hour is imposed on I by noting that median drop size does not continue to grow beyond this limit. For computation of EI values, the storm energy is expressed in hundreds of foot-tons per acre. Therefore, the value of E computed from Equation (22.3) must be divided by 100 before multiplying by I_{30} to compute EI.

Thus, EI can be determined for a given storm. For a specified period, the individual storm EI values can be summed up, which will provide a numerical measure of the erosive potential of the rainfall within that period. In this manner, the average annual total of the storm EI values in a particular area is the rainfall erosion index for that area. Other methods of computing EI have also been suggested (Foster et al., 1982; Richardson et al., 1983).

22.3.2.2 Soil erodibility factor, *K*

The factor K is determined experimentally. Representative values of K for most of the soil types and texture classes have been compiled by the Soil Conservation Service, as shown in Table 22.1. For a specific area, K can be obtained from the soil-erodibility monograph of Figure 22.2 presented by Wischmeier and Mannering (1969) and Wischmeier et al. (1971). For soils containing less than 70% of silt and very fine

TABLE 22.1 COMPUTED *K* VALUES FOR SOILS ON EROSION RESEARCH STATIONS (AFTER WISCHMEIER AND SMITH, 1978)

Soil	Source of Data	Computed *K*
Dunkirm silt loam	Geneva, NY	0.69[a]
Keene silt loam	Zanesville, OH	0.48
Shelby loam	Bethany, MO	0.41
Lodi loam	Blacksburg, VA	0.39
Fayette silt loam	LaCrosse, WI	0.38[a]
Cecil sandy clay loam	Watkinsville, GA	0.36
Marshall silt loam	Clarinda, IA	0.33
Ida silt loam	Castana, IA	0.33
Mansic clay loam	Hays, KS	0.32
Hagerstown silty clay loam	State College, PA	0.31[a]
Austin clay	Temple, TX	0.29
Mexico silt loam	McCredie, MO	0.28
Honeoye silt loam	Marcellus, NY	0.28[a]
Cecil sandy loam	Clemson, SC	0.28[a]
Ontario loam	Geneva, NY	0.27[a]
Cecil clay loam	Watkinsville, GA	0.26
Boswell fine sandy loam	Tyler, TX	0.25
Cecil sandy loam	Watkinsville, GA	0.23
Zaneis fine sandy loam	Guthrie, OK	0.22
Tifton loamy sand	Tifton, GA	0.10
Freehold loamy sand	Marlboro, NJ	0.08
Blath flaggy silt loam with surface stones > 2 inches removed	Arnot, NY	0.05[a]
Albia gravelly loam	Beemerville, NJ	0.03

[a] Evaluated from continuous fallow. All others were computed from rowcrop data.

sand, K can be expressed as

$$100K = 2.1\ M^{1.14}(10^{-4})(12 - a) + 3.25(b - 2) + 2.5(c - 3) \tag{22.4}$$

where M is the particle-size parameter defined as percent silt (si) plus very fine sand (vfs) (size 0.1–0.002 mm) times the quantity (100 − minus percent clay), a is the percent organic matter, b is the soil-texture code used in USDA soil classification, and c is the profile-permeability class. It may be noted that when the silt fraction does not exceed 70%, erodibility varies approximately as the 1.14 power of M, but addition of organic matter content, soil structure, and profile-permeability class as done in Equation (22.4) improves the prediction accuracy.

22.3.2.3 Topographic factor, *LS*

For practical purposes, the two factors, L and S, can be combined into a single factor called the topographic factor. LS is the ratio of soil loss per unit area from a field slope to that from a 72.6-ft length of uniform 9% slope under otherwise identical conditions. For a specified slope and its length, LS can be computed as

$$LS = (\lambda/72.6)^m(65.41\ \sin^2\theta + 4.56\ \sin\theta + 0.065) \tag{22.5a}$$

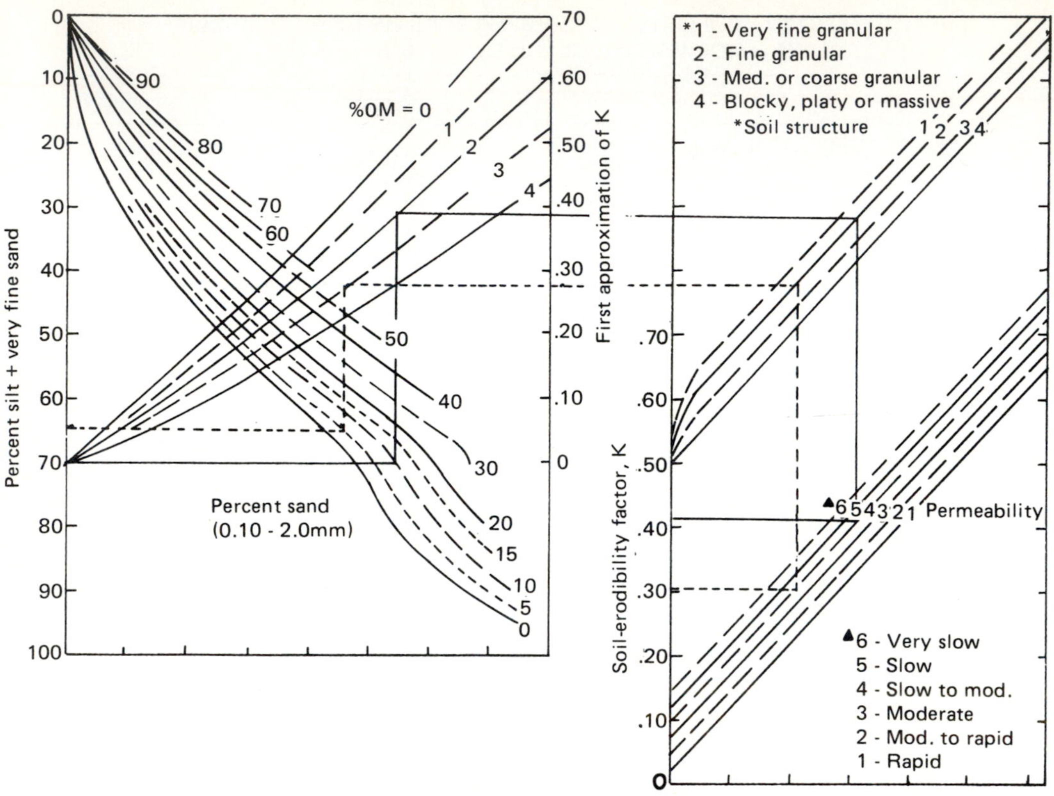

Figure 22.2 The soil-erodibility nomograph. Where the silt fraction does not exceed 70%, the equation is $100K = 2.1M^{1.14}(10^{-4})\ (12 - a) + 3.25(b - 2) + 2.5(c - 3)$, where M = (percent si + vfs) (100 − percent c), a = percent organic matter, b = structure code, c = profile permeability class, si = silt, and vfs = very fine sand.

where λ is the slope length in feet; θ is the angle of the slope; and $m = 0.5$ if the percent slope is 5 or more, $m = 0.4$ on slopes of 3.5 to 4.5%, $m = 0.3$ on slopes of 1 to 3%, and $m = 0.2$ on uniform gradients of less than 1%. A tabular solution of Equation (22.5a) is given in Table 22.2, and a graphical solution in Figure 22.3. Equation (22.5a) is based on slopes having essentially a uniform gradient. For slopes that are appreciably convex, concave, or complex, the *LS* values obtained must be modified as suggested by Wischmeier and Smith (1978) and Foster and Wischmeier (1974). The modification is based on the following: (1) Divide the irregular slope into a small number of equal-length segments such that the gradient within each segment can be considered practically uniform, (2) list the segment gradients in the order in which they occur on the slope from the upstream end to the downstream, (3) obtain *LS* for each segment from Figure 22.3 using the total length (not the segment length), (4) multiply these by the corresponding factors obtained as:

$$\text{Factor} = \text{soil-loss fraction} = \frac{J^{m+1} - (J - 1)^{m+1}}{N^{m+1}} \tag{22.5b}$$

TABLE 22.2 VALUES OF THE TOPOGRAPHIC FACTOR *LS* FOR SPECIFIC COMBINATIONS OF SLOPE LENGTH AND STEEPNESS (AFTER WISCHMEIER AND SMITH, 1978)[a]

Percent	Slope length (feet)											
Slope	25	50	75	100	150	200	300	400	500	600	800	1000
0.2	0.060	0.069	0.075	0.080	0.086	0.092	0.099	0.105	0.110	0.114	0.121	0.126
0.5	0.073	0.083	0.090	0.096	0.104	0.110	0.119	0.126	0.132	0.137	0.145	0.152
0.8	0.086	0.098	0.107	0.113	0.123	0.130	0.141	0.149	0.156	0.162	0.171	0.179
2	0.133	0.163	0.185	0.201	0.227	0.248	0.280	0.305	0.326	0.344	0.376	0.402
3	0.190	0.233	0.264	0.287	0.325	0.354	0.400	0.437	0.466	0.492	0.536	0.573
4	0.230	0.303	0.357	0.400	0.471	0.528	0.621	0.697	0.762	0.820	0.920	1.01
5	0.268	0.379	0.464	0.536	0.656	0.758	0.928	1.07	1.20	1.31	1.52	1.69
6	0.336	0.476	0.583	0.673	0.824	0.952	1.17	1.35	1.50	1.65	1.90	2.13
8	0.496	0.701	0.859	0.992	1.21	1.41	1.72	1.98	2.22	0.243	2.81	3.14
10	0.685	0.968	1.19	1.37	1.68	1.94	2.37	2.74	3.06	3.36	3.87	4.33
12	0.903	1.28	1.56	1.80	2.21	2.55	3.13	3.61	4.04	4.42	5.11	5.71
14	1.15	1.62	1.99	2.30	2.81	3.25	3.98	4.59	5.13	5.62	6.49	7.26
16	1.42	2.01	2.46	2.84	3.48	4.01	4.92	5.68	6.35	6.95	8.03	8.98
18	1.72	2.43	2.97	3.43	4.21	3.86	5.95	6.87	7.68	8.41	9.71	10.9
20	2.04	2.88	3.53	4.08	5.00	5.77	7.07	8.16	9.12	10.0	11.5	12.9

[a] $LS = (\lambda/72.6)^m(65.41 \sin^2 \theta + 4.56 \sin \theta + 0.065)$, where λ = slope length in feet; $m = 0.2$ for gradients $< 1\%$, 0.3 for 1 to 3% slopes, 0.4 for 3.5 to 4.5% slopes, 0.5 for 5% slopes and steeper; and θ = angle of slope. (For other combinations of length and gradient, interpolate between adjacent values or see Figure 22.3.)

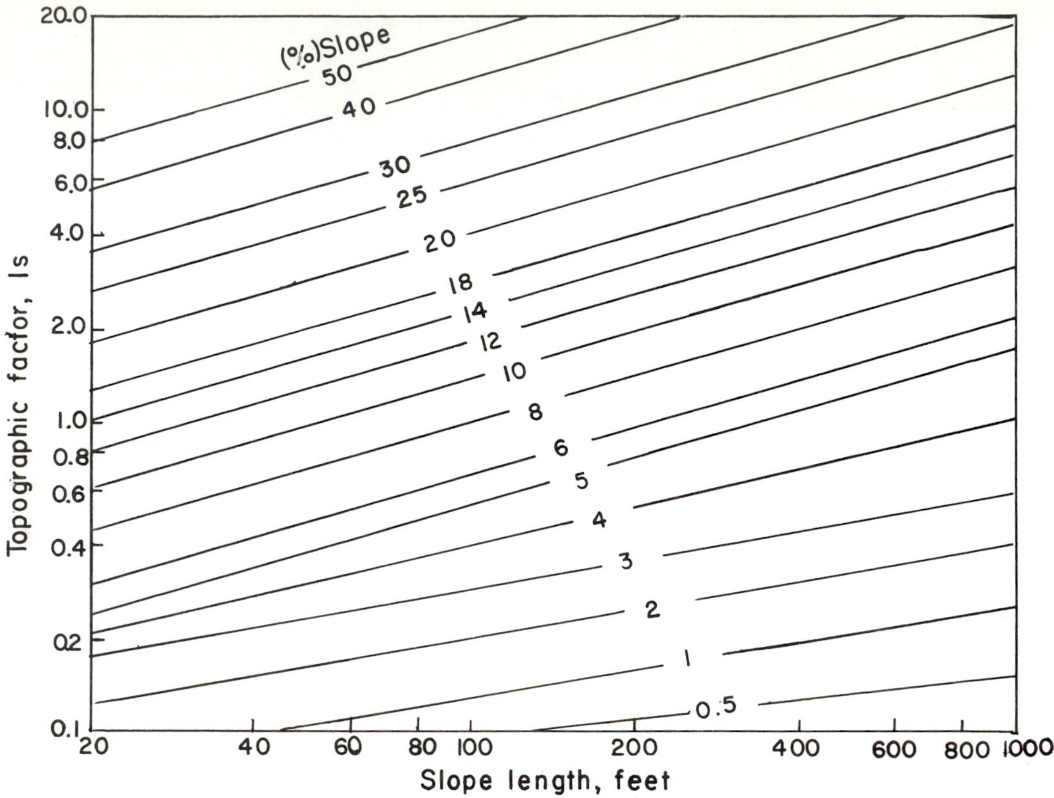

Figure 22.3 Slope-effect chart (topographic factor, *LS*). $LS = (\lambda/72.6)$ (65.41 $\sin^p \theta + 4.56 \sin \theta + 0.065$), where λ = slope length in feet; θ = angle of slope; and $m = 0.2$ for gradients $< 1\%$. 0.3 for 1 to 3% slopes, 0.4 for 3.5 to 4.5% slopes, and 0.5 for 5% or steeper slopes.

where J is the segment sequence number, m is the slope-length exponent as defined in Equation (22.5a), and N is the number of equal-length segments into which the slope was divided, and (5) then add the products to obtain LS for the entire slope.

22.3.2.4 Cover and management factor, *C*

This factor measures the combined effect of all the interrelated cover and management variables including the type of vegetation, plant spacing, the stand, the quality of growth, crop sequence, tillage practices, crop residues, incorporated residues, land-use residues, fertility treatments, etc. Wischmeier and Smith (1978) have given a detailed account of the quantitative evaluation of crop and management effects on C for agricultural lands, construction areas, pasture, range and idle land, and woodland. If C is to be obtained for a woodland, then three categories are distinguished: (1) undisturbed forest land; (2) woodland that is grazed, burned, or selectively harvested; and (3) forest lands that have experienced site-preparation treatments for reestablishment after harvest. For brevity of space, values of C for only pasture, range, idle, and woodland are given here. Given a combination of cover conditions on these types of lands, C can be obtained from Table 22.3 (Wischmeier, 1975).

22.3.2.5 Support practice factor, *P*

Important cropland practices are contour tillage, strip cropping on the contour, and terrace systems. Each of these practices must include stabilized waterways for disposal of the effective rainfall. Wischmeier and Smith (1978) have given values of P for each of these practices. Table 22.4 gives P for contouring. In general, tillage and planting on the contour reduce erosion. Contouring appears to be most effective on slopes of 3 to 8%. Since slope length influences effectiveness of contouring, the P values are based on maximum slope lengths (Stewart et al., 1975a, 1975b). The P values for contour strip cropping with respect to maximum strip widths and slope-length limits are given by Wischmeier and Smith (1978). For contour-farmed terraced fields, the P values apply to contour-farmed broadbase terraces most common on gently sloping land, to steep backslope terraces most common on steeper slopes, and to level terraces.

EXAMPLE 22.2 Compute the rainfall erosion index *EI* for the storms whose maximum 30-minute intensities in inches per hour are 1.0, 1.5, 2.75, 2.0, 2.25, 2.5, 2.75, and 3.0. How sensitive is *EI* to variation in rainfall intensity?

Solution From Equation (22.3):

I $\left(\frac{\text{inch}}{\text{hour}}\right)$	E $\left(\frac{\text{foot-tons}}{\text{acre-inch}} \div 100\right)$	EI $\left(\frac{\text{foot-tons}}{\text{acre-hour}}\right)$
1.0	916/100	9.16
1.5	974/100	14.61
1.75	996/100	17.43
2.0	1016/100	20.32
2.25	1032/100	23.22
2.5	1048/100	26.20
2.75	1061/100	29.18
3.0	1073/100	32.19

EXAMPLE 22.3 Compute the soil-erodibility factor K for the surface soil whose properties are:

PROPERTY	%
Sand	10
Silt	60
Clay	28
Organic matter	2

The soil structure is fine granular and the permeability is slow to moderate.

TABLE 22.3 FACTOR *C* FOR PERMANENT PASTURE, RANGE, AND IDLE LAND (AFTER WISCHMEIER AND SMITH, 1978)[a]

Vegetative canopy		Cover that contracts the soil surface						
			Percent ground cover					
Type and height[b]	Percent cover[c]	Type[d]	0	20	40	50	80	95+
No appreciable canopy		G	0.45	0.20	0.10	0.042	0.013	0.003
		W	0.45	0.24	0.15	0.091	0.043	0.011
Tall weeds or short brush with the average drop fall height of 20 in.	25	G	0.36	0.17	0.09	0.038	0.013	0.003
		W	0.36	0.20	0.13	0.083	0.041	0.011
	50	G	0.26	0.13	0.07	0.035	0.12	0.003
		W	0.26	0.16	0.11	0.076	0.039	0.001
	75	G	0.17	0.10	0.06	0.032	0.011	0.003
		W	0.17	0.12	0.09	0.068	0.038	0.011
Appreciable brush or bushes, with the average drop fall height of $6\frac{1}{2}$ ft	25	G	0.40	0.18	0.09	0.040	0.013	0.003
		W	0.40	0.22	0.14	0.087	0.042	0.011
	50	G	0.34	0.16	0.08	0.038	0.012	0.003
		W	0.34	0.19	0.13	0.082	0.042	0.011
	75	G	0.28	0.14	0.08	0.036	0.012	0.003
		W	0.28	0.17	0.12	0.078	0.040	0.011
Trees, but no appreciable low brush. Average drop fall height of 13 ft	25	G	0.42	0.19	0.10	0.041	0.013	0.003
		W	0.42	0.23	0.14	0.089	0.042	0.011
	50	G	0.39	0.18	0.09	0.040	0.013	0.003
		W	0.39	0.21	0.14	0.087	0.042	0.011
	75	G	0.36	0.17	0.09	0.039	0.012	0.003
		W	0.36	0.20	0.13	0.084	0.041	0.011

[a] The listed *C* values assume that the vegetation and mulch are randomly distributed over the entire area.

[b] Canopy height is measured as the average fall height of water drops falling from the canopy to the ground. Canopy effect is inversely proportional to the drop fall height and is negligible if all height exceeds 33 ft.

[c] Portion of total area surface that would be hidden from view by canopy in a vertical projection (a bird's-eye view).

[d] G: cover at the surface is grass, grasslike plants, decaying compacted duff, or litter at least 2 in. deep.
W: cover at the surface is mostly broadleaf herbaceous plants (as weeds with little lateral-root network near the surface) or undecayed residues or both.

TABLE 22.4 *P* VALUES AND SLOPE-LENGTH LIMITS FOR CONTOURING (AFTER WISCHMEIER AND SMITH, 1978)

Land slope	*P* value	Maximum length[a] (feet)
1 to 2	0.60	400
3 to 5	0.50	300
6 to 8	0.50	200
9 to 12	0.60	120
13 to 16	0.70	80
17 to 20	0.80	60
21 to 25	0.90	50

[a] Limit may be increased by 25% if the residue cover after crop seedlings will regularly exceed 50%.

Solution By using the left graph of Figure 22.2 for silt and sand 70% and organic matter 2%, an initial approximation of $K = 0.385$ is obtained. From the right graph of Figure 22.2 for fine granular soil structure and slow–moderate permeability, the final value of $K = 0.415$ is obtained.

EXAMPLE 22.4 Calculate the topographic factor for the following surfaces:

SLOPE SEGMENTS FROM UPSTREAM TO DOWNSTREAM

	Slope length (ft)	Slope (%)	Slope length (ft)	Slope (%)	Slope length (ft)	Slope (%)
(1) Convex	150	10	150	20	150	30
(2) Straight line	150	20	150	20	150	20
(3) Concave	150	30	150	20	150	10

Solution By using Figure 22.3 or Table 22.2 for segments from the upstream to the downstream, the following is obtained:

Segment (1)	Percent slope (2)	From figure 22.3 (3)	From equation (22.5b), table 22.2 (4)	Product = (3) × (4)
(a) Convex slope				
1	10	2.9	0.19	$(LS)_1 = 0.551$
2	20	8.6	0.35	$(LS)_2 = 3.01$
3	30	17.0	0.46	$(LS)_3 = 7.82$
				$(LS)_{total} = 11.38$

Segment (1)	Percent slope (2)	From figure 22.3 (3)	From equation (22.5b), table 22.2 (4)	Product = (3) × (4)
(b) Concave slope				
1	30	17.0	0.19	$(LS)_1$ = 3.23
2	20	8.6	0.35	$(LS)_2$ = 3.01
3	10	2.9	0.46	$(LS)_3$ = 1.334
				$(LS)_{\text{total}}$ = 7.57
(c) Straight slope				
1	20	8.6	0.19	$(LS)_1$ = 1.634
2	20	8.6	0.35	$(LS)_2$ = 3.01
3	20	8.9	0.46	$(LS)_3$ = 3.956
				$(LS)_{\text{total}}$ = 8.6

EXAMPLE 22.5 Watershed Y, located near Riesel (Waco), Texas, has an area of 309 acres. The average land slope is 2.41% and the slope length is 393 feet. It has Blackland Prairie soil with a high content of montmorillonite clay (Houston black clay 66%, Heiden clay 23%, Austin silt clay 10%, and Trinity clay 1%) and a soil-erodibility factor of $K = 0.33$. The average annual rainfall is 34 inches and the rainfall–runoff factor is 345. The watershed is a conservation watershed with well-maintained terraces and grassed waterways and has a crop-management factor of 0.3. The value of P is 0.2. Calculate the average annual erosion from this watershed.

Solution The topographic factor LS is computed from Equation (22.5a) for $\lambda = 393$ ft, $m = 0.3$, and $\theta = 2.17°$, or using Table 22.2:

$$LS \cong 0.35$$

The average annual soil loss per unit area, using Equation (22.2), is

$$A = 345 \times 0.33 \times 0.35 \times 0.30 \times 0.20 = 2.39 \text{ tons/acre/year}$$

Therefore, the average annual erosion for watershed Y is

$$309 \times 1.59 = 738.51 \text{ tons/year}$$

EXAMPLE 22.6 A watershed area of 600 acres above a proposed flood-retarding structure in Fountain County, Indiana, is shown in Figure 22.4. Compute the annual soil loss from sheet erosion for present conditions and for future conditions after recommended land treatment is applied on all land in the watershed. This example is due to Renfro (1975).

Solution

Present Conditions

Cropland, 280 acres

Continuous corn with residue removed, average yield 70 bu/acre, cultivated up and down slope; soil, Fayette silt loam: slope 8% and slope length 200 ft. The USLE parameters are $R = 185$, $K = 0.37$, $LS = 1.41$, $C = 0.43$, and $P = 1.0$. This yields soil loss as

$$A = 185 \times 0.37 \times 1.41 \times 0.43 \times 1.0 = 41.5 \text{ tons/acre/year}$$

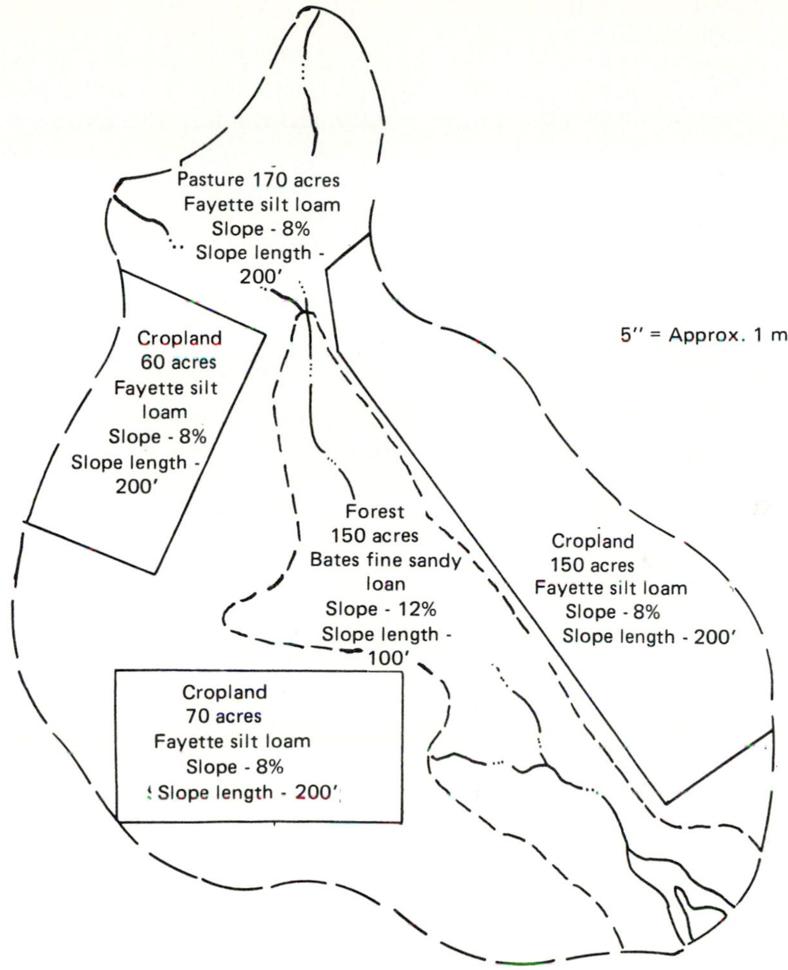

Figure 22.4 A 600-acre watershed.

Pasture, 1970 acres

Canopy of short brush, 0.5 m fall height; percent cover provided by canopy, 50% surface cover, grass and grasslike plants; percent of surface or ground cover, 8%; soil, Fayette silt loam; slope 8% and slope length 200 ft. The USLE parameters are $R = 185$, $K = 0.37$, $LS = 1.41$, and $C = 0.012$. This yields soil loss as

$$A = 185 \times 0.37 \times 1.41 \times 0.012 = 1.16 \text{ tons/acre/year}$$

Forest, 150 acres

Poorly stocked; percent of area covered by tree canopy, 80%; percent of area covered by litter, 50%; undergrowth, unmanaged; soil, Bates silt loam; slope 12% and slope length 100 ft. The USLE parameters are $R = 185$, $K = 0.32$, $LS = 1.8$, and $C = 0.05$. This yields soil loss as

$$A = 185 \times 0.32 \times 1.8 \times 0.05 = 5.3 \text{ tons/acre/year}$$

Future Conditions

Cropland, 280 acres

Rotation of wheat, meadow, corn, corn with residue left; contour strip cropped; soil, Fayette silt loam; slope 8% and slope length 200 ft. The USLE parameters are R = 185, K = 0.37, LS = 1.41, C = 0.119, and P = 0.3. This yields soil loss as

$$A = 185 \times 0.37 \times 1.41 \times 0.119 \times 0.3 = 3.4 \text{ tons/acre/year}$$

Pasture, 170 acres, with improved management

Canopy cover increased to 75% with 0.4 m fall height; ground cover increased to 95% (for area not predicted by canopy); soil, Fayette silt loam; slope 8% and slope length 200 ft. The USLE parameters are R = 185, K = 0.37, LS = 1.41, and C = 0.003. The soil loss is

$$A = 185 \times 0.37 \times 1.41 \times 0.003 = 0.29 \text{ tons/acre/year}$$

Forest, 150 acres, with improved management

Medium stocked; canopy cover increased to 80%; undergrowth, managed; soil, Bates silt loam; slope 12% and slope length 100 ft. The USLE parameters are R = 185, K = 0.32, LS = 1.8, and C = 0.003. The soil loss is

$$A = 185 \times 0.32 \times 1.8 \times 0.003 = 0.32 \text{ tons/acre/year}$$

To summarize:

Present conditions:		tons/year
Cropland:	280 acres × 41.5 tons/acre =	11,620
Pasture:	170 acres × 1.16 tons/acre =	197
Forest:	150 acres × 5.3 tons/acre =	795
	Total =	12,612
Future conditions;		
Cropland:	280 acres × 3.4 tons/acre =	952
Pasture:	170 acres × 0.29 tons/acre =	49
Forest:	150 acres × 0.32 tons/acre =	48
	Total =	1,049

22.3.2.6 Application to agricultural watersheds

Williams and Berndt (1972) extended the USLE for computing sediment yields from agricultural watersheds. For this purpose, all factors except the rainfall factor must be modified because it is impractical to compute erosion for each field in a watershed. However, if the land use or conservation practices vary widely in different parts of the watershed, source erosion should be computed separately for each part.

The soil-erodibility factor for a watershed is determined by weighting its value K_i of each soil in the watershed (Soil Conservation Service, 1969) according to the area A_{Di} encompassed by the soil as

$$K = \frac{1}{A_D} \sum_{i=1}^{N} K_i A_{Di} \tag{22.6}$$

where A_D is the watershed area, and N is the number of different soil areas in the watershed.

The slope length L is the average overland flow length for the watershed. This is obtained by considering the watershed to be rectangular with one channel in the center extending over the length of the watershed. The overland flow length is then half the width. Therefore,

$$L = 0.5A_D/L_c \tag{22.7}$$

where L_c represents the total length of the channels in the watershed.

The slope gradient S can be determined from a topographic map. The average slope between any two contours can be computed by

$$S_i = \frac{H}{2A_{Di}}(L_{cj} + L_{cj+1}) \times 100 \tag{22.8a}$$

where S_i is the average percent slope for the acre A_{Di} enclosed between contours j and $j + 1$, H is the difference in elevation between the contours, and L_{cj}, is the length of contour j.

The average watershed slope is determined by weighting the slopes between the contours given by Equation (22.8a) with the areas between them as

$$S = \frac{1}{A_D}\sum_{i=1}^{N} S_iA_{Di} \tag{22.8b}$$

where N is the number of areas between contours within the watershed.

The slope-length factor LS is determined by

$$LS = 8.52\left(\frac{L}{72.6}\right)^w(0.0076 + 0.0053S + 0.00076S^2) \tag{22.9}$$

where w is a watershed constant dependent upon slope gradient, and L is expressed in feet. If L is in meters,

$$LS = 4.705\left(\frac{L}{22.13}\right)^w(0.0076 + 0.0053S + 0.00076S^2) \tag{22.10}$$

If $w = 0.5$,

$$LS = L^{0.5}(0.0076 + 0.0053S + 0.00076S^2) \tag{22.11}$$

The crop-management factor C for a watershed is determined by weighting its value C_i of each crop and management level according to the area represented by this level:

$$C = \frac{1}{A_D}\sum_{i=1}^{N} C_iA_{Di} \tag{22.12}$$

where N is the number of crops grown multiplied by the number of management levels.

Williams and Berndt applied the USLE to the Texas Blacklands, where the erosion-control-practice factor is highly dependent upon two erosion-control practices: terraces and grassed waterways. To determine the P value for a watershed in this area, they considered only the cultivated area, which was divided into three categories according to the erosion-control practice: (1) straight rows, (2) straight rows with grassed waterways, and (3) terraces. The factor P was then computed as

$$P = 1.0A_r + 0.3A_s + P_iA_t \tag{22.13}$$

where A_r is the watershed area farmed with straight rows, A_s is the watershed area farmed with straight rows and grassed waterways, P_i is the erosion-control-practice factor for terracing, and A_t is the watershed area that is terraced.

EXAMPLE 22.7 Compute the average annual soil erosion from an area that is comprised of four parts.

	Parts			
	1	2	3	4
Drainage area (mi^2)	0.21	0.48	1.73	6.84
Average land slope (%)	2.6	2.4	2.10	2.06
Slope length (ft)	250	400	300	350
Channel slope (%)	1.20	1.13	0.51	0.40
Length/width	2.25	1.86	2.88	3.60

The parameters of the USLE for the four parts of the area are as follows:

Part	R	K	LS	C	P
1	380	0.33	0.32	0.22	0.10
2	300	0.33	0.35	0.20	0.10
3	400	0.36	0.28	0.11	0.12
4	350	0.34	0.30	0.12	0.11

The total length of the channels in the area is 3.24 miles.

Solution

1. Using Equation (22.6), compute the weighted soil-erodibility factor:

$$K = \frac{1}{9.26}(0.33 \times 0.21 + 0.33 \times 0.48 + 0.36 \times 1.73 + 0.34 \times 6.84)$$

$$= 0.343$$

2. Using Equation (22.7), compute the average overland flow length:

$$L = \frac{1}{2} \cdot \frac{9.26}{3.24} = 1.43$$

3. Using Equation (22.8b), compute the average watershed slope:

$$S = \frac{1}{9.26}(2.6 \times 0.21 + 2.4 \times 0.48 + 2.1 \times 1.73 + 2.06 \times 6.84)$$

$$= 2.1\%$$

4. Using Equation (22.9), compute the LS factor for $w = 0.3$ (adequate for slopes 1 to 3%):

$$LS = 8.52 \left(\frac{1.43}{72.6}\right)^{0.3} [0.0076 + 0.0053 \times 2.1 + 0.00076 \times (2.1)^2]$$

$$= 0.758$$

5. Using Equation (22.12), compute the weighted crop-management factor:

$$C = \frac{1}{9.26} (0.22 \times 0.21 + 0.2 \times 0.48 + 0.11 \times 1.73 + 0.12 \times 6.84)$$

$$= 0.125$$

Similarly, compute the weighted P and R:

$$P = \frac{\sum_{i=1}^{4} P_i A_{D_i}}{A_D} = 0.11$$

$$R = \frac{\sum_{i=1}^{4} R_i A_{D_i}}{A_D} = 357.4$$

Finally, the average annual erosion is

$$A = 357.4 \times 0.343 \times 0.758 \times 0.125 \times 0.11 = 1.28 \text{ tons/acre}$$

or

$$A = (5926 \text{ acres}) \times 1.28 = 5016.15 \text{ tons for the whole watershed}$$

22.3.3 Modified Universal Soil Loss Equation (MUSLE)

The USLE was modified by Williams (1975a) by replacing the rainfall–runoff factor with a runoff factor. The MUSLE, therefore, can be expressed as

$$A = B(V_Q Q_P)^{0.56} K(LS)CP \tag{22.14}$$

where V_Q is the volume of runoff in acre-feet, Q_p is the peak flow rate in cfs, B is a constant, and the other terms have the same meanings as in the USLE. Based on the data from small watersheds near Riesel, Texas, and Hastings, Nebraska, B was found to be 95. If V_Q is in m^3, Q_P in m^3/s, then B would be 11.8. Equation (22.14) explained 92% of the variation in sediment yield of these watersheds.

The MUSLE was developed to predict sediment yield for individual storms (Williams 1975b, 1978a, 1978b, 1979, 1980). Williams (1975a) argued that inclusion of the runoff factor in place of the rainfall energy factor in the USLE would be better because studies (Dragoun and Miller, 1964; Williams et al., 1971) have shown that runoff is the single best indicator of sediment yield. The quantities V_Q and Q_P can be determined using the methods discussed in Chapters 14 and 18, respectively. Therefore, the MUSLE may be more suitable for estimating sediment yield for individual storms on small watersheds.

EXAMPLE 22.8 Compute the soil erosion from an area subjected to a storm generating a volume of runoff of 10 acre-feet and a peak flow rate of 15 ft^3/s. The parameters of the USLE are $K = 0.32$, $LS = 0.30$, $C = 0.10$, and $P = 0.12$.

Solution Take $B = 95$. By using Equation (22.14),

$$A = 95 \times (10 \times 15)^{0.56} \times 0.32 \times 0.30 \times 0.10 \times 0.12$$

$$= 1.81 \text{ tons/acre/year}$$

22.3.4 Sediment Delivery Ratio Method (SDRM)

Sediment yield of a watershed is only a part of its gross erosion, and equals the gross erosion minus the sediment deposited en route to the point of reference. Sediments produced by sheet and rill erosion often move only short distances and may get deposited away from the stream system. They may remain in the areas of their origin or be deposited on more level slopes downstream. A sediment delivery ratio D_R is a fraction or percentage of the gross erosion that is transported off a given area of land as sediment yield:

$$D_R = \frac{S_y}{E_g} \tag{22.15}$$

where D_R is the sediment delivery ratio (SDR), E_g is the gross erosion, and S_y is the sediment yield. Commonly, D_R is used as the fraction of E_g that appears at some downstream point as S_y. Here S_y is the total sediment passing a reference point; it includes suspended load and bed load and is, thus, the accumulated load. D_R is then a composite of delivery rates of sediments of all sizes and from all sources. If D_R is known or can be approximated, then S_y can be determined by computing E_g and multiplying it by D_R.

22.3.4.1 Factors affecting D_R

Many factors, including watershed physiography, sediment source, proximity and magnitude of source, transport system, texture of eroded material and depositional areas, influence D_R. Eroded soil materials derived from rill and interrill erosion often move only short distances and may lodge in areas remote from the stream system. On the other hand, channel erosion produces sediment that is immediately available to the transport system, and much of it tends to be in motion as suspended sediment or bed load.

The magnitude and proximity of sediment source affect D_R. A large amount of sediment may be produced by severe erosion in an area, but if the area is remote from the stream, its D_R will be less than when a smaller amount of sediment is produced by moderate erosion close to the stream.

Runoff is the chief carrier of sediment. The ability to transport sediment depends upon the velocity and volume of discharge as well as the amount and type of eroded material. Frequency and duration of runoff affect the total volume of sediment transport. Furthermore, the stream system itself has considerable bearing on the amount of sediment it can transport. For example, the higher the drainage density, the higher the D_R. The condition of the channels, clogged or open, meandering or straight, sloping gently or steeply, affects the delivery ratio.

The texture of eroded material is another important consideration. If the material is fine silt and clay, it probably will stay in suspension, and much of it will be transported downstream. If it is sand, it requires efficient transport systems and high velocities. Much of it is deposited in the upstream areas when the velocity drops down significantly.

Depositional areas decrease D_R. Sometimes sediment gets deposited at the foot of the upland slopes, along the edges of large valleys, in valley flats, in and along mainstream channels, and at the heads of and in lakes, reservoirs, and ponds.

Watershed characteristics have considerable bearing on D_R. Drainage area, topography, slope, relief, length, and vegetation are important characteristics. High relief is often associated with high D_R. As drainage area increases, D_R decreases. The relief-length ratio has a pronounced effect on D_R. Settling basins, sediment traps, and vegetative filter strips across the lower end of an upland area trap sediments near their origin.

22.3.4.2 Determination of D_R

To determine D_R, it is clear from Equation (22.15) that S_y at a given point in the watershed and E_g must be known. However, measurements of these quantities are generally not available for most small watersheds. The USLE is currently the most accepted method for predicting sheet and rill erosion. The Soil Conservation Service (1971) has developed guidelines for estimating gully and channel erosion. Channel erosion can also be estimated by mapping channel enlargements and periodic measurements of channel cross-section. Gully erosion can similarly be estimated from map studies and field surveys. Sediment yield can be estimated by reservoir sedimentation surveys or by a sediment-load measurement. However, acquisition of pertinent data is often very expensive and, sometimes, even difficult to undertake. Therefore, simple procedures, which potentially can be extended to other areas, have been attempted.

It is generally considered that the sediment discharged to large rivers is usually less than one-fourth of that eroded from the land surface. Although D_R experiences a wide variation for a given area, it roughly varies inversely as the 0.2 power of the drainage area:

$$D_R = A_D^{-0.2} \tag{22.16}$$

Considering area A_D as the dominant factor, Roehl (1962) expressed S_y as a function of A_D by a curve. The curve can be expressed as a set of points. Thus, a D_R–A_D relationship for known E_g can be constructed.

Table 22.5 shows a set of values of D_R for various values of A_D. These values, however, should be used with judgment, taking into account such factors as texture, type of erosion, transport system, and depositional areas. For example, if a watershed has predominantly sandy soils, its D_R should be decreased from the value given in Table 22.5.

TABLE 22.5 RELATIONSHIP BETWEEN AREA AND SEDIMENT DELIVERY RATIO

Area (mi²)	Sediment delivery ratio
0.01	0.65
0.1	0.44
0.5	0.33
1.0	0.29
5	0.22
10	0.18
50	0.12
100	0.095
200	0.08
600	0.05

Equation (22.16) can be recast as

$$\frac{D_{R1}}{D_{R2}} = \left(\frac{A_{D2}}{A_{D1}}\right)^{0.2} \tag{22.17}$$

where subscripts 1 and 2 denote two points on the D_R–A_D curve. This relation can be used to interpolate between tabular values. Equation (22.17) can also be used to estimate S_y for a watershed, provided data on a similar nearby watershed are available. In such a case, E_g would be proportional to area. Therefore,

$$S_y = S_{y*}\left(\frac{A_D}{A_{D*}}\right)^{0.8} \tag{22.18}$$

where * signifies the watershed where measurements are available.

Other relationships of D_R and watershed characteristics have been formulated for specific geographic areas. For the Blackland Prairie in Texas, Maner (1958) used 14 watersheds ranging in size from 0.43 to 97.4 mi² and derived the following relationship:

$$\log D_R = 1.8768 - 0.14191 \log 10A_D \tag{22.19}$$

where D_R is the percentage of annual erosion, and A_D is in square miles. Its standard error of estimate was ±0.03007 log units, and the coefficient of correlation 0.96 expressed 92% of the variation in D_R between watersheds. Figure 22.5 graphs this relationship.

Maner (1958) also developed a sediment delivery ratio curve for the Red Hills physiographic area in Oklahoma and Texas. This area is also referred to as the Rolling Red Plains Land Resource Area by the Soil Conservation Service of the U.S. Department of Agriculture. The relationship, as shown in Figure 22.6, can be expressed as

$$\log D_R = 2.94259 - 0.82362 \text{ colog } (R_e/L) \tag{22.20}$$

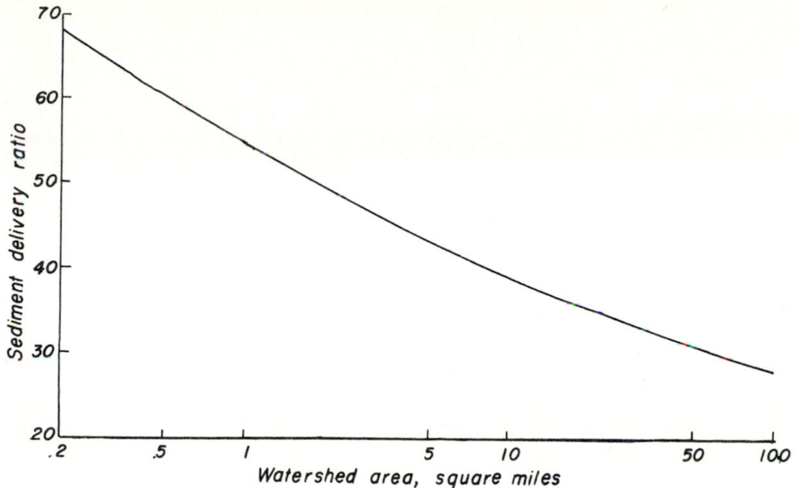

Figure 22.5 Sediment-delivery ratio curve, Blackland Prairie Land Resource Area (after Maner, 1962).

where R_e is the watershed relief defined as the difference in elevation between the average elevation of the watershed divide at the headwaters of the main channel and the elevation of the streambed at the point of sediment-yield measurement, and L is the length defined as the maximum valley length measured essentially parallel to the main channel from the point of yield measurement to the watershed divide. The

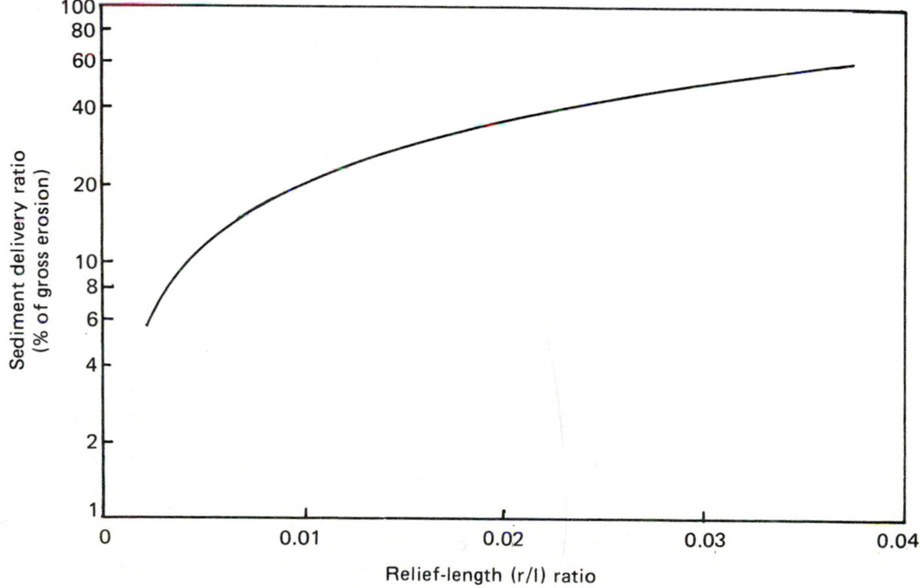

Figure 22.6 Sediment-delivery ratio curve, Rolling Red Plains Land Resource Area (after Maner, 1958).

standard error of estimate for this equation was ±0.04041 log units, and the coefficient of correlation was 0.987.

Williams (1977) computed sediment delivery ratios for 15 watersheds in Texas, with areas up to 65 km², using sediment and runoff models. These ratios were related to watershed characteristics as

$$D_R = 1.366 \times 10^{-11} A_D^{-0.0998} (R_e/L)^{0.3629} C_N^{5.444} \tag{22.21}$$

where A_D is in km², R_e is in m, L is in km, and C_N is the long-term average SCS curve number. Equation (22.21) explained about 93% of variation in D_R.

22.3.5 Dendy–Bolton (DB) Method

Dendy and Bolton (1976) used data from over 500 reservoirs throughout the United States to obtain measured sediment-yield values. They segregated the data into areas where runoff was either less than or greater than 2 inches per year. In areas where runoff is less than 2 inches, they derived a relation (for 175 reservoirs) as

$$S_y = 1280 V_Q^{0.46} (1.43 - 0.26 \log A_D) \tag{22.22}$$

and for the other areas (330 reservoirs):

$$S_y = 1958 \exp(-0.055 V_Q)(1.43 - 0.26 \log A_D) \tag{22.23}$$

where S_y is the annual sediment yield in tons/mi²/year, A_D is in mi², and V_Q is the annual runoff in inches. According to Williams (1981), the DB method may not be applicable to specific watersheds because it does not account for variations in many local factors such as slope, cover, and management. However, it can be useful in preliminary and large area planning. According to Renard (1980), Equation (22.22) is realistic for conditions encountered in the southwestern United States. For the Walnut Gulch Experimental Watershed in Arizona, Renard (1977) found

$$V_Q = 0.4501 A_D^{-0.1449} \tag{22.24}$$

By substituting Equation (22.24) into Equation (22.23),

$$S_y = 887 A_D^{-0.0667} (1.43 - 0.26 \log A_D) \tag{22.25}$$

To convert the annual sediment yield to acre-feet/mi²/year, he assumed that the sediment deposited weighed 90 lb/ft³.

EXAMPLE 22.9 Calculate the sediment yield, using the DB method, for the drainage areas: 2.25; 35.5; 110; 1,115; 11,283; and 27,395 square miles for which the average annual runoff values, respectively, are 2.61, 3.15, 6.37, 6.81, 1.45, and 1.25 inches.

Solution For the first four drainage areas, Equation (22.23) is used since $V_Q > 2$ inches/year.

DRAINAGE AREA (mi²)	RUNOFF (in./year)	SEDIMENT YIELD (tons/mi²/year)
2.25	2.61	2270
35.5	3.15	1690
110.0	6.37	1240
1115.0	6.81	859

For the last two drainage areas, Equation (22.22) is used since $V_Q < 2$ inches/year:

DRAINAGE AREA (mi^2)	RUNOFF (in./year)	SEDIMENT YIELD (tons/mi^2/year)
11,283	1.45	572
27,395	1.25	392

22.3.6 Regression Analysis

Many workers developed regression equations relating sediment yield or discharge to watershed and climatic characteristics. These equations were developed for specific locations and yielded satisfactory results for those locations. Flaxman (1972) developed a regression equation for reservoir design on rangeland watersheds in the western United States relating sediment yield to watershed conditions. Data from 27 watersheds ranging in size from 12 to 54 mi^2 in 10 western states were used. His equation is

$$\log (S_y + 100) = 6.21301 - 2.19113 \log (X_1 + 100) + 0.06034 \log (X_2 + 100) - 0.01644 \log (X_3 + 100) + 0.0425 \log (X_4 + 100) \quad (22.26)$$

where S_y is the average annual sediment yield (acre-feet/mi^2/year), X_1 is the ratio of average annual precipitation (inches) to average annual temperature (°F), X_2 is the average watershed slope (%), X_3 are the soil particles greater than 1.0 mm (%), and X_4 is the soil-aggregation index.

Parameters X_i, $i = 1, 2, 3$, and 4, take into account the effect on sediment yield of climate and vegetative growth through X_1, of topography through X_2, and of soil properties through X_3 and X_4. The correlation coefficient was 0.958, which explained about 91% of the variance in average annual sediment yield. In a discussion of the above method, Singh (1973) noted that S_y could be obtained also by a linear regression analysis:

$$S_y = 0.60854 - 0.76099X_1 + 0.03806X_2 - 0.01252X_3 + 0.04227X_4 \quad (22.27)$$

with a correlation coefficient of 0.8826. He compared Equations (22.26) and (22.27) using Flaxman's data and found that the two equations were comparable.

22.3.7. Runoff–Sediment Relation

Rendon-Herrero (1974) suggested a linear relation in log space between sediment yield (wash load) and surface-runoff volume due to rainfall from small upland watersheds. Furthermore, the slope of this straight line would remain approximately constant from one watershed to another. These hypotheses were further investigated by Rendon-Herrero et al. (1980) and Singh and Chen (1982). Figure 22.7 shows this relationship for three watersheds.

An investigation by Singh and Chen (1982) of 39 small watersheds from 14 states in the United States concluded the following: (1) The linear relationship is

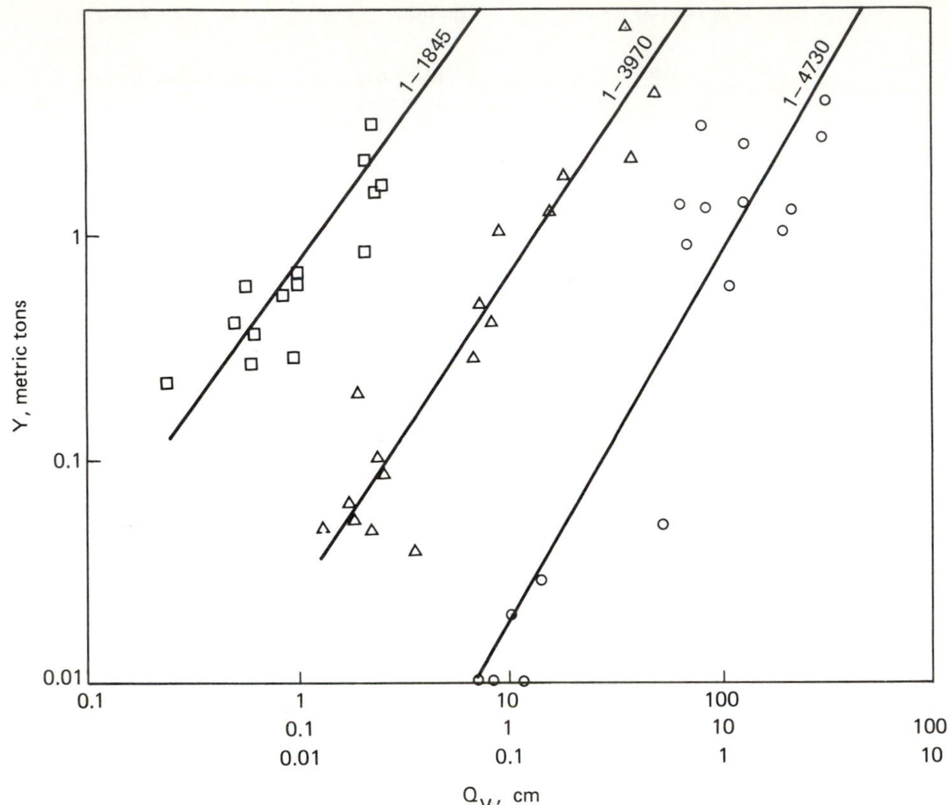

Figure 22.7 The relation between sediment yield and volume of direct runoff for some sample watersheds (summer and winter storms are combined).

satisfactory for a number of watersheds. (2) This relationship is improved if the storms are separated into winter and summer storms, excluding snow storms. (3) The amount of variance explained by this relationship can be as high as 95%, depending upon watershed characteristics. (4) The watersheds can be grouped in such a way that a family of parallel straight lines result, each representing a group. Thus, the only unknown that remains to be determined is the intercept.

This relationship can be expressed as

$$Y = aQ_v^b \tag{22.28}$$

or

$$\log Y = \log a + b \log Q_v \tag{22.29}$$

where log a is the intercept, and b is the slope of the line. Parameters a and b were both correlated with watershed characteristics, including watershed area A_D (km^2), main channel length L (km), main channel slope S_0 (m/km), mean basin elevation E (m), erodibility factor K, shape factor R, forest area A_f (km^2), and lake storage area

S_a (%). The regression equations obtained are

$$\log a = -2.5628 + 0.0006A_D - 0.0075L - 0.1543S_0 - 0.00055A_f + 0.0029E + 0.0004S_a + 8.6996K - 0.0033R \tag{22.30}$$

with a correlation coefficient of 0.9393 and a standard error of estimate of 0.2977. Equation (22.30) explained 88.23% of the variation in log a. With reduced geomorphic characteristics,

$$\log a = -2.337 - 0.003L - 0.1415S_0 - 0.0004A_f + 0.004E + 7.576K \tag{22.31}$$

with a correlation coefficient of 0.9107 and a standard error of estimate of 0.3205.

Similarly,

$$\log b = 0.2163 - 0.00002A_D + 0.00053L + 0.00019S_0 + 0.00008A_f - 0.00045E - 0.0238A_a + 0.0923K - 0.000003R \tag{22.32}$$

with a correlation coefficient of 0.7422 and a standard error of estimate of 0.0695. This explained a 55.08% of the variation in log b. With reduced geomorphic characteristics,

$$\log b = 0.2625 - 0.00001L + 0.00007A_f - 0.0004E - 0.0251S_0 - 0.0000007R \tag{22.33}$$

with a correlation coefficient of 0.7384 and a standard error of estimate of 0.06205.

EXAMPLE 22.10 Compute the sediment yield for a watershed that has the following characteristics: area = 45.6 km^2; main channel length = 12.9 km; main channel slope = 2.94 m/km; mean basin elevation = 171 m; erodibility factor = 0.36, shape factor = 0.28; and forest area = 8.7 km^2.

Solution By using Equations (22.31) and (22.33) for reduced geomorphic characteristics:

$$\log a = -2.337 - 0.003 \times 12.9 - 0.1415 \times 2.94 - 0.0004 \times 8.7 + 0.004 \times 171 + 7.576 \times 0.36 = 0.61617$$

$$\log b = 0.2625 - 0.00001 \times (12.9) + (0.00007) \times (8.7) - (0.0004) \times (171) - (0.0251) \times (2.94) - (0.0000007) \times (0.28) = 0.1207$$

The sediment yield Y can then be expressed as

$$\log Y = 0.61617 + 1.321Q_v$$

22.3.8 Sediment-Concentration Graph (SCG)

It is generally observed that suspended sediment concentration in a stream arises and declines in a manner similar to that of streamflow due to an isolated runoff event. Figure 22.8 shows typical variations of sediment concentration and streamflow. The maximum concentration may occur in advance of, or coincide with, the maximum streamflow, depending upon the stream discharge. The time difference between these two maxima is often termed as the "time of lead" (Johnson, 1943). Generally, the sediment concentration experiences a rapid rise from the beginning until about the peak stage. It declines with the streamflow recession, but usually at a less rapid rate than the increase during the rising stage and has a long tail. As the discharge and concentration continue to decrease, a point is reached where the sediment load becomes negligible in relation to that transported during the rise. In a small stream, this condition usually may occur a few hours after the peak. Furthermore, the concentration often fluctuates erratically during the rising limb, but decreases uniformly during the falling limb.

Based on these considerations, Johnson (1943) was perhaps the first to derive a distribution graph for suspended-sediment concentration employing a hypothesis analogous to that embodied in the unit hydrograph (Sherman, 1932). He examined streamflow hydrographs and corresponding suspended-sediment concentration

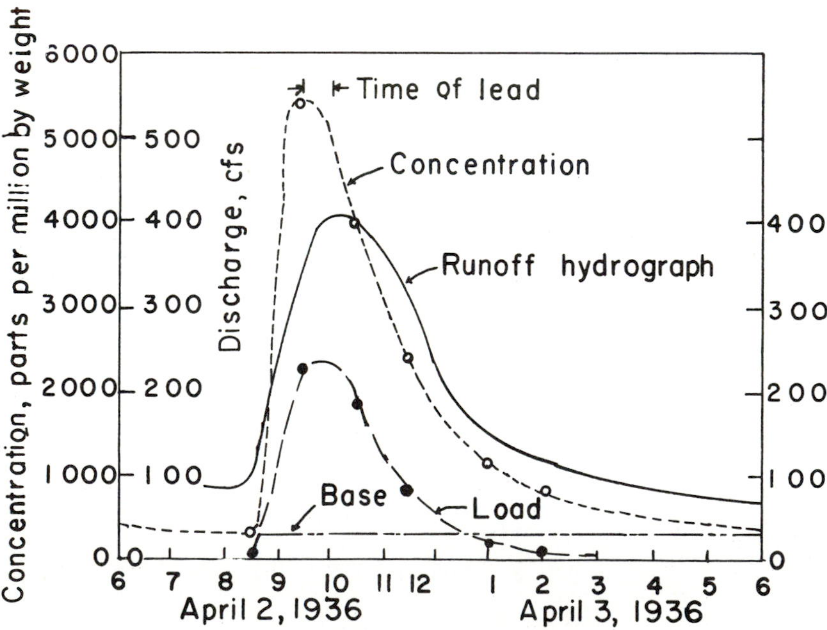

Figure 22.8 Graphs of discharge, suspended-matter concentration, and suspended load for a typical rise on the East Fork of the Deep River, North Carolina.

graphs for a number of isolated runoff events from the East Fork of Deep River, near High Point, North Carolina, and found that the distribution graph of concentration was analogous to the distribution graph of streamflow. The distribution graph for sediment concentration can be constructed in the same way as that for streamflow. To briefly summarize, the following steps are outlined:

1. Select an isolated runoff event for which a sufficient number of water samples are available to define the variation of sediment concentration, sediment load, and water discharge. Plot these three quantities on the same diagram as shown in Figure 22.8.
2. Select a horizontal baseline of concentration that gives the value of concentration prior to the rise, as shown in Figure 22.8.
3. Separate the base concentration from the observed concentration values. This, we hope, gives the concentration due to the runoff event.
4. Plot the concentration values, determined as before, against time. The time base of this graph extends from the beginning of the rise to the time when the concentration has receded so low that the sediment load can be considered negligible.
5. Divide the time base arbitrarily into a number of time intervals. One-half to an hour may be an appropriate length of the time interval. Determine the average concentration for each time interval.
6. Sum the average concentration values of all the time intervals.
7. Compute the percentage of the total concentration occurring during each of the time intervals.
8. Plot the percent concentration against the time interval. This is the distribution graph of the suspended sediment, corresponding to the period of stream rise (the time interval between the start of surface runoff and the occurrence of peak discharge) or the effective rainfall duration, which is the unit duration. This is the desired SCG. It is tacitly assumed that the effective rainfall is uniform in space and time.

Johnson (1943) found that the shape of the distribution graph was a function of the period of the stream rise, and somewhat of the intensity of the effective rainfall. Obviously, land use, vegetal cover, soil moisture, rainfall characteristics, and the time of the year will have a bearing on the graph. To apply the SCG method for computing suspended load, it is required that the streamflow hydrograph and one measurement of the suspended-sediment concentration during a rise be available. It is preferable that the concentration be known at, or shortly after, the peak discharge, because it fluctuates erratically during the rising limb. If the streamflow hydrograph has two peaks, then two observed concentrations near the peaks should be known. Given a concentration and a distribution graph oriented with respect to the peak discharge according to Figure 22.8, a graph of concentration can be obtained for each peak. If the effects of two peaks overlap, the individual concentration curves are calculated separately, and then added graphically to give the concentration curve for the entire hydrograph.

22.3.9 Unit Sediment Graph (USG)

Rendon-Herrero (1974, 1978) extended the unit hydrograph method to directly derive a unit sediment graph for a small watershed. The USG, thus, is based on hypotheses analogous to those in the unit hydrograph. The sediment load considered in the USG is the wash load only. Rendon-Herrero (1974) expressed the following to define the USG:

> A form of a unit sediment graph was indeed developed whose standard unit was 1.0 ton (910 kg) for a given duration, distributed over the watershed area, analogous in unit-hydrograph analysis to 1.00 in. (25 mm) of excess (effective) rainfall over the same area.

In light of this definition, the USG and UH are similar in their derivations. To discuss the derivation of the USG by Rendon-Herrero, the following steps are outlined:

1. Select an isolated rainfall–runoff event of a desired duration in accordance with the requirements of the UH, for which sediment-concentration graph C is known.
2. Separate the baseflow Q_b from the runoff hydrograph Q_T, using a standard hydrograph separation technique, to obtain the direct runoff hydrograph Q:
$$Q(t) = Q_T(t) - Q_b(t) \tag{22.34}$$
3. Using the same baseflow separation technique, separate out the concentration due to baseflow. It may be noted that Rendon-Herrero assumed that the maximum of runoff and sediment concentration occurred at the same time.
4. Compute the sediment discharge Q_s due to direct runoff by noting that sediment discharge is the product of water discharge and sediment concentration:
$$Q_s = Q_T C_T - Q_b C_b \tag{22.35}$$
where C_T and C_b are sediment concentrations corresponding to Q_T and Q_b, respectively.
5. Compute the volume of direct runoff, which is the area under the direct-runoff hydrograph (DRH):
$$V_Q = \int_0^\infty Q(t)\,\mathrm{d}t \tag{22.36}$$
6. Compute the sediment load, V_s, which is the area under the sediment graph (DSG) due to direct runoff:
$$V_s = \int_0^\infty Q_s\,\mathrm{d}t \tag{22.37}$$
7. Divide the ordinates of the sediment graph by the sediment load to obtain ordinates of the USG, h_s:
$$h_s = \frac{Q_s}{V_s} \tag{22.38}$$

The USG varies somewhat with the intensity of the effective rainfall (ER). It can be used to generate a sediment graph for a given storm if the wash load produced by that storm is known. A relationship between V_s and V_Q was proposed, as shown in Figure 22.7. Using this relation, V_s can be determined. Therefore, Q_s can be determined by multiplying h_s with V_s. It must be noted that the duration of the USG chosen to determine Q_s must be the same as that of the effective rainfall generating V_Q. The USG method was tested on a small wash-load-producing watershed, Bixler Run Watershed, 39 km^2 in area, near Loysville, Pennsylvania.

Rendon-Herrero (1974) proposed the use of the so-called "series" graph to determine the sediment hydrograph. This method has the advantage that the duration of the effective rainfall is neglected altogether, but requires construction of the series graphs beforehand. Thus, this method cannot be extended to ungaged basins.

22.3.10 Instantaneous Unit Sediment Graph (IUSG)

Williams (1978b) extended the concept of the instantaneous unit hydrograph (IUH) to determine sediment discharge from an agricultural watershed, using the instantaneous unit sediment graph (IUSG). The concept of the IUSG has been employed by Singh et al. (1982), Chen and Kuo (1984), and Srivastava et al. (1984). The IUSG is similar to the IUH and is, therefore, subject to similar limitations. It can be defined as the distribution of sediment, $h_s(t)$, from an instantaneous burst of rainfall producing one unit of runoff, and is considered to be the product of the IUH, $h(t)$, and the sediment-concentration distribution, $C(t)$.

$$h_s(t) = h(t)C(t) \tag{22.39}$$

A basic assumption of the IUH is that h varies linearly with the volume of the effective rainfall, V_Q. Likewise, IUSG assumes that C also varies linearly with V_Q. Thus, the storm-sediment discharge Q_s can be obtained by convolving $h_s(t)$ with the incremental source runoff squared. Chen and Kuo (1984) expressed

$$Q_s(t) = \sum_{k=1}^{n} h_s[D,t - (i - 1)D]S_e(t)D \tag{22.40}$$

where $S_e(t)$ is the effective sediment-erosion intensity (ESEI) (FT^{-1}), $h_s(D,t)$ is the USG corresponding to the duration D, and D is the time interval of each $S_e(t)$ pulse. The dimensions of Q_s and h_s, respectively, are FT^{-1} and T^{-1}. A value of one hour or less may be appropriate for D.

Equation (22.28) shows a relationship between $Y(F)$ and $Q_v(L)$. A similar relationship exists between $Q(t)$ (LT^{-1}) and $Q_s(t)$ (FT^{-1}). As $Q(t)$ varies with effective rainfall intensity (ERI), so does $Q_s(t)$ with what is called effective sediment-erosion intensity (ESEI), $S_e(t)$. In other words, effective rainfall, depth or intensity, produces the direct runoff and, in turn, the direct runoff produces the direct sediment yield. Conceptually, therefore, the effective rainfall intensity generates its counterpart, the effective sediment-erosion intensity, which when combined with the unit sediment graph produces the sediment graph. The relation between S_e and effective rainfall is of the same form with the same parameter values, as that between $Q_s(t)$ and

$Q(t)$. The sum of the S_e ordinates must equal Y as that of effective rainfall ordinates equals Q_v.

This method can be applied to compute $h_s(t)$ for any watershed as follows. For each pair of $Q(t)$ and $Q_s(t)$ graphs, Q_v and Y can be computed. By using a log-transformed linear regression relation, $Q_s = \alpha Q^\beta$ and Equation (22.29) can be established. By using values of α and β, $S_e(t)$ is computed from the effective rainfall intensity. By employing Equation (22.40), the USG, $h_s(t)$, is then obtained. One can also determine $h_s(t)$ from Equation (22.38).

For a fixed D, say, 1 hour, out of many direct sediment graphs for a watershed, only a few USGs will be available. Thus, an average USG can be computed using an appropriate averaging scheme. This can be done for a number of watersheds, and a regional one-hour mean USG can be obtained. Chen and Kuo (1984) related the peak sediment discharge, the time to peak, and the time base of the USG to the soil-erodibility factor, watershed area, main channel length, main channel slope, and mean watershed elevation. These relationships can be used to synthesize the USG for ungaged watersheds.

EXAMPLE 22.11 Data are given on rainfall, runoff, and sediment for watershed W5 (1.76 mi^2), Pigeon Roost Creek, Mississippi. Develop a unit sediment graph in both dimensional and dimensionless form.

RAINFALL

Time interval (h)	5:00	6:00	7:00	8:00	9:00	10:00	11:00
Intensity (in./h)	0.28	0.19	0.13	0.39	0.03	0.10	0.0

RUNOFF

Time interval (h)	5:00	6:00	7:00	8:00	8:25	9:00	10:00	11:00	12:00	13:00	14:00
Discharge (cfs)	1.4	25.6	53.4	102.5	313.4	266.9	114.7	105.0	75.7	66.0	47.4

SEDIMENT DISCHARGE

Time interval (h)	5:30	6:30	7:30	8:00	8:25	9:00	10:00	10:25	11:00	12:00	13:00	14:00
Discharge (tons/h)	3.6	46.8	71.6	106.2	159.3	158.0	118.8	50.4	41.9	30.0	18.0	7.2

Solution

1. Plot ordinates of the sediment discharge Q_s (tons/h) versus time (h), constructing a sediment discharge graph (SDG), as shown in Figure 22.9.
2. Compute the sediment load (tons) by interpreting the area under the DSG:

$$V_s = \int_{t_1=5}^{t_2=14} Q_s \, dt = 515.33 \text{ tons}$$

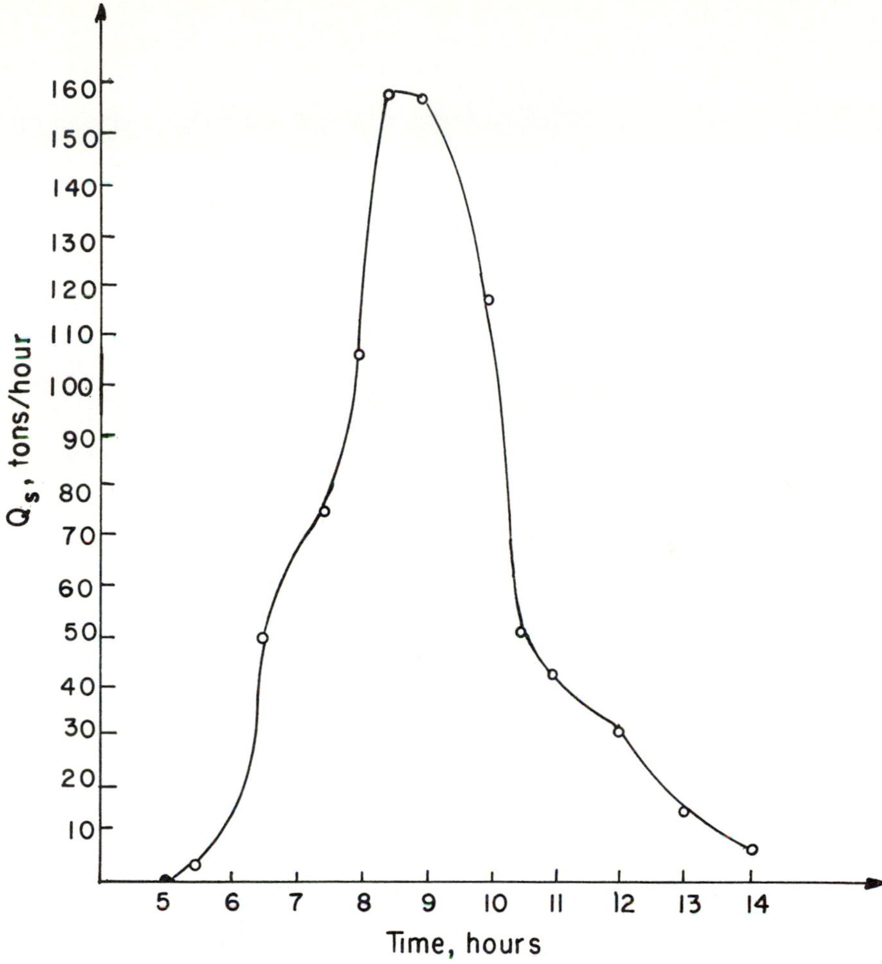

Figure 22.9 Sediment–discharge hydrograph.

3. Divide the ordinates of the sediment graph by the sediment load V_s and obtain the unit sediment graph (USG) in dimensional form, as shown in Figure 22.10.

$$h_s = \frac{Q_s}{V_s}\left[\frac{1}{\text{hour}}\right]$$

Construct the table:

(h–min)	5:30	6:30	7:30	8:00	8:25	9:00
h_s (1/h)	0.0069	0.0908	0.1389	0.2061	0.3091	0.3067
(h–min)	10:00	10:25	11:00	12:00	13:00	14:00
h_s (1/h)	0.2305	0.0978	0.0813	0.0592	0.035	0.014

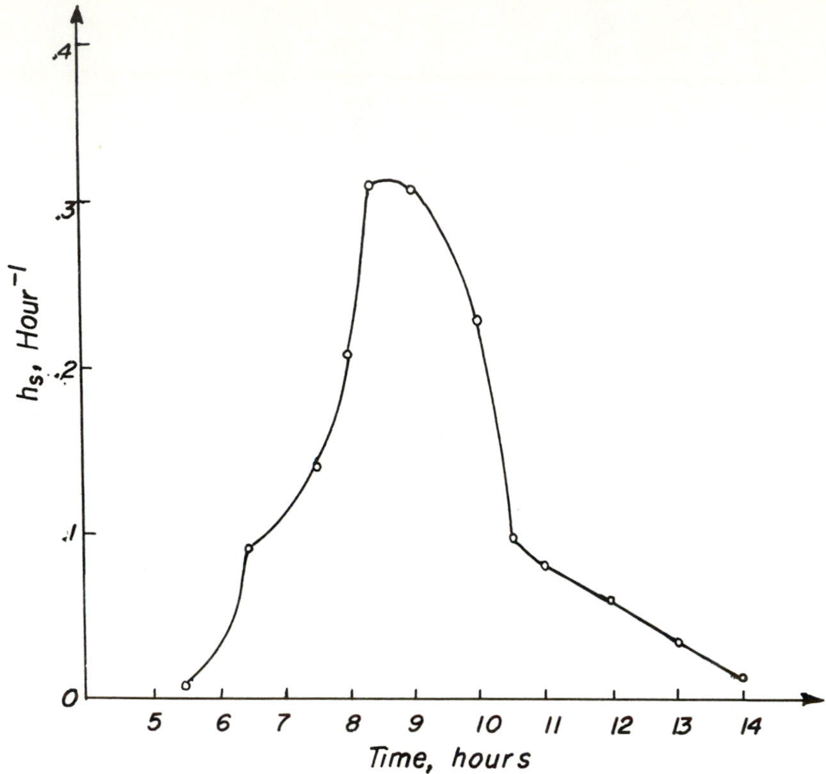

Figure 22.10 Unit-sediment graph.

The USG is nondimensionalized by dividing the sediment graph ordinates by the peak ($h_{sp} = 0.3091$ at 8:25) and its time axis by the lag time (the time interval between the centroid of the rainfall hyetograph and the UH peak: t_L = 1 hour, 25 minutes). Construct the table:

TIME

$t_s^* = \dfrac{t}{t_L}$	1.06	1.76	2.47	2.82	3.12	3.53	4.24	4.53	4.94
$h_s^* = \dfrac{h_s}{h_{sp}}$	0.022	0.294	0.449	0.667	1.00	0.99	0.746	0.316	0.263
$t_s^* = \dfrac{t}{t_L}$	5.65	6.35	7.06						
$h_s^* = \dfrac{h_s}{h_{sp}}$	0.1921	0.113	0.045						

The dimensionless USG is shown in Figure 22.11.

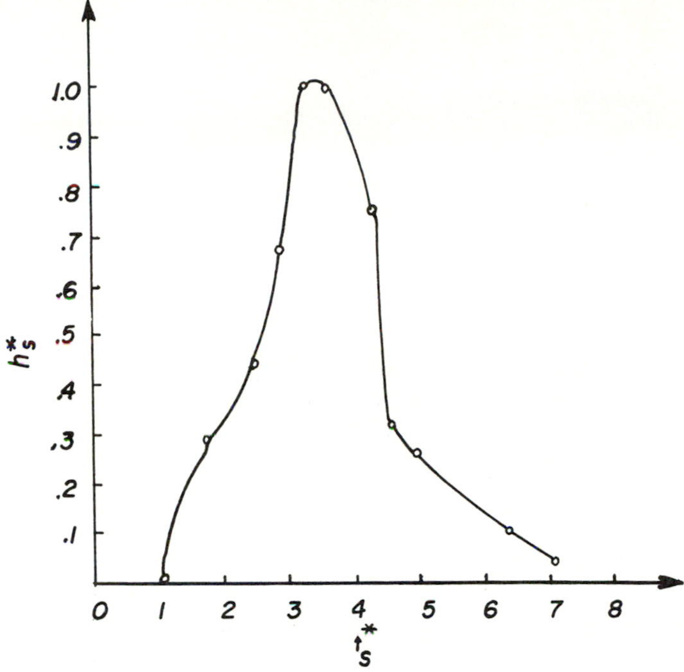

Figure 22.11 Dimensionless unit-sediment graph.

EXERCISES

22.1. Compute the average soil loss by the Musgrave equation for an area near Temple, Texas, where the soil is Austin clay. The erodibility of the soil is 0.18 inches/year. The land slope is 4%, the slope length is 75 feet, and the maximum 30-minute rainfall is 1.5 inches. The crop-cover factor is taken as 40% of that of continuous row crops.

22.2. Estimate soil erosion using the Musgrave equation for a corn-producing area. The soil is silt loam. The land slope is 3%, the slope length is 200 feet, and the maximum 30-minute rainfall is 1.5 inches. The erodibility of the soil is 0.3 inches per year. The corn crop factor is 20% of that of continuous row crop.

22.3. Compute the rainfall erosion index for maximum 30-minute rainfall intensities of 0.5, 0.75, 1.0, 1.25, 1.5, 1.75, and 2.0 inches per hour. Plot variation of this index with rainfall intensity.

22.4. Compute the soil-erodibility factor for three surface soils available in your area. Comment on the variation of this factor with soil composition.

22.5. Calculate the topographic factor for the following surfaces:

	SLOPE (%)	LENGTH (ft)	SLOPE (%)	LENGTH (ft)	SLOPE (%)	LENGTH (ft)
(a) Convex	10	100	30	100	50	100
(b) Straight line	30	100	30	100	30	100
(c) Concave	50	100	30	100	10	100

22.6. Calculate the average annual soil erosion from an area (5 mi^2) composed of three parts (0.5, 1, 2.5 mi^2) whose USLE parameters are as follows:

PART	R	K	LS	C	P
1	300	0.32	0.30	0.20	0.10
2	250	0.31	0.35	0.15	0.10
3	400	0.36	0.30	0.10	0.12

22.7. Calculate the soil erosion from an area subjected to a storm producing a runoff volume of 20 acre-feet and a peak flow rate of 20 ft^3/s. The parameters of the USLE are $K = 0.3$, $LS = 0.3$, $C = 0.1$, and $P = 0.15$.

22.8. Calculate sediment yield using the DB method for the drainage areas for which annual runoff values are given:

Drainage Area (mi^2)	2	10	100	1,000	10,000
Annual Runoff (in.)	1.5	2.0	3.5	5	3.75

22.9 Estimate sediment yield for a drainage basin having the following characteristics: area = 50 km^2, main channel length = 10 km; main channel slope = 3 m/km; mean basin elevation = 200 m; erodibility factor = 0.32, shape factor = 0.3, and forest area = 10 km^2.

CHAPTER 23

Streamflow Simulation

Streamflow at the outlet of a watershed is simulated either for individual storm events or continuously in time. Both the event-based streamflow simulation (EBSS) and the continuous streamflow simulation (CSS) are employed to address a wide array of environmental and water-resources problems. Excellent discussions on the use of streamflow simulation have been presented by Klemes (1973), Jackson (1982), James and Burges (1982), James et al. (1982), Office of Technology Assessment (1982), and Sorooshian (1983), amongst others. Table 23.1 lists some example applications. Since the development of the famous Stanford Watershed Model (SWM) IV (Crawford and Linsley, 1966), numerous CSS models have been developed for particular purposes; many of them are variants of the SWM. Similarly, many EBSS models have been developed during the last 25 years or so (Renard et al., 1982; Singh, 1982). The objectives of this chapter are to present general concepts perceived to be relevant to streamflow simulation and to make some comments about some of the models.

23.1 EVENT-BASED STREAMFLOW SIMULATION (EBSS)

23.1.1 Elements of EBSS

The EBSS models attempt to mimic rainfall–runoff processes. Main elements are presented here in the same order as needed to develop an EBSS model. Minor deviations may occur, depending upon availability of data and some other factors.

TABLE 23.1 EXAMPLE PROBLEMS FOR APPLICATION OF THE EVENT-BASED AND CONTINUOUS STREAMFLOW SIMULATION

Application of EBSS	Application of CSS
1. Design of hydraulic structures such as dams, culverts, bridges, and spillways	1. Extension of streamflow records
2. Design of urban and highway drainage	2. Flow forecasting
3. Planning of flood control works	3. Watershed experimentation
4. Urban planning and development	4. Supplementing of stream gaging program
5. Assessment of nonpoint source pollution	5. Evaluation of the effect of land-use practice on watershed response
6. Disposal of waste material	6. Design of urban drainage, highway culverts, reservoirs, etc.
7. Evaluation of environmental impacts of land use and management practices	7. Planning urban development, river training works
8. Planning of soil-conservation works	8. Water-quality modeling
9. Assessment of flood damage	9. Water-supply development
10. Erosion and sediment transport	10. Flood mitigation
11. Evaluation of hydrologic consequences of climatic change	11. Drought management
12. Determination of flood-insurance program	12. Irrigation planning and management

WATERSHED REPRESENTATION

A watershed is usually a complex and heterogeneous system. Its characteristics vary in space. Hydrologic processes vary both in space and time. One way to account, at least partly, for spatial variability of governing hydrologic factors is to divide the watershed into nearly homogeneous subbasins. As a minimum, there should generally be one subbasin for each rain gage that provides independent data. If rain gages are located too close together, then they would virtually record the same precipitation and hence should not be treated separately. Furthermore, differences in soil, vegetation, land use, or topography, which significantly affect streamflow, must be considered to demarcate the subbasins. Of course, the larger the number of subbasins, the greater the accuracy as well as the cost of simulation. There is, thus, a trade-off between the accuracy of simulated streamflow and the cost of simulation that must be determined. Once the watershed is decomposed, as shown in Figure 23.1, computations proceed from the most remote upstream subbasin in a downstream direction, as shown in Figure 23.2.

DETERMINATION OF EFFECTIVE RAINFALL AMOUNT

Fundamental to streamflow simulation is the determination of the amount of effective rainfall (ER) or the volume of direct runoff (DR) that a given rainfall event produces on each portion of the watershed. Determination of this quantity is most difficult and is a major cause of error. Methods for obtaining this quantity are presented in Chapter 14. For example, the SCS curve-number method is one such

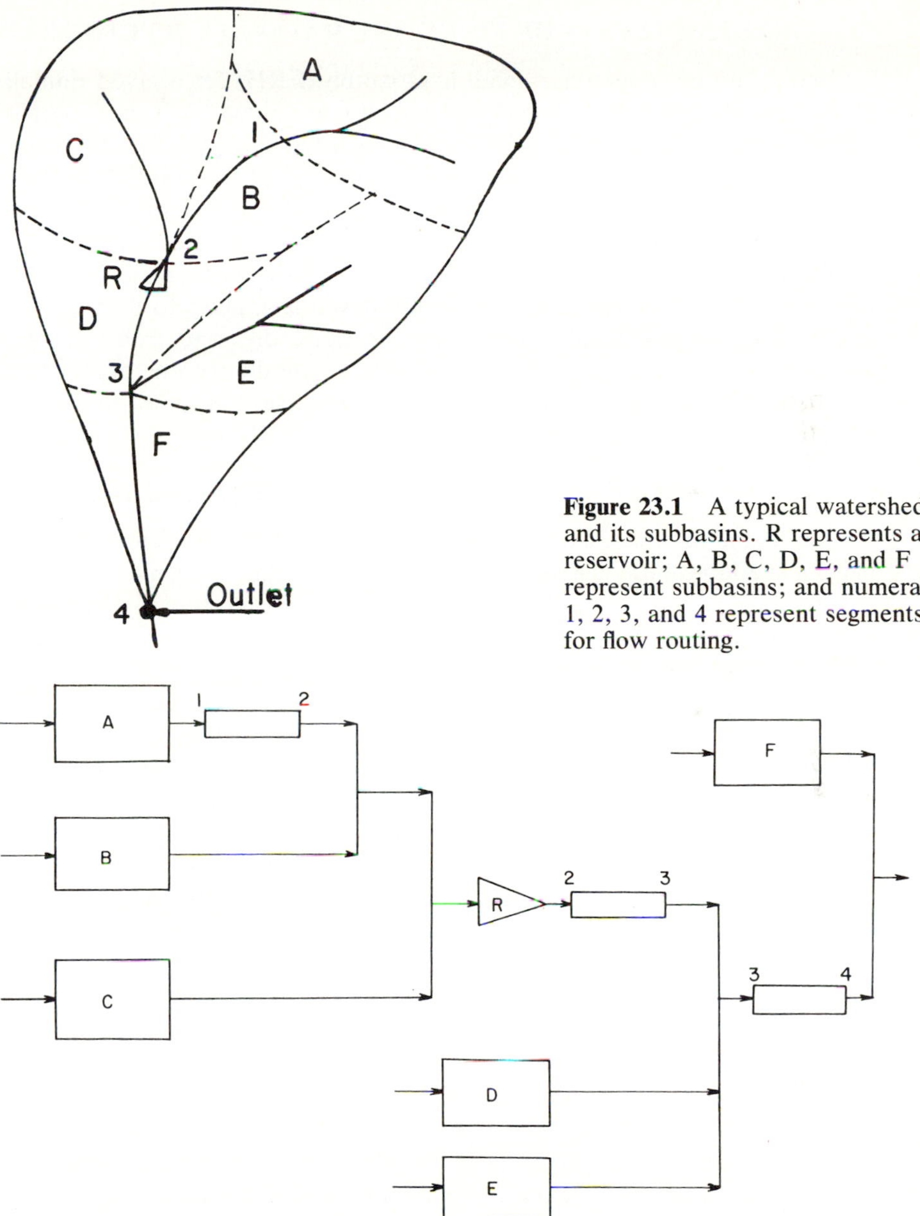

Figure 23.1 A typical watershed and its subbasins. R represents a reservoir; A, B, C, D, E, and F represent subbasins; and numerals 1, 2, 3, and 4 represent segments for flow routing.

Figure 23.2 Computation sequence for streamflow simulation for the watershed in Figure 23.1. Each block represents a subbasin; it receives effective rainfall and produces direct runoff which is then routed through a channel and/or reservoir.

method. If streamflow observations are available, then the DR volume can be estimated using hydrograph separation for each subbasin.

DETERMINATION OF EFFECTIVE RAINFALL HYETOGRAPH

To determine the effective rainfall hyetograph (ERH) for a given rainfall event requires knowledge of the ER amount, which is obtained as discussed before. Given this amount, the ERH is then determined for each subbasin by employing an appropriate infiltration model and satisfying the equality of ER and DR amounts or continuity equation. This step usually requires an iterative procedure. If the ER amount is not known, then the rainfall infiltration can be estimated explicitly and, consequently, the ERH, as discussed in Chapter 8. In this case, however, satisfaction of the continuity equation cannot be assured when applied to actual data. In other words, the computed ER amount may be quite different from the observed DR amount. It is, of course, implied that infiltration model parameters are known or can be estimated from given data. In the previous case, the infiltration parameters are adjusted for satisfying the equality of the ER and DR amounts.

COMPUTATION OF DIRECT-RUNOFF HYDROGRAPH

At this stage, a decision has to be made regarding the type of the model to be used in light of available data. The various types of models have been presented in Chapters 15, 16, and 17. Suppose a linear conceptual model is chosen for this purpose. Then the instantaneous unit hydrograph (IUH) for this model must be derived, as well as a method for estimating the IUH parameters. Thereafter, the IUH is convoluted with the ERH estimated already to obtain the direct-runoff hydrograph (DRH). To summarize, three tasks are performed for this choice for each subbasin: (a) computation of the IUH, (b) determination of the IUH parameters, and (c) convolution of the IUH with the ERH.

FLOW ROUTING

Depending upon the drainage pattern and the existence of dams or reservoirs, bridges, etc., within the basin, the DRH of each subbasin is to be routed during its journey to the watershed outlet (see Figure 23.2). Here again a decision has to be made about the routing method. Suppose that the Muskingum method is chosen for flow routing through channels, and the storage indication method for flow routing through reservoirs. Then the parameters of the Muskingum method (weighting factor and lag time) are estimated first for each channel reach. Because this method is linear, it will be sufficient to obtain its IUH and then convolute it with the reach inflow hydrograph, which is usually the DRH, to compute the reach outflow hydrograph. Alternatively, once the Muskingum parameters are estimated, the channel-flow routing can be accomplished using the discrete Muskingum formulation. For reservoir-flow routing, storage–elevation and outflow–elevation relationships must be established for each reservoir using topographic maps and empirical observations. These relationships are then combined to form a storage–outflow graph, which is then used to perform flow routing through the reservoir. Methods of channel-flow and reservoir-flow routing are discussed in Chapters 21 and 20, respectively.

PARAMETER ESTIMATION

Two approaches are used to estimate parameters of the EBSS models. The first approach estimates parameters from physical measurements of the watershed; the implicit assumption here is that the model is physically based and contains physical parameters. Strictly speaking, no EBSS model seems to satisfy this assumption. The second approach systematically optimizes model parameters corresponding to a chosen objective function. Most EBSS models have some parameters that can be estimated from physical measurements and have other parameters that are found by optimization using observed hydrologic data. Several optimization methods have been used for parameter estimation of EBSS models (Ibbitt and O'Donnell, 1971; Johnston and Pilgrim, 1976; Sorooshian and Gupta, 1983). A comprehensive discussion of potential problems encountered in parameter optimization has been given by Sorooshian (1983). Green and Stephenson (1986) have discussed a number of objective criteria that can be used in optimization as well as for comparing the models. James and Burges (1982) have presented a comprehensive discussion on selection, calibration, and testing of hydrologic models.

The parameter estimation is accomplished in two steps: model calibration and model verification. To this end, hydrologic data are split into calibration and verification sets. According to Sorooshian (1983), the purpose of calibration is twofold: (1) to obtain a conceptually realistic and unique parameter set that closely reflects our understanding of the watershed system; and (2) to obtain a parameter set that produces the best possible agreement between model-simulated and observed streamflows for the calibration data set. Sorooshian and Dracup (1980) have pointed out that most calibration methods tend to emphasize the second aspect. The consequence of so doing is that the model, using the calibrated parameters, produces a close reproduction of the observed streamflow (for the calibration data), but may predict streamflow poorly for the verification set. This phenomenon is sometimes referred to as model divergence. Thus, three requirements can be specified for a parameter estimation method: (1) a suitable estimation criterion, (2) an efficient optimization algorithm, and (3) appropriate calibration data.

The estimation criterion must describe both the stochastic elements of the data and the physical situation. Lacking that, the parameter estimates will be biased (Sorooshian and Dracup, 1980). The most commonly used least squares criterion provides efficient and unbiased parameter estimates only if the variance of data errors is constant and these errors are time-independent. Streamflow measurements seldom satisfy these restrictions. As a result, Williams and Yeh (1983), and Kuczera (1982) have suggested the use of the generalized least squares estimator to eliminate the effect of serially correlated errors associated with streamflow measurements.

Obtaining unique and realistic parameter estimates through optimization is complicated by the existence of multiple optima, parameter interaction, and parameters with varying degrees of sensitivity. According to Sorooshian and Gupta (1983), these complications arise from model structure representation and nonlinearity in the parameter space, imperfect representation of the physical process by the model, and data and their associated measurement errors. The performance of an optimization method can be significantly improved by employing such techniques as optimization in transformed parameter space, reparameterization of the model equation, and the use of an objective convergence criterion.

The calibration data play a vital role in the success of an optimization method. The data should be as representative of the watershed conditions as possible. Sorooshian et al. (1983) have noted that the information contained in the data (hydrologic variability) and the efficiency with which this information is extracted are more important than the length of the data. Such data should have as few errors as possible and should represent disparate climatic and geographical locations. Fiering and Kuczera (1982) have stressed the need for the use of three or more state variables, and not simply streamflow data, in parameter estimation.

The models should be validated under varied and diverse hydrologic conditions (Klemes, 1982). In other words, the verification record should be as remote from the calibration record as possible. The conventional statistical goodness-of-fit criteria such as root-mean-square error, coefficient of variation, bias, etc., are not necessarily accurate indicators of the best model fit (Sorooshian et al., 1983). Kitanidis and Bras (1980) have suggested the use of the coefficient of persistence. Other indicators suggested are a linear combination of the mean ratio and its standard deviation (Cunday and Brooks, 1981), the percent root-mean-square error by flow group (Restrepo-Posada and Bras, 1982), and the percent bias by flow group (Sorooshian et al., 1983).

23.2 DATA NEEDS FOR EBSS MODELS

Before undertaking streamflow simulation for a particular storm, it is advisable to assess the quantity and quality of available data. Quite often, the available data dictate the type of simulation model to be used more than the problem itself. A general inventory of data frequently available or needed is given in what follows.

23.2.1 Watershed Characteristics

The most commonly available is the topographic map from which many useful geomorphic parameters can be extracted, that is, watershed area, subbasin areas, elevations, slopes, channel lengths, channel profiles, centroid, etc. Many other geomorphic parameters can then be computed, as discussed in Chapter 5. Another useful map is the land-use map, which provides data on areas of land-use practice, soil types, vegetation, forest areas, lakes, urban development, etc.

23.2.2 Rainfall Characteristics

Rainfall hyetographs are needed for each subbasin. Some of the subbasins may not have a recording rain gage and may involve extrapolation of rainfall data from neighboring subbasins. If a subbasin has more than one rain gage, then the mean areal rainfall hyetograph is to be determined using one of the methods described in Chapter 7. Sometimes, only storage-type rain gages are available in some watersheds. The rainfall amounts then need to be properly distributed in time so that rainfall hyetographs can be prescribed.

23.2.3 Infiltration and Other Loss Characteristics

In a majority of cases, no data are available on soil infiltration, interception, depression storage, and antecedent soil moisture. If data do exist in part or full, maximum advantage must be taken to estimate infiltration (discussed in Chapter 8) and other loss functions (discussed in Chapters 10 and 11). If no information is available on antecedent soil moisture, then an antecedent precipitation index can be used to get an estimate of the antecedent soil moisture. Soil type and land-use vegetation complex can be used to estimate infiltration parameters.

23.2.4 Streamflow Characteristics

Streamflow may be available in terms of the stage at the watershed outlet and at some other gages within the watershed. Appropriate rating curves can be used to convert stages into discharges. Part of the streamflow data may be used for model calibration and the remaining data for model verification.

23.3 BUILDING AN EBSS MODEL

A number of steps are involved in building a streamflow simulation model, which are outlined in what follows. Two cases are distinguished: (a) the model is to be calibrated using an optimization method, and (b) the model calibration is not required.

1. Divide the watershed into subbasins depending upon the number of rain gages, land-use practices, stream gages, etc.
2. Assess the data that are available.
3. Decide the type of the model to be used.
4. Prepare the rainfall data for each subbasin.
5. Choose the infiltration model.
6. Choose the ER–DR model.
7. Choose the channel-routing model.
8. Choose the reservoir-routing model.
9. Decide if the model parameters are to be optimized. If the answer is no, go to step 18; otherwise, go to the next step.
10. Choose an objective function, for example, the sum of squares of deviations.
11. Choose a parameter-optimization method, for example, the Rosenbrock–Palmer algorithm.
12. Choose infiltration parameters.
13. Choose ER–DR model parameters.
14. Choose channel-flow routing parameters.

15. Choose reservoir-flow routing parameters.
16. Compute the ERH using the infiltration model.
17. Go to step 22.
18. Compute the ERH for each subbasin. This can be done in two ways. First, perform hydrograph separation if streamflow observations are available. Compute the DR amount. Using the infiltration model, compute the ERH such that the ER amount is the same as the DR amount. The equality between these two amounts is achieved by adjusting infiltration parameters. Second, estimate the infiltration parameters. Then estimate the rainfall infiltration and consequently the ERH.
19. Estimate the parameters of the ER–DR model for each subbasin.
20. Estimate the parameters of the channel-flow routing model for each channel reach.
21. Estimate the parameters of the reservoir-flow routing model for each model.
22. Compute the DRH for each subbasin.
23. Route the DRH through the appropriate channel.
24. Route the flow through the reservoir.
25. Combine the routed flows with the DRHs of the subbasins to produce the DRH at the watershed outlet.
26. Stop if the model parameters are not to be optimized. Otherwise, go to the next step.
27. Compute the objective function.
28. Check if the objective function is satisfied. If yes, stop. Otherwise, go to the next step.
29. Update the parameter values.
30. Go to step 18.

23.4 EBSS MODELS

A large number of EBSS models are available. Some of these models along with other types have been briefly described by Brown, et al. (1974). Some of these models have also been discussed by Fleming (1975), James et al. (1982), Larson et al. (1982), Singh (1989), and Viessman et al. (1989). Only a few representative models are summarized in Table 23.2. To complete the discussion, we briefly consider the HEC-1 model.

The HEC-1 flood hydrograph model was developed by the Hydrologic Engineering Center (1981) to simulate the DRH due to precipitation by representing the watershed with interconnected hydrologic and hydraulic components. In addition, the model has options for multiplan–multiflood analysis, dam-break simulation, economic assessment of flood damage, and optimal sizing of flood-control systems. This model has been extended to determining discharge-frequency relationships for ungaged watersheds (Hydrologic Engineering Center, 1982). This is perhaps the most comprehensive EBSS model. Many elements of simulation are modeled using several options. Infiltration is estimated using four options: (a) initial and uniform

TABLE 23.2 SOME EVENT-BASED STREAMFLOW SIMULATION MODELS

Model			Model components						
Name	Author(s)	Year	Baseflow separation	DR volume	Infiltration and loss	DR hydrograph	Channel routing	Reservoir routing	Parameter optimization
HEC-1	Hydrologic Engineering Center	1981, 1982	Yes	SCS curve-number and two other methods	Variable loss-rate method	Clark's and Snyder's unit hydrograph methods	Muskingum method and five other methods	Storage indication method	Automatic calibration capability
TR-20	Soil Conservation Service	1973	Constant-rate method	SCS curve-number method	SCS curve-number method	Unit hydrograph method	Convex method	Storage indication method	No
USGS	Dawdy, et al.	1972	Constant-rate method	Soil-moisture accounting	Philip equation	Clark's unit hydrograph	Translation method	No	Rosenbrock's method
HYMO	Williams and Hann	1973	No	SCS curve-number method	SCS curve-number method	Nash model	Variable storage-coefficient method	Storage indication method	No
SWMM	Metcalf and Eddy, Inc., et al.	1971	No	Loss accounting	Horton's equation	Hydraulic method	Hydraulic routing method	No	No
WAHS	Singh	1983	Recession equation	SCS curve-number method	Philip's equation	Geomorphological unit hydrograph method	Linear reservoir	No	Rosenbrock–Palmer method
RORB	Laurenson and Mein	1983	Two options	No	Constant and variable loss-rate methods	Nonlinear storage routing	Nonlinear storage routing	Yes	No
WBNM	Boyd et al.	1979	No	Yes	ϕ index	Linear as well as storage elements for routing	Storage routing	No	Yes
FHSM	Foroud and Broughton	1981	Yes	Yes	Modified Horton's equation	Time–area curve + a linear reservoir	No	No	Nonlinear least square curve fitting
XJM	Zhao et al.	1980	Yes	Yes	Storage-capacity curve	Unit hydrograph method	Muskingum method	No	No
GAWSER	Ghate and Whiteley	1982	Yes	Yes	Holtan's equation	Time–area curve + convolution	HYMO method	No	No
MIT	Maddaus and Eagleson	1969	No	No	Any suitable model	Linear channel and reservoir	Linear	No	Optimization
HM	Huggins and Monke	1968	No	Yes	Holtan's equation	Kinematic wave method	No	No	No
Kansas	Smith and Lumb	1966	Yes	Yes	Soil-moisture accounting	Lag and route method	No	No	No
IHM	Morris	1980	Yes	Yes	Richards equation	St. Venant equation	St. Venant equation	No	No

loss rate, (b) exponential loss rate, (c) SCS curve-number method, and (d) Holtan's infiltration equation. The DRH is estimated using the unit hydrograph method (with Clark, Snyder, and SCS dimensionless UH methods as options) and the kinematic wave method. The conic method, normal-depth storage and outflow, and the modified Puls method are used for storage–flow routing, whereas the lag and route and Muskingum methods are used for channel-flow routing. A univariate search technique is employed to determine optimal model parameters. This model is one of the most commonly used models in the United States, and can be used for hydrologic analyses under a wide variety of conditions (Feldman, 1981).

23.5 CONTINUOUS STREAMFLOW SIMULATION

23.5.1 Comparison with Event-Based Simulation

The CSS models allow simulation of streamflow for long periods of time, and thus more fully utilize the capability of the digital computer. These models maintain more or less a continuous accounting of the water in storage in the watershed. Because of the long periods of time, such hydrologic processes as evaporation and transpiration, infiltration, interception, depression storage, subsurface flow, and baseflow assume added significance. In EBSS, some of these processes are neglected, some are lumped, and some are considered with considerable approximation, for the period of simulation is usually as long as the duration of the direct-runoff hydrograph (DRH). The emphasis in CSS is on simulation of the entire land phase of the hydrologic cycle, whereas the emphasis in EBSS is on modeling the DRH or its peak characteristics. Thus, CSS models are models of the hydrologic cycle, whereas EBSS models are models of the rainfall–runoff cycle. It is logical to say that CSS models are more general and encompass EBSS models as their special cases. Naturally then, discussion of CSS models involves what has already been presented about EBSS models plus discussion of components not included in EBSS models. Reviews of CSS models have been reported by Fleming (1975), Larson et al. (1982), and Linsley (1982), among others. Some of these models are presented in Haan (1982) and Singh (1982). Renard et al. (1982) inventoried currently available models.

23.5.2 Elements of CSS

Two phases are simulated: the land phase and the channel phase. The land phase is much more complicated in CSS than in EBSS, but the channel phase is about the same.

WATERSHED REPRESENTATION

As in EBSS, the watershed is divided into homogeneous subbasins or units. Most watersheds contain some impervious area consisting of roads, rock outcrops, buildings, and impervious soils, as well as exposed surfaces of streams, ponds, lakes, reservoirs, and swamps. In most watersheds, the impervious area is usually less than 3% of the total area. In urban areas, this percentage will be much higher. In some watersheds, lakes and swamps occupy a significant portion of the total area. Proper allowance is made for inclusion of imperviousness and water surfaces in CSS models.

MEAN AREAL RAINFALL

Rainfall measured at a gage may not be equal to the average rainfall over the subbasin the gage is supposed to represent. Some subbasins may have more than one rain gage. Whatever the case, the mean areal rainfall is estimated with least bias for each subbasin. Chapter 7 presents methods of estimating the mean areal rainfall.

INTERCEPTION

Maximum interception storage is estimated for each subbasin. Some areas have more forest canopy and urban dwellings, and have higher interception storage than others. The intercepted water is usually disposed of through potential evapotranspiration. The maximum interception storage capacity is usually less than 3 mm. That is why it is negligible for large rainfall events, but may be significant over the period of a year in CSS models. A more complete discussion of interception is contained in Chapter 11.

DEPRESSION STORAGE

On most watersheds, depression storage is a small quantity, and can be lumped with soil-moisture storage, which is discussed in the following section. However, in urban areas or where the watersheds have significant depressions, ponds, etc., it will be significant. Moisture as depression storage is partly evapotranspired and partly infiltrates the soil. An appropriate model is provided to compute depression storage, and then to link with evapotranspiration and infiltration models. Depression storage is discussed in Chapter 11.

SOIL-MOISTURE STORAGE

Central to a CSS model is soil-moisture storage, for it factors in estimation of infiltration, interflow, and percolation to groundwater. Usually, the upper soil profile is considered for this storage; the thickness of the profile may be more than the plant root zone (up to 100 cm). This profile is divided into two or more zones whose maximum moisture-holding capacities are estimated from knowledge of soil and vegetation characteristics. The moisture in these zones is depleted by evapotranspiration and replenished by rainfall infiltration or snowmelt. A model taking into account these interactions is needed. Chapter 14 discusses soil-moisture accounting more fully.

INFILTRATION

When it rains, a portion of rainfall infiltrates into the soil. The rate at which water infiltrates depends on the moisture available in the soil. An infiltration model is chosen to specify the infiltration-capacity rate and then linked with the soil-moisture model. Infiltration models are discussed in Chapter 8. The initial infiltration capacity is estimated using available soil-moisture storage. Thus, infiltration and soil-moisture storage are treated interactively.

EVAPOTRANSPIRATION

Evapotranspiration is the mechanism to transport moisture from the land surface to the atmosphere. Intercepted water is evaporated. Moisture from the soil profile is evapotranspired. Evapotranspiration can take place even from shallow groundwater. The rate of evapotranspiration depends upon the availability of moisture. Thus, evapotranspiration and soil moisture interact with one another. An appropriate evapotranspiration model can be used to estimate the rate of evapotranspiration, and then to link with the soil-moisture model. Evapotranspiration models are discussed in Chapter 10.

INTERFLOW

Interflow depends on the moisture contents of the soil-moisture zones. Thus, soil-moisture storage is the key to determining interflow. A method is provided to compute interflow due to given rainfall. Interflow eventually merges with surface runoff.

BASEFLOW

Baseflow is derived from groundwater storage and sustains streamflow during dry periods. Part of the moisture in excess of soil-moisture storage recharges the groundwater aquifer. A method is provided to compute the baseflow contribution to streamflow during and subsequent to a rainfall–runoff event. Chapter 13 discusses separation of groundwater from surface runoff, and Chapter 9 presents models for baseflow recession.

SURFACE RUNOFF

The portion of rainfall, not lost to infiltration and other losses, is to be transformed to surface runoff (or overload flow). In reality, surface runoff, infiltration, and, in turn, soil moisture are interactive. An appropriate model is chosen to compute surface runoff from the land surface of the subbasin. If the subbasins are highly heterogeneous, it is advisable to have different surface-runoff models reflecting particular features of the subbasins. Surface-runoff models are presented in Chapters 15 to 17.

CHANNEL-FLOW ROUTING

The DRH obtained from the land surface defines the inflow hydrograph for the channel. Of course, the channel may have its own drainage area contributing lateral flow to it. An appropriate model is chosen to route flows in the channel. During the course of routing, smaller streams contribute flow to larger streams. Therefore, flow routing is to be performed in accordance with watershed partitioning. Channel-flow routing models are discussed in Chapter 21.

RESERVOIR-FLOW ROUTING

If the watershed has reservoirs, the flow is routed accordingly. Proper allowance must be made for the presence of highways, culverts, bridges, etc. The model to be chosen must reflect these features. Models for reservoir-flow routing are given in Chapter 20.

23.6 DATA NEEDS FOR CSS MODELS

In order for a CSS model to be widely usable, its data requirements must be as minimal as possible. Three types of data are usually available: (1) watershed characteristics (discussed in Chapter 7), (2) climatic characteristics, and (3) hydrological characteristics. Soil, land use, and topographic data reflect watershed characteristics. Rainfall and meteorological data such as temperature, radiation, humidity, pressure, etc., are normally the climatic data. Hydrological data may include not only observed hydrologic data such as streamflow, potential evapotranspiration, soil moisture, infiltration, etc., but also information on parameters of hydrologic models used in CSS. As far as possible, reliance on observed data should be minimized and that on hydrologic understanding be maximized.

23.7 PARAMETER ESTIMATION

Most CSS models require parameter optimization, for their parameters cannot be entirely determined from physically measurable physiographic characteristics. In some cases, some of the model parameters are related to watershed characteristics, but not all parameters. Many parameter-optimization methods have been used in hydrology, as indicated earlier. Suitability of an optimization method often depends on the structure of the CSS model whose parameters are to be optimized, data errors, and convergence criterion. For a comprehensive discussion on this subject, reference is made to James and Burges (1982) and Sorooshian (1983).

23.8 BUILDING A CSS MODEL

From the preceding discussion, it is clear that building a CSS model involves modeling the various components of the hydrologic cycle and maintaining a continuous water balance involving these components. Some of these components are interactive and involve iterative calculation. Larson et al. (1982) have presented a good discussion on assembling these components into a CSS model. Figure 23.3 shows a general conceptual framework for building a CSS model. Many of the models possess similar arrangements of components. These models are summarized in Table 23.3.

The Stanford water model IV (SWM), developed by Crawford and Linsley (1966), is perhaps the most widely accepted model for simulation of the land phase of the hydrologic cycle. The model has been applied to many watersheds throughout the world, and many modified versions of it have been developed. Its applications have encompassed data extension, flood forecasting, flood-frequency analysis, estimation of peak discharge, sediment transport, effect of urbanization and land-use practices, etc.

Hourly and daily precipitation, daily temperature, radiation, wind, monthly or daily evaporation, as well as a variety of watershed parameters constitute input to the SWM. Hourly or daily streamflow at the watershed outlet is the output. The time interval for calculation is 15 minutes. The model is a lumped-parameter representation with 34 parameters. Most of these parameters are physically based, but four of them are obtained by using an optimization scheme, automatic or otherwise. These four parameters pertain to infiltration, soil-moisture zones, and interflow. The remaining parameters (30) are evaluated from maps, surveys, or hydrometeorologic records. When snowmelt simulation is not needed, the number of model parameters reduces to 25.

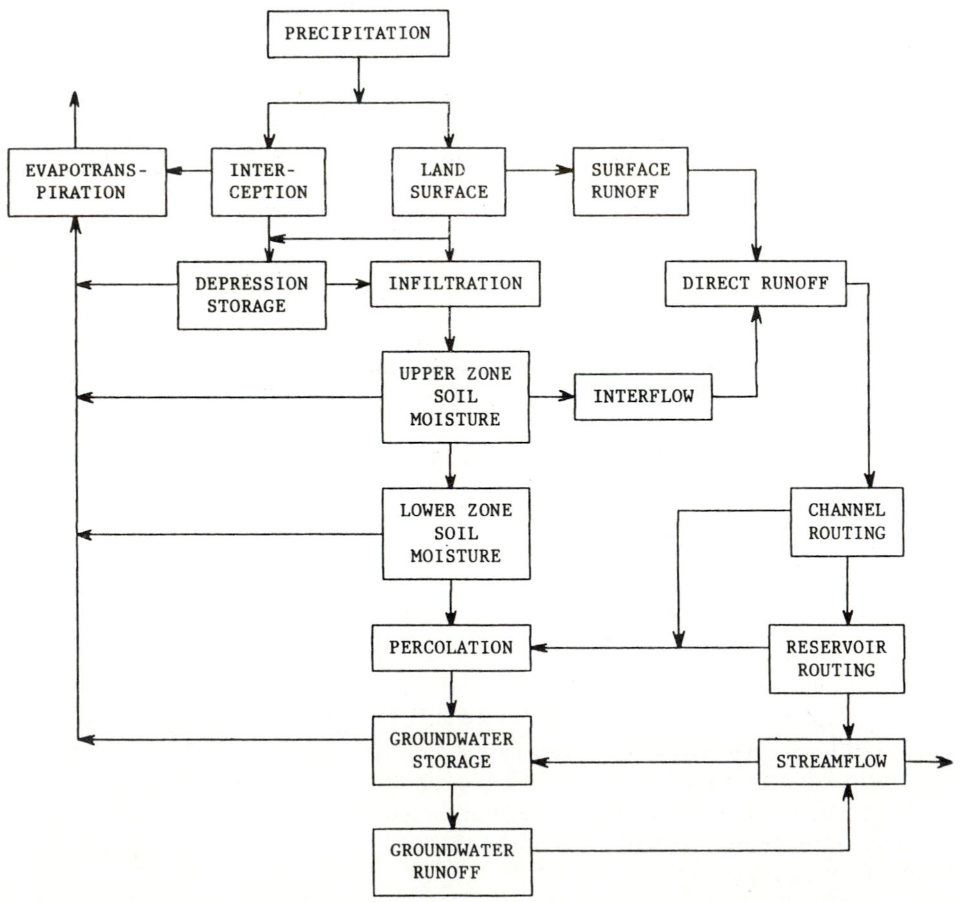

Figure 23.3 Components of a continuous streamflow simulation model.

TABLE 23.3 SOME CONTINUOUS STREAMFLOW SIMULATION MODELS

Model		Model components										
Name	Author	Inter-ception	Infiltra-tion	Soil-moisture storage	Evapo-trans-piration	Surface runoff	Snowmelt runoff	Inter-flow	Ground water runoff	Channel routing	Reservoir routing	Parameter optimiza-tion
SWM IV	Crawford & Linsley (1966)	Yes	Yes	Yes	Yes	Yes	Yes	Yes	Yes	Yes	Yes	No
KWM	Liou (1970)	Yes	Yes	Yes	Yes	Yes	No	Yes	Yes	Yes	Yes	Yes
OPSET	James (1972)	Yes	Yes	Yes	Yes	Yes	Yes	Yes	Yes	Yes	Yes	No
OSUM	Ricca (1972)	Yes	Yes	Yes	Yes	Yes	No	Yes	Yes	Yes	Yes	No
NWSRFS	Hydrologic Research Laboratory (1972)	Yes	Yes	Yes	Yes	Yes	No	Yes	Yes	Yes	Yes	Yes
SSARR	U.S. Army Engineer Division, North Pacific (1975)	No	No	Yes	Yes	Yes	Yes	Yes	Yes	Yes	Yes	Yes
API	Sittner et al. (1969)	No	No	Yes	No	Yes	No	No	Yes	No	No	No
USDA	Holtan et al. (1975)	No	Yes	Yes	Yes	Yes	Yes	Yes	Yes	Yes	No	No
TWM	Claborn and Moore (1970)	Yes	Yes	Yes	Yes	Yes	No	Yes	Yes	Yes	Yes	Yes
TANK	Sugawara et al. (1984)	No	Yes	Yes	Yes	Yes	Yes	Yes	Yes	Yes	No	Yes
HBV	Bergstrom (1976)	No	No	Yes	Yes	Yes	Yes	Yes	Yes	No	No	Yes
SHE	Abbott et al. (1986a, 1986b)	Yes	Yes	Yes	Yes	Yes	Yes	Yes	Yes	No	No	No
CEQUEAU	Charbonneau et al. (1977)	No	Yes	Yes	Yes	Yes	Yes	Yes	Yes	Yes	No	No
MC	Deschenes et al. (1985a, 1985b)	No	Yes	Yes	Yes	Yes	Yes	Yes	Yes	No	No	No
SCM	Refsgard (1981)	Yes	Yes	Yes	Yes	Yes	Yes	Yes	Yes	No	No	No
SRBM	Bultot and Dupriez (1976a, 1976b)	Yes	Yes	Yes	Yes	Yes	Yes	Yes	Yes	No	No	Yes
UBCWM	Quick and Pipes (1977)	No	Yes	Yes	Yes	Yes	Yes	Yes	Yes	Yes	Yes	No
MRM	Porter and McMahon (1971)	Yes	Yes	Yes	Yes	Yes	No	Yes	Yes	Yes	No	Yes
HYSIM	Manly (1978)	Yes	Yes	Yes	Yes	Yes	Yes	Yes	Yes	Yes	Yes	No
ARBM	Chapman (1968)	Yes	Yes	Yes	Yes	Yes	No	Yes	Yes	No	No	Yes
BM	Boughton (1966)	Yes	Yes	Yes	Yes	Yes	No	Yes	Yes	No	No	Yes
HHM	Ando et al. (1983)	No	Yes	Yes	Yes	Yes	No	Yes	Yes	No	No	Yes
TVA	Tennessee Valley Authority (1972)	Yes	Yes	Yes	Yes	Yes	No	Yes	Yes	No	No	Yes
USUWSM	Andrews et al. (1978)	Yes	Yes	Yes	Yes	Yes	Yes	Yes	Yes	Yes	No	Yes

EXERCISES

23.1. Consider a large watershed in your area to which an EBSS model is to be applied. Show partitioning of this watershed and then sketch the computation sequence for streamflow simulation.

23.2. Outline the data requirements of an EBSS model and the sources from which the data can be obtained.

23.3. What are the two most important elements for building an EBSS model? Discuss and present your answer in a logical manner.

23.4. Outline five criteria for selecting an EBSS model for practical use to solve a specified problem. Now select another problem. Will these criteria apply to this problem? If not, indicate the five criteria to be used.

23.5. Analyze the status of EBSS modeling. What elements are represented adequately and what elements not so adequately?

23.6. Draw a schematic showing the computation sequence for a CSS model. Indicate the components of the hydrologic cycle included in the model.

23.7. What are the three most important processes in CSS modeling? Present your answer in a logical manner.

23.8. Consider the following watersheds: (1) entirely agricultural, (2) entirely rural, (3) entirely urban, and (4) entirely forested. What will be the major differences to be considered in building EBSS models for these watersheds?

23.9. What will be the major differences to be considered in building CSS models for the watersheds of Exercise 23.8?

23.10. Indicate the components of the hydrologic cycle that are negligible in an EBSS model but become significant in a CSS model.

PART 9
Hydrologic Design

CHAPTER 24
Statistical Preliminaries

Hydrologic processes such as rainfall, snowfall, floods, droughts, etc., are usually investigated by analyzing their records of observations. Many characteristics of these processes seem to vary in a way not amenable to deterministic analysis. In other words, deterministic relationships, presented so far, do not seem to be applicable for analysis of these characteristics. For example, if one plots instantaneous peak discharges from each year for a river, as shown in Figure 24.1, a rather erratic graph is obtained, and there does not appear to be any deterministic relationship capable of explaining the variation of peak discharge from one year to another. For purposes of hydrologic analysis, the annual peak discharge is then considered to be a random variable. Other examples of random variables are annual maximum rainfall, maximum temperature, maximum wind speed, period of flooding, minimum annual flow, etc. Methods of probability and statistics are employed for analysis of random variables. In this chapter, some elementary concepts of probability and statistics are presented without derivations, which are used for frequency analysis in hydrology.

24.1 MOTIVATION FOR USING PROBABILITY

An engineering design always takes into account the extreme situations. When a highway bridge is built, it must be able to pass the design discharge of specified magnitude without being flooded during its life-span. The same holds when a dam is built, a culvert is built, or a drainage structure is built. When an urban area is developed, storm drainage is provided so that the area is not likely to be flooded. When a building is constructed, it must withstand design wind loads of given magnitude. When a water-supply scheme is designed, the scheme must be able to supply

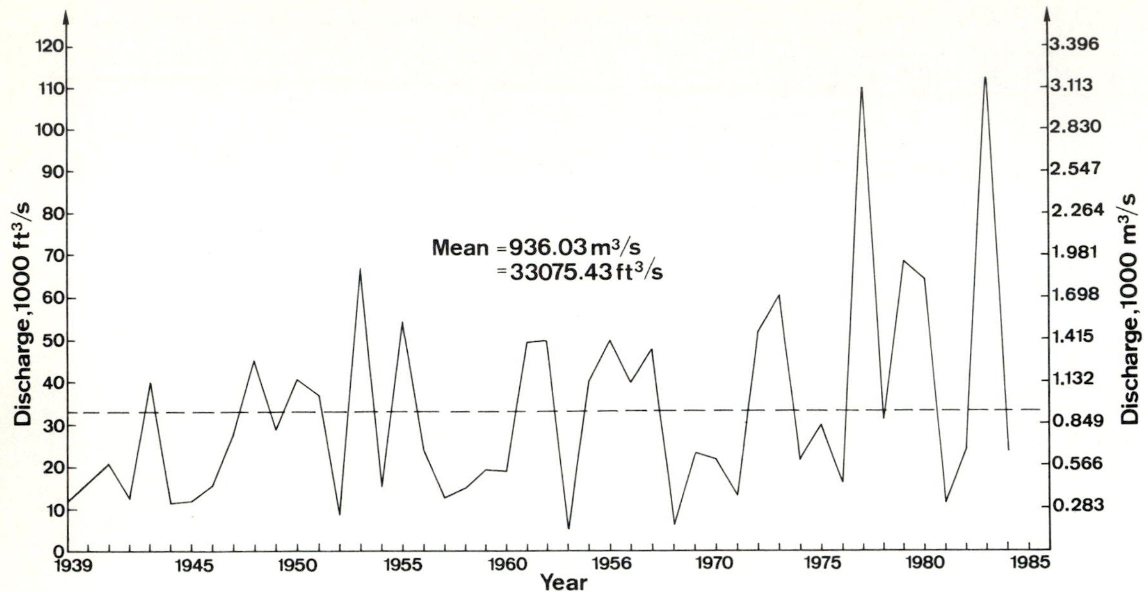

Figure 24.1 Instantaneous peak discharge for the Amite River at Denham Springs, Louisiana, for the period 1939–1984. Mean = 33,075.43 cfs and the standard deviation = 24,277.38 cfs.

water to the specified extent. Clearly, a design event is required for designing a water-resources project. The magnitude of this design event varies from one project to the other, that is, it will be different for different structures such as culverts, bridges, dams, etc. No matter how large the design event used, there is always some chance (or risk) that this design event will be exceeded during the intended useful life of the project. This raises the questions: How large or small should that risk be? How can this risk be quantified? This and similar questions can be answered using concepts of probability.

24.2 CONCEPTS OF PROBABILITY

Let us consider an experiment of tossing a coin, which has a head (H) and a tail (T). When the coin is tossed, we get either H or T, but cannot get both at the same time, for they are mutually exclusive. The term "mutually exclusive" implies that occurrence of H at any time automatically precludes the possibility of occurrence of T at that time. Since there are only two possibilities in this case, the probability of getting H is $\frac{1}{2}$ and the probability of getting T is also $\frac{1}{2}$. This can be verified by tossing the coin a large number of times, recording the appearance of H or T each time, and then dividing the number of times H or T appears by the number of tossings. If the coin is defect-free and the style of tossing is unbiased, the number of appearances of H will be half the number of tossings. Thus, one can say that in a single trial, the probability of an equally likely event is equal to the number of outcomes divided by the number of possible outcomes. In coin tossing, H is an event, the appearance of H is an outcome, and H and T are possible outcomes, In more than one trial, the probability of an event is equal to the number of successes divided by the number of trials. The probability of H or T is always 1, for success will be achieved every time the coin is tossed. This leads to formulation of the following rules.

Rule 1. The probability of an event (E) is nonnegative and less than or equal to 1.

$$0 \leq P(E) \leq 1 \tag{24.1}$$

Furthermore, the sum of probabilities of all possible outcomes in any trial is 1. For a coin tossing,

$$P(E_1) + P(E_2) = P(H) + P(T) = 1$$

More generally, for N possible outcomes,

$$\sum_{i=1}^{N} P(E_i) = 1 \tag{24.2}$$

Clearly, when a coin is tossed,

$$P(H) = 0.5, \qquad P(T) = 0.5 = 1 - P(H)$$

Rule 2. For two independent mutually exclusive events, E_1 and E_2, *the probability of* E_1 or E_2 is equal to the probability of E_1 plus the probability of E_2.

$$P(E_1 \text{ or } E_2) = P(E_1 \cup E_2) = P(E_1) + P(E_2) \tag{24.3}$$

$P(E_1 \cup E_2)$ signifies the union of probabilities and reads "probability of E_1 or E_2." Equation (24.3) can be generalized to any number of events. For example, in case of a die having six edges, there are six possibilities, or six events can occur.

The coin-tossing experiment can be further analyzed. Suppose the coin is tossed twice. What is the probability of obtaining head twice in succession? Four outcomes are possible—HH, HT, TT, TH—each with a probability of $\frac{1}{4}$. Two heads in a row is one of four possibilities. Thus, the probability of HH is $\frac{1}{4}$, and is equal to the probability of getting H the first time multiplied by the probability of getting H the second time.

Rule 3. For two independent events, E_1 and E_2, the probability of E_1 and E_2 is equal to the product of individual probabilities of E_1 and E_2.

$$P\,(E_1 \text{ and } E_2) = P(E_1 \cap E_2) = P(E_1) \times P(E_2) \tag{24.4}$$

$P(E_1 \cap E_2)$ signifies the intersection or joint probability and reads "probability of E_1 and E_2." The term "independent events" implies that the occurrence of one event does not influence the occurrence of the other event. The two events may or may not occur at the same time. Equation (24.4) can be extended to any number of events.

In the case of coin tossing, H and T are independent, mutually exclusive events. In hydrology, many events are independent, many are mutually exclusive, and many take place simultaneously and are not mutually exclusive and may not be even independent. For example, extreme rainfall and extreme yearly temperature can be considered as independent events. Floods and droughts are mutually exclusive events in a given watershed. Rainfall and floods are not independent events. For events that are not independent, the following rule applies.

Rule 4. For two events, E_1 and E_2, the probability of E_1 or E_2 is equal to the sum of individual probabilities of E_1 and E_2 minus the joint probability of E_1 and E_2.

$$P(E_1 \cup E_2) = P(E_1) + P(E_2) - P(E_1 \cap E_2) \tag{24.5}$$

If E_1 and E_2 are independent, then Equation (24.4) applies and Equation (24.5) can be written as

$$P(E_1 \cup E_2) = P(E_1) + P(E_2) - P(E_1)\, P(E_2) \tag{24.6}$$

If E_1 and E_2 are independent mutually exclusive events, then $P(E_1 \cap E_2) = 0$ and Equation (24.5) specializes in Equation (24.3). Thus, Equation (24.5) is a general equation embracing Equations (24.3) and (24.6) as special cases. The term $P(E_1 \cap E_2)$ can be expressed using Bayes' theorem as

$$P(E_1|E_2) = \frac{P(E_1 \cap E_2)}{P(E_2)} \tag{24.7}$$

where $P(E_1|E_2)$ is the probability of E_1 conditioned on E_2. If E_1 and E_2 are independent events, then $P(E_1|E_2) = P(E_1)$.

Let us carry the analysis of coin tossing further. Suppose the coin is biased such that $P(H) = \frac{2}{3}$ and $P(T) = \frac{1}{3}$. Consider tossing five such coins simultaneously. What is then the probability of obtaining three heads in one toss? Two outcomes are possible for each coin. Thus, the probability of obtaining one H for a single coin is $\frac{2}{3}$, and the probability of obtaining T is $\frac{1}{3}$. There can be many ways in which three Hs can be obtained. The order in which they are obtained is immaterial. One way to obtain three Hs is to get three Hs first and two Ts later, that is, the probability of this particular sequence is

$$\frac{2}{3} \cdot \frac{2}{3} \cdot \frac{2}{3} \cdot \frac{1}{3} \cdot \frac{1}{3} = \left(\frac{2}{3}\right)^3 \left(\frac{1}{3}\right)^2 \tag{24.8}$$

Since the order is immaterial, this is also the probability of any other sequence. According to Rule 2, the probability of obtaining exactly three Hs when tossing five coins is equal to the sum of probabilities of the individual combinations:

$$P(3\text{ H in 5 tosses}) = N\left[\left(\frac{2}{3}\right)^3 \left(\frac{1}{3}\right)^2\right] \tag{24.9}$$

where N is the number of possible combinations of 3 Hs, which can be obtained by listing all possible combinations, as shown in Figure 24.2.

If the order in which H or T appears is important, then the number of possible combinations is 2^5. If the order is immaterial, as in this case, then there are only six different combinations, or $(5 - 2)!$. The number of possible combinations with three Hs is 10 or $\binom{5}{3} = 5!/[3!(5 - 3)!]$. This leads to the following:

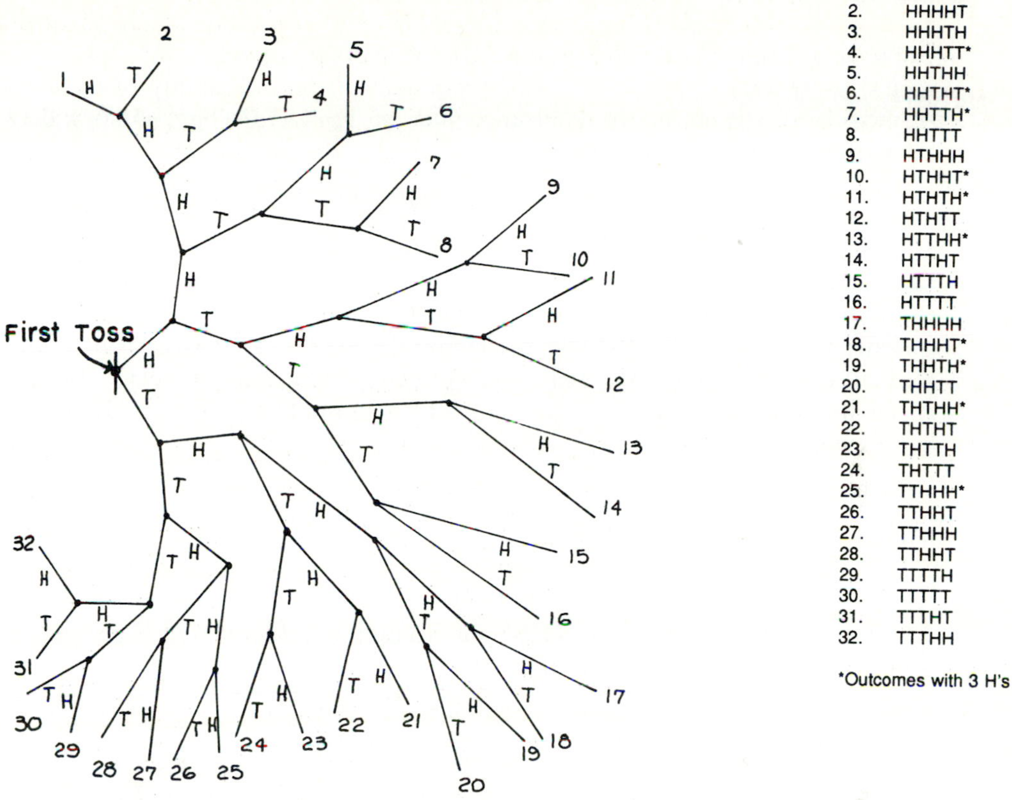

Figure 24.2 Branch diagram of possible outcomes of tossing five coins.

The number of different possible outcomes of k items out of a population of n items is $\binom{n}{k}$ expressed as

$$\binom{n}{k} = \frac{n!}{k!(n-k)!} \tag{24.10}$$

Thus, N in Equation (24.9) is

$$N = \frac{5!}{3!(5-3)!} = 10$$

and

$$(P(3\text{ H in 5 tosses}) = 10\left[\frac{8}{27} \cdot \frac{1}{9}\right] = \frac{80}{243} = 0.329$$

Rule 5. The probability of k successes in the n trials can be determined as

$$P(k\text{ successes in } n\text{ trials}) = \binom{n}{k} p^k(1-p)^{n-k} \tag{24.11}$$

where p is the probability of success in any one trial.

24.3 RETURN PERIOD, T

Suppose a coin is tossed once a year. On the average, its head will appear once in 2 years, or $1/P(H) = 1/0.5 = 2$. This reciprocal of the probability of occurrence is termed the return period, or recurrence interval, T, in hydrology, and is widely used in hydrologic frequency analysis.

$$T = 1/P \tag{24.12}$$

Clearly, T is an average value and not an actual value of the period of occurrence of the associated outcome. Thus, a storm that has been exceeded on the average once in 10 years has a probability of exceedance in any 1 year of 1/10, or 0.1. Stated another way, the storm that is exceeded on the average of 1 year in 10 years has a percent probability of $0.1 \times 100 = 10\%$ of being exceeded any year. This does not mean that every 10 years a storm of that magnitude will occur. Therefore, the probability that the storm will occur in any year is

$$p = 1/T$$

The probability that the storm will not occur, $\bar{p}$, in any year is

$$\bar{p} = 1 - p = 1 - 1/T$$

The probability that the storm will not occur for n successive years is given by application of Equation (24.4) as

$$(\bar{p})^n = (1 - p)^n = (1 - 1/T)^n$$

This is because the probability of storm occurrence is the same from year to year. The probability that the storm will occur at least once in n successive years, sometimes also called risk R, is

$$R = 1 - (\bar{p})^n = 1 - (1 - p)^n = 1 - (1 - 1/T)^n \tag{24.13}$$

Equation (24.13) can also be derived from application of Equation (24.11).

$$\begin{aligned} P(1 \text{ or more storms in } n \text{ years}) &= 1 - P(\text{no storms in } n \text{ years}) \\ &= 1 - \binom{n}{0} p^0 (1 - p)^{n-0} \\ &= 1 - (1 - p)^n = 1 - (1 - 1/T)^n \end{aligned}$$

Equation (24.13) can be used to calculate return periods for various degrees of risk and expected project life, as given in Table 24.1.

EXAMPLE 24.1 The Amite River near Denham Springs, Louisiana, reaches flood stage every year with a probability of 0.2. Parts of Denham Springs get flooded every year with a probability of 0.1. Experience shows that when the Amite River is in flood, the probability of Denham Springs getting flooded increases to 0.2. Compute the probability of flooding for either the Amite River or Denham Springs. What is the probability of both the Amite River being in flood and Denham Springs getting flooded?

Solution Let R denote the flooding of the Amite River and S the flooding of Denham Springs. Let $\bar{R}$ and $\bar{S}$ denote, respectively, opposites of R and S.

$$P(R) = 0.2 \qquad\qquad P(S) = 0.1$$

TABLE 24.1 RETURN PERIODS FOR VARIOUS VALUES OF RISK AND PROJECT LIFE

Risk (%)	Project life in years (N) / Return period in years (T)									
	1	2	5	10	15	20	25	50	75	100
1	100.00	199.50	497.99	995.44	1492.90	1990.42	2487.72	4975.45	7459.86	9945.00
2	50.00	99.50	247.99	495.47	742.95	990.45	1237.90	2475.25	3712.59	4949.03
5	20.00	39.49	97.98	195.46	292.93	390.41	487.88	975.25	1462.58	1949.93
10	10.00	19.49	47.96	95.41	142.87	190.32	237.78	475.06	712.32	949.58
15	6.67	12.81	31.27	62.03	92.80	123,56	154.33	308.16	461.98	615.81
20	5.00	9.47	22.91	45.32	67.72	90.13	112,54	224.57	336.60	448.64
25	4.00	7.46	17.89	35.26	52.64	70.02	87.40	174.30	261.20	348.10
30	3.33	6.12	14.52	28.54	42.56	56.57	70.59	140.68	210.77	280.87
35	2.86	5.16	12.11	23.72	35.32	46.93	58.54	116.57	174.60	232.64
40	2.50	4.44	10.30	20.08	29.87	39.65	49.44	98.38	147.32	196.26
45	2.22	3.87	8.87	17.23	25.59	33.96	42.32	84.14	125.95	167.77
50	2.00	3.41	7.73	14.93	22.14	29.36	36.57	72.64	108.70	144.77
60	1.67	2.72	5.97	11.42	16.88	22.33	27.79	55.07	82.35	109.64
70	1.43	2.21	4.67	8.82	12.97	17.12	21.27	42.03	62.80	83.56
80	1.25	1.81	3.63	6.73	9.83	12.93	16.04	31.57	47.10	62.63
90	11.11	1.46	2.71	4.86	7.03	9.20	11.37	22.22	33.07	43.93
100	1.00	1.00	1.00	1.00	1.00	1.00	1.00	1.00	1.00	1.00

$$P(\bar{R}) = 1 - 0.2 = 0.8 \qquad P(\bar{S}) = 0.9$$

$$P(S \text{ given } R) = P(S|R) = 0.2$$

By applying Equation (24.5),

$$P(S \cup R) = 0.1 + 0.2 - P\,(S \cap R)$$

where $P\,(S \cap R) = P(S|R)P(R) = 0.2 \times 0.2 = 0.04$. Therefore,

$$P\,(S \cup R) = 0.1 + 0.2 - 0.04 = 0.26$$

The probability of both getting flooded is

$$P(S \cap R) = 0.2 \times 0.2 = 0.04$$

as computed earlier.

Let us analyze this problem further. If S and R are mutually exclusive, then

$$P(S \cap R) = 0$$

and

$$P(S \cup R) = 0.1 + 0.2 = 0.3$$

If S and R are independent, then

$$P(S \cup R) = 0.1 + 0.2 - 0.1 \times 0.2 = 0.3 - 0.02 = 0.28$$

and

$$P(S \cap R) = 0.1 \times 0.2 = 0.02$$

EXAMPLE 24.2 Compute the return period of a design storm to be used for the design of a culvert. There is a 5% probability that the design storm will occur in the next 5 years.

Solution From Equation (24.13),

$$0.05 = 1 - (1 - 1/T)^5$$

or

$$T = 1/[1 - (0.95)^{0.2}] = 97.98 \text{ years}$$

EXAMPLE 24.3 Compute the probability that exactly two 10-year floods will occur in a single 30-year period.

Solution From Equation (24.11), where $p = 1/10$,

$$P(2 \text{ in } 30) = \binom{30}{2}\left(\frac{1}{10}\right)^2\left(\frac{9}{10}\right)^{28}$$

$$= 435\left(\frac{1}{10}\right)^2\left(\frac{9}{10}\right)^{28} = 0.2277$$

There is a 22.8% chance that exactly two 10-year floods will occur in 30 years.

EXAMPLE 24.4 Compute the probability that two or more 10-year floods will occur in 30 years.

Solution

$$P(2 \text{ or more in } 30) = P\ (2 \text{ in } 30) + P(3 \text{ in } 30) + P\ (4 \text{ in } 30) + \cdots$$

By taking advantage of Equation (24.2),

$$\begin{aligned} P(2 \text{ or more in } 30) &= 1 - P(0 \text{ in } 30) - P(1 \text{ in } 30) \\ &= 1 - \binom{30}{0}(0.1)^0\left(\frac{9}{10}\right)^{30} - \binom{30}{1}(0.1)\left(\frac{9}{10}\right)^{29} \\ &= 1 - \left(\frac{9}{10}\right)^{30} - 3\left(\frac{9}{10}\right)^{29} \\ &= 0.816 \end{aligned}$$

Thus, if a structure is designed for a 10-year storm, there is a 81.6% chance that the storm will be exceeded two or more times in 30 years.

EXAMPLE 24.5 A rainfall station has been in operation for 3 years. During this time, there has been only one rainy day in November. Compute the probability of having a rainy day in November.

Solution

November = 30 days

3 Novembers = 90 days

Probability of having rain on any one day (P) = 1/90

Probability of not having rain on any one day = 1 − 1/90

Therefore,

Probability of not having rain for the entire month = $(1 - 1/90)^{30}$

Therefore, the probability of having at least one rainy day in November is

$$\begin{aligned} P &= 1 - (89/90)^{30} \\ &= 1 - (0.7152) = 0.2848 = 28.48\% \end{aligned}$$

EXAMPLE 24.6 What is the probability of occurrence of a flood equal to or greater than a 20-year flood during the next 3 years; the next 20 years?

Solution

$P = 1/20 = 0.05$

Probability of not having a flood next year = 1 − 0.05 = 0.95

Probability of not having a flood in 3 years = $1 - (0.95)^3$

= 1 − 0.857 = 0.143 = 14.3%

Similarly, for the next 20 years,

$$\text{Probability} = 1 - (0.95)^{20} = 1 - 0.358 = 0.642 = 64.2\%$$

EXAMPLE 24.7 What is the probability of a flood equal to or greater than a 5-year flood occurring next year; next 3 years?

Solution

Probability of having a 5-year flood next year = 1/5 = 0.2

Probability of not having a 5-year flood next year = 1 − 0.2 = 0.8

Probability of not having a 5-year flood in the next 3 years = $(0.8)^3$

Probability of having a flood in the next 3 years:

$R = 1 - (0.8)^3 = 1 - (0.512) = 0.488 = 48.8\%$

24.4 DISCRETE AND CONTINUOUS RANDOM VARIABLES

In the case of coin tossing, it is noticed that there are only two probable events, H or T. Let us say that we assign a value of 0 when H appears and a value of 1 when T appears. If the coin tossing is a random variable, then this random variable takes on fixed values of 0 and 1. Consider now the case of counting the number of rainy days in the month of June in Baton Rouge, Louisiana. If the number of rainy days is a random variable, then this takes on specific integral values such as 0, 1, 2, 3, etc.; it most certainly cannot take on fractional values such as 0.15, or 2.135, etc. If the number of floods in a year is a random variable, then this too assumes specific integral values. Thus, a random variable is defined to be discrete if it assumes a finite set of values; usually, these values are integral. A discrete random variable is described by a discrete probability distribution.

Consider rainfall being measured at a given location. The amount of rainfall over a specified period of time can take on any value, such as 0.1, 0.11, 0.111, 0.12, . . ., 0.2, 0.21, 0.215, 0.217, . . . , 1.05, 1.15, . . . , 2.01, etc. These values form an infinite set. The rainfall amount is a continuous random variable. Similarly, consider the river discharge above a certain datum at a gaging station. Call this discharge as exceedance. Then the exceedance can assume any value, and the values of exceedance will constitute an infinite set. The exceedance is a continuous random variable. The same observation can be made about wind velocity, flow velocity, rainfall intensity, the time interval between two rainfall events, the interarrival time of floods, etc. Thus, if a random variable can assume any value, it is a continuous random variable; the set of these values is infinite. A continuous random variable is described by a continuous probability distribution. The usual method to denote a random variable is by a capital letter, say, X, and a specific value of this variable by a lowercase letter, say, x.

EXAMPLE 24.8 Give at least five examples of a discrete variable and five of a continuous variable.

Solution Discrete random variables: (1) the number of floods in a year, (2) the number of snowstorms in a week, (3) the number of dry days in the month of February, (4) the number of sunshine hours in a day, and (5) the number of houses damaged by a flood in Baton Rouge, Louisiana.

Continuous random variables: (1) the amount of daily rainfall, (2) the magnitude of flood peak in a year, (3) the time span between two rainy days, (4) the duration of flooding, and (5) the 5-minute intensity of thunderstorms.

24.5 DEFINITION OF PROBABILITY DISTRIBUTIONS

In the previous discussion, we have mentioned the rules for determining the probability of a random variable assuming a specific value in some simple cases. The random variable can assume any value and each value is associated with a probability. If for all values of the given random variable, the corresponding probabilities are known or found, then the relation between the values of the random variable and the probability values is described by what is called as probability distribution. If this distribution is known, then the probability of any value of the random variable can be easily determined.

As an illustration, first, consider the case of a discrete random variable. Suppose we wish to determine the probability of the number of wet days in the month of May in New Orleans, Louisiana. Climatological records are available for a long period of time that show wet and dry days in May. From an analysis of these records where the day had to be described as either wet or dry, it is found that the number of wet days is not more than 10 in the month of May. The number of wet days is a discrete random variable and takes on fixed values such as 0, 1, 2, . . . , 10. The corresponding probabilities obtained from the analysis are

$P(0) = 0.01$ $\quad P(5) = 0.20$
$P(1) = 0.05$ $\quad P(6) = 0.12$
$P(2) = 0.08$ $\quad P(7) = 0.08$
$P(3) = 0.15$ $\quad P(8) = 0.05$
$P(4) = 0.22$ $\quad P(9) = 0.03$
$\quad P(10) = 0.01$

The sum of probabilities $\Sigma_{i=0}^{10} P(x_i) = 1.0$.

These probabilities are plotted against the number of wet days in Figure 24.3 and show the probability distribution of wet days in May at New Orleans.

Another way to describe probabilities is to determine cumulative probabilities and their distribution. In the previous example, we may wish to determine the probability of the number of wet days equal to or less than a given value. This is determined by summing the individual probabilities, as shown in Table 24.2. When the cumulative probabilities are plotted against the random variable, the resulting relation is called the cumulative distribution function (CDF), as shown in Figure 24.4. This is a monotonically nondecreasing function with a lower limit of 0 and an upper limit of 1.

Consider now the case of a continuous random variable. The instantaneous annual peak discharge of the Amite River at Denham Springs in Louisiana is used to illustrate the point. Table 24.3 shows the discharge data for a period of 46 years. These data are grouped into eight class intervals and the number of values falling in

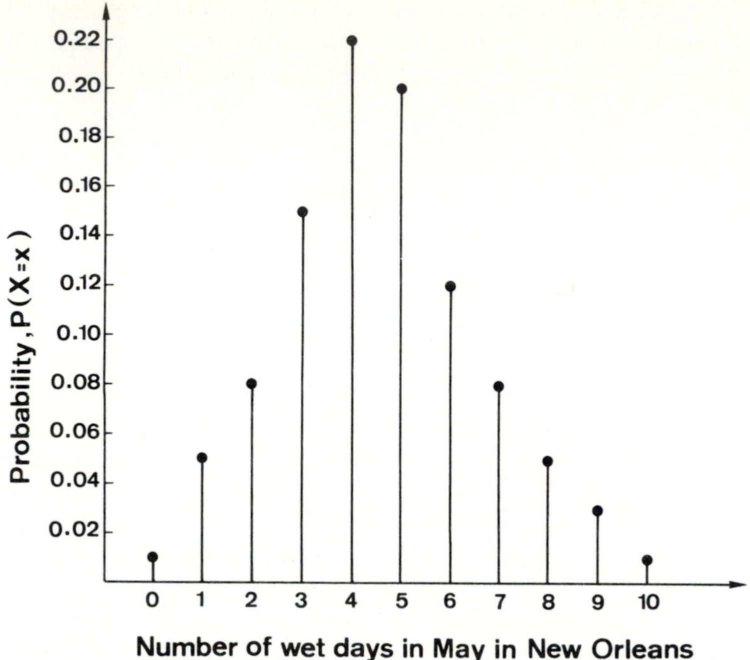

Figure 24.3 Probability distribution of wet days in the month of May at New Orleans, Louisiana.

each class interval is counted as shown in Table 24.4. This number defines the absolute frequency and the sum of absolute frequencies of all class intervals must equal the total number of data points (or observations). These frequencies are plotted in Figure 24.5. A more instructive way to plot, however, is to use the relative frequency, which for a given class interval is the absolute frequency of that class interval divided by the total number of data points. This is shown on the right

TABLE 24.2 CUMULATIVE DISTRIBUTION FUNCTION

No. of wet days, x	$P(X = x)$	$P(X \leq x)$
0	0.01	0.01
1	0.05	0.06
2	0.08	0.14
3	0.15	0.29
4	0.22	0.51
5	0.20	0.71
6	0.12	0.83
7	0.08	0.91
8	0.05	0.96
9	0.03	0.99
10	0.01	1.00

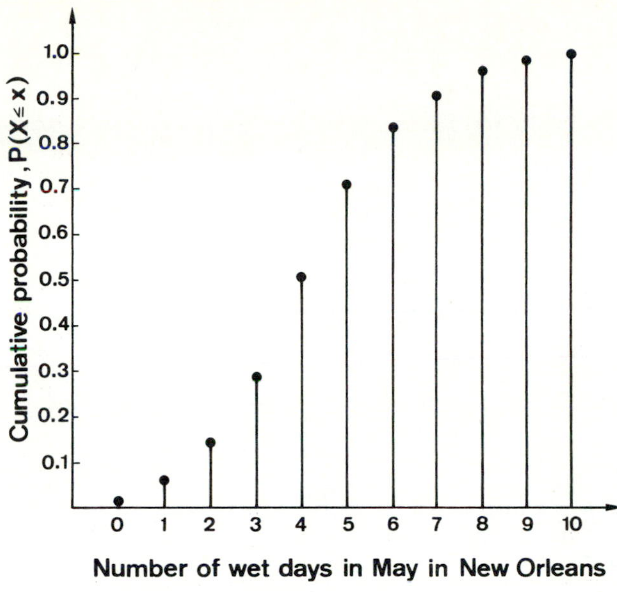

Figure 24.4 Cumulative probability distribution of wet days in the month of May at New Orleans, Louisiana.

TABLE 24.3 INSTANTANEOUS ANNUAL PEAK DISCHARGE OF THE AMITE RIVER AT DENHAM SPRINGS, LOUISIANA

Year	Discharge (cfs)	Discharge (m^3/s)	Year	Discharge (cfs)	Discharge (m^3/s)
1939	12,100	342.43	1961	49,100	1,398.53
1940	16,000	452.80	1962	49,700	1,406.51
1941	20,800	588.64	1963	5,150	145.75
1942	12,200	345.26	1964	40,500	1,146.15
1943	40,200	1,137.66	1965	49,900	1,412.17
1944	11,000	311.30	1966	39,700	1,123.51
1945	11,600	328.28	1967	47,800	1,352.74
1946	15,500	438.65	1968	6,290	178.06
1947	27,800	786.74	1969	23,000	650.90
1948	45,100	1,276.33	1970	21,700	614.11
1949	28,800	815.04	1971	12,600	356.58
1950	40,800	1,154.64	1972	51,800	1,465.94
1951	36,900	1,044.27	1973	60,200	1,703.66
1952	8,230	232.91	1974	21,300	602.79
1953	67,000	1,896.10	1975	29,900	846.17
1954	15,200	430.16	1976	16,100	455.63
1955	54,300	1,536.69	1977	110,000	3,113.000
1956	23,400	662.22	1978	31,300	885.79
1957	12,300	348.09	1979	68,600	1.941.38
1958	14,700	416.01	1980	64,200	1,816.86
1959	19,100	540.53	1981	11,300	319.79
1960	18,800	532.04	1982	23,900	676.37
			1983	112,000	3,169.60
			1984	23,600	667.88

TABLE 24.4 ARRANGEMENT OF DISCHARGE DATA OF TABLE 24.3 IN CLASS INTERVALS, AND ABSOLUTE, RELATIVE, AND CUMULATIVE FREQUENCIES[a]

Discharge (ft^3/s) class interval	Absolute frequency, n	Cumulative absolute frequency, Σn	Relative frequency, n/N	Relative cumulative frequency, $\Sigma(n/N)$	$\frac{n}{N\,\Delta x}$ ($\times$ 6.66666 $\times$ 10^{-5})
5,000– 20,000	17	17	0.370	0.370	0.370
20,000– 35,000	11	28	0.239	0.609	0.239
35,000– 50,000	10	38	0.217	0.826	0.217
50,000– 65,000	4	42	0.087	0.913	0.087
65,000– 80,000	2	44	0.043	0.956	0.843
80,000– 95,000	0	44	0	0.956	0
95,000–110,000	1	45	0.022	0.978	0.022
110,000–125,000	1	46	0.022	1.000	0.022

[a] Class size = 15,000 ft^3/s (Δx).

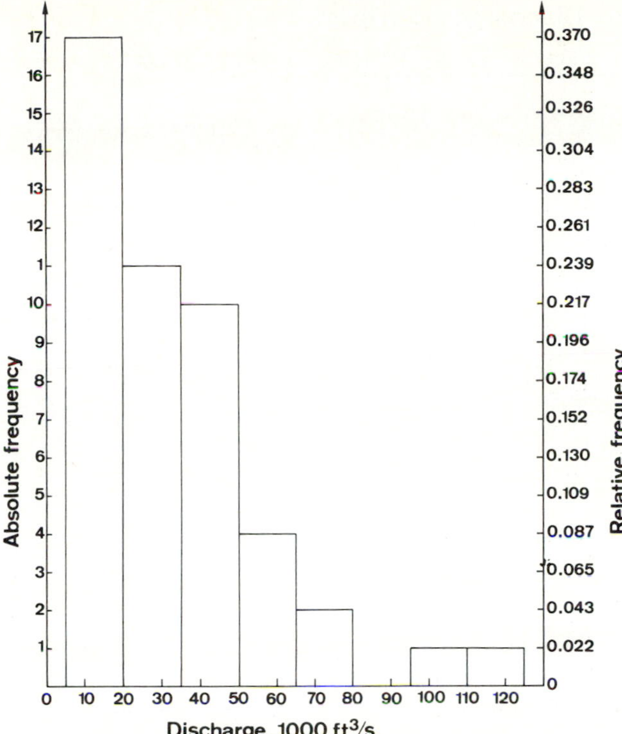

Figure 24.5 Histogram of peak-discharge data for the Amite River at Denham Springs, Louisiana.

ordinate of the figure. This diagram represents the frequency of the peak discharge of the Amite River. By summing the relative frequencies of class intervals, cumulative frequencies are obtained, which when plotted against the peak discharge give the cumulative frequency distribution, as shown in Figure 24.6.

The selection of a class interval can affect the appearance of a frequency histogram. Several criteria are used to determine the class-interval size. The range of data, the number of observations, and the behavior of data are important considerations. According to Sturges (1926), the number of classes can be estimated as

$$m = 1 + 3.3 \log n \tag{24.14}$$

where m is the number of classes, and n is the number of observations. Another criterion is given by Spiegel (1961) that limits the classes between 5 and 20. Steel and Torrie (1960) recommend that the class interval not exceed one-fourth to one-half the standard deviation. A good guide is to avoid using either too few or too many class intervals and to see that a smooth curve, representative of the data, is produced.

Because the continuous variable—the peak discharge—can take on an infinite number of values, its distribution will be a continuous curve, not a broken one, as shown in Figures 24.5 and 24.6. As the number of observations increases and the class-interval size is reduced, the discontinuous curves of these figures will approach, in the limit, the continuous curves of Figure 24.7, which are probability distributions. Then the relative frequency has the same connotation as the probability. To be consistent with the definition of probability, the area under the frequency

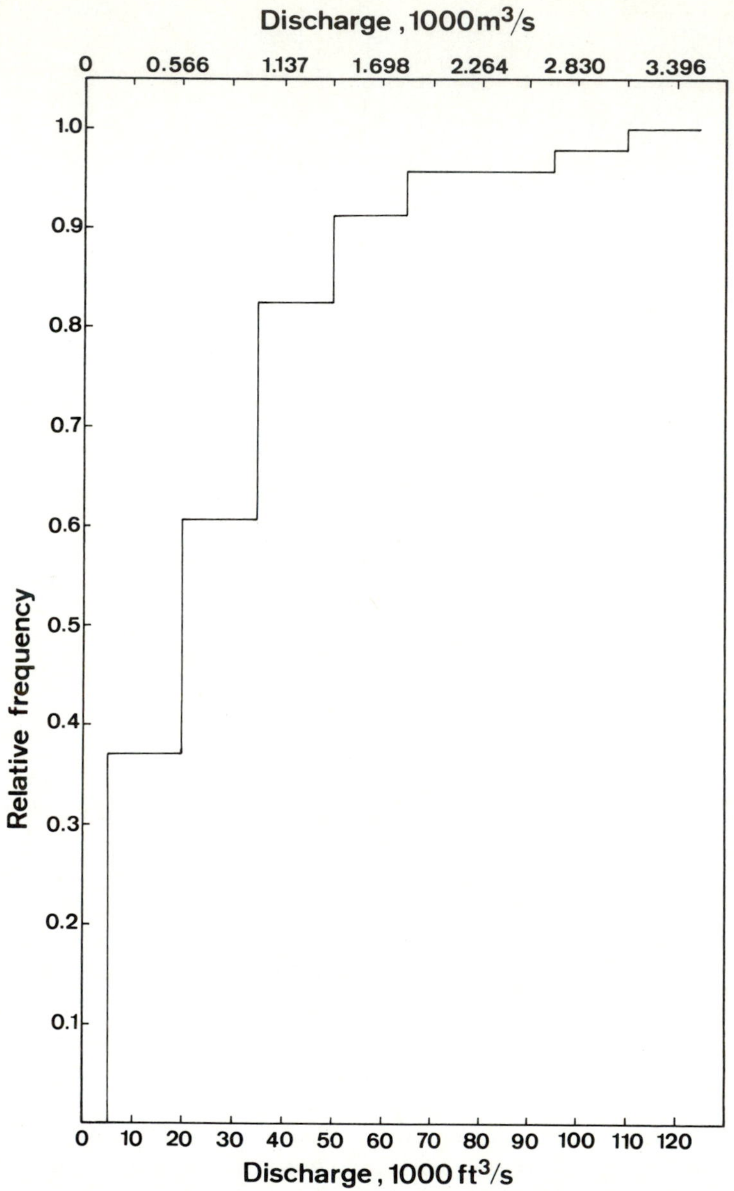

Figure 24.6 Cumulative histogram of peak-discharge data for the Amite River at Denham Springs, Louisiana.

histogram must also be unity. This requirement is satisfied by dividing the relative frequency (n/N) of each class interval by the interval width Δx. The result of this computation is shown in Table 24.4. This quantity ($n/(N\,\Delta x)$) represents the average density of probability, or, literally, the probability per unit width of the class interval, as shown in Figure 24.7. If $F(X \leq x) = F(x)$ represents the cumulative distribution function, then

$$\frac{n}{N} = \Delta F(x) = F\left(x + \frac{\Delta x}{2}\right) - F\left(x - \frac{\Delta x}{2}\right) \tag{24.15a}$$

Let $\Delta x \rightarrow 0$, then

$$f(x) = \lim_{\Delta x \rightarrow 0} \frac{\Delta F(x)}{\Delta x} = \frac{\mathrm{d}F(x)}{\mathrm{d}x} \tag{24.15b}$$

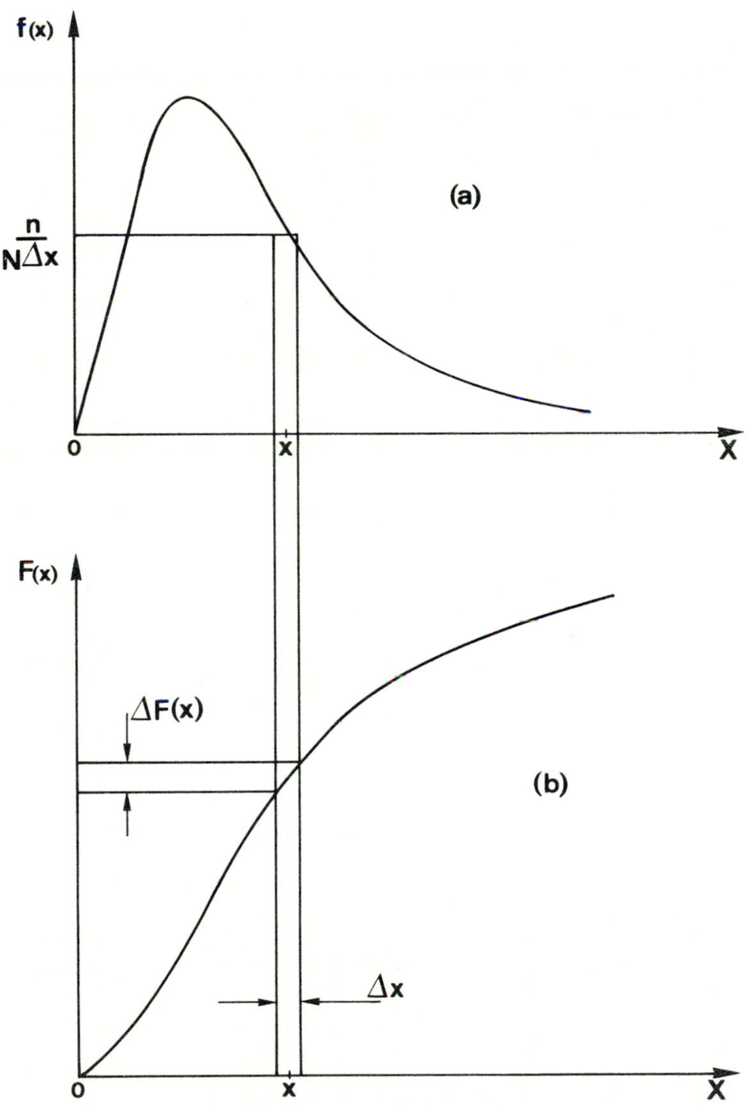

Figure 24.7 (a) Probability density function (PDF) and (b) cumulative distribution function (CDF) of annual peak discharge.

where $f(x)$ is called the probability density function (PDF) of the continuous random variable X, and, of course, $F(x)$ is the cumulative distribution function (CDF). The quantity $f(x) \cdot \Delta x$ defines the differential probability.

Now it may be worthwhile to state some of the properties of $f(x)$ and $F(x)$ for continuous random variables.

1. $f(x) \geq 0, \qquad -\infty < x < \infty$ (24.16)

This implies that the probability of a random variable cannot be negative.

2. $\int_{-\infty}^{\infty} f(x)\, \mathrm{d}x = 1$ (24.17)

This implies that the sum of probabilities of all possible outcomes is equal to 1. The area under the PDF is unity.

3. $P(X \leq x) = F(X \leq x) = F(x) = \int_{-\infty}^{x} f(x)\, \mathrm{d}x$ (24.18)

If a and b are any real numbers such that a < b, the events $X \leq a$ and $a < X \leq b$ will be mutually exclusive. Then $P(X \leq b) \geq P(X \leq a)$ or $F(X \leq b) \geq F(X \leq a)$, and

$$P(X \leq b) = P(X \leq a) + P(a < X \leq b)$$

$$= \int_{-\infty}^{a} f(x)\, \mathrm{d}x + \int_{-\infty}^{b} f(x)\, \mathrm{d}x = \int_{-\infty}^{b} f(x)\, \mathrm{d}x = F(X \leq b)$$

From this, we obtain:

$$P(a < X \leq b) = P(X \leq b) - P(X \leq a)$$

$$= \int_{-\infty}^{b} f(x)\, \mathrm{d}x - \int_{-\infty}^{a} f(x)\, \mathrm{d}x \tag{24.19}$$

$$= \int_{a}^{b} f(x)\, \mathrm{d}x, \qquad \text{for } a < b$$

4. The probability that X (continuous variable) assumes a particular value is zero, that is, $P(X = a) = F(x = a) = 0$,

$$\int_{a}^{a} f(x)\, \mathrm{d}x = F(a) - F(a) = 0 \tag{24.20}$$

5. $F(+\infty) = \lim_{x\to\infty} F(x) = 1$ (24.21)

Also

$$F(-\infty) = \lim_{x\to\infty} F(x) = 0 \tag{24.22}$$

This can be seen from the area under the PDF.

For discrete random variables, analogous statements can be made:

1. $\sum_{i} f(x_i) = 1$ (24.23)

where $f(x_i)$ represents the probability of $X = x_i$ in the sample space, if the observations are finite in the sample. Thus, this can be replaced by $p(x_i)$.

2. $$P(a \le x \le b) = \sum_{x_i \ge a}^{x_i \le b} p(x_i) \tag{24.24}$$

3. $$P(X \le x_k) = \sum_{i=1}^{k} p(x_i) \tag{24.25}$$

EXAMPLE 24.9 A coin is tossed until its head appears. The number of coin tossings N until a head appears can be considered as a random variable. Compute its cumulative distribution function and sketch it.

Solution From the previous discussion, $P(H) = P(T) = 0.5$, in any coin tossing. In the first tossing, $P(T) = 0.5$. In the second tossing, $P(T) = 0.5$. In both the first and second tossings, the probability of appearance of T is $P(T) = P(T^2) = P(T) \cdot P(T) = 0.5 \times 0.5 = (0.5)^2$. Similarly, in the $(n - 1)$th tossing, the probability of appearance of T is $P(T^{n-1}) = (0.5)^{n-1}$. The probability of H on the nth tossing is $P(H) \times P(T^{n-1}) = (0.5)^{n-1} \cdot (0.5)$. $P(N = n) = (0.5)^n$; thus, $P(N = n)$ and $P(N \le n)$ can be computed and sketched, as shown in Figure 24.8. Some sample values are:

N	$P(N = n)$	$P(N \le n)$
1	0.5	0.5
2	0.25	0.75
3	0.125	0.875
4	0.0625	0.9375
5	0.0313	0.9688
6	0.0156	0.9844
7	0.0078	0.9922

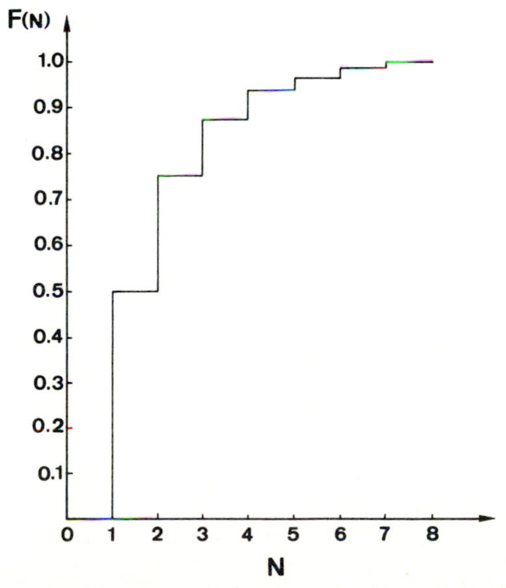

Figure 24.8 Cumulative distribution function for a number of tosses until a head appears.

EXAMPLE 24.10 A function is given as $f(x) = A \exp(-Ax)$, $A > 0$, $x \geq 0$. Show that this function can be a PDF.

Solution The area under the PDF is unity. Therefore,

$$\int_0^\infty A \exp(-Ax) = A\left[-\frac{\exp(-Ax)}{A}\right]\Bigg|_0^\infty = -\exp(-Ax)\Big|_0^\infty = 1$$

This shows that this function can be considered as a PDF.

24.6 DISTRIBUTION CHARACTERISTICS

There are four principal characteristics or moments of probability distributions: central tendency, dispersion, skewness, and kurtosis. These characteristics are expressed in terms of the distribution parameters, which themselves can be expressed in terms of moments. These parameters are estimated from the distribution of observed sample data, and are then used as estimates of the parameters of the population distribution.

24.6.1 Central Tendency

This describes the grouping or clustering of observations or probability about a central value. Several measures of central tendency are used.

ARITHMETIC MEAN

This is the most common and most reliable measure of central tendency. This is the first moment about the origin and can be expressed as

$$E[X] = \mu = \int_{-\infty}^{\infty} xf(x)\,\mathrm{d}x \tag{24.26}$$

and

$$E[X] = \mu = \sum_{i=1}^{n} x_1 p(x_i) \tag{24.27}$$

for continuous and discrete random variables, respectively. The notation $E[X]$ is called the expected value of the random variable X, and is denoted by μ. An estimate of the population mean is obtained from a sample as

$$\bar{x} = \frac{1}{n}\sum_{i=1}^{n} x_i \tag{24.28}$$

The sample mean $\bar{x}$ is the best estimate of the population mean μ. In the computation of $\bar{x}$, it is assumed that all values of X are equally important. In other words, each value has the same weight. In some instances, different weights may be assigned to different values. Then a weighted mean is calculated as

$$\bar{x} = \frac{1}{n} \sum_{i=1}^{n} w_i x_i, \qquad \sum_{i=1}^{n} x_i = 1, \qquad 0 \leq w_i \leq i, \qquad i = 1, 2, \ldots, n \tag{24.29}$$

OTHER MEASURES

Other types of mean are harmonic mean and geometric mean, but these are not commonly used. Two other measures of central tendency are the median and mode. The median represents the middle value of the observed data. In other words, half of the values lie on either side of the sample median. The population median, μ_{me}, can be expressed

$$\int_{-\infty}^{\mu_{me}} f(x)\,\mathrm{d}x = 0.5$$

and $\mu_{me} = x_n$, with n determined from

$$\sum_{i=1}^{n} p(x_i) = 0.5$$

Thus, the median divides the population distribution into two equal parts, respectively, for continuous and discrete random variables. The median may not exist for a sample or population.

The mode is the peak value of the PDF of a continuous random variable, or is the most frequently occurring value if the variable is discrete. A sample or population may not have or may have more than one peak.

24.6.2 Dispersion

The two most common measures of dispersion or variability are range and variance. For a sample, the range is the difference between the largest and smallest values, and conveys an idea about the spread of data. The range of many continuous hydrologic variables is from 0 to ∞.

The most commonly used measure of dispersion is the variance, or standard deviation. The variance is the second moment about the mean and is expressed for a continuous variable X as

$$\mathrm{Var}\,[X] = \sigma^2 = \mu_2 = E[(X - \mu)^2] = E[X^2] - (E[X])^2 \tag{24.30}$$

This expresses variance as the average squared deviations about the mean.

For a discrete population of size n,

$$\sigma^2 = \frac{1}{n} \sum_{i=1}^{n} (x_i - \mu)^2 \tag{24.31}$$

Because μ is not known precisely, an estimate of variance s^2 is computed from the observed sample as

$$s^2 = \frac{1}{n}\sum_{i=1}^{n}(x_i - \bar{x})^2$$
$$= \frac{1}{n}\sum_{i=1}^{n}x_i^2 - \left(\frac{1}{n}\sum_{i=1}^{n}x_i\right)^2 \tag{24.32}$$

If the sample data are grouped, then

$$s^2 = \sum_{i=1}^{k} f(x_i)(x_i - \bar{x})^2 \tag{24.33}$$

where k is the number of class intervals, and $f(x_i)$ is the relative frequency of the ith class interval. For $n < 30$, the best estimate of σ^2 is found by replacing n by $n - 1$ in Equation (24.32).

The positive square root of variance is known as the standard deviation, σ or s. The unit of s is the same as that of $\bar{x}$. Another statistic used to measure dispersion is the coefficient of variation, C_v, which is defined as

$$C_v = \sigma/\mu \qquad \text{for the population} \tag{24.34}$$

$$C_v = s/\bar{x} \qquad \text{for the sample} \tag{24.35}$$

It is dimensionless and is useful in comparing relative variability.

24.6.3 Skewness or Asymmetry

Most distributions encountered in hydrology are not symmetrical. They may be skewed to the left or skewed to the right (the more frequently occurring case). Whatever the case may be, they are skewed. If a distribution is symmetrical about the mean, then all its odd moments about the centroid vanish. The third moment, therefore, is defined to measure skewness, which can be expressed as

$$\mu_3 = E[(X - \mu)^3] = \int_{-\infty}^{\infty}(x - \mu)^3 f(x)\, dx \tag{24.36}$$

for a continuous population, and

$$\mu_3 = \frac{1}{n}\sum_{i=1}^{n}(x_i - \mu)^3 \tag{24.37}$$

for a discrete population. From a sample of n observations, an estimate of μ_3, M_3, is obtained as

$$M_3 = \frac{1}{n}\sum_{i=1}^{n}(x_i - \bar{x})^3 \tag{24.38}$$

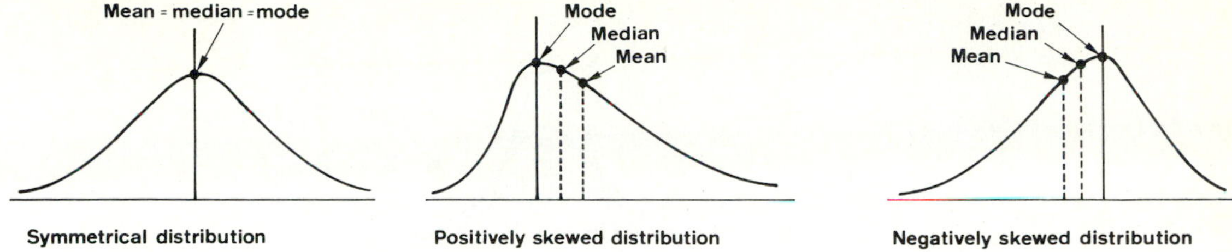

Figure 24.9 Symmetrical and skewed distributions.

However, the most commonly used measure of asymmetry is the coefficient of skewness (C_s), γ, which is defined as the ratio of the third central moment divided by the cube of standard deviation:

$$\gamma = \mu_3/\sigma^3 \tag{24.39}$$

An unbiased estimate of γ for a sample size n is obtained as

$$C_s = \frac{n^2 M_3}{(n-1)(n-2)s^3} \tag{24.40}$$

This coefficient is dimensionless and is useful in comparing distributions. For a symmetrical distribution, $\mu_3 = 0$ and, therefore, $\gamma = 0$. For a distribution that has a long tail to the right, $\gamma > 0$, the distribution is positively skewed. If the distribution has a long tail to the left, it has negative skewness. Figure 24.9 shows symmetrical, and positively and negatively skewed distributions. If the sample is small, $n < 30$, this measure of skewness may not be a reliable one.

24.6.4 Kurtosis

Kurtosis is a measure of peakedness or flatness of a distribution relative to the normal distribution. The coefficient of kurtosis, or simply kurtosis, is expressed as

$$K = \frac{\mu_4}{\mu_2^2} \tag{24.41}$$

where μ_4 is the fourth central moment of the population. Its sample estimate is

$$K = \frac{M_4}{s^4} \tag{24.42}$$

A less biased estimate of K is obtained (Yevjevich, 1972) as

$$\hat{K} = \frac{n^3}{(n-1)(n-2)(n-3)} \frac{M_4}{s^4} \tag{24.43}$$

where n is the sample size.

The coefficient of excess kurtosis, ε, is defined as $K - 3$. The normal distribution has a kurtosis of 3. Thus, a distribution, which is more peaked than normal, will

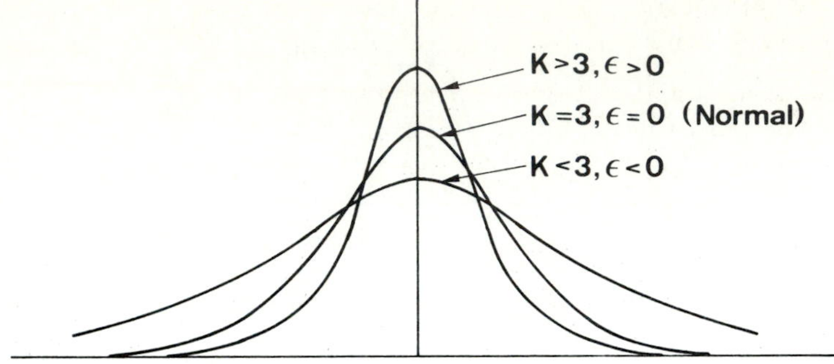

Figure 24.10 Distribution having more and less kurtosis than a normal distribution.

have a positive value of the coefficient; otherwise, the coefficient will be negative. Figure 24.10 shows kurtosis of distributions with $\varepsilon = 0$, $\varepsilon > 0$, and $\varepsilon < 0$.

EXAMPLE 24.11 Compute the arithmetic mean of the discharge values given in Table 24.3.

Solution

$$\bar{x} = \frac{1}{n}\sum_{i=1}^{46} x_i$$

The sum of all discharge values is obtained and is then divided by $n = 46$.

$$\sum_{i=1}^{46} x_i = 43{,}057.595 \text{ m}^3/\text{s}$$

$$\bar{x} = 936.035 \text{ m}^3/\text{s}$$

EXAMPLE 24.12 Determine the mean temperature from the following data:

Temperature (°C)	35	28	30	37	32	20	23
Number of days	1	5	10	2	5	2	5

Solution This is a case of grouped data. Therefore, Equation (24.29) can be used, where $w_i = n_i/N$, N = total number of days = 30. Therefore,

$$\bar{x} = 35\left(\frac{1}{30}\right) + 28\left(\frac{5}{30}\right) + 30\left(\frac{10}{30}\right) + 37\left(\frac{2}{30}\right)$$

$$+ 32\left(\frac{5}{30}\right) + 20\left(\frac{2}{30}\right) + 23\left(\frac{5}{30}\right)$$

$$= 1.167 + 4.667 + 10.000 + 2.467 + 5.333 + 1.333 + 3.833$$

$$= 28.8°\text{C}$$

EXAMPLE 24.13 The discharge data of Table 24.3 is arranged in class intervals as given in Table 24.4. Find the mean discharge.

Solution First, the midvalue of each class interval is found, as shown in column (2) of Table 24.5. The midvalue is multiplied by the relative frequency of the associated class interval, and then the products are summed up to yield the mean, as shown in Table 24.5. Then

$$\bar{x} = 33{,}264.75 \text{ ft}^3/\text{s}$$
$$= 941.39 \text{ m}^3/\text{s}$$

EXAMPLE 24.14 Find the median and mode class of the discharge data given in Table 24.3.

Solution The discharge data are arranged in descending or ascending order and the middle value is located. Since there are 46 observations, two middle values exist: 23,600 and 23,900. Therefore,

$$\text{median} = (23{,}600 + 23{,}900)/2 = 23{,}750 \text{ ft}^3/\text{s}$$
$$= 672{,}125 \text{ m}^3/\text{s}$$

From Table 24.4, it is seen that the mode class is 5,000–20,000, wherein the absolute frequency is 17 and the relative frequency is 0.375.

EXAMPLE 24.15 Compute the standard deviation, variance, and coefficient of variation for the data in Table 24.3. Tabulate the computations, as shown in Table 24.6.

$$\text{Variance, } s^2 = 540{,}697{,}318.3$$

$$\text{Standard deviation, s} = 23{,}252.89914$$

$$\text{Coefficient of variation, } C_v = 0.6990$$

TABLE 24.5 COMPUTATION OF THE MEAN BY ARRANGING DATA IN CLASS INTERVALS

Discharge (ft^3/s) class interval (1)	Midvalue (2)	Relative frequency (3)	$x_i \times f(x_i)$ (2) × (3) = (4)
[a]5,000– 20,000	12,575	0.370	4,652.75
20,000– 35,000	27,500	0.239	6,572.50
35,000– 50,000	42,500	0.217	9,222.50
50,000– 65,000	57,500	0.087	5,002.50
65,000– 80,000	72,500	0.043	3,117.50
80,000– 95,000	87,500	0.0	0.0
95,000–110,000	102,500	0.022	2,255.00
[b]110,000–125,000	111,000	0.022	2,442.00
			$\Sigma x_i f(x_i)$ = 33,264.75 ft^3/s

[a] The lowest discharge is 5,150 ft^3/s, so the midvalue for computation should be (5,150 + 20,000)/2 = 12,575 ft^3/s.

[b] The highest discharge is 112,000 ft^3/s, so the midvalue for computation should be (110,000 + 112,000)/2 = 111,000 ft^3/s.

TABLE 24.6 COMPUTATION FOR VARIANCE

Discharge (ft^3/s) class interval (1)	Midvalue, x_i (2)	$x_i - \bar{x}$ (3)	$(x_i - \bar{x})^2$ (3) × (3) (4)	Relative frequency, $f(x_i)$ (5)	$f(x_i)x(x_i - \bar{x})^2$ (4) × (5) = (6)
5,000– 20,000	[a]12,575	−20,689.75	428,065,755.1	0.370	158,384,329.4
20,000– 35,000	27,500	−5,764.75	33,232,342.56	0.239	7,942,529.87
35,000– 50,000	42,500	9,235.24	85,289,842.56	0.219	18,678,475.52
50,000– 65,000	57,500	24,235.25	587,347,324.6	0.087	51,099,218.8
65,000– 80,000	72,500	39,235.25	1,539,404,843.0	0.043	66,194,408.23
80,000– 95,000	87,500	54,235.25	2,941,412,343.0	0.0	0
95,000–110,000	102,500	69,235.25	4,793,518,943.0	0.022	105,457,436.5
110,000–125,000	*111,000	77,735.25	6,042,769,093.0	0.022	132,940.920.0
					$\Sigma f(x_i)(x_i - \bar{x})^2$ = 540,697,318.3

[a] As explained in Table 24.5:

$\bar{x}$ = 33,264.75 ft^3/s

Variance, s^2 = 540,697,318.3

Coefficient of variation, C_v = 0.6990

Standard deviation, s = 23,252.9

TABLE 24.7 COMPUTATION FOR SKEWNESS

Discharge (ft^3/s) class interval (1)	Midvalue, x_i (2)	$x_i - \bar{x}$ (3)	$(x_i - \bar{x})^3$ (3) × (3) × (3) (4)	Relative frequency, $f(x_i)$ (5)	$f(x_i) \times (x_i - \bar{x})^3$ (5) × (4) = (6)
5,000– 20,000	[a]12,575	−20,689.75	-8.856573×10^{12}	0.370	-3.276932×10^{12}
20,000– 35,000	27,500	−5,764.75	-1.915761×10^{11}	0.239	-4157867×10^{10}
35,000– 50,000	42,500	9,235.24	7.876705×10^{11}	0.219	1.724998×10^{11}
50,000– 65,000	57,500	24,235.25	1.423451×10^{13}	0.087	1.238402×10^{12}
65,000– 80,000	72,500	39,235.25	6.039893×10^{13}	0.043	2.597154×10^{12}
80,000– 95,000	87,500	54,235.25	1.595309×10^{14}	0.0	0
95,000–110,000	102,500	69,235.25	3.318805×10^{14}	0.022	7.301372×10^{12}
110,000–125,000	[a]111,000	77,735.25	4.697362×10^{14}	0.022	1.03342×10^{13}
					$\Sigma f(x_i)(x_i - \bar{x})^3 = 1.832091 \times 10^{13}$

[a] As explained in Table 24.5:

$\bar{x} = 33{,}264.75\ ft^3/s$

Third central moment, $M_3 = 1.832091 \times 10^{13}$

Standard deviation, $s = 23{,}252.89914$

Coefficient of skewness = 1.5573

TABLE 24.8 COMPUTATION FOR KURTOSIS

Discharge (ft^3/s) class interval (1)	Midvalue, x_i (2)	$x_i - \bar{x}$ (3)	$(x_i - \bar{x})^4$ $(3)^4$ (4)	Relative frequency, $f(x_i)$ (5)	$f(x_i) \times (x_i - \bar{x})^4$ (4) × (5) = (6)
5,000– 20,000	[a]12,575	−20,689.75	1.832403×10^{17}	0.370	6.779891×10^{16}
20,000– 35,000	27,500	−5,764.75	1.104389×10^{15}	0.239	2.639489×10^{14}
35,000– 50,000	42,500	9,235.24	7.274326×10^{15}	0.219	1.593077×10^{15}
50,000– 65,000	57,500	24,235.25	3.449769×10^{17}	0.087	3.001299×10^{16}
65,000– 80,000	72,500	39,235.25	2.369767×10^{18}	0.043	1.019×10^{17}
80,000– 95,000	87,500	54,235.25	8.652201×10^{18}	0.0	0
95,000–110,000	102,500	69,235.25	2.297783×10^{19}	0.022	5.055123×10^{17}
110,000–125,000	[a]111,000	77,735.25	3.651506×10^{19}	0.022	8.033313×10^{17}
					$\Sigma f(x_i)(x_i - \bar{x})^4 = 1.510413 \times 10^{18}$

[a] As explained in Table 24.5:

$\bar{x} = 33{,}264.75$ ft^3/s

Fourth central moment, $M_4 = 1.510413 \times 10^{18}$

Standard deviation, $s = 23{,}252.89914$

Kurtosis, $\hat{K} = 5.9065$

Coefficient of excess kurtosis, $\varepsilon = 2.9065$

EXAMPLE 24.16 Estimate the coefficient of skewness for the data in Table 24.3. Tabulate the computations, as shown in Table 24.7.

$$M_3 = 1.832091 \times 10^{13}$$

$$s = 23{,}252.89914, \text{ from the previous example}$$

$$C_s = 1.5573$$

EXAMPLE 24.17 Estimate kurtosis for the data in Table 24.3, and show the computations in tabular form. Table 24.8 shows the computations.

$$M_4 = 1.510413 \times 10^{18}$$

$$s = 23{,}252.89914, \text{ from the previous example}$$

$$\hat{K} = 5.9065$$

$$\varepsilon = \hat{K} - 3 = 2.9065$$

24.7 RELIABILITY OF ESTIMATES OF DISTRIBUTION CHARACTERISTICS

24.7.1 Central-Limit Theorem

This theorem can be stated as follows:

> If S_n is the sum of n independently and identically distributed random variables X_i each having a mean μ and variance σ^2, then in the limit, as n approaches infinity, the distribution of S_n approaches a normal distribution with mean $n\mu$ and variance $n\sigma^2$.

If a random variable satisfies the conditions of the theorem, it can be expected to follow a normal distribution. The underlying distribution of the random variable can be any distribution. In hydrology, these conditions (n is large, independent, and identical variable distributions) usually do not hold. Of greater application in hydrology is the generalized central-limit theorem (Thomas, 1971), which holds under some very general conditions. If X_i, for $i = 1, 2, \ldots, n$, is a random variable independent of X_j, for $j \neq i$, and $E[X_i] = \mu_i$ and Var $[X_i] = \sigma_i^2$, then the sum $S_n = X_1 + X_2 + \cdots + X_n$ approaches a normal distribution with $E[S_n] = \Sigma_{i=1}^n \mu_i$ and Var $[S_n] = \Sigma_{i=1}^n \sigma_i^2$ as n approaches infinity. One condition for this generalized theorem is that each X_i has a negligible effect on the distribution of S_n.

24.7.2 Standard Error

Consider a long discharge record of the Amite River at Denham Springs, Louisiana. Say this record is 500 years long and is divided into 25 samples, each 20 years long. The mean of each sample is computed and designated as $\bar{x}_i$, $i = 1, 3, \ldots, 25$. These values of the mean can be used to describe the sample distribution of the mean $\bar{x}$. Then basic statistics such as mean, standard deviation, etc., can be computed for the mean. The standard deviation of this mean (computed from 25 means of the samples, $\bar{x}_i$, 1, 2, . . . , 25) is called the standard error of the mean. Similarly, the standard error of other statistics can be derived. Then, according to the central-limit theorem, the distribution of the means will approach a normal distribution with mean μ and standard deviation $\sigma/\sqrt{n}$, as the sample size increases. The statistic $\sigma/\sqrt{n}$ is the standard error of the mean. As n increases, the standard error decreases, and the parameter estimate becomes more reliable. The standard error of some basic statistics is given in Table 24.9. Thus, the reliability of distribution-parameter estimates is a function of sample size. Using Monte Carlo simulation techniques, one can generate error-free samples from any population, and then evaluate reliability of parameter estimates for various sample sizes.

EXAMPLE 24.18 Compute the standard error of the mean, standard deviation, coefficient of variation, and coefficient of skewness of the discharge data of Table 24.3.

Solution

$$\text{Standard error of the mean} = \frac{\sigma}{n^{1/2}} \cong \frac{s}{n^{1/2}}$$

$$= \frac{23{,}252.89914}{\sqrt{46}} = 3428.4529 \text{ ft}^3/\text{s}$$

$$\text{Standard error of the standard deviation} = \frac{\sigma}{(2n)^{1/2}} \cong \frac{s}{(2n)^{1/2}}$$

$$= \frac{23{,}252.89914}{(2 \times 46)^{1/2}}$$

$$= \frac{23{,}252.89914}{(92)^{1/2}}$$

$$= 2424.2823 \text{ ft}^3/\text{s}$$

$$\text{Standard error of coefficient of variation} = \frac{C_v(1 + 2C_v^2)^{1/2}}{(2n)^{1/2}}$$

$$= \frac{0.6990[1 + 2(0.6990)^2]^{1/2}}{(92)^{1/2}}$$

$$= 0.1025$$

TABLE 24.9 STANDARD ERROR OF BASIC STATISTICS

Statistics	Standard error
Mean	$\dfrac{\sigma}{n^{1/2}}$
Standard deviation	$\dfrac{\sigma}{(2n)^{1/2}}$
Coefficient of variation	$\dfrac{C_v(1 + 2C_v^2)^{1/2}}{(2n)^{1/2}}$
Coefficient of skewness	$\left[\dfrac{6n(n-1)}{(n+1)(n+2)(n+3)}\right]^{1/2}$

$$\text{Standard error of coefficient of skewness} = \left[\frac{6n(n-1)}{(n+1)(n+2)(n+3)}\right]^{1/2}$$

$$= \left[\frac{6(46)(45)}{(47)(48)(49)}\right]^{1/2}$$

$$= 0.3352$$

24.8 FREQUENCY DISTRIBUTIONS

Water-resources projects often require frequency distributions of magnitudes, volumes, durations, or depths of hydrologic variables. For example, the frequency with which the flood of a particular magnitude will be equalled or exceeded is frequently needed. Observed data usually form the basis of deriving frequency distributions of hydrologic variables. First, we consider some of the commonly used distributions in hydrology. Table 24.10 summarizes these distributions.

24.8.1 Binomial Distribution

This distribution arises in Bernoulli processes, where in any trial, an event may or may not occur. The probability of occurrence of an event is the same from one trial to the other. The trial can correspond to a point in discrete time or space. For example, the probability of head appearing in the first trial is the same in every other trial. The binomial distribution usually occurs when we deal with complementary events. In the case of coin tossing, heads and tails are complementary events. Wet and dry days in a given time interval are complementary events. The binomial distribution is given by Equation (24.11), which derives from the binomial theorem:

$$\begin{aligned}(q+p)^x &= xp(q)^{x-1} + \frac{x(x-1)}{2!}p^2q^{x-2} + \cdots \\ &= \sum_{i=0}^{x}\binom{x}{i}p^iq^{x-i}\end{aligned} \qquad (24.44)$$

where $q = \bar{p} = 1 - p$, the probability that the event will not occur. Note that Equation (24.11) is given by the successive terms of this theorem. Here x represents a specific value of the random variable X. For example, if n represents the number of days in a time interval, then $X = x$ can be the number of wet days, dry days, number of floods, number of cold frosty days, etc. In this sense, x is a subsample and the probabilities of $x = 0, 1, 2, \ldots$, are given by Equation (24.44) or (24.11). Thus, the number of events in a given time interval follows the binomial distribution. As the sample size increases, the binomial distribution approaches the normal distribution, which is discussed later, more rapidly for values of p near 0.5 than for those near 0 or 1.

The cumulative distribution function is given by

$$F(x) = \sum_{i=0}^{x}\binom{n}{i}p^iq^{n-i}, \qquad x = 0, 1, 2, \ldots \qquad (24.45)$$

TABLE 24.10 SUMMARY OF DISTRIBUTIONS COMMONLY USED IN HYDROLOGY

Distribution	Probability density function	Range	Mean	Variance
Binomial	$P(x) = \binom{n}{x} p^x(1-p)^{n-x}$	$0 \le x \le n$	np	$np(1-p)$
Geometric	$P(x) = pq^{x-1},\ q = 1-p$	$1 \le x \le \ldots$	$1/p$	q/p^2
Poisson	$P(x) = \dfrac{\lambda^x \exp(-\lambda)}{x!}$	$0 \le x \ldots$	λ	λ
Exponential	$f(x) = \lambda \exp(-\lambda x)$	$0 \le x \le \infty$	$1/\lambda$	$1/\lambda^2$
Gamma	$f(x) = \dfrac{\lambda^n x^{n-1}}{(n-1)!} \exp(-\lambda x)$	$0 \le x \le \infty$	n/λ	n/λ^2
Normal	$f(x) = \dfrac{1}{\sigma\sqrt{2n}} \exp\left[-\dfrac{(x-\mu)^2}{2\sigma^2}\right]$	$-\infty < x < \infty$	μ	σ^2
Log-normal ($y = \ln x$)	$f(x) = \dfrac{1}{\sigma x\sqrt{2n}} \exp\left[-\dfrac{(\ln x-\mu_y)^2}{2\sigma_y^2}\right]$	$0 < x < \infty$	μ_y or $\exp(\mu_y + \sigma_y^2/2)$	σ_y^2 or $\mu_x^2 [\exp(\sigma_y^2) - 1]$
Gumbel	$f(x) = \alpha \exp\{-\alpha(x-\beta) - \exp[-\alpha(x-\beta)]\}$	$-\infty < x < \infty$	$\mu + \gamma/\alpha$ $\gamma = 0.5772$	$n^2/6\alpha^2$
Pearson Type III	$f(x) = \dfrac{1}{a\Gamma(b)} \left(\dfrac{x-c}{a}\right)^{b-1} \exp\left(-\dfrac{x-c}{a}\right)$	$-\infty < x < \infty$	$ab + c$	a^2b
Log Pearson Type III ($y = \ln x$)	$f(x) = \dfrac{1}{ax\Gamma(b)} \left(\dfrac{\ln x-c}{a}\right)^{b-1} \exp\left(-\dfrac{\ln x-c}{a}\right)$	$0 < x < \infty$	$\mu_y = c + ab$	$\sigma_y^2 = a^2b$

This gives the probability of x or fewer times an event occurs in n independent trials. The parameters of the binomial distribution are n and p.

The mean and variance of the binomial distribution are expressed as

$$E[X] = np \tag{24.46}$$

$$\text{Var}[X] = npq = np(1 - p) \tag{24.47}$$

This distribution is symmetrical if $p = q$, skewed to the right if $q > p$, and skewed to the left if $q < p$.

EXAMPLE 24.19 In the month of June in Baton Rouge, Louisiana, afternoon thunderstorms are quite common. Suppose that there is a 20% chance of rainfall on any given day in this month. A convention is being planned during this month that may last about 5 days. To make proper arrangements for the convention, a no-rainfall 5-day period is preferred. What is the probability of having no rainfall in these 5 days? What is the probability of one rainy day during this period?

Solution Here, $n = 5$, $p = 0.2$, $q = 0.8$, and $x = 0$. Therefore,

$$P(x = 0) = \frac{5!}{0!(5 - 0)!}(0.2)^0(0.8)^5$$

$$= (0.8)^5 = 0.3277$$

If one day is a rainy day, then $x = 1$. Therefore,

$$P(x = 1) = \frac{5!}{1!(5 - 1)!}(0.2)^1(0.8)^4$$

$$= 5 \times 0.2 \times (0.8)^4 = 0.4096$$

EXAMPLE 24.20 The probability of an event occurring every year is 0.4. What is the probability of this event occurring in year 2 and not in year 1?

Solution

Probability of this event not occurring in year 1 = 0.6

Probability of this event occurring in year 2 = 0.4

Probability of this event not occurring in year 1 and occurring in year 2 = 0.6 × 0.4 = 0.24

EXAMPLE 24.21 How many times will a 5-year flood occur, on the average, in a 10-year period? Compute the probability that exactly this number of floods will occur in a 10-year period.

Solution A 5-year flood occurs every year with the probability

$$p = 1/5 = 0.2$$

Number of years = 10

By using Equation (24.46),

$$E[X] = np = 10 \times 0.2 = 2$$

By using Equation (24.11),

$$P(X = 2) = \binom{10}{2}(0.2)^2(1 - 0.2)^{10-2} = \binom{10}{2}(0.2)^2(0.8)^8$$

$$P(2) = \frac{10!}{2!(10 - 2)!}(0.2)^2(0.8)^8 = \frac{10!}{2!(8)!}(0.2)^2(0.8)^8 = 0.3020$$

EXAMPLE 24.22 What is the probability that a 5-year flood will occur four times in a 10-year period?

Solution

$$p = 1/5 = 0.2, \qquad n = 10, \qquad x = 4$$

By using Equation (24.11),

$$P(X = 4) = \binom{10}{4}(0.2)^4(0.8)^6 = \frac{10!}{4!(6)!}(0.2)^4(0.8)^6 = 0.0881$$

EXAMPLE 24.23 What is the probability that a 5-year flood will not occur at all in a 10-year period?

Solution

$$p = 1/5 = 0.2, \qquad n = 10, \qquad x = 0$$

By using Equation (24.11),

$$P(X = 0) = \binom{10}{0}(0.2)^0(0.8)^{10} = \frac{10!}{0!(10 - 0)!}(0.2)^0(0.8)^{10}$$

$$= (0.8)^{10} = 0.1074$$

24.8.2 Geometric Distribution

This distribution also arises in Bernoulli processes, and is expressed as

$$P(x) = pq^{x-1}, \qquad x = 1, 2, 3, \ldots \tag{24.48}$$

where $P(x)$ denotes the probability of an event to occur the first time on the xth trial. This means that the event has not occurred on any one of the first $x - 1$ trials. This should be expected because the event occurs with probability p on any trial. The mean and variance of the geometric distribution can be expressed as

$$E[X] = 1/p \tag{24.49}$$

$$\text{Var}\,[X] = q/p^2 \tag{24.50}$$

EXAMPLE 24.24 What is the probability that a 5-year flood will occur for the first time in the ith year, where $i = 1, 2, 3, 4$, and 5.

Solution

$$p = 1/5 = 0.2$$

$$P(X = 1) = (0.2)(0.8)^{1-1} = 0.2$$

$$P(X = 2) = (0.2)(0.8) = 0.16$$

$$P(X = 3) = (0.2)(0.8)^2 = 0.2 \times 0.64 = 0.128$$

$$P(X = 4) = (0.2)(0.8)^3 = 0.1024$$

$$P(X = 5) = (0.2)(0.8)^4 = 0.0819$$

EXAMPLE 24.25 What is the probability that there will be at least 2 years before the 5-year flood occurs? Also compute for 4 years.

Solution

$$p = 0.2$$

$$q = 0.8$$

The probability of the 5-year flood not occurring in 2 years $= 0.8 \times 0.8 = 0.64$.
The probability of the 5-year flood not occurring in 4 years $= (0.8)^4 = 0.41$.

24.8.3 Poisson Distribution

The Poisson distribution is a limiting form of the binomial distribution when p is very small and n is very large, and np tends to a constant λ. This may happen when the interval over which the Bernoulli process is defined gets smaller and smaller so that p of the event occurring in the interval gets smaller and smaller and the number of trials becomes greater and greater, keeping np constant. The quantity $\lambda = np$ is a parameter of the Poisson distribution, and is the only parameter in the distribution, in contrast to binomial and normal distributions, which have two parameters. This parameter denotes the expected mean frequency of occurrence of some event in a given time t. For the Poisson distribution, the probability of the event occurring 0, 1, 2, 3, . . . times in a given time t is equal to the successive terms of the following sequence:

$$\frac{1}{\exp(\lambda)}, \frac{\lambda}{\exp(\lambda)}, \frac{\lambda^2}{2\exp(\lambda)}, \frac{\lambda^3}{(2)(3)\exp(\lambda)}, \frac{\lambda^4}{(2)(3)(4)\exp(\lambda)}, \ldots$$

in which exp is the base of the natural logarithm, is an irrational number whose value is about 2.718, and whose logarithm (base 10) is about 0.4343. The number of events in any interval of time is independent of the number of events in any other nonoverlapping time interval.

The Poisson distribution can be expressed as

$$P(X = x) = \frac{\lambda^x \exp(-\lambda)}{x!}, \qquad \lambda > 0, \qquad x = 0, 1, 2, \ldots \tag{24.51}$$

This gives the probability distribution of the number of events in a given time, say, a year. The cumulative distribution function (CDF) of the Poisson distribution is

$$P(X \leq x) = \sum_{i=0}^{x} \frac{\lambda^i \exp(-\lambda)}{i!} \tag{24.52}$$

The mean and variance of the Poisson distribution are

$$E[X] = \lambda \tag{24.53}$$

$$\text{Var}\,[X] = \lambda \tag{24.54}$$

Equation (24.51) can be expressed for the number of events in time t as

$$P(x) = \frac{(\lambda t)^x \exp(-\lambda t)}{x!} \tag{24.55}$$

where $P(x)$ gives the probability of x events in time t. λ is interpreted as the average rate of occurrence of the event. The distribution parameter then becomes λt. The mean and variance of X then are λt.

EXAMPLE 24.26 What is the probability that a 5-year flood will occur three times in a period of 10 years? What is the probability that it will not occur at all? How many floods of this magnitude will occur on the average during 10 years?

Solution

$$E[X] = np = 10 \times 1/5 = 2$$

On the average, two floods will occur in 10 years. Thus, $\lambda = 2$.

$$P[X = 3] = \frac{(2)^3 \exp(-2)}{3!} = \frac{4}{3} \exp(-2) = 0.18$$

The probability that the 5-year flood will occur three times = 0.18. The probability that it will not occur at all is given as

$$P[X = 0] = \frac{(2)^0 \exp(-2)}{0!} = \exp(-2) = 0.135$$

EXAMPLE 24.27 What is the probability that a 5-year flood will occur fewer than four times in a period of 10 years?

Solution

$$\lambda = 2$$

By using Equation (24.52),

$$P[X \le 4] = \sum_{i=1}^{4} \frac{(2)^i \exp(-2)}{i!}$$

$$= 2 \exp(-2) + 2 \exp(-2) + (4/3) \exp(-2) + (2/3) \exp(-2)$$

$$= 6 \exp(-2) = 0.812$$

24.8.4 Exponential Distribution

The exponential distribution is given as

$$f(x) = \lambda \exp(-\lambda x) \tag{24.56}$$

Its CDF is

$$F(x) = 1 - \exp(-\lambda x) \tag{24.57}$$

This distribution is used to determine the length of the time interval between occurrences of the event, say, a flood. Its mean and variance are

$$E(X) = 1/\lambda \tag{24.58}$$

$$\text{Var}\,[X] = 1/\lambda^2 \tag{24.59}$$

EXAMPLE 24.28 From observations of floods in the Baton Rouge area, it is seen that the average time interval between floods is approximately 2 years. What is the probability that there will be a lapse greater than 10 years before a next flood arrives?

$$\lambda = 1/2 = 0.5$$

$$F(x \leq 10) = 1 - \exp(-0.5 \times 10) = 1 - \exp(-5) = 1 - 0.0067 = 0.9933$$

$$F(x \geq 10) = 0.0067$$

This shows that there is a very small chance that 10 years or more will pass before the next flood.

24.8.5 Gamma Distribution

The probability density function of the gamma distribution, with λ and n as parameters, is given as

$$f(x) = \frac{\lambda^n x^{n-1}}{(n-1)!} \exp(-\lambda x), \qquad x > 0, \qquad \lambda > 0, \qquad n = 1, 2, 3, \ldots \tag{24.60}$$

Extensive use of this distribution is found in hydrologic literature, although comparatively less in flood-frequency analysis. It can be used to determine the time to the nth event. The time to the nth event is the time to the first event T_1 plus the time interval between the first and second events T_2 plus the time interval between the second and third events T_3, and so on, or $T_1 + T_2 + T_3 + \cdots + T_n$. Because the time interval between events is described by the exponential distribution, the gamma distribution then results from sum of exponentials.

The mean and the variance of the gamma distribution are

$$E[X] = n/\lambda \tag{24.61}$$

$$\text{Var}\,[X] = n/\lambda^2 \tag{24.62}$$

EXAMPLE 24.29 The average time interval between floods in some parts of Louisiana is 2 years. Compute the probability that there will be a period less than or equal to 10 years for occurrence of five floods.

Solution

$$\lambda = \tfrac{1}{2} = 0.5$$

$$F(x \leq 10) = \int_0^{10} \frac{(0.5)^5 x^4}{4!} \exp(-0.5x)\,dx$$

$$= \frac{1}{768} \int_0^{10} x^4 \exp(0.5x)\,dx$$

$$= \left[\frac{\exp(-0.5x)}{768}(-2x^4 - 16x^3 - 96x^2 - 384x - 768)\right]\Bigg|_0^{10}$$

$$= \frac{1}{768} \{\exp(-5)\,[-2(10)^4 - 16(10)^3 - 96(10)^2 - 384(10)$$

$$- 768] - \exp(0)(0 - 0 - 0 - 0 - 768)\}$$

$$= -0.175 - 0.140 - 0.084 - 0.034 - 0.007 + 1$$

$$= -0.433 + 1 = 0.567$$

24.8.6 Normal Distribution

The normal distribution has a symmetrical, bell-shaped probability density function, and is also known as the Gaussian distribution, or the natural law of errors. It has two parameters, the mean μ and the standard σ, and can be written as

$$f(x) = \frac{1}{\sigma\sqrt{2\pi}} \exp\left[-\frac{(x-\mu)^2}{2\sigma^2}\right], \qquad -\infty < x < \infty \tag{24.63}$$

or

$$f(z) = \frac{1}{\sqrt{2\pi}} \exp(-z^2/2) \tag{24.64}$$

where $z = (x - \mu)/\sigma$, called the standard normal variate, and $f(x)$ is the probability density function (PDF) of the variable X. The parameters μ and σ^2 are sometimes called location and scale parameters, respectively. From Equation (24.63), it is seen that $f(x) \to 0$ as $x \to \pm\infty$, and $f(x)$ is maximum at $x = \mu$. Therefore, $f(x)$ is symmetrical about μ. By integrating Equation (24.63), we can write the CDF as

$$F(x) = \frac{1}{\sigma\sqrt{2\pi}} \int_{-\infty}^{x} \exp\left[-\frac{(u-\mu)^2}{2\sigma^2}\right] du \tag{24.65}$$

or

$$F(z) = \frac{1}{\sqrt{2\pi}} \int_{-\infty}^{z} \exp(-w^2/2)\, dw \tag{24.66}$$

The standard variate z has 0 mean and unit variance. Equations (24.64) and (24.66) represent the standard normal distribution. Table D.1 in Appendix D gives values of $F(z)$—the area under the standard normal curve—usually available in standard statistical textbooks. Figure 24.11 shows areas for three ranges of z. Corresponding to a given cumulative probability, the value of z is found from tables of area and then x is found from the inverse transform as

$$x = \mu + \sigma z \tag{24.67}$$

or

$$x = \bar{x} + sz \tag{24.68}$$

The normal distribution is the most widely used distribution. It is used in analysis of variance, testing of hypotheses, estimation of random errors of hydro-

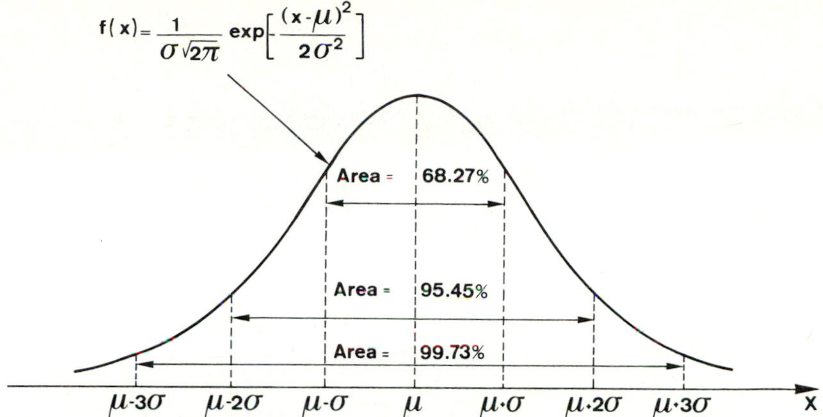

Figure 24.11 Areas under the normal probability distribution for three values of the standard variate.

logic measurements, comparison of distributions, derivation of distributions of statistical parameters, generation of random numbers, etc. A random variable is expected to follow a normal distribution if it is made up of the sum of many small independent effects.

Many hydrologic variables usually are not normally distributed, but transformations can, in many cases, make them approximately normally distributed. When the time interval over which a hydrologic variable is measured is large, say, a year or more, then the variable approximately follows a normal distribution. For example, yearly runoff at a gaging station approximates a normal distribution. This same is true of 2-year, 5-year, or 10-year runoff. This is because the number of causative effects increases. However, this is not true for daily, weekly, or monthly runoff.

24.8.7 Log-Normal Distribution

When a random variable is made up of the product of many small effects, then its logarithm is made up of the sum of logarithms of these small effects. This suggests that the logarithm of the random variable can be expected to follow a normal distribution. In other words, the logarithmic transformation reduces the random variable to be approximately normally distributed. Let $Y = \ln X$. If Y is normally distributed, then X is log-normally distributed. The PDF of a log-normal distribution can be expressed as

$$f(x) = \frac{1}{\sqrt{2\pi}x\sigma} \exp\left[-\frac{(\ln x - \mu_y)^2}{2\sigma_y^2}\right], \qquad x > 0 \tag{24.69}$$

The parameters are μ_y and σ_y^2, which can be estimated by first transforming all of x_i's to y_i's as

$$y_i = \ln x_i \tag{24.70}$$

and then

$$\bar{y} = \frac{1}{n}\sum_{i=1}^{n} y_i \tag{24.71}$$

$$S_y^2 = \frac{1}{n-1}\sum_{i=1}^{n} (y_i^2 - n\bar{y}^2) \tag{24.72}$$

The mean, variance, coefficient of variation, and coefficient of skewness of the log-normal distribution are

$$E[X] = \exp\left(\mu_y + \frac{\sigma_y^2}{2}\right)$$

$$\text{Var}[X] = \sigma_x^2 = \mu_x^2[\exp(\sigma_y^2) - 1]$$

$$C_v = [\exp(\sigma_y^2) - 1]^{0.5}$$

$$C_s = 3C_v + C_v^2$$

The log-normal distribution can be evaluated using tables of standard normal distribution.

EXAMPLE 24.30 Compute the area under the standard normal curve for the following cases: (a) between $z = 0$ and $z = 1.5$, (b) between $z = -1.5$ and $z = 0$, and (c) between $z = -0.5$ and $z = 2.5$.

Solution From Table D.1 in Appendix D, the area between $z = 0$ and $z = 1.5$ is $F(0 \leq z \leq 1.5) = 0.4332$.

The area between $z = -1.5$ and $z = 0$ is the same as the area between $z = 0$ and $z = 1.5$ by virtue of symmetry of the normal PDF. This area is 0.4332. Thus, $F(-1.5 \leq z \leq 0) = 0.4332$.

The area between $z = -0.5$ and $z = 2.5$ is the same as the area between $z = 0$ and $z = -0.5$ plus the area between $z = 0$ and $2.5 = 0.1915 + 0.4938 = 0.6853$. Thus, $F(-0.5 \leq z \leq 2.5) = 0.6853$.

24.8.8 Extreme Value Type I Distribution

Consider a series of N observations of a random variable, where N is quite large. If the series is divided into n subsamples of size m, such that $N = nm$, each sample contains a largest value and a smallest value. The largest and smallest values are commonly referred to as extremes corresponding, respectively, to floods and droughts. Gumbel (1958) has shown that the n largest values of subsamples asymptotically follow an extreme value (EV) type I distribution, with PDF and CDF given as

$$f(x) = \alpha \exp\{-\alpha(x - \beta) - \exp[-\alpha(x - \beta)]\}, \qquad -\infty < x < \infty, \quad -\infty < \beta < \infty, \ \alpha > 0 \tag{24.73}$$

$$F(x) = \exp\{-\exp[-\alpha(x - \beta)]\}$$

or

$$F(y) = \exp[-\exp(-y)], \qquad -\infty < y < \infty \tag{24.74}$$

where α and β are parameters, scale and location, with β being the mode of the distribution, and $y = (x - \beta)\alpha$.

Equation (24.73) can also be expressed as

$$y = -\ln[-\ln F(y)] \tag{24.75a}$$

$$y = -\ln\left(\ln \frac{T}{T-1}\right) \tag{24.75b}$$

or

$$y = -\left(0.834 + 2.303 \log \log \frac{T}{T-1}\right) \tag{24.75c}$$

Gumbel (1954) was the first to employ the theory of extreme values for flood-frequency analysis. This distribution is frequently referred to as the Gumbel distribution, and also as the double negative exponential distribution. It is one of the most widely used distributions for frequency analysis of floods, maximum rainfall, maximum wind speed, etc.

The mean and variance of the distribution are

$$E[X] = \mu = \beta + \gamma/\alpha \tag{24.76a}$$

$$\text{Var}[X] = \sigma^2 = \pi/\alpha^2\sqrt{6} = 1.645/\alpha^2 \tag{24.76b}$$

Gumbel has shown that $\bar{y}$ is a function of sample size and is shown in Table D.2 in Appendix D. When $n \to \infty$, $\bar{y} \to 0.577$. The standard deviation of y is also a function of sample size, and is given in Table D.3 in Appendix D. As $n \to \infty$, $s_y \to 1.2825$. This simplified procedure can be used to perform flood-frequency analysis using the Gumbel distribution.

Chow (1951) has shown that the Gumbel distribution is essentially a log-normal distribution with constant skewness. The coefficient of skewness is approximately 1.14 for the Gumbel distribution.

The parameters of the Gumbel distribution have been estimated in a number of ways. Lowery and Nash (1970) and Jain and Singh (1986) compared several methods and found that the method of moments was comparable to other methods. The method of moments estimates are

$$\hat{\alpha} = \frac{1.283}{s} \tag{24.77}$$

$$\hat{\beta} = \bar{x} - 0.45s \tag{24.78}$$

The "mean annual flood," used in hydrology, usually refers to a flood with a return period of 2.33 years. This return period is actually computed using the Gumbel distribution. This can be shown as follows. Equation (24.74) gives the nonexceedance probability of $y = (x - \beta)\alpha$. Since $\mu = \beta + \gamma/\alpha$, $y = [\beta + \gamma/\alpha - \beta]\alpha = \gamma = 0.5772$, where γ is Euler's constant. This gives $F(y) = 0.5703$, and $1 - F(y) = 0.4297$. Therefore, $T = 1/0.4297 = 2.33$ years.

24.8.9 Pearson Type (PT) III Distribution

The Pearson type (PT) III distribution is a three-parameter gamma distribution and is widely used in hydrology. Its PDF can be written as

$$f(x) = \frac{1}{a\Gamma(b)} \left(\frac{x - c}{a}\right)^{b-1} \exp\left(-\frac{x - c}{a}\right) \tag{24.79}$$

where a, b, and c are scale, shape, and location parameters, respectively, and $\Gamma(b)$ is a gamma function. If $c = 0$, this distribution becomes a two-parameter gamma distribution. When $z = (x - c)/a$, Equation (24.79) reduces to a one-parameter gamma distribution:

$$f(z) = \frac{1}{\Gamma(b)} z^{b-1} \exp(-z) \tag{24.80}$$

Parameters a, b, and c are related to mean, standard deviation, and coefficient of skewness as

$$a = \sigma/\sqrt{b} \tag{24.81}$$

$$b = (2/\gamma_s)^2 \tag{24.82}$$

$$c = \mu - \sigma\sqrt{b} \tag{24.83}$$

24.8.10 Log Pearson Type III Distribution

The log Pearson type (LPT) III distribution is widely used in the United States since the recommendation in 1967 by the U.S. Water Resources Council for adoption of this distribution as the standard flood-frequency distribution by all U.S. government agencies. If the random variable $Y = \ln X$ has a PT III distribution, then X has LPT III distribution with PDF given as

$$f(x) = \frac{1}{ax\Gamma(b)} \left(\frac{\ln x - c}{a}\right)^{b-1} \exp\left[-\frac{(\ln x - c)}{a}\right] \tag{24.84}$$

where a, b, and c are scale, shape, and location parameters, respectively. This distribution is very versatile and accommodates a variety of shapes.

The mean, standard deviation, and the coefficient of skewness of LPT III distribution are expressed as

$$\mu_y = c + ab \tag{24.85}$$

$$\sigma_y = a(b)^{1/2} \tag{24.86}$$

$$\gamma_y = \frac{2}{b^{1/2}} \tag{24.87}$$

Parameters a, b, and c can be computed from the sample data by replacing population statistics with sample statistics in Equations (24.85) to (24.87).

24.8.11 Transformations

Transformations are used for two purposes. First, data can be transformed such that they can be represented by a normal distribution. Second, the transformed data can be better represented by a known distribution. For example, if flood data are transformed logarithmically, then the Pearson type III distribution may better fit the transformed data than the original data. Of the several transformations (Jain and Singh, 1986), the power transformation is the most frequently used.

$$y = \frac{x^{\lambda} - 1}{\lambda} \tag{24.88}$$

in which $\lambda \neq 0$, and

$$y = \log x \tag{24.89}$$

in which $\lambda = 0$ and is a constant of transformation. The reciprocal and square-root transformations can be obtained as special cases of Equation (24.88).

EXERCISES

24.1. A residential area located near Amite River has been flooded in 5 of the past 20 years. What is the return period of this flood? What is the probability that the residential area will not be flooded next year? What is the probability that the area will be flooded at least once in the next 3 years?

24.2. A hydrologist wants to design a small dam for a flood whose return period is 50 years. What risk is he willing to accept for the flooding to occur in the next 2 years?

24.3. A temporary dam is to be built to protect the 10-year construction activity for a major cross-valley project. If the dam is designed to withstand the 30-year flood, determine the risk that the structure will be overtopped (1) in the first year, (2) in the third year exactly, (3) at least once in the 10-year construction period, and (4) not at all during the 10-year period.

24.4. Two rain gages, which are of the same type, are inspected weekly. Experience shows that these gages are each 20% of the times out of order at the end of a week. In 10% of the times, both gages are out of order.

(a) Compute the probability that at least one gage is out of order at the end of a week.

(b) Compute the probability that at the end of a week both gages are still working.

(c) Denote one gage as A and the other gage as B. Then $P(\mathrm{A}) \neq 0$ and $P(\mathrm{B}) \neq 0$. Show that

$$P(\mathrm{A}|\mathrm{B}) > P(\mathrm{A})$$

$$P(\mathrm{A} \cap \mathrm{B}) > P(\mathrm{A}) \cdot P(\mathrm{B})$$

What do these statements mean?

(d) Find lower and upper bounds on the probability that at the end of a week at least one rain gage is broken down when information on the proportion of weeks both gages are broken down is not available, but when it is known that there are some common factors causing breakdown of the rain gages. Assume that the conditional probability of one rain gage breaking down when the other rain gage has broken down is larger than the unconditional probability of the same event.

(e) Find an upper bound on the probability that at least one rain gage is in order in the same circumstances as in (d).

24.5. If A and B are two events, then show that

$$\overline{(A \cup B)} = \bar{A} \cap \bar{B}$$

where the overbar symbol denotes the complement.

24.6. Show that $\overline{(A \cap B)} = \bar{A} \cup \bar{B}$, where A and B are events, and the overbar symbol denotes the complement.

24.7. Consider two tributaries A and B joining to form a river C. River A has a yearly discharge smaller than its mean μ_A in 50% of the years and river B has a yearly discharge smaller than its mean μ_B in 60% of the time. In 70% of the years in which river A has a discharge smaller than μ_A, river B also has a discharge smaller than its mean.

(a) Calculate the probability that both rivers have a discharge smaller than their mean values.

(b) Calculate the probability that at least one river has a discharge smaller than its mean.

(c) Calculate the probability that river A has a discharge smaller than its mean value given that river B has a discharge smaller than its mean.

(d) Calculate the probability that at least one river has a discharge higher than its mean.

(e) Calculate the probability of no shortage in river C if no shortage occurs when both rivers have a discharge higher than their means.

24.8. The probability of at least one flood in a year is 0.2. The occurrence of floods in different years is considered independent.

(a) Let X be the number of years up to and including the year in which flooding occurs. Find the probability function of X.

(b) What is the mean return period of a flood?

(c) Calculate the probability function of the number of floods in 7 years.

24.9. The yearly peak flows X of a river section follow a log-normal probability distribution, that is, $\ln X \sim N(4.4, 0.4)$ if X is expressed in m^3/s.

(a) Calculate the median peak value; this is the peak flow that is not exceeded in 50% of the years.

(b) Calculate the flow, up to which the river section must be protected, for the return period of floods to be 50 years.

(c) If the number of floods in 50 years in (b) is denoted by N, calculate $P(N = 0)$, $P(N = 1)$, and $P(N > 1)$.

(d) If the river section is protected for flow up to 106.7 m^3/s, what is the return period of this flood? What is the probability of more than 40 floods in 100 years and the probability of more than one flood in 50 years?

24.10. During winters in an area, showers follow a Poisson process. The mean number of showers in 4 days is 3.

(a) Calculate the probability of zero showers in 3 days.

(b) Calculate the probability of one shower in 3 days.

(c) Calculate the probability of two or three showers in 3 days.

(d) What is the probability distribution of the number of showers in 60 days?

(e) Calculate the probability of more than 54 showers in 60 days.

(f) Calculate the mean time between two showers.

(g) Calculate the probability that the time to the next shower is less than 1 day.

24.11. Assume that the yearly mean discharge of a river is a random variable X, such that $\ln X$ is normally distributed with a mean of 5 and a standard deviation of 0.5 for X expressed in m^3/s. A shortage arises when the yearly mean discharge is smaller than a critical value, Q_c. Assume that discharges in different years are independent.

(a) If the mean return period of a shortage is 25 years, then what is Q_c?

(b) If the mean return period of a shortage is 15 years, then what is the probability of at least two shortages in 10 years?

(c) If the mean return period of a shortage is 15 years, then what are the mean and standard deviation of the number of shortages in 100 years?

(d) If the mean return period of a shortage is 30 years and if each shortage causes a damage of $2000 due to an agricultural shortfall, then what is the mean cost in 30 years and what is the standard deviation of this cost?

24.12. Assume that floods occur following a Poisson process with the mean of three floods per 4 years.

(a) What is the probability of at least one flood in a year?

(b) What is the probability that the next flood occurs within 2.5 years given that the latest flood was 3 years ago?

(c) What is the standard deviation of the number of floods in 10 years?

24.13. If $F(Q)$ is the probability distribution function of yearly peak flows and $T(Q)$ is the return period of floods when river banks are protected for flows below Q, then what is the relation between $T(Q)$ and $F(Q)$?

24.14. Flood banks are designed for a return period of T years. Calculate the probability of one or more floods in $\bar{T}$ years for all combinations of T and $\bar{T}$ given:

$$T = 2, 5, 25, 50, \text{ and } 100 \text{ years}$$

$$\bar{T} = 2, 5, 25, 50, \text{ and } 100 \text{ years}$$

24.15. A rain gage has been in operation for 3 years. During this time, there has been only one rainy day in June. What is the probability of having a rainy day at this gage in June?

24.16. Analysis of an annual flood series covering the period 1910 to 1959 at a gaging station on a certain river shows that the 100-year flood has a magnitude of 310,000 cfs and the 10-year flood has a magnitude of 140,000 cfs. Assume that the floods follow the Gumbel distribution.

(a) What is the probability of having as large as or larger than 200,000 cfs next year?

(b) What is the magnitude of a flood having a return period of 40 years?

(c) What is the probability of having at least a 100-year flood in the next 8 years?

(d) Find the mean annual flood.

(e) Find the standard deviation of the annual floods.

24.17. What is the probability of occurrence of a flood equal to or greater than a 20-year flood during the next 5 years?

24.18. What is the probability of having a flood equal to or greater than a 5-year flood next year? What is the probability of having at least one flood equal to or greater than a 5-year flood during the next 5 years?

24.19. What is the probability that a 50-year flood will occur in the next 10 years?

24.20. Compute the probability that a 10-year flood will occur in a single 20-year period.

24.21. Compute the probability that two or more 10-year floods will occur in 20 years.

24.22. Instantaneous annual peak-discharge data for the Comite River near Comite, Louisiana, are available for a 38-year period:

S.N.	DISCHARGE (m^3/s)	S.N.	DISCHARGE (m^3/s)	S.N.	DISCHARGE (m^3/s)
1	66.8	2	68.5	3	94.6
4	102.8	5	122.3	6	136.5
7	141.6	8	170.2	9	180.1
10	196.8	11	202.4	12	205.3
13	207.0	14	241.5	15	260.8
16	267.6	17	273.5	18	283.1
19	286.0	20	291.6	21	300.1
22	305.8	23	308.6	24	325.6
25	359.6	26	373.7	27	390.7
28	430.4	29	436.0	30	444.5
31	464.3	32	467.2	33	498.3
34	569.1	35	580.4	36	586.1
37	676.7	38	682.3		

Determine the mean, variance, coefficient of variation, skewness, coefficient of skewness, kurtosis, and coefficient of excess kurtosis for the discharge data.

24.23. The relation between risk R and return period T (years) for independent events can be defined as

$$R = 1 - \left(1 - \frac{1}{T}\right)^n$$

where n is the number of years. Plot R versus T for $n = 5$, 10, 20, 50, and 100. Note that you will have a family of curves. What is the risk that a 50-year flood will occur at least once in the next 5 years?

24.24. Suppose there is a 10% chance of rainfall on any given day in the month of February in Baton Rouge, Louisiana. What is the probability of having no rainfall in a 10-day period? What is the probability of having one rainy day and of having two rainy days during this period?

24.25. The probability of a tornado striking Baton Rouge every year is 0.3. What is the probability that the tornado will occur in year 2 and not in year 1?

24.26. How many times will a 2-year flood occur, on an average, in a 5-year period? Compute the probability that exactly this number of floods will occur in a 10-year period.

24.27. What is the probability that a 2-year flood will occur four times in an 8-year period?

24.28. What is the probability that a 2-year flood will not occur at all in a 5-year period?

24.29. What is the probability that a 2-year flood will occur for the first year in the ith year, where $i = 1$, 2, 3, 4, and 5?

24.30. What is the probability that there will be at least a year before the 2-year flood occurs?

24.31. What is the probability that a 2-year flood will occur five times in a period of 8 years? What is the probability that it will not occur at all? How many floods of this magnitude will occur on the average during 8 years?

24.32. What is the probability that a 2-year flood will occur fewer than five times in a period of 8 years?

24.33. The average time interval between tornados striking Baton Rouge is approximately 4 years. What is the probability that there will be a lapse of less than or equal to 8 years before the next tornado strikes?

24.34. Compute the probability that there will be three tornados in a period less than or equal to 8 years. The average time interval is 4 years.

24.35. Compute the area under the standard normal curve for the following cases:

(a) between $z = 0$ and $z = 1.0$

(b) between $z = -1$ and $z = 0$

(c) between $z = -1.5$ and $z = 1.5$

24.36. The distribution of mean annual rainfall at 35 stations in thc James River basin, Virginia, is given in the following summary:

INTERVAL (2-IN. GROUPINGS)	NO. OF OBSERVATIONS
36 or 37	2
38–39	4
40–41	7
42–43	9
44–45	5
46–47	4
48–49	2
50–51	2

(a) Plot the relative frequency distribution.

(b) Plot the cumulative frequency distribution.

(c) Estimate the probability that the mean annual rainfall will (1) exceed 40 in., (2) exceed 50 in., and (3) will be between 40 and 50 in.

CHAPTER 25
Frequency Analysis

Frequency analysis is performed to determine the frequency of the likely occurrence of hydrologic events. This information is required in the solution of a variety of water-resource problems. Some pertinent examples include design of reservoirs, floodways, bridges, culverts, highways, levees, urban drainage systems, airfield drainage, irrigation systems, stream-control works, water-supply systems, and hydroelectric power plants; zoning of floodplain land for industrial, residential, and recreational use; setting of flood-insurance premiums; economic evaluation of flood-protection projects; drought-mitigation programs; etc. Although the frequency analysis of virtually every component of the hydrologic cycle is required, our emphasis here will be on frequency analyses of streamflow extremes and rainfall only.

25.1 TREATMENT OF DATA

The hydrologic data to be analyzed for frequency analysis must be treated in light of the objectives of the analysis, length of record, completeness of record, randomness of data, and homogeneity. The length of record should be more than 25 years for the derived distribution to be acceptable. The hydrologic data must have been controlled by a uniform set of hydrologic and operational factors. For example, the factors causing a winter rainflood are quite different from those during a spring snowmelt flood or a local cloudburst flood. These two types of floods should not be combined into a single record. Sometimes a hydrologic record may be interrupted by a period of 1 or more years. This interruption should be investigated carefully. Missing data may sometimes be estimated using regional analysis or by correlation with other hydrologic data in the region.

The hydrologic data are generally presented in chronological order. These data constitute the complete duration series (CDS). For frequency analysis, the CDS is seldom used because the hydrologic design of a project is normally dictated by only a few critical events. Therefore, the hydrologic data can be selected in two ways: (1) partial duration series (PDS) and (2) annual duration series (ADS). The PDS is comprised of the data exceeding a specified base level. In the ADS, one value (usually the highest) is selected from each year. For all practical purposes, the two series are comparable if the record is longer than 10 years, and either, therefore, can be used.

25.2 POINT FREQUENCY ANALYSIS

The frequency distributions presented in the preceding chapter can be fitted to the data. Two commonly used methods of fitting are discussed here: (1) the graphical method and (2) frequency factors.

25.2.1 Graphical Method

This method involves fitting of an assumed probability distribution to observed data. The sample data are arranged in either ascending or descending order of magnitude. Each data point is assigned a rank starting with 1. If these are arranged in descending order of magnitude, then the highest value will be assigned the rank of 1, the second highest the rank of 2, and so on; the lowest value will have the rank of n, where n also signifies the number of data points in the sample. This arrangement gives an estimate of the exceedance probability, that is, the probability of a value being equal to or greater than the ranked value. If the values are arranged in ascending order, then an estimate of the non-exceedance probability, that is, the probability of a value being less than or equal to the ranked value, is obtained. These data points are plotted on probability paper, with their positions determined from a plotting-position formula.

Many plotting-position formulas are available; some of the more commonly used ones are given in Table 25.1. Adamowski (1981) has shown that all of these formulas can be expressed as special cases of

$$P_m = \frac{m - a}{N + b} \tag{25.1}$$

where a and b are constants, and P_m is the probability of the mth observation, and m is the mth value of N ordered observations such as $P_1 < P_2 < \cdots < P_N$. Perhaps one of the most commonly used plotting-position formulas in hydrology is

$$p_m = \frac{m}{N + 1} \tag{25.2}$$

where p_m is the exceedance probability of the mth data point (observed value) in the sample arranged in descending order. Clearly, the return period of the mth data

TABLE 25.1 SOME COMMONLY USED PLOTTING—POSITION FORMULAS

Method	Formula $P_m (X > X_m)$	$P_m = \frac{m-a}{N+b}$		Values of P_m and T_m for	
		a	b	$m = 1$	$N = 10$
Hazen (1914)	$\frac{m-0.5}{N}$	0.5	0.0	0.05	20.0
California (1923)	$\frac{m}{N}$	1.0	0.0	0.10	10.0
Weibull (1939)	$\frac{m}{N+1}$	0.0	1.0	0.091	11.0
Beard (1943)	$\frac{M-0.31}{N+0.38}$	0.31	0.38	0.066	15.0
Chegodayev (1955)	$\frac{m-0.3}{N+0.4}$	0.30	0.40	0.067	14.9
Blom (1958)	$\frac{m-0.375}{N+0.25}$	0.375	0.25	0.0609	16.4
Gringorten (1963)	$\frac{m-0.44}{N+0.12}$	0.44	0.12	0.055	18.1
Cunnane (1978)	$\frac{m-0.4}{N+0.2}$	0.4	0.2	0.58	17.2
Adamowski (1981)	$\frac{m-0.25}{N+0.5}$	0.25	0.50	0.071	14.1

point, T_m, is

$$T_m = \frac{N+1}{m} \tag{25.3}$$

The observed values and their exceedance probabilities are then plotted on the probability paper corresponding to the assumed probability distribution. This probability paper is commonly available or else can be constructed. On the ordinate of the graph paper are observed values and on the abscissa the probabilities or return periods. The objective of using the probability paper is to linearize the distribution so that plotted data can be represented by a straight line. A best-fit straight line is then fitted through the plotted points. The line is assumed to give the probabilities of all values beyond the observed range.

25.2.2 Frequency-Factor Method

Chow (1951, 1954) proposed the use of a frequency factor in hydrologic frequency analysis. If a hydrologic variable X is plotted chronologically in time, then a particular value x is found to be composed of two parts: namely, the mean, $\bar{x}$, and the departure from the mean, Δx:

$$x = \bar{x} + \Delta x \tag{25.4}$$

The value of Δx can be positive or negative, large or small, irregular and variable, and can be expressed as the product of the standard deviation S and the frequency factor K. Therefore,

$$x = \bar{x} + SK \tag{25.5}$$

where K depends on the return period T and the PDF of X; K literally means the number of standard deviations above and below the mean to achieve the desired quantile. For an assumed distribution, a relation between K and T can be derived. For two-parameter distributions, the value of K varies with the probability or T. For skewed distributions, it varies with the coefficient of skewness and is very sensitive to the length of record.

NORMAL DISTRIBUTION

Recall the definition of the standard normal variate, $Z = (X - \mu)/\sigma$, where μ is the population mean, and σ is population standard deviation of the variable X. Its observed values are then expressed as

$$z = \frac{x - \bar{x}}{S}$$

This can be written as

$$x = \bar{x} + Sz \tag{25.6}$$

Thus, for the normal distribution, K is the standard normal variate, which can be obtained from Table D.1 of standard normal distributions given in Appendix D.

LOG-NORMAL DISTRIBUTION

Chow (1964) has derived the frequency factor for the lognormal distribution, which is expressed as

$$K = \frac{\exp\left(\sigma_y K_y - \dfrac{\sigma_y^2}{2}\right) - 1}{[\exp(\sigma^2) - 1]^{1/2}} \tag{25.7}$$

where $Y = \ln X$, and $K_y = (Y - \mu_y)/\sigma_y$. Table D.4 gives the values of K.

GUMBEL DISTRIBUTION

Equation (24.73) can be rearranged as

$$\frac{1}{T} = 1 - F(x) = 1 - \exp\{-\exp[\alpha(x - \beta)]\} \tag{25.8}$$

This can be written as

$$x = \beta - \frac{1}{\alpha}\ln[\ln T - \ln(T - 1)] \tag{25.9}$$

By substituting Equations (24.77) and (24.78) for α and β into Equation (24.98),

$$x = \bar{x} - \frac{S\sqrt{6}}{\pi}\left[0.5772 + \ln\left(\ln\frac{T}{T-1}\right)\right] \tag{25.10}$$

On comparing this with Equation (25.5), the frequency factor K for the Gumbel distribution can be derived as

$$K = -\frac{\sqrt{6}}{\pi}\left[0.5772 + \ln\left(\ln\frac{T}{T-1}\right)\right] \tag{25.11}$$

Note that the expression for the reduced variate Y can be rearranged as

$$y = \frac{1.283(x - \bar{x})}{S} + 0.577 \tag{25.12}$$

On comparing with Equation (25.5),

$$K = \frac{y - 0.577}{1.283} \tag{25.13}$$

Equation (25.11) is valid only when N approaches infinity. For a finite sample, K varies with the length of the record. For various sample sizes, Chow (1964) has derived K versus T curves, as given in Table D.5 in Appendix D. When $K = 0$ in Equation (25.11), $T = 2.33$ years. Thus, in flood-frequency analysis, the recurrence interval of the mean annual flood is commonly designated as 2.33 years.

LOG-PEARSON TYPE 3 DISTRIBUTION

For the Log-Pearson Type III Distribution, K is a function of both the return period and the coefficient of skewness. Values of K for log-transformed values are given by the Water Resources Council (1967), as shown in Table D.6 in Appendix D. The procedure for fitting this distribution is as follows. Transform the data, x_i (annual floods), to their logarithmic values, y_i. Compute the mean, standard deviation, and coefficient of skewness for the log values. Get the value of K for the desired return period from the tabulated values. Compute y from $y = \bar{y} + KS_y$. Then compute $x = \exp(y)$ for the T value considered.

Sometimes the coefficient of skewness is adjusted to account for the effect of sample size N by using the relation proposed by Hazen (1930):

$$\hat{C}_s = C_s\left(1 + \frac{8.5}{N}\right) \tag{25.14}$$

where $\hat{C}_s$ is the adjusted coefficient of skewness. Beard (1962, 1974), Benson (1968), and others recommend using a regionalized coefficient of skewness in flood-frequency analysis if the record of a gaging station is less than 100 years.

EXAMPLE 25.1 Peak-discharge values of the Kentucky River near Salvisa, Kentucky, are given in Table 25.2. Fit the normal distribution to this data on a normal probability curve. Determine graphically the distribution parameters. Show the computations in tabular form.

TABLE 25.2 INSTANTANEOUS ANNUAL PEAK-DISCHARGE VALUES OF THE KENTUCKY RIVER NEAR SALVISA, KENTUCKY (MCCABE, 1962)

Year	Peak discharge (cfs)	Peak discharge (m^3/s)
1895	47,300	1,338.59
1896	54,400	1,539.52
1897	87,200	2,467.76
1898	65,700	1,859.31
1899	91,500	2,598.45
1900	53,500	1,514,05
1901	67,800	1,918.74
1902	70,000	1,981.00
1903	66,900	1,893.27
1904	34,700	982.01
1905	58,000	1,641.14
1906	47,000	1,330.10
1907	66,300	1,876.29
1908	80,900	2,289.47
1909	80,000	2,264.09
1910	52,300	1,480.09
1911	58,000	1,641.40
1912	67,200	1,901.76
1913	115,000	3,254.50
1914	46,100	1,304.63
1915	52,400	1,482.92
1916	94,300	2,668.69
1917	111,000	3,141.30
1918	71,700	2,029.11
1919	96,100	2,719.63
1920	92,500	2,617.75
1921	34,100	965.03
1922	69,000	1,952.70
1923	73,400	2,077.22
1924	99,100	2,804.53
1925	79,200	2,241.36
1926	62,600	1,771.58
1927	93,700	2.651.71
1928	68,700	1,944.21
1929	80,100	2,266.83
1930	32,300	914.89
1931	43,100	1,219.73
1932	77,000	2,179.91
1933	53,600	1,516.88
1934	70,800	2,003.64
1935	89,400	2,530.02
1936	62,600	1,771.58
1937	112,000	3,169.60
1938	44,000	1,245.20
1939	84,300	2,385.69
1940	45,000	1,273.50
1941	28,400	803.72
1942	46,000	1,301.80
1943	80,400	2,275.32

TABLE 25.2 *(Continued)*

Year	Peak discharge (cfs)	Peak discharge (m^3/s)
1944	55,000	1,556.50
1945	72,900	2,063.07
1946	71,200	2,014.96
1947	46,800	1,324,44
1948	84,100	2,380.03
1949	61,300	1,734.79
1950	87,100	2,464.93
1951	70,500	1,995.15
1952	77,700	2,198.91
1953	44,200	1,250.86
1954	20,600	582.98
1955	85,000	2,405.50
1956	82,900	2,346.07
1957	88,700	2,510.21
1958	60,200	1,703.66
1959	40,300	1,140.49
1960	50,500	1,429.15

Solution The data are arranged in ascending order and each value is assigned a rank starting from $m = 1$ for the lowest value, as shown in Table 25.3. Relative frequency of each data point is computed using Equation (25.2). On a normal probability paper, peak-discharge values are plotted against their relative frequencies, as shown in Figure 25.1. A straight line is drawn such that there are about equal numbers of points on each side of the line.

Discharge corresponding to relative frequency of 50% = 67,500 cfs

Therefore,

Mean $\bar{x}$ = 67,500 cfs

Corresponding to relative frequency of 84.13%, discharge $(x + S)$ = 88,000 cfs

Corresponding to relative frequency of 15.87%, discharge $(x - S)$ = 40,500 cfs

Therefore, the standard deviation is S = 20,500 cfs.

EXAMPLE 25.2 Fit the log-normal distribution to the discharge data of Table 25.4. Determine the distribution parameters graphically as well as from the data. Show the computations in tabular form.

Solution The data are arranged in descending order and each value is assigned a rank starting from $m = 1$ for the highest value, as shown in Table 25.4. The probability of each data point is computed using Equation (25.2). On log-normal probability paper, peak-discharge values are plotted against their relative frequencies, as shown in Figure 25.2. A straight line is drawn through the data points. From the figure, the distribution parameters are obtained as follows:

TABLE 25.3 PROBABILITIES AND RETURN PERIODS OF ANNUAL PEAK DISCHARGE VALUES OF THE KENTUCKY RIVER NEAR SALVISA, KENTUCKY

Rank, m	Discharge (cfs)	Nonexceedance probability (%), $(m/n + 1) \times 100$	Return period, T (years) $(n + 1)/m$
1	115,000	1.49	67
2	112,000	2.98	33.5
3	111,000	4.48	22.3
4	99,100	5.97	16.75
5	96,100	7.46	13.4
6	94,300	8.96	11.17
7	93,700	10.45	9.57
8	92,500	11.94	8.38
9	91,500	13.43	7.44
10	89,400	14.93	6.70
11	88,700	16.42	6.09
12	87,200	17.91	5.58
13	87,100	19.40	5.15
14	85,000	20.89	4.79
15	84,300	22.39	4.47
16	84,100	23.88	4.19
17	82,900	25.37	3.94
18	80,900	26.87	3.72
19	80,400	28.36	3.53
20	80,100	29.85	3.35
21	80,000	31.34	3.19
22	79,200	32.84	3.05
23	77,700	34.33	2.91
24	77,000	35.82	2.79
25	73,400	37.31	2.68
26	72,900	38.81	2.58
27	71,700	40.3	2.48
28	71,200	41.79	2.39
29	70,800	43.28	2.31
30	70,500	44.78	2.23
31	70,000	46.27	2.16
32	69,000	47.76	2.09
33	68,700	49.25	2.03
34	67,800	50.75	1.97
35	67,200	52.24	1.91
36	66,900	53.73	1.86
37	66,300	55.22	1.81
38	65,700	56.72	1.76
39	62.600	58.21	1.72
40	62,600	59.70	1.68
41	61,300	61.19	1.63
42	60,200	62.69	1.60
43	58,000	64.18	1.56
44	58,000	65.67	1.52
45	55,000	67.16	1.49
46	54,400	68.66	1.46
47	53,600	70.15	1.43
48	53,500	71.64	1.40

TABLE 25.3 *(Continued)*

Rank, m	Discharge (cfs)	Nonexceedance probability (%), $(m/n+1)\times 100$	Return period, T (years) $(n+1)/m$
49	52,400	73.13	1.37
50	52,300	74.63	1.34
51	50,500	76.12	1.31
52	47,300	77.61	1.29
53	47,000	79.10	1.26
54	46,800	80.60	1.24
55	46,100	82.09	1.22
56	46,000	83.58	1.20
57	45,000	85.08	1.18
58	44,200	86.57	1.16
59	44,000	88.06	1.14
60	43,100	89.55	1.12
61	40,300	91.05	1.10
62	34,700	92.54	1.08
63	34,100	94.03	1.06
64	32,300	95.52	1.05
65	28,400	97.02	1.03
66	20,600	98.51	1.02

At 50% probability, $x = \bar{x} =$ mean discharge $= 26{,}000$ cfs

Hence, mean $\ln x = \overline{\ln x} = \bar{y} = 10.166$ cfs

At 84.10% probability, $x_1 = 12{,}000$ cfs

At 15.9% probability, $x_2 = 57{,}000$ cfs

Hence, the standard deviation of $\ln x$, S_n,

$$S_n \cong \frac{\ln x_2 - \ln x_1}{2} = \frac{\ln(57{,}000) - \ln(12{,}000)}{2}$$

or

$$S_n \cong 0.779$$

In terms of statistics of x, we have

$$\bar{x} = \exp\left(\bar{y} + \frac{S_n^2}{2}\right) = \exp\left[10.166 + (0.779/2)\right] = 35{,}221.8 \text{ cfs}$$

$$S^2 = \bar{x}^2\left[\exp(S_n^2) - 1\right] = (35{,}221.8)^2\{\exp\left[(0.779)^2\right] - 1\}$$

$$S^2 = 1{,}035{,}417{,}119$$

or

$$S = 32{,}177.9 \text{ cfs}$$

Coefficient of variation, C_V:

$$C_V = \frac{S}{\bar{x}} = \frac{32{,}177.9}{35{,}221.8} = 0.9136$$

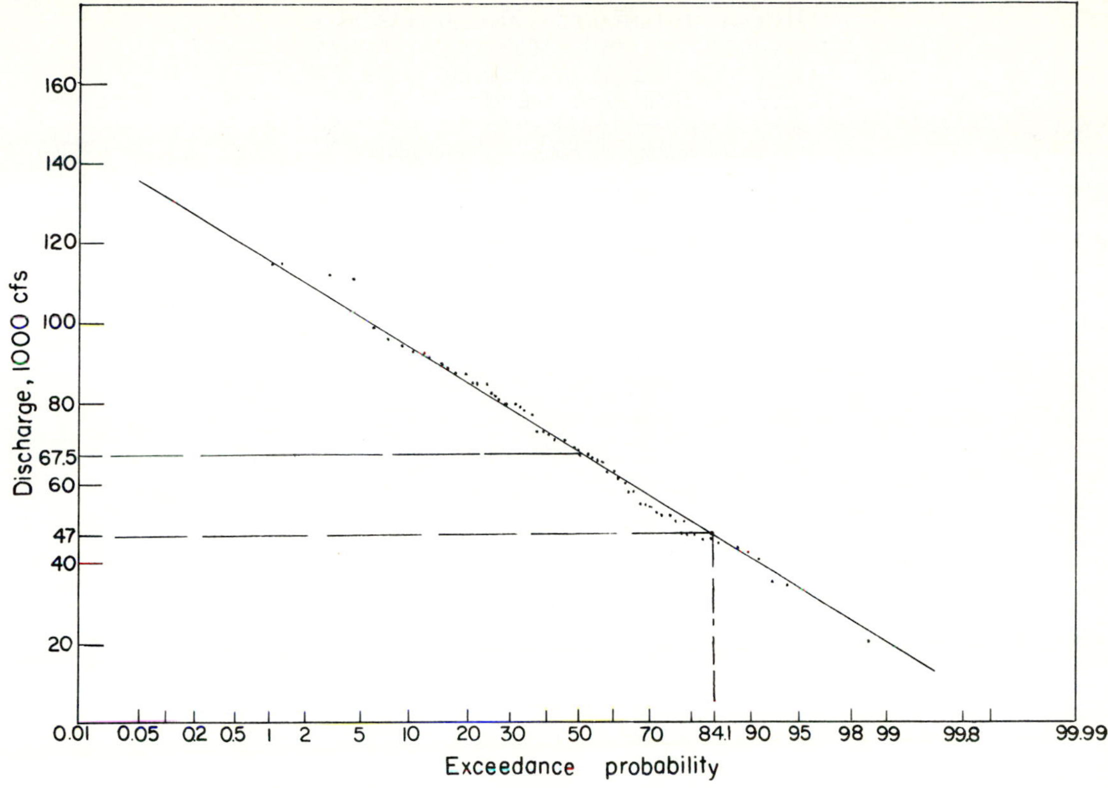

Figure 25.1 Normal fit for the Kentucky River discharges.

Coefficient of skewness, C_s:

$$C_s = (C_V)^3 + 3(C_V) = (-0.9136)^3 + 3(0.9136) = 3.5$$

Sample statistics calculated from the data are

$$\bar{x} = \frac{\Sigma x_i}{n} = \frac{1{,}521{,}470}{46} = 33{,}075.4 \text{ cfs}$$

$$S = \left[\frac{\Sigma x_i^2 - \dfrac{(\Sigma x_i)^2}{n}}{n - 1}\right]^{0.5} = \left[\frac{7.685 \times 10^{10} - \dfrac{(1.52147 \times 10^6)^2}{46}}{45}\right]^{0.5}$$

$$= 24{,}277.4 \text{ cfs}$$

$$C_V = \frac{S}{\bar{x}} = \frac{24{,}277.4}{33{,}075.4} = 0.734$$

$$C_s = \frac{n\,\Sigma(x_i^2 - \bar{x})^3}{(n - 1)(n - 2)S^3} = \frac{46(9.521 \times 10^{14})}{(45)(44)(24{,}277.4)^3} \cong 1.55$$

EXAMPLE 25.3 Fit the Gumbel distribution to the discharge data of Table 25.4. Determine the distribution parameters. Use Gumbel probability paper.

TABLE 25.4 INSTANTANEOUS ANNUAL PEAK DISCHARGE OF THE AMITE RIVER AT DENHAM SPRINGS, LOUISIANA; RANKS ASSIGNED TO DISCHARGE VALUES; AND EXCEEDANCE PROBABILITIES AND RETURN PERIODS

Year 1	Discharge (cfs) 2	Rank, M 3	Probability $M/(N + 1)$ 4	Return period 5 (= 1/4)
1939	12,100.00	40	0.851	1.17
1940	16,000.00	33	0.702	1.42
1941	20,800.00	29	0.617	1.62
1942	12,200.00	39	0.830	1.21
1943	40,200.00	16	0.340	2.94
1944	11,000.00	43	0.915	1.09
1945	11,600.00	41	0.872	1.15
1946	15,500.00	34	0.723	1.38
1947	27,800.00	22	0.468	2.14
1948	45,100.00	13	0.277	3.62
1949	28,800.00	21	0.447	2.24
1950	40,800.00	14	0.298	3.36
1951	36,900.00	18	0.383	2.61
1952	8,230.00	44	0.936	1.07
1953	67,000.00	4	0.085	11.75
1954	15,200.00	35	0.745	1.34
1955	54,300.00	7	0.149	6.71
1956	23,400.00	25	0.532	1.88
1957	12,300.00	38	0.809	1.24
1958	14,700.00	36	0.766	1.31
1959	19,100.00	30	0.638	1.57
1960	18,800.00	31	0.660	1.52
1961	49,100.00	11	0.234	4.27
1962	49,700.00	10	0.213	4.70
1963	5,150.00	46	0.979	1.02
1964	40,500.00	15	0.319	3.13
1965	49,900.00	9	0.191	5.22
1966	39,700.00	17	0.362	2.76
1967	47,800.00	12	0.255	3.92
1968	6,290.00	45	0.957	1.04
1969	23,000.00	26	0.553	1.81
1970	21,700.00	27	0.574	1.74
1971	12,600.00	37	0.787	1.27
1972	51,800.00	8	0.170	5.88
1973	60,200.00	6	0.128	7.83
1974	21,300.00	28	0.596	1.68
1975	29,900.00	20	0.426	2.35
1976	16,100.00	32	0.681	1.47
1977	110,000.00	2	0.043	23.50
1978	31,300.00	19	0.404	2.47
1979	68,600.00	3	0.064	15.67
1980	64,200.00	5	0.106	9.40
1981	11,300.00	42	0.894	1.12
1982	23,900.00	23	0.489	2.04
1983	112,000.00	1	0.021	47.00
1984	23,600.00	24	0.511	1.96

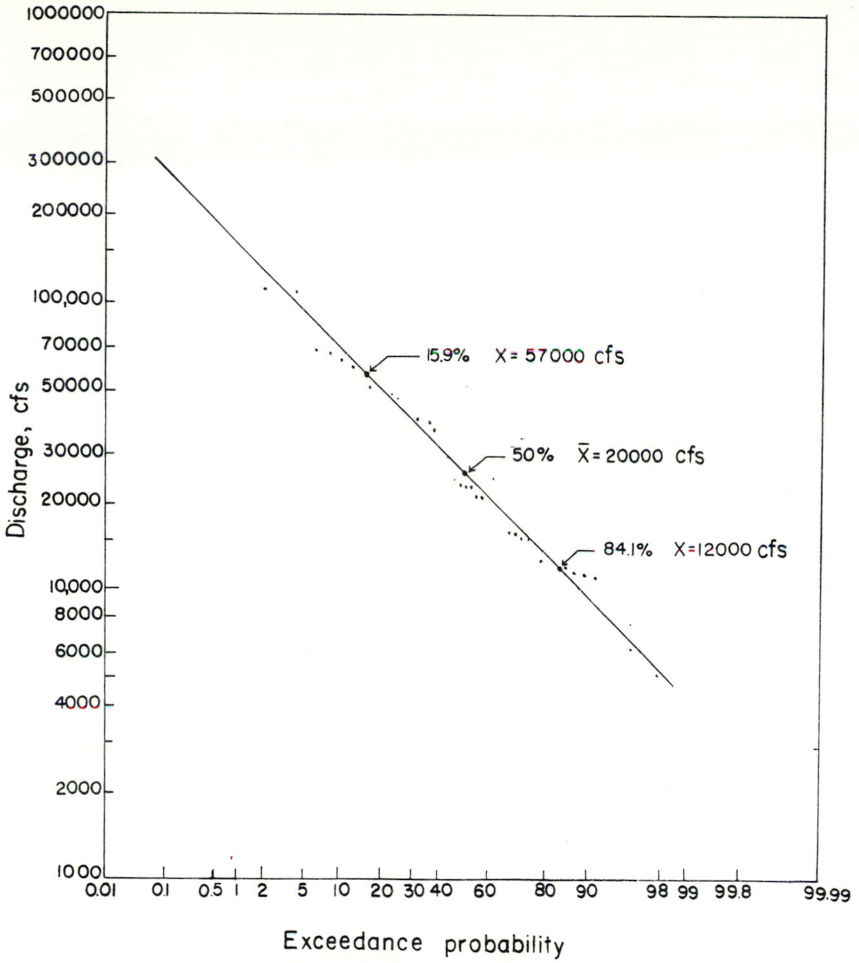

Figure 25.2 Log-normal fit for the Amite River discharges.

Solution The arrangement of data and the computation of probabilities are shown in Table 25.4. Discharge values are plotted against the recurrence interval on the Gumbel probability paper, as shown in Figure 25.3. The best-fit line is drawn through the points. From the graph, the mode is at

$$m = \hat{\beta} \cong 25{,}000 \text{ cfs}$$

By using the 25% and 75% quantiles,

$$P(x_1) = 0.25; \text{ hence, } x_1 = 18{,}500 \text{ cfs}$$

$$P(x_2) = 0.75; \text{ hence, } x_2 = 45{,}500 \text{ cfs}$$

α is estimated as follows:

$$\frac{1}{\alpha} = 0.6359(x_2 - x_1)$$

$$= 0.6359(45{,}500 - 18{,}500)$$

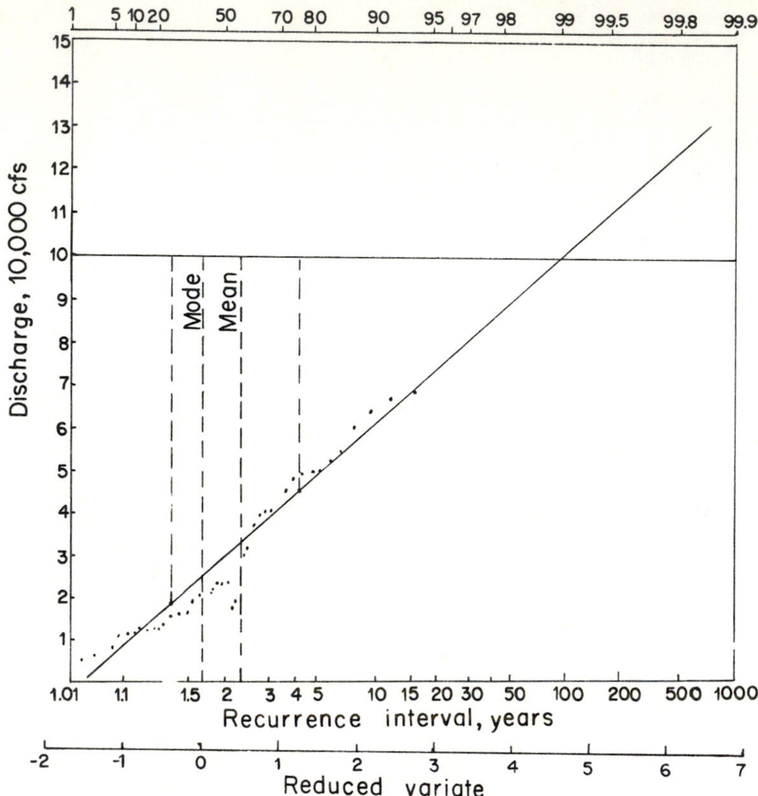

Figure 25.3 Gumbel fit for the Amite River discharges.

Therefore,

$$\alpha = 0.0000582$$

Hence,

$$\bar{x} = \beta + \frac{0.5772}{\alpha} = 25{,}000 + \frac{0.5772}{0.0000582} = 34{,}910.1 \text{ cfs}$$

$$S = \frac{1.281}{\alpha} = \frac{1.281}{0.0000582} = 22{,}010.3 \text{ cfs}$$

EXAMPLE 25.4 Fit the Pearson type III distribution to the discharge data of Table 25.4. Use probability paper.

Solution Calculations for plotting positions and recurrence intervals are given in Table 25.4. Discharge values are plotted against exceedance probabilities on the normal probability paper, as shown in Figure 25.4. A best-fit curve is drawn through the plotted points. Estimates of the 100-, 150-, and 200-year floods based on the fitted frequency curve are

T (year)	DISCHARGE (cfs)
100	126,000
150	139,000
200	146,000

EXAMPLE 25.5 Fit the log Pearson type III distribution to the discharge data of Table 25.4. Use probability paper.

Solution Calculations for plotting positions and recurrence intervals are given in Table 25.4. Discharge values versus probabilities are graphed on the probability paper, as

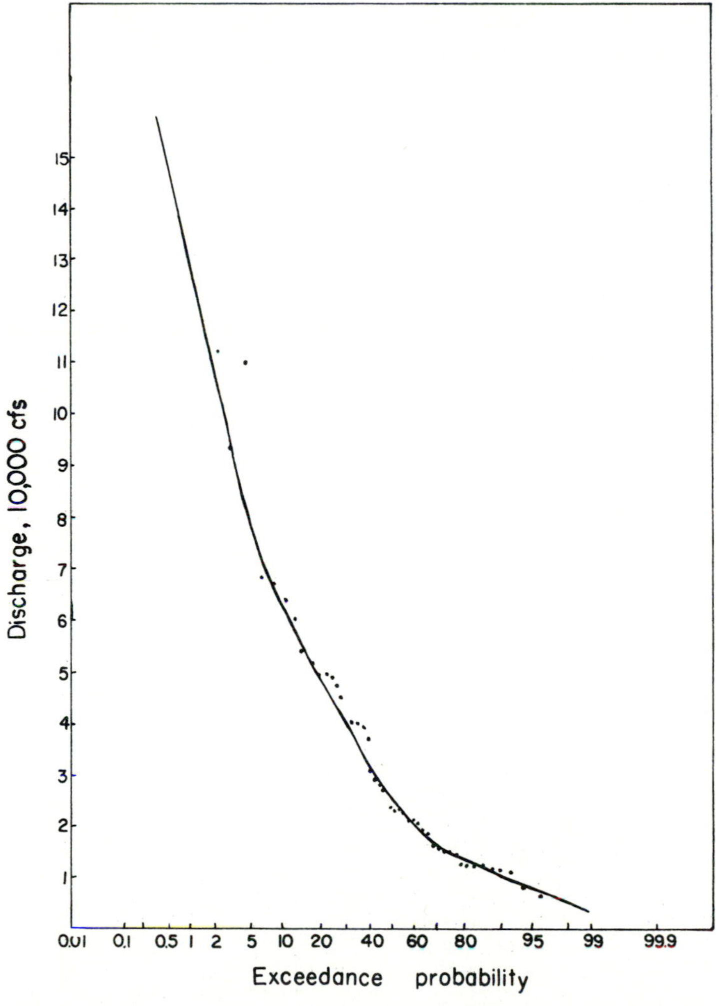

Figure 25.4 Pearson type III fit for the Amite River discharges.

shown in Figure 25.5. A best-fit line is drawn through the plotted points. Estimates of the 100-, 150-, and 200-year floods based on the fitted line are

T (year)	DISCHARGE (cfs)
100	145,000
150	168,000
200	178,000

EXAMPLE 25.6 Use power transformation to transform the data of Table 25.4 and then fit the normal distribution.

Solution On normal probability paper, the discharge values and their plotting positions are graphed, as shown in Figure 25.4. The scatter of points tends to follow a curve, which indicates nonnormality of the data. A trial-and-error calculation suggested that the logarithmic transformation was adequate in this case. The log values are then plotted using the same plotting positions on a normal probability paper, as shown in

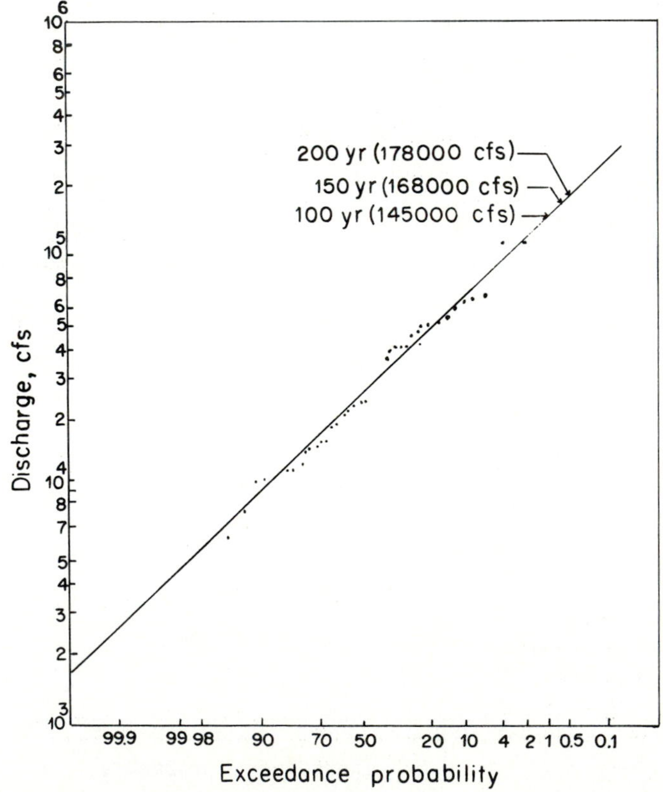

Figure 25.5 Log Pearson type III fit for the Amite River discharges.

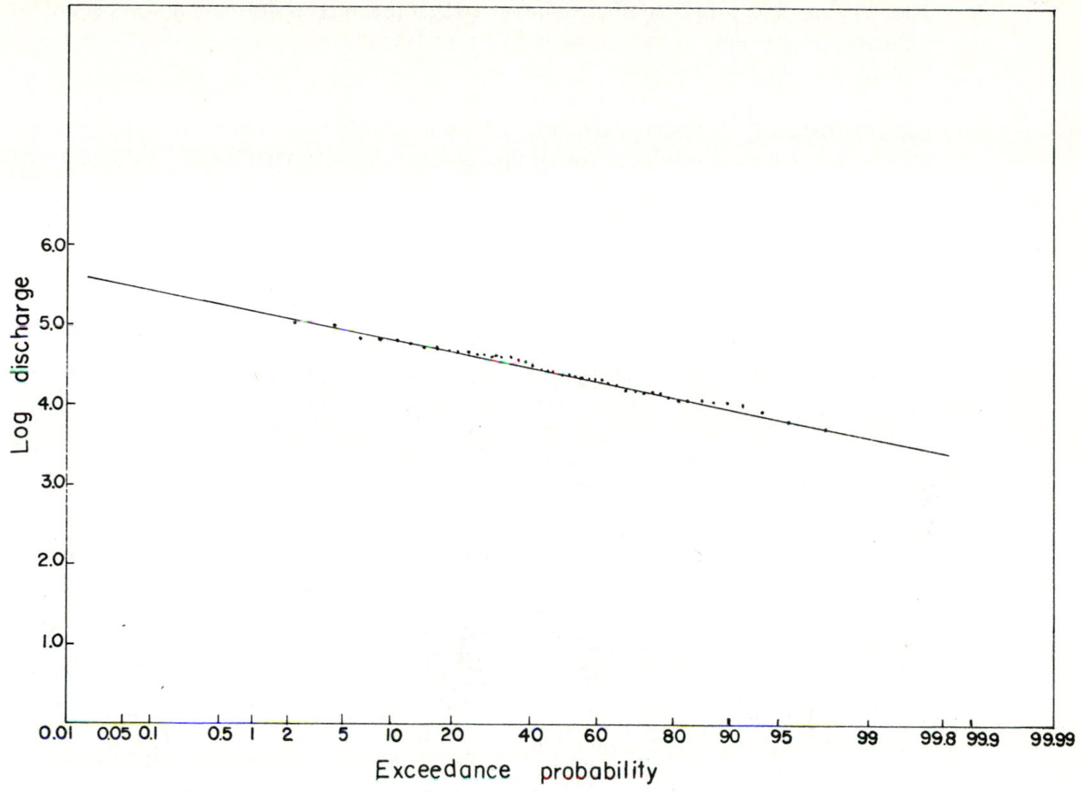

Figure 25.6 Normal fit to the log of the discharges of the Amite River.

Figure 25.6. The calculated statistics of the log values ($y = \log x$) are

$$\text{Mean, } \bar{y} = \overline{\log x} = \frac{\Sigma \log x_i}{n} = \frac{202.986}{46}$$

$$\overline{\log x} = 4.413 \text{ cfs}$$

$$\text{Standard deviation, } S_y = S(\log x) = \left[\frac{\Sigma (\log x_i)^2 - \dfrac{(\Sigma \log x_i)^2}{n}}{n - 1} \right]^{0.5}$$

$$= \left[\frac{900.153 - \dfrac{202.986)^2}{46}}{45} \right]^{0.5} = 0.3102 \text{ cfs}$$

$$\text{Coefficient of skewness, } C_s = \frac{n \, \Sigma (\log x_i = \overline{\log x_i})^3}{(n - 1)(n - 2)S_y^3}$$

$$= \frac{46(-0.0921)}{(45)(44)(0.3102)^3} \cong -0.072$$

The very small skewness coefficient indicates that the log transformation is adequate for reducing original data to normalized data. A log transformation is equivalent to a power transformation, where $\lambda = 0$.

EXAMPLE 25.7 Predict the 100-year, 150-year, and 200-year floods from the previous fits and compare the results. How comparable are these results?

Solution Estimates of 100-, 150-, and 200-year floods are

Distribution	100 year	150 year	200 year
Log normal	160,000	180,000	195,000
Gumbel	101,000	106,500	112,500
Pearson III	126,000	139,000	146,000
Log Pearson III	145,000	168,000	178,000
Log-transformed normal	151,400	166,000	177,800

Based on the goodness of fit of the different distributions to data, the log-transformed and the log Pearson type III distributions appear to better fit the empirical distribution of the Amite River discharges. The original data is nonnormal. Logarithmic transformation reduces skewness of the data substantially and, hence, adequately normalizes the data.

EXAMPLE 25.8 The mean annual flood peak of a given river is 32,100 cfs and the standard deviation of the flood peaks is 6,000 cfs. What is the probability of a flood of the magnitude of 40,000 cfs occurring in the river within the next 5 years?

Solution

$$\bar{x} = 32{,}000 \text{ cfs}$$

$$S = 6{,}000 \text{ cfs}$$

$$x = 40{,}000 \text{ cfs}$$

Assume that the flood peaks are Gumbel distributed. From Equation (24.74), $F(y) = \exp[-\exp(-y)]$, $y = (x - \beta)/\alpha$. By using Equations (24.77) and (24.78),

$$y = \frac{1.283}{S}(x - \bar{x} + 0.45\,S)$$

$$= \frac{1.283}{6{,}000}(40{,}000 - 32{,}100 + 0.45 \times 6{,}000) = 3.28$$

$$F(y) = \exp[-\exp(-3.28)]$$

Therefore,

$$P = 1 - F(y) = 1 - \exp[-\exp(-3.28)] = 1 - \exp(-0.036)$$

$$= 1 - 0.97 = 0.03$$

The return period of this flood event $= 1/0.03 = 33.33$ years. The probability of a 40,000-cfs flood occurring in the next 5 years $= 1 - (0.97)^5 = 1 - 0.859 = 0.141 =$ 14.1%.

EXAMPLE 25.9 An analysis of annual flood peaks on a stream with 60 years of record indicates that the mean value of the annual peaks is 190,000 cfs. The standard deviation of flood peaks is 32,000 cfs. Assume that the flood peaks follow the Gumbel distribution. Find the probability of a flood of the magnitude equal to or greater than 300,000 cfs occurring next year. What is the magnitude of a flood with a recurrence interval of 5 years?

Solution

$$x = 300{,}000 \text{ cfs}$$

$$S = 32{,}000 \text{ cfs}$$

$$\bar{x} = 190{,}000 \text{ cfs}$$

By following the procedure of Example 25.8,

$$y = \frac{1.283}{32{,}000}(300{,}000 - 190{,}000 + 0.45 \times 32{,}000)$$

$$= \frac{124.4}{25} \cong 5.00$$

Hence,

$$P = 1 - \exp[-\exp(-5.0)] = 1 - \exp(-0.0067)$$

$$= 1 - 0.9901 = 0.0099 = 0.99\%$$

The probability of a flood equal to or greater than 300,000 cfs next year $= 1 - (1 - 0.0099)^1 = 0.0099 \cong 1\%$.

The recurrence interval = 5 years. The probability of flood occurrence once a year = 1/5 = 0.2.

$$P = 1 - \exp[-\exp(-y)] = 0.2$$

This yields

$$y = 0.079$$

Therefore,

$$0.079 = \frac{1.283}{32{,}000}(x - 190{,}000 + 0.45 \times 32{,}000)$$

This produces

$$x = 177{,}580 \text{ cfs}$$

EXAMPLE 25.10 An analysis of an annual flood series covering the period 1890 to 1966 (76 years) on a certain river shows that the 80-year flood has a magnitude of 620,000 cfs and 1.4-year flood has a magnitude of 215,000 cfs. Assume the annual floods are Gumbel distributed.

(a) What is the probability of having a flood as great as or greater than 440,000 cfs?
(b) What is the magnitude of flood having a recurrence interval of 40 years?
(c) What is the probability of having 575,000 cfs flood or a greater flood in the coming 25 years time?

(d) Find the mean, $\bar{x}$, of the annual floods and its recurrence interval.
(e) Find the standard deviation of the annual floods.

Solution

(a) 76-year records:

$$80\text{-year flood} = 620{,}000 \text{ cfs}$$

$$1.4\text{-year flood} = 215{,}000 \text{ cfs}$$

$$\text{Probability of 80-year flood} = 1/80 = 0.0125$$

$$\text{Probability of 1.4-year flood} = 1/1.4 = 0.7142$$

$$P = 1 - \exp\,[-\exp\,(-y)]$$

For an 80-year flood, $P = 0.0125$. Therefore,

$$0.0125 = 1 - \exp\,[-\exp\,(-y)]$$

or

$$\exp\,[-\exp\,(-y)] = 1 - 0.0125 = 0.9875$$

This yields

$$y = 4.4$$

$$y = \frac{1.283}{S}\,(x - \bar{x} + 0.45S)$$

Therefore,

$$4.4 = \frac{1.283}{S}\,(620{,}000 - \bar{x} + 0.45S)$$

$$3.52S = 620{,}000 - \bar{x} + 0.45S$$

Therefore,

$$\bar{x} + 3.07S = 620{,}000 \text{ cfs}$$

Similar calculations are done for the 1.4-year flood.

$$y = -0.222$$

This leads to

$$-0.22 = \frac{1.283}{S}\,(215{,}000 - \bar{x} + 0.45S)$$

$$-0.1715S = 215{,}000 - \bar{x} + 0.45S$$

$$\bar{x} - 0.2785S = 215{,}000$$

By solving this equation and the one prior for S and $\bar{x}$, that is,

$$\bar{x} + 3.07S = 620{,}000$$

$$\bar{x} - 0.2785S = 215{,}000$$

Solution of these equations yields

$$S = 145{,}000 \text{ cfs}$$

$$\bar{x} = 175{,}000 \text{ cfs}$$

Now the probability of having a flood greater than or equal to 440,000 cfs is computed.

$$y = \frac{1.283}{S}(x - \bar{x} + 0.45S)$$

Here $x = 440{,}000$ cfs, $\bar{x} = 175{,}000$ cfs, and $S = 145{,}000$ cfs. Hence,

$$y = \frac{1.283}{145{,}000}(440{,}000 - 175{,}000 + 0.45 \times 145{,}000) = \frac{330}{112.5} = 2.93$$

Therefore,

$$P = 1 - \exp\,[-\exp\,(-2.93)]$$

$$P = 1 - 0.9462 = 0.0538 = 5.38\%$$

(b) The magnitude of a flood having a recurrence interval of 40 years is obtained as follows:

$$P = \frac{1}{40} = 0.25$$

$$P = 1 - \exp\,[-\exp\,(-y)] = 0.025$$

Hence,

$$y = 1.245$$

Therefore,

$$1.245 = \frac{1.283}{S}(x - \bar{x} + 0.45 \times S)$$

$$= \frac{1.283}{145{,}000}(x - 175{,}000 + 0.45 \times 145{,}000)$$

or

$$x = 90{,}000 + 112{,}500 \times 1.245 = 90{,}000 + 140{,}000 = 230{,}000 \text{ cfs}$$

(c) The probability of a flood 575,000 cfs or a greater flood in coming 25 years is obtained as follows:

$$x = 575{,}000$$

$$y = \frac{1.283}{145{,}000}(575{,}000 - 175{,}000 + 0.45 \times 145{,}000)$$

$$y = \frac{1}{112{,}500}(465{,}000) = 4.13$$

Therefore,

$$P = 1 - \exp\,[-\exp\,(-4.13)] = 1 - 0.942 = 0.058 = 5.8\%$$

$$R = 1 - (1 - P)^n = 1 - (0.942)^{25} = 1 - 0.2188 = 0.7812 = 78.12\%$$

EXAMPLE 25.11 Construct a probability paper for the following values corresponding to the Gumbel distribution:

Values of recurrence interval in years	y
1.07	−1.00
1.2	−0.70
1.3	−0.50
1.4	−0.30
1.5	−0.10
1.6	−0.0
2	0.37
3	0.99
4	1.33
5	1.50
10	2.25
20	2.97
30	3.45
40	3.80
50	3.90
75	4.30
100	4.60
150	5.00
200	5.30
250	5.55
300	5.75
400	6.00

Solution For drawing the probability graph paper, the recurrence interval is taken on x-axis. The recurrence interval is spread in proportion to y. Values of y for different recurrence intervals are shown in the table.

Take a scale of 1″ = 1 for y and mark values of y. Against y on each vertical line, prescribe the recurrence interval. Ordinates may be taken on the usual scale of 1″ = 100,000 cfs. Figure 25.7 shows the construction of the probability paper.

EXAMPLE 25.12 Solve Example 25.10 using Figure 25.2 of Example 25.11.

Solution From Figure 25.2:

(a) Probability of having a flood as great as or greater than 440,000 cfs:

$$\text{Recurrence interval} = 10 \text{ years}$$

$$\text{Hence, probability} = 1/10 = 10\%$$

(b) Magnitude of flood having a recurrence interval of 20 years = 570,000 cfs.

(c) Probability of having a 575,000-cfs flood or greater for the coming 25 years:

$$\text{Recurrence interval} = 48 \text{ years}$$

$$P = \frac{1}{48} = 0.028 = 2.8\%$$

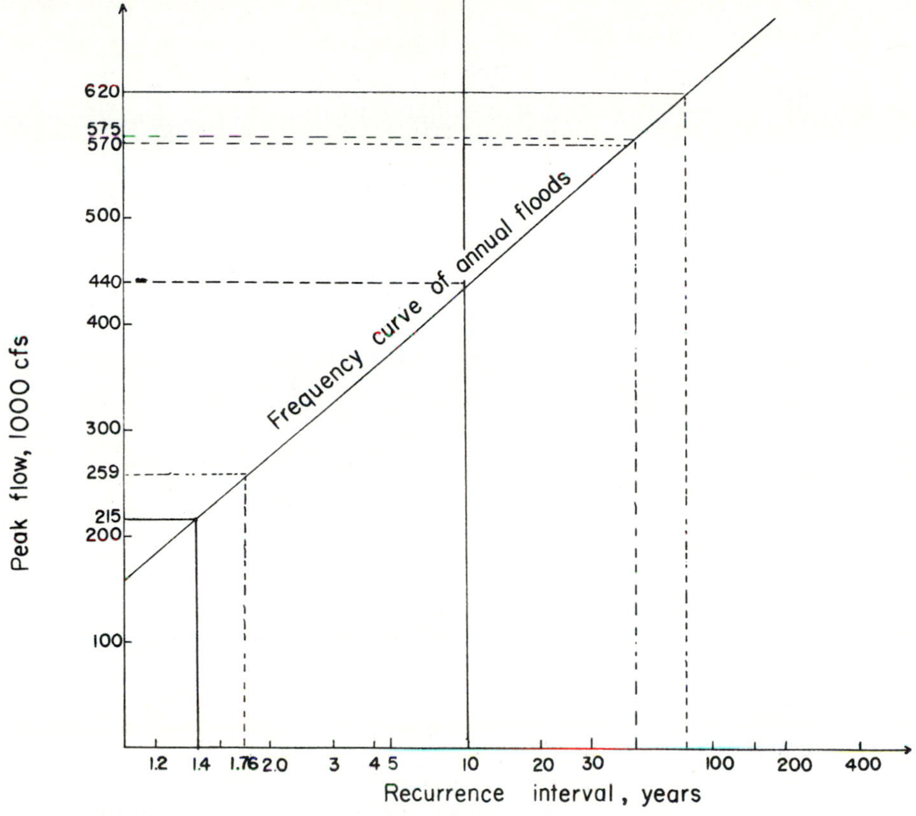

Figure 25.7 Construction of probability paper for the Gumbel distribution.

$$P \text{ of not having the flood} = 1 - 0.028 = 0.972$$

$$\text{Probability } R = 1 - (1 - P)^n = 1 - (0.972)^{25}$$

Therefore,

$$P = 1 - 0.4732 = 0.5268 = 52.68\%$$

(d) $\bar{x}$ and its recurrence interval:

$$80\text{-year flood} = 620{,}000 \text{ cfs}$$

$$1.4\text{-year flood} = 215{,}000 \text{ cfs}$$

From Figure 25.2, y for the 80-year flood = 4.36.

$$y \text{ for the 1.4-year flood} = -0.3$$

Now

$$y = \frac{1.283}{S}(x - \bar{x} + 0.45S)$$

$$= \frac{1.283}{S}(620{,}000 - \bar{x} + 0.45S)$$

or

$$\bar{x} + 2.95S = 620{,}000$$

Similarly,

$$-0.3 \times 0.7797S = 215{,}000 - \bar{x} + 0.45S$$

or

$$\bar{x} - 6.8391S = 215{,}000$$

Therefore,

$$\bar{x} + 2.95S = 620{,}000$$

$$\bar{x} - 0.683S = 215{,}000$$

Solving the equations, we get:

$$S = 122{,}500 \text{ cfs}$$

$$\bar{x} = 259{,}000 \text{ cfs}$$

Its recurrence interval from Figure 25.2 is

$$T = 1.76 \text{ years}$$

(e) The standard deviation as determined before is 122,500 cfs.

25.3 CONFIDENCE LIMITS

A value of the variate estimated from a probability distribution for a given return period is usually in error due to limited sample size. Therefore, a statement is needed indicating the limits about the estimated value within which the true value is contained with a specific probability. This statement is made by constructing confidence limits, which are also called the confidence intervals, confidence bands, error limits, or control curves. Thus, the confidence interval indicates the limits about the estimated value and the probability with which the true value will lie between those limits. This statement accounts for the sampling errors only.

Let the confidence probability be α. The confidence interval of the variate x corresponding to a return period T is bounded by values x_1 and x_2 (Nemec, 1973) as

$$x_{1,2} = x \pm G(\alpha)S_e \tag{25.14}$$

where $G(\alpha)$ is a function of the confidence probability α and can be determined by using the table of normal variates. As an example,

α(%)	50	68	80	90	95	99
$G(\alpha)$	0.674	1.00	1.282	1.645	1.96	2.58

S_e is the probable error expressed as

$$S_e = b\frac{S_{N-1}}{\sqrt{N}} \tag{25.15}$$

where

$$b = (1 + 1.3K + 1.1K^2)^{0.5} \tag{25.16}$$

in which K is the frequency factor of the given distribution under consideration, S_N is the standard deviation of the sample, and N is the sample size. By using this method, confidence limits can be placed above and below the fitted distribution curve. If the Gumbel distribution is considered, then for a given sample and T, 80% confidence limits are about twice as big as 50% ones, and 95% confidence limits are about thrice as big as 50% limits.

Beard (1962) proposed a method for constructing approximate error limits for frequency curves. His method involves computing standard deviations of the variate, and factors by which these deviations must be multiplied to mark off a given confidence band above and below the frequency curve. Table 25.5 shows errors limits for a 90% confidence band. Given in the table are the multiplying factors for the standard deviations of the variate. The 5% level indicated in the table means that only 5% of the variate values should be higher and only 5% should be lower than the 95% limit. Nine out of ten values should lie within the confidence band.

EXAMPLE 25.13 Compute the 90% confidence limits for the fitted curve of Example 25.5.

Solution Statistical calculations using these discharge values are given in Chapter 24 (Examples 24.13 to 24.17).

Mean of discharge values, $\bar{Q}$ = 33,264.8 cfs ≅ 33,265 cfs

Standard deviation of discharge values, S = 23,252.9 cfs ≅ 23,253 cfs

TABLE 25.5 ERROR LIMITS FOR FLOOD-FREQUENCY CURVES (AFTER BEARD, 1962)

Years of record, n	Exceedance frequency (%, at 5% level)						
	99.9	99	90	50	10	1	0.1
5	1.22	1.00	0.76	0.95	2.12	3.41	4.41
10	0.94	0.76	0.57	0.58	1.07	1.65	2.11
15	0.80	0.65	0.48	0.46	0.79	1.19	1.52
20	0.71	0.58	0.42	0.39	0.64	0.97	1.23
30	0.60	0.49	0.35	0.31	0.50	0.74	0.93
40	0.53	0.43	0.31	0.27	0.42	0.61	0.77
50	0.49	0.39	0.28	0.24	0.36	0.54	0.67
70	0.42	0.34	0.24	0.20	0.30	0.44	0.55
100	0.37	0.29	0.21	0.17	0.25	0.36	0.45
	0.1	1	10	50	90	99	99.9
	Exceedance frequency (%, at 95% level)						

Note: Tabular values are multiples of the standard deviation of the variate. Five percent error limits are added to the flood value from the fitted curve at the same exceedance frequency and the sum plotted. Ninety-five percent limits are subtracted from the flood value at the same exceedance frequency. Log values are added or subtracted before antilogging and plotting.

90% confidence limits are constructed using Table 25.5, and are shown in Table 25.6.

Years of record = 46

The lower- (95%) and upper- (5%) limit curves can be constructed by plotting their corresponding discharges against exceedence probabilities.

25.4 REGIONAL FREQUENCY ANALYSIS

For many watersheds, streamflow data are either insufficient or non-existent at the sites of interest. The methods of frequency analysis using data from a single site will have then limited predictive value because of large sampling errors. As a result, a regional frequency analysis is performed. By defining a region that is hydrologically similar in terms of the parameter or variable to be studied, data from several gaging sites within this homogeneous region are pooled together into a single regional frequency analysis. Examples of regional frequency analysis are estimation of design flood from rainfall–runoff relationship, prediction of flood peaks from the relation between observed values and drainage-basin characteristics, estimation of rainfall depths and frequencies in ungaged areas from characteristics at well-gaged sites in the same area.

Several methods are available to perform a regional analysis. One of the first steps in a regional analysis is to define the region itself. The definition of a region depends on the quantities to be estimated. Many methods are available to define a region that is homogeneous. For mean annual precipitation, large physiographic regions within the United States can be used, whereas for peak flow, the regions may be confined to drainage basins of certain sizes. Regional boundaries can be defined in terms of similarity of flood-frequency curves or flow curves. Homogeneity tests (Dalrymple, 1960) have been used to test if flood-frequency curves in a region can be considered homogeneous.

25.4.1 Index-Flood (IF) Method

The index-flood (IF) method developed by the U.S. Geological Survey (Dalrymple, 1960; Benson, 1962) is widely used to perform regional flood-frequency analysis. The basic premise of this method is that a combination of streamflow records maintained at a number of gaging stations will produce a more reliable, not a longer, record, and thus will increase the reliability of frequency analysis within a region. There are two major parts of the IF method. The first is the development of basic dimensionless frequency curves representing the ratio of the flood of any frequency to an index flood (the mean annual flood). The second is the development of relations between geomorphologic characteristics of drainage areas and the mean annual flood by which to predict the mean annual flood at any point within the region. By combining the mean annual flood with the basic frequency curve, a regional frequency curve is produced.

TABLE 25.6 CONSTRUCTION OF CONFIDENCE LIMITS

Exceedance probability (1)	Discharge (cfs) from the curve (Figure 25.6) (2)	Multiplying factor for 0.05 error limit from Table 25.5 (3)	Multiplying factor for 0.95 error limit from Table 25.5 (4)	Error (cfs) for 0.05 limit curve (5) − (3) × S	Error (cfs) for 0.95 limit curve (6) = (4) × S	Discharge (cfs) for 0.05 limit curve (7) = (2) + (5)	Discharge (cfs) for 0.95 limit curve (8) = (2) − (6)
0.99	1,650	0.514	0.730	11,952.04	16,974.69	13,602.04	−15,324.69
99	4,700	0.414	0.582	9,626.74	12,277.58	14,324.74	−7,577.58
90	10,200	0.298	0.396	6,929.39	9,208.19	17,129.39	991.81
50	25,300	0.258	0.258	5,999.27	5,999.27	31,299.27	19,300.73
10	69,000	0.396	0.298	9,208.19	6,929.39	78,208.19	62,070.61
1	145,000	0.582	0.414	12,277.58	9,626.74	157,277.58	135,373.26
0.1	260,000	0.730	0.514	16,974.69	11,952.04	276,974.69	248,047.96

BASIC FREQUENCY CURVE

In large regions that are homogeneous with respect to flood-producing characteristics, individual streams whose drainage areas vastly differ in size have frequency curves of approximately equal slope if the discharge is expressed as a ratio of the mean. The flood peaks at each gaging station are divided by an index flood (which is often taken as the mean annual flood at the station) and are thus reduced to dimensionless ratios. The individual curves plotted using the flood ratios can be superimposed and will nearly coincide.

These curves will pass through the recurrence interval of 2.33 years at the ratio of 1.0, but may have different slopes. The variation of these slopes may be used to test the homogeneity of the region. Within the homogeneous region, the ratios are compiled for all stations and then the median ratio for each is obtained. By plotting the median ratios against recurrence interval, a regional frequency curve is obtained. This is supposedly the best representation of a flood-frequency relation obtained by combining all dimensionless curves.

HOMOGENEITY TEST

The homogeneity test was developed by Langbein (Dalrymple, 1960) and can be used to test if a region is homogeneous to permit combining individual frequency curves with confidence to form a regional curve. The question is: Do the records differ from one another by amounts attributable to chance? The answer to this question lies in computing these differences and setting limits that will be acceptable statistically.

The standard error of estimate S_e of the reduced variate y for the Gumbel distribution (EV1) can be written as

$$S_e = \exp(y)\{1/[(T - 1)N]\}^{0.5} \tag{25.17}$$

where T is the recurrence interval, and N is the number of years of record. If a normal distribution of the estimates is assumed, then 95% of the estimates of the T-year flood will lie with $\pm 2S_e$ of their most probable value of T. The test employs the 10-year flood because this is the longest recurrence interval for which most records will give dependable estimates. For $T = 10$ years,

$$2S_e = 0.666 \exp(y)/N^{0.5} \tag{25.18}$$

and y for the EV1 distribution is 2.25. Therefore, the confidence limits are specified by

$$2.25 \pm 6.33/N^{0.5}$$

Dalrymple (1960) listed values of y corresponding to T, and lower and upper confidence limits with the corresponding T for various values of N, as shown in Table 25.7.

APPLICATION OF HOMOGENEITY TEST

The homogeneity test uses the 10-year floods estimated from the frequency curves of each gaging station within the region. These 10-year floods are divided by mean annual floods that have a recurrence interval of 2.33 years in accordance with

TABLE 25.7 VALUES OF y CORRESPONDING TO T AND CONFIDENCE LIMITS (AFTER DALRYMPLE, 1960)

n (years)	y	$2S_e = 6.33/n^{0.5}$	Lower confidence limit		Upper confidence limit	
			$y - 2S_e$	T_L	$y + 2S_e$	T_U
5	2.25	2.84	−0.59	1.2	5.09	160
10	2.25	2.00	0.25	1.8	4.25	70
20	2.25	1.42	0.83	2.8	3.67	40
50	2.25	0.90	1.53	4.4	3.15	24
100	2.25	0.63	1.62	5.6	2.88	18
200	2.26	0.45	1.80	6.5	2.70	15
500	2.25	0.28	1.97	7.7	2.53	13
1000	2.25	0.20	2.05	8.3	2.45	12

the EV1 distribution. These ratios are averaged to obtain the mean 10-year ratio for the region. The mean annual flood for each station is multiplied by the mean 10-year ratio to yield a modified 10-year flood for that station. The recurrence intervals of these modified floods are obtained for all stations from their corresponding frequency curves. These are then plotted against the periods of record of the gaging stations on the test graph. Any station that plots outside the 95% confidence limits is excluded from the homogeneous region. When this happens, partitioning of the region into two or more homogeneous regions is made. The period of record of a gaging station is taken as the number of recorded annual floods plus one-half the number of floods computed by interstation correlation.

The 95% confidence limits are found as follows. For each station, the upper and lower limits of a 10-year flood are determined corresponding to the 95% confidence interval as shown before. The return periods of these upper and lower limits are obtained for the station and are then plotted against the years of record used for obtaining the return periods.

MEAN ANNUAL FLOOD (MAF)

The mean annual flood is defined here as the value of the graphical frequency curve at the recurrence interval of 2.33 years. Benson (1960, 1962) has confirmed experimentally that the MAF has a magnitude equivalent to the flood of a 2.33-year recurrence interval. It is dependent on many factors that can be classified as either physiographic or meteorologic.

The physiographic factors influencing the MAF at a given point are (1) drainage area, (2) stream slopes, (3) natural storage in lakes, swamps, or channels, (4) land slope, (5) land use, (6) geology, (7) stream density, (8) stream pattern, (9) elevation, (10) aspect, (11) orographic position, (12) basin shape, (13) basin relief, and (14) soil cover. Langbein and others (1947) have described practical methods of computing some of the important physiographic factors.

The meterologic factors are connected with the magnitude and distribution of precipitation received by a drainage area, and include: (1) storm types, (2) type of region, whether humid or arid, (3) storm pattern, (4) storm direction, (5) precipitation intensities, (6) snowmelt, (7) storm volume, and (8) extent of ice jams. The evaluation, treatment, and use of some of these factors are difficult. The most commonly used factor is the mean annual precipitation.

There are other factors, representing the composite effect of some of these factors, that have been used in flood-frequency studies. The mean annual runoff is one such factor reflecting the effect of precipitation and basin characteristics. Another factor is the lag time, which is the time difference between the centers of rainfall and runoff. It represents the composite effect of most or all of the topographic factors.

MEDIAN FLOOD RATIOS

The ratios of several floods of different recurrence intervals to the MAF are tabulated for each station. Enough recurrence intervals are selected to define the curve appropriately. By tabulating the flood ratios, the median ratio for each recurrence interval is computed. This median of a recurrence interval is the midvalue of its flood ratios if their number is odd or the mean of the two central ratios if their number is even.

REGIONAL FREQUENCY CURVE

Each median flood ratio is plotted against its recurrence interval on probability paper. An average frequency curve is then drawn. This is the regional frequency curve, showing flood discharge in a ratio to the MAF, is based on all significant discharge records available, and represents the most likely flood-frequency values for all areas in the region.

WEAKNESSES

Some deficiencies of the IF method are:

1. The flood ratios for comparable streams may differ due to large differences in the index flood. If the index flood is not typical and is used as obtained from a short period of record, the remainder of the frequency curve may be faulty.

2. The homogeneity test cannot be applied at a level much higher than that of the 10-year flood because many individual records are too short to adequately define the frequency curve at higher levels. In many cases, individual curves show wide and sometimes systematic differences at higher levels.

3. Within a flood-frequency region, frequency curves are combined for all sizes of drainage areas, excluding the largest. Recent studies using ratios of less frequency floods have shown in all cases that the ratio of any specified flood to the MAF vary inversely with the drainage area. In general, the larger the drainage area, the flatter the frequency curve. The effect of drainage area is relatively greater for floods of higher recurrence intervals.

25.4.2 Multiple-Regression Method

The relation of flood peaks of selected recurrence intervals to basin and climatic parameters is determined by multiple-regression methods. The resulting relation is of the form

$$Q_T = b \prod_{i=1}^{N} x_i^{a_i} \tag{25.19}$$

where Q_T denotes the T-year flood; b is the regression constant; x_i, $i = 1, 2, \ldots, N$, are independent basin and climatic parameters; a_i, $i = 1, 2, \ldots, N$, are regression coefficients. Many studies throughout the United States have used models similar to Equation (25.19) for estimating the flood magnitude of recurrence interval T. The basin and climatic characteristics are normally evaluated from topographic, geologic, and climatic maps. Although the basin and climatic factors to be used in regression analysis vary from one region to another, some of the most important of these factors are drainage area, main-channel slope, and mean annual precipitation. Many of these factors are interrelated. For example, in general, as drainage area increases, slope and rainfall intensity decrease.

Some of the advantages of using a multiple-regression analysis are as follows: (1) It provides a mathematical relation between Q of a specified value of T and the independent variables. (2) It provides an evaluation of the independent variables that best define the dependent variable. (3) It provides a measure of the accuracy of the equation in terms of the standard error of estimate, and tests the significance of the coefficients of each independent variable. (4) It evaluates the relative significance of each independent variable by indicating those variables that have a coefficient that is significantly different from zero at a particular percent confidence level. (5) It provides an easy evaluation of the coefficients when the dependent and independent variables are transformed to their logarithms and used in a linear regression.

EXAMPLE 25.14 Develop a regional flood-frequency curve from the flood-frequency data given in Table 25.8 for southwestern Louisiana. Check if the gaging stations can be considered to belong to a homogeneous region.

Solution The index-flood method is used to develop the regional flood-frequency curve. Flood ratios are obtained by taking the ratio of the T-year flood to the 2.33 year flood for the 33 stations in southwestern Louisiana, as shown in Table 25.8. For each T-year flood, the median ratio is obtained, as shown in Table 25.9. Flood ratios versus recurrence intervals are plotted, as shown in Figure 25.8, and slopes of the curves are inspected. Except for a minor deviation for station 33, no severe deviations due to inhomogeneity can be detected in the figure. Thus, the stations can be considered homogeneous. The discharge ratio (median value) is plotted on probability paper and a curve is fitted through the points, which is the regional flood-frequency curve, as shown in Figure 25.9.

EXAMPLE 25.15 Basin and climatic characteristics and weighted flood-frequency relations for 48 small watersheds in Louisiana are given in Table 25.10. Establish the relations between basin and climatic characteristics and peak discharge of a specified recurrence interval. How good are these relations?

Solution Regression analysis is performed between basin and climatic characteristics and peak discharge for a specified recurrence interval for 48 small watersheds in Louisiana. To that effect, a regression equation is formulated as

$$Q_p = b_0 A^{b_1} S^{b_2} R^{b_3} \tag{25.20}$$

where Q_p is the peak discharge in cfs for a given T, A is the basin area in mi^2, S is the basin slope in ft/mi, R is the rainfall in in., and b_i's are regression parameters.

For fitting, Equation (25.20) is transformed as

$$\log Q_p = \log b_0 + b_1 \log A + b_2 \log S + b_3 \log R \tag{25.21}$$

The least squares analysis is used to estimate b_0, b_1, b_2, and b_3, and the following relations are obtained:

$T = 2$ years: $\quad Q_p = 0.0327A^{0.586}S^{0.306}R^{1.981}$; $R^2 = 62.91\%$

$T = 5$ years: $\quad Q_p = 0.0869A^{0.627}S^{0.342}R^{1.829}$; $R^2 = 75.00\%$

$T = 10$ years: $\quad Q_p = 0.1168A^{0.646}S^{0.354}R^{1.806}$; $R^2 = 79.37\%$

$T = 25$ years: $\quad Q_p = 0.1599A^{0.664}S^{0.364}R^{1.788}$; $R^2 = 82.64\%$

TABLE 25.8 REGIONAL AT-SITE FLOOD QUANTILES (CFS) BY GUMBEL DISTRIBUTION BASED ON OBSERVED DATA FOR SOUTHWESTERN LOUISIANA

Serial no.	Station no.	Flood quantiles for return period (cfs)						Obs. mean flood (cfs)	$Q_{2.33}$ (cfs)
		2-year	10-year	25-year	50-year	100-year	200-year		
1	07386500	927	2,336	3,392	4,384	5,589	7,056	1,223	1,332
2	07381800	2,165	5,452	7,917	10,232	13,045	16,469	2,855	2,524
3	08012000	7,054	17,763	25,792	33,336	42,498	53,654	9,303	7,991
4	08010000	3,863	9,728	14,125	18,256	23,273	29,383	5,095	5,349
5	08011800	2,246	5,657	8,214	10,617	13,535	17,088	2,963	2,785
6	08015500	25,282	63,916	92,805	119,949	152,913	193,053	33,472	29,806
7	08013500	13,305	33,504	48,648	62,877	80,156	101,198	17,546	16,222
8	08014500	12,432	31,307	45,457	58,753	74,899	94,561	16,395	12,564
9	08014000	4,387	11,048	16,042	20,734	26,432	33,370	5,786	4,750
10	08014200	3,963	9,980	14,491	18,729	23,876	30,144	5,227	4,258
11	08013000	12,586	31,694	46,091	59,479	75,826	95,730	16,597	15,448
12	08016800	3,368	8,483	12,317	15,919	20,294	25,622	4,442	3,959
13	08016400	3,709	92,342	13,564	17,531	22,349	28,216	4,892	4,331
14	08016600	3,839	9,668	14,038	18,144	23,130	29,202	5,063	4,554
15	08015000	6,529	16,442	23,874	30,856	39,336	49,662	8,611	6,580
16	08014800	3,915	9,858	14,314	18,501	23,585	29,777	5,163	4,721
17	08014600	1,938	4,881	7,088	9,161	11,678	14,744	2,556	2,122
18	08013800	1,001	2,520	3,660	4,730	6,030	7,613	1,320	1,227
19	08031000	1,290	3,248	4,716	6,096	7,771	9,812	1,701	1,570
20	08030000	1,929	4,858	7,053	9,117	11,622	14,673	2,544	2,420
21	08028700	709	1,785	2,592	3,350	4,271	5,393	935	832
22	08029500	2,913	7,337	10,654	13,770	17,554	22,162	3,842	2,637
23	08028000	10,395	26,176	38,007	49,123	62,623	79,062	13,708	8,985
24	08025850	598	1,507	2,188	2,828	3,605	4,551	789	601
25	08025500	4,889	12,312	17,877	23,106	29,457	37,189	6,448	4,261
26	08023000	1,866	4,749	6,896	8,913	11,362	14,345	2,487	2,252
27	07353500	2,323	5,851	8,495	10,980	13,997	17,672	3,064	2,508
28	07354000	2,243	5,648	8,201	10,600	13,514	17,061	2,958	3,037
29	07353990	3,631	9,143	13,276	17,159	21,874	27,617	4,788	3,616
30	07351700	1,138	2,866	4,161	5,378	6,856	8,656	1,501	1,066
31	07351500	4,510	11,358	16,491	21,315	27,172	34,305	5,948	5,923
32	07351000	3,212	8,090	11,747	15,183	19,356	24,437	4,237	4,253
33	07344450	3,133	7,891	11,457	14,008	18,878	23,834	4,132	3,267

TABLE 25.9 FLOOD RATIOS FOR SOUTHWESTERN LOUISIANA STATIONS

Station	2-year	10-year	25-year	50-year	100-year	200-year	Mean
1	0.6959	1.7538	2.5465	3.2913	4.1959	5.2973	0.9182
2	0.8578	2.1601	3.1367	4.0539	5.1684	6.5250	1.1311
3	0.8827	2.2229	3.2276	4.1717	5.3182	6.7143	1.1642
4	0.7222	1.8187	2.6407	3.4130	4.3509	5.4932	0.9525
5	0.8065	2.0312	2.9494	3.8122	4.8600	6.1357	1.0639
6	0.8482	2.1444	3.1136	4.0243	5.1303	6.4770	1.1230
7	0.8202	2.0653	2.9989	3.8760	4.9412	6.2383	1.0816
8	0.9895	2.4918	3.6180	4.6763	5.9614	7.5263	1.3049
9	0.9236	2.3259	3.3773	4.3651	5.5646	7.0253	1.2181
10	0.9307	2.3438	3.4032	4.3985	5.6073	7.0794	1.2276
11	0.8147	2.0517	2.9836	3.8503	4.9085	6.1969	1.0744
12	0.8507	2.1427	3.1111	4.0210	5.1260	6.4718	1.1220
13	0.8564	2.1570	3.1318	4.0478	5.1602	6.5149	1.1295
14	0.8430	2.1230	3.0826	3.9842	5.0791	6.4124	1.1118
15	0.9922	2.4988	3.6283	4.6894	5.9781	7.5474	1.3087
16	0.8293	2.0881	3.0320	3.9189	4.9958	6.3074	1.0936
17	0.9133	2.3002	3.3402	4.3172	5.5033	6.9482	1.2045
18	0.8158	2.0538	2.9829	3.8549	4.9144	6.2046	1.0758
19	0.8217	2.0688	3.0038	3.8828	4.9497	6.2497	1.0834
20	0.7971	2.0074	2.9145	3.7674	4.8025	6.0632	1.0512
21	0.8522	2.1454	3.1154	4.0264	5.1334	6.4820	1.1238
22	1.1047	2.7823	4.0402	5.2218	6.6568	8.4042	1.4570
23	1.1569	2.9133	4.2301	5.4672	6.9697	8.7993	1.5257
24	0.9950	2.5075	3.6406	4.7055	5.9983	7.5724	1.3128
25	1.1474	2.8895	4.1955	5.4227	6.9132	8.7278	1.5133
26	0.8375	2.1088	3.0622	3.9578	5.0453	6.3699	1.1044
27	0.9262	2.3329	3.3872	4.3780	5.5809	7.0463	1.2217
28	0.7386	1.8597	2.7004	3.4903	4.4498	5.6177	0.9740
29	1.0041	2.5285	3.6715	4.7453	6.0492	7.6374	1.3241
30	1.0675	2.6886	3.9034	5.0450	6.4315	8.1201	1.4081
31	0.7613	1.9173	2.7838	3.5981	4.5868	5.7909	1.0041
32	0.7552	1.9022	2.7621	3.5700	4.551	5.7458	0.9962
33	0.9590	2.4154	3.5069	4.2877	5.7784	7.2954	1.2648
Median	0.8522	2.1454	3.1154	4.0264	5.1334	6.4820	1.1238

$$T = 50 \text{ years:} \qquad Q_p = 0.1028A^{0.677}S^{0.370}R^{1.924};\ R^2 = 84.34\%$$

$$T = 100 \text{ years:} \qquad Q_p = 0.0829A^{0.689}S^{0.375}R^{2.007};\ R^2 = 85.48\%$$

The R^2 obtained in the regression analysis ranges from 64% to 86%. A fairly good fit is obtained, with at least 63% of the variation of Q_p for $T = 2$ years and at most 86% of variation for Q_p for 100 years accounted for by the regression model. R^2 seems to increase with an increase in T value.

EXAMPLE 25.16 Determine the 2-, 10-, 25-, 50-, and 100-year discharge quantiles for the following watersheds using equations developed in Example 25.15.

1. Watershed area = 4 mi^2, slope = 50 ft/mi, and rainfall = 50 in.
2. Watershed area = 7 mi^2, slope = 50 ft/mi, and rainfall = 50 in.

3. Watershed area = 5 mi², slope = 50 ft/mi, and rainfall = 50 in.
4. Watershed area = 1210 mi², slope = 4 ft/mi, and rainfall = 62 in.
5. Watershed area = 900 mi², slope = 4.3 ft/mi, and rainfall = 61 in.
6. Watershed area = 4 mi², slope = 15 ft/mi, and rainfall = 55 in.
7. Watershed area = 158 mi², slope = 12 ft/mi, and rainfall = 61 in.
8. Watershed area = 804 mi², slope = 2.8 ft/mi, and rainfall = 50 in.
9. Watershed area = 35.1 mi², slope = 4.1 ft/mi, and rainfall = 49 in.
10. Watershed area = 1097 mi², slope = 2.2 ft/mi, and slope = 49 in.

Solution The watershed area, slope, and rainfall values are substituted in the equations for their corresponding recurrence intervals. For example, for $A = 4.0$ mi², $S = 50$ ft/mi, and rainfall = 50 in.:

$$Q_{2\text{ year}} = 0.0327(4)^{0.586}(50)^{0.306}(50)^{1.981}$$
$$= 566.1 \text{ cfs}$$

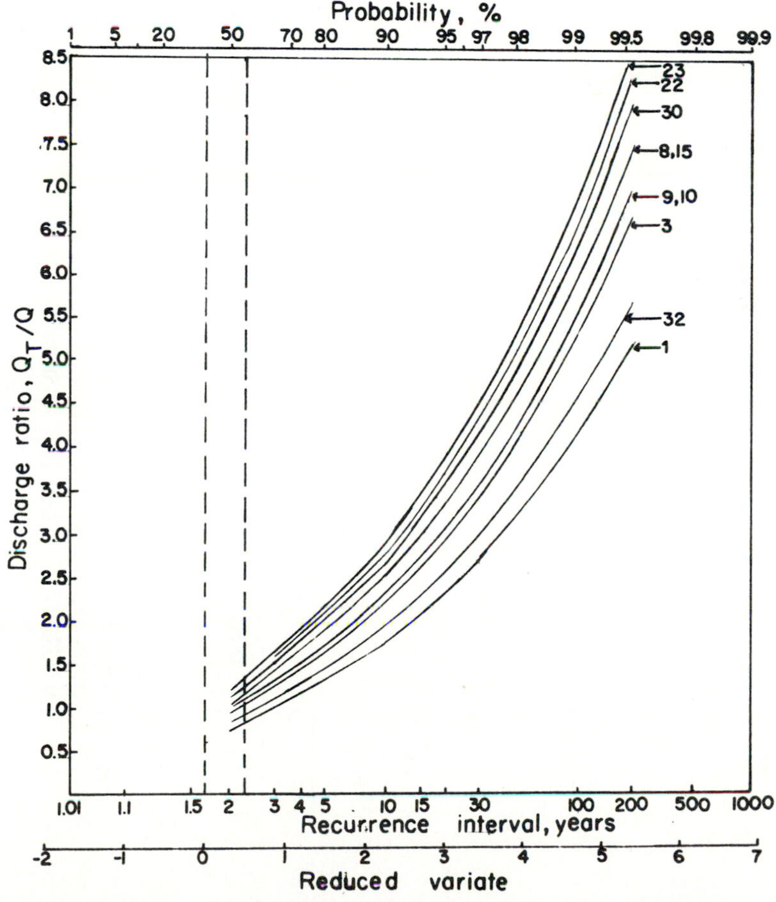

Figure 25.8 Flood ratios versus T_r for southwestern Louisiana.

TABLE 25.10 BASIN AND CLIMATIC CHARACTERISTICS AND WEIGHTED FLOOD-FREQUENCY RELATIONS FOR 48 SMALL WATERSHEDS IN LOUISIANA

Serial no.	Station no.	Basin and climatic characteristics: Drainage area (mi^2)	Channel slope (ft/mi)	Mean annual precipitation (in.)	Peak discharge (cfs) for the indicated period					
		Recurrence interval (years)			2	5	20	25	50	100
1	02489990	0.41	40.6	61	148	242	314	416	497	585
2	07344467	2.60	25.6	47	568	765	893	1,050	1,170	1,290
3	07349794	7.39	12.7	48	321	728	1,060	1,500	1,870	2,250
4	07349870	0.44	26.0	48	124	193	242	305	342	381
5	07350980	0.26	96.8	46	71	119	152	195	222	250
6	07351725	0.79	74.2	47	460	645	761	900	996	1,090
7	07351990	7.05	13.5	52	459	748	942	1,190	1,400	1,600
8	07352265	2.90	50.1	52	379	747	1,050	1,470	1,760	2,080
9	07352580	0.91	20.1	49	215	329	402	492	561	628
10	07352723	0.79	43.3	49	231	370	464	582	663	743
11	07353780	0.65	36.9	51	244	384	478	594	676	757
12	07354420	0.82	59.5	52	306	510	643	800	928	1,050
13	07355456	0.57	30.8	56	183	291	371	480	556	637
14	07355618	0.22	18.0	59	80	132	166	208	242	276
15	07364108	0.49	52.7	50	90	174	234	307	367	426
16	07364507	2.05	20.1	50	722	1,060	1,290	1,580	1,750	1,940

17	07366403	0.54	47.8	51	116	214	286	379	453	526
18	07367250	8.76	18.7	51	562	1,090	1,500	2,040	2,480	2,940
19	07368300	0.42	4.7	52	35	64	89	122	145	170
20	07368558	2.78	1.9	52	127	180	222	280	314	352
21	07369950	0.53	6.5	52	138	216	272	348	397	450
22	07370950	0.46	24.3	53	82	177	251	351	431	512
23	07372113	0.14	38.2	56	55	96	124	159	189	216
24	07372766	0.30	58.0	56	141	246	321	416	492	567
25	07373640	0.20	64.8	56	152	204	239	284	317	351
26	07374590	0.22	74.1	63	110	191	254	343	415	491
27	07375185	0.91	28.6	62	266	408	514	661	776	898
28	07375345	1.74	27.6	64	539	848	1,080	1,400	1,650	1,910
29	07375790	0.76	36.6	64	274	487	649	868	1,060	1,250
30	07375920	0.78	25.4	64	235	399	516	673	812	956
31	07376250	1.88	5.2	57	195	308	392	505	592	684
32	07376627	6.79	4.2	59	559	812	1,010	1,290	1,480	1,690
33	07377405	0.29	81.9	59	237	356	437	439	622	706
34	07377650	0.73	24.0	57	416	649	815	1,030	1,200	1,370
35	07379085	0.13	12.2	54	94	117	134	157	171	187
36	07381750	0.18	58.4	57	97	140	169	206	231	257
37	07381850	0.95	42.2	59	437	644	784	963	1,090	1,230
38	07386503	1.46	14.4	55	388	592	729	906	1,050	1,200
39	08011700	0.15	29.0	61	116	165	197	238	266	295
40	08011760	1.25	6.9	60	202	320	405	515	596	682
41	08013350	1.80	5.6	60	217	328	405	505	579	656
42	08013610	0.32	37.4	55	335	409	458	523	561	601
43	08014900	0.25	6.0	57	180	242	283	334	370	405
44	08015687	1.37	24.6	58	528	808	994	1,230	1,410	1,600
45	08022680	1.27	20.9	46	103	194	265	366	434	506
46	08023424	0.89	43.7	50	172	293	377	480	557	632
47	08024160	0.92	42.0	51	273	491	650	853	997	1,140
48	08027600	0.19	56.8	55	128	190	233	285	321	357

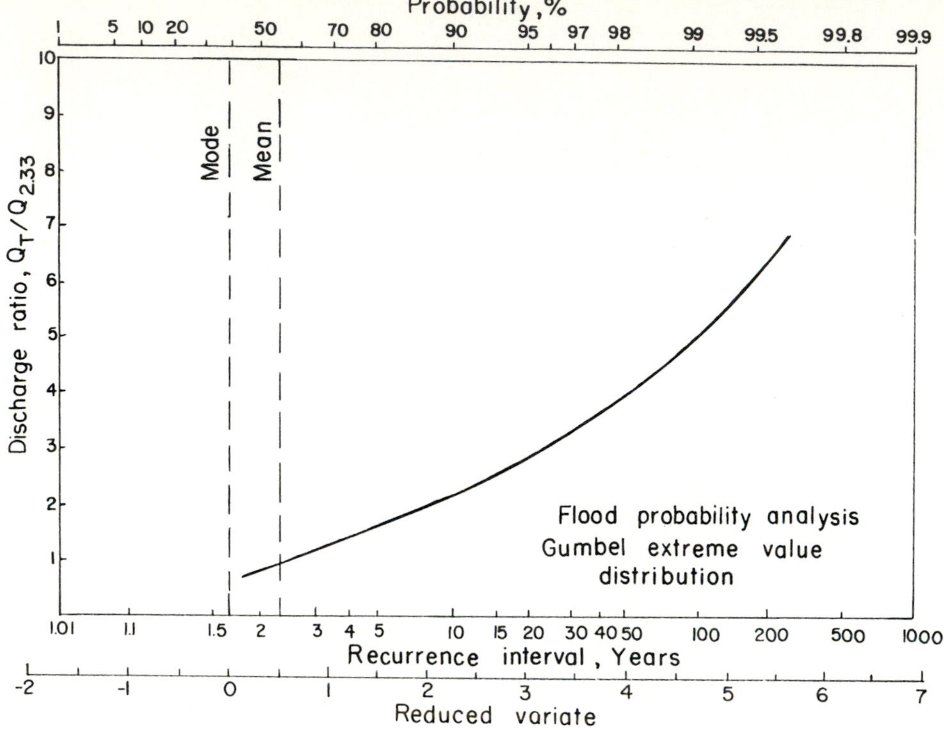

Figure 25.9 Regional frequency curve for southwestern Louisiana.

Similarly, discharge is computed for other recurrence intervals, and is tabulated as follows:

Area (mi²)	Slope (ft/mi)	Rain (in.)	Discharge (cfs)				
			Recurrence interval (years)				
			2	10	25	50	100
4.0	50	50	566.1	1,337.1	1,749.1	2,075.0	2,400.6
7.0	50	50	785.8	1,919.3	2,536.2	3,030.8	3,530.0
5.0	50	50	645.2	1,544.4	2,028.4	2,413.8	2,799.6
1,210.0	4	62	11,376.8	32,293.2	46,638.4	58,933.6	73,400.0
900.0	4.3	61	9,469.2	26,572.8	38,184.1	48,014.6	59,529.9
4.0	15.0	55	473.0	1,037.1	1,354.3	1,596.6	1,850.6
158.0	12.0	61	4,676.6	12,419.7	17,296.4	21,615.0	26,380.3
804.0	2.8	50	5,242.3	14,820.8	21,330.0	25,889.2	31,462.5
35.1	41	49	903.4	2,163.4	2,493.9	3,442.2	4,029.8
1,097.0	2.2	49	5,612.4	16,037.2	23,218.5	28,109.2	34,189.2

EXAMPLE 25.17 Determine the 30-, 50-, 100-, and 200-year discharges from both the multiple-regression analysis and the regional frequency curve.

(a) Watershed area = 96.3 mi², slope = 7.2 ft/mi, and rainfall = 48 in., and $Q_{2.33}$ = 2252 cfs (SW Louisiana).
(b) Watershed area = 154 mi², slope = 5.3 ft/mi, and rainfall = 51 in., and $Q_{2.33}$ = 2750 cfs (SW Louisiana).

How comparable are the two methods?

Solution

(a) Watershed area = 96.3 mi², slope = 7.2 ft/mi, rainfall = 48 in., and $Q_{2.33}$ = 2252 cfs. From the regional frequency curve in Figure 25.9, the discharges are computed as

T (year)	Ratio	Q (cfs)
30	3.35	7,544.2
50	4.03	9,075.6
100	5.13	11,552.8
200	6.48	14,593.0

From multiple-regression analysis, the discharges are obtained as:

T (years)	2	5	10	25	50	100
Q (cfs)	1861.4	3556	4883	6769.1	8069.2	9572.5

These discharge values are plotted against their T values on probability paper, and the frequency curve is fitted, as shown in Figure 25.10. The 30- and 200-year discharge quantiles are obtained as

$$Q_{30} = 7{,}100 \text{ cfs}$$

$$Q_{200} = 10{,}950 \text{ cfs}$$

Comparison of regional frequency and regression results in the following:

T (year)	Regional curve Q (cfs)	Regression Q (cfs)	% Difference (based on regional curve)
30	7,544.2	7,100	−5.9
50	9,075.6	8,069.2	−11.1
100	11,552.8	9,572.5	−17.1
200	14,593.0	10,950	−25.0

The difference between the two methods increases with T value. Discharges obtained from the regression equation are lower than those obtained from the regional frequency curve. The large difference can result because in regression analysis the equations developed were for relatively small watersheds.

(b) Watershed area = 154 mi², slope = 5.3 ft/mi, rainfall = 51 in., and $Q_{2.33}$ = 2750 cfs. From the regional frequency curve in Figure 25.9, the discharges are computed as

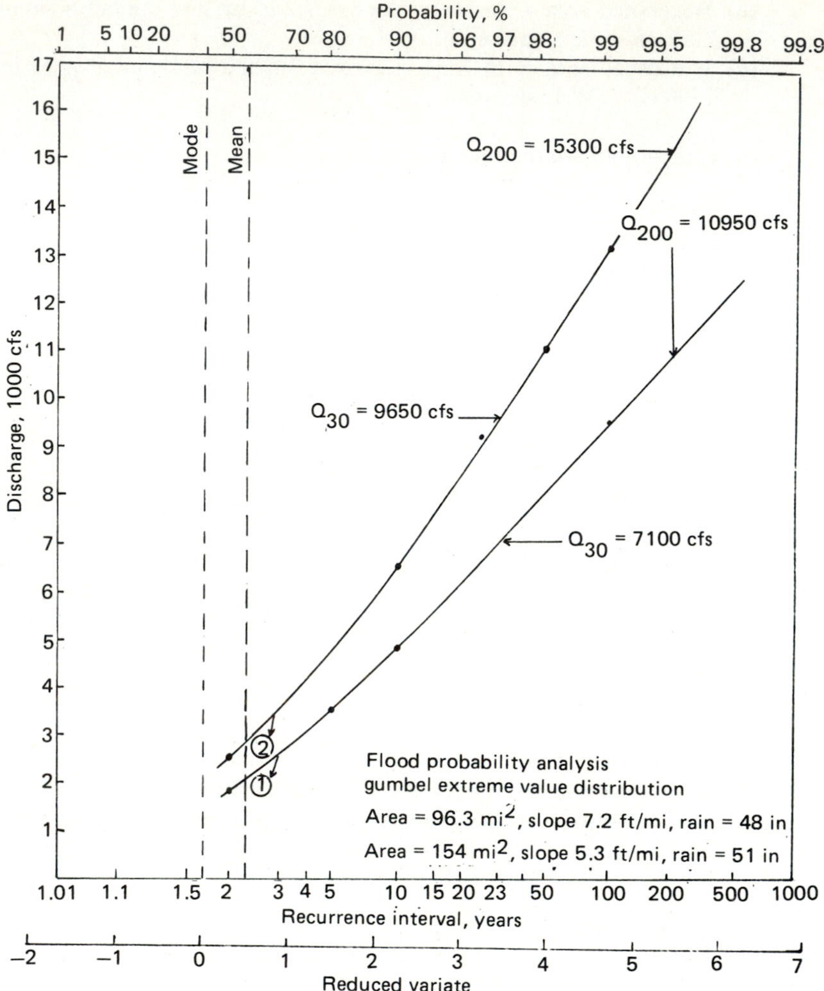

Figure 25.10 Frequency curves for watershed 3 and watershed 4.

T (year)	Ratio	Q (cfs)
30	3.35	9,212.5
50	4.03	11,082.5
100	5.13	14,107.5
200	6.48	17,820.0

From multiple-regression analysis, the discharges are obtained as:

T (years)	2 yr	5 yr	10 yr	25 yr	50 yr	100 yr
Q (cfs)	2,516.3	4,802.2	6,620	9,245	11,125	13,318

These discharge values are plotted against their T values on probability paper, and the frequency curve is fitted, as shown in Figure 25.10. The 30- and 200-year discharge quantiles are obtained as

$$Q_{30} = 9{,}650 \text{ cfs}$$

$$Q_{200} = 15{,}300 \text{ cfs}$$

Comparison of regional frequency and regression results in the following:

T (year)	Regional curve Q (cfs)	Regression Q (cfs)	% Difference (based on regional curve)
30	9,212.5	9,650	4.7
50	11,082.5	11,125	0.4
100	14,107.5	13,318	5.6
200	17,820.0	15,300	−14.1

The differences are attributed to the fact that the regression equations were developed for small watersheds.

EXERCISES

25.1. Perform a frequency analysis for the annual peak-discharge data of Table 25.11 by fitting the extreme value type I distribution. Plot the upper and lower confidence limits (90% reliability band) for the frequency curve.

TABLE 25.11 PEAK STAGES AND DISCHARGES OF THE COMITE RIVER NEAR COMITE, LOUISIANA (USGS GAGE 07378000)

Water year	Date	Gage height (ft)	Discharge (cfs)
1944	Apr. 25	21.12	3,440
1945	May 1	22.80	4,820
1946	May 16	23.61	6,010
1947	Mar. 14	25.16	10,600
1948	Mar. 4	25.00	10,000
1949	Mar. 25	25.14	10,300
1950	Jan. 8	25.07	11,500
1951	Mar. 30	25.62	11,500
1952	Apr. 5	21.83	3,630
1953	May 19	28.64	20,500
1954	Dec. 10	24.02	7,150
1955	Apr. 15	26.07	10,900
1956	Mar. 13	25.54	9,450
1957	Apr. 5	23.07	4,320

TABLE 25.11 *(Continued)*

Water year	Date	Gage height (ft)	Discharge (cfs)
1958	Mar. 24	23.18	5,000
1959	Feb. 3	24.15	6,360
1960	Dec. 18	24.46	6,950
1961	Mar. 19	26.72	15,200
1962	Apr. 29	19.03	20,900
1963	Jan. 21	8.23	2,420
1964	Mar. 4	16.92	15,400
1965	Oct. 6	18.82	20,100
1966	Feb. 12	16.94	13,200
1967	Apr. 14	21.20	17,600
1968	Mar. 23	7.67	2,360
1969	Apr. 13	17.06	12,700
1970	Oct. 7	17. 9	13,800
1971	Sep. 17	12. 9	7,310
1972	Dec. 8	20.51	16,500
1973	Apr. 29	19.58	14,500
1974	Jan. 20	15.14	9,210
1975	Jan. 8	16.49	10,800
1976	Mar. 26	15.52	9,660
1977	Apr. 23	25.52	23,700

Note: Datum of gage is 25.85 ft above msl.

25.2. Estimate the magnitudes of 50-year, 70-year, and 100-year floods from analysis of Exercise 25.1.

25.3. Determine the return periods for floods of 400, 500, 600, and 800 m^3/s from the analysis in Exercise 25.1.

25.4. The following peak discharges represent the annual maximum flows for the Appomattox River near Mattoax, Virginia, for the years 1961 through 1970:

YEAR	FLOW
1961	4510
1962	7060
1963	4550
1964	3500
1965	3420
1966	3880
1967	2740
1968	3650
1969	4350
1970	2660

Based on this recorded data, determine the 2-, 10-, 50-, and 100-year discharges assuming a log Pearson distribution. Use a station skew coefficient of 0.67. Plot this frequency curve on normal probability paper. Plot the three largest values at their proper plotting positions. Place a 90% confidence interval on the estimate of the 100-year event.

25.5. Derive K–T relationships for normal, log-normal, and Gumbel distributions, where K is the frequency factor, and T is the recurrence time. Using these relationships, construct theoretical K–T curves for these distributions for various values of the coefficient of variation.

25.6. Annual flood peaks for Greenbrier River at Alderson, West Virginia, are given in Table 25.12. Construct the K–T curves for normal, log-normal, and Gumbel distributions, and compare them with theoretical curves for each of the distributions. Comment on the differences between the two types of curves.

TABLE 25.12 ANNUAL FLOOD PEAKS FOR THE GREENBRIER RIVER AT ALDERSON, WEST VIRGINIA

Year	Flood (cfs)	Year	Flood (cfs)
1896	28,800	1932	50,100
1897	54,000	1933	26,400
1898	52,500	1934	32,300
1899	48,900	1935	49,600
1900	17,100	1936	58,600
1901	56,800	1937	36,600
1902	43,800	1938	32,800
1903	48,900	1939	41,600
1904	25,700	1940	29,900
1905	37,600	1941	50
1906	26,000	1942	35,300
1907	52,500	1943	36,200
1908	52,500	1944	25,200
1909	20,000	1945	19,000
1910	45,900	1946	43,600
1911	43,800	1947	24,400
1912	35,500	1948	40,300
1913	64,000	1949	37,100
1914	40,800	1950	31,500
1915	27,200	1951	29,300
1916	10,200	1952	27,600
1917	26,000	1953	47,100
1918	60,500	1954	29,700
1919	32,000	1955	44,400
1920	21,000	1956	18,200
1921	1,000	1957	28,900
1922	5,200	1958	26,700
1923	2,500	1959	23,900
1924	36,200	1960	35,500
1925	500	1961	31,400
1926	3,700	1962	34,700
1927	23,200	1963	47,200
1928	1,000	1964	39,600
1929	15,700	1965	28,400
1930	19,600	1966	26,400
1931	100	1967	54,500

25.7. There has been considerable controversy regarding the topic of plotting position for cumulative probability distribution curves. Gumbel has suggested five conditions that a plotting-position formula should satisfy. Making reference to Gumbel (1960, pp. 29–34), state these conditions. Show graphically the difference between the cumulative distribution curves for the plotting-position formulas in Table 25.1. Discuss the differences with particular regard to where they occur and the effect of sample size.

25.8. Take the appropriate probability papers (e.g., normal probability, log-normal probability, and Gumbel probability). Plot the annual flood peaks of Table 25.12 on these papers and fit straight lines on them. Estimate the parameters of the distributions from these plots. Discuss which distribution fits better and why.

25.9. Construct a 95% confidence band on each of the plots in Exercise 25.8. Are there any points lying outside of the band? If so, what do they indicate?

25.10. Estimate the recurrence interval for the mean annual flood (from the sample) from

TABLE 25.13 ANNUAL PEAK DISCHARGE FOR THE PECATONICA RIVER AT FREEPORT, ILLINOIS

Rank	Discharge (cfs)	Rank	Discharge (cfs)
1	18,400	34	5,700
2	17,000	35	5,650
3	15,700	36	5,600
4	14,500	37	5,400
5	13,900	38	5,380
6	13,700	39	5,370
7	13,100	40	5,170
8	13,000	41	4,840
9	12,300	42	4,830
10	12,000	43	4,800
11	11,000	44	4,600
12	11,000	45	4,570
13	10,500	46	4,480
14	10,000	47	4,400
15	9,070	48	4,190
16	8,820	49	4,180
17	8,640	50	4,160
18	8,600	51	4,040
19	8,450	52	3,650
20	8,070	53	3,500
21	7,890	54	3,350
22	7,820	55	3,080
23	7,600	56	3,000
24	7,220	57	2,730
25	7,110	58	2,720
26	6,880	59	2,660
27	6,700	60	2,600
28	6,310	61	2,600
29	6,220	62	2,280
30	6,150	63	2,180
31	6,070	64	2,120
32	6,000	65	2,040
33	5,820	66	1,190

normal, log-normal, and Gumbel distributions fitted in Exercise 25.8. Find the 20-year flood from each of the distributions.

25.11. What do you understand by a 10-year flood of 250,000 cfs? What is the probability of having as much as or a greater flood next year? What is the probability of having as much as or a greater flood continuously for 2 years? What is the probability of having as much as or a greater flood in the coming 10 years? How many floods as much as or greater do you anticipate in 100 years? What is the recurrence interval of a flood of 200,000 cfs if the mean annual flood is 115,500 cfs and the standard deviation is 45,300 cfs?

25.12. The available data for the Pecatonica River at Freeport, Illinois, are given in Tables 25.13 to 25.16 for a proposed bridge at Pecatonica. Estimate the expected 100-year discharge. Assume all storms covered the entire drainage basin uniformly for the same duration of the precipitation excess. The drainage area is 1326 square miles.

TABLE 25.14 A DISCHARGE RECORD FOR THE PECATONICA RIVER AT FREEPORT, ILLINOIS

Date	Time	Discharge (cfs)
7–18–77	0000	292
	1200	546
7–19–77	0000	1223
	1200	1562
7–20–77	0000	1763
	1200	1864
7–21–77	0000	1900
	1200	1927
7–22–77	0000	1959
	1200	1998
7–23–77	0000	2046
	1200	2093
7–24–77	0000	2145
	1200	2195
7–25–77	0000	2236
	1200	2277
7–26–77	0000	2169
	1200	1847
7–27–77	0000	1443
	1200	1107
7–28–77	0000	883
	1200	736
7–29–77	0000	643
	1200	587
7–30–77	0000	538
	1200	504
7–31–77	0000	478
	1200	457
8–01–77	0000	434
	1200	417

TABLE 25.15 A DISCHARGE RECORD FOR THE PECATONICA RIVER AT FREEPORT, ILLINOIS

Date	Time	Discharge (cfs)
8–06–77	0000	400
	1200	512
8–07–77	0000	557
	1200	531
8–08–77	0000	508
	1200	649
8–09–77	0000	1295
	1200	1448
8–10–77	0000	1477
	1200	1436
8–11–77	0000	1352
	1200	1200
8–12–77	0000	1027
	1200	857
8–13–77	0000	732
	1200	637
8–14–77	0000	572
	1200	523
8–15–77	0000	493
	1200	472
8–16–77	0000	455

(a) Estimate the 100-year runoff volume.
(b) Estimate the peak discharge using at least two methods.
(c) Compare the peak discharges associated with this volume.
(d) Discuss how you would decide which peak discharge to recommend for the design discharge.

TABLE 25.16 A DISCHARGE RECORD FOR THE PECATONICA RIVER AT FREEPORT, ILLINOIS

Date	Time	Discharge (cfs)
3–18–79	1200	1012
3–19–79	0000	1192
	1200	1961
3–20–79	0000	2580
	1200	2903
3–21–79	0000	3613
	1200	4636
3–22–79	0000	5044
	1200	4806

TABLE 25.16 (*Continued*)

Date	Time	Discharge (cfs)
3–23–79	0000	4853
	1200	4964
3–24–79	0000	5461
	1200	5849
3–25–79	0000	5995
	1200	5940
3–26–79	0000	5831
	1200	5813
3–28–79	0000	5444
	1200	5274
3–29–79	0000	5093
	1200	4884
3–30–79	0000	4652
	1200	4516
3–31–79	0000	4472
	1200	4398
4–01–79	0000	4369
	1200	4297
4–02–79	0000	4128
	1200	3924
4–03–79	0000	3818
	1200	3727
4–04–79	0000	3688
	1200	3663
4–05–79	0000	3613
	1200	3580
4–06–79	0000	3510
	1200	3422
4–07–79	0000	3297
	1200	3138
4–08–79	0000	3043
	1200	2940
4–09–79	0000	2832
	1200	2740
4–10–79	0000	2631
	1200	2525
4–11–79	0000	2420
	1200	2327
4–12–79	0000	2241
	1200	2157
4–13–79	0000	2137
	1200	2130
4–14–79	0000	2107
	1200	2061
4–15–79	0000	2017
	1200	1967

TABLE 25.16 (*Continued*)

Date	Time	Discharge (cfs)
4–16–79	0000	1908
	1200	1837
4–17–79	0000	1776
	1200	1708
4–18–79	0000	1652
	1200	1600
4–19–79	0000	1560
	1200	1517
4–20–79	0000	1480
	1200	1444
4–21–79	0000	1414
	1200	1389

25.13. Fit an appropriate frequency distribution to the data given in Table 25.17.

25.14. In Chapter 7, Table 7.4 lists the annual precipitation in inches for part of the Tana River basin of East Africa.

(a) Estimate the area mean (5-station) annual rainfall value for which the probability of the occurrence of a value equal to or less is $P = 0.01$. Would one have more confidence in the value determined from the actual short record or from a longer record, which was partially synthesized in Example 7.4?

(b) What is the probability of an area (5-station) mean rainfall of 60 inches or more occurring twice in any 10-year period?

(c) Compare the coefficient of variation for the single station of Fort Hall with the area mean. Comment upon the difference if any.

25.15. The 26-hour maximum rainfall values (in mm) over a 15-year period are tabulated. Find the highest rainfall that may be expected once in 5, 20, and 100 years.

Serial number	24-Hour maximum rainfall (mm)	Serial number	24-Hour maximum rainfall (mm)
1	99	15	94
2	76	16	67
3	113	17	52
4	102	18	37
5	82	19	45
6	56	20	79
7	47	21	104
8	38	22	116
9	92	23	121
10	98	24	131
11	118	25	147
12	142	26	136
13	135	27	133
14	109		

TABLE 25.17 ANNUAL PRECIPITATION IN INCHES AT THE UNIVERSITY OF ARIZONA WEATHER STATION AND AT THE ATTERBURY EXPERIMENTAL WATERSHED

Year	Tucson	Year	Tucson	Atterbury
1905	24.17	1940	14.75	
1906	11.75	1941	15.63	
1907	14.09	1942	8.05	
1908	14.04	1943	9.91	
1909	10.21	1944	13.32	
1910	9.80	1945	7.63	
1911	11.25	1946	13.67	
1912	9.84	1947	5.72	
1913	9.32	1948	6.81	
1914	19.90	1949	7.04	
1915	12.62	1950	8.51	
1916	13.15	1951	12.49	
1917	10.70	1952	10.62	
1918	9.12	1953	6.47	
1919	18.01	1954	11.31	
1920	10.72	1955	10.93	
1921	13.78	1956	5.82	7.18
1922	9.39	1957	13.35	13.52
1923	15.22	1958	12.63	13.49
1924	5.07	1959	12.16	11.24
1925	9.80	1960	9.34	8.76
1926	12.15	1961	10.25	10.59
1927	9.74	1962	10.17	8.78
1928	6.50	1963	9.61	10.59
1929	9.33	1964	12.42	12.63
1930	11.27	1965	15.80	14.92
1931	16.26			
1932	10.94	Mean	10.47	11.17
1933	9.60			
1934	8.59			
1935	15.77			
1936	12.24			
1937	8.43			
1938	8.89			
1939	7.05			

25.16. The 20-year data of mean annual discharges and peak discharges beginning from 1951 up to 1970 are tabulated.

DISCHARGES FROM THE LITTLE BLUE RIVER BASIN

Year	1951	1952	1953	1954	1955	1956	1957	1958	1959	1960
Peak discharges (cfs)	7,350 6,580 7,740 31,000 6,200 6,800	8,380	1,760	4,650	6,360 8,560	1,270	6,710 14,300	9,180 6,160 6,640 21,700 7,900	5,120	7,980 7,900
Mean annual discharge (cfs)	290	133	22	23	91	13	56	231	65	98

Year	1961	1962	1963	1964	1965	1966	1967	1968	1969	1970
Peak discharges (cfs)	8,200 7,780 7,430	9,140 7,090	4,390	7,090 8,130	12,100 9,050	5,520	14,600 10,100	4,880	9,160 14,200 4,100	8,760 6,630 8,280 6,630 15,400
Mean annual discharge (cfs)	285	194	64	113	178	80	186	138	184	217

Find the statistical parameters of the data and verify whether the mean annual discharges obey the normal distribution. Find that discharge that will be equalled or exceeded with a 20% probability (value of K for an exceedence probability of 20% is 0.842 for a normal distribution).

25.17. Annual rainfall values at a given station over a period of 20 years are given. Make the ranking of these rainfalls in ascending order and find the probability of occurrence and return periods of these events. Use the following three methods as options for calculating the probability of occurrence: California, Hazen, and Weibull methods.

Year	Annual rainfall (mm)	Year	Annual rainfall (mm)
1	805	11	925
2	797	12	911
3	1125	13	822
4	841	14	897
5	969	15	952
6	947	16	1093
7	918	17	975
8	783	18	956
9	875	19	863
10	834	20	870

25.18. The mean monthly rainfall values over a 10-year period that occurred at Campina Grande, Brazil, are given. Applying Weibull's formula, perform the frequency analysis.

Month	J	F	M	A	M	J	J	A	S	O	N	D
Mean Rainfall	39.4	54.9	95.2	108	109.8	111.7	108.1	61	26.9	10.7	12.6	20.5

25.19. The daily mean flood peak of a catchment is related to basin characteristic variables as:

$$Q_{\text{peak}} = K + a \ln A + b \ln \text{MAR} + c \ln \text{LONG} + d \ln \text{DF}$$

where

Q_{peak} = daily mean flood peak

A = area of catchment (km^2)

DF = drainage factor (number of confluences/km^2)

MAR = mean annual runoff

LONG = longitude of the place (degrees)

TABLE 25.18 REGIONAL AT-SITE FLOOD QUANTILES (CFS) BY THE GUMBEL DISTRIBUTION BASED ON OBSERVED DATA FOR SOUTHEASTERN LOUISIANA

Station no.	Flood quantiles for return period (cfs)						Obs. mean flood (cfs)	$Q_{2.33}$ (cfs)
	2-year	10-year	25-year	50-year	100-year	200-year		
02492000	19,649	46,483	63,662	78,346	94,791	113,265	23,203	2,451
02492360	5,917	13,998	19,171	23,593	28,546	34,109	7,289	6,847
02490105	2,440	5,772	7,906	9,729	11,772	14,066	3,006	2,517
02491500	21,423	50,670	69,408	85,418	103,348	123,489	26,387	24,331
02491700	3,489	8,255	11,306	13,914	16,835	20,116	4,298	3,561
02491350	2,494	5,902	8,082	9,946	12,033	14,379	3,073	2,523
02490000	1,958	4,633	6,345	7,808	9,448	11,290	2,412	1,963
07378500	26,892	63,617	87,128	107,225	129,732	155,016	33,124	30,180
07375222	2,128	5,036	6,897	8,488	10,269	12,271	2,622	2,626
07380160	969	2,292	3,140	3,864	4,665	5,586	1,194	1,192
07375170	3,804	9,000	12,327	15,170	18,355	21,932	4,687	4,238
07376000	5,487	12,980	17,778	21,878	26,471	31,630	6,759	6,075
07376500	3,026	7,158	9,804	12,065	14,598	17,443	3,727	3,550
07375500	14,876	35,192	48,197	59,315	71,765	85,752	18,323	16,092
07377300	24,321	57,535	78,798	96,974	117,329	140,196	29,957	27,723
07376600	1,114	2,635	3,610	4,442	5,375	6,423	1,372	1,451
07375480	6,948	16,436	22,510	27,703	33,518	40,050	8,558	7,280
07375000	4,974	11,768	16,117	19,835	23,998	28,675	6,127	4,848
07377000	21,208	50,170	68,711	84,560	102,310	122,249	26,122	23,616
07375800	4,664	11,035	15,113	18,599	22,503	26,889	5,746	4,170
07375307	4,272	10,107	13,842	17,035	20,611	24,627	5,263	3,727
03778000	9,826	23,246	31,837	39,181	47,405	56,644	12,104	12,038
07377500	6,967	16,482	22,574	22,780	33,612	40,162	8,582	7,721
07373500	6,120	14,478	19,829	24,403	29,525	35,279	7,539	7,230

These values for a 100-year recurrence interval flood are as follows:

$$K = 56.51, \quad A = 0.91, \quad b = -1.1, \quad c = 15.05, \quad d = 0.21,$$

$$A = 75 \text{ km}^2, \quad \text{MAR} = 750 \text{ mm}, \quad \text{DF} = 3, \quad \text{LONG} = 65$$

Calculate the surface runoff of the flood for this case.

25.20. Develop a regional flood-frequency curve from the flood-frequency data given in Table 25.18. Check if the gaging stations can be considered to belong to a homogeneous region.

25.21. Determine the 30-, 50-, 100-, and 200-year discharges from the regional frequency curve developed in Exercise 25.20 for the following watersheds in southeastern Louisiana:

(a) Watershed area = 12.1 mi^2, slope = 15.9 ft/mi, rainfall = 61 in., and $Q_{2.33}$ = 1963 cfs.

(b) Watershed area = 502 mi^2, slope = 4.7 ft/mi, rainfall = 60 in., and $Q_{2.33}$ = 14,000 cfs.

Also estimate these discharges from the multiple-regression equations derived in Example 25.14. How comparable are these two methods?

25.22. Develop a regional flood-frequency curve from flood-frequency data given in Table 25.19. Check if the gaging stations can be considered to belong to a homogeneous region.

TABLE 25.19 REGIONAL AT-SITE FLOOD QUANTILES (CFS) BY THE GUMBEL DISTRIBUTION BASED ON OBSERVED DATA FOR SOUTHEASTERN LOUISIANA

Station no.	Flood quantiles for return period (cfs) 2-year	10-year	25-year	50-year	100-year	200-year	Obs. mean flood (cfs)	$Q_{2.33}$ (cfs)
07373000	3,372	9,301	14,063	18,728	24,599	32,008	4,732	3,240
07372500	3,403	9,386	14,191	18,898	23,823	32,300	4,775	3,475
07372200	17,223	47,499	71,815	95,635	125,618	163,451	24,164	21,242
07370750	1,840	5,075	7,674	10,291	13,423	17,466	2,582	2,136
07373110	2,152	5,935	8,973	11,949	15,696	20,423	3,019	19,278
07372000	6,334	18,298	27,665	36,842	48,392	62,967	9,309	9,442
07370500	4,561	15,792	19,018	25,327	33,267	43,286	6,399	6,175
07371500	6,117	16,870	25,507	33,967	44,616	58,053	8,582	7,580
07366420	3,332	9,181	13,894	18,502	24,303	31,623	4,675	3,418
07365000	5,485	15,128	22,873	30,459	40,009	52,058	7,696	6,959
07365870	1,832	5,053	7,640	10,174	13,364	17,389	2,571	2,639
07365500	2,851	7,864	11,891	15,835	20,799	27,064	4,001	2,821
07366000	5,924	16,338	24,701	32,895	43,208	56,221	8,311	6,674
07366200	2,434	9,444	14,278	19,015	24,976	32,498	4,804	3,922
07364700	3,196	8,816	13,329	17,751	23,316	30,338	4,485	2,542
07362100	6,217	17,295	26,149	34,823	45,740	59,516	8,799	8,800
07365800	5,394	14,877	22,493	29,953	39,344	51,194	7,568	7,570
07352000	2,420	6,676	10,093	13,441	17,655	22,973	3,396	2,934
07352500	3,488	9,620	14,545	19,369	24,441	34,104	4,894	4,325
07348700	6,134	16,917	25,578	34,062	44,741	58,216	8,606	7,168
07349500	3,626	10,000	15,120	20,135	26,448	34,414	5,088	4,805
07348725	1,315	3,626	5,483	7,302	9,591	12,480	1,845	2,049
07348800	1,814	5,003	7,564	10,074	13,232	17,217	2,547	2,154

25.23 Determine the 30-, 50-, 100-, and 200-year discharges from the regional frequency curve developed in Exercise 25.21 for the following watersheds in northwestern Louisiana:

(a) Watershed area = 271 mi², slope = 3 ft/mi, rainfall = 54 in., and $Q_{2.33}$ = 6175 cfs.

(b) Watershed area = 47.6 mi², slope = 8.1 ft/mi², rainfall = 56 in., and $Q_{2.33}$ = 2136 cfs.

How comparable are the two methods?

25.24. From a given streamflow, what flow would you choose as part of an annual series if you desired to calculate the 7-day low flow with a return period of 10 years?

25.25. For a period of record on a stream, the average discharge = 26,000 cfs, and the standard deviation of discharge = 5600 cfs. Determine the flood discharge for a return period of 50 years.

25.26. If the mean $\bar{x}$ and the standard deviation S are 53,020 cfs and 16,470 cfs, respectively, determine the 25-year flood if the number of samples of annual data is 30.

25.27. Construct a frequency curve for the Tombigbee River at Aliceville, Alabama, from the daily discharge data at the Columbus gaging station.

(a) How many days during the year would you expect the flow to be greater than 10,000 cfs?

(b) How many days during the year would you expect the flow to be greater than 60,000 cfs?

CHAPTER 26
Design Storm and Design Flood

Hydrologic design of many water-resources projects is based on either peak discharge or complete discharge hydrograph. The design project must be capable of safely passing this peak discharge or the discharge hydrograph, which has a certain frequency of occurrence. It is this discharge that is called the design flood and its frequency the design frequency. The term design flood is the peak discharge or the flood hydrograph that is adopted for design of water-resources projects. The design flood is the maximum flood against which the project is protected. The connotation of a flood ranges from a particular value of discharge to the entire discharge hydrograph.

The selection of a design flood for a water-resources project such as dams and spillways involves (1) selecting safety criteria and (2) estimating the flood that meets these criteria. The safety criteria for dams and spillways are selected in two ways. First, the dam failure must be prevented at any cost if such a failure would cause loss of human life or would result in catastrophic economic and social consequences. This is a "no-risk" criterion. Second, because the society cannot afford to prevent all dams from failure, some probability of failure must be tolerated. This is referred to as "probability-based" criterion.

Depending upon the size of a water-resources project, three types of design floods are recognized: (1) probable maximum flood (PMF), (2) standard project flood (SPF), and (3) frequency-based flood (FBF). With proper consideration of flood characteristics, flood frequency, flood-damage potential, socioeconomic factors, land-use change, and so on, the selected design flood may either be greater or smaller than the SPF.

A design flood can be estimated using a frequency-based method or a rainfall–runoff method. The former involves a frequency analysis of long-term streamflow data at the site of interest. Or a frequency analysis of rainfall data is performed and coupled with a rainfall–runoff model to get the design flood. However, such data are not always available. Then the design flood may be estimated from a rainfall–runoff method that may be either data-based or hypothetical. Here rainfall data are analyzed to obtain the critical depth of rainfall. This critical depth of rainfall is designated as design storm. Corresponding to the three flood designations, design storms are classified as (1) probable maximum precipitation (PMP), (2) standard project storm (SPS), and (3) frequency-based storm (FBS). The design storm, when used with an appropriate rainfall–runoff model, leads to the design flood.

According to National Weather Service (1982), the PMP is defined as the theoretically greatest depth of precipitation for a given duration that is physically possible over a given size storm at a particular geographic location at a certain time of the year. The PMF is the flood associated with the PMP. Thus, the PMF, according to the U.S. Army Corps of Engineers (1979), is the flood that may be expected from the most severe combination of critical meteorological and hydrologic conditions that are reasonably possible in the region under consideration. It is a hypothetical flood estimated by transforming the PMP. The probability of occurrence of the PMP is unknown, but is very low. The PMP can be exceeded in the future, and does not, therefore, represent the maximum possible depth at the design site.

The SPS of a given duration is the most severe flood-producing rainstorm of the same duration that has occurred in the region of the watershed. This storm then corresponds to the most critical depth–area–duration relationship. The SPF is the flood derived from the SPS.

Water-resources projects are classified as small, medium, and large. Examples of each type are given in Table 26.1. Based on size, classification of dams is given in

TABLE 26.1 SOME EXAMPLES OF WATER-RESOURCES PROJECTS

Type of project	Examples
Small	Levees, drainage ditches, irrigation tanks, minor road bridges, urban storm drains, airport drainage systems, small crossroad culverts, spillway appurtenances of small dams, detention storage reservoirs, etc.
Medium	Hydropower plants, irrigation canals, medium-size dams and reservoirs, navigation in medium-size rivers, urban flood-control reservoirs, drainage systems, state highways, thermal power plants, barrages, railway bridges, etc.
Large	Multipurpose water-resources projects, large dams, levees, spillways, major river bridges, major irrigation canals, interstate highways, nuclear power plants, large dams, large navigation channels, big hydropower plants, etc.

TABLE 26.2 CLASSIFICATION OF DAMS BASED ON SIZE (AFTER THE U.S. ARMY CORPS OF ENGINEERS, 1979)

	Impoundment			
	Storage		Height	
Size	$\times 10^3$ m^2	a-ft	m	ft
Small	<1230 and >61.5	<1000 and >50	<12.2 and >7.5	<40 and >25
Intermediate	>1230 and <61,5000	>1000 and <50,000	>12.2 and <30.5	>40 and <100
Large	>61,500	>50,000	>30.5	>100

Table 26.2. No universally accepted safety criteria have been established as yet. However, a general consensus seems to be emerging in favor of the PMF for design of new large dams of high hazard potential. A similar consensus does not appear to be forming for existing dams of high hazard potential or for new and existing dams of intermediate and low hazard potential. A classification of hazard potential based on loss of life and property has been given by the U.S. Army Corps of Engineers (1979), as shown in Table 26.3. The hazard potential is also a function of the hydraulic head of the dam and can also be classified accordingly.

According to the Central Water Commission (1972) for design of spillways for major and medium-sized dams with storage capacity of more than 50,000 acre-feet, the PMF is to be used. For design of barrages and minor dams with a storage capacity of less than 50,000 acre-feet, the SPF or a 100-year flood, whichever is greater, is to be used. For design of minor structures, an FBF is used; depending upon the importance of a structure, a flood of 50- or 100-year frequency is used. The U.S. Army Corps of Engineers (1979) recommends selection of a design flood as some percentage of PMF, based on the size of a dam and its hazard potential, as shown in Table 26.4.

Estimation of a design storm and design flood is needed for the design of a broad spectrum of civil works. Dams, reservoirs, bridges, culverts, levees and em-

TABLE 26.3 CLASSIFICATION OF HAZARD POTENTIAL (AFTER THE U.S. ARMY CORPS OF ENGINEERS, 1979)

Hazard	Loss of life (extent of development)	Economic loss (extent of development)
Low	None expected (no permanent structures for human habitation)	Minimal (undeveloped to occasional structures or agriculture)
Significant	Few (no urban development and no more than a small number of inhabitable structures)	Appreciable (notably agriculture, industry, or structures)
High	More than a few	Excessive (extensive community, industry, or agriculture

TABLE 26.4 RECOMMENDED SAFETY STANDARDS (AFTER U.S. ARMY CORPS OF ENGINEERS, 1979; WANG, 1988a)

Hazard	Size	Safety standard
Low	Small	50- to 100-year flood
	Intermediate	100-year flood to SPF
	Large	SPF to PMF[a]
Significant	Small	100-year flood to SPF
	Intermediate	SPF to PMF[a]
	Large	PMF
High	Small	SPF to PMF[a]
	Intermediate	PMF
	Large	PMF

[a] Replacing the PMF by the SPF may be considered for concrete dams.

bankments, detention ponds, spillways, water-supply schemes, power-generation schemes, navigation channels, culverts, drainage systems, flood insurance, urban development, flood-damage assessment, dam-breach mitigation, flood warning systems, etc., typify such works.

26.1 METHODS OF ESTIMATING DESIGN FLOODS

The methods of estimating a design flood can be grouped, depending upon the data requirements, into three classes: (1) rainfall–runoff methods, (2) frequency-based methods, and (3) risk-based methods.

26.1.1 Rainfall–Runoff Methods

Four methods can be distinguished in this class. These methods are either data-based or hypothetical.

- **(a)** *Greatest Storm of Record*. The design flood is computed by converting the greatest storm of record at the site under consideration. This storm is the design storm.
- **(b)** *Transposition*. This involves transposing a severe historical storm from another similar watershed in the region to the watershed under consideration, and then converting the transposed storm to runoff. The transposed storm is the design storm and the computed runoff is the design flood.
- **(c)** *Standard Project Storm (SPS)*. From the rainfall record, the SPS is derived, which constitutes the design storm. The SPS value is then converted to runoff, which is the design flood.
- **(d)** *Probable Maximum Precipitation (PMP)*. Based on meteorological analysis, a theoretical PMP value is derived, which constitutes the design storm. The PMP value is then converted to runoff, which is the design flood.

26.1.2 Frequency-Based Methods

Three methods can be distinguished in this class.

(a) *Precipitation Frequency Analysis.* A frequency analysis of rainfall data is performed to estimate the design storm, which is then converted to the design flood. Precipitation records are more plentifully available and are, therefore, used in a majority of designs, especially for small and large watersheds.

(b) *Flood-Frequency Analysis.* A frequency analysis of flood flows recorded at the site is performed to estimate the design flood. If the flow data are not available at the design location, then flow data from a similar watershed in the region may be used for frequency analysis. These methods are preferable to precipitation frequency analysis, for the flood flow is ultimately to be obtained. However, flood data are not, in general, plentifully available, especially on small and large watersheds. Flood frequency is more commonly used for medium-size watersheds, and provides reliable estimates of 2-, 5-, 10-, 15-, 20-, and 25-year floods.

(c) *Regional Frequency Analysis.* A regional frequency analysis is performed to estimate the design flood. This may involve flood flows at several locations or regression equations developed from recorded flood data.

26.1.3 Risk-Based Methods

Actually, most projects have two levels of risk, that is, functional and structural safety. The functional risk involves a cost–benefit analysis, whereas structural safety is lower, usually for a 100-year design flood or SPF. Thus, the risk-based methods combine economic risk analysis with frequency-based engineering analysis. Cost estimates are made for the project as well as environmental and other consequences resulting from its failure. Such consequences may be exemplified by structural damages, road and bridge losses, traffic interruptions, loss of human and

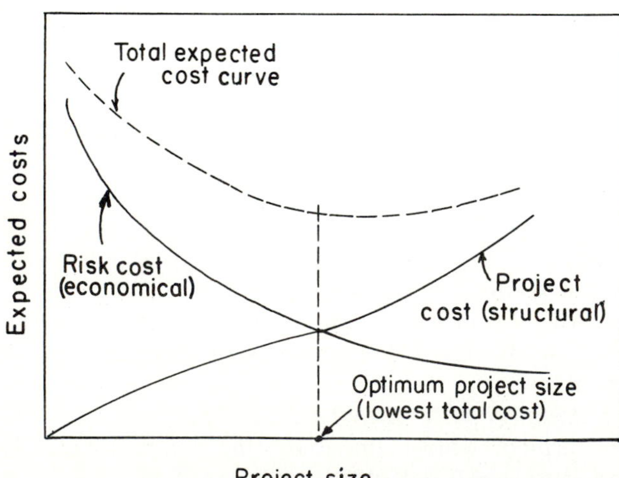

Figure 26.1 Optimum project-size selection based on economic-risk costs.

animal life, crop damage, land loss, damage to dwellings, clean-up costs, etc. All these costs are summed up to compute the total cost of the project. The optimal project size may be the one that corresponds to the lowest total cost. Or the ratio of the cost to the benefit is computed, and the project size corresponding to the lowest value of this ratio is selected. Figure 26.1 illustrates the concept of a risk-based project size selection.

26.2 CONSIDERATIONS IN SELECTION OF A METHOD

The selection of a method for estimation of a design flood can be based on the following considerations.

(a) *Project Objective*. The purpose of the project determines if the peak discharge, volume, or the entire discharge hydrograph is needed. The estimation method follows from this need.

(b) *Project Type*. The size of the project to be constructed determines the type of design flood to be adopted. The hydrologic design of a major water-resources project, say, a large dam, is quite different from that of a small project, say, a drainage channel. The estimation method varies accordingly. A major water-resources project is usually a multipurpose project providing protection from floods, and storage for irrigation, power, water supply, navigation, low-flow augmentation, recreation, commerce, etc. To allocate storage properly to these uses requires the frequency of occurrence of low flows, and average monthly, seasonal, and yearly flows, in addition to historical and design floods. This means that the entire streamflow history is needed. For a major structure, the design hydrograph is routed to evaluate the adequacy of spillways and outlet works operated in conjunction with reservoir storage.

(c) *Availability of Data*. The amount, quality, type, and completeness of hydrologic data may be the principal factors in selecting a method for estimating the design flood. Procedures for ungaged or data-scarce watersheds are quite different from those for gaged watersheds. Similarly, procedures may vary from long records to short records.

TABLE 26.5 METHODS USED FOR COMPUTING PEAK DISCHARGE FROM WATERSHEDS OF DIFFERENT AREAS

Watershed area (km^2)	Methods commonly used
Less than 2.5	Rational method; infiltration approach; time–area method; overland flow hydrograph
Less than 250	Time–area method; overland flow hydrograph; unit hydrograph; flood-frequency analysis; regional regression relating peak discharge to drainage characteristics
250 to 5000	Unit hydrograph; flood-frequency analysis; regional regression equations; watershed models
Greater than 5000	Flood routing; flood-frequency analysis; watershed models; regional regression equations

(d) *Watershed Characteristics*. The size and type of the watershed, and its soil–land-use–vegetation complex determine the method of conversion of design storms to design floods. Table 26.5 shows methods that can be used for estimating peak discharge from a watershed of given size.

(e) *Accuracy of Results*. The accuracy desired of flood estimates depends upon the sophistication of analysis and the computing facility. The training level of personnel involved, availability of funds, time available for analysis, etc., are also factors closely associated with the accuracy of results desired.

(f) *Other Considerations*. In addition to the amount of risk to be accepted in providing the degree of protection to downstream life and property, social and economic considerations, legal and environmental implications, etc., also influence the hydrologic design of water-resources projects.

26.3 STEPS IN ESTIMATION OF A DESIGN FLOOD

Several steps are common to some methods of estimating design floods for the design of minor water-resources projects. These steps are as follows:

1. Determine the duration of the design storm. For most watersheds this duration is not equal to the watershed time of concentration. A critical period within the storm is established. This duration is approximately taken as equal to the time of concentration of the watershed.
2. Select the design frequency. This depends on the size of the project.
3. Estimate the storm depth for the selected frequency and duration. This may involve areal adjustment of the point depth value.
4. Estimate the abstractions and the amount of direct runoff.
5. Select the probability level and the corresponding storm pattern.
6. Compute the time distribution and the intensity of the effective rainfall.
7. Determine the unit hydrograph for the watershed.
8. Determine the design hydrograph by convoluting the unit hydrograph with the time distribution of the effective rainfall.

These steps will be discussed in detail in the ensuing section.

26.4 ANALYSIS OF DESIGN STORMS

Once the design frequency has been established, the design storm is to be analyzed next. This analysis involves storm selection, and determining rainfall duration, point rainfall depth, areal depth adjustment, rainfall intensity, time distribution of rainfall, and areal distribution pattern.

26.4.1 Storm Duration

The duration of design rainfall depends on the watershed size and the scale of water-resources project. This duration is usually taken as approximately equal to the time of concentration of the watershed, for this duration permits the whole watershed area to contribute to peak discharge if rainfall is occurring uniformly over the area. When rainfall is of variable intensity during the period of its occurrence, a duration greater than the time of concentration is used. This longer duration should be such that any higher duration does not increase the peak discharge but only sustains it. For small watersheds, the ratio of volume of runoff to peak discharge may be used, assuming a triangular discharge hydrograph, to yield an appropriate value of storm duration.

The Soil Conservation Service (1971) uses the greater of 6-hour duration and the time of concentration in design of emergency spillways and freeboard for small dams. The Soil Conservation Service (1975) uses 24-hour durations for all urban watershed studies. The distribution of rainfall intensity is not uniform for this long duration; it is very high during the middle of the storm period.

Storm durations of less than or equal to 6 hours are satisfactory for small watersheds. For large watersheds, these durations may vary up to 10 days. Frequency-based rainfall values can be obtained from maps in Weather Bureau technical papers (e.g., U.S. Weather Bureau, 1964):

For duration up to 1 day:

TP-40	48 contiguous United States
TP-42	Puerto Rico and the Virgin Islands
TP-43	Hawaii
TP-47	Alaska

For durations from 2 to 10 days:

TP-49	48 contiguous United States
TP-41	Hawaii
TP-52	Alaska
TP-53	Puerto Rico and the Virgin Islands

26.4.2 Storm Selection

For determining design storm depths, only pertinent storms are selected for analysis. The storm selection is facilitated by a detailed meteorologic study of major storms in the region of the project watershed. For a preliminary selection, a review is made of such relevant records as daily rainfall data, storm tracks, cyclones, low-pressure depressions, history of storms, and flood data. Then an appropriate threshold value of rainfall depth is fixed to screen storms. As a guideline, a value of 2.5 cm may be appropriate for watersheds in semiarid regions, and a value of 5 cm for watersheds in humid regions. The threshold value varies with the watershed size—a higher value for small watersheds and a lower value for larger watersheds. An adequate rainfall depth for different watershed sizes can be 10 cm for watersheds up to 5000 km^2, 5 to 8 cm for watersheds between 5000 and 10,000 km^2, and 5 cm for watersheds greater than 10,000 km^2. Only the storms with a daily rainfall equal to or greater than the threshold value are selected for further analysis. Sometimes a heavy storm may be preceded or followed by another storm just short of the threshold value. If the total rainfall depth of the two storms on 2 days equals or exceeds the 2-day threshold value, then both storms are included in the analysis.

26.4.3 Rainfall Depth

An estimate of the rainfall amount and its distribution over the watershed is needed for the design storm. Different procedures are available to make this estimate, depending upon the size and the type of the water resources project. For example, for small dams, small bridges, and other minor structures, the standard project storm may be used. Frequency-based estimates of annual maximum D-day average rainfall depth (D = 1 day, 2 day, and so on), using point-frequency analysis or regional frequency analysis, are obtained for the watershed for design of medium and small projects. For large projects, either the probable maximum precipitation or rainfall frequency analysis is used.

DEPTH–DURATION (DD) ANALYSIS

The DD analysis is performed to determine rainfall depths over the watershed for different durations. Both in-situ and transposed storms are included in the analysis. The average watershed rainfall depth is computed using the isohyetal method or another standard method of computing mean areal rainfall. This average depth corresponds to a specific duration. The values of the areal average depth corresponding to different durations are tabulated and plotted on a graph that shows the relation between rainfall depth and duration.

DEPTH–AREA–DURATION (DAD) ANALYSIS

The distribution of rainfall amounts for various area sizes and durations is called depth–area–duration (DAD) analysis (World Meteorological Organization, 1969). The purpose of this analysis is to determine the maximum precipitation amounts that occur over various sizes of drainage area during standard passages of time or storm periods in hours or days, such as 6 hours, 12 hours, 24 hours, 2 days, 6 days, etc. For example, the DAD analysis can produce the largest rainfall amount over 500 km^2 in 6 hours. This analysis uses recorded rainfall data and is prepared for different regions and for different seasons. Such data are then used to determine design storms. Figure 26.2 illustrates the depth–area–duration relations for a large storm in the southern United States. This figure shows the relation between depth, area, and duration for various areas up to 100,000 km^2. Although the relation is not linear, it is clear that there is a decrease in the depth of precipitation as the area increases. Small drainage basins have a greater average depth of precipitation than do large drainage basins. This is consistent with the relation between rainfall from smaller convective storms and rainfall from larger cyclonic storms. Cyclonic storms cover large areas with less rainfall and convective storms cover small areas with greater rainfall.

Two methods are used in the DAD analysis: (1) the mass-curve method and (2) the incremental isohyetal method. Each method portrays the time division of rainfall differently. The mass-curve method involves construction of mass curves for individual stations, an average mass curve for the entire area, and then construction of

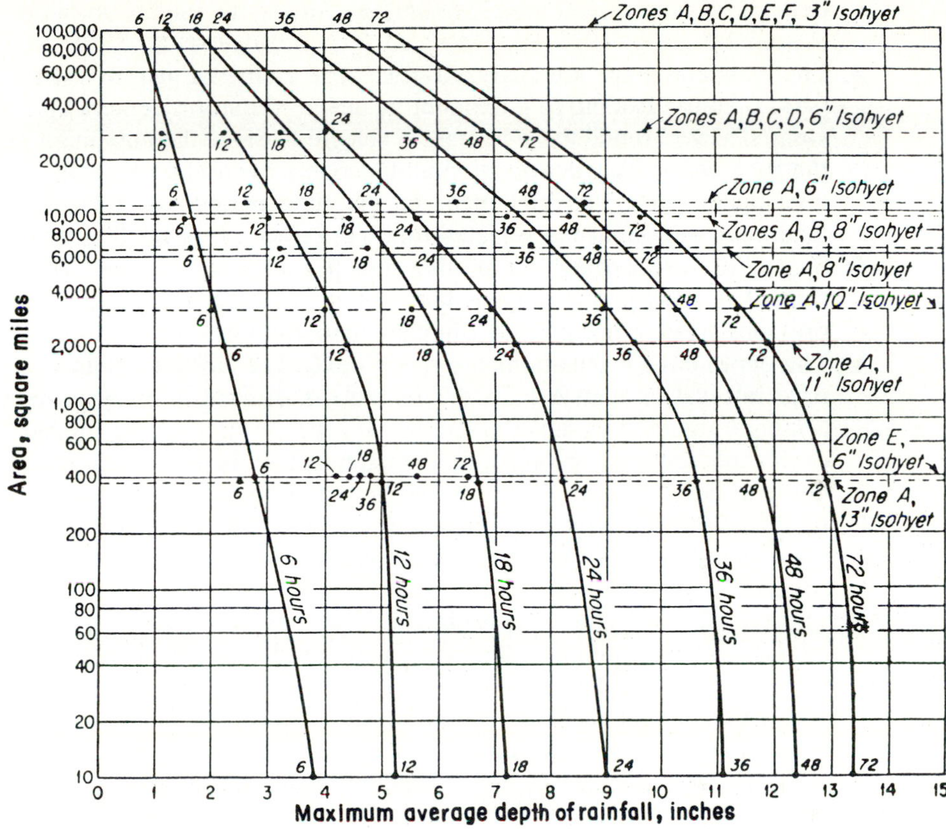

Figure 26.2 Maximum depth–area–duration curves for the storm of January 18–21, 1935, centered near Bolivar, Tennessee, and Hernando, Mississippi (after U.S. Army Corps of Engineers, 1956).

one isohyetal map for the total storm rainfall. When rainfall readings are available at different times of the day, as is the case in the United States, construction of mass curves is fairly easy.

The incremental isohyetal method involves construction of a number of isohyetal maps. If the isohyetal pattern is simple and rainfall readings are available at a uniform time each day, then this method is relatively simple. In many cases, especially in developing countries, rainfall is recorded and reported once a day, and the standard durations are, therefore, integral multiples of a day such as 2 days, 3 days, and so on; hence, the incremental isohyetal method may be preferable.

The steps involved in carrying out a DAD analysis are as follows:

1. The storm to be analyzed is assigned a definite beginning and a definite ending to ensure the same time interval for all rain gages.
2. For each time period during the total storm duration, rainfall amounts are computed for each rain gage and tabulated.
3. Separate isohyetal maps are prepared for the area for each time period, for

example, 6 hours, 12 hours, 24 hours, etc., using the rainfall amounts obtained in step 2.

4. The isohyetal maps are analyzed next. The isohyetal map of the time period (say, 6 hours) leading to maximum values at the storm centers and over substantial areas around these centers is selected. Since the maximum rainfalls of the time period may not occur simultaneously over all areas, it is advisable to investigate other isohyetal maps of the time period and then establish the maximum rainfall of the selected time period for areas of all sizes.
5. Steps 2 to 4 are repeated for another time period, say, 12 hours. The maximum rainfall for this period may or may not include the maximum rainfall of the previous smaller time period. Similar calculations are done for other durations.
6. The depth–area calculations are made next. The isohyetal map of the lowest time interval is analyzed first. This map is divided into zones to represent principal rainfall centers. Beginning with the central isohyet (higher value) in each zone, the area encompassed by each isohyet is planimetered and then the net area between each pair of isohyets is determined.
7. In the storm-centered DAD analysis, the last enclosing isohyet is taken as boundary, whereas the watershed boundary is used in watershed-centered analysis.
8. The average isohyetal value is obtained for consecutive pairs of isohyets. This average value is multiplied by the area enclosed between them, and the volume is then computed. The volumes are computed for all successive isohyetal pairs and then summed up. The volumes are divided by the corresponding areas to obtain maximum rainfall depths over these areas, which are of various sizes.
9. The maximum average depths corresponding to various area sizes are plotted on semilog paper using the log scale for area, and a smooth curve is drawn enveloping all the points on the graph. This is the depth–area curve of the maximum rainfall for the selected time period.

   Similarly, depth–area curves are prepared for other time periods. Taken together, the depth–area curves of the various durations are the depth–area–duration curves, and the whole procedure to construct these DAD curves is the DAD analysis.

NUMERICAL METHOD FOR DAD ANALYSIS

Nicks and Igo (1980) developed a computer-optimized depth–area–duration model for areas up to 2600 km^2 and durations from 1 to 24 hours. Data analyzed were from 10 years of records (1962–1971) collected from 168 gages of a 228-gage network operated by the Southern Great Plains Research Watershed located in the Washita River basin in central Oklahoma. The network consisted of recording gages spaced on a 5- by 5-km uniform grid. Storms were selected that produced surface runoff at one or more of the 18 runoff-measuring stations on the subdrainage basins. By using this criterion, 118 runoff-producing storms ranging in amounts from 25.4 to 233 mm with durations from 0.5 to 24 hours were selected.

Storm data for depth–area calculations included longitude, latitude, Thiessen area of each gage, storm duration, and a list of gage rainfall amounts for each storm selected. From these input data, the method selected the maximum amount of each storm and assigned the associated gage location as the storm center from which area–depth calculations were made as follows. Radial distances from the center gage to all other gages were calculated using the longitude and latitude of each station. Then areas, for which average rainfall depths were estimated, were computed by summing the Thiessen polygon areas of gages falling within 5-km radial increments from the storm center. Thus, the area-weighted mean rainfall $\bar{P}$ was computed as

$$\bar{P} = \frac{\sum_{i=1}^{n} P_j A_j}{\sum_{j=1}^{n} A_j} \tag{26.1}$$

where P_j is the jth station rainfall amount in mm, A_j is the jth station Thiessen area in km^2, and n is the number of gages included in each radius.

By using linear interpolation, the depth–area data generated before were converted to uniform area depths. Rainfall amounts were computed for successive 130-km^2 increments of area, starting with a point amount and ending with the depth for 2600 km^2. The uniform area–depth data so generated were used to develop a model for storm rainfall.

The mean rainfall covering a given area for a given duration was calculated as

$$P_A = P_c - \frac{P_c A D^m}{a + bA} \tag{26.2}$$

where P_A is the mean rainfall depth in millimeters for area A in km^2; P_c is the storm center point amount in millimeters; D is the duration of the storm center rainfall in hours; and m, a, and b are parameters. Equation (26.2) can also be written as

$$R = \frac{P_A}{P_c} = 1 - \frac{AD^m}{a + bA} \tag{26.3}$$

A nonlinear least squares optimization procedure, developed by DeCoursey and Snyder (1969), was used to estimate parameters m, a, and b. For the 138 storms used, the values of the parameters were $a = 337.48$, $b = 1.094$, and $m = -0.148$, and the multiple correlation coefficient was 0.88. Verification of the model showed that the mean error between computed and observed rainfall depths increased with area. The mean errors ranged from 8% on 130 km^2 to 27% on 3100 km^2.

EXAMPLE 26.1 Hourly rainfall for a storm in July 1964 at several stations in Western Ontario, Canada, are given in Table 26.6. Determine the maximum depth–area curve for 6-, 12-, and 24-h duration for the portion of the storm covering the South Branch of the Thames River watershed shown in Figure 26.3.

Solution The maximum depth–area curves for durations of 6, 12, and 24 hours for the portion of the storm covering the south branch of the Thames River watershed are determined as follows:

1. Mass curves of rainfall are constructed for the given stations for July 12 and 13, as shown in Figure 26.4.
2. The calculations for the mass curves are shown in Table 26.7.

TABLE 26.6 HOURLY PRECIPITATION IN INCHES FOR WESTERN ONTARIO, JULY 12–13, 1964

	Hour ending											
July 12	0100	0200	0300	0400	0500	0600	0700	0800	0900	1000	1100	1200
Brantford OWRC	0.00	0.00	0.00	0.11	0.08	0.00	0.01	0.02	0.05	0.05	0.00	0.01
Woodstock	0.00	0.00	0.05	0.01	0.03	0.03	0.01	0.05	0.06	0.09	0.05	0.04
London A	0.07	0.00	0.01	0.03	0.07	0.01	0.06	0.16	0.06	0.02	0.00	0.02
Glen Allan	0.00	0.01	0.01	0.01	0.02	0.03	0.00	0.00	0.00	0.00	0.00	0.00
Fullarton	0.01	0.03	0.05	0.00	0.01	0.02	0.02	0.02	0.00	0.00	0.00	0.00
Delhi C.D.A.	0.00	0.00	0.10	0.07	0.00	0.00	0.05	0.03	0.08	0.02	0.00	0.01
Guelph W-1	0.00	0.00	0.05	0.05	0.00	0.04	0.04	0.04	0.09	0.06	0.00	0.08
July 12	1300	1400	1500	1600	1700	1800	1900	2000	2100	2200	2300	2400
Brantford	0.04	0.05	0.06	0.02	0.05	0.08	0.14	0.16	0.10	0.08	0.11	0.18
Woodstock	0.04	0.02	0.03	0.01	0.03	0.08	0.04	0.03	0.15	0.12	0.11	0.03
London A	0.01	0.03	0.02	0.07	0.04	0.04	0.23	0.05	0.08	0.14	0.19	0.07
Glen Allan	0.00	0.00	0.00	0.00	0.01	0.01	0.01	0.09	0.06	0.08	0.17	0.15
Fullarton	0.00	0.00	0.00	0.00	0.00	0.00	0.00	0.00	0.07	0.01	0.05	0.26
Delhi C.D.A.	0.01	0.04	0.03	0.03	0.06	0.06	0.07	0.11	0.15	0.10	0.08	0.07
Guelph W-1	0.03	0.02	0.01	0.04	0.04	0.10	0.05	0.15	0.11	0.10	0.09	0.22
July 13	0100	0200	0300	0400	0500	0600						
Brantford	0.46	0.27	0.07	0.05								
Woodstock	0.18	0.30	0.20	0.02								
London A	0.09	0.7										
Glen Allan	0.05	0.12	0.34	0.30	0.10	0.05						
Fullarton	0.10	0.08	0.12	0.02								
Delhi C.D.A.	0.39	0.33	0.04									
Guelph W-1	0.34	0.22	0.11									

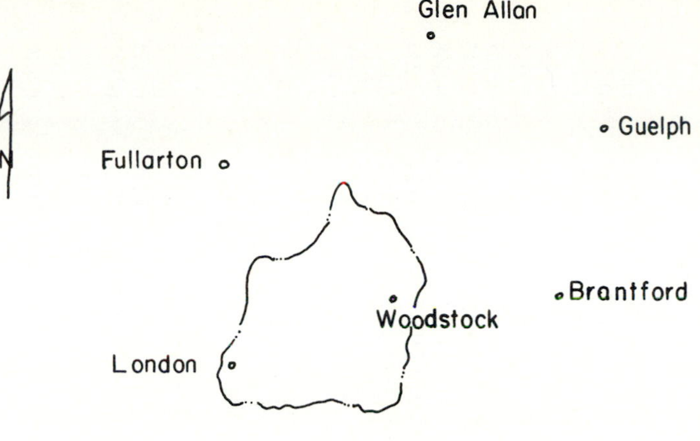

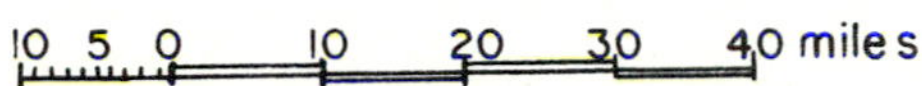

Figure 26.3 Drainage-area map of the south branch of the Thames River.

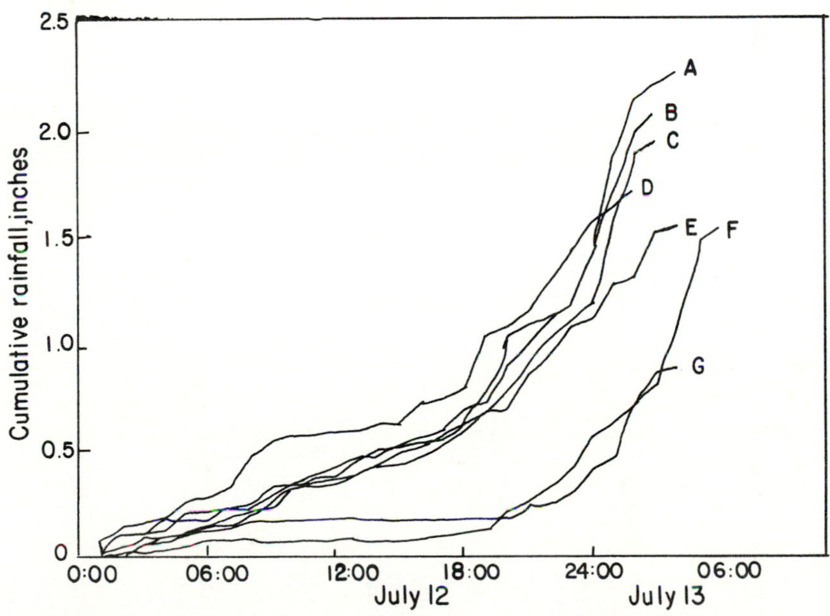

Figure 26.4 Mass curves of rainfall for stations in the Thames River basin.

TABLE 26.7 MASS-CURVE COMPUTATIONS

Time	Cumulative precipitation (in.)							Seven-station precipitation (in.)	
(hour ending) (1)	Brantford (2)	Woodstock (3)	London A (4)	Glen Allan (5)	Fullarton (6)	Delhi C.D.A. (7)	Guelph W-1 (8)	Total (9)	Average (10)
July 12									
0100	0.00	0.00	0.07	0.00	0.01	0.00	0.00	0.08	0.011
0200	0.01	0.00	0.14	0.01	0.04	0.00	0.00	0.12	0.017
0300	0.01	0.05	0.15	0.02	0.09	0.10	0.05	0.27	0.038
0400	0.12	0.06	0.18	0.03	0.09	0.17	0.10	0.28	0.040
0500	0.20	0.09	0.25	0.05	0.10	0.17	0.10	0.21	0.030
0600	0.20	0.12	0.26	0.08	0.12	0.17	0.14	0.13	0.018
0700	0.21	0.13	0.32	0.08	0.14	0.22	0.18	0.19	0.027
0800	0.25	0.18	0.48	0.08	0.16	0.25	0.22	0.32	0.045
0900	0.28	0.24	0.54	0.08	0.16	0.33	0.31	0.34	0.048
1000	0.33	0.33	0.56	0.08	0.16	0.35	0.37	0.24	0.034
1100	0.33	0.38	0.56	0.08	0.16	0.35	0.37	0.05	0.071
1200	0.34	0.42	0.58	0.08	0.16	0.36	0.45	0.16	0.0228
1300	0.38	0.46	0.59	0.08	0.16	0.37	0.48	0.13	0.018
1400	0.43	0.48	0.62	0.08	0.16	0.41	0.50	0.16	0.0228
1500	0.49	0.51	0.64	0.08	0.16	0.44	0.51	0.15	0.021
1600	0.51	0.52	0.71	0.08	0.16	0.47	0.55	0.17	0.024
1700	0.56	0.55	0.75	0.09	0.16	0.53	0.59	0.23	0.032
1800	0.64	0.63	0.79	0.10	0.16	0.59	0.69	0.37	0.0528
1900	0.78	0.67	1.02	0.11	0.16	0.66	0.74	0.44	0.062
2000	0.94	0.70	1.07	0.20	0.16	0.77	0.89	0.59	0.084
2100	1.04	0.85	1.15	0.26	0.23	0.92	1.00	0.72	0.1028
2200	1.12	0.97	1.29	0.34	0.24	1.02	1.12	0.63	0.09
2300	1.23	1.08	1.48	0.51	0.29	1.17	1.20	0.80	0.14
2400	1.41	1.11	1.55	0.66	0.55		1.29	0.98	0.140
July 13									
0100	1.87	1.29	1.64	0.71	0.65	1.56	1.85	1.61	0.23
0200	2.14	1.59	1.71	0.83	0.73	1.89	2.07	1.40	0.20
0300	2.21	1.79	1.79	1.17	0.85	1.93	2.18	0.88	0.125
0400	2.26	1.81	1.79	1.47	0.87	1.93	2.18	0.39	0.055
0500	2.26	1.81	1.79	1.57	0.87	1.93	2.18	0.10	0.014
0600	2.26	1.81	1.79	1.62	0.87	1.93	2.18	0.05	0.007

3. Determine those 6-hour, 12-hour, and 24-hour periods from the given data for which the precipitation will be maximum over the watershed. This can be determined in two ways.

   (a) Plot the average mass curve for the seven stations and then pick up the 6-hour, 12-hour, and 24-hour periods of maximum rainfall. This average mass curve (of course, unweighted) is shown in Figure 26.5. Computations are shown in Table 26.7 for this unweighted average mass curve. The 6-, 12-, and 24-hour periods of maximum precipitation are shown in the figure. These are as follows:

   6-hour period of maximum precipitation = after 2100, July 12, to 0300, July 13
   12-hour period of maximum precipitation = after 1600, July 12, to 0400, July 13
   24-hour period of maximum precipitation = 0400, July 12, to 0400, July 13

   (b) Determine the 6-, 12-, and 24-hour periods of maximum precipitation as follows. Hourly precipitation values are summed up for the seven stations to yield seven-station hourly totals. These totals are summed up and cumulated for every 6-, 12-, and 24-hour period throughout the period of precipitation, as shown in Table 26.8. Then mark the periods of maximum precipitation. The value of maximum precipitation for 6 h = 6.29 in.; the corresponding period of 6 hours follows from 2100, July 12, to 0300, July 13. Maximum precipitation for 12 h = 9.13 in.; the corresponding period follows from 1600, July 12, to 0400, July 13, 1964. Maximum precipitation for 24 hours = 11.38 in.; the corresponding period is from 0400, July 12, to 0400, July 13, 1964.

4. For these periods, the maximum precipitation is found for each station. Calculated values are indicated in Table 26.9. Maximum precipitation values for these are shown in Figure 26.6.

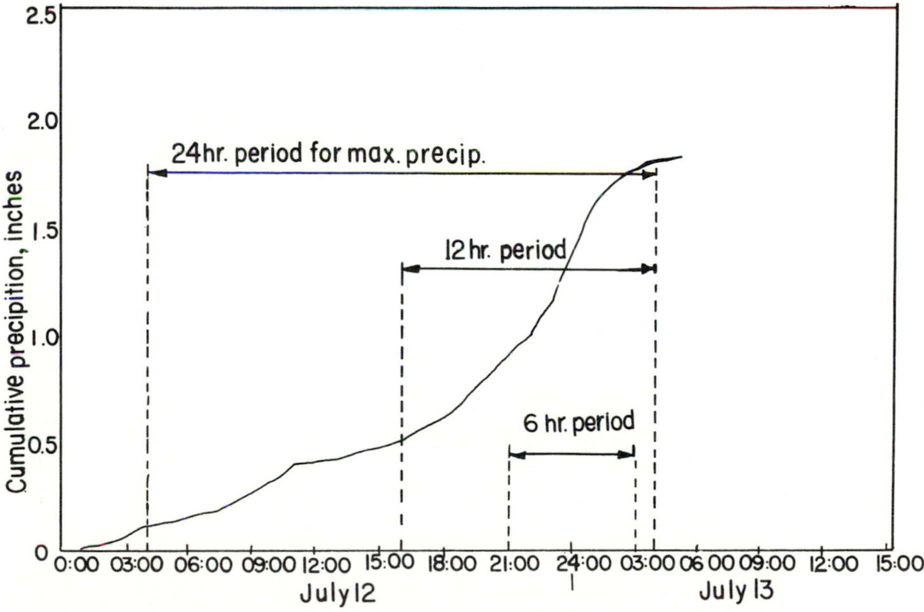

Figure 26.5 Average unweighted mass curve showing periods of maximum precipitation.

TABLE 26.8 DETERMINATION OF 6-h, 12-h, AND 24-h PERIODS OF MAXIMUM PRECIPITATION

Time (hour ending)	Seven-station hourly totals (in.)	6-Hour precip. (in.)	12-Hour precip. (in.)	24-Hour precip. (in.)
0.00	0.08			
0200	0.12			
0300	0.27			
0400	0.28			
0500	0.21			
0600	0.13	1.09		
0700	0.19	1.20		
0800	0.32	1.40		
0900	0.34	1.47		
1000	0.24	1.43		
1100	0.05	1.27		
1200	0.16	1.30	2.39	
1300	0.13	1.24	2.44	
1400	0.16	1.08	2.48	
1500	0.15	0.89	2.36	
1600	0.17	0.82	2.25	
1700	0.23	1.00	2.27	
1800	0.37	1.21	2.51	
1900	0.44	1.66	2.76	
2000	0.59	2.05	2.93	
2100	0.72	2.62	3.31	
2200	0.63	3.08	3.76	
2300	0.80	3.65	4.45	
2400	0.98	4.26	5.27	7.76
July 13				
0100	1.61	5.33	6.75	9.29
0200	1.40	6.13	7.99	10.57
0300	0.88	6.29	8.72	11.18
0400	0.39	6.05	9.13	11.38
0500	0.10	5.35	9.00	11.29
0600	0.05	4.42	8.68	11.19

TABLE 26.9 MAXIMUM PRECIPITATION VALUES FOR 6-h, 12-h, AND 24-h PERIODS

Station	Max 6-h precip. 2100 to 0300 h July 12–13 (in.)	Max 12-h precip. 1600 to 0400 h July 12–13 (in.)	Max 24-h precip. 0400 to 0400 h July 12–13 (in.)
Brant Ford	1.17	1.75	2.14
Woodstock	0.94	1.20	1.75
London A	0.56	1.00	1.53
Glen Allan	0.91	1.39	1.44
Fullarton	0.62	0.71	0.78
Delhi C.D.A.	1.01	1.46	1.76
Guelph W-1	1.08	1.53	1.98
Σ Precipitation	6.29	9.13	11.38

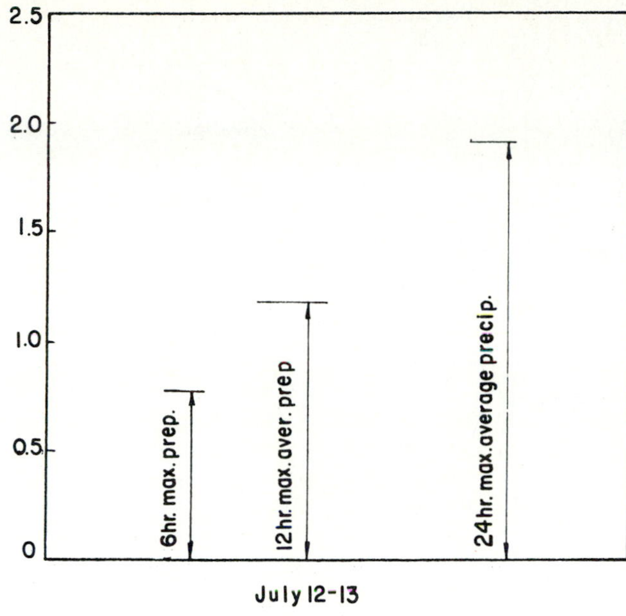

Figure 26.6 Maximum average precipitation for 6-hour, 12-hour, and 24-hour periods.

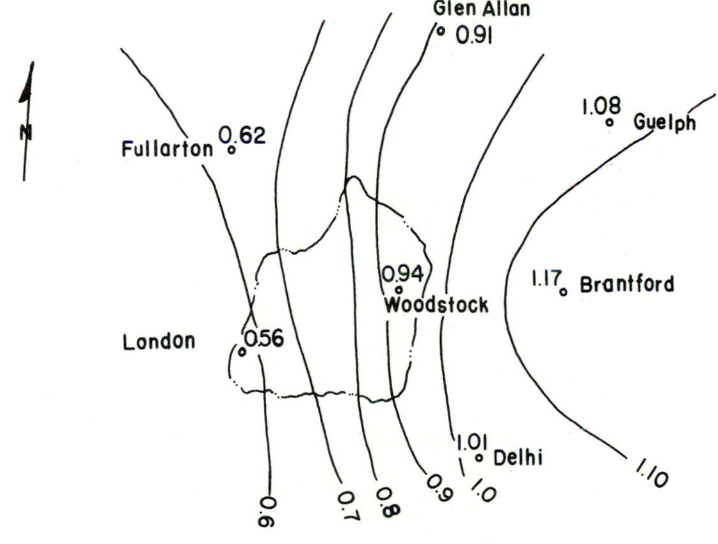

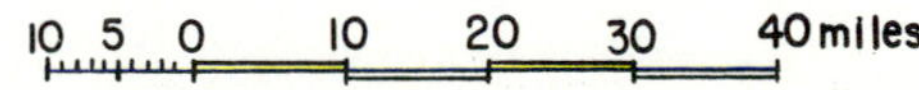

Figure 26.7 Isohyetal map for 6-hour maximum precipitation.

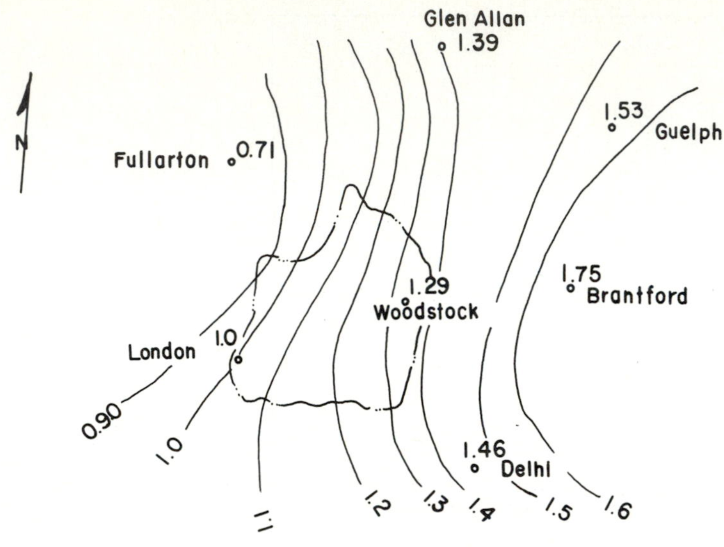

Figure 26.8 Isohyetal map for 12-hour maximum precipitation.

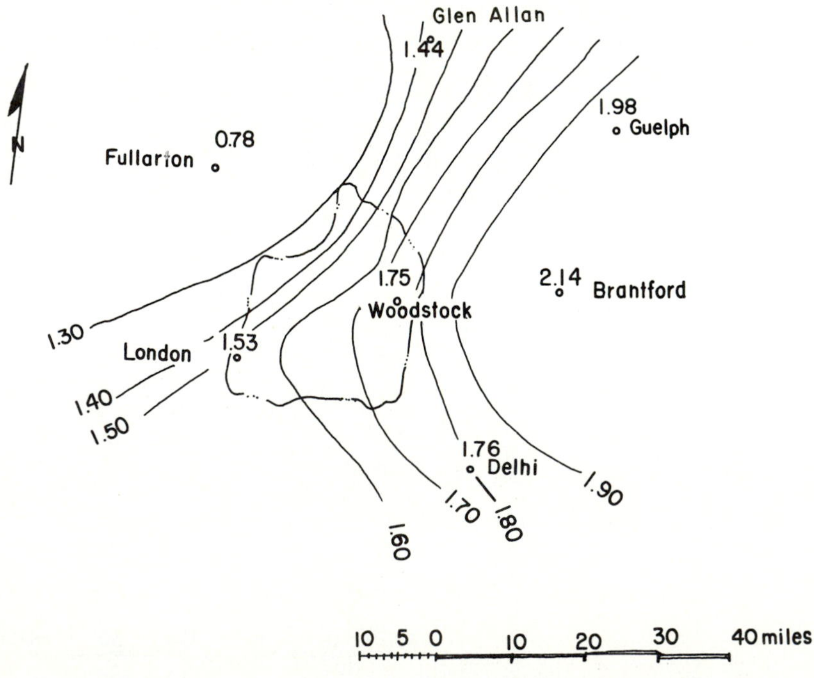

Figure 26.9 Isohyetal map for 24-hour maximum precipitation.

TABLE 26.10 COMPUTATION FOR AVERAGE MAXIMUM PRECIPITATION (6-h PERIOD)

SN (1)	Isohyetal line (2)	Net area (mi^2) (3)	Mean precip. (in.) (4)	Precip. volume (5) = (3) × (4)	Average weighted precip. (in.) (6)
1	0.5	33	0.58	19.2	0.037
2	0.6	92	0.65	60.0	0.115
3	0.7	178	0.75	133.5	0.260
4	0.8	115	0.85	97.8	0.188
5	0.9	102	0.95	96.0	0.184
6	1.0				
				Σ Precip. Volume = 406.5	Σ Precip. = 0.78 in.
Average maximum precipitation = 406.5/520 = 0.78 in.					

5. The maximum precipitation values for 6, 12, and 24 hours, given in Table 26.9, are then used for drawing isohyets.
6. For isohyetal maps corresponding to these periods for the watershed area, a road map is used. The relevant portion of the complete road map is traced on a separate sample paper. Figure 26.3 shows the drainage area for the south branch of the Thames River watershed.
7. Isohyets are drawn at the interval of 0.1 in. on the drainage-area map for 6-, 12-, and 24-hour periods separately, as shown in Figures 26.7 to 26.9.
8. The drainage area of the South Branch of the Thames River watershed is planimetered and is found to be 520 square miles.
9. The average precipitation over the watershed for these periods is determined by the isohyetal method. Relevant calculations are shown in Tables 26.10 to 26.12.

TABLE 26.11 COMPUTATION FOR AVERAGE MAXIMUM PRECIPITATION (12-h PERIOD)

SN (1)	Isohyetal line (2)	Net area (mi^2) (3)	Mean precip. (in.) (4)	Precip. volume (5) = (3) × (4)	Average weighted precip. (in.) (6)
1	0.90	41	0.95	39.0	0.075
2	1.0	130	1.05	136.2	0.261
3	1.1	154	1.15	187.0	0.357
4	1.2	155.5	1.25	194.0	0.372
5	1.3	39.5	1.35	53.0	0.102
6	1.4				
				Σ Precip. Volume = 609.4	Σ Precip. = 1.17 in.
Average maximum precipitation = 609.4/520 = 1.17 in.					

TABLE 26.12 COMPUTATION FOR AVERAGE MAXIMUM PRECIPITATION (24-h PERIOD)

SN (1)	Isohyetal line (2)	Net area (mi^2) (3)	Mean precip. (in.) (4)	Precip. volume (5) = (3) × (4)	Average weighted precip. (in.) (6)
1	1.30	50.0	1.35	67.5	0.13
2	1.40	60.0	1.45	87.0	0.168
3	1.50	150.0	1.55	233.0	0.45
4	1.60	150.0	1.65	247.0	0.475
5	1.70	110.0	1.75	193.0	0.37
6	1.80				
				Σ Precip. Volume = 827.5	Σ Precip. = 1.59 in.

Average maximum precipitation = 827.5/520 = 1.59 in.

10. The maximum average precipitation over the watershed for the given period is thus determined.
11. The maximum depths of precipitation are determined over areas of various sizes. These values are tabulated in Tables 26.13 to 26.15 respectively for durations of 6, 12 and 24 h. The maximum depth–area–duration curves are shown in Figure 26.10.

STORM TRANSPOSITION

The probable maximum storm (PMS) or PMP can be estimated using storm transposition and maximization. The purpose of storm transposition is to enhance the storm experience of a watershed by considering not only the storms that have occurred over or near the watershed, but also the storms that have resulted in heavy rainfall on meteorologically homogeneous adjacent areas. Storm transposition is used in watersheds that have inadequate rainfall data or have experienced no severe storms. Furthermore, the watershed is of highly irregular shape, which does not

TABLE 26.13 MAXIMUM DEPTH–AREA CALCULATIONS FOR THE 6-h PERIOD

SN (1)	Isohyets (2)	Net area (mi^2) (3)	Cumulative area (4)	Average precip. (in.) (5)	Average weighted precip. (in.) (6)
1	>0.90	102	102.0	0.95	0.95
2	0.80	115	217.0	0.85	0.90
3	0.70	178	395.0	0.75	0.84
4	0.60	92	487.0	0.65	0.825
5	<0.50	33	520.0	0.58	0.79

TABLE 26.14 MAXIMUM DEPTH–AREA CALCULATIONS FOR THE 12-h PERIOD

SN (1)	Isohyets (2)	Net area (mi^2) (3)	Cumulative area (4)	Average precip. (in.) (5)	Average weighted precip. (in.) (6)
1	>1.3	39.5	39.5	1.35	1.35
2	1.2	155.5	195.0	1.25	1.27
3	1.1	154.0	349.0	1.15	1.23
4	1.0	130.0	479.0	1.05	1.17
5	<0.90	41.0	520.0	0.95	1.102

conform to the isohyetal pattern of major storms in the region containing the watershed. Transposition of storms from one watershed to the other is based on the assumption that these storms could occur on the watershed in question. Two main steps are involved in storm transposition: (1) assurance of meteorological homogeneity and (2) selection and analysis of major recorded storms.

1. *Meteorological Homogeneity*. Two watersheds can be considered meteorologically homogeneous if they are influenced by the same moisture source, experience similar types of storms, have similar topographic features, and exhibit similar orientation to seasonal winds. The factors affecting the homogeneity of an area are discussed in Chapter 6. To summarize, chief among these factors are topography, distance from the sea, direction of the prevailing winds, and mean annual temperature. The factors that can be considered to ascertain homogeneity may include (a) mean seasonal and annual rainfall, (b) storm tracks over the watershed and the region, (c) weather systems generating heavy storms, (d) average annual runoff and peak discharge of streams in the region, and (e) watershed characteristics such as land form, soil vegetation, etc.

The areal limits for transposition of a storm can be delineated as follows: (1) Identify the area where the maximum storm rain occurred. This is accomplished by constructing an isohyetal map prepared from records of rain gages in the area. (2) Analyze the synoptic meteorology of the storm, and identify the causes of storm rainfall, etc. (3) Examine all major recorded storms and identify those comparable

TABLE 26.15 MAXIMUM DEPTH–AREA CALCULATIONS FOR THE 24-h PERIOD

SN (1)	Isohyets (2)	Net area (mi^2) (3)	Cumulative area (4)	Average precip. (in.) (5)	Average weighted precip. (in.) (6)
1	>1.70	110.0	110.0	1.75	1.75
2	1.60	150.0	260.0	1.65	1.69
3	1.50	150.0	410.0	1.55	1.64
4	1.40	60.0	470.0	1.45	1.62
5	<1.3				

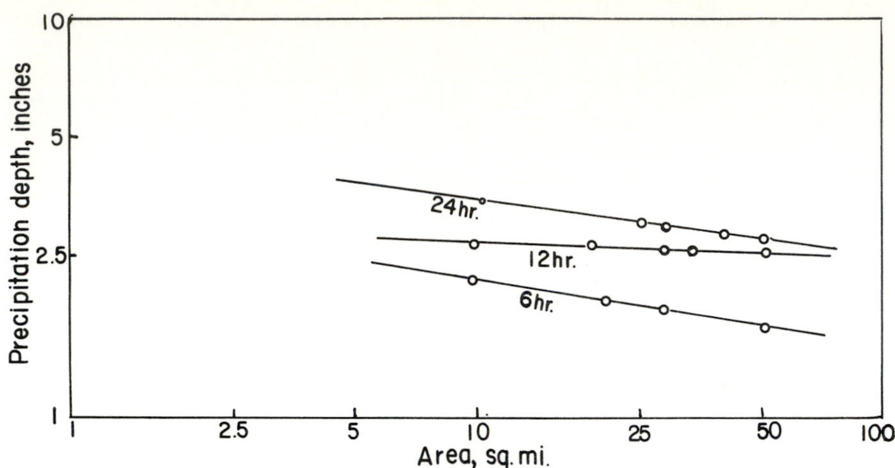

Figure 26.10 Maximum depth area curves for 6, 12, and 24 hours.

with the storm under consideration. Mark the locations of these storms on a map. Draw a boundary around these locations, which is the approximate area in which the storm under consideration can be expected to occur. (4) The approximate area is adjusted by considering the effects of topography, the general location of the storm with respect to the moisture, and inflow barriers.

2. *Method of Transportation.* Having defined the meteorologically homogeneous region of which the project watershed constitutes a part, the next step is to determine the permissible change in orientation of a storm isohyetal pattern when the storm is transposed to the watershed. The following steps can be followed to accomplish storm transposition.

(1) Identify where and when the heaviest storm rain fell in the meteorologically homogeneous region. This is done by analyzing the synoptic meteorology of the storm.
(2) Prepare the isohyetal pattern of the selected rainstorm.
(3) Prepare a layout of the storm pattern on a transparent sheet. Superimpose this layout on the watershed in the most critical manner that yields the maximum depths of rainfall over the watershed.
(4) Use the depth–duration analysis to calculate the average rainfall depth of the transposed storm over the watershed.
(5) Carry out the previous steps for the selected storms. Select the storm producing the maximum depths of rainfall over the watershed. This storm is then considered for further adjustment to obtain the design storm.

3. *Precautionary Measures for Storm Transposition.* When transposing storms, the following considerations should be taken into account: (1) Storm transposition is not recommended for mountainous areas where necessary adjustments for topography cannot be made satisfactorily. Therefore, when the terrain is highly orographic, rainstorms are neither transposed to nor away from these areas. Similarly, it would not be advisable to transpose cyclonic storms in coastal areas to areas

far inland. (2) Areal limits for transposition must be defined for each individual storm. A storm does not necessarily have the same probability of occurrence over all parts of the transposition region. The synoptic features of a particular storm are likely to have varying frequencies over different portions of this region. (3) Storm transposition over large changes in latitude should be avoided, for a high latitudinal shift may involve significant changes in storm air-mass characteristics. (4) The axis of isohyetal pattern with respect to the axis of the watershed should be taken into account. Generally, the storm axis is not rotated by more than 15° to 20° and the direction of rotation must be toward the axis of normal isohyetal pattern.

STORM MAXIMIZATION

The purpose of storm maximization is to determine the amount of increase in rainfall of a storm due to physically possible alterations in the meteorological factors producing the storm. The physical factors can be distinguished as: (1) mechanisms causing atmospheric moisture to precipitate and (2) the moisture content of the air mass responsible for the storm.

In general, it is difficult to manipulate physical mechanisms for increased precipitation or increase mechanical efficiency of a storm. In practice, storm maximization, therefore, is not carried out in this way. Furthermore, the mechanisms producing intense rainfall can be reasonably assumed to be highly efficient and may likely be near maximum efficiency. Storms are, therefore, maximized for moisture content of the air mass.

Moisture and Precipitable Water. The atmospheric moisture content can be expressed as precipitable water. This refers to the amount of liquid water that would occur in an air column of unit cross-section if all of its water vapor were condensed, which, of course, is seldom the case. The vapor content is available in varying amounts in atmospheric layers close to the earth's surface. The bulk of the vapor content of the atmosphere is present in the lower layers. For practical purposes, it is satisfactory to consider the vapor content up to the 300-mb level in the atmosphere.

The amount of precipitable water, W, can be computed by considering an air column extending up from the earth's surface (Solot, 1939). Let this column be a 1-cm square. If ρ_w is the absolute humidity, then

$$dW = \rho_w \, dz \tag{26.4}$$

where dz is the incremental height of the air column, and W is precipitable water. If the height is z, then

$$\int_0^W dW = \int_0^z \rho_w \, dz$$

or

$$W = \int_0^z \rho_w \, dz, \qquad W(z = 0) = 0 \tag{26.5}$$

From the hydrostatic equation,

$$dp = -\rho g \, dz \tag{26.6}$$

where ρ is the total air density, p is the total pressure of moist air, g is the acceleration due to gravity. Note that $\rho = \rho_w + \rho_d$, where ρ_d is the density of dry air. Equation (26.6) can be written as

$$dp = -(\rho_w + \rho_d)g\, dz \qquad \text{or} \qquad dz = -\frac{dp}{g(\rho_w + \rho_d)} \tag{26.7}$$

Substituting Equation (26.7) into Equation (26.5) gives

$$W = \frac{1}{g}\int_p^{p_0} \frac{\rho_w}{\rho_w + \rho_d}\, dp, \qquad p(z = 0) = p_0 \tag{26.8}$$

The term $\rho_w/(\rho_w + \rho_d)$ is the specific humidity q, which is approximately equal to

$$q = 0.622\,\frac{e}{p} = \frac{\rho_w}{\rho_w + \rho_d} \tag{26.9}$$

where e is the partial pressure of the water vapor. Inserting Equation (26.9) into Equation (26.8) yields

$$W = \frac{0.622}{g}\int_p^{p_0} e\,\frac{dp}{p} \tag{26.10a}$$

in which W is in gm/cm^2. Inasmuch as the density of precipitable water is equal to 1 gm/cm^3, W is the depth of precipitable water in centimeters. By using the appropriate conversion factor, W can be expressed in inches as

$$W = 0.0004\int_p^{p_0} q\, dp \tag{26.10b}$$

with q in g/kg, and p in mb.

By knowing the atmospheric pressure and relative humidity to compute vapor pressure, the depth of precipitable water can be computed.

Since pressure and temperature are related, the moisture content is also related to temperature. The moisture content can also be calculated from temperature considerations. When the air is cooled at a constant atmospheric pressure, the temperature at which the air becomes saturated is called the dew-point temperature. The dew-point depression, which is the difference between the air temperature near the ground and the dew-point temperature, is an indication of the degree of saturation. The lesser the dew-point depression is, the nearer the air is to saturation. For an atmospheric column, the dew-point is usually known at the ground surface. It can be assumed that the dew-point in the column varies with altitude and has a pseudoadiabatic lapse rate. With this assumption, the precipitable water can be computed for an atmospheric column. Figure 26.11 shows the relation for depths of precipitable water in a column of air of given height above the mean sea level and 1000-millibar pressure as a function of dew point, assuming saturation and pseudoadiabatic lapse rate (U.S. Weather Bureau, 1941). The dew point is replaced by the surface temperature when the air is not saturated. The dew point observed at any elevation can be transformed to the 1000-mb pressure, assuming saturation with the pseudoadiabatic lapse rate, as shown in Figure 26.12. The transformation of dew

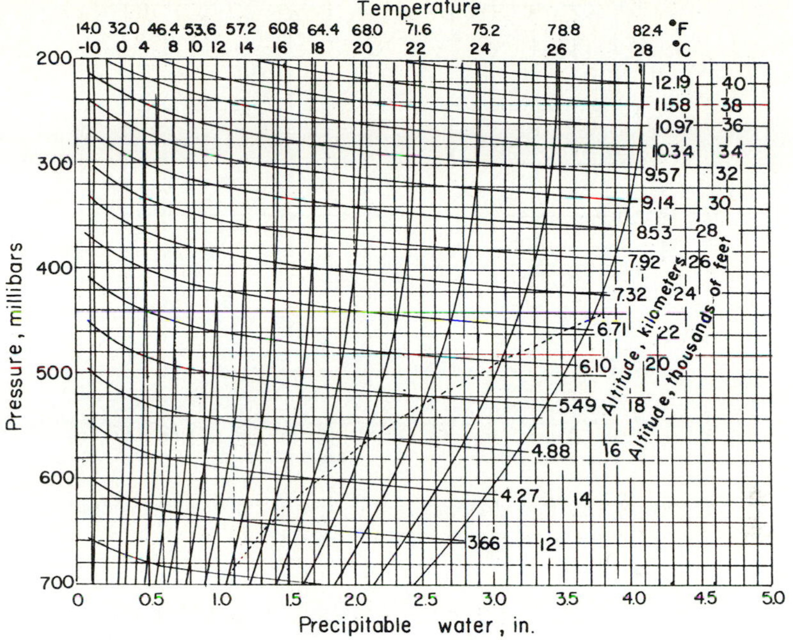

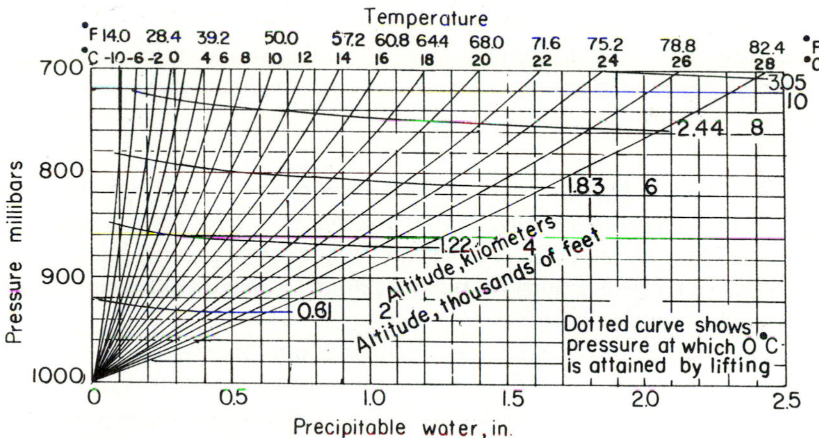

Figure 26.11 Depths of precipitable water in a column of air given the height above 1000 mb as a function of dew point, assuming saturation and pseudoadiabatic lapse rate (U.S. Weather Bureau, 1941).

point can also be obtained analytically. Based on thermodynamic considerations and assuming adiabatic cooling, the following can be expressed:

$$TP^{(1-r)/r} = \text{constant} \tag{26.11}$$

where T is the absolute temperature, P is the pressure, and r is a constant equal to

$$r = \frac{c_p}{c_v} \tag{26.12}$$

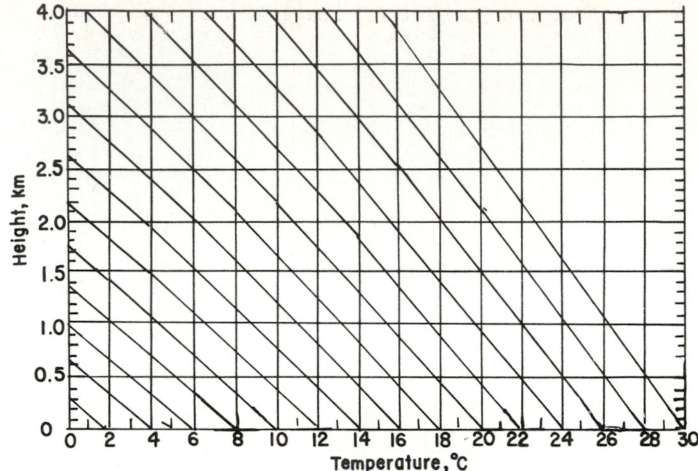

Figure 26.12 Pseudoadiabatic diagram for dew-point reduction to 1000 mb at the mean sea level (height zero).

where c_p is the specific heat of air at constant pressure, and c_v is the specific heat of air at constant volume. For air, the value of r is approximately 1.4. Thus,

$$T_1 P_1^{(1-r)/r} = T_2 P_2^{(1-r)/r} \tag{26.13}$$

where subscripts 1 and 2 refer to points 1 and 2 on the temperature–pressure curve. If P_1, P_2, and T_1 are known, then T_2 can be obtained from Equation (26.13).

Storm Dew Point and Maximum Dew Point. Because the storm occurs for several hours, it is desirable to compute the most persistent value of the dew point from observations made at regular intervals of time. In practice, the highest storm dew point persisting over 6 hours or 12 hours is used; this could also be an average value over that time period.

The maximum dew point is obtained from an analysis of historical data. The calendar period of 15 days during which the storm has occurred is marked. The dew-point data corresponding to this period is gathered for each of 25 to 30 years. The highest persisting dew point of each year for the selected period is noted. Then the highest of these dew point values is chosen for storm maximization.

EXAMPLE 26.2 Select the highest persistent 12-h dew point for the following dew-point data in May 1988:

	MAY 1	MAY 2		MAY 3		MAY 4	
TIME	1 P.M.	1 A.M.	1 P.M.	1 A.M.	1 P.M.	1 A.M.	1 P.M.
Dew-point temperature (°C)	22	23	24	25	24	22	22

Solution The most persistent 12-h dew point for this data is 24 °C. Another value could be the average of 12-h consecutively highest dew-point values = (24 + 25 + 24)/3 = 24.33°C.

EXAMPLE 26.3 The dew-point temperature at 2.5-km elevation is about 6°C. What will be the dew-point temperature at the mean sea level?

Solution

Dewpoint temperature = 6°C

Elevation = 2.5 km

From Figure 26.12, the dew point reduced to the mean sea level is 18°C.

EXAMPLE 26.4 The dew-point temperature of a storm is known to be 24°C, and the watershed elevation for inflow barriers is 0.5 km. Calculate the depth of precipitable water.

Solution

Dew-point temperature = 24°C

Elevation = 0.5 km

Dew-point temperature reduced to the mean sea level (from Figure 26.12) = 25.8°C

Depth of precipitable water between 1000-mb and 200-mb pressure levels (from Figure 26.11) = 3.4 in.

Depth of precipitable water at the inflow barrier between the 1000-mb pressure level and an elevation of 0.5 km (Figure 26.11) = 0.4 in.

Precipitable water (between the watershed level and the 200-mb pressure level) = 3.4 − 0.4 = 3.0 in.

Moisture Adjustment. The moisture adjustment for an observed storm is made to determine the rainfall that would result if the moisture available to the storm were maximum over the project watershed, assuming the mechanical efficiency to be the same. The moisture-adjustment factor (MAF) is defined as the ratio of the maximum total precipitable water over the watershed, W_T, to the storm precipitable water, W_S,

$$\text{MAF} = \frac{W_T}{W_S} \tag{26.14}$$

The standard project storm depths, obtained from transposition or DAD analysis, are multiplied by the MAF.

The definition of MAF is influenced by use of transposition. When there is no transposition of storm from one location to the other, the MAF is defined as the ratio of the precipitable water corresponding to the maximum dew point, W_M, to the precipitable water corresponding to the storm dew point, W_S:

$$\text{MAF} = \frac{W_M}{W_S} \tag{26.15}$$

When storm transposition is involved, it is necessary to adjust the storm depths obtained by transposition for change in the moisture regime due to relocation of the storm. If W_S is the precipitable water associated with the storm in its original location and W_{M1} is the precipitable water corresponding to the maximum dew-point temperature in the original location of the storm, then

$$\text{MAF} = \frac{W_{M1}}{W_S} \tag{26.16}$$

If W_{M2} is defined as the precipitable water corresponding to the maximum dew-point temperature in the new location, then

$$\text{MAF} = \frac{W_{M2}}{W_{M1}} \tag{26.17}$$

The moisture-maximization factor (MMF) is then defined as

$$\text{MMF} = \frac{W_{M1}}{W_S} \cdot \frac{W_{M2}}{W_{M1}} = \frac{W_{M2}}{W_S} \tag{26.18}$$

Barrier Adjustments. If storm transposition is made across mountains less than 600 m high, then proper allowance must be made to account for the mountains acting as barriers. When the moist air encounters a barrier, it may lose moisture during its ascension to the windward slopes of the mountain. The MAF is defined as

$$\text{MAF} = \frac{W_c}{W_1} \tag{26.19}$$

where W_1 is the precipitable water at the foothills, and W_c is the precipitable water at the crest of the barrier. The value of MAF would be less than the one for windward slopes and greater than the one for leeward slopes.

ENVELOPMENT OF TRANSPOSED, ADJUSTED STORMS

For envelopment of storms, two cases are considered. First, the storm isohyetal pattern is assumed to perfectly fit the watershed. The storms are then subject to transposition and moisture maximization. For each storm, the maximum adjusted average rainfall depths for various durations are obtained for an area equal to the project watershed. These values are plotted on graph paper and a smooth enveloping curve is drawn through the highest values. This curve is supposed to yield the probable maximum rainfall depths for the watershed.

Second, the isohyetal pattern does not exactly conform to the project watershed, owing to its irregular shape, large size, and peculiar orientation. This means that the maximum rainfall occurring in the watershed may be less than that would fall in an area of the same size for a severe storm. One way to account for this lack of conformity is to superimpose the isohyetal pattern of the transposed storm over the watershed with an appropriate orientation that produces the maximum areal average rainfall depth. The maximum average depth is then obtained using the isohyetal method. This is done for each storm considered suitable for transposition. After maximizing each transposed storm, the maximum average rainfall depths for various durations are obtained and plotted on graph paper. A smooth enveloping curve through the largest values is constructed.

Underlying the envelopment procedure is the assumption that over the entire watershed, the maximum rainfall for all durations occur as the result of a single storm. Because two storms may be dynamically quite different from one another, envelopment of both storms is questionable. The effect of this envelopment is one of maximization. One way to eliminate this maximization effect is to use adjusted storms separately. Another way is to consider the isohyetal pattern of a critical storm observed over the watershed and multiply it by the lowest value of the adjustment factors. The adjusted isohyetal pattern is then used for further computations.

EXAMPLE 26.5 Atmospheric conditions at Guelph, Ontario, Canada, preceding, during, and following the storm of July 1964, are indicated in Table 26.16. Assuming the meteorological conditions as recorded at Guelph were representative of the air mass producing precipitation on the Thames River watershed of Example 26.1, maximize the precipitation to the 12-hour persistent dew point of 74°F considered appropriate for July in Ontario.

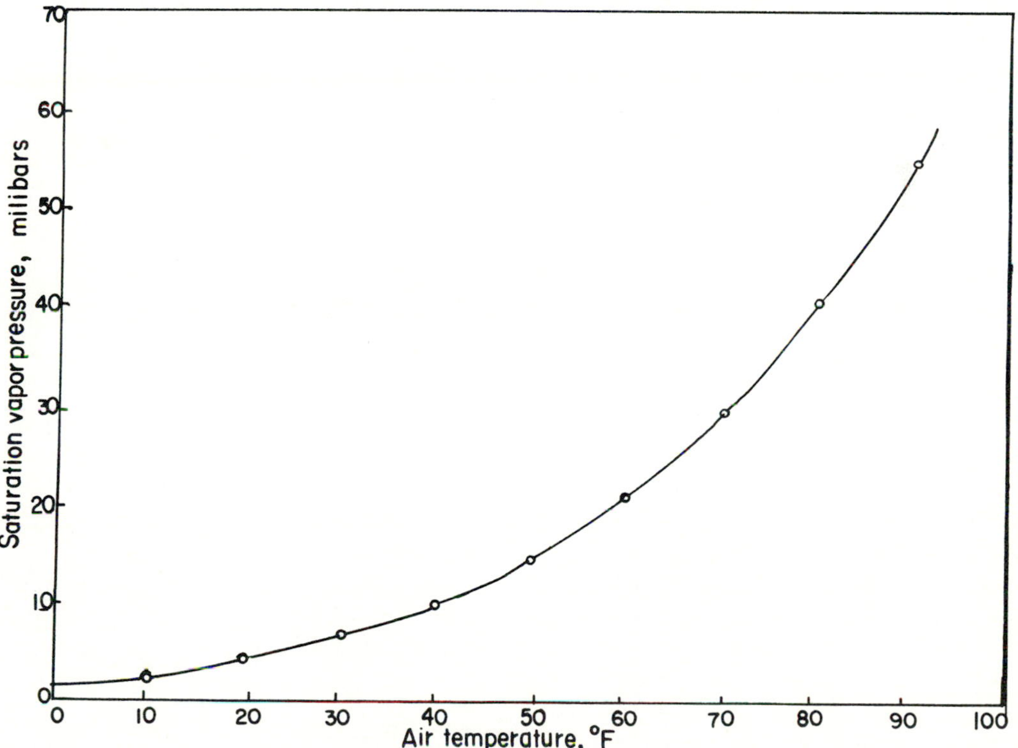

Figure 26.13 Saturated vapor-pressure curve.

TABLE 26.16 DRY-BULB TEMPERATURE, HUMIDITY, AND PRESSURE FOR THE GUELPH BENCHMARK STATION

	0100 Hours			0700 Hours			1300 Hours			1900 Hours		
Date	Temp. (°F)	Rel. hum. (%)	Press. (mb)	Temp. (°F)	Rel. hum. (%)	Press. (mb)	Temp. (°F)	Rel. hum. (%)	Press. (mb)	Temp. (°F)	Rel. hum. (%)	Press. (mb)
July 9	62.0	100	974	64.5	88	974	80.0	61	974	76.0	68	973
10	58.5	93	974	64.0	81	975	83.0	63	975	77.5	75	974
11	55.0	96	975	62.5	85	975	81.5	69	974	71.0	80	973
12	62.5	93	974	60.0	100	973	60.5	99	972	59.5	100	970
13	60.5	100	962	61.5	98	964	70.0	80	967	65.0	71	969
14	56.0	95	971	62.0	98	972	67.0	76	972	62.5	92	971
15	59.5	92	973	60.5	95	976	70.0	73	979	69.0	73	980

Solution Maximization of precipitation to the 12-hour persistent dew point of 74°F, considered appropriate for July in Ontario, Canada, is achieved as follows:

1. From given atmospheric conditions, the dew-point temperature is determined for the period of every 6 hours. From these dew-point values, a maximum value of the dew point, which must persist for at least 12 hours, is selected as follows:
   (a) Determine the saturation vapor-pressure curve.
   (b) Determine the saturated vapor pressure corresponding to each value of the given air temperature.
   (c) The saturation vapor-pressure curve is shown in Figure 26.13.

TABLE 26.17 RESULTS OF CALCULATIONS FOR DEW-POINT TEMPERATURE

Time	Temp. (°F)	RH (%)	Pressure (mb)	Saturation vapor pressure (mb)	Vapor pressure (mb)	Dew-point temp. (°F)
July 9						
0100	62.0	100	974	18.8	18.8	62.0
0700	64.5	88	974	20.5	18.03	61.0
1300	80.0	61	974	34.61	21.15	65.3
1900	76.0	68	973	30.40	20.65	64.8
July 10						
0100	58.5	93	974	16.55	15.4	56.3
0700	64.0	81	975	20.1	16.3	58.0
1300	83.0	63	975	38.20	24.05	69.0
1900	77.5	75	974	32.0	24.0	68.0
July 11						
0100	55.0	96	975	14.63	14.06	54.0
0700	62.5	85	975	19.15	16.28	58.0
1300	81.5	69	974	36.40	25.10	70.4
1900	71.0	60	974	25.35	20.40	64.5
July 12						
0100	62.5	93	974	19.15	17.80	60.2
0700	60.5	100	973	17.55	17.51	60.0
1300	60.5	99	972	17.53	17.37	60.0
1900	59.5	100	970	17.10	17.20	59.5
July 13						
0100	59.5	100	962	17.53	17.53	60.5
0700	61.5	98	964	11.50	18.11	61.2
1300	70.0	80	967	24.79	19.81	64.2
1900	65.0	71	969	20.86	14.81	55.5
July 14						
0100	56	95	971	15.20	14.42	54.8
0700	62	98	972	18.80	18.41	61.5
1300	67	76	972	22.40	14.90	59.0
1900	62.5	92	971	19.15	17.07	60.0
July 15						
0100	59.5	92	973	17.20	15.81	57.0
0700	60.5	95	976	17.53	16.68	59.5
1300	70.0	73	929	24.79	18.10	61.0
1900	69.0	73	980	23.90	17.48	60.0

(d) Determine the vapor pressure at the given temperature.
(e) Making use of the psychrometric chart, enter the values of the dry-bulb temperature and relative humidity, and determine the dew-point temperature.
(f) The dew-point temperature can also be found using an appropriate formula.
(g) The results of computations are shown in Table 26.17. The maximum value of dew point persisting for 12 hours is 68°F.
(h) The maximum dew-point value is reduced to the 1000-millibar pressure level. In fact, all the dew-point values should be reduced to the 1000-millibar pressure level before selecting the maximum dew-point value persisting for 12 hours. Because the pressure variation is very small over the area, the maximum value of the dew point is chosen, assuming that this value would remain maximum even after being reduced to the 1000-millibar pressure level. This is done for purposes of simplicity.

1. The dew-point values of 68°F and 74°F are reduced to the 1000-millibar pressure level. Assume the process to be adiabatic. $r = 1.4$ for air.

$$T_1 = 68 + 460 = 528°\text{F}$$

$$P_1 = 974$$

$$P_2 = 1000$$

By using Equation (26.13),

$$T_2 = 528\left(\frac{974}{1000}\right)^{(1-1.4)/1.4} = 528 \times 1.0075 = 531.96°\text{F absolute}$$

$$T_2 = 531.96 - 460 = 71.96°\text{F}$$

Thus, the dew-point temperature = 71.96°F. Similarly, the given dew-point temperature of 74°F is reduced to the 1000-millibar pressure level. The air pressure corresponding to this value is 974 millibars. The dew-point temperature after reduction is 78°F.

2. Moisture adjustment is based on the highest topographic barrier and the 200-millibar base pressure. The highest topographic barrier in this case is the 1200-ft elevation. Hence, maximization of the precipitation is calculated as follows:

Two sets of conditions are available. Condition A consists of 12-hour persistent dew-point temperature = 68°F; 68°F reduced to the 1000-millibar pressure = 71.96°F, ≅ 72°F; 68°F dew-point temperature corresponds to a pressure level = 974 millibars; and the elevation of the watershed = 1200 ft above the mean sea level.

Condition B consists of 12-hour persistent dew-point temperature = 74°F; 74°F reduced to the 1000-millibar pressure level = 78°F; and the elevation = 1200 ft above the mean sea level. Figure 26.11 shows curves for maximum precipitable water at given conditions of dew-point temperature, elevation, and pressure level. From these curves, the values of precipitable water are obtained.

Condition A

(a) Storm precipitable water (1000-millibar to 200-millibar pressure level) at 71.96°F = 2.55 in.
(b) Storm precipitable water (1000-millibar pressure level to 1200-ft elevation) at 71.96°F = 0.22 in.

Therefore, the net precipitable water = (a) − (b) = 2.55 − 0.22 = 3.33 in.

Condition B

(a) Maximum storm precipitable water (1000-millibar to 200-millibar pressure level at 78°F dew-point temperature = 3.25 in.

(b) Maximum storm precipitable water (1000-millibar pressure level to 1200-ft. elevation) at 78°F dew-point temperature = 0.28 in.

Therefore, the net precipitable water = 3.25 − 0.28 = 2.97 in. Hence, maximization factor = 2.97/2.33 = 1.27. In this manner, storm precipitation can be maximized under a given set of conditions.

The average maximum precipitation for 6, 12, and 24 hours, calculated in Example 26.1, can be maximized by multiplying the values of the average maximum depth over the watershed area for a given duration with the maximization factor; for example, the maximum average depth over the watershed area for 6 hours = 0.78 in.; the maximum precipitation depth for that duration = 0.78 × 1.27 = 0.99 in. The maximized precipitation values are as follows:

S.N.	Area of watershed	Avg. max. depth (in.)	Duration (h)	Maximization factor	Maximized precip. depth (in.)
1	520	0.78	6	1.27	0.99
2	520	1.17	12	1.27	1.49
3	520	1.59	24	1.27	2.02

EXAMPLE 26.6 Considering precipitation from the storm of Example 26.5 further east (see daily values in the monthly record), compute the appropriate 24-hour maximum probable precipitation on the watershed using storm transposition. Use may be made of a road map for plotting isohyets.

Solution Consider precipitation from the same storm further east. An approximate 24-hour maximum probable precipitation on the watershed of the transposed storm is calculated as follows:

1. Take only that portion of the road map that contains a watershed area, south branch and eastern side of the watershed catch (in this case, toward Toronto, Oshawa, and Lindsay).
2. From the monthly records, select rainfall values of 24-hour duration for the month of July 1964. These values should be maximum. The 24-hour rainfall values are from 8:00 A.M., July 12, to 8:00 A.M., July 13. The rainfall values for this period are found for all the stations, as shown in the areal map of Figure 26.14.
3. From the 24-hour precipitation values, the isohyets are drawn, starting with 2 in. of rainfall to 3.5 in. of rainfall at intervals of 0.25 in. The isohyetal map is shown in Figure 26.15.
4. On this isohyetal map, the watershed is located, using trial and error, in such a way that the maximum precipitation is enclosed within the watershed area. This is shown in Figure 26.15. After two or three trials, the watershed location is shifted onto the sites corresponding to the maximum precipitation from the storm over the watershed. There appears to be two sites that can give maximum probable precipitation. One site appears to be toward 44°N. Another site appears to be

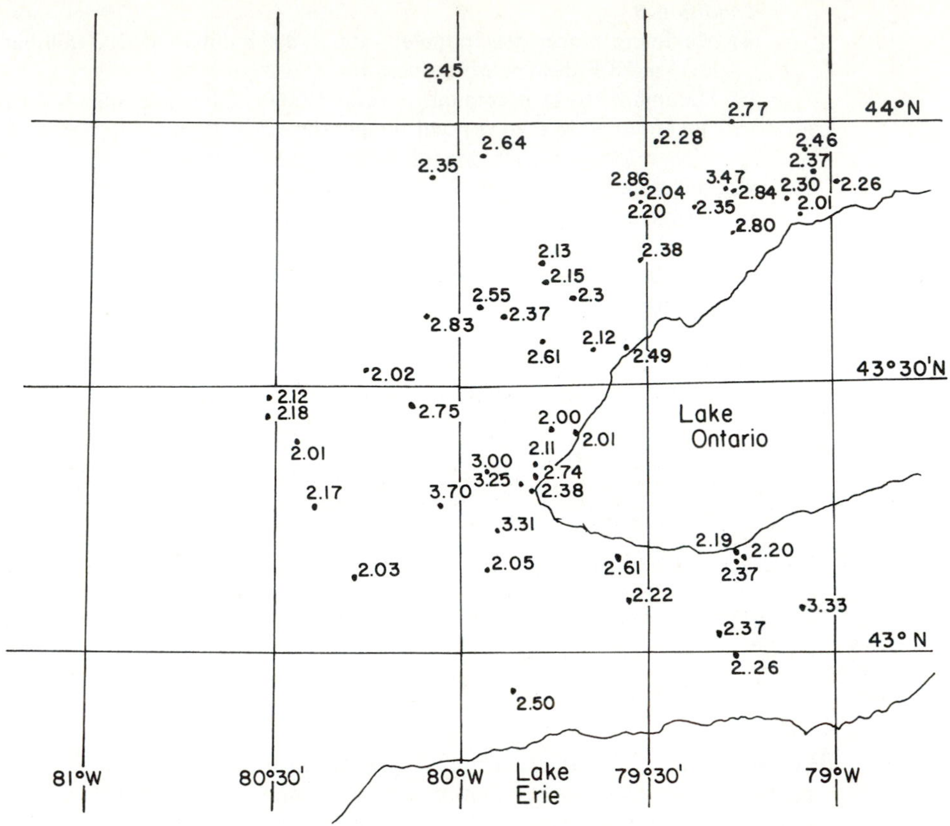

Figure 26.14 Areal map showing precipitation records on July 12, 1964.

between 43°N and 43°30′N. By trial and error, it is found that the second site yields the probable maximum precipitation and, hence, the watershed is relocated to this site so as to trap the maximum amount of precipitation. The transposed location of the watershed is depicted by the superimposed boundaries of the drainage basin.

5. The average maximum precipitation is then determined for the watershed location by the isohyetal method. Calculations are indicated in Table 26.18. The maximum depth of 24-hour precipitation by virtue of the transposed storm over the watershed is found to be 2.82 in.
6. The 24-hour PMP of the transposed storm over the watershed is determined. Transposition should also consider the effects of changes in elevation, slope, topography, and atmospheric conditions, for the watershed has been shifted to a new location. This means that the probable maximum depth of 24-hour precipitation must be corrected for these changes.
7. From storm transposition calculations, it is seen that variations between the conditions of the given site of the watershed and the conditions of the newly transposed site are minor. Transposition does not seem to have much affect in this particular case. This case can be characterized as a case of maximization only, and not that of transposition.

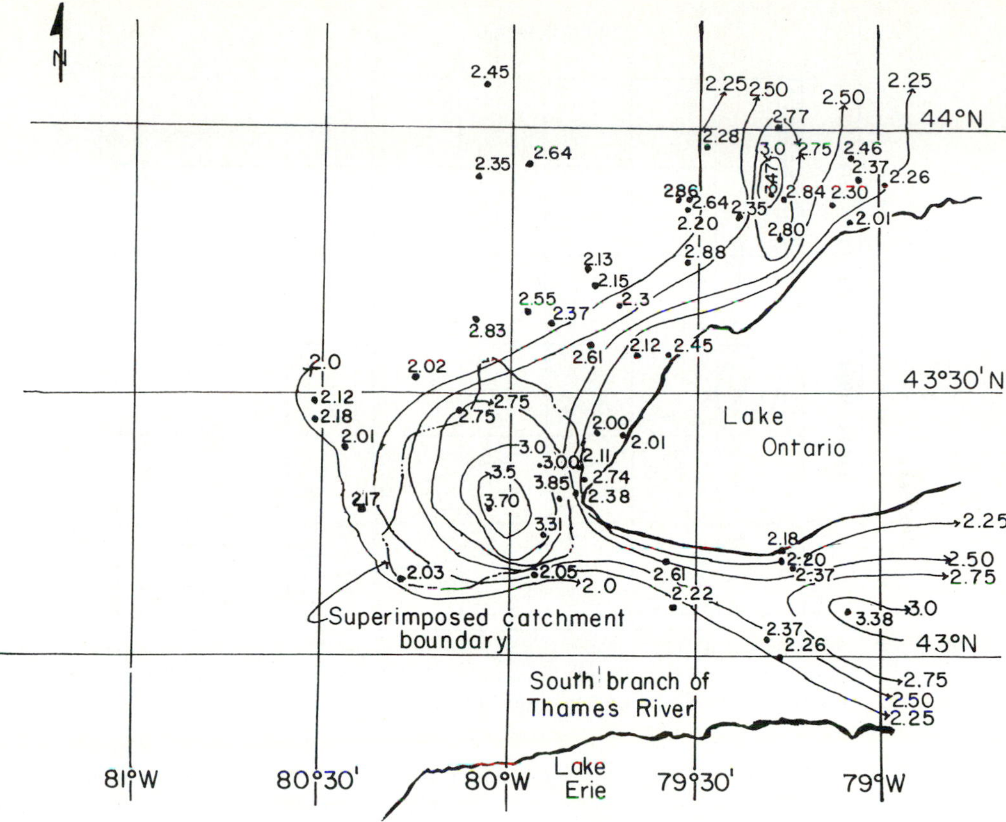

Figure 26.15 Isohyetal map for 24-hour rainfall of July 12, 1964.

Storm transposition over the watershed involves two sites. Site A consists of the old location of the watershed transposition of the storm: dew-point temperature = 68°F; reduced to 1000 millibars = 71.96°F; elevation = 1200 ft; atmospheric pressure corresponding to 68°F = 974 millibars. Site B consists of the

TABLE 26.18 COMPUTATIONS FOR MAXIMUM PROBABLE PRECIPITATION (24 h)

S.N. (1)	Isohyetal line (2)	Total area (mi^2) (3)	Net area enclosed, a (4)	Mean precip., $\bar{p}$ (5)	Precip. volume, $a \times \bar{p}$ (6) = (4) × (5)
1	>3.5	44.40	44.40	3.60	163.0
2	3.0	143.50	99.10	3.25	321.0
3	2.75	292.50	149.0	2.875	433.3
4	2.50	415.50	123.0	2.63	322.5
5	2.25	490.5	75.0	2.38	179.0
6	2.0	520.0	27.5	2.13	59.3

Σ precipitation volume = 1478.9

Average maximum 24-h precipitation = 1478.1/520 = 2.82 in.

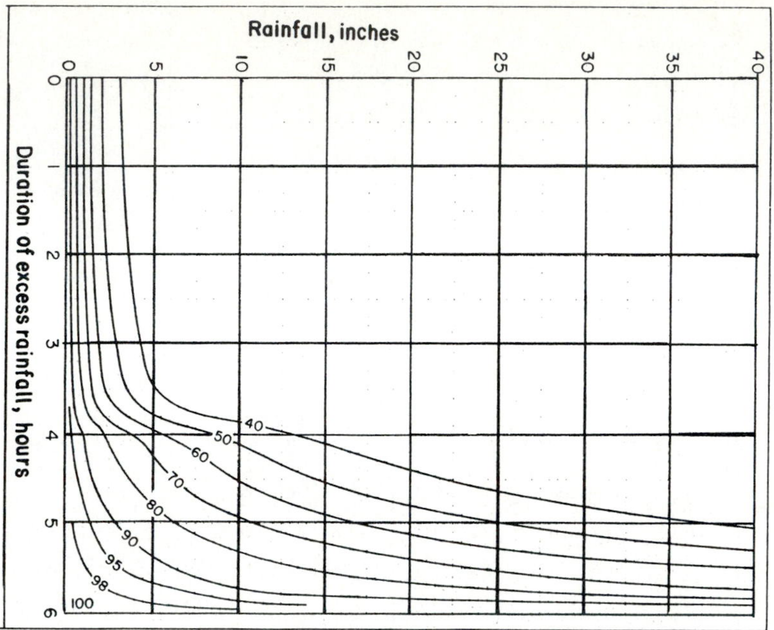

Figure 26.16 Duration of effective rainfall for a range of runoff curve numbers (after Soil Conservation Service, 1971).

new location of the watershed: dew-point temperature = 68°F; reduced to 1000 millibars = 71.96°F; elevation = 1200 ft approximately; atmospheric pressure corresponding to 68°F = 974 millibars.

From this data, we observe that even after transposition of the storm over the watershed, the conditions do not change significantly. The effect of transposition is insignificant in this case. The probable maximum precipitation over the watershed of the transposed storm is 2.82 in.

26.4.4 Duration of Effective Rainfall

If the storm duration is denoted as D, then the duration of the effective rainfall $D_0 < D$, for the initial rainfall is usually abstracted by infiltration, interception, depression storage, and so on. The Soil Conservation Service (1971) has presented a family of curves to estimate D_0 for a 6-hour storm as a function of curve number, CN, and storm precipitation amount, P, as shown in Figure 26.16. When CN = 100, $D_0 = D = 6$ h, for there is no abstraction.

To determine D_0 for any storm of duration greater than 6 h, the following steps are involved: (1) Determine the amount of rainfall P_* prior to the effective rainfall as a function of CN. The SCS has developed a table of P_* values, as given in Table 26.19. (2) Compute the rainfall ratio, P^*/P. (3) Obtain the time ratio, D_0/D. The values of this ratio are given in Table 26.20. (4) Multiply the time ratio by the rainfall duration to obtain D_0.

EXAMPLE 26.7 A 10-hour storm produced 5 in. of rainfall on a small mixed-use watershed whose curve number is 70. Determine the duration of effective rainfall of this storm.

Solution

$$\text{CN} = 70$$

$$P_* \text{ from } \textit{Table } 26.19 = 0.86 \text{ in.}$$

$$P^*/P = \frac{0.86}{5.0} = 0.172$$

$$D_0/D \text{ from Table } 26.20 = 0.717$$

Therefore, $D_0 = 0.717 \times 10 = 7.17$ h

26.4.5 Time Distribution

A representative time distribution or hyetograph is selected for the storm depth and duration estimated before. This distribution is needed for estimating the design flood hydrograph. When a watershed does not have a network of recording rain gages, rainfall data from a nearby network located in a meteorologically homogeneous area can be used to establish the time distribution of the design storm. It can be based on an analysis of recorded storm distribution patterns or the worst-possible storm pattern. The time distribution significantly affects the shape and peak of the resulting discharge hydrograph.

Derivation by Analysis of Records. The following steps are involved: (1) Select all storms whose 24-hour totals exceed 150 mm and 48-hour totals exceed 200 mm. Depending upon the climatic region, the base level for storm totals may vary

TABLE 26.19 RAINFALL PRIOR TO EXCESS OR EFFECTIVE RAINFALL (AFTER SOIL CONSERVATION SERVICE, 1971)

CN	*P** (in.)	CN	*P** (in.)	CN	*P** (in.)	CN	*P** (in.)	CN	*P** (in.)
100	0.00	86	0.33	72	0.78	58	1.45	44	2.54
99	0.02	85	0.35	71	0.82	57	1.51	43	2.64
98	0.04	84	0.38	70	0.86	56	1.57	42	2.76
97	0.06	83	0.41	69	0.90	55	1.64	41	2.88
96	0.08	82	0.44	68	0.94	53	1.70	40	3.00
95	0.11	81	0.47	67	0.98	53	1.77	39	3.12
94	0.13	80	0.50	66	1.03	52	1.85	38	3.26
93	0.15	79	0.53	65	1.08	51	1.92	37	3.40
92	0.17	78	0.56	64	1.12	50	2.00	36	3.56
91	0.20	77	0.60	63	1.17	49	2.08	35	3.72
90	0.22	76	0.63	62	1.23	48	2.16	34	3.88
89	0.25	75	0.67	61	1.28	47	2.26	33	4.06
88	0.27	74	0.70	60	1.33	46	2.34	32	4.24
87	0.30	73	0.74	59	1.39	45	2.44	31	4.44

TABLE 26.20 RAINFALL AND TIME RATIOS FOR DETERMINING D_0 WHEN THE STORM DURATION IS GREATER THAN 6 HOURS (AFTER SOIL CONSERVATION SERVICE, 1971)

Rainfall Ratio	Time Ratio	Rainfall Ratio	Time Ratio	Rainfall Ratio	Time Ratio	Rainfall Ratio	Time Ratio
0.000	1.000	0.070	0.852	0.140	0.746	0.210	0.684
0.002	0.995	0.072	0.848	0.142	0.744	0.212	0.682
0.004	0.990	0.074	0.844	0.144	0.742	0.214	0.680
0.006	0.985	0.076	0.841	0.146	0.740	0.216	0.679
0.008	0.981	0.078	0.837	0.148	0.739	0.218	0.677
0.010	0.976	0.080	0.833	0.150	0.737	0.220	0.675
0.012	0.971	0.082	0.830	0.152	0.735	0.222	0.673
0.014	0.967	0.084	0.827	0.154	0.733	0.224	0.672
0.016	0.962	0.086	0.824	0.156	0.732	0.226	0.670
0.018	0.957	0.088	0.821	0.158	0.730	0.228	0.668
0.020	0.952	0.090	0.818	0.160	0.728	0.230	0.667
0.022	0.948	0.092	0.815	0.162	0.726	0.232	0.666
0.024	0.943	0.094	0.812	0.164	0.724	0.234	0.666
0.026	0.938	0.096	0.809	0.166	0.723	0.236	0.665
0.028	0.933	0.098	0.806	0.168	0.721	0.238	0.665
0.030	0.929	0.100	0.803	0.170	0.719	0.240	0.664
0.032	0.924	0.102	0.800	0.172	0.717		
0.034	0.919	0.104	0.797	0.174	0.716	(Change in tabulation increment)	
0.036	0.915	0.106	0.794	0.176	0.714		
0.038	0.911	0.108	0.791	0.178	0.712		
0.040	0.908	0.110	0.788	0.180	0.710	0.250	0.662
0.042	0.904	0.112	0.785	0.182	0.709	0.300	0.651
0.044	0.900	0.114	0.782	0.184	0.707	0.350	0.640
0.046	0.896	0.116	0.779	0.186	0.705	0.400	0.628
0.048	0.893	0.118	0.776	0.188	0.703	0.450	0.617
0.050	0.889	0.120	0.773	0.190	0.702	0.500	0.606
0.052	0.885	0.122	0.770	0.192	0.700	0.550	0.595
0.054	0.882	0.124	0.767	0.194	0.698	0.600	0.583
0.056	0.878	0.126	0.764	0.196	0.696	0.650	0.542
0.058	0.874	0.128	0.761	0.198	0.695	0.700	0.500
0.060	0.870	0.130	0.758	0.200	0.693	0.750	0.447
0.062	0.867	0.132	0.755	0.202	0.691	0.800	0.386
0.064	0.863	0.134	0.751	0.204	0.689	0.850	0.310
0.066	0.859	0.136	0.749	0.206	0.687	0.900	0.220
0.068	0.856	0.138	0.747	0.208	0.686	0.950	0.116

from one watershed to the other. (2) Compute the hourly rainfall totals for 1, 2, 3, 6, 9, 12, 15, 18, 21, 24, . . . , 48 hours using consecutive hourly rainfall values. (3) Divide the rainfall totals by the total rainfall amount of 24-hour or 48-hour duration, and multiply the ratios by 100. (4) Plot the percentages of different durations on graph paper. The plot should monotonically increase. (5) Repeat steps (2) to (4) for each selected storm. Use the same graph paper for plotting; use the same graph paper for the 24-hour duration and another for the 48-hour duration or a different one. (6) Draw an enveloping curve through the maximum percentages of each duration from amongst the different storms.

Huff's Distributions. Huff (1967) analyzed storm patterns for midwestern watersheds. The storms were divided into four equal probability groups: first quartile, second quartile, third quartile, and fourth quartile. The first quartile represented the most severe storms, the fourth quartile represented the mildest storms, and the second and third quartiles represented storms in order of decreasing severity between these two limits. The first and second quartile storms made up 66% of the total number of storms analyzed.

Huff (1967) constructed the time distribution of the first quartile storms as well as for other quartile storms for various probability levels. For the first and second quartile storms, the time distributions are shown in Figures 26.17 and 26.18. An 80% probability level in the figures corresponds to the distribution occurring in 20% or less of the storms. On the other hand, the 20% level is the distribution occurring in 80% or less of the storms.

Huff also constructed histograms for different quartile storms. Figure 26.19 shows selected histograms for first quartile storms for three probability levels. The bulk of the total rainfall (nearly 80%) occurs in the first 20% of storm duration for the 10% probability histogram. This type of storm may be typical of thunderstorms. The 90% probability-level histogram represents more or less a steady rainstorm or a chain of rainfall storms. The 50% probability level or medium histogram is used most commonly. The median time distribution of the first quartile storms is shown in Figure 26.20.

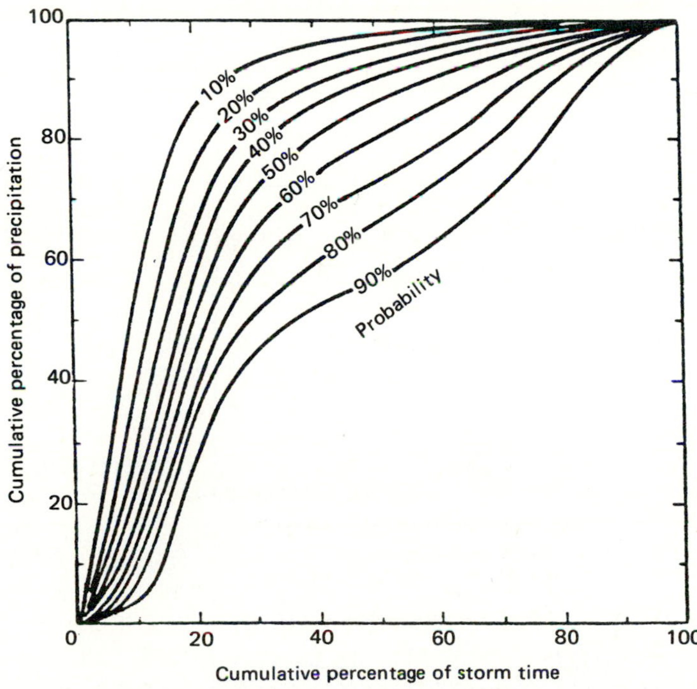

Figure 26.17 The distribution of first-quartile storms (after Huff, 1967).

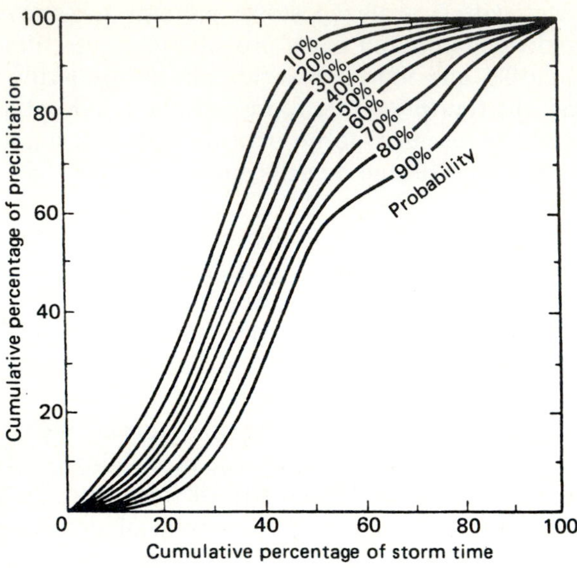

Figure 26.18 Time distribution of second-quartile storms (after Huff, 1967).

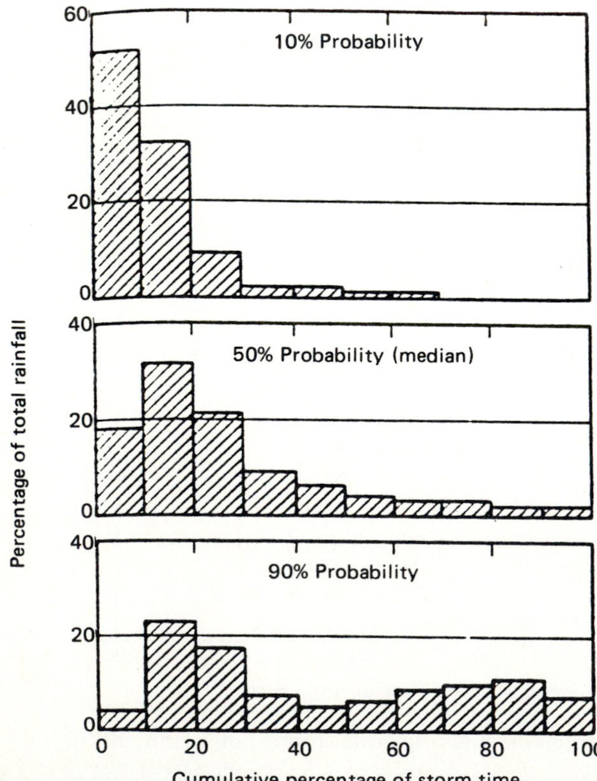

Figure 26.19 Selected histographs for first-quartile storms (after Huff, 1967).

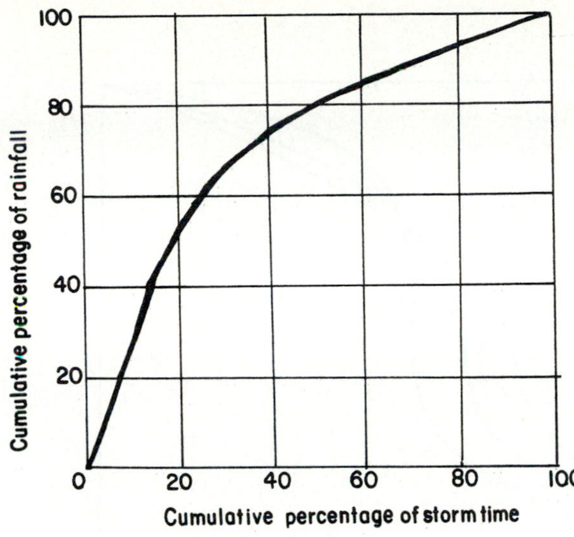

Figure 26.20 Time distribution of storm rainfall, median first-quartile curve for point rainfall (after Huff, 1967).

SCS Distributions. The Soil Conservation Service (1975) uses a uniform distribution for short-duration storms. It also uses a time distribution for a 6-hour storm, as shown in Figure 26.21. For 24-hour storms, the SCS developed four types of distributions designated as type I, IA, II, and III, as shown in Figure 26.22. The United States was divided in four regions, as shown in Figure 26.23. Different distribution curves are applicable to different regions.

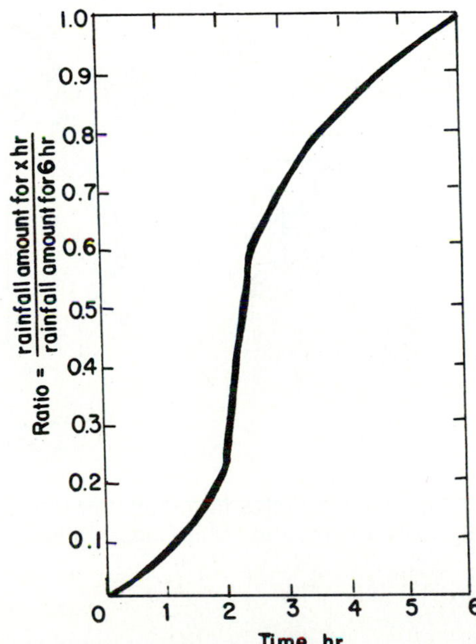

Figure 26.21 A 6-hour design storm distribution (after Soil Conservation Service, 1975).

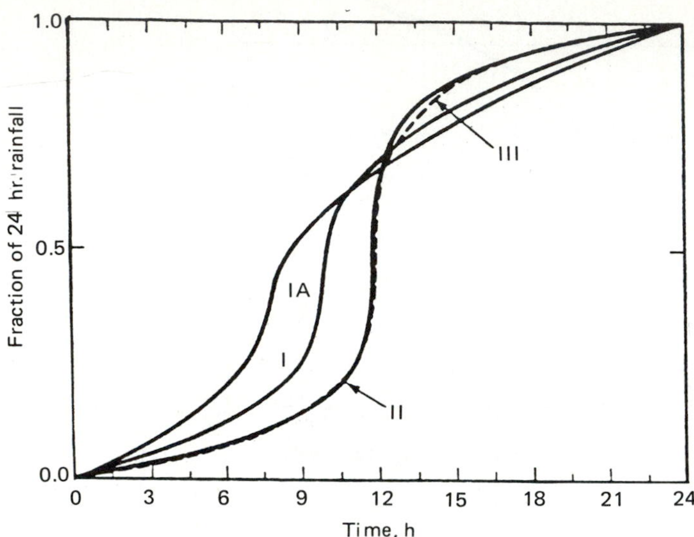

Figure 26.22 The 24-hour rainfall durations for zones I, IA, II, and III (after Soil Conservation Service, 1986).

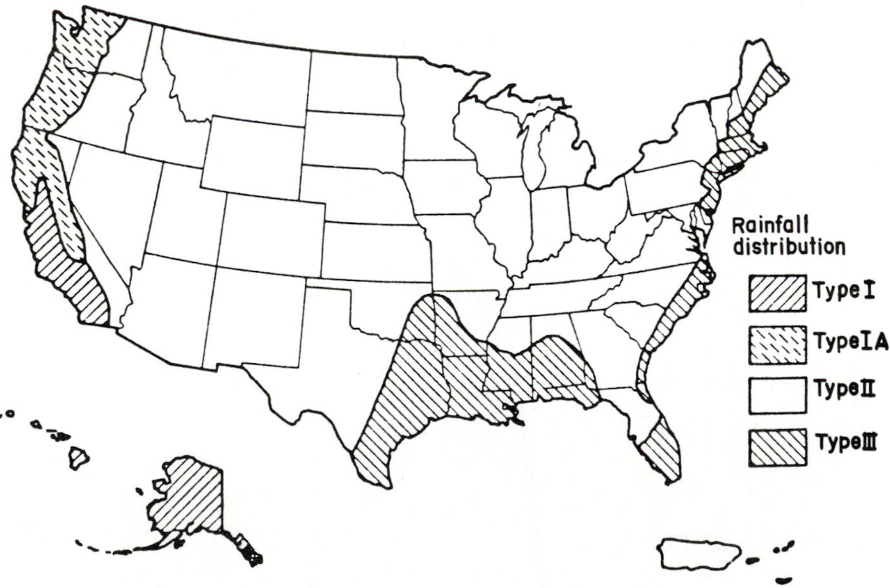

Figure 26.23 Division of the United States into four zones for application of the SCS rainfall distributions (after Soil Conservation Service, 1986).

26.4.6 Critical Storm Distribution

Several high-intensity bursts of 3 or 6 hours can occur within a storm duration of 48 or 72 hours. Some of these bursts can occur in succession. For a 48-hour distribution, the 6-hour time increments can be averaged to compute the critical time distribution as follows: (1) Arrange the 6-hour rainfall increments in descending order. (2) Select the four largest 6-hour rainfall increments and make a 24-hour sequence by grouping them. Keep the second largest increment next to the right of the largest. Put the third largest before the largest value, and put the remaining fourth on either end. (3) Arrange the next four largest increments (numbers 5, 6, 7, and 8) in the same way as in step (2) to make another 24-hour sequence. (4) Form a 48-hour critical distribution by placing the 24-hour sequence of step (3) to the right of the 24-hour sequence of step (2).

EXAMPLE 26.8 Determine the time distribution (or hyetograph) of a 5-cm rainfall resulting from a 6-hour storm. Use the SCS method.

Solution Determine the ratio of the rainfall amount for a given hour to the total rainfall amount from Figure 26.21. Then compute the rainfall amount for that hour. Calculations are given as follows:

TIME	RATIO	CUMULATIVE RAINFALL AMOUNT (cm)	INCREMENTAL RAINFALL AMOUNT (cm)	RAINFALL INTENSITY (cm/h)
0	0	0	0	
				0.41
1	0.082	0.41	0.41	
				0.715
2	0.225	1.125	0.715	
				2.35
3	0.695	3.475	2.35	
				0.70
4	0.835	4.175	0.70	
				0.45
5	0.925	4.625	0.45	
				0.375
6	1.0	5.00	0.375	
				0.0

26.4.7 Spatial Distribution

Spatial variability of design storm rainfall is usually considered when designing major water resources projects which involve an estimate of PMP or SPS.

U.S. ARMY CORPS OF ENGINEERS' METHOD

The U.S. Army Corps of Engineers (1965) uses an SPS isohyetal pattern for a 96-hour storm as shown in Figure 26.24. This pattern is somewhat elliptical in shape, may be oriented in any appropriate direction and may correspond to the depth-area

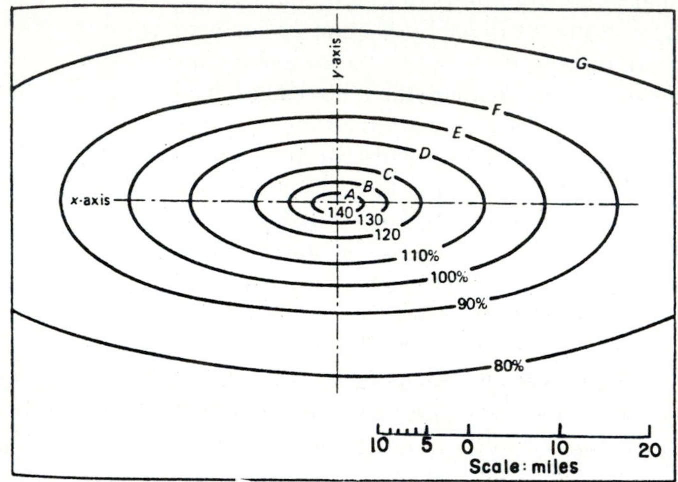

Figure 26.24 General SPS isohyetal pattern for a 96-hour storm (after U.S. Army Corps of Engineers, 1965).

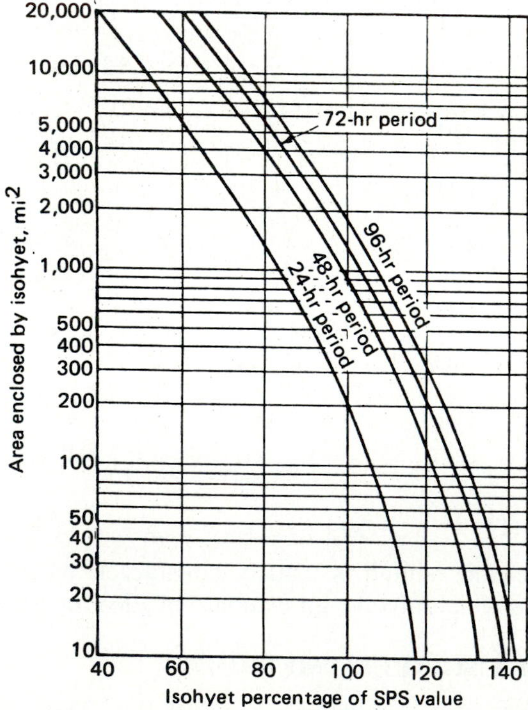

Figure 26.25 The SPS DAD curves by 24-hour increments (after U.S. Army Corps of Engineers, 1965).

relation represented by a 96 h storm. The isohyets, designated as A, B, C, D, E, F and G, encompass a watershed area as:

ISOHYETS

	A	B	C	D	E	F	G
Area (mi^2)	16	100	320	800	1800	3700	7100

These isohyets, as indicated in Figure 26.24, also correspond to certain percentages which distribute a 96 h SPS storm elliptically without altering the average storm depth. For example, let us say that a 96-hour storm is known and its depth is 40 in. (102 cm). Then the isohyet A will correspond to 56 in. (142 cm), B to 52 in. (132 cm), C to 48 in. (122 cm), D to 44 in. (112 cm), E to 40 in. (102 cm), F to 36 in. (91 cm), and G to 32 in. (81 cm) of rainfall depth. In this manner, the 96 h SPS storm depth of 40 in. (102 cm) is distributed in an elliptical pattern.

If the storm duration is less than 96 hours and is 72, 48 or 24 hours, then the isohyetal patterns of these duration storms can be developed by making use of the DAD curves shown in Figure 26.25. These DAD curves involve isohyetal depths as percentages of the index storm depth. For a given storm, the isohyetal depths in percent corresponding to isohyetal areas in Figure 26.24 are obtained from Figure 26.25. For example, for a 24-hour storm, the isohyet B encloses an area of 100 mi^2, its corresponding isohyet percentage is approximately 106 from Figure 26.24, and not 130 as used with a 96-hour storm. If a storm duration is not included in the DAD curves, then calculations can be done using linear interpolation.

TVA METHOD

The Tennessee Valley Authority (TVA) uses an elliptical type isohyetal pattern, as shown in Figure 26.26, in the computation of PMP and TVA precipitation for watersheds of 100 to 3000 mi^2. A TVA precipitation is the precipitation resulting from transposition and adjustment (without maximization) of extreme storms that have occurred in the Tennessee Valley. A few of the most important extreme events are excluded from TVA precipitation.

EXAMPLE 26.9 A 96-hour storm produced 80 cm of rainfall on a watershed. The areas enclosed by isohyets A to G are: 10 km^2 by A, 15 km^2 by B, 30 km^2 by C, 70 km^2 by D, 120 km^2 by E, 320 km^2 by F, and 500 km^2 by G. Determine the areal distribution of rainfall using the U.S. Army Corps of Engineers' method.

Solution The rainfall amount of 80 cm is distributed according to percentages assigned to different isohyets.

ISOHYET	PERCENTAGE	AREA ENCLOSED (km^2)	RAINFALL VALUE (cm)
A	140	10	112
B	130	15	104
C	120	30	96
D	110	70	88
E	100	120	80
F	90	320	72
G	80	500	64

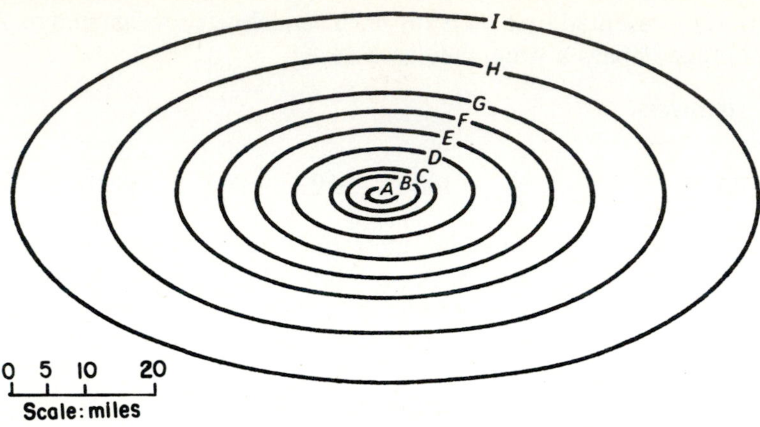

Figure 26.26 Generalized storm pattern (after National Weather Service, 1969).

26.5 PROBABLE MAXIMUM PRECIPITATION

26.5.1 Generalized PMP Estimates

Generalized estimates of PMP for the United States have been mapped by the National Weather Service, U.S. Department of Commerce, as reported in its NOAA Hydrometeorological Report (HMR) Series. For example, for watersheds located east of the 105° meridian, the PMP curves are available in HMR 51 and 52, and for those located west of the 105° meridian, these are presented in HMR 36, 43, and 49 (National Weather Service, 1961, 1966, 1977, 1978, 1982). For the area between the 105° meridian and the Continental Divide, the PMP curves are reported in HMR 55 (National Weather Service, 1984). Figures 26.27 to 26.29 are samples that show all-season PMP for 200-mi^2 (518-km^2) rainfall (National Weather Service, 1978). The generalized estimates are obtained by areal averaging of PMP contours of equal values. Procedures for estimating PMP from areal rainfall values involve three steps: (1) transposition of storms, (2) moisture maximization, and (3) envelopment of the transposed, adjusted storms. The envelopment involves interpolation between precipitation maxima for different durations and areas, and is intended to account for the random occurrence of large storms. These three steps are described in the previous section.

26.5.2 Enveloping Curves

The National Weather Service (1969) has presented an enveloping curve of the world's greatest rainfalls, as shown in Figure 26.30. The probable rainfall depth for an area can be approximated from this curve.

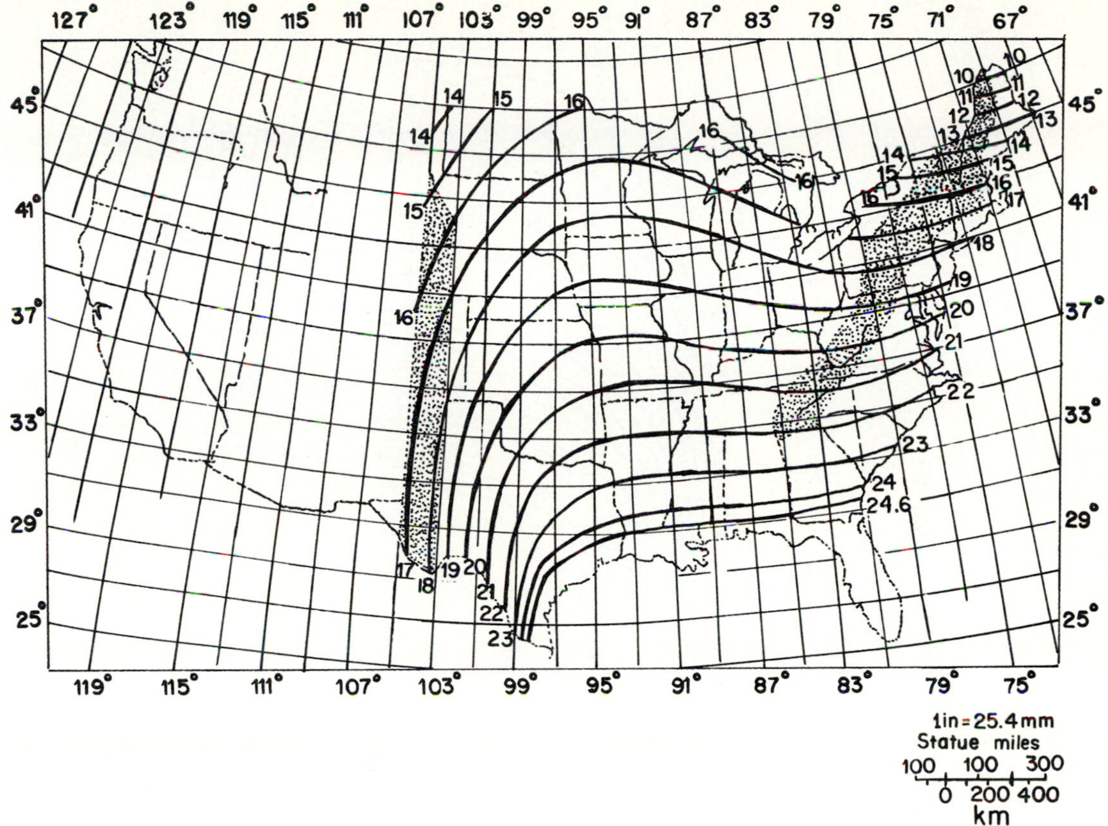

Figure 26.27 All-season PMP (in.) for 6-hour 200-mi^p (515 km^p) rainfall (after National Weather Service, 1978).

26.5.3 Statistical Method

The statistical frequency equation (as discussed in Chapter 24) can be used to estimate the PMP as

$$P_M = \bar{P} + K_p S_p \tag{26.20}$$

where P_M is the maximum rainfall of a given duration D in hours for a return period T, $\bar{P}$ is the mean of the D-hour annual rainfall maxima over the period of record, K_p is a constant equal to the frequency factor associated with the return period T, and S_p is the standard deviation of the D-hour rainfall maxima. Hershfield (1961) applied this method to 24-hour maximum rainfall data from 2600 stations, 90% of which were located in the United States. His analysis found K_p to be 15. He noted adjustments to the values of $\bar{P}$ and S_p for the record length. However, these adjustments do not alter the results by more than 5 to 10%. In a later study, Hershfield (1965) showed that K_p actually varies with $\bar{P}$ and D, as shown in Figure 26.31. This procedure

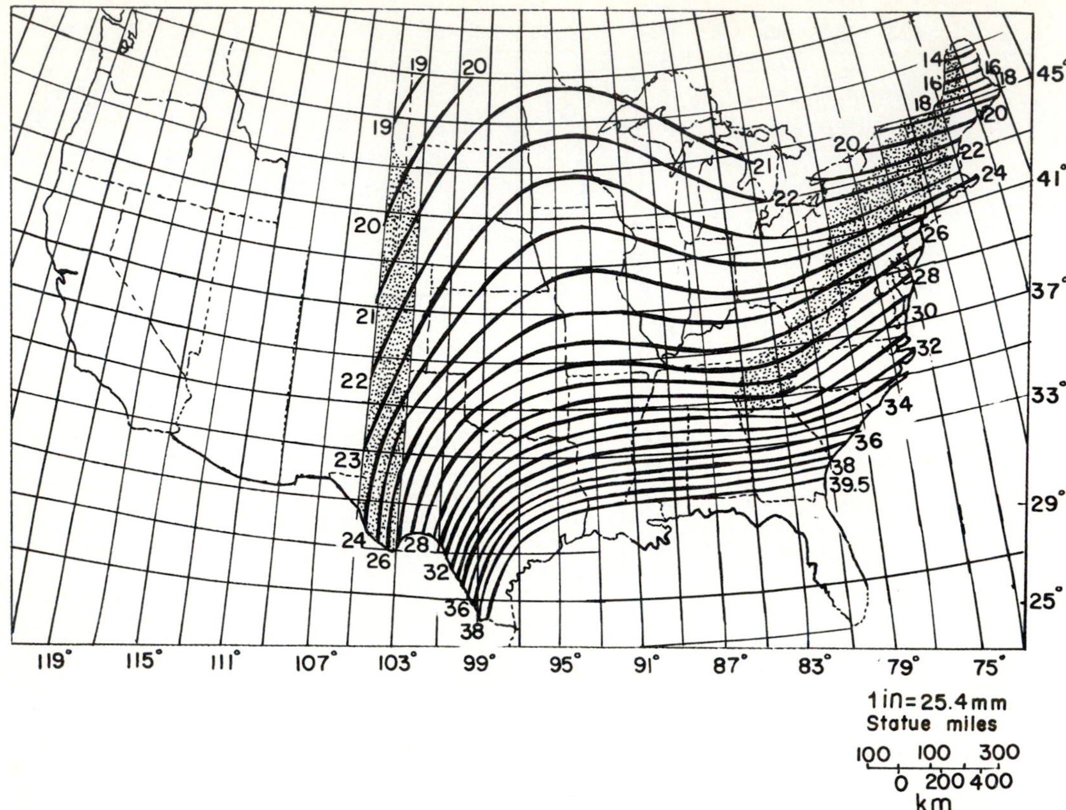

Figure 26.28 All-season PMP (in.) for 24-hour 200-mi^p (515 km^p) rainfall (after National Weather Service, 1978).

yields point estimates of PMP, which can then be converted to areal estimates by the use of a suitable depth–area relation.

26.6 STANDARD PROJECT STORM (SPS)

The SPS, like PMP, is used in design of large water-resources projects. Unlike PMP, an SPS value can be obtained from a survey of severe storms that have occurred in the watershed and its vicinity, and then selecting the storm causing the most severe depth–area–duration relation. The selected storm may be oriented appropriately to produce the SPS. Generally, the SPS rainfall is about half of the PMP value. A reasonable value of SPS can be gotten from critically examining four or five of the largest storms. Alternatively, severe storms from meteorologically homogeneous areas can be transposed to the project watershed and the storms then analyzed to produce the SPS value.

26.7 FREQUENCY-BASED METHOD

In 1985, the National Research Council Committee on Safety Criteria for Dams published its report on dam safety. Two of the recommendations were as follows: (1) The use of PMF can be continued as the general design standard for proposed high-hazard potential dams. (2) For existing high-hazard dams, the design should include an estimate of flood probabilities, expected project performance, and damages resulting from dam failure for a range of floods up to and including the PMF. The second recommendation of a frequency-based approach is in response to the concern over the need and costs to modify existing federally licensed dams so that they safely satisfy the PMF criteria. It represents a marked departure from a long tradition of no-risk design based on PMF.

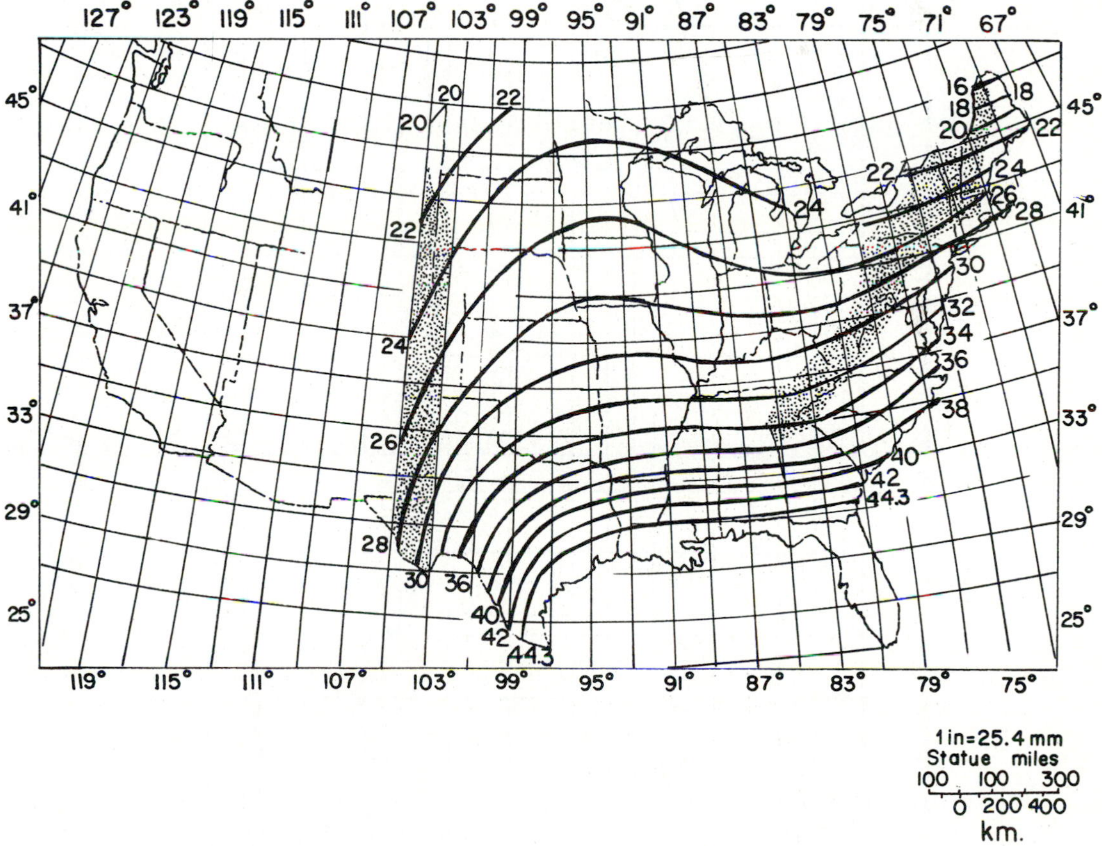

Figure 26.29 All-season PMP (in.) for 48-hour 200-mi^p (515 km^p) rainfall (after National Weather Service, 1978).

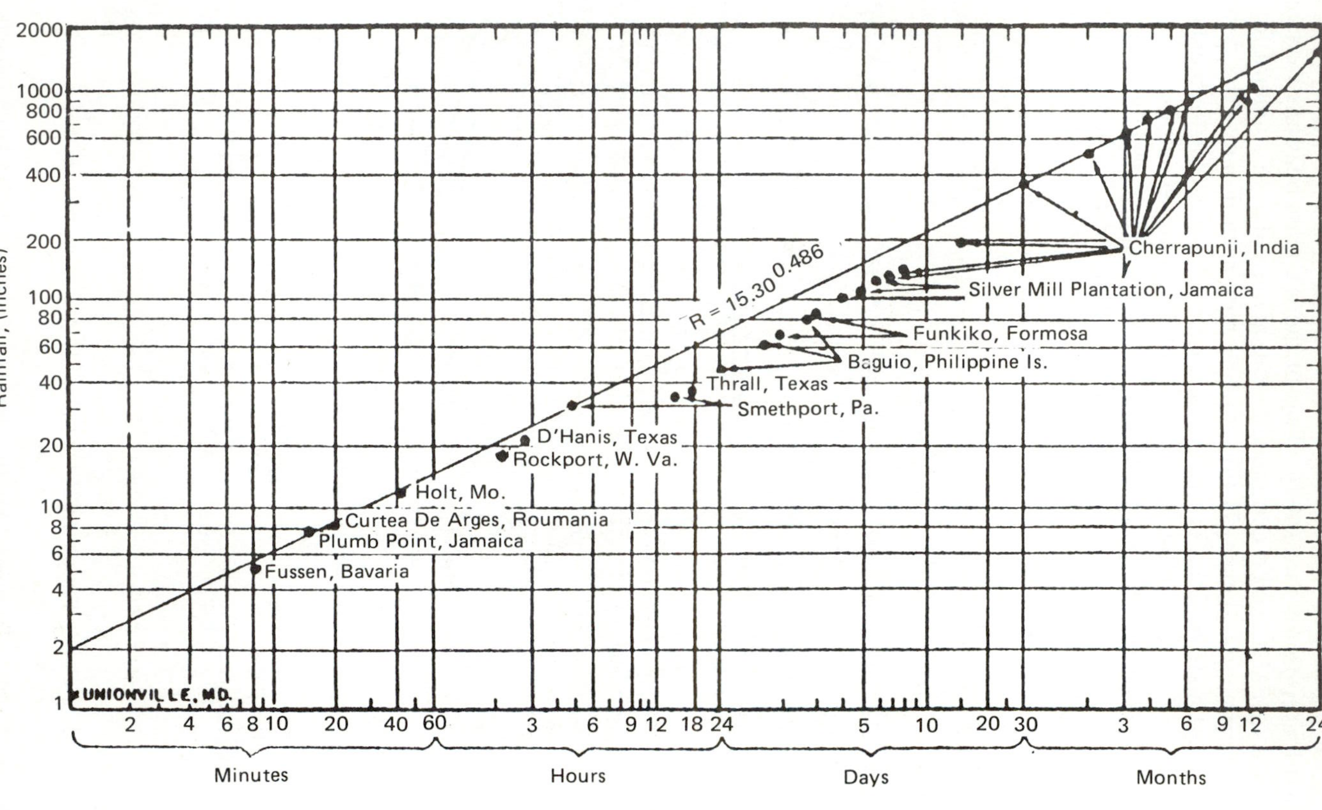

Figure 26.30 Creager curves of world's greatest rainfall (after National Weather Service, 1969).

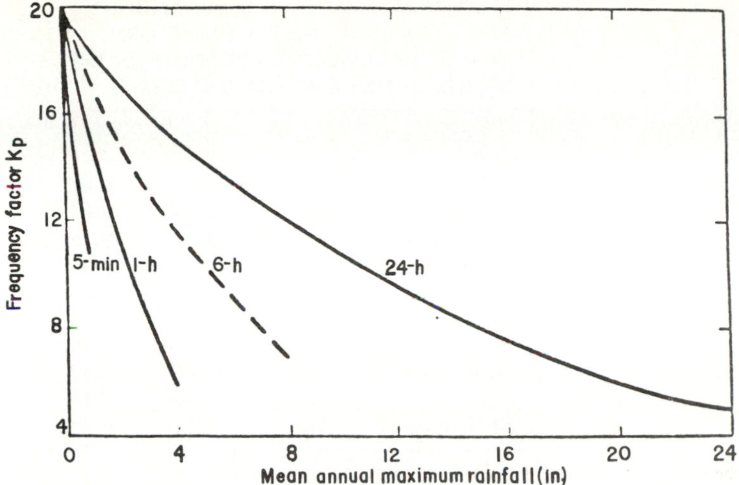

Figure 26.31 Frequency factor as a function of mean annual maximum rainfall and storm duration (after Hershfield, 1965).

26.7.1 Intensity–Duration Frequency (IDF) Analysis

Many water-resources projects such as urban drainage, airport drainage, design of detention storage, etc., are designed on the basis of the rainfall intensity–duration–frequency (IDF) relationship. Sometimes rainfall intensity is replaced by depth and frequency by return period in an IDF relationship. The rainfall intensity is usually the average intensity over the rainfall duration. It is known that the greatest point rainfall intensity occurs during short time intervals. Also, higher rainfall intensities occur less frequently than lower intensities.

A mathematical relation between rainfall intensity and duration, developed by the National Weather Service, is of the form

$$i = \frac{a}{(D^b + c)^n} \tag{26.21}$$

where i is the average intensity; D is the duration; and a, b, and c are coefficients. This equation can be written in logarithmic linear form as

$$\log i = \log a - n \log (D^b + c) \tag{26.22}$$

This intensity–duration relation plots as a straight line (for $b = 1$ and $c = 0$) on log–log paper with n being the slope of the line. Nearly every municipality has prepared a set of such curves for various frequencies as guidelines to be used for hydrologic design within their jurisdiction.

Wenzel (1982) has given values of the coefficients a, b, and c with $n = 1$ from a number of cities in the United States, as shown in Table 26.21, where i is in in./h, and

TABLE 26.21 VALUES OF CONSTANTS a, b AND c FOR n = 1 AND T = 10 yr STORM INTENSITIES AT VARIOUS LOCATIONS IN THE UNITED STATES (AFTER WENZEL, 1982)

Location	a	b	c
Atlanta	97.5	0.83	6.88
Chicago	94.9	0.88	9.04
Cleveland	73.7	0.86	8.25
Denver	96.6	0.97	13.90
Houston	97.4	0.77	4.80
Los Angeles	20.3	0.63	2.06
Miami	124.2	0.81	6.19
New York	78.1	0.82	6.57
Santa Fe	62.5	0.89	9.10
St. Louis	104.7	0.89	9.44

Rainfall in in./h and D in minutes

D is in min. Equation (26.21) is modified with the return period T included as

$$i = \frac{aT^e}{D + c} \quad \text{or} \quad i = \frac{aT^e}{D^b + c} \tag{26.23}$$

where e is a coefficient.

The IDF curves can be constructed using frequency analysis of recorded rainfall data. The following steps are outlined for construction of these curves: (1) Select the duration of rainfall, such as 5, 10, 20, 30, 60, or 100 min. (2) For the duration selected, compute the annual maximum rainfall intensity or depth for each year of record. If there are 30 years of data, then there will be 30 annual maximum depths associated with the selected duration. (3) Fit an appropriate frequency curve to these values. The methods of frequency analysis are described in Chapters 24 and 25. (4) Obtain from the fitted frequency curve the values of rainfall intensity for the selected return periods such as 5, 10, 20, 30, 50, 80, and 100 years. Tabulate these values. (5) Repeat steps (2) and (3) for different durations. (6) Rearrange the data as rainfall intensity against duration for various return periods or frequencies. (7) For a selected frequency, plot on semilog graph paper rainfall intensity values on the ordinate versus duration on the abscissa, and fit a smooth curve or connect the points linearly. Do the same for other frequencies. These plots will be a family of IDF curves, as shown in Figure 26.32 for Omaha, Nebraska.

EXAMPLE 26.10 Compute the 10-year, 30-min design rainfall intensities for Atlanta, Houston, and Miami.

Solution Equation (26.21) is used to compute the 10-year, 30-min design rainfall intensities.

Atlanta: $a = 97.5$, $b = 0.83$, $n = 1$, and $c = 6.88$. Therefore,

$$i = \frac{97.5}{(20)^{0.83} + 6.88} = \frac{97.5}{12.02 + 6.88} = \frac{97.5}{18.9} = 5.16 \text{ in./h}$$

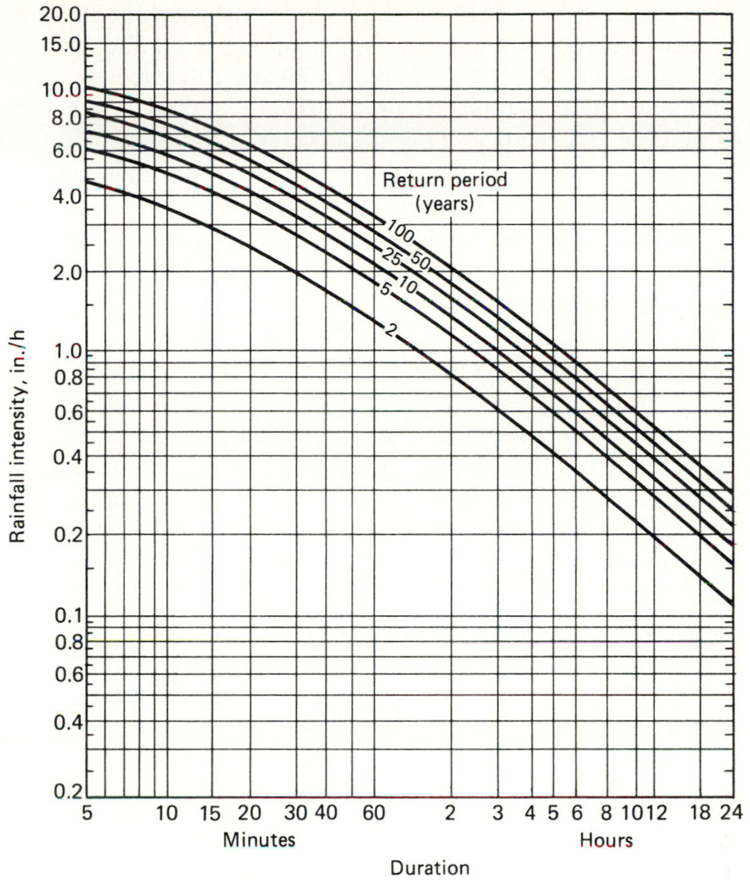

Figure 26.32 Rainfall intensity, duration, and frequency curves for Omaha, Nebraska (after Weather Bureau, 1956).

Houston: $a = 97.4$, $b = 0.77$, $n = 1$, and $c = 4.8$.

$$i = \frac{97.4}{(20)^{0.77} + 4.8} = \frac{97.4}{10.04 + 4.8} = \frac{97.4}{14.84} = 6.56 \text{ in./h}$$

Miami: $a = 124.2$, $b = 0.81$, $n = 1$, and $c = 6.19$.

$$i = \frac{124.2}{(20)^{0.81} + 6.19} = \frac{124.2}{11.32 + 6.19} = \frac{124.2}{17.51} = 7.09 \text{ in./h}$$

26.7.2 Areal Adjustment

A point-rainfall measurement does not accurately represent the mean areal rainfall. Eagleson (1970) has pointed out that the difference between the mean rainfall over an area about the storm center and the point rainfall measured at the storm center increases with a decrease in the total rainfall depth, decreases with increasing duration, is greater for convective and orographic precipitation than for cyclonic storms, and increases with increasing area.

Point-rainfall measurements have been and continue to be utilized in deriving rainfall depth–frequency curves. For example, a number of depth–duration–frequency atlases published by the National Weather Service are based on point-rainfall measurements. For watersheds up to 10 mi², rainfall depths obtained from such atlases may be satisfactory. For larger watersheds, these depths, however, should be adjusted. The National Weather Service (1963) has developed a set of area-depth curves for various storm durations for watersheds up to 400 mi², as shown in Figure 26.33. These curves can be used with frequency curves to transform the point-rainfall depth to areal average depth. Leclerc and Schaake (1972) have expressed the figure algebraically as

$$K = \frac{P_A}{P} = 1 - \exp(-1.1D^{1/4}) + \exp(-1.1D^{1/4} - 0.01A) \tag{26.24}$$

where D is the rainfall duration in hours, and A is the area in square miles. For smaller watersheds, 10 to 100 mi², the Soil Conservation Service (1971) modifies, by using Figure 26.34, the 6-hour storm depths obtained from a frequency atlas. The U.S. Bureau of Reclamation (1965) has developed a curve, as shown in Figure 26.35, to convert a 6-hour point PMP rainfall to a 6-hour areal rainfall for areas west of the 105° meridian.

A multitude of equations have been derived to express mean areal depth of a storm (Court, 1961). A stochastic methodology has been proposed by Rodriguez-Iturbe and Mejia (1974). Simply put, the average rainfall depth over the area for a given duration and return period P_A is a fraction of the point-rainfall value (or mean of the point-rainfall values) within the same area for the same duration and return period P, or

$$P_A = KP, \qquad 0 < K \leq 1 \tag{26.25}$$

where K is the areal reduction factor, varying with the duration and size of the area. Values of K for areas up to 30,000 km² and durations up to 25 days are given by the

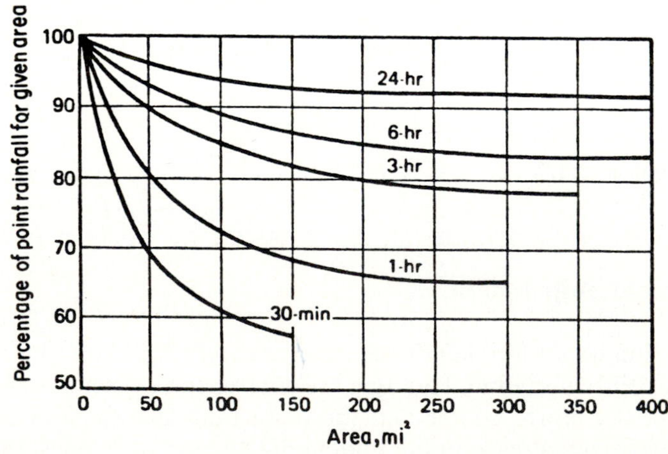

Figure 26.33 Area–depth curves (after National Weather Service, 1963).

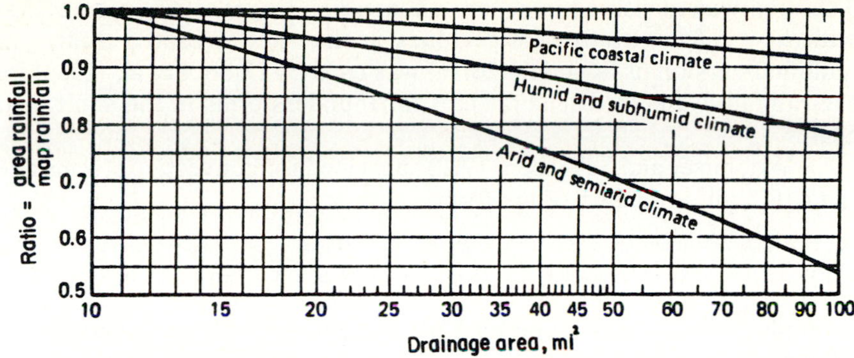

Figure 26.34 Rainfall ratios for 10- to 100-mi^p watersheds (after Soil Conservation Service, 1971).

Natural Environment Research Council (1975). Bell (1976) investigated the adequacy of these K values and found them to be adequate for the United Kingdom.

The thrust of the many equations is to express K as a function of area, and possibly rainfall duration. Horton (1924) expressed storm areal mean depth P_A as

$$P_A = P \exp(-aA^b) \tag{26.26}$$

where P is the maximum point-rainfall depth at the center of the storm, A is the watershed area, and a and b are empirical constants. Dhar and Bhattacharya (1977) found, using the least squares method, the values of a and b for the north India plains as

DURATION	VALUE OF a	VALUE OF b
1 day	0.0016	0.6614
2 day	0.0018	0.6306
3 day	0.003	0.5691

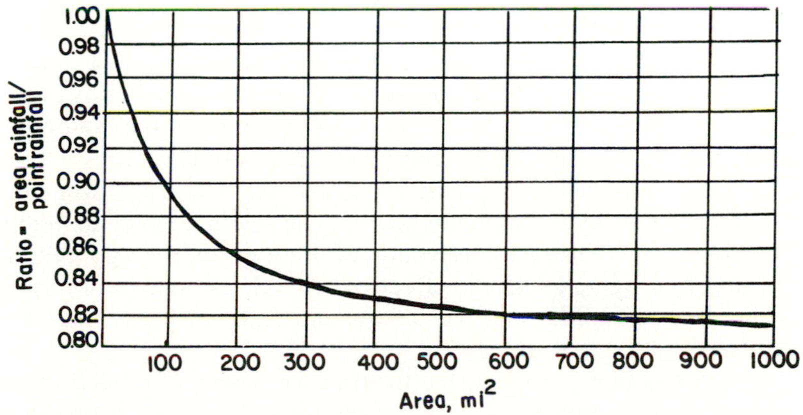

Figure 26.35 Area–depth curve for 6-hour PMP for areas west of the 105° meridian (U.S. Bureau of Reclamation, 1965).

The area was measured in square kilometers. Boyer (1957) used a similar relation with $b = 1$ for extratropical cyclones in the central United States. A somewhat modified version of Equation (26.26) was used by Osborn et al. (1980) for describing the rainfall–watershed relationships for thunderstorms in the southwestern United States.

Woolhiser and Schwalen (1960) suggested for convective storms over areas less than 18.5 mi^2 a relation of the form

$$\frac{P_A}{P} = 1 - \frac{0.14}{P} A^{0.6} \tag{26.27}$$

A similar relation was found by Huff and Stout (1952) for thunderstorm rainfall in Illinois.

Using digitized radar data, Frederick et al. (1977) found for 30-min and 60-min durations:

$$\frac{P_A}{P} = (a + bA)^{-1} \tag{26.28}$$

where a and b are constants, which can be determined by least squares regression. A correlation coefficient of 0.95 or better was reported.

EXAMPLE 26.11 A 3-h storm occurred over a 200 mi^2 watershed. The rain gage located in this watershed measured 5 in. of rainfall. What was the areal average rainfall for this storm?

Solution From Figure 26.33, the percentage of point rainfall for a 3-h storm on a 200 mi^2 watershed is 80%. Therefore, the areal average rainfall $= 5 \times 0.8 = 4$ in.

26.8 DESIGN FLOOD

A design flood is estimated in two ways: (1) transformation of the design storm to design flood and (2) flood-frequency analysis. The first method may use a unit-hydrograph procedure or a simulation model. The first step in applying either method is to make a careful reconnaissance of the project watershed and divide the watershed into a number of subwatersheds in order to account for spatial variability in rainfall and watershed characteristics. Each subwatershed is more or less homogeneous. The number of subwatersheds depends upon the watershed size, rainfall pattern, vegetation, soil, and topographic characteristics. As a general guideline, five to fifteen subwatersheds may be sufficient. The transformation of rainfall to runoff using a unit-hydrograph method involves estimation of abstractions, computation of the effective rainfall, derivation of the unit-hydrograph method, convolution of the unit hydrograph with the effective rainfall, estimation of baseflow, routing of the flood hydrograph through the stream and reservoir system, and combining with floods from intervening areas.

If a simulation model is used for transformation of rainfall to runoff, then it is necessary to ascertain that sufficient data are available to properly calibrate the model on the project watershed. Any of the available simulation models can be employed.

It is prudent to evaluate reasonableness of the PMF before it is adopted for design of a project. One way is to compare the PMP and PMF estimates with observed rainfalls and floods. Observed rainfall, such as given in Table 26.22, can be used as a guide, with proper allowance made for geographical, meteorological, and topographical differences between the watershed and the area used for comparison. Similarly, observed floods, such as shown in Figure 26.30, may be used for comparison, with proper allowance given for length of record.

TABLE 26.22 MAXIMUM OBSERVED DEPTH–AREA–DURATION DATA FOR THE UNITED STATES (AFTER U.S. ARMY CORPS OF ENGINEERS, 1946)[a]

	Duration (hours)						
Area	6	12	18	24	36	48	72
10 mi^2	24.7a	29.8b	36.3c	38.7c	41.8c	43.1c	45.2c
26 km^2	(627)	(757)	(922)	(983)	(1062)	(1095)	(1148)
100 mi^2	19.6b	26.3c	32.5c	32.5c	35.2c	37.9c	40.6c
259 km^2	(498)	(668)	(826)	(894)	(963)	(988)	(1031)
200 mi^2	17.9b	25.6c	31.4c	34.2c	36.7c	37.7c	39.2c
518 km^2	(455)	(650)	(798)	(869)	(932)	(958)	(996)
500 mi^2	15.4b	24.6c	29.7c	32.7c	35.0c	36.0c	37.3c
1,295 km^2	(391)	(625)	(754)	(831)	(889)	(914)	(947)
1,000 mi^2	13.4b	22.6c	27.4c	30.2c	32.9c	33.7c	34.9c
2,590 km^2	(340)	(574)	(696)	(767)	(836)	(856)	(886)
2,000 mi^2	11.2b	17.7c	22.5c	24.8c	27.3c	28.4c	29.7c
5,180 km^2	(282)	(450)	(572)	(630)	(693)	(721)	(754)
5,000 mi^2	8.1bj	11.1b	14.1b	15.5c	18.7d	20.7d	24.4d
12,950 km^2	(206)	(282)	(358)	(394)	(475)	(526)	(620)
10,000 mi^2	5.7j	7.9k	10.1e	12.1e	15.1d	17.4d	21.3d
25,900 km^2	(145)	(201)	(307)	(384)	(384)	(442)	(541)
20,000 mi^2	4.0j	6.0k	7.9e	9.6e	11.6d	13.8d	17.6d
51,800 km^2	(102)	(152)	(201)	(244)	(295)	(351)	(447)
50,000 mi^2	2.5eh	4.2g	5.3e	6.3e	7.9e	8.9e	11.5f
129,500 km^2	(64)	(107)	(135)	(160)	(201)	(226)	(292)
100,000 mi^2	1.7h	2.5ih	3.5e	4.3e	5.6e	6.6f	8.9f
259,000 km^2	(43)	(64)	(89)	(109)	(142)	(168)	(226)

Storm	Date	Location of Center	
a	July 17–18, 1942	Smethport, PA	
b	Sept. 8–10, 1921	Thrall, TX	
c	Sept. 3–7, 1950	Yankeetown, FL	Hurricane
d	June 27–July 1, 1899	Hearne, TX	
e	Mar. 13–15, 1929	Elba, AL	
f	July 5–10, 1916	Bonifay, FL	Hurricane
g	Apr. 15–18, 1900	Eutaw, AL	
h	May 22–26, 1908	Chattanooga, OK	
i	Nov. 19–22, 1934	Millry, AL	
j	June 27–July 4, 1936	Bebe, TX	
k	Apr. 12–16, 1927	Jefferson Parish, LA	

[a] Average rainfall in inches and (millimeters).

For a selected recurrence interval, the design flood can be estimated using the flood data recorded at gaging stations located at or near the dam site. Transposition of flood peaks from the stations to the dam site is often at a 0.5 power of the drainage-area ratio unless another functional relationship is known more precisely. Methods of frequency analysis discussed in Chapters 24 and 25 can be employed to compute the design flood. It should be noted that floods are often caused by more than one mechanism and, as a result, the probability distributions of differing mechanisms can be quite different. For example, flood data for a station in a tropical area may contain floods caused by hurricanes, also called typhoons or tropical cyclones, and floods caused by frontal rains or thunderstorms. Flood data from areas of winter snow may contain floods caused by snowmelt and floods caused by rainfall. In such cases, more than one frequency distribution may be required to adequately describe the flood data. Recent advances made in dating of paleofloods and in estimating the magnitude of such floods (Stedinger and Baker, 1987) show promise in enhancing the data base for frequency analysis.

EXAMPLE 26.12 An interior area of 100 acres houses an apartment complex and a shopping center. The land slope is 5% and there is a levee further down toward the downstream side of the area, as shown in Figure 26.36. The length of flow for this area is 3500 ft. The curve number for the area is 85. Design a drainage ditch along the levee to carry off the runoff from this area. Also design a detention basin for a 50-year flood. The detention basin is to be emptied in 24 hours.

Solution First, the value of potential maximum retention S is computed from Equation (14.6):

$$S = \frac{1000}{85} - 10 = 1.765$$

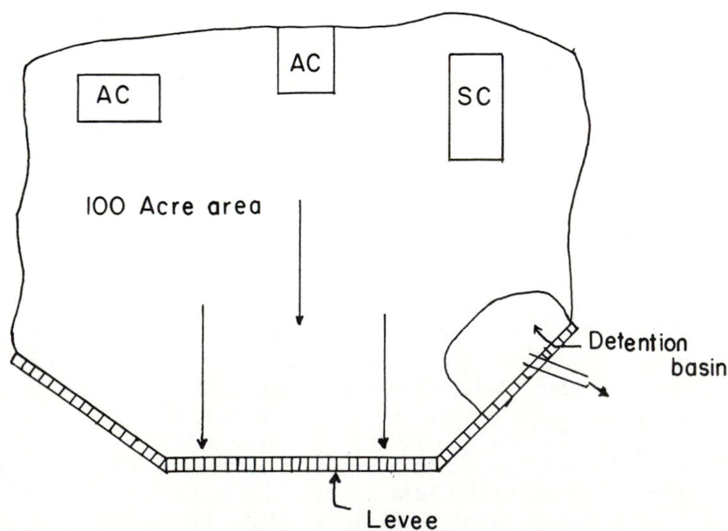

Figure 26.36 An interior area housing an apartment complex (AC) and a shopping center (SC).

Then the watershed lag is computed from Equation (13.9):

$$t_L \text{ (h)} = \frac{L^{0.8}(S + 1)^{0.7}}{1900s^{0.5}} = \frac{(3500)^{0.8}(1.765 + 1)^{0.7}}{1900 \times 5^{0.5}} = 0.328 \text{ h}$$

$$\cong 20 \text{ min.}$$

The time of concentration T_c is computed from $t_L = 0.6T_c$, which yields

$$T_c = \frac{20}{0.6} = 33 \text{ min.}$$

Corresponding to the return period of 50 years and $T_c = 33$ min., the maximum rainfall intensity can be obtained from Technical Paper No. TP-25 of the U.S. Department of Commerce, Weather Bureau. For this area, it is taken to be 5.5 in./h.

The value of runoff coefficient C for this drainage area is $0.5 \le C \le 0.7$. The average value is, therefore, selected as $C = 0.6$.

By using the rational method,

$$Q_p = CIA = (0.6)(5.5)(100) = 330 \text{ cfs}$$

Now we want to design a drainage ditch to carry 330 cfs at normal depth. The ditch is trapezoidal in cross-section, as shown in Figure 26.37, with a depth D, a side slope $1 : z$, and a bottom width b. The channel bottom slope is 0.001 and Manning's $n = 0.035$. Then,

$$\text{cross-sectional area } A \text{ of the ditch} = D(b + zD)$$

$$\text{hydraulic radius } R \text{ of the ditch} = \frac{D(b + zD)}{b + 2D(1 + z^2)^{0.5}}$$

By applying Manning's equation,

$$Q_p = \frac{1.49}{n} AR^{2/3}s^{1/2} = \frac{1.49}{n} [D(b + zD)] \left[\frac{D(b + zD)}{b + 2D(1 + z^2)^{0.5}}\right]^{2/3} \times (0.001)^{1/2}$$

Here D, z, and b are unknown. The depth is taken to be 5 ft and the side slope as $1:2$. Then the remaining unknown is b.

$$\frac{0.035(330)}{1.49(0.001)^{0.5}} = 5(b + 10) \left[\frac{5(b + 10)}{b + 10(1 + 4)^{0.5}}\right]^{2/3}$$

or

$$245.13 = 5(b + 10) \left[\frac{5(b + 10)}{b + 22.36}\right]^{2/3}$$

By trial and error, $b \cong 12.5$ ft. Therefore, $A = 112.5 \text{ ft}^2$. The velocity v is

$$v = \frac{330}{112.5} = 2.93 \text{ ft/s}$$

Top width = 10 + 12.5 + 10 = 32.5 ft

Right of way = 32.5 + (60; it is 50 to 110 ft) = 92.5 ft

This completes the design of the drainage channel, whose cross-section is shown in Figure 26.37.

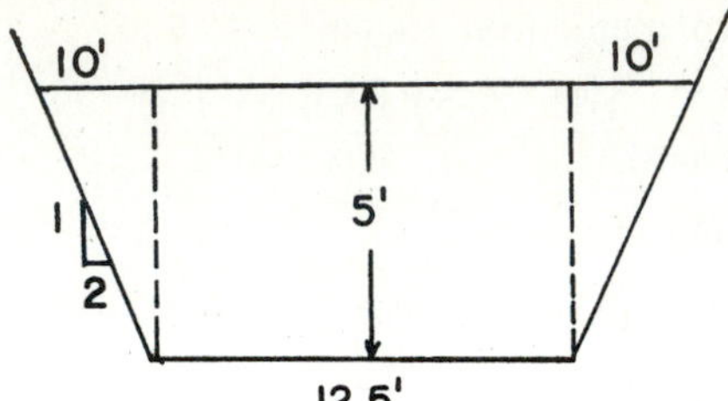

Figure 26.37 Cross-section of designed trapezoidal channel.

For design of the detention basin or storage pond, the rainfall intensity $I = 5.5$ in./h and its duration is 0.55 h.

$$\text{Amount of rain} = 5.5 \times 0.55 = 3.025 \text{ in.}$$

$$\text{Volume of water over the drainage area} = \frac{3.025}{12} \times 100 = 25.2 \text{ a-ft}$$

This amount does not include losses (abstractions). From Technical Release No. TR-55 of the U.S. Department of Agriculture, Soil Conservation Service, the amount of runoff V_Q for CN = 85 and rainfall of 3.025 in. is

$$V_Q = 1.59 \text{ in.} = \frac{1.59}{12} \times 100 = 13.25 \text{ a-ft}$$

The discharge hydrograph having the volume of 1.59 in. can be obtained using the SCS triangular hydrograph method or Snyder's unit hydrograph method. For the SCS method,

$$\text{Time base of the hydrograph} = t_p + t_r$$

$$t_p = \frac{D}{2} + t_L = \frac{1}{4}\text{ h} + \frac{1}{3}\text{ h} = 0.58 \text{ h}$$

$$t_r = 1.67 t_p = 1.67 \times 0.58 \text{ h} = 0.97 \text{ h}$$

Therefore,

$$\text{Time base} = 1.55 \text{ h}$$

The discharge hydrograph, as shown in Figure 26.38, is thus a triangle with a base of 1.55 h, a volume of 1.59 in., and a peak of 330 cfs located 52 min. away from the origin.

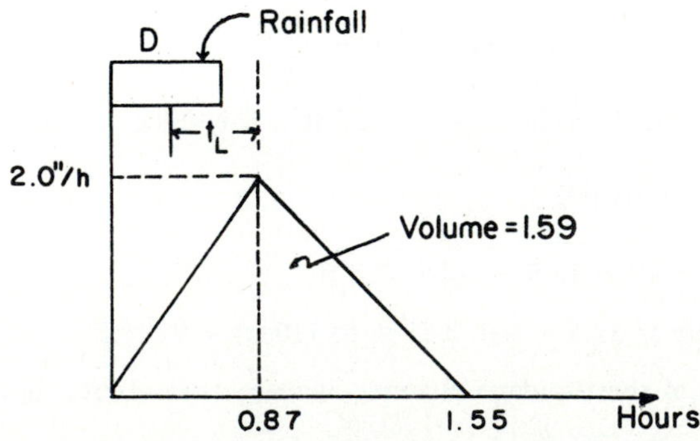

Figure 26.38 Triangular discharge hydrograph.

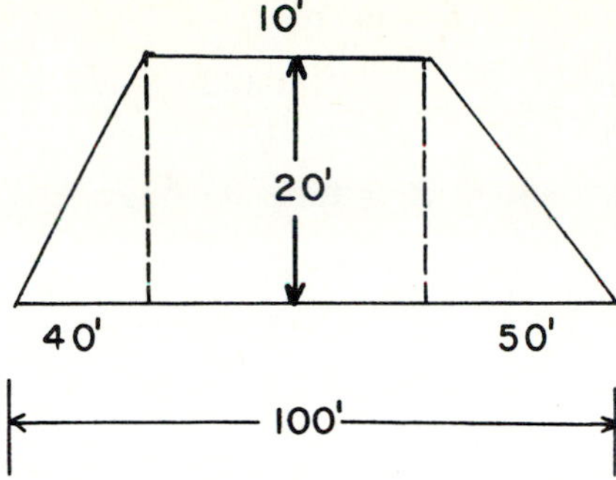

Figure 26.39 Levee cross-section.

The storage area is needed to contain the volume of runoff of 13.25 a-ft. This volume of water is to be emptied in 24 hours. The detention storage is assumed to have the depth of 2.65 ft over an area of 5 acres.

The average discharge

$$Q = 13.25/24 = 0.552 \text{ a-ft/h}$$

$$= 0.552 \times \frac{43{,}560}{3600} = 6.68 \text{ cfs}$$

Say, the levee cross-section, as shown in Figure 26.39, has a 100-ft base width, a 10-ft top width, a 20-ft height, a slope of 1 : 2 on the left side and 1 : 2.5 on the right side of the levee.

A pipe is run from the detention storage through the levee at a slope of 0.01; the length of the pipe is 100 ft. The maximum head is 2.65 + 0.5 (= 100 × 0.01/2) = 3.15 ft. at the pipe outlet, as shown in Figure 26.40. The pipe is assumed to be 24 in. RCP with Manning's $n = 0.012$.

$$\text{Cross-section area of the pipe} = \pi r^2 = 3.14 \times 1 \times 1 = 3.14 \text{ ft}^2$$

$$\text{Wetted perimeter} = \pi D = 3.14 \times 2 = 6.28 \text{ ft}$$

$$\text{Hydraulic radius} = R = \frac{3.14}{6.28} = 0.5 \text{ ft}$$

$$Q_{\max} = A\left(\frac{2gH}{K}\right)^{0.5} = 3.14\left(\frac{2 \times 32.17 \times H}{K}\right)^{0.5}$$

where K is the total loss = entry loss (K_e) + friction loss (K_f) + exit loss (K_x).

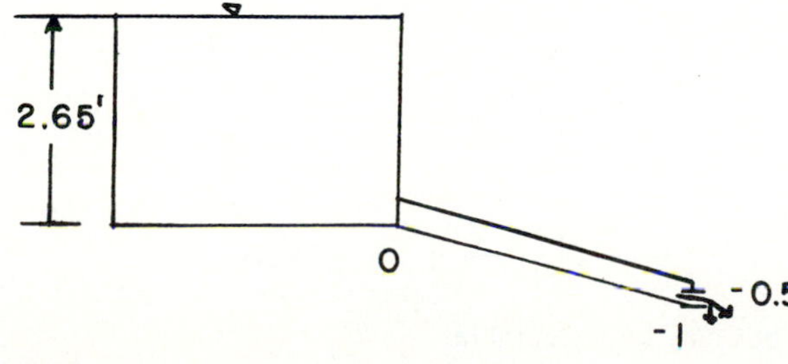

Figure 26.40 Discharge pipe extending from the detention storage.

$$H = 3.15 \text{ ft}, \qquad L = 100 \text{ ft}$$

$$K_f = \frac{2gn^2L}{R^{4/3}} = \frac{2 \times 32.17 \times (0.012)^2 \times 100}{(0.5)^{4/3}} = 1.05$$

$$K_e = 1 \qquad \text{and} \qquad K_x = 0.1$$

(These are standard values of exit and entry losses.)

$$K = 1.0 + 1.05 + 0.1 = 2.15$$

Therefore,

$$Q_{\max} = 3.14 \left(\frac{2 \times 32.17 \times 3.15}{2.15}\right)^{0.5} = 30.5 \text{ cfs}$$

The minimum Q, say, at elevation of 0.5 ft, is

$$Q_{\min} = 3.14 \left(\frac{2 \times 32.17 \times 1.0}{2.15}\right)^{0.5} = 17.2 \text{ cfs}$$

This is an overdesign of the pipe.

Assume the diameter of the pipe is 1 ft. Then

$$A = 0.785 \text{ ft}^2$$

$$P = 3.14 \text{ ft}$$

$$R = 0.25 \text{ ft}$$

$$K_f = \frac{2 \times 32.17 \times (0.012)(100)}{(0.25)^{4/3}} = 2.65$$

$$K = 3.75$$

Therefore,

$$Q_{\max} = 0.785 \left(\frac{2 \times 32.17 \times 3.15}{3.75}\right)^{0.5} = 5.77 \text{ cfs}$$

This is an underdesign.

Assume the diameter of the pipe is 18 in. Then

$$A = 1.77 \text{ ft}^2$$

$$P = 4.21 \text{ ft}$$

$$R = 0.376$$

$$K_f = \frac{2 \times 32.17 \times (0.012)^2 (100)}{(0.376)^{4/3}} = 1.54$$

$$K = 2.64$$

$$Q_{\max} = 1.77 \left(\frac{2 \times 32.17 \times 3.15}{2.64}\right)^{0.5} = 15.5 \text{ cfs}$$

$$Q_{\min} = 1.77 \left(\frac{2 \times 32.17 \times 1.0}{2.64}\right)^{0.5} = 8.75 \text{ cfs}$$

This is still an overdesign, but may be acceptable.

The drawdown schedule as well as rating-curve computations need to be made for this pipe. The detention basin will have a sort of parabolic storage shape, as shown in Figure 26.41. The results of the rating-curve computations are as follows:

ELEVATION (ft)	HEAD (ft)	FLOW (cfs)
4.8	5.05	19.6
4.5	4.75	19.0
4.0	4.25	18.0
3.5	3.75	16.9
3.0	3.25	15.7
2.5	2.75	14.5
2.0	2.25	13.1
1.5	1.75	11.5
1.0	1.25	9.8
0.5	0.75	7.5
0.1	0.35	5.1

The drawdown schedule is as follows:

BEGINNING TIME (h)	BEGINNING ELEVATION (ft)	FLOW (cfs)	CHANGE IN VOLUME (a-ft)	VOLUME (ft)	END ELEVATION (ft)
0	2.65	—	—	13.25	—
1	2.65	15.5	1.28	11.97	2.40
2	2.40	14.88	1.23	10.74	2.10

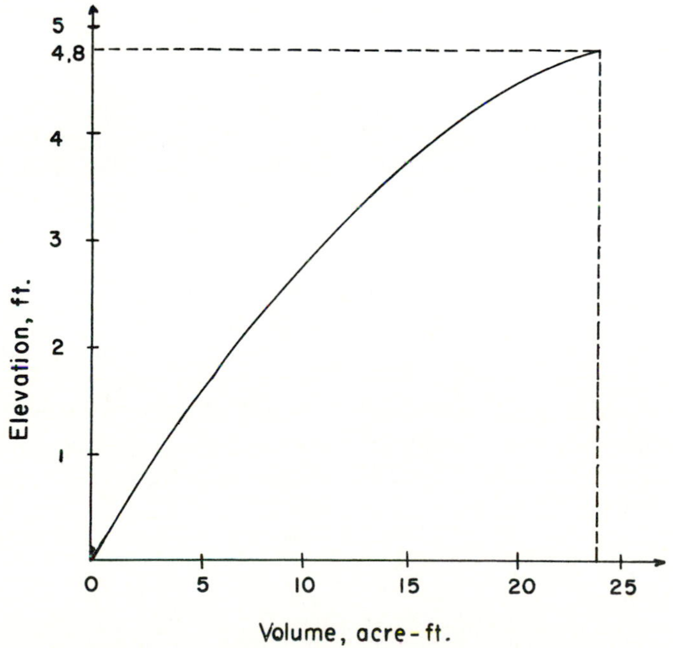

Figure 26.41 Storage–elevation relation.

BEGINNING TIME (h)	BEGINNING ELEVATION (ft)	FLOW (cfs)	CHANGE IN VOLUME (a-ft)	VOLUME (ft)	END ELEVATION (ft)
3	2.10	14.10	1.16	9.58	1.90
4	1.90	13.50	1.12	8.46	1.60
5	1.60	12.70	1.05	7.41	1.30
6	1.30	11.70	0.97	6.44	1.0
7	1.0	10.70	0.88	5.56	0.8
8	0.8	9.97	0.82	4.74	0.6
9	0.6	9.17	0.76	3.98	0.3
10					

EXERCISES

26.1. The dew-point temperature at an elevation of 1000 m has been observed to be 10°C. What is the corresponding dew-point temperature at the mean sea level?

26.2. The dew-point temperature at the meteorological station in the watershed at an elevation of 500 m is 15°C. Calculate the depth of precipitable water between the 1000-mb pressure level and the watershed elevation level.

26.3. The dew-point temperature at the measurement station (500-m elevation) in the watershed is 15°C. The elevation of the inflow barrier of the watershed is about 200 m. Calculate the depth of precipitable water between 200 m and the 200-mb pressure level.

26.4. A storm has occurred during July 10–15, 1980, over a watershed. The persisting dew-point temperature of the storm is 20°C. The maximum dew-point temperature from historical data on dew-point temperature for the period of storm (July 10–15) is found to be 25°C. The watershed elevation is 400 m. Compute the moisture-adjustment factor (without maximization).

26.5. The dew-point temperature of a storm in its original location reduced to the 1000-mb pressure level at the mean sea level is 20°C. The storm is transposed to a new location and the maximum dew-point temperature for this location reduced to the 1000-mb pressure level is 23°C. The maximum dew-point temperature in the region of storm occurrence reduced to the 1000-mb pressure level is 25°C. Compute the moisture-adjustment factor.

26.6. A storm has occurred over a watershed with an average elevation of 200 m. Its dew-point temperature reduced to the 1000-mb pressure level is 20°C. This storm is transposed to a watershed with an average elevation of 300 m in the same meteorologically homogeneous region. The maximum dew-point temperature during the season (based on historical data) at the transposed site is 23°C and at the original location is 25°C. Compute the moisture-adjustment factor.

26.7. The daily rainfall record at a given station is as follows:

Day	1	2	3	4	5	6	7	8	9	10	11	12	13	14
Rainfall (mm)	2	5	3	8	7	10	8	7	7	5	7	4	3	2

Plot the depth–duration and intensity–duration curves.

26.8. Compute the depth of precipitable water in cm for a 5000-m high atmospheric column. The pressure and specific humidity at the ground are, respectively, 1015 mb and 4.5 gm/kg and at 5000-m elevation 400 mb and 0.5 gm/kg. Assume a linear variation between specific humidity and pressure.

26.9. Compute the depth of precipitable water in cm for 6,000 m high atmospheric column. The following information is available:

ELEVATION (m)	TEMPERATURE (°F)	PRESSURE (mb)	VAPOR PRESSURE (mb)
0	58	1013	7.0
600	52	942	5.0
1200	44	875	3.8
1800	38	812	3.2
2400	30	753	2.0
3000	23	697	1.6
3600	16	644	1.1
4200	8	595	0.8
4800	2	550	0.6
5400	−6	500	0.4
6000	−13	450	0.2

26.10. Construct depth–area–duration curves for a storm observed over a drainage area, as shown in Figure 26.42. Choose time intervals of 2, 4, 6, 12, and 18 h, which occurred on July 12–13, 1974. The data are given in Table 26.23.

26.11. Table 26.24 indicates the hourly rainfall values (cm) for a storm at several stations that occurred on August 15, 1955, in a watershed shown in Figure 26.43. Determine the maximum depth–area curves for 6-, 12-, 18-, and 24-h durations.

26.12. Perform a depth–area–duration analysis for the storm of January 24–25, 1978, that occurred in the state of Mississippi, as shown in Figure 26.44. Select 2-, 4-, and 6-h time intervals for computation of the maximum depth. Table 26.25 gives the data.

26.13. You have been retained as a consulting engineer to assist in the design of a channelization project for flood-control purposes. The level of protection (channel capacity) of this project is to be for a flood event with a recurrence interval of 50 years. (This means that there is a 2% chance that the channel capacity will be exceeded in any given year.) Accordingly, you must calculate the maximum rate of runoff from the project watershed that would result from a storm of that magnitude. The topographic, physical, soil, and land-use characteristics of the watershed are given in what follows. In deriving this design discharge, you will be using the Snyder method. The basic methodology for converting rainfall excess to runoff will be the Soil Conservation Service method. The hydrologic flood routing will be by the Muskingum method. The design storm rainfall will be obtained from the National Weather Service TP-40, *Rainfall-Frequency Atlas of the United States.*

The Watershed

The total watershed area is 10.0 mi^2. This area is subdivided into three subbasins corresponding to two tributaries and a main routing reach. The characteristics of each subarea are as follows:

TABLE 26.23 HOURLY PRECIPITATION (cm) FOR A STORM THAT OCCURRED ON JULY 12–13, 1974

	Hour ending											
July 12	0100	0200	0300	0400	0500	0600	0700	0800	0900	1000	1100	1200
Station 1	0.00	0.03	0.00	0.35	0.30	0.00	0.03	0.08	0.15	0.15	0.00	0.03
2	0.00	0.00	0.15	0.03	0.10	0.10	0.03	0.20	0.20	0.30	0.15	0.12
3	0.25	0.22	0.03	0.10	0.22	0.05	0.18	0.50	0.20	0.06	0.00	0.06
4	0.00	0.03	0.03	0.05	0.05	0.12	0.00	0.00	0.00	0.00	0.00	0.00
5	0.03	0.10	0.15	0.00	0.03	0.06	0.05	0.06	0.00	0.00	0.00	0.00
6	0.00	0.00	0.30	0.25	0.00	0.00	0.15	0.10	0.25	0.08	0.00	0.03
7	0.00	0.00	0.15	0.20	0.00	0.12	0.20	0.15	0.30	0.25	0.00	0.15
July 12	1300	1400	1500	1600	1700	1800	1900	2000	2100	2200	2300	2400
Station 1	0.10	0.15	0.20	0.07	0.15	0.25	0.45	0.50	0.30	0.25	0.30	0.55
2	0.15	0.06	0.05	0.03	0.10	0.22	0.15	0.10	0.45	0.40	0.35	0.10
3	0.05	0.10	0.06	0.20	0.12	0.12	0.70	0.15	0.25	0.45	0.55	0.22
4	0.00	0.00	0.00	0.00	0.03	0.03	0.03	0.27	0.18	0.25	0.50	0.50
5	0.00	0.00	0.00	0.01	0.01	0.00	0.00	0.03	0.21	0.03	0.10	0.80
6	0.03	0.12	0.10	0.15	0.20	0.18	0.21	0.35	0.45	0.30	0.20	0.25
7	0.10	0.06	0.03	0.15	0.10	0.30	0.15	0.45	0.30	0.30	0.30	0.70
July 13	0100	0200	0300	0400	0500	0600						
Station 1	1.50	1.00	0.21	0.10	0.05	0.05						
2	0.55	0.90	0.60	0.06	0.02	0.01						
3	0.30	0.20	0.05	0.05	0.05	0.05						
4	0.15	0.35	1.10	0.50	0.30	0.20						
5	0.15	0.25	0.30	0.60	0.10	0.10						
6	1.20	1.00	0.12	0.00	0.00	0.00						
7	1.25	0.70	0.30	0.00	0.00	0.15						

TABLE 26.24 HOURLY RAINFALL VALUES (cm) FOR A STORM THAT OCCURRED ON AUGUST 15, 1955

	Hour ending											
August 15	0100	0200	0300	0400	0500	0600	0700	0800	0900	1000	1100	1200
Station A	0.5	0.2	1.5	2.0	3.0	5.0	0.5	0.0	0.0	1.0	2.0	0.5
B	0.1	0.0	0.5	1.5	2.0	3.0	0.0	0.0	0.0	0.0	1.0	0.5
C	0.2	0.2	1.0	1.5	2.5	4.0	0.5	0.5	0.5	0.5	1.0	0.5
D	0.0	0.0	1.5	2.0	4.0	4.0	3.0	2.5	2.0	1.0	0.5	1.0
E	0.5	1.0	1.0	1.5	2.0	4.5	3.0	2.0	3.0	0.0	0.0	0.5
F	0.5	0.5	1.0	1.0	2.0	2.0	2.0	3.0	0.0	1.5	1.5	1.5
G	0.3	0.5	0.5	0.5	5.0	5.0	5.0	2.0	3.5	0.0	0.0	0.0
August 15	1300	1400	1500	1600	1700	1800	1900	2000	2100	2200	2300	2400
Station A	0.0	0.0	0.5	1.0	1.5	2.0	3.0	4.0	2.0	1.5	0.5	0.0
B	0.5	0.1	0.5	1.0	2.0	3.0	1.5	1.0	1.5	0.5	0.5	0.5
C	1.0	1.5	2.0	2.5	3.0	2.5	2.0	1.5	1.0	0.5	0.0	0.0
D	0.5	0.5	0.5	1.0	1.0	1.0	2.0	1.0	0.5	0.5	0.5	0.0
E	0.0	0.0	0.0	0.0	1.5	2.0	2.5	3.0	3.5	4.0	2.5	0.0
F	1.0	1.0	2.0	2.0	2.5	3.0	3.5	4.0	2.0	1.5	1.0	0.0
G	0.0	0.0	0.5	1.0	1.5	2.0	2.5	3.0	3.5	2.5	2.0	0.0
August 15	0100	0200	0300	0400	0500	0600						
Station A	0.0	0.5	0.2	0.1	0.0	0.0						
B	0.5	0.5	0.0	0.0	0.0	0.0						
C	0.0	0.5	0.5	0.5	0.0	0.0						
D	0.0	0.5	0.5	0.0	0.0	0.0						
E	0.0	0.5	0.0	0.0	0.5	0.0						
F	0.0	0.5	0.5	0.5	0.0	0.0						
G	0.5	0.5	0.0	0.0	0.5	0.0						

TABLE 26.25 STORM OF JANUARY 24–25, 1978, IN MISSISSIPPI

Station	Date	Hourly precipitation amounts (cm)											
		A.M. Hour ending											
		1	2	3	4	5	6	7	8	9	10	11	12
Bude Fire Tower	24	0.0	0.0	0.33	0.5	0.3	0.71	0.41	0.48	0.56	0.36	0.18	0.03
	25	0.0	0.0	0.03	0.0	0.0	0.0	0.03	0.13	0.91	0.46	0.28	0.0
Cleveland 3N	24	0.0	0.0	0.0	0.76	0.25	0.51	0.51	0.76	0.0	0.25	0.0	0.0
	25	0.25	0.0	0.0	0.0	0.0	0.0	0.0	0.0	0.0	0.25	0.0	0.51
Collins	24	0.0	0.0	0.0	0.0	0.5	0.2	0.3	0.33	0.43	0.43	1.17	0.71
	25	0.0	0.0	0.0	0.0	0.0	0.0	0.0	0.15	0.46	0.30	0.66	0.25
Canton	24	0.0	0.0	0.0	0.0	0.25	0.25	0.76	0.25	0.51	0.76	0.0	0.25
	25	0.0	0.0	0.0	0.0	0.0	0.0	0.0	0.0	0.01	0.51	0.51	0.25
Lexington 2NNW	24	0.0	0.0	0.0	0.0	0.51	0.25	0.25	0.25	0.25	0.25	0.25	0.25
	25	0.0	0.25	0.0	0.0	0.0	0.0	0.0	0.0	0.0	0.25	0.25	0.0
Meridian WSOAP	24	0.0	0.0	0.0	0.0	0.05	0.13	0.18	0.28	0.38	0.66	0.36	0.33
	25	0.0	0.0	0.0	0.0	0.0	0.0	0.0	0.0	0.05	0.38	0.89	1.22
Saucier Exp	24	0.0	0.0	0.0	0.0	0.0	0.0	0.0	0.0	0.51	1.24	0.61	0.43
Forest	25	0.0	0.0	0.0	0.0	0.03	0.03	1.45	1.42	1.7	0.91	0.28	0.38
State Univ.	24	0.0	0.0	0.0	0.0	0.0	0.25	0.25	0.25	0.25	0.25	0.25	0.0
	25	0.0	0.0	0.0	0.0	0.0	0.0	0.0	0.0	0.0	0.0	0.25	0.51
University	24	0.25	0.0	0.0	0.0	0.51	0.51	0.25	0.25	0.25	0.25	0.25	0.25
	25	0.51	0.25	0.0	0.0	0.0	0.0	0.25	0.0	0.0	0.0	0.0	0.25
Vicksburg Water	24	0.0	0.0	0.25	0.51	0.25	0.75	0.51	0.25	0.51	0.51	0.0	0.0
Ways Exp. Stat.	25	0.0	0.0	0.0	0.0	0.0	0.0	0.0	0.0	0.0	0.51	0.0	0.25

Hourly precipitation amounts (cm)											
P.M. Hour ending											
1	2	3	4	5	6	7	8	9	10	11	12
0.05	0.03	0.10	0.10	0.15	0.41	0.33	0.30	0.0	0.0	0.03	0.03
0.03	0.03	0.0	0.0	0.0	0.0	0.0	0.0	0.0	0.0	0.0	0.0
0.0	0.0	0.0	0.0	0.0	0.0	0.25	0.0	0.0	0.0	0.0	0.0
0.25	0.0	0.25	0.0	0.25	0.25	0.0	0.0	0.0	0.0	0.0	0.0
0.25	0.03	0.03	0.1	0.25	0.13	0.15	0.1	0.03	0.0	0.03	0.0
0.03	0.05	0.0	0.03	0.03	0.0	0.0	0.0	0.0	0.0	0.0	0.0
0.0	0.0	0.0	0.0	0.25	0.25	0.25	0.25	0.0	0.0	0.0	0.0
0.0	0.25	0.0	0.0	0.0	0.25	0.0	0.0	0.0	0.0	0.0	0.0
0.0	0.0	0.0	0.0	0.0	0.25	0.0	0.0	0.0	0.0	0.0	0.0
0.0	0.25	0.0	0.0	0.0	0.0	0.0	0.0	0.0	0.0	0.0	0.0
0.38	0.18	0.08	0.05	0.10	0.18	0.25	0.25	0.15	0.05	0.05	0.03
0.56	0.18	0.13	0.03	0.0	0.03	0.0	0.0	0.0	0.0	0.0	0.0
0.99	1.04	1.17	0.86	0.53	0.79	0.74	0.58	0.20	0.08	0.08	0.03
0.10	0.03	0.0	0.0	0.0	0.0	0.0	0.0	0.0	0.0	0.0	0.0
0.0	0.0	0.0	0.0	0.0	0.0	0.0	0.25	0.0	0.0	0.0	0.0
0.51	0.25	0.0	0.25	0.0	0.0	0.0	0.0	0.0	0.0	0.0	0.0
0.25	0.0	0.0	0.0	0.0	0.0	0.0	0.25	0.0	0.0	0.0	0.0
0.51	0.0	0.25	0.25	0.0	0.0	0.0	0.0	0.0	0.0	0.0	0.0
0.0	0.0	0.25	0.25	0.25	0.0	0.25	0.25	0.0	0.0	0.0	0.0
0.0	0.25	0.0	0.0	0.0	0.25	0.0	0.0	0.0	0.0	0.0	0.0

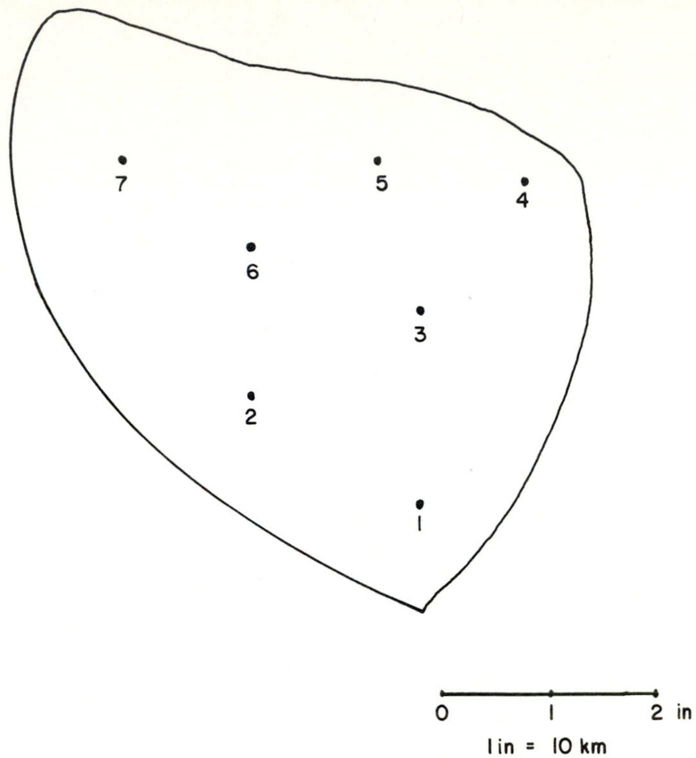

Figure 26.42 A drainage basin.

Subarea 1

Drainage area = 3.1 mi^2
Water-course length (L) = 3.6 mi
Average land slope = 65.0 ft/mi
Soil types = WgB (20%), HoA (40%), and Qu (40%)
Land use = commercial (10%), meadow (30%), medium-density residential (30%), and high-density residential (30%)

Subarea 2

Drainage area = 3.4 mi^2
Water-course length (L) = 3.9 mi
Average land slope = 71.0 ft/mi
Soil types = HoA (40%), Po (30%), and Qu (30%)
Land use = high-density residential (25%), medium-density residential (45%), industrial (30%)

Subarea 3

Drainage area = 3.5 mi^2
Water-course length (L) = 4.1 mi
Average land slope = 59.0 ft/mi

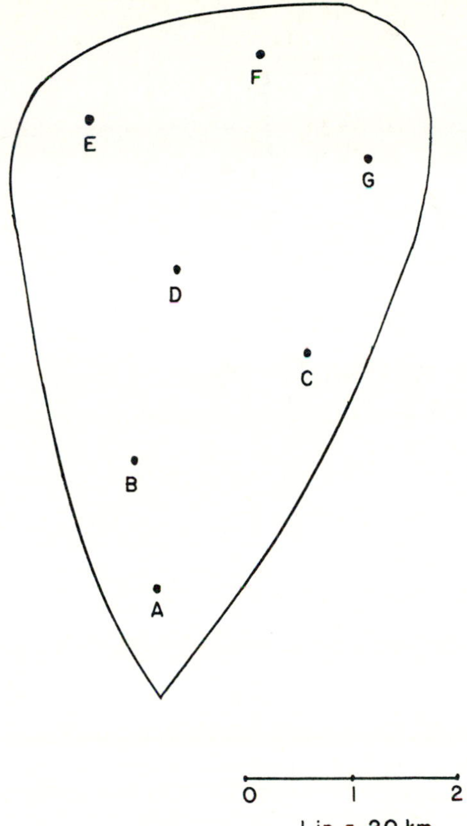

Figure 26.43 A drainage basin.

Soil types = HoA (35%), Qu (30%), and Po (35%)
Land use = high-density residential (20%), parking lots (20%), and commercial (60%)

where: HoA = Hockley loamy fine sand
Po = Portsmouth fine sandy loam
Qu = Quitman loamy sand
WgB = Wagram loamy fine sand

Routing Parameters

Muskingum weighting factor (x) value = 0.25
Reach travel time, k = 3.8 h
Number of steps = 1

26.14. Compute and compare the 10-year, 20-min design rainfall intensities for Chicago, Cleveland, and St. Louis.

26.15. **(a)** Construct a schematic of a watershed model for the drainage basin shown in Figure 26.45. Show subareas, combining points, and routing reaches.

(b) It is desired to construct a highway embankment just downstream of subarea 1. The following data describe this subarea: drainage area = 2.5 mi^2, length = 3.2 mi, slope = 2.1%, land use: medium-density residential ($\frac{1}{4}$-acre lots), and soils: 50% D and 50% C.

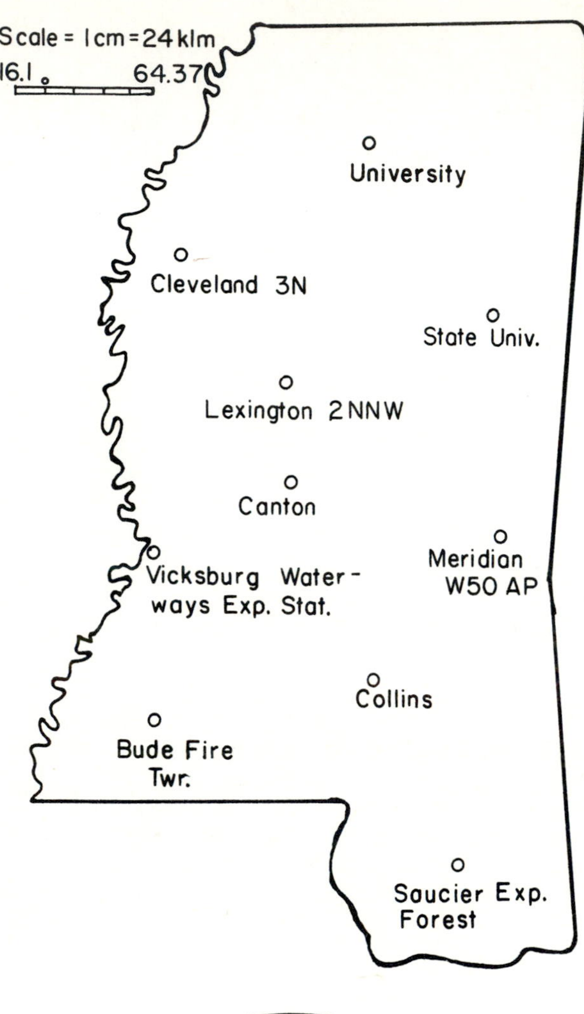

Figure 26.44 The state of Mississippi.

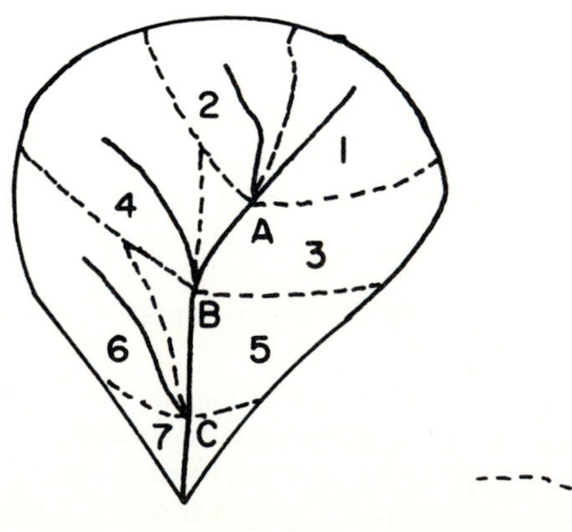

Figure 26.45 A drainage basin.

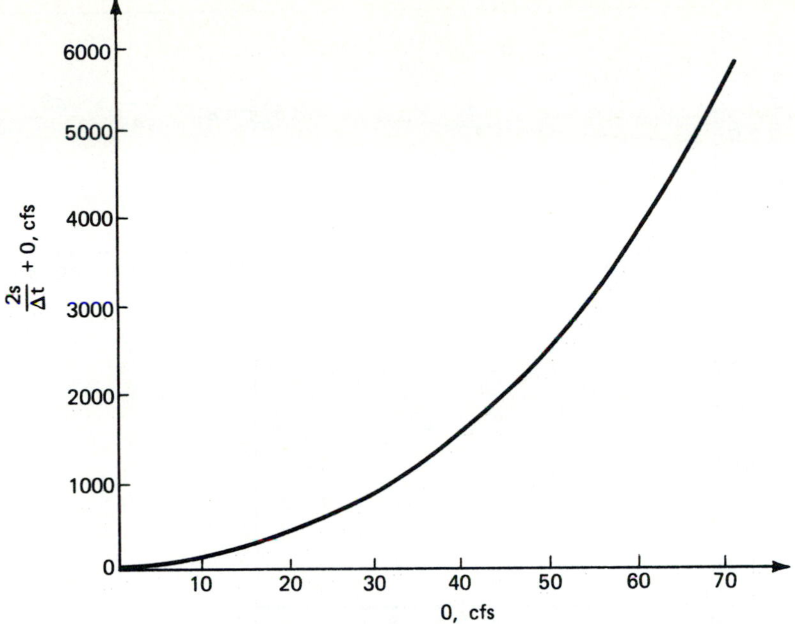

Figure 26.46 $(2S/\Delta t) + 0$ versus 0 curve.

(i) Derive a 1-h unit hydrograph for this basin by the SCS method.

(ii) Derive a design hydrograph for this basin assuming it will have 2.0 in. of runoff. Use 1 cfs/mi^2 for baseflow.

(iii) The embankment will be traversed by the well-known 36-in. CMP. The $(2S/\Delta T) + 0$ curve for the site is shown in Figure 26.46, and the rating curve

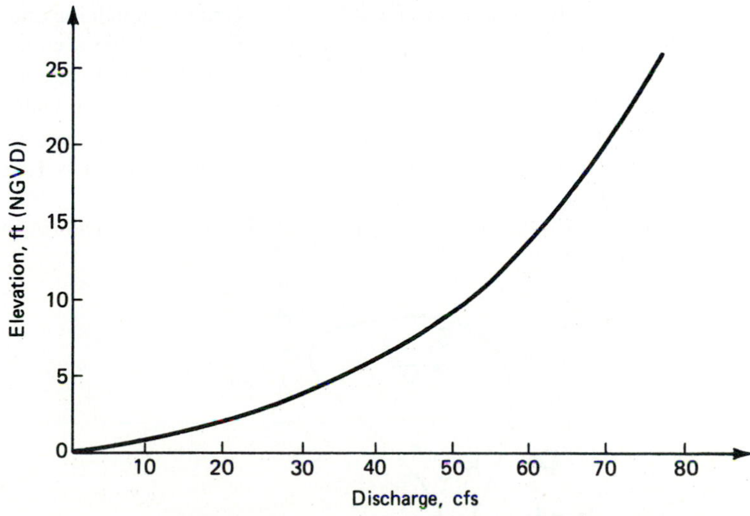

Figure 26.47 Discharge–elevation curve.

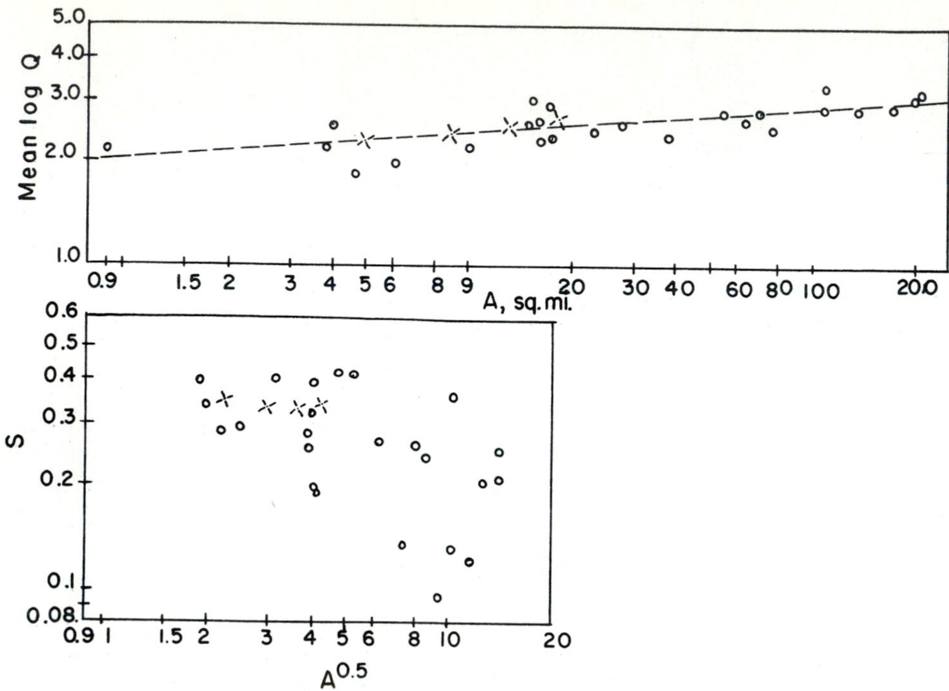

Figure 26.48 (a) Mean logarithm of discharge versus drainage area. (b) Standard deviation of logarithms of discharge versus the square root of the drainage area.

for the culvert is shown in Figure 26.47. Route the design hydrograph through the site by storage indication and determine the minimum allowable embankment height.

(c) Using the plots shown in Figure 26.48, estimate a mean log and standard deviation of the logs for this area. If the skew coefficient of the area is 0:

(i) What is the recurrence interval of the design hydrograph?

(ii) What is the risk that the embankment will fail in 10 years?

26.16. **(a)** A drainage basin is shown in Figure 26.49. The drainage area = 2.5 mi², the mainstream length = 2.6 mi, the length to the centroid = 1.4 mi, C_t = 1.95, and C_p = 0.62.

(i) Determine the basin lag time and the time of concentration.

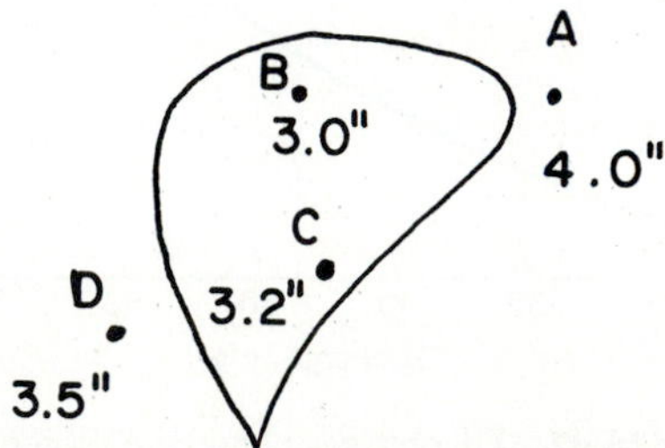

Figure 26.49 A drainage basin.

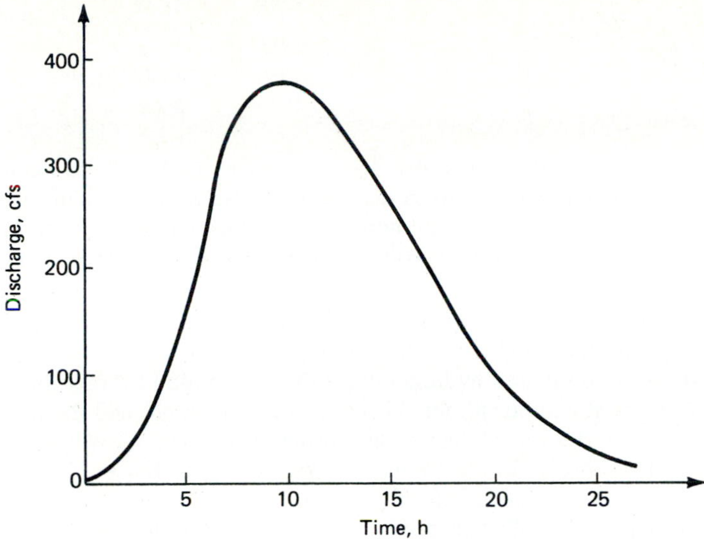

Figure 26.50 A discharge hydrograph.

(ii) Determine the peak of Snyder's unit hydrograph, the 50% width, and 75% width, and the duration.

(iii) Select a proper duration for the drainage area (other than the one calculated before) and the computation interval.

(iv) Convert the unit hydrograph to a 3-h duration (just convert the lag and peak).

(b) Estimate the basin average precipitation for the area.

(c) A runoff hydrograph for this basin is shown in Figure 26.50.

(i) Draw the separation line to determine the direct runoff.

(ii) What is the significance of the intersection point of this line with the falling side of the hydrograph?

(iii) If the sum of the ordinates of the direct-runoff hydrograph at a 1-h computation interval is 3812 cfs-h, what is the ϕ index for this area? Assume the total rainfall amount fell in a 6-h period.

(d) A lake with a surface area of 500 acres has an average weekly inflow of 30 cfs and an average outflow of 25 cfs. If the area received 2.0 in. of rainfall during the week and the evaporation losses totaled 1 in., what is the change in storage in the lake for the week?

(e) The following peak discharges represent the annual maximum flows for the Appomattox River near Mattoax, Virginia, for the years 1961 through 1970:

YEAR	FLOW
1961	4510
1962	7060
1963	4550
1964	3500
1965	3420
1966	3880
1967	2740

YEAR	FLOW
1968	3650
1969	4350
1970	2660

Based on these recorded data, determine the 2-, 10-, 50-, and 100-year discharges assuming a log Pearson distribution. Use a station skew coefficient of 0.67. Plot its frequency curve on normal probability curve. Plot the three largest values at their proper plotting positions. Place a 90% confidence interval on the estimate of the 100-year event.

26.17. The storm sequence, given in what follows, was recorded over a 8.26 mi^2 basin.

(a) Find the ϕ index for this storm.

(b) Derive the unit hydrograph for this storm (don't forget the duration).

(c) The stream length for this basin is $L = 6.56$ mi, and the length to the centroid is $L_c = 3.5$ mi. Obtain C_t and using an average C_p of 0.60, derive Snyder's unit hydrograph for this basin. Use $W_{50} = 9$ h and $W_{75} = 5$ h. Convert Snyder's UHG to a 2-h duration before plotting.

(d) Apply the following rainfall excess pattern to this unit hydrograph. Compare this hydrograph with the original hydrograph.

Time	1	2	3	4	5	6
Excess	0.10	0.06	0.11	0.60	0.29	0

(e) Assume that the original hydrograph is to be used as a design hydrograph for a detention reservoir. The capacity curve of the site is shown in Figure 1.10 in Chapter 1. Outflow is controlled by the ubiquitous 36-in. CMP with a downstream invert of 50.0. Using the orifice equation with a C value of 0.52, rate the pipe up to an elevation 65.0 NGVD. (This corresponds to a head of 15 ft.) Then route the inflow hydrograph through the reservoir to determine the maximum amount of storage used and thus the required height of the embankment. Use a routing interval of 1 hour. What is the maximum outflow from the pipe?

Time (h)	Q (cfs)	Time (h)	Q (cfs)
1	5	17	178
2	13	18	131
3	37	19	97
4	64	20	80
5	131	21	68
6	388	22	57
7	744	23	54
8	723	24	48
9	649	25	44
10	624	26	40
11	582	27	36
12	366	28	33
13	302	29	32
14	332	30	31
15	296	31	29
16	258	32	28

Time (h)	Precipitation (in.)	Time (h)	Precipitation (in.)
1	0.04	8	0.01
2	0.38	9	0.01
3	0.38	10	0
4	0.24	11	0.07
5	0.37	12	0.23
6	0.80	13	0.19
7	0.74	14	0.03

26.18. (a) The following storm sequence and the hydrograph shown in Figure 26.51 were observed over a 1.5 mi^2 area:

Time	0	1	2	3
Precipitation	0	0.4	1.2	0.5

Determine the ϕ index for this storm.

(b) Calculate the unit hydrograph from the observed storm (don't forget the duration). What would be the estimate of the basin lag time?

(c) If rain fell at an average rate of 0.2 in./h for 3 hours over this basin, what would be the ordinate of the resulting hydrograph at a time of 12 hours based on the previous unit hydrograph?

(d) It is desired to fill a reservoir for a hydropower project by means of a cut-and-fill operation through the coffer dam using a 36-in. diameter CMP. The tailwater rating curve, shown in Figure 1.10 in Chapter 1, gives the flow through the pipe for

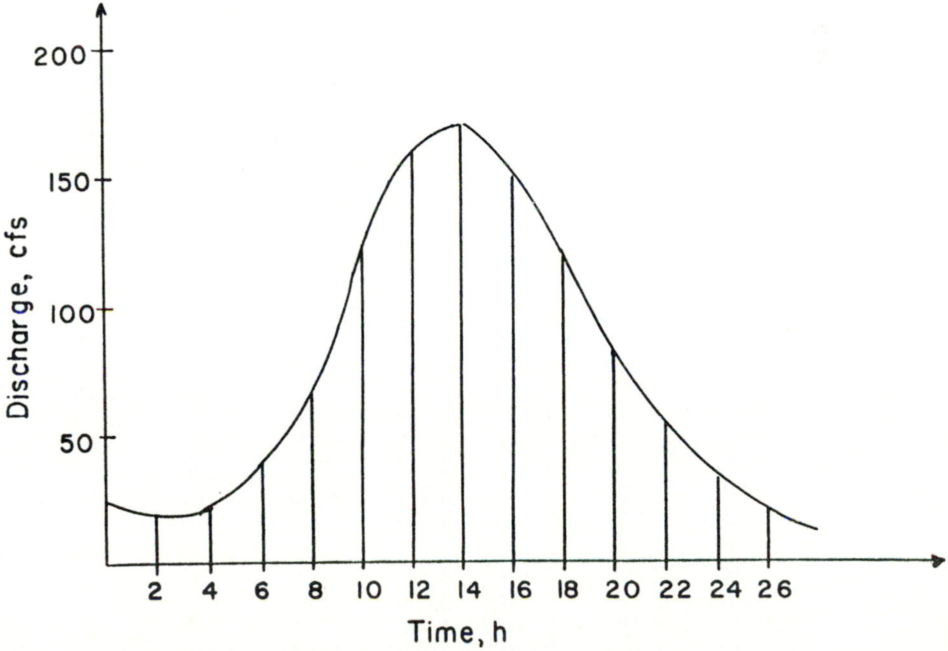

Figure 26.51 A discharge hydrograph.

various reservoir levels. If the volume in the reservoir between elevations 60 and 62 ft NGVD is 160 a-ft, how long will it take to fill this area? Use the average pipe discharge between these elevations in the calculations.

26.19. Construct rainfall-frequency-duration curves for Shreveport, Louisiana, using the data given in Table 26.26.

26.20. Construct rainfall-frequency-duration curves for New Orleans, Louisiana, using the data given in Table 26.27.

26.21. Determine the maximum 2-hour rainfall for DAD analysis for the following record:

TIME	HOURLY RAINFALL (cm)
12:00	0
1:00	5
2:00	5
3:00	2
4:00	4
5:00	0
6:00	2
7:00	4
8:00	5
9:00	3
10:00	2
11:00	1
12:00	0

This record should be read such that rainfall at the end of an hour implies rainfall for this entire hour. For example, at 4:00, the hourly rainfall is 4 cm. This means that the rainfall that occurred between 3:00 and 4:00 is 4 cm.

TABLE 26.26 ANNUAL MAXIMUM RAINFALL FOR DIFFERENT DURATIONS OBSERVED AT SHREVEPORT, LOUISIANA (RAIN-GAGE STATION 0022)

	Rainfall amount (in.)						
Year	3 h	4 h	12 h	24 h	48 h	72 h	96 h
1948	2.190	2.200	2.200	2.830	3.470	3.470	3.470
1949	1.850	3.220	5.540	6.830	7.820	8.170	8.190
1950	2.100	2.780	3.360	3.400	3.530	4.690	5.460
1951	1.470	2.170	2.300	2.320	2.330	2.330	2.890
1952	1.910	2.810	3.260	3.260	3.260	3.740	3.740
1953	3.680	4.470	5.350	5.550	5.550	5.550	5.550
1954	2.210	2.870	3.500	3.500	4.660	4.660	4.660
1955	2.030	3.270	4.230	5.570	5.800	5.800	6.990
1956	2.500	3.750	4.580	4.620	4.620	4.620	4.620
1957	3.320	3.370	4.350	4.750	5.380	5.380	5.410
1958	2.230	2.580	4.270	4.490	5.750	6.200	6.260
1959	1.840	2.410	2.470	3.010	3.360	3.430	3.650
1960	2.700	3.010	3.500	3.670	4.720	4.840	5.080
1961	2.540	3.130	3.650	5.390	5.670	7.260	8.170
1962	1.520	1.780	1.990	2.290	2.610	2.610	2.610
1963	1.560	2.610	3.160	4.010	4.060	4.780	5.160
1964	1.930	2.330	2.400	2.410	3.410	4.630	4.630

TABLE 26.26 *(Continued)*

	Rainfall amount (in.)						
Year	3 h	4 h	12 h	24 h	48 h	72 h	96 h
1965	2.780	2.800	2.880	3.530	3.760	4.930	4.930
1966	1.550	1.920	1.960	2.300	3.740	3.890	3.890
1967	3.200	4.050	4.410	4.420	6.480	7.260	7.270
1968	2.500	4.130	4.810	5.140	5.140	5.140	5.140
1969	1.820	2.960	3.510	3.830	7.300	7.300	7.300
1970	1.960	2.510	2.950	3.300	3.300	3.300	3.760
1971	2.160	2.620	3.100	3.100	3.260	3.650	4.360
1972	2.850	4.230	4.300	4.300	4.340	4.340	4.440
1973	2.090	2.760	3.460	3.460	4.710	4.710	4.720
1974	2.280	2.620	4.030	4.080	4.500	4.550	5.800
1975	1.380	1.980	2.140	2.710	2.960	3.710	3.750
1976	3.020	3.080	3.080	3.080	3.080	3.390	5.380
1977	1.780	1.870	2.480	3.150	3.150	3.150	3.150
1978	4.650	5.220	5.220	5.270	5.280	5.280	5.440
1979	2.470	3.270	3.610	3.800	4.580	4.940	5.350
1980	1.310	1.520	1.910	2.100	3.590	3.750	3.750
1981	3.230	3.230	3.320	5.230	5.230	5.230	5.230
1982	3.520	3.440	3.440	3.450	3.450	3.550	3.770
1983	3.650	3.650	4.120	4.340	4.340	4.460	4.460
1984	3.010	3.490	3.490	3.490	5.060	6.510	6.540
1985	3.160	3.170	3.260	3.490	3.490	4.280	4.540
1986	4.590	5.950	7.000	7.060	7.060	7.080	7.130
1987	2.570	3.690	4.630	6.510	6.510	6.510	6.510

TABLE 26.27 ANNUAL MAXIMUM RAINFALL FOR DIFFERENT DURATIONS OBSERVED AT NEW ORLEANS, LOUISIANA (RAIN-GAGE STATION 0022)

	Rainfall amount (in.)						
Year	3 h	4 h	12 h	24 h	48 h	72 h	96 h
1948	3.940	7.030	9.320	10.930	11.900	13.100	15.770
1949	3.810	5.800	6.420	6.420	6.830	7.440	7.460
1950	1.990	1.990	2.180	2.550	3.560	3.660	4.150
1951	2.060	3.720	5.000	5.000	5.010	5.010	5.010
1952	2.330	2.720	3.640	4.550	4.640	4.650	4.660
1953	5.890	6.520	7.340	7.770	7.770	9.710	9.710
1954	1.880	2.590	2.850	4.620	4.980	5.010	5.010
1955	3.680	4.050	4.670	4.680	5.570	6.910	7.140
1956	3.820	5.520	5.920	7.650	7.930	7.930	8.030
1957	3.090	4.150	5.720	6.620	6.680	7.260	7.310
1958	2.070	2.270	2.670	2.770	3.720	4.670	4.870
1959	3.480	3.830	3.830	4.520	6.310	6.310	6.710
1960	3.860	4.050	4.050	4.050	5.860	5.860	5.860
1961	3.830	4.040	4.270	4.270	5.180	5.950	6.320
1962	3.040	3.180	3.180	3.420	3.760	3.780	3.780
1963	3.040	3.900	4.030	4.030	6.310	6.330	6.330
1964	3.320	3.720	3.880	3.880	3.880	5.210	5.210

TABLE 26.27 *(Continued)*

Year	Rainfall amount (in.)						
	3 h	4 h	12 h	24 h	48 h	72 h	96 h
1965	3.020	3.900	5.440	6.750	6.800	6.850	6.860
1966	3.170	3.590	3.590	3.980	4.930	5.960	6.110
1967	3.340	3.890	5.500	6.900	7.150	7.150	7.280
1968	1.650	1.810	2.050	2.570	2.940	3.260	4.180
1969	2.240	2.240	2.740	3.400	4.760	5.160	5.160
1970	2.380	3.140	3.400	3.400	3.400	3.400	3.990
1971	2.920	3.050	3.050	5.440	6.070	7.800	8.090
1972	2.710	3.220	3.400	3.470	3.470	3.470	3.470
1973	3.050	3.370	4.200	5.590	7.750	7.750	7.750
1974	1.750	2.480	3.210	3.750	4.000	5.140	5.140
1975	2.370	2.380	2.600	3.410	4.290	4.390	5.240
1976	2.500	3.040	3.630	3.660	3.660	3.660	3.800
1977	2.700	3.000	3.100	4.440	6.280	7.250	8.880
1978	5.900	8.800	9.200	10.400	10.500	10.500	10.500
1979	2.300	2.900	4.200	5.600	7.000	7.300	7.800
1980	2.720	3.500	6.500	8.400	8.800	8.800	9.500
1981	4.900	5.900	5.900	5.900	6.700	6.700	7.400
1982	4.100	5.600	6.800	6.900	6.900	7.600	7.700
1983	5.000	7.300	9.000	9.400	9.400	9.400	9.400
1984	2.890	2.890	2.890	3.020	4.370	4.840	5.110
1985	2.400	2.600	3.100	5.000	7.400	9.700	10.000
1986	2.600	2.900	3.400	4.060	4.460	4.860	4.960
1987	2.720	3.100	3.780	4.150	4.350	4.380	4.700

APPENDIX A

Conversion from One System of Units to Another

A.1 COMMONLY USED ABBREVIATIONS

ITEM	ABBREVIATION	ITEM	ABBREVIATION
Acre	a	Liter	l
Acre-foot	a-ft	Langley	ly
British thermal unit	Btu	Log to the base *e*	ln
Cubic feet per second	cfs	Log to the base 10	log
Cubic meters per second	cms	Meter	m
Centimeter	cm	Mile	mi
Degrees Celsius	°C	Millibar	mb
Degrees Fahrenheit	°F	Minute	min
Foot or feet	ft	Millimeter	mm
Gallon per minute	gpm	Newton	N
Gram	g	Parts per million by weight	ppm
Hectare	ha	Pound	lb
Hour	h	Second	s
Inch	in.	Second foot day	sfd
Joule	J	Yard	yd
Kilogram	kg	Watt	W
Kilometer	km		

A.2 COMMONLY USED UNITS OF MEASUREMENT

A.2.1 Length

Meter = 39.37 in. = 3.2808 ft
Inch = 2.54 cm
Foot = 12 in.
Mile = 5280 ft
Kilometer = 1000 m

A.2.2 Area

Square meter
Square foot
Acre = 43,560 ft^2 = 0.4047 ha
Hectare = 0.01 km^2 = 2.471 acres
Square kilometer = 0.3861 mi^2 = 247.1 acres
= 100 ha = 10^6 m^2
Square yard = 9 ft^2

A.2.3 Volume

Cubic meter = 1000 liters = 1.308 yd^3 = 35.31 ft^3
Liter = 0.001 m^3
Gallon
Acre-foot = 1233 m^3
Acre-inch
Cubic foot
Cubic yard

A.2.4 Weight

Kilogram = 2.205 lb
Pound
Long ton = 2240 lb
Short ton = 2000 lb
Metric ton = 1000 kg = 2204.62 lb

A.2.5 Pressure

Bar
Pound per square inch
Newton per square meter

A.2.6 Force

Newton, N (kg-m/s)
Dyne
Pound = 4.448 N
Kilogram = 9.807 N

A.2.7 Power

Watt = 1 joule per second
Horsepower = 745.7 W
Kilowatt = 1.341 horsepower = 737 ft-lb/s
= 102 kg-m/s

A.2.8 Energy

Joule (Newton × meter)
Erg = 10^{-7} J
Calorie = 4.184 J = 0.00397 Btu
Btu = 1055 J
Kilowatt-hour = 3.6×10^6 J

A.2.9 Temperature

Degrees Celsius (°C)
Degrees Fahrenheit (°F)
Degrees Kelvin (°K)

A.3 COMMONLY USED CONVERSION FACTORS

$e = 2.71828$
$\log_{10} e = 0.43429$
$\ln 10 = \log_e 10 = 2.30259$
Acceleration due to gravity,
$g = 32.17$ ft/s^2
= 980.6 cm/s^2
1 cubic foot per second
= 0.9917 acre-inch per hour
= 0.000214 cubic mile per year
= 1 s-ft
= 1 cfs

1 horsepower
= 0.746 kilowatt
= 550 foot-pounds per second
1 inch of runoff per square mile
= 26.9 second-foot-days
= 53.3 acre-feet
= 2,323,200 cubic feet
1 second-foot-day per square mile
= 0.03719 inch
1 cubic inch of water = 0.0361 lb
Heat of fusion = 79.7 cal/g = 144 Btu/lb
Heat of vaporization of water
= 540 cal/g = 970 Btu/lb
1 calorie/cm^2 = 3.69 Btu/ft^2

A.4 CONVERSION OF UNITS

Many of the hydrologic variables are expressed in either the length, the area, the volume, or the flow rate units. Accordingly, tables are prepared for easy conversion of these units. Table A.1 includes conversion of length units, Table A.2 does conversion of area units, Table A.3 does conversion of volume units, and Table A.4 does conversion of flow units.

TABLE A.1 CONVERSION OF LENGTH UNITS (AFTER SCHULZ, 1973)

1 unit	Equivalent of unit value, i.e., 1 m = 3.28 ft									
	mm	cm	dm	m	km	in.	ft	yd	mi	nt mi
millimeters (mm)	1	0.1	0.1	0.001	10^{-6}	0.03937	0.00328	0.00109	6.21×10^{-7}	5.39×10^{-7}
centimeters (cm)	10	1	0.1	0.01	0.00001	0.3937	0.0328	0.0109	6.21×10^{-6}	5.39×10^{-6}
decimeters (dm)	100	10	1	0.1	0.0001	3.937	0.3281	0.1094	6.21×10^{-5}	5.39×10^{-5}
meters (m)	1000	100	10	1	0.001	39.37	3.281	1.094	6.21×10^{-4}	5.39×10^{-4}
kilometers (km)	10^6	10^5	10^4	10^3	1	39,370	3281	1093.6	0.621	0.5396
inches (in.)	25.4	2.54	0.254	0.0254	2.54×10^{-5}	1	0.0833	0.0278	1.58×10^{-5}	1.37×10^{-5}
feet (ft)	304.8	30.48	3.048	0.3048	3.05×10^{-4}	12	1	0.333	1.89×10^{-4}	1.64×10^{-4}
yards (yd)	914.4	91.44	9.144	0.9144	9.14×10^{-4}	36	3	1	5.68×10^{-4}	4.93×10^{-4}
miles (mi)	1.61×10^6	1.61×10^5	1.61×10^4	1.61×10^3	1.6093	63,360	5280	1760	1	0.8684
nautical mile (nt mi)	1.85×10^6	1.85×10^5	1.85×10^4	1853	1.853	72,963	6076	2025	1.151	1

TABLE A.2 CONVERSION OF AREA UNITS (AFTER SCHULZ, 1973)

1 unit	Equivalent of unit value, i.e., 1 mi^2 = 640 acres										
	cm^2	m^2	km^2	hectares	$in.^2$	ft^2	yd^2	mi^2	a	morgen	rai
square centimeters (cm^2)	1	0.0001	10^{-10}	10^{-8}	0.155	1.08×10^{-3}	1.2×10^{-4}	3.861×10^{-11}	2.471×10^{-8}	—	—
square meters (m^2)	10^4	1	10^{-6}	10^{-4}	1550	10.76	1.196	3.861×10^{-7}	2.47×10^{-4}	1.17×10^{-4}	6.25×10^{-4}
square kilometers (km^2)	10^{10}	10^6	1	100	1.55×10^9	10.76×10^6	1.196×10^6	0.3861	247.1	117	625
hectares (ha)	10^8	10^4	0.01	1	1.55×10^7	10.76×10^4	1.196×10^4	3.861×10^{-3}	2.471	1.17	6.25
square inches ($in.^2$)	6.452	6.4×10^{-4}	6.45×10^{-10}	6.45×10^{-8}	1	6.94×10^{-3}	7.7×10^{-4}	2.49×10^{-10}	1.574×10^{-7}	7.55×10^{-8}	4.03×10^{-7}
square feet (ft^2)	929	0.0929	9.29×10^{-8}	9.29×10^{-6}	144	1	0.1111	3.587×10^{-8}	2.3×10^{-5}	1.09×10^{-5}	5.80×10^{-5}
square yards (yd^2)	8361	0.8361	8.36×10^{-7}	8.36×10^{-5}	1296	9	1	3.23×10^{-7}	2.07×10^{-4}	9.82×10^{-5}	5.22×10^{-4}
square miles (mi^2)	—	2,589,998	2.59	259	—	27.87×10^6	3.098×10^6	1	640	302.5	1619
acre (a)	40.4×10^6	4047	4.047×10^{-3}	0.4047	6.27×10^6	43,560	4840	1.562×10^{-3}	1	0.473	2.53
morgen	85.5×10^6	8550	8.55×10^{-3}	0.855	1.322×10^7	92,000	10,220	3.3×10^{-3}	2.111	1	5.34
rai	16×10^6	1600	1.600	0.1600	2.48×10^6	17,250	1917	6.18×10^{-4}	0.395	0.187	1

TABLE A.3 CONVERSION OF VOLUME UNITS (AFTER SCHULZ, 1973)

1 unit	Equivalent of unit value, i.e., 1 a-ft = 43,560 ft^3									
	cc or cm^3	l	m^3	$in.^3$	ft^3	U.S. gallons	Imperial gallons	a-ft	sfd	mil gallons
cubic centimeters (cc or cm^3)	1	0.001	10^{-6}	0.06102	3.53×10^{-5}	2.64×10^{-4}	2.2×10^{-4}	8.1×10^{-10}	4.09×10^{-10}	2.64×10^{-10}
liters (l)	1000	1	0.001	61.023	0.0353	0.26417	0.22008	8.1×10^{-7}	4.09×10^{-7}	2.64×10^{-7}
cubic meters (m^3)	10^6	1000	1	61,023	35.314	264.17	220.08	8.107×10^{-4}	4.09×10^{-4}	2.642×10^{-4}
cubic inches ($in.^3$)	16.39	1.64×10^{-2}	1.64×10^{-5}	1	5.79×10^{-4}	4.33×10^{-3}	3.61×10^{-3}	1.218×10^{-8}	6.698×10^{-9}	4.329×10^{-9}
cubic feet (ft^3)	28,317	28.317	0.02832	1728	1	7.481	6.232	2.296×10^{-5}	1.157×10^{-5}	7.48×10^{-6}
U.S. gallons	3785.4	3.785	3.785×10^{-3}	231	0.13368	1	0.83311	3.069×10^{-6}	1.547×10^{-6}	10^{-6}
Imperial gallons	4542.5	4.542	4.54×10^{-3}	277.3	0.16046	1.200	1	3.684×10^{-6}	1.86×10^{-6}	1.2×10^{-6}
acre-feet (a-ft)	1.233×10^9	1.233×10^6	1233.5	75.27×10^6	43,560	3.26×10^5	2.715×10^5	1	0.5042	0.3260
second-foot-day (sfd)	2.45×10^9	2.45×10^6	2451.3	1.493×10^8	86,400	6.46×10^5	5.38×10^5	1.9835	1	0.6463
mil gallons	3.785×10^9	3.785×10^6	3785	2.31×10^8	1.338×10^5	10^6	8.33×10^5	3.0684	1.5472	1

TABLE A.4 CONVERSION OF FLOW UNITS (AFTER SCHULZ, 1973)

1 unit	Equivalent of unit value, i.e., 1 cfs = 28.317 liters/second										
	cms or m^3/s	cmd or m^3/day	l/s	cu ft/sec (cfs)	cfd	ac ft/day	gpm	US gal/day	mgd	Igpm	Imp gal/day
cubic meters/second (cms or m^3/s)	1	86,400	1000	35.314	3.051×10^6	70.045	15,850	22.82×10^6	22.824	13,208	19.02×10^6
cubic meters/day (cmd or m^3/day)	1.157×10^{-5}	1	0.116	4.09×10^{-4}	35.314	8.1×10^{-4}	0.1835	264.17	2.64×10^{-4}	0.1528	220.14
liters/second (l/s)	0.001	86.4	1	0.0353	3051.2	0.070	15.85	22824	2.28×10^{-2}	13.208	19020
cubic feet/second (cfs or ft^3/s)	0.0283	2446.6	28.317	1	86,400	1.9835	448.83	646,323	0.6463	374.03	538.603
cubic feet/day (cfd or ft^3/day)	3.28×10^{-7}	0.02832	3.28×10^{-4}	1.16×10^{-5}	1	2.3×10^{-5}	5.19×10^{-3}	7.48	7.48×10^{-6}	4.33×10^{-3}	6.233
acre-feet/day (afd or a-ft/day)	0.0143	1233.5	14.276	0.5042	43.560	1	226.28	325,850	0.3258	188.57	271,542
US gallons/minute (gpm)	6.309×10^{-5}	5.451	0.0631	2.23×10^{-3}	192.5	4.42×10^{-3}	1	1440	1.44×10^{-3}	0.8333	1200
US gallons/day (gpd)	4.38×10^{-8}	3.785×10^{-3}	4.382×10^{-5}	1.55×10^{-6}	0.1337	3.07×10^{-6}	6.94×10^{-4}	1	10^{-6}	5.79×10^{-4}	0.833
mil gallons/day (mgd)	4.38×10^{-2}	3785	43.82	1.55	1.337×10^5	3.07	694	10^6	1	579	8.33×10^5
Imperial gallons/minute (Igpm)	7.57×10^{-5}	6.541	0.0757	2.67×10^{-3}	231.12	5.3×10^{-3}	1.2	1728	1.7×10^{-3}	1	1440
Imperial gallons/day (Igpd)	5.26×10^{-8}	4.54×10^{-3}	5.26×10^{-5}	1.86×10^{-6}	0.160	3.68×10^{-6}	8.33×10^{-4}	1.2	1.2×10^{-6}	6.94×10^{-4}	1

APPENDIX B

Analytical Solutions of the Muskingum Method

Analytical solutions of the Muskingum equation have been obtained for various initial conditions. We will derive these solutions here. By combining Equations (21.11) and (21.12),

$$k(1 - x)\frac{dQ}{dt} + kx\frac{dI}{dt} = I - Q \tag{B.1}$$

Let us define the Laplace transform of I and Q:

$$L[I] = P = \int_0^\infty e^{-st} I \, dt$$

$$L[Q] = q = \int_0^\infty e^{-st} Q \, dt$$

By taking the Laplace transform of Equation (B.1),

$$a[sq - Q(0)] + b[sP - I(0)] = P - q$$

where $a = k(1 - x)$ and $b = kx$. $Q(0)$ and $I(0)$ are the initial conditions. By rearranging the terms,

$$q = \frac{aQ(0) + bI(0)}{1 + as} + \frac{P}{1 + as} - \frac{bsP}{1 + as}$$

The last term in this equation can be arranged as

$$\frac{bsP}{1 + as} = P\frac{b}{a}\left(-\frac{1}{1 + as} + 1\right)$$

By inserting this and rearranging,

$$q = \frac{qQ(0) + bI(0)}{1 + as} - \frac{b}{a}P + \left(a + \frac{b}{a}\right)\frac{P}{1 + as}$$

By taking the inverse Laplace transform,

$$Q(t) = \frac{qQ(0) + bI(0)}{a} e^{-t/a} - \frac{b}{a} I(t) + \left(1 + \frac{b}{a}\right)\frac{1}{a}\int_0^\infty I(u)\, e^{-(t-u)/a}\, du$$

$$= \frac{k(1 - x)q(0) + kxI(0)}{k(1 - x)} e^{-t/[k(1-x)]} - \frac{x}{1 - x} I(t) + \frac{1}{k(1 - x)^2} \tag{B.2}$$

$$\times \int_0^t e^{-(t-u)/[k(1-x)]} I(u)\, du$$

Equation (B.2) is the solution of Equation (B.1) subject to the given initial conditions. This equation can be utilized to study the effect of initial conditions on the Muskingum solution as well as to determine the instantaneous unit hydrograph (IUH), unit hydrograph (UH), or summation hydrograph (SH). If initially $Q(0) = I(0)$,

$$Q(t) = \frac{I(0)}{1-x} e^{-t/[k(1-x)]} - \frac{x}{1-x} I(t) + \frac{1}{k(1-x)^2} \times \int_0^t e^{-(t-u)/[k(1-x)]} I(u)\, \mathrm{d}u \tag{B.3}$$

This equation was also obtained by Kulandaiswamy (1966) and Diskin (1967). If, on the other hand, $I(0) = Q(0) = 0$,

$$Q(t) = \frac{1}{k(1-x)^2} \int_0^t e^{-(t-u)/[k(1-x)]} I(u)\, \mathrm{d}u - \frac{x}{1-x} I(t) \tag{B.4}$$

B.1 INSTANTANEOUS UNIT HYDROGRAPH (IUH)

Because the Muskingum method is linear, it may suffice to determine its IUH. If $I(0)$ is interpreted as $\lim_{t\to 0^+} I(t)$, then in the case $I(t) = \delta(t)$, the IUH can be determined. Because $\delta(t)$ is not defined at $t = 0$, we can determine the IUH meaningfully only for initial conditions used in Equation (B.4). Therefore, replacing $I(t)$ there by $\delta(t)$,

$$\begin{aligned} h(0, t) &= -\frac{x}{1-x}\delta(t) + \frac{1}{k(1-x)^2}\int_0^t e^{-(t-s)/[k(1-x)]}\,\delta(s)\,\mathrm{d}s \\ h(0, t) &= -\frac{x}{1-x}\delta(t) + \frac{1}{k(1-x)^2} e^{-t/[k(1-x)]} \end{aligned} \tag{B.5}$$

where $h(0, \mathrm{t})$ is the IUH. This equation was obtained by Venetis (1969). This shows a negative impulse at the origin and a positive logarithmic recession.

However, Morel-Seytoux (1979) has argued that it is preferable to obtain the SH because it is better behaved; then obtain $h(0, t)$ and $h(D, t)$ therefrom. To obtain the SH, we combine Equations (15.1) and (15.6) with S as a dependent variable and solve

$$k(1-x)\frac{\mathrm{d}S}{\mathrm{d}t} + S = kI \tag{B.6}$$

The solution of Equation (B.6) with the right side equal to 1, subject to $S(0) = 0$, is

$$U_s(t) = 1 - \exp\left[\frac{-t}{k(1-x)}\right] \tag{B.7}$$

where $U_s(t)$ is the Muskingum storage-step response function. The Muskingum storage IUH is obtained by simply differentiating $U_s(t)$:

$$h_s(0, t) = \frac{1}{k(1-x)} \exp\left[\frac{-t}{k(1-x)}\right] \tag{B.8}$$

We can express $S(t)$ as

$$S(t) = \int_0^t kI(\tau) \frac{1}{k(1-x)} \exp\left[\frac{-t+\tau}{k(1-x)}\right] d\tau$$

Substituting it into Equation (15.6), we obtain Equation (B.4):

$$Q(t) = \frac{1}{k(1-x)^2} \int_0^t \exp\left[\frac{-(t-\tau)}{k(1-x)}\right] I(\tau)d\tau - \frac{\mathrm{x}}{1-x} I(t)$$

The UH of the Muskingum model can be derived from Equation (15.5). If the period of the UH is 1, then

$$h(1, t) = \int_0^t h(0, t-\tau)\, d\tau, \qquad t \leq 1$$

and

$$h(1, t) = \int_0^1 h(0, t-\tau)\, d\tau, \qquad t > 1$$

Therefore, by using Equation (B.5),

$$h(1, t) = -\frac{x}{1-x} \int_0^t \delta(t-\tau)\, d\tau + \frac{1}{k(1-x)^2}$$
$$\times \int_0^t \exp\left[\frac{-(t-\tau)}{k(1-x)}\right] d\tau, \qquad t \leq 1$$

which yields

$$\begin{aligned} h(1, t) &= -\frac{x}{1-x} + \frac{\exp[-t/k(1-x)]}{1-x} \left\{\exp\left[\frac{t}{k(1-x)}\right] - 1\right\}, \qquad t \leq 1 \\ &= 1 - \frac{1}{1-x} \exp\left[\frac{-t}{k(1-x)}\right], \qquad t \leq 1 \end{aligned} \tag{B.9}$$

Similarly, for $t > 1$,

$$\begin{aligned} h(1, t) &= -\frac{x}{1-x} \int_0^1 \delta(\mathrm{t}-\tau)d\tau + \frac{1}{k(1-x)^2} \int_0^1 \exp\left[\frac{-(t-\tau)}{k(1-x)}\right] d\tau \\ &= 0 + \frac{\exp[-t/k(1-x)]}{1-x} \left\{\exp\left[\frac{1}{k(1-x)}\right] - 1\right\} \\ &= \frac{\exp[1/k(1-x)] - 1}{1-x} \exp\left[\frac{-t}{k(1-x)}\right], \qquad t > 1 \end{aligned} \tag{B.10}$$

The ordinates of the UH, $h(1, t)$, can be computed by replacing t by j, $j = 1, 2, \ldots,$ in Equations (B.9) and (B.10).

APPENDIX C

Least Squares Method for Estimation of the Muskingum Parameters

The method is based on minimizing the sum of squares of deviations between observed storage and estimated storage for a given inflow and outflow sequence. That is,

$$E = \sum_{j=1}^{N} [S_0(j) - S_e(j)]^2 \rightarrow \min \tag{C.1}$$

where E is the error function to be minimized, $S_0(j)$ is the observed storage for the jth time interval, $S_e(j)$ is the estimated storage for the jth time interval, and N is the number of observations or data.

The estimated storage should be given by Equation (21.12). It should be noted that in actual practice, the storage that is given is a relative storage, not an absolute one. Therefore, Equation (21.12) can be modified to incorporate the difference C between relative and absolute storages:

$$S = k[xI + (1 - x)Q] + C \tag{C.2}$$

If the flood wave is starting on a dry river bed, parameter C will vanish. First, consider the case $C \neq 0$. By substituting Equation (C.2) into Equation (C.1) and dropping the index j for brevity,

$$E = \sum_{1}^{N} [S_0 - kxI - k(1 - x)Q - C]^2 \rightarrow \min \tag{C.3}$$

Let $A = kx$ and $B = k(1 - x)$. By differentiating Equation (C.3) with respect to A, B, and C, equating each time to zero, and rearranging,

$$\Sigma S_0 I = A\Sigma I^2 + B\Sigma QI + C\Sigma I$$

$$\Sigma S_0 Q = A\Sigma IQ + B\Sigma Q^2 + C\Sigma Q$$

$$\Sigma S_0 = A\Sigma I + B\Sigma Q + CN$$

Solving for A, B, and C,

$$A = \frac{y_1}{y_2} - \frac{y_3}{y_2}$$

$$B = \frac{y_1 z_2 - z_1 y_2}{y_3 z_2 - y_2 z_3} \tag{C.4}$$

$$C = (\Sigma S_0 - A\Sigma I - B\Sigma Q)/N$$

where

$$y_1 = \Sigma S_0 I - \Sigma S_0 \Sigma I / N$$

$$y_2 = \Sigma I^2 - (\Sigma I)^2 / N$$

$$y_3 = \Sigma IQ - \Sigma Q \Sigma I / N$$

$$z_1 = \Sigma S_0 Q - \Sigma S_0 \Sigma Q / N$$

$$z_2 = \Sigma IQ - \Sigma I \Sigma Q / N$$

$$z_3 = \Sigma Q^2 - (\Sigma Q)^2 / N$$

Therefore,

$$k = A + B \tag{C.5}$$

$$x = A/k \tag{C.6}$$

Equations (C.4) and (C.5) give parameters k, x, and C for a known inflow–outflow sequence.

If $C = 0$, then the error function E follows:

$$E = \sum_{1}^{N} [S_0 - kxI - (1 - x)kQ]^2 \rightarrow \min \tag{C.7}$$

By using A and B as defined before, differentiating E with respect to A and B, and equating each time to zero,

$$\Sigma S_0 I = A \Sigma I^2 + B \Sigma IQ$$

$$\Sigma S_0 Q = A \Sigma IQ + B \Sigma Q^2$$

Solving for A and B,

$$A = \frac{S_0 I \Sigma Q^2 - \Sigma S_0 Q \Sigma IQ}{\Sigma I^2 \Sigma Q^2 - (\Sigma IQ)^2} \tag{C.8}$$

$$B = \frac{\Sigma S_0 Q \Sigma I^2 - \Sigma S_0 I \Sigma IQ}{\Sigma I^2 \Sigma Q^2 - (\Sigma IQ)^2} \tag{C.9}$$

Therefore,

$$k = A + B \tag{C.10}$$

$$x = A/k \tag{C.11}$$

Thus, k and x can be obtained objectively and conveniently. The computation involved in the method can be easily carried out on a desk calculator.

APPENDIX D
Statistical Tables

The last section of the book deals with hydrologic design that involves frequency analysis. The hydrologic frequency analyses employ normal, log-normal, the Gumbel and the Pearson type III distributions. To facilitate application of these distributions, pertinent tables are prepared. Table D.1 includes the area under the standard normal curve (or probability value) for various values of the standard normal variate. Tables D.2 and D.3 are for the Gumbel distribution; for various sample sizes these tables give values of the reduced mean and the reduced standard deviation, respectively. Table D.4 lists the values of the frequency factor for log-normal distribution corresponding to various values of the coefficient of variation c_v, the coefficient of skewness c_s and the exceedance probability. Table D.5 gives the values of the frequency factor for the Gumbel distribution for different sample sizes and return periods. Tables D.6 and D.7 list the values of the frequency factor for the Pearson type III distribution with positive as well as negative coefficients of skewness.

TABLE D.1 AREA UNDER THE STANDARD NORMAL CURVE. $Z = (X - \mu)/\sigma$

Z	0.00	0.01	0.02	0.03	0.04	0.05	0.06	0.07	0.08	0.09
0.0	0.0000	0.0040	0.0080	0.0120	0.0159	0.0199	0.0239	0.0279	0.0319	0.0359
0.1	0.0398	0.0438	0.0478	0.0517	0.0557	0.0596	0.0636	0.0675	0.0714	0.0753
0.2	0.0793	0.0832	0.0871	0.0910	0.0948	0.0987	0.1026	0.1064	0.1103	0.1141
0.3	0.1179	0.1217	0.1255	0.1293	0.1331	0.1368	0.1406	0.1443	0.1480	0.1517
0.4	0.1554	0.1591	0.1628	0.1664	0.1700	0.1736	0.1772	0.1808	0.1844	0.1879
0.5	0.1915	0.1950	0.1985	0.2019	0.2054	0.2088	0.2123	0.2157	0.2190	0.2224
0.6	0.2257	0.2291	0.2324	0.2357	0.2389	0.2422	0.2454	0.2486	0.2518	0.2549
0.7	0.2580	0.2611	0.2642	0.2673	0.2704	0.2734	0.2764	0.2794	0.2823	0.2852
0.8	0.2881	0.2910	0.2939	0.2967	0.2995	0.3023	0.3051	0.3078	0.3106	0.3133
0.9	0.3159	0.3186	0.3212	0.3238	0.3264	0.3289	0.3315	0.3340	0.3365	0.3389
1.0	0.3413	0.3438	0.3461	0.3485	0.3508	0.3531	0.3554	0.3577	0.3599	0.3621
1.1	0.3643	0.3665	0.3686	0.3708	0.3729	0.3749	0.3770	0.3790	0.3810	0.3830
1.2	0.3849	0.3869	0.3888	0.3907	0.3925	0.3944	0.3962	0.3980	0.3997	0.4015
1.3	0.4032	0.4049	0.4066	0.4082	0.4099	0.4115	0.4131	0.4147	0.4162	0.4177
1.4	0.4192	0.4207	0.4222	0.4236	0.4251	0.4265	0.4279	0.4292	0.4306	0.4319
1.5	0.4332	0.4345	0.4357	0.4370	0.4382	0.4394	0.4406	0.4418	0.4430	0.4441
1.6	0.4452	0.4463	0.4474	0.4485	0.4495	0.4505	0.4515	0.4525	0.4535	0.4545
1.7	0.4554	0.4564	0.4573	0.4582	0.4591	0.4599	0.4608	0.4616	0.4625	0.4633
1.8	0.4641	0.4649	0.4656	0.4664	0.4671	0.4678	0.4686	0.4693	0.4699	0.4706
1.9	0.4713	0.4719	0.4726	0.4732	0.4738	0.4744	0.4750	0.4756	0.4762	0.4767
2.0	0.4772	0.4778	0.4783	0.4788	0.4793	0.4798	0.4803	0.4808	0.4812	0.4817
2.1	0.4821	0.4826	0.4830	0.4835	0.4838	0.4842	0.4846	0.4850	0.4854	0.4857
2.2	0.4861	0.4865	0.4868	0.4871	0.4875	0.4878	0.4881	0.4884	0.4887	0.4890
2.3	0.4893	0.4896	0.4898	0.4901	0.4904	0.4906	0.4909	0.4911	0.4913	0.4916
2.4	0.4918	0.4920	0.4922	0.4925	0.4927	0.4929	0.4931	0.4932	0.4934	0.4936
2.5	0.4938	0.4940	0.4941	0.4943	0.4945	0.4946	0.4948	0.4949	0.4951	0.4952
2.6	0.4953	0.4955	0.4956	0.4957	0.4959	0.4960	0.4961	0.4962	0.4963	0.4964
2.7	0.4965	0.4966	0.4967	0.4968	0.4969	0.4970	0.4971	0.4972	0.4973	0.4974
2.8	0.4974	0.4975	0.4976	0.4977	0.4977	0.4978	0.4979	0.4980	0.4980	0.4981
2.9	0.4981	0.4982	0.4983	0.4983	0.4984	0.4984	0.4985	0.4985	0.4986	0.4986
3.0	0.4986	0.4987	0.4987	0.4988	0.4988	0.4989	0.4989	0.4989	0.4990	0.4990

TABLE D.2 REDUCED MEAN $\bar{y}_N$ FOR THE GUMBEL DISTRIBUTION

N^a	0	1	2	3	4	5	6	7	8	9
10	0.4952	0.4996	0.5035	0.5070	0.5100	0.5128	0.5157	0.5181	0.5202	0.5220
20	0.5236	0.5252	0.5268	0.5283	0.5296	0.5309	0.5320	0.5332	0.5343	0.5353
30	0.5362	0.5371	0.5380	0.5388	0.5396	0.5402	0.5410	0.5418	0.5424	0.5330
40	0.5436	0.5442	0.5448	0.5453	0.5458	0.5463	0.5468	0.5473	0.5477	0.5481
50	0.5485	0.5489	0.5493	0.5497	0.5501	0.5504	0.5508	0.5511	0.5515	0.5518
60	0.5521	0.5524	0.5527	0.5530	0.5533	0.5535	0.5538	0.5540	0.5543	0.5545
70	0.5548	0.5550	0.5552	0.5555	0.5557	0.5559	0.5561	0.5563	0.5565	0.5567
80	0.5589	0.5570	0.5572	0.5574	0.5576	0.5578	0.5580	0.5581	0.5583	0.5585
90	0.5586	0.5587	0.5589	0.5591	0.5592	0.5593	0.5595	0.5596	0.5598	0.5599
100	0.5600									

[a] N = sample size.

TABLE D.3 REDUCED STANDARD DEVIATION s_y FOR THE GUMBEL DISTRIBUTION

N^a	0	1	2	3	4	5	6	7	8	9
10	0.9496	0.9676	0.9833	0.9971	1.0095	1.0206	1.0316	1.0411	1.0493	1.0565
20	1.0628	1.0696	1.0754	1.0811	1.0864	1.0915	1.0961	1.1004	1.1047	1.1086
30	1.1124	1.1159	1.1193	1.1226	1.1255	1.1285	1.1313	1.1339	1.1363	1.1388
40	1.1413	1.1436	1.1458	1.1480	1.1499	1.1519	1.1538	1.1557	1.1574	1.1590
50	1.1607	1.1623	1.1638	1.1658	1.1667	1.1681	1.1696	1.1708	1.1721	1.1734
60	1.1747	1.1759	1.1770	1.1782	1.1793	1.1803	1.1814	1.1824	1.1834	1.1844
70	1.1854	1.1863	1.1873	1.1881	1.1890	1.1898	1.1906	1.1915	1.1923	1.1930
80	1.1938	1.1945	1.1953	1.1959	1.1967	1.1973	1.1980	1.1987	1.1994	1.2001
90	1.2007	1.2013	1.2020	1.2026	1.2032	1.2038	1.2044	1.2049	1.2055	1.2060
100	1.2065									

[a] N = sample size.

TABLE D.4 FREQUENCY FACTOR FOR LOG-NORMAL DISTRIBUTION (AFTER CHOW, 1964)

	Probability in percent equal to or greater than the given variate								
	99	95	80	50	20	5	1	0.1	
c_s	−	−	−	−	+	+	+	+	c_v
0	2.33	1.65	0.84	0	0.84	1.64	2.33	3.09	0
0.1	2.25	1.62	0.85	0.02	0.84	1.67	2.40	3.22	0.033
0.2	2.18	1.59	0.85	0.04	0.83	1.70	2.47	3.39	0.067
0.3	2.11	1.56	0.85	0.06	0.82	1.72	2.55	3.56	0.100
0.4	2.04	1.53	0.85	0.07	0.81	1.75	2.62	3.72	0.136
0.5	1.98	1.49	0.86	0.09	0.80	1.77	2.70	3.88	0.166
0.6	1.91	1.46	0.85	0.10	0.79	1.79	2.77	4.05	0.197
0.7	1.85	1.43	0.85	0.11	0.78	1.81	2.84	4.21	0.230
0.8	1.79	1.40	0.84	0.13	0.77	1.82	2.90	4.37	0.262
0.9	1.74	1.37	0.84	0.14	0.76	1.84	2.97	4.55	0.292
1.0	1.68	1.34	0.84	0.15	0.75	1.85	3.03	4.72	0.324
1.1	1.63	1.31	0.83	0.16	0.73	1.86	3.09	4.87	0.351
1.2	1.58	1.29	0.82	0.17	0.72	1.87	3.15	5.04	0.381
1.3	1.54	1.26	0.82	0.18	0.71	1.88	3.21	5.19	0.409
1.4	1.49	1.23	0.81	0.19	0.69	1.88	3.26	5.35	0.436
1.5	1.45	1.21	0.81	0.20	0.68	1.89	3.31	5.51	0.462
1.6	1.41	1.18	0.80	0.21	0.67	1.89	3.36	5.66	0.490
1.7	1.38	1.16	0.79	0.22	0.65	1.89	3.40	5.80	0.517
1.8	1.34	1.14	0.78	0.22	0.64	1.89	3.44	5.96	0.544
1.9	1.31	1.12	0.78	0.23	0.63	1.89	3.48	6.10	0.570
2.0	1.28	1.10	0.77	0.24	0.61	1.89	3.52	6.25	0.596
2.1	1.25	1.08	0.76	0.24	0.60	1.89	3.55	6.39	0.620
2.2	1.22	1.06	0.76	0.25	0.59	1.89	3.59	6.51	0.643
2.3	1.20	1.04	0.75	0.25	0.58	1.88	3.62	6.65	0.667
2.4	1.17	1.02	0.74	0.26	0.57	1.88	3.65	6.77	0.691
2.5	1.15	1.00	0.74	0.26	0.56	1.88	3.67	6.90	0.713
2.6	1.12	0.99	0.73	0.26	0.55	1.87	3.70	7.02	0.734
2.7	1.10	0.97	0.72	0.27	0.54	1.87	3.72	7.13	0.755
2.8	1.08	0.96	0.72	0.27	0.53	1.86	3.74	7.25	0.776
2.9	1.06	0.95	0.71	0.27	0.52	1.86	3.76	7.36	0.796
3.0	1.04	0.93	0.71	0.28	0.51	1.85	3.78	7.47	0.818
3.2	1.01	0.90	0.69	0.28	0.49	1.84	3.81	7.65	0.857
3.4	0.98	0.88	0.68	0.29	0.47	1.83	3.84	7.84	0.895
3.6	0.95	0.86	0.67	0.29	0.46	1.81	3.87	8.00	0.930
3.8	0.92	0.84	0.66	0.29	0.44	1.80	3.89	8.16	0.966
4.0	0.90	0.82	0.65	0.29	0.42	1.78	3.91	8.30	1.000
4.5	0.84	0.78	0.63	0.30	0.39	1.75	3.93	8.60	1.081
5.0	0.80	0.74	0.62	0.30	0.37	1.71	3.95	8.86	1.155

TABLE D.5 FREQUENCY FACTOR FOR THE GUMBEL DISTRIBUTION

Sample size n	Return period								
	5	10	15	20	25	50	75	100	1000
15	0.967	1.703	2.117	2.410	2.632	3.321	3.721	4.005	6.265
20	0.919	1.625	2.023	2.302	2.517	3.179	3.563	3.836	6.006
25	0.888	1.575	1.963	2.235	2.444	3.088	3.463	3.729	5.842
30	0.866	1.541	1.922	2.188	2.393	3.026	3.393	3.653	5.727
35	0.851	1.516	1.891	2.152	2.354	2.979	3.341	3.598	
40	0.838	1.495	1.866	2.126	2.326	2.943	3.301	3.554	5.576
45	0.829	1.478	1.847	2.104	2.303	2.913	3.268	3.520	
50	0.820	1.466	1.831	2.086	2.283	2.889	3.241	3.491	5.478
55	0.813	1.455	1.818	2.071	2.267	2.869	3.219	3.467	
60	0.807	1.446	1.806	2.059	2.253	2.852	3.200	3.446	
65	0.801	1.437	1.796	2.048	2.241	2.837	3.183	3.429	
70	0.797	1.430	1.788	2.038	2.230	2.824	3.169	3.413	5.359
75	0.792	1.423	1.780	2.029	2.220	2.812	3.155	3.400	
80	0.788	1.417	1.773	2.020	2.212	2.802	3.145	3.387	
85	0.785	1.413	1.767	2.013	2.205	2.793	3.135	3.376	
90	0.782	1.409	1.762	2.007	2.198	2.785	3.125	3.367	
95	0.780	1.405	1.757	2.002	2.193	2.777	3.116	3.357	
100	0.779	1.401	1.752	1.998	2.187	2.770	3.109	3.349	5.261
∞	0.719	1.305	1.635	1.866	2.044	2.592	2.911	3.137	4.936

TABLE D.6 FREQUENCY FACTOR FOR THE PEARSON TYPE III DISTRIBUTION WITH POSITIVE SKEW COEFFICIENTS (AFTER WATER RESOURCES COUNCIL, 1967)

Skew coef.	Recurrence interval (years)							
	1.0101	2	5	10	25	50	100	200
	Percent chance (≥)							
c_S	99	50	20	10	4	2	1	0.5
3.0	−0.667	−0.396	0.420	1.180	2.278	3.152	4.051	4.970
2.9	−0.690	−0.390	0.440	1.195	2.277	3.134	4.013	4.904
2.8	−0.714	−0.384	0.460	1.210	2.275	3.114	3.973	4.847
2.7	−0.740	−0.376	0.479	1.224	2.272	3.093	3.932	4.783
2.6	−0.769	−0.368	0.499	1.238	2.267	3.071	3.889	4.718
2.5	−0.799	−0.360	0.518	1.250	2.262	3.048	3.845	4.652
2.4	−0.832	−0.351	0.537	1.262	2.256	3.023	3.800	4.584
2.3	−0.867	−0.341	0.555	1.274	2.248	2.997	3.753	4.515
2.2	−0.905	−0.330	0.574	1.284	2.240	2.970	3.705	4.444
2.1	−0.946	−0.319	0.592	1.294	2.230	2.942	3.656	4.372
2.0	−0.990	−0.307	0.609	1.302	2.219	2.912	3.605	4.298
1.9	−1.037	−0.294	0.627	1.310	2.207	2.881	3.553	4.223
1.8	−1.087	−0.282	0.643	1.318	2.193	2.848	3.499	4.147
1.7	−1.140	−0.268	0.660	1.324	2.179	2.815	3.444	4.069
1.6	−1.197	−0.254	0.675	1.329	2.163	2.780	3.388	3.990
1.5	−1.256	−0.240	0.690	1.333	2.146	2.743	3.330	3.910
1.4	−1.318	−0.225	0.705	1.337	2.128	2.706	3.271	3.828
1.3	−1.383	−0.210	0.719	1.339	2.108	2.666	3.211	3.745
1.2	−1.449	−0.195	0.732	1.340	2.087	2.626	3.149	3.661
1.1	−1.518	−0.180	0.745	1.341	2.066	2.585	3.087	3.575
1.0	−1.588	−0.164	0.758	1.340	2.043	2.542	3.022	3.489
0.9	−1.660	−0.148	0.769	1.339	2.018	2.498	2.957	3.401
0.8	−1.733	−0.132	0.780	1.336	1.993	2.453	2.891	3.312
0.7	−1.806	−0.116	0.790	1.333	1.967	2.407	2.824	3.223
0.6	−1.880	−0.099	0.800	1.328	1.939	2.359	2.755	3.132
0.5	−1.955	−0.083	0.808	1.323	1.910	2.311	2.686	3.041
0.4	−2.029	−0.066	0.816	1.317	1.880	2.261	2.615	2.949
0.3	−2.104	−0.050	0.824	1.309	1.849	2.211	2.544	2.856
0.2	−2.178	−0.033	0.830	1.301	1.818	2.159	2.472	2.763
0.1	−2.252	−0.017	0.836	1.292	1.785	2.107	2.400	2.670
0	−2.326	0	0.842	1.282	1.751	2.054	2.326	2.576

TABLE D.7 FREQUENCY FACTOR FOR THE PEARSON TYPE III DISTRIBUTION WITH NEGATIVE SKEW COEFFICIENTS (AFTER WATER RESOURCES COUNCIL, 1967)

	Recurrence interval (years)							
	1.0101	2	5	10	25	50	100	200
Skew coef.	Percent chance (≥)							
c_S	99	50	20	10	4	2	1	0.5
0.0	−2.326	0	0.842	1.282	1.751	2.054	2.326	2.576
−0.1	−2.400	0.017	0.846	1.270	1.716	2.000	2.252	2.482
−0.2	−2.472	0.033	0.850	1.258	1.680	1.945	2.178	2.388
−0.3	−2.544	0.050	0.853	1.245	1.643	1.890	2.104	2.294
−0.4	−2.615	0.066	0.855	1.231	1.606	1.834	2.029	2.201
−0.5	−2.686	0.083	0.856	1.216	1.567	1.777	1.955	2.108
−0.6	−2.755	0.099	0.857	1.200	1.528	1.720	1.880	2.016
−0.7	−2.824	0.116	0.857	1.183	1.488	1.663	1.806	1.926
−0.8	−2.891	0.132	0.856	1.166	1.448	1.606	1.733	1.837
−0.9	−2.957	0.148	0.854	1.147	1.407	1.549	1.660	1.749
−1.0	−3.022	0.164	0.852	1.128	1.366	1.492	1.588	1.664
−1.1	−3.087	0.180	0.848	1.107	1.324	1.435	1.518	1.581
−1.2	−3.149	0.195	0.844	1.086	1.282	1.379	1.449	1.501
−1.3	−3.211	0.210	0.838	1.064	1.240	1.324	1.383	1.424
−1.4	−3.271	0.225	0.832	1.041	1.198	1.270	1.318	1.351
−1.5	−3.330	0.240	0.825	1.018	1.157	1.217	1.256	1.282
−1.6	−3.388	0.254	0.817	0.994	1.116	1.166	1.197	1.216
−1.7	−3.444	0.268	0.808	0.970	1.075	1.116	1.140	1.155
−1.8	−3.499	0.282	0.799	0.945	1.035	1.069	1.087	1.097
−1.9	−3.553	0.294	0.788	0.920	0.996	1.023	1.037	1.044
−2.0	−3.605	0.307	0.777	0.895	0.959	0.980	0.990	0.995
−2.1	−3.656	0.319	0.765	0.869	0.923	0.939	0.946	0.949
−2.2	−3.705	0.330	0.752	0.844	0.888	0.900	0.905	0.907
−2.3	−3.753	0.341	0.739	0.819	0.855	0.864	0.867	0.869
−2.4	−3.800	0.351	0.725	0.795	0.823	0.830	0.832	0.833
−2.5	−3.845	0.360	0.711	0.771	0.793	0.798	0.799	0.800
−2.6	−3.889	0.368	0.696	0.747	0.764	0.768	0.769	0.769
−2.7	−3.932	0.376	0.681	0.724	0.738	0.740	0.740	0.741
−2.8	−3.973	0.384	0.666	0.702	0.712	0.714	0.714	0.714
−2.9	−4.013	0.390	0.651	0.681	0.683	0.689	0.690	0.690
−3.0	−4.051	0.396	0.636	0.660	0.666	0.666	0.667	0.667

References

CHAPTER 1

ACKERMANN, W. C., E. A. COLMAN, AND H. O. OGROSKY. 1955. "From Ocean to Sky to Land to Ocean." In *U.S. Department of Agriculture Yearbook 1955*. Washington, D.C.: U.S. Department of Agriculture, pp. 41–51.

BISWAS, A. K. 1972. *History of Hydrology*. New York: American Elsevier.

CHOW, V. T., Ed. 1964. *Handbook of Applied Hydrology: A Compendium of Water Resources Technology*. New York: McGraw-Hill.

DOOGE, J. C. I. 1973. *Linear Theory of Hydrologic Systems,* Technical Bulletin No. 1468. Washington, D.C.: Agricultural Research Service, U.S. Department of Agriculture.

———. 1984. "The Waters of the Earth." *Hydrological Sciences Journal* 29(2/6): 149–176.

JONES, P. B., D. G. WALKER, R. W. HARDEN, AND L. L. MCDANIELS. 1963. *The Development of the Science of Hydrology,* Circular No. 63-03. Austin: Texas Water Commission.

KLEMES, V. 1983. "Conceptualization and Scale in Hydrology." *Journal of Hydrology* 65: 1–23.

MEINZER, O. E., Ed. 1949. *Hydrology, Vol. 9 of Physics of the Earth*. New York: Dover.

PRICE, W. E., AND L. A. HEINDL. 1968. "What is Hydrology?" *Transactions, American Geophysical Union* 49(2): 529–533.

ROUSE, H., AND S. INCE. 1957. *History of Hydraulics*. New York: Dover.

SINGH, V. P. 1989. *Hydrologic Systems, Vol. II, Watershed Modeling*. Englewood Cliffs, NJ: Prentice Hall.

CHAPTER 2

DAVIS, R. M. 1976. "What Is Left of Our Cropland 'Frontier'?" *Soil Conservation* 42(1): 1.

DOOGE, J. C. I. 1988. "Hydrology in Perspective." *Hydrological Sciences Journal* 33(1/2): 61–85.

JOURDAN, M. R., AND G. J. SULLIVAN. 1987. "Tactical Uses of Induced Flooding." *The Military Engineer* 79(516): 433–444.

LANGBEIN, W. B., AND OTHERS. 1949. *Annual Runoff in the United States,* Geological Survey Circular 52. Washington, D.C.: U.S. Department of the Interior.

LEEDEN, V. 1975. *Water Resources of the World.* Port Washington, NY: Water Information Center.

L'VOVICH, M. I. 1979. *World Water Resources and Their Future,* English translation from the Russian edited by R. L. Nace. Washington, D.C.: American Geophysical Union.

MAIONE, U. 1988. "Present and Future Perspectives on Water Resources in Developed Countries." *Hydrological Sciences Journal* 33(1/2): 87–102.

Military Engineer. 1958. "Principal Rivers of the World." *The Military Engineer* 50(337): 386–387.

MORRIS, W. V. 1968. *Water.* Ottawa: Department of Energy, Mines and Resources.

NACE, R. L., Ed. 1971. *Scientific Framework of World Water Balance,* UNESCO Technical Papers Hydrology, No. 7. Paris: UNESCO Press.

SCHENDEL, V. 1975. *The World's Water Resources and Water Balance. Natural Resources and Development,* Vol. 1. Hanover, West Germany: Institute for Scientific Cooperation.

U.S. Geological Survey. 1949. *Large Rivers of the United States,* Geological Survey Circular 44. Washington, D.C.: U.S. Department of the Interior.

———. 1969. *Water of the World,* Leaflet GOP:1969-0-348-604. Washington, D.C.: U.S. Department of the Interior.

Water Resources Council. 1968. *The Nation's Water Resources*. Washington, D.C.: U.S. Government Printing Office.

WICKHAM, M. P., AND M. R. JOURDAN. 1985. "Military Hydrology Program: Tactical Environment Applications." *The Military Engineer* 77(502): 468–469.

CHAPTER 3

ANDERSON, H. W., M. D. HOOVER, AND K. G. REINHART. 1976. *Forests and Water: Effects of Forest Management on Floods, Sedimentation, and Water Supply,* Forest Service Technical Report RSW-18. Berkeley: U.S. Department of Agriculture.

CHERKAUER, D. S. 1975. "Urbanization's Impact on Water Quality During a Flood in Small Watersheds." *Water Resources Bulletin* 11(5): 987–998.

FLEMING, G. 1969. "Design Curves for Suspended Load Estimation." *Proceedings of the Institution of Civil Engineers* 43: 1–9.

HARROLD, L. L., G. O. SCHWAB, AND B. L. BONDURANT. 1974. *Agricultural and Forest Hydrology*. Columbus, OH: Agricultural Engineering Department, Ohio State University.

LANDSBERG, H. E. 1970. *Climates and Urban Plan-*

ning in Urban Climates, WMO Technical Note 108. Geneva: World Meteorological Organization.

LEE, R. 1980. *Forest Hydrology.* New York: Columbia University Press.

PONCE, V. M. 1989. *Engineering Hydrology: Principles and Practices.* Englewood Cliffs, NJ: Prentice Hall.

Soil Conservation Service. 1971. "Hydrology." In *SCS National Engineering Handbook.* Washington, D.C.: U.S. Department of Agriculture, Section 4.

STALL, J. B., M. L. TERSTRIEP, AND F. A. HUFF. 1970. "Storm Effects of Urbanization on Floods." Paper presented at the American Society of Civil Engineers National Water Resources Engineering and Management Conference, January 26–30, 1970, Memphis, Tennessee, preprint 1130.

CHAPTER 4

DOOGE, J. C. I. 1973. *Linear Theory of Hydrologic Systems,* Technical Bulletin No. 1468. Washington, D.C.: Agricultural Research Service, U.S. Department of Agriculture.

DRAPER, N. R., AND H. SMITH. 1966. *Applied Regression Analysis.* New York: Wiley.

SINGH, V. P. 1988. *Hydrologic Systems, Vol. 1, Rainfall-Runoff Modeling.* Englewood Cliffs, NJ: Prentice Hall.

CHAPTER 5

BENSON, M. A. 1962. *Factors Influencing the Occurrence of Flood in a Humid Region of Diverse Terrain,* USGS Water Supply Paper 1580-B. Washington, D.C.: U.S. Department of the Interior.

CARLSTON, C. W. 1963. *Drainage Density and Stream Flow,* U.S. Geological Survey Professional Paper 422-C. Washington, D.C.: U.S. Department of the Interior.

GRAY, D. M. 1961. "Interrelationships of Watershed Characteristics." *Journal of Geophysical Research* 66(4): 1215–1223.

GRAY, D. M., AND J. M. WIGHAM. 1970. "Peak Flow-Rainfall Events." In *Handbook on the Principles of Hydrology,* D. M. Gray, Ed. Ottawa: National Research Council of Canada, Section VIII.

GUPTA, V. K., I. RODRIGUEZ-ITURBE, AND E. F. WOOD, Eds. 1986. *Scale Problems in Hydrology: Runoff Generation and Basin Response.* Boston: Reidel.

HACK, J. T. 1957. *Studies of Longitudinal Stream Profiles in Virginia and Maryland,* U.S. Geological Survey Professional Paper 294-B. Washington, D.C.: U.S. Department of the Interior.

HORTON, R. E. 1932. "Drainage Basin Characteristics." *Transactions, American Geophysical Union* 13: 350–361.

———. 1945. "Erosional Development of Streams and their Drainage Basins: A Hydrophysical Approach to Quantitative Morphology." *Geological Society of America Bulletin* 56: 275–370.

JOHNSTONE, D., AND W. P. CROSS. 1949. *Elements of Applied Hydrology.* New York: Ronald Press.

LANE, L. J. 1975. *Influence of Simplifications of Watershed Geometry in Simulation of Surface Runoff.* Unpublished Ph.D. dissertation, Colorado State University, Fort Collins, Colorado.

LAURENSON, E. M. 1962. *Hydrograph Synthesis by Runoff Routing,* Report No. 66. New South Wales: Water Research Laboratory, The University of New South Wales.

LEOPOLD, L. B., AND J. P. MILLER. 1956. *Ephemeral Streams: Hydraulic Factors and Their Relation to the Drainage Net,* U.S. Geological Survey Professional Paper 282-A. Washington, D.C.: U.S. Department of the Interior.

MELTON, M. A. 1957. *An Analysis of the Relations Among Elements of Climate, Surface Properties, and Geomorphology,* Technical Report 11. New York: Department of Geology, Columbia University.

MILLER, V. C. 1959. *A Quantitative Geomorphic Study of Drainage Basin Characteristics in the Clinch Mountain Area, Virginia and Tennessee,* Project NR 389-042, Technical Report 3. New York: Department of Geology, ONR Geography Branch, Columbia University.

MORISAWA, M. E. 1959. "Measurement of Drainage Basin Outline Form." *Journal of Geology* 66: 587–591.

———. 1967. "Relation of Discharge and Stream Length in Eastern United States." In *Proceedings of the First International Hydrology Symposium,* Vol. 1. Fort Collins, CO: Colorado State University.

RODRIGUEZ-ITURBE, I. 1982. "The Coupling of Climate and Geomorphology in Rainfall–Runoff Analysis." In *Rainfall–Runoff Relationship,* V. P. Singh, Ed. Littleton, CO: Water Resources Publications, pp. 431–438.

ROGERS, W. J. 1971. "Hydrograph Analysis and Some Related Variables." In *Quantitative Geomorphology: Some Aspects and Applications,* M. Morisawa, Ed. Binghamton, NY: State University of New York, pp. 245–257.

———. 1972. "A New Concept in Hydrograph Analysis." *Water Resources Research* 8(4): 973–981.

SCHUMM, S. A. 1954. *Evolution of Drainage Systems and Slopes in Badlands at Perth Amboy, New Jersey,* Technical Report No. 8. New York: Department of Geology, Columbia University.

———. 1956. "Evolution of Drainage Systems and Slopes in Badlands at Perth Amboy, New Jersey." *Geological Society of America Bulletin* 67: 597–646.

SMART, J. S., AND A. J. SURKAN. 1967. "The Relation between Mainstream Length and Area in Drainage Basins." *Water Resources Research* 3: 963–974.

SMITH, K. G. 1950. "Standards for Grading Texture of Erosional Topography." *American Journal of Science* 248: 655–668.

SNYDER, F. F. 1938. "Synthetic Unit-Graphs." *Transactions, American Geophysical Union* 19: 447–454.

SRIBNYI, M. F. 1961. "Geomorphological Characteristics of Catchment Basins (Drainage Areas)." In *Problems of River Runoff Control,* translated by Israeli Program for Scientific Translation. Moscow: U.S.S.R. Academy of Sciences.

STRAHLER, A. N. 1957. "Quantitative Analysis of Watershed Geomorphology." *Transactions, American Geophysical Union* 38: 913–920.

———. 1964. "Quantitative Geomorphology of Drainage Basins and Channel Networks." In *Handbook of Applied Hydrology,* V. T. Chow, Ed. New York: McGraw-Hill, Section 4-II, p. 4-40–4-76.

TAYLOR, A. B., AND H. E. SCHWARZ. 1952. "Unit-Hydrograph Lag and Peak Flow Related to Basin Characteristics." *Transactions, American Geophysical Union* 23: 235–246.

U.S. Army Corps of Engineers. 1954. *Unit Hydrograph Compilations,* Civil Works Investigations, Project CW 153 (three volumes in 1949; one volume in 1954). Washington, D.C.: Department of the Army, Washington District.

U.S. Department of Agriculture. 1961. *Hydrologic Data for Experimental Agricultural Watersheds in the United States, 1956–59,* USDA Miscellaneous Publication No. 945. Washington, D.C.: U.S. Department of Agriculture.

CHAPTER 6

BERGERON, T. 1933. *On the Physics of Clouds.* Lisbon: Mem. Meteorology Association, International Union of Geodesy, Geophysics, pp. 156–175.

BOSEN, J. F. 1958. "An Appropriate Formula to Compute Relative Humidity from Drybulb and Dewpoint Temperatures." *Monthly Weather Review* 86: 486.

———. 1960. "A Formula for Approximation of the Saturation Vapor Pressure Over Water." *Monthly Weather Review* 88: 275.

———. 1961. "Formula for Approximation of the Ratio of the Saturation Vapor Pressure Over Ice to That Over Water at the Same Temperature." *Monthly Weather Review* 92: 508.

BUDYKO, M. I. 1974. *Climate and Man.* New York: Academic Press.

GEIGER, R. 1957. *The Climate Near the Ground.* Cambridge, MA: Harvard University Press.

JOHNSON, J. C. 1960. *Physical Meteorology.* New York: Wiley.

LANDSBERG, H. E. 1958. *Physical Climatology.* State College, PA: Pennsylvania State University.

LANDSBERG, H. E., AND W. C. JACOBS. 1951. *Applied Climatology—Compendium of Meteorology.* Boston: American Meteorological Society.

PETTERSSEN, S. 1958. *Introduction to Meteorology.* New York: McGraw-Hill.

———. 1964. "Meteorology." In *Handbook of Applied Hydrology,* V. T. Chow, Ed. New York: McGraw-Hill.

ROSSBY, C. G. 1941. "The Scientific Basis of Modern Meteorology." In *Climate and Man, U.S. Department of Agriculture Yearbook.* Washington, D.C.: U.S. Department of Agriculture, pp. 599–655.

SUTTON, O. G. 1953. *Micrometeorology.* New York: McGraw-Hill.

CHAPTER 7

ALVAREZ, F., AND W. K. HENRY. 1970. "Rain Gage Spacing and Reported Rainfall." *Bulletin of the International Association of Scientific Hydrology* 15(3): 97–107.

ARON, G., AND T. M. RACHFORD. 1974. "Procedures for Filling Gaps in Hydrologic Event Series." *Water Resources Bulletin* 10(4): 719–727.

BROWN, M. J., AND E. L. PECK. 1962. "Reliability of Precipitation Measurements as Related to Exposure." *Journal of the Applied Meteorology* 1(2): 203–207.

BUISHAND, T. A. 1982. "Some Methods for Testing the Homogeneity of Rainfall Records." *Journal of Hydrology* 58: 11–27.

CAUSEY, O. Y. 1953. "The Distribution of Summer Showers Over Small Drainage Area." *Monthly Weather Review* 81: 111–114.

CHANGNAN, S. A., JR., R. G. SOMONIN, AND F. A. HUFF. 1976. "A Hypothesis for Urban Rainfall Anomalies." *Journal of Applied Meteorology* 16(6): 544–560.

EAGLESON, P. S. 1967. "Optimum Density of Rainfall Networks." *Water Resources Research* 3(4): 1021–1033.

HORTON, R. E. 1924. "Estimates on Determining the Mean Precipitation of a Drainage Basin." *New England Water Works Association* 38: 1–47.

HUFF, F. A. 1970. "Sampling Errors in Measurement of Mean Precipitation." *Journal of Applied Meteorology* 9(1): 35–44.

HUFF, F. A., AND J. C. NEILL. 1957. "Area Representativeness of Point Rainfall." *Transactions, American Geophysical Union* 38(3): 341–345.

HUFF, F. A., AND J. L. VOGEL. 1978. "Urban, Topographic and Diurnal Effects on Rainfall in the St. Louis Region." *Journal of Applied Meteorology* 17(5): 565–577.

JOHANSON, R. C. 1971. *Precipitation Network Requirements for Streamflow Estimation,* Technical Report No. 147. Stanford, CA: Department of Civil Engineering, Stanford University.

KOHLER, M. A. 1949. "Double-Mass Analysis for Testing Consistency of Records and for Making Required Adjustments." *Bulletin of the American Meteorological Society* 30(5): 188–189.

LARSON, L. W., AND E. L. PECK. 1974. "Accuracy of Precipitation Measurements for Hydrologic Modeling." *Water Resources Research* 10(4): 857–863.

LINSLEY, R. K., AND M. A. KOHLER. 1951. "Variations in Storm Rainfalls Over Small Areas." *Transactions, American Geophysical Union* 32: 215–220.

MASTERS, G. M. 1974. *Introduction to Environmental Science and Technology.* New York: Wiley.

MITCHELL, J. M., JR. 1964. *A Critical Appraisal of Periodicities in Climate.* Cent. Agric. Econ. Dev. Report 20. Ames, IA: Iowa State University, pp. 189–227.

PAULHUS, J. L. H., AND M. A. KOHLER. 1952. "Interpolation of Missing Precipitation Records." *Monthly Weather Review* 80: 129–133.

SHAW, N. 1942. *Manual of Meteorology,* Vol. 2, 2d ed. London: Cambridge University Press.

SINGH, V. P., AND Y. K. BIRSOY. 1975a. *Studies on Rainfall-Runoff Modeling: 1. Estimation of Mean Areal Rainfall,* WRRI Report 065. Las Cruces, NM: New Mexico Water Resources Research Institute, New Mexico State University.

———. 1975b. "Comparison of the Methods of Estimating Mean Areal Rainfall." *Nordic Hydrology* 6(4): 222–241.

SINGH, V. P., AND P. K. CHOWDHURY. 1986. "Comparing Some Methods of Estimating Mean Areal Rainfall." *Water Resources Bulletin* 22(2): 275–282.

TABIOS III, G. Q., AND J. D. SALAS. 1985. "A Comparative Analysis of Techniques for Spatial Interpolation of Precipitation." *Water Resources Bulletin* 21(3): 365–380.

TUNG, Y. K. 1983. "Point Rainfall Estimation for a Mountainous Region." *ASCE Journal of Hydraulics Engineering* 109(10): 1386–1393.

U.S. Army Corps of Engineers. 1967. *Review Report for Papillion Creek and Tributaries, Nebraska,* Vol. II, Appendices III–VIII. Omaha, NB: U.S. Army Corps of Engineers.

WEISS, L. L. 1963. "Securing More Nearly True Precipitation Measurements." *Journal of the Hydraulic Division, ASCE* 89(HY3): 11–18.

WHITMORE, J. S., F. J. VAN EFDEN, AND K. J. HARVEY. 1961. *Assessment of Average Annual Rainfall Over Large Catchments,* C.C.T.A. Publication 66. Pretoria: Inter-African Conference on Hydrology, pp. 100–107.

World Meteorological Organization. 1969. *Guide to Hydrometeorological Practices,* 2d ed., Technical Report No. 82, WMO NO. 168. Geneva: World Meteorological Association.

CHAPTER 8

BAUER, S. W. 1974. "A Modified Horton Equation During Intermittent Rainfall." *Hydrological Sciences Bulletin* 19(2/6): 219–224.

BOUWERS, H. 1966. "Rapid Field Measurement of Air Entry Value and Hydraulic Conductivity of Soil as Significant Parameters in Flow System Analysis." *Water Resources Research* 2: 729–738.

BROOKS, R. H., AND A. T. COREY. 1964. *Hydraulic Properties of Porous Media,* Hydrology Papers No. 3. Fort Collins, CO: Colorado State University.

DOOGE, J. C. I. 1973. *Linear Theory of Hydrologic Systems,* Technical Bulletin No. 1468. Washington, D.C.: Agricultural Research Service, U.S. Department of Agriculture, pp. 267–291.

DULEY, F. L. 1939. "Surface Factors Affecting the Rate of Intake of Water by Soils." *Soil Science Society of American Proceedings* 4: 60–64.

FOK, Y. S. 1975. "A Comparison of the Green–Ampt and Philip Two-Term Infiltration Equations." *Transactions, American Society of Agricultural Engineers* 18(6): 1073–1075.

GRAY, D. M., D. I. NORUM, AND J. M. WIGHAM. 1970. "Infiltration and Physics of Flow of Water Through Media." In *Handbook on the Principles of Hydrology,* D. M. Gray, Ed. Ottawa: National Research Council of Canada, Section V.

GREEN, W. H., AND C. A. AMPT. 1911. "Studies on Soil Physics, I. Flow of Water and Air Through Soils." *Journal of Agricultural Sciences* 4: 1–24.

HJELMFELT, A. T., JR. 1979. "Curve Number Procedures as Infiltration Method." *Journal of the Hydraulics Division, Proceedings of the American Society of Civil Engineers* 106(HY6): 1107–1111.

HOLTAN, H. N. 1961. *A Concept of Infiltration Estimates in Watershed Engineering,* Paper 41-51. Washington, D.C.: Agricultural Research Service, U.S. Department of Agriculture.

HOLTAN, H. N., G. F. STILTNER, W. H. HENSON, AND N. C. LOPEZ. 1975. *USDA-74 Revised Model of Watershed Hydrology,* Technical Bulletin No. 1518, USDA-ARS. Washington, D.C.: Agricultural Research Service, U.S. Department of Agriculture.

HORTON, R. E. 1933. "The Role of Infiltration in Hydrologic Cycle." *Transactions, American Geophysical Union* 14: 446–460.

———. 1939. "Analysis of Runoff-Plot Experiments with Varying Infiltration Capacities." *Transactions, American Geophysical Union* 20(Part IV): 693–694.

———. 1940. "An Approach Toward a Physical Interpretation of Infiltration Capacity." *Soil Science Society of America* 5: 399–417.

HUGGINS, L. F., AND E. J. MONKE. 1966. *The Mathematical Simulation of the Hydrology of Small Watersheds,* Technical Report No. 1. West Lafayette, IN: Indiana Water Resources Research Center, Purdue University.

KOSTIAKOV, A. M. 1932. "On the Dynamics of the Coefficient of Water Percolation in Soils and of the Necessity of Studying It from a Dynamic Point of View for Purposes of Amelioration." In *Transactions, Sixth Comm. International Soil Science Society, Russian, Part A.* pp. 17–29.

MEIN R. G., AND C. L. LARSON. 1971. *Modeling the Infiltration Component of the Rainfall-Runoff Process,* WRC Bulletin 43. Minneapolis: Water Resources Research Institute, University of Minnesota.

MOREL-SEYTOUX, H. J. 1981. "Application of Infiltration Theory for the Determination of Excess Rainfall Hyetograph." *Water Resources Bulletin* 17(6): 1012–1022.

MUSGRAVE, G. W. 1955. "How Much of the Rain Enters the Soil?" In *Water—The Yearbook of Agriculture, 1955.* Washington, D.C.: U.S. Department of Agriculture, pp. 151–155.

OVERTON, D. E. 1964. *Mathematical Refinement of an Infiltration Equation for Watershed Engineering,* ARS41-99. Washington, D.C.: Agricultural Research Service, U.S. Department of Agriculture.

PESCHKE, G., AND M. KUTILEK. 1982. "Infiltration Model in Simulated Hydrographs." *Journal of Hydrology* 56: 369–379.

PHILIP, J. R. 1954. "An Infiltration Equation with Physical Significance." *Soil Science* 77(2): 153–157.

———. 1957. "The Theory of Infiltration: 4. Sorptivity and Algebraic Equations." *Soil Science* 84: 257–265.

RAUZI, F., C. L. FLY, AND E. J. DYKSTERHERES. 1968. *Water Intake on Midcontinental Rangelands as Influenced by Soil and Plant Cover,* Technical Bulletin 1390. Washington, D.C.: Agricultural Research Service, U.S. Department of Agriculture.

RAWLS, W. J., D. L. BRAKENSIEK, AND N. MILLER. 1983. "Green–Ampt Infiltration Parameters from Soils Data." *Journal of Hydraulics Division, Proceedings of the American Society of Civil Engineers* 109(HY1): 62–70.

ROGERS, W. F. 1970. "A Portable Infiltrometer." *Agricultural Engineering* 51: 469–471.

———. 1971. "Hydrograph Analysis and Some Related Geomorphic Variables." In *Quantitative Geomorphology: Some Aspects and Applications, Proceedings of Second Annual Geomorphology Symposium.* Binghamton, NY: State University of New York at Binghampton, pp. 249–253.

SINGH, V. P., AND F. X. YU. 1990. "Derivation of an Infiltration Equation Using a Systems Approach." *Journal of Irrigation and Drainage Engineering, ASCE* 116(6): 837–858.

SMILES, D. E., AND J. H. KNIGHT. 1976. "A Note on the Use of the Philip Infiltration Equation." *Australian Journal of Soil Research* 14: 103–108.

Soil Consevation Service. 1972. "Hydrology." In *SCS National Engineering Handbook.* Washington, D.C.: U.S. Department of Agriculture, Section 4.

VERMA, S. C. 1982. "Modified Horton's Infiltration Equation." *Journal of Hydrology* 58: 383–388.

YOUNGS, E. G. 1968. "An Estimation of Sorptivity for Infiltration Studies from Moisture Movement Considerations." *Soil Science* 106(3): 157–163.

CHAPTER 9

ANDERSON, M. G., AND T. P. BURT. 1980. "Interpretation of Recession Flow." *Journal of Hydrology* 46: 89–101.

BARNES, B. S. 1939. "The Structure of Discharge Recession Curves." *Transactions, American Geophysical Union* 20: 721–725.

———. 1940. "Discussion of 'Analysis of Runoff Characteristics.'" *Transactions, American Society of Civil Engineers* 105: 104–108.

BEAR, J. 1972. *Dynamics of Fluids in Porous Media.* New York: American Elsevier.

BOUSSINESQ, J. 1904. "Recherches theoriques sur le coulement des nappes d'eau infiltrees den le sol et sur le debit des sources." *Journal de Mathematiques Pure et Applied* 10: 5–70.

CARLSTON, C. W. 1963. *Drainage Density and Streamflow,* U.S. Geological Survey Professional Paper 422-C. Washington, D.C.: U.S. Department of the Interior.

COOPER, H. H., JR., AND C. E. JACOB. 1946. "A Generalized Graphical Method for Evaluating Formation Coefficients and Summarizing Field Well History." *Transactions, American Geophysical Union* 27: 526–534.

DAHL, N. J. 1985. "Baseflow and Water Supply." *Nordic Hydrology* 16: 309–324.

DARCY, H. 1856. Les Fontaines Publiques de la Ville de Dijon. Victor Dalmont, Paris, France.

DOOGE, J. C. I. 1960. *The Routing of Groundwater Recharge Through Typical Elements of Linear Storage,* Publication No. 52. Wallingford, England: International Association of Scientific Hydrology, pp. 286–300.

GLOVER, R. E. 1977. *Transient Ground Water Hydraulics.* Fort Collins, CO: Water Resources Publications.

HALL, F. R. 1968. "Base-Flow Recessions—A Review." *Water Resources Research* 4(5): 973–983.

———. 1982. "Subsurface Water Contributions to Streamflow." In *Rainfall-Runoff Relationship,* V. P. Singh, Ed. Littleton, CO: Water Resources Publications, pp. 237–244.

HANTUSH, M. S. 1964. "Hydraulics of Wells." In *Advances in Hydroscience,* Vol. 1, V. T. Chow, Ed. New York: Academic Press, pp. 281–432.

HORTON, R. E. 1933. "The Role of Infiltration in the Hydrologic Cycle." *Transactions, American Geophysical Union* 14: 446–460.

———. 1935. *Surface Runoff Phenomena, Part I, Analysis of the Hydrograph,* Publication 101. Voorheesville, NY: Horton Hydrological Laboratory.

JACOB, C. E. 1943. "Correlation of Ground-Water Levels and Precipitation on Long Island, New York: 1. Theory." *Transactions, American Geophysical Union* 24: 564–573.

———. 1944. "Correlation of Ground-Water Levels and Precipitation on Long Island, New York: 2. Correlation of Data." *Transactions, American Geophysical Union* 24: 928–939.

———. 1949. "Flow of Ground Water." In *Engineering Hydraulics,* H. Rouse, Ed. New York: Wiley, pp. 321–386.

JAMES, L. D., AND W. O. THOMPSON. 1970. "Least Squares Estimation of Constants in a Linear Recession Model." *Water Resources Research* 6(4): 1062–1069.

JONES, P. N., AND C. A. MCGILCHRIST. 1978. "Analysis of Hydrological Recession Curves." *Journal of Hydrology* 36: 365–374.

KNISEL, W. G. 1963. "Baseflow Recession Analysis for Comparison of Drainage Basins and Geology." *Journal of Geophysical Research* 68: 3649–3653.

KRAIJENHOFF VAN DE LEUR, D. A. 1958. "A Study of Non-Steady Groundwater Flow with Special Reference to a Reservoir-Coefficient." *Ingenieur* 19: 87–94.

KUNKLE, G. R. 1962. "The Baseflow-Duration Curve, a Technique for the Study of Groundwater Discharge from a Drainage Basin." *Journal of Geophysical Research* 67(4): 1543–1554.

LANGBEIN, W. B. 1940. "Some Channel Storage and Unit Hydrograph Studies." *Transactions, American Geophysical Union* 21: 620–627.

LAURENSON, E. M. 1961. "A Study of the Hydrograph Recession Curves of an Experimental Catchment." *Journal of the Institute of Engineers* (Australia) 33: 253–358.

LOHMAN, S. W. 1972. *Ground-Water Hydraulics,* U.S. Geological Survey Professional Paper 708. Washington, D.C.: U.S. Department of the Interior.

MEINZER, O. E. 1923. *Outline of Ground-Water Hydrology with Definitions,* U.S. Geological Survey Water-Supply Paper 494. Washington, D.C.: U.S. Department of the Interior.

MEYBOOM, P. 1961. "Estimating Ground-Water Recharge from Streamflow Hydrographs." *Journal of Geophysical Research* 66(4): 1203–1214.

MITCHELL, W. D. 1948. *Unit Hydrographs in Illinois.* Springfield, IL: Illinois Division of Waterways, pp. 26–28.

SHIRMOHAMMADI, A., W. G. KNISEL, AND J. M. SHERIDAN. 1984. "An Approximate Method for Partitioning Daily Streamflow Data." *Journal of Hydrology* 74: 335–354.

SINGH, V. P., AND Y. K. BIRSOY. 1977. "Some Statistical Relationships between Rainfall and Runoff." *Journal of Hydrology* 34: 251–268.

THEIS, C. V. 1935. "The Relation between the Lowering of the Piezometric Surface and the Rate and Duration of Discharge of a Well Using Ground-Water Discharge." *Transactions, American Geophysical Union* 16: 519–524.

TJOMSLAND, T., E. RUUD, AND K. NORDSETH. 1978. "The Physiographic Influence on Recession Runoff in Small Norwegian Rivers." *Nordic Hydrology* 9: 17–30.

TODD, D. K. 1980. *Ground Water Hydrology.* New York: Wiley.

TOEBES, C., W. B. MORRISSEY, R. SHORTER, AND M. HENDY. 1969. "Baseflow Recession Curves." In *Handbook of Hydrological Procedures: Procedure No. 9.* Wellington, New Zealand: Soil and Water Division, Ministry of Works, pp. 1–8.

TOEBES, C., AND D. D. STRANG. 1964. "On Recession Curves, I. Recession Equations." *Journal of Hydrology* (New Zealand) 3(3): 2–14.

VENETIS, C. 1969. "A Study of the Recession of Unconfirmed Aquifers." *International Association of Hydrological Sciences* 14(4/12): 119–125.

WALTON, W. C. 1970. *Ground Water Resource Evaluation.* New York: McGraw-Hill.

WEISMAN, R. N. 1977. "The Effect of Evapotranspiration on Streamflow Recession." *Hydrological Science Bulletin* 22(3/9): 371–377.

WHITE, E. L. 1977. "Sustained Flow in Small Appalachian Watersheds Underlain by Carbonate Rocks." *Journal of Hydrology* 32: 71–86.

YEVJEVICH, V. M. 1963. *Fluctuations of Wet and Dry Years,* Hydrology Paper 1. Fort Collins, CO: Colorado State University, pp. 18–20.

CHAPTER 10

BLACK, J. N., C. W. BONYTHON, AND J. A. PRESCOTT. 1954. "Solar Radiation and Duration of Sunshine." *Proceedings, Royal Meteorological Society* 80: 231–235.

BLANEY, H. F., AND W.D. CRIDDLE. 1962. *Determining Consumptive Use and Irrigation Water Requirements,* Technical Bulletin 1275. Washington, D.C.: U.S. Department of Agriculture.

BOWEN, I. S. 1926. "The Ratio of Heat Losses by Conduction and by Evaporation from Any Water Surface." *Physical Review* 27: 779–787.

California Department of Public Works. 1947. *Evaporation from Water Surfaces in California,* Bulletin 54. Sacramento: California Department of Public Works.

CHOW, V. T., Ed. 1964. *Handbook of Applied Hydrology.* New York: McGraw-Hill, pp. 11-25 to 11-33.

CRIDDLE, W. D. 1958. "Methods of Computing Consumptive Use of Water." *Journal of Irrigation and Drainage Division, Proceedings of the American Society of Civil Engineers* 84(IR1): 1–27.

———. 1966. "Empirical Methods of Predicting Evapotranspiration Using Air Temperature as the Primary Variable." In *Proceedings, Evapotranspiration Conference, American Society of Civil Engineers.* New York: American Society of Civil Engineers, pp. 54–66.

CUMMINGS, N. W. 1935. "Evaporation from Water Surfaces: Status of Present Knowledge and Need for Further Investigations." *Transactions, American Geophysical Union* 16(Part 2): 507–510.

FARNSWORTH, R. K., AND E. S. THOMPSON. 1982. *Mean Monthly, Seasonal, and Annual Pan Evaporation for the United States,* NOAA Technical Report NWS 34. Washington, D.C.: Office of Hydrology, National Weather Service, U.S. Department of Commerce.

FORTIER, S. 1907. "Evaporation Losses in Irrigation." *Engineering News* 58: 304–307.

GRAY, D. M., G. A. MCKAY, AND J. M. WIGHAN. 1970. "Energy, Evaporation and Evapotranspiration." In *Handbook on the Principles of Hydrology,* D. M. Gray, Ed. Ottawa: National Research Council of Canada, Section III, pp. 3.1 to 3.66.

HARBECK, G. E., JR., et al. 1954. *Water-Loss Investigations, Vol. 1, Lake Hefner Studies,* U.S. Geological Survey Paper 2610. Washington, D.C.: U.S. Department of the Interior.

HORTON, R. E. 1943a. "Evaporation Maps of the United States." *Transactions, American Geophysical Union* 24: 750.

———. 1943b. "Hydrologic Interrelations between Lands and Oceans." *Transactions, American Geophysical Union* 23(Part 2): 753–764.

JENSEN, M. E. 1966. "Empirical Methods of Estimating or Predicting Evapotranspiration Using Radiation." In *Proceedings, Conference on Evapotranspiration.* Chicago: St. Joseph: American Society of Agricultural Engineers, pp. 57–61.

———, Ed. 1973. *Consumptive Use of Water and Irrigation Requirements.* New York: American Society of Civil Engineers.

JENSEN, M. E., AND H. R. HAISE. 1963. "Estimating

Evapotranspiration from Solar Radiation.'' *Journal of Irrigation and Drainage Division. Proceedings of the American Society of Civil Engineers* 89: 15–41.

JENSEN, M. E., D. C. N. ROBB, AND C. E. FRANZOY. 1970. ''Scheduling Irrigations Using Climate–Crop–Soil Data.'' *Journal of Irrigation and Drainage Division, Proceedings of the American Society of Civil Engineers* 96: 25–28.

KOHLER, M. A., T. J. NORDENSEN, AND D. R. BAKER. 1959. *Evaporation Maps for the United States,* Technical Paper 37. Washington, D.C.: U.S. Weather Bureau.

MEYER, A. F. 1944. *Evaporation from Lakes and Reservoirs*. St. Paul: Minnesota Resources Commission.

PENMAN, H. L. 1948. ''Natural Evaporation from Open Water, Bare Soil and Grass.'' *Proceedings, Royal Society* (London) A193: 120–146.

RICHARDSON, B. 1931. ''Evaporation as a Function of Insolation.'' *Transactions, American Society of Civil Engineers* 95: 996–1011.

THORNWAITE, C. W. 1948. ''An Approach Toward a Rational Classification of Climate.'' *Geographical Review* 38: 55–94.

TURC, L. 1961. ''Estimation of Irrigation Water Requirements, Potential Evapotranspiration: A Simple Climatic Formula Evolved Up to Date.'' *Annals of Agronomy* 12: 13–14.

U.S. GEOLOGICAL SURVEY. 1954. *Water-Loss Investigations, Vol. 1, Lake Hefner Studies,* U.S. Geological Survey Professional Paper No. 269. Washington, D.C.: U.S. Department of the Interior.

VEIHMEYER, F. J. 1964. ''Evapotranspiration.'' In *Handbook of Applied Hydrology,* V. T. Chow, Ed. New York: McGraw-Hill, Section 11.

CHAPTER 11

HELVEY, J. D. 1971. ''A Summary of Rainfall Interception by Certain Conifers of North America.'' In *Biological Effects of the Hydrological Cycle,* E. J. Monke, Ed. West Lafayette: Purdue University.

HORTON, R. E: 1919. ''Rainfall Interception.'' *Monthly Weather Review* 147: 603–623.

LINSLEY, R. K., JR., M. A. KOHLER, AND J. L. H. PAULHUS. 1949. *Applied Hydrology*. New York: McGraw-Hill.

SOIL CONSERVATION SERVICE. 1971. ''Hydrology.'' In *SCS National Engineering Handbook*. Washington, D.C.: U.S. Department of Agriculture, Section 4.

TRIMBLE, G. R., JR. 1959. *A Problem Analysis and Program for Watershed Management Research, Northeast Forest Experiment Station,* Station Paper 116. Washington, D.C.: U.S. Forest Service.

VIESSMAN, W., JR., G. L. LEWIS, AND J. W. KNAPP. 1989. *Introduction to Hydrology*. New York: Harper and Row.

CHAPTER 12

ACKERS, P., W. R. WHITE, J. A. PERKINS, AND A. J. M. HARRISON. 1978. *Weirs and Flumes for Flow Measurement*. New York: Wiley.

BOS, M. G., Ed. 1976. *Discharge Measurement Structures,* Publication No. 20. Wagengen, The Netherlands: International Institute for Land Reclamation and Improvement.

BOYER, M. C. 1964. ''Streamflow Measurement.'' In *Handbook of Applied Hydrology,* V. T. Chow, Ed. New York: McGraw-Hill, Section 15.

BRATER, E. F., AND H. W. KING. 1976. *Handbook of Hydraulics*. New York: McGraw-Hill.

CHOW, V. T. 1959. *Open-Channel Hydraulics*. New York: McGraw-Hill.

GROVER, N. C., AND A. W. HARRINGTON. 1966. *Stream Flow Measurement, Records and Their Uses*. New York: Dover.

HERSCHY, R. W., Ed. 1978. *Hydrometry*. New York: Wiley.

KENNEDY, E. J. 1984. ''Discharge Ratings at Gaging Stations.'' In *Applications of Hydraulics, Techniques of Water Resources Investigation of the United States Geological Survey*. Washington, D.C.: U.S. Department of the Interior.

KOLUPAILA, S. 1960. *Bibliography of Hydrometry*. South Bend, IN: University of Notre Dame.

LANGBEIN, W. B. 1968. *Hydrological Bench Marks,* WMO/IHD Project, Report No. 8. Geneva: World Meteorological Organization.

LEOPOLD, L. B., AND T. MADDOCK, JR. 1953. *The Hydraulic Geometry of Stream Channels and Some Physiographic Implications,* U.S. Geological Survey Professional Paper 252. Washington, D.C.: U.S. Department of the Interior.

LINSLEY, R. K., M. A. KOHLER, AND J. L. H. PAULHUS. 1975. *Hydrology for Engineers*. New York: McGraw-Hill.

RANTZ, S. E., AND OTHERS. 1983a. *Measurement and Computation of Streamflow: Vol. 1. Measurement of Stage and Discharge,* U.S. Geological Surgey Water-Supply Paper 2175. Washington, D.C.: U.S. Department of the Interior.

RANTZ, S. E., AND OTHERS. 1983b. *Measurement and Computation of Streamflow: Vol. 2. Measurement of Stage and Discharge,* U.S. Geological Survey Water-Supply Paper 2175. Washington, D.C.: U.S. Department of the Interior.

U.S. GEOLOGICAL SURVEY. 1979. *Water Resources Data for Nebraska,* U.S. Geological Survey Water-Data Report NE-79-1, Water Year 1979, Lincoln, Nebraska. Washington, D.C.: U.S. Department of the Interior.

WISLER, C. O., AND E. F. BRATER. 1949. *Hydrology*. New York: Wiley.

WORLD METEOROLOGICAL ORGANIZAITON. 1965. *Guide to Hydrometeorological Practices,* WMO Publication No. 168. Geneva: World Meteorological Organization.

CHAPTER 13

BARNES, B. S. 1940. ''Discussion of 'Analysis of Runoff Characteristics.''' *Transactions, American Society of Civil Engineers* 105: 104–108.

BOYD, M. J. 1978. ''A Storage Routing Model Relating Drainage Basin Hydrology and Geomorphology." *Water Resources Research* 14(5): 921–928.

CHOW, V. T., Ed. 1964. *Handbook of Applied Hydrology*. New York: McGraw-Hill, Chapter 14.

KIRPICH, T. P. 1940. "Time of Concentration of Small Agricultural Watersheds." *Civil Engineering* 10(6): 362.

LINSLEY, R. K., M. A. KOHLER, AND J. L. H. PAULHUS. 1958. *Hydrology for Engineers*. New York: McGraw-Hill.

MCCUEN, R. H., S. L. WONG, AND W. J. RAWLS. 1984. "Estimating Urban Time of Concentration." *Journal of Hydraulic Engineering, ASCE* 110(7): 887–904.

PANU, U. S., AND V. P. SINGH. 1981. *Basin Lag*, Technical Report MSSU-EIRS-CE-81-4. Mississippi State. MS: Engineering and Industrial Research Station, Mississippi State University.

SINGH, V. P. 1988. *Hydrologic Systems, Vol. 1: Rainfall–Runoff Modeling*. Englewood Cliffs, NJ: Prentice Hall.

SINGH, V. P., AND H. AMINIAN. 1986. "An Empirical Relation between Volume and Peak of Direct Runoff." *Water Resources Bulletin* 22(5): 725–730.

Soil Conservation Service. 1972. "Hydrology." In *SCS National Engineering Handbook*. Washington, D.C.: U.S. Department of Agriculture, Section 4.

———. 1975. *Urban Hydrology for Small Watersheds*, Technical Release No. 55. Washington, D.C.: U.S. Department of Agriculture.

CHAPTER 14

AYERS, H. D. 1962. *A Survey of Watershed Yield*, Report No. 63. New South Wales: Water Research Laboratory, University of New South Wales, Kensington.

BUDYKO, M. I. 1948. *Evaporation Under Natural Conditions*, translated from the Russian by the Israeli Program for Scientific Translation. Washington, D.C.: National Science Foundation and U.S. Department of the Interior.

HAMON, W. R. 1963. *Computation of Direct Runoff Amounts from Storm Rainfall*, Publication No. 63. Wallingford, England: International Association of Scientific Hydrology, pp. 52–62.

HAWKINS, R. H. 1978. "Effects of the Rainfall Intensity on Runoff Curve Numbers." In *Proceedings of the 1978 Meetings of the Arizona Section of the American Water Resources Association and the Hydrology Section of the Arizona Academy of Science*, Vol. 8. Flagstaff, Arizona: Arizona Academy of Sciences, pp. 53–64.

LANGBEIN, W. B., AND OTHERS. 1949. *Annual Runoff in the United States*, U.S. Geological Survey Circular 52. Washington, D.C.: U.S. Department of the Interior.

RIPPL, W. 1882. "The Capacity of Storage Reservoirs for Water Supply." *Proceedings, Institute of Civil Engineers* 71: 270–278.

SINGH, V. P., AND W. T. DICKINSON. 1975. "A Simple Runoff Model Utilizing Soil Moisture Parameters." In *Proceedings, Second World Congress*, Vol. V. New Delhi: Central Board of Irrigation and Power, pp. 111–116.

SMITH, R. E. 1978. *A Proposed Infiltration Model for Use in Simulation of Field Scale Watershed Hydrology*. Paper presented at the U.S. Department of Agriculture, Agricultural Research Service, Nonpoint Pollution Modeling Workshop, Arlington, Texas.

SOIL CONSERVATION SERVICE. 1969. "Hydrology." In *SCS National Engineering Handbook*. Washington, D.C.: U.S. Department of Agriculture, Section 4.

CHAPTER 15

BARNES, B. S. 1959. "Consistency in Unit Hydrographs." *Journal of the Hydraulics Division, Proceedings of the American Society of Civil Engineers* 85(HY8): 39–61.

BERNARD, M. 1935. "An Approach to Determinate Streamflow." *Transactions, American Society of Civil Engineers* 100: 347–395.

COLLINS, W. T. 1939. "Runoff Distribution Graphs from Precipitation Occurring in More Than One Time Unit." *Civil Engineering* 9(9): 559–561.

DICKINSON, T. W., AND H. D. AYERS. 1965. "The Effect of Storm Characteristics on the Unit Hydrograph." *Transactions of the Engineering Institute of Canada* 8(A-15): 3–7.

DISKIN, M. H. 1979. "Some Dimensional Considerations in the Unit Hydrograph Theory." *Journal of Hydrology* 42: 199–208.

JOHNSTONE, D., AND W. P. CROSS. 1949. *Elements of Applied Hydrology*. New York: Ronald Press.

MINSHALL, N. E. 1960. "Predicting Storm Runoff on Small Agricultural Watersheds." *Journal of the Hydraulics Division, Proceedings of the American Society of Civil Engineers* 86(HY8): 17–38.

RENDON-HERRERO, O. 1978. "Unit Sediment Graph." *Water Resources Research* 14(5): 889–901.

SHERMAN, L. K. 1932. "Streamflow from Rainfall by the Unit Graph Method." *Engineering News-Record* 108: 501–505.

SINGH, V. P., A. BANIUKIEWICZ, AND V. J. CHEN. 1982. "An Instantaneous Unit Sediment Graph Study for Small Upland Watersheds." In *Modeling Components of the Hydrologic Cycle*, V. P. Singh, Ed. Littleton, CO: Water Resources Publications, pp. 539–554.

WILLIAMS, J. R. 1978. "A Sediment Graph Model Based on Instantaneous Unit Sediment Graph." *Water Resources Research* 14(4): 659–664.

CHAPTER 16

BULL, J. A. 1968. "Unit Graphs for Nonuniform Rainfall Distribution." *Journal of the Hydraulics Division, Proceedings of the American Society of Civil Engineers* 94(HY1): 235–257.

CHU, S. T., AND W. F. LYTLE. 1972. "Time Base for Watersheds in South Dakota." *Transactions of the American Society of Agricultural Engineers* 15: 276–279.

Cordery, I. 1968. "Synthetic Unit Graphs for Small Catchments in Eastern New South Wales." *Civil Engineering Transactions, Institution of Engineers* (Australia) 10: 47–58.

Dooge, J. C. I. 1973. *Linear Theory of Hydrologic Systems,* Technical Bulletin No. 1468. Washington, D.C.: Agricultural Research Service, U.S. Department of Agriculture.

Hudlow, M. D., and R. A. Clark. 1969. "Hydrological Synthesis by Digital Computer." *Journal of the Hydraulics Division, Proceedings of the American Society of Civil Engineers* 95(HY3): 839–860.

Laurenson, E. M. 1964. "A Catchment Storage Model for Runoff Routing." *Journal of Hydrology* 2: 141–163.

Linsley, R. K. 1943. "Application to the Synthetic Unit Graph in the Western Mountain States." *Transactions, American Geophysical Union* 24: 581–587.

Mathur, B. S. 1974. "Natural Catchment Representation by a Series of Linear Channels." In *Proceedings of the Warsaw Symposium on Mathematical Models in Hydrology,* Vol. 2. Warsaw, Poland: UNESCO/IAHS, pp. 634–642.

Miller, A. C., S. N. Kerr, and D. J. Spaeder. 1983. "Calibration of Snyder Coefficients for Pennsylvania." *Water Resources Bulletin* 19(4): 625–630.

Morgan, P. E., and S. M. Johnson. 1962. "Analysis of Synthetic Unit Graph Methods." *Journal of the Hydraulics Division, Proceedings of the American Society of Civil Engineers* 88(HY5): 199–220.

Nash, J. E. 1958. "Determining Runoff from Rainfall." *Proceedings of the Institution of Civil Engineers* (Ireland) 10: 163–184.

Snydre, F. F. 1938. "Synthetic Unit Graphs." *Transactions, American Geophysical Union* 19: 447–454.

Soil Conservation Service. 1971. "Hydrology." In *SCS National Engineering Handbook.* Washington, D.C.: U.S. Department of Agriculture, Section 4.

U.S. Army Corps of Engineers. 1940. *Engineering Construction—Flood Control.* Fort Belvoir, VA: Engineering School, U.S. Army Corps of Engineers.

Viessman, W., Jr., J. W. Knapp, G. L. Lewis, and T. E. Harbaugh. 1977. *Introduction to Hydrology.* New York: Harper and Row.

CHAPTER 17

Dooge, J. C. I. 1959. "A General Theory of the Unit Hydrograph." *Journal of Geophysical Research* 64(2): 241–256.

Kalinin, G. P., and P. I. Milyukov. 1958. "On the Computation of Unsteady Flow in Open Channels." *Meteorologiya Gidrologiya Zhuzurnal* (Leningrad) 10: 10–18.

Mathur, B. S. 1972a. "Natural Catchment Representation by a Series of Linear Channels, Part I." In *Proceedings of the Warsaw Symposium on Mathematical Models in Hydrology,* Vol. 2. Warsaw, Poland: UNESCO/IAHS, pp. 634–642.

Mathur, B. S. 1972b. "Natural Catchment Representation by a Series of Linear Channels, Part II." In *Proceedings of the Warsaw Symposium on Mathematical Models in Hydrology,* Vol. 2. Warsaw, Poland: UNESCO/IAHS, pp. 643–652.

Nash, J. E. 1957. *The Form of the Instantaneous Unit Hydrograph,* Publication 42. Wallingford, England: International Association of Scientific Hydrology, pp. 114–112.

———. 1958. "Determining Runoff from Rainfall." *Proceedings of the Institution of Civil Engineers* (Ireland) 10: 163–184.

———. 1959. "Systematic Determination of Unit Hydrograph Parameters." *Journal of Geophysical Research* 64(1): 111–115.

———. 1960. "A Unit Hydrograph Study, with Particular Reference to British Catchments." *Proceedings of the Institution of Civil Engineers* (London) 17: 249–282.

Soil Conservation Service. 1972. "Hydrology." In *SCS National Engineering Handbook.* Washington, D.C.: U.S. Department of Agriculture, Section 4.

CHAPTER 18

Chow, V. T. 1962. *Hydrologic Determination of Waterway Areas for the Design of Drainage Structures in Small Drainage Basins,* Engineering Experiment Station Bulletin No. 462. Urbana: University of Illinois, Engineering Experiment Station.

Kuichling, E. 1889. "The Relation between the Rainfall and the Discharge of Sewers in Populous Districts." *Transactions, American Society of Civil Engineers* 20: 37–40.

Rogers, W. F. 1980. "A Practical Model for Linear and Nonlinear Runoff." *Journal of Hydrology* 46: 51–78.

Rogers, W. F., and V. P. Singh. 1986. "Evaluating Flood Retarding Structures." *Advances in Water Resources* 9: 236–244.

———. 1988. "Drainage Basin Peak Discharge Rating Curve." *Hydrological Processes* 2: 245–253.

Rogers, W. F., and H. A. Zia. 1982. "Linear and Nonlinear Runoff from Large Drainage Basins." *Journal of Hydrology* 55: 267–278.

Singh, V. P., and H. Aminian. 1986. "An Empirical Relation between Volume and Peak of Direct Runoff." *Water Resources Bulletin* 22(5): 725–730.

CHAPTER 19

Anderson, E. A., and N. H. Crawford. 1964. *The Synthesis of Continuous Snowmelt Runoff Hydrographs on a Digital Computer,* Technical Report No. 36. Stanford, CA: Department of Civil Engineering, Stanford University.

Garstka, W. U., L. D. Love, B. C. Goodell, and F. A. Bertle. 1958. *Factors Affecting Snowmelt and Streamflow.* Washington, D.C.: Bureau of Reclamation and U.S. Forest Service.

Gray, D. M., and D. H. Male. 1991. *Handbook of Snow*. Toronto: Pergamon Press.

Kondrat'ev, K. 1954. *Radiant Energy of the Sun*. Leningrad: Gidromet.

Light, P. 1941. "Analysis of High Rates of Snow-Melting." *Transactions, American Geophysical Union* 22 (Part I): 195–205.

Pysklywec, D. W. 1966. *Correlation of Snowmelt with the Controlling Meteorological Parameters*. Masters thesis, Department of Civil Engineering, University of New Brunswick, Fredericton.

Schneider, S. R., D. R. Wiesnet, and D. F. McGinnis. 1976. *River Basin Snow Mapping at the National Environmental Satellite Service*, NOAA Technical Memorandum NESS 83. Washington, D.C.: U.S. National Weather Service.

Seligman, G. 1962. *Snow Structure and Ski Fields*. Edinburgh: R. R. Clarke.

Soil Conservation Service. 1972. "Snow Survey and Water Supply Forecasting." In *SCS National Engineering Handbook*. Washington, D.C.: U.S. Department of Agriculture, Section 22.

Sverdrup, H. U. 1936. "The Eddy Conductivity of the Air Over a Smooth Snowfield." *Geophysical Publication* 11(7):1–69.

U.S. Army Corps of Engineers. 1956. *Snow Hydrology—Summary Report of Snow Investigations*. Portland, OR: North Pacific Division, Department of the Army.

U.S. Army Corps of Engineers. 1960. *Engineering and Design Manuals*, EM1110-2-1406. Washington, D.C.: U.S. Department of the Army.

U.S. Weather Bureau. 1951. *Operation and Maintenance of Storage Precipitation Gages*. National Weather Service, NOAA, U.S. Department of Commerce.

Warnick, C. C., and V. E. Penton. 1971. "New Methods of Measuring Equivalent of Snow Pack for Automatic Recording at Remote Mountain Locations." *Journal of Hydrology* 13: 201–213

Wilson, W. T. 1941. "An Outline of the Thermodynamics of Snowmelt." *Transactions, American Geophysical Union* 22: 182–195.

Work, R. A. 1953. *Stream-Flow Forecasting from Snow Surveys*, Circular No. 914. Washington, D.C.: U.S. Department of Agriculture.

CHAPTER 20

Chow, V. T. 1959. *Open-Channel Hydraulics*. New York: McGraw-Hill.

Dooge, J. C. I. 1980. *Flood Routing in Channels*. Lecture Notes, Department of Civil Engineering, University College, Dublin.

Fread, D. L. 1982. "Flood Routing: A Synopsis of Past, Present and Future Capability." In *Rainfall–Runoff Relationship*, V. P. Singh, Ed. Littleton, CO: Water Resources Publications, pp. 521–542.

Goodrich, R. D. 1931. "Rapid Calculation of Reservoir Discharge." *Civil Engineering* 1: 417–418.

Linsley, R. K., M. A. Kohler, and J. L. H. Paulhus. 1975. *Applied Hydrology*. New York: McGraw-Hill.

Singh, V. P. 1988. *Hydrologic Systems, Vol. 1: Rainfall–Runoff Modeling*. Englewood Cliffs, NJ: Prentice Hall.

Weinmann, P. E., and E. M. Laurenson. 1979. "Approximate Flood Routing Methods: A Review." *Journal of the Hydraulics Division, Proceedings of the American Society of Civil Engineers* 105(HY12): 1521–1536.

CHAPTER 21

Dooge, J. C. I. 1980. *Flood Routing in Channels*. Lecture Notes, Department of Civil Engineering, University College, Dublin.

Fread, D. L. 1982. "Flood Routing: A Synopsis of Past, Present and Future Capability." In *Rainfall–Runoff Relationship*, V. P. Singh, Ed. Littleton, CO: Water Resources Publications, pp. 521–542.

Linsley, R. K., M. A. Kohler, and J. L. H. Paulhus. 1958. *Hydrology for Engineers*. New York: McGraw-Hill.

Singh, V. P. 1988. *Hydrologic Systems, Vol. 1: Rainfall–Runoff Modeling*. Englewood Cliffs, NJ: Prentice Hall.

Soil Conservation Service. 1964. "Hydrology." In *SCS National Engineering Handbook*. Washington, D.C.: U.S. Department of Agriculture, Section 4, Chapters 17 and 21.

CHAPTER 22

American Society of Civil Engineers. 1970. "Sediment Sources and Sediment Yields." *Journal of the Hydraulics Division, Proceedings of the American Society of Civil Engineers* 96(HY6): 1283–1329.

Bennett, J. P. 1974. "Concepts of Mathematical Modeling of Sediment Yield." *Water Resources Research* 10(3): 485–492.

Carter, C. E., J. D. Greer, H. J. Braud, and J. M. Floyd. 1974. "Raindrop Characteristics in South Central United States." *Transactions, American Society of Agricultural Engineers* 17: 1033–1037.

Chen, V. J., and C. Y. Kuo. 1984. "A Study of Synthetic Sediment Graphs for Ungaged Watersheds." *Journal of Hydrology* 84: 35–54.

Dendy, F. E., G. C. Bolton. 1976. "Sediment Yield–Runoff Drainage Area Relationships in the United States." *Journal of Soil and Water Consideration* 31(6): 264–266.

Dragoun, F. J., and C. R. Miller. 1964. *Sediment Characteristics of Two Small Agricultural Watersheds in Central Nebraska*. Paper presented at the 1964 Summer Meeting of the American Society of Agricultural Engineers, June 21–24, Fort Collins, Colorado.

Ellison, W. D. 1946. "Soil Detachment and Transportation." *Soil Conservation* 11(8): 179.

———. 1947a. "Soil Erosion Studies—Part I." *Agricultural Engineering* 28: 145–146.

———. 1947b. "Soil Erosion Studies—Part II, Soil

Detachment Hazard by Raindrop Splash." *Agricultural Engineering* 28: 197–201.

———. 1947c. "Soil Erosion Studies—Part III, Some Effects of Soil Erosion on Infiltration and Surface Runoff." *Agricultural Engineering* 28:245–248.

———. 1947d. "Soil Erosion Studies—Part IV, Soil Erosion, Soil Loss, and Some Effects of Soil Erosion." *Agricultural Engineering* 28: 297–300.

———. 1947e. "Soil Erosion Studies—Part V, Soil Transportation in the Splash Process." *Agricultural Engineering* 28: 349–351, 353.

FLAXMAN, E. M. 1972. "Predicting Sediment Yield in Western United States." *Journal of the Hydraulics Division, Proceedings of the American Society of Civil Engineers* 98(HY12): 2073–2085.

FOSTER, G. R. 1971. "The Overland Flow Process Under Natural Conditions." In *Biological Effects in the Hydrological Cycle, Proceedings of the Third International Seminar for Hydrology Professors*. West Lafayette, In: Purdue University, pp. 173–185.

FOSTER, G. R., J. LAMBARDI, AND W. C. MOLDENHAUER. 1982. "Evaluation of Rainfall-Runoff Erosivity Factors for Individual Storms." *Transactions of the ASAE* 25: 124–129.

FOSTER, G. R., AND L. D. MEYER. 1975. "Mathematical Simulation of Upland Erosion by Fundamental Erosion Mechanics." In *Present and Prospective Technology for Predicting Sediment Yields and Sources*, ARS-S-40. Washington, D.C.: Agricultural Research Service, U.S. Department of Agriculture, pp. 190–207.

———. 1977. "Soil Erosion and Sedimentation by Water—An Overview." In *Proceedings of the National Symposium on Soil Erosion and Sedimentation by Water*. St. Joseph, MI: American Society of Agricultural Engineers, pp. 1–13.

FOSTER, G. R., AND W. H . WISCHMEIER. 1974. "Evaluating Irregular Slopes for Soil Erosion Prediction." *Transactions, American Society of Agricultural Engineers* 17: 305–309.

GUNN, R., AND G. D. KINZER. 1949. "The Terminal Velocity of Fall for Water Droplets." *Journal of Meteorology* 6: 243–248.

GYR, A. 1983. "Towards a Better Definition of Three Types of Sediment Transport." *Journal of Hydraulic Research* 1(1): 1–15.

HARTLEY, D. M. 1984. "Runoff and Erosion Response of Reclaimed Surfaces." *Journal of Hydraulic Engineering, ASCE* 110(9): 1181–1199.

HEEDE, B. H. 1984. "Overland Flow and Sediment Delivery: An Experiment with Small Subdrainage in Southeastern Ponderosa Pine Forests (Colorado, USA)." *Journal of Hydrology* 72: 261–273.

HUDSON, N. W. 1971. "Raindrop Size." In *Soil Conversation*. Ithaca, NY: Cornell University Press, pp. 50–56.

JOHNSON, J. W. 1943. "Distribution Graphs of Suspended-Matter Concentration." *Transactions, American Society of Civil Engineers* 108:941–964.

LLOYD, C. H., AND G. W. ELEY. 1952. "Graphical Solutions of Probable Soil Loss Formula for Northeastern Region." *Journal of Soil and Water Conservation* 7: 189–191.

LOUGHRAN, R. J., B. L. CAMPBELL, AND G. L. ELLIOTT. 1986. "Sediment Dynamics in Partially Cultivated Catchment in New South Wales, Australia." *Journal of Hydrology* 83: 285–297.

MANER, S. B. 1958. "Factors affecting Sediment Delivery Ratios in the Red Hills Physiographic Area." *Transactions, American Geophysical Union* 39(4): 669–675.

MEYER, L. D., G. R. FOSTER, AND S. MIKOLOV. 1975. "Effect of Flow Rate and Canopy on Rill Erosion." *Transactions of the American Society of Agricultural Engineers* 12(6): 754–758, 762.

MEYER, L. D., AND W. H. WISCHMEIER. 1969. "Mathematical Simulation of the Process of Soil Erosion by Water." *Transactions of the American Society of Agricultural Engineers* 12(6): 754–758.

MUSGRAVE, G. W. 1947. "The Quantitative Evaluation of Factors in Water Erosion: A First Approximation." *Journal of Soil and Water Conservation* 2(3): 133–138.

MUTCHLER, C. K., AND J. D. GREER. 1980. "Effect of Slope Length on Erosion from Low Slopes." *Transactions of the American Society of Agricultural Engineers* 23: 866–869, 876.

MUTCHLER, C. K., AND K. C. MCGREGOR. 1983. "Erosion from Low Slopes." *Water Resources Research* 19(5): 1323–1326.

PARK, S. W., J. K. MITCHELL, AND G. D. BUBENZER. 1982. "Splash Erosion Modeling: Physical Analyses." *Transactions of the American Society of Agricultural Engineers* 25: 357–361.

REID, L. M., AND T. DUNNE. 1984. "Sediment Production from Forest Streams." *Transactions, American Society of Agricultural Engineers* 20(11): 1753–1761.

RENARD, K. G. 1977. "Past, Present, and Future Water Resources Research in Arid and Semi-Arid Areas of the Southwestern United States." In *Proceedings of the 1977 Hydrology Symposium*. Canberra: Australian Institution of Engineers, pp. 1–29.

———. 1980. "Estimating Erosion and Sediment Yield from Rangeland." In *Proceedings, ASCE Symposium on Watershed Management*. Australia: Institution of Engineers, pp. 162–175.

RENDON-HERRERO, O. 1974. "Estimation of Washload Produced by Certain Small Watersheds." *Journal of the Hydraulics Division, Proceedings of the American Society of Civil Engineers* 109(HY7): 835–848.

———. 1978. "Unit Sediment Graph." *Water Resources Research* 14(5): 889–901.

RENDON-HERRERO, O., V. P. SINGH, AND V. J. CHEN. 1980. "ER-ES Watershed Relationship." In *Proceedings, International Symposium on Water Resources Systems*, Vol. 1. Roorkee, India: University of Rourkee, pp. II-8-41 to II-8-47.

RENFRO, G. W. 1975. "Use of Erosion Equations and Sediment Delivery Ratios for Predicting Sediment Yield." In *Present and Prospective Technol-*

ogy for Predicting Sediment Yields and Sources, ARS-S-40. Washington, D.C.: Agricultural Research Service, U.S. Department of Agriculture, pp. 33–45.

RICHARDSON, C. W., G. R. FOSTER, AND D. A. WRIGHT. 1983. "Estimation of Erosion Index from Daily Rainfall Amount." *Transactions of the American Society of Agricultural Engineers* 26:153–156, 160.

RITCHIE, J. C. 1972. "Sediment, Fish and Fish Habitat." *Journal of Soil and Water Conservation* 27: 124–125.

ROEHL, J. R. 1962. *Sediment Source Areas, Delivery Ratios and Influencing Morphological Factors,* Publication No. 59. Wallingford, England: International Association of Scientific Hydrology, pp. 202–213.

SHERMAN, L. K. 1932. "Streamflow from Rainfall by Unit-Graph Method." *Engineering News-Record* 108(4): 501–505.

SINGH, V. P. 1973. "Predicting Sediment Yield in Western United States—A Discussion." *Journal of the Hydraulics Division, Proceedings of the American Society of Civil Engineers* 99(HY10): 1891–1894.

SINGH, V. P., A. BANIUKIEWICZ, AND V. J. CHEN. 1982. "An Instantaneous Unit Sediment Graph Study for Small Upland Watersheds." In *Modeling Components of the Hydrologic Cycle,* V. P. Singh, Ed. Littleton, CO: Water Resources Publications, pp. 539–554.

SINGH, V. P., AND V. J. CHEN. 1982. "On the Relation between Sediment Yield and Runoff Volume." In *Modeling Components of the Hydrologic Cycle,* V. P. Singh, Ed. Littleton, CO: Water Resources Publications, pp. 555–570.

Soil Conservation Service. 1969. *Hydrologic Group K and T Factors of Soils Having Type Locations in the South Region. USDA: South Regional Technical Service, U.S. Department of Agriculture.*

——. 1971. "Sediment Sources, Yields, and Delivery Ratios." In *SCS National Engineering Handbook.* Washington, D.C.: U.S. Department of Agriculture, Section 3.

SRIVASTAVA, P. K., R. A. RASTOGI, AND H. S. CHAUHAN. 1984. "Prediction of Storm Sediment Yields from a Small Watershed." *Journal of Agricultural Engineering, ISAE* 21(1–2): 121–126.

STEWART, B. A., W. H. WISCHMEIER, D. A. WOOLHISER, et al. 1975a. *Control of Water Pollution from Cropland: A Manual for Guideline Development,* Vol. 1, ARS-H-5-1. Washington, D.C.: Agricultural Research Service, U.S. Department of Agriculture.

——. 1975b. *Control of Water Pollution from Cropland: An Overview,* Vol. 2, ARS-H-5-2. Washington, D.C.: Agricultural Research Service, U.S. Department of Agriculture.

WILLIAMS, J. R. 1975a. "Sediment Yield Prediction with Universal Equation Using Runoff Energy Factor." In *Present and Prospective Technology for Predicting Sediment Yields and Sources,* ARS-S-40. Washington, D.C.: Agricultural Research Service, U.S. Department of Agriculture, pp. 244–252.

——. 1975b. "Sediment Routing for Agricultural Watersheds." *Water Resources Bulletin* 11(5): 965–974.

——. 1977. *Sediment Delivery Ratios Determined with Sediment and Runoff Models,* Publication No. 122. Wallingford, England: International Association of Hydrological Sciences, pp. 168–178.

——. 1978a. "A Sediment Yield Routing Model." In *Proceedings of the Speciality Conference on Verification of Mathematical and Physical Models in Hydraulic Engineering, ASCE.* New York: American Society of Civil Engineers, pp. 662–670.

——. 1978b. "A Sediment Graph Model Based on an Instantaneous Unit Sediment Graph." *Water Resources Research* 14(4): 659–664.

——. 1979. "Model for Predicting Sediment, Phosphorus, and Nitrogen Yields from Rural Basins." In *Proceedings, XVII Congress of the International Association for Hydraulic Research,* Vol. 5. Delft, The Netherlands: International Association of Hydraulic Research, pp. 107–116.

——. 1980. "SPNM, A Model for Predicting Sediment, Phosphorus, Nitrogen Yields from Agricultural Basins." *Water Resources Bulletin* 16(5): 843–848.

——. 1981. *Mathematical Modeling of Watershed Sediment Yield.* Paper presented at the International Symposium on Rainfall–Runoff Modeling, May 18–21, Mississippi State University, Mississippi State, Mississippi.

WILLIAMS, J. R., AND H. D. BERNDT. 1972. "Sediment Yield Computed with Universal Equation." *Journal of the Hydraulics Division, Proceedings of the American Society of Civil Engineers* 98(HY12): 2087–2098.

WILLIAMS, J. R., E. A. HILER, AND R. W. BAIRD. 1971. "Prediction of Sediment Yields from Small Watersheds." *Transactions, American Society of Agricultural Engineers* 14(6): 1158–1162.

WISCHMEIER, W. H. 1959. "A Rainfall–Erosion Index for Universal Soil Loss Equation." *Soil Science Society of America Proceedings* 23: 246–249.

——. 1962. "Storms and Soil Conservation." *Journal of Soil and Water Conservation* 17(2): 55–59.

——. 1975. "Estimating the Soil Loss Equation's Cover and Management Factor for Undisturbed Areas." In *Present and Prospective Technology for Predicting Sediment Yields and Sources,* ARS-S-40. Washington, D.C.: Agricultural Research Service, U.S. Department of Agriculture, pp. 118–124.

WISCHMEIER, W. H., C. B. JOHNSON, AND B. V. CROSS. 1971. "A Soil Erodibility Nomograph for Farm Land and Construction Sites." *Journal of Soil and Water Conservation* 26:189–193.

WISCHMEIER, W. H., AND J. V. MANNERING. 1969. "Relation of Soil Properties to Its Erodibility." *Soil Science Society of America Proceedings* 23:131–137.

WISCHMEIER, W. H., AND D. D. SMITH. 1958. "Rainfall Energy and Its Relationship to Soil Loss." *Transactions, American Geophysical Union* 36:285–291.

———. 1978. *Predicting Rainfall Erosion Losses—A Guide to Conservation Planning,* Agricultural Handbook No. 537. Washington, D.C.: Science and Education Administration, U.S. Department of Agriculture.

WOO, H. S., P. Y. JULIEN, AND E. V. RICHARDSON. 1986. "Washload and Fine Sediment Load." *Journal of Hydraulic Engineering, ASCE* 112(6) 541–545.

CHAPTER 23

ABBOTT, M. B., J. C. BATHURST, J. A. CUNGE, P. E. O'CONNELL, AND J. RASMUSSEN. 1986a. "An Introduction to the European Hydrological System—Systeme Hydrologique Europeen, 'SHE,' 1: History and Philosophy of a Physically Based Distributed Modelling System." *Journal of Hydrology* 87: 45–59.

———. 1986b. "An Introduction to the European Hydrological System—Systeme Hydrologique Europeen, 'SHE,' 2: Structure of a Physically Based Distributed Modelling System." *Journal of Hydrology* 87: 61–77.

ANDO, Y., K. MUSIAKE, AND Y. TAKAHASI. 1983. "Modeling of Hydrologic Processes in a Small Natural Hillslope Basin, Based on the Synthesis of Partial Hydrological Relationships." *Journal of Hydrology* 64: 311–337.

ANDREWS, W. H., P. H. RILEY, AND M. B. MASTELLER. 1978. *Mathematical Modeling of a Sociological and Hydrologic Decision System,* Water Resources Planning Series Report P-78-004. Logan, UT: Utah Water Research Laboratory, Utah State University.

BERGSTROM, S. 1976. *Development and Application of a Conceptual Model for Scandinavian Catchments,* SMHI Report No. RH07. Norrkopping, Sweden: Swedish Meteorological and Hydrological Institute.

BOUGHTON, M. E. 1966. "A Mathematical Model for Relating Runoff to Rainfall with Daily Data." *Civil Engineering Transactions* (Institution of Engineers, Australia) CE 8(1): 83–97.

BOYD, J. M., D. H. PILGRIM, AND I. CORDERY. 1979. "A Storage Routing Model Based on Catchment Geomorphology." *Journal of Hydrology* 42: 209–230.

BROWN, J. W., M. R. WALSH, R. W. MCCARLEY, A. J. GREEN, AND H. W. WEST. 1974. Models and Methods Applicable to Corps of Engineers Urban Studies. Miscellaneous Paper H-74-8, Hydraulics Laboratory, Vicksburg: U.S. Army Engineer Waterways Experiment Station.

BULTOT, F., AND G. L. DUPRIEZ. 1976a. "Conceptual Hydrological Model for an Average-Sized Catchment Area, I. Concepts and Relationships." *Journal of Hydrology* 29: 251–272.

———. 1976b. "Conceptual Hydrological Model for an Average-Sized Catchment Area, II. Estimate of Parameters, Validity of Model, Applications." *Journal of Hydrology* 29: 273–292.

CHAPMAN, T. G. 1968. "Catchment Parameters for a Deterministic Rainfall-Runoff Model." In *Land Evaluation,* G. A. Stewart, Ed. Melbourne: Macmillan.

CHARBONNEAU, R., J. P. FORTIN, AND G. MORIN. 1977. "The CEQUEAU Model: Description and Examples of Its Use in Problems Related to Water Resource Management." *Hydrological Science Bulletin* 22(1/3): 193–202.

CLABORN, B. J., AND W. MOORE. 1970. *Numerical Simulation in Watershed Hydrology,* Report No. HYD14-7001. Austin, TX: Hydraulic Engineering Laboratory, University of Texas.

CRAWFORD, N. H., AND R. K. LINSLEY. 1966. *Digital Simulation in Hydrology: Stanford Watershed Model IV,* Technical Report No. 39. Stanford, CA: Department of Civil Engineering, Stanford University.

CUNDAY, T. W., AND K. N. BROOKS, 1981. "Calibrating and Verifying the SSARR Model—Missouri River Watersheds Study." *Water Resources Bulletin* 17(5): 775–782.

DAWDY, D. R., R. W. LITCHY, AND J. M. BERGMANN. 1972. *A Rainfall–Runoff Simulation Model for Estimation of Flood Peaks for Small Drainage Basins,* U.S. Geological Survey Professional Paper 506-B. Washington, D.C.: U.S. Department of the Interior.

DESCHESNES, J., J. P. VILLENEUVE, E. LEDOUX, AND G. GIRARD. 1985a. "Modeling the Hydrologic Cycle: The MC Model, Part I—Principles and Description." *Nordic Hydrology* 16: 257–272.

———. 1985b. "Modeling the Hydrologic Cycle: The MC Model, Part II—Modeling Applications." *Nordic Hydrology* 16: 273–290.

FELDMAN, A. D. 1981. "HEC Models for Water Resources System Simulation: Theory and Experience." In *Advances in Hydroscience,* V. T. Chow, Ed. New York: Academic Press, pp. 297–423.

FIERING, M. B., AND G. KUCZERA. 1982. "Robust Estimators in Hydrology." In *Scientific Basis of Water Resources Management.* Washington, D.C.: National Academy Press, pp. 85–94.

FLEMING, G. 1975. *Computer Simulation Techniques in Hydrology.* New York: Elsevier.

FOROUD, N., AND R. S. BROUGHTON. 1981. "Flood Hydrograph Simulation Model." *Journal of Hydrology* 49: 139–172.

GHATE, S. R., AND H. R. WHITELEY. 1982. "GAWSER—A Modified HYMO Model Incorporating Areally Variable Infiltration." *Transactions of the American Society of Agricultural Engineers* 25: 134–142, 149.

GREEN, I. R. A., AND D. STEPHENSON. 1986. "Criteria for Comparison of Single Event Models." *Hydrological Sciences Journal* 31(3/9): 395–411.

HAAN, C. T., Ed. 1982. *Hydrologic Modeling of Ag-*

ricultural Watersheds, ASAE Monograph No. 5. St. Joseph, MI: American Society of Agricultural Engineers.

HOLTAN, H. N., G. J. STILTNER, W. H. HENSEN, AND N. C. LOPEZ. 1975. *USDA HL-74 Revised Model of Watershed Hydrology,* Technical Bulletin No. 1518. Washington, D.C.: Agricultural Research Service, U.S. Department of Agriculture.

HUGGINS, L. F., AND E. J. MONKE. 1968. "A Mathematical Model for Simulating the Hydrologic Response of a Watershed." *Water Resources Research* 4(3): 529–539.

Hydrologic Engineering Center. 1981. *HEC-1 Flood Hydrograph Package, User's Manual.* Davis, CA: U.S. Army Corps of Engineers.

———. 1982. *Hydrologic Analysis of Ungaged Watersheds Using HEC-1,* Training Document No. 15. Davis, CA: U.S. Army Corps of Engineers.

Hydrologic Research Laboratory. 1972. *National Weather Service River Forecast System: Forecast Procedures,* NOAA Technical Memorandum NWS-HYDRO-14. Washington, D.C.: National Weather Service, National Oceanic and Atmospheric Administration, U.S. Department of Commerce.

IBBITT, R. P., AND T. O'DONNELL. 1971. "Fitting Methods for Conceptual Catchment Models. *Journal of the Hydraulics Division, Proceedings of the American Society of Civil Engineers* 97(HY9): 1331–1342.

JACKSON, T. J. 1982. "Application and Selection of Hydrologic Models." In *Hydrologic Modeling of Agricultural Watersheds,* ASAE Monograph No. 5, C. T. Haan, Ed. St. Joseph, MI: American Society of Agricultural Engineers, pp. 475–504.

JAMES, L. D. 1972. "Hydrologic Modeling, Parameter Estimation and Watershed Characteristics." *Journal of Hydrology* 17: 283–307.

JAMES, L. D., D. S. BOWLES, AND R. H. HAWKINS. 1982. "A Taxonomy for Evaluating Surface Water Quantity Model Reliability." In *Applied Modeling in Catchment Hydrology,* V. P. Singh, Ed. Littleton, CO: Water Resources Publications, pp. 189–228.

JAMES, L. D., AND S. J. BURGES. 1982. "Selection, Calibration, and Testing of Hydrologic Models." In *Hydrologic Modeling of Agricultural Watersheds,* ASAE Monograph No. 5, C. T. Haan, Ed. St. Joseph, MI: American Society of Agricultural Engineers, pp. 437–472.

JOHNSTON, P. R., AND D. H. PILGRIM. 1976. "Parameter Estimation for Watershed Models." *Water Resources Research* 12(3): 477–486.

KITANIDIS, P. K., AND R. L. BRAS. 1980. "Real-Time Forecasting with a Conceptual Hydrologic Model: 1. Analysis of Uncertainty." *Water Resources Research* 16(6): 1025–1033.

KLEMES, V. 1973. *Applications of Hydrology to Water Resources Management,* Operational Hydrology Report No. 4. Geneva: World Meteorological Organization.

———. 1982. "Empirical and Causal Models in Hydrology." In *Scientific Basis of Water Resources Management*. Washington, D.C.: National Academy Press, 95–104.

KUCZERA, G. 1982. "On the Relationship between the Reliability of Parameter Estimates and Hydrologic Time Series Data Used in Calibration. *Water Resources Research* 18(1): 146–154.

LARSON, C. L., C. A. ONSTAD, H. H. RICHARDSON, AND K. N. BROOKS. 1982. "Some Particular Watershed Models." In *Hydrologic Modeling of Agricultural Watersheds,* ASAE Monograph No. 5, C. T. Haan, Ed. St. Joseph, MI: American Society of Agricultural Engineers, pp. 409–434.

LAURENSON, E. M., AND R. G. MEIN. 1983. *RORB—Version 3: Runoff Routing Program—User Manual,* 2d ed. Monash, Australia: Department of Civil Engineering, Monash University.

LINSLEY, R. K. 1982. "Rainfall–Runoff Models: An Overview." In *Rainfall–Runoff Relationship,* V. P. Singh, Ed. Littleton, CO: Water Resources Publications, pp. 3–22.

LIOU, E. Y. 1970. *OPSET: Program for Computerized Selection of Watershed Parameter Values for the Stanford Watershed Model,* Research Report No. 34. Lexington, KY: Water Resources Institute, University of Kentucky.

MADDAUS, W. O., AND P. S. EAGLESON. 1969. *A Distributed Linear Representation of Surface Runoff,* Report No. 115. Cambridge, MA: Hydrodynamics Laboratory, Massachusetts Institute of Technology.

MANLY, R. E. 1978. "Simulation of Flows in Ungaged Basins." *Hydrological Sciences Bulletin* 23(1/3): 85–101.

Metcalf and Eddy, Inc., University of Florida, and Water Resources Engineers, Inc. 1971. *Storm Water Management Model, Vol. I—Final Report,* Water Pollution Control Research Series 11024 DOC 07/71. Washington, D.C.: Environmental Protection Agency.

MORRIS, E. M. 1980. *Forecasting Flood Flows in Grassy and Forested Basins Using a Deterministic Distributed Mathematical Model,* Publication No. 129. Wallingford, England: International Association of Scientific Hydrology, pp. 247–255.

Office of Technology Assessment. 1982. *Use of Models for Water Resources Management, Planning and Policy*. Washington, D.C.: Congress of the United States.

PECK, E. L. 1976. *Catchment Modeling and Initial Parameter Estimation for the National Weather Service River Forecast System,* NOAA Technical Memo NWS HYDRO-31. Silver Spring, MD: National Weather Service, U.S. Department of Commerce.

PORTER, J. W., AND T. A. MCMAHON. 1971. "A Model for the Simulation of Streamflow Data from Climatic Records." *Journal of Hydrology* 13: 297–324.

QUCK, M. C., AND A. PIPES. 1977. "U. B. C. Watershed Model." *Hydrological Sciences Bulletin* 22(1/3): 153–161.

REFSGARD, J. C. 1981. *The Surface Component of an*

Integrated Hydrological Model, Susa Hydrology Report 12. Lyngby, Denmark: Technical University of Denmark.

RENARD, K. G., W. J. RAWLS, AND M. M. FOGEL. 1982. "Currently Available Models." In *Hydrologic Modeling of Agricultural Watersheds,* ASAE Monograph No. 5, C. T. Haan, Ed. St. Joseph, MI: American Society of Agricultural Engineers, pp. 507–522.

RESTREPO-POSADA, J., AND R. L. BRAS. 1982. *Automatic Parameter Estimation of a Large Conceptual Rainfall–Runoff Model: A Maximum Likelihood Approach,* Report No. 267. Cambridge, MA: R. M. Parsons Laboratory of Hydrodynamics and Water Resources, Massachusetts Institute of Technology.

RICCA, V. T. 1972. *The Ohio State University Version of the Stanford Streamflow Simulation Model, Part I—Technical Aspects.* Columbus, OH: Ohio State University.

SINGH, V. P., Ed. 1982. *Applied Modeling in Catchment Hydrology.* Littleton, CO: Water Resources Publications.

SINGH, V. P. 1983. *A Geomorphic Approach to Hydrograph Synthesis with Potential for Application to Ungaged Watersheds.* Technical Completion Report. Baton Rouge: Louisiana Water Resources Research Institute, Louisiana State University.

———. 1989. *Hydrologic Systems, Vol. 2, Watershed Modeling.* Englewood Cliffs, NJ: Prentice Hall.

SITTNER, W. C., C. E. SCHAUSS, AND J. C. MONRO. 1969. "Continuous Hydrograph Synthesis with an API-Type Hydrologic Model." *Water Resources Research* 5(5): 1007–1022.

SMITH, R. L., AND A. M. LUMB. 1966. *Derivation of Basin Hydrographs,* Contribution No. 19. Manhattan, KS: Kansas Water Resources Research Institute, Kansas State University.

Soil Conservation Service. 1973. *Computer Program for Project Formulation Hydrology,* Technical Release No. 20. Washington, D.C.: U.S. Department of Agriculture.

SOROOSHIAN, S. 1983. "Surface Water Hydrology: On-Line Estimation." *Reviews of Geophysics and Space Physics* 21(3): 706–721.

SOROOSHIAN, S., AND J. M. DRACUP. 1980. "Stochastic Parameter Estimation Procedures for Hydrologic Rainfall–Runoff Models: Correlated and Heteroscedastic Error Cases." *Water Resources Research* 16(2): 430–442.

SOROOSHIAN, S., AND V. K. GUPTA. 1983. "Automatic Calibration of Conceptual Rainfall–Runoff Models: The Question of Parameter Observability and Uniqueness." *Water Resources Research* 19(1): 260–268.

SOROOSHIAN, S., V. K. GUPTA, AND J. L. FULTON. 1983. "Evaluation of Maximum Likelihood Parameter Estimation Techniques for Conceptual Rainfall–Runoff Models: Influence of Calibration Data Variability and Length on Model Credibility." *Water Resources Research* 19(1): 251–259.

SUGAWARA, M., I. WATANABE, E. OZAKI, AND Y. KATSUYAMA. 1984. *Tank Model with Snow Component,* Research Notes of the National Research Center for Disaster Prevention No. 65. Ibaraki-Ken, Japan: Science and Technology Agency.

Tennessee Valley Authority. 1972. *Upper Bear Creek Experimental Project—A Continuous Daily-Streamflow Model,* Research Paper No. 8. Knoxville: Tennessee Valley Authority.

U.S. Army Engineer Division, North Pacific. 1972. *Program Description and User Manual for SSARR Model—Streamflow Synthesis and Reservoir Regulation Model.* Portland, Oregon: U.S. Army Corps of Engineers.

VIESSMAN, W., JR., G. L. LEWIS, AND J. W. KNAPP. 1989. *Introduction to Hydrology.* New York: Harper and Row.

WILLIAMS, B. J., AND W. W.-G. YEH. 1983. Paper Estimation in Rainfall–Runoff Models." *Journal of Hydrology* 63: 373–393.

WILLIAMS, J. R., AND R. W. HANN. 1973. *HYMO: Problem-Oriented Computer Language for Hydrologic Modeling.* Washington, D.C.: Agricultural Research Service, U.S. Department of Agriculture.

ZHAO, R. J., Y. L. ZUANG, L. R. FANG, X. R. LIU, AND Q. S. ZHANG. 1980. "The Zinanjiang Model." In *Proceedings of the Oxford Symposium on Hydrological Forecasting,* Publication No. 129. Wallingford, England: International Association of Hydrological Sciences, pp. 351–356.

CHAPTER 24

BOBEE, B. 1975. "The Log Pearson Type III Distribution and Its Application in Hydrology." *Water Resources Research* 11(5): 681–689.

CHOW, V. T. 1951. "A General Formula for Hydrologic Frequency Analysis." *Transactions, American Geophysical Union* 32: 231–237.

GUMBEL, E. J. 1954. *Statistical Theory of Extreme Values and Some Practical Applications,* Applied Mathematics Series No. 33. Washington, D.C.: U.S. National Bureau of Standards.

———. 1958. *Statistics of Extremes.* New York: Columbia University Press.

JAIN, D., AND V. P. SINGH. 1986. "A Comparison of Transformation Methods for Flood Frequency Analysis." *Water Resources Bulletin* 22(6): 903–912.

LOWERY, M. D., AND J. E. NASH. 1970. "A Comparison of Methods of Fitting the Double Exponential Distribution." *Journal of Hydrology* 10: 259–275.

SPEIGEL, M. R. 1961. *Theory and Problems of Statistics.* New York: Schaum.

STEEL, R. G. D., AND J. H. TORRIE. 1960. *Principles and Procedures of Statistics.* New York: McGraw-Hill.

STURGES, H. A. 1926. "The Choice of a Class Interval." *Journal of the American Statistical Association* 21: 65–66.

THOMAS, J. B. 1971. *An Introduction to Applied Probability and Random Processes.* New York: Wiley.

Water Resources Council. 1967. *Guidelines for Determining Flood Flow Frequency,* Hydrology Committee Bulletin 15. Washington, D.C.: Water Resources Council.

YEVJEVICH, V. 1972. *Probability and Statistics in Hydrology*. Fort Collins, CO: Water Resources Publications.

CHAPTER 25

ADAMOWSKI, K. 1981. "Plotting Formula for Flood Frequency." *Water Resources Bulletin* 17(2): 197–202.

BEARD, L. R. 1943. "Statistical Analysis in Hydrology." *Transactions, American Society of Civil Engineers* 108: 1110–1160.

———. 1962. *Statistical Methods in Hydrology*. Sacramento: Engineer District, U.S. Army Corps of Engineers.

———. 1974. *Flood Flow Frequency Techniques*. CRWR-119. Austin, TX: Center for Research in Water Resources, University of Texas.

BENSON, M. A. 1960. *Characteristics of Frequency Curves Based on a Theoretical 1,000 Year Record,* U.S. Geological Survey Water Supply Paper 1543-A. Washington, D.C.: U.S. Department of the Interior, pp. 51–77.

———. 1962. *Evaluation of Methods for Evaluation of the Occurrence of Floods,* U.S. Geological Survey Water Supply Paper 1580-A. Washington, D.C.: U.S. Department of the Interior.

———. 1968. "Uniform Flood Frequency Estimating Methods for Federal Agencies." *Water Resources Research* 4(4): 891–908.

BLOM, G. 1958. *Statistical Estimates and Transformed Beta Variables*. New York: Wiley.

California. 1923. *Flow in California Streams,* CDPW Bulletin No. 5. Sacramento: California State Department of Public Works.

CHEGODAYEV. 1955. "Quoted in 'Formulas for the Calculation of the Confidence of Hydrologic Quantities,' by A. G. Alekseyev," in *Handbook of Applied Hydrology,* V. T. Chow, Ed., New York: McGraw-Hill.

CHOW, V. T. 1951. "A General Formula for Hydrologic Frequency Analysis." *Transactions, American Geophysical Union* 32: 231–237.

———. 1954. "The Log-Probability Law and Its Engineering Applications." *Proceedings, American Society of Civil Engineers* 80: 1–25.

———, Ed. 1964. *Handbook of Applied Hydrology*. New York: McGraw-Hill, Section 8-1.

CUNNANE, C. 1978. "Unbiased Plotting Positions—A Review." *Journal of Hydrology* 37: 205–222.

DALRYMPLE, T. 1960. "Flood Frequency Analysis," Water Supply Paper 1543-A. In *Manual of Hydrology,* Part 3. Washington, D.C.: U.S. Government Printing Office.

GRINGORTEN, I. I. 1963. "A Plotting Rule for Extreme Probability Paper." *Journal of Geophysical Research* 68(3): 813–814.

GUMBEL, E. J. 1960. *Statistics of Extremes*. New York: Columbia University Press.

HAZEN, A. 1914. "Storage to be Provided in Improving Reservoirs for Municipal Water Supply." *Transactions, American Society of Civil Engineers* 77: 1547–1550.

———. 1930. *Flood Flows: A Study of Frequencies and Magnitudes*. New York: Wiley.

LANGBEIN, W. B., and Others. 1947. "Topographical Characteristics of Drainage Basins." U.S. Geological Survey Water Supply Paper 968-C. Washington, D.C.: U.S. Government Printing Office.

MCCABE, J. A. 1962. *Floods in Kentucky—Magnitude and Frequency,* Information Circular 9. Lexington, KY: Kentucky Geological Survey.

NEMEC, J. 1973. *Engineering Hydrology*. London: McGraw-Hill.

Water Resources Council. 1967. *Guidelines for Determining Flood Flow Frequency,* Hydrology Committee Bulletin 15. Washington, D.C.: Water Resources Council.

WEIBULL, W. 1939. "A Statistical Theory of the Strength of Materials." *Ing. Vetenskaps Akad. Handl* (*Stockholm*) 151: 15

CHAPTER 26

BELL, F. C. 1976. *The Areal Reduction Factor in Rainfall Frequency Estimation,* Report No. 35. Wallingford, England: Institute of Hydrology.

BOYER, M. C. 1957. "A Correlation of the Characteristics of Great Storms." Transactions of the American Geophysical Union 38(2): 233–236.

Central Water Commission. 1972. *Estimation of Design Flood Recommended Procedures*. New Delhi: Ministry of Irrigation and Power.

COURT, A. 1961. "Area–Depth Rainfall Formulas." *Journal of Geophysical Research* 66(6): 1823–1831.

DECOURSEY, D. G., AND W. M. SNYDER. 1969. "Computer-Oriented Method of Optimizing Hydrologic Model Parameters." *Journal of Hydrology* 9:34–56.

DHAR, O. N., AND B. K. BHATTACHARYA. 1977. "Relationship between Central Rainfall and its Areal Extent." Journal of Irrigation and Power 34(2): 245–250.

EAGLESON, P. S. 1970. *Dynamic Hydrology*. New York: McGraw-Hill.

FREDERICK, R. H., V. A. MYERS, AND E. P. AUCIELLO. 1977. "Storm Depth–Area Relations from Digitized Radar Returns." *Water Resources Research* 13(3): 675–679.

HERSHFIELD, D. M. 1961. "Estimating the Probable Maximum Precipitation." *Journal of Hydraulics Division, Proceedings of the American Society of Civil Engineers* 87(HY5): 99–116.

———. 1965. "Method for Estimating Probable Maximum Precipitation." *Journal of the American Water Works Association,* 57(8): 965–972.

HORTON, R. E. 1924. "Discussion of 'The Distribution of Intense Rainfall and Some Other Factors in the Design of Storm-Water Drains,' by F. A. Marston." Proceedings of the American Society of Civil Engineers 50: 660–667.

HUFF, F. A. 1967. "Time Distribution of Rainfall in Heavy Storms." *Water Resources Research* 3(4): 1007–1019.

Huff, F. A., and G. E. Stout. 1952. "Area–Depth Studies for Thunderstorm Rainfall in Illinois." *Transactions, American Geophysical Union* 33(4): 495–498.

Leclerc, G., and J. C. Schaake. 1972. *Derivation of Hydrologic Frequency Curves,* Report No. 142. Cambridge, MA: R. M. Parsons Laboratory of Hydrodynamics and Water Resources, Massachusetts Institute of Technology.

National Research Council. 1985. *Safety of Dams—Flood and Earthquake Criteria.* Washington, D.C.: Committee on Safety Criteria for Dams, National Academy Press.

National Weather Service. 1961. *Probable Maximum Precipitation in California,* NOAA Hydrometeorological Report No. 36. Washington, D.C.: U.S. Department of Commerce.

———. 1963. *Rainfall Frequency Atlas of the United States for Durations from 30 Minutes to 24 Hours and Return Periods from 1 to 100 Years,* U.S. Weather Bureau Technical Paper No. 4. Washington, D.C.: U.S. Department of Commerce.

———. 1966. *Probable Maximum Precipitation, Northwest States,* NOAA Hydrometeorological Report No. 43. Washington, D.C.: U.S. Department of Commerce.

———. 1969. *Probable Maximum TVA Precipitation for Tennessee River Basins Up to 3000 Square Miles in Area and Durations to 72 Hours,* Weather Bureau Hydrometeorological Report No. 45. Washington, D.C.: U.S. Department of Commerce.

———. 1977. *Probable Maximum Precipitation Estimates, Colorado River and Great Basin Drainages,* NOAA Hydrometeorological Report No. 49. Washington, D.C.: U.S. Department of Commerce.

———. 1978. *Probable Maximum Precipitation, United States East of the 105th Meridian,* NOAA Hydrometeorological Report No. 51. Washington, D.C.: U.S. Department of Commerce.

———. 1982. *Application of Probable Maximum Precipitation Estimates, United States East of the 105th Meridian,* NOAA Hydrometeorological Report No. 52. Washington, D.C.: U.S. Department of Commerce.

———. 1984. *Probable Maximum Precipitation Estimates—United States between the Continental Divide and 103rd Meridian,* NOAA Hydrometeorological Report No. 55. Washington, D.C.: U.S. Department of Commerce.

Natural Environment Research Council. 1975. *Flood Studies Report,* Vol. 2. London: Natural Environment Research Council.

Nicks, A. D., and F. A. Igo. 1980. "A Depth–Area–Duration Model of Storm Rainfall in the Southern Great Plains." *Water Resources Research* 16(5): 939–945.

Osborn, H. B., L. J. Lane, and V. A. Myers. 1980. "Rainfall/Watershed Relationships for Southwestern Thunderstorms." *Transactions of the American Society of Agricultural Engineers* 23(1): 82–87, 91.

Rodriguez-Iturbe, I., and J. M. Mejia. 1974. "On the Transformation of Point Rainfall to Areal Rainfall." *Water Resources Research* 10(4): 729–735.

Soil Conservation Service. 1971. "Hydrology." In *SCS National Engineering Handbook.* Washington, D.C.: U.S. Department of Agriculture, Section 4.

———. 1975. *Urban Hydrology for Small Watersheds,* Technical Release No. 55. Washington, D.C.: U.S. Department of Agriculture.

Solot, S. 1939. "Computation of Depth of Precipitable Water in a Column of Air." *Monthly Weather Review* 67: 100–103.

Stedinger, J. R., and V. R. Baker. 1987. "Surface Water Hydrology—Historical and Paleoflood Information," AGU US National Report, 1983–1986. *Reviews of Geophysics and Space Physics* 00: 119–124.

U.S. Army Corps of Engineers. 1946. *Storm Rainfall in the United States: Depth–Area–Duration Data.* Washington, D.C.: U.S. Army Corps of Engineers.

———. 1956. *Seasonal Variations of the Probable Maximum Precipitation East of the 105th Meridian,* Hydrometeorological Report No. 33. Washington, D.C.: U.S. Department of the Army.

———. 1979. *Engineering and Design, National Program for Inspection of Non-Federal Dams,* ER NO-1110-2-106. Washington, D.C.: U.S. Department of the Army.

U.S. Bureau of Reclamation. 1965. *Design of Small Dams.* Washington, D.C.: U.S. Department of the Interior.

U.S. Weather Bureau. 1941. *Psychrometric Tables,* Report No. 235. Washington, D.C.: U.S. Department of Commerce.

———. 1964. *Two-to-Ten-Day Precipitation for Return Periods of 2 to 100 Years in the Contiguous United States,* Technical Paper No. 49. Washington, D.C.: U.S. Department of Commerce.

Wang, B. H. 1988a. "Determination of Design Flood for Spillways." *Transactions, 16th Congress on Large Dams,* Vol. IV, Q. 63, R. 39. Paris, France: International Commission on Large Dams, pp. 647–666.

———. 1988b. *Hydrology of Extreme Events.* Lecture presented at the International Seminar on Hydrology of Extremes, Roorkie, India.

Wenzel, H. G. 1982. "Rainfall for Urban Stormwater Design." In *Urban Storm Water Hydrology,* Water Resources Monograph, D. F. Kibler, Ed. Washington, D.C.: American Geophysical Union.

Woolhiser, D. A., and H. C. Schwalen. 1960. *Area–Depth–Frequency Relations for Thunderstorm Rainfall in Southern Arizona,* Technical Paper 527. Tucson: Arizona Agricultural Experiment Station, The University of Arizona.

World Meteorological Organization. 1969. *Manual for Depth–Area–Duration Analysis of Storm Precipitation,* WMO No. 237, Technical Paper 129. Geneva: World Meteorological Organization, pp. 1–81.

APPENDIX A

SCHULZ, E. F. 1973. *Problems in Applied Hydrology*. Fort Collins, CO: Water Resources Publications.

APPENDIX B

DISKIN, M. H. 1967. "On the Solution of the Muskingum Method of Flood Routing Equation." *Journal of Hydrology* 5: 286–289.

KULANDAISWAMY, V. C. 1966. "A Note on the Muskingum Method of Flood Routing." *Journal of Hydrology* 4: 273–276.

MOREL-SEYTOUX, H. J. 1979. "Flow Forecasting Based on Preseason Conditions." In *Modeling of Rivers,* H. W. Shen, Ed. New York: Wiley, pp. 4-1 to 4-46.

VENETIS, C. 1969. "The IUH of the Muskingum Channel Reach." *Journal of Hydrology* 4: 185–200.

APPENDIX C

CHOW, V. T., Ed. 1964. *Handbook of Applied Hydrology: A Compendium of Water Resources Technology*. New York: McGraw-Hill.

Water Resources Council. 1967. "A Uniform Technique for Determining Flood Flow Frequencies," Bulletin No. 15. Washington, D.C.: Water Resources Council.

Index

AUTHOR INDEX

SUBJECT INDEX